Fundamentals of Osteoporosis

Fundamentals of Osteoporosis

ROBERT MARCUS

DAVID FELDMAN

DOROTHY A. NELSON

CLIFFORD J. ROSEN

AMSTERDAM • BOSTON • HEIDELBERG • LONDON
NEW YORK • OXFORD • PARIS • SAN DIEGO
SAN FRANCISCO • SINGAPORE • SYDNEY • TOKYO

Academic Press is an imprint of Elsevier

Academic Press is an imprint of Elsevier
30 Corporate Drive, Suite 400, Burlington, MA 01803, USA
525 B Street, Suite 1900, San Diego, California 92101-4495, USA
84 Theobald's Road, London WC1X 8RR, UK

Library of Congress Cataloging-in-Publication Data
APPLICATION SUBMITTED

British Library Cataloguing-in-Publication Data
A catalogue record for this book is available from the British Library.

ISBN: 978-0-12-375098-3

For information on all Academic Press publications
visit our Web site at www.elsevierdirect.com

Printed in the United States of America
09 10 11 9 8 7 6 5 4 3 2 1

Working together to grow
libraries in developing countries

www.elsevier.com | www.bookaid.org | www.sabre.org

ELSEVIER   BOOK AID International   Sabre Foundation

Contents

CHAPTER 1   The Bone Organ System: Form and Function

ELISE F. MORGAN, GEORGE L. BARNES, AND

THOMAS A. EINHORN

CHAPTER 2   The Nature of Osteoporosis

ROBERT MARCUS AND MARY L. BOUXSEIN

CHAPTER 3   Skeletal Heterogeneity and the Purposes of Bone Remodeling: Implications for the Understanding of Osteoporosis

A. M. PARFITT

CHAPTER 4   Osteoblast Biology

JANE B. LIAN AND GARY S. STEIN

CHAPTER 5   Osteoclast Biology

HARRY C. BLAIR, SCOTT SIMONET, DAVID L. LACEY,

AND MONE ZAIDI

CHAPTER 12   Regulation of Bone Cell Function by Estrogens

BARRY S. KOMM, BORIS CHESKIS, AND PETER V.N. BODINE

CHAPTER 13   Androgens and Skeletal Biology: Basic Mechanisms

KRISTINE M. WIREN

CHAPTER 14   Phosphatonins

PETER J. TEBBEN, THERESA J. BERNDT, AND RAJIV KUMAR

CHAPTER 15   Wnt Signaling in Bone

MARK L. JOHNSON ND ROBERT R. RECKER

CHAPTER 16   Cytokines and Bone Remodeling

GREGORY R. MUNDY, BABATUNDE OYAJOBI, GLORIA GUTIERREZ, JULIE STERLING, SUSAN PADALECKI, FLORENT ELEFTERIOU, AND MING ZHAO

CHAPTER 17   Skeletal Growth Factors

ERNESTO CANALIS

CHAPTER 18   Intercellular Communication during Bone Remodeling

T. JOHN MARTIN AND GIDEON A. RODAN

Contributors

Eva Balint
Endocrinology Division, Stanford University School of Medicine, Stanford, CA

George L. Barnes
Department of Orthopaedic Surgery, Boston University School of Medicine, Boston, MA

Wesley G. Beamer
The Jackson Laboratory, Bar Harbor, ME

Theresa J. Berndt
Division of Nephrology, Mayo Clinic, Rochester, MN

Harry C. Blair
Department of Pathology, Section of Laboratory Medicine, McGowan Institute for Regenerative Medicine, University of Pittsburgh, School of Medicine, Pittsburgh, PA

Peter V. N. Bodine
Project Management, Wyeth Research, Collegeville, PA

Lynda F. Bonewald
Department of Oral Biology, University of Missouri at Kansas City School of Dentistry, Kansas City, MO

Adele L. Boskey
Hospital for Special Surgery, Weill Medical College of Cornell University, New York, NY

Mary L. Bouxsein
Department of Orthopaedic Surgery, Beth Israel Deaconess Medical Center and Harvard Medical School, Boston, MA

Ernesto Canalis
Department of Research, Saint Francis Hospital and Medical Center, Hartford, CT; University of Connecticut School of Medicine, Farmington, CT

Boris Cheskis
Women's Health Musculoskeletal Biology, Wyeth-Ayerst Research, Collegeville, PA

Thomas A. Einhorn
Department of Orthopaedic Surgery, Boston University Medical Center, Boston, MA

Florent Elefteriou
Center for Bone Biology, Vanderbilt Medical Center, Nashville, TN

David Feldman
Division of Endocrinology, Gerontology, and Metabolism, Stanford University School of Medicine, Stanford, CA

Gloria Gutierrez
Center for Bone Biology, Vanderbilt Medical Center, Nashville, TN

Mark L. Johnson
Department of Oral Biology, University of Missouri at Kansas City School of Dentistry, Kansas City, MO

Barry S. Komm
Women's Health Musculoskeletal Biology, Wyeth-Ayerst Research, Collegeville, PA

Aruna V. Krishnan
Division of Endocrinology, Gerontology, and Metabolism, Stanford University School of Medicine, Stanford, CA

Henry Kronenberg
Endocrine Unit, Massachusetts General Hospital and Harvard Medical School, Boston, MA

Rajiv Kumar
Departments of Internal Medicine, Biochemistry, and Molecular Biology, Mayo Clinic, Rochester, MN

David L. Lacey
Amgen, Thousand Oaks, CA

Jane B. Lian
Department of Cell Biology, University of Massachusetts Medical School, Worcester, MA

Peter J. Malloy
Division of Endocrinology, Gerontology, and Metabolism, Stanford University School of Medicine, Stanford, CA

Robert Marcus
Senior Medical Fellow, Eli Lilly & Company, Indianapolis, IN; Professor Emeritus, Department of Medicine, Stanford University, Stanford, CA

T. John Martin
Saint Vincent's Institute of Medical Research, Victoria, Australia

Elise F. Morgan
Department of Aerospace and Mechanical Engineering, Boston University; Department of Orthopaedic Surgery, Boston University School of Medicine, Boston, MA

Gregory R. Mundy
Department of Cellular and Structural Biology, University of Texas Health Science Center, San Antonio, TX

Robert A. Nissenson
Endocrine Unit, San Francisco Veterans Affairs Medical Center and Departments of Medicine and Physiology, University of California, San Francisco, CA

Babatunde Oyajobi
Department of Cellular and Structural Biology, University of Texas Health Science Center, San Antonio, TX

Susan Padalecki
Department of Urology and Cellular and Structural Biology, University of Texas Health Science Center, San Antonio, TX

A. M. Parfitt
Division of Endocrinology, University of Arkansas for Medical Sciences, Little Rock, AR

Sylvain Provot
Department of Anatomy, University of California at San Francisco, San Francisco, CA

Robert R. Recker
Osteoporosis Research Center, Creighton University School of Medicine, Omaha, NE

Pamela Gehron Robey
Bone Research Branch, National Institute of Dental Research, National Institutes of Health, Bethesda, MD

Gideon A. Rodan (deceased)
University of Pennsylvania, Philadelphia, PA

Clifford J. Rosen
Maine Center for Osteoporosis Research and Education, Bangor, ME

Ernestina Schipani
Endocrine Unit, Massachusetts General Hospital and Harvard Medical School, Boston, MA

Scott Simonet
Amgen, Thousand Oaks, CA

Gary S. Stein
Department of Cell Biology, University of Massachusetts Medical School, Worcester, MA

Julie Sterling
Center for Bone Biology, Vanderbilt University, Nashville, TN

Peter J. Tebben
Division of Endocrinology, Diabetes, Metabolism, and Nutrition, Department of Internal Medicine, Mayo Clinic, Rochester, MN

Kristine M. Wiren
Departments of Medicine and Behavioral Neuroscience, Oregon Health & Science University; and VA Medical Center, Portland, OR

Joy Wu
Endocrine Unit, Massachusetts General Hospital and Harvard Medical School, Boston, MA

Mone Zaidi
Mount Sinai School of Medicine, New York, NY

Ming Zhao
Department of Cellular and Structural Biology, University of Texas Health Science Center, San Antonio, TX

Wei Zhu
Hospital for Special Surgery, New York NY

Preface

This volume consists of 18 basic science chapters taken from the 80-chapter comprehensive reference work, *Osteoporosis, 3rd Edition*. The rationale for this separate publication is the thought that many of the scientists and laboratories in the bone field that have a strictly basic science focus might find utility in a volume devoted exclusively to skeletal science without requiring a huge collection of material on epidemiology, diagnosis, clinical manifestations, and therapeutics. This volume opens with three introductory chapters on the fundamentals of skeletal structure and organization, the nature of osteoporosis, and the purposes of bone remodeling, and is followed by detailed chapters on all major bone cell types: osteoblasts, osteoclasts, and osteocytes; and extensive discussions of bone matrix proteins. Next are chapters on skeletal development and insights gained from mouse genetics to the study of bone development and physiology. The last set of chapters deals with skeletal regulatory hormones: parathyroid hormone-PTHrP, vitamin D, estrogens, and androgens; cytokines and growth factors active in bone; and the exciting new fields of phosphatonins and Wnt signaling. The final chapter brings together novel thoughts on intercellular communication during bone remodeling. We hope that these chapters, written by outstanding scholars in their fields, will offer readers information of substantial relevance to their own work, as well as providing a great deal of enjoyable reading.

ROBERT MARCUS, M.D.
DAVID FELDMAN, M.D.
DOROTHY A. NELSON, Ph.D.
CLIFFORD J. ROSEN, M.D.

The Bone Organ System: Form and Function

ELISE F. MORGAN, GEORGE L. BARNES, AND THOMAS A. EINHORN

I. INTRODUCTION

Bone is a vital, dynamic connective tissue whose structure and composition reflect a balance between its two major functions: provision of mechanical integrity for locomotion and protection, and involvement in the metabolic pathways associated with mineral homeostasis. In addition, bone is the primary site of hematopoiesis, and a rich picture of the complex interplay between the bone organ system and the immune system continues to emerge [1–3].

Beginning with the observations of Galileo, it has been assumed that the shape and internal structure of bone are influenced by the mechanical loads associated with normal function. The 19th century saw active development of this concept, particularly with respect to the cross-sectional geometry of whole bones [4] and to the structure of trabecular bone (see [5, 6] for a review). The most well known of the published works from this time period is by Julius Wolff, who synthesized many others' observations in postulating that the structure of trabecular bone is aligned with the principal stress directions that occur in this tissue during normal skeletal function [7]. In this hypothesis, known by the misnomer "Wolff's Law," Wolff further proposed, as others before him had [8], that this alignment results from a self-regulating functional adaptation process. Although errors in various components of Wolff's writings have been identified [9, 10], what is generally thought of today as Wolff's Law is the overall concept that, in bone, form follows function. This concept underlies much of the scientific investigation of relationships between bone structure and its mechanical and metabolic functions.

In maintaining these structure–function relationships, bone tissue is constantly being broken down and rebuilt in a process called *remodeling*. The cellular link between bone resorbing cells or osteoclasts, and bone forming cells or osteoblasts, is known as coupling. With age, remodeling tends to result in a negative bone balance, in that at each remodeling site slightly less bone is deposited than is resorbed. This negative balance leads to osteopenia and osteoporosis, thus predisposing the bone to fracture during even minimal trauma. However, in normal states, the remodeling activities in bone serve to reduce bone mass where the mechanical demands of the skeleton are low and to add mass at those sites where the demands are repeatedly high. It is worth emphasizing that, were the removal and deposition of bone tissue to occur independently of mechanical considerations, fluctuations in systemic needs for calcium and magnesium could very well be disastrous for the integrity of the skeleton. Hence, bone is a well-designed organ system whose homeostasis depends on processing of external mechanical input and physiological signals from the systemic environment and the transduction of these signals into cellular and chemical events.

II. COMPOSITION AND ORGANIZATION OF BONE

Bone is a composite material consisting of an inorganic and an organic phase. By weight, approximately 60% of the tissue is inorganic matter, 8–10% is water, and the remainder is organic matter [11]. By volume, these proportions are approximately 40%, 25%, and 35%, respectively. The inorganic phase is an impure form of hydroxyapatite ($Ca_{10}[PO_4]_6[OH]_2$), which is a naturally occurring calcium phosphate. The organic phase is composed predominantly (98% by weight) of type I collagen and a variety of noncollagenous proteins, and cells make up the remaining 2% of this phase [12].

A. Organic Phase

The organic phase of bone plays a wide variety of roles, influencing profoundly the structure and also the mechanical and biochemical properties of the tissue. Growth factors and cytokines, and extracellular matrix proteins such as osteonectin, osteopontin, bone sialoprotein, osteocalcin, proteoglycans, and other phosphoproteins and proteolipids, make small contributions to the overall volume of bone but major contributions to its biologic function.

Type I collagen is a ubiquitous protein of extremely low solubility, and it is the major structural component of the bone matrix. The type I collagen molecule consists of three polypeptide chains composed of approximately 1000 amino acids each. These chains take the form of a triple helix of two identical 1(I) chains and one unique 2(I) chain cross-linked by hydrogen bonding between hydroxyproline and other charged residues. This produces a very rigid linear molecule that is approximately 300 nm in length. Each molecule is aligned with the next in a parallel fashion and in a quarter-staggered array to produce a collagen fibril. The collagen fibrils are then grouped in bundles to form the collagen fiber. Within the collagen fibril, gaps known as "hole zones" are present between the ends of the molecules. In addition, pores exist between the sides of parallel molecules (Figure 1-1). Noncollagenous proteins or mineral deposits can be found within these spaces, and mineralization of the matrix is thought to be initiated in the hole zones.

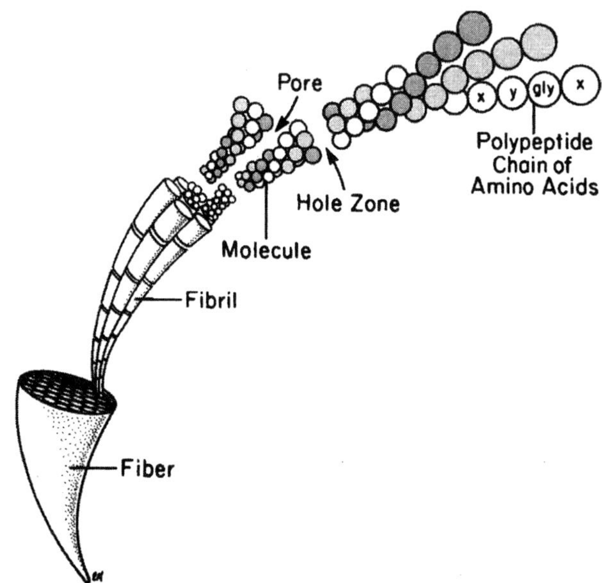

FIGURE 1-1   Collagen fiber and fibril structure with putative locations of pores and hole zones shown. Reprinted with permission from [12].

Several noncollagenous proteins have been identified in bone. One of the more extensively studied of these in bone is osteocalcin (OC) or bone-carboxyglutamic acid-containing protein (bone Gla protein). This is a small (5.8 kDa) protein in which three glutamic acid residues are carboxylated as a result of a vitamin K dependent, post-translational modification. The carboxylation of these residues confers on this protein calcium and mineral binding properties. Osteocalcin is one of the most abundant noncollagenous proteins in bone, accounting for 10–20% of the noncollagenous protein content, and it is closely associated with the mineral phase. Evidence suggests that this bone-specific protein may regulate activities of osteoclasts and osteoclast precursors. However, through characterization of the phenotype of osteocalcin-deficient mice, it was also found that osteocalcin has an important role in inhibiting bone formation and in mineral maturation [13].

Other noncollagenous proteins found in bone may also be important in mineral binding, including nucleation and crystal growth. In addition, several of the bone matrix proteins, such as osteopontin, bone sialoprotein, bone acidic glycoprotein, thrombospondin, and fibronectin, contain arginine–glycine–aspartic acid (RGD) sequences. These amino acid sequences, which are characteristic of cell-binding proteins, are recognized by a family of cell membrane proteins known as integrins. The integrins span the cell membrane and provide a link between the extracellular matrix and the cytoskeleton of the cell. Integrins on osteoblasts, osteoclasts, and fibroblasts provide a means for anchoring these cells to the extracellular matrix. Once anchored, the cells are then enabled to express their phenotype and conduct the types of activities that characterize their functions [14].

Growth factors and cytokines such as transforming growth factor-β (TGF-β), insulin-like growth factor (IGF), osteoprotegerin (OPG), interferon-γ, the tumor necrosis factors (TNFs), the interleukins, and the bone morphogenetic proteins (BMPs 2–10) are present in very small quantities in bone matrix. Such proteins have important effects regulating bone cell differentiation, activation, growth, and turnover (see Chapter 12, Komm). It is also likely that these growth factors serve as coupling factors that link the processes of bone formation and bone resorption (Table 1-1).

B. Inorganic Phase

Bone mineral is not pure hydroxyapatite. The small plate-shaped (20–50 nm long, 15 nm wide, and 2–5 nm thick) apatite crystals contain impurities, most notably carbonate in place of the phosphate groups.

TABLE 1-1   Noncollagenous Proteins of the Extracellular Matrix

Structural matrix proteins

Osteocalcin	Restricted to the osteoblast lineage. Vitamin K dependent. May regulate osteoclasts and their precursors.
Osteopontin	Expressed by a variety of cells. Highly expressed in bone and inflammatory tissue. Contains an RGD sequence. Supports osteoblast attachment to bone. Member of the small integrin-binding ligand N-linked glycoprotein (sibling) family. Binds and activates MMP-3.
Bone sialoprotein	Made by osteoblasts and hypertrophic chondrocytes. May initiate mineralization. Supports cell attachment. Binds Ca+ with a high affinity. Member of the sibling family. Binds and activates MMP-2.
Decorin	Also known as chondroitin sulfate proteoglycan I. Regulates collagen fibrillogenesis and TGFβ1 activity. Binds to fibrinogen.
Biglycan	Also known as chondroitin sulfate proteoglycan II. Involved in the regulation of fibrillogenesis. Modulates BMP-2 induced osteogenesis.
Osteonectin	Expressed in a variety of connective tissues. Strong affinity for Ca+. May play a role in matrix mineralization.

Enzymatic matrix modifiers

MMPs	The matrix metalloproteinases (MMPs) includes collagenases (MMP-1 and -13) and gelatinases (MMP-2 and -9). MMPs are required for collagen degradation. Most are expressed in mature chondrocytes and osteoblasts.
TIMPS	Tissue inhibitors of MMPs (TIMPs) are the inhibitors of MMP activity.
Lysyl oxidase	Copper-dependent extracellular enzyme that catalyzes oxidative deamination of elastin and collagen precursors leading to the formation of a mature ECM.
Stromelysin	Member of the MMP family (MMP-3). Degrades most components of the ECM. Activates other MMPs.

Bone morphogens

TGFβ superfamily	The transforming growth factor β (TGFβ) superfamily of morphogens include TGFβ1-3, the bone morphogenic proteins (BMPs), and the growth and differentiation factors (GDFs). This family of morphogens regulates most steps in chondrogenic, osteogenic, and osteoclastogenic cellular differentiation.
FGFs	Fibroblast growth factors 1 and 2 have angiogenic properties. FGFs promote cellular proliferation.
PDGFs	Platelet-derived growth factors exist in three forms (AA, AB, BB). PDGF is associated with mesenchymal cell chemotaxis and proliferation.

The concentration of carbonate (4–6%) makes bone mineral similar to a carbonate apatite known as dahllite. Other documented substitutions are potassium, magnesium, strontium, and sodium in place of the calcium ions and chloride and fluoride in place of the hydroxyl groups [15]. These impurities reduce the crystallinity of the apatite [16], and in doing so may alter certain properties such as solubility [17]. The solubility of bone mineral is critical for mineral homeostasis and bone adaptation.

The crystal size and crystallinity of bone mineral are altered with certain diseases and therapies. For example, crystal size is decreased with Paget's disease [18] and diabetes [19], but increased in osteopetrotic individuals [20] and with bisphosphonate treatment [21]. Whether osteoporosis is associated with abnormal crystal size or crystallinity is the subject of some controversy [22].

C. Organization of Bone

The skeleton is composed of two parts: the axial skeleton, which includes the bones of the head and trunk, and the appendicular skeleton, which includes all of the bones of the limbs and pelvic girdle. The standard example used in discussions of the macroscale structure of whole bones is the long bone. Long bones such as the tibia, femur, and humerus are divided into three parts: the epiphysis, metaphysis, and diaphysis (Figure 1-2). The *epiphysis* is found at either end of the bone and develops from a center of ossification that is distinct from the rest of the long bone shaft. It is separated from the rest of the bone by a layer of growth cartilage known as the *physis*. The *metaphysis* is the region between the physis and the central portion of the long bone (known as the *diaphysis*). From a structural perspective, the metaphysis is the region of transition from the wider epiphysis to the more slender diaphysis.

Membranes on both the outer and inner surface of the whole bone play important roles in bone modeling and remodeling, as well as in fracture healing. The periosteum lines the outer surface of nearly the entire long bone. It is not present on the articulating surfaces and at ligament and tendon insertion points. The periosteum is composed of two layers: an outer fibrous layer that is in direct contact with muscle and other

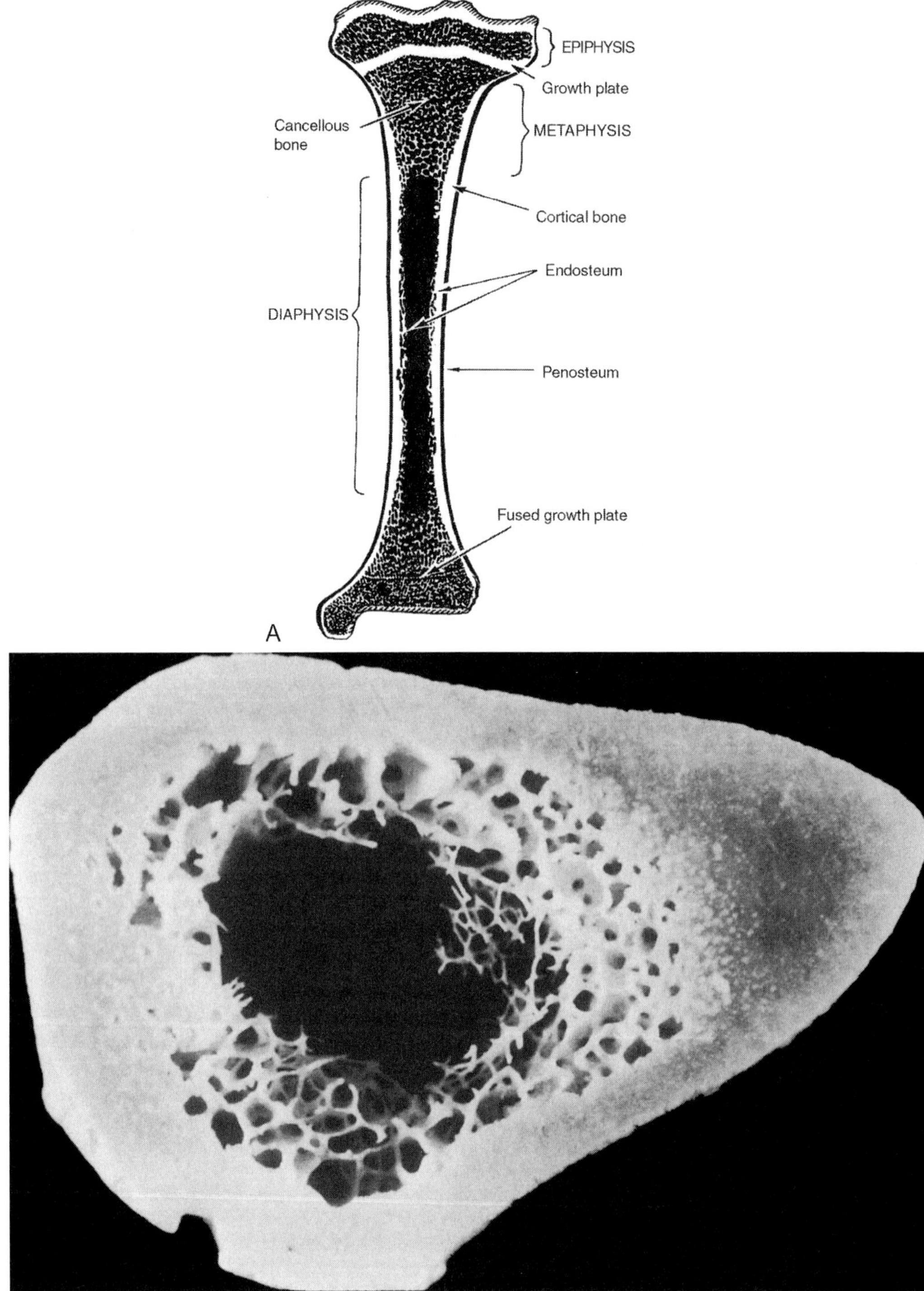

FIGURE 1-2   (A) Sketch of a longitudinal cross-section of a long bone. Reprinted with permission from [102]. (B) Cross-section of the mid-diaphysis of the tibia.

soft tissues, and an inner layer, known as the cambium layer. Whereas the outer layer is much like a sheath of fibrous connective tissue, the cambium layer is populated by uncommitted progenitors of osteoblasts and chondrocytes (Figure 1-3). Through this pool of precursor cells, the periosteum contributes to appositional bone growth during bone development and is responsible for the expansion of the diameters of the

FIGURE 1-3   Magnified view of the periosteum of a long bone. The darker staining tissue at the lower portion of the figure is mineralized cortical bone. Above this is the periosteum, which consists of two layers. The outer layer contains elongated fibroblast-like cells embedded in a fibrous-like tissue. The inner layer, known as the cambium layer, is a loose connective tissue populated by osteoblast and chondrocyte precursors.

long bones with aging. The endosteum lines the inner surfaces of the long bone and consists of bone surface cells, including osteoblasts and bone lining cells.

The building block of bone tissue is the mineralized collagen fibril ($\sim$0.1–3 µm in diameter). These fibrils are arranged either as a collection of randomly oriented fibrils known as woven bone (Figure 1-4) or as aligned in thin sheets called lamellae, which are then stacked in a plywood-type arrangement known as lamellar bone (Figure 1-5). Woven bone is considered immature or primitive bone and is normally found in the embryonic and newborn skeletons, in fracture callus, and in some metaphyseal regions of the growing skeleton. Given that fracture healing and skeletal growth are scenarios in which rapid deposition of bone tissue is advantageous, it is perhaps not surprising that woven bone is laid down relatively quickly (as much as 4 µm per day compared to 1 µm per day for lamellar bone). Woven bone is also found in certain bone tumors, in patients with osteogenesis imperfecta, and in patients with Paget's disease. Lamellar bone is the more mature form of bone tissue that results from the remodeling of woven bone or preexisting lamellar tissue. Lamellar bone begins to develop in the human skeleton at approximately 1 month of age, and by the age of 4, most of the bone in the body is lamellar.

In addition to the difference in fibril arrangement, woven and lamellar bone differ somewhat in composition. As compared to lamellar bone, woven bone has a smaller average apatite crystal size and higher cell density, and the distribution of osteocytes appears random rather than closely associated with the mineralized fibril structure (Figures 1-4 and 1-5). Newly formed woven bone is not as highly mineralized as lamellar bone, although the opposite is true when comparing the final degree of mineralization in these two types of tissues. The differences in composition and structure lead to differences in the mechanical behavior. Due to the random orientation of the fibrils, woven bone is more isotropic than lamellar bone; i.e., its mechanical properties such as stiffness and strength do not depend on the direction in which the forces are applied. In contrast, the stiffness and strength of individual lamellae are greatest in the direction of the fibrils. Depending on the distribution of fibril orientation throughout a region of lamellar bone, however, the stiffness and strength of lamellar bone can range from anisotropic (direction-dependent) to nearly isotropic.

In both woven and lamellar bone, the osteocytes reside in small ellipsoidal holes (5 µm minor diameter; 7–8 µm major diameter) called lacunae (Figure 1-6). In lamellar bone, the lacunae are located along the interfaces between lamellae. There are about 25,000 lacunae

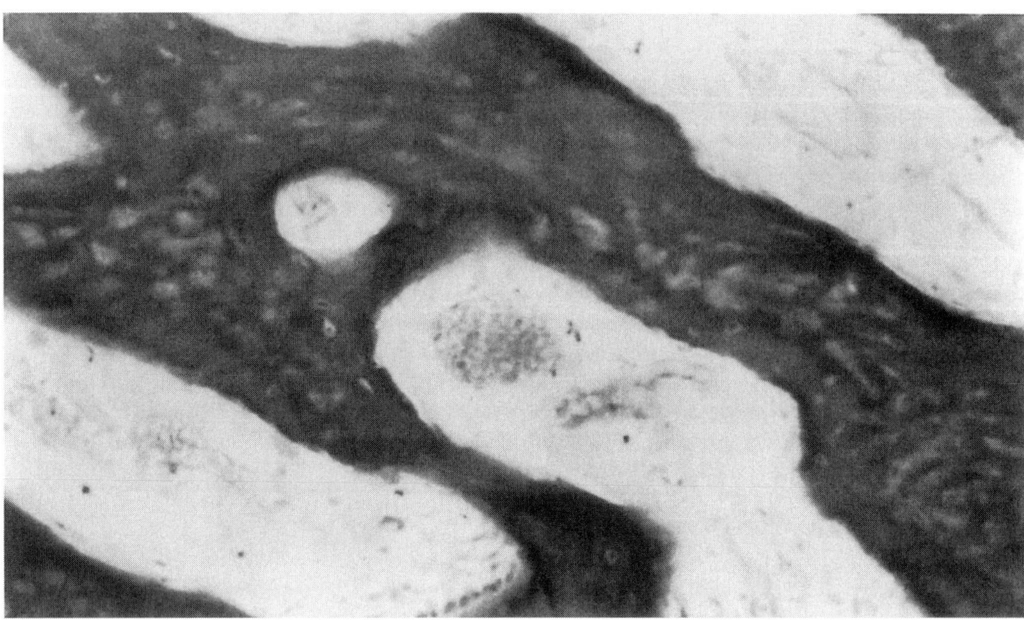

FIGURE 1-4   Woven bone. Note the area of active bone formation (top) and the lack of any particular alignment of the collagen fibrils.

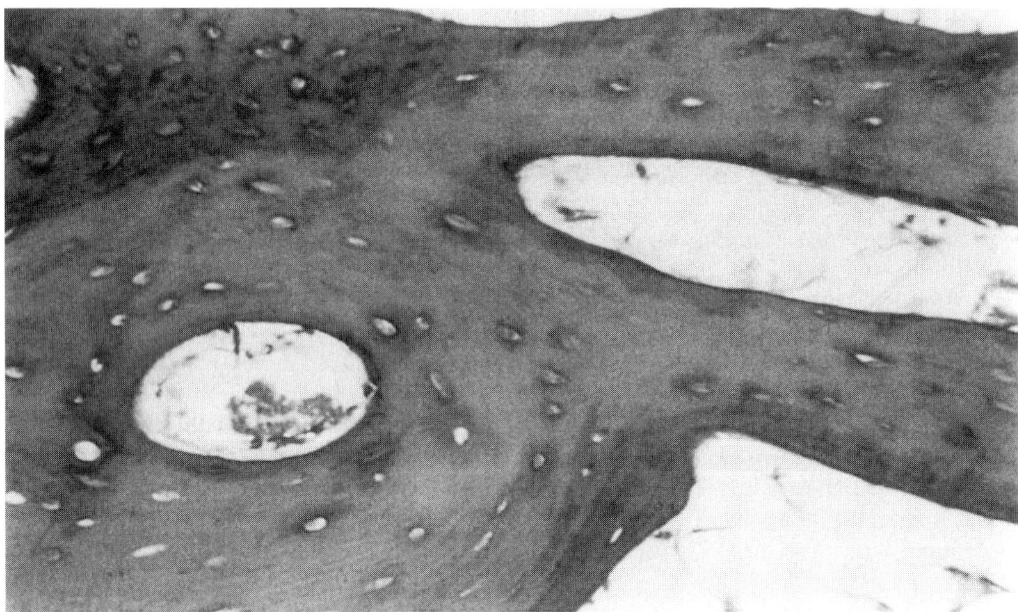

FIGURE 1-5   Lamellar bone. Note the well-delineated orientation of the collagen fibrils and coordinated arrangement of the cells.

per mm^3 in bone tissue, and this number decreases with age [23, 24], although it is not clear if it is further altered with diseases such as osteoporosis [25]. Each osteocyte has dendritic processes that extend from the cell through tiny ($\approx 0.5\,\mu m$ diameter, $3–7\,\mu m$ long) channels called canaliculi, to meet at cellular gap junctions with the processes of surrounding cells. There are about 50–100 canaliculi per single lacuna and about one million per mm^3 of bone tissue. The lacunar-canalicular network may play a central role in bone mechanotransduction.

Both woven and lamellar bone can occupy fairly large volumes, extending uniformly throughout volumes as large as several cubic millimeters. In particular, lamellar bone is found in the long bone diaphysis as large concentric rings of lamellae in the outer 2–3 mm of the circumference. However, lamellar bone is also commonly

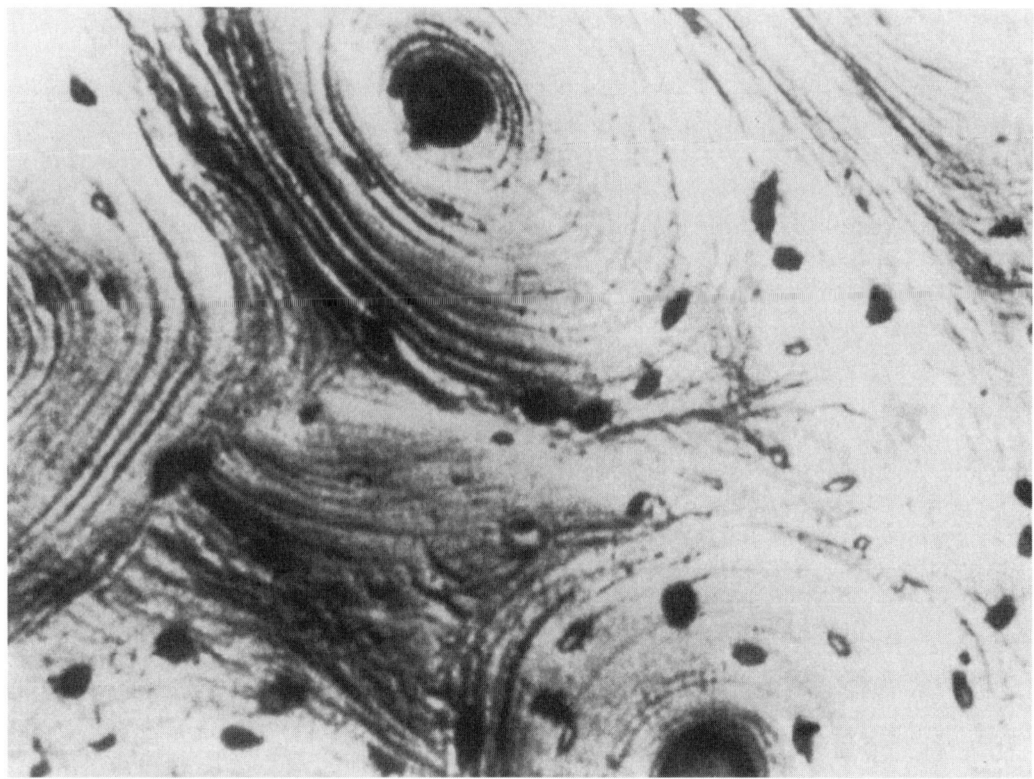

FIGURE 1-6   Scanning electron micrograph of cortical bone showing individual secondary osteons, surrounded by lamellar bone. Osteocytes are housed in the small ellipsoidal lacunae, whose locations are closely associated with the lamellar interfaces.

arranged in smaller cylindrical structures called *secondary osteons* or Haversian systems. These osteons are termed secondary because they are formed through bone remodeling, replacing the previous primary bone tissue. Their diameter and length (typically 200 µm and 1–3 mm, respectively) are determined by the diameter and length of the cutting cone, which is described in the next section on bone remodeling. Each osteon consists of 10–30 concentric rings of lamellae that surround a central cavity, the *Haversian canal*, containing one or more blood vessels and nerves [26] (Figure 1-6). A second type of canal, the Volkmann's canal, runs transverse to the osteonal axis, providing a radial path for blood flow through the whole bone. The outer surface of the osteon is lined with a thin (1–2 µm) layer, known as the cement line, consisting of calcified mucopolysaccharides and very little collagen [27]. In the diaphysis, secondary osteons are typically oriented such that their longitudinal axis is aligned with the diaphyseal axis, although evidence exists that in some bones, the osteons loosely spiral around the diaphyseal axis [28, 29]. Although these osteons are often viewed in cross-section, it is important to note that in three dimensions, the osteon is an irregular, anastomosing cylinder.

Most vessels in Haversian and Volkmann's canals have the ultrastructural features of capillaries, although some smaller-sized vessels may resemble lymphatic vessels. When examined histologically, these small vessels contain only precipitated protein; their endothelial walls are not surrounded by a basement membrane. The basement membrane of capillary walls may function as a rate-limiting or selective ion-limiting transport barrier, because all material traversing the vessel wall must go through the basement membrane. The presence of this barrier is particularly important in calcium and phosphorous ion transport to and from bone. The capillaries in the central canals are derived from the principal nutrient arteries of the bone: the epiphyseal and metaphyseal arteries. The vascular system is critical for bone function, not only with respect to nutrient supply but also as a source of cells of both the osteoclast and osteoblast lineage [30, 31].

At the scale of 1–10 mm, there are two types of bone: trabecular bone (also known as cancellous or spongy bone) and cortical bone (also known as compact or dense bone). Trabecular bone is found principally in the axial skeleton and in the metaphyses and epiphyses of long bones (Figure 1-2). It is a highly porous structure consisting of a network of rod- and plate-shaped trabeculae surrounding an interconnected pore space that is filled with bone marrow (Figure 1-7).

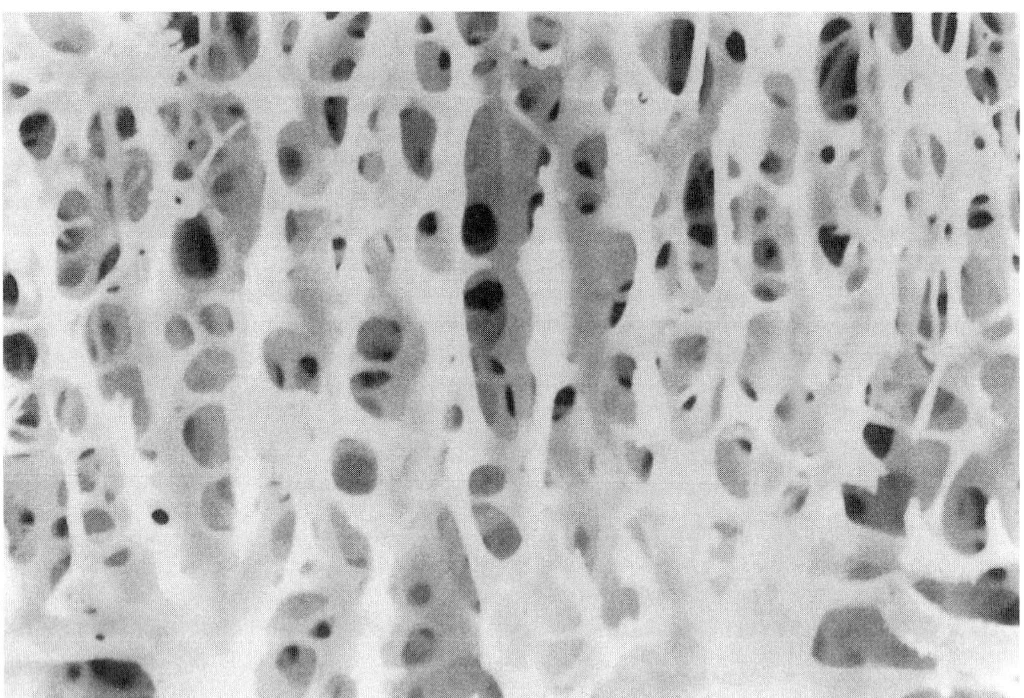

FIGURE 1-7   Trabecular bone. The field of view is approximately 15 mm in width.

Trabeculae range in thickness from 50 to 300 µm and are composed almost exclusively of lamellar bone arranged in packets that are sometimes referred to as hemiosteons. However, the thicker trabeculae can contain secondary osteons, presumably because their thickness is such that nutrient transport via the lacunar-canalicular network alone is insufficient. In the mature human skeleton, cortical bone consists largely of secondary osteons and, to a lesser extent, circumferential lamellae that ring the outer surface of the diaphysis and a type of lamellar bone known as interstitial bone (Figure 1-8). Interstitial bone is merely composed of portions of secondary osteons that were not removed by a cutting cone during remodeling. Both the metaphyses and epiphyses of long bones have a thin shell of cortical bone surrounding the trabecular compartment, and the diaphyses are entirely cortical (Figure 1-2).

The distinction between cortical and trabecular bone can be made largely on the basis of porosity. The porosity of cortical bone ranges only 5–20% and is due to the Haversian and Volkmann's canals and, to a lesser extent, the lacunar and canalicular spaces. Trabecular bone has another scale of porosity due to the marrow space; typical spacing between trabeculae ranges from 100 to 500 µm. The porosity of trabecular bone can range from 40% in the primary compressive group of the femoral neck to more than 95% in the elderly spine.

Porosity is the major determinant of the stiffness and strength of trabecular bone [32, 33].

In addition to porosity, the three-dimensional structure of trabecular bone, known as the *trabecular architecture*, can vary tremendously among anatomic sites and with age. Trabecular bone from the vertebral body tends to be predominantly rod-like, while that from the proximal femur contains a more balanced mixture of rods and plates (Figure 1-9). Quantitative descriptors of trabecular architecture such as trabecular thickness and trabecular spacing contribute somewhat independently of porosity to trabecular bone stiffness and strength [34].

With age and also with disuse, trabeculae become progressively thinner and can become perforated by resorption cavities. In certain anatomic sites such as the vertebral body and proximal tibia, age-related changes in trabecular architecture include an increase in the anisotropy of the trabecular structure (Figure 1-10) [35, 36]. With the overall decrease in bone mass with age, this increase in anisotropy helps to preserve the load-carrying capacity of trabecular bone along its main "grain" axis, but at the necessary expense of the load-carrying capacity in other directions. Nonhabitual loading conditions such as impact after a fall can subject trabecular bone to such off-axis loads. Thus, the risk of fracture due to off-axis loads can increase with age to

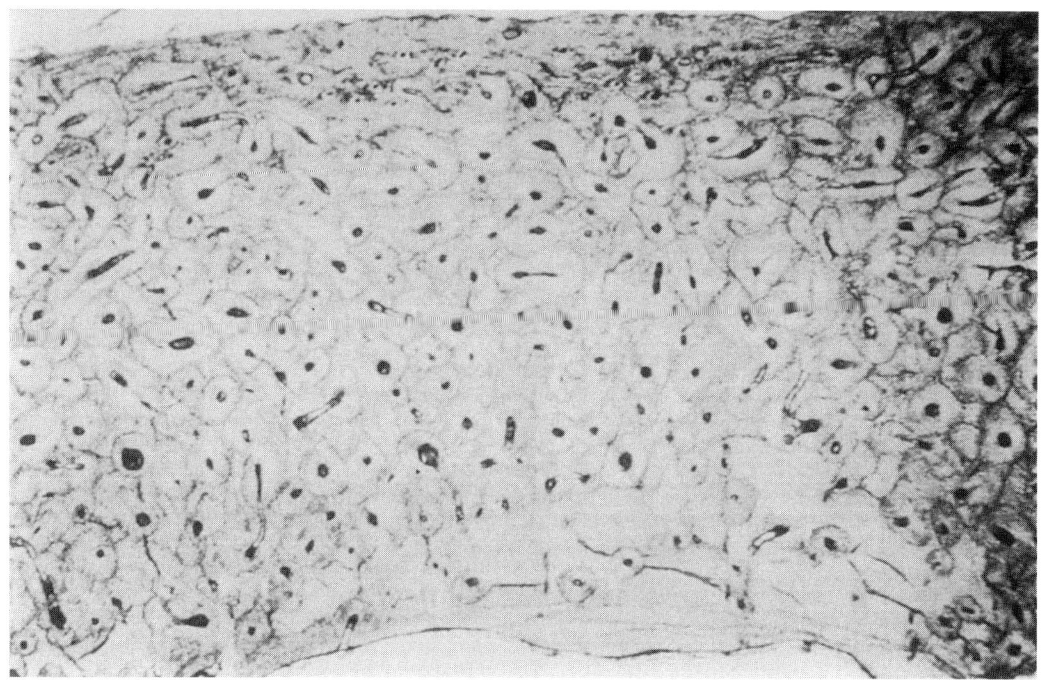

FIGURE 1-8   A transverse section of a long bone diaphysis showing circumferential lamellar bone, secondary osteons, and interstitial bone.

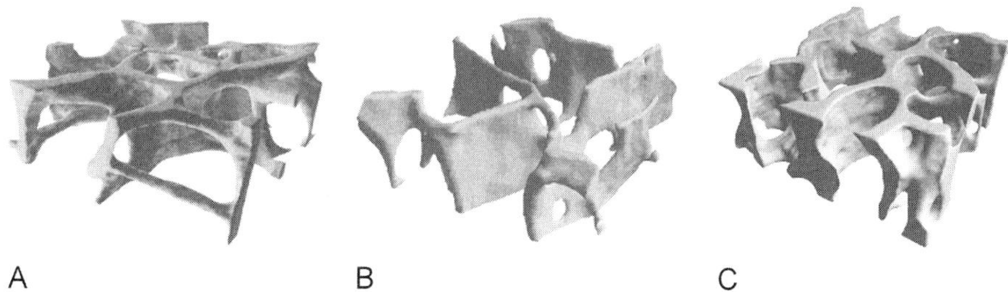

FIGURE 1-9   High-magnification, three-dimensional renderings of trabecular bone from the human (A) vertebra, (B) femoral greater trochanter, and (C) femoral neck. Each volume is $3 \times 3 \times 1 \, mm^3$. From [33].

a greater extent than the decrease in bone mass alone would suggest.

III. CELLULAR COMPONENTS OF BONE

A. Bone Cells

Bone metabolism is regulated by multiple environmental signals including chemical, mechanical, electrical, and magnetic. The local cellular compartment of the bone responds to these environmental signals by modulating the balance between new bone formation and the local resorption of older bone (i.e., remodeling). Three cell types are typically associated with bone homeostasis: osteoblasts, osteocytes, and osteoclasts. These three cell types are derived from two separate stem cell lineages—the mesenchymal lineage and the hematopoietic lineage—underscoring the unique regulation of bone homeostasis and the intimate interactions between the immune system and bone.

B. Mesenchymal Lineage Cells

Bone formation, both embryonic and postnatal, is carried out by the mesenchymal lineage osteoblast. As noted previously, osteoblasts produce the protein matrix of bone made up of type I collagen and several noncollagenous proteins. This protein matrix, referred to as the osteoid, creates a template for mineralization and production of the mature bone. In addition to

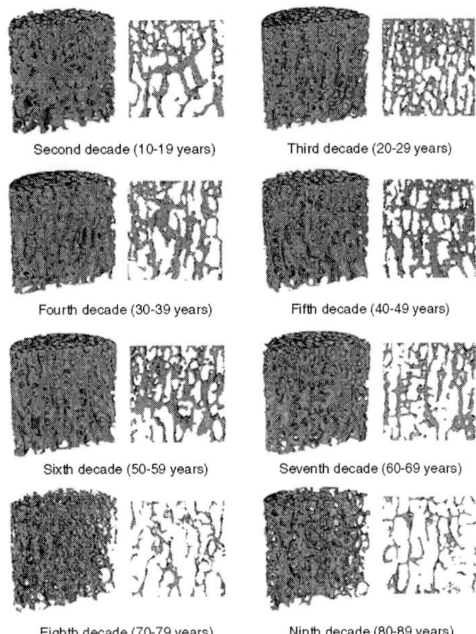

FIGURE 1-10   High-resolution, three-dimensional renderings and two-dimensional cross-sections of trabecular bone from the human proximal tibia in the 2nd–9th decades of life. With age, bone density decreases and an overall deterioration of the trabecular structure occurs. In addition, the tissue becomes more preferentially aligned with the diaphyseal axis of the tibia (here, the vertical direction). This preferential alignment results in anisotropy, or directional dependence, of the structure. The main direction of alignment in the structure is often referred to as the "grain" axis. Reprinted with permission from [36].

bone formation, osteoblasts assist with the initiation of bone resorption by secreting factors that recruit and promote the differentiation of monocytic lineage cells into mature osteoclasts and also by producing neutral proteases that degrade the osteoid and prepare the bone surface for osteoclast-mediated remodeling.

Osteoblasts are derived from mesenchymal stem cells, pluripotent cells that can differentiate into a variety of cell types including myoblasts, adipocytes, chondrocytes, osteoblasts, and osteocytes. The specific lineage selection of an individual mesenchymal stem cell involves a number of coordinated lineage selection steps and the actions of a number of transcriptional regulators whose activities are modulated in response to the local microenvironment (Figure 1-11). Two transcription factors have been demonstrated to be required for osteoblast formation and differentiation: Runx2 and Osterix [37]. The regulatory activity of these central osteoblast regulators is modified by cofactors including members of the Dlx (distaless), Msx, and Hox homeodomain gene families and downstream signal transduction mediators such as the TGF-β superfamily-related SMADs. Runx2 is a member of the runt homology domain transcription factors and acts as a scaffolding protein organizing nuclear complexes at discrete sites on the nuclear matrix associated with active gene transcription. Transgenic knockout studies have clearly demonstrated the requirement for

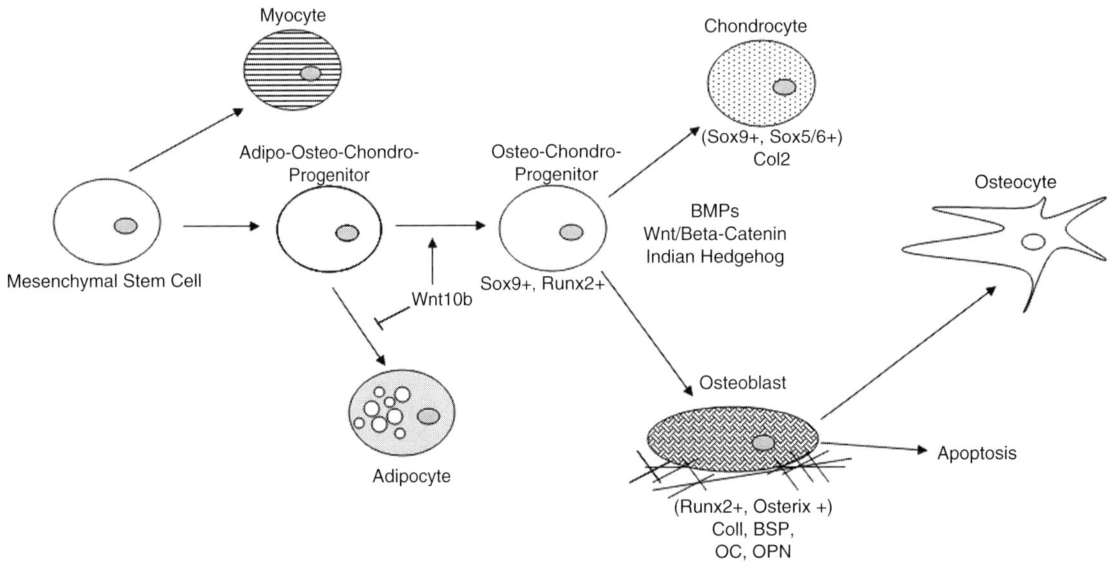

FIGURE 1-11   Graphic representation of the steps involved in osteoblast differentiation from mesenchymal stem cell to matrix expressing mature osteoblast and on to the osteocyte stage.

Runx2 activity for osteoblast differentiation, as these knockout mice produce no bone during embryogenesis [38]. These Runx2 knockout mice lack osteoblasts and display defects in chondrocyte hypertrophy demonstrating the role of Runx2 in both osteoblast differentiation and chondrocyte maturation. Runx2 regulates the expression of many mature osteoblast-related genes including osteocalcin, bone sialoprotein, osteopontin, and collagen type I. The second required transcription factor for osteoblast differentiation is the zinc finger motif containing factor Osterix. Like Runx2 knockout mice, the Osterix knockouts lack embryonic bone formation and osteoblast differentiation [39]. Unlike the Runx2 deficient animals, Osterix knockouts do not display the defects in chondrocyte hypertrophy, and Runx2 expression levels are comparable to controls. Osterix functions downstream of Runx2 activity as Runx2-/- cells express no Osterix. While the mechanism through which Osterix regulates osteoblast differentiation is poorly understood, it has been noted that in the Osterix knockout mice, the pool of Runx2-expressing pre-osteoblasts express several genes associated with chondrogenesis, suggesting Osterix plays a role in stabilizing osteogenic commitment and osteoblast maturation.

The relative expression and activity of Runx2 and Osterix are regulated by the local microenvironment and, more specifically, the locally produced morphogens to which the cells are exposed. Growth factors including members of the fibroblast growth factors (FGFs), insulin-like growth factors (IGFs), transforming growth factor-beta (TGF-β), bone morphogenetic proteins (BMPs), and Wnts have all been demonstrated to play important roles in regulating embryonic osteoblast differentiation. While each of these morphogens is likely to play some role in postnatal osteoblast differentiation, their role in bone homeostasis is less clear. One exception is the recent data demonstrating that Wnt signaling is an important component of the regulation of bone mineral density (BMD) recognized as a result of mutations in humans. The autosomal recessive disorder osteoporosis pseudoglioma (OPPG), characterized by low bone mass, frequent deformations and fractures, and defects in eye vascularization, has been linked to mutations in lipoprotein-related peptide 5, LRP5 [40–42]. LRP5 is a Wnt co-receptor that, along with the Wnt receptor, frizzled, activates canonical Wnt signaling in cells. Children with OPPG have normal endochondral growth and bone turnover, but their trabecular bone volume is significantly decreased [42a]. Furthermore, gain of function experiments in mature mouse models has shown that organisms with a constitutively activated LRP5 mutation exhibit a high bone mass (HBM) [43]. Thus, these data support the conclusion that canonical Wnt signaling is important in the regulation of postnatal bone mass.

The other mesenchymal lineage cell type found in bone is the osteocyte. Osteocytes are predominantly associated with a mechanosensory function in bone and potentially also a role in Ca+ homeostasis. Osteocytes are a type of osteoblast and thus differentiate from the same mesenchymal lineage under the regulation of the same transcription factors discussed previously [44, 45]. Osteocytes, however, escape apoptosis, reduce their production of matrix molecules, and eventually end up encapsulated in the bone matrix. In the bone they are characterized by their long processes that extend through the lacunocanalicular system of the bone. Osteocytes are in fact the most abundant cellular component of mammalian bones, making up 95% of all bone cells. Relative to the other bone cells, osteocytes are long lived, with estimates running as high as 25 years, as compared to osteoblasts, which are estimated in humans to live approximately an average of 3 months [44]. Osteocytes create an interconnected network in bone allowing for intercellular communications between both neighboring osteocytes and the surface-lining osteoblasts. This interconnection between osteocytes allows for the transmission of mechanical and chemical signals across the network through direct transmission of mechanical forces either through the triggering of integrin force receptors, changes in membrane conformation, chemical signals via the gap junctions, or secreted factors that travel through the extracellular fluid of the lacunocanalicular system [44]. This interconnected signaling allows for the adaptation of bone to the external mechanical and chemical inputs that regulate bone homeostasis.

C. The Hematopoietic Cell Lineage

Bone homeostasis involves the constant remodeling and rebuilding of bone, a process that leads to the replacement of 4–10% of bone each year in humans. While the bone formation side of the equation is carried out by the mesenchymal lineage-derived osteoblasts, the remodeling side of the homeostasis equation in bone is carried out by the hematopoietic lineage osteoclast. Osteoclasts play a role in balancing calcium homeostasis with skeletal remodeling. Histologically, osteoclasts are found at the apex of the classical "cutting cones" in cortical bone and in the resorptive cavities known as Howship's lacunae on trabecular bone surfaces undergoing active remodeling. Osteoclasts are multinucleated cells derived from hematopoietic mononuclear cells [46, 47]. In order to remove bone, newly formed osteoclasts become polarized, form a ruffled membrane, and adhere tightly to the bone matrix via an $\alpha_v\beta_3$ integrin mediated binding to the bone surface to form the "sealing zone." The osteoclast then secretes acid via H^+-ATPase (for hydroxyapatite dissolution) and proteases including cathepsin K (for matrix protein digestion) into this closed microcompartment along the bone surface referred to as the hemivacuole, thereby removing the underlying bone. By focusing the secretion of these acids and enzymes, osteoclasts are able to move along a bone surface or into a cutting cone slowly solubilizing bone in a defined area without disrupting the surrounding local microenvironment.

Osteoclasts are members of the hematopoietic cell lineage and are derived from mononuclear/macrophage cells (Figure 1-12). A mature multinucleated osteoclast forms by fusion of cells from the hematopoietic and myelomonocytic origin and is therefore a member of the mononuclear phagocyte series and may be thought of as a specialized type of macrophage [46]. Indeed, the bone resorption process employs some of the same cellular machinery as phagocytosis. The early differentiation stages of osteoclast formation depend on the transcription factor PU.1, which regulates c-fms expression along with the transcription factor src [46, 47]. The expression of c-fms, the M-CSF receptor, is a central component of early osteoclast formation as M-CSF responsiveness is required for both monocyte progenitor proliferation and the expression of the receptor activator of NF-κB (RANK), a critical receptor for osteoclast differentiation. The ligand for RANK (RANKL) is the critical cytokine for the final stages in osteoclast differentiation and a member of the TNF-α family of cytokines. The binding of RANKL to the RANK receptor activates NF-κB signaling leading to the formation of mature multinucleated osteoclasts [48]. The activity of RANKL is balanced by the level of expression of its inhibitor osteoprotogerin (OPG), a soluble RANK decoy receptor. It is the local ratio of RANKL to OPG that ultimately determines if osteoclast formation will occur by regulating the amount of available RANKL. In addition to the regulation of osteoclast formation, osteoclast activity can be regulated as can the life span of an osteoclast. Various cytokines

have been demonstrated to play a role in enhancing osteoclast activity (IL-1 and RANKL itself) and prolong the life span of an osteoclast (IL-1, IL-6, M-CSF, TNF-α, LPS) [46–48]. Thus osteoclast-mediated bone resorption is regulated by many cytokines associated with inflammation that can regulate osteoclast formation, activity, and apoptosis.

IV.   BONE HOMEOSTASIS

A.   Osteoblast-Osteoclast Coupling

Bone homeostasis is maintained by the coordinated actions of osteoblast-mediated formation and osteoclast-mediated bone removal. This coordination is referred to as "coupling." The concept of coupling is based on the idea that osteoblasts influence osteoclast formation and activity, and likewise osteoclasts influence osteoblast differentiation and activity (Figure 1-13). Currently, the majority of our understanding of coupling revolves around the influence of osteoblasts on osteoclast formation. Osteoblasts express the majority of cytokines that regulate osteoclast progenitor differentiation including M-CSF, RANKL, and OPG in bone, the primary cytokines that regulate osteoclast formation [48]. During osteoblast differentiation, the level of expression of these cytokines changes with the immature osteoblast producing the highest levels of M-CSF and RANKL. Thus, as an osteoblast begins to mature into a matrix-producing bone cell, it signals to local osteoclast precursors with RANKL to differentiate, thereby coupling the new bone formation with the recruitment of new osteoclasts for its subsequent remodeling. By coordinating osteoclast differentiation with osteoblast differentiation, the system stays in balance. Conversely, many researchers believe that osteoclasts signal back to osteoblast progenitors through the release of BMPs and other growth factors that promote osteogenesis from the bone matrix as a part of the bone removal process completing the circle [49].

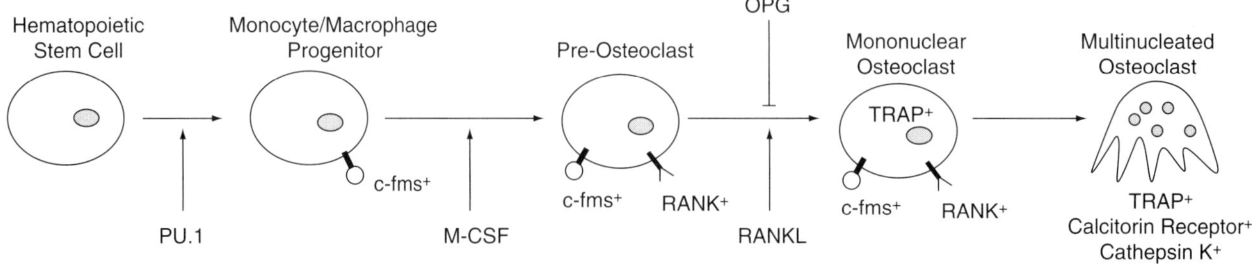

FIGURE 1-12   Graphic representation of the steps involved in osteoclast differentiation from a hematopoietic stem cell to a mature multinucleated osteoclast.

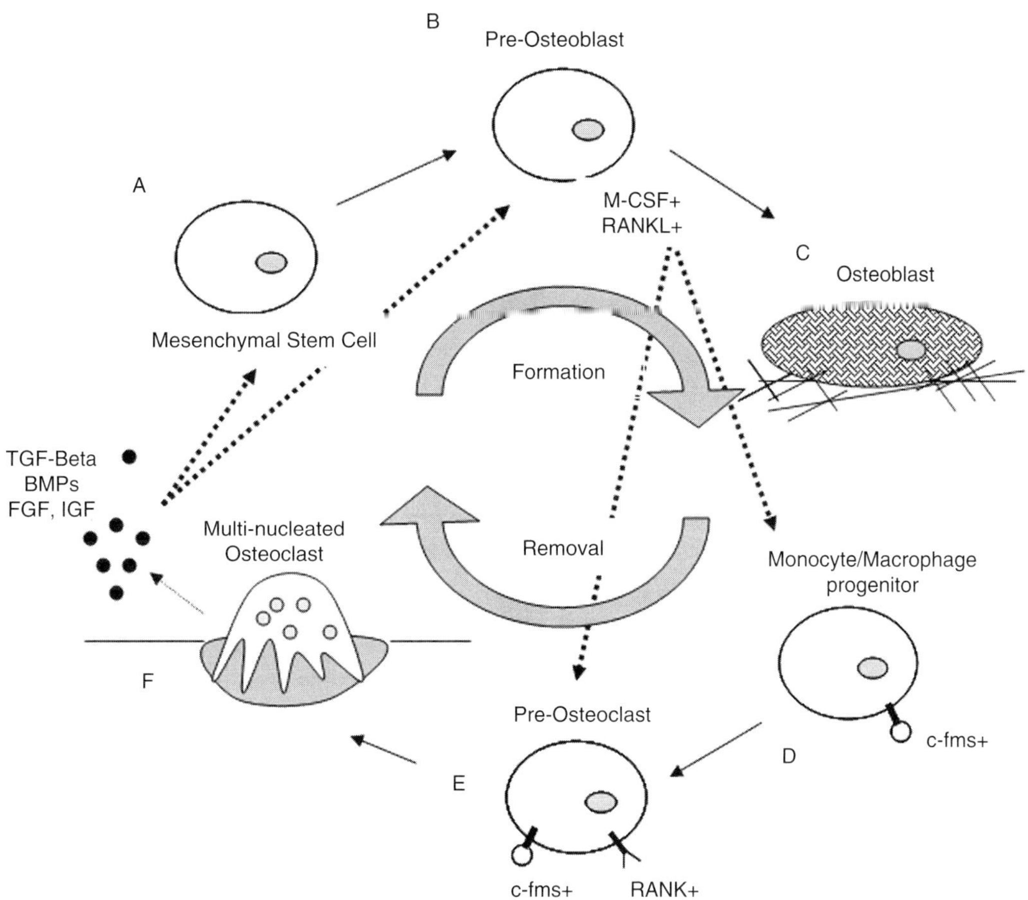

FIGURE 1-13   Graphic representation of the cellular interactions between osteoblast lineage cells (A–C) and osteoclast lineage cells (D–F). Dashed lines indicate cell signaling events important in the coupled differentiation of the respective lineages. Pre-osteoblasts (B) express the majority of M-CSF and RANKL that induce osteoclast differentiation (dashed lines indicating effects on progenitor cells and pre-osteoclast). Conversely, morphogens released from bone during osteoclast-mediated removal (F) influence the differentiation of the mesenchymal stem cell and pre-osteoblast (indicated with dashed line)

While the local interactions between osteoblast and osteoclast regulate the local balance of formation and removal, this system was evolutionarily adapted to provide a means of responding to more global mechanical forces and systemic metabolic requirements. The skeleton supports two major functions for the organism as a whole, including load bearing and mineral storage. Bones adapt to the mechanical forces placed upon them. The interconnected osteocyte network is widely perceived to provide mechanosensory feedback that is communicated to the lining osteoblasts [44]. While a multitude of studies have demonstrated that intracellular responses to mechanical input can include increased cAMP, IP3, intracellular calcium, and activation of MAPK pathway, exactly how the mechanical response is transmitted between cells remains unclear [50].

The second primary function of bone is as a mineral store, and bone remodeling plays an important role in systemic mineral homeostasis, with Ca+ being the primary mineral stored in bone. Systemic Ca+ levels are monitored by Ca+ sensors in the parathyroid gland. As Ca+ levels drop, the parathyroid releases parathyroid hormone (PTH). Systemic PTH leads to increased remodeling and the release of Ca+, bringing levels back up into the optimal range. PTH achieves this increase in remodeling primarily through its actions on the osteoblast. PTH increases the expression of the Notch ligand Jagged1 in osteoblasts [51]. It has been demonstrated that osteoblasts can regulate the expansion of the hematopoietic stem cell niche in bone marrow through a Notch-mediated mechanism, and by increasing Jagged1 expression on osteoblasts, PTH leads to an expansion of the hematopoietic lineage from which the osteoclasts are derived. In addition osteoblasts respond to PTH, as well as interleukin-11, prostaglandin E2 (PGE2), and 1,25(OH)2D, by increasing the expression of RANKL and other osteoclast regulatory cytokines leading to increased osteoclast differentiation and activity and decreased osteoclast apoptosis [52]. Finally, PTH induces increased neutral protease expression by osteoblasts and causes osteoblasts to contract away from the bone surface, exposing the bone and providing the osteoclasts access to the surface. Consequently, systemic release of PTH can induce increased bone resorption and Ca+ release by enhancing osteoclast formation and activity, by increasing osteoblast-mediated preparation of the bone surface by neutral protease secretion, and by providing the osteoclasts access to the bone surface by causing contraction of lining osteoblasts away from the bone.

B. Bone Remodeling

Cortical bone constitutes approximately 80% of the skeletal mass and trabecular bone approximately 20%. Bone surfaces may be undergoing formation or resorption, or they may be inactive. These processes occur throughout life in both cortical and trabecular bone. Bone remodeling is a surface phenomenon, and it occurs on periosteal, endosteal, Haversian canal, and trabecular surfaces. The rate of cortical bone remodeling, which may be as high as 50% per year in the midshaft of the femur during the first 2 years of life, eventually declines to a rate of 2–5% per year in the elderly. Rates of remodeling in trabecular bone are proportionally higher throughout life and may normally be 5–10 times higher than cortical bone remodeling rates in the adult [53].

Historically, bone histologists have described the skeleton as being composed of individual structural units or bone metabolic units (BMU) [17]. The BMU of cortical bone is the osteon or Haversian system. As described previously, the canals are connected to each other by transverse Volkmann's canals and periodically either divide or reunite to form a branching network. Osteons form approximately two-thirds of cortical bone volume, a proportion that falls with age, with the remainder consisting of interstitial bone representing the previous generation of osteons. There are also subperiosteal and subendosteal circumferential lamellae.

In trabecular bone, the BMU is the hemiosteon. In two-dimensional sections, these are shaped like thin crescents about $600\,\mu m$ long and about $60\,\mu m$ in depth. Three-dimensionally, these BMUs are actually larger than they appear in two-dimensional histological sections with prolongations in different directions that interlock with adjacent BMUs [54]. These BMUs follow the same shape as the trabecular surface, most of which are concave toward the marrow.

Under normal conditions, the remodeling process of resorption followed by formation is closely coupled and results in no net change in bone mass. As such,

the BMU consists of a group of cells that participate in remodeling in a concerted and coordinated fashion. Cortical bone remodeling proceeds via cutting cones and is similar to processes in other hard biological tissues. Cuttings cones, or sheets of osteoclasts, bore holes through the hard bone, leaving tunnels, which appear in cross-section as cavities. The head of the cutting cone consists of osteoclasts that resorb the bone. Following closely behind the osteoclast is a capillary loop and a population of endothelial cells and perivascular mesenchymal cells that are progenitors for osteoblasts and soon begin to lay down the osteoid and refill the resorption cavity. By the end of the process, a new osteon will have been formed. Trabecular bone remodeling occurs on the surface of bone at specific sites. These areas are then filled in with newly formed osteoid. The mechanisms that control the activity and site specificity of this process are unknown.

According to the model proposed by Parfitt, the normal remodeling sequence in bone follows a scheme of quiescence, activation, resorption, reversal, formation, and return to quiescence. In the adult, approximately 80% of trabecular and approximately 95% of intracortical bone surfaces are inactive with respect to bone remodeling [55, 56]. The surface of bone is covered by a layer of thin, flattened lining cells approximately $15\,\mu m$ in diameter, which arise by terminal transformation of osteoblasts. Between these lining cells and bone is a layer of unmineralized osteoid. These lining cells have receptors for a variety of substances, which are important for initiating bone resorption (PTH, PGE2), and may respond to such substances by resorbing this surface osteoid, which is covering the bone. In doing so, mineralized bone will be exposed, and the activation sequence of bone remodeling may be initiated.

The conversion of a small area of bone surface from quiescence to activity is referred to as activation. The cycle of this response begins with the recruitment of osteoclasts, followed by the initiation of mechanisms for their attraction (chemotaxis) and attachment to the bone surfaces. Several known growth factors may be active in promoting chemotaxis. In addition, several proteins are known to be attachment factors for osteoclasts, such as those that contain the RGD amino acid sequences as noted earlier. Osteopontin, osteocalcin, and osteonectin may be important proteins in this process. In the adult skeleton, activation occurs about every 10 seconds. For intracortical remodeling, osteoclast precursors travel to the site of activation via the circulation, gaining access to the site by either a Volkmann or Haversian canal. In trabecular remodeling, activation occurs at sites that are apposed to bone marrow cells.

In cortical bone, the osteoclast and the cutting cone travel at a speed of about 20 or $40\,\mu m$ per day, roughly parallel to the long axis of the bone and about 5–10 µm per day perpendicular to the main direction of advance [57]. In trabecular bone, osteoblasts erode to a depth of about two-thirds of the final cavity; the remainder of the cavity is eroded more slowly by mononuclear cells [58].

The reversal phase is a time interval between the completion of resorption and the initiation of bone formation at a particular skeletal site. Under normal conditions, it lasts about 1–2 weeks. The appearance of new osteoblasts at the base of the resorption cavity depends on chemotaxis for these osteoblasts and their progenitors, as well as conditions that stimulate proliferation. Hence, chemotaxis, attachment, proliferation, and differentiation occur in a stepwise and concerted fashion in order for new bone formation ultimately to take place.

V. BONE MECHANICS

The hierarchical structure of bone, together with evidence that changes in structure can occur with age and disease at many different levels of this hierarchy, renders bone a classic subject for study of mechanical behavior at multiple length scales. In answering a given research question, one may be interested in measuring the mechanical properties of a whole bone, trabecular or cortical bone, single osteons or lamellae, individual mineralized collagen fibrils, or several of the above. Tests performed at each of these length scales can provide insight into bone mechanical properties and, in particular, effects of various age-, disease-, and treatment-related changes in these properties. However, because of the hierarchical complexity of bone structure, it is at best difficult and sometimes impossible to extrapolate across different length scales based only on results from one type of test. For example, a whole bone may be stronger simply because it is larger, not because the tissue itself is any stronger. Similarly, a higher degree of mineralization of the collagen fibrils may not produce a stiffer tissue if those fibrils are not particularly well organized. These examples are just two of the many that motivate consideration of structure–function relationships in bone from the macroscale to microscale to nanoscale.

A. Mechanical Behavior of Whole Bones

The principal advantage of mechanical tests performed on whole bones is that these tests are highly relevant clinically, provided that the manner in which the loads are applied during the test approximates well the *in vivo* loads in the clinical situation of interest. For long bone diaphyses, common *in vivo* loading conditions and *in vitro* mechanical testing configurations include compression, torsion, and bending. Less common is tension. Each of these loading modes results in a characteristic fracture pattern (Figure 1-14). For studies focused on hip fractures, loads are applied *in vitro* in order to simulate gait or fall loading conditions. For the vertebrae, common loading modes include compression and compression combined with bending (specifically, anterior or posterior flexion). Although simple compression and flexion are likely simplified representations of the loads to which vertebrae and motion segments are subjected *in vivo*, these idealized loading conditions do produce clinically observed fracture patterns, including crush, endplate, and wedge fractures (Figure 1-15).

The stiffness and strength of a whole bone are structural properties, not material properties. Structural properties depend on the size and shape of the whole bone as well as on the mechanical properties of the bone tissue itself (material properties). Therefore, quantifying the size and shape of the whole bone can provide some insight into the respective contributions of geometry versus material properties. Principles of engineering mechanics stipulate that the axial stiffness, either in compression or tension, of a structure is proportional to the cross-sectional area, while the bending and torsional stiffnesses of beam-like structures (such as diaphyses) depend on how the material (tissue) is distributed around the axis of bending or torsion (Figure 1-16). Material distributed farther away from these

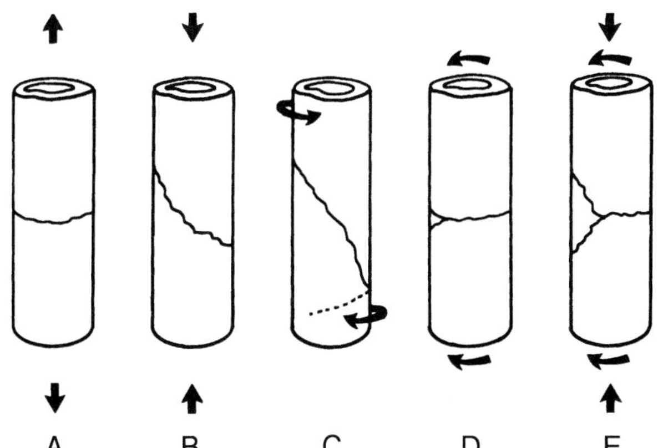

FIGURE 1-14   Fracture patterns in a cylindrical section of bone subjected to different loading configurations. (A) Pure tensile loading produces a transverse fracture. (B) Pure compressive loading produces an oblique fracture. (C) Torsional loading produces a spiral fracture. (D) Bending produces a transverse fracture with a small fragment on the compressive side. (E) Bending superimposed with compression produces a transverse fracture with a larger fragment on the concave side.

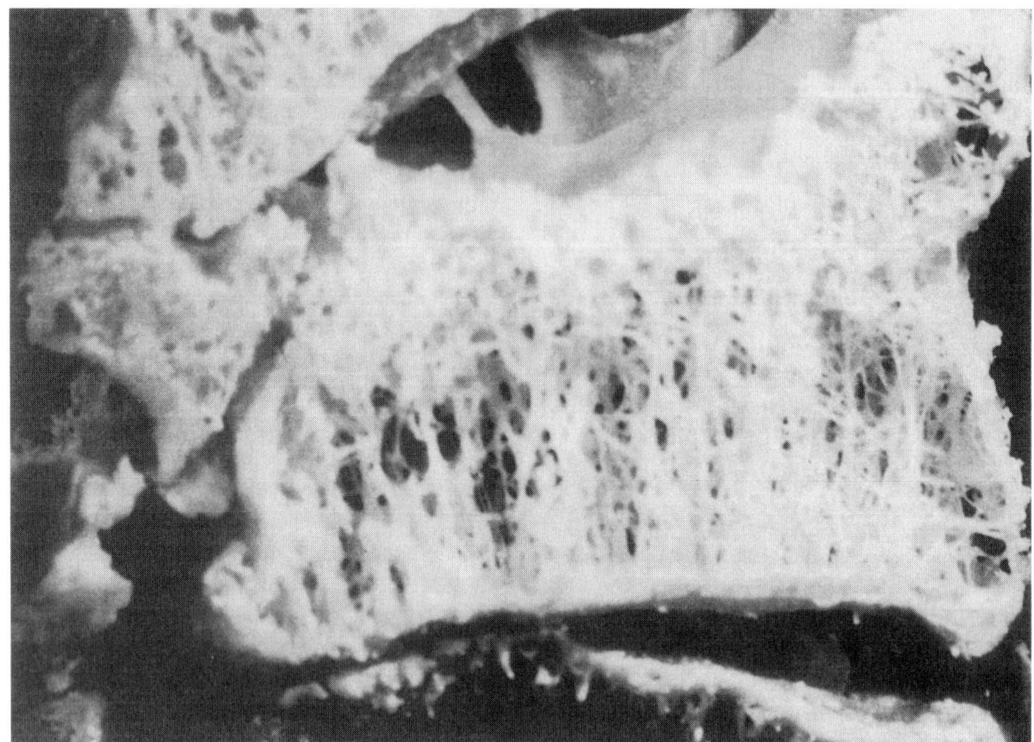

FIGURE 1-15   Sagittal section of a vertebral compression fracture.

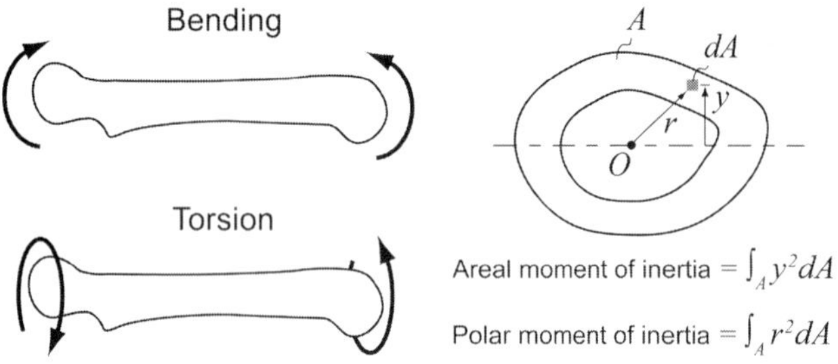

FIGURE 1-16   The bending stiffness of a structure such as a long bone diaphysis is proportional to the areal moment of inertia. If the diaphyseal cross-section is circular, then the torsional stiffness is proportional to the polar moment of inertia; otherwise, this proportionality is only approximate. These moments of inertia are geometric quantities that describe how the tissue is distributed with respect to the axis of bending (shown here as the dotted line on the diaphyseal cross-section) or the axis of torsion (the line that passes through point O and that is directed out of the plane of the figure).

axes contributes more to resisting the applied bending and torsional loads than does material near the axes. Two geometric properties, the *areal moment of inertia* (also known as the cross-sectional moment of inertia) and *polar moment of inertia*, quantify this distribution in manners relevant for bending and torsion, respectively. These geometric properties can change with physical activity and with aging. For example, with age, both the outer and inner diameter of the diaphysis increase due to a combination of endosteal resorption and periosteal bone formation. The net result is a thinner cortex and smaller cross-sectional area, but also an increase—or at least less of a decrease—in areal moment of inertia and polar moment of inertia [59, 60]. The changes in moment of inertia can serve to mitigate the mechanical consequences of the age-related decline in bone mass. Comparisons of cross-sectional geometry in femoral diaphyses of different inbred mouse strains provide a powerful illustration of the independent contributions of tissue properties and bone size and structure to the mechanical properties of whole bones [61–63].

If the bone is straight, prismatic (the cross-sectional geometry does not change along the length of the structure), and if it is of uniform composition, it is straightforward to calculate the Young's modulus or shear modulus (defined in the next section) of the bone tissue from the results of a test performed on the whole bone [64]. Of course, none of these three descriptors is accurate for vertebral bodies and diaphyses. For the latter, one can calculate an effective elastic modulus of the tissue if the true cross-sectional geometry and its variation along the diaphyseal axis are included in the calculations. However, without accounting for the true geometry of the specimen, substantial errors in the modulus can result [65].

B.   Mechanical Behavior of Bone Tissue

Bone tissue is subjected to a wide variety of mechanical demands during activities of daily living and during nonhabitual scenarios such as trauma. Experiments on the mechanical behavior of bone tissue determine the ability of the tissue to meet those demands. In working with bone tissue, one can avoid the confounding influences of specimen size and shape by preparing tissue samples of regular geometry such that the geometry can be easily accounted for. With this approach, the applied loads can be expressed easily in terms of *stress* rather than force, and the deformation that the specimen undergoes as a result of the applied loads can be expressed in terms of *strain* rather than displacement. Stress is the force per unit area acting on a specimen and thus quantifies the intensity of the force. For a specimen of regular geometry, it is easily calculated by dividing the applied force by the cross-sectional area (Figure 1-17). There are two kinds of stresses: normal stresses and shear stresses. Normal stresses act either to pull the specimen apart (tensile stress) or to shorten or compact it (compressive stress), and shear stresses act to slide one part of the specimen relative to another part. In general, regions of bone tissue are subjected to both normal and shear stresses during normal skeletal function (Figure 1-18).

Strain is a measure of how the specimen deforms, but unlike displacement, the deformation is expressed in terms of a *relative* change in the size or shape of

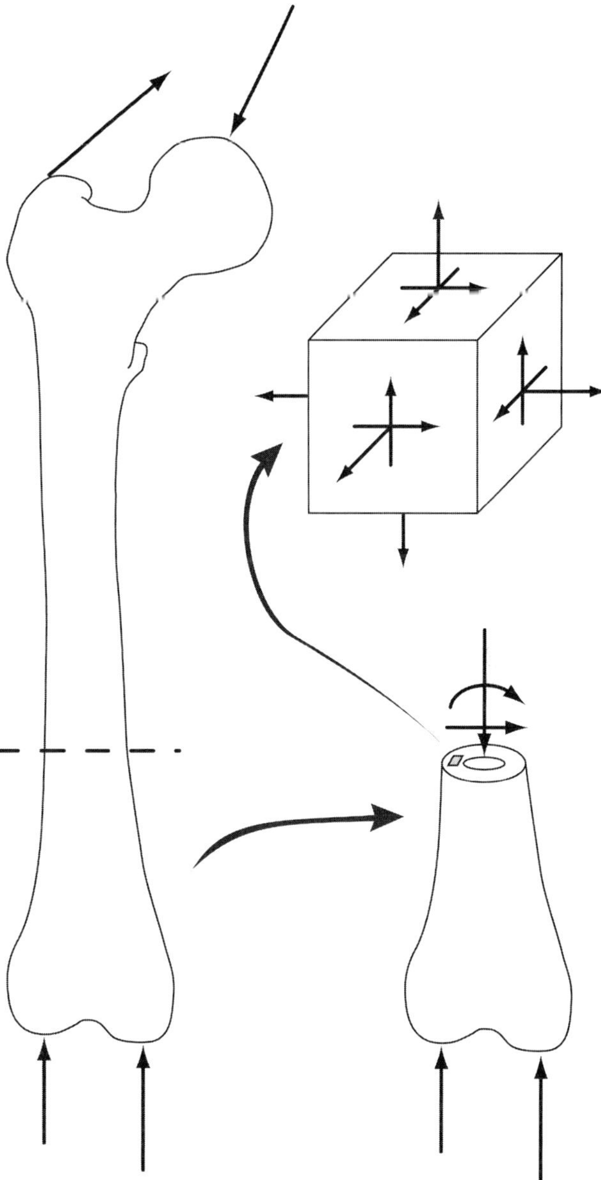

FIGURE 1-18   During normal skeletal function, including gait, regions of bone tissue are subjected to a combination of normal and shear stresses. In the most general case, a region of tissue is subjected to normal and shear stresses on each face. The state of stress shown for this specimen is a multiaxial stress state.

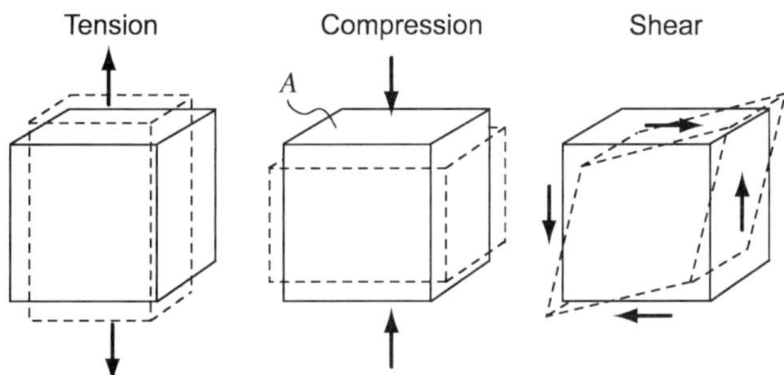

FIGURE 1-17   Normal and shear stresses acting on a specimen of tissue produce normal and shear strains. The dotted lines represent the specimen that is deformed under the action of the applied forces. Whether the applied force is tensile, compressive, or shear, the stress is calculated by dividing the magnitude of the force by the area over which the force is applied (denoted here by *A*). Tensile and compressive stresses cause tensile and compressive strains, respectively, along the direction of the applied force; however, they also cause contraction and expansion, respectively, in the perpendicular directions. The latter effect is quantified by the Poisson's ratio, which is defined as the ratio of transverse to longitudinal strain. Shear strain represents the deformation of the specimen that consists of a change of angle between two lines that were originally perpendicular to each other.

the specimen (Figure 1-17). Normal strains, whether tensile or compressive, quantify the change in length of the specimen relative to its original length. Shear strain quantifies the change in angle of two lines in the material that were originally perpendicular to each other. Strain is dimensionless and is often expressed in microstrain (10^{-6} mm/mm) or percent (10^{-2} mm/mm).

How much strain a specimen of bone tissue will undergo in response to an applied stress depends on the stiffness of the tissue. The material property that describes stiffness is the elastic modulus or *Young's modulus*. The Young's modulus is defined from a uniaxial test (stress applied along one direction only); it is the slope of the initial portion of the *stress–strain curve*, which is a plot of the applied stress against the normal strain in the direction of applied stress (Figure 1-19). Similarly, the shear modulus is defined as the slope of the initial portion of the shear stress–(shear) strain curve. For cortical bone, the stress–strain curves are fairly linear at low values of stress [66], making reproducible measurement of the modulus straightforward. In contrast, trabecular bone exhibits nonlinearity even at low stresses, and care must be taken to calculate the curve's slope in a manner that is standardized across specimens and experiments [67].

As mentioned briefly in Section II, most types of bone tissue exhibit elastic anisotropy in that the elastic modulus differs depending on the direction of applied load. In the most general case, the type of anisotropy exhibited by bone tissue is orthotropy [68, 69], which means that there is a different elastic modulus along each of three mutually perpendicular directions (Figure 1-20). Some types of bone tissue (e.g., woven bone) are isotropic in that the elastic modulus is the same in all directions. Finally, some types of bone tissue (e.g., cortical bone with a secondary osteon structure and trabecular bone from the vertebral body) exhibit an intermediate class of anisotropy, known as transverse isotropy. For transversely isotropic materials, the elastic modulus is distinct along the direction of the main grain of the tissue but is the same in all directions perpendicular to the grain axis.

In the context of osteoporosis, it is clearly of interest to determine the strength of a specimen of bone tissue. For a uniaxial test, strength is defined either as the ultimate stress (the maximum value of stress that the specimen can bear) or the yield stress. The latter is technically the stress above which the tissue no longer behaves elastically; that is, if the specimen is loaded above the yield stress and then unloaded to zero stress, the specimen will show some permanent deformation and/or a reduction in stiffness upon reloading. In practice, the yield stress and yield strain are defined from the stress–strain curve using an offset method (Figure 1-19).

Determining the strength of a specimen when it is subjected to a multiaxial stress state (a combination of normal and/or shear stresses acting along multiple directions) is more challenging with respect to the experimental methods, but this type of test is clinically relevant, given the complexity of the tissue's mechanical environment *in vivo*. This task is further complicated by the fact that strength, like elastic modulus, is anisotropic, being higher along the grain axis than along a direction oblique to this axis. Thus, whether a specimen will fail depends not only on the magnitudes and types of the applied stresses, but also on the orientation of these stresses with respect to the specimen microstructure. Development of multiaxial failure criteria for bone tissue is the subject of ongoing research [70–72].

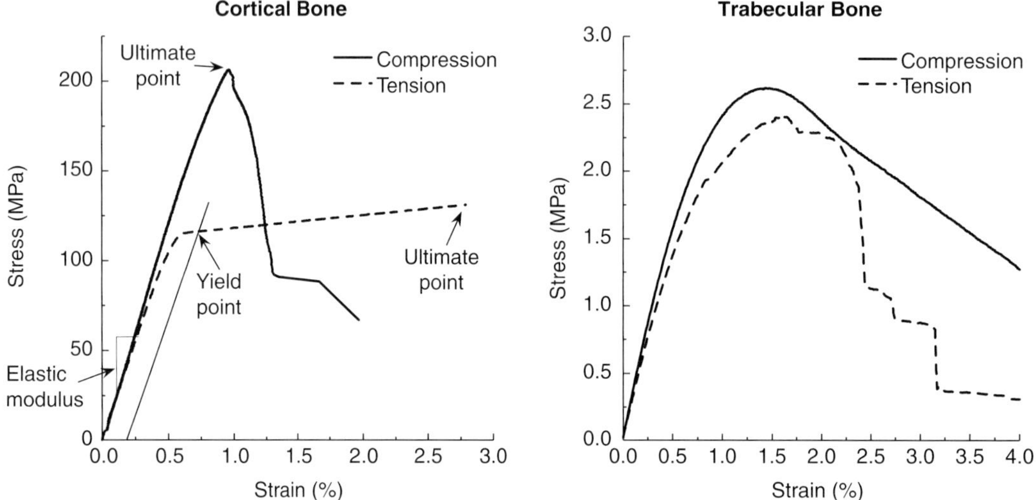

FIGURE 1-19   Stress–strain curves for cortical and trabecular bone in both compression and tension. The elastic modulus is the slope of the initial portion of the curve. Two measures of strength, the yield stress and ultimate stress, are the values of stress at the yield and ultimate points, respectively. In practice the yield point is defined using an offset method: This point is the intersection of the stress–strain curve with a line that has a slope equal to the elastic modulus but that is offset along the strain axis by a certain amount (typically, 0.2%). Data from [33, 103].

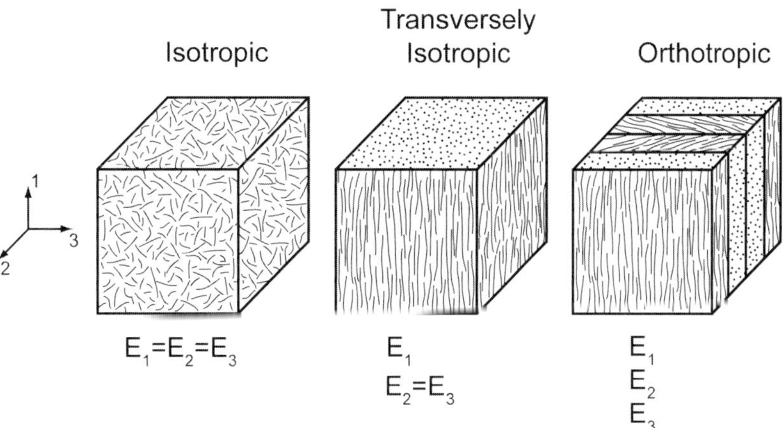

FIGURE 1-20   Three types of anisotropy are typically encountered in bone tissue. If the mineralized collagen fibrils have no particular orientation (such as in woven bone), the tissue is isotropic, and the elastic modulus measured in each of the three directions shown is the same. If the fibrils all have a single, consistent orientation, the tissue is transversely isotropic. The elastic modulus is higher along the direction of the fibrils (the grain axis) but is the same in all directions perpendicular to this axis. Cortical bone composed of secondary osteons is nearly transversely isotropic; in this case the osteons are the "fibrils." Finally, if there are several preferred orientations of the fibrils, such as shown here in a schematic of lamellar bone, the tissue is orthotropic. In this case, the elastic modulus is different along each of the three directions shown. In general, trabecular bone is also orthotropic.

Measures of strength provide a quantitative but essentially dichotomous description of failure, and it is helpful to supplement these measures with investigations of what the mechanisms of failure are for a given type of bone tissue. Failure mechanisms are dictated by not only the nature of the applied loads, but also the composition and microstructure of the tissue. Both cortical bone and trabecular bone are stronger in compression than tension, which reflects the fact that the inorganic phase is stronger in compression than tension. The organic phase contributes to the ductility and toughness of the tissues. *Ductility* is quantified by the amount of strain that the specimen can withstand before complete fracture. *Toughness* is defined in one of two ways, either as the amount of energy that the specimen can absorb prior to complete fracture (also known as the *work to failure* or energy to failure) or as the resistance of the tissue to the initiation and propagation of cracks. The latter is often referred to as the *fracture toughness*. Both the ductility and toughness of intact bone tissue are substantially higher than that of apatite and heat-treated bone tissue [73, 74], suggesting that the organic phase is indeed critical in these two aspects of bone failure. However, the microstructure of bone tissue also plays a role. In cortical tissue, crack growth often initiates at pores such as lacunae and Volkmann's canals and appears to arrest, at least temporarily, at cement lines, leaving secondary osteons intact [27, 75, 76]. For tensile loading along the grain axis, continued crack growth results in debonding of osteons from the interstitial bone and transverse fracture of the osteons themselves, giving the appearance on the fracture surfaces of the specimen that the osteons have "pulled out" of the surrounding tissue [77, 78]. For compressive loading, the osteons tend to fracture oblique to their longitudinal axis, and little pull-out is observed [78].

Crack initiation and propagation are also observed in trabeculae prior to complete, macroscopic failure of trabecular bone, and the extent of microcracking in a given region of trabecular bone appears to be related to the magnitudes of the strains that the region has experienced during loading [79]. As a consequence of the highly porous structure of trabecular bone, even simple loading conditions such as uniaxial compression applied to the entire specimen can produce a wide distribution of stresses and strains in the tissue comprising the trabeculae. Even at low magnitudes of applied stress, some tissue-level strains can be high enough to cause local yielding of the tissue and a concomitant decline in the mechanical properties of the entire specimen of trabecular bone [80].

Although the bulk of the work to date on the mechanical properties of bone tissue has been performed on specimens 1–10 mm in scale, a rapidly growing body of research has focused on micro- and nanoscale properties. Micromechanical tests on osteons and single trabeculae seek to characterize the elastic and failure properties of these small structures within cortical and trabecular bone [81–88]. Osteon push-out tests have been developed to quantify the shear strength of cement lines [89, 90]. In addition, several techniques, including acoustic microscopy and nanoindentation, allow measurement of mechanical properties of regions of bone tissue composed of just one or several lamellae [91–101]. When combined with other high-resolution imaging and compositional measurement techniques such as x-ray tomography, Raman microspectroscopy, backscattered electron imaging, and infrared (IR) spectroscopy, these nanoscale testing methods enable investigation of relationships among composition, structure, and mechanical function at a very fine level of detail.

VI. SUMMARY

Bone is a complex, hierarchically organized organ system whose composition and structure are closely related to, and in many ways controlled by, the functional demands made upon it. Bone tissue is a composite material composed of a proteinaceous extracellular matrix impregnated with impure apatite crystals. In this sense, the structure and mechanical properties of bone tissue are similar to engineering composite materials such as fiberglass. However, bone tissue is a living tissue that is constantly undergoing turnover via coordinated activities by osteoblasts, osteoclasts, osteocytes, and their precursors. Through this process of bone remodeling, bone is an organ system that can respond relatively quickly to changes in metabolic and structural needs.

Recent and ongoing research has continued to enhance understanding of cellular and hormonal control of bone remodeling. In particular, knockout studies have played essential roles in identifying key transcription factors and signaling pathways involved in bone formation, resorption, and mechanotransduction. Several of these studies and others have linked abnormalities in signaling with changes in bone mechanical properties. Although the picture is by no means complete, it is clear that diseases such as osteoporosis can involve deficits in bone mechanical properties at multiple length scales and that the underlying causes of these deficits can be associated with multiple aspects of bone homeostasis. The concepts presented in this chapter provide a framework for further elucidation of the biological and biomechanical mechanisms underlying the close relationship between form and function in bone.

REFERENCES

1. I. A. Bab and T. A. Einhorn, Polypeptide factors regulating osteogenesis and bone marrow repair. *J Cell Biochem.* **55**, 358–365 (1994).
2. M. Horowitz and R. L. Jilka, Colony stimulating factors in bone remodeling. In *Cytokines and Bone Metabolism* (M. Gowen, ed.). CRC Press, Boca Raton, FL, 185–227 (1992).
3. M. N. Wein, D. C. Jones, and L. H. Glimcher, Turning down the system: Counter-regulatory mechanisms in bone and adaptive immunity. *Immunol Rev.* **208**, 66–79 (2005).
4. C. Bell, *Animal Mechanics, or Proofs of Design in the Animal Frame.* Morrill Wyman, Cambridge, MA, 1827.
5. J. C. Koch, The laws of bone architecture. *Am J Anat.* **21**, 177–298 (1917).
6. H. Roesler, The history of some fundamental concepts in bone biomechanics. *J Biomech.* **20**, 1025–1034 (1987).
7. J. Wolff, *The Law of Bone Remodelling.* Springer-Verlag, Berlin, New York, 1986.
8. W. Roux, *Der Zuchtende Kampf der Teile, oder die "Teilauslese" im Organismus ("Theorie der Funktionellen Anpassung").* Wilhelm Engelmann, Leipzig, 1881.
9. J. E. Bertram and S. M. Swartz, The "law of bone transformation": A case of crying Wolff? *Biol Rev Camb Philos Soc.* **66**, 245–273 (1991).
10. S. C. Cowin, The false premise of Wolff's law. *Forma.* **12**, 247–262 (1997).
11. J. K. Gong, J. S. Arnold, and S. H. Cohn, Composition of trabecular and cortical bone. *Anat Rec.* **149**, 325–332 (1964).
12. T. A. Einhorn, Bone metabolism and metabolic bone disease. In *Orthopaedic Knowledge Update 4 Home Study Syllabus* (J. W. Frymoyer, ed.). Am Acad Orthop Surg., Rosemont, 69–88 (1994).
13. A. L. Boskey, S. Gadaleta, C. Gundberg, S. B. Doty, P. Ducy, and G. Karsenty, Fourier transform infrared microspectroscopic analysis of bones of osteocalcin-deficient mice provides insight into the function of osteocalcin. *Bone.* **23**, 187–196 (1998).
14. E. Ruoslahti, Integrins. *J Clin Invest.* **87**, 1–5 (1991).
15. D. McConnell, The crystal structure of bone. *Clin Orthop.* **23**, 253–268 (1962).
16. H. Ou-Yang, E. P. Paschalis, W. E. Mayo, A. L. Boskey, and R. Mendelsohn, Infrared microscopic imaging of bone: Spatial distribution of $CO3(2-)$. *J Bone Miner Res.* **16**, 893–900 (2001).
17. F. S. Kaplan, W. C. Hayes, T. M. Keaveny, A. L. Boskey, T. A. Einhorn, and J. P. Iannotti, Form and function of bone, In *Orthopaedic Basic Science* (S. R. Simon, ed.). Am. Acad. Orthop. Surg., 127–184 (1994).
18. E. S. Siris, Paget's disease of bone. *J Bone Miner Res.* **13**, 1061–1065 (1998).
19. T. A. Einhorn, A. L. Boskey, C. M. Gundberg, V. J. Vigorita, V. J. Devlin, and M. M. Beyer, The mineral and mechanical properties of bone in chronic experimental diabetes. *J Orthop Res.* **6**, 317–323 (1988).
20. A. L. Boskey and S. C. Marks, Jr., Mineral and matrix alterations in the bones of incisors-absent (ia/ia) osteopetrotic rats. *Calcif Tissue Int.* **37**, 287–292 (1985).
21. P. Fratzl, S. Schreiber, P. Roschger, M. H. Lafage, G. Rodan, and K. Klaushofer, Effects of sodium fluoride and alendronate on the bone mineral in minipigs: A small-angle X-ray scattering and backscattered electron imaging study. *J Bone Miner Res.* **11**, 248–253 (1996).
22. A. L. Boskey, Bone mineral and matrix. Are they altered in osteoporosis? *Orthop Clin North Am.* **21**, 19–29 (1990).
23. M. G. Mullender, D. D. Vandermeer, R. Huiskes, and P. Lips, Osteocyte density changes in aging and osteoporosis. *Bone.* **18**, 109–113 (1996).
24. D. Vashishth, O. Verborgt, G. Divine, M. B. Schaffler, and D. P. Fyhrie, Decline in osteocyte lacunar density in human cortical bone is associated with accumulation of microcracks with age. *Bone.* **26**, 375–380 (2000).

25. M. G. Mullender, S. D. Tan, L. Vico, C. Alexandre, and J. Klein-Nulend, Differences in osteocyte density and bone histomorphometry between men and women and between healthy and osteoporotic subjects. *Calcif Tissue Int.* **77**, 291–296 (2005).

26. G. Marotti and A. Z. Zallone, Changes in the vascular network during the formation of Haversian systems. *Acta Anat (Basel).* **106**, 84–100 (1980).

27. D. B. Burr, M. B. Schaffler, and R. G. Frederickson, Composition of the cement line and its possible mechanical role as a local interface in human compact bone. *J Biomech.* **21**, 939–945 (1988).

28. J. Cohen and W. H. Harris, The three-dimensional anatomy of the Haversian system. *J Bone Joint Surg.* **40A**, 419–434 (1958).

29. S. Mohsin, D. Taylor, and T. C. Lee, Three-dimensional reconstruction of Haversian systems in ovine compact bone. *Eur J Morphol.* **40**, 309–315 (2002).

30. S. Shi and S. Gronthos, Perivascular niche of postnatal mesenchymal stem cells in human bone marrow and dental pulp. *J Bone Miner Res.* **18**, 696–704 (2003).

31. I. M. Shapiro, E. E. Golub, B. Chance, C. Piddington, O. Oshima, O. C. Tuncay, P. Frasca, and J. C. Haselgrove, Linkage between energy status of perivascular cells and mineralization of the chick growth cartilage. *Dev Biol.* **129**, 372–379 (1988).

32. D. R. Carter and W. C. Hayes, The compressive behavior of bone as a two-phase porous structure. *J Bone Joint Surg.* **59–A**, 954–962 (1977).

33. E. F. Morgan and T. M. Keaveny, Dependence of yield strain of human trabecular bone on anatomic site. *J Biomech.* 34, 569–577 (2001).

34. D. Ulrich, B. Van Rietbergen, A. Laib, and P. Rueegsegger, The ability of three-dimensional structural indices to reflect mechanical aspects of trabecular bone. *Bone.* **25**, 55–60 (1999).

35. L. Mosekilde, Sex differences in age-related loss of vertebral trabecular bone mass and structure—Biomechanical consequences. *Bone.* **10**, 425–432 (1989).

36. M. Ding, A. Odgaard, F. Linde, and I. Hvid, Age-related variations in the microstructure of human tibial cancellous bone. *J Orthop Res.* **20**, 615–621 (2002).

37. T. Kobayashi and H. Kronenberg, Minireview: Transcriptional regulation in development of bone. *Endocrinology.* **146**, 1012–1017 (2005).

38. T. Komori, H. Yagi, S. Nomura, A. Yamaguchi, K. Sasaki, K. Deguchi, Y. Shimizu, R. T. Bronson, Y. H. Gao, M. Inada, M. Sato, R. Okamoto, Y. Kitamura, S. Yoshiki, and T. Kishimoto, Targeted disruption of Cbfa1 results in a complete lack of bone formation owing to maturational arrest of osteoblasts. *Cell.* **89**, 755–764 (1997).

39. K. Nakashima, X. Zhou, G. Kunkel, Z. Zhang, J. M. Deng, R. R. Behringer, and B. de Crombrugghe, The novel zinc finger-containing transcription factor osterix is required for osteoblast differentiation and bone formation. *Cell.* 108, 17–29 (2002).

40. Y. Gong, R. B. Slee, N. Fukai, G. Rawadi, S. Roman-Roman, A. M. Reginato, H. Wang, T. Cundy, F. H. Glorieux, D. Lev, M. Zacharin, K. Oexle, J. Marcelino, W. Suwairi, S. Heeger, G. Sabatakos, S. Apte, W. N. Adkins, J. Allgrove, M. Arslan-Kirchner, J. A. Batch, P. Beighton, G. C. Black, R. G. Boles, L. M. Boon, C. Borrone, H. G. Brunner, G. F. Carle, B. Dallapiccola, A. De Paepe, B. Floege, M. L. Halfhide, B. Hall, R. C. Hennekam, T. Hirose, A. Jans, H. Juppner, C. A. Kim, K. Keppler-Noreuil, A. Kohlschuetter, D. LaCombe, M. Lambert, E. Lemyre, T. Letteboer, L. Peltonen, R. S. Ramesar, M. Romanengo, H. Somer, E. Steichen-Gersdorf, B. Steinmann, B. Sullivan, A. Superti-Furga, W. Swoboda, M. J. van den Boogaard, W. Van Hul, M. Vikkula, M. Votruba, B. Zabel, T. Garcia, R. Baron, B. R. Olsen, and M. L. Warman, LDL receptor-related protein 5 (LRP5) affects bone accrual and eye development. *Cell.* **107**, 513–523 (2001).

41. W. M. Cheung, L. Y. Jin, D. K. Smith, P. T. Cheung, E. Y. Kwan, L. Low, and A. W. Kung, A family with osteoporosis pseudoglioma syndrome due to compound heterozygosity of two novel mutations in the LRP5 gene. *Bone.* **39**, 470–476 (2006).

42. H. Hartikka, O. Makitie, M. Mannikko, A. S. Doria, A. Daneman, W. G. Cole, L. Ala-Kokko, and E. B. Sochett, Heterozygous mutations in the LDL receptor-related protein 5 (LRP5) gene are associated with primary osteoporosis in children. *J Bone Miner Res.* **20**, 783–789 (2005).

42a. M. Kato, M. S. Patel, R. Levasseur, I. Lobov, B. H. Chang, D. A. Glass, 2nd, C. Hartmann, L. Li, T. H. Hwang, C. F. Brayton, R. A. Lang, G. Karsenty, L. Chan. Cbfa1-independent decrease in osteoblast proliferation, osteopenia, and persistent embryonic eye vascularization in mice deficient in Lrp5, a Wnt coreceptor. *J Cell Biol.* **157**, 303–314 (2002).

43. P. Babij, W. Zhao, C. Small, Y. Kharode, P. J. Yaworsky, M. L. Bouxsein, P. S. Reddy, P. V. Bodine, J. A. Robinson, B. Bhat, J. Marzolf, R. A. Moran, and F. Bex, High bone mass in mice expressing a mutant LRP5 gene. *J Bone Miner Res.* **18**, 960–974 (2003).

44. M. Tate, T. A. Adamson Jr, and T. W. Bauer, Cells in focus. The osteocyte. *IJBCB.* **36**, 1–8 (2004).

45. T. A. Franz-Odendaal, B. K. Hall, and P. E. Witten, Buried alive: How osteoblasts become osteocytes. *Dev Dyn.* **235**, 176–190 (2006).

46. J. M. Quinn and M. T. Gillespie, Modulation of osteoclast formation. *Biochem Biophys Res Commun.* **328**, 739–745 (2005).

47. M. Zaidi, H. C. Blair, B. S. Moonga, E. Abe, and C. L. Huang, Osteoclastogenesis, bone resorption, and osteoclast-based therapeutics. *J Bone Miner Res.* **18**, 599–609 (2003).

48. T. Wada, T. Nakashima, N. Hiroshi, and J. M. Penninger, RANKL-RANK signaling in osteoclastogenesis and bone disease. *Trends Mol Med.* **12**, 17–25 (2006).

49. T. J. Martin and N. A. Sims, Osteoclast-derived activity in the coupling of bone formation to resorption. *Trends Mol Med.* **11**, 76–81 (2005).

50. J. Rubin, C. Rubin, and C. R. Jacobs, Molecular pathways mediating mechanical signaling in bone. *Gene.* **367**, 1–16 (2006).

51. J. M. Weber, S. R. Forsythe, C. A. Christianson, B. J. Frisch, B. J. Gigliotti, C. T. Jordan, L. A. Milner, M. L. Guzman, and L. M. Calvi, Parathyroid hormone stimulates expression of the Notch ligand Jagged1 in osteoblastic cells. *Bone.* **39**, 485–493 (2006).

52. G. J. Atkins, P. Kostakis, B. Pan, A. Farrugia, S. Gronthos, A. Evdokiou, K. Harrison, D. M. Findlay, and A. C. Zannettino, RANKL expression is related to the differentiation state of human osteoblasts. *J Bone Miner Res.* **18**, 1088–1098 (2003).

53. A. M. Parfitt, Bone remodeling. *Henry Ford Hosp Med J.* **36**, 143–144 (1988).

54. J. Kragstrup and F. Melsen, Three-dimensional morphology of trabecular bone osteons reconstructed from serial sections. *Metab Bone Dis Relat Res* **5**, 127–130 (1983).

55. A. M. Parfitt, The cellular basis of bone remodeling: The quantum concept reexamined in light of recent advances in the cell biology of bone. *Calcif Tissue Int.* **36** (Suppl 1), S37–45 (1984).

56. A. M. Parfitt, D. S. Rao, J. Stanciu, A. R. Villanueva, M. Kleerekoper, and B. Frame, Irreversible bone loss in osteomalacia. Comparison of radial photon absorptiometry with iliac bone histomorphometry during treatment. *J Clin Invest.* **76**, 2403–2412 (1985).

57. T. A. Einhorn and R. J. Majeska, Neutral proteases in regenerating bone. *Clin Orthop Relat Res.* 286–297 (1991).

58. E. F. Eriksen, *et al.*, Reconstruction of the resorptive site in iliac trabecular bone: A kinetic model for bone resorption in 20 normal individuals. *Metab Bone Dis Rel Res.* **5**, 235–242 (1984).

59. R. W. Smith, Jr. and R. R. Walker, Femoral expansion in aging women. Implications for osteoporosis and fractures. *Henry Ford Hosp Med J.* **28**, 168–170 (1980).

60. C. B. Ruff and W. C. Hayes, Subperiosteal expansion and cortical remodeling of the human femur and tibia with aging. *Science.* **217**, 945–948 (1982).

61. M. P. Akhter, U. T. Iwaniec, M. A. Covey, D. M. Cullen, D. B. Kimmel, and R. R. Recker, Genetic variations in bone density, histomorphometry, and strength in mice. *Calcif Tissue Int.* **67**, 337–344 (2000).

62. M. C. van der Meulen, K. J. Jepsen, and B. Mikic, Understanding bone strength: Size isn't everything. *Bone.* **29**, 101–104 (2001).

63. J. E. Wergedal, M. H. Sheng, C. L. Ackert-Bicknell, W. G. Beamer, and D. J. Baylink, Genetic variation in femur extrinsic strength in 29 different inbred strains of mice is dependent on variations in femur cross-sectional geometry and bone density. *Bone.* **36**, 111–122 (2005).

64. J. M. Geer and S. P. Timoshenko, *Mechanics of Materials.* PWS Publishing Company, Boston, 1990.

65. M. E. Levenston, G. S. Beaupre, and M. C. van der Meulen, Improved method for analysis of whole bone torsion tests. *J Bone Miner Res.* **9**, 1459–1465 (1994).

66. D. T. Reilly and A. H. Burstein, The elastic and ultimate properties of compact bone tissue. *J Biomech.* **8**, 393–405 (1975).

67. E. F. Morgan, O. C. Yeh, W. C. Chang, and T. M. Keaveny, Non-linear behavior of trabecular bone at small strains. *J Biomech Eng.* **123**, 1–9 (2001).

68. A. Odgaard, J. Kabel, B. Van Rietbergen, M. Dalstra, and R. Huiskes, Fabric and elastic principal directions of cancellous bone are closely related. *J Biomech.* **30**, 487–495 (1997).

69. G. Yang, J. Kabel, B. Van Rietbergen, A. Odgaard, R. Huiskes, and S. Cowin, The anisotropic Hooke's law for cancellous bone and wood. *J Elasticity.* **53**, 125–146 (1999).

70. H. H. Bayraktar, A. Gupta, R. Y. Kwon, P. Papadopoulos, and T. M. Keaveny, The modified super-ellipsoid yield criterion for human trabecular bone. *J Biomech Eng.* **126**, 677–684 (2004).

71. T. M. Keaveny, E. F. Wachtel, S. P. Zadesky, and Y. P. Arramon, Application of the Tsai-Wu quadratic multiaxial failure criterion to bovine trabecular bone. *J Biomech Eng.* **121**, 99–107 (1999).

72. P. K. Zysset, M. S. Ominsky, and S. A. Goldstein, A novel 3D microstructural model for trabecular bone: II. The relationship between fabric and the yield surface. *Comput Methods Biomech Biomed Engin.* **2**, 1–11 (1999).

73. E. F. Morgan, D. N. Yetkinler, B. R. Constantz, and R. H. Dauskardt, Mechanical properties of carbonated apatite bone mineral substitute: Strength, fracture and fatigue behaviour. *J Mater Sci Mater Med.* **8**, 559–570 (1997).

74. J. C. I. Catanese, J. D. B. Featherstone, and T. M. Keaveny, Characterization of the mechanical and ultrastructural properties of heat-treated cortical bone for use as a bone substitute. *J Biomed Mater Res.* **45**, 327–336 (1999).

75. G. C. Reilly and J. D. Currey, The development of microcracking and failure in bone depends on the loading mode to which it is adapted. *J Exp Biol.* **202**, 543–552 (1999).

76. G. C. Reilly, Observations of microdamage around osteocyte lacunae in bone. *J Biomech.* **33**, 1131–1134 (2000).

77. K. Piekarski, Fracture of bone. *J Appl Phys.* **41**, 215–223 (1970).

78. W. E. Caler and D. R. Carter, Bone creep-fatigue damage accumulation. *J Biomech.* **22**, 625–635 (1989).

79. S. Nagaraja, T. L. Couse, and R. E. Guldberg, Trabecular bone microdamage and microstructural stresses under uniaxial compression. *J Biomech.* **38**, 707–716 (2005).

80. E. F. Morgan, O. C. Yeh, and T. M. Keaveny, Damage in trabecular bone at small strains. *Eur J Morphol.* **42**, 13–21 (2005).

81. A. Ascenzi and E. Bonucci, The tensile properties of single osteons. *Anat Rec.* **158**, 375–386 (1967).

82. A. Ascenzi and E. Bonucci, The compressive properties of single osteons. *Anat Rec.* **161**, 377–391 (1968).

83. A. Ascenzi, A. Benvenuti, F. Mango, and R. Simili, Mechanical hysteresis loops from single osteons: Technical devices and preliminary results. *J Biomech.* **18**, 391–398 (1985).

84. A. Ascenzi, P. Baschieri, and A. Benvenuti, The torsional properties of single selected osteons. *J Biomech.* **27**, 875–884 (1994).

85. P. R. Townsend, R. M. Rose, and E. L. Radin, Buckling studies of single human trabeculae. *J Biomech.* **8**, 199–201 (1975).

86. K. Choi and S. A. Goldstein, A comparison of the fatigue behavior of human trabecular and cortical bone tissue. *J Biomech.* **25**, 1371–1381 (1992).

87. J. Y. Rho, R. B. Ashman, and C. H. Turner, Young's modulus of trabecular and cortical bone material: Ultrasonic and microtensile measurements. *J Biomech.* **26**, 111–119 (1993).

88. C. J. Hernandez, S. Y. Tang, B. M. Baumbach, P. B. Hwu, A. N. Sakkee, F. van der Ham, J. DeGroot, R. A. Bank, and T. M. Keaveny, Trabecular microfracture and the influence of pyridinium and non-enzymatic glycation-mediated collagen cross-links. *Bone.* **37**, 825–832 (2005).

89. A. Ascenzi and E. Bonucci, The shearing properties of single osteons. *Anat Rec.* **172**, 499–510 (1972).

90. X. N. Dong and X. E. Guo, Geometric determinants to cement line debonding and osteonal lamellae failure in osteon pushout tests. *J Biomech Eng.* **126**, 387–390 (2004).

91. J. L. Katz and A. Meunier, Scanning acoustic microscope studies of the elastic properties of osteons and osteon lamellae. *J Biomech Eng.* 115, 543–548 (1993).

92. P. K. Zysset, X. E. Guo, C. E. Hoffler, K. E. Moore, and S. A. Goldstein, Mechanical properties of human trabecular bone lamellae quantified by nanoindentation. *Technol Health Care.* **6**, 429–432 (1998).

93. J. Y. Rho, P. Zioupos, J. D. Currey, and G. M. Pharr, Variations in the individual thick lamellar properties within osteons by nanoindentation. *Bone.* **25**, 295–300 (1999).

94. C. H. Turner, J. Rho, Y. Takano, T. Y. Tsui, and G. M. Pharr, The elastic properties of trabecular and cortical bone tissues are similar: Results from two microscopic measurement techniques. *J Biomech.* **32**, 437–441 (1999).

95. S. Hengsberger, A. Kulik, and P. Zysset, Nanoindentation discriminates the elastic properties of individual human bone lamellae under dry and physiological conditions. *Bone.* **30**, 178–184 (2002).

96. Z. Fan, J. G. Swadener, J. Y. Rho, M. E. Roy, and G. M. Pharr, Anisotropic properties of human tibial cortical bone as measured by nanoindentation. *J Orthop Res.* **20**, 806–810 (2002).

97. B. Busa, L. M. Miller, C. T. Rubin, Y. X. Qin, and S. Judex, Rapid establishment of chemical and mechanical properties during lamellar bone formation. *Calcif Tissue Int.* **77**, 386–394 (2005).

98. J. Litniewski, Determination of the elasticity coefficient for a single trabecula of a cancellous bone: Scanning acoustic microscopy approach. *Ultrasound Med Biol.* **31**, 1361–1366 (2005).

99. T. Hofmann, F. Heyroth, H. Meinhard, W. Franzel, and K. Raum, Assessment of composition and anisotropic elastic properties of secondary osteon lamellae. *J Biomech.* **39**, 2282–2294 (2005).

100. G. Balooch, M. Balooch, R. K. Nalla, S. Schilling, E. H. Filvaroff, G. W. Marshall, S. J. Marshall, R. O. Ritchie, R. Derynck, and T. Alliston, TGF-beta regulates the mechanical properties and composition of bone matrix. *Proc Natl Acad Sci U S A.* **102**, 18813–18818 (2005).

101. T. Hoc, L. Henry, M. Verdier, D. Aubry, L. Sedel, and A. Meunier, Effect of microstructure on the mechanical properties of Haversian cortical bone. *Bone.* **38**, 466–474 (2006).

102. L. Weiss, *Cell and Tissue Biology, A Textbook of Histology.* Urban and Schwarzenberg, Baltimore, 1988.

103. E. F. Morgan, J. J. Lee, and T. M. Keaveny, Sensitivity of multiple damage parameters to compressive overload in cortical bone. *J Biomech Eng.* 127, 557–562 (2005).

The Nature of Osteoporosis

ROBERT MARCUS AND MARY L. BOUXSEIN

I. DEFINING OSTEOPOROSIS

This chapter introduces the topic of osteoporosis from the perspective of the bone. Its purpose is to consider the definition of osteoporosis and to discuss the nature of osteoporotic bone, including the characteristics that affect its ability to resist fracture. Osteoporosis is a condition of generalized skeletal fragility in which bone strength is sufficiently weak that fractures occur with minimal trauma, often no more than is applied by routine daily activity. Albright and Reifenstein [1] proposed in 1948 that primary osteoporosis consists of two separate entities: one related to menopausal estrogen loss and the other to aging. This concept was elaborated upon by Riggs and associates [2], who suggested the terms "Type I osteoporosis," to signify a loss of trabecular bone after menopause, and "Type II osteoporosis," to represent a loss of cortical and trabecular bone in men and women as the end result of age-related bone loss. By this formulation, the Type I disorder directly results from lack of endogenous estrogen, while Type II osteoporosis reflects the composite influences of long-term remodeling inefficiency, adequacy of dietary calcium and vitamin D, intestinal mineral absorption, renal mineral handling, and parathyroid hormone (PTH) secretion. Although there may be heuristic value to defining subsets of patients in this manner, the model suffers by not accounting for the complex and multifactorial nature of a disease that defies rigid categorization.

Bone mass at any time in adult life reflects the peak investment in bone mineral at skeletal maturity minus that which has been subsequently lost. A woman who experienced interruption of menses, extended bed rest, eating disorder, or systemic illness during her adolescent growth years might enter adult life having failed to achieve the bone mass that would have been predicted from her genetic or constitutional profile. If she then underwent a perfectly normal rate of bone loss, her skeleton would still be in jeopardy simply due to the deficit in peak bone mass. Thus, it seems most

appropriate to consider osteoporosis the consequence of a stochastic process, that is, multiple genetic, physical, hormonal, and nutritional factors acting alone or in concert to diminish skeletal integrity.

Historical artifacts show that characteristic deformities of vertebral osteoporosis were recognized in antiquity [3], although broad awareness of this condition has come about only during the past few decades. Unfortunately, because traditional radiographic techniques cannot distinguish osteoporosis until it is severe, confirmation of the diagnosis remained problematic until recently. Diagnosis was by necessity clinical, requiring a history of one or more low-trauma fractures. Although highly specific, such a grossly insensitive diagnostic criterion offered no assistance to physicians who hope to identify and treat affected individuals who have been fortunate not yet to have sustained a fracture.

The introduction of accurate noninvasive bone mass measurements afforded the opportunity to estimate a person's fracture risk and to make an early diagnosis of osteoporosis. Briefly stated, large prospective studies have shown that a reduction in BMD of 1 standard deviation from the mean value for an age-specific population confers a 2- to 3-fold increase in long-term fracture risk [4–9]. In a manner similar to that by which serum cholesterol concentration predicts risk for heart attack or blood pressure predicts risk for stroke, BMD measurements can successfully identify subjects at risk of fracture and can help physicians select those individuals who will derive greatest benefit for initiation of therapy.

Several factors limit the ability of BMD measurements to predict an individual's fracture risk with great accuracy. The normative data against which BMD comparisons are most often made have been determined for Caucasian men and women, and do not necessarily apply to other ethnic groups. BMD is clearly related to body weight, yet routine clinical bone mass assessments are not weight adjusted. Various features of bone geometry that affect bone strength and fracture risk are not generally considered in the clinical interpretation of bone mass measurements, including bone size as well as the spatial distribution of bone mass. Moreover, bone mass determinations cannot distinguish individuals with low mass and intact microarchitecture from those with equal mass who have trabecular disruption and cortical porosity [10].

In 1994, a group of senior investigators in this field offered a working definition of osteoporosis based exclusively on bone mass [11]. The reasoning behind this proposal, made on behalf of the World Health Organization (WHO), was that the clinical significance of osteoporosis lies exclusively in the occurrence of fracture, that bone mass predicts long-term fracture risk, and that selection of rigorous diagnostic criteria would minimize the number of patients who are incorrectly diagnosed. The authors suggested a cutoff BMD value of 2.5 standard deviations below the average for healthy young adult women. Using this value, approximately 30% of postmenopausal women would be designated as osteoporotic, which gives a realistic projection of lifetime fracture rates. In addition, Kanis *et al.* [11] proposed that BMD values of 1–2 standard deviations below the young adult mean be designated as "osteopenic." Such values identify individuals at increased risk for fracture, but for whom a diagnosis of osteoporosis would not be justified since it would mislabel far more individuals than would actually be expected ever to fracture.

This approach has proven useful for clinical management, but has several limitations. The applicability of this criterion to young people prior to the completion of peak bone acquisition would be inappropriate; and it remains unclear exactly what the best means to assess fracture risk in men may be. The BMD measurement is itself subject to several confounding factors, including bone size and geometry [12]. As BMD correlations among skeletal sites are not strong, designating a person "normal" based on a single site, for example, the lumbar spine, necessarily overlooks individuals with low bone density elsewhere, such as the hip. It seems reasonable to suppose that adjustment of bone density readings for such factors as body size, bone geometry, and ethnic background might improve the accuracy of this technique. Finally, recent studies indicate that, although individuals with low BMD are at greater relative risk to fracture, many fractures in the population are experienced by individuals with bone mass measurements in the normal to osteopenic range by WHO criteria [13–15]. Altogether, it should be evident that whereas the WHO guidelines provide an operational definition of osteoporosis to facilitate clinical diagnosis, the BMD-based guidelines are of limited use to investigators whose interest is the nature and causes of osteoporosis. Knowledge of a low bone density at a particular point in time offers no information regarding the adequacy of peak bone mass attained, the amount of bone that may have been lost, the rate of bone loss, or the quality of bone that remains.

II.   MATERIAL AND STRUCTURAL BASIS OF SKELETAL FRAGILITY

The need to understand more fully the nature of skeletal fragility and overcome the limitations of BMD measurements has brought renewed attention to the broader array of factors that influence skeletal fragility [16, 17]. In support of this view, osteoporosis

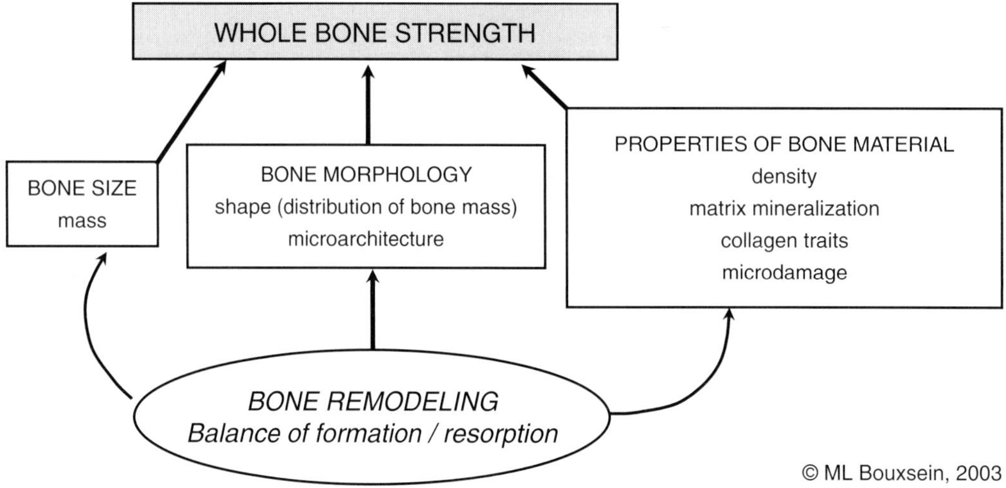

FIGURE 2-1   Determinants of whole bone strength.

was defined at a recent NIH Consensus Development Conference as "a disease characterized by low bone strength, leading to enhanced bone fragility and a consequent increase in fracture risk" [18]. This definition underscores the role of bone strength, and implies that understanding bone strength is key to understanding fracture risk.

The enhanced fragility associated with osteoporotic fractures has been attributed to several factors, chief among them low bone mass and microarchitectural deterioration. Implicit in this view is that osteoporosis results from deficits in the amount and structure of bone, but that the residual bone is not, in contrast to osteomalacia, grossly undermineralized. However, recent data challenge this long-held view, indicating that subtle changes in bone matrix properties such as the degree of mineralization and extent of collagen cross-linking may contribute to skeletal fragility.

For many years the prevailing view has been that osteoporosis develops through excessive loss of bone. Only recently has attention been drawn to abnormalities in bone acquisition as a basis for subsequent bone fragility. This latter issue notwithstanding, the dominant model of osteoporosis among workers in the field has, until recently, emphasized only the amount and distribution of bone substance. However, the great overlap in bone density between individuals with and without fracture indicates the limitations of such a model to account adequately for individual differences in fracture susceptibility. In other words, additional properties of bone likely contribute to skeletal fragility.

The ability of a bone to resist fracture (or "whole bone strength") depends on the amount of bone (i.e., mass), its spatial distribution (i.e., shape and microarchitecture), and the intrinsic properties of the materials that comprise it [19] (Figure 2-1). Bone remodeling, specifically the *balance* between formation and resorption, is the biologic process that mediates changes in the traits that influence bone strength. Thus, diseases and drugs that have an impact on bone remodeling will influence bone's resistance to fracture. Due to a combination of changes in the structural and material properties of bone, whole bone strength declines markedly with age. For instance, laboratory studies of human cadaveric specimens have shown that the strength of the proximal femur and vertebral body is 2- to 10-fold lower in older persons than in young individuals [20, 21].

In considering these determinants of bone strength, one must keep in mind several important concepts. First, unlike most engineering materials, bone is continually adapting to changes in its mechanical and hormonal environment, and is capable of self-renewal and

repair via the process of remodeling. Thus, in response to increased mechanical loading, bone may adapt by altering its size, shape, and/or matrix properties. This type of adaptation is readily seen by the greater size of the bones in the dominant versus nondominant arm of tennis players [22]. In addition, favorable changes in bone geometry may occur in response to deleterious changes in bone matrix properties. For example, in a mouse model of osteogenesis imperfecta, a defect in the collagen that leads to increased bone fragility can be compensated for by a favorable change in bone geometry to preserve whole bone strength [23]. Thus, the loss of bone strength with age likely reflects the ongoing skeletal response to changes in its hormonal (i.e., a decline in gonadal steroids) and mechanical environments (i.e., decreased physical activity).

A second important concept concerns the hierarchical nature of the factors that influence whole bone strength. Thus, properties at the cellular, matrix, microarchitectural, and macroarchitectural levels may all impact bone mechanical properties [16]. Importantly, though, these various factors are interrelated, and therefore one cannot expect that changes in a single property will be solely predictive of changes in bone mechanical behavior.

In any discussion of bone strength, it is important to distinguish between the material and structural properties of bone. During any activity, a complex distribution of forces (or loads) is applied to the skeleton. With the imposition of these forces, bones undergo deformations. The relationship between the forces applied to the bone and the resulting deformations characterizes the *structural behavior*, or *structural properties*, of the whole bone. Thus, structural properties are influenced by the size and shape of the bone, as well as the properties of the bone tissue. In contrast to the structural behavior, the *material behavior*, or *material properties*, of bone tissue is independent of the specimen geometry. Thus, the material properties reflect the intrinsic biomechanical characteristics of cortical and trabecular bone.

The material properties of trabecular bone are influenced by many factors; however, the strongest determinants are apparent density (or volume fraction, the fraction of bone actually occupied by bone tissue) and the microstructural arrangement of the trabecular network. Sampled over a wide range of densities, the stiffness and strength of trabecular bone are related to density in a nonlinear fashion, such that the change in strength is disproportionate to (i.e., greater than) the change in density [24–27]. For example, a 25% decrease in density, approximately equivalent to 15 years of age related bone loss, would be predicted to cause a 44% decrease in the stiffness and strength of trabecular bone. However, given the heterogeneous nature of trabecular bone, it is clear that density alone cannot explain all of the variation in trabecular bone mechanical properties. Both empirical observations and theoretical analyses indicate that trabecular microarchitecture plays an important role (see "Role of Bone Microarchitecture" below).

The primary determinants of the biomechanical properties of cortical bone include porosity and the mineralization density of the bone matrix (or ash content). Indeed, over 80% of the variation in cortical bone stiffness and strength is explained by a power–law relationship with mineralization and porosity as explanatory variables [28–31]. Other properties that influence cortical bone mechanical behavior include, but are not limited to, its histologic structure (primary, lamellar vs. osteonal bone), the collagen content and orientation of collagen fibers, the extent and nature of collagen cross-linking, the number and composition of cement lines, and the presence of fatigue-induced microdamage [32–37]. A few of the factors that influence both the structural and material behavior of bone will be briefly presented in the sections that follow.

A. Role of Bone Microarchitecture

Although bone density is among the strongest predictors of the mechanical behavior of trabecular bone, both empirical observations and theoretical analyses show that aspects of the trabecular microarchitecture influence trabecular bone strength as well [26, 27, 38]. Trabecular architecture can be described by the shape of the basic structural elements and their orientation. The trabecular structure is generally characterized by the number of trabeculae in a given volume, their average thickness, the average distance between adjacent trabeculae, and the degree to which trabeculae are connected to each other. Previously, assessment of trabecular microarchitecture was possible only by two-dimensional histomorphometry (for discussion of this topic based on 2D studies, the reader should consult previous editions of this book). However, newer imaging modalities such as high-resolution microcomputed tomography and magnetic resonance imaging allow for three-dimensional assessment of trabecular structure on excised bone specimens [39–41] and *in vivo* [10, 42–44].

Laboratory studies have demonstrated moderate to strong correlations between trabecular bone architecture and biomechanical properties of trabecular bone [45–49]. Generally, however, trabecular bone microarchitecture is strongly correlated with trabecular bone volume [39, 45, 46], and therefore discerning the independent effects of specific architectural features on

bone mechanical properties has proven challenging. Nonetheless, Ulrich *et al.* reported that including indices of trabecular architecture assessed by 3D microcomputed tomography enhanced prediction of the biomechanical properties of human trabecular bone [49]. To further address this issue, analytical studies have investigated how specific changes in trabecular architecture may influence trabecular bone mechanical behavior [50–52]. For example, an analytical model of vertebral trabecular bone was used to demonstrate that for the same decline in bone mass, loss of trabecular elements was 2 to 5 times more deleterious to bone strength than thinning of the trabecular struts, implying that maintaining connectivity of the trabecular network is critical [50]. Their finding may be explained by examining one potential mechanism by which individual trabecular elements may fail.

Bell [53] proposed that isolated trabeculae may fail by buckling, which describes the failure mode of a long, slender column. In this case, the critical buckling load (or buckling strength) is proportional to the cross-sectional area of the column and to its elastic modulus, and is inversely proportional to the square of unsupported length of the column. Therefore, loss of horizontal trabecular elements leads to a marked increase in the unsupported length of a trabecular strut, markedly decreasing its buckling strength. Inversely, preservation of one or more horizontal struts can profoundly influence trabecular bone buckling strength with very little change in bone mass. This concept is illustrated in Figure 2-2, which shows the theoretical effect of adding one or more horizontal struts on trabecular bone buckling strength.

Another potential mechanism whereby trabecular bone properties decline with increased bone resorptive activity is the hypothesis that the presence of resorption cavities themselves serves as a site of local weakness where cracks in the trabeculae may initiate [54]. van der Linden and colleagues evaluated this possibility

# Horizontal Trabeculae	Effective Length	Buckling Strength
0	L	S
1	1/2 L	4 × S
3	1/4 L	16 × S

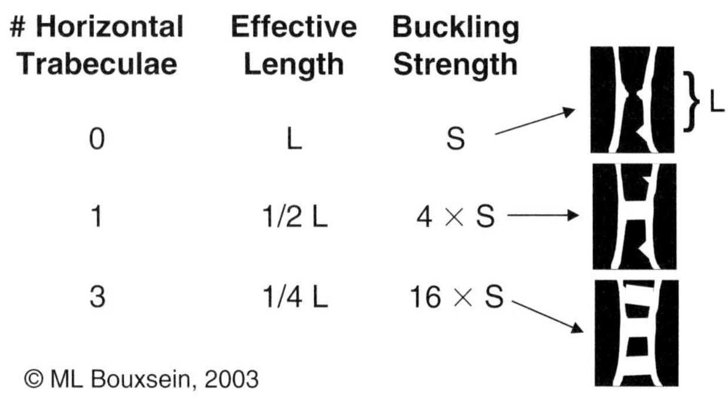

© ML Bouxsein, 2003

FIGURE 2-2 Influence of trabecular cross-struts on buckling strength.

using an analytical model of vertebral trabecular bone, wherein they induced a 20% decline in bone mass either by thinning the entire trabecular structure or by randomly introducing resorption cavities [51]. They made two important observations. First, in both cases the predicted decline in vertebral trabecular bone strength was larger (30% for trabecular thinning and 50% for introduction of resorption cavities) than the decline in bone mass. Second, the reduction in bone strength was greater when bone loss occurred by introduction of resorption cavities than by trabecular thinning. Altogether these observations confirm the deleterious impact of high bone resorption in the absence of increased bone formation on trabecular bone strength and provide a partial explanation for why small changes in bone mass due to therapy can have marked effects on vertebral fracture risk.

The importance of trabecular bone microarchitecture has since been supported by clinical studies showing altered trabecular microarchitecture in subjects with fragility fractures compared to age-matched controls with no fractures [55–58]. For example, after adjusting for bone volume, trabecular bone from the femoral head of individuals who suffered hip fracture was more oriented in a single direction than bone from unfractured individuals [55]. One interpretation of this finding is that the femoral trabecular bone from individuals with hip fracture was less able to withstand unusual loading conditions, such as would be expected during a sideways fall. Other studies have also shown altered trabecular microarchitecture among those with vertebral fracture and that the extent of microarchitectural deterioration is related to vertebral fracture severity [56, 58]. A recent study of individuals undergoing organ transplant showed that changes in trabecular architecture distinguished individuals with vertebral fracture, even after adjustment for BMD [57, 59]. Moreover, data from iliac crest biopsies obtained during clinical trials suggest that maintenance of trabecular architecture with bisphosphonate therapy [60–62] or improvement of trabecular architecture with teriparatide [63, 64] may contribute to the antifracture efficacy of these agents. Altogether these clinical observations point to an important role of trabecular architecture in fragility fractures, particularly at skeletal sites rich in trabecular bone such as the spine.

B. Role of Bone Matrix Properties

In addition to macro- and microarchitecture, features of the bone matrix itself influence bone mechanical properties. Characteristics that affect bone mechanical properties which involve the composition of the matrix include (but are not limited to) the relative ratio of inorganic (i.e., mineral) to organic (i.e., water, collagen, and noncollagenous proteins); the degree of matrix mineralization; mineral crystal size and maturation; the extent and nature of collagen cross-links; and the amount and nature of matrix microdamage [16].

I. MATRIX MINERALIZATION

During the course of bone remodeling, the initial wave of resorption removes both matrix and mineral. The subsequent bone formation phase involves an initial laying down of organic matrix, with an initial component of mineralization occurring after the new matrix reaches a thickness of about 20 microns. Initially, mineralization proceeds at a rapid pace, the new bone achieving most of its ultimate mineral content within a few weeks. After perhaps 2 months, however, the rate of mineralization slows substantially and continues thereafter at a linear rate. It appears that the bone never actually becomes saturated with mineral and that mineralization continues essentially forever, being interrupted only when a new wave of resorption occurs to remove that bone and start the process over again. Thus, the rate at which new remodeling units are brought into play, referred to as the "birthrate" of new remodeling osteons (estimated in biopsy material as the "activation frequency"), constitutes a primary mechanism by which bone mineralization is influenced [65].

It is well established that the degree of matrix mineralization, or ash content, strongly influences the mechanical behavior of cortical [28, 66, 67] and trabecular bone [68, 69]. The elastic modulus and strength of cortical bone are positively related to the degree of matrix mineralization. In fact, a modest 7% increase in bone mineral content is associated with a 3-fold increase in bone stiffness and a doubling in breaking strength [66]. Thus, it seems inescapable that undermineralization would promote bone fragility. However, the ability of cortical bone to absorb energy may either increase (if the bone is relatively undermineralized to begin with) or decrease (if the bone is already fully mineralized) with increasing mineral content [70].

Among the first efforts to assess the composition of human osteoporotic bone was that of Burnell *et al.* [71], who compared iliac crest biopsies from osteoporotic postmenopausal women with vertebral compression fractures to biopsies from normal controls. As expected, osteoporotic bone was less dense. However, the fraction of mineral per gram of bone tissue was also reduced. Moreover, within the mineral phase, carbonate and the calcium-to-phosphorus ratio were decreased, while sodium and magnesium content were increased, yet the same biopsies gave no hint of osteomalacia. Although these results describe average values for the entire study cohort, they reveal considerable heterogeneity in bone composition, even within this group of clinically homogeneous patients. Most patients had normal results; one-quarter showed undermineralized matrix, and only a few showed decreased matrix but normal mineralization. The subjects with decreased mineral fraction were those who also had an increased content of sodium and magnesium in the mineral phase, suggesting the presence of skeletal calcium deficiency.

Drug therapies that decrease bone turnover will eventually increase the degree of matrix mineralization by prolonging the period of secondary mineralization [65, 72]. In contrast, agents that increase bone turnover may lead to a transient decrease in the degree of matrix mineralization as new remodeling units are initiated and new bone laid down. Thus, iliac crest biopsies from postmenopausal women treated with antiresorptive therapy (calcium + vitamin D, raloxifene, risedronate, and alendronate) show an increase in the degree of mineralization that mirrors the suppression of bone turnover [73–76], whereas iliac crest biopsies from men treated with teriparatide show a slight decrease in the degree of mineralization [77]. These effects on matrix mineralization will be reflected in BMD measurements, and likely contribute to the antifracture efficacy of these agents [78, 79].

Another aspect of matrix mineralization that may influence skeletal fragility is the spatial distribution and heterogeneity of mineralization. Individuals with vertebral fractures have a more heterogeneous distribution of mineralization density values than individuals of similar age without fractures [80]. Individuals with fractures had regions of very low mineralization and regions of extremely high mineralization. This finding suggests that the fracture group may have an impaired capacity to regulate bone remodeling to avoid these extremes of tissue mineralization that are likely to be sites of mechanical weakness. Additional data regarding heterogeneity of mineralization density are provided by evaluation of iliac crest biopsy specimens after osteoporosis therapy. In these studies, the heterogeneity of mineralization density values *increases* following intermittent PTH therapy [77] and *decreases* following bisphosphonate therapy [74], yet both treatments are associated with reduced fracture risk. Thus, although theoretical arguments suggest that increasing material homogeneity may negatively impact bone's resistance to fracture, empirical evidence contradicts this view. Clearly, further studies are needed to unravel the complex relationships between material heterogeneity, skeletal fragility, and fracture risk.

2. COLLAGEN CHARACTERISTICS

Bone is a composite material with two primary constituents: mineral and collagen. Although collagen has long taken a back seat to mineral with regards to concepts about skeletal fragility, mounting evidence indicates an important role for age- and disease-related changes in collagen content and structure [37]. The majority of evidence suggests that in normal bone, the mineral provides stiffness and strength, whereas collagen affords bone its ductility and ability to absorb energy before fracturing [81]. The dramatic fragility seen in osteogenesis imperfecta underscores the potential for collagen abnormalities to influence bone strength. However, more subtle alterations in collagen, as noted by polymorphisms in the COL1A1 gene, have also been associated with fracture risk independent of BMD status [82, 83]. Post-translational modifications of collagen have also been shown to influence bone mechanical properties [37, 84, 85], although their specific contribution to age-related skeletal fragility remains to be defined [86–88].

3. MICRODAMAGE

Throughout life, physiologic loading of the skeleton produces fatigue damage in bone. Although the optimal methods to quantify microdamage in bone are under debate, numerous studies show that the accumulation of damage weakens bone (reviewed by Burr [36]). Moreover, it appears that microdamage initiates activation of remodeling, presumably to repair the damaged tissue [89]. This intriguing observation suggests that one important role of bone remodeling is to repair fatigue-induced microdamage in bone. It has been hypothesized that excessive suppression of bone turnover may reduce the capacity of bone to repair microdamage, and eventually lead to reduced mechanical properties [90–93]. Debate regarding the optimal level of bone turnover to prevent architectural deterioration while preserving the ability of bone to maintain calcium homeostasis, respond to altered mechanical loading, and repair microdamage is ongoing [89, 94, 95]. It is interesting to note that whereas the accumulation of microdamage is associated with reduced mechanical properties, the ability of a material to undergo "microcracking" may actually *increase* its toughness [96–99]. As a simple explanation for this latter phenomenon, consider that when a material with a crack in it is loaded, energy is accumulated at the tip of the crack. This energy can either be dissipated by growth of the crack, or by the generation of microcracks near the tip of the larger crack. In this latter case, growth of the larger crack is inhibited, and the material can absorb more energy (i.e., making it tougher) before this larger crack eventually progresses through the material to cause failure. The specific characteristics of bone that confer "good" microcracking versus "bad" microdamage remain to be elucidated.

III. CONCLUSIONS

At the beginning of this chapter we discussed the limitations of a bone mass based diagnosis of osteoporosis. A primary difficulty with such a definition is that its sensitivity to factors known collectively as "bone quality" has not been clarified, and it is tempting to attribute the diagnostic ambiguities of BMD measurements to their failure to account for these features. Although these concerns persist, the fact that information contained in the BMD estimate accounts in part for some of the important geometric, material, and microarchitectural properties solidifies its rationale as a diagnostic criterion. Certainly, any substantial degree of matrix undermineralization would be reflected in a lower BMD, and trabecular disruption of sufficient magnitude to be mechanically important would also register as a bone mineral deficit, and therefore as a lower BMD. Qualitative features that would not be included in a BMD assessment include collagen characteristics, ultrastructural morphology such as cement lines, and the extent and type of accumulated fatigue damage.

The question remains whether osteoporosis should be viewed as one or more unique diagnostic entities, as is the case for Paget's disease, or whether it is more useful to consider it a condition of skeletal fragility resulting from a stochastic process, in which contributory factors include age, body size, adequacy of peak bone mass, degree of adult bone loss, and accumulation of qualitative impairments. Since the overall trajectory over time of adolescent bone acquisition and adult bone loss appears to be universal, the only basis for considering osteoporosis as one or more distinct entities would be a demonstration that its qualitative abnormalities, such as those discussed in this chapter, are restricted to those patients who have suffered a fragility fracture. Although evidence remains incomplete, it seems unlikely that such specificity will be validated for most of these abnormalities.

REFERENCES

1. F. Albright, and E. C. Reifenstein, Jr. *The Parathyroid Glands and Metabolic Bone Disease: Selected Studies*. Williams and Wilkins, Baltimore (1948).

2. B. L. Riggs, H. W. Wahner, E. Seeman, K. P. Offord, W. L. Dunn, R. B. Mazess, K. A. Johnson, and L. J. Melton, 3rd. Changes in bone mineral density of the proximal femur and spine with aging. Differences between the postmenopausal and senile osteoporosis syndromes. *J Clin Invest*, **70**(4), 716–723 (1982).

3. T. Appelboom, and J. J. Body. The antiquity of osteoporosis: More questions than answers. *Calcif Tissue Int*, **53**(6), 367–369 (1993).

4. S. Hui, C. Slemenda, and C. J. Johnston. Baseline measurement of bone mass predicts fracture in White women. *Ann Intern Med*, **111**, 355–361 (1989).

5. L. Melton, E. Atkinson, W. O'Fallon, H. Wahner, and B. Riggs. Long-term fracture prediction by bone mineral assessed at different skeletal sites. *J Bone Miner Res*, **8**, 1227–1233 (1993).

6. S. R. Cummings, D. M. Black, M. C. Nevitt, W. Browner, J. Cauley, K. Ensrud, H. K. Genant, L. Palermo, J. Scott, and T. M. Vogt. Bone density at various sites for prediction of hip fractures. The Study of Osteoporotic Fractures Research Group. *Lancet*, **341**(8837), 72–75 (1993).

7. D. Marshall, O. Johnell, and H. Wedel. Meta-analysis of how well measures of bone mineral density predict occurrence of osteoporotic fractures. *BMJ*, **312**(7041), 1254–1259 (1996).

8. S. C. Schuit, M. van der Klift, A. E. Weel, C. E. de Laet, H. Burger, E. Seeman, A. Hofman, A. G. Uitterlinden, J. P. van Leeuwen, and H. A. Pols. Fracture incidence and association with bone mineral density in elderly men and women: The Rotterdam Study. *Bone*, **34**(1), 195–202 (2004).

9. O. Johnell, J. A. Kanis, A. Oden, H. Johansson, C. De Laet, P. Delmas, J. A. Eisman, S. Fujiwara, H. Kroger, D. Mellstrom, P. J. Meunier, L. J. Melton, 3rd, T. O'Neill, H. Pols, J. Reeve, A. Silman, and A. Tenenhouse. Predictive value of BMD for hip and other fractures. *J Bone Miner Res*, **20**(7), 1185–1194 (2005).

10. S. Boutroy, M. L. Bouxsein, F. Munoz, and P. D. Delmas. In vivo assessment of trabecular bone microarchitecture by high-resolution peripheral quantitative computed tomography. *J Clin Endocrinol Metab*, **90**(12), 6508–6515 (2005).

11. J. A. Kanis, L. J. Melton, 3rd, C. Christiansen, C. C. Johnston, and N. Khaltaev. The diagnosis of osteoporosis. *J Bone Miner Res*, **9**(8), 1137–1141 (1994).

12. D. R. Carter, M. L. Bouxsein, and R. Marcus. New approaches for interpreting projected bone densitometry. *J Bone Miner Res*, **7**, 137–145 (1992).

13. E. S. Siris, Y. T. Chen, T. A. Abbott, E. Barrett-Connor, P. D. Miller, L. E. Wehren, and M. L. Berger. Bone mineral density thresholds for pharmacological intervention to prevent fractures. *Arch Intern Med*, **164**(10), 1108–1112 (2004).

14. E. Sornay-Rendu, F. Munoz, P. Garnero, F. Duboeuf, and P. D. Delmas. Identification of osteopenic women at high risk of fracture: The OFELY study. *J Bone Miner Res*, **20**(10), 1813–1819 (2005).

15. S. A. Wainwright, L. M. Marshall, K. E. Ensrud, J. A. Cauley, D. M. Black, T. A. Hillier, M. C. Hochberg, M. T. Vogt, and E. S. Orwoll. Hip fracture in women without osteoporosis. *J Clin Endocrinol Metab*, **90**(5), 2787–2793 (2005).

16. M. L. Bouxsein. Bone quality: Where do we go from here? *Osteoporos Int*, **14**(Suppl 5), 118–127 (2003).

17. E. Seeman, and P. D. Delmas. Bone quality—The material and structural basis of bone strength and fragility. *N Engl J Med*, **354**(21), 2250–2261 (2006).

18. NIH consensus statement. Osteoporosis prevention, diagnosis, and therapy. *JAMA*, **285**(6), 785–795 (2001).

19. M. Bouxsein. Biomechanics of age-related fractures. In *Osteoporosis*, 2nd ed. (R. Marcus, D. Feldman and J. Kelsey eds.), pp. 509–534. Academic Press, San Diego (2001).

20. A. Courtney, E. F. Wachtel, E. R. Myers, and W. C. Hayes. Age-related reductions in the strength of the femur tested

in a fall loading configuration. *J Bone Jt Surg*, **77**, 387–395 (1995).

21. L. Mosekilde. The effect of modelling and remodelling on human vertebral body architecture. *Technol Health Care*, **6**(5–6), 287–297 (1998).

22. H. H. Jones, J. D. Priest, W. C. Hayes, C. C. Tichenor, and D. A. Nagel. Humeral hypertrophy in response to exercise. *J Bone Jt Surg*, **59–A**, 204–208 (1977).

23. J. Bonadio, K. J. Jepsen, M. K. Mansoura, R. Jaenisch, J. L. Kuhn, and S. A. Goldstein. A murine skeletal adaptation that significantly increases cortical bone mechanical properties. Implications for human skeletal fragility. *J Clin Invest*, **92**(4), 1697–1705 (1993).

24. D. R. Carter, and W. C. Hayes. Bone compressive strength: The influence of density and strain rate. *Science*, **194**, 1174–1176 (1976).

25. D. R. Carter, and W. C. Hayes. The compressive behavior of bone as a two-phase porous structure. *J Bone Jt Surg*, **59–A**(7), 954–962 (1977).

26. J. C. Rice, S. C. Cowin, and J. A. Bowman. On the dependence of the elasticity and strength of cancellous bone on apparent density. *J Biomech*, **21**(2), 155–168 (1988).

27. T. M. Keaveny, E. F. Morgan, G. L. Niebur, and O. C. Yeh. Biomechanics of trabecular bone. *Annu Rev Biomed Eng*, **3**, 307–333 (2001).

28. J. Currey. Effects of porosity and mineral content on the Young's modulus of bone. *Eur Soc Biomech*, **5**, 104 (1986).

29. J. Currey. Physical characteristics affecting the tensile failure properties of compact bone. *J Biomech*, **23**, 837–844 (1990).

30. M. Schaffler, and D. Burr. Stiffness of compact bone: Effects of porosity and density. *J Biomech*, **21**, 13–16 (1988).

31. R. McCalden, J. McGeough, M. Barker, and C. Court-Brown. Age-related changes in the tensile properties of cortical bone. *J Bone Jt Surg*, **75–A**, 1193–1205 (1993).

32. D. B. Burr, M. B. Schaffler, and R. G. Frederickson. Composition of the cement line and its possible mechanical role as a local interface in human compact bone. *J Biomechanics*, **21**(11), 939–945 (1988).

33. D. B. Burr, M. R. Forwood, D. P. Fyhrie, R. B. Martin, M. B. Schaffler, and C. H. Turner. Bone microdamage and skeletal fragility in osteoporotic and stress fractures. *J Bone Miner Res*, **12**(1), 6–15 (1997).

34. S. Hoshaw, D. Cody, A. Saad, and D. Fhyrie. Decrease in canine proximal femoral ultimate strength and stiffness due to fatigue damage. *J Biomech*, **30**, 323–329 (1997).

35. M. Schaffler, K. Choi, and C. Milgrom. Aging and matrix microdamage accumulation in human compact bone. *Bone*, **17**, 521–525 (1995).

36. D. Burr. Microdamage and bone strength. *Osteop Int*, **14**, 67–72 (2003).

37. D. B. Burr. The contribution of the organic matrix to bone's material properties. *Bone*, **31**(1), 8–11 (2002).

38. L. Gibson. The mechanical behaviour of cancellous bone. *J Biomech*, **18**, 317–328 (1985).

39. T. Hildebrand, A. Laib, R. Muller, J. Dequeker, and P. Ruegsegger. Direct three-dimensional morphometric analysis of human cancellous bone: Microstructural data from spine, femur, iliac crest, and calcaneus. *J Bone Miner Res*, **14**(7), 1167–1174 (1999).

40. S. Majumdar, M. Kothari, P. Augat, D. C. Newitt, T. M. Link, J. C. Lin, T. Lang, Y. Lu, and H. K. Genant. High-resolution magnetic resonance imaging: Three-dimensional trabecular bone architecture and biomechanical properties. *Bone*, **22**(5), 445–454 (1998).

41. H. K. Genant, C. Gordon, Y. Jiang, T. M. Link, D. Hans, S. Majumdar, and T. F. Lang. Advanced imaging of the macrostructure and microstructure of bone. *Horm Res*, **54**(Suppl 1), 24–30 (2000).

42. T. M. Link, and S. Majumdar. Current diagnostic techniques in the evaluation of bone architecture. *Cur Osteoporosis Rep*, **2**(2), 47–52 (2004).

43. F. W. Wehrli, P. K. Saha, B. R. Gomberg, H. K. Song, P. J. Snyder, M. Benito, A. Wright, and R. Weening. Role of magnetic resonance for assessing structure and function of trabecular bone. *Top Magn Reson Imaging*, **13**(5), 335–355 (2002).

44. S. Khosla, B. L. Riggs, E. J. Atkinson, A. L. Oberg, L. J. McDaniel, M. Holets, J. M. Peterson, and L. J. Melton, 3rd. Effects of sex and age on bone microstructure at the ultradistal radius: A population-based noninvasive in vivo assessment. *J Bone Miner Res*, **21**(1), 124–131 (2006).

45. R. Goulet, S. Goldstein, M. Ciarelli, J. Kuhn, M. Brown, and L. Feldkamp. The relationship between the structural and orthogonal compressive properties of trabecular bone. *J Biomech*, **27**, 375–389 (1994).

46. M. Bouxsein, and S. Radloff. Quantitative ultrasound of the calcaneus reflects the material properties of calcaneal trabecular bone. *J Bone Miner Res*, **12**, 839–846 (1997).

47. S. Goldstein, R. Goulet, and D. McCubbrey. Measurement and significance of three-dimensional architecture to the mechanical integrity of trabecular bone. *Calcif Tissue Int*, **53**(Suppl 1), S127–S133 (1993).

48. B. D. Snyder, and W. C. Hayes. Multiaxial structure-property relations in trabecular bone. In *Biomechanics of Diarthrodial Joints* (V. C. Mow, A. Ratcliffe, and SL-Y Woo, eds.), pp. 31–59. Springer-Verlag, New York (1990).

49. D. Ulrich, B. van Rietbergen, A. Laib, and P. Ruegsegger. The ability of three-dimensional structural indices to reflect mechanical aspects of trabecular bone. *Bone*, **25**(1), 55–60 (1999).

50. M. J. Silva, and L. J. Gibson. Modeling the mechanical behavior of vertebral trabecular bone: Effects of age-related changes in microstructure. *Bone*, **21**(2), 191–199 (1997).

51. J. C. van der Linden, J. Homminga, J. A. Verhaar, and H. Weinans. Mechanical consequences of bone loss in cancellous bone. *J Bone Miner Res*, **16**(3), 457–465 (2001).

52. O. C. Yeh, and T. M. Keaveny. Biomechanical effects of intraspecimen variations in trabecular architecture: A three-dimensional finite element study. *Bone*, **25**(2), 223–228 (1999).

53. G. H. Bell, O. Dunbar, and J. S. Beck. Variations in strength of vertebrae with age and their relation to osteoporosis. *Calc Tiss Res*, **1**, 75–86 (1967).

54. A. Parfitt. Age-related structural changes in trabecular and cortical bone: Cellular mechanisms and biomechanical consequences. *Calcif Tissue Int*, **36(S)**, 123–128 (1984).

55. T. E. Ciarelli, D. P. Fyhrie, M. B. Schaffler, and S. A. Goldstein. Variations in three-dimensional cancellous bone architecture of the proximal femur in female hip fractures and in controls. *J Bone Miner Res*, **15**(1), 32–40 (2000).

56. E. Legrand, D. Chappard, C. Pascaretti, M. Duquenne, S. Krebs, V. Rohmer, M. F. Basle, and M. Audran. Trabecular bone microarchitecture, bone mineral density, and vertebral fractures in male osteoporosis. *J Bone Miner Res*, **15**(1), 13–19 (2000).

57. T. M. Link, A. Lotter, F. Beyer, S. Christiansen, D. Newitt, Y. Lu, C. Schmid, and S. Majumdar. Changes in calcaneal trabecular bone structure after heart transplantation: An MR imaging study. *Radiology*, **217**(3), 855–862 (2000).

58. J. E. Aaron, P. A. Shore, R. C. Shore, M. Beneton, and J. A. Kanis. Trabecular architecture in women and men of similar bone mass with and without vertebral fracture: II. Three-dimensional histology [In Process Citation]. *Bone*, **27**(2), 277–282 (2000).

59. T. M. Link, K. Kisters Saborowski, M. Kempkes, M. Kosch, D. Newitt, Y. Lu, S. Waldt, and S. Majumdar. Changes in calcaneal trabecular bone structure assessed with high-resolution MR imaging in patients with kidney transplantation. *Osteoporos Int*, **13**(2), 119–129 (2002).

60. T. E. Dufresne, P. A. Chmielewski, M. D. Manhart, T. D. Johnson, and B. Borah. Risedronate preserves bone architecture in early postmenopausal women in 1 year as measured by three-dimensional microcomputed tomography. *Calcif Tissue Int*, **73**(5), 423–432 (2003).

61. B. Borah, T. E. Dufresne, P. A. Chmielewski, T. D. Johnson, A. Chines, and M. D. Manhart. Risedronate preserves bone architecture in postmenopausal women with osteoporosis as measured by three-dimensional microcomputed tomography. *Bone*, **34**(4), 736–746 (2004).

62. R. Recker, P. Masarachia, A. Santora, T. Howard, P. Chavassieux, M. Arlot, G. Rodan, L. Wehren, and D. Kimmel. Trabecular bone microarchitecture after alendronate treatment of osteoporotic women. *Curr Med Res Opin*, **21**(2), 185–194 (2005).

63. D. W. Dempster, F. Cosman, E. S. Kurland, H. Zhou, J. Nieves, L. Woelfert, E. Shane, K. Plavetic, R. Muller, J. Bilezikian, and R. Lindsay. Effects of daily treatment with parathyroid hormone on bone microarchitecture and turnover in patients with osteoporosis: A paired biopsy study. *J Bone Miner Res*. **16**(10), 1846–1853 (2001).

64. Y. Jiang, J. J. Zhao, B. H. Mitlak, O. Wang, H. K. Genant, and E. F. Eriksen. Recombinant human parathyroid hormone (1–34) [teriparatide] improves both cortical and cancellous bone structure. *J Bone Miner Res*, **18**(11), 1932–1941 (2003).

65. P. J. Meunier, and G. Boivin. Bone mineral density reflects bone mass but also the degree of mineralization of bone: Therapeutic implications. *Bone*, **21**(5), 373–377 (1997).

66. J. Currey. The mechanical consequences of variation in the mineral content of bone. *J Biomech*, **2**, 1–11 (1969).

67. J. D. Currey. What determines the bending strength of compact bone? *J Exp Biol*, **202**(Pt 18), 2495–2503 (1999).

68. C. J. Hernandez, G. S. Beaupre, T. S. Keller, and D. R. Carter. The influence of bone volume fraction and ash fraction on bone strength and modulus. *Bone*, **29**(1), 74–78 (2001).

69. H. Follet, G. Boivin, C. Rumelhart, and P. J. Meunier. The degree of mineralization is a determinant of bone strength: A study on human calcanei. *Bone*, **34**(5), 783–789 (2004).

70. C. H. Turner. Biomechanics of bone: Determinants of skeletal fragility and bone quality. *Osteoporos Int*, **13**(2), 97–104 (2002).

71. J. M. Burnell, D. J. Baylink, C. H. Chestnut, 3rd, M. W. Mathews, and E. J. Teubner. Bone matrix and mineral abnormalities in postmenopausal osteoporosis. *Metabolism*, **31**(11), 1113–1120 (1982).

72. P. J. Meunier, M. Arlot, P. Chavassieux, and A. J. Yates. The effects of alendronate on bone turnover and bone quality. *Int J Clin Pract Suppl*, **101**, 14–17 (1999).

73. G. Y. Boivin, P. M. Chavassieux, A. C. Santora, J. Yates, and P. J. Meunier. Alendronate increases bone strength by increasing the mean degree of mineralization of bone tissue in osteoporotic women. *Bone*, **27**(5), 687–694 (2000).

74. P. Roschger, S. Rinnerthaler, J. Yates, G. A. Rodan, P. Fratzl, and K. Klaushofer. Alendronate increases degree and uniformity of mineralization in cancellous bone and decreases the porosity in cortical bone of osteoporotic women. *Bone*, **29**(2), 185–191 (2001).

75. G. Boivin, P. Lips, S. M. Ott, K. D. Harper, S. Sarkar, K. V. Pinette, and P. J. Meunier. Contribution of raloxifene and calcium and vitamin D3 supplementation to the increase of the degree of mineralization of bone in postmenopausal women. *J Clin Endocrinol Metab*, **88**(9), 4199–4205 (2003).

76. G. Boivin, and P. J. Meunier. Methodological considerations in measurement of bone mineral content. *Osteoporos Int*, **14**, 22 28 (2003).

77. B. M. Misof, P. Roschger, F. Cosman, E. S. Kurland, W. Tesch, P. Messmer, D. W. Dempster, J. Nieves, E. Shane, P. Fratzl, K. Klaushofer, J. Bilezikian, and R. Lindsay. Effects of intermittent parathyroid hormone administration on bone mineralization density in iliac crest biopsies from patients with osteoporosis: A paired study before and after treatment. *J Clin Endocrinol Metab*, **88**(3), 1150–1156 (2003).

78. P. D. Delmas. How does antiresorptive therapy decrease the risk of fracture in women with osteoporosis? *Bone*, **27**(1), 1–3 (2000).

79. D. W. Dempster. The impact of bone turnover and bone-active agents on bone quality: Focus on the hip. *Osteoporos Int*, **13**(5), 349–352 (2002).

80. T. E. Ciarelli, D. P. Fyhrie, and A. M. Parfitt. Effects of vertebral bone fragility and bone formation rate on the mineralization levels of cancellous bone from white females. *Bone*, **32**(3), 311–315 (2003).

81. J. Currey. Role of collagen and other organics in the mechanical properties of bone. *Osteop Int*, **14**, 29–36 (2003).

82. V. Mann, E. E. Hobson, B. Li, T. L. Stewart, S. F. Grant, S. P. Robins, R. M. Aspden, and S. H. Ralston. A COL1A1 Sp1 binding site polymorphism predisposes to osteoporotic fracture by affecting bone density and quality. *J Clin Invest*, **107**(7), 899–907 (2001).

83. Z. Efstathiadou, A. Tsatsoulis, and J. P. Ioannidis. Association of collagen ialpha 1 Sp1 polymorphism with the risk of prevalent fractures: A meta-analysis. *J Bone Miner Res*, **16**(9), 1586–1592 (2001).

84. P. Garnero, O. Borel, E. Gineyts, F. Duboeuf, H. Solberg, M. L. Bouxsein, C. Christiansen, and P. D. Delmas. Extracellular post-translational modifications of collagen are major determinants of biomechanical properties of fetal bovine cortical bone. *Bone*, **38**(3), 300–309 (2006).

85. S. Viguet-Carrin, J. P. Roux, M. E. Arlot, Z. Merabet, D. J. Leeming, I. Byrjalsen, P. D. Delmas, and M. L. Bouxsein. Contribution of the advanced glycation end product pentosidine and of maturation of type I collagen to compressive biomechanical properties of human lumbar vertebrae. *Bone*, **39**, 1073–1079 (2006).

86. P. Garnero, P. Cloos, E. Sornay-Rendu, P. Qvist, and P. Delmas. Type I collagen racemization and isomerization and the risk of fracture in postmenopausal women: The OFELY Prospective Study. *J Bone Miner Res*, **17**(5), 826–833 (2002).

87. M. Saito, K. Fujii, and K. Marumo. Degree of mineralization-related collagen crosslinking in the femoral neck cancellous bone in cases of hip fracture and controls. *Calcif Tissue Int*, **79**(3), 160–168 (2006).

88. M. Saito, K. Fujii, S. Soshi, and T. Tanaka. Reductions in degree of mineralization and enzymatic collagen cross-links and increases in glycation induced pentosidine in the femoral neck cortex in cases of femoral neck fracture. *Osteoporos Int*, **17**(7), 986–995 (2006).

89. M. Schaffler. Role of bone turnover in microdamage. *Osteop Int*, **14**, 73–80 (2003).

90. T. Mashiba, T. Hirano, C. H. Turner, M. R. Forwood, C. C. Johnston, and D. B. Burr. Suppressed bone turnover by bisphosphonates increases microdamage accumulation and reduces some biomechanical properties in dog rib [see comments]. *J Bone Miner Res*, **15**(4), 613–620 (2000).

91. T. Mashiba, C. H. Turner, T. Hirano, M. R. Forwood, C. C. Johnston, and D. B. Burr. Effects of suppressed bone turnover by bisphosphonates on microdamage accumulation and biomechanical properties in clinically relevant skeletal sites in beagles. *Bone*, **28**(5), 524–531 (2001).

92. S. Komatsubara, S. Mori, T. Mashiba, M. Ito, J. Li, Y. Kaji, T. Akiyama, K. Miyamoto, Y. Cao, J. Kawanishi, and H. Norimatsu. Long-term treatment of incadronate disodium accumulates microdamage but improves the trabecular bone microarchitecture in dog vertebra. *J Bone Miner Res*, **18**(3), 512–520 (2003).

93. S. Komatsubara, S. Mori, T. Mashiba, J. Li, K. Nonaka, Y. Kaji, T. Akiyama, K. Miyamoto, Y. Cao, J. Kawanishi, and H. Norimatsu. Suppressed bone turnover by long-term bisphosphonate treatment accumulates microdamage but maintains intrinsic material properties in cortical bone of dog rib. *J Bone Miner Res*, **19**(6), 999–1005 (2004).

94. A. M. Parfitt. Targeted and nontargeted bone remodeling: Relationship to basic multicellular unit origination and progression. *Bone*, **30**(1), 5–7 (2002).

95. D. B. Burr. Targeted and nontargeted remodeling. *Bone*, **30**(1), 2–4 (2002).

96. D. Vashishth, J. C. Behiri, and W. Bonfield. Crack growth resistance in cortical bone: Concept of microcrack toughening. *J Biomech*, **30**(8), 763–769 (1997).

97. D. Vashishth, K. E. Tanner, and W. Bonfield. Experimental validation of a microcracking-based toughening mechanism for cortical bone. *J Biomech*, **36**(1), 121–124 (2003).

98. P. Zioupos. Recent developments in the study of failure of solid biomaterials and bone: "Fracture" and "pre-fracture" toughness. *Math Sci and Eng*, **6**, 33–40 (1998).

99. P. Zioupos. On microcracks, microcracking, in-vivo, in-vitro, in-situ and other issues. *J Biomech*, **32**(2), 209–211, 213–259 (1999).

Skeletal Heterogeneity and the Purposes of Bone Remodeling: Implications for the Understanding of Osteoporosis

A. M. PARFITT

I. INTRODUCTION

The cells of bone influence its structure by means of four processes: growth, repair, modeling, and remodeling, the last being the basis of bone tissue turnover in the adult skeleton. The purposes of growth and repair are obvious. Modeling serves to adapt bones to changes in mechanical loading [1], and remodeling serves to thicken trabeculae in the growing skeleton [2], processes that are most effective during adolescence [3]. But why does a tissue that can survive for thousands of years after death need to be maintained by periodic replacement during life? Most of those interested in bone, whether as physicians, as clinical investigators, or as basic scientists, show remarkably little interest in this fundamental question. Many articles and book chapters discuss the *regulation* of bone remodeling, but regulation, at least in the physiologic sense, implies a target [4]. The target value of any regulatory process in biology has been optimized by natural selection. Mechanisms have evolved which ensure that deviations from the target are detected and that corrective measures to restore the target value are carried out. In this sense, body temperature, extracellular fluid osmolality, tissue oxygen tension, and countless other physiologic quantities are regulated, but the mechanisms of regulation could not be determined until the existence of the target had been recognized and its precise nature defined. Is there a target for bone remodeling or for some characteristic of bone that is influenced by remodeling?

The piecemeal, quantal nature of bone remodeling is well known. The process is carried out by temporary anatomic structures known as basic multicellular units, or BMUs [5–8], which excavate and replace tunnels through cortical bone (osteonal remodeling) or trenches across the surface of cancellous bone (hemiosteonal remodeling). Each BMU includes two teams of executive cells (osteoclasts and osteoblasts), supported by blood vessels, nerves, and loose connective tissue. The life span of the BMU is measured in months, but the life span of osteoblasts while they are making bone is measured in weeks, and the life span of osteoclast nuclei is measured in days. During progression of the BMU through or across the surface of bone, the spatial and temporal relationships between its components are maintained by the continued growth of the central capillary in cortical bone [9], and extension of the remodeling compartment in cancellous bone [10], together with recruitment of new cells [9–11]. These cells, like the formed elements of the blood, originate from stem cells in the bone marrow [12] except that in the peripheral skeleton osteoblasts are derived from local precursors [9]. For blood cells, as for other short-lived cells, control of cell production and survival is more important than control of differentiated cell function; although the details are less clear, the same applies also to bone cells [12].

Each type of blood cell is normally produced at a basal rate that is sufficient for ordinary purposes but that can be increased when needed [13]. For each cell type, the circumstances under which demand is increased are well known, and are related to the function of the particular cell, although the cell types differ with respect to the time scale of this response, its specificity, the relative importance of reactive and anticipatory homeostasis [14], and the extent to which the control mechanisms have been elucidated. The importance of these relationships between supply and demand, and between demand and function, applies also to bone cells. For osteoblasts in the adult nongrowing skeleton, the demand is created by bone resorption, since the function of osteoblasts is to replace the bone removed by osteoclasts. However, the circumstances that create a demand for osteoclasts are much less well defined, since these circumstances are dictated by the purposes of bone remodeling. Indeed, the questions "What are the purposes of bone remodeling and how are they achieved?" are essentially equivalent to the questions "Where and when are osteoclasts needed, and how is this need recognized and satisfied?" The answers to these questions are different in different types of bone and in different regions of the skeleton.

II. SKELETAL HETEROGENEITY

A. Structure and Function

The structural differences between cortical bone, in which porosity and surface-to-volume ratio are low, and cancellous bone, in which these geometric quantities are high [15], are now widely recognized. All intermediate values for these quantities can occur, but they are infrequent, implying that transitional structures tend to be temporary and short-lived [16]. Less often noted are the differences between the axial and appendicular subdivisions of the skeleton (Table 3-1); the pelvis, defined anatomically as appendicular, behaves functionally as part of the axial skeleton, so that it is more accurate to contrast central with peripheral regions. This distinction is important because the different functions of the skeleton are divided differently between the central and peripheral components. The primary function is load-bearing—to support posture, permit movement (including locomotion), and provide protection for the soft tissues. Subsidiary functions are to participate in mineral homeostasis and to provide a favorable microenvironment for hematopoiesis. For convenience the former functions will be referred to as "mechanical" and the latter as "metabolic" [13].

It is commonly believed that the mechanical functions are carried out mainly by cortical bone and the metabolic functions mainly by cancellous bone, regardless of their central or peripheral locations. In fact, the functions of the peripheral skeleton, cancellous as well as cortical, are mainly mechanical, whereas the central skeleton, cortical as well as cancellous, in addition to its mechanical function, participates to a much greater extent in the metabolic functions of bone. This revision in functional attribution is most striking for peripheral cancellous bone, such as in the metaphyses of the long bones [17]. As is evident from the orientation of the trabeculae (Figure 3-1A), metaphyseal cancellous bone transmits loads from the joint surfaces to diaphyseal cortical bone. Indeed, the metaphyses are flared in shape precisely to make such load transmission possible. Similar functional and architectural considerations apply to the cancellous bone in the small bones of the hands and feet (Figure 3-1B). As will subsequently be discussed in detail, there is no evidence that such peripheral cancellous bone participates to a significant extent in the metabolic functions of the skeleton, whether related to mineral homeostasis or to hematopoiesis.

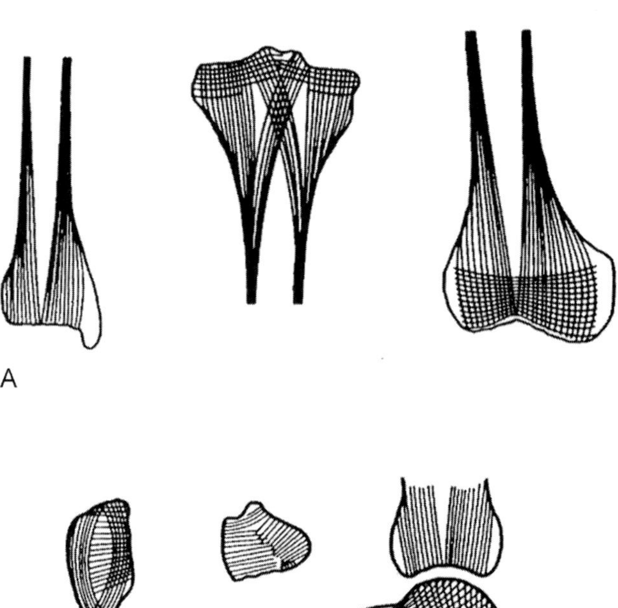

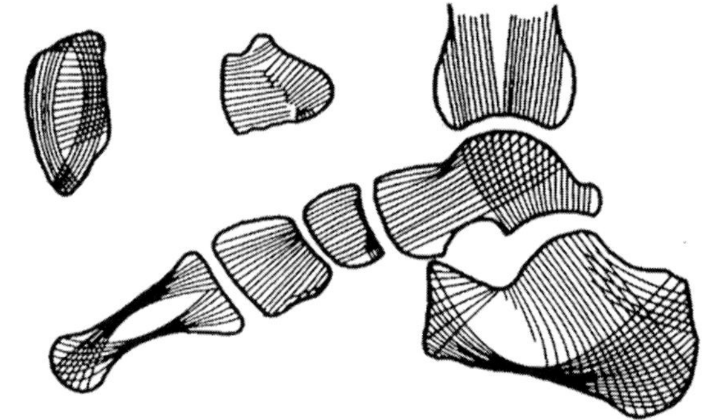

Figure 3-1 (A) Examples of trabecular orientation in metaphyseal cancellous bone in the appendicular skeleton. The alignment with stress trajectories facilitates transmission of loads from the joints to diaphyseal cortical bone. (B) Examples of trabecular orientation in the small bones of the feet. The alignment with stress trajectories facilitates transmission of loads during locomotion to the ankle joint and thence to diaphyseal cortical bone in the tibia. Modified from [17].

Table 3-1 Subdivisions of the Skeleton

Feature	Central	Peripheral
Main bone tissue	Cancellous	Cortical
Main soft tissue	Viscera	Muscle
Main joint type	Various	Synovial
Cortices	Thin	Thick
Marrow	Hematopoietic	Fatty
Turnover	High	Low

B. Remodeling and Turnover

The frequent assertion that cancellous bone has higher turnover than cortical bone is usually supported by comparing central cancellous with peripheral cortical bone, but this is to confuse the geometrical and biological factors that influence turnover [18]. The remodeling process occurs only on bone surfaces, and the intensity of remodeling is expressed by the activation frequency, which is the reciprocal of the average time interval between the initiation of consecutive cycles of remodeling at the same surface location, referred to as the regeneration period [19]. Turnover refers to volume replacement, which depends not only on the surface-defined activation frequency but on the surface-to-volume ratio. This geometrical property is about four to five times higher in typical cancellous bone than in typical cortical bone [15, 20]. Consequently, the former could have higher turnover despite a lower intensity of remodeling. Systematic site-specific measurements of turnover in the human skeleton are available only for the rib [5] and for the ilium. In the latter, activation frequency is similar on the cancellous, endocortical, and intracortical subdivisions of the endosteal envelope [21, 22], so that the difference in turnover between cortical and cancellous bone at this site depends entirely on the difference in surface-to-volume ratio. Turnover depends also on distance from the surface; iliac interstitial bone has much lower turnover than inner cortical bone close to the marrow [18]. Unfortunately, the ilium, although probably representative of the central skeleton in general, and of the vertebral bodies in particular [23], is quite unrepresentative of the peripheral skeleton [23, 24].

In peripheral cortical bone, turnover is lower by about half than in the ribs (around 2% per year vs. 4% per year), based on a variety of indirect methods [18, 25]. For peripheral cancellous bone, estimates of turnover are based on fewer data, but they suggest that the central-peripheral difference is greater than for cortical bone. During the treatment of osteomalacia the increase in cancellous bone mineral was about 35% in the ilium, measured histologically, but only 1–2% in the distal radius, measured by single photon absorptiometry [26]. On the reasonable assumption that unmineralized osteoid tissue accumulates during the evolution of osteomalacia in proportion to the initial rate of turnover, this rate in the cancellous bone of the distal radius is normally only about 2% per year. This is similar to the estimate for peripheral cortical bone; since the surface-to-volume ratio would be higher in cancellous bone, the activation frequency would be even lower than on the intracortical surfaces. Direct measurements of turnover in the beagle con-

firm a much lower value for peripheral than for central cancellous bone, even though the absolute values for both were higher than in human subjects [27].

C. Relationship to Marrow Composition

In the embryo, hematopoietic marrow appears first in the yolk sac and subsequently migrates to the liver and spleen, and then to the marrow cavities. At birth, hematopoiesis is active in cancellous bone throughout the skeleton but has virtually ceased at extramedullary sites [28, 29]. During growth, there is gradual conversion of red to yellow marrow, a process that begins in the distal extremities and proceeds centripetally. By age 25, hematopoiesis has disappeared from the peripheral skeleton, except to a limited extent in the upper femora [28]. Macroscopically visible hematopoiesis continues in the central skeleton throughout life, although there is a gradual increase in the number of fat cells at the expense of hematopoietic cells [30]. At any age, a sustained increase in demand can lead to reappearance of hematopoiesis in the extremities [28, 31]. Whether this results from reactivation of dormant local stem cells or from recolonization of fatty marrow by circulating stem cells is not known, but the latter seems more likely since it is now certain that hematopoietic stem cells do circulate [32].

The data presented, although incomplete, indicate that in the adult human skeleton central cancellous bone has persistent hematopoiesis and high bone turnover (except for interstitial bone), whereas peripheral cancellous bone has absent hematopoiesis and low bone turnover (Table 3-2). Furthermore, based on external radionuclide counting, there is a close correlation between the extent of hematopoiesis and bone blood flow [33]. When different bones sampled at autopsy were compared, there was a good relationship between the proportion of cancellous bone surface in contact with hematopoietic cells and the proportion engaged in

TABLE 3-2 Cancellous Bone and Its Marrow

Feature	Red Marrow	Yellow Marrow
Bone type	Metabolic	Mechanical
Location	Central	Peripheral
Main functions	Calcium homeostasis	Transmit loads
	Support hematopoiesis	Absorb energy
Cellularity	High	Low
Blood flow	High	Low
Turnover	High	Low

bone remodeling [34]. In adult beagles, there is an even more striking correspondence between marrow composition and bone remodeling (Table 3-3). Adjacent to red marrow, there is a 15% higher mineral apposition rate and an almost 10-fold higher bone formation rate than adjacent to yellow marrow, with corresponding differences in the uptake of plutonium [35, 36]. If there are no resident hematopoietic stem cells in yellow marrow, all osteoclasts in the peripheral skeleton, cancellous as well as cortical, must be derived from circulating mononuclear precursor cells [11, 13]. In the central skeleton also, participation of the local microcirculation has now been established for normal bone remodeling [10, 37]. In pathologic bone resorption, as in neoplastic bone disease [38] or osteoprotegerin deficiency [39], osteoclast precursors produced in much larger numbers than needed might be able to migrate directly to the bone surface.

The relationship between marrow composition and remodeling can be disturbed in pathologic conditions. For example, after ovariectomy, bone turnover and amount of fat in the marrow both increase [40]. No relationship between marrow composition and bone remodeling was found in a single patient with osteoporosis who died from an unrelated cause after administration of tetracycline labels in preparation for bone biopsy [41]. The relationship is also disturbed by proximity to synovial joints; turnover is higher within 1 mm of the articular surface than at more distant locations [26]. Nevertheless, the spatial association between hematopoiesis and active remodeling appears to be characteristic of the healthy skeleton. To most observers, this is simply the expected consequence of the presence or absence of precursor cells in close proximity to the bone surface, but this is a superficial view, since circulating osteoclast precursors can be made available anywhere in the skeleton. Does cancellous bone *need* to turn over so much faster in some locations than in others, and if so, why?

TABLE 3-3　Cancellous Bone Turnover in Normal Beagles

Site	Marrow	MAR[a] (μm/day)	BFR[b] (%/year)
Lumbar vertebra	Red	1.29 ± 0.10	106 ± 9
Proximal humerus	Red	1.23 ± 0.10	89 ± 18
Pelvis	Red	1.26 ± 0.10	83 ± 25
Proximal ulna	Yellow	0.90 ± 0.06	13 ± 6
Distal ulna	Yellow	0.97 ± 0.07	7 ± 3

Note: Data expressed as mean ± SE. From [35] and [36].
[a] Mineral apposition rate: n = 8.
[b] Bone formation rate: n = 4.

III.　THE PURPOSES OF BONE REMODELING

There is probably no physiologic function other than bone remodeling that has attracted so much study in the face of so much uncertainty about why it occurs. Many in the field act as if they believed that the only purpose of remodeling was to cause osteoporosis and thus provide employment for scientists and business opportunities for the pharmaceutical industry! In the analysis of this problem, it seems reasonable to make two assumptions. First, periodic replacement of bone serves to maintain its ability to carry out its functions, as previously summarized. Second, since the most obvious difference between the old bone removed and the new bone put in its place is in their ages, excessive age of bone in some way compromises its functional capacity. *Bone* age must be carefully distinguished from *subject* age. Different regions of bone have widely different ages, since some bone was made yesterday and some was made decades ago, but *mean* bone age is the reciprocal of the mean rate of turnover [19]; if turnover increases with subject age for any reason, mean bone age will be lower in an older than in a younger person. Only at sites where turnover is extremely low such as the auditory ossicles [42] or deep interstitial cancellous bone [18] does bone age necessarily increase with chronological age.

As bone gets older, its true density increases as secondary mineralization progresses and water is displaced; consequently, it becomes more brittle [43]. There are also changes in matrix constituents such as accumulation of products of advanced glycation [44] and increased cross-linking of collagen [45]. Osteocytes have a finite life span, eventually dying by apoptosis leaving an apparently empty lacuna that may eventually become occluded by mineralized debris [46]. As will later be discussed in more detail, fatigue microcracks increase in number with bone age and are spatially associated with missing osteocytes [47]. The adverse effects of increased bone age have been studied mainly in cortical bone, but central cancellous bone regions of bone of widely different ages are in close proximity [18]. The mean age of surface bone close to the marrow varies from about 0.5 to 4 years, but mean age increases progressively with increasing distance from the surface, and beyond 75 μm the bone is essentially isolated from surface remodeling so that its age is close to the age of the subject [48]. Differences between interstitial and surface bone, due entirely to the difference in age, are listed in Table 3-4. The remodeling of interstitial cancellous bone carries a high risk for trabecular plate perforation [8, 19] so that at this location excessive bone age is the price that has to be paid for the preservation of cancellous architecture [48].

TABLE 3-4   The Effect of Location on Iliac Cancellous Bone Age and Age-Dependent Properties

Features	Surface (superficial)	Interstitial (deep)
Turnover	High	Low
Age	Low	High
Density	Low	High[a]
Microdamage	Low	High[b]
Osteocytes[c]		
Density	High	Low
Effect of age	No change	Fall
Determinant	Initial density	Life span

[a][50].
[b]By analogy with cortical bone [47].
[c][46].

The salutary effects of remodeling in preventing excessive aging of bone will differ in different regions, depending on which effects of age are more important. Since the primary function of bone is mechanical, the primary purpose of remodeling of bone is to maintain its load-bearing capacity. This is accomplished both by preventing the adverse effects of excessive bone age at the microscopic and submicroscopic levels, and by repairing damage after it occurs. The role of remodeling in maintaining the metabolic functions of bone will be considered later, but first the remodeling apparatus and how it behaves over time must be examined in greater detail.

A.   The Life History of a BMU

The stages of a BMU are commonly depicted as quiescence, activation, resorption, reversal, formation, and back to quiescence [8]. These terms refer to successive states of the bone surface at a single location; it is the *surface* that becomes activated. This simple down and up model has been useful in describing temporal relationships and in analyzing the cellular basis of bone loss, but it conceals the three-dimensional reality, already briefly mentioned, that the BMU moves through tissue space, so that it has to *begin* somewhere and *end* somewhere else [6–8, 49]. The beginning of a new BMU, termed *origination*, occurs in response to the identification of a target—a region of bone that needs to be replaced [51]. The need is recognized by osteocytes, which communicate in some way with the cells that line the nearest bone surface, which in turn communicate with the nearest small blood vessel. In cortical bone the vessel of origin is the central capillary of a Haversian or Volkmann canal from which a new capillary grows [9]; neoangiogenesis is an essential component of bone remodeling. The new capillary passes between the lining cells, which have digested the thin layer of unmineralized matrix beneath them and then retracted [8], and progresses through the bone in the direction of the target, in the wake of the cutting cone of new osteoclasts [6, 7]. The relationship between the capillary and the osteoclasts is symbiotic: The capillary cannot advance until room is made for it by resorption, and the osteoclasts cannot continue to advance unless their dying nuclei are replaced by the diapedesis of monocytic osteoclast precursors from the capillary, which requires the right area code to be turned on [7, 11]. In cancellous bone, the new capillary sprout penetrates, in the same manner as in cortical bone, between lining cells, which then form a canopy over the new remodeling site, now in direct contact with a temporary extension of the circulation [37]. The new structure, termed the bone remodeling compartment, represents a form of vasculogenic mimicry [10].

In order for the new BMU to reach its target, it excavates a tunnel through cortical bone—osteonal remodeling—or a trench across the surface of cancellous bone—hemiosteonal remodeling [6]. Progression of the BMU requires continued access to the circulation and arrival of new osteoclast precursor cells and their replacement from the bone marrow [9, 11]. Directional information must somehow be provided by the osteocytic lacunar–canalicular system; the limited available knowledge will be summarized later. During *longitudinal* progression of the BMU, successive *transverse* cycles of remodeling are generated, each new cycle slightly out of step with the one before [8]. The total number of such cycles per unit time is the activation frequency, usually reported for a specific surface or region, but which conceptually can be aggregated for the whole skeleton and estimated rather crudely by biochemical indices of bone turnover. After the target has been reached and replaced by new bone, the BMU will continue to progress for some distance beyond the target because it has acquired some biological momentum, but will eventually come to a stop. Such post-targeted remodeling has been described as "redundant," "surplus," "spare," "nontargeted," or "stochastic." Each of these terms has some merit, but "spare" is probably the least inaccurate. Origination is such an intricate process that it could not occur by chance [49]. All remodeling is either targeted, requiring a new BMU, or post-targeted, requiring progression of an existing BMU; the distinction between them will be discussed in greater detail in subsequent sections.

B. Fatigue Damage and Mechanical Competence

All structural materials that undergo repetitive cyclical loading are subject to fatigue, a phenomenon that has been most extensively studied in fabricated materials such as steel [42]. After a certain number of load cycles, tiny cracks appear that are detectable at first at the ultramicroscopic level, but were probably preceded by damage at the submicroscopic and molecular levels. If cyclical loading continues, the cracks extend and accumulate into microscopic and then macroscopic damage and eventually into overt structural failure. The essence of fatigue is that in each cycle, the load-induced strain (relative deformation) is far below the instantaneous breaking strain of the intact material. Biological materials such as bone also undergo fatigue damage but differ from man-made materials in their capacity for self-repair [52]. The occurrence of fatigue damage has been demonstrated unequivocally in cortical bone [53, 54], and there is compelling evidence that experimentally induced fatigue damage in cortical bone induces repair by remodeling, so that the damaged bone is removed and replaced by new undamaged bone [55, 56]. It is reasonable to assume that the same applies to load-bearing cancellous bone, which also develops fatigue damage with repetitive cyclical loading [57, 58]. Various degrees of microdamage can be identified in human cancellous bone [59], including microcracks that closely resemble those observed in cortical bone [60, 61]. Unlike those in femoral cortical bone [62] such cracks do not increase significantly with age in the vertebral body [59] and do not increase until after age 60 in the femoral head [60], so that for the most part they must be repaired by remodeling. However, it is not certain that such lesions are due to fatigue, since identical lesions can be produced experimentally by compression [63]. Microfractures in cancellous bone heal by callus formation rather than by remodeling [64], and although often called fatigue fractures, most of them (at least in the vertebrae) can be explained, not by fatigue, but by instantaneous overload, leading to failure by buckling [65].

Evidently, a major function of remodeling is to provide a means for replacing load-bearing bone that has undergone fatigue microdamage; indeed, it is quite possible that all BMU origination events in the peripheral skeleton are triggered by microdamage [66], and that this mechanism has evolved to allow large long-lived vertebrates to maintain a light skeleton [52, 67]. But repair of microdamage may not be the only way in which remodeling maintains the mechanical competence of bone. The similarity between different members of the same species in the spatial distribution of remodeling activity at different skeletal sites [68] is difficult to explain by a mechanism that is purely reparative. One of the most striking aspects of such remodeling maps is their bilateral symmetry, such that cross-sections at the same level of bones on opposite sides of the body are virtually mirror images of one another [42, 69]. It seems unlikely that such consistent symmetry could be the expression of fatigue damage *repair*, but it might be an expression of fatigue damage *prevention*.

Because of bilateral symmetry in local bone geometry and mass, there will be bilateral symmetry in the local strains engendered by mechanical loading. The relationship between strain and remodeling rate is "U" shaped [70]; increasing strain is accompanied by increased remodeling before the occurrence of strain-induced damage [71], presumably by prolonging the post-targeted progression of existing BMUs. For material of the same mechanical properties, the major determinants of fatigue damage are the number of load cycles and the average change in strain in each cycle, and for the same level of physical activity, the major determinant of the number of load cycles is the age of the structure. The customary pattern and intensity of physical activity are species specific and so are genetically determined [3]. Consequently, it seems possible that the remodeling map is the expression of a genetic program to prevent bone age from exceeding some critical level, a level that is different in different regions of the skeleton [68–71]. This would be consistent with the notion that remodeling evolved as a means to prolong the fatigue life of bone [52, 67].

The contrast between the *prevention* of fatigue and other forms of damage by keeping bone age below some critical value and the *repair* of such damage by removal of the bone involved is analogous to the contrast between anticipatory and reactive homeostasis [14], except that the basis of the anticipation is genetic rather than physiologic. More specifically, it exemplifies the distinction between targeted and post-targeted remodeling, a distinction that establishes an order of priority for different remodeling projects. There is a wide range of turnover rates consistent with skeletal health [5, 72], and the low rates that occur in hypothyroidism [73] and hypoparathyroidism [74] do not appear to increase fracture risk. Presumably, the reason is that spare remodeling to prevent excessive bone age provides a substantial margin of safety. Consequently, curtailing the post-targeted progression of a particular BMU is unlikely to have any harmful effects. However, targeted remodeling to remove fatigued bone before the damage escalates from microscopic to macroscopic, which requires new BMU origination, must be carried out promptly, or else it will fail in its purpose.

The existence of such a temporal hierarchy has an important impact on the therapeutic reduction of bone turnover, a point that will subsequently be discussed in more detail.

The mechanism of targeted remodeling in load-bearing bone is now much clearer. The only cell that is in the right location to detect microscopic damage is the osteocyte. This cell can be activated by mechanically induced strain to increase the synthesis of various proteins and prostaglandins, nitrous oxide, and no doubt other signaling molecules [75, 76], effects that probably mediate the addition of bone to the nearest bone surface during growth [3, 5], but microdamage repair requires the origination of a new BMU as previously described. In the adult rat ulna, there is a close relationship, both spatial and temporal, between experimentally induced fatigue damage, osteocytes undergoing DNA fragmentation during apoptosis, and resorption spaces containing osteoclasts [77, 78], but osteocyte death is preceded by increased expression of Bax, a pro-apoptotic gene [79]. Osteocytes more than 1 to 2 mm from the damaged bone show increased expression of Bcl2, an anti-apoptotic gene [79]. Osteocytes exert a general suppressant effect on bone remodeling [70, 80], but BMU origination requires a positive signal, either from dying osteocytes or from surrounding Bcl2 expressing osteocytes, which serves also as a beacon or homing signal for the advancing BMU [79]. Whether the signal is biochemical, electrical, hydraulic, or neural is unknown. Many other factors can influence one or more steps in this complex process, but their role is permissive, not regulatory [4]. In unloaded bone also, osteocyte apoptosis serves as a beacon for osteoclastic removal of bone perceived as no longer needed [81], but the bone removed for damage repair is completely replaced, whereas the bone removed in response to unloading is replaced incompletely or not at all.

C.   Metabolic Functions of Remodeling

The foregoing argument has established three interconnected facts. First, the primary function of metaphyseal cancellous bone in the extremities is mechanical load bearing. Second, the reason why load-bearing bone must be remodeled is to maintain its mechanical competence. Third, the rate of turnover of load-bearing bone adjacent to fatty marrow, whether cortical or cancellous, is low. Clearly, a low rate of turnover, of the order of 2–5% per year, is sufficient to maintain the mechanical competence of bone, regardless of its location in the skeleton or its geometric features. Consequently, the rate of turnover of axial cancellous bone adjacent to hematopoietic marrow (15–35% per year) is much higher (by a factor of at least 5) than is necessary to maintain mechanical competence [82]. Remodeling rates were higher in the past because of changes in nutrition and physical activity [72], but even the lower rates in pre-agricultural humans were much higher than needed for maintenance [82]. Unless this mechanically surplus or spare remodeling is simply a form of occupational therapy for cells with nothing better to do, it must serve an entirely different purpose. This conclusion will not surprise the many endocrinologists who have always believed that the main purpose of bone remodeling was to support calcium homeostasis, but the restriction of this function mainly to cancellous bone adjacent to red marrow has not previously been emphasized. The relative importance of the mechanical and metabolic aspects of remodeling, debated inconclusively for many years [69, 70], is evidently different in different regions of the skeleton, although both are essential to the organism as a whole.

The most important nonmechanical function of bone remodeling concerns the regulation of calcium homeostasis. Bone is involved in both determining the steady-state target value for plasma-free calcium and correcting deviations from the target value [83]. Both of these processes depend on a relatively high rate of bone remodeling, but in quite different ways. Bone mineral also functions as a reservoir for sodium and as a buffer for hydrogen ion regulation. Bone remodeling may also provide biochemical support for hematopoiesis as well as the mechanical support provided by the bone itself. Both the number and the proliferative activity of stem cells are greatest adjacent to the endosteal surface, where they are segregated in microenvironmental niches [84], and for this reason bone lining cells may need timely replacement. Bone matrix contains growth factors and other regulatory molecules, some of which may act on blood-forming cells rather than on bone cells. For several reasons, it could be advantageous for such molecules to be released into the bone marrow during bone resorption rather than directly from the cells involved in their biosynthesis. Possible reasons include cell polarization, with osteoblasts transporting substances away from, and osteoclasts toward, the marrow; the high proton concentration within the ruffled border of osteoclasts; and a need for intermittent rapid release rather than more continuous slow release. However, this is speculative, and the remainder of the discussion will focus on the relationship between bone remodeling and calcium homeostasis.

Except under conditions of extreme calcium deprivation, the calcium homeostatic function of remodeling is not antagonistic to the mechanical function, since normally calcium homeostasis does not depend

42   A. M. PARFITT

on continued net loss of calcium from bone [83]. Steady-state levels of plasma-free calcium can be high, normal, or low, regardless of the directional changes in osteoclastic bone resorption or in calcium balance [85]. Plasma-free calcium is regulated by the joint effects of parathyroid hormone (PTH) on the renal tubular reabsorption of calcium and on the blood–bone equilibrium. This equilibrium is achieved when the inward and outward fluxes of calcium at quiescent bone surfaces are equal, and the calcium level at which this occurs is determined by some effect of PTH on bone lining cells [86, 87]. For this mechanism to be effective, several conditions must be met. First, there must be a high blood flow, which is ensured by the proximity of hematopoietic marrow. Second, the bone at the surface must retain enough water to permit rapid diffusion of minerals, which is ensured by a high rate of remodeling. As previously indicated, as bone ages, secondary mineralization proceeds slowly to completion by crystal enlargement and displacement of water, with a progressive decline in its ability to support the rapid mineral exchanges on which plasma-calcium homeostasis depends [83]. Spare, post-targeted remodeling could prevent excessive aging of surface bone, but as for fatigue damage, from time to time targeted remodeling will be needed to remove bone that has become hypermineralized. The mechanism of targeting is less well understood than for fatigue damage but should be simpler, since the bone to be removed is on rather than beneath the surface. One signal to surface remodeling is loss of osteocytes, a mechanism that serves to maintain osteocyte density, probably in the interests of mineral exchange [80].

In addition to determining the steady-state target level of plasma-free calcium, the bone also participates in the correction of deviations from the target value. A fall in plasma-free calcium stimulates PTH secretion, which increases the outflow of calcium from bone, not only by shifting the balance of exchange at quiescent bone surfaces but also by increasing the resorptive activity of existing osteoclasts. This acute effect is quite separate from the long-term effect of PTH to increase activation frequency, osteoclast recruitment, and bone turnover in primary and secondary hyperparathyroidism [88]. Obviously, the rapidity of the correction depends on the number of osteoclasts available, which is determined by the number of BMUs present, and by the efficiency of the local circulation. The most important use for this mechanism is to accommodate the circadian changes in the supply of calcium from intestinal absorption, with an approximately 12- to 16-hour period of eating, followed by an 8- to 12-hour period of fasting, during which both PTH secretion and bone resorption increase [89]. In each BMU, the cutting

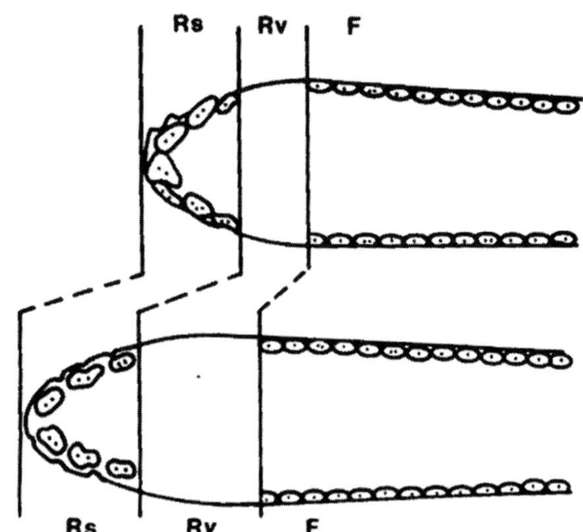

FIGURE 3-2   Contribution of BMU-based remodeling to short-term demands for calcium. During the night the osteoclasts of the cutting cone (Rs) advance more quickly than the osteoblasts of the closing cone (F), increasing the extent of the reversal zone (Rv). During the day, the cutting cone slows down, and the closing cone catches up. The same concertina-like action can occur with cancellous BMUs (hemiosteonal remodeling). Reprinted with permission from [83].

cone (in osteonal remodeling) or hemicone (in hemiosteonal remodeling) advances more rapidly at night and slows down to allow the closing cone (or hemicone) to catch up during the day. This concertina-like action (Figure 3-2) allows the skeleton to supply calcium at night when it is needed, without affecting the terminal balance of the BMUs and so without causing an irreversible loss of bone.

A final aspect of the relationship between remodeling and calcium homeostasis is that the remodeling apparatus can supply a temporary but sustained demand for calcium lasting for many months by a temporary increase in BMU progression and activation frequency and a corresponding increase in the remodeling-dependent reversible mineral deficit [91]. The best known example is cyclic physiologic osteoporosis in deer, in which a seasonal increase in cortical porosity is entrained to the antler growth cycle [91]. The phenomenon has been demonstrated only in ribs; whether it is confined to the central skeleton or affects the peripheral skeleton as well is not known. The same phenomenon can satisfy the increased demand for calcium that occurs during growth, pregnancy, and lactation; based on densitometric data, in these circumstances the peripheral skeleton is also involved [91]. The bone loss of lactation is generalized, accompanied by high bone turnover and completely reversible [92]. During the adolescent growth spurt, some of the calcium needed for endochondral ossification and subperiosteal apposition is provided by a further increase in the already high cortical porosity that subsides after cessation of longitudinal growth [3].

IV. IMPLICATIONS FOR UNDERSTANDING OSTEOPOROSIS

"Osteoporosis" is a convenient term with which to cover the health implications of two related phenomena. First, bone mass in *individuals* falls with age. Second, partly as a result, the incidence of fractures in the *population* rises with age. Regrettably, for a variety of nonmedical and nonscientific reasons, it has become fashionable to define "osteoporosis" as a disease that is either present or absent, but in this text the term is used only in the former sense.

A. Pathogenesis of Fractures

The relationship of bone remodeling to bone loss and to bone fragility will be considered separately, since bone loss is not the only cause of increased bone fragility.

I. MECHANISMS OF BONE LOSS

The most remarkable feature of age-related bone loss is its universality. There are useful analogies between osteoporosis and hypertension [93, 94], but there are also differences. In some communities remote from Western civilization, mean blood pressure does not rise with age. However, there is no subset of the human species in which mean bone mass does not fall with age, although the rate and magnitude of loss may differ between individuals and between groups [95]. Bone loss not only affects almost all persons but almost every bone, and it is of interest to compare the observed rates of loss at different skeletal sites with those predicted from remodeling theory. There are many problems in comparing rates of loss between different sites [96], including differences in methodology, instrumentation, and units. Rates of bone loss are usually expressed as percentages of the initial value per year. This is not the best way of comparing measurements at the same site between individuals or groups [97], but in the absence of a better mathematical model, it is the most practical way of comparing different sites.

For a few years after menopause, the rate of loss is substantially faster for vertebral cancellous bone than for either cancellous bone at other sites or cortical bone [94, 98, 99], but the wider the age range over which data are collected, the more similar the rates become.

For example, 25 years after menopause the average amount of bone that has been lost in healthy women is about 35% of the initial value (or about 1.4% per year), in both the vertebral bodies and the distal forearm [98]. In the ilium, loss of cancellous bone, measured histologically in autopsy specimens, is about 1% per year in women between the ages of 25 and 75 [100, 101]. About the same rate of loss is found in biopsy specimens and the proportional loss of cancellous and cortical bone is very similar [15, 20]. Likewise, in healthy women studied between the ages of 55 and 75 years, the average rates of loss (% per year) were 1.0 in the distal radius, 1.2 in the calcaneum, and 1.4 in the proximal radius [102]. Thus, at both central and peripheral sites, comprising various proportions of cortical and cancellous bone, the long-term rates of bone loss measured cross-sectionally are in the range of 1–1.5% per year. In cross-sectional studies the subjects differ not only in age but in year of birth and so may have been subject to different environmental influences [103]. This generational or cohort effect could increase the apparent rate of loss compared to longitudinal studies, but would apply to every site, so that the real differences between sites could be even smaller than they appear. Furthermore, more recent longitudinal studies support the same general conclusion. Forearm bone loss in women was 1.25% per year from age 42 to 72 years [104], spinal bone loss in postmenopausal women was 0.5% per year over 6 years [105], and upper femur bone loss in women aged over 65 years was 0.2–0.8% per year [106].

All bone loss occurs from one of the internal surfaces of bone, and the rate of loss from any surface location depends on the average bone deficit at the end of each cycle of remodeling and the frequency with which cycles occur on that surface. Thus, for the same focal imbalance, the rate of bone loss from a surface is proportional to the rate of remodeling on that surface [8, 19]. It is impossible to measure remodeling rates at individual surface locations noninvasively, but biochemical indices of bone turnover reflect the aggregate of the separate contributions of each BMU currently present in the skeleton, although each index is also influenced by several other factors [107]. In accordance with remodeling theory, differences in these indices between persons are significantly correlated with differences in the subsequent rate of bone loss [107–110]. However, when different sites are compared, a serious paradox emerges. Remodeling theory predicts that for the same focal imbalance, the average rate of loss will be about five times higher from cancellous bone adjacent to red marrow than from cancellous bone adjacent to yellow marrow, because of their difference in turnover, but sustained differences of even half this magnitude have never been demonstrated. The inescapable conclusion is that the degree of focal remodeling imbalance in, for example, the calcaneum, is much greater than in the ilium, the only site where such imbalance has so far been measured [21, 22].

For the same absolute rates of bone loss from a surface, the fractional loss depends on the thickness of bone beneath the surface, and hence is proportional to the surface-to-volume ratio [97]. Accordingly, it would be expected that for the same degree of remodeling imbalance and the same frequency of remodeling activation, the average fractional rate of bone loss would be about five times higher in cancellous than in cortical bone, because of their difference in surface-to-volume ratio. However, again, sustained differences in rates of bone loss of even half this magnitude have never been demonstrated. Only in the ilium have rates of both bone remodeling and bone loss been measured at both cortical and cancellous sites in the same bone. As previously mentioned, the results indicated similar rates of surface remodeling, similar fractional rates of bone loss, much larger absolute rates of loss from the endocortical surface, and by inference much greater remodeling imbalance on this surface [15, 21, 22]. In primary hyperparathyroidism, in normal age and menopause-related bone loss, and in patients with vertebral fracture, cortical thinning is mainly the result of increased resorption depth [21, 22], which is the two-dimensional reflection of deeper penetration by endocortical BMUs. The same phenomenon has been demonstrated in the rib [111] and inferred for the metacarpal [94] and is presumably a universal feature of cortical bone loss throughout the skeleton. Furthermore, the similarity in fractional rates of bone loss indicates that the increase in resorption depth at different sites is inversely related to the customary rate of turnover, and positively related to the usual thickness of cortical bone, at each site.

This is a remarkable and unexpected conclusion. When bone loss is both generalized and sustained, as in normal aging, it appears that resorption depth at different sites increases to the extent necessary to bring about roughly the same rates of fractional bone loss and, as it were, "compensates" for differences in bone turnover contingent on differences in marrow composition and for differences in local bone structure and geometry. The only conceivable kind of explanation for such a phenomenon is biomechanical [5, 42]. All mechanical influences on bone remodeling are mediated by strain, the technical term for relative deformation of a structural material as the result of load bearing. Similar fractional rates of bone loss throughout the skeleton will produce similar proportional changes in the strains that occur in different bones as a result of the same pattern and intensity of physical activity. Frost [1], building on earlier work by others [42, 112], has proposed the existence of the "mechanostat," which orchestrates the recruitment and activity of osteoclasts and osteoblasts in such a way that strain is maintained within an acceptable range [113].

The primary function of the mechanostat is to ensure that during growth each bone acquires the strength it needs to support the species-specific pattern and intensity of physical activity customary during adult life [3]. After growth has ceased, the mechanostat is much less effective in adapting the bones to an *increase* in mechanical demand, but is highly effective in adapting them to a *decrease*, accounting for the rapidity, severity, and usual irreversibility of bone loss consequent on disuse [114]. As a result of the sedentary life-style made possible by economic development, aging is in most persons accompanied by a progressive reduction in physical activity and muscle strength of earlier onset and greater severity than is biologically mandated [115]. According to biomechanical theory, this should not increase the risk of fracture, since the reduced bone mass would remain appropriate to the reduced level of activity, but this does not take account of the age-related increase in liability to fall, to which the mechanostat is blind. Frost postulated that as a result of estrogen deficiency the mechanostat is reset, so that the skeleton responds not so much to actual but to erroneously perceived disuse [1]. Could a universal resetting of the mechanostat account for disproportionately rapid loss of central cancellous bone in the first few years after menopause? Possibly, if the distribution of estrogen receptors α and β differs between surfaces [116, 117], but the mechanostat set point could also be influenced more directly by some aspect of the aging process.

2. Mechanisms of Bone Fragility

Bone mass is inversely related to fracture risk, both current and future, but there are also qualitative abnormalities in bone that contribute to its fragility [118, 119, Table 3-5]. The best known and most well established of these nonmass factors relates to cancellous bone architecture. When cancellous bone is lost as a result of estrogen deficiency, whole structural elements are removed, leaving those that remain more widely separated and less well connected [100]. As a result, vertebral fracture risk is increased to a greater extent than would be expected for the reduction in bone mass [120]. This could be why the presence of at least one vertebral fracture is an independent risk factor for further vertebral fractures [121], but the increased risk applies also to other fractures [122] so that some nonarchitectural factor is also involved. The structural changes are the result of perforation of trabecular plates because the cutting hemicones of individual BMUs penetrate more deeply

TABLE 3-5   Qualitative Aspects of Bone Strength

1. Microarchitectural disorganization
2. Accumulation of unrepaired fatigue damage
3. Abnormal mineral density distribution
4. Unnecessarily high bone turnover
5. Osteocyte deficiency

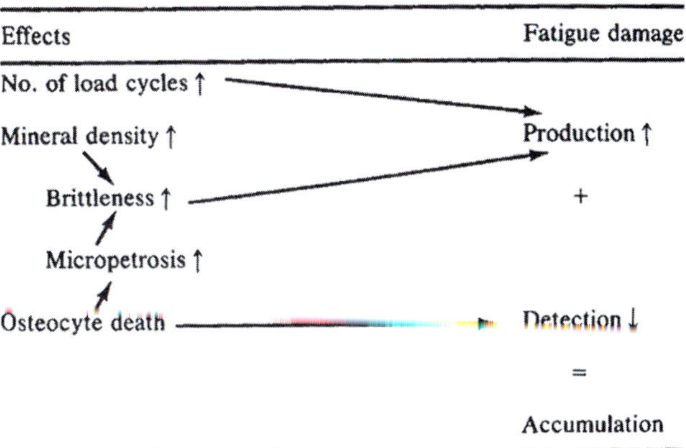

FIGURE 3-3   Mechanisms whereby increased bone age could lead to accumulation of fatigue damage. Some effects would increase fatigue damage production, and some effects would decrease its detection and repair. Reproduced with permission from [129].

into the bone away from the surface [19, 123]. This qualitative abnormality is due to delayed apoptosis [124], and consequent prolonged life span of osteoclasts, not to excessively rapid resorption by individual osteoclasts. However, a more fundamental problem may be loss of BMU directional control [125]. Although these various changes could be due to resetting of the mechanostat set point [1], the occurrence of severe vertebral osteopenia in elite athletes with exercise-associated amenorrhea [126] indicates that the effects of estrogen deficiency are not prevented by increased physical activity.

The second qualitative factor in bone fragility is accumulation of fatigue microdamage (Table 3-5). Frost [127] proposed that normally there is such a wide margin of safety that the adverse effect of bone loss on bone fragility is mediated, not by a reduction in instantaneous breaking strength, but by fatigue damage accumulation due to increased strain in the bone that remains. However, most investigators believe that the margin of safety is not as great as Frost claimed [128]. Frost further proposed that a defective damage repair mechanism could be overwhelmed by even normal damage production. As already mentioned, there is a close relationship between fatigue damage accumulation and bone age [47, 129, Figure 3-3]. Increased bone age would increase susceptibility to fatigue damage, both directly (by exceeding the fatigue life) and indirectly. Osteocyte death, which can occur spontaneously when bone age exceeds about 20 years [46, 130], leads to perilacunar hypermineralization (or micropetrosis), which would make the bone more brittle and more susceptible to fatigue damage [43]. Osteocyte death would also impair detection of fatigue damage, since the process of death is part of the signaling pathway, and osteocytes can die only once. The repair of microdamage by a new BMU could be delayed by an age-related decline in any of the intervening steps previously outlined, or by loss of the directional control needed for the new BMU to find its target [125], another likely consequence of osteocyte death [77]. This reasoning is plausible, but it has been difficult to prove unequivocally that defective microdamage repair is important in the pathogenesis of clinically significant fractures [53]. Perhaps the strongest evidence for the concept is the occurrence of spontane-

ous fractures in patients with radiation necrosis of bone [131] and in dogs in which bone remodeling has been completely abrogated [132].

In these circumstances, the fractures occur predominantly in the long bones of the extremities, where turnover is generally low. Whether such a mechanism also operates in the pathogenesis of vertebral compression fractures is still uncertain; I have equivocated about this point for 20 years [48, 65, first and second editions of this book]. In iliac cancellous bone, it is much more difficult than in cortical bone to tell whether a microcrack was present *in vivo* or was produced during biopsy or section preparation (Qiu, personal communication). Among patients with true vertebral fracture (not just radiographic deformation), there is greater than normal variability of bone formation rate, and the mean value is lower than normal [133]. Consequently, the proportion of patients with a large amount of very old cancellous bone is greater than normal [48], although increased susceptibility to fatigue damage has not been demonstrated. Microcracks occur in central cancellous bone, but they have not been shown to be more common than expected in patients with vertebral fracture. In the vertebral body, the perforations and loss of structural elements previously mentioned occur preferentially in horizontal rather than in vertical trabeculae. The compressive strength of a vertical trabecula will decline in proportion to the square of the unsupported length, so that a 50% reduction in the number of horizontal trabeculae will lead to a 4-fold increase in the susceptibility to buckling [134]. Based on estimates of *in vivo* stresses during normal activity [135], and on the production of microcracks by experimental compression [63], vertebral microfractures can be explained by instantaneous overload as a result of the architectural changes previously mentioned without the need to invoke a fatigue-based mechanism [65, 135, 136]. Nevertheless, a role for defective microdamage repair has not been ruled out.

Hip fractures share with vertebral fractures the inverse relationship of risk to bone mass, but differ from vertebral fractures with respect to the qualitative contribution to bone fragility. Loss of cancellous bone connectivity due to estrogen deficiency is less important, whereas fatigue damage accumulation is more important; although small islands of hematopoietic tissue can persist in the upper femur much longer than at more distal sites, particularly in the femoral head, the proportion of red marrow is much lower than in the ilium [24]. There are no tetracycline-based measurements of bone remodeling in the upper femur, but other indices of bone remodeling are lower than in the ilium or vertebral body [34, 137], and this difference is exaggerated in patients with hip fracture [24]. The proportion of osteocytes that are viable declines progressively with increasing subject age in the femoral neck [138], and the age-related decline in osteocyte lacunar density in the femoral mid shaft is associated with microcrack accumulation [139]; it seems very likely that osteocyte death is involved in hip fracture pathogenesis [140]. True bone mineral density increases with age in the femoral shaft cortex but not in the spine [141]. Fatigue microdamage occurs in the cortical bone of both the femoral neck and the femoral shaft, and in the latter, crack density increases exponentially with age, more so in women than in men [63]. Cancellous microfractures in the femoral head increase in number with age and with reduction in mineral density [142, 143] and are significantly more frequent in hip fracture patients than in controls, despite a statement by the authors to the contrary [144]; because of the lower bone turnover and differences in architecture, there is greater reason to invoke a fatigue-based mechanism than in the spine [24, 143]. All these data indicate that increased bone age and its adverse effects on bone fragility (Figure 3-3) are likely to be of major importance in the pathogenesis of hip fracture [129] (Table 3-6).

Although not shown to enhance fatigue damage, the other adverse effects of low bone turnover and increased bone age would be expected in some patients with true vertebral fracture (Table 3-5). In these patients the mineral density of iliac bone varies over a wider range than in normal subjects. The frequency distribution is bimodal, different subsets having higher than or lower than normal mean mineral density. In some patients with osteoporotic vertebral fracture, there appears to be a substantial delay in secondary mineralization, the process whereby mineral crystals enlarge at the expense of water [83]. This would remove much of the need for bone

Table 3-6   Fracture Pathogenesis at Different Sites

	Vertebra	Femoral neck
Function of cancellous bone	Metabolic	Mechanical
Marrow/turnover	Red/high	Yellow/low
Osteocyte death	Yes	Yes
Increase with age	Small	Large
Fatigue damage	?[a]	Yes[b]
Hypermineralization	No	Yes[b]
Main qualitative factor	Architecture	Bone age

[a]Microdamage, not shown to be due to fatigue.
[b]In femoral cortical bone, not necessarily at fracture site.

remodeling to prevent hypermineralization, but there was no relationship to surface bone formation rate [50]. Nevertheless, hypomineralization would be expected to reduce the stiffness and strength of bone as a material [42] and to be an independent risk factor for bone fragility, so that it is important to discover its pathogenesis. In other patients, abnormally high mineral density would increase brittleness and reduce fracture toughness; these would be expected consequences of low bone formation rate, but such a relationship could not be demonstrated [50]. The factors that influence true bone density and how these factors may be altered to produce change in either direction in patients with vertebral fracture merits more attention than they have received.

During the treatment of osteoporosis with so-called antiresorptive drugs, improvement in spinal bone mineral density accounts for only a small part of the observed reduction in vertebral fracture rates [145], and the reason for this discrepancy has received much attention [119]. High bone turnover, assessed by biochemical indices, contributes independently to subsequent fracture risk [146, 147], and reduction in bone turnover contributes independently to the beneficial effect of estrogen therapy, and could account for the discrepancy just mentioned [148]. The adverse effect of high turnover on bone strength has been attributed to increased perforative resorption [148], but the mechanism is actually more subtle [65, 149]. For high turnover to be a mechanical threat, some horizontally oriented trabeculae must have been removed [100], which withdraws lateral support from the remaining vertically oriented trabeculae that bear the compressive loads. As previously mentioned, the resistance to buckling decreases as the *square* of the increase in unsupported length [134]. A contributory factor is that residual vertical trabeculae slowly become thinner with increasing age [100].

Each episode of bone remodeling that occurs on a thin unsupported vertical trabecula, as found in most women more than 5 years post menopause [150], acts as a stress concentrator [151] and represents a focal weakness that poses a small risk of buckling. In iliac cancellous bone, the usual mean depth of resorption is about one-third of the usual mean trabecular thickness [8, 14, 22]. Because of the wide frequency distribution of these measurements [152], even a normal size resorption cavity may penetrate halfway or more through a trabecula [20, 22]. For this effect, it is not necessary for the resorptive process to *perforate* the trabecula; being present is enough. When turnover increases, the risk of buckling will increase within only a few weeks, and it is not necessary for additional irreversible bone destruction to occur [149]. The relationship between bone fragility and bone turnover is U-shaped, both abnormally low and unnecessarily high rates increasing fracture risk [72]. A localized increase in bone turnover and cortical porosity during fatigue damage repair may also temporarily increase fracture risk [53].

The most recently discovered qualitative aspect of bone strength is osteocyte deficiency (Table 3-5). Osteocytes are necessary for the detection of fatigue damage and initiating its repair, but they also contribute directly to bone strength. In mice, the prevalence of osteocytes undergoing apoptosis is an independent predictor of vertebral compressive strength [153], and when osteocytes are protected from the adverse effects of glucocorticoids, compressive strength is preserved even though bone is lost [154]. Women with genuine vertebral fractures have about 30% fewer osteocytes and lacunae in iliac cancellous bone than controls [155]. In normal women, osteocyte density declines with age in deep interstitial bone because of death by apoptosis, but not in surface bone, which is renewed by remodeling (Table 3-4) [46]. In patients with vertebral fracture, osteocyte and lacunar density are low in superficial bone because of reduced incorporation of osteocytes while the bone is being made, and also low in deep bone, indicating that this defect was present many years before the fractures occurred [155]. These data, together with the data on bone formation rate and true bone density previously mentioned, indicate that some patients with vertebral fracture have a real disease of unknown etiology that is not just a consequence of age-related bone loss. Why osteocyte deficiency impairs bone strength is unknown, but disruption of the canalicular circulation may be involved [156].

B. Prevention of Fractures

It is customary to discuss the "prevention" and "treatment" of osteoporosis separately, but this is a misleading distinction, since the only therapeutic goal is to prevent fractures; whether one's aim is to prevent the first fracture or a subsequent fracture does not alter this principle. Of the several aspects of fracture prevention, the theme of this chapter relates most clearly to the prevention and restoration of bone loss. Agents that accomplish these aims are usually referred to respectively as "inhibitors of bone resorption" and "stimulators of bone formation," but these vague terms betray a serious lack of comprehension of bone remodeling. They ignore the indivisible unity of the BMU as a structural and functional entity, obscure the crucial distinction between effects on cell recruitment and effects on differentiated cell function, and engender the absurd notions that all bone resorption is bad and all bone formation is good. The former error is potentially more dangerous than the latter, so this aspect of therapy will be the focus of subsequent discussion. A reduction in activation frequency and consequent reduction in bone turnover will have several salutary effects. The mechanical threat of remodeling will fall within a few weeks before there has been a detectable change in bone mass, contraction of the remodeling space will lead to reversal of temporary bone loss within a few months, and there will be a long-term reduction in the rate of irreversible bone loss. How can these benefits be obtained without frustrating the purposes of bone remodeling?

Activation frequency is the best histologic index of the intensity of bone remodeling on a surface and is the main determinant of the rate of bone turnover, but it is not a measure of the frequency of BMU *origination*, since it depends also on the mean distance of BMU *progression* [7, 49, 157–159]. The effects on all histologic, biochemical, and radiokinetic indices of bone turnover would be the same whether, for example, one BMU traveled for 9 units of distance through or across the surface of bone, or each of 3 BMUs traveled for 3 units of distance (Figure 3-4). But the biological significance would be different, since each new BMU represents a separate remodeling project. Approximately 90% of new mononuclear osteoclast precursor cells are used to sustain the progression of existing BMUs, and only 10% are used to originate a new BMU [7]. Consequently, substantial changes in activation frequency and bone turnover can be brought about by manipulating the distance and duration of BMU progression without changing the frequency of BMU origination. Each episode of targeted remodeling requires a new BMU, but spare remodeling could be accomplished if each BMU progressed for a variable distance beyond its target [7].

It is this arrangement that makes it possible for therapeutic agents to reduce activation frequency and bone turnover by curtailing BMU progression, without inhibiting BMU origination and so to reduce spare post-targeted remodeling without interfering with targeted remodeling. Obviously, the ability to prioritize different remodeling

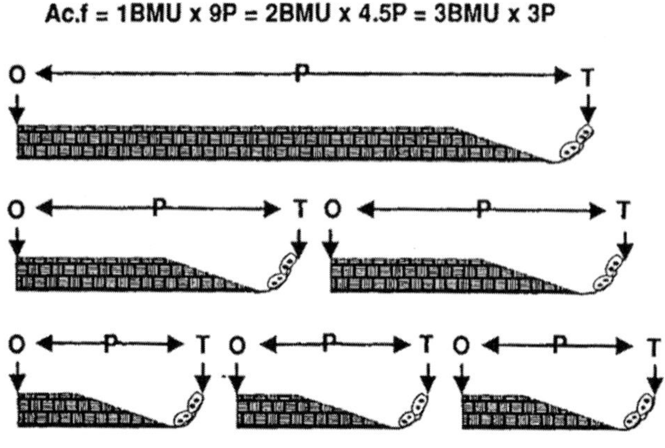

FIGURE 3-4   Relationship between BMU origination and remodeling activation. Activation frequency represents the product of frequency of BMU origination and the average distance of BMU progression. In this example, activation frequency would be the same with 1 BMU that progresses for 9 units of distance, or 2 BMUs that each progress for 4.5 units of distance, or 3 BMUs that each progress for 3 units of distance. However, the biological significance would be different, because each BMU represents a separate remodeling project. Copyright 1995, A. M. Parfitt, used with permission.

tasks is a feature of the remodeling system itself, not of the individual therapeutic agents. It must be assumed that the signals for osteoclast precursors to arrive at a particular location are more compelling for BMU origination than for BMU progression, more compelling for BMU progression *toward* its target than *beyond* its target, and more compelling for the peripheral than for the central skeleton, because of the difference in margin of safety. These hierarchies could reflect differences in the types as well as the amounts of signal molecules. However, therapeutic agents may differ in their ability to exploit these differences in signal strength. Agents that act directly on osteoclasts to reduce their resorptive activity are more likely to act indiscriminately on all osteoclasts throughout the skeleton, and in some locations this is likely to negate their purpose; consequently, the net outcome of the intervention could be harmful rather than beneficial. However, agents that reduce the supply of osteoclast precursor cells leave the remodeling system able to deploy its more limited resources to the best advantage.

Not surprisingly, hormone replacement therapy (HRT) is the most effective means of preventing the adverse effects on bone of the hormone deficiency that results from menopausal ovarian failure. This remains true despite uncertainty about the overall long-term safety and efficacy of HRT [160] and whether there are important differences between regimens [161]. Estrogen deficiency increases the availability of osteoclast precursor cells [12] and so increases the spare component

of bone remodeling by removing a constraint on post-targeted BMU progression, particularly in the central skeleton. However, the most destructive consequence of estrogen deficiency is delayed osteoclast apoptosis [7, 124], leading to deeper BMU penetration (reflected in two-dimensional histologic sections as increased resorption depth), trabecular plate perforation, and loss of connectivity. Both of these effects—increased osteoclast recruitment and delayed osteoclast apoptosis—are prevented by HRT, and ideally both of them should be prevented by any agent that is used as a substitute for HRT. Until recently, the most widely used substitute was calcitonin, but this agent prevents rather than promotes osteoclast apoptosis [162], and its effects on resorption depth are unknown. The newer bisphosphonates appear to be more complete substitutes for HRT. Although their best known effect is to acutely inhibit the function of existing osteoclasts, in the long term, they reduce osteoclast recruitment by mechanisms that remain uncertain [7, 163], promote earlier osteoclast apoptosis [164], and reduce resorption depth [165].

The safety of reducing bone turnover depends on the ability to limit spare post-targeted remodeling preferentially in the central skeleton without interfering with targeted remodeling at any skeletal site. Obviously, there is a lower limit to osteoclast precursor cell recruitment below which the purposes of remodeling will be frustrated. As would be predicted from the earlier discussion, complete suppression of remodeling in beagles leads after a few months to the occurrence of spontaneous fractures [132]; this occurred with etidronate, which causes osteomalacia, but also with clodronate, which does not. A dangerous reduction in bone turnover could never occur with physiological agents such as estrogen or calcitonin, but can readily be produced by bisphosphonates if given in excessive dose. Regrettably, there is very little information on what lower limit is safe. The safe level will be different in different regions of the skeleton, which reduces the value of biochemical indices of turnover to determine safety, since these are necessarily blind to regional differences. Quite low levels of whole body bone turnover are consistent with skeletal health when they occur naturally, but might conceal regional ill health when induced by therapeutic intervention. Reducing osteoclast recruitment to a level just sufficient to allow the completion of targeted remodeling in most cases but leaving no room for spare remodeling would also lead eventually to spontaneous fractures, but the time required would probably be measured in years rather than in months, which limits the use of animal models to determine long-term safety.

As explained earlier, when vertical trabeculae have lost their horizontal supports, even normal remodeling may constitute a mechanical threat. In this situation, reducing turnover even within the normal range (defined by biochemical indices) may be useful in the prevention of vertebral fractures [65]. But the long-term effects of reducing turnover on hip fracture risk are less easily predictable. For reasons given previously, the adverse effects of prolonged bone age on bone fragility (Figure 3-3) are likely to be more serious in the upper femur than in the spine (Table 3-6). Indeed, in several parts of the femoral neck, regions of hypermineralization and reduced fracture toughness become more extensive with age [166]. The large increase in the use of bisphosphonates that followed the approval of alendronate by the FDA has reduced the incidence of vertebral fractures [167], and may also reduce the incidence of other fractures, but the data available today do not exclude the possibility that 10–20 years from now there will be an epidemic of hip fractures. By then, a large proportion of the elderly population will have levels of bisphosphonate of one kind or another in the femoral heads and necks that are possibly dangerous, and there will be nothing that can be done about it. Both risedronate and alendronate increase microdamage accumulation in canine rib [168]; on a body weight basis, the doses were much higher than used clinically, but smaller doses for a much longer time could have the same effect. Vertebral compressive strength was increased because of increased bone mass, but fracture toughness was reduced [169].

It is not too late to find out what is really going on in the bones of hip fracture patients, but only if we abandon the exclusive reliance on biochemical and densitometric methods and on histologic examination at a site chosen for its convenience rather than its relevance to the problem of greatest importance. An important recent discovery from direct examination is that clustered remodeling and giant resorption cavities due to confluence of clusters are more common in hip fracture patients than age-matched control subjects [170, 171]. A reduction in the frequency of such clusters could account for the early reduction in fracture risk by correction of vitamin D deficiency [172] and also for the beneficial effect of bisphosphonate administration—another example of the danger of high bone turnover. On the other hand, if the clusters are an effect of defective directional control of BMUs, then reducing their number could compromise microdamage repair. Obviously, a great deal more research will be needed to resolve these uncertainties.

REFERENCES

1. H. M. Frost, Bone's mechanostat: A 2003 update. *Anat Rec*, **275**, 1081–1101 (2003).
2. A. M. Parfitt, R. Travers, F. Rauch, F. H. Glorieux, Structural and cellular changes during bone growth in healthy children. *Bone*, **27**, 487–494 (2000).
3. A. M. Parfitt, The two faces of growth—Benefits and risks to bone integrity. *Osteoporosis Int*, **4**, 382–398 (1994).
4. A. M. Parfitt, Regulation or permission? *J Bone Miner Res*, **21**, 659–660 (2006).
5. H. M. Frost, *Intermediary Organization of the Skeleton*. CRC Press, Boca Raton, FL (1986).
6. A. M. Parfitt, Osteonal and hemi-osteonal remodeling. The spatial and temporal framework for signal traffic in adult human bone. *J Cell Biochem*, **55**, 273–286 (1994).
7. A. M. Parfitt, G. R. Mundy, G. D. Roodman, D. E. Hughes, and B. Boyce, A new model for the regulation of bone resorption. *J Bone Miner Res*, **11**, 150–159 (1996).
8. A. M. Parfitt, New concepts of bone remodeling: A unified spatial and temporal model with physiological and pathophysiologic implications. In *Bone Loss and Osteoporosis: An Anthropological Perspective* (S. C. Agarwal and S. D. Stout, eds.), pp. 3–17. Kluwer Academic/Plenum Publishers, New York (2003).
9. A. M. Parfitt, Mini-review—The mechanism of coupling—A role for the vasculature. *Bone*, **26**, 319–323 (2000).
10. A. M. Parfitt, The bone remodeling compartment: A circulatory function for bone lining cells. *J Bone Miner Res*, **16**, 1583–1585 (2001).
11. A. M. Parfitt, Mini-review—Osteoclast precursors as leucocytes: Importance of the area code. *Bone*, **23**, 491–494 (1998).
12. S. C. Manolagas, Editorial: Cell number versus cell vigor. What really matters to a regenerating skeleton? *Endocrinology*, **140**, 4377–4381 (1999).
13. A. M. Parfitt, Problems in the application of *in vitro* systems to the study of human bone remodeling. Proceedings from the Workshop on Human Models of Skeletal Aging, NIH. *Calcif Tissue Int*, **56**(Suppl. 1), S5–S7 (1995).
14. M. C. Moore-Ede, Physiology of the circadian timing system: Predictive versus reactive homeostasis. *Am J Physiol*, **250**, R735–R752 (1986).
15. J. Foldes, A. M. Parfitt, M-S. Shih, D. S. Rao, and M. Kleerekoper, Structural and geometric changes in iliac bone: Relationship to normal aging and osteoporosis. *J Bone Miner Res*, **6**, 759–766 (1991).
16. D. P. Fyhrie, S. M. Lang, and A. M. Parfitt, *Cortical and Cancellous Bone Structure in the Ilia of Normals and Osteoporotics*, p. 443 [Abstract]. Orthopaedic Research Society, 40th Annual Meeting (1994).
17. G. H. Meyer, Die Architectur der Spongiosa. *Arch Anat Physiol Wissensch*, **34**, 615–628 (1987).
18. A. M. Parfitt, Misconceptions (2), Bone turnover is always higher in cancellous than in cortical bone. *Bone*, **30**, 807–809 (2002).
19. A. M. Parfitt, The physiologic and pathogenetic significance of bone histomorphometric data. In *Disorders of Bone and Mineral Metabolism*, 2nd ed. (F. L. Coe and M. J. Favus, eds.), pp. 469–485. Lippincott Williams and Wilkins, Philadelphia (2002).
20. Z. H. Han, S. Palnitkar, D. S. Rao, D. Nelson, and A. M. Parfitt, Effect of ethnicity and age or menopause on the structure and geometry of iliac bone. *J Bone Miner Res*, **11**, 1967–1975. (1996).
21. A. M. Parfitt, Surface specific bone remodeling in health and disease. In *Clinical Disorders of Bone and Mineral Metabolism* (M. Kleerekoper and S. Krane, eds.), pp. 7–14. Mary Ann Liebert, New York (1989).
22. Z. H. Han, S. Palnitkar, D. S. Rao, D. Nelson, and A. M. Parfitt, Effects of ethnicity and age or menopause on the remodeling and turnover of iliac bone: Implications for mechanisms of bone loss. *J Bone Miner Res*, **12**, 498–508 (1997).
23. D. W. Dempster, Relationship between the iliac crest bone biopsy and other skeletal sites. In *Clinical Disorders of Bone*

and Mineral Metabolism (M. N. Kleerekoper and S. M. Krane, eds.), pp. 247–252. Mary Ann Liebert, New York (1989).

24. I. Eventov, B. Frisch, Z. Cohen, and I. Hammel, Osteopenia, hematopoiesis, and bone remodelling in iliac crest and femoral biopsies: A prospective study of 102 cases of femoral neck fractures. *Bone*, **12**, 1–6 (1991).

25. A. M. Parfitt, The physiologic and clinical significance of bone histomorphometric data. In *Bone Histomorphometry: Techniques and Interpretations* (R. Recker, ed.), pp. 142–223. CRC Press, Boca Raton, FL (1983).

26. A. M. Parfitt, D. S. Rao, J. Stanciu, A. R. Villenueva, M. Kleerekoper, and B. Frame, Irreversible bone loss in osteo-malacia: Comparison of radial photon absorptiometry with iliac bone histomorphometry during treatment. *J Clin Invest*, **76**, 2403–2412 (1985).

27. D. B. Kimmel, and W. S. S. Jee, A quantitative histologic study of bone turnover in young adult beagles. *Anat Rec*, **203**, 31–45 (1982).

28. S. N. Wickramasinghe, *Human Bone Marrow*. Blackwell Scientific, Oxford (1975).

29. M. Tavassoli, Embryonic and fetal hemopoiesis: An overview. *Blood Cells*, **17**, 269–281 (1991).

30. P. Meunier, J. Aaron, C. Edouard, and G. Vignon. Osteoporosis and the replacement of cell populations of the marrow by adipose tissue. A quantitative study of 84 iliac bone biopsies. *Clin Ortho & Related Res*, **80**, 147–154 (1971).

31. R. P. Custer, *An Atlas of the Blood and Bone Marrow*. Saunders, Philadelphia (1949).

32. S. McKinney-Freeman, and M. A. Goodell, Circulating hematopoietic stem cells do not efficiently home to bone marrow during homeostasis. *Exp Hematol*, **32**, 868–876 (2004).

33. D. Van Dyke, Similarity in distribution of skeletal blood flow and erythropoietic marrow. *Clin Orthop*, **52**, 37–51 (1967).

34. B. Krempien, F. M. Lemminger, E. Ritz, and E. Weber, The reaction of different skeletal sites to metabolic bone disease— A micromorphometric study. *Klin Wochenschr*, **56**, 755–759 (1978).

35. T. J. Wronski, J. M. Smith, and W. S. S. Jee, The microdis-tribution and retention of injected [239]Pu on trabecular bone surfaces of the beagle: Implications for the induction of osteo-sarcoma. *Radiat Res*, **83**, 74–89 (1980).

36. T. J. Wronski, J. M. Smith, and W. S. S. Jee, Variations in mineral apposition rate of trabecular bone within the beagle skeleton. *Calcif Tissue Int*, **33**, 583–586 (1981).

37. E. M. Hauge, D. Qvesel, E. F. Eriksen, L. Mosekilde, and F. Melsen, Cancellous bone remodeling occurs in specialized compartments lined by cells expressing osteoblastic markers. *J Bone Miner Res*, **16**, 1575–1582 (2001).

38. A. M. Parfitt, Bone remodeling, normal and abnormal: A biological basis for the understanding of cancer-related bone disease and its treatment. *Can J Oncol*, **5**(Suppl 1), 1–10 (1996).

39. M. P. Whyte, S. E. Obrecht, P. M. Finnegan, J. L. Jones, M. N. Podgornik, W. H. McAlister, and S. Mumm, Osteoprotegerin deficiency and juvenile Paget's disease. *New Engl J Med*, **347**, 175–184 (2002).

40. R. B. Martin, B. D. Chow, and P. A. Lucas, Bone marrow fat content in relation to bone remodeling and serum chemistry in intact and ovariectomized dogs. *Calcif Tissue Int*, **46**, 189–194 (1990).

41. J. Podenphant, and U. Engel, Regional variations in histomor-phometric bone dynamics from the skeleton of an osteoporotic woman. *Calcif Tissue Int*, **40**, 184–188 (1987).

42. J. Currey, *Bones: Structure and Mechanics*. Princeton University Press, Princeton, NJ (2002).

43. H. M. Frost. Micropetrosis. *J Bone Jt Surg*, **42**, 138–143 (1980).

44. D. Vashishth, G. J. Gibson, J. L. Khoury, M. B. Schaffler, J. Kimura, and D. P. Fyhrie, Influence of nonenzymatic glyca-tion on biomechanical properties of cortical bone. *Bone*, **28**, 195–201 (2001).

45. A. J. Bailey, and L. Knott, Molecular changes in bone collagen in osteoporosis and osteoarthritis in the elderly. *Exp Gerontol*, **34**, 337–351 (1999).

46. S. Qiu, S. Palnitkar, D. S. Rao, and A. M. Parfitt, Age and distance from the surface but not menopause reduce osteocyte viability in human cancellous bone. *Bone*, **31**, 313–318 (2002).

47. S. Qiu, D. S. Rao, D. P. Fyhrie, S. Palnitkar, and A. M. Parfitt, The morphological association between microcracks and osteocyte lacunae in human cortical bone. *Bone*, **37**, 10–15 (2005).

48. A. M. Parfitt, M. Kleerekoper, and A. R. Villanueva, Increased bone age: Mechanisms and consequences. In *Osteoporosis* (C. Christiansen, C. Johansen, and B. J. Riis, eds.), pp. 301–308. Osteopress ApS, Copenhagen (1987).

49. A. M. Parfitt, Misconceptions V—Activation of osteoclasts is the first step in the bone remodeling cycle. *Bone*, **39**, 1170–1172 (2006).

50. T. E. Ciarelli, D. P. Fyhrie, and A. M. Parfitt, Effects of verte-bral bone fragility and bone formation rate on the mineraliza-tion levels of cancellous bone from white females. *Bone*, **32**, 311–315 (2003).

51. A. M. Parfitt, Targeted and non-targeted bone remodeling: Relationship to BMU origination and progression. *Bone*, **30**, 5–7 (2002).

52. R. B. Martin, Fatigue microdamage as an essential element of bone mechanics and biology. *Calcif Tissue Int*, **73**, 101–107 (2003).

53. D. B. Burr, M. R. Forwood, D. P. Fyhrie, R. B. Martin, M. B. Schaffler, and C. H. Turner, Bone microdamage and skeletal fragility in osteoporotic and stress fractures. *J Bone Miner Res*, **12**, 6–15 (1997).

54. F. J. O'Brien, O. Brennan, O. D. Kennedy, and T. C. Lee, Microcracks in cortical bone. How do they affect bone biol-ogy? *Cur Osteoporosis Rep*, **3**, 39–45 (2005).

55. S. Mori, and D. B. Burr, Increased intracortical remodeling following fatigue damage. *Bone*, **14**, 103–109 (1993).

56. V. Bentolila, T. M. Boyce, D. P. Fyhrie, R. Drumb, T. M. Skerry, and M. B. Schaffler, Intracortical remodeling in adult rat long bones after fatigue loading. *Bone*, **23**, 275–281 (1998).

57. D. R. Carter, D. P. Fyhrie, and R. T. Whalen. Trabecular bone density and loading history: Regulation of connective tissue biology by mechanical energy. *J Biomechanics*, **20**, 785–794 (1987).

58. S. M. Bowman, X. E. Guo, D. W. Cheng, T. M. Keaveny, L. J. Gibson, W. C. Hayes, and T. A. McMahon, Creep con-tributes to the fatigue behavior of bovine trabecular bone. *J Biomechanical Engineering*, **120**, 647–654 (1998).

59. T. E. Wenzel, M. B. Schaffler, and D. P. Fyhrie, In vivo trabecular microcracks in human vertebral bone. *Bone*, **19**, 89–95 (1996).

60. S. Mori, R. Harruff, W. Ambrosius, and D. B. Burr, Trabecular bone volume and microdamage accumulation in the femoral heads of women with and without femoral neck fractures. *Bone*, **20**, 521–526 (1997).

61. N. H. Fazzalari, M. R. Forwood, B. A. Manthey, K. Smith, and P. Kolesik. Three-dimensional confocal images of microdam-age in cancellous bone. *Bone*, **23**, 373–378 (1998).

62. M. B. Schaffler, K. Choi, and C. Milgrom, Aging and matrix microdamage accumulation in human compact bone. *Bone*, **17**, 521–525 (1995).

63. D. P. Fyhrie, and M. B. Schaffler, Failure mechanisms in human vertebral cancellous bone. *Bone*, **15**, 105–109 (1994).

64. E. P. M. Urovitz, V. L. Fornasier, M. I. Risen, and I. MacNab, Etiological factors in the pathogenesis of femoral trabecular fatigue fractures. *Clin Orthop*, **127**, 275–280 (1977).

65. A. M. Parfitt, Pathophysiology of bone fragility. In *Proceedings of the Fourth International Symposium on Osteoporosis* (C. Christiansen and B. Riis, eds.), pp. 164–166. Hong Kong, Handelstrykkeriet Aalborg ApS, Aalborg, Denmark (1993).

66. R. B. Martin, Is all cortical bone remodeling initiated by microdamage? *Bone*, **30**, 8–13 (2002).

67. R. B. Martin, Fatigue damage, remodeling, and the minimization of skeletal weight. *J Theoretical Biology*, **220**, 271–276 (2003).

68. G. Marotti, Map of bone formation rate values recorded throughout the skeleton of the dog. In *Bone Morphometry: Proceedings of the First Workshop* (Z. F. G. Jaworski, ed.), pp. 202–207. University of Ottawa Press, Ottawa (1976).

69. D. Dempster, Bone remodeling. In *Disorders of Bone and Mineral Metabolism* (F. Coe and M. Z. Favus, eds.), pp. 355–380. Raven Press, New York (1992).

70. R. B. Martin, Toward a unifying theory of bone remodeling. *Bone*, **26**, 1–6 (2000).

71. Z. F. G. Jaworski, Haversian systems and Haversian bone. In *Bone—A Treatise: Bone Metabolism and Mineralization* (B. K. Hall, ed.), Vol. 4, pp. 21–45. CRC Press, Boca Raton, FL (1992).

72. R. P. Heaney, Is the paradigm shifting? *Bone*, **33**, 457–465 (2003).

73. E. F. Eriksen, T. Steiniche, L. Mosekilde, and F. Melsen, Histomorphometric analysis of bone in metabolic bone disease. *Endocrinol Metabol Clin North Am*, **18**, 919–954 (1989).

74. A. M. Parfitt, Idiopathic, surgical and other varieties of parathyroid hormone deficient hypoparathyroidism. In *Endocrinology*, 2nd ed. (L. DeGroot, ed.), pp. 1049–1064. Saunders, Philadelphia (1989).

75. C. H. Turner, and M. R. Forwood, What role does the osteocyte network play in bone adaptation? *Bone*, **16**, 283–285 (1995).

76. B. S. Noble, and J. Reeve, Osteocyte function, osteocyte death and bone fracture resistance. *Mol & Cell Endocrinol*, **159**, 7–13 (2000).

77. O. Verborgt, G. J. Gibson, and M. B. Schaffler, Loss of osteocyte integrity in association with microdamage and bone remodeling after fatigue in vivo. *J Bone Miner Res*, **15**, 60–67 (2000).

78. B. S. Noble, N. Peet, H. Y. Stevens, A. Brabbs, J. R. Mosley, G. C. Reilly, J. Reeve, T. M. Skerry, and L. E. Lanyon, Mechanical loading: Biphasic osteocyte survival and targeting of osteoclasts for bone destruction in rat cortical bone. *Amer J Physiology Cell Physiology*, **284**, C934–943 (2003).

79. O. Verborgt, N. A. Tatton, R. J. Majeska, and M. B. Schaffler, Spatial distribution of Bax and Bcl-2 in osteocytes after bone fatigue: Complementary roles in bone remodeling regulation? *J Bone Miner Res*, **17**, 907–914 (2002).

80. S. Qiu, S. Palnitkar, D. S. Rao, and A. M. Parfitt, Relationship between osteocyte density and bone formation rate in human cancellous bone, *Bone*, **31**, 709–711 (2002).

81. J. I. Aguirre, L. I. Plotkin, S. A. Stewart, R. S. Weinstein, A. M. Parfitt, S. C. Manolagas, and T. M. Bellido, Osteocyte apoptosis is induced by weightlessness in mice and precedes osteoclast recruitment and bone loss. *J Bone Miner Res*, **21**, 605–615 (2006).

82. A. M. Parfitt, What is the normal rate of bone remodeling. *Bone*, **35**, 1–3 (2004).

83. A. M. Parfitt, Calcium homeostasis. In *Handbook of Experimental Pharmacology: Physiology and Pharmacology of Bone* (G. R. Mundy and T. J. Martin, eds.), Vol. 107, pp. 1–65. Springer-Verlag, Heidelberg (1993).

84. G. B. Adams, K. T. Chabner, I. R. Alley, D. P. Olson, Z. M. Szczepiorkowski, M. C. Poznansky, C. H. Kos, M. R. Pollak, E. M. Brown, and D. T. Scadden, Stem cell engraftment at the endosteal niche is specified by the calcium-sensing receptor. *Nature*, **437**, 519–603 (2006).

85. A. M. Parfitt, Equilibrium and disequilibrium hypercalcemia: New light on an old concept. *Metab Bone Dis Rel Res*, **1**, 279–293 (1979).

86. A. M. Parfitt, Misconceptions (3): Calcium leaves bone only by resorption and enters only by formation. *Bone*, **33**, 259–263 (2003).

87. A. M. Parfitt, Calcium homeostasis. *J Musculoskel Neuron Interact*, **4**, 109–110 (2004).

88. A. M. Parfitt, Renal bone disease: A new conceptual framework for the interpretation of bone histomorphometry. *Curr Opin Nephrol Hypertens*, **12**, 387–408 (2003).

89. A. Blumsohn, K. Herrington, R. A. Hannon, P. Shao, D. R. Eyre, and R. Eastell, The effect of calcium supplementation on the circadian rhythm of bone resorption. *J Clin Endocrinol Metab*, **79**, 730–735 (1994).

90. A. M. Parfitt, Morphologic basis of bone mineral measurements. Transient and steady state effects of treatment in osteoporosis. *Miner Electrolyte Metab*, **4**, 273–287 (1980). [Editorial].

91. A. M. Parfitt, Integration of skeletal and mineral homeostasis. In *Osteoporosis: Recent Advances in Pathogenesis and Treatment* (H. F. DeLuca, H. Frost, W. Jee, C. Johnston, and A. M. Parfitt, eds.), pp. 115–126. University Park Press, Baltimore (1981).

92. H. J. Kalkwarf, Lactation and maternal bone health. *Adv Exp Med Biol*, **554**, 101–114 (2004).

93. M. Kleerekoper, and D. Nelson, Osteoporosis as a community health problem: Lessons learned from studying hypertension. *HFH Med J*, **36**, 113–116 (1988).

94. A. M. Parfitt, Osteoporosis: 50 years of change, mostly in the right direction. In *Osteoporosis and Bone Biology: The State of the Art* (J. Compston and S. Ralston, eds.), pp. 1–13. International Medical Press, London (2000).

95. S. M. Garn, *The Earlier Gain and the Later Loss of Cortical Bone*. Thomas, Springfield, IL (1970).

96. A. M. Parfitt, Interpretation of bone densitometry measurements. The disadvantages of a percentage scale and a discussion of available alternatives. *J Bone Miner Res*, **5**, 537–540 (1990).

97. A. M. Parfitt, Is the rate of bone loss influenced by the initial value? Biological and statistical issues. Abstract, Perth International Bone Meeting, Fremantle, Australia. *Osteoporosis Int*, **5**, 309–310 (1995).

98. B. E. C. Nordin, A. G. Need, R. L. Prince, M. Horowitz, D. H. Gutteridge, and S. E. Papapoulos, Osteoporosis. In *Metabolic Bone and Stone Disease* (B. E. C. Nordin, A. G. Need, and H. A. Morris, eds.), pp. 1–82. Churchill Livingstone, London (1993).

99. C. W. Slemenda, Adult bone loss. In *Osteoporosis* (R. Marcus, ed.), pp. 101–124. Blackwell Scientific, London (1994).

100. A. M. Parfitt, C. H. E. Mathews, A. R. Villanueva, M. Kleerekoper, B. Frame, and D. S. Rao, Relationship between surface, volume and thickness of iliac trabecular bone in aging and in osteoporosis. Implications for the microanatomic and cellular mechanism of bone loss. *J Clin Invest*, **72**, 1396–1409 (1983).

101. P. J. Meunier, Assessment of bone turnover by histomorphometry in osteoporosis. In *Osteoporosis: Etiology, Diagnosis, and Management* (B. L. Riggs and L. J. Melton, eds.), pp. 317–332. Raven Press, New York (1988).

102. J. W. Davis, J. S. Grove, P. D. Ross, J. M. Vogel, and R. D. Wasnich, Relationship between bone mass and rates of bone change at appendicular measurement sites. *J Bone Miner Res*, **7**, 719–725 (1992).

103. K. M. Davies, R. R. Recker, M. R. Stegman, and R. P. Heaney, Tallness versus shrinkage: Do women shrink with age or grow taller with recent birth date? *J Bone Miner Res*, **6**, 1115, 1120 (1991).

104. H. G. Ahlborg, O. Johnell, and M. K. Karlsson, Long term effects of oestrogen therapy on bone loss in postmenopausal women: A 23 year prospective study. *BJOG*, **111**, 335–339 (2004).

105. Y. Z. Bagger, L. B. Tanko, P. Alexandersen, P. Pava, and C. Christiansen, Alendronate has a residual effect on bone mass in postmenopausal Danish women up to 7 years after treatment withdrawal. *Bone*, **33**, 301–307 (2003).

106. D. E. Sellimeyer, K. L. Stone, A. Sebastian, and S. R. Cummings, A high ratio of dietary animal to vegetable protein increases the rate of bone loss and the risk of fracture in postmenopausal women. *Am J Clin Nutr*, **73**, 118–122 (2001).

107. P. D. Delmas, R. Eastell, P. Garnero, M. J. Seibel, and J. Stepan, The use of biochemical markers of bone turnover in osteoporosis. *Osteoporosis Intern*, **11**(Suppl 6), S2–17 (2000).

108. D. N. Cruz, J. J. Wysolmerski, H. M. Brickel, C. G. Gundberg, C. A. Simpson, M. A. Mitnick, A. S. Kliger, M. I. Lorber, G. P. Basadonna, A. L. Friedman, K. L. Insogna, and M. J. Bia, Parameters of high bone-turnover predict bone loss in renal transplant patients: A longitudinal study. *Transplantation*, **72**, 83–88 (2001).

109. O. Lofman, P. Magnusson, G. Toss, and L. Larsson, Common biochemical markers of bone turnover predict future bone loss: A 5-year follow-up study. *Clin Chim Acta*, **356**, 67–75 (2005).

110. A. K. Srivastava, E. L. Vliet, E. M. Lewiecki, M. Maricic, A. Abdelmalek, O. Gluck, and D. J. Baylink, Clinical use of serum and urine bone markers in the management of osteoporosis. *Cur Med Res Opinion*, **21**, 1015–1026 (2005).

111. K. Wu, S. Jett, and H. M. Frost, Bone resorption rates in rib in physiological, senile, and postmenopausal osteoporoses. *J Lab Clin Med*, **69**, 810–818 (1967).

112. L. E. Lanyon, A. E. Goodship, C. J. Pye, and J. H. MacFie, Mechanically adaptive bone remodeling. *J Biomech*, **15**, 141, 154 (1982).

113. H. M. Frost, Structural adaptations to mechanical usage (SATMU). *Anat Rec*, **226**, 414–422 (1990).

114. Z. F. G. Jaworski, and H. K. Uhthoff, Disuse osteoporosis: Current status and problems. In *Current Concepts of Bone Fragility* (H. K. Uhthoff, ed.), p. 181. Springer-Verlag, Berlin (1986).

115. H. M. Frost, On our age-related bone loss: Insights from a new paradigm. *J Bone Miner Res*, **12**, 1539–1546 (1997).

116. L. Lanyon, V. Armstrong, D. Ong, G. Zaman, and J. Price, Is estrogen receptor alpha key to controlling bones' resistance to fracture? *J Endocrin*, **182**, 183–191 (2004).

117. L. K. Saxton, and C. H. Turner, Estrogen receptor beta: The antimechanostat? *Bone*, **36**, 185–192 (2005).

118. A. M. Parfitt, Bone quality: Definition, history, current overview. Webcast from the meeting, Bone Quality: What Is It and Can We Measure It? At www.asbmr.org/bonequality.cfm#WebCast.

119. E. Seeman, and P. D. Delmas, Bone quality—The material and structural basis of bone strength and fragility. *N Eng J Med*, **354**, 2250–2261 (2006).

120. M. Kleerekoper, A. R. Villanueva, J. Stanciu, D. S. Rao, and A. M. Parfitt, The role of three dimensional trabecular microstructure in the pathogenesis of vertebral compression fractures. *Calcif Tissue Int*, **37**, 594–597 (1985).

121. P. D. Ross, J. W. Davis, R. S. Epstein, and R. D. Wasnich, Pre-existing fractures and bone mass predict vertebral fracture incidence in women. *Ann Intern Med*, **114**, 919–923 (1991).

122. C. M. Klotzbuecher, P. D. Ross, P. B. Landsman, T. A. Abbott, and M. Berger, Patients with prior fractures have an increased risk of future fractures: A summary of the literature and statistical synthesis. *J Bone Miner Res*, **15**, 721–739 (2000).

123. E. F. Eriksen, B. Langdahl, A. Vesterby, J. Rungby, and M. Kassem, Hormone replacement therapy prevents osteoclastic hyperactivity: A histomorphometric study in early postmenopausal women. *J Bone Miner Res*, **14**, 1217–1221 (1999).

124. D. E. Hughes, A. Dai, J. C. Tiffee, H. H. Li, G. R. Mundy, and B. F. Boyce, Estrogen promotes apoptosis of murine osteoclasts mediated by TGF-β. *Nature Medicine*, **2**, 1132–1135 (1996).

125. A. M. Parfitt, Abnormal structure and fragility of bone as expressions of disordered remodeling. Abstract XXIV, European Symposium on Calcified Tissues, Aarhus. *Calcif Tissue Int*, **56**, 423 (1995).

126. B. L. Drinkwater, K. Nilson, C. H. Chesnut, W. J. Bremner, S. Shainholtz, and M. B. Southworth, Bone mineral content of amenorrheic and eumenorrheic athletes. *N Engl J Med*, **311**, 277–281 (1984).

127. H. M. Frost, The pathomechanics of osteoporosis. *Clin Orthop*, **200**, 198–225 (1985).

128. R. M. Alexander, Optimum strengths for bones liable to fatigue and accidental fracture. *J Theor Biol*, **109**, 621–636 (1984).

129. A. M. Parfitt, Bone age, mineral density and fatigue damage. *Calcif Tissue Int*, **53**(Suppl 1), S82–S86 (1993).

130. H. M. Frost, *In vivo* osteocyte death. *J Bone Jt Surg*, **42A**, 138–143 (1960).

131. J. Vaughn, The effects of skeletal irradiation. *Clin Orthop*, **56**, 283–303 (1968).

132. L. Flora, G. S. Hassing, A. M. Parfitt, and A. R. Villanueva, Comparative skeletal effects of two diphosphonates in dogs. In *Bone Histomorphometry: Third International Workshop* (W. S. S. Jee and A. M. Parfitt, eds.), pp. 389–407. ArmourMontagu, Paris (1981).

133. A. M. Parfitt, A. R. Villanueva, J. Foldes, D. S. Rao, Relations between histologic indices of bone formation: Implications for the pathogenesis of spinal osteoporosis. *J Bone Miner Res*, **10**, 466–473 (1995).

134. G. H. Bell, O. Dunbar, J. S. Beck, and A. Gibb. Variations in strength of vertebrae with age and their relation to osteoporosis. *Calcif Tissue Res*, **1**, 75–86 (1967).

135. B. D. Snyder, S. Piazza, W. T. Edwards, and W. C. Hayes, Role of trabecular morphology in the etiology of age-related vertebral fractures. *Calcif Tissue Int*, **53**(Suppl. 1), S14–S22 (1993).

136. B. Vernon-Roberts, and C. J. Pirie, Healing trabecular microfractures in the bodies of lumbar vertebrae. *Ann Rheum Dis*, **32**, 406–412 (1973).

137. J. S. Wand, T. Smith, J. R. Green, R. Hesp, J. N. Bradbeer, and J. Reeve, Whole-body and site-specific bone remodeling in patients with previous femoral fractures: Relationships between reduced physical activity, reduced bone mass and increased bone resorption. *Clinical Sciences*, **83**, 665–675 (1992).

138. S. Y. P. Wong, J. Kariks, R. A. Evans, C. R. Dunstan, and E. Hills, The effect of age on bone composition and viability in the femoral head. *J Bone Jt Surg*, **67A**, 274–283 (1985).

139. D. Vashishth, O. Verborgt, G. Divine, M. B. Schaffler, and D. P. Pyhrie, Decline in osteocyte lacunar density in human cortical bone is associated with accumulation of microcracks with age. *Bone*, **26**, 375–380 (2000).

140. C. R. Dunstan, N. M. Somers, and R. A. Evans, Osteocyte death and hip fracture. *Calcif Tissue Int*, **53**(Suppl 1), S113–S117 (1993).

141. M. Grynpas, Age and disease-related changes in the mineral of bone. *Calcif Tissue Int*, **53**(Suppl 1), S57–S64 (1993).

142. M. A. R. Freeman, R. C. Todd, and C. J. Pirie, The role of fatigue in the pathogenesis of senile femoral neck fractures. *J Bone Jt Surg*, **56B**, 698–702 (1974).

143. N. L. Fazzalari, Trabecular microfracture. *Calcif Tissue Int*, **53**(Suppl 1), S143–S147 (1993).

144. M. Wicks, R. Garrett, B. Vernon-Roberts, and N. Fazzalari, Absence of metabolic bone disease in the proximal femur in patients with fracture of the femoral neck. *J Bone Jt Surg*, **64B**, 319–322 (1982).

145. S. R. Cummings, D. B. Karpf, F. Harris, H. K. Genant, K. Ensrud, A. Z. LaCroix, and D. M. Black, Improvement in spine bone density and reduction in risk of vertebral fractures during treatment with antiresorptive drugs. *Am J Med*, **112**, 281–289 (2002).

146. C. Meier, T. V. Nguyen, J. R. Center, M. J. Seibel, and J. A. Eisman, Bone resorption and osteoporotic fractures in elderly men: The Dubbo osteoporosis epidemiology study. *J Bone Miner Res*, **20**, 579–587 (2005).

147. E. Sornay-Rendu, F. Munoz, P. Garnero, F. Dubboeuf, and P. D. Delmas, Identification of osteopenic women at high risk of fracture: The OFELY study. *J Bone Mine Res*, **20**, 1813–1819 (2005).

148. B. L. Riggs, and L. J. Melton, Bone turnover matters: The raloxifene treatment paradox of dramatic decreases in vertebral fracture without commensurate increases in bone density. *J Bone Mine Res*, **17**, 11–14 (2002).

149. A. M. Parfitt, High bone turnover is intrinsically harmful: Two paths to a similar conclusion. *J Bone Miner Res*, **17**, 1558–1559 (2002).

150. L. Mosekilde, Age-related changes in vertebral trabecular bone architecture—Assessed by a new method. Bone, 9, 247–250 (1988).

151. T. A. Einhorn, Bone strength: The bottom line. *Calcif Tissue Int*, **51**, 333–339 (1992).

152. J. Reeve, A stochastic analysis of iliac trabecular bone dynamics. *Clin Orthop Rel Res*, **213**, 264–278 (1986).

153. R. S. Weinstein, C. C. Powers, A. M. Parfitt, and S. C. Manolagas, Preservation of osteocyte viability by bisphosphonate contributes to bone strength in glucocorticoid-treated mice independently of BMD: An unappreciated determinant of bone strength. *J Bone Miner Res*, Suppl 1, S156 (2002).

154. C. A. O'Brien, D. Jia, L. I. Plotkin, T. Belido, C. C. Powers, S. A. Stewart, S. C. Manolagas, and R. S. Weinstein, Glucocorticoids act directly on osteoblasts and osteocytes to induce their apoptosis and reduce bone formation and strength. *Endocrinology*, **145**, 1835–1841 (2004)

155. S. Qiu, D. S. Rao, S. Palnitkar, and A. M. Parfitt, Reduced iliac cancellous osteocyte density in patients with osteoporotic vertebral fracture. *J Bone Miner Res*, **18**, 1657–1663 (2003).

156. R. S. Weinstein, J. J. Goellner, D. Jia, R. S. Shelton, A. M. Parfitt, and S. C. Manolagas, Glucocorticoid excess disrupts the canalicular circulation: Potential mechanism of the disparity between bone density and strength in glucocorticoid-induced osteoporosis and osteonecrosis. *J Bone Miner Res*, **20**(Suppl 1), S50 (2005).

157. A. M. Parfitt, The actions of parathyroid hormone on bone. Relation to bone remodeling and turnover, calcium homeostasis and metabolic bone disease. I. Mechanisms of calcium transfer between blood and bone and their cellular basis. Morphologic and kinetic approaches to bone turnover. *Metabolism*, **25**, 809–844 (1976).

158. R. B. Martin, On the histologic measurement of osteonal BMU activation frequency. *Bone*, **15**, 547–549 (1994).

159. C. J. Hernandez, S. J. Hazelwood, and R. B. Martin, The relationship between basic multicellular unit activation and origination in cancellous bone. *Bone*, **25**, 583–587 (1999).

160. J. C. Stevenson, Hormone replacement therapy. *Cur Osteoporosis Rep*, **2**, 12–16 (2004).

161. A. W. Popp, C. Bodmer, C. Senn, G. Fuchs, M. E. Kraenzlin, H. Wyss, M. H. Birkhaeuser, and K. Lippuner. Prevention of postmenopausal bone loss with long-cycle hormone replacement therapy. *Maturitas*, **53**, 191–200 (2006).

162. K. Kanaoka, Y. Kobayashi, F. Hashimoto, T. Nakashima, M. Shibata, K. Kobayashi, Y. Kato, and H. Sakai. A common downstream signaling activity of osteoclast survival factors that prevent nitric oxide-promoted osteoclast apoptosis. *Endocrinology*, **141**, 2995–3005 (2000).

163. H. Fleisch, *Bisphosphonates in Bone Disease. From the Laboratory to the Patient,* 4th ed. Academic Press, San Diego CA (2000).

164. D. E. Hughes, K. R. Wright, H. L. Uy, S. Saski, T. Yoneds, G. D. Roodman, G. R. Mundy, and B. F. Boyce, Bisphosphonates promote apoptosis in murine osteoclasts *in vitro* and *in vivo*. *J Bone Miner Res*, **10**, 1478–1487 (1995).

165. R. W. Boyce, C. L. Paddock, J. R. Gleason, W. K. Sletsema, and E. F. Eriksen, The effects of risedronate on canine cancellous bone remodeling: Three-dimensional kinetic reconstruction of the remodeling site. *J Bone Miner Res*, **10**, 211–221 (1995).

166. T. M. Boyce, and R. D. Bloebaum, Cortical aging differences and fracture implications for the human femoral neck. *Bone*, **14**, 769–778 (1993).

167. U. A. Liberman, S. R. Weiss, J. Broll, *et al.*, Effect of oral alendronate on bone mineral density and the incidence of fractures in postmenopausal osteoporosis. *N Engl J Med*, **333**, 1437–1443 (1995).

168. J. Li, T. Mashiba, and D. B. Burr, Bisphosphonate treatment suppresses not only stochastic remodeling but also the targeted repair of microdamage. *Calcif Tiss Intl*, **69**, 281–286 (2001).

169. T. Mashiba, C. H. Turner, T. Hirano, M. R. Forwood, C. C. Johnston, and D. B. Burr, Effects of suppressed bone turnover by bisphosphonates on microdamage accumulation and biomechanical properties in clinically relevant skeletal sites in beagles. *Bone*, **28**, 524–531 (2001).

170. G. R. Jordan, N. Loveridge, K. L. Bell, J. Power, N. Rushton, and J. Reeve, Spatial clustering of remodeling osteons in the femoral neck cortex: A cause of weakness in hip fracture? *Bone*, **26**, 305–313 (2000).

171. K. L. Bell, N. Loverdige, G. R. Jordan, J. Power, C. R. Constant, and J. Reeve, A novel mechanism for induction of increased cortical porosity in cases of intracapsular hip fracture. *Bone*, **27**, 297–304 (2000).

172. M. C. Chapuy, M. E. Arlot, P. D. Delmas, and P. J. Meunier, Effect of calcium and cholecalciferol treatment for three years on hip fractures in elderly women. *BMJ*, **308**, 1081–1082 (1994).

CHAPTER 4

Osteoblast Biology

JANE B. LIAN AND GARY S. STEIN

I. OVERVIEW

Bone formation takes place throughout life to support growth, mechanical forces, bone turnover to meet metabolic needs, and the reparative process. The requirement for continuous renewal of bone through the remodeling process necessitates recruitment, proliferation, and differentiation of osteoblast-lineage cells. A contributing factor to bone loss in the aging skeleton is the decreased ability of bone-forming osteoblasts to replace bone removed by the activity of the bone-resorbing osteoclasts. It is now appreciated that numerous developmental, growth factor, cytokine, and hormone responsive regulatory signals mediate competency for expression of genes associated with bone matrix synthesis and metabolic responses as a function of the stages of osteoblast growth and differentiation.

Osteoblast differentiation is a multistep series of events modulated by an integrated cascade of regulatory factors and that initially supports proliferation and the sequential expression of genes associated with the biosynthesis, organization, and mineralization of the bone extracellular matrix. This chapter discusses the current understanding of the phenotypic definition of the spectrum of bone-forming cells with respect to their functional properties and responses. It is becoming apparent that signaling pathways, important for early skeletal patterning during embryonic bone development, have postnatal functions in supporting bone formation and maintaining bone mass. Recent advances in the identification of obligatory factors that contribute to osteoblast growth and differentiation in the adult skeleton are presented. Knowledge of unique properties and definition of the molecular mechanisms that control progression through the osteoblast cell lineage will allow a rational intervention for stimulating bone formation in the aging skeleton, fracture repair, pathologies of metabolic bone diseases, and implant stability.

II. EMBRYONIC DEVELOPMENT OF THE OSTEOBLAST PHENOTYPE: LESSONS FOR BONE FORMATION IN THE POSTNATAL SKELETON

The complexities of bone formation are immediately apparent in the embryo where different regions of the skeleton arise from specific primordial structures. Skeletogenesis involves two different processes that continue in the adult skeleton. Intramembranous bone formation, as occurs in the development of the flat bones of the skull, results from the differentiation of mesenchymal cell condensations directly to osteoblasts. The replacement of resorbed bone in the adult skeleton is essentially an intramembranous process in which mesenchymal stromal cells are recruited from the marrow for differentiation to osteoblasts. The endochondral sequence of bone formation (EBF), as occurs for all long bones, involves the differentiation of mesenchymal progenitors first to form a cartilage template of the bone, which undergoes maturation to calcified cartilage that is then replaced by bone. The EBF sequence occurs during fracture repair of the adult skeleton. Our understanding of skeletal patterning and limb development has been expanded significantly by characterization of the signaling factors and transcription factors that serve as morphogenic determinants of bone formation [1–3]. See Chapter 8 (Provot) for a discussion of the development of the skeleton.

A. Regulatory Factors for Mesenchyme Organization

Progenitors of the bone-forming cells for all osseous tissues derive from the mesodermal germ cell layer. Secreted signaling proteins induce skeletal elements in the embryo from different regions of the mesoderm and also function in the adult. Cells of the paraxial mesoderm undergo condensation and segmentation to form somites under the regulation of the cell surface receptor Notch1 and its ligands, delta and serrate [4]. Notch, a transmembrane domain that functions in determination of cell differentiation pathways, has multiple effects on early stages of osteoblast differentiation, both enhancing maturation through bone morphogenetic protein-2 (BMP-2) and impairing differentiation by suppressing Wnt/β-catenin signaling, as well as inhibiting the mineralization stage through activity of the Notch1 target gene Hey 1 [5–7]. Sclerotome cells from somites are induced to form cartilage by the cytokine sonic hedgehog (Shh) secreted by the notochord, producing the axial skeleton (spine, sternum, and ribs) [8]. Shh was shown to regulate mesenchymal cell recruitment into the osteogenic lineage and is involved in osteoblast differentiation by mediating the effects of BMP-2 [9–11]. The Gli family of transcription factors has critical regulatory roles in mediating Shh signaling for sclerotome development and BMP-2 response to hedgehog signaling [11, 12], as well as for regulation of limb development [13].

The lateral plate mesoderm gives rise to the appendicular skeleton (limbs), and the cephalic mesoderm gives rise to the neural crest, which provides progenitor cells for facial skeletal structures. Homeodomain proteins, including Msh homeobox (Msx1 and Msx2) and the distal-less (Dlx) family, specify the spatial and temporal formation of the craniofacial skeleton [2, 14, 15], and these proteins regulate osteoblast differentiation throughout life. Considering these different embryonic developmental programs of the mesoderm to form intramembranous bone and subtypes of endochondral bone (e.g., limbs and vertebrae), an early osteoprogenitor may divert from a stem cell at these specific skeletal sites [16]. In the adult skeleton, it has been shown that axial and appendicular-derived osteoblasts exhibit different responses to hormones. It remains to be determined whether this selective activity reflects the tissue environment or inherent properties of the cells selected at an early stage during osteoblast differentiation.

Knowledge of how mesenchymal condensations are initiated and grow, and how their sizes and boundaries are regulated, is being accrued through genetic studies in mice and the characterization of molecular defects in skeletal development. The neural crest–mesoderm boundary is but one example in which the boundary between two compartments serves as a signaling center that is most strikingly reflected by the craniosynostosis disorders in which premature fusion of the parietal and frontal bones is caused by mutations in homeodomain protein Msx2 and the transcription factor Twist, which cooperate in development of the neural crest–mesenchyme tissue [17]. This boundary is critically maintained by

the ephrinB1–EphB interactions [18–20]. Extracellular matrix molecules, cell surface receptors, and cell adhesion molecules, such as fibronectin, tenascin, syndecan, and N-CAM, initiate condensation and set boundaries for the forming mesenchyme. The major signaling proteins involved in segmentation of the vertebrate body are Hox genes, which define the positions where bone structures will develop [21, 22]. Signals essential for limb patterning arise from the zone of polarizing activity, which resides in the posterior limb mesoderm. The formation of skeletal tissues from the condensations of the mesenchymal progenitor cells at a specific site is determined by epithelial–mesenchymal interactions that control shape and size of the limbs through secretion of regulatory factors and target transcription factors. A group of epithelial cells, the apical ectodermal ridge (AER), caps the limb buds and secretes growth factors that pattern the limb. Formation and activities of the AER are regulated by Hox genes, Wnt and fibroblast growth factor (FGF) signaling. For example, Hox genes modulate the proliferation of cells within condensations [1], and the clustered Dlx gene family of homeobox-related genes, which are expressed in the AER of the limb bud, regulates the proximal–distal pattern of outgrowth [23]. In the postnatal skeleton, Hox genes are observed to be expressed during fracture repair, which may contribute to reforming the bone in a specific orientation [24].

B. Signaling Pathways: FGF, BMP/TGF-β, Wnt, and Indian Hedgehog

Regulatory factors that induce and control development of skeletal structures remain as key signals for induction of bone renewal after turnover, maintenance of bone structure, and bone repair in the adult skeleton. These factors include members of major signaling pathways, the transforming growth factor-β (TGF-β)/ bone morphogenic protein (BMP) superfamily, which induce mesenchyme condensation for bone formation; fibroblast growth factors (FGFs), which are essential for the earlier stages of limb bud outgrowth; and Wnt proteins, which contribute to the formation of the bone axes [25–28]. Thus, it is instructive to understand how these pathways regulate embryonic bone formation for potential therapeutic strategies to treat disorders of the skeleton after birth.

The fibroblast growth factor receptors, FGFR1, FGFR2, and FGFR3, and the FGF ligands are expressed throughout skeletal cell populations and contribute to the regulation of progenitor and differentiated populations of cells during both intramembranous and endochondral bone formation [29]. FGFR1 is expressed in limb mesenchyme and in osteoprogenitor cells at the osteogenic front separating the nonosseous suture tissue between the ossification plates of the calvarial bone tissue. An FGFR1 mutation leads to premature fusion of craniofacial structure (craniosynostoses), whereas dominant-negative forms of FGFR1 will inhibit calvarial suture fusion [30, 31]. Mutations in FGFR2 are responsible for Apert syndrome, a severe craniosynostosis [30], and Pfeiffer and Crouzon disorders [32]. FGFR2 is expressed as two variants: Fgfr2b is required for limb outgrowth, whereas Fgfr2c is required for osteoblast maturation [33]. Interestingly, different point mutations result in distinct phenotypic alterations in gene expression of the osteoblasts conveying the mutation [34, 35]. FGFR3 expression is initiated as chondrocytes differentiate in long bones and various knock-in mutations of this gene lead to severe dwarfism [36]. The FGFR3 and parathyroid hormone–related peptide (PTHrP) signals coordinate cartilage and bone formation [37]. Complete ablation of FGFR3 leads to embryonic skeletal overgrowth [38] but an osteopenia phenotype in the adult mouse [39]. The FGFR1, FGFR2, and FGFR3 activities appear to be linked with respect to their positive and negative regulator of endochondral bone formation and osteoblast growth and differentiation [40].

Specific FGF ligands control limb outgrowth by increasing proliferation of mesenchymal cells [29, 41–43]. Among these, FGF2 and FGF18 contribute to bone formation. Mice lacking FGF18 have defects in both chondrogenesis and osteogenesis [44, 45]. FGF2 was initially isolated from the cartilage matrix [46] and later identified in periosteal cells and osteoblasts [47]. FGF2 activates several signaling factors [48], including Wnt genes [49], Notch ligand expression [50, 51], Hedgehog factors, and transcriptional regulators such as helix–loop–helix proteins [52]. FGF2 supports osteoblast growth and differentiation and contributes to osteoblast survival [53–55]. Mice lacking FGF2 have decreased bone mass [56]. These examples illustrate how FGF signaling through multiple receptors and ligands and with specific activities on different cell populations controls expansion of embryonic cartilaginous tissue for endochondral bone formation.

Bone morphogenetic proteins of the TGF-β superfamily are multifunctional growth and differentiation factors (GDFs) that support the development of many tissues, including cartilage and bone [25, 57, 58]. They are actively involved in determining parameters of size and shape during mesenchymal cell condensation [59]. Specificity of the activities of TGF-β and BMPs with target cells is regulated by their activation of distinct type I and type II serine/threonine kinase receptors that phosphorylate intracellular receptor–regulated (R) Smad proteins (R-Smads). R-Smads are anchored to the cell membrane by SARA (the Smad anchor for receptor activation). Smad2 and -3 mediate TGF-β responses, whereas Smad1, -5, and -8 are activated by BMP receptors. Interactions between R-Smads and Smad4, the common DNA-binding Smad, result in translocation of the complex to the nucleus for transcription of target genes [27, 60]. Antagonistic or inhibitory Smads (I-Smads) contribute to regulating this pathway. Genetic studies of receptors and Smad components in the mouse reveal phenotypes that affect many tissues [61–63]. In the skeleton, constitutively active forms of BMP-1A and BMP-1B promote chondrogenesis and osteogenesis, but only the dominant-negative form of BMPR-1B inhibits these events [64]. Complete null mutation of the BMPR-1B type I receptor revealed defects mainly in the appendicular skeleton, with marked reduction in proliferation of the prechondrocytic population and subsequent chondrocyte differentiation [65]. Conditional deletion of the BMP-1A receptor in GDF5-expressing cells in developing joints resulted in an osteoarthritis phenotype in the mouse [66]. Mice lacking Smad3, which mediates TGF-β signaling, develop degenerative joint disease [67].

BMP signaling is required at an early stage of skeletal development for formation of the AER and dorsal–ventral patterning of the limb (mediated by BMP-2, -4, and -7) [68]. Selective expression of BMPs regulates mesenchymal condensations and contributes to restriction of options for lineages. BMP-5 is expressed in condensing mesenchyme, perichondral periosteum. BMP-2 and BMP-6 can stimulate cartilage differentiation to the hypertrophic phenotype [69]. BMP-2, -4, -6, and -7 are potent osteoinductive growth factors for bone formation, but BMP-6 appears to be detected in bone marrow mesenchymal cells prior to differentiation [70]. BMP-4 overexpression is linked to fibrodysplasia ossificans progressive, and the gene mutation causing the phenotype is the BMP-4 I receptor activin (ACVR1) [71–73]. BMP activities are highly regulated by inhibitors including Sclerostin, Chordin, Noggin, and Gremlin, which play a critical role in skeletal development by inhibiting BMP-2 signaling [74, 75]. These requirements for regulated BMP-2 activity are illustrated by disruption of Noggin control, which results in early skeletal malformations in the mouse [76, 77]. Transducer of ErbB2 (Tob) antagonizes BMP-2-mediated bone formation by forming a complex with Smad1. The Tob null mouse has increased bone mass resulting from osteoblasts hyperresponsive to BMP-2 [78].

A question is how the BMP/TGF-β signal is transduced to specific cell differentiation programs when the same receptors are present in all cells. One mechanism appears to be through tissue-specific transcriptional control. BMPs induce a spectrum of transcription factors essential for differentiation of skeletal tissues [79–82]. Among these are Sox genes, which are required for chondrogenesis [83, 84]; homeodomain transcription factors, which pattern the skeleton and promote bone formation; and Runx2 as well as Osterix, which are both essential for osteoblast differentiation and formation of a mineralized skeleton [85–89]. Runx2 is expressed early in embryogenesis and upregulated in late stages of bone development, suggesting that this factor may be important in early specification of the phenotype, as well as having an essential role for osteoblast differentiation [89–91]. Studies indicate an important positive regulatory loop between BMP-2 and the Runx2 transcription factor [92–94]. BMP-2 and Runx2 together have synergistic effects in promoting osteogenesis [95, 96]. Such combinatorial regulation provides a "feed forward" mechanism necessary to support development of the skeleton. These examples show how chondrogenic and the osteogenic activity of BMPs are related to induction of specialized transcriptional regulators of cell differentiation. Figure 4-1 illustrates an example of the developmental signaling pathways that converge to regulate bone formation through the transcription factor Runx2, as well as pathways independent of Runx2 [97].

I. WNT SIGNALING AND SKELETAL DEVELOPMENT

The Wnt family of 19 secreted cysteine-rich glycoproteins regulates numerous developmental processes, including cell polarity, cell differentiation, and migration [98]. Wnt signaling is mediated through several pathways, but the canonical β-catenin pathway is a well-recognized regulator of early embryogenesis, skeletal development, as well as maintenance of bone mass in the adult [99–101]. Wnt/β-catenin signaling has been identified as a normal physiological response to mechanical loading [102]. Wnt proteins bind to and activate receptor complexes consisting of the Frizzled family of G protein–coupled receptors and the low-density lipoprotein (LDL) receptor-related proteins (LRP5/6). Activation of the canonical pathway results in stabilization of β-catenin by inhibiting its phosphorylation involving casein kinase 1 and glycogen synthase kinase 3 (GSK3) within a protein complex (with Axin and APC), which prevents the targeting of β-catenin for ubiquitination and proteasome degradation. As a result, β-catenin is translocated into the nucleus to form heterodimers with the TCF1 or LEF transcription factors for expression of Wnt-responsive genes. In the absence of nuclear β-catenin, TCF/LEF is associated with transcriptional co-repressors and

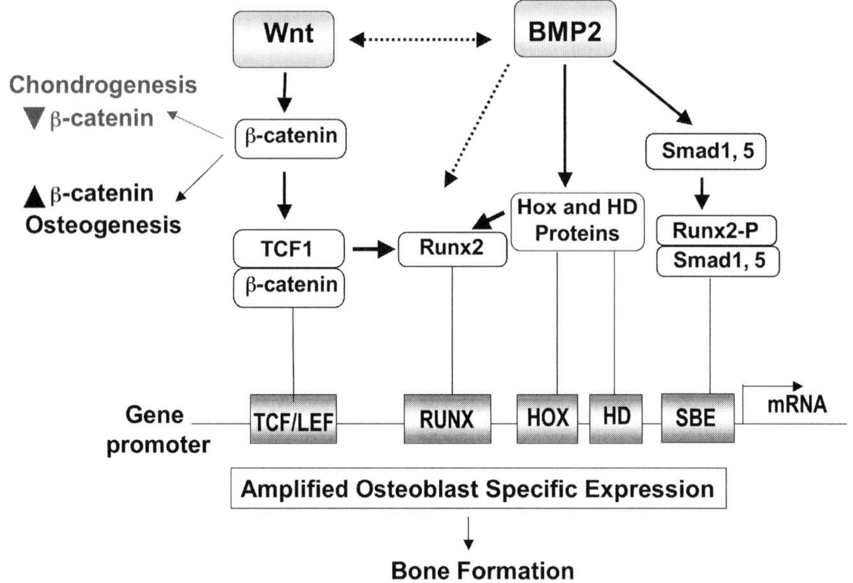

FIGURE 4-1   Integration of developmental signaling pathways through Runx2. Wnt and BMP-2 promote osteogenesis through multiple signaling cascades. The canonical Wnt/β-catenin pathway modulates the switches between chondrogenesis and osteogenesis in early mesenchymal cells through cellular levels of β-catenin. The osteogenic effects of Wnt are in part mediated through TCF1/β-catenin activation of Runx2. BMP-2 signaling is mediated through receptor Smads and MAPK signaling, pathways that can phosphorylate Runx2. BMP-2 can also induce several classes of transcription factors as early response genes, including Runx2, several Hox and homeodomain proteins, and Osterix (not shown). Hoxa10, Dlx3, and Dlx5 also contribute to activation of Runx2 in mesenchymal cells [97]. These transcription factors target gene promoters through protein–DNA interactions. In addition, Runx2 forms protein–protein complexes, as illustrated for +Smad and Runx2, which can regulate transcription through both Runx sites and Smad binding elements (SBEs). Runx, SBEs, and homeodomain response elements are abundant in osteoblast genes and the activity of multiple developmental regulators together may represent an osteogenic signature for gene expression and bone formation.

suppresses Wnt target genes. Each component of the canonical β-catenin pathway has identified roles in skeletogenesis [100]. The other two pathways, the planar cell polarity and the calcium pathways, are not as well defined in the mammalian tissues in the skeleton.

The Wnt signaling pathway is regulated by several antagonists. Dickkopf (Dkk1 and -2) and their receptor Kremen [103] interact with the frizzled receptor/LRP5/6 complex preventing transduction of the Wnt signal. Dkk1 negatively regulates bone formation [104], whereas the Dkk2 Wnt antagonist functions at a late stage of osteogenesis and is required for mineralization of bone. Dkk2$^{-/-}$ are osteopenic [105, 106]. A class of secreted frizzled-related proteins (sFRPs) interact with Wnt proteins, sequestering them from interaction with frizzled receptors, as does WIF (Wnt inhibitory factor-1) and Cerberus [98]. Inactivation of sFRP1 in the mouse results in a high bone phenotype in older mice, increased bone mineral density in young mice, and accelerated hypertrophic chondrocyte and osteoblast maturation [107–109]. Axin is an intracellular inhibitor of canonical Wnt signaling, and its absence leads to increased β-catenin and osteoblast proliferation and differentiation [110]. Specific Wnt proteins that activate either the canonical β-catenin pathway or noncanonical pathways have been identified, although further studies are necessary to clarify the specific roles of Wnt factors that may operate through multiple pathways. Both gain- and loss-of-function mutations in components of the Wnt signaling pathway, including specific Wnt proteins, agonists, and anti-agonists, have revealed the significance of Wnt signaling in regulating skeletal development. By conditional ablation in mice, β-catenin has been identified as a key regulator of formation of the AER and of the dorsal–ventral axis of the limb [111]. Wnt10a misexpression in the developing chick limb identified its importance for AER formation [112]. Wnt3a knockout mice have a skeletal phenotype since Wnt3a is required for somite formation [113]. A rare human genetic disorder, tetra-amelia, characterized by the absence of all the limbs, has been linked to mutation in Wnt3 [114]. The expression of several Wnts (Wnt4, -5a, -5b, -6, -11, and -14) during limb development in the chick suggests key functions in initial stages of skeletal development [115, 116]. Misexpression of Wnt14 identified its role in induction of joint interzone [117]. Wnt5a and Wnt5b coordinate the pace and transitions between chondrocyte zones [118–120]. Wnt5b, which is expressed in the prehypertrophic chondrocyte zone as well as in joints and perichondrium, delays hypertrophy, whereas Wnt4 blocks initiation of chondrogenesis but accelerates hypertrophy [49, 121, 122]. Wnt10b is a potent osteogenic factor, both enhancing bone formation and inhibiting adipogenesis [123, 124]. Therefore, the developing limb and endochondral bone formation,

which both rely on the regulation of the proliferation, maturation, and spatial organization of chondrocytes, are processes highly dependent on Wnt signals.

The importance of canonical Wnt/β-catenin signaling in skeletal development and postnatal bone formation is established from numerous genetic studies and was initially identified by two human mutations in the Wnt co-receptor LRP5. Direct effects on formation and turnover of the mature skeleton have been revealed by an activating mutation (gain of function) in the Wnt co-receptor LRP5, resulting in the high bone mass trait in humans [125, 126], a phenotype reproduced in the mouse model [127, 128]. The LRP5 loss-of-function mutation leads to osteopenia accompanied by fractures in humans causing osteoporosis pseudoglioma syndrome [129, 130]. This low bone mass phenotype is also recapitulated in the mouse [131]. The developmental significance of Wnt/β-catenin signaling was revealed by genetic studies of inactivation of β-catenin in mesenchymal lineage cells, which resulted in severe loss of bone from inhibited osteoblast maturation and increases in osteoclast differentiation [132–134]. Increasing Wnt signaling (e.g., by ectopic expression in a cell type or by expressing a stabilized form of β-catenin) produced enhanced ossification and suppression of chondrogenesis [133–135]. One contributing molecular mechanism for the β-catenin cellular levels regulating the switch between chrondrogenesis and osteogenesis is the positive regulation of Runx2 by β-catenin/TCF1 in mesenchymal cells [108]. Enhanced canonical Wnt signaling in the sFRP1 mouse model shows a seven- or eightfold higher level of Runx2 in osteoblasts of young and old mice exhibiting increased hypertrophic chondrocyte activity and a high bone mass phenotype [109]. Runx2 also cooperates with the TCF/LEF transcription on the FGF18 target gene, another key regulator of bone formation [136]. Osteoblast cells can also negatively regulate Wnt signaling for control of normal bone formation. Sclerostin, a secreted protein from osteocytes, is a marker of the Dan family of glycoproteins. Sclerostin functions as a ligand for LRP5 to inhibit Wnt activity [137, 138]. Taken together, these findings underscore the significance of regulating the Wnt signaling pathway for bone formation and in maintaining bone mass, and they provide new avenues for potential therapeutic targets for decreasing bone loss in the aging skeleton. These findings also indicate that Wnt signaling is not only necessary for bone development but also supports maintenance of bone tissue functions in the adult skeleton through several mechanisms. Further discussion of the Wnt pathway is provided in Chapter 15 (Johnson).

Indian hedgehog signaling has emerged as a significant regulator of bone collar formation during embryonic development. The coordination of chondrocyte maturation during endochondral bone formation with bone formation events is tightly regulated. Both selective expression of secreted factors in chondrogenic subpopulations and feedback loops control the proliferation and maturation of chondrocytes and the pace of endochondral ossification [139, 140]. An essential regulatory protein is PTHrP, with specialized autocrine and paracrine activities for regulating endochondral bone formation. PTHrP activities are regulated by Indian hedgehog (Ihh), the other key factor in the normal development of the growth plate and endochondral bone formation [141]. Together, Ihh and PTHrP regulate the proportions of proliferating and hypertrophic chondrocytes and, hence, rate of cartilage differentiation [139, 142–144]. They are expressed abundantly in the mature and hypertrophic chondrocyte zones [145] for regulation of chondrocyte maturation. Ihh signaling is mediated through its receptor, Patch, and serves multiple functions in coordinating the events of expansion of the proliferative zone and regulating osteoprogenitor differentiation for maturation of the bone collar [146].

The importance of coordinated activities among these pathways is the focus of new investigations. Interactions between FGF, BMP, Wnt, and the PTH/PTHrP/Ihh signals occur primarily at the level of regulating secreted factors for coordinating the timing of developmental events. For example, the BMP pathway targets Ihh [147] during embryogenesis, and the BMP and Notch pathways network with Wnt. At appropriate levels, PTH is an anabolic factor of bone, and studies show that PTH increases β-catenin cellular levels in osteoblasts [148]. Sclerostin, an osteocyte-derived negative regulator of bone formation, functions by inhibiting Wnt signaling that is required for BMP-stimulated osteoblast differentiation [137, 138].

C.　Runx Factors during Embryogenesis

1.　THE RUNX FAMILY IS EXPRESSED IN THE SKELETON

It is important to recognize that the molecular mechanisms underlying the signaling pathways induced by secreted factors for skeletal development point to specific transcriptional regulators that control pattern formation and/or guide the mesenchymal cell to the chondrogenic and/or the osteoblast lineage. The *runt* homology domain-related core binding factor family of transcription factors (RUNX/CBFA) comprises three related genes that each support tissue specification and organogenesis together with the Runx DNA-binding partner protein CBFβ. In human, these factors were first designated as AML because they were identified as

the protein encoded by a gene locus rearranged in acute myelogenous leukemia (AML). Other names include core-binding factor (CBFA) and polyoma enhancer-binding protein (PEBP2). The Human Genome Nomenclature Committee has referred to this family as RUNX. RUNX1 (CBFA2/AML-1B/PEBP2B) is critical for hematopoietic cell differentiation, RUNX3 (CBFA3/AML-2/PEBP2C) is required for gut development [149, 150], and RUNX2 (CBFA1/AML-3/PEBP2A) is essential for the differentiation of osteoblastic cells for formation of the mineralized skeleton [88, 89, 151–153]. The obligatory role of Runx2 for the formation of mature bone in the developing skeleton has been shown by the absence of a calcified skeleton and bone formation in Runx2 null mouse models that die at birth [154–157]. The human cleidocranial dysplasia abnormalities are all derived from various mutations in runx2 and include intramembranous bone and supernumerary teeth defects [158–161]. Runx2 is expressed not only in abundance in osteoblasts and hypertrophic chondrocytes but also in cartilage, thymus, testis, and tooth.

It is now appreciated that all RUNX factors are expressed in condensing mesenchyme and skeletal tissues, but the roles of RUNX1 and RUNX3 are not established. Runx1 is highly expressed in calvarial sutures, periochondrium and periosteum, and epithelia of many organs [90, 162–164]. Runx1 expression partly overlaps Runx2 in periosteal tissue, whereas Runx2 and Runx3 overlap in the hypertrophic zone [165]. In zebrafish, knockdown of Runx2 and Runx3 compromises craniofacial formation [166]. The roles of Runx factors in tooth development have been explored (reviewed by Ryoo and Wang [167] and Yamashiro *et al.* [168]). Runx3 is selectively expressed in upper molars, whereas Runx2 expressed in the lower jaw revealed functions related to the bud stage of tooth development through regulation of FGF3 and SHH signaling [169, 170].

2.　RUNX2 FUNCTIONS AT STAGES OF EMBRYONIC DEVELOPMENT

Runx2 is involved in epithelial–mesenchymal interactions and is a target of three signaling pathways—FGF, BMP-2/TGF-β, and Wnt signaling. Runx2 is activated by BMP-2-induced homeodomain proteins [97, 171], and Runx2 can both negatively and positively regulate BMP or TGF-β target genes through formation of Smad–Runx2 complexes [92, 172]. Runx2 becomes phosphorylated in response to FGF signaling, which stimulates its transcriptional activity for bone and tooth development [55, 173]. Canonical/β-catenin Wnt signaling also activates Runx2 in mesenchymal cells [108]. Studies that have defined Runx2 levels during mesenchyme differentiation to cartilage and bone indicate Runx2 functions as a molecular switch.

Runx2 is expressed in prechondrogenic mesenchyme, several days prior to bone, and is retained in perichondrium [89, 91, 174]. Indeed, Runx2 function in the perichondrium must be downregulated for cells to enter the chondrogenic lineage [175]. A key mechanism for directing the bipotential cell through the chondrogenic pathway involves suppression of Runx2 gene expression in prechondrocytes by Nkx3.2, the gene associated with the bagpipe mutation in mouse [176] and Runx2/Cbfa1 [177].

Nkx3.2 is a strong repressor transcription factor and one of the earliest mediators of chondrocyte commitment [178, 179]. Repression of Nkx3.2 is critical for development of the axial skeleton and position of the jaw joint [176, 180, 181]. Through a series of de-repression events, Nkx3.2 allows for activation of Sox9, a requirement for chondrocyte differentiation [182]. As the mesenchymal cells are recruited into the chondrogenic lineage by Nkx3.2, Runx2 becomes downregulated through direct transcriptional control mediated by an Nkx3.2 response element in the Runx2 gene [175]. Stein and colleagues [175] showed that this mechanism of Runx2 repression is required for mesenchymal cells to enter the chondrogenic lineage. Thus, Nkx3.2 initiates a cascade of events for both suppressing osteogenesis and activating chondrogenesis. The activation of Sox9 by Nkx3.2 further promotes decreased Runx2 function. Sox9 directs chondrogenesis in part by directly interacting with Runx2 to repress Runx2 activity [183]. Downregulation of Nkx3.2 occurs in the hypertrophic zone of the growth plate, which may allow Runx2 to be re-expressed to high levels for endochondral bone formation to progress. Runx2 functional activity is also repressed by the transcription factor Twist in proliferating chondrocytes, which binds to Runx2 and interferes with Runx2 DNA binding [184]. It is anticipated that other factors will be identified that contribute to complete repression of Runx2 for commitment of cells to chondrocytes. In a complementary manner, Runx2 regulates NFATc2, a repressor of cartilage growth [185]. Mechanisms by which Runx2 is reactivated in the hypertrophic chondrocyte need to be addressed, but here BMP-2 and Wnt signaling are viable candidates, as described previously. Runx2 activates vascular endothelial growth factor and matrix metalloproteinase 9, both essential factors for vascular invasion and recruitment of osteoblasts to absorb calcified cartilage matrix and osteoprogenitors for bone formation [91, 186–188].

3. RUNX2 CONTROLS CELL FATE DETERMINATION

A compelling question is the function of Runx2 in neural crest cells that migrate to form the cranio-

facial skeleton and in mesenchyme at E9.5 prior to bone formation. The involvement of Runx2 supporting osteogenesis includes a central role in cell growth control. The proliferative expansion of mesenchymal cells, osteoprogenitor cells, and immature osteoblasts in response to mitotic growth factors is critical for normal skeletal development and bone formation. Cell growth control is mediated in part at the transcriptional level because there are cell cycle stage–specific demands for *de novo* synthesis of proteins (e.g., histones and cyclins) [189]. However, transcription factors that control proliferation of osteoblasts are minimally understood. Runx2 is expressed in mesenchymal cells and is required for growth suppression to support the transition stage from proliferation to exit from the cell cycle for phenotype commitment [188]. Calvarial-derived progenitors from Runx2 null and Runx2 ΔC mice, which express a protein lacking the normal C-terminus of Runx2, exhibit increased cell growth. Reintroduction of Runx2 into Runx2-deficient cells by adenoviral delivery restores physiological control of proliferation in osteoblasts, suggesting that Runx2 contributes to transcriptional control in immature osteoblasts to regulate proliferation. In agreement with the cell growth regulatory function of Runx2, the levels of Runx2 are tightly regulated upon entry and exit from the cell cycle in osteoblasts. Runx2 is upregulated at the onset of quiescence in contact-inhibited or serum-deprived immature osteoblasts, whereas Runx2 levels are diminished to low levels upon re-entry into the cell cycle [190]. Such findings, together with genetic data, indicate a cell growth–suppressive function for Runx2 in mesenchymal bone cell progenitors. This cell growth regulatory activity of Runx2 is distinct from the genetic requirement for Runx2 in the final stages of osteoblast maturation and osteogenesis. We have observed elevated levels of cyclin E in Runx2-deficient mice. Runx2-dependent control of cyclins, CDK inhibitors, growth factors, and growth factor receptors is particularly relevant because together they function as components of cell signaling pathways that control cell cycle entry and/or the subsequent transitions between different cell cycle stages.

Runx2 protein has been shown to be retained in the cell during mitosis, when most proteins are downregulated or degraded while the cell is engaged in DNA synthesis for cell division [191–193]. During cell division, cessation of transcription is coupled with mitotic chromosome condensation. A fundamental biological question is how patterns in gene expression are retained during mitosis to ensure the phenotype of progeny cells. Findings suggest that Runx transcription factors are determinants of cell fate and provide a genetic mechanism for the retention of gene expression patterns during cell division. Runx2 proteins are stable during cell division and are associated with chromosomes during mitosis through sequence-specific DNA binding (Figure 4-2). During mitosis, Runx2 transcriptionally regulates ribosomal RNA genes, which support protein synthesis after cell division [194]. Using siRNA-mediated silencing, mitotic synchronization, and expression profiling, our laboratory also identified Runx2-regulated genes that are modulated postmitotically [192]. Novel target genes involved in cell growth and differentiation in bone cells were validated by chromatin immunoprecipitation studies [192, 194]. These findings indicate that Runx proteins have an active role in retaining a cell's genotype during cell division to support lineage-specific gene expression in progeny cells. Runx2 is equally partitioned between the two daughter nuclei and may provide a mechanism for regulating genes that retain the chondro- and osteogenic lineage properties of dividing progenitor cells [191, 192]. Importantly, after exit from the cell cycle, Runx2 protein increases several-fold, and evidence from Runx2 null cells indicates that Runx2 may function as an inhibitor of proliferation of progenitors, thus providing a mechanism for regulating the transition from growth to a postproliferative

stage as a component of cellular commitment to the osteogenic lineage [195]. Runx2 may therefore function as a lineage determinant in several capacities by (1) regulating protein synthesis and gene transcription during mitosis that "bookmark" cells for the chondro-osseous lineage upon exit from the cell cycle and (2) by serving as a transcriptional mediator of BMP and Wnt osteogenic signals.

III. DEVELOPMENTAL SEQUENCE OF OSTEOBLAST PHENOTYPE DEVELOPMENT

Whereas subpopulations of osteoblasts are recognizable *in vivo* morphologically in relation to tissue organization, *in vitro* phenotypic differences with respect to expression of genes reflecting their maturational stages, functions, and responses to physiologic mediators of bone formation can be demonstrated. Primary cell cultures from calvaria and trabecular bone tissue and marrow stromal cells, as well as cultures of established lines that produce an organized bonelike matrix, provide a basis for studies that map the temporal expression of cell growth and tissue-specific genes during

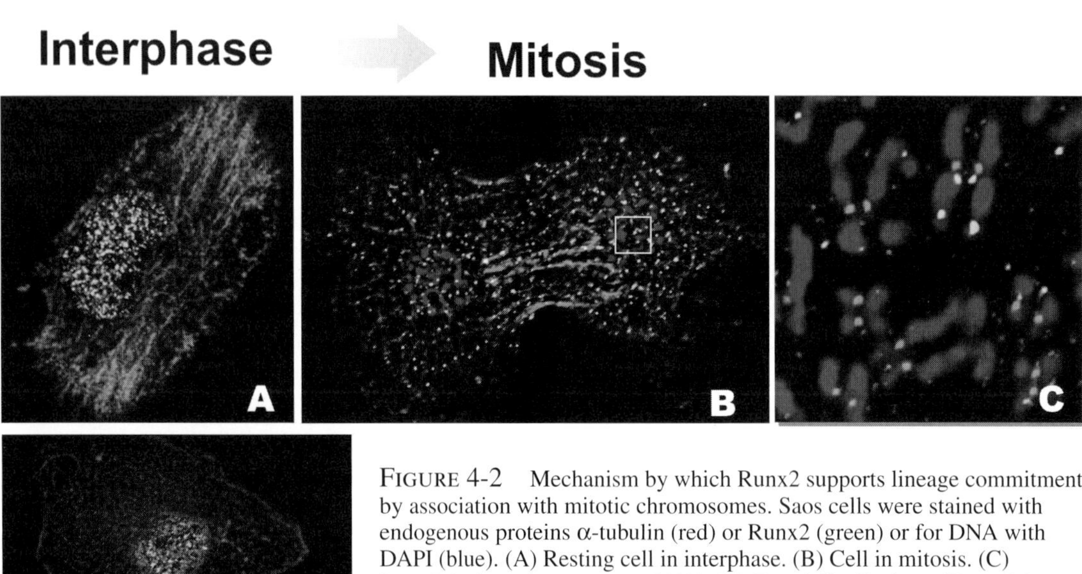

FIGURE 4-2   Mechanism by which Runx2 supports lineage commitment by association with mitotic chromosomes. Saos cells were stained with endogenous proteins α-tubulin (red) or Runx2 (green) or for DNA with DAPI (blue). (A) Resting cell in interphase. (B) Cell in mitosis. (C) Runx2 foci on chromosomes [192]. (D) Equal distribution of Runx2 in the two daughter cells [191]. This association of Runx2 may function in bookmarking target genes for postmitotic osteogenic lineage determination.

the progressive establishment of the osteoblast phenotype [196]. Profiles of gene expression have defined developmental stages of the osteoblast phenotype and allowed for investigating regulatory mechanisms that support the progression of osteoblast growth and differentiation and maturation-specific responses to physiological mediators of bone formation and remodeling (Figure 4-3).

A. Markers of Osteoblast Maturation Stages

The sequential expression of cell growth and tissue-specific genes that are useful markers for progressive development of the bone cell phenotype are presented in Figure 4-3. Four principal developmental periods can be defined by expression of the major functional bone matrix proteins, often designated "phenotypic markers." Initially, proliferation supports expansion of the proliferating preosteoblast cell population to form a multilayered cellular nodule. Genes requisite for the activation of proliferation (e.g., c-myc, c-fos, and c-jun) and cell cycle progression (e.g., histones and cyclins) are expressed together with the expression of genes encoding growth factors (e.g., FGF and insulin-like growth factor-1 [IGF-1]), TGF-β, BMPs, cell adhesion proteins (e.g., fibronectin), and type I collagen, the major component of the bone extracellular matrix to support osteoblast growth and differentiation. BMP-2, which is expressed primarily at early stages, enhances BMP-3 and BMP-4 expression for the later mineralization stage.

Following the initial proliferation period, a second stage of gene expression is associated with the maturation and organization of the bone extracellular matrix (ECM). Collagen synthesis continues and undergoes cross-link maturation [197]. Genes that contribute to rendering the extracellular matrix competent for mineralization (e.g., alkaline phosphatase) are upregulated. Two principal transition points are key elements of this

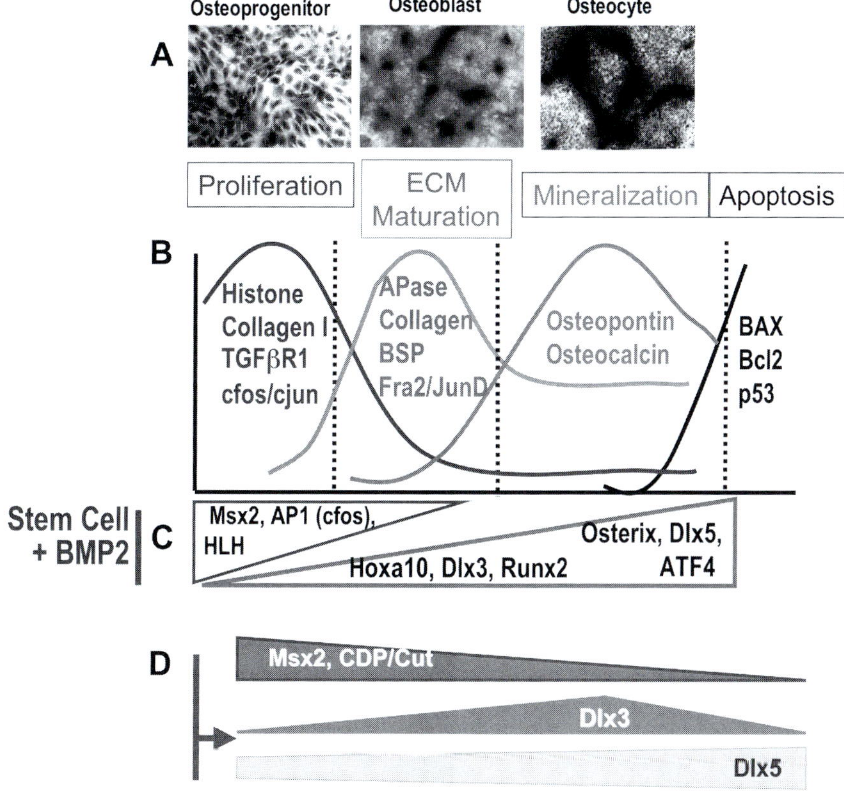

FIGURE 4-3   Stages of osteoblast differentiation *in vitro*. (A) Histologic staining by toluidine blue (left), alkaline phosphatase (middle), and von Kossa silver stain (right) to reflect the major stages of osteoblast maturation. (B) Expression of marker genes reaching peak expression that is characteristic of each stage. (C) Transcription factor expression is represented. Several factors are rapidly induced in pluripotent stem cells in response to osteogenic BMPs. Runx2 continuously increases during osteoblast differentiation. Dlx5, Osterix, and ATF4 are functionally linked to the mineralization stage [97, 221, 222, 594]. (D) Binding of homeodomain proteins to gene promoters during differentiation [97, 226]. The association and dissociation of these factors at the TAAT core motif in genes form a regulatory network to support transcription.

temporal expression of genes that support the progression of differentiation. These transitions have been established experimentally and defined functionally as restriction points during osteoblast differentiation to which developmental expression of genes can proceed but cannot pass without additional cellular signaling [198]. The first transition point is at the completion of the proliferation period when genes for cell cycle and cell growth control are downregulated and expression of genes encoding proteins for extracellular matrix maturation and organization is initiated. During the proliferation period, the absence of gene expression observed in postproliferative mature osteoblasts is called "phenotype suppression" [199]. The model is supported by the binding of repressor-type regulatory factors abundant in proliferating osteoblasts (e.g., oncogene-encoded factors, helix–loop–helix proteins, and homeobox suppressor proteins) to genes that are expressed in postproliferative cells. This suppression is reversed when proliferating cells exit the cell cycle for differentiation.

The second transition is at the onset of extracellular matrix mineralization. Signals for the third developmental period involve gene expression related to the accumulation of hydroxyapatite in the ECM. Genes encoding several proteins with mineral-binding proteins (e.g., osteopontin, osteocalcin, and bone sialoprotein) exhibit maximal expression at this time, when mineralization of the bone tissue-like organized matrix is ongoing. This profile suggests functional roles for these proteins in the regulation of the ordered deposition of hydroxyapatite. A fourth developmental period follows in mature mineralized cultures during which collagenases are elevated, apoptotic activity occurs, and compensatory proliferative activity is evident *in vitro* [200, 201]. This stage appears to serve an editing/remodeling function for modifications in the bone ECM, not unlike *in vivo* activities that sustain the structural and functional properties of the tissue.

B.   Regulatory Networks for Osteoblast Differentiation

Both expression profiling (mRNA and protein levels) and analysis of the association of transcription factors with target gene promoters during osteoblast differentiation have revealed how transcription families can temporally regulate genes. Regulatory factors that directly engage in protein–DNA as well as protein–protein interactions are important mechanisms for both the activation and the suppression of genes, reflecting stages of osteoblast maturation. It is now appreciated that transcription factors that contribute to position and pattern formation in the developing embryo (e.g., Hox

and homeodomain genes) can provide mechanisms for regulating the progression of osteoblast differentiation in the adult. The selective representation of these factors during osteoblast differentiation and family members within a class of transcription factors, as well as evidence for their functional consequences (e.g., by forced expression, antisense, or antibody blocking studies) in osteoblasts, provides compelling evidence for regulatory effects in driving osteoblast maturation. The developmental expression of the bone-specific osteocalcin gene has provided a paradigm for defining osteoblast-restricted gene expression. Homeodomain proteins, Fos/Jun family members in response to growth factors, helix–loop–helix factors, and RUNX2/ CBFA1, Osterix, C/EBP, and ATF proteins and steroid hormone receptors are among the well-characterized transcription factors that are obligatory for osteoblast differentiation.

Modifications in the representation of classes of transcription factors at different stages of osteoblast differentiation (Figure 4-3C) reflect linkage to the transcriptional control of the osteoblast phenotype. Helix–loop–helix factors, which are negative regulators of osteogenesis, illustrate this point. Id (inhibitor of differentiation), twist, and scleraxis are expressed in mesoderm of the developing embryo [202]. Scleraxis is expressed in cells that form the skeleton and is not detected at the onset of ossification [203]. Twist is a key regulator of epithelial–mesenchymal interactions [204]. Id and Twist expression must be downregulated for osteoblast differentiation to proceed, and overexpression of these factors inhibits osteogenesis *in vitro* [205] through multiple mechanisms [184]. This complexity ensures developmental, tissue-specific regulated expression of the postproliferative bone-specific genes in osteoblasts.

The fos and jun family members for heterodimers at AP-1 motifs on gene promoters are responsive to numerous growth regulators and essential for bone development [206]. They exhibit developmental stage-specific expression and activities during osteoblast differentiation *in vivo* [207] and *in vitro* [208]. AP-1 factors also cooperate with other factors for transcriptional control. This is exemplified by AP-1 and Runx2 sites in the MMP13 collagenase gene essential for activation of this collagenase by PTH [209, 210]. The importance of c-fos in establishing the osteoblast phenotype was first revealed by the null mouse [211]. *In vivo* immunohistochemical staining reveals that c-fos is expressed in osteoprogenitor cells, in the perichondrium and periosteal tissues, but not in mature osteoblasts. The *in vitro* temporal profile of c-Fos and c-Jun expressed maximally in proliferating preosteoblasts, while Fra2 and JunD are upregulated in differentiated osteoblasts, is consistent with *in vivo* functional studies [212]. Mouse models have revealed that Fra1, Jun, and JunB are essential in the adult organism, in addition to development of bone. JunB null mice exhibit osteopenia with defects in osteoblasts and osteoclasts [213]. Fra1 is an activator of bone formation [214]. Thus, AP-1 factors, although ubiquitous, have specialized activities in bone.

Sp1 family members regulate skeletal development. Sp3 is an activator of a chondrocyte-specific gene, chondromodulin [215]. The Sp1 factor also supports bone formation, mainly through activation of collagen type I synthesis. Several studies have identified polymorphisms in the Sp1 promoter element of ColIα1, but the linkage to osteoporosis or bone mineral density changes is not clear [216, 217]. Osterix (SP-7) is an osteoblast-restricted member of the SP-1 family of zinc finger transcription factor discovered as a BMP-2-inducible protein. The Osterix null mouse lacks a mineralized skeleton and is therefore considered essential for osteogenesis since Runx2 is expressed in Osterix null mice [87]. Osterix appears to require cooperation with nuclear factor of activated T cells (NFAT) through formation of an NFAT–Osterix complex for bone formation [218, 219].

Specific ATF factors contribute to either cartilage development or the late stage of osteoblast maturation. Chondrodysplasias are observed in ATF2-deficient mice [220]. ATF4 was identified through a mouse model with a null mutation in a growth factor–regulated kinase (RSK2). ATF4 is a substrate of RSK2 and mice deficient in this kinase have exhibited defects in terminal differentiation of osteoblasts [221]. ATF4 must cooperate with Runx2 for expression of osteoblast-specific genes [222]. An inhibitor of ATF4 called FIAT, a leucine zipper protein, when expressed in transgenic mice reduced osteocalcin expression, bone mineral density, volume, and trabecular thickness. Thus, ATF4 functions as a downstream component of the Runx2 and Osterix transcriptional network of osteoblast essential factors that contribute to the final stage of osteoblast differentiation.

Homeodomain (HD) protein binding sites contribute to bone-specific expression of several genes—collagen type I, osteocalcin, bone sialoprotein, and Runx. During osteoblast differentiation *in vitro*, homeobox proteins are temporally expressed (Figure 4-3D) and bound to gene promoters at different levels during osteoblast differentiation. Msx-2 is expressed maximally in the preosteoblast and is subsequently downregulated in the mature bone cell. Msx2 in proliferating osteoblasts represses phenotypic genes. However, in aortic cells, Msx2 has procalcific actions indirectly through activation of Wnt signals [223]. *In situ* hybridization studies confirm the reciprocal expression of osteocalcin (in osteoblasts) and Msx-2 in preosseous cells of developing bone [224]. In contrast, Dlx3 and Dlx5 increase during differentiation to promote expression of phenotypic genes and drive osteoblast maturation [97, 225, 226]. Regulation of osteoblast differentiation by HD proteins not only occurs by protein–DNA binding but also involves protein–protein interactions among HD factors. The formation of heterodimers with each other or other transcription factors further amplifies or attenuates gene transcription and cell differentiation.

C. Regulation of Osteoblast Activity

In vitro models of osteoblast differentiation provide a better understanding of the properties and physiologic responses of osteoblast lineage cells at their individual stages of differentiation. This is best exemplified in bone marrow stromal cell cultures and calvarial-derived osteoblasts that produce a mineralizing matrix. Parathyroid hormone will promote the differentiation of preosteoblasts but suppress late stages of maturation through mechanisms related to PTH repression of Runx2 and Osterix transcription factors [227–229]. Although caution should be exercised in translation from *in vitro* to *in vivo* effects of PTH on bone formation, these studies indicate that even pulsed PTH administration may increase bone formation by stimulating the proliferation of progenitors, and not by anabolic effects of differentiated osteoblasts [230]. It is established that TGF-β stimulates the replication of progenitor cells and directly increases collagen synthesis. When proliferating calvarial osteoblasts are exposed to TGF-β, a block in differentiation is observed [231–234]. The mitogenic effects of TGF-β are not apparent on mature postproliferative osteoblasts.

The anabolic activities of osteoblasts are regulated in large part by growth factors stimulating osteoprogenitor proliferation and hormones that promote differentiation, as described previously. The IGF system has significant control of bone formation. IGF-1 and IGF-2 are synthesized in many tissues, and both are highly expressed in active osteoblasts. IGFs stimulate cell proliferation and collagen synthesis and, at the same time, inhibit matrix collagen degradation by decreasing collagenase 3 transcription [235]. The synthesis of IGF-1 is regulated by physiologic mediators of bone formation. PTH stimulates [236] whereas glucocorticoids [237] are inhibitory to IGF-1 expression. IGF activities are regulated by a family of IGF-binding proteins, designated IGFBP-1 through IGFBP-6. These binding proteins have either stimulatory effects (e.g., IGFBP-5) or inhibitory activity (e.g., IGFBP-4)

[238]. Both clinical studies and mouse genetics reveal associations of IGF-1 with bone formation and turnover [239–244]. Notable is that IGF-1 is required for anabolic properties of PTH [243, 245]. However, the finding that IGF-1 mediates expression of Osterix [246] and the expression activity of Runx2 [247, 248], the two transcription factors essential for bone formation, underscores the central control of IGF-1 in osteoblast growth and differentiation.

The various steroid hormones, including glucocorticoids [249, 250], $1,25(OH)_2D_3$ [251, 252], and estrogen [253], also have selective effects, either promoting differentiation of the cells at early stages of maturation or inhibiting anabolic activities and promoting resorptive properties of the osteoblast at later stages. In general, growth factors and steroid hormones have the most robust responses in immature osteoblasts and can radically modify the differentiation program when added to proliferating cells.

Glucocorticoid effects on bone metabolism and induced osteoporosis are presented elsewhere [254]. In the osteoblast, glucocorticoids directly regulate expressed genes that contribute to bone formation, including cytokines, growth factors, and bone matrix proteins. Increases in alkaline phosphatase, osteocalcin, and collagen are observed at early maturation stages, but inhibition of these genes occurs in differentiated osteoblasts. The molecular mechanisms by which glucocorticoids exert selective effects on a particular gene are complex, but numerous examples have been documented. Both transcriptional and post-transcriptional gene regulation will be affected by glucocorticoids, as shown for osteocalcin and collagen [249, 255]. Dexamethasone will interfere with the binding of transcription factors (e.g., TCF/LEF and Egr2/Krox20), mRNA stability, and BMP signaling [256, 257]. Glucocorticoids promote osteoblast colony formation in human and rat marrow–derived cells and accelerate osteoblast differentiation in proliferating calvarial-derived cells, reflected by both increased number and size of the bone nodules and early mineralization [258–260]. In contrast, dexamethasone exerts an antiproliferative effect on mouse osteoblasts [261, 262] and blocks their maturation. Because postproliferative cultures cannot be stimulated to produce more mineralizing nodules, the mature osteoblast is refractory to growth stimulation by dexamethasone. However, glucocorticoids *in vivo* and *in vitro* also induce apoptosis of osteoblast populations [254, 263]. These findings, together with glucocorticoid effects on osteoclast activity, contribute to glucocorticoid-induced osteopenia observed *in vivo* when pharmacologic doses of glucocorticoids are required.

The active metabolite of vitamin D, $1,25(OH)_2D_3$, has complex effects on the skeletal system related to targeting of many cell types, dose, and timing. Vitamin D_3 is a biphasic regulator of osteoblast activity for bone formation and bone resorption. The active hormone regulates the expression of genes in osteoblasts that form the bone ECM or provides signals for osteoclast differentiation. Upregulation by $1,25(OH)_2D_3$ of numerous osteoblast parameters related to bone matrix formation and mineralization (e.g., collagen, alkaline phosphatase, osteopontin, osteocalcin, and matrix Gla protein), and to bone resorption (e.g., cytokines), reflects influences of the hormone on osteoblast function and regulation of bone turnover. In addition, the hormone, when administered at high doses or when endogenously produced at high levels, stimulates RANKL production in osteoclasts. Thus, either depletion of the hormone or pharmacological doses and long-term exposure to the hormone can result in abnormalities of bone formation, rickets, osteomalacia, as well as osteopenia. The general anabolic effect of $1,25(OH)_2D_3$ on the skeleton has been shown by increasing mineral ion homeostasis in mice deficient in 1α-hydroxylase (an enzyme needed for $1,25[OH]_2D_3$ synthesis) and the PTH null mouse [264]. *In vitro* analysis of $1,25(OH)_2D_3$ in primary cultures of osteoblasts shows stage-dependent effects [251, 265, 266]. The steroid is antiproliferative in the growth period and can block formation of the mineralized nodule when introduced during the growth period or stimulate differentiation-related gene expression in mature osteoblasts [252, 267–269]. Because of these properties, acute versus continuous exposure of cells to $1,25(OH)_2D_3$ can lead to opposing results.

The sex steroids have diverse effects on osteoblast cell population, in addition to their effects on osteoclasts. In general, androgens and estrogens have proapoptotic effects on osteoclasts and antiapoptotic effects on osteoblasts. These hormones exhibit anabolic effects on bone through very distinct pathways, and the reader is referred to more comprehensive reviews [270, 271] and Chapters 12 and 13. Estrogen contributes to anabolic activities by suppression of resorptive cytokines, whereas androgens have direct effects on osteoprogenitor proliferation. There is a preference for androgen stimulation of periosteal osteoprogenitors, whereas estrogen stimulates endosteal osteoblasts. New concepts are emerging related to sex steroid control of bone formation with the knowledge that both hormones together are needed to support expansion of osteoprogenitors [272]. The very different expression profiles of the receptors, with androgen receptor (AR) increasing and estrogen receptor (ERα) declining during mineralization stages, imply that there are specialized activities of the hormones on osteoblast populations [273]. Furthermore, both ERα and ERβ isoforms appear to regulate different genes in osteoblasts, adding to the complexity of estrogen responses [274].

Osteoblast activities and survival are highly regulated by the mediators of cell–ECM and cell–cell interactions. A spectrum of integrins has been shown to be expressed by osteoblasts and adhere to the full range of RGD-containing bone matrix proteins [275]. It is well established that growth of osteoprogenitors on collagen promotes differentiation, whereas disruption of collagen–integrin interactions suppresses expression of the osteoblast phenotype [276, 277]. A nonintegrin adhesion receptor, CD44, the hyaluronate receptor, is linked to the cytoskeleton. CD44 has been identified as a useful marker for osteocyte differentiation [278, 279]. Several members of the cadherin family of cell-adhesion proteins are expressed in osteoblasts, including cadherin-11, cadherin-4, N-cadherin, and OB-cadherin [280, 281]. N-cadherin is present in proliferative preosteoblastic cells and may support osteoblast differentiation [282], but it is lost as they become osteocytic [283]. In contrast, OB-cadherin is barely detected in osteoprogenitor cells and is upregulated in alkaline phosphatase–expressing cells [284]. Indeed, the relative abundance of different cadherins defines the differentiation pathway of mesenchymal precursors to specific lineages; for example, R-cadherin is downregulated and cadherin-11 upregulated in response to BMP-2-induced osteogenesis [281, 285]. Signaling pathways from the extracellular matrix through the cytoskeleton and, finally, to the nucleus, which allow expression and upregulation of bone-specific and bone-related genes, are being investigated.

Cell–cell communication is important for the differentiation and maturation of osteoblasts. Cytoplasmic processes on the secreting side of the surface osteoblast extend deep into the osteoid matrix and are in contact with the extended cellular processes of osteocytes. Junctional complexes (gap junctions) are often found between the osteoblasts on the surface as well as between cellular processes. In this manner, surface osteoblasts establish cell–cell communication with neighboring cells in the mineralized matrix. Gap junctions are a structure of six multiple protein units (connexins) that couple with an identical unit in a neighboring cell to form a channel connecting the two cytoplasms. Studies in osteoblasts suggest that the selective utilization of connexin proteins contributes to the modulation of molecular permeability [286]. Connexin-43 is the major gap junction protein in osteoblasts, and decreased expression reduces cell–cell communication and expression of osteoblastic

genes [287]. In connexin-43 null mice, craniofacial abnormalities are observed, and although the axial and appendicular developed normally, impaired function was reflected by delayed ossification [288].

Studies from many groups using different osteoblast models have reinforced two important concepts: (1) that the stage of osteoblast maturation influences the selective responsiveness of specific genes to hormones or growth factors and (2) that there is a window of responsiveness of a cell during which the factor can alter development and maintenance of the bone cell phenotype. These analyses have provided clinically relevant information toward an understanding of the consequences incurred by the osteoblast when exposed to therapeutic agents that may stimulate or inhibit cell proliferation or differentiation.

D.   Osteoblast Culture Models

Characterizing cell phenotypes from genetic studies, identifying molecular mechanisms or novel markers of osteogenic cell populations, and determining the activities of a therapeutic agent must take into consideration the cell model being used. Primary cell cultures offer several advantages, particularly for studying cell growth control mechanisms and differentiation in the context of a mineralizing matrix. Primary cultures from genetically altered mouse models often reflect the *in vivo* status of the defects in osteoblast and osteoclast phenotypic properties. For example, cultured marrow-derived cells from osteopetrotic rats that exhibit precocious and intensified mineralization and failure to form osteoclasts retain these defects *ex vivo* [289, 290]. Modifications in osteoblast and chondrocyte differentiation in mouse models characterizing components of the Wnt/β-catenin pathway or cell cycle regulators are other examples in which the *in vivo* phenotype is reflected in cell culture models [108, 109, 123, 134].

Calvarial-derived cells from the newborn and marrow-derived mesenchymal cells are readily induced into the osteogenic lineages by media supplemented with ascorbate to amplify matrix production and β-glycerolphosphate to promote mineralization. The cells progress through a normal sequence of osteoblast differentiation for evaluating exogenous factors. Mouse embryo fibroblasts (isolated at E12.5) can be differentiated into several mesenchymal lineages, including chondrocytes and osteoblasts. Osteoprogenitor cells that differentiate have also been isolated from the periosteum. Primary cell cultures may not always be appropriate or practical for some lines of experimentation. They are limited in their ability to maintain phenotypic properties with passaging. Observations from primary cultures of fetal and adult bone, marrow, and periosteum reported during the past decade reveal considerations for the following in interpretation of these studies. The age of isolation influences the growth properties and representations of subpopulations of bone-forming cells. The expression of osteogenic and other phenotypic responses appears to also be related to bone sites and cell passages. Thus, studies of osteoblast activities must be controlled carefully. Cells closer to the progenitor/preosteoblast stage are differentiated more readily *in vitro*.

Results from studies of isolated human bone cells require consideration of many variables, including the site of tissue origin, age, sex, hormonal status, underlying bone pathology, and influence of medications [291, 292]. Nonetheless, the effective use of cultured human osteoblasts in assessing functional activity related to aging skeleton and disease is being validated, for example, with cells from patients with Paget's disease, osteogenesis imperfecta, and vitamin D–related disorders [293]. However, there is still a need for rigorous studies of normal subjects and osteoporotic patients to understand age-related responses of osteogenic lineage cells.

Osteosarcoma cell lines, immortalized or selected cells representing different stages of osteoblast maturation, have been characterized and are particularly useful for addressing molecular mechanisms in more or less homogeneous populations. Nonosseous cell lines provide tools for evaluating determinants of the osteoblast phenotype. One example shows how the NIH3T3 fibroblast cell line can be induced to express alkaline phosphatase. However, caution should be maintained when using only alkaline phosphatase as a marker of osteoblast properties in such cells [294]. Variability in biological responses may occur with respect to passage number, and cells of mixed morphology may appear if they are not maintained under appropriate conditions. Human cell lines from young and aged subjects have been established by immortalization with a temperature-sensitive large T-antigen and have been advantageous in evaluating hormone responses [295–297]. The telomerase reverse transcriptase immortalizes cells and maintains the properties of human bone marrow mesenchymal stem cells (MSCs), including their ability to differentiate [298]. Established mouse cell lines representing either the phenotype of a genetically modified mouse such as Runx2 null cells [299] or an osteoblast *in vivo* population such as the osteocyte cell line [300] have provided excellent models for identifying biological mechanisms controlled by the target gene and signal transduction pathways mediated by the specific cell phenotype.

IV. PHENOTYPIC PROPERTIES OF OSTEOGENIC LINEAGE CELLS

A. Stem Cells and Mesenchymal Osteoprogenitors

I. PROPERTIES FOR ISOLATION

Adult stem cells are being harvested from many tissues, including bone marrow [301, 302]. Maintaining the stem cell–like properties *in vitro* has been challenging. Stem cells by their nature are generally in a noncycling (G_0) stage of the cell cycle. Embryonic stem (ES) cells can be propagated in culture on a feeder layer of mouse embryonic fibroblasts or without feeders in the presence of leukemia inhibitory factor (LIF), required for maintenance of mouse but not human ES cells [303]. Oct4 (a POU homeodomain protein) and nanog (a new homeodomain protein) are also requirements for a self-renewal of ES cells [304, 305]. The progression of the most primitive pluripotent cell to the undifferentiated multipotential mesenchymal cell and presumed osteoprogenitor is not understood. Progenitor cells must be responsive to a broad spectrum of regulatory signals that mediate their proliferation, commitment, and progression of phenotype development, as well as sustain their structural and functional properties. In fully developed bone, there is a requirement for utilization of the same factors that can mediate the growth and differentiation of osteoprogenitor cells during skeletal development, as well as for osteoblast differentiation during bone remodeling and fracture healing in the adult.

From a bone developmental perspective, mesenchymal-derived osteoprogenitor cells arise/reside in the periosteal tissue or the bone marrow stroma. The marrow and its stromal "bedding" give rise to multipotential cells of both hematopoietic lineage (origin of osteoclasts) and nonhematopoietic lineage cells, designated MSCs, from which many tissue-specific cells derive, such as riboblasts, chondrocytes, myoblasts, and adipocytes. When suspensions of marrow cells are plated *in vitro*, clonal colonies of adherent fibroblasts are formed, each derived from the single cell that has been designated as the colony-forming fibroblastic unit or CFU/F. Formation of CFUs requires the presence of hematopoietic cells [306]. A proportion of these cells have a high proliferative and differentiation capacity and exhibit characteristics of stem cells when transplanted in the closed environment of a diffusion chamber or transplanted into the circulation [307–309].

A key obstacle in understanding the origin of osteoblast lineage cells is the inability to identify passage of the MSC to osteoprogenitors prior to the expression of bone phenotypic properties. Using characterization of hematopoietic stem cells as a paradigm, several groups have developed antibodies to cell surface proteins using presumptive marrow stromal cell populations. The antigen to the cell surface marker antibody (SB-10) produced in response to MSCs is the activated leukocyte cell adhesion molecule ALCAM [310]. Expression of ALCAM becomes downregulated in concert with changes in morphology and detection of alkaline phosphatase activity of the periosteal osteoprogenitors as they migrate and develop into osteoblasts. These reagents have the potential for both recognition and purification of skeletal stem cells. STRO-1-positive cells are well documented to have osteoprogenitor properties. With the advent of many cell surface markers that distinguish hematopoietic lineage and mesenchymal multipotential cells [311, 312], better defined populations can be studied for their differentiation potential. Multiple markers are needed to identify subpopulations of a cell phenotype through their lineage from growth to differentiation.

The osteoprogenitor appears to have limited self-renewal capacity compared to the stem cell. In contrast, a key feature of the osteoprogenitor/preosteoblast population is its capacity to divide and increase the size of bone. Labeling studies ([^{3}H]thymidine and autoradiography) indicate that the proliferating cells are principally confined to progenitor cells and preosteoblasts, with very few osteoblasts labeled. The determined osteoprogenitor is recognizable in bone as a preosteoblast. Proliferation and differentiation of the osteoprogenitor and preosteoblast pool are influenced by many growth factors (TGF-β1, BMPs, FGFs, endothelial growth factor, nerve growth factor, platelet-derived growth factor, and stromal cell–derived factor-1) that have been identified as stimulating expansion of MSCs or the CFU/F. LIF maintains stem cell populations and osteoprogenitors and inhibits their differentiation *in vitro*, but it will have osteogenic activity *in vivo* [313, 314]. FGF signaling has both negative and positive effects on proliferation of osteoprogenitors [40, 315, 316]. The plethora of growth factors expressed and produced by osteoblast lineage cells are stored in the bone ECM [317, 318]. A local mechanism for stimulating the proliferation of progenitors in the bone microenvironment is thereby provided.

2. LINEAGE ALLOCATION

Much attention has been given to lineage allocation of the mesenchymal stem between the osteoblast and adipocyte in marrow bone. Although debate regarding whether the inherent osteogenic potential of the MSC in marrow declines with aging is ongoing, locally secreted and systemic factors, as well as nuclear factors

influencing lineage direction, have been clearly defined. PTHrP, by enhancing BMP-1A receptor expression and BMP-2 responsiveness, promotes osteoblastogenesis but decreases adipogenesis [319]. Menin, a product of the multiple endocrine neoplasia type 1 (MEN1) gene, was identified in the null mouse as a requirement for MSC commitment to osteoblasts [320]. The orphan receptor tyrosine kinase ROR2 promotes osteoblast differentiation by shifting MSC cell fate to the osteoblast through induction of Osterix and suppression of adipogenic factors C/EBPα and PPARγ [321]. Commitment of a stem cell to a phenotype is regulated by cell shape and cytoskeleton changes that involve Rho GTPase activity. A dominant-negative RhoA promotes a round shape, leading to adipocyte differentiation, whereas a constitutively active RhoA induces the osteogenic phenotype independent of cell shape [322, 323]. Physical forces on the MSC appear to be a significant component for osteoblast allocation because microgravity inhibits osteoblast colony formation of human MSCs and increases adipocytes [324, 325]. Finally, transcriptional regulators of gene expression have potent and direct effects on modifying cellular phenotypes.

Commitment of stem cells to specific mesenchymal lineages occurs early in development of the limb. Transcription factors, which function as "master switches," mediate cell differentiation by induction of a set of phenotypic genes that characterize the muscle, adipocyte, chondrocyte, or osteoblast cells (Figure 4-4). A number of studies have defined master genes that direct a pluripotent cell to different lineages (Figure 4-4). Adipogenesis is promoted through the activities of PPARγ and CEBPα [326, 327], chondrogenesis requires Sox9 [328], and *in vivo* osteogenesis requires Runx2 [88, 89, 154] and Osterix [87, 329] (Figure 4-4). Inhibitory transcription factors, such as GILZ or retinoic acid, can block adipogenesis [330], thereby increasing a pool of progenitors for osteoblast differentiation.

The plasticity of these lineages is indicated by several lines of evidence. Forced expression of the transcription factors that function as master switches (Figure 4-4B) in phenotype commitment can transdifferentiate a cell to a different phenotype. The reciprocal relationship between adipocyte and osteoblast differentiation is indicated by numerous such studies [331, 332]. Forced expression of PPARγ in marrow stromal cell lines results in the inhibition of terminal osteoblast differentiation with concomitant downregulation of Runx2. The bipotential property of the late-stage osteoprogenitor or preadipocyte is markedly sensitive to biological regulatory signals influencing master switch transcription factor expression. Regulatory signals

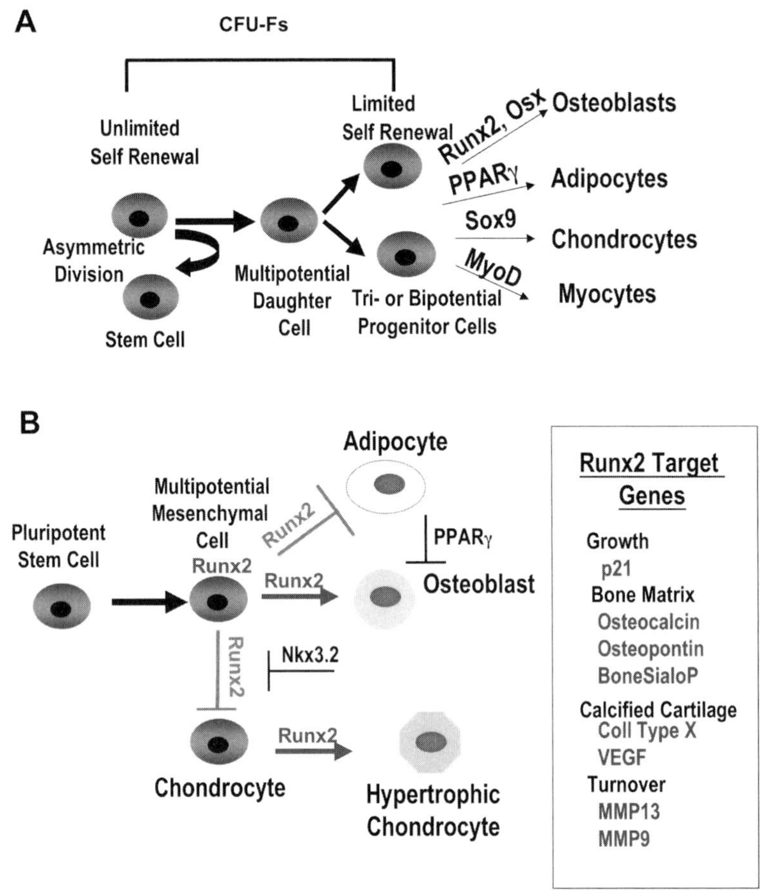

FIGURE 4-4   Lineage allocation of stem cells. (A) Representation of stem cell renewal and maturation to the mesenchymal stromal cell with limited pluripotency. The transcription factors proven through genetic studies to function as master regulatory genes required for the indicated phenotypes are shown. (B) Transcriptional regulation of lineage determination and the role of Runx2 expressed in the undifferentiated mesenchymal cell are indicated. Runx2 will inhibit other cell phenotypes including the myoblasts (not shown). For cells to enter the chondrogenic lineage, Runx2 must be downregulated, and several transcription factors, including Sox9 and Twist, are negative regulators of Runx2. The downregulation of Nkx3.2 permits reactivation of Runx2 expression in the hypertrophic chondrocyte. Adipogenesis and osteoblastogenesis can be regulated by expression of either Runx2 or PPARγ. A sampling of Runx2 target genes that reflect the different cell phenotypes and Runx2 functions for bone formation is shown.

influencing osteogenesis in preference to adipogenesis include $1,25(OH)_2D_3$, BMP-2, and Runx2.

The expression of Runx2 in early embryogenesis, followed by an upregulation in late stages of bone development, suggests that Runx2 may be important in both early specification of the mesenchymal stromal phenotype and for supporting the final stages of osteoblast differentiation. The potency of Runx2 in directing osteogenic commitment is provided by numerous studies that show Runx2 expression can activate bone phenotypic genes in pluripotent cells and redirect a committed premuscle cell into the osteoblast lineage [333] or inhibit the adipogenic phenotype [331]. Conversely, activation of PPARγ in osteoblasts will downregulate Runx2-mediated transcription of bone phenotypic genes [334]. More significant, PPARγ-deficient ES cells not only failed to become adipocytes but also spontane-

ously differentiated to osteoblasts [335]. The *in vivo* significance of Runx2 in early commitment to the osteoblast lineage is indicated by evidence that mesenchymal progenitor cells from Runx2 null mice differentiate more toward chondrocytes and adipocytes, consistent with the requirement for inhibition of Runx2 in normal cells for cartilage and fat tissue to develop from mesenchymal cells [336]. In normal skeletal development, the Osterix transcription factor functions to drive Runx2-expressing cells farther through the osteoblast lineage. From these studies, it is clear that tissue-specific transcription factors control cell fate, but questions remain regarding how expression of these master regulators of cell programs is controlled and how a hierarchy of cell selection is established.

We are currently beginning to reach an understanding of the complexity of factors required to support expansion of a progenitor cell and the signals that must be initiated for stem cells to acquire an osteogenic property. With new discoveries, considerations for how the different regulatory proteins can be applied for a therapeutic strategy must take into account their effects on a spectrum of diverse activities from different pathways.

B. Osteoblasts

I. *In Vivo* Morphology

Based on morphological and histological studies, osteoblastic cells are categorized in a presumed linear sequence progressing from osteoprogenitor cells to pre-osteoblasts, which mature to osteoblasts and then to lining cells or osteocytes (Figure 4-5). There is a gradient of differentiation that can be observed morphologically either in the periosteum or in the marrow as the osteoprogenitor cell reaches the bone surface and the osteoblast phenotype becomes fully expressed. Preosteoblasts

are usually observed as one or two layers of cells behind the osteoblast near bone-forming surfaces; that is, they are usually present where active mature osteoblasts are laying down a bone matrix. They appear elongated, fibroblastic, or spindle shaped with an oval or elongated nucleus and with notable glycogen content (Figure 4-5). Preosteoblasts may express a few phenotypic markers of the osteoblast (e.g., alkaline phosphatase activity), but less than mature osteoblasts. The preosteoblast, however, has not yet acquired many of the differentiated characteristics of mature osteoblasts; for example, there is no evidence of a well-developed rough endoplasmic reticulum. Osteoblasts that are derived from proliferating osteoprogenitors can be observed in clusters at the bone surface (Figure 4-5). These cells synthesize the bone ECM, designated osteoid (Figure 4-3). In metabolic bone disorders leading to decreased calcium or phosphate deposition in bone, as in vitamin D deficiency, wide osteoid seams are evident. Mineralization leads to the final stage of osteoblast differentiation. When the bone-forming osteoblast becomes encased in its own mineralized matrix, it is an

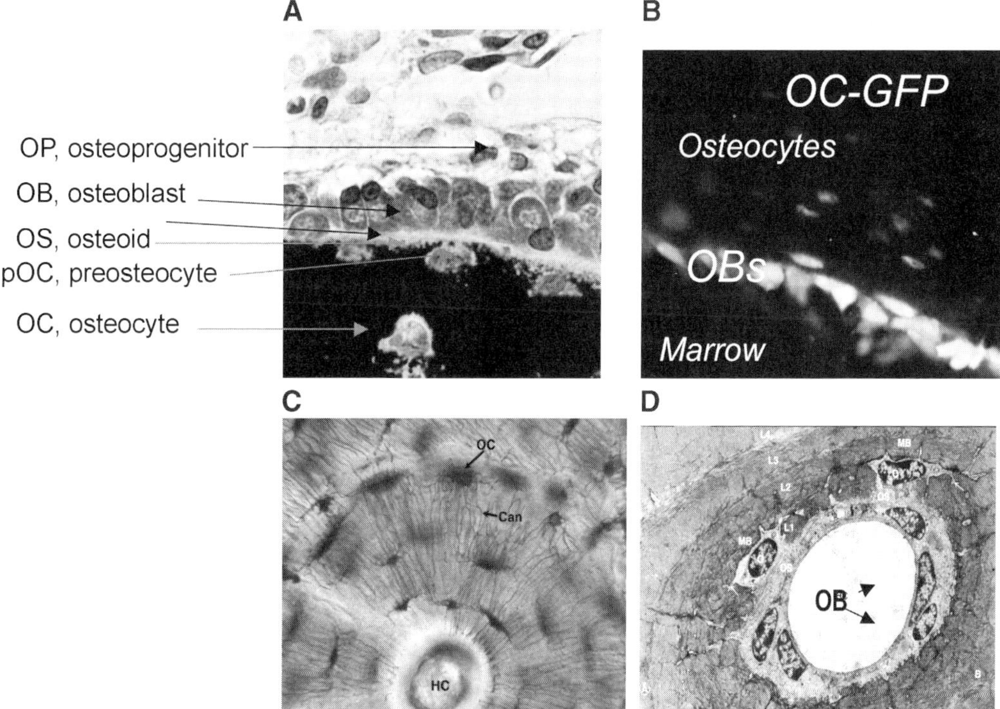

FIGURE 4-5   Osteoblast lineage cells. (A) Stages of osteoblast maturation are visualized on the surface of this bone trabeculae, Goldner trichrome stain. (B) Mouse cortical bone from a transgenic mouse expressing green fluorescent protein under control of the osteocalcin promoter is shown to illustrate that this bone-specific marker is expressed in osteoblasts (OBs) and osteocytes. (C) An osteon of human bone shows circumferential layers of cells and tissue around the Haversian canal (HC). The osteocyte cell body (OC) in lacunae with dendritic process in canaliculi (Can) are visualized. (D) Low magnification of electron micrograph of demineralized osteon showing the lamellar organization of the matrix (L1, L2, and L3 layers) with active osteoblasts on the surface.

osteocyte. On a quiescent bone surface, the osteoblast flattens to a lining cell, forming an endosteum. Bone lining cells are in direct communication with the osteocytes within the mineralized matrix through cellular processes that lie within the canaliculi. They are considered to provide a selective barrier between bone and other extracellular fluid compartments and contribute to mineral homeostasis by regulating the fluxes of calcium and phosphate in and out of bone fluids [337]. Four forms of the osteoblast cell lineage are thus recognized *in vivo*. They are the committed progenitors: preosteoblasts, mature osteoblasts, osteocytes, and the bone-lining cell.

When the preosteoblast ceases to proliferate, a key signaling event occurs for development of the mature osteoblast from the spindle-shaped osteoprogenitor. The osteoblast expresses all of the differentiated functions required to synthesize bone. Osteoblasts are defined *in vivo* by their appearance along the bone surface as large cuboidal cells actively producing matrix (Figure 4-5), which is not yet calcified (osteoid tissue). Several structural features characterize this osteoblast, including its size and cuboidal morphology, a round distinguishing nucleus at the base of the cell (opposite the bone surface), a strongly basophilic cytoplasm, and a prominent Golgi complex located between the nucleus and the apex of the cell [338]. At the ultrastructural level, one observes an extremely well-developed rough endoplasmic reticulum with dilated cisternae and a dense granular content, and also a large circular Golgi complex consisting of multiple Golgi stacks. These are typical characteristics of a secretory cell.

2. SECRETED MATRIX PROTEINS

The primary functional activity of the active surface osteoblast is production of an ECM with competency for mineralization. In this regard, the high level of tissue nonspecific alkaline phosphatase (TNAP) (bone, kidney, and liver isoform) and the ability to synthesize a number of noncollagenous proteins that are in either representative or restricted abundance in mineralized tissues are important features. Alkaline phosphatase activity, a hallmark of the osteoblast phenotype, is a widely accepted marker of new bone formation and early osteoblast activity. Gradations of enzyme intensity and mRNA expression are found in bone, with lowest levels (or absence) in osteocytes and osteoprogenitors and maximal levels in surface osteoblasts and hypertrophic chondrocytes at the mineralization front.

The osteoblast synthesizes and vectorially secretes most of the bone ECM protein; others are accumulated in bone as a result of their binding to bone mineral. Fetal bone is enriched in type III collagen and several minor collagens such as type V are found. The major matrix component synthesized by osteoblasts is collagen type I, which compromises nearly 90% of bone protein matrix and provides the essential substrate for mineral deposition. Collagen forms a fibrillar network stabilized by unique cross-links to maintain structural integrity of the tissue upon mineralization [339–341]. The fibrils organize with precise spacing that can accommodate deposited mineral. Discrete sites in the collagenous matrix serve as initial sites of mineral deposition in the hole regions between end-to-end collagen fibrils [342], accommodating small specialized bone proteins that interact with collagen and function as nucleators of hydroxyapatite. Collagen determines the structural organization of trabecular and cortical bone (woven, lamellar, and Haversian architecture) and supports the flexibility of mineralized tissues. Collagen and the highly specialized noncollagenous proteins that are either synthesized by the osteoblast or derive from other tissues and accumulate in bone bound to collagen and/or mineral contribute to mineralization of the osteoid, bone structure, and bone tissue metabolic functions.

The noncollagenous groups of proteins that represent components of the bone ECM function in mediating cell signaling from the ECM, cell adhesion/matrix attachment, protein–protein interactions by binding to collagen to regulate fibrillogenesis, as well as controlling mineral deposition through nucleation and inhibitor activities. Noncollagenous proteins have been classified by functional protein domains and post-translational modifications; they include proteoglycans, RGD-containing proteins, leucine-rich repeat proteins, glycoproteins, γ-carboxyglutamic acid (Gla-containing proteins), and the small integrin binding ligand (sibling) proteins and phosphoproteins. The most abundant noncollagenous proteins include osteonectin, osteocalcin, bone sialoprotein, osteopontin, and dentin matrix protein-1.

Ablation of the genes encoding some of the more abundant and bone-restricted noncollagenous proteins (osteocalcin [343, 344], osteopontin [345], and biglycan [346]) has resulted in only subtle changes in the bone matrix and mineral phase of bone that were not anticipated from *in vitro* studies and the calcium and phosphate binding properties. However, the phenotypes are revealing their functions and structural components for bone tissue integrity. The glycosylaminoglycan chains of decorin and biglycan facilitate their strong association to hydroxyapatite [347]. Deletions of the decorin and biglycan genes in mice disturb collagen fibril organization. Biglycan null mice have progressively diminished bone mass with age, whereas decorin-deficient mice have normal bone mass. However, biglycan/decorin double-knockout mice have severe osteopenia [348, 349]. The importance of biglycan in bone was shown by forced expression, which accelerated osteoblast differentiation *in vitro* and *in vivo* following transplantation of biglycan expressing cells, resulting in large areas of lamellar bone [350]. Osteonectin is a glycoprotein and binds to collagen. The null mice exhibit osteoporosis and decreased bone formation, consistent with a decrease in collagen accumulation [351–354].

Several of the SIBLING noncollagenous proteins synthesized by the osteoblast and osteocyte are upregulated during osteoblast differentiation and participate in ECM mineralization. Bone sialoprotein (BSP), a phosphorylated glycoprotein with a hydrophobic domain that binds to collagen, is expressed almost exclusively in bone [355]. BSP binds to hydroxyapatite through polyglutamic acid regions required for its functional activity as a nucleator of hydroxyapatite [356]. To date, a null mouse mutant has not been characterized. Matrix extracellular phosphoglycoprotein (MEPE) functions as a regulator of Pi concentration and the null mouse exhibits increased bone mass and density [357], indicating an inhibitory role of MEPE in osteoblast activity [358]. Osteopontin (OPN) is a protein with a wide distribution with many functions, including as an inhibitor of bone mineralization and ectopic calcification [359–361]. This protein tends to be enriched on surfaces undergoing bone turnover and provides an interacting protein module for adherence and activity of the bone-resorbing osteoclast. Thus, an important OPN function is facilitating osteoclastic bone resorption [345, 362]. OPN null mice have a subtle phenotype [360], but technologies for resolution of crystal size and maturity in null mutant mouse models reveal defects in bone mineral and quality. Increased mineral content and maturity (i.e., perfection) was found throughout all anatomical regions of the OPN-deficient mouse bone, consistent with osteopontin function in bone resorption/turnover [363]. Interestingly, the multiphosphorylated proteins characterized by stretches of serines, including OPN, BSP, and MEPE, map to the q arm of chromosome 4 [364, 365].

Dentin matrix protein 1 (DMP1) is another SIBLING protein highly expressed in osteocytes. It is an acidic (glutamic and aspartic rich) phosphoprotein that functions in stimulating osteoblast differentiation, as well as responding to mechanical loading [366]. The DMP1 knockout mouse exhibits a hypomineralization phenotype in tooth [367] and bone [368]. The nonphosphorylated protein is a hydroxyapatite nucleator *in vitro*, but it exhibits inhibitory properties when phosphorylated. However, phosphorylated DMP1 peptides that are isolated from bone and teeth behave as nucleators [369]. These findings support mechanisms for controlling hydroxyapatite formation, not only by protein phosphorylation but also through protein cleavage at specific aspartic acid sites. Through molecular approaches,

it is now appreciated that multiple functional groups in NC proteins have nucleating and inhibitory activities. For example, by generating a chimeric protein that included the collagen-binding domain of decorin and the apatite nucleating domain of BSP, the deposition of large needle crystals was far greater on collagen than on each of the control proteins [370]. In other studies, the inhibitory activity of MEPE was localized to the C-terminal Asp–Ser with peptide (ASARM) [371], but a peptide fragment of MEPE containing the integrin-binding RGD and the glycosaminoglycan attachment sequence SGDS supported increased bone formation in *in vitro* and *in vivo* models.

Osteocalcin is one of the most abundant bone ECM proteins accumulated in relation to mineral deposition. It is a vitamin K–dependent protein necessary for synthesis of its three calcium-binding Gla residues. Gla residues promote osteocalcin binding to hydroxyapatite. This property, as well as its upregulation by $1,25(OH)_2D_3$, suggests a dynamic role in calcium deposition and mobilization. Inactivation of the osteocalcin gene did not result in a major phenotype during development and growth, but after 4 months, a higher mass was observed compared to WT without a change in osteoblast number [344]. The crystal properties of osteocalcin-deficient mice differ from WT [343]. Although osteocalcin's precise function remains obscure, more than 5000 papers have documented osteocalcin as a valued serum marker of bone turnover and a marker of the mature osteoblasts whose expression correlates with matrix mineralization.

From genetic studies of bone matrix protein, it appears that inactivation of genes representing the abundant noncollagenous proteins suggests that no one protein is a major determinant of mineralization and bone. The implication of genetic studies is that the noncollagenous proteins may have redundant or coordinated functions and that each of their specialized functions is contributing to the properties of the mineral phase.

3. MATRIX MINERALIZATION PATHWAYS

Mechanisms for facilitating apatite deposition specifically in bone matrix are operative. Although substrate requirements for mineralization of bone cannot be underestimated, initiation of hydroxyapatite formation must also be considered in the context of (1) the organization of ECM components, (2) enzymes required to support an environment for nucleation, and (3) mineral homeostasis for appropriate mineral composition. Early inductive events for nucleation involve (1) removal of inhibitors of mineralization (ATP, pyrophosphate, citrate, and proteins) by enzyme activities; (2) mechanisms for raising local calcium and phosphate ion concentrations; and (3) propagation of hydroxyapatite crystals from initial deposits, which occurs through epitaxy of initial crystallites mediated by the matrix.

The removal of inhibitors involves two enzymes that have been identified as central regulators of the mineralization inhibitor pyrophosphate (PPi): (1) the nucleotide pyrophosphatase phosphodiesterase 1 (NPP1), which produces PPi from ATP and the nucleoside triphosphates, and (2) the TNAP, which hydrolyzes PPi to Pi. In addition, a transmembrane protein, ANK, transports intracellular PPi functions as a calcification inhibitor by increasing extracellular pyrophosphate [372]. The Ank gene is associated with ankylosis [373].

Alkaline phosphatase activity is still considered critical to the initiation of mineralization, a concept supported by characterization of the genetic defect in hypophosphatasia [374]. Generating the TNAP null mouse [375] demonstrated that the mechanism of impaired mineralization of cartilage and bone in this mouse represents the defect of infantile hypophosphatasia [376, 377]. The ability of TNAP to cleave pyrophosphate, removing the inhibitor, is an essential function [378–381]. As a consequence, Pi is generated, providing a local environment for nucleation and growth of the mineral phase as proven by *in vitro* studies of TNAP$^{-/-}$ osteoblasts, which cannot initiate mineralization [382]. Consistent with deficiencies in alkaline phosphatase that inhibit mineralization due to a rise in pyrophosphate levels, inactivating mutations in enzymes that produce pyrophosphate, the family of ectonucleotide pyrophosphatase/phosphodiesterase (Enpp1) or nucleoside triphosphate pyrophosphohydrolases as plasma cell glycoprotein 1 (PC-1), results in hypermineralization defects [378, 383–385]. Inactivation of the ANK gene, which transports PPi, also led to a hypermineralization phenotype [372]. An elegant series of studies of the genetic crosses of the TNAP$^{-/-}$, Enpp1$^{-/-}$, and ANK$^{-/-}$ mouse provided several lines of *in vivo* evidence that pyrophosphate is an inhibitor of mineralization and that mineralization occurs in bone as a result of the ability of TNAP to cleave pyrophosphate [378, 381].

Modifications in calcium and phosphate homeostatic mechanisms must be considered for an understanding of the mineralization pathologies that are associated with metabolic bone disease. Maintaining serum calcium levels through calcitrophic hormone axis (parathyroid hormone, calcitonin, and 1,25[OH]$_2$D$_3$) impacts on the bone reservoir. Bone will mineralize in a normal physiologic manner when serum calcium is maintained through physiologic dietary absorption. Vitamin D deficiency or metabolic bone diseases associated with enzymes or receptors for the hormone 1,25(OH)$_2$D$_3$ will lead to osteomalacia in adults and rickets in children and impaired bone formation in the mouse [386–388]. Transgenic mice expressing two- or threefold higher levels of the vitamin D receptor expressed in osteoblasts

had bone with a higher calcium content compared to wild-type mice, with decreased bone resorption and increased homogeneity of the mineral deposits and collagen maturity [389]. These findings are consistent with the importance of vitamin D for bone structural integrity and the anabolic effects of 1,25(OH)$_2$D$_3$ on bone and are leading the way to better therapeutic approaches by ligand-specific modulation of the VDR/RXR receptor [390].

Novel factors for regulation of calcium and phosphate ion levels are being identified. Mediators of ion levels include a G protein–coupled calcium-sensing receptor that is found on many cells [391]. An activating mutation of the calcium-sensing receptor *in vivo*, the NUF mouse, resulted in ectopic calcification, hypocalcemia, hyperphosphatemia, and inappropriately reduced PTH levels [392]. Inorganic phosphate is essential with calcium for mineralization in bone and formation of the hydroxyapatite crystals. The majority of homeostatic regulation of inorganic phosphate occurs through actions of renal Pi handling by PTH and its regulation of the 25-hydroxyvitamin D1 α-hydroxylase enzyme, which increases levels of the active hormone, 1,25(OH)$_2$D$_3$. In addition, Pi regulatory proteins, called phosphatonins, have been identified through rare genetic disorders in humans—X-linked autosomal dominant hypophosphatemic rickets (XLH) and autosomal dominant hypophosphatemic rickets (ADHR) [358, 393]. The hyp mouse, representing the syndrome of XLH genetic defect, was found to be an inactivating mutation in an endopeptidase called PHEX, proposed to be functionally linked to a phosphatonin. Transgenic expression of PHEX in osteoblasts improved the defective bone mineralization in the hyp mouse [394, 395] but did not fully rescue the metabolic phenotype [396]. The genetic basis of ADHR was identified to be a mutation in FGF23, which appears to have phosphatonin properties in that increased secretion will induce phosphaturia and hypophosphatemia [397]. In tumors inducing osteomalacia, FGF23 is expressed at abnormally high levels, as are two other proteins with apparent phosphaturic action: an MEPE and frizzled-related protein 4 [398]. Mouse models of FGF23 defined a key role in Pi metabolism, with the demonstration of an osteomalacia phenotype, implicating FGF23 in bone mineralization, or an indirect effect through the hyperphosphatemia and high vitamin D levels [397, 399, 400]. Although it is not certain if the effects of FGF23 are linked to a phosphatonin pathway (because FGF is not identified as a substrate for effects) or a direct physiologic role of a phosphatonin in handling phosphate homeostasis or tissue-specific aspects [401], clearly these disturbances impact on bone mineralization. Until questions relating to specific pathway mechanisms and indirect versus direct effects on how phosphate ion concentrations are regulated through bone cells can be elucidated, we can only conclude from these significant studies that phosphate levels are critical for normal mineralization.

C. Osteocytes

1. OSTEOCYTE MORPHOLOGY

As the active matrix-forming osteoblast becomes encased in the mineralized matrix, the cell differentiates further into osteocytes, the cells comprising 90–95% of bone tissue [402]. The osteocyte is considered the mechanosensor of bone tissue that impacts on its primary function to maintain bone as a viable tissue supporting physiological needs and structural requirements. Labeling studies suggest that the transition from an osteoblast to an osteocyte lasts approximately 3–5 days [403]. Mechanisms that induce the osteocyte morphology to a smaller cell body with numerous cytoplasmic extensions are not understood [404, 405], but transitional stages are recognized *in vivo* [406]. The osteocyte is considered the most mature or terminally differentiated cell of the osteoblast lineage, not capable of cell division *in vivo*. Osteocytes are embedded in bone matrix-occupying spaces (lacunae) in the interior of bone and are connected to adjacent cells by long cytoplasmic projections radiating from the cell body. These dendritic processes are enriched in microfilaments and lie within channels (canaliculi) through the mineralized matrix and form gap junctions with processes of neighboring cells and cells lining the bone surface. *In vitro*, markers of the osteocyte associated with dendritic extensions have identified cell lines with preosteocyte and mature osteocyte properties [297, 407]. In isolated cultures, mature osteocytes retain their cellular projections [408, 409]. Through different gap junction proteins called connexins (described previously), osteoblasts and osteocytes are coupled metabolically and electrically. Rapid fluxes of bone calcium across these junctions facilitate the transmission of information between osteoblasts on the bone surface and osteocytes within the structure of bone [410]. The osteocytes and surface-lining cells form a continuum, or syncytium, by connection of their cytoplasmic projections through gap junctions that facilitate the exchange of both mechanical and metabolic signals for responsiveness to physiologic demands on the skeleton. The role of osteocytes is discussed in detail in Chapter 6 (Bonewald).

2. MECHANOTRANSDUCER FUNCTION OF OSTEOCYTES

Osteocytes in their lacunae are now being appreciated for their dynamic functions in homeostatic adaptation of bone to mechanical forces [411]. Osteocytes maintain bone mass through anabolic activities, and even dying osteocytes promote bone repair through recruitment of osteoclast-mediated turnover. Some, but not all, of the biochemical features of the osteoblast are expressed in the osteocyte. There is a decrease in the volume of the cell. An older osteocyte, located deeper within the calcified bone, shows fewer of these features; in addition, glycogen stores become evident in its cytoplasm. Osteocytes have been shown to synthesize new bone matrix at the surface of the lacunae, and there is evidence for their ability to resorb calcified bone from the same surface [412]. With accumulating evidence that reduced mechanical forces on bone (e.g., weightlessness) promote osteocyte apoptosis [413, 414], a concept has emerged that the necrotic state of an osteocyte recruits osteoclasts for bone repair [415, 416]. This structural organization and the direct contact of the active osteoblast or surface lining cells with the osteocyte are consistent with the concept that bone cells, responding to varying physiological signals, can communicate their responses and transmit regulatory signals.

Mechanisms by which osteocytes function as mechanotransducers are being defined. Bone-lining cells receive the majority of systemic and local signals and can transmit these to osteocytes. However, mechanical forces on the bone produce stress-generated signals that are perceived by osteocytes, which then transmit the regulatory information to surface osteoblasts. Stress-generated electric potentials experienced by bone are either produced by strain in the organic components (piezoelectric potential) or result from electrolyte fluid flow produced by deformation of the bone (streaming potential). Mechanical strain induces factors for the proliferation, differentiation, and anabolic activities of osteoblasts [417]. Evidence that osteocytes sense mechanical loading includes the following: rapid changes in metabolic activity by [^{3}H]uridine uptake, increased metabolic activity (e.g., glucose-6-phosphate dehydrogenase), activation of several channels, periosteal gene expression, and rapid induction of small signaling molecules [418–422]. In response to mechanical strain, a volume-sensitive calcium influx pathway is activated [423], potentiated by PTH and the connexin-43 hemichannels, which is a component of gap junctions.

The phenotype of P2X7R reveals a direct anabolic role for this receptor in bone formation and an indirect role in limiting osteoclast activity in trabecular bone [424]. Gap junction–mediated signaling in response to mechanical strain requires PGE$_2$, L-type calcium channels, and P2Y receptor activation [425, 426]. L-type (long-lasting) voltage-sensitive calcium channels [427] and the P2X7 nucleotide receptor, an ATP gated ion channel, are involved in mechanotransduction. Mechanical loading sensitivity was reduced up to 73% in P2X7R null mice [428]. Among the rapidly induced signals (within seconds) are the prostaglandin PGE_2, cAMP, ATP, and nitric oxide (NO). PGE_2 promotes bone formation, whereas NO inhibits resorption [429–431]. Numerous anabolic pathways are activated in response to mechanical loads, including IGF-1, BMPs, and Wnt canonical/β-catenin signaling [127, 432–434]. Transcription factors essential for osteoblast activity, such as Runx2, TCF/LEF1, Osterix, and AP-1, are increased, as is expression of their target genes representing constituents of the bone ECM. The DMP1 matrix protein, which has been functionally linked to osteocyte maturation and mineral metabolism, responds to mechanical loading *in vitro* and *in vivo* [366, 422, 435].

The majority of the evidence to date suggests that mechanical tension can trigger bone remodeling and favor bone formation. Increased expression and synthesis of bone matrix proteins are documented; for example, osteopontin may facilitate bone remodeling by osteoclasts. However, it has been reported that mechanical strain inhibits expression of the RANKL/ TRANCE osteoclast differentiation factor [436]. Thus, the osteocyte is a mechanosensor that responds to loading and fluid shear forces in a manner that supports bone mass and viability. These exquisite mechanisms provide bone with the ability to act as a tissue responding to physiological homeostatic demands and functioning as a structural connective tissue organ that depends on communication among its resident cells.

Understanding how osteocytes sense load is an area of active investigation with respect to identity of the mechanoreceptor(s) [437, 438]. Extracellular matrix receptors, such as the integrins and CD44 receptors, appear to mediate cellular sensing of mechanical forces. The integrin cytoskeleton complex is affected by changes in cell shape induced by mechanical strain and facilitates the transduction of signals that may ultimately lead to modifications in gene expression [439]. Thickening of actin stress fibers and increased synthesis of cytoskeleton components in osteoblasts in response to mechanical strain have been documented [440]. The osteoblast and osteocyte cell surface glycocalyx is a primary sensor, and primary cilia, long known as a sensor of cell matrix [441], has been identified in association with bone abnormalities in mice deficient in polycystin-1, a protein component of cilia and a mechanosensory protein in kidney and present in osteoblasts [442].

The life span of osteoblast and osteocyte lineage cells is dependent on several factors. Because more osteoblasts are recruited to bone remodeling sites than can be organized on the bone surface for further differentiation by mineralizing osteoid, a high percentage of surface osteoblasts will die [443]. Apoptosis of preosteoblast clusters may be triggered by the lack of an adequate ECM and appropriate cell–matrix interactions for survival [444]. Apoptosis is a general mechanism for limiting organ size in embryonic development and in the adult when there is a need to regenerate tissue. In contrast to osteoblasts, osteocytes are very long-lived in their lacunae but will undergo apoptosis in response to systemic metabolic factors and when the structural integrity of bone is compromised. A dynamic function of osteocytes is in the repair of normal bone injured by microcracks. Such disruption of bone integrity and osteocyte apoptosis provides a signal for recruitment of osteoclasts for bone turnover [445]. For normal bone homeostasis, physiologic, weight bearing decreases apoptosis of osteocytes [446]. Glucocorticoid excess and estrogen or androgen deficiencies are well established to provoke osteocyte apoptosis [263, 271]. Microfracture in bones [447] and disruption of cell–cell contacts with the consequent inability to receive stimulatory signals and cell nutrients will lead to apoptosis. Increased empty lacunae and apoptotic cells (detected by DNA fragmentation using the TUNEL assay) are observed during bone turnover in aged human bone [448, 449], in glucocorticoid-treated mouse models [450], and following estrogen withdrawal [451]. Parathyroid hormone [452], bisphosphonates [453], and estrogen treatments can prevent/reduce osteocyte apoptosis. These studies all demonstrate that the osteocyte, like the surface osteoblast, is responsive to a broad spectrum of physiological mediators of bone metabolism.

D.   Cellular Cross-Talk and Osteoblast Function

Biological functions of osteoblast lineage cells extend beyond their role in bone growth (stromal osteoprogenitors), matrix production (osteoblasts), and structural integrity of bone (osteocytes). Importantly, all these cells respond to endocrine factors, such as PTH/PTHrP and $1,25(OH)_2D_3$, released to meet physiologic needs for osteoclastic resorption of bone. The bone-forming cell populations then produce cytokines and coupling factors that are essential for the sequelae of events mediating the growth and differentiation of osteoclasts. Interestingly, even the osteocyte cell line MLO-Y4 was shown to support osteoclastogenesis in the absence of exogenous factors [445], reminiscent of earlier studies requiring stromal cells for osteoclast differentiation prior to the discovery of RANKL [454]. The mechanism coupling osteoblast and osteoclast activities for regulated bone turnover is well established by the knowledge of several signaling pathways defined during the past decade [455]. Induction of bone resorption and turnover is initiated through the osteoblast, mediated by two key pathways: (1) indirect mechanisms by which calciotrophic hormones stimulate stromal cells and osteoblasts to secrete macrophage colony-stimulating factor (M-CSF/CSF-1) that will promote growth of hematopoietic precursors and activate osteoclastogenesis [456, 457], and (2) the RANK–RANKL (receptor activator of NF-κB), also know as TRANCE (tumor necrosis factor-α–related activated induced cytokine), system that involves direct interactions between a ligand on osteoblast lineage cells and its receptor on preosteoclasts to activate intracellular signaling cascades for osteoclast differentiation. The interleukin (IL)-6 family of cytokines is also secreted by osteoblasts in response to hormones. These potent stimulators of bone resorption also participate in osteoclastogenesis at early and later stages. Cytokine production by human bone marrow stromal cells can be affected by age and estrogen status [458–460]. RANKL on osteoprogenitor stromal cells interacts directly with RANK on osteoclast precursors and was demonstrated to have competency for inducing osteoclast formation from hematopoietic cells in the absence of stromal cells [461, 462]. Mice with a disrupted RANKL gene completely lack osteoclasts because of the inability of osteoblasts to support their differentiation [463]. Activating mutations in RANK have been identified as the cause of the bone disorder familial expansile osteolysis [464]. Thus, both null mutations and transgenic expression of RANKL proved the *in vivo* requirement for the RANK–RANKL system. However, costimulatory factors, such as the immunoreceptor tyrosine-based activation motif (ITAM) adaptor proteins, cooperate with RANKL to activate osteoclast differentiation. Mice lacking two ITAM adaptor proteins (DAP12 and Fc receptor gamma chain) are severely osteopetrotic [465, 466]. Together, RANK–RANKL and M-CSF/Cfms receptors represent essential factors required for coupling stromal/osteoblastic cells to the formation of osteoclasts and are appropriately controlled by cytokine and hormonal mediators of bone resorption for regulated bone turnover.

In the adult, the resorption and formation of bone at a single site is designated the bone remodeling unit. Reversal from resorption to formation is regulated by calcitonin, which inhibits osteoclast resorption when serum Ca/P is normalized. However, a key negative regulator of osteoclast differentiation, also mediated by crosstalk from osteoblasts, is through secreted osteoprotegrin (OPG), formerly designated osteoclastogenesis inhibitory factor, a secreted protein with strong homology to the TNF receptor family. OPG is expressed in several tissues, including bone, cartilage, kidney, and blood vessels [467, 468]. This soluble inhibitor of the RANKL–RANK interaction ensures that bone formation predominates when required. Several experimental approaches established OPG as a soluble factor competent to inhibit osteoclast differentiation by blocking the RANKL–RANK interaction [469–471]. Expression of the OPG gene in osteoblast lineage cells is upregulated by calcium and is downregulated by the glucocorticoid dexamethasone [471]. In addition to the RANK–RANKL–OPG system, the Toll-like receptor 9 on osteoclasts and osteoblast mediates CpG oligodeoxynucleotide signaling for regulation of osteoclastogenesis [472–474]. Also, *in vitro* and *in vivo* studies show that P2Y nucleotide receptors mediate intercellular calcium signaling between osteoblasts and osteoclasts to regulate bone formation and bone resorption [424, 475].

Newly identified factors, secreted from osteocytes and osteoclasts, appear to function in maintaining a balance between resorption and formation. For example, osteoclasts express ephrinB2, whereas osteoblasts express its receptor ephrinB4. This signaling from ephrinB2 suppresses osteoclast differentiation, whereas ephrinB4-initiated signaling enhances osteogenic differentiation [476]. Following the activation and resorption phases of the bone remodeling sequence, the recruitment, proliferation, and differentiation of osteoprogenitors and osteoblasts on the resorbed surface are accomplished in part by the bone microenvironment. Stored growth factors in the bone matrix are released to provide a local concentration of factors that initiate the formation phase by recruitment of osteoprogenitors to the resorbed bone surface. Thus, the ephrin signal system appears to be essential for bone homeostasis [476].

Although the interrelationship of bone tissue cells with the hematopoietic lineage cells for regulating bone resorption is well established, cross-talk of osteoblasts with other systems is emerging. Cell–cell interactions have been recognized between early hematopoietic cells and osteoblasts via integrins on CD34-positive cells and various cell adhesion molecules on bone marrow stromal cells [477]. The chemokine SDF-1 (CSCL12) and its receptor (CXCR4) are expressed in CD34$^+$/CD38$^-$ cells and STRO-1$^+$ stromal cells [478]. Dynamic levels of SDF-1 and CXCR4 expression induce proliferation of hematopoietic and mesenchymal progenitors and recruitment of bone-resorbing osteoclasts, osteoblasts, neutrophils, and other myeloid cells, leading to leukocyte mobilization. The expression of ephrinB2 in hematopoietic cells is regulated by interaction with stromal cells. Interaction of ephrinB2 with EphB4 receptor modulates the migration and colonization of the hematopoietic cells in the local stromal microenvironment. Ephrin signaling is active in both osteoblasts and osteoclasts, with bidirectional effects enhancing osteoblast differentiation through ephrinB4 receptor and inhibiting osteoclast differentiation through ephrin B2 on the nuclear factor of activated T cells (NFATC1) target gene [476, 479]. NFATC1 in osteoblasts controls expression of chemoattractant for monocytic osteoclast precursors. Inhibitors of the calcineurin/NFAT pathway (known immunosuppressants) impair bone formation by decreasing NFATC1 in osteoblasts that is necessary for activity of the bone essential Osterix [218, 480].

Studies have raised provocative implications of a direct influence of immune cells in contributing to osteogenic differentiation [481]. The immune and bone organ systems are linked by the production of multiple cytokines from T lymphocytes regulating bone turnover by the modulation of both osteoblast and osteoclast activities. ICAM-1 and VCAM-1 have been reported on the osteoblast surface in response to inflammatory cytokines, thereby providing a potential mechanism for T cell interactions that contribute to the regulation of bone turnover. Aside from bone turnover activities, osteoblasts produce a number of immune molecules, including induction of the Toll-like receptor 5 on osteoblasts, which is upregulated in response to bacterial pathogens. This defines an important function of osteoblasts shared with immune cells [482]. Cross-talk between osteoblasts and the endothelial cell is beginning to be investigated; this communication is likely important for vascular invasion into the bone matrix. Osteoblasts secrete paracrine factors that regulate endothelial cell (EC) function [483], including vascular endothelial growth factor (VEGF) and its receptors [484]. VEGF secreted by ECs has been reported to enhance the anabolic effects of $1,25(OH)_2$ vitamin D_3 on osteoblasts [485] and to be necessary for angiogenesis during endochondral bone formation *in vivo* [486]. Of note, osteoblasts influence the expression of E-selectin on EC cells, and bone sialoprotein, which is upregulated in osteogenic tumors and mediates cell attachment via $\alpha_v\beta_3$ integrins, can directly promote adhesion of endothelial cells [487, 488]. In a reciprocal manner, EC cells can promote osteoblast differentiation via gap junction communication [489, 490]. From these reports, it can be predicted that osteoblast lineage cells would interact with different cell systems to support the general systemic properties of bone as a tissue responsive to many physiologic activities.

V. MOLECULAR MECHANISMS MEDIATING PROGRESSION OF OSTEOBLAST GROWTH AND DIFFERENTIATION

With recognition of decreased osteoblast surfaces in osteoporotic bone and reports of decreased marrow osteoprogenitors with age [491–493], defining mechanisms contributing to the regulation of proliferative activity and differentiation in osteoblast lineage cells is increasing in importance. In this section, two fundamental parameters are presented that have identified (1) how osteogenic factors establish control of bone growth through modification of regulatory events in the cell cycle and (2) how differentiation of osteoblasts is established by control of gene expression through modification in nuclear architecture. Both these parameters contribute to the determination, differentiation, and biological functions of osteogenic lineage cells.

A. Cell Cycle Control

To understand regulatory parameters of proliferation, one must consider mechanisms that support the requisite responsiveness to growth factors through signaling pathways and the consequent induction of proliferation. To explain the induction, synthesis, activation, and suppression of the complex and interrelated regulatory factors associated with the growth control of osteoprogenitor cell proliferation *in vivo*, an understanding of mechanisms that control cell proliferation is required. Proliferation is controlled through the cell cycle by the activity of regulatory proteins that support progression of cells that have responded to a mitogenic stimulus through DNA replication and cell division. The cell cycle is a stringent growth-regulated series of sequential biochemical and molecular events that support genome replication and mitotic division [494]. Stages of the cell cycle regulated by specific cyclin and cyclin-dependent kinase complexes and checkpoints that monitor competency of cells to progress through DNA replication and mitotic division illustrate some of the requirements for growth control (Figure 4-6). Suppression of certain cell cycle–regulated genes is requisite for the cessation of proliferation and upregulation of phenotypic genes.

When quiescent cells (G_0) are stimulated to proliferate and divide, they enter G_1, the first phase of the cell cycle in which the enzymes required for DNA replication are synthesized. Before a cell can progress through G_1 and begin DNA synthesis (S phase), it must pass through a checkpoint in late G_1, which is known as the restriction point [495]. At this cell cycle restriction point, both positive and negative external growth signals are integrated. If conditions are appropriate, the cell proceeds through the remainder of G_1 and enters the S phase. Once the cell passes the restriction point, it is refractory to withdrawal of mitogens or to growth inhibitory signals and is committed to progressing through the remainder of the cell cycle unless it is subjected to DNA damage or metabolic disturbance [495].

In mammalian cells, progression through the cell cycle is regulated by a cascade of complexes containing

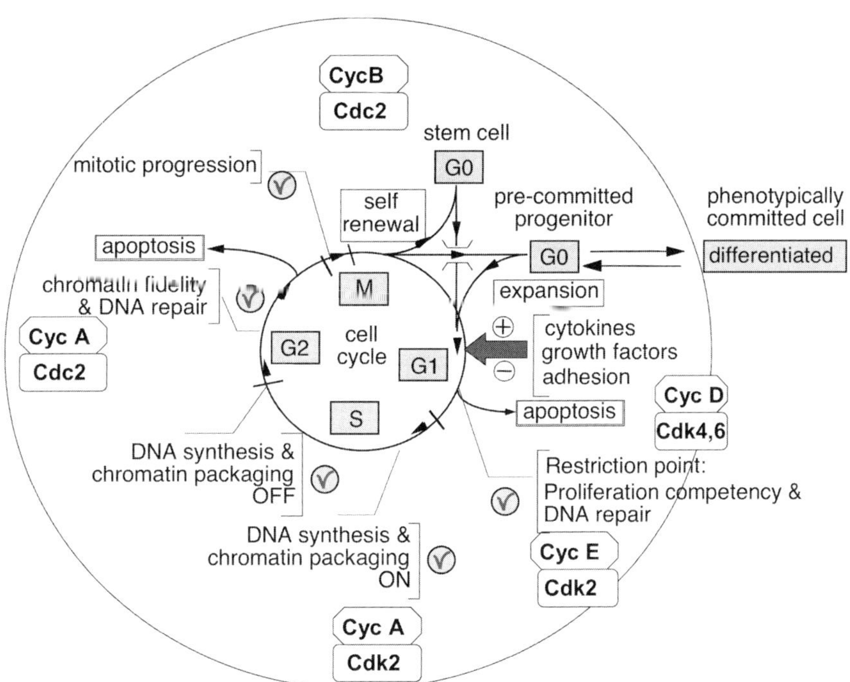

FIGURE 4-6   Control of cell cycle progression in bone cells. Progression through the cell cycle is controlled by formation of cyclin and cyclin-dependent kinase (cdk) complexes at each stage (M, G_1, S, and G_2). Activities associated with each stage are indicated. Entry into G_0 and exit from the cell cycle are controlled by growth-regulatory factors (e.g., cytokines, growth factors, cell adhesion, and/or cell–cell contact) that determine self-renewal of stem cells and expansion of precommitted progenitor cells. The cell cycle is regulated by several critical cell cycle checkpoints (checkmarks), at which competency for cell cycle progression is monitored. The biochemical parameters associated with each cell cycle checkpoint are indicated. Options for defaulting to apoptosis during G_1 and G_2 are evaluated by surveillance mechanisms that assess fidelity of structural and regulatory parameters of cell cycle control. Apoptosis also occurs in mature differentiated bone cells.

cyclins and a family of threonine/serine kinases designated cyclin-dependent kinases (cdks) that transduce growth factor–mediated signals into discrete phosphorylation events. Cyclin activity is modulated by the formation of complexes distinct at each stage of the cell cycle. In general, the levels of cdk proteins remain relatively constant during the cell cycle, whereas the expression of specific cyclins is confined to distinct phases of the cell cycle in which they are degraded quickly after having completed their function. The activity of cyclin–cdk complexes is regulated by a number of mechanisms: (1) positive and negative phosphorylation events for formation and reversible association of the proteins and (2) inhibitory proteins designated CKIs of the Cip/Kip and INK families.

Cyclins and cdks are responsive to regulation by the phosphorylation-dependent signaling pathways associated with activities of the early response genes, which are upregulated following the mitogen stimulation of cell proliferation [496]. Cyclin-dependent phosphorylation activity is functionally linked to activation and suppression of both p53 and RB-related tumor-suppressor genes [497]. p53 accumulates in response to stress, inducing arrest at G_1 or G_2. The retinoblastoma

protein (Rb), a tumor suppressor, is a member of a family of related proteins that includes p105, p107, and p130. Rb has been shown to have a critical role in the regulation of cell proliferation, particularly in progression through G_1. Rb functions as a signal transducer, receiving both growth-promoting and -inhibitory signals and linking them to the transcriptional machinery required for cell cycle progression or cell cycle arrest. In quiescent cells or cells reentering G_1 from mitosis, Rb exists in an underphosphorylated or dephosphorylated state. Phosphorylation of Rb occurs late in G_1 and modifies the activities of regulatory complexes that are required for gene expression linked to the onset of S phase [498].

The activities of the cdk are downregulated by a series of inhibitors (designated CDIs) and mediators of ubiquitination, which signal destabilization and/or destruction of these regulatory complexes in a cell cycle–dependent manner. The cyclin inhibitory protein (CIP) class of CDIs includes the proteins p21, p27, and p57. Growth arrest is, in part, due to induction of the cyclin-dependent kinase inhibitor (cdki) protein p21, which can interact with multiple cyclin–cdk complexes. The INK class is represented by proteins p15, p16, p18, and p19, which are linked to apoptosis control mechanisms.

I. GROWTH CONTROL IN SKELETAL CELLS

Expression of cell cycle regulatory proteins, cyclins, and cyclin-dependent kinases appears not to be solely confined to control of proliferation but, for example, associated with differentiation in bone osteoblasts and nonosseous cells [248–250]. The cell cyclin cycle regulatory factors are targets of many signaling cascades that contribute to skeletal development. Studies have shown that the cyclin B/Cdk1 complex phosphorylates Runx2 in mitosis, while cyclin B/Cdki converts Runx2 to a hypophosphorylated form by PP1/PP2A to support postmitotic regulation of Runx2 target genes. The cyclin D1/Cdk complex can ubiquinate Runx2, degrading it to inhibit Runx2-mediated differentiation [499, 500]. Secreted osteogenetic factors, which include high calcium (via the calcium-sensing receptor), β-catenin, PTHrP, TGF-β [501–504], Ihh [505], Wnt 5B [119], and c-Fos [506], will affect cell growth through cyclin D1. Numerous transcription factors that control gene expression in chondro- and osteogenic lineage cells exert their effects on cyclin B1, including STAT, ETS, CREB, nuclear receptors (estrogen, glucocorticoid, and vitamin D receptor), c-Jun, JunB, c-Fos, PPARγ, and INI1/SNF5 [507, 508]. Cell cycle regulatory factors, particularly cyclin E, have been noted in several systems involved in the regulation of differentiation, in myoblasts [509], in osteoblasts [510], and in promyeloid cell differentiation into macrophages [511]. JunB, a target of both cyclin D1 and cyclin A [512], is a key regulator of osteoblast growth and differentiation. Mice lacking JunB exhibit an osteopenia phenotype with decreased proliferation and decreased expression of cyclin D1 and cyclin A and increased expression of p60 (INK4A) [213].

During osteoblast differentiation, cdki are also developmentally expressed. The cdki p21 (CIP/WAF1) is expressed in the growth period and contributes to cell cycle exit and differentiation, with dramatic increases in p21 observed in postmitotic chondrocytes [513, 514]. In contrast, p27 (KIP-1) is expressed in the immediate postproliferative period and is upregulated again during differentiation [515]. Thus, p21 has multiple effects in regulating the growth and differentiation of skeletal lineage cells, possibly by (1) responding to signaling factors that regulate chondrogenic and osteogenic activities, including FGF, Sox9, thyroid hormone, and BMPs; (2) promoting cell cycle exit; and (3) functioning to attenuate osteoblast maturation.

Deregulation of the cell cycle regulatory factors in skeletal disorders are being understood through genetic studies. The overexpression of STAT1, STAT5, and p21 correlates with the phenotypic severity of chondrodysplasias associated with activating mutations of FGFR3 [516]. Overexpression of cyclin D1 is associated with the development of parathyroid adenomas [517]. Cyclin D1 and cyclin A genes are the targets of activated PTH/PTHrP receptors in Jansen's metaphysiochondrodysplasia [503]. Studies characterizing bone abnormalities associated with null mutations of cell cycle and cell growth regulatory factors have revealed their significance in providing signals for the control of both the number and the differentiation of bone-related cells. For example, marrow harvested from p27–/–mice shows a three- or fourfold increase in osteogenic nodule formation compared to wild type. Thus, the absence of this cdki allows the marrow population to extend its growth phase, increasing cell numbers. This expansion of the osteoprogenitor population is consistent with the larger size of the animals and the proportionally increased cortical width of the long bones [515].

Most significantly, the p53–MDM pathway regulates bone formation and osteoblast differentiation [518, 519]. A major role of p53 is to promote cell cycle arrest and programmed cell death. The p53 tumor suppressor has a critical role in preventing cancers in the most commonly mutated gene in human cancers [520]. p53 activity is negatively controlled by MDM2, which encodes an E3 ubiquitin ligase that becomes induced as p53 cellular levels increase [521] and targets p53 proteosomal degradation. This autoregulatory negative feedback loop between p53 and MDM2 to keep p53 activities under control has been established through several mouse models. Mice deleted for p53 will form tumors with 100% penetrance, but do undergo normal development [522]. In contrast, a mouse model carrying a mutated p53 allele that increased p53 activity showed early aging-like phenotypes in several organs and osteoporosis [523]. The results suggested that negative regulation of p53 might be important to maintain proper tissue homeostasis in adult mice. Two studies in the mouse using different experimental approaches provided genetic evidence that p53 blocks osteoblast differentiation during bone development [518, 519]. In general, proteins that suppress cell proliferation would be expected to promote differentiation. In p53 knockout mice, elevated levels of osterix, a transcription factor essential for osteogenesis, also promoted the differentiation of osteoclasts but with a net anabolic effect. The study by Lengner et al. [518] showed that p53 null osteoprogenitor cells have increased expression of Runx2, increased osteoblast maturation, and increased osteogenic potential. Runx2 is also an activator of osteogenic differentiation and functions upstream of osterix. Thus, both studies establish that p53 suppresses osteoblast differentiation by repressing the expression of two transcription factors essential for bone formation, either Runx2 or osterix. Importantly, the results have been confirmed by deletion of MDM2 in osteoblast progenitor cells, which resulted in elevated p53 activity, reduced proliferation, and reduced expression of Runx2 and differentiation. Both phenotypes were rescued by crossing MDM2 and p53 null mice. Thus, the p53–MDM2 regulatory link for control of cell proliferation regulates the number of proliferating osteoprogenitor cells for normal bone development by modifying expression levels of transcription factors essential for osteogenesis; however, the development of osteosarcomas by deregulation of this pathway is not necessarily linked to Runx2 or Osterix.

pRb regulates cell cycle progression through its interaction with E2F transcription factors and inhibits the G_1-to-S phase in cell cycle transition. The pRb-related p130 and p107 proteins' overlapping roles and genetic deletion of these in mice identified their importance in regulating chondrocyte growth. Mice that exhibited defective endochondral-bone development shortened limbs died soon after birth [524]. FGF signaling targets these two pRb proteins to induce chondrocyte growth arrest [525].

Investigations of the effects of growth factors and osteogenic hormones on cell cycle target genes are increasing our understanding of their precise molecular mechanisms in the regulation of growth, differentiation, and apoptosis of osteoprogenitor cells and osteoblasts. Several studies have reported BMP-2 and BMP-4 induction of cell cycle arrest in the G_1 phase that is mediated by enhanced expression of the p21 cyclin inhibitor [526] and rapid induction of cyclin G, a cyclin that is increased after the induction of p53 by DNA damage [527]. Both of these events are linked to the induction of apoptosis, and in the developing tooth, p21 and BMP-4 are co-expressed in cells destined to undergo apoptosis in a transitional epithelial structure known as the enamel knot [528]. The apoptotic-promoting effects of BMP-2 have been reported to oppose the estradiol-induced growth of human breast cancer cells. Where estradiol stimulates cyclins and cyclin-dependent kinases, the BMP induction of the cyclin kinase inhibitor p21 leads to the inactivation of cyclin D1 [529]. The abundance of TGF-β and BMPs in the early stages of osteoblast maturation and the targeting of BMP action to p21 may provide a mechanism not only for promoting osteogenic differentiation but also for apoptosis of proliferating cells that are recruited to the bone surface and may not progress to the mature osteocyte.

The effects of other cytokines and growth factors that target the proliferation phase are coupled through p21. IL-6 promotes differentiation and exhibits anti-apoptotic effects on human osteoblasts [530]. The effects of IL-6 on the p21 promoter are mediated by STAT-binding proteins and a STAT response element in the p21 promoter. FGFs are classic mitogens of the osteoprogenitor pool as well as modulators of osteoblast differentiation [53, 56, 531–533]. FGF signaling also activates STAT1 and p21, a mechanism that accounts for the ability of FGF-2 to induce both mitogenic responses and growth arrest in cancer cells [534, 535]. TGF-β also inhibits cell cycle progression in part through the upregulation of p21 gene expression [515, 536]. Regulation of the p21 promoter is mediated by TGF-β induction of Smad3 and Smad4 [536, 537]. The steroid hormone $1,25(OH)_2D_3$ exerts antiproliferative effects in undifferentiated cells also mediated by the enhanced expression of p21 [538] and p27 [515]. This finding is consistent with the high levels of p27 in mature osteoblasts and $1,25(OH)_2D_3$ induction of markers of the mature osteoblast phenotype. Osteoblast responses to regulators of bone formation that involve cell cycle control are summarized in Table 4-1.

It is becoming increasingly evident that each step in the regulatory cycles (cell cycle, cyclin/cdk cycle, and cdki cycle) governing proliferation is responsive to multiple signaling pathways and has multiple regulatory options. The diversity in cyclin–cyclin-dependent kinase complexes accommodates the control of proliferation under multiple biological circumstances and provides functional redundancy as a compensatory mechanism. Similarly, the inhibitors of cyclin–cdk complexes bind to and regulate multiple cyclin–cdk-containing complexes at several checkpoints [539–541]. The regulatory events associated with these proliferation-related cycles support control within the contexts of (1) responsiveness to a broad spectrum of positive and negative mitogenic factors, (2) cell–cell and cell–ECM interactions, (3) monitoring genome integrity and invoking DNA repair and/or apoptotic mechanisms if required, and (4) competency for differentiation. Perturbation of any of these cell cycle regulatory mechanisms can result in unregulated or neoplastic growth.

B. Nuclear Architectural Control of Regulatory Machinery: The Runx2 Paradigm

It is becoming increasingly apparent that nuclear architecture provides a basis for support of the stringently regulated modulation of cell growth and tissue-specific transcription necessary for the onset and progression of osteoblast differentiation. Here, multiple lines of evidence point to contributions by three levels of nuclear organization: (1) the DNA regulatory elements for gene

TABLE 4-1   Osteogenic Responses Mediated by Cell Cycle Regulation

Regulator	Response	Reference
PTHrP	Induces G_1 growth arrest by inhibiting cyclin D1/cdk4/cdk6	Datta *et al.* [596]
	JunB increased	
PTH	Arrests cells in G_1 via induction of MAPK phosphatase and p21 and decreases cyclin D1/mRNA	Qin *et al.* [597] Onishi and Hruska [598]
Glucocorticoid	Exerts antimitotic and antiphenotypic effects on postconfluent growth by reduction of cyclin A/cdk level and interferes with growth-permissive axis by GSK3β activation via c-myc downregulation and inhibition of G_1/S cell cycle transition	Smith *et al.* [599, 600]
$1,25(OH)_2D_3$	Antiproliferative effects at multiple levels, including appearance of growth-suppressing hypophosphorylated pRb and decreased cdk activities	Jensen *et al.* [601] Sinkkonen *et al.* [602]
	Exerts direct effects on gene regulation of cyclin C and p21 genes	Saramaki *et al.* [603]
		Liu *et al.* [538]
Estrogen	Regulates expression and function of c-myc and cyclin D1, inhibits p21, stimulates growth through increases in cyclin D2 inducing G_1 to S progression	Doisneau-Sixou *et al.* [604] Fujita *et al.* [605] Kanda and Watanabe [606]
FGF2 FGFR3	Induces growth arrest by a cascade initiated by disruption of cyclin D3/cdk6 complexes; increases in p21 and p27, and underphosphorylation of p107 and p130	Aikawa *et al.* [607] Krejci *et al.* [608]
	Activates STAT, which is mitogenic in normal cells and results in growth arrest in cancer cells	Dailey *et al.* [609] Laplantine *et al.* [610]
BMP-2/4	Differentiation mediated by p27 p21 and BMP4 are co-expressed in the enamel not limited to apoptosis induces cell cycle arrest by increasing p21	Thomas *et al.* [611] Jernvall *et al.* [528]
	Rapid induction of cyclin G after DNA damage	Yamato *et al.* [526] Okamoto and Prives [527]
TGF-β	Inhibits cell cycle progression via increase of p21	Paradali 2000 Drissi *et al.* [515]
IL-6	Increases p21 to promote differentiation and has anti-apoptotic effects via increases in STAT1	Bellido *et al.* [530]
Heparin sulfate	Negatively regulates cell cycle resulting in growth arrest	Manton *et al.* [612]
Clock genes	Antiproliferative effects by inhibiting G_1 cyclins	Fu *et al.* [613]
	PER1 and PER2 null mice exhibit increased proliferation, increases in c-myc and G_1 cyclins resulting in osteoblast proliferation and increased bone mass	
c-fos	Overexpression accelerates cell cycle progression via induction of cyclins A and E in osteoblasts but not fibroblast	Sunters *et al.* [614]
	Cdk activity is increased by dissociation of P27 through cdk2 complexes	
Runx2	Runx2 null mice exhibit increased proliferation and promote cell cycle exit, induces growth arrest by increasing P27	Pratap *et al.* [195]
		Galindo *et al.* [190] Galindo *et al.* [190] Thomas *et al.* [611]

transcription by specific protein–DNA interactions; (2) the chromatin structure and nucleosome organization that establishes competency for activation of a silent gene; and (3) the nuclear matrix scaffold, which accommodates the organization of functional domains within the nucleus (Figures 4-7A and 4-7B). During the past several years, there has been a focus on contributions of higher order nuclear organization to architecturally supporting compartmentalization of regulatory machinery in subnuclear microenvironments that are functionally coupled to regulatory events for *in vivo* transcriptional control (reviewed in Zaidi [542]). Examples of the transcriptional regulatory machinery organized in functional domains associated with the nuclear matrix (NM) scaffold include nucleoli, chromosomes, and promyelocytic leukemia protein bodies (Figure 4-7C). Regulatory functions of the NM include, but are by no means restricted to, DNA replication; gene location; physical constraints on chromatin structure that support the formation of loop domains; concentration and targeting of transcription factors; RNA synthesis; processing and transport of gene transcripts; and post-translational modifications of chromosomal proteins, as well as imprinting and modifications of chromatin structure. Among the transcription factors that organize regulatory complexes in NM-associated subnuclear domains are Runx factors (Figures 4-7D and 4-7E).

It is apparent that local nuclear environments generated by the multiple aspects of nuclear structure are intimately tied to the developmental expression of cell growth and tissue-specific genes. During osteoblast differentiation, nuclear matrix protein profiles are changing, suggesting dynamic changes in factors associated with subsets of genes representing each stage of maturation [543]. Osteoblasts receive physiologic cues that initiate signaling pathways that ultimately influence transcription. Here, the mechanisms that sense, amplify, dampen, and/or integrate regulatory signals involve structural as well as functional components of cellular membranes. Extending the structure–regulation paradigm to nuclear architecture expands the cellular context in which cell structure–gene expression interrelationships are operative. Modifications in cell structure by mechanical forces or physiologic mediators that affect cell shape will influence nuclear architecture and change gene expression to accommodate the biological signal. Nuclear structure is a primary determinant of transcriptional control. Thus, the power of addressing gene expression within the three-dimensional context of nuclear structure would be difficult to overestimate. The levels of nuclear architecture will be explained using as paradigms the bone-specific and Runx2 genes.

I. CONTEXT OF GENE REGULATORY ELEMENTS

The primary level of gene organization establishes a linear ordering of promoter regulatory elements. This representation of regulatory sequences reflects competency for the responsiveness to physiological regulatory signals as discussed previously. The organization of the Runx2 and osteocalcin promoters is shown in Figure 4-8A, and common features include multiple Runx elements and protein regulatory motifs. The well-studied osteocalcin gene provides a paradigm for the involvement of nuclear organization in transcriptional control that is linked to bone formation, homeostatic regulation, and bone remodeling. The regulatory elements of the bone-specific osteocalcin gene are organized in a manner that supports developmental expression in relation to bone cell differentiation and responsiveness to physiologic mediators. Characterized regulatory elements and cognate transcription factors can support both osteocalcin suppression in nonosseous cells and proliferating osteoblasts, as well as transcriptional activation in mature osteoblasts and steroid hormone enhancement (Figure 4-8). A bipartite element in the proximal promoter confers responsiveness to growth factors FGF-2, TGF-β, and cAMP [544, 545]. Two motifs confer bone-specific expression. The OC box (99 to 76 bp) with a core homeodomain protein element binds factors that can repress (Msx2) or activate OC (Dlx3 and Dlx5) [226, 546, 547]. Chromatin immunoprecipitation studies have identified association of different HD proteins with the OC promoter at specific stages of development [226]. These findings illustrate how one regulatory element can function in either repression or activation of gene transcription.

Multiple Runx regulatory elements are strategically positioned in many gene promoters. Runx sites contribute to chromatin structure of active genes and the integration of physiologic signals. These functions were identified in the OC gene. Microarray profiling studies are identifying hundreds of Runx target genes that can function in many capacities [548, 549]. Two Runx sites, A and B, flanking the vitamin D response element, and the Runx2 sites B and C contribute to positioning of a nucleosome in the actively transcribed gene [550]. By mutation studies, all three sites were found to be required for maximal basal expression of OC. Strikingly, mutation of the three Runx sites leads to abrogation of responsiveness to vitamin D, glucocorticoids, and TGF-β. Direct interactions between Runx2 and the VDR for transcription have been established [550–552]. Runx2 is also a positive regulator of estrogen activity and functions with the ER, possibly in a manner analogous to the VDR [553]. Cooperative interactions between Runx2 and C/EBP elements, first reported for OC synergistic transcription in mature osteoblasts, occur in other genes [554, 555]. These findings strongly support multifunctional roles for Runx2 factors in regulating gene expression, not only as a simple transactivator but also by facilitating modifications in promoter architecture and chromatin organization.

The vitamin D responsive element (VDRE) functions as an enhancer of the osteocalcin gene by binding the transcriptionally active VDR/RXR heterodimer complex. The core motif of the VDRE, two steroid half elements with a three-nucleotide spacer, is highly conserved. However, subtle variations, both within the core domain and within the flanking sequences, render VDRE promoter elements of various genes selectively ligand

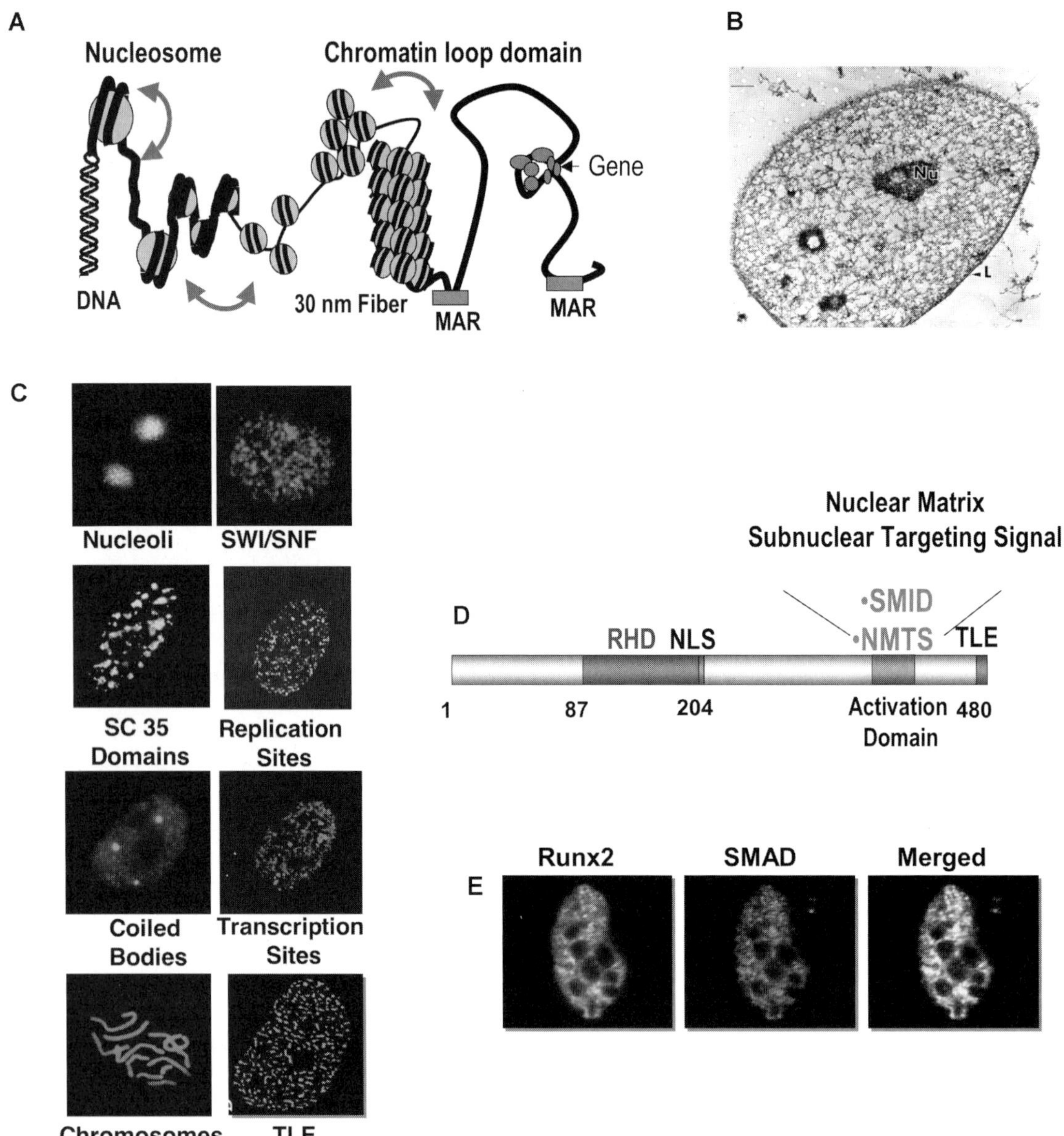

FIGURE 4-7   Nuclear architecture contributes to bone-specific gene regulation. (A) Levels of chromatin organization. Chromatin organization and the nucleosome of core histone protein for binding DNA (left). Post-translational modifications of histone proteins regulate active (open) chromatin and inactive (condensed) chromatin. Chromatin loop domains (10–100 kb) are tethered to components of the nuclear matrix through matrix attachment region (MAR) sequences. An individual gene with a positioned nucleosome is illustrated within the loop (right). (B) Electron micrograph of the filamentous structure of the nuclear matrix scaffold [566]. (C) Organization of functional activities in domains associated with the nuclear matrix scaffold [542]. Antibodies to markers of the indicated functional domains reveal the organization of structures and transcriptional foci. (D) Domain organization of Runx2 showing a nuclear localization signal (NLS) contiguous to the runt homology DNA binding domain (RHD) and a second intranuclear trafficking signal designated the nuclear matrix targeting signal (NMTS) located in the C-terminus. The C-terminal Groucho/TLE interacting protein is also nuclear matrix associated with its own distinct targeting signal [595]. The Runx2-Smad interacting domain (SMID) overlaps the NMTS [93]. (E) Runx2 recruits co-regulatory proteins to Runx2 domains in the nuclear matrix compartment. Shown is the interaction of Runx2 and the BMP-2-induced Smad1 *in situ* in HeLa cells transfected with XPRESS tag Runx2 and flag tagged Smad [92].

responsive in a developmental and tissue-specific manner. This is particularly significant for the OC gene in which a contiguous Runx2 site to the VDRE forms a bridge complex with the VDR/RXR complex [551]. Specificity of VDRE utilization is further conferred by protein–DNA and/or protein–protein interactions in addition to the VDR/RXR complexes. Interacting co-regulatory proteins with VDR/RXR include AP1 factors, YY1, Runx2, and several coactivators (DRIP205, SRCs, and p160/CBP), as well as components of the RNA polymerase II complex.

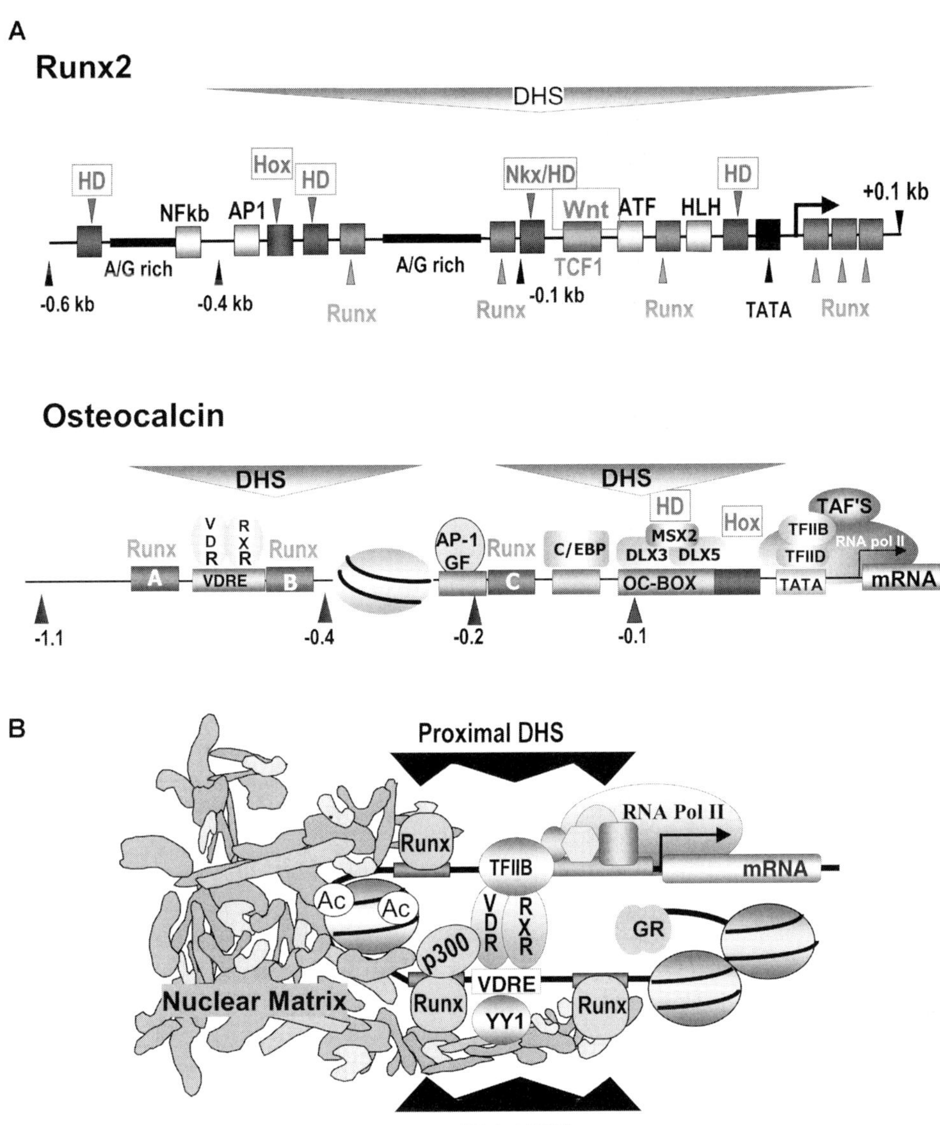

FIGURE 4-8   Regulatory elements in the osteocalcin and Runx2 promoters. (A) The transcriptionally active osteocalcin gene with the positioned nucleosome between the proximal and distal domains. Runx2 is shown in the lower panel, in which the first 600 kb are transcriptionally active, indicated by strong DNase hypersensitivity (DHS) across this domain. Note that multiple homeodomain (HD) sites occur in the Runx2 gene promoter, in contrast to a signal tissue-specific HD element in the designated osteocalcin box. Both genes have multiple Runx sites. (B) Three-dimensional model of OC promoter structure is based on experimental evidence, including a positioned nucleosome in the transcribed gene between the proximal and distal domains; direct physical interactions between the VDR and Runx2, as well as the VDR and TFIIB; and mutation of the Runx sites, which decreases DNase I hypersensitivity (DHS).

$1,25(OH)_2D_3$ enhancement of OC transcription is highly dependent on basal levels of expression [251]. This finding and knowledge of the cross-talk between the TATA box in the proximal promoter and the distal promoter VDRE are examples of evidence that the linear organization of gene regulatory sequences is necessary but insufficient to accommodate the requirements for physiological responsiveness to homeostatic, developmental, and tissue-related regulatory signals. The three-dimensional model of OC promoter structure accommodating protein–protein interactions between regulatory elements is based on OC chromatin of the actively transcribed gene (Figure 4-8B). The three-dimensional conformation of the OC promoter is facilitated by Runx2 association with the nuclear matrix stabilizing the transcriptional complexes (protein–protein interactions).

2. Epigenetic Control and Regulation of Chromatin Modifications

Parameters of chromatin structure and nucleosome organization are a second level of genome architecture. There is a requirement to render promoter regulatory elements competent for protein–DNA and protein–protein interactions that mediate positive and negative controls. Additionally, activities of regulatory complexes at the proximal and distal promoter must be integrated. Modifications in chromatin reduce the distance between promoter elements, thereby supporting interactions between the modular components of transcriptional control. Each nucleosome (approximately 140 nucleotide base pairs wound around a core complex of two each of H3, H4, H2, and H2B histone proteins) contracts linear spacing of the DNA. Folding of nucleosome arrays into solenoid-type structures provides a potential for interactions that support synergism between promoter elements and responsiveness to multiple signaling pathways (Figure 4-7A).

The molecular mechanisms that mediate chromatin remodeling are being defined. A family of proteins comprising multimeric protein complexes has been described in yeast (SWI/SNF complex) and in mammalian cells that promote transcription by altering chromatin structure. Chromatin remodeling for activation of genes is initiated by the large ATPase-containing SWI/SNF complex, which is required for induction of phenotype programs such as myogenesis and osteogenesis [96, 556–558]. Alterations in chromatin render DNA sequences containing regulatory elements accessible for binding cognate transcription factors and mediate protein–protein interactions that influence the structural and functional properties of chromatin. The remodeling of nucleosomal structure involves alterations in histone–DNA and/or histone–histone interactions. These "epigenetic" mechanisms that do involve DNA sequences contribute to heritable changes in gene expression. DNA hypermethylation at a specific lysine 9 in H4 histone and histone hypoacetylation are characteristics associated with gene silencing, whereas H3 histone acetylation, methylation of lysine 4 in H3 histone, and phosphorylation post-translational modifications have been functionally linked with changes in nucleosomal structure that alter the accessibility to specific regulatory elements and hence gene activation [559]. Core histone hyperacetylation mediated by co-regulatory factors such as p300 and CBP enhances the binding of most transcription factors to nucleosomes [560]. Histone deacetylation (HDAC) enzymes reverse the transcriptionally active chromatin structure and promote a condensed nucleosome configuration, inactivating genes. Indeed, HDAC inhibition promotes osteogenic differentiation by increasing expression of osteogenic transcription factors [561, 562] and genetic disruption of HDACs results in skeletal defects.

Alterations in the chromatin organization of the OC gene promoter during osteoblast differentiation provide a paradigm for remodeling chromatin structure and nucleosome organization that is linked to a long-term commitment to phenotype-specific gene expression (Figure 4-9). In nonosseous cells, the packing of chromatin contributes to the extent that promoter elements are accessible to transcriptional activation complexes. An array of nucleosomes on the OC promoter in nonosseous cells contributes to maintaining the suppression of gene transcription. Figure 4-8 schematically depicts modifications in chromatin structure and nucleosome organization that parallel competency for gene activation. When the OC gene is activated in osteoblasts, there is a rearrangement in nucleosome placement, with a single nucleosome becoming positioned between proximal regulatory elements and distal domains [563]. DNase I hypersensitivity is detected in two promoter regulatory domains (proximal and distal VDRE) and is enhanced by vitamin D treatment. Thus, structural properties of the chromatin, reflected by DNase I hypersensitivity, describe the extent to which the osteocalcin gene is transcribed in bone cells. The Runx sites, as described previously, are essential for chromatin remodeling that leads to active transcription. Mutation of the three Runx sites in the osteocalcin gene promoter near the VDRE results in a complete loss of DNase hypersensitivity, reflecting a closed chromatin configuration and inaccessibility of transcription factors to the promoter [564]. Mutation of the OC distal domain VDRE also affects nuclease sensitivity in the OC proximal promoter domain [552]. The mechanisms for vitamin D enhancement of osteocalcin dependent on Runx2 are related to formation of the Runx2/p300 complex on the OC promoters, which interacts with the VDR/RXR. Not only Runx2 but also many other transcription factors, including the VDR/RXR, homeodomain proteins, and C/EBP, have the ability to interact with co-regulatory proteins that can acetylate or deacetylate histone proteins and thereby modify transcriptional levels. From the many studies of Runx2 on the OC gene, three key functions of Runx2 have been documented: (1) induction of chromatin remodeling of tissue-specific activation, (2) facilitation of the recruitment of other transcription factors to the OC gene promoter through chromatin modifications, and (3) conformation of the promoter organization of transcription factor complexes through association with the nuclear matrix scaffold.

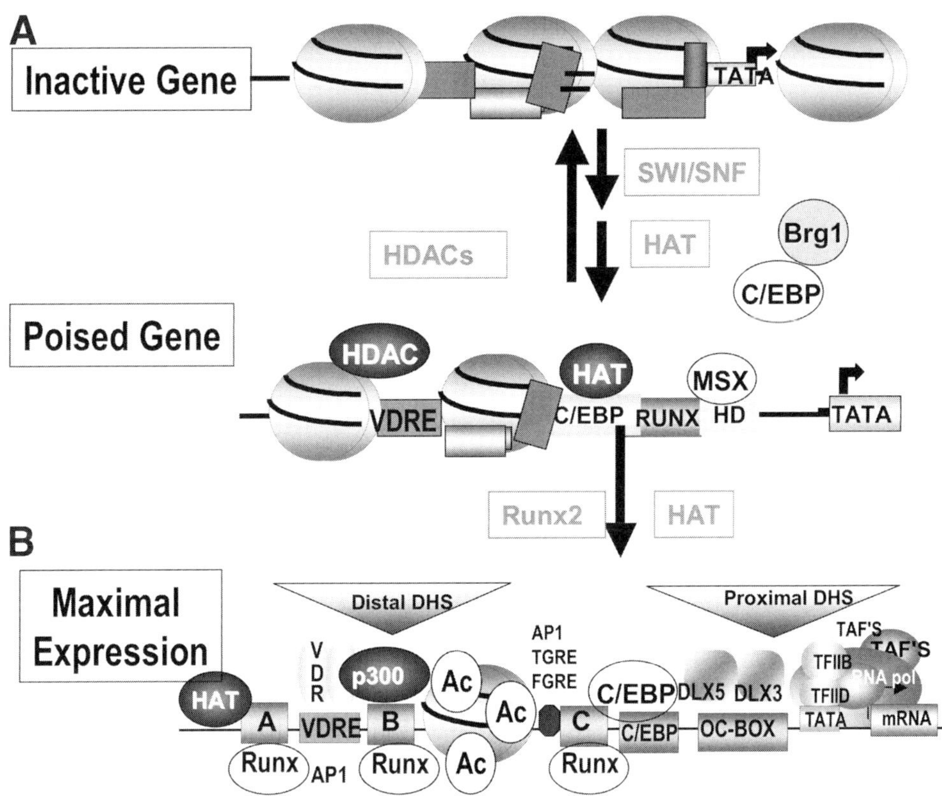

FIGURE 4-9   (A) Chromatin organization of the osteocalcin gene promoter. Representation of the inactive OC gene in a closed chromatin configuration that is remodeled by SWI/SNF complexes and histone acetyltransferases (HAT), to prepare the promoter for transcription factor–mediated activation of gene expression. Brg1 is a subunit of a SWI/SNF complex that interacts with C/EBP and the p300 activator. Here, the OC gene is under repression by Msx2. Upon binding of Runx2, the OC gene promoter is further remodeled to accommodate binding of all regulators and is maximally expressed by vitamin D stimulation. A positioned acetylated (Ac) nucleosome separating the proximal and distal regulatory elements is found in the transcribed gene. Mutation of all Runx2 sites results in inactive chromatin and loss of the DNase hypersensitive sites I and II [564]. (B) Three-dimensional organization of the OC gene facilitated by the association of Runx with the nuclear matrix scaffold that facilitates interaction between the proximal RNA polymerase complex interactions with the vitamin D receptor complex. This mechanism allows for physiologic upregulation of osteocalcin by vitamin D coordinated with basal transcriptional levels. Runx2 association with the nuclear matrix supports gene promoter conformation and regulatory element cross-talk.

3. SUBNUCLEAR TARGETING OF TRANSCRIPTIONAL REGULATORY COMPLEXES

Transcriptional control is provided by a third level of nuclear architecture, the NM scaffold. The anastomosing network of fibers and filaments that constitute the nuclear matrix accommodates structural modifications of the nucleus associated with proliferation, differentiation, and changes necessary to sustain phenotypic requirements of specialized cells [565, 566]. Many functional activities of the NM support the structural basis and necessary modifications in chromatin for accessibility of transactivation factors to regulate gene expression. As the intricacies of gene organization and regulation are elucidated, the implications of a fundamental biological paradox become strikingly evident. With a limited representation of gene-specific regulatory elements and a low abundance of cognate transactivation factors, how can sequence-specific interactions occur to support a threshold for the initiation of transcription within nuclei of intact cells? Viewed from a quantitative perspective, the *in vivo* regulatory challenge is to account for the formation of functional transcription initiation complexes. A number of NM-associated proteins have identified functions in osteoblasts; the most significant of which is Runx2 (NMP2), carrying a unique nuclear matrix targeting signal (NMTS) conserved only in Runx factors [542]. Others include YY1 (NMP1), C1Z (NMP4), and an uncharacterized protein interacting with the collagen gene (NMP3) [567]. Another class of NM proteins, the special AT-rich sequence-binding proteins SATB1 and SATB2, are DNA helix destabilizing factors that are primarily localized at the base of large loop domains, designated matrix attachment regions. They are important for their functions in recruiting chromatin remodeling factors [568] (Figure 4-7A). SATB1 regulates gene expression in several cell phenotypes (e.g., hematopoietic cells, B cells, and differentiating neurons). SATB2, identified as the cleft palate gene [569], has an essential role in jaw development [570, 571]. Notably, SATB2 acts in conjunction with other regulatory factors organizing complexes to regulate transcription. During embryonic development, SATB2 interacts with Runx2 and ATF4 to promote cooperative binding to Runx2 target genes and thereby promote osteoblast differentiation [571].

The Runx2 transcription factor serves as a paradigm for the obligatory relationships between nuclear structure and the control of skeletogenesis. An essential feature of Runx proteins is their targeting to subnuclear domains through a specific sequence in the C-terminus, designated the nuclear matrix targeting signal (NMTS) [572, 573] (Figure 4-10). The *in vivo* relevance of subnuclear targeting of Runx2 for biological activity was shown by several studies. Deletion mutants or point mutations that prevent or decrease the association of Runx factors with the nuclear matrix scaffold, result in compromised expression of target genes in hematopoietic cells by Runx1 and in osteoblasts by Runx2 [157, 574]. The phenotype of Runx2 ΔC (lacking the C-terminus NMTS domain) mouse is a complete absence of a mineralized skeleton and lethality before birth, analogous to the Runx2 null mouse models [157] (Figures 4-10A–4-10C). Rescue of the phenotype requires both Runx2 and a BMP–Smad constituent with the finding that the Smad-interacting domain overlaps the NMTS signal [93]. In studies of breast and prostate cancer cells metastatic to bone, Runx2 is highly expressed [575–578]. Significantly, Runx2 upregulates genes related to vascularization (VEGF), tissue invasion (MMP9 and MMP13), and adhesion (osteopontin and bone sialoprotein) [578]. These genes are linked to metastasis. The metastatic breast cancer cell line MDA-MB-231, expressing either a Runx2 dominant-negative or a subnuclear targeting-deficient point mutant protein, suppressed expression of Runx2 target genes. These modified cell lines exhibited suppressed tumor growth and prevented osteolytic disease when directly injected into the bone microenvironment [579, 580]. Thus, disrupting the subnuclear association of Runx2 through the mutation of the NMTS blocks expression of Runx2-dependent target genes that promote bone metastasis and osteolysis. In conclusion, the subnuclear targeting of Runx factors is a unique property of Runx factors that contributes to their tissue-specific and master gene regulatory activities.

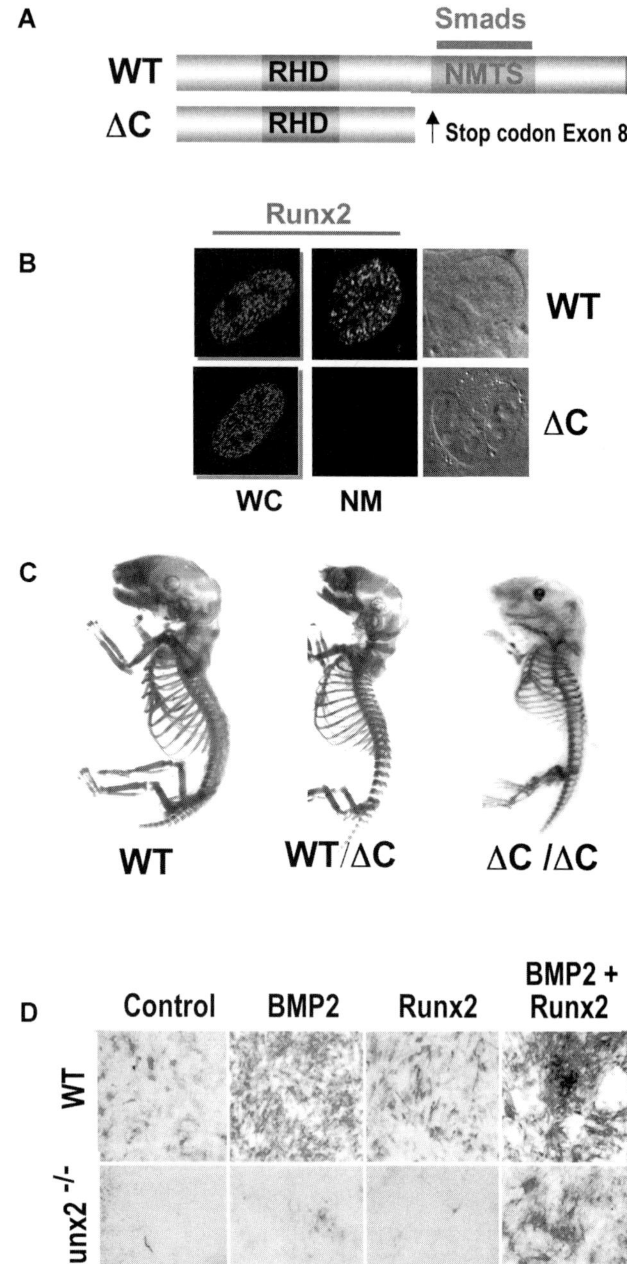

FIGURE 4-10 Knock-in mutation of the Runx2 C-terminal domain results in a lethal phenotype. (A) Schematic of stop codon mutation to eliminate translation of the last exon 8 encoding the NMTS domain, which overlaps the Smad interacting domain [93]. Mice died just before birth [157]. (B) Cells isolated from the calvarium show that wild-type and ΔC Runx2 proteins are synthesized, enter the nucleus, and bind to DNA; however, the ΔC Runx2 cannot associate with the nuclear matrix scaffold. (C) Phenotype of the heterozygote mouse with missing clavicle and delayed intramembranous bone formation and the homozygote mouse completely devoid of mineralized tissue. (D) *Ex vivo* rescue of Runx2 null cells isolated from the cranium of ΔC Runx2 mice only occurs in response to a combination of both BMP2 and Runx2 repletion [96].

An important property of Runx factors is that they are scaffolding proteins interacting with numerous classes of co-regulatory proteins for rendering Runx transcription factors competent to function as master switches of cell differentiation [581, 582] (Figure 4-11). Runx2 co-regulatory protein complexes are visualized as punctuate foci in the nucleus. Multiple lines of evidence support the essential requirement for Runx2 to be targeted to specific domain interaction with co-regulatory proteins for regulating osteogenesis. Although the mediators of Src, BMP, and TGF-β signaling are competent to interact with a Runx2 point mutant protein (Y428A) that cannot associate with the nuclear matrix, their signals will not be transduced to target genes [92, 583]. For example, the nuclear import of YAP, the mediator of Src signaling, and Smad transducers of BMP/TGF-β signaling, enter the nucleus in response to Src and BMP/TGF-β signal, but there is a stringent requirement for fidelity of Runx2 location in subnuclear domains for recruitment of these signaling proteins to Runx2 transcriptionally active foci for execution of the signal (Figure 4-10D).

Among the Runx2 co-regulatory interacting proteins [581, 584] are chromatin remodeling factors that function to alter nucleosomal organization within the confines of nuclear architecture (Figure 4-11). Runx interacting factors that have associated histone acety-

lase transferase (HAT) activity include the coactivator p300/CBP, which functions as a transcriptional adaptor [585], and the MYST family of HATS, MOZ, and MORF [586]. Runx factors also have the ability to repress gene transcription through interaction with several HDAC enzymes. HDAC3 binds to the Runx2 NH2-terminus, whereas the HDAC6 interacting domain is in the C-terminus of Runx2. Functional activities of several HDACs on osteoblast genes are well documented [587, 588]. The significance of HDAC4 repressor activity of Runx2 during endochondral bone formation was demonstrated by the phenotype of the HDAC4 null mouse involving disruption of hypertrophic chondrocyte maturation. Runx2 is a potent activator of VEGFs, which are required for vascular invasion for endochondral bone formation [589]. Thus, Runx2 interacts with numerous chromatin remodeling factors competent to modify chromatin for regulation of gene transcription in a Runx2-dependent manner and provide physiologic levels of target gene expression.

Distinct protein modules in Runx2 are targets for modification in Runx2 transcriptional control of osteoblast differentiation. Many transcription factors can interfere with Runx2 DNA binding to target genes and therefore inhibit bone formation (Figure 4-11). The PPXY motif in Runx2 interacts with WW domain proteins. In response to TGF-β, Smurf1, a WW domain

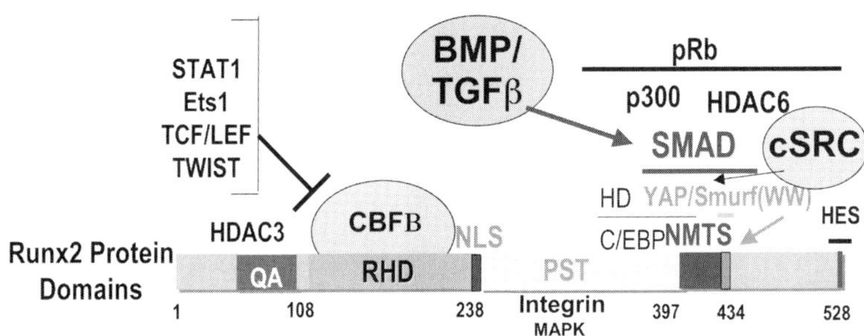

Runx2 Interacting Proteins
- DNA binding partner CBFβ
- Chromatin related factors (HAT(p300), HDAC 1,3,4,6)
- Transcription factors (homeodomain proteins, C/EBP,Twist, TCF, Nrf2)
- Signal transducers (Smads, YAP, pRB)
- WW domain proteins (YAP, TAZ)
- E3Ubiquitin ligases (Smurf1, Schnurri-WWP1)
- Co-repressors (mSin3a, STAT1, YAP)
- Co-activators (TAZ, Hes1, pRB)

FIGURE 4-11   Runx2 interacting proteins. Illustrated are examples of proteins that form complexes with Runx2 altering Runx2 transcriptional activity, providing mechanisms for positive and negative Runx2-mediated gene expression as a cell progresses through stages of differentiation or in response to physiological signals that affect bone metabolism. Indicated are numerous classes of interacting proteins that function as positive and negative regulators of Runx2 activity. PST, proline–serine–threonine; RHD, runt homology domain.

protein, targets Runx2 for proteasomal degradation. Src signaling negatively regulates osteoblast differentiation at the PPXY motif by a mechanism in which YAP, a cytoplasmic shuttling protein, in response to Src nonreceptor kinase activity complexes with Runx2 to inhibit expression of Runx-regulated genes [583]. Yet another WW domain protein, TAZ, interacts with Runx2 to increase its activation potential on target genes [590, 591]. Another Runx2 protein sequence, the C-terminal VWRPY motif, interacts with TLE/Groucho (a co-repressor), a nuclear matrix protein, and the TLE dominant-negative protein Grg5, which is a co-activator [592]. HES-1, a basic HLH factor (and target of Notch signaling) that associates with the nuclear matrix as does Groucho, can antagonize the Runx2–Groucho interaction and promote the transactivation function of Runx2 [593]. These distinct protein–protein interactions provide an exquisite example of the modification of Runx2 activities dependent on the interacting co-regulatory protein for facilitating its function as a master regulatory gene throughout the course of osteogenesis. Table 4-2 summarizes the many functions of Runx2 that are accommodated by the organization of multimeric complexes in Runx subnuclear domains. Such dynamic changes of co-regulatory protein interactions in nuclear microenvironments are consistent with the concept that the nuclear matrix is functionally involved in gene localization and in the concentration of subnuclear localization of regulatory factors. Thus, the 150–300 punctate Runx2 foci observed in osteoblasts (Figure 4-7E) represent a spectrum of multimeric functional complexes of Runx2 with different co-regulatory proteins on gene promoters. This discovery

of a Runx family–specific functional protein module that targets Runx complexes to sites within the nucleus is the basis of an important concept for tissue-specific control of gene expression. A significant conclusion from all these studies is that a key property of Runx2 protein is the ability to integrate within nuclear microenvironments the signaling of numerous pathways that contribute to the control of osteoblast differentiation (Figure 4-12 summarizes those known to date).

VI. CONCLUSION

This chapter presented the cell biology of osteoblasts within the context of our current understanding of the regulatory controls operative in promoting osteoblast differentiation. We have attempted to address how physiologic parameters of gene expression are integrated to support the requirements of bone development and functional integrity of the tissue. During osteoblast phenotype development and bone formation, stages of maturation are defined by levels of expression of subsets of osteoblast genes. A cohort of tissue-specific, developmental, steroid hormone and growth factor–related transcription factor complexes impinge on gene transcription, providing a complex and integrated series of regulatory signals for the selective activation and repression of genes related to activity. We presented a growing body of evidence for the molecular mechanisms that contribute to the effects of a hormone or growth factor on expression of a specific gene, which are related to the osteoblast phenotype (i.e., the stage of cellular maturation). These selective effects are the result of proteins regulatory factors and

TABLE 4-2   Runx2 Identified Functions

Bone formation

Osteoblast differentiation

Lineage determinant

Neural crest and craniofacial development

Regulates chondrogenesis

Tissue specification

Tooth morphogenesis

Bone turnover through RANKL

Mediates BMP2 osteogenic effects

Promotes Wnt/β-catenin switch from chondrogenesis to osteogenesis

Integrates ECM and growth factor signaling responses

Autoregulates its own transcription

Responsive to hormones

Chromatin remodeling for gene activation and regulation

Growth regulators

Cooperative oncogene

Cell migration and invasion

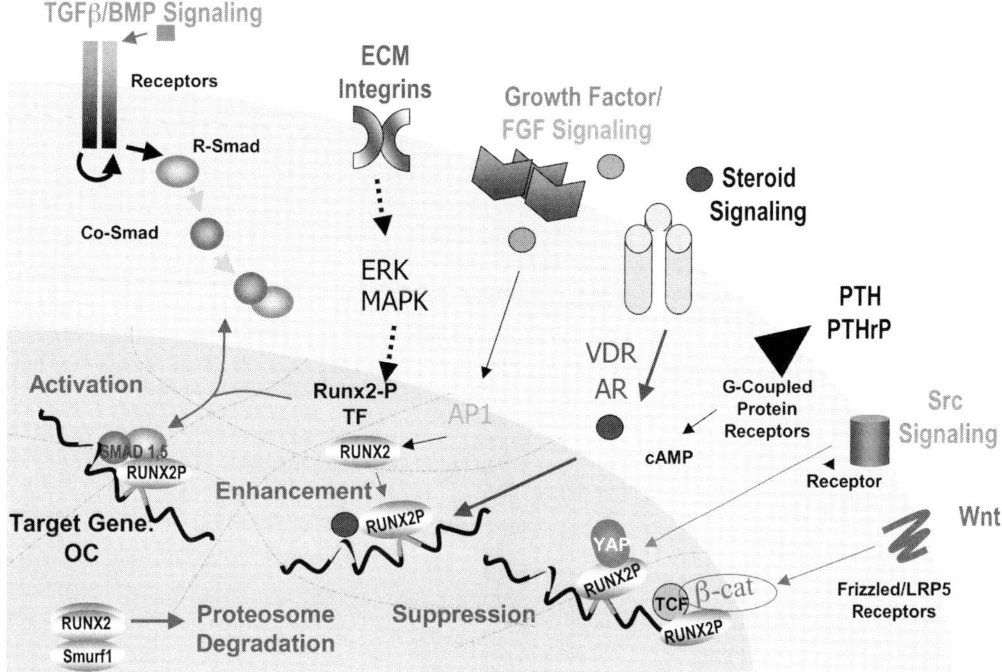

FIGURE 4-12   Runx2 is responsive to multiple osteogenic signaling pathways. Signaling pathways that transduce their signal to genes through interactions of the intracellular receptor or shuttling protein with Runx2 in transcriptionally active domains. Runx2 can be phosphorylated in response to integrin and growth factor signaling and complexes with intracellular transducers of signaling pathways (e.g., TGF-β/BMP/Smads and the YAP mediator of Src signaling). The complexes are targeted to gene promoters in Runx2 subnuclear domains. Some factors, such as TGF-β signaling, can induce proteosomal degradation of Runx2 through Smurf1 interaction with Runx2.

transcriptional complexes associated with a gene that contributes to its expression. Thus, clinical consideration for treatment and therapeutic regimens can be approached with greater knowledge of the consequential effects at the level of gene-regulating responses.

ACKNOWLEDGMENTS

We gratefully appreciate preparation of the manuscript by Judy Rask and thank colleagues Janet Stein, André van Wijnen, Amjad Javed, Kaleem Zaidi, Mohammad Hassan, Jitesh Pratap, and Tripti Gaur for helpful discussions and also members of our research group. The National Institutes of Health grants supporting the research program related to this chapter include AR45688, AR45689, AR39588, and DE12528. The contents of this chapter are solely the responsibility of the authors and do not necessarily represent the official views of the National Institutes of Health.

REFERENCES

1. B. K. Hall and T. Miyake, All for one and one for all: Condensations and the initiation of skeletal development. *Bioessays* **22**, 138–147 (2000).

2. S. A. Brugmann, M. D. Tapadia, and J. A. Helms, The molecular origins of species-specific facial pattern. *Curr Top Dev Biol* **73**, 1–42 (2006).

3. I. Thesleff, The genetic basis of tooth development and dental defects. *Am J Med Genet A* **140**, 2530–2535 (2006).

4. N. Vargesson, K. Patel, J. Lewis, and C. Tickle, Expression patterns of Notch1, Serrate1, Serrate2, and Delta1 in tissues of the developing chick limb. *Mech Dev* **77**, 197–199 (1998).

5. M. Nobta, T. Tsukazaki, Y. Shibata, C. Xin, T. Moriishi, S. Sakano, H. Shindo, and A. Yamaguchi, Critical regulation of bone morphogenetic protein-induced osteoblastic differentiation by Delta1/Jagged1-activated Notch1 signaling. *J Biol Chem* **280**, 15842–15848 (2005).

6. V. Deregowski, E. Gazzerro, L. Priest, S. Rydziel, and E. Canalis, Notch 1 overexpression inhibits osteoblastogenesis by suppressing Wnt/beta-catenin but not bone morphogenetic protein signaling. *J Biol Chem* **281**, 6203–6210 (2006).

7. N. Zamurovic, D. Cappellen, D. Rohner, and M. Susa, Coordinated activation of Notch Wnt and TGF-beta signaling pathways in BMP-2 induced osteogenesis: Notch target gene Hey1 inhibits mineralization and Runx2 transcriptional activity. *J Biol Chem* **279**, 37704–37715 (2004).

8. B. D. Harfe, P. J. Scherz, S. Nissim, H. Tian, A. P. McMahon, and C. J. Tabin, Evidence for an expansion-based temporal Shh gradient in specifying vertebrate digit identities. *Cell* **118**, 517–528 (2004).

9. S. Spinella-Jaegle, G. Rawadi, S. Kawai, S. Gallea, C. Faucheu, P. Mollat, B. Courtois, B. Bergaud, V. Ramez, A. M. Blanchet, G. Adelmant, R. Baron, and S. Roman-Roman, Sonic hedgehog increases the commitment of pluripotent mesenchymal cells into the osteoblastic lineage and abolishes adipocytic differentiation. *J Cell Sci* **114**, 2085–2094 (2001).

10. T. Yuasa, H. Kataoka, N. Kinto, M. Iwamoto, M. Enomoto-Iwamoto, S. Iemura, N. Ueno, Y. Shibata, H. Kurosawa, and A. Yamaguchi, Sonic hedgehog is involved in osteoblast differentiation by cooperating with BMP-2. *J Cell Physiol* **193**, 225–232 (2002).

11. M. Zhao, M. Qiao, S. E. Harris, D. Chen, B. O. Oyajobi, and G. R. Mundy, The zinc finger transcription factor Gli2 mediates bone morphogenetic protein 2 expression in osteoblasts in response to hedgehog signaling. *Mol Cell Biol* **26**, 6197–6208 (2006).

12. D. Miao, H. Liu, P. Plut, M. Niu, R. Huo, D. Goltzman, and J. E. Henderson, Impaired endochondral bone development and osteopenia in Gli2-deficient mice. *Exp Cell Res* **294**, 210–222 (2004).

13. M. Barna, P. P. Pandolfi, and L. Niswander, Gli3 and Plzf cooperate in proximal limb patterning at early stages of limb development. *Nature* **436**, 277–281 (2005).

14. M. J. Depew, T. Lufkin, and J. L. Rubenstein, Specification of jaw subdivisions by Dlx genes. *Science* **298**, 381–385 (2002).

15. R. F. Robledo, L. Rajan, X. Li, and T. Lufkin, The Dlx5 and Dlx6 homeobox genes are essential for craniofacial axial and appendicular skeletal development. *Genes Dev* **16**, 1089–1101 (2002).

16. S. O. Akintoye, T. Lam, S. Shi, J. Brahim, M. T. Collins, and P. G. Robey, Skeletal site-specific characterization of orofacial and iliac crest human bone marrow stromal cells in same individuals. *Bone* **38**, 758–768 (2006).

17. M. Ishii, A. E. Merrill, Y. S. Chan, I. Gitelman, D. P. Rice, H. M. Sucov, and R. E. Maxson Jr., Msx2 and Twist cooperatively control the development of the neural crest-derived skeletogenic mesenchyme of the murine skull vault. *Development* **130**, 6131–6142 (2003).

18. A. Compagni, M. Logan, R. Klein, and R. H. Adams, Control of skeletal patterning by ephrinb1–Ephb interactions. *Dev Cell* **5**, 217–230 (2003).

19. A. Davy, J. Aubin, and P. Soriano, Ephrin-B1 forward and reverse signaling are required during mouse development. *Genes Dev* **18**, 572–583 (2004).

20. A. E. Merrill, E. G. Bochukova, S. M. Brugger, M. Ishii, D. T. Pilz, S. A. Wall, K. M. Lyons, A. O. Wilkie, and R. E. Maxson Jr., Cell mixing at a neural crest–mesoderm boundary and deficient ephrin–Eph signaling in the pathogenesis of craniosynostosis. *Hum Mol Genet* **15**, 1319–1328 (2006).

21. J. Zakany, M. Kmita, and D. Duboule, A dual role for Hox genes in limb anterior–posterior asymmetry. *Science* **304**, 1669–1672 (2004).

22. F. R. Goodman, Limb malformations and the human HOX genes. *Am J Med Genet* **112**, 256–265 (2002).

23. M. J. Depew, C. A. Simpson, M. Morasso, and J. L. Rubenstein, Reassessing the Dlx code: The genetic regulation of branchial arch skeletal pattern and development. *J Anat* **207**, 501–561 (2005).

24. R. P. Gersch, F. Lombardo, S. C. McGovern, and M. Hadjiargyrou, Reactivation of Hox gene expression during bone regeneration. *J Orthop Res* **23**, 882–890 (2005).

25. B. S. Yoon and K. M. Lyons, Multiple functions of BMPs in chondrogenesis. *J Cell Biochem* **93**, 93–103 (2004).

26. P. J. Marie, Fibroblast growth factor signaling controlling osteoblast differentiation. *Gene* **316**, 23–32 (2003).

27. P. ten Dijke, J. Fu, P. Schaap, and B. A. Roelen, Signal transduction of bone morphogenetic proteins in osteoblast differentiation. *J Bone Joint Surg Am* **85-A**(Suppl. 3), 34–38 (2003).

28. J. J. Westendorf, R. A. Kahler, and T. M. Schroeder, Wnt signaling in osteoblasts and bone diseases. *Gene* **341**, 19–39 (2004).

29. X. Sun, F. V. Mariani, and G. R. Martin, Functions of FGF signalling from the apical ectodermal ridge in limb development. *Nature* **418**, 501–508 (2002).

30. Y. X. Zhou, X. Xu, L. Chen, C. Li, S. G. Brodie, and C. X. Deng, A Pro250Arg substitution in mouse Fgfr1 causes increased expression of Cbfa1 and premature fusion of calvarial sutures. *Hum Mol Genet* **9**, 2001–2008 (2000).

31. M. K. Hajihosseini, M. D. Lalioti, S. Arthaud, H. R. Burgar, J. M. Brown, S. R. Twigg, A. O. Wilkie, and J. K. Heath, Skeletal development is regulated by fibroblast growth factor receptor 1 signalling dynamics. *Development* **131**, 325–335 (2004).

32. P. Rutland, L. J. Pulleyn, W. Reardon, M. Baraitser, R. Hayward, B. Jones, S. Malcolm, R. M. Winter, M. Oldridge, S. F. Slaney, M. D. Poole, and A. O. Wilkie, Identical mutations in the FGFR2 gene cause both Pfeiffer and Crouzon syndrome phenotypes. *Nat Genet* **9**, 173–176.

33. V. P. Eswarakumar, M. C. Horowitz, R. Locklin, G. M. Morriss-Kay, and P. Lonai, A gain-of-function mutation of Fgfr2c demonstrates the roles of this receptor variant in osteogenesis. *Proc Natl Acad Sci USA* **101**, 12555–12560 (2004).

34. T. Baroni, P. Carinci, C. Lilli, C. Bellucci, M. C. Aisa, L. Scapoli, S. Volinia, F. Carinci, F. Pezzetti, M. Calvitti, A. Farina, C. Conte, and M. Bodo, P253R fibroblast growth factor receptor-2 mutation induces RUNX2 transcript variants and calvarial osteoblast differentiation. *J Cell Physiol* **202**, 524–535 (2005).

35. Y. Tanimoto, M. Yokozeki, K. Hiura, K. Matsumoto, H. Nakanishi, T. Matsumoto, P. J. Marie, and K. Moriyama, A soluble form of fibroblast growth factor receptor 2 (FGFR2) with S252W mutation acts as an efficient inhibitor for the enhanced osteoblastic differentiation caused by FGFR2 activation in Apert syndrome. *J Biol Chem* **279**, 45926–45934 (2004).

36. D. M. Ornitz and P. J. Marie, FGF signaling pathways in endochondral and intramembranous bone development and human genetic disease. *Genes Dev* **16**, 1446–1465 (2002).

37. N. Amizuka, D. Davidson, H. Liu, G. Valverde-Franco, S. Chai, T. Maeda, H. Ozawa, V. Hammond, D. M. Ornitz, D. Goltzman, and J. E. Henderson, Signalling by fibroblast growth factor receptor 3 and parathyroid hormone-related peptide coordinate cartilage and bone development. *Bone* **34**, 13–25 (2004).

38. C. Deng, A. Wynshaw-Boris, F. Zhou, A. Kuo, and P. Leder, Fibroblast growth factor receptor 3 is a negative regulator of bone growth. *Cell* **84**, 911–921 (1996).

39. G. Valverde-Franco, H. Liu, D. Davidson, S. Chai, H. Valderrama-Carvajal, D. Goltzman, D. M. Ornitz, and J. E. Henderson, Defective bone mineralization and osteopenia in young adult FGFR3-/- mice. *Hum Mol Genet* **13**, 271–284 (2004).

40. A. L. Jacob, C. Smith, J. Partanen, and D. M. Ornitz, Fibroblast growth factor receptor 1 signaling in the osteo-chondrogenic cell lineage regulates sequential steps of osteoblast maturation. *Dev Biol* **296**, 315–328 (2006).

41. A. M. Boulet, A. M. Moon, B. R. Arenkiel, and M. R. Capecchi, The roles of Fgf4 and Fgf8 in limb bud initiation and outgrowth. *Dev Biol* **273**, 361–372 (2004).

42. A. Abzhanov and C. J. Tabin, Shh and Fgf8 act synergistically to drive cartilage outgrowth during cranial development. *Dev Biol* **273**, 134–148 (2004).

43. E. Laufer, C. E. Nelson, R. L. Johnson, B. A. Morgan, and C. Tabin, Sonic hedgehog and Fgf-4 act through a signaling cascade and feedback loop to integrate growth and patterning of the developing limb bud. *Cell* **79**, 993–1003 (1994).

44. Z. Liu, K. J. Lavine, I. H. Hung, and D. M. Ornitz, FGF18 is required for early chondrocyte proliferation hypertrophy and vascular invasion of the growth plate. *Dev Biol* **302,** 80–91 (2006).

45. N. Ohbayashi, M. Shibayama, Y. Kurotaki, M. Imanishi, T. Fujimori, N. Itoh, and S. Takada, FGF18 is required for normal cell proliferation and differentiation during osteogenesis and chondrogenesis. *Genes Dev* **16,** 870–879 (2002).

46. R. Sullivan and M. Klagsbrun, Purification of cartilage-derived growth factor by heparin affinity chromatography. *J Biol Chem* **260,** 2399–2403 (1985).

47. M. M. Hurley, C. Abreu, G. Gronowicz, H. Kawaguchi, and J. Lorenzo, Expression and regulation of basic fibroblast growth factor mRNA levels in mouse osteoblastic MC3T3-E1 cells. *J Biol Chem* **269,** 9392–9396 (1994).

48. H. J. Kim, D. P. Rice, P. J. Kettunen, and I. Thesleff, FGF-, BMP- and Shh-mediated signalling pathways in the regulation of cranial suture morphogenesis and calvarial bone development. *Development* **125,** 1241–1251 (1998).

49. C. Hartmann and C. J. Tabin, Dual roles of Wnt signaling during chondrogenesis in the chicken limb. *Development* **127,** 3141–3159 (2000).

50. A. B. Zimrin, M. S. Pepper, G. A. McMahon, F. Nguyen, R. Montesano, and T. Maciag, An antisense oligonucleotide to the notch ligand jagged enhances fibroblast growth factor-induced angiogenesis *in vitro*. *J Biol Chem* **271,** 32499–32502 (1996).

51. S. Artavanis-Tsakonas, K. Matsuno, and M. E. Fortini, Notch signaling. *Science* **268,** 225–232 (1995).

52. D. P. Rice, T. Aberg, Y. Chan, Z. Tang, P. J. Kettunen, L. Pakarinen, R. E. Maxson, and I. Thesleff, Integration of FGF and TWIST in calvarial bone and suture development. *Development* **127,** 1845–1855 (2000).

53. H. Tanaka, H. Ogasa, J. Barnes, and C. T. Liang, Actions of bFGF on mitogenic activity and lineage expression in rat osteoprogenitor cells: Effect of age. *Mol Cell Endocrinol* **150,** 1–10 (1999).

54. F. Debiais, G. Lefevre, J. Lemonnier, S. Le Mee, F. Lasmoles, F. Mascarelli, and P. J. Marie, Fibroblast growth factor-2 induces osteoblast survival through a phosphatidylinositol 3-kinase-dependent, -beta-catenin-independent signaling pathway. *Exp Cell Res* **297,** 235–246 (2004).

55. L. Xiao, P. Liu, T. Sobue, A. Lichtler, J. D. Coffin, and M. M. Hurley, Effect of overexpressing fibroblast growth factor 2 protein isoforms in osteoblastic ROS 17/2.8 cells. *J Cell Biochem* **89,** 1291–1301 (2003).

56. A. Montero, Y. Okada, M. Tomita, M. Ito, H. Tsurukami, T. Nakamura, T. Doetschman, J. D. Coffin, and M. M. Hurley, Disruption of the fibroblast growth factor-2 gene results in decreased bone mass and bone formation. *J Clin Invest* **105,** 1085–1093.

57. D. Chen, M. Zhao, and G. R. Mundy, Bone morphogenetic proteins. *Growth Factors* **22,** 233–241 (2004).

58. H. M. Ryoo, M. H. Lee, and Y. J. Kim, Critical molecular switches involved in BMP-2-induced osteogenic differentiation of mesenchymal cells. *Gene* **366,** 51–57 (2006).

59. D. M. Kingsley, A. E. Bland, J. M. Grubber, P. C. Marker, L. B. Russell, N. G. Copeland, and N. A. Jenkins, The mouse short ear skeletal morphogenesis locus is associated with defects in a bone morphogenetic member of the TGFb superfamily. *Cell* **71,** 399–410 (1992).

60. R. Derynck and Y. E. Zhang, Smad-dependent and Smad-independent pathways in TGF-beta family signalling. *Nature* **425,** 577–584 (2003).

61. T. T. Phan, I. J. Lim, O. Aalami, F. Lorget, A. Khoo, E. K. Tan, A. Mukhopadhyay, and M. T. Longaker, Smad3 signalling plays an important role in keloid pathogenesis via epithelial–mesenchymal interactions. *J Pathol* **207,** 232–242 (2005).

62. J. Aubin, A. Davy, and P. Soriano, *In vivo* convergence of BMP and MAPK signaling pathways: Impact of differential Smad1 phosphorylation on development and homeostasis. *Genes Dev* **18,** 1482–1494 (2004).

63. G. Han, A. G. Li, Y. Y. Liang, P. Owens, W. He, S. Lu, Y. Yoshimatsu, D. Wang, D. P. Ten, X. Lin, and X. J. Wang, Smad7-induced beta-catenin degradation alters epidermal appendage development. *Dev Cell* **11,** 301–312 (2006).

64. M. Zhao, S. E. Harris, D. Horn, Z. Geng, R. Nishimura, G. R. Mundy, and D. Chen, Bone morphogenetic protein receptor signaling is necessary for normal murine postnatal bone formation. *J Cell Biol* **157,** 1049–1060 (2002).

65. S. E. Yi, A. Daluiski, R. Pederson, V. Rosen, and K. M. Lyons, The type I BMP receptor BMPRIB is required for chondrogenesis in the mouse limb. *Development* **127,** 621–630 (2000).

66. R. B. Rountree, M. Schoor, H. Chen, M. E. Marks, V. Harley, Y. Mishina, and D. M. Kingsley, BMP receptor signaling is required for postnatal maintenance of articular cartilage. *Plos Biol* **2,** E355 (2004).

67. X. Yang, L. Chen, X. Xu, C. Li, C. Huang, and C. X. Deng, TGF-beta/Smad3 signals repress chondrocyte hypertrophic differentiation and are required for maintaining articular cartilage. *J Cell Biol* **153,** 35–46 (2001).

68. K. Ahn, Y. Mishina, M. C. Hanks, R. R. Behringer, and E. B. Crenshaw III, BMPR-IA signaling is required for the formation of the apical ectodermal ridge and dorsal–ventral patterning of the limb. *Development* **128,** 4449–4461 (2001).

69. C. D. Grimsrud, P. R. Romano, M. D'Souza, J. E. Puzas, P. R. Reynolds, R. N. Rosier, and R. J. O'Keefe, BMP-6 is an autocrine stimulator of chondrocyte differentiation. *J Bone Miner Res* **14,** 475–482 (1999).

70. M. S. Friedman, M. W. Long, and K. D. Hankenson, Osteogenic differentiation of human mesenchymal stem cells is regulated by bone morphogenetic protein-6. *J Cell Biochem* **98,** 538–554 (2006).

71. E. M. Shore, M. Xu, G. J. Feldman, D. A. Fenstermacher, M. A. Brown, and F. S. Kaplan, A recurrent mutation in the BMP type I receptor ACVR1 causes inherited and sporadic fibrodysplasia ossificans progressiva. *Nat Genet* **38,** 525–527 (2006).

72. F. S. Kaplan, J. Fiori, L. S. De La Pena, J. Ahn, P. C. Billings, and E. M. Shore, Dysregulation of the BMP-4 signaling pathway in fibrodysplasia ossificans progressiva. *Ann N Y Acad Sci* **1068,** 54–65 (2006).

73. L. Kan, M. Hu, W. A. Gomes, and J. A. Kessler, Transgenic mice overexpressing BMP4 develop a fibrodysplasia ossificans progressiva (FOP)-like phenotype. *Am J Pathol* **165,** 1107–1115 (2004).

74. V. Rosen, BMP and BMP inhibitors in bone. *Ann N Y Acad Sci* **1068,** 19–25 (2006).

75. E. Abe, Function of BMPs and BMP antagonists in adult bone. *Ann N Y Acad Sci* **1068,** 41–53 (2006).

76. C. K. Wang, M. Omi, D. Ferrari, H. C. Cheng, G. Lizarraga, H. J. Chin, W. B. Upholt, C. N. Dealy, and R. A. Kosher, Function of BMPs in the apical ectoderm of the developing mouse limb. *Dev Biol* **269,** 109–122 (2004).

77. C. W. Archer, G. P. Dowthwaite, and P. Francis-West, Development of synovial joints. *Birth Defects Res C Embryo Today* **69,** 144–155 (2003).

78. Y. Yoshida, S. Tanaka, H. Umemori, O. Minowa, M. Usui, N. Ikematsu, E. Hosoda, T. Imamura, J. Kuno, T. Yamashita, K. Miyazono, M. Noda, T. Noda, and T. Yamamoto, Negative

regulation of BMP/Smad signaling by Tob in osteoblasts. *Cell* **103,** 1085–1097 (2000).

79. E. Balint, D. Lapointe, H. Drissi, C. van der Meijden, D. W. Young, A. J. van Wijnen, J. L. Stein, G. S. Stein, and J. B. Lian, Phenotype discovery by gene expression profiling: Mapping of biological processes linked to BMP-2-mediated osteoblast differentiation. *J Cell Biochem* **89,** 401–426 (2003).

80. C. Banerjee, A. Javed, J.-Y. Choi, J. Green, V. Rosen, A. J. van Wijnen, J. L. Stein, J. B. Lian, and G. S. Stein, Differential regulation of the two principal Runx2/Cbfa1 N-terminal isoforms in response to bone morphogenetic protein-2 during development of the osteoblast phenotype. *Endocrinology* **142,** 4026–4039 (2001).

81. S. E. Harris, D. Guo, M. A. Harris, A. Krishnaswamy, and A. Lichtler, Transcriptional regulation of BMP-2 activated genes in osteoblasts using gene expression microarray analysis: Role of Dlx2 and Dlx5 transcription factors. *Frontiers Biosci* **8,** S1249–S1265 (2003).

82. K. S. Lee, H. J. Kim, Q. L. Li, X. Z. Chi, C. Ueta, T. Komori, J. M. Wozney, E. G. Kim, J. Y. Choi, H. M. Ryoo, and S. C. Bae, Runx2 is a common target of transforming growth factor beta1 and bone morphogenetic protein 2 and cooperation between Runx2 and Smad5 induces osteoblast-specific gene expression in the pluripotent mesenchymal precursor cell line C2C12. *Mol Cell Biol* **20,** 8783–8792 (2000).

83. J. Chimal-Monroy, J. Rodriguez-Leon, J. A. Montero, Y. Ganan, D. Macias, R. Merino, and J. M. Hurle, Analysis of the molecular cascade responsible for mesodermal limb chondrogenesis: Sox genes and BMP signaling. *Dev Biol* **257,** 292–301 (2003).

84. Y. Mori-Akiyama, H. Akiyama, D. H. Rowitch, and B. de Crombrugghe, Sox9 is required for determination of the chondrogenic cell lineage in the cranial neural crest. *Proc Natl Acad Sci USA* **100,** 9360–9365 (2003).

85. I. Semba, K. Nonaka, I. Takahashi, K. Takahashi, R. Dashner, L. Shum, G. H. Nuckolls, and H. C. Slavkin, Positionally-dependent chondrogenesis induced by BMP4 is co-regulated by Sox9 and Msx2. *Dev Dyn* **217,** 401–414 (2000).

86. Z. Zhang, Y. Song, X. Zhang, J. Tang, J. Chen, and Y. Chen, Msx1/Bmp4 genetic pathway regulates mammalian alveolar bone formation via induction of Dlx5 and Cbfa1. *Mech Dev* **120,** 1469–1479 (2003).

87. K. Nakashima, X. Zhou, G. Kunkel, Z. Zhang, J. M. Deng, R. R. Behringer, and B. de Crombrugghe, The novel zinc finger-containing transcription factor osterix is required for osteoblast differentiation and bone formation. *Cell* **108,** 17–29 (2002).

88. C. Banerjee, L. R. McCabe, J.-Y. Choi, S. W. Hiebert, J. L. Stein, G. S. Stein, and J. B. Lian, Runt homology domain proteins in osteoblast differentiation: AML-3/CBFA1 is a major component of a bone specific complex. *J Cell Biochem* **66,** 1–8 (1997).

89. P. Ducy, R. Zhang, V. Geoffroy, A. L. Ridall, and G. Karsenty, Osf2/Cbfa1: A transcriptional activator of osteoblast differentiation. *Cell* **89,** 747–754 (1997).

90. N. Smith, Y. Dong, J. Pratap, J. B. Lian, P. Kingsley, A. J. van Wijnen, J. L. Stein, E. M. Schwarz, R. J. O'Keefe, G. S. Stein, and M. H. Drissi, Overlapping expression of Runx1(Cbfa2) and Runx2(Cbfa1) transcription factors supports cooperative induction of skeletal development. *J Cell Physiol* **203,** 133–143 (2005).

91. C. J. Lengner, H. Drissi, J.-Y. Choi, A. J. van Wijnen, J. L. Stein, G. S. Stein, and J. B. Lian, Activation of the bone related *Runx2/Cbfa1* promoter in mesenchymal condensations and developing chondrocytes of the axial skeleton. *Mech Dev* **114,** 167–170 (2002).

92. S. K. Zaidi, A. J. Sullivan, A. J. van Wijnen, J. L. Stein, G. S. Stein, and J. B. Lian, Integration of Runx and Smad regulatory signals at transcriptionally active subnuclear sites. *Proc Natl Acad Sci USA* **99,** 8048–8053 (2002).

93. F. Afzal, J. Pratap, K. Ito, Y. Ito, J. L. Stein, A. J. van Wijnen, G. S. Stein, J. B. Lian, and A. Javed, Smad function and intranuclear targeting share a Runx2 motif required for osteogenic lineage induction and BMP2 responsive transcription. *J Cell Physiol* **204,** 63–72.

94. Y. W. Zhang, N. Yasui, K. Ito, G. Huang, M. Fujii, J. Hanai, H. Nogami, T. Ochi, K. Miyazono, and Y. Ito, A RUNX2/PEBP2aA/CBFA1 mutation displaying impaired transactivation and Smad interaction in cleidocranial dysplasia. *Proc Natl Acad Sci USA* **97,** 10549–10554 (2000).

95. S. Yang, D. Wei, D. Wang, M. Phimphilai, P. H. Krebsbach, and R. T. Franceschi, *In vitro* and *in vivo* synergistic interactions between the Runx2/Cbfa1 transcription factor and bone morphogenetic protein-2 in stimulating osteoblast differentiation. *J Bone Miner Res* **18,** 705–715 (2003).

96. D. W. Young, J. Pratap, A. Javed, B. Weiner, Y. Ohkawa, A. van Wijnen, M. Montecino, G. S. Stein, J. L. Stein, A. N. Imbalzano, and J. B. Lian, SWI/SNF chromatin remodeling complex is obligatory for BMP2-induced Runx2-dependent skeletal gene expression that controls osteoblast differentiation. *J Cell Biochem* **94,** 720–730 (2005).

97. M. Q. Hassan, R. S. Tare, S. Lee, M. Mandeville, M. I. Morasso, A. Javed, A. J. van Wijnen, J. L. Stein, G. S. Stein, and J. B. Lian, BMP2 commitment to the osteogenic lineage involves activation of Runx2 by Dlx3 and a homeodomain transcriptional network. *J Biol Chem* **281,** 40515–40526 (2006).

98. C. Y. Logan and R. Nusse, The Wnt signaling pathway in development and disease. *Annu Rev Cell Dev Biol* **20,** 781–810 (2004).

99. C. Hartmann, A Wnt canon orchestrating osteoblastogenesis. *Trends Cell Biol* **16,** 151–158 (2006).

100. P. V. Bodine and B. S. Komm, Wnt signaling and osteoblastogenesis. *Rev Endocr Metab Disord* **7,** 33–39 (2006).

101. V. Krishnan, H. U. Bryant, and O. A. MacDougald, Regulation of bone mass by Wnt signaling. *J Clin Invest* **116,** 1202–1209 (2006).

102. J. A. Robinson, M. Chatterjee-Kishore, P. J. Yaworsky, D. M. Cullen, W. Zhao, C. Li, Y. Kharode, L. Sauter, P. Babij, E. L. Brown, A. A. Hill, M. P. Akhter, M. L. Johnson, R. R. Recker, B. S. Komm, and F. J. Bex, Wnt/beta-catenin signaling is a normal physiological response to mechanical loading in bone. *J Biol Chem* **281,** 31720–31728 (2006).

103. B. Mao, W. Wu, G. Davidson, J. Marhold, M. Li, B. M. Mechler, H. Delius, D. Hoppe, P. Stannek, C. Walter, A. Glinka, and C. Niehrs, Kremen proteins are Dickkopf receptors that regulate Wnt/beta-catenin signalling. *Nature* **417,** 664–667 (2002).

104. F. Morvan, K. Boulukos, P. Clement-Lacroix, R. S. Roman, I. Suc-Royer, B. Vayssiere, P. Ammann, P. Martin, S. Pinho, P. Pognonec, P. Mollat, C. Niehrs, R. Baron, and G. Rawadi, Deletion of a single allele of the Dkk1 gene leads to an increase in bone formation and bone mass. *J Bone Miner Res* **21,** 934–945 (2006).

105. X. Li, P. Liu, W. Liu, P. Maye, J. Zhang, Y. Zhang, M. Hurley, C. Guo, A. Boskey, L. Sun, S. E. Harris, D. W. Rowe, H. Z. Ke, and D. Wu, Dkk2 has a role in terminal osteoblast differentiation and mineralized matrix formation. *Nat Genet* **37,** 945–952 (2005).

106. J. Li, I. Sarosi, R. C. Cattley, J. Pretorius, F. Asuncion, M. Grisanti, S. Morony, S. Adamu, Z. Geng, W. Qiu,

P. Kostenuik, D. L. Lacey, W. S. Simonet, B. Bolon, X. Qian, V. Shalhoub, M. S. Ominsky, K. H. Zhu, X. Li, and W. G. Richards, Dkk1-mediated inhibition of Wnt signaling in bone results in osteopenia. *Bone* **39**, 754–766 (2006).

107. P. V. Bodine, W. Zhao, Y. P. Kharode, F. J. Bex, A. J. Lambert, M. B. Goad, T. Gaur, G. S. Stein, J. B. Lian, and B. S. Komm, The Wnt antagonist secreted frizzled-related protein-1 is a negative regulator of trabecular bone formation in adult mice. *Mol Endocrinol* **18**, 1222–1237 (2004).

108. T. Gaur, C. J. Lengner, H. Hovhannisyan, R. A. Bhat, P. V. N. Bodine, B. S. Komm, A. Javed, A. J. van Wijnen, J. L. Stein, G. S. Stein, and J. B. Lian, Canonical WNT signaling promotes osteogenesis by directly stimulating RUNX2 gene expression. *J Biol Chem* **280**, 33132–33140 (2005).

109. T. Gaur, C. J. Lengner, S. Hussain, B. Trevant, D. Ayers, J. L. Stein, P. V. N. Bodine, B. S. Komm, G. S. Stein, and J. B. Lian, Secreted frizzled protein 1 regulates Wnt signaling for BMP2 induced chondrocyte differentiation. *J Cell Physiol* **208**, 87–96 (2006).

110. H. M. Yu, B. Jerchow, T. J. Sheu, B. Liu, F. Costantini, J. E. Puzas, W. Birchmeier, and W. Hsu, The role of Axin2 in calvarial morphogenesis and craniosynostosis. *Development* **132**, 1995–2005 (2005).

111. B. A. Parr and A. P. McMahon, Dorsalizing signal Wnt-7a required for normal polarity of D-V and A-P axes of mouse limb. *Nature* **374**, 350–353 (1995).

112. T. Narita, S. Sasaoka, K. Udagawa, T. Ohyama, N. Wada, S. Nishimatsu, S. Takada, and T. Nohno, Wnt10a is involved in AER formation during chick limb development. *Dev Dyn* **233**, 282–287 (2005).

113. M. Ikeya and S. Takada, Wnt-3a is required for somite specification along the anteroposterior axis of the mouse embryo and for regulation of Cdx-1 expression. *Mech Dev* **103**, 27–33 (2001).

114. S. Niemann, C. Zhao, F. Pascu, U. Stahl, U. Aulepp, L. Niswander, J. L. Weber, and U. Muller, Homozygous WNT3 mutation causes tetra-amelia in a large consanguineous family. *Am J Hum Genet* **74**, 558–563 (2004).

115. M. Lako, T. Strachan, P. Bullen, D. I. Wilson, S. C. Robson, and S. Lindsay, Isolation characterisation and embryonic expression of WNT11, a gene which maps to 11q13.5 and has possible roles in the development of skeleton, kidney and lung. *Gene* **219**, 101–110 (1998).

116. P. G. Loganathan, S. Nimmagadda, R. Huang, M. Scaal, and B. Christ, Comparative analysis of the expression patterns of Wnts during chick limb development. *Histochem Cell Biol* **123**, 195–201 (2005).

117. C. Hartmann and C. J. Tabin, Wnt-14 plays a pivotal role in inducing synovial joint formation in the developing appendicular skeleton. *Cell* **104**, 341–351 (2001).

118. T. P. Yamaguchi, A. Bradley, A. P. McMahon, and S. Jones, A Wnt5a pathway underlies outgrowth of multiple structures in the vertebrate embryo. *Development* **126**, 1211–1223 (1999).

119. Y. Yang, L. Topol, H. Lee, and J. Wu, Wnt5a and Wnt5b exhibit distinct activities in coordinating chondrocyte proliferation and differentiation. *Development* **130**, 1003–1015 (2003).

120. L. Topol, X. Jiang, H. Choi, L. Garrett-Beal, P. J. Carolan, and Y. Yang, Wnt-5a inhibits the canonical Wnt pathway by promoting GSK-3-independent {beta}-catenin degradation. *J Cell Biol* **162**, 899–908 (2003).

121. Y. Kawakami, N. Wada, S. I. Nishimatsu, T. Ishikawa, S. Noji, and T. Nohno, Involvement of Wnt-5a in chondrogenic pattern formation in the chick limb bud. *Dev Growth Differ* **41**, 29–40 (1999).

122. V. Church, T. Nohno, C. Linker, C. Marcelle, and P. Francis-West, Wnt regulation of chondrocyte differentiation. *J Cell Sci* **115**, 4809–4818 (2002).

123. C. N. Bennett, K. A. Longo, W. S. Wright, L. J. Suva, T. F. Lane, K. D. Hankenson, and O. A. MacDougald, Regulation of osteoblastogenesis and bone mass by Wnt10b. *Proc Natl Acad Sci USA* **102**, 3324–3329 (2005).

124. K. A. Longo, W. S. Wright, S. Kang, I. Gerin, S. H. Chiang, P. C. Lucas, M. R. Opp, and O. A. MacDougald, Wnt10b inhibits development of white and brown adipose tissues. *J Biol Chem* **279**, 35503–35509 (2004).

125. L. M. Boyden, J. Mao, J. Belsky, L. Mitzner, A. Farhi, M. A. Mitnick, D. Wu, K. Insogna, and R. P. Lifton, High bone density due to a mutation in LDL-receptor-related protein 5. *N Engl J Med* **346**, 1513–1521 (2002).

126. R. D. Little, R. R. Recker, and M. L. Johnson, High bone density due to a mutation in LDL-receptor-related protein 5. *N Engl J Med* **347**, 943–944 (2002).

127. M. P. Akhter, D. J. Wells, S. J. Short, D. M. Cullen, M. L. Johnson, G. R. Haynatzki, P. Babij, K. M. Allen, P. J. Yaworsky, F. Bex, and R. R. Recker, Bone biomechanical properties in LRP5 mutant mice. *Bone* **35**, 162–169 (2004).

128. P. Babij, W. Zhao, C. Small, Y. Kharode, P. J. Yaworsky, M. L. Bouxsein, P. S. Reddy, P. V. Bodine, J. A. Robinson, B. Bhat, J. Marzolf, R. A. Moran, and F. Bex, High bone mass in mice expressing a mutant LRP5 gene. *J Bone Miner Res* **18**, 960–974 (2003).

129. Y. Gong, R. B. Slee, N. Fukai, G. Rawadi, S. Roman-Roman, A. M. Reginato, H. Wang, T. Cundy, F. H. Glorieux, D. Lev, M. Zacharin, K. Oexle, J. Marcelino, W. Suwairi, S. Heeger, G. Sabatakos, S. Apte, W. N. Adkins, J. Allgrove, M. Arslan-Kirchner, J. A. Batch, P. Beighton, G. C. Black, R. G. Boles, L. M. Boon, C. Borrone, H. G. Brunner, G. F. Carle, B. Dallapiccola, A. De Paepe, B. Floege, M. L. Halfhide, B. Hall, R. C. Hennekam, T. Hirose, A. Jans, H. Juppner, C. A. Kim, C. Keppler-Noreuil, A. Kohlschuetter, D. Lacombe, M. Lambert, E. Lemyre, T. Letteboer, L. Peltonen, R. S. Ramesar, H. Romanengo, H. Somer, E. Steichen-Gersdorf, B. Steinmann, B. Sullivan, A. Superti-Furga, W. Swoboda, M. J. van den Boogaard, W. van Hul, M. Vikkula, M. Votruba, B. Zabel, T. Garcia, R. Baron, B. R. Olsen, and M. L. Warman, LDL receptor-related protein 5 (LRP5) affects bone accrual and eye development. *Cell* **107**, 513–523 (2001).

130. L. van Wesenbeeck, E. Cleiren, J. Gram, R. K. Beals, O. Benichou, D. Scopelliti, L. Key, T. Renton, C. Bartels, Y. Gong, M. L. Warman, M. C. de Vernejoul, J. Bollerslev, and W. van Hul, Six novel missense mutations in the LDL receptor-related protein 5 (LRP5) gene in different conditions with an increased bone density. *Am J Hum Genet* **72**, 763–771 (2003).

131. M. Kato, M. S. Patel, R. Levasseur, I. Lobov, B. H. Chang, D. A. Glass, C. Hartmann, L. Li, T. H. Hwang, C. F. Brayton, R. A. Lang, G. Karsenty, and L. Chan, Cbfa1-independent decrease in osteoblast proliferation osteopenia and persistent embryonic eye vascularization in mice deficient in Lrp5 A Wnt coreceptor. *J Cell Biol* **157**, 303–314 (2002).

132. S. L. Holmen, C. R. Zylstra, A. Mukherjee, R. E. Sigler, M. C. Faugere, M. L. Bouxsein, L. Deng, T. L. Clemens, and B. O. Williams, Essential role of beta-catenin in postnatal bone acquisition. *J Biol Chem* **280**, 21162–21168 (2005).

133. D. A. Glass, P. Bialek, J. D. Ahn, M. Starbuck, M. S. Patel, H. Clevers, M. M. Taketo, F. Long, A. P. McMahon, R. A. Lang,

and G. Karsenty, Canonical wnt signaling in differentiated osteoblasts controls osteoclast differentiation. *Dev Cell* **8,** 751–764 (2005).

134. T. F. Day, X. Guo, L. Garrett-Beal, and Y. Yang, Wnt/beta-catenin signaling in mesenchymal progenitors controls osteoblast and chondrocyte differentiation during vertebrate skeletogenesis. *Dev Cell* **8,** 739–750 (2005).

135. T. P. Hill, D. Spater, M. M. Taketo, W. Birchmeier, and C. Hartmann, Canonical Wnt/beta-catenin signaling prevents osteoblasts from differentiating into chondrocytes. *Dev Cell* **8,** 727–738 (2005).

136. M. I. Reinhold and M. C. Naski, Direct interactions of Runx2 and canonical Wnt signaling induce FGF18. *J Biol Chem* **282,** 3653–3663 (2007).

137. D. L. Ellies, B. Viviano, J. Mccarthy, J. P. Rey, N. Itasaki, S. Saunders, and R. Krumlauf, Bone density ligand, Sclerostin, directly interacts with LRP5 but not LRP5G171V to modulate Wnt activity. *J Bone Miner Res* **21,** 1738–1749 (2006).

138. R. L. Van Bezooijen, J. P. Svensson, D. Eefting, A. Visser, G. van der Horst, M. Karperien, P. H. Quax, H. Vrieling, S. E. Papapoulos, P. ten Dijke, and C. W. Lowik, Wnt but not BMP signaling is involved in the inhibitory action of Sclerostin on BMP-stimulated bone formation. *J Bone Miner Res* **22,** 19–28 (2007).

139. E. Minina, C. Kreschel, M. C. Naski, D. M. Ornitz, and A. Vortkamp, Interaction of FGF Ihh/Pthlh and BMP signaling integrates chondrocyte proliferation and hypertrophic differentiation. *Dev Cell* **3,** 439–449 (2002).

140. E. Schipani and S. Provot, Pthrp PTH and the PTH/Pthrp receptor in endochondral bone development. *Birth Defects Res C Embryo Today* **69,** 352–362 (2003).

141. B. St. Jacques, M. Hammerschmidt, and A. P. McMahon, Indian hedgehog signaling regulates proliferation and differentiation of chondrocytes and is essential for bone formation. *Genes Dev* **13,** 2072–2086 (1999).

142. B. St. Jacques, M. Hammerschmidt, and A. P. McMahon, Indian hedgehog signaling regulates proliferation and differentiation of chondrocytes and is essential for bone formation. *Genes Dev* **13,** 2072–2086 (1999).

143. T. Kobayashi, U. I. Chung, E. Schipani, M. Starbuck, G. Karsenty, T. Katagiri, D. L. Goad, B. Lanske, and H. M. Kronenberg, Pthrp and Indian hedgehog control differentiation of growth plate chondrocytes at multiple steps. *Development* **129,** 2977–2986 (2002).

144. B. Lanske, A. C. Karaplis, K. Lee, A. Luz, A. Vortkamp, A. Pirro, M. Karperien, L. H. K. Defize, C. Ho, R. C. Mulligan, A. B. Abou-Samra, H. Juppner, G. V. Segre, and H. M. Kronenberg, PTH/PTHrP receptor in early development and Indian hedgehog-regulated bone growth. *Science* **273,** 663–666 (1996).

145. A. Vortkamp, K. Lee, B. Lanske, G. V. Segre, H. M. Kronenberg, and C. J. Tabin, Regulation of rate of cartilage differentiation by Indian hedgehog and PTH-related protein. *Science* **273,** 613–622 (1996).

146. F. Long, U. I. Chung, S. Ohba, J. McMahon, H. M. Kronenberg, and A. P. McMahon, Ihh signaling is directly required for the osteoblast lineage in the endochondral skeleton. *Development* **131,** 1309–1318 (2004).

147. K. Seki and A. Hata, Indian hedgehog gene is a target of the bone morphogenetic protein signaling pathway. *J Biol Chem* **279,** 18544–18549 (2004).

148. T. Tobimatsu, H. Kaji, H. Sowa, J. Naito, L. Canaff, G. N. Hendy, T. Sugimoto, and K. Chihara, Parathyroid hormone increases beta-catenin levels through Smad3 in mouse osteoblastic cells. *Endocrinology* **147,** 2583–2590 (2006).

149. C. A. Yoshida and T. Komori, Role of Runx proteins in chondrogenesis. *Crit Rev Eukaryot Gene Expr* **15,** 243–254 (2005).

150. K. Blyth, E. R. Cameron, and J. C. Neil, The Runx genes: Gain or loss of function in cancer. *Nat Rev Cancer* **5,** 376–387 (2005).

151. H. L. Merriman, A. J. van Wijnen, S. Hiebert, J. P. Bidwell, E. Fey, J. Lian, J. Stein, and G. S. Stein, The tissue-specific nuclear matrix protein NMP-2 is a member of the AML/CBF/PEBP2/Runt domain transcription factor family: Interactions with the osteocalcin gene promoter. *Biochemistry* **34,** 13125–13132 (1995).

152. C. Banerjee, S. W. Hiebert, J. L. Stein, J. B. Lian, and G. S. Stein, An AML-1 consensus sequence binds an osteoblast-specific complex and transcriptionally activates the osteocalcin gene. *Proc Natl Acad Sci USA* **93,** 4968–4973 (1996).

153. P. Ducy and G. Karsenty, Two distinct osteoblast-specific cis-acting elements control expression of a mouse osteocalcin gene. *Mol Cell Biol* **15,** 1858–1869 (1995).

154. T. Komori, H. Yagi, S. Nomura, A. Yamaguchi, K. Sasaki, K. Deguchi, Y. Shimizu, R. T. Bronson, Y.-H. Gao, M. Inada, M. Sato, R. Okamoto, Y. Kitamura, S. Yoshiki, and T. Kishimoto, Targeted disruption of *Cbfa1* results in a complete lack of bone formation owing to maturational arrest of osteoblasts. *Cell* **89,** 755–764 (1997).

155. F. Otto, A. P. Thornell, T. Crompton, A. Denzel, K. C. Gilmour, I. R. Rosewell, G. W. H. Stamp, R. S. P. Beddington, S. Mundlos, B. R. Olsen, P. B. Selby, and M. J. Owen, *Cbfa1,* a candidate gene for cleidocranial dysplasia syndrome, is essential for osteoblast differentiation and bone development. *Cell* **89,** 765–771 (1997).

156. K. Y. Choi, S. W. Lee, M. H. Park, Y. C. Bae, H. I. Shin, S. Nam, Y. J. Kim, H. J. Kim, and H. M. Ryoo, Spatio-temporal expression patterns of Runx2 isoforms in early skeletogenesis. *Exp Mol Med* **34,** 426–433 (2002).

157. J.-Y. Choi, J. Pratap, A. Javed, S. K. Zaidi, L. Xing, E. Balint, S. Dalamangas, B. Boyce, A. J. van Wijnen, J. B. Lian, J. L. Stein, S. N. Jones, and G. S. Stein, Subnuclear targeting of Runx/Cbfa/AML factors is essential for tissue-specific differentiation during embryonic development. *Proc Natl Acad Sci USA* **98,** 8650–8655 (2001).

158. S. Mundlos, F. Otto, C. Mundlos, J. B. Mulliken, A. S. Aylsworth, S. Albright, D. Lindhout, W. G. Cole, W. Henn, J. H. M. Knoll, M. J. Owen, R. Mertelsmann, B. U. Zabel, and B. R. Olsen, Mutations involving the transcription factor CBFA1 cause cleidocranial dysplasia. *Cell* **89,** 773–779 (1997).

159. S. Mundlos, Cleidocranial dysplasia: Clinical and molecular genetics. *J Med Genet* **36,** 177–182 (1999).

160. B. Lee, K. Thirunavukkarasu, L. Zhou, L. Pastore, A. Baldini, J. Hecht, V. Geoffroy, P. Ducy, and G. Karsenty, Missense mutations abolishing DNA binding of the osteoblast-specific transcription factor OSF2/CBFA1 in cleidocranial dysplasia. *Nat Genet* **16,** 307–310 (1997).

161. Y. Zhang, N. Yasui, N. Kakazu, T. Abe, K. Takada, S. Imai, M. Sato, S. Nomura, T. Ochi, S. Okuzumi, H. Nogami, T. Nagai, H. Ohashi, and Y. Ito, PEBP2alphaA/CBFA1 mutations in Japanese cleidocranial dysplasia patients. *Gene* **244,** 21–28 (2000).

162. Y. Wang, R. M. Belflower, Y. F. Dong, E. M. Schwarz, R. J. O'Keefe, and H. Drissi, Runx1/AML1/Cbfa2 mediates onset of mesenchymal cell differentiation toward chondrogenesis. *J Bone Miner Res* **20,** 1624–1636 (2005).

163. J. B. Lian, E. Balint, A. Javed, H. Drissi, R. Vitti, E. J. Quinlan, L. Zhang, A. J. van Wijnen, J. L. Stein, N. Speck, and

G. S. Stein, Runx1/AML1 hematopoietic transcription factor contributes to skeletal development *in vivo*. *J Cell Physiol* **196,** 301–311 (2003).

164. D. Levanon, O. Brenner, V. Negreanu, D. Bettoun, E. Woolf, R. Eilam, J. Lotem, U. Gat, F. Otto, N. Speck, and Y. Groner, Spatial and temporal expression pattern of Runx3 (Aml2) and Runx1 (Aml1) indicates non-redundant functions during mouse embryogenesis. *Mech Dev* **109,** 413–417 (2001).

165. C. A. Yoshida, H. Yamamoto, T. Fujita, T. Furuichi, K. Ito, K. Inoue, K. Yamana, A. Zanma, K. Takada, Y. Ito, and T. Komori, Runx2 and Runx3 are essential for chondrocyte maturation and Runx2 regulates limb growth through induction of Indian hedgehog. *Genes Dev* **18,** 952–963 (2004).

166. M. V. Flores, E. Y. Lam, P. Crosier, and K. Crosier, A hierarchy of Runx transcription factors modulate the onset of chondrogenesis in craniofacial endochondral bones in zebrafish. *Dev Dyn* **235,** 3166–3176 (2006).

167. H. M. Ryoo and X. P. Wang, Control of tooth morphogenesis by Runx2. *Crit Rev Eukaryot Gene Expr* **16,** 143–154 (2006).

168. T. Yamashiro, T. Aberg, D. Levanon, Y. Groner, and I. Thesleff, Expression of Runx1, -2, and -3 during tooth, palate, and craniofacial bone development. *Mech Dev* **119,** S107–S110 (2002).

169. T. Aberg, X. P. Wang, J. H. Kim, T. Yamashiro, M. Bei, R. Rice, H. M. Ryoo, and I. Thesleff, Runx2 mediates FGF signaling from epithelium to mesenchyme during tooth morphogenesis. *Dev Biol* **270,** 76–93 (2004).

170. X. P. Wang, T. Aberg, M. J. James, D. Levanon, Y. Groner, and I. Thesleff, Runx2 (Cbfa1) inhibits Shh signaling in the lower but not upper molars of mouse embryos and prevents the budding of putative successional teeth. *J Dent Res* **84,** 138–143 (2005).

171. M. H. Lee, Y. J. Kim, W. J. Yoon, J. I. Kim, B. G. Kim, Y. S. Hwang, J. M. Wozney, X. Z. Chi, S. C. Bae, K. Y. Choi, J. Y. Cho, J. Y. Choi, and H. M. Ryoo, Dlx5 specifically regulates Runx2-II expression by binding to homeodomain response elements in the Runx2 distal promoter. *J Biol Chem* **280,** 35579–35587 (2005).

172. J. S. Kang, T. Alliston, R. Delston, and R. Derynck, Repression of Runx2 function by TGF-beta through recruitment of class II histone deacetylases by Smad3. *EMBO J* **24,** 2543–2555 (2005).

173. H. J. Kim, J. H. Kim, S. C. Bae, J. Y. Choi, H. J. Kim, and H. M. Ryoo, The protein kinase C pathway plays a central role in the fibroblast growth factor-stimulated expression and transactivation activity of Runx2. *J Biol Chem* **278,** 319–326 (2003).

174. E. Hinoi, P. Bialek, Y. T. Chen, M. T. Rached, Y. Groner, R. R. Behringer, D. M. Ornitz, and G. Karsenty, Runx2 inhibits chondrocyte proliferation and hypertrophy through its expression in the perichondrium. *Genes Dev* **20,** 2937–2942 (2006).

175. C. J. Lengner, M. Q. Hassan, R. W. Serra, C. Lepper, A. J. van Wijnen, J. L. Stein, J. B. Lian, and G. S. Stein, Nkx3.2 mediated repression of RUNX2 promotes chondrogenic differentiation. *J Biol Chem* **280,** 15872–15879 (2005).

176. C. Triboli and T. Lufkin, The murine Bapx1 homeobox gene plays a critical role in embryonic development of the axial skeleton and spleen. *Development* **126,** 5699–5711 (1999).

177. H. Akazawa, I. Komuro, Y. Sugitani, Y. Yazaki, R. Nagai, and T. Noda, Targeted disruption of the homeobox transcription factor Bapx1 results in lethal skeletal dysplasia with asplenia and gastroduodenal malformation. *Genes Cells* **5,** 499–513 (2000).

178. L. Zeng, H. Kempf, L. C. Murtaugh, M. E. Sato, and A. B. Lassar, Shh establishes an Nkx3.2/Sox9 autoregulatory loop that is maintained by BMP signals to induce somitic chondrogenesis. *Genes Dev* **16,** 1990–2005 (2002).

179. L. C. Murtaugh, L. Zeng, J. H. Chyung, and A. B. Lassar, The chick transcriptional repressor Nkx3.2 acts downstream of Shh to promote BMP-dependent axial chondrogenesis. *Dev Cell* **1,** 411–422 (2001).

180. J. Wilson and A. S. Tucker, Fgf and Bmp signals repress the expression of Bapx1 in the mandibular mesenchyme and control the position of the developing jaw joint. *Dev Biol* **266,** 138–150 (2004).

181. L. A. Lettice, L. A. Purdie, G. J. Carlson, F. Kilanowski, J. Dorin, and R. E. Hill, The mouse bagpipe gene controls development of axial skeleton skull and spleen. *Proc Natl Acad Sci USA* **96,** 9695–9700 (1999).

182. H. Akiyama, M. C. Chaboissier, J. F. Martin, A. Schedl, and B. de Crombrugghe, The transcription factor Sox9 has essential roles in successive steps of the chondrocyte differentiation pathway and is required for expression of Sox5 and Sox6. *Genes Dev* **16,** 2813–2828 (2002).

183. G. Zhou, Q. Zheng, F. Engin, E. Munivez, Y. Chen, E. Sebald, D. Krakow, and B. Lee, Dominance of SOX9 function over RUNX2 during skeletogenesis. *Proc Natl Acad Sci USA* **103,** 19004–19009 (2006).

184. P. Bialek, B. Kern, X. Yang, M. Schrock, D. Sosic, N. Hong, H. Wu, K. Yu, D. M. Ornitz, E. N. Olson, M. J. Justice, and G. Karsenty, A twist code determines the onset of osteoblast differentiation. *Dev Cell* **6,** 423–435 (2004).

185. K. Thirunavukkarasu, Y. Pei, T. L. Moore, T. Wei, H. Wang, and S. Chandrasekhar, Regulation of NFATC2 gene expression by the transcription factor Runx2. *Mol Biol Rep* **34,** 1–10 (2007).

186. M. J. Jimenez, M. Balbin, J. Alvarez, T. Komori, P. Bianco, K. Holmbeck, H. Birkedal-Hansen, J. M. Lopez, and C. Lopez-Otin, A regulatory cascade involving retinoic acid, Cbfa1, and matrix metalloproteinases is coupled to the development of a process of perichondrial invasion and osteogenic differentiation during bone formation. *J Cell Biol* **155,** 1333–1344 (2001).

187. E. Zelzer, W. Mclean, Y. S. Ng, N. Fukai, A. M. Reginato, S. Lovejoy, P. A. d'Amore, and B. R. Olsen, Skeletal defects in VEGF(120/120) mice reveal multiple roles for VEGF in skeletogenesis. *Development* **129,** 1893–1904 (2002).

188. J. Pratap, S. Vashi, J. Zhang, L. Languino, A. J. van Wijnen, J. L. Stein, J. B. Lian, and G. S. Stein, Runx2 regulates transcription of gelatinases (MMP9) in metastatic cancer cell lines and is functionally related to cell migration. *J Bone Miner Res* **19,** S123 (2004) [Abstract].

189. A. J. van Wijnen, G. S. Stein, J. L. Stein, and J. B. Lian, Cell cycle control of transcription at the G_1/S phase transition. In *Cell Cycle Inhibitors in Cancer Therapy: Current Strategies* (A. Giordano and K. J. Soprano, eds.), pp. 31–48. Humana Press, Totowa, NJ (2003).

190. M. Galindo, J. Pratap, D. W. Young, H. Hovhannisyan, H. J. Im, J. Y. Choi, J. B. Lian, J. L. Stein, G. S. Stein, and A. J. van Wijnen, The bone-specific expression of RUNX2 oscillates during the cell cycle to support a G_1 related antiproliferative function in osteoblasts. *J Biol Chem* **280,** 20274–20285 (2005).

191. S. K. Zaidi, D. W. Young, S. H. Pockwinse, A. Javed, J. B. Lian, J. L. Stein, A. J. van Wijnen, and G. S. Stein, Mitotic partitioning and selective reorganization of tissue specific transcription factors in progeny cells. *Proc Natl Acad Sci USA* **100,** 14852–14857 (2003).

192. D. W. Young, M. Q. Hassan, X.-Q. Yang, M. Galindo, A. Javed, S. K. Zaidi, P. Furcinitti, D. Lapointe, M. Montecino, J. B. Lian, J. L. Stein, A. J. van Wijnen, and G. S. Stein, Mitotic retention of gene expression patterns by the cell fate determining transcription factor Runx2. *Proc Natl Acad Sci USA* **104**, 3189–3194 (2007).

193. M. Parisien, S. J. Silverberg, E. Shane, C. L. De La Cruz, R. Lindsay, J. P. Bilezikian, and D. W. Dempster, The histomorphometry of bone in primary hyperparathyroidism: Preservation of cancellous bone structure. *J Clin Endocrinol Metab* **70**, 930–938 (1990).

194. D. W. Young, M. Q. Hassan, J. Pratap, M. Galindo, S. K. Zaidi, S. Lee, X. Yang, R. Xie, J. Underwood, P. Furcinitti, A. N. Imbalzano, S. Penman, J. A. Nickerson, M. A. Montecino, J. B. Lian, J. L. Stein, A. J. van Wijnen, and G. S. Stein, Mitotic occupancy and lineage-specific transcriptional control of ribosomal RNA genes by Runx2. *Nature* **445**, 442–446 (2007).

195. J. Pratap, M. Galindo, S. K. Zaidi, D. Vradii, B. M. Bhat, J. A. Robinson, J.-Y. Choi, T. Komori, J. L. Stein, J. B. Lian, G. S. Stein, and A. J. van Wijnen, Cell growth regulatory role of Runx2 during proliferative expansion of pre-osteoblasts. *Cancer Res* **63**, 5357–5362 (2003).

196. G. S. Stein and J. B. Lian, Molecular mechanisms mediating proliferation/differentiation interrelationships during progressive development of the osteoblast phenotype. *Endocr Rev* **14**, 424–442 (1993).

197. L. C. Gerstenfeld, A. Riva, K. Hodgens, D. R. Eyre, and W. J. Landis, Post-translational control of collagen fibrillogenesis in mineralizing cultures of chick osteoblasts. *J Bone Miner Res* **8**, 1031–1043 (1993).

198. T. A. Owen, M. Aronow, V. Shalhoub, L. M. Barone, L. Wilming, M. S. Tassinari, M. B. Kennedy, S. Pockwinse, J. B. Lian, and G. S. Stein, Progressive development of the rat osteoblast phenotype *in vitro*: Reciprocal relationships in expression of genes associated with osteoblast proliferation and differentiation during formation of the bone extracellular matrix. *J Cell Physiol* **143**, 420–430 (1990).

199. J. B. Lian, G. S. Stein, R. Bortell, and T. A. Owen, Phenotype suppression: A postulated molecular mechanism for mediating the relationship of proliferation and differentiation by Fos/Jun interactions at AP-1 sites in steroid responsive promoter elements of tissue-specific genes. *J Cell Biochem* **45**, 9–14 (1991).

200. M. P. Lynch, C. Capparelli, J. L. Stein, G. S. Stein, and J. B. Lian, Apoptosis during bone-like tissue development *in vitro*. *J Cell Biochem* **68**, 31–49 (1998).

201. S. K. Winchester, N. Selvamurugan, R. C. D'Alonzo, and N. C. Partridge, Developmental regulation of collagenase-3 mRNA in normal differentiating osteoblasts through the activator protein-1 and the runt domain binding sites. *J Biol Chem* **275**, 23310–23318 (2000).

202. Z. F. Chen and R. R. Behringer, Twist is required in head mesenchyme for cranial neural tube morphogenesis. *Genes Dev* **9**, 686–699 (1995).

203. P. Cserjesi, D. Brown, K. L. Ligon, G. E. Lyons, N. G. Copeland, D. J. Gilbert, N. A. Jenkins, and E. N. Olson, Scleraxis: A basic helix–loop–helix protein that prefigures skeletal formation during mouse embryogenesis. *Development* **121**, 1099–1110 (1995).

204. J. Yang, S. A. Mani, and R. A. Weinberg, Exploring a new twist on tumor metastasis. *Cancer Res* **66**, 4549–4552 (2006).

205. M. S. Lee, G. N. Lowe, D. D. Strong, J. E. Wergedal, and C. A. Glackin, TWIST, a basic helix–loop–helix transcription factor, can regulate the human osteogenic lineage. *J Cell Biochem* **75**, 566–577 (1999).

206. E. F. Wagner, Functions of AP1 (Fos/Jun) in bone development. *Ann Rheum Dis* **61**(Suppl. 2), ii40–ii42 (2002).

207. M. Machwate, A. Jullienne, M. Moukhtar, and P. J. Marie, Temporal variation of c-fos proto-oncogene expression during osteoblast differentiation and osteogenesis in developing bone. *J Cell Biochem* **57**, 62–70 (1995).

208. L. R. McCabe, M. Kockx, J. Lian, J. Stein, and G. Stein, Selective expression of fos- and jun-related genes during osteoblast proliferation and differentiation. *Exp Cell Res* **218**, 255–262 (1995).

209. J. Hess, D. Porte, C. Munz, and P. Angel, AP-1 and Cbfa/Runt physically interact and regulate PTH-dependent MMP13 expression in osteoblasts through a new OSE2/AP-1 composite element. *J Biol Chem* **276**, 20029–20038 (2001).

210. R. C. D'Alonzo, N. Selvamurugan, G. Karsenty, and N. C. Partridge, Physical interaction of the activator protein-1 factors c-Fos and c-Jun with Cbfa1 for collagenase-3 promoter activation. *J Biol Chem* **277**, 816–822 (2002).

211. A. E. Grigoriadis, K. Schellander, Z.-Q. Wang, and E. F. Wagner, Osteoblasts are target cells for transformation in *c-fos* transgenic mice. *J Cell Biol* **122**, 685–701 (1993).

212. L. R. McCabe, C. Banerjee, R. Kundu, R. J. Harrison, P. R. Dobner, J. L. Stein, J. B. Lian, and G. S. Stein, Developmental expression and activities of specific fos and jun proteins are functionally related to osteoblast maturation: Role of fra-2 and jun D during differentiation. *Endocrinology* **137**, 4398–4408 (1996).

213. L. Kenner, A. Hoebertz, T. Beil, N. Keon, F. Karreth, R. Eferl, H. Scheuch, A. Szremska, M. Amling, M. Schorpp-Kistner, P. Angel, and E. F. Wagner, Mice lacking Junb are osteopenic due to cell-autonomous osteoblast and osteoclast defects. *J Cell Biol* **164**, 613–623 (2004).

214. R. Eferl, A. Hoebertz, A. F. Schilling, M. Rath, F. Karreth, L. Kenner, M. Amling, and E. F. Wagner, The Fos-related antigen Fra-1 is an activator of bone matrix formation. *EMBO J* **23**, 2789–2799 (2004).

215. T. Aoyama, T. Okamoto, S. Nagayama, K. Nishijo, T. Ishibe, K. Yasura, T. Nakayama, T. Nakamura, and J. Toguchida, Methylation in the core-promoter region of the chondromodulin-I gene determines the cell-specific expression by regulating the binding of transcriptional activator Sp3. *J Biol Chem* **279**, 28789–28797 (2004).

216. V. Mann, E. E. Hobson, B. Li, T. L. Stewart, S. F. Grant, S. P. Robins, R. M. Aspden, and S. H. Ralston, A COL1A1 Sp1 binding site polymorphism predisposes to osteoporotic fracture by affecting bone density and quality. *J Clin Invest* **107**, 899–907 (2001).

217. S. M. Pluijm, H. W. van Essen, N. Bravenboer, A. G. Uitterlinden, J. H. Smit, H. A. Pols, and P. Lips, Collagen type I alpha1 Sp1 polymorphism, osteoporosis, and intervertebral disc degeneration in older men and women. *Ann Rheum Dis* **63**, 71–77 (2004).

218. T. Koga, Y. Matsui, M. Asagiri, T. Kodama, C. B. de Crombrugghe, K. Nakashima, and H. Takayanagi, NFAT and osterix cooperatively regulate bone formation. *Nat Med* **11**, 880–885 (2005).

219. J. Liu, H. Yang, W. Liu, X. Cao, and X. Feng, Sp1 and Sp3 regulate the basal transcription of receptor activator of nuclear factor kappa B ligand gene in osteoblasts and bone marrow stromal cells. *J Cell Biochem* **96**, 716–727 (2005).

220. A. M. Reimold, M. J. Grusby, B. Kosaras, J. W. Fries, R. Mori, S. Maniwa, I. M. Clauss, T. Collins, S. L. Sidman,

M. J. Glimcher, and L. H. Glimcher, Chondrodysplasia and neurological abnormalities in ATF-2-deficient mice. *Nature* **379,** 262–265 (1996).

221. X. Yang, K. Matsuda, P. Bialek, S. Jacquot, H. C. Masuoka, T. Schinke, L. Li, S. Brancorsini, P. Sassone-Corsi, T. M. Townes, A. Hanauer, and G. Karsenty, ATF4 is a substrate of RSK2 and an essential regulator of osteoblast biology: Implication for Coffin–Lowry syndrome. *Cell* **117,** 387–398 (2004).

222. G. Xiao, D. Jiang, C. Ge, Z. Zhao, Y. Lai, H. Boules, M. Phimphilai, X. Yang, G. Karsenty, and R. T. Franceschi, Cooperative interactions between activating transcription factor 4 and Runx2/Cbfa1 stimulate osteoblast-specific osteocalcin gene expression. *J Biol Chem* **280,** 30689–30696 (2005).

223. J. S. Shao, S. L. Cheng, J. M. Pingsterhaus, N. Charlton-Kachigian, A. P. Loewy, and D. A. Towler, Msx2 promotes cardiovascular calcification by activating paracrine Wnt signals. *J Clin Invest* **115,** 1210–1220 (2005).

224. M. Bidder, T. Latifi, and D. A. Towler, Reciprocal temporospatial patterns of Msx2 and osteocalcin gene expression during murine odontogenesis. *J Bone Miner Res* **13,** 609–619 (1998).

225. H.-M. Ryoo, H. M. Hoffmann, T. L. Beumer, B. Frenkel, D. A. Towler, G. S. Stein, J. L. Stein, A. J. van Wijnen, and J. B. Lian, Stage-specific expression of Dlx-5 during osteoblast differentiation: Involvement in regulation of osteocalcin gene expression. *Mol Endocrinol* **11,** 1681–1694 (1997).

226. M. Q. Hassan, A. Javed, M. I. Morasso, J. Karlin, M. Montecino, A. J. van Wijnen, G. S. Stein, J. L. Stein, and J. B. Lian, Dlx3 transcriptional regulation of osteoblast differentiation: Temporal recruitment of Msx2, Dlx3, and Dlx5 homeodomain proteins to chromatin of the osteocalcin gene. *Mol Cell Biol* **24,** 9248–9261 (2004).

227. G. van der Horst, H. Farih-Sips, C. W. Lowik, and M. Karperien, Multiple mechanisms are involved in inhibition of osteoblast differentiation by PTHrP and PTH in KS483 cells. *J Bone Miner Res* **20,** 2233–2244 (2005).

228. A. Hollnagel, M. Ahrens, and G. Gross, Parathyroid hormone enhances early and suppresses late stages of osteogenic and chondrogenic development in a BMP-dependent mesenchymal differentiation system (C3H10T1/2). *J Bone Miner Res* **12,** 1993–2004 (1997).

229. C. Gray and S. J. Jones, PTH(1–34) suppresses appositional bone formation by cultured rat cranial osteoblasts. *Bone* **23,** 453–457 (1998).

230. M. H. Kroll, Parathyroid hormone temporal effects on bone formation and resorption. *Bull Math Biol* **62,** 163–188 (2000).

231. M. E. Antosz, C. G. Bellows, and J. E. Aubin, Effects of transforming growth factor beta and epidermal growth factor on cell proliferation and the formation of bone nodules in isolated fetal rat calvaria cells. *J Cell Physiol* **140,** 386–395 (1989).

232. E. C. Breen, R. A. Ignotz, L. McCabe, J. L. Stein, G. S. Stein, and J. B. Lian, TGF beta alters growth and differentiation related gene expression in proliferating osteoblasts *in vitro,* preventing development of the mature bone phenotype. *J Cell Physiol* **160,** 323–335 (1994).

233. S. E. Harris, L. F. Bonewald, M. A. Harris, M. Sabatini, S. Dallas, J. Q. Feng, N. Ghosh-Choudhury, J. Wozney, and G. R. Mundy, Effects of transforming growth factor beta on bone nodule formation and expression of bone morphogenetic protein 2, osteocalcin, osteopontin, alkaline phosphatase, and type I collagen mRNA in long-term cultures of fetal rat calvarial osteoblasts. *J Bone Miner Res* **9,** 855–863 (1994).

234. M. Iwasaki, K. Nakata, H. Nakahara, T. Nakase, T. Kimura, K. Kimata, A. I. Caplan, and K. Ono, Transforming growth factor-beta 1 stimulates chondrogenesis and inhibits osteogenesis in high density culture of periosteum-derived cells. *Endocrinology* **132,** 1603–1608 (1993).

235. E. Canalis, S. Rydziel, A. M. Delany, S. Varghese, and J. J. Jeffrey, Insulin-like growth factors inhibit interstitial collagenase synthesis in bone cell cultures. *Endocrinology* **136,** 1348–1354 (1995).

236. R. C. Pereira and E. Canalis, Parathyroid hormone increases mac25/insulin-like growth factor-binding protein-related protein-1 expression in cultured osteoblasts. *Endocrinology* **140,** 1998–2003 (1999).

237. R. C. Pereira, F. Blanquaert, and E. Canalis, Cortisol enhances the expression of mac25/insulin-like growth factor-binding protein-related protein-1 in cultured osteoblasts. *Endocrinology* **140,** 228–232 (1999).

238. F. Minuto, C. Palermo, M. Arvigo, and A. M. Barreca, The IGF system and bone. *J Endocrinol Invest* **28,** 8–10 (2005).

239. D. Faibish, S. M. Ott, and A. L. Boskey, Mineral changes in osteoporosis: A review. *Clin Orthop Relat Res* **443,** 28–38 (2006).

240. J. He, C. J. Rosen, D. J. Adams, and B. E. Kream, Postnatal growth and bone mass in mice with IGF-I haploinsufficiency. *Bone* **38,** 826–835 (2006).

241. T. Niu and C. J. Rosen, The insulin-like growth factor-I gene and osteoporosis: A critical appraisal. *Gene* **361,** 38–56 (2005).

242. K. M. Delahunty, K. L. Shultz, G. A. Gronowicz, B. Koczon-Jaremko, M. L. Adamo, L. G. Horton, J. Lorenzo, L. R. Donahue, C. Ckert-Bicknell, B. E. Kream, W. G. Beamer, and C. J. Rosen, Congenic mice provide *in vivo* evidence for a genetic locus that modulates serum insulin-like growth factor-I and bone acquisition. *Endocrinology* **147,** 3915–3923 (2006).

243. S. Yakar, M. L. Bouxsein, E. Canalis, H. Sun, V. Glatt, C. Gundberg, P. Cohen, D. Hwang, Y. Boisclair, D. Leroith, and C. J. Rosen, The ternary IGF complex influences postnatal bone acquisition and the skeletal response to intermittent parathyroid hormone. *J Endocrinol* **189,** 289–299 (2006).

244. J. Jiang, A. C. Lichtler, G. A. Gronowicz, D. J. Adams, S. H. Clark, C. J. Rosen, and B. E. Kream, Transgenic mice with osteoblast-targeted insulin-like growth factor-I show increased bone remodeling. *Bone* **39,** 494–504 (2006).

245. D. D. Bikle, T. Sakata, C. Leary, H. Elalieh, D. Ginzinger, C. J. Rosen, W. Beamer, S. Majumdar, and B. P. Halloran, Insulin-like growth factor I is required for the anabolic actions of parathyroid hormone on mouse bone. *J Bone Miner Res* **17,** 1570–1578 (2002).

246. A. B. Celil and P. G. Campbell, BMP-2 and insulin-like growth factor-I mediate Osterix (Osx) expression in human mesenchymal stem cells via the MAPK and protein kinase D signaling pathways. *J Biol Chem* **280,** 31353–31359 (2005).

247. M. Qiao, P. Shapiro, R. Kumar, and A. Passaniti, Insulin-like growth factor-1 regulates endogenous RUNX2 activity in endothelial cells through a phosphatidylinositol 3-kinase/ERK-dependent and Akt-independent signaling pathway. *J Biol Chem* **279,** 42709–42718 (2004).

248. H. Koch, J. A. Jadlowiec, and P. G. Campbell, Insulin-like growth factor-I induces early osteoblast gene expression in human mesenchymal stem cells. *Stem Cells Dev* **14,** 621–631 (2005).

249. V. Shalhoub, F. Aslam, E. Breen, A. van Wijnen, R. Bortell, G. S. Stein, J. L. Stein, and J. B. Lian, Multiple levels of steroid hormone-dependent control of osteocalcin during osteoblast differentiation: Glucocorticoid regulation of basal and

vitamin D stimulated gene expression. *J Cell Biochem* **69**, 154–168 (1998).

250. Y. Ishida and J. N. Heersche, Glucocorticoid-induced osteoporosis: Both *in vivo* and *in vitro* concentrations of glucocorticoids higher than physiological levels attenuate osteoblast differentiation. *J Bone Miner Res* **13**, 1822–1826 (1998).

251. T. A. Owen, M. S. Aronow, L. M. Barone, B. Bettencourt, G. S. Stein, and J. B. Lian, Pleiotropic effects of vitamin D on osteoblast gene expression are related to the proliferative and differentiated state of the bone cell phenotype: Dependency upon basal levels of gene expression, duration of exposure, and bone matrix competency in normal rat osteoblast cultures. *Endocrinology* **128**, 1496–1504.

252. L. C. Gerstenfeld, D. Zurakowski, J. L. Schaffer, D. P. Nichols, C. D. Toma, M. Broess, S. P. Bruder, and A. I. Caplan, Variable hormone responsiveness of osteoblast populations isolated at different stages of embryogenesis and its relationship to the osteogenic lineage. *Endocrinology* **137**, 3957–3968 (1996).

253. P. V. N. Bodine, R. A. Henderson, J. Green, M. Aronow, T. Owen, G. S. Stein, J. B. Lian, and B. S. Komm, Estrogen receptor-A is developmentally regulated during osteoblast differentiation and contributes to selective responsiveness of gene expression. *Endocrinology* **139**, 2048–2057 (1998).

254. G. Mazziotti, A. Angeli, J. P. Bilezikian, E. Canalis, and A. Giustina, Glucocorticoid-induced osteoporosis: An update. *Trends Endocrinol Metab* **17**, 144–149 (2006).

255. A. M. Delany, B. Y. Gabbitas, and E. Canalis, Cortisol downregulates osteoblast alpha 1 (I) procollagen mRNA by transcriptional and posttranscriptional mechanisms. *J Cell Biochem* **57**, 488–494 (1995).

256. E. Smith and B. Frenkel, Glucocorticoids inhibit the transcriptional activity of LEF/TCF in differentiating osteoblasts in a glycogen synthase kinase-3beta-dependent and -independent manner. *J Biol Chem* **280**, 2388–2394 (2005).

257. N. Leclerc, C. A. Luppen, V. V. Ho, S. Nagpal, J. G. Hacia, E. Smith, and B. Frenkel, Gene expression profiling of glucocorticoid-inhibited osteoblasts. *J Mol Endocrinol* **33**, 175–193 (2004).

258. P. S. Leboy, J. N. Beresford, C. Devlin, and M. E. Owen, Dexamethasone induction of osteoblast mRNAs in rat marrow stromal cell cultures. *J Cell Physiol* **146**, 370–378.

259. C. G. Bellows, J. E. Aubin, and J. N. M. Heersche, Physiological concentrations of glucocorticoids stimulate formation of bone nodules from isolated rat calvaria cells *in vitro*. *Endocrinology* **121**, 1985–1992 (1987).

260. V. Shalhoub, D. Conlon, M. Tassinari, C. Quinn, N. Partridge, G. S. Stein, and J. B. Lian, Glucocorticoids promote development of the osteoblast phenotype by selectively modulating expression of cell growth and differentiation associated genes. *J Cell Biochem* **50**, 425–440 (1992).

261. C. G. Bellows, A. Ciaccia, and J. N. Heersche, Osteoprogenitor cells in cell populations derived from mouse and rat calvaria differ in their response to corticosterone, cortisol, and cortisone. *Bone* **23**, 119–125 (1998).

262. J. B. Lian, V. Shalhoub, F. Aslam, B. Frenkel, J. Green, M. Hamrah, G. S. Stein, and J. L. Stein, Species-specific glucocorticoid and 1,25-dihydroxyvitamin D responsiveness in mouse MC3T3-E1 osteoblasts: Dexamethasone inhibits osteoblast differentiation and vitamin D downregulates osteocalcin gene expression. *Endocrinology* **138**, 2117–2127 (1997).

263. C. A. O'Brien, D. Jia, L. I. Plotkin, T. Bellido, C. C. Powers, S. A. Stewart, S. C. Manolagas, and R. S. Weinstein, Glucocorticoids act directly on osteoblasts and osteocytes to induce their apoptosis and reduce bone formation and strength. *Endocrinology* **145**, 1835–1841 (2004).

264. Y. Xue, A. C. Karaplis, G. N. Hendy, D. Goltzman, and D. Miao, Exogenous 1,25-dihydroxyvitamin D3 exerts a skeletal anabolic effect and improves mineral ion homeostasis in mice that are homozygous for both the 1alpha-hydroxylase and parathyroid hormone null alleles. *Endocrinology* **147**, 4801–4810 (2006).

265. M. Broess, A. Riva, and L. C. Gerstenfeld, Inhibitory effects of 1,25(OH)2 vitamin D$_3$ on collagen type I osteopontin and osteocalcin gene expression in chicken osteoblasts. *J Cell Biochem* **57**, 440–451 (1995).

266. V. Viereck, H. Siggelkow, S. Tauber, D. Raddatz, N. Schutze, and M. Hufner, Differential regulation of Cbfa1/Runx2 and osteocalcin gene expression by vitamin-D3 dexamethasone and local growth factors in primary human osteoblasts. *J Cell Biochem* **86**, 348–356 (2002).

267. H. Ishida, C. G. Bellows, J. E. Aubin, and J. N. Heersche, Characterization of the 1,25-(OH)2D3-induced inhibition of bone nodule formation in long-term cultures of fetal rat calvaria cells. *Endocrinology* **132**, 61–66 (1993).

268. H. Siggelkow, H. Schulz, S. Kaesler, K. Benzler, M. J. Atkinson, and M. Hufner, 1,25 Dihydroxyvitamin-D3 attenuates the confluence-dependent differences in the osteoblast characteristic proteins alkaline phosphatase, procollagen I peptide, and osteocalcin. *Calcif Tissue Int* **64**, 414–421 (1999).

269. C. G. Bellows, S. M. Reimers, and J. N. Heersche, Expression of mRNAs for type-I collagen, bone sialoprotein, osteocalcin, and osteopontin at different stages of osteoblastic differentiation and their regulation by 1,25 dihydroxyvitamin D3. *Cell Tissue Res* **297**, 249–259 (1999).

270. K. M. Wiren, Androgens and bone growth: It's location, location, location. *Curr Opin Pharmacol* **5**, 626–632 (2005).

271. S. C. Manolagas, S. Kousteni, and R. L. Jilka, Sex steroids and bone. *Recent Prog Horm Res* **57**, 385–409 (2002).

272. D. Vanderschueren, K. Venken, J. Ophoff, R. Bouillon, and S. Boonen, Clinical review: Sex steroids and the periosteum—Reconsidering the roles of androgens and estrogens in periosteal expansion. *J Clin Endocrinol Metab* **91**, 378–382 (2006).

273. K. M. Wiren, E. A. Chapman, and X. W. Zhang, Osteoblast differentiation influences androgen and estrogen receptor-alpha and -beta expression. *J Endocrinol* **175**, 683–694 (2002).

274. D. G. Monroe, B. J. Getz, S. A. Johnsen, B. L. Riggs, S. Khosla, and T. C. Spelsberg, Estrogen receptor isoform-specific regulation of endogenous gene expression in human osteoblastic cell lines expressing either ERalpha or ERbeta. *J Cell Biochem* **90**, 315–326 (2003).

275. J. H. Bennett, S. Moffatt, and M. Horton, Cell adhesion molecules in human osteoblasts: Structure and function. *Histol Histopathol* **16**, 603–611 (2001).

276. C. H. Damsky, Extracellular matrix–integrin interactions in osteoblast function and tissue remodeling. *Bone* **25**, 95–96 (1999).

277. J. Heino, The collagen receptor integrins have distinct ligand recognition and signaling functions. *Matrix Biol* **19**, 319–323 (2000).

278. D. E. Hughes, D. M. Salter, and R. Simpson, CD44 expression in human bone: A novel marker of osteocytic differentiation. *J Bone Miner Res* **9**, 39–44 (1994).

279. H. H. Jamal and J. E. Aubin, CD44 expression in fetal rat bone: *In vivo* and *in vitro* analysis. *Exp Cell Res* **223**, 467–477 (1996).

280. S. L. Cheng, F. Lecanda, M. K. Davidson, P. M. Warlow, S. F. Zhang, L. Zhang, S. Suzuki, T. St. John, and R. Civitelli,

Human osteoblasts express a repertoire of cadherins which are critical for BMP-2-induced osteogenic differentiation. *J Bone Miner Res* **13**, 633–644 (1998).

281. E. Hay, J. Lemonnier, D. Modrowski, A. Lomri, F. Lasmoles, and P. J. Marie, N- and E-cadherin mediate early human calvaria osteoblast differentiation promoted by bone morphogenetic protein-2. *J Cell Physiol* **183**, 117–128 (2000).

282. S. L. Ferrari, K. Traianedes, M. Thorne, M. H. Lafage-Proust, P. Genever, M. G. Cecchini, V. Behar, A. Bisello, M. Chorev, M. Rosenblatt, and L. J. Suva, A role for N-cadherin in the development of the differentiated osteoblastic phenotype. *J Bone Miner Res* **15**, 198–208 (2000).

283. Y. S. Lee and C. M. Chuong, Adhesion molecules in skeletogenesis: I. Transient expression of neural cell adhesion molecules (NCAM) in osteoblasts during endochondral and intramembranous ossification. *J Bone Miner Res* **7**, 1435–1446 (1992).

284. M. Okazaki, S. Takeshita, S. Kawai, R. Kikuno, A. Tsujimura, A. Kudo, and E. Amann, Molecular cloning and characterization of OB-cadherin, a new member of cadherin family expressed in osteoblasts. *J Biol Chem* **269**, 12092–12098 (1994).

285. C. S. Shin, F. Lecanda, S. Sheikh, L. Weitzmann, S. L. Cheng, and R. Civitelli, Relative abundance of different cadherins defines differentiation of mesenchymal precursors into osteogenic myogenic or adipogenic pathways. *J Cell Biochem* **78**, 566–577 (2000).

286. T. H. Steinberg, R. Civitelli, S. T. Geist, A. J. Robertson, E. Hick, R. D. Veenstra, H. Z. Wang, P. M. Warlow, E. M. Westphale, and J. G. Laing, Connexin43 and Connexin45 form gap junctions with different molecular permeabilities in osteoblastic cells. *EMBO J* **13**, 744–750 (1994).

287. J. P. Stains and R. Civitelli, Cell–cell interactions in regulating osteogenesis and osteoblast function. *Birth Defects Res C Embryo Today* **75**, 72–80 (2005).

288. F. Lecanda, P. M. Warlow, S. Sheikh, F. Furlan, T. H. Steinberg, and R. Civitelli, Connexin43 deficiency causes delayed ossification, craniofacial abnormalities, and osteoblast dysfunction. *J Cell Biol* **151**, 931–944 (2000).

289. M. E. Jackson, V. Shalhoub, J. B. Lian, G. S. Stein, and S. C. Marks Jr., Aberrant gene expression in cultured mammalian bone cells demonstrates an osteoblast defect in osteopetrosis. *J Cell Biochem* **55**, 366–372 (1994).

290. V. Shalhoub, M. E. Jackson, C. Paradise, G. S. Stein, J. B. Lian, and S. C. Marks Jr., Heterogeneity of colony stimulating factor-1 gene expression in the skeleton of four osteopetrotic mutations in rats and mice. *J Cell Physiol* **166**, 340–350 (1996).

291. E. J. Moerman, K. Teng, D. A. Lipschitz, and B. Lecka-Czernik, Aging activates adipogenic and suppresses osteogenic programs in mesenchymal marrow stroma/stem cells: The role of PPAR-gamma2 transcription factor and TGF-beta/BMP signaling pathways. *Aging Cell* **3**, 379–389 (2004).

292. S. M. Mueller and J. Glowacki, Age-related decline in the osteogenic potential of human bone marrow cells cultured in three-dimensional collagen sponges. *J Cell Biochem* **82**, 583–590 (2001).

293. C. Silve, B. Grosse, C. Tau, M. Garabedian, J. Fritsch, P. D. Delmas, G. Cournot-Witmer, and S. Balsan, Response to parathyroid hormone and 1,25-dihydroxyvitamin D3 of bone-derived cells isolated from normal children and children with abnormalities in skeletal development. *J Clin Endocrinol Metab* **62**, 583–590 (1986).

294. C. Shui and A. M. Scutt, Mouse embryo-derived NIH3T3 fibroblasts adopt an osteoblast-like phenotype when treated with 1alpha,25-dihydroxyvitamin D(3) and dexamethasone *in vitro*. *J Cell Physiol* **193**, 164–172 (2002).

295. M. Prince, C. Banerjee, A. Javed, J. Green, J. B. Lian, G. S. Stein, P. V. Bodine, and B. S. Komm, Expression and regulation of Runx2/Cbfa1 and osteoblast phenotypic markers during the growth and differentiation of human osteoblasts. *J Cell Biochem* **80**, 424–440 (2001).

296. K. M. Waters, D. J. Rickard, B. L. Riggs, S. Khosla, J. A. Katzenellenbogen, B. S. Katzenellenbogen, J. Moore, and T. C. Spelsberg, Estrogen regulation of human osteoblast function is determined by the stage of differentiation and the estrogen receptor isoform. *J Cell Biochem* **83**, 448–462 (2001).

297. P. V. N. Bodine, S. K. Vernon, and B. S. Komm, Establishment and hormonal regulation of a conditionally-transformed preosteocytic cell line from adult human bone. *Endocrinology* **137**, 4592–4604 (1996).

298. B. M. Abdallah, M. Haack-Sorensen, J. S. Burns, B. Elsnab, F. Jakob, P. Hokland, and M. Kassem, Maintenance of differentiation potential of human bone marrow mesenchymal stem cells immortalized by human telomerase reverse transcriptase gene despite [corrected] extensive proliferation. *Biochem Biophys Res Commun* **326**, 527–538 (2005).

299. J.-S. Bae, S. Gutierrez, R. Narla, J. Pratap, R. Devados, A. J. van Wijnen, J. L. Stein, G. S. Stein, J. B. Lian, and A. Javed, Reconstitution of Runx2/Cbfa1 null cells identifies a requirement for BMP2 signaling through a Runx2 functional domain during osteoblast differentiation. *J Cell Biochem* **100**, 434–449 (2006).

300. Y. Kato, A. Boskey, L. Spevak, M. Dallas, M. Hori, and L. F. Bonewald, Establishment of an osteoid preosteocyte-like cell MLO-A5 that spontaneously mineralizes in culture. *J Bone Miner Res* **16**, 1622–1633 (2001).

301. M. Serafini and C. M. Verfaillie, Pluripotency in adult stem cells: State of the art. *Semin Reprod Med* **24**, 379–388 (2006).

302. D. S. Vieyra, K. A. Jackson, and M. A. Goodell, Plasticity and tissue regenerative potential of bone marrow-derived cells. *Stem Cell Rev* **1**, 65–70 (2005).

303. I. Ginis, Y. Luo, T. Miura, S. Thies, R. Brandenberger, S. Gerecht-Nir, M. Amit, A. Hoke, M. K. Carpenter, J. Itskovitz-Eldor, and M. S. Rao, Differences between human and mouse embryonic stem cells. *Dev Biol* **269**, 360–380 (2004).

304. I. Chambers and A. Smith, Self-renewal of teratocarcinoma and embryonic stem cells. *Oncogene* **23**, 7150–7160 (2004).

305. B. C. Heng, T. Cao, L. W. Stanton, P. Robson, and B. Olsen, Strategies for directing the differentiation of stem cells into the osteogenic lineage *in vitro*. *J Bone Miner Res* **19**, 1379–1394 (2004).

306. A. J. Friedenstein, N. V. Latzinik, Gorskaya, E. A. Luria, and I. L. Moskvina, Bone marrow stromal colony formation requires stimulation by haemopoietic cells. *Bone Miner* **18**, 199–213 (1992).

307. Z. Hou, Q. Nguyen, B. Frenkel, S. K. Nilsson, M. Milne, A. J. van Wijnen, J. L. Stein, P. Quesenberry, J. B. Lian, and G. S. Stein, Osteoblast-specific gene expression after transplantation of marrow cells: Implications for skeletal gene therapy. *Proc Natl Acad Sci USA* **96**, 7294–7299 (1999).

308. P. H. Krebsbach, S. A. Kuznetsov, K. Satomura, R. V. Emmons, D. W. Rowe, and P. G. Robey, Bone formation *in vivo*: Comparison of osteogenesis by transplanted mouse and human marrow stromal fibroblasts. *Transplantation* **63**, 1059–1069 (1997).

309. A. J. Friedenstein, R. K. Chailakhyan, and U. V. Gerasimov, Bone marrow osteogenic stem cells: *In vitro* cultivation and

transplantation in diffusion chambers. *Cell Tissue Kinet* **20**, 263–272 (1987).

310. S. P. Bruder, N. S. Ricalton, R. E. Boynton, T. J. Connolly, N. Jaiswal, J. Zaia, and F. P. Barry, Mesenchymal stem cell surface antigen SB-10 corresponds to activated leukocyte cell adhesion molecule and is involved in osteogenic differentiation. *J Bone Miner Res* **13**, 655–663 (1998).

311. E. A. Jones, S. E. Kinsey, A. English, R. A. Jones, L. Straszynski, D. M. Meredith, A. F. Markham, A. Jack, P. Emery, and D. McGonagle, Isolation and characterization of bone marrow multipotential mesenchymal progenitor cells. *Arthritis Rheum* **46**, 3349–3360 (2002).

312. K. Stewart, P. Monk, S. Walsh, C. M. Jefferiss, J. Letchford, and J. N. Beresford, STRO-1, HOP-26 (CD63), CD49a, and SB-10 (CD166) as markers of primitive human marrow stromal cells and their more differentiated progeny: A comparative investigation *in vitro*. *Cell Tissue Res* **313**, 281–290 (2003).

313. L. Malaval, F. Liu, A. B. Vernallis, and J. E. Aubin, GP130/OSMR is the only LIF/IL-6 family receptor complex to promote osteoblast differentiation of calvaria progenitors. *J Cell Physiol* **204**, 585–593 (2005).

314. L. Malaval and J. E. Aubin, Biphasic effects of leukemia inhibitory factor on osteoblastic differentiation. *J Cell Biochem* **81**, 63–70 (2001).

315. K. Yu, J. Xu, Z. Liu, D. Sosic, J. Shao, E. N. Olson, D. A. Towler, and D. M. Ornitz, Conditional inactivation of FGF receptor 2 reveals an essential role for FGF signaling in the regulation of osteoblast function and bone growth. *Development* **130**, 3063–3074 (2003).

316. A. Fakhry, C. Ratisoontorn, C. Vedhachalam, I. Salhab, E. Koyama, P. Leboy, M. Pacifici, R. E. Kirschner, and H. D. Nah, Effects of FGF-2/-9 in calvarial bone cell cultures: Differentiation stage-dependent mitogenic effect, inverse regulation of BMP-2 and noggin, and enhancement of osteogenic potential. *Bone* **36**, 254–266 (2005).

317. F. Liu, L. Malaval, and J. E. Aubin, Global amplification polymerase chain reaction reveals novel transitional stages during osteoprogenitor differentiation. *J Cell Sci* **116**, 1787–1796 (2003).

318. D. H. Kim, K. H. Yoo, K. S. Choi, J. Choi, S. Y. Choi, S. E. Yang, Y. S. Yang, H. J. Im, K. H. Kim, H. L. Jung, K. W. Sung, and H. H. Koo, Gene expression profile of cytokine and growth factor during differentiation of bone marrow-derived mesenchymal stem cell. *Cytokine* **31**, 119–126 (2005).

319. G. K. Chan, D. Miao, R. Deckelbaum, I. Bolivar, A. Karaplis, and D. Goltzman, Parathyroid hormone-related peptide interacts with bone morphogenetic protein 2 to increase osteoblastogenesis and decrease adipogenesis in pluripotent C3H10T 1/2 mesenchymal cells. *Endocrinology* **144**, 5511–5520 (2003).

320. H. Sowa, H. Kaji, G. N. Hendy, L. Canaff, T. Komori, T. Sugimoto, and K. Chihara, Menin is required for bone morphogenetic protein 2- and transforming growth factor beta-regulated osteoblastic differentiation through interaction with Smads and Runx2. *J Biol Chem* **279**, 40267–40275 (2004).

321. Y. Liu, R. A. Bhat, L. M. Seestaller-Wehr, S. Fukayama, A. Mangine, R. A. Moran, B. S. Komm, P. V. Bodine, and J. Billiard, The orphan receptor tyrosine kinase Ror2 promotes osteoblast differentiation and enhances *ex vivo* bone formation. *Mol Endocrinol* **21**, 376–387 (2007).

322. J. Settleman, Tension precedes commitment—Even for a stem cell. *Mol Cell* **14**, 148–150 (2004).

323. R. McBeath, D. M. Pirone, C. M. Nelson, K. Bhadriraju, and C. S. Chen, Cell shape, cytoskeletal tension, and RhoA regulate stem cell lineage commitment. *Dev Cell* **6**, 483–495 (2004).

324. M. Zayzafoon, W. E. Gathings, and J. M. McDonald, Modeled microgravity inhibits osteogenic differentiation of human mesenchymal stem cells and increases adipogenesis. *Endocrinology* **145**, 2421–2432 (2004).

325. C. Ontiveros, R. Irwin, R. W. Wiseman, and L. R. McCabe, Hypoxia suppresses Runx2 independent of modeled microgravity. *J Cell Physiol* **200**, 169–176 (2004).

326. Q. Q. Tang, T. C. Otto, and M. D. Lane, CCAAT/enhancer-binding protein beta is required for mitotic clonal expansion during adipogenesis. *Proc Natl Acad Sci USA* **100**, 850–855 (2003).

327. E. Mueller, S. Drori, A. Aiyer, J. Yie, P. Sarraf, H. Chen, S. Hauser, E. D. Rosen, K. Ge, R. G. Roeder, and B. M. Spiegelman, Genetic analysis of adipogenesis through peroxisome proliferator–activated receptor gamma isoforms. *J Biol Chem* **277**, 41925–41930 (2002).

328. B. de Crombrugghe, V. Lefebvre, R. R. Behringer, W. Bi, S. Murakami, and W. Huang, Transcriptional mechanisms of chondrocyte differentiation. *Matrix Biol* **19**, 389–394 (2000).

329. J. Skillington, L. Choy, and R. Derynck, Bone morphogenetic protein and retinoic acid signaling cooperate to induce osteoblast differentiation of preadipocytes. *J Cell Biol* **159**, 135–146 (2002).

330. X. Shi, W. Shi, Q. Li, B. Song, M. Wan, S. Bai, and X. Cao, A glucocorticoid-induced leucine-zipper protein GILZ inhibits adipogenesis of mesenchymal cells. *EMBO Rep* **4**, 374–380 (2003).

331. F. Gori, T. Thomas, K. C. Hicok, T. C. Spelsberg, and B. L. Riggs, Differentiation of human marrow stromal precursor cells: Bone morphogenetic protein-2 increases OSF2/CBFA1, enhances osteoblast commitment, and inhibits late adipocyte maturation. *J Bone Miner Res* **14**, 1522–1535 (1999).

332. B. Lecka-Czernik, I. Gubrij, E. J. Moerman, O. Kajkenova, D. A. Lipschitz, S. C. Manolagas, and R. L. Jilka, Inhibition of Osf2/Cbfa1 expression and terminal osteoblast differentiation by PPARgamma2. *J Cell Biochem* **74**, 357–371 (1999).

333. M. H. Lee, A. Javed, H. J. Kim, H. I. Shin, S. Gutierrez, J. Y. Choi, V. Rosen, J. L. Stein, A. J. van Wijnen, G. S. Stein, J. B. Lian, and H. M. Ryoo, Transient upregulation of CBFA1 in response to bone morphogenetic protein-2 and transforming growth factor B1 in C2C12 myogenic cells coincides with suppression of the myogenic phenotype but is not sufficient for osteoblast differentiation. *J Cell Biochem* **73**, 114–125 (1999).

334. M. J. Jeon, J. A. Kim, S. H. Kwon, S. W. Kim, K. S. Park, S. W. Park, S. Y. Kim, and C. S. Shin, Activation of peroxisome proliferator-activated receptor-gamma inhibits the Runx2-mediated transcription of osteocalcin in osteoblasts. *J Biol Chem* **278**, 23270–23277 (2003).

335. T. Akune, S. Ohba, S. Kamekura, M. Yamaguchi, U. I. Chung, N. Kubota, Y. Terauchi, Y. Harada, Y. Azuma, K. Nakamura, T. Kadowaki, and H. Kawaguchi, PPARgamma insufficiency enhances osteogenesis through osteoblast formation from bone marrow progenitors. *J Clin Invest* **113**, 846–855 (2004).

336. H. Kobayashi, Y. Gao, C. Ueta, A. Yamaguchi, and T. Komori, Multilineage differentiation of Cbfa1-deficient calvarial cells *in vitro*. *Biochem Biophys Res Commun* **273**, 630–636 (2000).

337. S. C. Miller and W. S. Jee, The bone lining cell: A distinct phenotype? *Calcif Tissue Int* **41**, 1–5 (1987).

338. D. A. Cameron, The Golgi apparatus in bone and cartilage cells. *Clin Orthop* **58,** 191–211 (1968).

339. D. A. Hanson and D. R. Eyre, Molecular site specificity of pyridinoline and pyrrole cross-links in type I collagen of human bone. *J Biol Chem* **271,** 26508–26516 (1996).

340. M. J. Glimcher, Mechanism of calcification: Role of collagen fibrils and collagen–phosphoprotein complexes *in vitro* and *in vivo. Anat Rec* **224,** 139–153 (1989).

341. E. P. Paschalis, K. Verdelis, S. B. Doty, A. L. Boskey, R. Mendelsohn, and M. Yamauchi, Spectroscopic characterization of collagen cross-links in bone. *J Bone Miner Res* **16,** 1821–1828 (2001).

342. M. J. Glimcher, The nature of mineral phase in bone, biological and clinical implications. In *Metabolic Bone Disease and Clinically Related Disorders* (L. V. Avioli and S. M. Krane, eds.), pp. 23–46. Academic Press, London (1998).

343. A. L. Boskey, S. Gadaleta, C. Gundberg, S. B. Doty, P. Ducy, and G. Karsenty, Fourier transformed infrared microspectroscopic analysis of bones of osteocalcin deficient mice provides insight into the function of osteocalcin. *Bone* **23,** 187 (1998).

344. P. Ducy, C. Desbois, B. Boyce, G. Pinero, B. Story, C. Dunstan, E. Smith, J. Bonadio, S. Goldstein, C. Gundberg, A. Bradley, and G. Karsenty, Increased bone formation in osteocalcin-deficient mice. *Nature* **382,** 448–452 (1996).

345. H. Yoshitake, S. R. Rittling, D. T. Denhardt, and M. Noda, Osteopontin-deficient mice are resistant to ovariectomy-induced bone resorption. *Proc Natl Acad Sci USA* **96,** 8156–8160 (1999). [Published erratum appears in *Proc Natl Acad Sci USA* 1999 Sep 14;**96**(19):10944.]

346. T. Xu, P. Bianco, L. W. Fisher, G. Longenecker, E. Smith, S. Goldstein, J. Bonadio, A. Boskey, A. M. Heegaard, B. Sommer, K. Satomura, P. Dominguez, C. Zhao, A. B. Kulkarni, P. G. Robey, and M. F. Young, Targeted disruption of the biglycan gene leads to an osteoporosis-like phenotype in mice. *Nat Genet* **20,** 78–82 (1998).

347. R. V. Sugars, A. M. Milan, J. O. Brown, R. J. Waddington, R. C. Hall, and G. Embery, Molecular interaction of recombinant decorin and biglycan with type I collagen influences crystal growth. *Connect Tissue Res* **44**(Suppl. 1), 189–195 (2003).

348. M. F. Young, Y. Bi, L. Ameye, and X. D. Chen, Biglycan knockout mice: New models for musculoskeletal diseases. *Glycoconj J* **19,** 257–262 (2002).

349. L. Ameye and M. F. Young, Mice deficient in small leucine-rich proteoglycans: Novel *in vivo* models for osteoporosis, osteoarthritis, Ehlers–Danlos syndrome, muscular dystrophy, and corneal diseases. *Glycobiology* **12,** 107R–116R (2002).

350. D. Parisuthiman, Y. Mochida, W. R. Duarte, and M. Yamauchi, Biglycan modulates osteoblast differentiation and matrix mineralization. *J Bone Miner Res* **20,** 1878–1886 (2005).

351. T. P. Strandjord, D. K. Madtes, D. J. Weiss, and E. H. Sage, Collagen accumulation is decreased in SPARC-null mice with bleomycin-induced pulmonary fibrosis. *Am J Physiol* **277,** L628–L635 (1999).

352. A. M. Delany, M. Amling, M. Priemel, C. Howe, R. Baron, and E. Canalis, Osteopenia and decreased bone formation in osteonectin-deficient mice. *J Clin Invest* **105,** 915–923 (2000).

353. A. M. Delany, I. Kalajzic, A. D. Bradshaw, E. H. Sage, and F. Canalis, Osteonectin-null mutation compromises osteoblast formation, maturation, and survival. *Endocrinology* **144,** 2588–2596 (2003).

354. A. L. Boskey, D. J. Moore, M. Amling, E. Canalis, and A. M. Delany, Infrared analysis of the mineral and matrix in bones of osteonectin-null mice and their wildtype controls. *J Bone Miner Res* **18,** 1005–1011 (2003).

355. C. E. Tye, G. K. Hunter, and H. A. Goldberg, Identification of the type I collagen-binding domain of bone sialoprotein and characterization of the mechanism of interaction. *J Biol Chem* **280,** 13487–13492 (2005).

356. C. E. Tye, K. R. Rattray, K. J. Warner, J. A. Gordon, J. Sodek, G. K. Hunter, and H. A. Goldberg, Delineation of the hydroxyapatite-nucleating domains of bone sialoprotein. *J Biol Chem* **278,** 7949–7955 (2003).

357. L. C. Gowen, D. N. Petersen, A. L. Mansolf, H. Qi, J. L. Stock, G. T. Tkalcevic, H. A. Simmons, D. T. Crawford, K. L. Chidsey-Frink, H. Z. Ke, J. D. McNeish, and T. A. Brown, Targeted disruption of the osteoblast/osteocyte factor 45 gene (OF45) results in increased bone formation and bone mass. *J Biol Chem* **278,** 1998–2007 (2003).

358. P. S. Rowe, The wrickkened pathways of FGF23, MEPE, and PHEX. *Crit Rev Oral Biol Med* **15,** 264–281 (2004).

359. S. A. Steitz, M. Y. Speer, M. D. McKee, L. Liaw, M. Almeida, H. Yang, and C. M. Giachelli, Osteopontin inhibits mineral deposition and promotes regression of ectopic calcification. *Am J Pathol* **161,** 2035–2046 (2002).

360. S. R. Rittling, H. N. Matsumoto, M. D. McKee, A. Nanci, X. R. An, K. E. Novick, A. J. Kowalski, M. Noda, and D. T. Denhardt, Mice lacking osteopontin show normal development and bone structure but display altered osteoclast formation *in vitro. J Bone Miner Res* **13,** 1101–1111 (1998).

361. M. Y. Speer, M. D. McKee, R. E. Guldberg, L. Liaw, H. Y. Yang, E. Tung, G. Karsenty, and C. M. Giachelli, Inactivation of the osteopontin gene enhances vascular calcification of matrix Gla protein-deficient mice: Evidence for osteopontin as an inducible inhibitor of vascular calcification *in vivo. J Exp Med* **196,** 1047–1055 (2002).

362. S. A. Shapses, M. Cifuentes, L. Spevak, H. Chowdhury, J. Brittingham, A. L. Boskey, and D. T. Denhardt, Osteopontin facilitates bone resorption, decreasing bone mineral crystallinity and content during calcium deficiency. *Calcif Tissue Int* **73,** 86–92 (2003).

363. A. L. Boskey, L. Spevak, E. Paschalis, S. B. Doty, and M. D. McKee, Osteopontin deficiency increases mineral content and mineral crystallinity in mouse bone. *Calcif Tissue Int* **71,** 145–154 (2002).

364. N. L. Huq, K. J. Cross, M. Ung, and E. C. Reynolds, A review of protein structure and gene organisation for proteins associated with mineralised tissue and calcium phosphate stabilisation encoded on human chromosome 4. *Arch Oral Biol* **50,** 599–609 (2005).

365. L. W. Fisher and N. S. Fedarko, Six genes expressed in bones and teeth encode the current members of the SIBLING family of proteins. *Connect Tissue Res* **44**(Suppl. 1), 33–40 (2003).

366. W. Yang, Y. Lu, I. Kalajzic, D. Guo, M. A. Harris, J. Gluhak-Heinrich, S. Kotha, L. F. Bonewald, J. Q. Feng, D. W. Rowe, C. H. Turner, A. G. Robling, and S. E. Harris, Dentin matrix protein 1 gene cis-regulation: Use in osteocytes to characterize local responses to mechanical loading *in vitro* and *in vivo. J Biol Chem* **280,** 20680–20690 (2005).

367. L. Ye, M. MacDougall, S. Zhang, Y. Xie, J. Zhang, Z. Li, Y. Lu, Y. Mishina, and J. Q. Feng, Deletion of dentin matrix protein-1 leads to a partial failure of maturation of predentin into dentin hypomineralization and expanded cavities of pulp and root canal during postnatal tooth development. *J Biol Chem* **279,** 19141–19148 (2004).

368. Y. Ling, H. F. Rios, E. R. Myers, Y. Lu, J. Q. Feng, and A. L. Boskey, DMP1 depletion decreases bone mineralization *in vivo*: An FTIR imaging analysis. *J Bone Miner Res* **20**, 2169–2177 (2005).

369. P. H. Tartaix, M. Doulaverakis, A. George, L. W. Fisher, W. T. Butler, C. Qin, E. Salih, M. Tan, Y. Fujimoto, L. Spevak, and A. L. Boskey, *In vitro* effects of dentin matrix protein-1 on hydroxyapatite formation provide insights into *in vivo* functions. *J Biol Chem* **279**, 18115–18120 (2004).

370. G. K. Hunter, M. S. Poitras, T. M. Underhill, M. D. Grynpas, and H. A. Goldberg, Induction of collagen mineralization by a bone sialoprotein–decorin chimeric protein. *J Biomed Mater Res* **55**, 496–502 (2001).

371. P. S. Rowe, Y. Kumagai, G. Gutierrez, I. R. Garrett, R. Blacher, D. Rosen, J. Cundy, S. Navvab, D. Chen, M. K. Drezner, L. D. Quarles, and G. R. Mundy, MEPE has the properties of an osteoblastic phosphatonin and minhibin. *Bone* **34**, 303–319 (2004).

372. A. M. Ho, M. D. Johnson, and D. M. Kingsley, Role of the mouse ank gene in control of tissue calcification and arthritis. *Science* **289**, 265–270 (2000).

373. F. T. Hakim, R. Cranley, K. S. Brown, E. D. Eanes, L. Harne, and J. J. Oppenheim, Hereditary joint disorder in progressive ankylosis (ank/ank) mice: I. Association of calcium hydroxyapatite deposition with inflammatory arthropathy. *Arthritis Rheum* **27**, 1411–1420 (1984).

374. P. S. Henthorn, M. Raducha, K. N. Fedde, M. A. Lafferty, and M. P. Whyte, Different missense mutations at the tissue-nonspecific alkaline phosphatase gene locus in autosomal recessively inherited forms of mild and severe hypophosphatasia. *Proc Natl Acad Sci USA* **89**, 9924–9928 (1992).

375. K. G. Waymire, J. D. Mahuren, J. M. Jaje, T. R. Guilarte, S. P. Coburn, and G. R. MacGregor, Mice lacking tissue non-specific alkaline phosphatase die from seizures due to defective metabolism of vitamin B-6. *Nat Genet* **11**, 45–51 (1995).

376. S. Narisawa, N. Frohlander, and J. L. Millan, Inactivation of two mouse alkaline phosphatase genes and establishment of a model of infantile hypophosphatasia. *Dev Dyn* **208**, 432–446 (1997).

377. K. N. Fedde, L. Blair, J. Silverstein, S. P. Coburn, L. M. Ryan, R. S. Weinstein, K. Waymire, S. Narisawa, J. L. Millan, G. R. MacGregor, and M. P. Whyte, Alkaline phosphatase knockout mice recapitulate the metabolic and skeletal defects of infantile hypophosphatasia. *J Bone Miner Res* **14**, 2015–2026 (1999).

378. D. Harmey, L. Hessle, S. Narisawa, K. A. Johnson, R. Terkeltaub, and J. L. Millan, Concerted regulation of inorganic pyrophosphate and osteopontin by akp2, enpp1, and ank: An integrated model of the pathogenesis of mineralization disorders. *Am J Pathol* **164**, 1199–1209 (2004).

379. H. C. Anderson, J. B. Sipe, L. Hessle, R. Dhanyamraju, E. Atti, N. P. Camacho, and J. L. Millan, Impaired calcification around matrix vesicles of growth plate and bone in alkaline phosphatase-deficient mice. *Am J Pathol* **164**, 841–847 (2004).

380. H. C. Anderson, R. Garimella, and S. E. Tague, The role of matrix vesicles in growth plate development and biomineralization. *Frontiers Biosci* **10**, 822–837 (2005).

381. M. Murshed, D. Harmey, J. L. Millan, M. D. McKee, and G. Karsenty, Unique coexpression in osteoblasts of broadly expressed genes accounts for the spatial restriction of ECM mineralization to bone. *Genes Dev* **19**, 1093–1104 (2005).

382. C. Wennberg, L. Hessle, P. Lundberg, S. Mauro, S. Narisawa, U. H. Lerner, and J. L. Millan, Functional characterization of osteoblasts and osteoclasts from alkaline phosphatase knockout mice. *J Bone Miner Res* **15**, 1879–1888 (2000).

383. H. C. Anderson, D. Harmey, N. P. Camacho, R. Garimella, J. B. Sipe, S. Tague, X. Bi, K. Johnson, R. Terkeltaub, and J. L. Millan, Sustained osteomalacia of long bones despite major improvement in other hypophosphatasia-related mineral deficits in tissue nonspecific alkaline phosphatase/nucleotide pyrophosphatase phosphodiesterase 1 double-deficient mice. *Am J Pathol* **166**, 1711–1720 (2005).

384. A. Okawa, I. Nakamura, S. Goto, H. Moriya, Y. Nakamura, and S. Ikegawa, Mutation in Npps in a mouse model of ossification of the posterior longitudinal ligament of the spine. *Nat Genet* **19**, 271–273 (1998).

385. L. Hessle, K. A. Johnson, H. C. Anderson, S. Narisawa, A. Sali, J. W. Goding, R. Terkeltaub, and J. L. Millan, Tissue-nonspecific alkaline phosphatase and plasma cell membrane glycoprotein-1 are central antagonistic regulators of bone mineralization. *Proc Natl Acad Sci USA* **99**, 9445–9449 (2002).

386. T. Yoshizawa, Y. Handa, Y. Uematsu, S. Takeda, K. Sekine, Y. Yoshihara, T. Kawakami, K. Arioka, H. Sato, Y. Uchiyama, S. Masushige, A. Fukamizu, T. Matsumoto, and S. Kato, Mice lacking the vitamin D receptor exhibit impaired bone formation, uterine hypoplasia, and growth retardation after weaning. *Nat Genet* **16**, 391–396 (1997).

387. M. Amling, M. Priemel, T. Holzmann, K. Chapin, J. M. Rueger, R. Baron, and M. B. Demay, Rescue of the skeletal phenotype of vitamin D receptor-ablated mice in the setting of normal mineral ion homeostasis: Formal histomorphometric and biomechanical analyses. *Endocrinology* **140**, 4982–4987 (1999).

388. Y. C. Li, A. E. Pirro, M. Amling, G. Delling, R. Baron, R. Bronson, and M. B. Demay, Targeted ablation of the vitamin D receptor: An animal model of vitamin D-dependent rickets type II with alopecia. *Proc Natl Acad Sci USA* **94**, 9831–9835 (1997).

389. B. M. Misof, P. Roschger, W. Tesch, P. A. Baldock, A. Valenta, P. Messmer, J. A. Eisman, A. L. Boskey, E. M. Gardiner, P. Fratzl, and K. Klaushofer, Targeted overexpression of vitamin D receptor in osteoblasts increases calcium concentration without affecting structural properties of bone mineral crystals. *Calcif Tissue Int* **73**, 251–257 (2003).

390. Y. K. Yee, S. R. Chintalacharuvu, J. Lu, and S. Nagpal, Vitamin D receptor modulators for inflammation and cancer. *Mini Rev Med Chem* **5**, 761–778 (2005).

391. E. M. Brown and R. J. MacLeod, Extracellular calcium sensing and extracellular calcium signaling. *Physiol Rev* **81**, 239–297 (2001).

392. T. A. Hough, D. Bogani, M. T. Cheeseman, J. Favor, M. A. Nesbit, R. V. Thakker, and M. F. Lyon, Activating calcium-sensing receptor mutation in the mouse is associated with cataracts and ectopic calcification. *Proc Natl Acad Sci USA* **101**, 13566–13571 (2004).

393. B. Bielesz, K. Klaushofer, and R. Oberbauer, Renal phosphate loss in hereditary and acquired disorders of bone mineralization. *Bone* **35**, 1229–1239 (2004).

394. X. Bai, D. Miao, D. Panda, S. Grady, M. D. McKee, D. Goltzman, and A. C. Karaplis, Partial rescue of the Hyp phenotype by osteoblast-targeted PHEX (phosphate-regulating gene with homologies to endopeptidases on the X chromosome) expression. *Mol Endocrinol* **16**, 2913–2925 (2002).

395. D. Miao, X. Bai, D. Panda, M. McKee, A. Karaplis, and D. Goltzman, Osteomalacia in Hyp mice is associated with abnormal phex expression and with altered bone matrix protein expression and deposition. *Endocrinology* **142**, 926–939 (2001).

396. R. G. Erben, D. Mayer, K. Weber, K. Jonsson, H. Juppner, and B. Lanske, Overexpression of human PHEX under the human beta-actin promoter does not fully rescue the Hyp mouse phenotype. *J Bone Miner Res* **20,** 1149–1160 (2005).

397. T. Shimada, T. Muto, I. Urakawa, T. Yoneya, Y. Yamazaki, K. Okawa, Y. Takeuchi, T. Fujita, S. Fukumoto, and T. Yamashita, Mutant FGF-23 responsible for autosomal dominant hypophosphatemic rickets is resistant to proteolytic cleavage and causes hypophosphatemia *in vivo*. *Endocrinology* **143,** 3179–3182 (2002).

398. T. Berndt, T. A. Craig, A. E. Bowe, J. Vassiliadis, D. Reczek, R. Finnegan, S. M. Jan De Beur, S. C. Schiavi, and R. Kumar, Secreted frizzled-related protein 4 is a potent tumor-derived phosphaturic agent. *J Clin Invest* **112,** 785–794 (2003).

399. T. Larsson, R. Marsell, E. Schipani, C. Ohlsson, O. Ljunggren, H. S. Tenenhouse, H. Juppner, and K. B. Jonsson, Transgenic mice expressing fibroblast growth factor 23 under the control of the alpha1(I) collagen promoter exhibit growth retardation, osteomalacia, and disturbed phosphate homeostasis. *Endocrinology* **145,** 3087–3094 (2004).

400. T. Shimada, Y. Yamazaki, M. Takahashi, H. Hasegawa, I. Urakawa, T. Oshima, K. Ono, M. Kakitani, K. Tomizuka, T. Fujita, S. Fukumoto, and T. Yamashita, Vitamin D receptor-independent FGF23 actions in regulating phosphate and vitamin D metabolism. *Am J Physiol Renal Physiol* **289,** F1088–F1095 (2005).

401. C. R. Dunstan, H. Zhou, and M. J. Seibel, Fibroblast growth factor 23: A phosphatonin regulating phosphate homeostasis? *Endocrinology* **145,** 3084–3086 (2004).

402. G. Marotti, The structure of bone tissues and the cellular control of their deposition. *Ital J Anat Embryol* **101,** 25–79 (1996).

403. M. Owen, Marrow stromal stem cells. *J Cell Sci* **10,** 63–76 (1988).

404. T. A. Franz-Odendaal, B. K. Hall, and P. E. Witten, Buried alive: How osteoblasts become osteocytes. *Dev Dyn* **235,** 176–190 (2006).

405. C. Palumbo, M. Ferretti, and G. Marotti, Osteocyte dendrogenesis in static and dynamic bone formation: An ultrastructural study. *Anat Rec A Discov Mol Cell Evol Biol* **278,** 474–480 (2004).

406. C. Palumbo, S. Palazzini, and G. Marotti, Morphological study of intercellular junctions during osteocyte differentiation. *Bone* **11,** 401–406 (1990).

407. C. Barragan-Adjemian, D. Nicolella, V. Dusevich, M. R. Dallas, J. D. Eick, and L. F. Bonewald, Mechanism by which MLO-A5 late osteoblasts/early osteocytes mineralize in culture: Similarities with mineralization of lamellar bone. *Calcif Tissue Int* **79,** 340–353 (2006).

408. A. van der Plas and P. J. Nijweide, Isolation and purification of osteocytes. *J Bone Miner Res* **7,** 389–396 (1992).

409. L. F. Bonewald, Mechanosensation and transduction in osteocytes. *IBMS BoneKEy-Osteovision* **3,** 7–15 (2006).

410. C. T. Rubin and L. E. Lanyon, Osteoregulatory nature of mechanical stimuli: Function as a determinant for adaptive remodeling in bone. *J Orthop Res* **5,** 300–310 (1987).

411. S. C. Manolagas, Choreography from the tomb: An emerging role of dying osteocytes in the purposeful, and perhaps not so purposeful, targeting of bone remodeling. *IBMS BoneKEy-Osteovision* **3,** 5–14 (2006).

412. E. M. Aarden, E. H. Burger, and P. J. Nijweide, Function of osteocytes in bone. *J Cell Biochem* **55,** 287–299 (1994).

413. J. I. Aguirre, L. I. Plotkin, S. A. Stewart, R. S. Weinstein, A. M. Parfitt, S. C. Manolagas, and T. Bellido, Osteocyte apoptosis is induced by weightlessness in mice and precedes osteoclast recruitment and bone loss. *J Bone Miner Res* **21,** 605–615 (2006).

414. A. Bakker, J. Klein-Nulend, and E. Burger, Shear stress inhibits while disuse promotes osteocyte apoptosis. *Biochem Biophys Res Commun* **320,** 1163–1168 (2004).

415. B. S. Noble, N. Peet, H. Y. Stevens, A. Brabbs, J. R. Mosley, G. C. Reilly, J. Reeve, T. M. Skerry, and L. E. Lanyon, Mechanical loading: Biphasic osteocyte survival and targeting of osteoclasts for bone destruction in rat cortical bone. *Am J Physiol Cell Physiol* **284,** C934–C943 (2003).

416. E. H. Burger, J. Klein-Nulend, and T. H. Smit, Strain-derived canalicular fluid flow regulates osteoclast activity in a remodelling osteon—A proposal. *J Biomech* **36,** 1453–1459 (2003).

417. T. S. Gross, J. L. Edwards, K. J. McLeod, and C. T. Rubin, Strain gradients correlate with sites of periosteal bone formation. *J Bone Miner Res* **12,** 982–988 (1997).

418. D. B. Burr, A. G. Robling, and C. H. Turner, Effects of biomechanical stress on bones in animals. *Bone* **30,** 781–786 (2002).

419. P. J. Ehrlich, B. S. Noble, H. L. Jessop, H. Y. Stevens, J. R. Mosley, and L. E. Lanyon, The effect of *in vivo* mechanical loading on estrogen receptor alpha expression in rat ulnar osteocytes. *J Bone Miner Res* **17,** 1646–1655 (2002).

420. T. M. Skerry, L. Bitensky, J. Chayen, and L. E. Lanyon, Early strain-related changes in enzyme activity in osteocytes following bone loading *in vivo*. *J Bone Miner Res* **4,** 783–788 (1989).

421. D. M. Raab-Cullen, M. A. Thiede, D. N. Petersen, D. B. Kimmel, and R. R. Recker, Mechanical loading stimulates rapid changes in periosteal gene expression. *Calcif Tissue Int* **55,** 473–478 (1994).

422. J. Gluhak-Heinrich, L. Ye, L. F. Bonewald, J. Q. Feng, M. MacDougall, S. E. Harris, and D. Pavlin, Mechanical loading stimulates dentin matrix protein 1 (DMP1) expression in osteocytes *in vivo*. *J Bone Miner Res* **18,** 807–817 (2003).

423. P. P. Cherian, A. J. Siller-Jackson, S. Gu, X. Wang, L. F. Bonewald, E. Sprague, and J. X. Jiang, Mechanical strain opens connexin 43 hemichannels in osteocytes: A novel mechanism for the release of prostaglandin. *Mol Biol Cell* **16,** 3100–3106 (2005).

424. H. Z. Ke, H. Qi, A. F. Weidema, Q. Zhang, N. Panupinthu, D. T. Crawford, W. A. Grasser, V. M. Paralkar, M. Li, L. P. Audoly, C. A. Gabel, W. S. Jee, S. J. Dixon, S. M. Sims, and D. D. Thompson, Deletion of the P2X7 nucleotide receptor reveals its regulatory roles in bone formation and resorption. *Mol Endocrinol* **17,** 1356–1367 (2003).

425. N. R. Jorgensen, S. C. Teilmann, Z. Henriksen, R. Civitelli, O. H. Sorensen, and T. H. Steinberg, Activation of L-type calcium channels is required for gap junction-mediated intercellular calcium signaling in osteoblastic cells. *J Biol Chem* **278,** 4082–4086 (2003).

426. B. Cheng, Y. Kato, S. Zhao, J. Luo, E. Sprague, L. F. Bonewald, and J. X. Jiang, PGE(2) is essential for gap junction-mediated intercellular communication between osteocyte-like MLO-Y4 cells in response to mechanical strain. *Endocrinology* **142,** 3464–3473 (2001).

427. J. Li, R. L. Duncan, D. B. Burr, and C. H. Turner, L-type calcium channels mediate mechanically induced bone formation *in vivo*. *J Bone Miner Res* **17,** 1795–1800 (2002).

428. J. Li, D. Liu, H. Z. Ke, R. L. Duncan, and C. H. Turner, The P2X7 nucleotide receptor mediates skeletal mechanotransduction. *J Biol Chem* **280,** 42952–42959 (2005).

429. T. N. McAllister, T. Du, and J. A. Frangos, Fluid shear stress stimulates prostaglandin and nitric oxide release in bone marrow-derived preosteoclast-like cells. *Biochem Biophys Res Commun* **270,** 643–648 (2000).

430. A. A. Pitsillides, S. C. Rawlinson, R. F. Suswillo, S. Bourrin, G. Zaman, and L. E. Lanyon, Mechanical strain-induced NO production by bone cells: A possible role in adaptive bone (re)modeling? *FASEB J* **9,** 1614–1622 (1995).

431. E. H. Burger, J. Klein-Nulend, C. M. Semeins, N. E. Ajubi, and P. J. Nijweide, Osteocytes but not periosteal fibroblasts produce nitric oxide (NO) in response to pulsatile fluid flow. *Trans Orthop Res Soc* **21,** 531 (1996). [Abstract]

432. K. H. Lau, S. Kapur, C. Kesavan, and D. J. Baylink, Upregulation of the Wnt, estrogen receptor, insulin-like growth factor-I, and bone morphogenetic protein pathways in C57BL/6J osteoblasts as opposed to C3H/HeJ osteoblasts in part contributes to the differential anabolic response to fluid shear. *J Biol Chem* **281,** 9576–9588 (2006).

433. A. J. Cheline, A. H. Reddi, and R. B. Martin, Bone morphogenetic protein-7 selectively enhances mechanically induced bone formation. *Bone* **31,** 570–574 (2002).

434. K. Sawakami, A. G. Robling, M. Ai, N. D. Pitner, D. Liu, S. J. Warden, J. Li, P. Maye, D. W. Rowe, R. L. Duncan, M. L. Warman, and C. H. Turner, The Wnt co-receptor LRP5 is essential for skeletal mechanotransduction but not for the anabolic bone response to parathyroid hormone treatment. *J Biol Chem* **281,** 23698–23711 (2006).

435. J. Q. Feng, L. M. Ward, S. Liu, Y. Lu, Y. Xie, B. Yuan, X. Yu, F. Rauch, S. I. Davis, S. Zhang, H. Rios, M. K. Drezner, L. D. Quarles, L. F. Bonewald, and K. E. White, Loss of DMP1 causes rickets and osteomalacia and identifies a role for osteocytes in mineral metabolism. *Nat Genet* **38,** 1310–1315 (2006).

436. J. Rubin, T. Murphy, M. S. Nanes, and X. Fan, Mechanical strain inhibits expression of osteoclast differentiation factor by murine stromal cells. *Am J Physiol Cell Physiol* **278,** C1126–C1132 (2000).

437. A. Liedert, D. Kaspar, R. Blakytny, L. Claes, and A. Ignatius, Signal transduction pathways involved in mechanotransduction in bone cells. *Biochem Biophys Res Commun* **349,** 1–5 (2006).

438. J. Rubin, C. Rubin, and C. R. Jacobs, Molecular pathways mediating mechanical signaling in bone. *Gene* **367,** 1–16 (2006).

439. R. O. Hynes, Integrins: Bidirectional allosteric signaling machines. *Cell* **110,** 673–687 (2002).

440. M. C. Meazzini, C. D. Toma, J. L. Schaffer, M. L. Gray, and L. C. Gerstenfeld, Osteoblast cytoskeletal modulation in response to mechanical strain *in vitro*. *J Orthop Res* **16,** 170–180 (1998).

441. V. Singla and J. F. Reiter, The primary cilium as the cell's antenna: Signaling at a sensory organelle. *Science* **313,** 629–633 (2006).

442. Z. Xiao, S. Zhang, J. Mahlios, G. Zhou, B. S. Magenheimer, D. Guo, S. L. Dallas, R. Maser, J. P. Calvet, L. Bonewald, and L. D. Quarles, Cilia-like structures and polycystin-1 in osteoblasts/osteocytes and associated abnormalities in skeletogenesis and Runx2 expression. *J Biol Chem* **281,** 30884–30895 (2006).

443. R. L. Jilka, R. S. Weinstein, T. Bellido, A. M. Parfitt, and S. C. Manolagas, Osteoblast programmed cell death (apoptosis): Modulation by growth factors and cytokines. *J Bone Miner Res* **13,** 793–802 (1998).

444. R. K. Globus, S. B. Doty, J. C. Lull, E. Holmuhamedov, M. J. Humphries, and C. H. Damsky, Fibronectin is a survival factor for differentiated osteoblasts. *J Cell Sci* **111,** 1385–1393 (1998).

445. S. Zhao, Y. K. Zhang, S. Harris, S. S. Ahuja, and L. F. Bonewald, MLO-Y4 osteocyte-like cells support osteoclast formation and activation. *J Bone Miner Res* **17,** 2068–2079 (2002).

446. B. Noble, Microdamage and apoptosis. *Eur J Morphol* **42,** 91–98 (2005).

447. O. Verborgt, G. J. Gibson, and M. B. Schaffler, Loss of osteocyte integrity in association with microdamage and bone remodeling after fatigue *in vivo*. *J Bone Miner Res* **15,** 60–67 (2000).

448. A. L. J. J. Bronckers, W. Goei, G. Luo, G. Karsenty, R. N. D'Souza, D. M. Lyaruu, and E. H. Burger, DNA fragmentation during bone formation in neonatal rodents assessed by transferase-mediated end labeling. *J Bone Miner Res* **11,** 1281–1291 (1996).

449. B. S. Noble, H. Stevens, N. Loveridge, and J. Reeve, Identification of apoptotic changes in osteocytes in normal and pathological human bone. *Bone* **20,** 273–282 (1997).

450. R. S. Weinstein, R. L. Jilka, A. M. Parfitt, and S. C. Manolagas, Inhibition of osteoblastogenesis and promotion of apoptosis of osteoblasts and osteocytes by glucocorticoids. Potential mechanisms of their deleterious effects on bone. *J Clin Invest* **102,** 274–282 (1998).

451. A. Tomkinson, J. Reeve, R. W. Shaw, and B. S. Noble, The death of osteocytes via apoptosis accompanies estrogen withdrawal in human bone. *J Clin Endocrinol Metab* **82,** 3128–3135 (1997).

452. R. L. Jilka, R. S. Weinstein, T. Bellido, P. Roberson, A. M. Parfitt, and S. C. Manolagas, Increased bone formation by prevention of osteoblast apoptosis with parathyroid hormone. *J Clin Invest* **104,** 439–446 (1999).

453. L. I. Plotkin, R. S. Weinstein, A. M. Parfitt, P. K. Roberson, S. C. Manolagas, and T. Bellido, Prevention of osteocyte and osteoblast apoptosis by bisphosphonates and calcitonin. *J Clin Invest* **104,** 1363–1374 (1999).

454. N. Udagawa, N. Takahashi, T. Akatsu, H. Tanaka, T. Sasaki, T. Nishihara, T. Koga, T. J. Martin, and T. Suda, Origin of osteoclasts: Mature monocytes and macrophages are capable of differentiating into osteoclasts under a suitable microenvironment prepared by bone marrow-derived stromal cells. *Proc Natl Acad Sci USA* **87,** 7260–7264 (1990).

455. A. Bruzzaniti and R. Baron, Molecular regulation of osteoclast activity. *Rev Endocr Metab Disord* **7,** 123–139 (2006).

456. H. Kodama, A. Yamasaki, M. Abe, S. Niida, Y. Hakeda, and H. Kawashima, Transient recruitment of osteoclasts and expression of their function in osteopetrotic (op/op) mice by a single injection of macrophage colony-stimulating factor. *J Bone Miner Res* **8,** 45–50 (1993).

457. H. Yoshida, S. Hayashi, T. Kunisada, M. Ogawa, S. Nishikawa, H. Okamura, T. Sudo, and L. D. Shultz, The murine mutation osteopetrosis is in the coding region of the macrophage colony stimulating factor gene. *Nature* **345,** 442–444 (1990).

458. D. Cheleuitte, S. Mizuno, and J. Glowacki, *In vitro* secretion of cytokines by human bone marrow: Effects of age and estrogen status. *J Clin Endocrinol Metab* **83,** 2043–2051 (1998).

459. A. Zallone, Direct and indirect estrogen actions on osteoblasts and osteoclasts. *Ann N Y Acad Sci* **1068,** 173–179 (2006).

460. J. J. Cao, T. J. Wronski, U. Iwaniec, L. Phleger, P. Kurimoto, B. Boudignon, and B. P. Halloran, Aging increases stromal/osteoblastic cell-induced osteoclastogenesis and alters the osteoclast precursor pool in the mouse. *J Bone Miner Res* **20,** 1659–1668 (2005).

461. H. Yasuda, N. Shima, N. Nakagawa, K. Yamaguchi, M. Kinosaki, S. Mochizuki, A. Tomoyasu, K. Yano, M. Goto, A. Murakami, E. Tsuda, T. Morinaga, K. Higashio, N. Udagawa, N. Takahashi, and T. Suda, Osteoclast differentiation factor is a ligand for osteoprotegerin/osteoclastogenesis-inhibitory factor and is identical to TRANCE/RANKL. *Proc Natl Acad Sci USA* **95,** 3597–3602 (1998).

462. D. L. Lacey, E. Timms, H. L. Tan, M. J. Kelley, C. R. Dunstan, T. Burgess, R. Elliott, A. Colombero, G. Elliott, S. Scully, H. Hsu, J. Sullivan, N. Hawkins, E. Davy, C. Capparelli, A. Eli, Y. X. Qian, S. Kaufman, I. Sarosi, V. Shalhoub, G. Senaldi, J. Guo, J. Delaney, and W. J. Boyle, Osteoprotegerin ligand is a cytokine that regulates osteoclast differentiation and activation. *Cell* **93,** 165–176 (1998).

463. Y. Y. Kong, H. Yoshida, I. Sarosi, H. L. Tan, E. Timms, C. Capparelli, S. Morony, A. J. Oliveira-Dos-Santos, G. Van, A. Itie, W. Khoo, A. Wakeham, C. R. Dunstan, D. L. Lacey, T. W. Mak, W. J. Boyle, and J. M. Penninger, OPGL is a key regulator of osteoclastogenesis lymphocyte development and lymph-node organogenesis. *Nature* **397,** 315–323 (1999).

464. A. E. Hughes, S. H. Ralston, J. Marken, C. Bell, H. Macpherson, R. G. Wallace, W. van Hul, M. P. Whyte, K. Nakatsuka, L. Hovy, and D. M. Anderson, Mutations in TNFRSF11A affecting the signal peptide of RANK cause familial expansile osteolysis. *Nat Genet* **24,** 45–48 (2000).

465. T. Koga, M. Inui, K. Inoue, S. Kim, A. Suematsu, E. Kobayashi, T. Iwata, H. Ohnishi, T. Matozaki, T. Kodama, T. Taniguchi, H. Takayanagi, and T. Takai, Costimulatory signals mediated by the ITAM motif cooperate with RANKL for bone homeostasis. *Nature* **428,** 758–763 (2004).

466. A. Mocsai, M. B. Humphrey, J. A. Van Ziffle, Y. Hu, A. Burghardt, S. C. Spusta, S. Majumdar, L. L. Lanier, C. A. Lowell, and M. C. Nakamura, The immunomodulatory adapter proteins DAP12 and Fc receptor gamma-chain (FcRgamma) regulate development of functional osteoclasts through the Syk tyrosine kinase. *Proc Natl Acad Sci USA* **101,** 6158–6163 (2004).

467. W. S. Simonet, D. L. Lacey, C. R. Dunstan, M. Kelley, M. S. Chang, R. Luthy, H. Q. Nguyen, S. Wooden, L. Bennett, T. Boone, G. Shimamoto, M. Derose, R. Elliott, A. Colombero, H. L. Tan, G. Trail, J. Sullivan, E. Davy, N. Bucay, L. Renshaw-Gegg, T. M. Hughes, D. Hill, W. Pattison, P. Campbell, and W. J. Boyle, Osteoprotegerin: A novel secreted protein involved in the regulation of bone density. *Cell* **89,** 309–319 (1997).

468. N. Yamamoto, S. Akiyama, T. Katagiri, M. Namiki, T. Kurokawa, and T. Suda, Smad1 and Smad5 act downstream of intracellular signalings of BMP-2 that inhibits myogenic differentiation and induces osteoblast differentiation in C2C12 myoblasts. *Biochem Biophys Res Commun* **238,** 574–580 (1997).

469. B. S. Kwon, S. Wang, N. Udagawa, V. Haridas, Z. H. Lee, K. K. Kim, K. O. Oh, J. Greene, Y. Li, J. Su, R. Gentz, B. B. Aggarwal, and J. Ni, TR1, a new member of the tumor necrosis factor receptor superfamily, induces fibroblast proliferation and inhibits osteoclastogenesis and bone resorption. *FASEB J* **12,** 845–854 (1998).

470. E. Tsuda, M. Goto, S. Mochizuki, K. Yano, F. Kobayashi, T. Morinaga, and K. Higashio, Isolation of a novel cytokine from human fibroblasts that specifically inhibits osteoclastogenesis. *Biochem Biophys Res Commun* **234,** 137–142 (1997).

471. H. Yasuda, N. Shima, N. Nakagawa, S. I. Mochizuki, K. Yano, N. Fujise, Y. Sato, M. Goto, K. Yamaguchi, M. Kuriyama, T. Kanno, A. Murakami, E. Tsuda, T. Morinaga, and K. Higashio, Identity of osteoclastogenesis inhibitory factor (OCIF) and osteoprotegerin (OPG): A mechanism by which OPG/OCIF inhibits osteoclastogenesis *in vitro. Endocrinology* **139,** 1329–1337 (1998).

472. W. Zou, A. Amcheslavsky, and Z. Bar-Shavit, CpG oligodeoxynucleotides modulate the osteoclastogenic activity of osteoblasts via Toll-like receptor 9. *J Biol Chem* **278,** 16732–16740 (2003).

473. A. Amcheslavsky, H. Hemmi, S. Akira, and Z. Bar-Shavit, Differential contribution of osteoclast- and osteoblast-lineage cells to CpG-oligodeoxynucleotide (CpG-ODN) modulation of osteoclastogenesis. *J Bone Miner Res* **20,** 1692–1699 (2005).

474. D. R. Madrazo, S. L. Tranguch, and I. Marriott, Signaling via Toll-like receptor 5 can initiate inflammatory mediator production by murine osteoblasts. *Infect Immun* **71,** 5418–5421 (2003).

475. N. R. Jorgensen, Z. Henriksen, O. H. Sorensen, E. F. Eriksen, R. Civitelli, and T. H. Steinberg, Intercellular calcium signaling occurs between human osteoblasts and osteoclasts and requires activation of osteoclast P2X7 receptors. *J Biol Chem* **277,** 7574–7580 (2002).

476. C. Zhao, N. Irie, Y. Takada, K. Shimoda, T. Miyamoto, T. Nishiwaki, T. Suda, and K. Matsuo, Bidirectional ephrinB2–EphB4 signaling controls bone homeostasis. *Cell Metab* **4,** 111–121 (2006).

477. R. S. Taichman and S. G. Emerson, The role of osteoblasts in the hematopoietic microenvironment. *Stem Cells* **16,** 7–15 (1998).

478. A. Dar, O. Kollet, and T. Lapidot, Mutual reciprocal SDF-1/CXCR4 interactions between hematopoietic and bone marrow stromal cells regulate human stem cell migration and development in NOD/SCID chimeric mice. *Exp Hematol* **34,** 967–975 (2006).

479. T. Okubo, N. Yanai, and M. Obinata, Stromal cells modulate ephrinB2 expression and transmigration of hematopoietic cells. *Exp Hematol* **34,** 330–338 (2006).

480. M. M. Winslow, M. Pan, M. Starbuck, E. M. Gallo, L. Deng, G. Karsenty, and G. R. Crabtree, Calcineurin/NFAT signaling in osteoblasts regulates bone mass. *Dev Cell* **10,** 771–782 (2006).

481. M. C. Horowitz and J. A. Lorenzo, The origins of osteoclasts. *Curr Opin Rheumatol* **16,** 464–468 (2004).

482. I. Marriott, D. M. Rati, S. H. McCall, and S. L. Tranguch, Induction of Nod1 and Nod2 intracellular pattern recognition receptors in murine osteoblasts following bacterial challenge. *Infect Immun* **73,** 2967–2973 (2005).

483. G. Fiorelli, C. Orlando, S. Benvenuti, F. Franceschelli, S. Bianchi, P. Pioli, A. Tanini, M. Serio, F. Bartucci, and M. L. Brandi, Characterization, regulation, and function of specific cell membrane receptors for insulin-like growth factor I on bone endothelial cells. *J Bone Miner Res* **9,** 329–337 (1994).

484. M. M. Deckers, M. Karperien, C. van der Bent, T. Yamashita, S. E. Papapoulos, and C. W. Lowik, Expression of vascular endothelial growth factors and their receptors during osteoblast differentiation. *Endocrinology* **141,** 1667–1674 (2000).

485. D. S. Wang, M. Miura, H. Demura, and K. Sato, Anabolic effects of 1,25-dihydroxyvitamin D3 on osteoblasts are enhanced by vascular endothelial growth factor produced by osteoblasts and by growth factors produced by endothelial cells. *Endocrinology* **138,** 2953–2962.

486. H. P. Gerber, T. H. Vu, A. M. Ryan, J. Kowalski, Z. Werb, and N. Ferrara, VEGF couples hypertrophic cartilage remodeling

ossification and angiogenesis during endochondral bone formation. *Nat Med* **5,** 623–628 (1999).

487. A. Bellahcene, K. Bonjean, B. Fohr, N. S. Fedarko, F. A. Robey, M. F. Young, L. W. Fisher, and V. Castronovo, Bone sialoprotein mediates human endothelial cell attachment and migration and promotes angiogenesis. *Circ Res* **86,** 885–891 (2000).

488. L. A. Makuch, D. M. Sosnoski, and C. V. Gay, Osteoblast-conditioned media influence the expression of E-selectin on bone-derived vascular endothelial cells. *J Cell Biochem* **98,** 1221–1229 (2006).

489. F. Villars, B. Guillotin, T. Amedee, S. Dutoya, L. Bordenave, R. Bareille, and J. Amedee, Effect of HUVEC on human osteoprogenitor cell differentiation needs heterotypic gap junction communication. *Am J Physiol Cell Physiol* **282,** C775–C785 (2002).

490. B. Guillotin, C. Bourget, M. Remy-Zolgadri, R. Bareille, P. Fernandez, V. Conrad, and J. Medee-Vilamitjana, Human primary endothelial cells stimulate human osteoprogenitor cell differentiation. *Cell Physiol Biochem* **14,** 325–332 (2004).

491. E. F. Eriksen, S. F. Hodgson, R. Eastell, S. L. Cedel, W. M. O'Fallon, and B. L. Riggs, Cancellous bone remodeling in type I (postmenopausal) osteoporosis: Quantitative assessment of rates of formation, resorption, and bone loss at tissue and cellular levels. *J Bone Miner Res* **5,** 311–319 (1990).

492. J. Glowacki, Influence of age on human marrow. *Calcif Tissue Int* **56,** S50–S51 (1995).

493. G. D'Ippolito, P. C. Schiller, C. Ricordi, B. A. Roos, and G. A. Howard, Age-related osteogenic potential of mesenchymal stromal stem cells from human vertebral bone marrow. *J Bone Miner Res* **14,** 1115–1122 (1999).

494. W. E. Mercer, Checking on the cell cycle. *J Cell Biochem Suppl* **30–31,** 50–54 (1998).

495. A. B. Pardee, G$_1$ events and regulation of cell proliferation. *Science* **246,** 603–608 (1989).

496. C. J. Sherr and J. M. Roberts, Living with or without cyclins and cyclin-dependent kinases. *Genes Dev* **18,** 2699–2711 (2004).

497. M. Macaluso, M. Montanari, C. Cinti, and A. Giordano, Modulation of cell cycle components by epigenetic and genetic events. *Semin Oncol* **32,** 452–457 (2005).

498. C. Giacinti and A. Giordano, RB and cell cycle progression. *Oncogene* **25,** 5220–5227 (2006).

499. R. Shen, X. Wang, H. Drissi, F. Liu, R. J. O'Keefe, and D. Chen, Cyclin D1–Cdk4 induce Runx2 ubiquitination and degradation. *J Biol Chem* **281,** 16347–16353 (2006).

500. A. Rajgopal, D. W. Young, K. A. Mujeeb, J. L. Stein, J. B. Lian, A. J. van Wijnen, and G. S. Stein, Mitotic control of RUNX2 phosphorylation by both CDK1/cyclin B kinase and PP1/PP2A phosphatase in osteoblastic cells. *J Cell Biochem* **100,** 1509–1517 (2007).

501. T. F. Li, D. Chen, Q. Wu, M. Chen, T. J. Sheu, E. M. Schwarz, H. Drissi, M. Zuscik, and R. J. O'Keefe, Transforming growth factor-beta stimulates cyclin D1 expression through activation of beta-catenin signaling in chondrocytes. *J Biol Chem* **281,** 21296–21304 (2006).

502. F. Beier, Z. Ali, D. Mok, A. C. Taylor, T. Leask, C. Albanese, R. G. Pestell, and P. Luvalle, TGFbeta and PTHrP control chondrocyte proliferation by activating cyclin D1 expression. *Mol Biol Cell* **12,** 3852–3863 (2001).

503. F. Beier and P. Luvalle, The cyclin D1 and cyclin A genes are targets of activated PTH/PTHrP receptors in Jansen's metaphyseal chondrodysplasia. *Mol Endocrinol* **16,** 2163–2173 (2002).

504. N. Chattopadhyay, S. Yano, J. Tfelt-Hansen, P. Rooney, D. Kanuparthi, S. Bandyopadhyay, X. Ren, E. Terwilliger, and E. M. Brown, Mitogenic action of calcium-sensing receptor on rat calvarial osteoblasts. *Endocrinology* **145,** 3451–3462 (2004).

505. F. Long, X. M. Zhang, S. Karp, Y. Yang, and A. P. McMahon, Genetic manipulation of hedgehog signaling in the endochondral skeleton reveals a direct role in the regulation of chondrocyte proliferation. *Development* **128,** 5099–5108 (2001).

506. A. Sunters. J. McCluskey, and A. E. Grigoriadis, Control of cell cycle gene expression in bone development and during c-Fos-induced osteosarcoma formation. *Dev Genet* **22,** 386–397 (1998).

507. O. Coqueret, Linking cyclins to transcriptional control. *Gene* **299,** 35–55 (2002).

508. F. Beier, Cell-cycle control and the cartilage growth plate. *J Cell Physiol* **202,** 1–8 (2005).

509. E. G. Reynaud, K. Pelpel, M. Guillier, M. P. Leibovitch, and S. A. Leibovitch, p57(Kip2) stabilizes the MyoD protein by inhibiting cyclin E-Cdk2 kinase activity in growing myoblasts. *Mol Cell Biol* **19,** 7621–7629 (1999).

510. E. Smith, B. Frenkel, R. Schlegel, A. Giordano, J. B. Lian, J. L. Stein, and G. S. Stein, Expression of cell cycle regulatory factors in differentiating osteoblasts: Postproliferative up-regulation of cyclins B and E. *Cancer Res* **55,** 5019–5024 (1995).

511. Q. Liu, R. W. VanHoy, J. H. Zhou, R. Dantzer, G. G. Freund, and K. W. Kelley, Elevated cyclin E levels, inactive retinoblastoma protein, and suppression of the p27(KIP1) inhibitor characterize early development of promyeloid cells into macrophages. *Mol Cell Biol* **19,** 6229–6239 (1999).

512. J. Hess, B. Hartenstein, S. Teurich, D. Schmidt, M. Schorpp-Kistner, and P. Angel, Defective endochondral ossification in mice with strongly compromised expression of JunB. *J Cell Sci* **116,** 4587–4596 (2003).

513. M. Stewart, A. Terry, M. Hu, M. O'Hara, K. Blyth, E. Baxter, E. Cameron, D. E. Onions, and J. C. Neil, Proviral insertions induce the expression of bone-specific isoforms of PEBP2alphaA (CBFA1): Evidence for a new myc collaborating oncogene. *Proc Natl Acad Sci USA* **94,** 8646–8651 (1997).

514. M. Zenmyo, S. Komiya, T. Hamada, K. Hiraoka, R. Suzuki, and A. Inoue, p21 and parathyroid hormone-related peptide in the growth plate. *Calcif Tissue Int* **67,** 378–381 (2000).

515. H. Drissi, D. Hushka, F. Aslam, Q. Nguyen, E. Buffone, A. Koff, A. van Wijnen, J. B. Lian, J. L. Stein, and G. S. Stein, The cell cycle regulator p27kip1 contributes to growth and differentiation of osteoblasts. *Cancer Res* **59,** 3705–3711 (1999).

516. L. Legeai-Mallet, C. Oist-Lasselin, A. Munnich, and J. Bonaventure, Overexpression of FGFR3, Stat1, Stat5, and p21Cip1 correlates with phenotypic severity and defective chondrocyte differentiation in FGFR3-related chondrodysplasias. *Bone* **34,** 26–36 (2004).

517. S. M. Mallya and A. Arnold, Cyclin D1 in parathyroid disease. *Frontiers Biosci* **5,** D367–D371 (2000).

518. C. J. Lengner, H. A. Steinman, J. Gagnon, B. E. Kream, G. S. Stein, J. B. Lian, and S. N. Jones, Osteoblast differentiation and skeletal development are regulated by Mdm2-p53 signaling. *J Cell Biol* **172,** 909–921 (2006).

519. X. Wang, H. Y. Kua, Y. Hu, K. Guo, Q. Zeng, Q. Wu, H. H. Ng, G. Karsenty, C. B. De, J. Yeh, and B. Li, p53 functions as a negative regulator of osteoblastogenesis, osteoblast-dependent osteoclastogenesis, and bone remodeling. *J Cell Biol* **172,** 115–125 (2006).

520. B. Vogelstein, D. Lane, and A. J. Levine, Surfing the p53 network. *Nature* **408,** 307–310 (2000).

521. D. Michael and M. Oren, The p53 and Mdm2 families in cancer. *Curr Opin Genet Dev* **12,** 53–59 (2002).

522. S. N. Jones, A. T. Sands, A. R. Hancock, H. Vogel, L. A. Donehower, S. P. Linke, G. M. Wahl, and A. Bradley, The tamorigenic potential and cell growth characteristics of p53-deficient cells are equivalent on the presence or absence of Mdm2. *Proc Natl Acad Sci USA* **93,** 14106–14111 (1996).

523. S. D. Tyner, S. Venkatachalam, J. Choi, S. Jones, N. Ghebranious, H. Igelmann, X. Lu, G. Soron, B. Cooper, C. Brayton, P. S. Hee, T. Thompson, G. Karsenty, A. Bradley, and L. A. Donehower, p53 mutant mice that display early ageing-associated phenotypes. *Nature* **415,** 45–53 (2002).

524. D. Cobrinik, M. H. Lee, G. Hannon, G. Mulligan, R. T. Bronson, N. Dyson, E. Harlow, D. Beach, R. A. Weinberg, and T. Jacks, Shared role of the pRB-related p130 and p107 proteins in limb development. *Genes Dev* **10,** 1633–1644 (1996).

525. F. Rossi, H. E. MacLean, W. Yuan, R. O. Francis, E. Semenova, C. S. Lin, H. M. Kronenberg, and D. Cobrinik, p107 and p130 coordinately regulate proliferation, Cbfa1 expression, and hypertrophic differentiation during endochondral bone development. *Dev Biol* **247,** 271–285 (2002).

526. K. Yamato, S. Hashimoto, N. Okahashi, A. Ishisaki, K. Nonaka, T. Koseki, M. Kizaki, Y. Ikeda, and T. Nishihara, Dissociation of bone morphogenetic protein-mediated growth arrest and apoptosis of mouse B cells by HPV-16 E6/E7. *Exp Cell Res* **257,** 198–205 (2000).

527. K. Okamoto and C. Prives, A role of cyclin G in the process of apoptosis. *Oncogene* **18,** 4606–4615 (1999).

528. J. Jernvall, T. Aberg, P. Kettunen, S. Keranen, and I. Thesleff, The life history of an embryonic signaling center: BMP-4 induces p21 and is associated with apoptosis in the mouse tooth enamel knot. *Development* **125,** 161–169 (1998).

529. N. Ghosh-Choudhury, G. Ghosh-Choudhury, A. Celeste, P. M. Ghosh, M. Moyer, S. L. Abboud, and J. Kreisberg, Bone morphogenetic protein-2 induces cyclin kinase inhibitor p21 and hypophosphorylation of retinoblastoma protein in estradiol-treated MCF-7 human breast cancer cells. *Biochim Biophys Acta* **1497,** 186–196 (2000).

530. T. Bellido, C. A. O'Brien, P. K. Roberson, and S. C. Manolagas, Transcriptional activation of the p21(WAF1, CIP1, SDI1) gene by interleukin-6 type cytokines. A prerequisite for their pro-differentiating and anti-apoptotic effects on human osteoblastic cells. *J Biol Chem* **273,** 21137–21144 (1998).

531. S. Walsh, C. Jefferiss, K. Stewart, G. R. Jordan, J. Screen, and J. N. Beresford, Expression of the developmental markers STRO-1 and alkaline phosphatase in cultures of human marrow stromal cells: Regulation by fibroblast growth factor (FGF)-2 and relationship to the expression of FGF receptors 1–4. *Bone* **27,** 185–195 (2000).

532. A. Mansukhani, P. Bellosta, M. Sahni, and C. Basilico, Signaling by fibroblast growth factors (FGF) and fibroblast growth factor receptor 2 (FGFR2)-activating mutations blocks mineralization and induces apoptosis in osteoblasts. *J Cell Biol* **149,** 1297–1308 (2000).

533. F. Debiais, M. Hott, A. M. Graulet, and P. J. Marie, The effects of fibroblast growth factor-2 on human neonatal calvaria osteoblastic cells are differentiation stage specific. *J Bone Miner Res* **13,** 645–654 (1998).

534. M. R. Johnson, C. Valentine, C. Basilico, and A. Mansukhani, FGF signaling activates STAT1 and p21 and inhibits the estrogen response and proliferation of MCF-7 cells. *Oncogene* **16,** 2647–2656 (1998).

535. H. Wang, M. Rubin, E. Fenig, A. Deblasio, J. Mendelsohn, J. Yahalom, and R. Wieder, Basic fibroblast growth factor causes growth arrest in MCF-7 human breast cancer cells while inducing both mitogenic and inhibitory G_1 events. *Cancer Res* **57,** 1750–1757 (1997).

536. K. Pardali, A. Kurisaki, A. Moren, P. ten Dijke, D. Kardassis, and A. Moustakas, Role of Smad proteins and transcription factor Sp1 in p21WAF1/Cip1 regulation by transforming growth factor-beta. *J Biol Chem* **275,** 29244–29256 (2000).

537. M. B. Datto, Y. Yu, and X. F. Wang, Functional analysis of the transforming growth factor beta responsive elements in the WAF1/Cip1/p21 promoter. *J Biol Chem* **270,** 28623–28628 (1995).

538. M. Liu, M. H. Lee, M. Cohen, M. Bommakanti, and L. P. Freedman, Transcriptional activation of the Cdk inhibitor p21 by vitamin D3 leads to the induced differentiation of the myelomonocytic cell line U937. *Genes Dev* **10,** 142–153 (1996).

539. G. S. Stein, T. A. Owen, A. J. van Wijnen, B. Cho, J. L. Stein, and J. B. Lian, Mechanisms of action of skeletal growth factors in osteoblasts. In *Skeletal Growth Factors* (E. Canalis, ed.), pp. 51–65. Lippincott Williams & Wilkins, Philadelphia (2000).

540. J. W. Harper, G. R. Adami, N. Wei, K. Keyomarsi, and S. J. Elledge, The p21 Cdk-interacting protein Cip1 is a potent inhibitor of G_1 cyclin-dependent kinases. *Cell* **75,** 805–816 (1993).

541. Y. Xiong, G. J. Hannon, H. Zhang, D. Casso, R. Kobayashi, and D. Beach, p21 is a universal inhibitor of cyclin kinases. *Nature* **366,** 701–704 (1993).

542. S. K. Zaidi, D. W. Young, J. Y. Choi, J. Pratap, A. Javed, M. Montecino, J. L. Stein, A. J. van Wijnen, J. B. Lian, and G. S. Stein, The dynamic organization of gene-regulatory machinery in nuclear microenvironments. *EMBO Rep* **6,** 128–133 (2005).

543. S. I. Dworetzky, K. L. Wright, E. G. Fey, S. Penman, J. B. Lian, J. L. Stein, and G. S. Stein, Sequence-specific DNA-binding proteins are components of a nuclear matrix-attachment site. *Proc Natl Acad Sci USA* **89,** 4178–4182 (1992).

544. J. M. Boudreaux and D. A. Towler, Synergistic induction of osteocalcin gene expression: Identification of a bipartite element conferring fibroblast growth factor 2 and cyclic AMP responsiveness in the rat osteocalcin promoter. *J Biol Chem* **271,** 7508–7515 (1996).

545. C. Banerjee, J. L. Stein, A. J. van Wijnen, B. Frenkel, J. B. Lian, and G. S. Stein, Transforming growth factor-beta 1 responsiveness of the rat osteocalcin gene is mediated by an activator protein-1 binding site. *Endocrinology* **137,** 1991–2000 (1996).

546. H. M. Hoffmann, K. M. Catron, A. J. van Wijnen, L. R. McCabe, J. B. Lian, G. S. Stein, and J. L. Stein, Transcriptional control of the tissue-specific developmentally regulated osteocalcin gene requires a binding motif for the Msx family of homeodomain proteins. *Proc Natl Acad Sci USA* **91,** 12887–12891 (1994).

547. D. A. Towler, S. J. Rutledge, and G. A. Rodan, Msx-2/Hox 8.1: A transcriptional regulator of the rat osteocalcin promoter. *Mol Endocrinol* **8,** 1484–1493 (1994).

548. J. Hecht, V. Seitz, M. Urban, F. Wagner, P. N. Robinson, A. Stiege, C. Dieterich, U. Kornak, U. Wilkening, N. Brieske, C. Zwingman, A. Kidess, S. Stricker, and S. Mundlos, Detection of novel skeletogenesis target genes by comprehensive analysis of a Runx2(-/-) mouse model. *Gene Expr Patterns* **7,** 102–112 (2007).

549. B. L. Vaes, P. Ducy, A. M. Sijbers, J. M. Hendriks, E. P. van Someren, N. G. de Jong, E. R. van den Heuvel. W. Olijve, F. T. van Zoelen, and K. J. Dechering, Microarray analysis on Runx2-deficient mouse embryos reveals novel Runx2

functions and target genes during intramembranous and endochondral bone formation. *Bone* **39**, 724–738 (2006).

550. C. R. Paredes, J. Gutierrez, S. Gutierrez, L. Allison, M. Puchi, M. Imschenetzky, A. van Wijnen, J. Lian, G. Stein, J. Stein, and M. Montecino, Interaction of the 1alpha,25-dihydroxy vitamin D3 receptor at the distal promoter region of the bone-specific osteocalcin gene requires nucleosomal remodeling. *Biochemistry* **363**, 667–676 (2002).

551. J. Sierra, A. Villagra, R. Paredes, F. Cruzat, S. Gutierrez, A. Javed, G. Arriagada, J. Olate, M. Imschenetzky, A. J. van Wijnen, J. B. Lian, G. S. Stein, and J. L. Stein, Regulation of the bone-specific osteocalcin gene by p300 requires Runx2/Cbfa1 and the vitamin D3 receptor but not p300 intrinsic histone acetyltransferase activity. *Mol Cell Biol* **23**, 3339–3351 (2003).

552. S. Gutierrez, J. Liu, A. Javed, M. Montecino, G. S. Stein, J. B. Lian, and J. L. Stein, The vitamin D response element in the distal osteocalcin promoter contributes to chromatin organization of the proximal regulatory domain. *J Biol Chem* **279**, 43581–43588 (2004).

553. T. L. McCarthy, W. Z. Chang, Y. Liu, and M. Centrella, Runx2 integrates estrogen activity in osteoblasts. *J Biol Chem* **278**, 43121–43129 (2003).

554. S. Gutierrez, A. Javed, D. Tennant, M. van Rees, M. Montecino, G. S. Stein, J. L. Stein, and J. B. Lian, CCAAT/enhancer-binding proteins (C/EBP) b and d activate osteocalcin gene transcription and synergize with Runx2 at the C/EBP element to regulate bone-specific expression. *J Biol Chem* **277**, 1316–1323 (2002).

555. S. Christakos, P. Dhawan, Y. Liu, X. Peng, and A. Porta, New insights into the mechanisms of vitamin D action. *J Cell Biochem* **88**, 695–705 (2003).

556. I. de la Serna, Y. Ohkawa, and A. N. Imbalzano, Chromatin remodelling in mammalian differentiation: Lessons from ATP-dependent remodellers. *Nat Rev Genet* **7**, 461–473 (2006).

557. Y. Ohkawa, C. G. Marfella, and A. N. Imbalzano, Skeletal muscle specification by myogenin and Mef2D via the SWI/SNF ATPase Brg1. *EMBO J* **25**, 490–501 (2006).

558. I. de la Serna, K. A. Carlson, and A. N. Imbalzano, Mammalian SWI/SNF complexes promote MyoD-mediated muscle differentiation. *Nat Genet* **27**, 187–190 (2001).

559. M. G. Rosenfeld, V. V. Lunyak, and C. K. Glass, Sensors and signals: A coactivator/corepressor/epigenetic code for integrating signal-dependent programs of transcriptional response. *Genes Dev* **20**, 1405–1428 (2006).

560. A. L. Clayton, C. A. Hazzalin, and L. C. Mahadevan, Enhanced histone acetylation and transcription: A dynamic perspective. *Mol Cell* **23**, 289–296 (2006).

561. H. H. Cho, H. T. Park, Y. J. Kim, Y. C. Bae, K. T. Suh, and J. S. Jung, Induction of osteogenic differentiation of human mesenchymal stem cells by histone deacetylase inhibitors. *J Cell Biochem* **96**, 533–542 (2005).

562. H. W. Lee, J. H. Suh, A. Y. Kim, Y. S. Lee, S. Y. Park, and J. B. Kim, Histone deacetylase 1-mediated histone modification regulates osteoblast differentiation. *Mol Endocrinol* **20**, 2432–2443 (2006).

563. M. Montecino, J. Lian, G. Stein, and J. Stein, Changes in chromatin structure support constitutive and developmentally regulated transcription of the bone-specific osteocalcin gene in osteoblastic cells. *Biochemistry* **35**, 5093–5102 (1996).

564. A. Javed, S. Gutierrez, M. Montecino, A. J. van Wijnen, J. L. Stein, G. S. Stein, and J. B. Lian, Multiple Cbfa/AML sites in the rat osteocalcin promoter are required for basal and vita-

min D responsive transcription and contribute to chromatin organization. *Mol Cell Biol* **19**, 7491–7500 (1999).

565. G. S. Stein, J. B. Lian, A. J. van Wijnen, J. L. Stein, A. Javed, M. Montecino, S. K. Zaidi, D. Young, J. Y. Choi, S. Gutierrez, and S. Pockwinse, Nuclear microenvironments support assembly and organization of the transcriptional regulatory machinery for cell proliferation and differentiation. *J Cell Biochem* **91**, 287–302 (2004).

566. J. A. Nickerson, Experimental observations of a nuclear matrix. *J Cell Sci* **114**, 463–474 (2001).

567. J. P. Bidwell, K. Torrungruang, M. Alvarez, S. J. Rhodes, R. Shah, D. R. Jones, K. Charoonpatrapong, J. M. Hock, and A. J. Watt, Involvement of the nuclear matrix in the control of skeletal genes: The NMP1 (YY1), NMP2 (Cbfa1), and NMP4 (Nmp4/CIZ) transcription factors. *Crit Rev Eukaryot Gene Expr* **11**, 279–297 (2001).

568. S. Cai, C. C. Lee, and T. Kohwi-Shigematsu, SATB1 packages densely looped transcriptionally active chromatin for coordinated expression of cytokine genes. *Nat Genet* **38**, 1278–1288 (2006).

569. D. R. Fitzpatrick, I. M. Carr, L. McLaren, J. P. Leek, P. Wightman, K. Williamson, P. Gautier, N. McGill, C. Hayward, H. Firth, A. F. Markham, J. A. Fantes, and D. T. Bonthron, Identification of SATB2 as the cleft palate gene on 2q32–q33. *Hum Mol Genet* **12**, 2491–2501 (2003).

570. O. Britanova, M. J. Depew, M. Schwark, B. L. Thomas, I. Miletich, P. Sharpe, and V. Tarabykin, Satb2 haploinsufficiency phenocopies 2q32–q33 deletions, whereas loss suggests a fundamental role in the coordination of jaw development. *Am J Hum Genet* **79**, 668–678 (2006).

571. G. Dobreva, M. Chahrour, M. Dautzenberg, L. Chirivella, B. Kanzler, I. Farinas, G. Karsenty, and R. Grosschedl, SATB2 is a multifunctional determinant of craniofacial patterning and osteoblast differentiation. *Cell* **125**, 971–986 (2006).

572. C. Zeng, A. J. van Wijnen, J. L. Stein, S. Meyers, W. Sun, L. Shopland, J. B. Lawrence, S. Penman, J. B. Lian, G. S. Stein, and S. W. Hiebert, Identification of a nuclear matrix targeting signal in the leukemia and bone-related AML/CBFa transcription factors. *Proc Natl Acad Sci USA* **94**, 6746–6751 (1997).

573. S. K. Zaidi, A. Javed, J.-Y. Choi, A. J. van Wijnen, J. L. Stein, J. B. Lian, and G. S. Stein, A specific targeting signal directs Runx2/Cbfa1 to subnuclear domains and contributes to transactivation of the osteocalcin gene. *J Cell Sci* **114**, 3093–3102 (2001).

574. D. Vradii, S. K. Zaidi, J. B. Lian, A. J. van Wijnen, J. L. Stein, and G. S. Stein, A point mutation in AML1 disrupts subnuclear targeting, prevents myeloid differentiation, and results in a transformation-like phenotype. *Proc Natl Acad Sci USA* **102**, 7174–7179 (2005).

575. N. Selvamurugan, S. Kwok, and N. C. Partridge, Smad3 interacts with JunB and Cbfa1/Runx2 for transforming growth factor-beta1-stimulated collagenase-3 expression in human breast cancer cells. *J Biol Chem* **279**, 27764–27773 (2004).

576. G. L. Barnes, A. Javed, S. M. Waller, M. H. Kamal, K. E. Hebert, M. Q. Hassan, A. Bellahcene, A. J. van Wijnen, M. F. Young, J. B. Lian, G. S. Stein, and L. C. Gerstenfeld, Osteoblast-related transcription factors Runx2 (Cbfa1/AML3) and MSX2 mediate the expression of bone sialoprotein in human metastatic breast cancer cells. *Cancer Res* **63**, 2631–2637 (2003).

577. P. Shore, A role for Runx2 in normal mammary gland and breast cancer bone metastasis. *J Cell Biochem* **96**, 484–489 (2005).

578. J. Pratap, A. Javed, L. R. Languino, A. J. van Wijnen, J. L. Stein, G. S. Stein, and J. B. Lian, The Runx2 osteogenic transcription factor regulates matrix metalloproteinase 9 in bone metastatic cancer cells and controls cell invasion. *Mol Cell Biol* **25,** 8581–8591 (2005).

579. G. L. Barnes, K. E. Hebert, M. Kamal, A. Javed, T. A. Einhorn, J. B. Lian, G. S. Stein, and L. C. Gerstenfeld, Fidelity of Runx2 activity in breast cancer cells is required for the generation of metastases associated osteolytic disease. *Cancer Res* **64,** 4506–4513 (2004).

580. A. Javed, G. L. Barnes, J. Pratap, T. Antkowiak, L. C. Gerstenfeld, A. J. van Wijnen, J. L. Stein, J. B. Lian, and G. S. Stein, Impaired intranuclear trafficking of Runx2 (AML3/CBFA1) transcription factors in breast cancer cells inhibits osteolysis *in vivo. Proc Natl Acad Sci USA* **102,** 1454–1459 (2005).

581. T. M. Schroeder, E. D. Jensen, and J. J. Westendorf, Runx2: A master organizer of gene transcription in developing and maturing osteoblasts. *Birth Defects Res C Embryo Today* **75,** 213–225 (2005).

582. G. S. Stein, J. B. Lian, A. J. van Wijnen, J. L. Stein, M. Montecino, A. Javed, S. K. Zaidi, D. W. Young, J. Y. Choi, and S. M. Pockwinse, Runx2 control of organization, assembly and activity of the regulatory machinery for skeletal gene expression. *Oncogene* **23,** 4315–4329 (2004).

583. S. K. Zaidi, A. J. Sullivan, R. Medina, Y. Ito, A. J. van Wijnen, J. L. Stein, J. B. Lian, and G. S. Stein, (2004). Tyrosine phosphorylation controls Runx2-mediated subnuclear targeting of YAP to repress transcription. *EMBO J* **23,** 790–799 (2004).

584. J. B. Lian, A. Javed, S. K. Zaidi, C. Lengner, M. Montecino, A. J. van Wijnen, J. L. Stein, and G. S. Stein, Regulatory controls for osteoblast growth and differentiation: Role of Runx/Cbfa/AML factors. *Crit Rev Eukaryot Gene Expr* **14,** 1–41 (2004).

585. V. A. Spencer and J. R. Davie, Role of covalent modifications of histones in regulating gene expression. *Gene* **240,** 1–12 (1999).

586. N. Pelletier, N. Champagne, S. Stifani, and X. J. Yang, MOZ and MORF histone acetyltransferases interact with the Runt-domain transcription factor Runx2. *Oncogene* **21,** 2729–2740 (2002).

587. J. J. Westendorf, S. K. Zaidi, J. E. Cascino, R. Kahler, A. J. van Wijnen, J. B. Lian, M. Yoshida, G. S. Stein, and X. Li, Runx2 (Cbfa1 AML-3) interacts with histone deacetylase 6 and represses the p21(CIP1/WAF1) promoter. *Mol Cell Biol* **22,** 7982–7992 (2002).

588. T. M. Schroeder, R. A. Kahler, X. Li, and J. J. Westendorf, Histone deacetylase 3 interacts with runx2 to repress the osteocalcin promoter and regulate osteoblast differentiation. *J Biol Chem* **279,** 41998–42007 (2004).

589. R. B. Vega, K. Matsuda, J. Oh, A. C. Barbosa, X. Yang, E. Meadows, J. McAnally, C. Pomajzl, J. M. Shelton, J. A. Richardson, G. Karsenty, and E. N. Olson, Histone deacetylase 4 controls chondrocyte hypertrophy during skeletogenesis. *Cell* **119,** 555–566 (2004).

590. C. B. Cui, L. F. Cooper, X. Yang, G. Karsenty, and I. Aukhil, Transcriptional coactivation of bone-specific transcription factor Cbfa1 by TAZ. *Mol Cell Biol* **23,** 1004–1013 (2003).

591. J. H. Hong, E. S. Hwang, M. T. McManus, A. Amsterdam, Y. Tian, R. Kalmukova, E. Mueller, T. Benjamin, B. M. Spiegelman, P. A. Sharp, N. Hopkins, and M. B. Yaffe, TAZ, a transcriptional modulator of mesenchymal stem cell differentiation. *Science* **309,** 1074–1078 (2005).

592. W. Wang, Y. G. Wang, A. M. Reginato, D. J. Glotzer, N. Fukai, S. Plotkina, G. Karsenty, and B. R. Olsen, Groucho homologue Grg5 interacts with the transcription factor Runx2-Cbfa1 and modulates its activity during postnatal growth in mice. *Dev Biol* **270,** 364–381 (2004).

593. K. W. McLarren, R. Lo, D. Grbavec, K. Thirunavukkarasu, G. Karsenty, and S. Stifani, The mammalian basic helix loop helix protein HES-1 binds to and modulates the transactivating function of the runt-related factor cbfa1. *J Biol Chem* **275,** 530–538 (2000).

594. V. W. Yu, G. Ambartsoumian, L. Verlinden, J. M. Moir, J. Prud'homme, C. Gauthier, P. J. Roughley, and R. St.-Arnaud, FIAT represses ATF4-mediated transcription to regulate bone mass in transgenic mice. *J Cell Biol* **169,** 591–601 (2005).

595. A. Javed, B. Guo, S. Hiebert, J.-Y. Choi, J. Green, S.-C. Zhao, M. A. Osborne, S. Stifani, J. L. Stein, J. B. Lian, A. J. van Wijnen, and G. S. Stein, Groucho/TLE/R-Esp proteins associate with the nuclear matrix and repress RUNX (Cbfa/AML/PEBP2a) dependent activation of tissue-specific gene transcription. *J Cell Sci* **113,** 2221–2231 (2000).

596. N. S. Datta, C. Chen, J. E. Berry, and L. K. McCauley, PTHrP signaling targets cyclin D1 and induces osteoblastic cell growth arrest. *J Bone Miner Res* **20,** 1051–1064 (2005).

597. L. Qin, X. Li, J. K. Ko, and N. C. Partridge, Parathyroid hormone uses multiple mechanisms to arrest the cell cycle progression of osteoblastic cells from G_1 to S phase. *J Biol Chem* **280,** 3104–3111 (2005).

598. T. Onishi and K. Hruska, Expression of p27Kip1 in osteoblast-like cells during differentiation with parathyroid hormone. *Endocrinology* **138,** 1995–2004 (1997).

599. E. Smith, R. A. Redman, C. R. Logg, G. A. Coetzee, N. Kasahara, and B. Frenkel, Glucocorticoids inhibit developmental stage-specific osteoblast cell cycle. Dissociation of cyclin A–cyclin-dependent kinase 2 from E2F4–p130 complexes. *J Biol Chem* **275,** 19992–20001 (2000).

600. E. Smith, G. A. Coetzee, and B. Frenkel, Glucocorticoids inhibit cell cycle progression in differentiating osteoblasts via glycogen synthase kinase-3beta. *J Biol Chem* **277,** 18191–18197 (2002).

601. S. S. Jensen, M. W. Madsen, J. Lukas, L. Binderup, and J. Bartek, Inhibitory effects of 1alpha,25-dihydroxyvitamin D(3) on the G(1)–S phase-controlling machinery. *Mol Endocrinol* **15,** 1370–1380 (2001).

602. L. Sinkkonen, M. Malinen, K. Saavalainen, S. Vaisanen, and C. Carlberg, Regulation of the human cyclin C gene via multiple vitamin D3-responsive regions in its promoter. *Nucleic Acids Res* **33,** 2440–2451 (2005).

603. A. Saramaki, C. M. Banwell, M. J. Campbell, and C. Carlberg, Regulation of the human p21(waf1/cip1) gene promoter via multiple binding sites for p53 and the vitamin D3 receptor. *Nucleic Acids Res* **34,** 543–554 (2006).

604. S. F. Doisneau-Sixou, C. M. Sergio, J. S. Carroll, R. Hui, E. A. Musgrove, and R. L. Sutherland, Estrogen and antiestrogen regulation of cell cycle progression in breast cancer cells. *Endocr Relat Cancer* **10,** 179–186 (2003).

605. M. Fujita, T. Urano, K. Horie, K. Ikeda, T. Tsukui, H. Fukuoka, O. Tsutsumi, Y. Ouchi, and S. Inoue, Estrogen activates cyclin-dependent kinases 4 and 6 through induction of cyclin D in rat primary osteoblasts. *Biochem Biophys Res Commun* **299,** 222–228 (2002).

606. N. Kanda and S. Watanabe, 17Beta-estradiol stimulates the growth of human keratinocytes by inducing cyclin D2 expression. *J Invest Dermatol* **123,** 319–328 (2004).

607. T. Aikawa, G. V. Segre, and K. Lee, Fibroblast growth factor inhibits chondrocytic growth through induction of p21 and subsequent inactivation of cyclin E–Cdk2. *J Biol Chem* **276,** 29347–29352 (2001).

608. P. Krejci, V. Bryja, J. Pachernik, A. Hampl, R. Pogue, P. Mekikian, and W. R. Wilcox, FGF2 inhibits proliferation and alters the cartilage-like phenotype of RCS Cells. *Exp Cell Res* **297,** 152–164 (2004).

609. L. Dailey, E. Laplantine, R. Priore, and C. Basilico, A network of transcriptional and signaling events is activated by FGF to induce chondrocyte growth arrest and differentiation. *J Cell Biol* **161,** 1053–1066 (2003).

610. E. Laplantine, F. Rossi, M. Sahni, C. Basilico, and D. Cobrinik, FGF signaling targets the pRb-related p107 and p130 proteins to induce chondrocyte growth arrest. *J Cell Biol* **158,** 741–750 (2002).

611. D. M. Thomas, S. A. Johnson, N. A. Sims, M. K. Trivett, J. L. Slavin, B. P. Rubin, P. Waring, G. A. McArthur, C. R. Walkley, A. J. Holloway, D. Diyagama, J. E. Grim, B. E. Clurman, D. D. Bowtell, J. S. Lee, G. M. Gutierrez, D. M. Piscopo, S. A. Carty, and P. W. Hinds, Terminal osteoblast differentiation, mediated by runx2 and p27KIP1, is disrupted in osteosarcoma. *J Cell Biol* **167,** 925–934 (2004).

612. K. J. Manton, M. Sadasivam, S. M. Cool, and V. Nurcombe, Bone-specific heparan sulfates induce osteoblast growth arrest and downregulation of retinoblastoma protein. *J Cell Physiol* **209,** 219–229 (2006).

613. L. Fu, M. S. Patel, A. Bradley, E. F. Wagner, and G. Karsenty, The molecular clock mediates leptin-regulated bone formation. *Cell* **122,** 803–815 (2005).

614. A. Sunters, D. P. Thomas, W. A. Yeudall, and A. E. Grigoriadis, Accelerated cell cycle progression in osteoblasts overexpressing the c-fos proto-oncogene: Induction of cyclin A and enhanced CDK2 activity. *J Biol Chem* **279,** 9882–9891 (2004).

Osteoclast Biology

HARRY C. BLAIR, SCOTT SIMONET, DAVID L. LACEY, AND MONE ZAIDI

I. INTRODUCTION

The osteoclast is a monocyte-derived cell responsible for degradation of mineralized connective tissue, cartilage, or bone. Osteoclasts appear early in evolution in bony fishes that inhabit both calcium-rich salt water and fresh water, which use the skeleton to maintain serum calcium while in fresh water. An ancient seven transmembrane-pass receptor, the parathyroid hormone (PTH) receptor, is adapted for regulation of this skeletal resorption via release of a soluble form of PTH in branchial (gill) organs, the parathyroid glands. The key function of the osteoclast is to acidify its substrate, which dissolves the bone mineral.

The air-breathing vertebrates maintain this system but have adapted the skeleton extensively, with vascularization of the developing skeleton and formation of a new type of lightweight, hollow bone based on dense cross-linked type I collagen. The osteoclast is essential both in the formation of this advanced skeleton and in the use of the skeleton for calcium homeostasis, as a sink for excess acid in the circulation, as well as the central mechanical support for the body. The complexity of skeletal modeling and turnover in terrestrial vertebrates is associated with many osteoclast-related diseases. Some of them, such as osteoporosis, are common causes of morbidity and mortality. Both bone formation and bone resorption are often highly active, with large quantities of bone made and destroyed. Excess of resorption over formation can destroy large portions of the skeleton in months to years (Figure 5-1).

New genomic and biochemical tools, together with knock-out and transgenic animals, have clarified the differentiation and regulation of the osteoclast. If knocked out, genes required for osteoclast development or function cause osteopetrosis, a rare disease in which mineralized cartilage cannot be removed. Complete osteopetrosis is fatal in the neonatal period. In humans this is called infantile malignant osteopetrosis, and it manifests as hepatosplenomegaly, failure to thrive, and blindness within weeks of birth; the only effective treatment is bone marrow transplant. Fortunately, it is also rare, occurring in humans on the order of 1 in 30,000–300,000 births depending on the population [1].

On the other hand, common diseases, including osteoporosis and bone damage in arthritis, involve abnormal or excessive osteoclastic activity. Many challenges remain in the treatment of these diseases due to complex regulatory pathways that are not fully understood. In addition, difficulties arise in modifying osteoclastic differentiation and activity using pathways that are clear because the same pathways in most cases are important in organs other than bone.

As with all monocyte derivatives, the osteoclast is dependent on tyrosine kinase signals for survival and differentiation, chiefly through monocyte colony-stimulating factor (M-CSF, also called CSF-1), which activates the receptor Fms. Specialized osteoclast differentiation is controlled largely by tumor necrosis factor (TNF) family receptors, of which RANK (receptor activator of nuclear factor-$\kappa\beta$) is of major importance. However, secondary regulation of the osteoclast involves steroid hormones, attachment proteins, and receptors for a number of cytokines, including inflammatory cytokines. These additional and subsidiary signals utilize a balanced web of intermediate proteins within the osteoclast and its precursor cells.

II. KEY OSTEOCLAST DIFFERENTIATION PATHWAYS

The earliest steps in osteoclast development are indistinguishable from immune cell differentiation, and PU.1, a B cell transcription factor, is essential for early

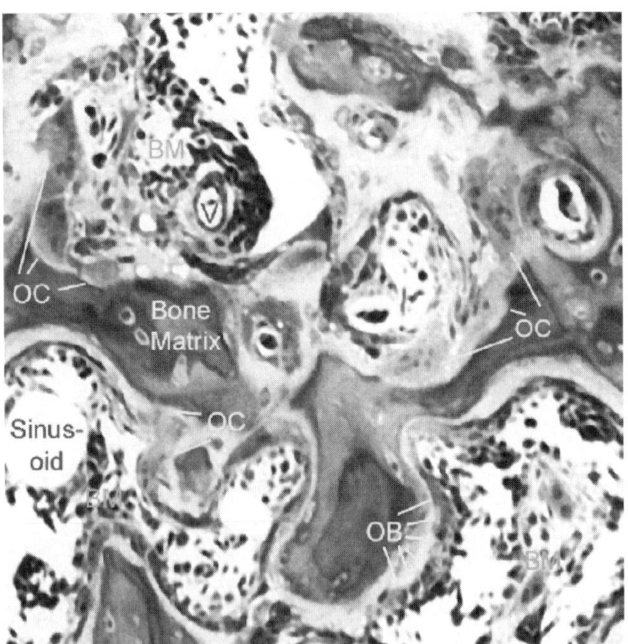

FIGURE 5-1   A thin section (1 μm) of undecalcified avian bone stained with methylene blue. Note the intimate relationship of bone cells to the marrow; osteoblasts and osteoclasts derive from marrow stem cells, although this relationship is usually not seen clearly. Osteoclasts (OC) are multinucleated cells that in the region shown occupy approximately 35% of the surface. Osteoblasts are rows of cells (OB) creating new matrix. Both mineralized (dark) and nonmineralized matrices are seen. The section is from an animal administered a low-calcium diet to cause high bone turnover. The field is 800 μm². BM, bone marrow.

osteoclast development [2]. Downstream, pluripotent monocytic stem cells produce macrophages, lymphocytes, dendritic cells, and osteoclasts. At this stage, osteoclast progenitors acquire Fms, the receptor for M-CSF. Low levels of M-CSF are required for survival and high levels permit monocyte proliferation. Activity of several related and downstream kinases, including Src, Grb2, and PI3-kinase, regulate this proliferation by activating cyclin D. M-CSF also activates c-Cbl, allowing it to ubiquitylate the proapoptotic gene *Bim*75, leading to degradation of the protein [3]. This mechanism ensures the survival of the formed precursors, as is evident from removing a key negative regulator of M-CSF, SHIP, which results in abundant osteoclasts.

Changes in the phenotype of mononuclear osteoclast precursors during osteoclast differentiation manifest as stepwise loss and acquisition of specific phenotypic markers [4]. Several stages preceding osteoclast specialization can be distinguished by cell surface antigens, including integrin family receptors. Human mononuclear osteoclast precursors circulating in peripheral blood express the monocyte–macrophage integrins CD11b-c and the lipopolysaccharide receptor CD14 [5, 6] but are negative for specialized osteoclast proteins. The integrin of the mature osteoclast is the vitronectin receptor (VNR, or $\alpha v \beta 3$) [7]; this integrin is also expressed on other monocyte derivatives including foreign body giant cells. Although the osteoclast precursors do not represent a distinct subpopulation by mononuclear surface markers, only 2–5% of human circulating mononuclear cells appear capable of osteoclast differentiation under typical conditions [6].

With activation of the TNF family receptor RANK, osteoclast precursors are committed. The RANK ligand (RANKL) is produced by osteoblasts as well as by bone marrow stromal cells and many other mesenchymal cells. RANKL is required for development of osteoclasts [8]; when it is knocked out in mice, severe osteopetrosis occurs.

A. The Role of TNF-α in Pathological Bone Resorption

Focal degradation of exposed mineralized matrix has long been known to occur due to macrophage activity and under some conditions, including stimulation of human macrophages by TNF-α (which is closely related to RANKL), lacunar resorption is reported [9, 10]. This level of resorption involves amounts of bone degradation insufficient to correct osteopetrosis, but the mechanism may be involved in some types of pathological bone resorption. In the presence of low levels of RANKL, TNF-α is a strong costimulus for bone resorption [11]. There is a large literature on TNF-α and it is clear that there are also indirect mechanisms affecting osteoclast differentiation, but a full review of the point is beyond the scope of this chapter. However, in some pathological states with bone loss, anti-TNF-α therapy may improve bone density at some sites [12, 13], although the mechanisms involved are unclear. It is noteworthy that whereas most TNF family ligands are mainly cell surface molecules, TNF-α circulates, particularly in pathological conditions in significant quantities. Thus, contributions of TNF-α to bone resorption in pathological states are likely to be of clinical significance. TNF-α may also mediate changes of glycoprotein hormone receptors not traditionally associated with bone resorption, including thyroid-stimulating hormone receptor (TSH-R) and follicle-stimulating hormone receptor (FSH-R), which have been shown by molecular approaches to occur and function in bone turnover due to the presence of their receptors in bone marrow cells [14–17].

In vivo, RANKL is essentially membrane bound, thereby limiting its effects to cell–cell-mediated activities. This stands in obvious contrast to *in vivo* pharmacologic studies that employ recombinantly derived, soluble RANKL. However, it is likely that in pathological conditions, soluble RANKL released into the circulation may in fact play a role in augmenting osteoclast activity systemically. For instance, serum RANKL is elevated in both the collagen-induced and adjuvant-induced arthritis models in rats [18]. As with other TNF family ligands, RANKL forms complexes, typically trimers, and this oligomerization is required for receptor activation. In addition to RANKL being a cell surface signal, there are scavenger receptors, importantly the soluble TNF receptor osteoprotegerin (OPG), that ensure that any RANKL released by proteinases does not remain in circulation [19]. Work indicates that costimulatory signals from immune-related receptors mediating signals by the ITAM motif cooperate with RANKL for osteoclast differentiation [20]. It is likely that this co-signaling pathway is the reason why, particularly with human cells, RANKL and CSF-1 produce osteoclasts only when supported by selected serum-containing media, the serum potentially supplying the necessary level of immune receptor costimulus. Main intracellular pathways involved in osteoclast differentiation are summarized in Figure 5-2.

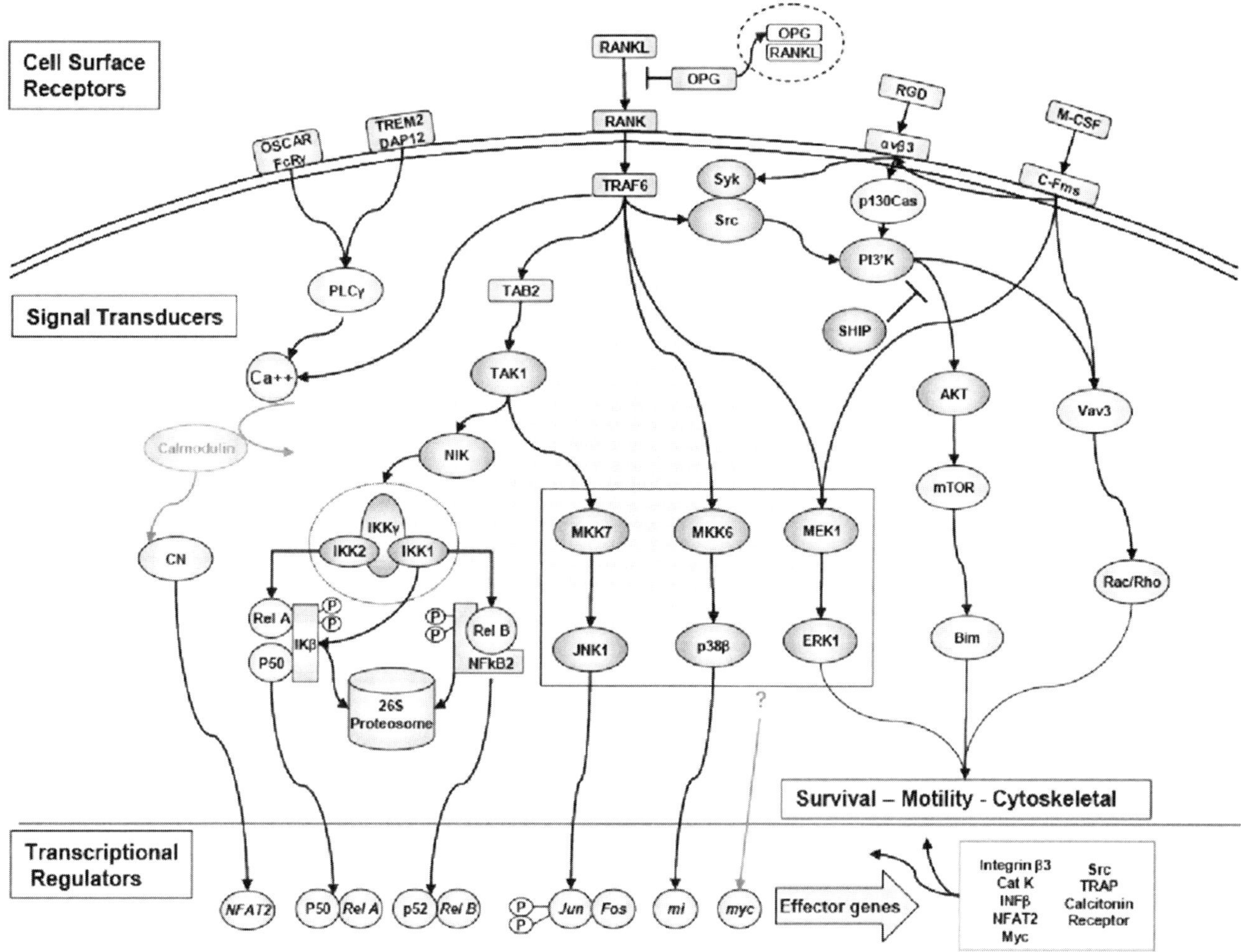

FIGURE 5-2   Key signaling pathways for the differentiation, survival, and activation of osteoclasts. RANK receptor ligation is followed by the recruitment of adaptor molecules, including TRAF6, which, interacting with c-src, stimulates the PI-3 kinase/Akt pathway. Additionally, the NF-κB and AP-1 families of transcription factors, key elements in osteoclast formation and function, are activated. Akt plays a role in phosphorylation of the IKK complex. Binding of M-CSF to c-FMS, its receptor, leads to activation of the $\alpha_v\beta_3$ integrin and recruitment of adapter proteins and cytosolic kinases, thus activating a variety of intracellular pathways required for osteoclast differentiation, survival, and activation of bone resorption. Reproduced from W. J. Boyle, W. S. Simonet, and D. L. Lacey, Osteoclast differentiation and activation. *Nature* **423,** 337–342 (2003).

RANK is activated by aggregation of two or more identical subunits. When activated, it recruits a second messenger, TRAF-6, to its intracellular domain. TRAF-6 mediates NF-βB and MAP kinase activation. In parallel, ITAM-harboring adapters Dap12 or FcRγ recruit Syk kinases that activate phospholipase Cγ and release Ca^{2+} from intracellular stores [21]. Syk and Dap12 are essential for normal osteoclastogenesis [22]. The periodic release of Ca^{2+} activates a calmodulin-dependent phosphatase, calcineurin. Calcineurin dephosphorylates the transcription factor NFAT2 [23]. Nuclear translocation of NFAT2, together with c-fos, mediates expression of osteoclast-specific genes, including further amplification of NFAT2 [24–26]. Knock-out of any of these pathways causes osteopetrosis and constitutive activation of NFAT2, which is sufficient to produce osteoclasts even without RANKL [27].

B. Activating Mutations in the RANK Pathway Cause Osteolytic Diseases, Whereas Defects in RANK Signaling Cause Osteopetrosis

Inactivating mutations in RANK or its receptor would also affect the immune system, and they have not been identified in humans. On the other hand, a stop codon mutation in the IKKγ gene impairs, but does not eliminate, NF-κB signaling and produces anhidrotic ectodermal dysplasia, immunodeficiency, and osteopetrosis [28]. It is likely that other incomplete defects in the RANK pathway will also cause defects in osteoclastic activity. Rare forms of osteopetrosis remain that are in search of an assigned gene. Most forms of human osteopetrosis are caused by mutations in molecules that mediate bone resorption (discussed later). On the other hand, constitutive activating mutations of TNFSF11a (RANK) cause rare autosomal dominant systemic osteolysis, and autosomal recessive inactivating mutations of the TNFSF11B (osteoprotegerin) gene cause juvenile Paget's disease. Bone loss, destruction of teeth, focal lesions in appendicular bones, and deafness occur in these diseases beginning during early childhood [29].

In addition to these central regulatory pathways, which are required for osteoclast formation, a number of additional stimuli modify osteoclast formation, survival, and activity. These include estrogen, inflammatory cytokines, and stretch. These are considered separately, after discussion of terminal osteoclastic differentiation and cellular function.

C. The Role of Parathyroid Hormone and Vitamin D

In this discussion of osteoclastic differentiation, PTH and vitamin D have not been considered. *In vitro* work has shown that there is little, if any, direct influence of PTH on bone resorption; this is secondary to signals in other cells that respond to PTH, including by varying the production of factors such as RANKL and OPG [30, 31]. Low-dose pulsatile PTH is anabolic in osteoblasts, but this activity is outside of the scope of this chapter. Osteoblasts and osteoclasts, and osteoclast precursors, all express vitamin D receptors. 1,25-dihydroxyvitamin D is important both in skeletal mineralization and, particularly at pharmacological levels, in stimulating maturation and activity of osteoclasts [32]. On the other hand, it is not clear that vitamin D is essential for osteoclast differentiation, although it may be supported by minimal levels of vitamin D in serum of osteoclast differentiation medium. In the development of vitamin D analogs for use in osteoporosis, a concern is the possibility of promoting high levels of osteoclast differentiation [33].

III. THE FULLY DIFFERENTIATED OSTEOCLAST: MECHANISMS OF BONE DEGRADATION

Osteoclasts are uniquely able to degrade marked quantities of bone matrix. The mineral is, for practical purposes, hydroxyapatite. To bring hydroxyapatite into solution at pH 7.4 requires acid secretion on a massive scale. Equation 1 shows the interconversion of calcium and phosphate in solution with hydroxyapatite:

$$14H^+ + Ca_{10}(PO_4)_6(OH)_2 \leftrightarrow 10Ca^{2+} + 6\,H_2PO_4^- + 2H_2O \quad (Eq.1)$$

The osteoclast moves a lot of acid since bone mineral requires the addition of ~1.5 moles of H^+ per mole of calcium removed at pH 7.4, and the osteoclast degrades in a day approximately its own volume in bone mineral. Indeed, activity of cultured osteoclasts *in vitro* can be judged by direct observation of the degree of acidification of their medium [34]. In keeping, osteoclasts are rich in mitochondria. With the necessity of maintaining an extracellular tight compartment, large cell diameter is an advantage, and the osteoclast is a giant cell with multiple nuclei. Normal human osteoclasts typically have 5–10 total nuclei; hyperactive cells found in pathological states, particularly Paget's disease, can contain many more nuclei and be much larger in size.

For acidification, the osteoclast produces a specialized microcompartment on the bone surface. This requires close apposition of an annulus of the osteoclast cell membrane to the matrix. Adhesion is via β_v integrins to matrix RGD peptides. This annulus, when osteoclasts are attached to bone, is a narrow ring of integrin binding with an associated actin ring (Figure 5-3A), which in the older literature is called the "clear zone" because of its appearance on electron microscopy. The major complementary subunit of α_v is β_3 [35].

A. Defects in Osteoclast Attachment

There are redundancies in integrin expression so that patients who have β_3 defects usually do not have osteopetrosis, although osteoclast attachment defects can cause osteopetrosis in humans [36]. The osteoclast's cytoskeleton, vesicular, and acid transport activities are reorganized to support this resorption compartment [37–39]. Attachment defects may be caused by intermediate proteins required to organize the osteoclast attachment ring and allow the expression of the acid-secreting apparatus. In particular, the Wiskott–Aldrich syndrome protein (WASp), a phosphoinositide-binding protein that regulates actin ring organization in podosomes and lamellipodia, is required for membrane ruffling in osteoclasts. Osteoclasts from WASp null mice fail to form actin rings at sealing zones on the bone surface, resulting in defects in bone resorption [40, 41].

B. Osteoclast Acid Secretion

The central activity in the isolated bone-attached compartment of the osteoclast is acid transport (Eq. 1). Because massive acid secretion is necessary, the membrane at the site of the acid transport is expanded to a loose, folded curtain-like structure, which on cross section appears mazelike and is called the "ruffled membrane" (Figure 5-3B).

C. Mutations Affecting HCl Secretion Are the Major Causes of Osteopetrosis

Acid transport is driven by a vacuolar H^+-ATPase [42]. This structure is a nano-motor. The ATPase is composed of membrane (V_o) and cytoplasmic (V_1) subassemblies. The V_o consists of a 17-kDa hydrogen channel and a large 116-kDa protein with multiple transmembrane domains that are essential for membrane insertion. Four homologous genes encode variants of this large membrane component [42, 43], one of which, TCIRG1 (ATP6i; A3), is amplified in osteoclasts [44, 45]. Defects in TCIRG1 are common causes of human osteopetrosis [36]. Variation in the amount and activity of TCIRG1 may underlie differences in bone density [46]. The V_1 assembly is common to all vacuolar-type H^+-ATPases and defects would presumably be embryonic lethal. The mechanism of the proton pump is believed to parallel that of the mitochondrial F-ATPase, which uses a proton gradient to produce ATP rather than

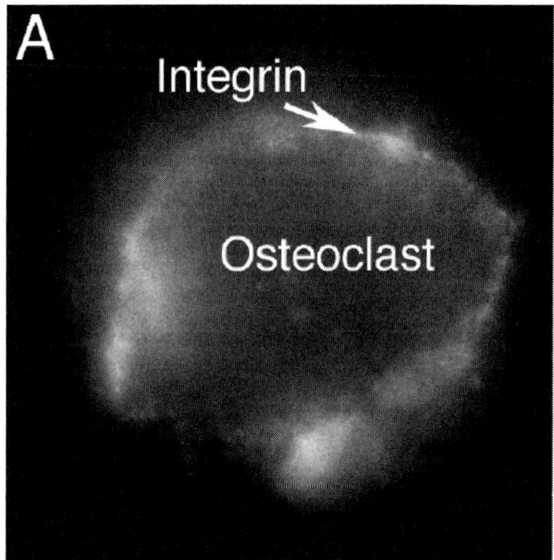

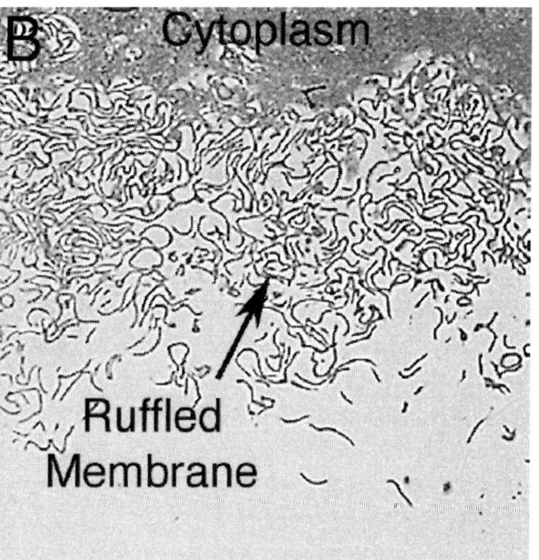

FIGURE 5-3   Osteoclast attachment features. (A) A human osteoclast on bone labeled for $\alpha_v\beta_3$ integrin (green) and actin (red) [91]. Note that there is a dense linear integrin attachment associated with a heavy actin ring. The photograph is 25 µm square. (B) An electron micrograph of the osteoclast's membrane at the attachment site showing the ruffled membrane. The membrane folds are ~20 nm thick and cannot be resolved by light microscopy. The section is orthogonal to the bone attachment and approximately 500 nm across.

consuming ATP to produce acid. Its structure is known in detail [47]. The F_1 or V_1 assemblies rotate with coupling of ATP hydrolysis and H^+ transport [48]. These ATPases are electrogenic; that is, they translocate H^+ only without any mechanism of charge balance. Hence, either cation countertransport or anion cotransport is required for the ATPase to function.

Studies of isolated osteoclast vesicles showed that chloride alone will support ATP-dependent acid transport in osteoclast membranes [49], and a Cl^- channel isolated from the avian osteoclast ruffled border is a homologue of a human intracellular chloride channel 5 (CLIC5) [50, 51]. CLIC5 is a member of a family of proteins that form chloride channels in membranes [52–54]. CLIC proteins are related to the omega family of glutathione *S*-transferases. They are required for development in *Caenorhabditis elegans* [55]. Furthermore, CLIC5 has been directly implicated in osteoclast bone resorption and H^+ transport [56]. However, it was also discovered that mice deficient in a widely expressed and unrelated chloride transporter, CLCN7 [57], are osteopetrotic [58, 59]. Furthermore, polymorphisms in CLCN7 are associated with many cases of human osteopetrosis [36]. On the other hand, CLCN7 is almost certainly a chloride–proton antiporter rather than a chloride channel, based on the properties of homologs of the same family [60, 61]. This requires at least two chloride transporters for osteoclast acid secretion since a chloride–proton antiporter will function without an H^+ gradient [62, 63]. Indeed, this is not unique to the osteoclast's acid secreting membrane since correcting CLCN7 expression in osteoclasts rescued bone metabolism but uncovered an underlying lysosomal defect [64]. Thus, the CLCN7 exchanger and CLIC5 provide charge neutralization by a mixed mechanism (Figure 5-4) that is important in acidification [65]. Proper distribution of ion transporters to their subcellular locations in the osteoclast depends on cytoskeletal interactions and on the intracellular tyrosine kinase Src [66–69]. Actin-directed insertion of CLIC proteins is also reported in other contexts [70].

D. Carbonic Anhydrase, Renal Tubular Acidosis, and Mild Osteopetrosis

The central metabolic pathway illustrated in Figure 5-4 excludes some elements that are of interest relative to human disease. Particularly, the osteoclast is highly metabolically active so that interconversion of CO_2 with carbonate is limited by the rate of hydration. Normally, this reaction in osteoclasts is accelerated by carbonic anhydrase II. The absence of carbonic anhydrase II

causes a mild form of osteopetrosis as well as renal tubular acidosis [1]. Other subsidiary mechanisms active in the osteoclast include chloride–bicarbonate exchange, which maintains the osteoclast's internal pH during acid secretion [71].

E. Osteoclastic Proteinases and Osteosclerotic Diseases

The acidic environment within the osteoclast attachment zone allows acid-optimal proteinases, principally the thiol proteinase cathepsin K [72], to cleave collagen and release peptides that are transcytosed and extruded at the osteoclast's dorsolateral surface. Lack of functional cathepsin K causes the disease pycnodysostosis (or Toulouse-Lautrec disease), a sclerotic disease that is less severe than osteopetrosis, presumably because there are other acid proteinases that are expressed in lesser quantities in the osteoclast and because there are neutral proteinases, such as matrix metalloproteinase-9, that are also expressed by the osteoclast at high levels [73]. Among the products of collagen degradation are post-translational lysine-

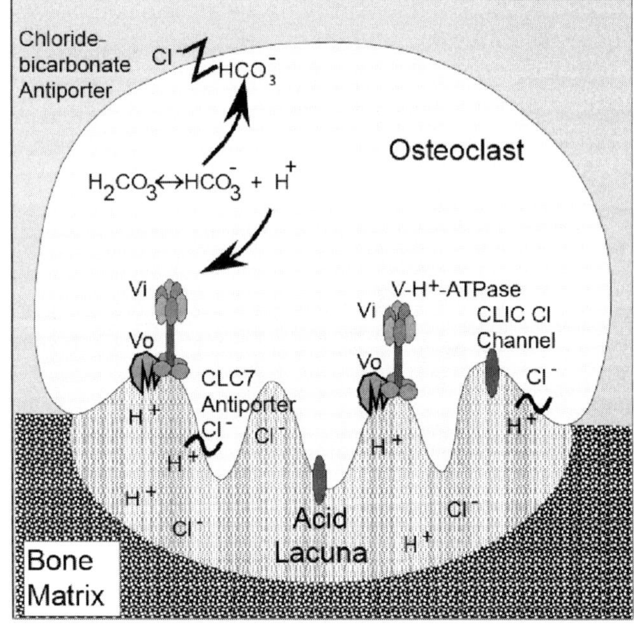

FIGURE 5-4   Ion transport by the osteoclast. Acid transport is powered by the vacuolar H^+-ATPase. Transport is balanced by chloride transport involving chloride channel (CLIC5) and chloride–proton antiporter (CLCN7) activity. Acid and base equivalents are derived ultimately from CO_2. Supporting transport processes include chloride–bicarbonate exchange in the basolateral membrane. Insertion of transporters in specific subcellular locations involves interaction of transporters with cytoskeletal components including actin.

derived collagen cross-link fragments, which are used as clinical markers for bone resorption, replacing the older measurement of urinary hydroxyproline. These assays are used particularly in evaluating the effects of novel therapeutic modalities on bone resorption [74]. In following an individual patient, bone resorption markers have been of relatively limited clinical utility in the overall management of osteoporosis due to the sensitivity and specificity of radiologic measurements of bone mass. However, they are extremely useful in the context of the development of novel therapeutics where there is a need to establish sensitive pharmacokinetic/pharmacodynamic relationships between drug exposures and target tissue responses.

To complete bone degradation, the high calcium solution [75] produced by osteoclast action must be moved to the extracellular space. Calcium may be released with osteoclast detachment, which occurs, in the absence of stimulated motility, at approximately 1-day intervals. However, studies by confocal imaging and labeled matrix show that bulk transport of degraded bone components occurs by vacuolar transcytosis through the osteoclast [76, 77]. The massive calcium movement in bone resorption suggests that the osteoclast may experience significant cytoplasmic calcium loading, even if the bulk transport is vesicular. The osteoclast highly expresses a Ca^{2+} ATPase [78]. Alternative calcium transport mechanisms have been proposed, including a calcium ferry involving the epithelial calcium channel, TRPV5, and calcium-binding proteins [79]. However, knock-out of TRPV5 does not cause osteopetrosis, although defects in bone mineralization occur [80].

IV. BONE RESORPTION COORDINATED BY INTERMEDIATE PROTEINS

There is a sequence of promoter activation that is more complex and better ordered than a linear pathway that releases large quantities of NF-κB or AP-1 transcription complexes for osteoclast differentiation. Since there is no osteoclast-specific promoter element, a mixture of nuclear cofactors, each regulated by specific interactions of adaptor and scaffolding proteins, integrates the mixed primary signals to which the osteoclast is exposed. This integration is required for the subtle and complex regulation of bone resorption that is observed *in vivo* (Figure 5-2).

The cell surface receptors that mediate osteoclast differentiation and attachment interact via a variety of intermediate proteins, including kinases and adaptor proteins. These interactions affect cellular activity, including motility, and may regulate cell survival. The adaptor proteins are particularly important in mediating signals that balance integrin receptor, PI-3-kinase, and small tyrosine kinase signals. Small tyrosine kinase activity is associated with receptors, including Fms, but also occurs in many other contexts. The adaptor proteins are proteins that have domains, which associate multiple proteins and often also cytoskeletal scaffolding components. This association places proteins in proximity that allows direct interaction of the effector signals, which are often kinases but also may include proteins such as trafs that transmit signals by other mechanisms, including regulation of degradation of targets. Adaptor proteins that are important in osteoclasts and osteoclast precursors include grb, shc, Gab2, and p130Cas [81–84], most of which have not been investigated thoroughly in the osteoclasts.

Although many of the associations of these proteins have not been completely elucidated, some worthy of specific mention include Gab2, which is required for normal RANKL signaling [85]. In regard to cell attachment–related signaling, p130Cas has been implicated by multiple studies [86, 87] and may also be involved in nongenomic estrogen signaling [88]. Adaptor and intermediate proteins are particularly important for signals that participate in many pathways that must be biologically separated in the cell, such as Fms. In some late regulatory signals, such as response to nitric oxide (NO) and calcium (discussed later), cytoplasmic targets have been proposed that may point to mechanisms of cross-regulation. However, it will be necessary to show that docking not only involves the same receptors and proteins but also occurs at the same time and place since adaptor proteins may also be involved in multiple discrete sequences of reactions within the cell.

Many of the regulatory interactions of adaptor proteins are mediated, in part, by cytoskeletal association, and interactions frequently cause cytoskeletal rearrangement. For example, $\alpha_v\beta_3$ activates c-src and Pyk-2, which recruit the adaptors c-Cbl and Cbl-b (which act at multiple sites) followed by PI3-kinase and the GTPase dynamin [89]. Outside-in signaling via attachment leads to the formation of a variety of complexes that include the kinases c-src and Syk, and the guanine nucleotide–binding factor Vav-3 [90]. Likewise, gelsolin and integrin-associated proteins occur, and a diverse array of proteins including VASPs [91], ITAM-harboring proteins [92], and c-src/Syk adaptors cooperate [93], although with unclear mechanisms.

V. OSTEOCLAST DEATH

Although the osteoclast number declines precipitously with changing hormonal conditions, such as with an increase in calcium, the direct mechanisms are not well characterized. It is likely that most osteoclast removal is, as with related immune cell death, by apoptotic mechanisms. In any case, it is clear that withdrawal of stimuli that support osteoclastic survival leads to apoptosis [94]. This also appears to be the case for removal of either RANKL or CSF-1 *in vitro* [95]. In this case, it appears that withdrawal of either leads to caspase activation. In mice, osteoclasts disappear rapidly following exposure to a single dose of OPG, the endogenous RANKL-binding inhibitor [95].

As with other macrophage family cells, death receptors including the Fas receptor directly cause osteoclast apoptosis [96]. Intermediates in osteoclast apoptosis include, as expected, caspases and calcium release [97].

Other initiating events that can kill osteoclasts include pharmacological doses of agents including bisphosphonates [98], reactive oxygen species, and nitric oxide [99, 100] or activation of osteoclast cell surface calcium receptors [101, 102]. Generally, these effects are balanced by survival effects of cytokines, and the response of the cell under survival conditions may be quite different, such as regulation of cell attachment or motility rather than death [91].

Lastly, there are a number of reports of osteoclast apoptosis *in vitro* in unexpected conditions, such as with exposure to estrogen. Indeed, there is robust bone resorption in the third trimester of pregnancy, when estrogen concentrations are at their highest physiological levels. These reports may reflect *in vitro* conditions in mixed cell populations with secondary cytokine production because these effects cannot be reproduced under most conditions. Work shows down-regulation of estrogen receptors with osteoclast differentiation [103].

VI. MECHANISTIC APPROACHES TO MODIFICATION OF OSTEOCLASTIC ACTIVITY *IN VIVO*

The common syndromes and diseases with bone loss have led to extensive interest in the pharmacological management of bone loss. There is a large literature on the subject, and a full discussion would overlap other chapters of this book. Thus, we limit this discussion to a brief comparison of the main approaches in the context of osteoclastic biology. Key approaches are compared in Table 5-1 for this purpose. Genetic defects in bone resorption are currently treatable only by bone

TABLE 5-1   Major Strategies for Pharmacological Inhibition of Bone Resorption Classified by the Target Osteoclast Biological Mechanism[a]

Biochemical mechanism	Example(s)	Key clinical concerns	Clinical use	Reference
Mineral-bound antimetabolites	Bisphosphonates	Osteonecrosis, long biological half-life	Many bone-losing states, cancer	Roelofs *et al.* [104]
RANKL inhibition	Denosumab (RANKL mAb)	Infection (theoretic)	Under investigation in PMO, oncology, RA settings	McClung *et al.* [105]
TNF-α inhibition	Etanercept, infliximab, adalimumab	Infection	RA, psoriatic arthritis, ankylosing spondylitis	Seriolo *et al.* [106]
Calcitonin receptor	Salmon calcitonin	Efficacy in bone loss	Bone pain; usually with additional therapy for bone loss	Munoz-Torrez *et al.* [108]
Thiol proteinases	SB-462795 (relacatib)	Bone quality	Preclinical trials	Kumare *et al.* [114]
H+-ATPase	Bafilomycin A	Toxicity at effective dose for bone loss	Uncertain	Warrell [110]
	Gallium nitrate	Toxicity	Cancer	Farina and Gagliardi [112]

[a]The list is not comprehensive; it summarizes major approaches in use or in development in rough inverse order relative to clinical utility. Endocrine and nutritional approaches are not included.

PMO, postmenopausal osteoporosis; RA, rheumatoid arthritis.

marrow transplant; these will not be further discussed. Deficiencies in vitamins and bone resorption with endocrine imbalance are also important in treatment; these are subjects addressed in detail elsewhere in this book and will not be included here.

A. Bone-Binding Metabolic Inhibitors

The major group in this class is the bisphosphonates, analogues of pyrophosphate with high affinity for hydroxyapatite. Bisphosphonates with nitrogen-containing side groups are highly effective, by mechanisms that may include interference with pathways including small GTPase signaling [104]. There is a large literature concerning bisphosphonate actions that is impractical to summarize. Key issues in evaluating the literature include the fact that concentrations of bisphosphonates in circulation are extremely low (picomolar to low nanomolar), other than during transient periods of administration. Thus, although the bisphosphonates will inhibit many processes *in vitro*, the importance of mechanisms inhibited by micromolar concentrations is difficult to assess. Nonetheless, these compounds are currently the most important antiresorptive drugs. They have shown excellent specificity, therapeutic response, and overall safety. Problems with administration have included osteonecrosis, particularly in the jaw, renal damage associated with the use of intravenous bisphosphonates, and effects of administration such as gastrointestinal damage for oral bisphosphonates. Another concern is that the compounds have extremely long half-lives *in vivo* (years to decades).

B. TNF Superfamily Signaling

In this class, there are two major types of candidate drugs: inhibitors of RANKL signaling and inhibitors of TNF-α signaling. The mechanism of inhibitors of RANKL signaling is obvious in that this would reduce osteoclastic differentiation, activation, and survival. The potential side effect of concern is interference with immune responses that could also depend potentially on RANKL, leading to the risk of infections. However, the data to date obtained in mature preclinical species using RANKL inhibitors have been reassuring in this regard. A monoclonal antibody to RANKL (denosumab) is in advanced trials and shows encouraging phase II clinical results [105]. Whether infection will emerge as a risk using RANKL inhibition in humans is

being closely monitored in ongoing clinical trials. As discussed previously, TNF-α accelerates bone loss, so inhibiting it may be of use in pathological conditions in which TNF-α is elevated [106]. Concerns with inhibitors of TNF-α are also mainly related to interference with immune function. The widespread use of TNF-α inhibitors has shown that concern regarding serious infections is well founded [107].

C. Calcitonin

Calcitonin receptors are present on osteoclasts and in the central nervous system. In fish, calcitonin is a potent antagonist of osteoclastic activity, but it has proven to be of limited activity in humans. However, in management of bone loss with bone pain, a trial of calcitonin is justified because many patients find rapid and significant relief, despite the limited effect on bone loss [108]. The mechanism for this analgesic response is believed to be dependent on central nervous system receptors.

D. Inhibitors of the H⁺-ATPase and Related Processes

Since the osteoclast is dependent on vacuolar-type H⁺-ATPase activity, inhibition of the proton pump may reduce bone resorption. Group IIA metals inhibit the H⁺-ATPase and accumulate on bone [109] but are potentially highly toxic and have been used clinically only in the management of cancer. They have not gone beyond clinical trials [110], largely due to the introduction of highly effective and less toxic agents. Bafilomycin A [111] inhibits all V-ATPases and at high concentrations will eliminate bone resorption, but it has not found clinical utility. It has shown promise in preclinical studies in rodents [112] but has not been used clinically in humans. Its broad spectrum of action and the fact that V-ATPases are involved in many vital non-bone mechanisms have led to skepticism regarding its future as a treatment for osteoporosis and related diseases. There are also numerous inhibitors of chloride transporters, but for the obvious reason that all cells require chloride transporters to maintain viability, these have not been considered as specific osteoclastic inhibitors. On the other hand, there are relatively specific inhibitors of individual thiol proteinases [113], including cathepsin K inhibitors in clinical trials [114].

VII. REGULATION OF OSTEOCLASTIC DIFFERENTIATION AND ACTIVITY *IN VIVO*

The production of the ligands that activate receptors which mediate osteoclast differentiation is highly regulated and critical to bone turnover. Centrally important factors are RANKL, OPG, and M-CSF (CSF-1) (Figure 5-2). RANKL is produced by mesenchymal cells in the bone marrow, as well as by osteoblasts, T cells, and other mesenchymal cells. RANKL is critical for osteoclastogenesis [115]. Both RANKL and CSF-1 are produced in major part as membrane proteins. Cell membrane proteins are of key importance in osteoclast differentiation, which limits osteoclast formation to the bone surface despite the presence of circulating osteoclast precursors and expression of RANKL and CSF-1 at many sites. Whereas RANKL can be released by proteolysis, OPG is a secreted soluble TNF receptor that binds directly to RANKL, blocking it from activating RANK [115, 116]. As with other "decoy receptors" for TNF family proteins, this system keeps osteoclast production localized to its intended locations on the bone surface.

Thus, the regulated expression of RANKL and OPG is coordinated to control bone resorption. Costimuli, particularly the inflammatory interleukin (IL)-1, TNF-α, and IL-6, can regulate osteoclast differentiation in secondary ways (Figure 5-2), and these cytokines in some cases regulate the capacity of stromal cells to produce RANKL and OPG [30]. A more controversial hypothesis holds that TNF-α not only promotes macrophage activity and is a costimulus for osteoclast formation but also may allow at least partial expression of osteoclast-specific proteins in a manner not requiring RANKL. A number of studies support this concept, although generally the lacunar bone resorption found in the absence of RANKL is weak relative to cells differentiating in the presence of RANKL [117]. Since osteoclasts are found *in vivo* only under conditions in which RANKL is expressed, this mechanism is likely to be mainly a laboratory artifact, although there appears to be no doubt that TNF-α is an important costimulator of osteoclast formation in pathological conditions (Table 5-1).

A large part of the effect of osteoclast-targeting cytokines and hormones is due to their influence on CSF-1, OPG, and RANKL expression. Thus, if CSF-1 is increased, the number of osteoclast precursors, and their ability to differentiate, is augmented. Increased RANKL activity—as a result of decreased OPG secretion, increased RANKL expression, or both—stimulates osteoclast recruitment, survival, and activity. Conversely, deficiency of RANKL activity will arrest osteoclast activity. Table 5-2 summarizes the role of the various cytokines regulating production of M-CSF, OPG, and RANKL. For brevity, individual cytokine associations are discussed. Note, however, that there is an apparent conundrum, namely that an osteoclast agonist may increase both OPG and RANKL. In this regard, most osteoclast differentiation is mediated by cell–cell signaling, and increased OPG may, in these circumstances, prevent unintended distant signaling by RANKL released by proteolytic activity. Thus, it is not a matter of summing the protein concentrations but, rather, of considering where the proteins are active at a highly localized level.

TABLE 5-2   Factors That Modulate Expression of RANKL, OPG, and CSF-1[a]

Class	Agent	RANKL	OPG	CSF-1 (M-CSF)
Hormone	1,25(OH)$_2$ vitamin D	↑	↑	↑
Estrogen		↑	↓	
PTH	↑	↓		
Glucocorticoids		↑	↓	↓
Cytokine	TNF-α	↑	↑	↑
IL-1	↑	↑		
IL-6	↑			
IL-1	↑			
Growth factors		TGF-β ↓	↑	
BMP-2		↑		
Canonical WNTs		↓	↑	
Prostaglandin	PGE$_2$	↑	↑	
Sclerostin	Sclerostin	↑	↓	

[a]Data from Hofbauer *et al.* [30], Lee and Lorenzo [31], and Canalis [124].

VIII. INTERACTION OF HORMONAL AND LOCAL SIGNALS WITH OSTEOCLAST ACTIVITY

Here, we consider direct and indirect interactions of hormones with osteoclast formation and activity. PTH, vitamin D, and TSH were discussed in Section II. In brief, PTH acts, in major part, on osteoblasts; 1,25-dihydroxyvitamin D affects osteoclast precursor differentiation. TSH increases osteoclast formation directly and via TNF-α. Calcitonin is discussed in Section VI; it directly causes osteoclast retraction and halts bone resorption in some species, but it has limited effects on bone degradation in humans, although it is useful for bone pain due to secondary effects.

A. Estrogen and Testosterone

Sex hormones are critical to maintenance of skeletal mass. The physiology is discussed elsewhere in this book. Regarding osteoclast differentiation and activity, there are three key estrogen-related mechanisms and a secondary mechanism. Briefly, estrogen acts through endothelial nitric oxide synthase (eNOS) to produce NO [118], which regulates osteoclastic activity and can lead to osteoclast apoptosis, depending on the cell context. These effects are probably mediated mainly by estrogen-dependent nongenomic interactions of the ERα with eNOS [119]. In osteoblasts and related cells, estrogen directly modifies the synthesis of osteoclast regulating cytokines, including OPG and CSF-1; in this regard, effects of testosterone are dissimilar [120]. Estrogen has small, but important, direct effects on osteoclast formation, which are also probably mediated, at least in part, by nongenomic mechanisms [103]. Finally, when estrogen (or testosterone) synthesis declines due to gonadal failure, there is a compensatory up-regulation of FSH, which has direct effects on the bone metabolism mediated directly by nongonadal FSH-R expression and by secondary TNF-α production [15, 16].

B. Prostaglandin E$_2$

Osteoclasts express prostaglandin receptors, and there are several reports of osteoclast differentiation either being negatively or positively regulated by prostaglandin E$_2$ [121, 122]. Most reports suggest positive regulation, and differences in response may be due to dose and costimuli. Osteoclasts express cyclooxygenase-2, and there is some suggestion that prosta-

glandin E$_2$ may be involved in normal differentiation [123]. Prolonged high-dose prostaglandin therapy in neonates causes a curious hyperostosis whose biological origin and possible relation to osteoclast function are unknown. Thus, there may be a role for prostaglandins, especially prostaglandin E$_2$, in osteoclast differentiation, but the mechanism and clinical importance are unclear.

C. Glucocorticoids

Glucocorticoid-induced osteopenia and osteonecrosis are major clinical problems. As with many other hormonal effects, the major effects on osteoclast formation are via changes in osteoblast and stromal cell production of RANKL, CSF-1, and OPG [124]. The mechanism for the dramatic bone resorption preceding femoral head collapse is probably also indirect, in that massive apoptosis of bone-forming units, osteons, composed of gap junction connected osteoblasts and osteocytes, precedes the formation of osteoclasts and osteolysis that leads to collapse of the bone [125].

D. Calcium

The osteoclast has a surface Ca^{2+} sensor, which mobilizes Ca^{2+} release from intracellular stores. This activates inducible nitric oxide synthase (iNOS), which in turn allows osteoclast detachment and retraction [102]. Detailed discussion of calcium and nitric oxide signaling in the osteoclast is beyond the scope of this chapter. However, this sensor is at a critical position to affect resorption by detachment and to initiate motility via NO or initiate apoptosis, depending on the context of the cell. Based on molecular and electrophysiological evidence, the sensor appears to be a type 2 ryanodine receptor, located uniquely in the osteoclast membrane. Activity of the sensor may be increased in high-resorption states, increasing the sensitivity of the osteoclast to down-regulation by calcium [126, 127].

E. Superoxide

Superoxide is produced by osteoclasts [128]. This acts to increase RANKL production [129], and since superoxide diffusion distance is very short, this is a possible mechanism for developing groups of adjacent osteoclasts in an area where one osteoclast has become active.

F. Nitric Oxide

As discussed previously, NO is a second messenger in bone for estrogen receptor activity in osteoblasts, and NO is also produced by iNOS, probably in response to a calcium signal. NO response in the osteoclast is largely via the NO-dependent guanosyl cyclase/cGMP-dependent protein kinase I (PKG I) pathway, which regulates osteoclast detachment or motility [91]. This involves regulation of cytoskeletal rearrangement. A central protein required for cytoskeletal rearrangement is VASP, the vasodilator-stimulated phosphoprotein that is an intermediate protein target of PKG I, which was discovered to mediate response to NO in other cells. High concentrations of NO mediate apoptosis in osteoclasts or their progenitors [91, 130].

IX. DISEASES WITH ALTERED BONE RESORPTION

A. Decreased or Absent Osteoclastic Activity

The causes of decreased osteoclastic activity leading to osteopetrosis and related disorders such as pycnodysostosis in humans were discussed previously, mainly in Section III. These include defects in cellular attachment, acid secretion, and acid proteinases. Defects in central differentiation pathways, such as RANK, have been observed mainly in genetically modified animals. These probably do not occur in humans because the defects would be lethal, although osteolytic diseases, some forms of Paget's disease, and some types of osteopetrosis occur with activating mutations in the RANK pathway or with partial defects in the RANK pathway, as discussed in Section II. Here, we briefly discuss the role of osteoclasts in other clinical diseases. These are also the subject of chapters elsewhere in this book, so we confine this discussion to factors related to osteoclast biology.

B. Osteoporosis

Osteoporosis is prevalent in aged people of either sex, and bone loss occurs rapidly after menopause in women. There are a number of other causes of osteoporosis, which are reviewed elsewhere in this book. Generally, there is nothing unique about osteoclasts in osteoporosis; there is simply more bone degradation than bone formation over a long period of time. This may be due to any of a number of hormonal and physical factors, but osteoclast abnormalities rarely, if ever, contribute to the development of osteoporosis that occurs in older adults.

C. Inflammatory-Related Bone Loss

Bone loss in periodontal disease occurs frequently. It is directly related to inflammatory infiltrates enhancing osteoclast formation and activity by the "usual suspects" of increased RANKL and TNF-α activity, attributed to immune cell infiltrates [131]. Bone loss also occurs related to all sorts of artificial implants cemented or inserted into bone at any site. In this case, the pathology is more complex, involving increased osteoclast formation due to immune infiltrates [132], but other factors are also involved. These include stress-shielding, in which an implant prevents flexion of surrounding bone, leading to induction of osteoclastic resorption at the shielded site [133]. Bone that is not stretched loses a key anabolic stimulus and is resorbed. In addition, there is stimulation of inflammation by debris from wear and tear of the implant, which can include increased reactive oxygen species [134]. Bacterial degradation fragments may also stimulate osteoclastic activity via Toll-like receptors [135].

D. Paget's Disease

Paget's disease is initiated by hyper-resorption with increased numbers of osteoclasts that are poorly regulated and have many nuclei. The only form of Paget's disease in which the etiology is clear is recessive inactivating mutations of TNFSF11B (osteoprotegerin) that cause juvenile Paget's disease (see Section II). Paget's disease in its general form occurs mainly during middle age and has a higher rate of occurrence in northern climates. Some forms of Paget's disease may be due to viral infection of osteoclasts [136].

E. Metastatic Cancer in Bone

Some types of cancers induce rapid osteolysis related to metastases, including multiple myeloma and breast cancer. Breast cancer often produces lytic lesions, and multiple myeloma always does so. Myeloma is a problematic and consistent producer of almost purely osteolytic bone lesions. Mechanisms of tumor osteolysis center around growth factors produced by tumor cells, including PTHrP, interleukins, TNF family proteins, and tyrosine kinase ligands including CSF-1 [137, 138].

Specific tumor cytokines produce additional dramatic specific effects on bone loss [139, 140]. Myeloma cells produce high levels of macrophage inflammatory peptide-1, which is linked to bone destruction. Other mechanisms include increased RANKL, tumor cell adhesion effects via VCAM, VEGF produced by tumor cells, and effects of bone cytokines on tumor cell survival. Tumor cells may also suppress bone formation.

REFERENCES

1. J. Tolar, S. L. Teitelbaum, and P. J. Orchard, Osteopetrosis. *N Engl J Med* **351**, 2839–2849 (2004).

2. M. M. Tondravi, S. R. McKercher, K. Anderson, J. M. Erdmann, M. Quiroz, R. Maki, and S. L. Teitelbaum, Osteopetrosis in mice lacking haematopoietic transcription factor PU.1. *Nature* **386**, 81–84 (1997).

3. T. Akiyama, P. Bouillet, T. Miyazaki, Y. Kadono, H. Chikuda, U. I. Chung, A. Fukuda, A. Hikita, H. Seto, T. Okada, T. Inaba, A. Sanjay, R. Baron, H. Kawaguchi, H. Oda, K. Nakamura, A. Strasser, and S. Tanaka, Regulation of osteoclast apoptosis by ubiquitylation of proapoptotic BH3-only Bcl-2 family member Bim. *EMBO J* **22**, 6653–6664 (2003).

4. J. Faust, D. L. Lacey, P. Hunt, T. Burgess, S. Scully, G. Van, A. Eli, Y. X. Qian, and V. Shalhoub, Osteoclast markers accumulate on cells developing from human peripheral blood mononuclear precursors. *J Cell Biochem* **72**, 67–80 (1999).

5. N. A. Athanasou and J. Quinn, Immunophenotypic differences between osteoclasts and macrophage polykaryons: Immunohistological distinction and implications for osteoclast ontogeny and function. *J Clin Pathol* **43**, 997–1003 (1990).

6. Y. Fujikawa, J. M. Quinn, A. Sabokbar, J. O. McGee, and N. A. Athanasou, The human osteoclast precursor circulates in the monocyte fraction. *Endocrinology* **137**, 4058–4060 (1996).

7. F. P. Ross, J. Chappel, J. I. Alvarez, D. Sander, W. T. Butler, M. C. Farach-Carson, K. A. Mintz, P. G. Robey, S. L. Teitelbaum, and D. A. Cheresh, Interactions between the bone matrix proteins osteopontin and bone sialoprotein and the osteoclast integrin alpha v beta 3 potentiate bone resorption. *J Biol Chem* **268**, 9901–9907 (1993).

8. Y. Y. Kong, H. Yoshida, I. Sarosi, H. L. Tan, E. Timms, C. Capparelli, S. Morony, A. J. Oliveira-dos-Santos, G. Van, A. Itie, W. Khoo, A. Wakeham, C. Dunstan, D. L. Lacey, T. W. Mak, W. J. Boyle, and J. M. Penninger, OPGL is a key regulator of osteoclastogenesis, lymphocyte development and lymph node organogenesis. *Nature* **397**, 315–323 (1999).

9. A. J. Kahn, C. C. Stewart, and S. L. Teitelbaum, Contact-mediated bone resorption by human monocytes *in vitro*. *Science* **199**, 988–990 (1978).

10. A. Sabokbar, O. Kudo, and N. A. Athanasou, Two distinct cellular mechanisms of osteoclast formation and bone resorption in periprosthetic osteolysis. *J Orthop Res* **21**, 73–80 (2003).

11. K. Fuller, C. Murphy, B. Kirstein, S. W. Fox, and T. J. Chambers, TNFα potently activates osteoclasts, through a direct action independent of and strongly synergistic with RANKL. *Endocrinology* **143**, 1108–1118 (2002).

12. L. Demis, C. Pony, M. Brehan, and M. Dougados, Infliximab in spondylarthropathy—Influence on bone density. *Clin Exp Rheumatol* **20**, S185–S186 (2002).

13. M. Vis, E. A. Havaardsholm, G. Haugeberg, T. Uhlig, A. E. Voskuyl, R. J. van de Stadt, B. A. Dijkmans, A. D. Woolf, T. K. Kvien, and W. F. Lems, Evaluation of bone mineral density, bone metabolism, osteoprotegerin and receptor activator of the NFkappaB ligand serum levels during treatment with infliximab in patients with rheumatoid arthritis. *Ann Rheum Dis* **65**, 1495–1499 (2006).

14. E. Abe, R. C. Marians, W. Yu, X. B. Wu, T. Ando, Y. Li, J. Iqbal, L. Eldeiry, G. Rajendren, H. C. Blair, T. F. Davies, and M. Zaidi, TSH is a negative regulator of skeletal remodeling. *Cell* **115**, 151–162 (2003).

15. J. Iqbal, L. Sun, T. R. Kumar, H. C. Blair, and M. Zaidi, Follicle-stimulating hormone stimulates TNF production from immune cells to enhance osteoblast and osteoclast formation. *Proc Natl Acad Sci USA* **103**, 14925–14930 (2006).

16. L. Sun, Y. Peng, A. C. Sharrow, J. Iqbal, Z. Zhang, D. J. Papachristou, S. Zaidi, L. L. Zhu, B. B. Yaroslavskiy, H. Zhou, A. Zallone, M. R. Sairam, T. R. Kumar, W. Bo, J. Braun, L. Cardoso-Landa, M. B. Schaffler, B. S. Moonga, H. C. Blair, and M. Zaidi, FSH directly regulates bone mass. *Cell* **125**, 247–260 (2006).

17. H. C. Wang, J. Dragoo, Q. Zhou, and J. R. Klein, An intrinsic thyrotropin-mediated pathway of TNF-alpha production by bone marrow cells. *Blood* **101**, 119–123 (2003).

18. M. Stolina, S. Adamu, M. Ominsky, D. Dwyer, F. Asuncion, Z. Geng, S. Middleton, H. Brown, J. Pretorius, G. Schett, B. Bolon, U. Feige, D. Zack, and P. J. Kostenuik, RANKL is a marker and mediator of local and systemic bone loss in two rat models of inflammatory arthritis. *J Bone Miner Res* **20**, 1756–1765 (2005).

19. W. S. Simonet, D. L. Lacey, C. R. Dunstan, M. Kelley, M. S. Chang, R. Luthy, H. Q. Nguyen, S. Wooden, L. Bennett, T. Boone, G. Shimamoto, M. DeRose, R. Elliott, A. Colombero, H. L. Tan, G. Trail, J. Sullivan, E. Davy, N. Bucay, L. Renshaw-Gegg, T. M. Hughes, D. Hill, W. Pattison, P. Campbell, S. Sander, G. Van, J. Tarpley, P. Derby, R. Lee, and W. J. Boyle, Osteoprotegerin: A novel secreted protein involved in the regulation of bone density. *Cell* **89**, 309–319 (1997).

20. T. Koga, M. Inui, K. Inoue, S. Kim, A. Suematsu, E. Kobayashi, T. Iwata, H. Ohnishi, T. Matozaki, T. Kodama, T. Taniguchi, H. Takayanagi, and T. Takai, Costimulatory signals mediated by the ITAM motif cooperate with RANKL for bone homeostasis. *Nature* **428**, 758–763 (2004).

21. Y. Katayama, M. Battista, W. M. Kao, A. Hidalgo, A. J. Peired, S. A. Thomas, and P. S. Frenette, Signals from the sympathetic nervous system regulate hematopoietic stem cell egress from bone marrow. *Cell* **124**, 407–421 (2006).

22. A. Mocsai, M. B. Humphrey, J. A. Van Ziffle, Y. Hu, A. Burghardt, S. C. Spusta, S. Majumdar, L. L. Lanier, C. A. Lowell, and M. C. Nakamura, The immunomodulatory adapter proteins DAP12 and Fc receptor gamma-chain (FcRgamma) regulate development of functional osteoclasts through the Syk tyrosine kinase. *Proc Natl Acad Sci USA* **101**, 6158–6163 (2004).

23. H. Takayanagi, S. Kim, T. Koga, H. Nishina, M. Isshiki, H. Yoshida, A. Saiura, M. Isobe, T. Yokochi, J. Inoue, E. F. Wagner, T. W. Mak, T. Kodama, and T. Taniguchi, Induction and activation of the transcription factor NFATc1 (NFAT2) integrate RANKL signaling in terminal differentiation of osteoclasts. *Dev Cell* **3**, 889–901 (2002).

24. M. Asagiri, K. Sato, T. Usami, S. Ochi, H. Nishina, H. Yoshida, I. Morita, E. F. Wagner, T. W. Mak, E. Serfling, and H. Takayanagi, Autoamplification of NFATc1 expression determines its essential role in bone homeostasis. *J Exp Med* **202**, 1261–1269 (2005).

25. J. Gohda, T. Akiyama, T. Koga, H. Takayanagi, S. Tanaka, and J. Inoue, RANK-mediated amplification of TRAF6 signaling leads to NFATc1 induction during osteoclastogenesis. *EMBO J* **24**, 790–799 (2005).

26. Y. Kim, K. Sato, M. Asagiri, I. Morita, K. Soma, and H. Takayanagi, Contribution of nuclear factor of activated T cells c1 to the transcriptional control of immunoreceptor osteoclast-associated receptor but not triggering receptor expressed by myeloid cells-2 during osteoclastogenesis. *J Biol Chem* **280**, 32905–32913 (2005).

27. K. Matsuo, D. L. Galson, C. Zhao, L. Peng, C. Laplace, K. Z. Wang, M. A. Bachler, H. Amano, H. Aburatani, H. Ishikawa, and E. F. Wagner, Nuclear factor of activated T-cells (NFAT) rescues osteoclastogenesis in precursors lacking c-Fos. *J Biol Chem* **279**, 26475–26480 (2004).

28. S. Dupuis-Girod, N. Corradini, S. Hadj-Rabia, J. C. Fournet, L. Faivre, F. Le Deist, P. Durand, R. Doffinger, A. Smahi, A. Israel, G. Courtois, N. Brousse, S. Blanche, A. Munnich, A. Fischer, J. L. Casanova, and C. Bodemer, Osteopetrosis, lymphedema, anhidrotic ectodermal dysplasia, and immunodeficiency in a boy and incontinentia pigmenti in his mother. *Pediatrics* **109**, e97 (2002).

29. A. Daroszewska and S. H. Ralston, Mechanisms of disease: Genetics of Paget's disease of bone and related disorders. *Nat Clin Pract Rheumatol* **2**, 270–277 (2006).

30. L. C. Hofbauer, S. Khosla, C. R. Dunstan, D. L. Lacey, W. J. Boyle, and B. L. Riggs, The roles of osteoprotegerin and osteoprotegerin ligand in the paracrine regulation of bone resorption. *J Bone Miner Res* **15**, 2–12 (2000).

31. S. K. Lee and J. A. Lorenzo, Regulation of receptor activator of nuclear factor-kappa B ligand and osteoprotegerin mRNA expression by parathyroid hormone is predominantly mediated by the protein kinase A pathway in murine bone marrow cultures. *Bone* **31**, 252–259 (2002).

32. T. Suda, Y. Ueno, K. Fujii, and T. Shinki, Vitamin D and bone. *J Cell Biochem* **88**, 259–266 (2003).

33. M. Shimazaki, Y. Miyamoto, K. Yamamoto, S. Yamada, M. Takami, T. Shinki, N. Udagawa, and M. Shimizu, Analogs of 1alpha,25-dihydroxyvitamin D3 with high potency in induction of osteoclastogenesis and prevention of dendritic cell differentiation: Synthesis and biological evaluation of 2-substituted 19-norvitamin D analogs. *Bioorg Med Chem* **14**, 4645–4656 (2006).

34. A. Carano, P. H. Schlesinger, N. A. Athanasou, S. L. Teitelbaum, and H. C. Blair, Acid and base effects on avian osteoclast activity. *Am J Physiol* **264**, C694–C701 (1993).

35. A. Miyauchi, J. Alvarez, E. M. Greenfield, A. Teti, M. Grano, S. Colucci, A. Zambonin-Zallone, F. P. Ross, S. L. Teitelbaum, D. Cheresh, and K. A. Hruska, Recognition of osteopontin and related peptides by an alpha v beta 3 integrin stimulates immediate cell signals in osteoclasts. *J Biol Chem* **266**, 20369–20374 (1991).

36. H. C. Blair, C. W. Borysenko, A. Villa, P. H. Schlesinger, S. E. Kalla, B. B. Yaroslavskiy, V. Garcia-Palacios, J. I. Oakley, and P. J. Orchard, *In vitro* differentiation of CD14 cells from osteopetrotic subjects: Contrasting phenotypes with TCIRG1, CLCN7, and attachment defects. *J Bone Miner Res* **19**, 1329–1338 (2004).

37. T. Akisaka, H. Yoshida, and R. Suzuki, The ruffled border and attachment regions of the apposing membrane of resorbing osteoclasts as visualized from the cytoplasmic face of the membrane. *J Electron Microsc (Tokyo)* **55**, 53–61 (2006).

38. P. H. Schlesinger, J. P. Mattsson, and H. C. Blair, Osteoclastic acid transport: Mechanism and implications for physiological and pharmacological regulation. *Miner Electrolyte Metab* **20**, 31–39 (1994).

39. H. K. Vaananen, H. Zhao, M. Mulari, and J. M. Halleen, The cell biology of osteoclast function. *J Cell Sci* **113**(Pt. 3), 377–381 (2000).

40. Y. Calle, G. E. Jones, C. Jagger, K. Fuller, M. P. Blundell, J. Chow, T. Chambers, and A. J. Thrasher, WASp deficiency in mice results in failure to form osteoclast sealing zones and defects in bone resorption. *Blood* **103**, 3552–3561 (2004).

41. M. A. Chellaiah, D. Kuppuswamy, L. Lasky, and S. Linder, Phosphorylation of a Wiscott-Aldrich syndrome protein-associated signal complex is critical in osteoclast bone resorption. *J Biol Chem* (2007).

42. H. C. Blair, S. L. Teitelbaum, R. Ghiselli, and S. Gluck, Osteoclastic bone resorption by a polarized vacuolar proton pump. *Science* **245**, 855–857 (1989).

43. T. Nishi and M. Forgac, The vacuolar (H+)-ATPases—Nature's most versatile proton pumps. *Nat Rev Mol Cell Biol* **3**, 94–103 (2002).

44. Y. P. Li, W. Chen, Y. Liang, E. Li, and P. Stashenko, Atp6i-deficient mice exhibit severe osteopetrosis due to loss of osteoclast-mediated extracellular acidification. *Nat Genet* **23**, 447–451 (1999).

45. J. P. Mattsson, X. Li, S. B. Peng, F. Nilsson, P. Andersen, L. G. Lundberg, D. K. Stone, and D. J. Keeling, Properties of three isoforms of the 116-kDa subunit of vacuolar H+-ATPase from a single vertebrate species. Cloning, gene expression and protein characterization of functionally distinct isoforms in *Gallus gallus*. *Eur J Biochem* **267**, 4115–4126 (2000).

46. G. Carn, D. L. Koller, M. Peacock, S. L. Hui, W. E. Evans, P. M. Conneally, C. C. Johnston Jr., T. Foroud, and M. J. Econs, Sibling pair linkage and association studies between peak bone mineral density and the gene locus for the osteoclast-specific subunit (OC116) of the vacuolar proton pump on chromosome 11p12–13. *J Clin Endocrinol Metab* **87**, 3819–3824 (2002).

47. P. D. Boyer, The ATP synthase—A splendid molecular machine. *Annu Rev Biochem* **66**, 717–749 (1997).

48. M. E. Finbow and M. A. Harrison, The vacuolar H+-ATPase: A universal proton pump of eukaryotes. *Biochem J* **324**(Pt. 3), 697–712 (1997).

49. H. C. Blair, S. L. Teitelbaum, H. L. Tan, C. M. Koziol, and P. H. Schlesinger, Passive chloride permeability charge coupled to H(+)-ATPase of avian osteoclast ruffled membrane. *Am J Physiol* **260**, C1315–C1324 (1991).

50. H. C. Blair and P. H. Schlesinger, Purification of a stilbene sensitive chloride channel and reconstitution of chloride conductivity into phospholipid vesicles. *Biochem Biophys Res Commun* **171**, 920–925 (1990).

51. P. H. Schlesinger, H. C. Blair, S. L. Teitelbaum, and J. C. Edwards, Characterization of the osteoclast ruffled border chloride channel and its role in bone resorption. *J Biol Chem* **272**, 18636–18643 (1997).

52. R. H. Ashley, Challenging accepted ion channel biology: p64 and the CLIC family of putative intracellular anion channel proteins. *Mol Membr Biol* **20**, 1–11 (2003). [Review]

53. N. S. Heiss and A. Poustka, Genomic structure of a novel chloride channel gene, CLIC2, in Xq28. *Genomics* **45**, 224–228 (1997).

54. T. J. Jentsch, M. Poet, J. C. Fuhrmann, and A. A. Zdebik, Physiological functions of CLC Cl− channels gleaned from human genetic disease and mouse models. *Annu Rev Physiol* **67**, 779–807 (2005).

55. K. L. Berry, H. E. Bulow, D. H. Hall, and O. Hobert, A *C. elegans* CLIC-like protein required for intracellular tube formation and maintenance. *Science* **302**, 2134–2137 (2003).

56. J. C. Edwards, C. Cohen, W. Xu, and P. H. Schlesinger, c-Src control of chloride channel support for osteoclast HCl

transport and bone resorption. *J Biol Chem* **281**, 28011–28022 (2006).

57. S. Brandt and T. J. Jentsch, ClC-6 and ClC-7 are two novel broadly expressed members of the CLC chloride channel family. *FEBS Lett* **377**, 15–20 (1995).

58. E. Cleiren, O. Benichou, H. E. Van, J. Gram, J. Bollerslev, F. R. Singer, K. Beaverson, A. Aledo, M. P. Whyte, T. Yoneyama, M. C. deVernejoul, and H. W. Van, Albers–Schonberg disease (autosomal dominant osteopetrosis, type II) results from mutations in the ClCN7 chloride channel gene. *Hum Mol Genet* **10**, 2861–2867 (2001).

59. U. Kornak, D. Kasper, M. R. Bosl, E. Kaiser, M. Schweizer, A. Schulz, W. Friedrich, G. Delling, and T. J. Jentsch. Loss of the ClC-7 chloride channel leads to osteopetrosis in mice and man. *Cell* **104**, 205–215 (2001).

60. A. Picollo and M. Pusch, Chloride/proton antiporter activity of mammalian CLC proteins ClC-4 and ClC-5. *Nature* **436**, 420–423 (2005).

61. O. Scheel, A. A. Zdebik, S. Lourdel, and T. J. Jentsch, Voltage-dependent electrogenic chloride/proton exchange by endosomal CLC proteins. *Nature* **436**, 424–427 (2005).

62. A. Accardi, M. Walden, W. Nguitragool, H. Jayaram, C. Williams, and C. Miller, Separate ion pathways in a Cl-/H+ exchanger. *J Gen Physiol* **126**, 563–570 (2005).

63. L. Diewald, J. Rupp, M. Dreger, F. Hucho, C. Gillen, and H. Nawrath, Activation by acidic pH of CLC-7 expressed in oocytes from *Xenopus laevis. Biochem Biophys Res Commun* **291**, 421–424 (2002).

64. D. Kasper, R. Planells-Cases, J. C. Fuhrmann, O. Scheel, O. Zeitz, K. Ruether, A. Schmitt, M. Poet, R. Steinfeld, M. Schweizer, U. Kornak, and T. J. Jentsch, Loss of the chloride channel ClC-7 leads to lysosomal storage disease and neurodegeneration. *EMBO J* **24**, 1079–1091 (2005).

65. M. Grabe and G. Oster, Regulation of organelle acidity. *J Gen Physiol* **117**, 329–344 (2001).

66. Y. Abu-Amer, F. P. Ross, P. Schlesinger, M. M. Tondravi, and S. L. Teitelbaum, Substrate recognition by osteoclast precursors induces C-src/microtubule association. *J Cell Biol* **137**, 247–258 (1997).

67. P. Soriano, C. Montgomery, R. Geske, and A. Bradley, Targeted disruption of the c-src proto-oncogene leads to osteopetrosis in mice. *Cell* **64**, 693–702 (1991).

68. S. Tehrani, R. Faccio, I. Chandrasekar, F. P. Ross, and J. A. Cooper, Cortactin has an essential and specific role in osteoclast actin assembly. *Mol Biol Cell* **17**, 2882–2895 (2006).

69. J. Zuo, J. Jiang, S. H. Chen, S. Vergara, Y. Gong, J. Xue, H. Huang, M. Kaku, and L. S. Holliday, Actin binding activity of subunit B of vacuolar H+-ATPase is involved in its targeting to ruffled membranes of osteoclasts. *J Bone Miner Res* **21**, 714–721 (2006).

70. M. Berryman, J. Bruno, J. Price, and J. C. Edwards, CLIC-5A functions as a chloride channel *in vitro* and associates with the cortical actin cytoskeleton *in vitro* and *in vivo. J Biol Chem* **279**, 34794–34801 (2004).

71. A. Teti, H. C. Blair, S. L. Teitelbaum, A. J. Kahn, C. Koziol, J. Konsek, A. Zambonin-Zallone, and P. H. Schlesinger, Cytoplasmic pH regulation and chloride/bicarbonate exchange in avian osteoclasts. *J Clin Invest* **83**, 227–233 (1989).

72. B. D. Gelb, G. P. Shi, H. A. Chapman, and R. J. Desnick, Pycnodysostosis, a lysosomal disease caused by cathepsin K deficiency. *Science* **273**, 1236–1238 (1996).

73. K. Sundaram, R. Nishimura, J. Senn, R. F. Youssef, S. D. London, and S. V. Reddy, RANK ligand signaling modulates the matrix metalloproteinase-9 gene expression during osteoclast differentiation. *Exp Cell Res* **313**, 168–178 (2007).

74. L. Ostanek, A. Pawlik, I. Brzosko, M. Brzosko, R. Sterna, M. Drozdzik, and B. Gawronska-Szklarz, The urinary excretion of pyridinoline and deoxypyridinoline during rheumatoid arthritis therapy with infliximab. *Clin Rheumatol* **23**, 214–217 (2004).

75. I. A. Silver, R. J. Murrills, and D. J. Etherington, Microelectrode studies on acid microenvironment beneath adherent macrophages and osteoclasts. *Exp Cell Res* **175**, 266–276 (1988).

76. S. A. Nesbitt and M. A. Horton, Trafficking of matrix collagens through bone-resorbing osteoclasts. *Science* **276**, 266–269 (1997).

77. J. Salo, P. Lehenkari, M. Mulari, K. Metsikko, and H. K. Vaananen, Removal of osteoclast bone resorption products by transcytosis. *Science* **276**, 270–273 (1997).

78. P. J. Bekker and C. V. Gay, Characterization of a Ca2(+)-ATPase in osteoclast plasma membrane. *J Bone Miner Res* **5**, 557–567 (1990).

79. B. C. van der Eerden, J. G. Hoenderop, T. J. de Vries, T. Schoenmaker, C. J. Buurman, A. G. Uitterlinden, H. A. Pols, R. J. Bindels, and J. P. van Leeuwen, The epithelial Ca^{2+} channel TRPV5 is essential for proper osteoclastic bone resorption. *Proc Natl Acad Sci USA* **102**, 17507–17512 (2005).

80. K. Y. Renkema, T. Nijenhuis, B. C. van der Eerden, A. W. van der Kemp, H. Weinans, J. P. van Leeuwen, R. J. Bindels, and J. G. Hoenderop, Hypervitaminosis D mediates compensatory Ca^{2+} hyperabsorption in TRPV5 knockout mice. *J Am Soc Nephrol* **16**, 3188–3195 (2005).

81. R. Rottapel, C. W. Turck, N. Casteran, X. Liu, D. Birnbaum, T. Pawson, and P. Dubreuil, Substrate specificities and identification of a putative binding site for PI3K in the carboxy tail of the murine Flt3 receptor tyrosine kinase. *Oncogene* **9**, 1755–1765 (1994).

82. A. Teti, S. Migliaccio, and R. Baron, The role of the alphaV-beta3 integrin in the development of osteolytic bone metastases: A pharmacological target for alternative therapy? *Calcif Tissue Int* **71**, 293–299 (2002).

83. I. Wolf, B. J. Jenkins, Y. Liu, M. Seiffert, J. M. Custodio, P. Young, and L. R. Rohrschneider, Gab3, a new DOS/Gab family member, facilitates macrophage differentiation. *Mol Cell Biol* **22**, 231–244 (2002).

84. M. Wolfson, C. P. Yang, and S. B. Horwitz, Taxol induces tyrosine phosphorylation of Shc and its association with Grb2 in murine RAW 264.7 cells. *Int J Cancer* **70**, 248–252 (1997).

85. T. Wada, T. Nakashima, A. J. Oliveira-dos-Santos, J. Gasser, H. Hara, G. Schett, and J. M. Penninger, The molecular scaffold Gab2 is a crucial component of RANK signaling and osteoclastogenesis. *Nat Med* **11**, 394–399 (2005).

86. I. Nakamura, E. Jimi, L. T. Duong, T. Sasaki, N. Takahashi, G. A. Rodan, and T. Suda, Tyrosine phosphorylation of p130Cas is involved in actin organization in osteoclasts. *J Biol Chem* **273**, 11144–11149 (1998).

87. I. Nakamura, G. A. Rodan, and L. T. Duong, Distinct roles of p130Cas and c-Cbl in adhesion-induced or macrophage colony-stimulating factor-mediated signaling pathways in pre-fusion osteoclasts. *Endocrinology* **144**, 4739–4741 (2003).

88. S. Cabodi, L. Moro, G. Baj, M Smeriglio, P. Di Stefano, S. Gippone, N. Surico, L. Silengo, E. Turco, G. Tarone, and P. Defilippi, p130Cas interacts with estrogen receptor alpha and modulates non-genomic estrogen signaling in breast cancer cells. *J Cell Sci* **117**, 1603–1611 (2004).

89. A. Bruzzaniti, L. Neff, A. Sanjay, W. C. Horne, P. De Camilli, and R. Baron, Dynamin forms a Src kinase sensitive complex with Cbl and regulates podosomes and osteoclast activity. *Mol Biol Cell* **16**, 3301–3313 (2005).

90. R. Faccio, S. L. Teitelbaum, K. Fujikawa, J. Chappel, A. Zallone, V. L. Tybulewicz, F. P. Ross, and W. Swat, Vav3 regulates osteoclast function and bone mass. *Nat Med* **11**, 284–290 (2005).

91. B. B. Yaroslavskiy, Y. Zhang, S. E. Kalla, P. Garcia, V. A. C. Sharrow, Y. Li, M. Zaidi, C. Wu, and H. C. Blair, NO-dependent osteoclast motility: Reliance on cGMP-dependent protein kinase I and VASP. *J Cell Sci* **118**, 5479–5487 (2005).

92. D. Mao, H. Epple, B. Uthgenannt, D. V. Novack, and R. Faccio, PLCgamma2 regulates osteoclastogenesis via its interaction with ITAM proteins and GAB2. *J Clin Invest* **116**, 2869–2879 (2006).

93. S. Fodor, Z. Jakus, and A. Mocsai, ITAM-based signaling beyond the adaptive immune response. *Immunol Lett* **104**, 29–37 (2006).

94. H. Glantschnig, J. E. Fisher, G. Wesolowski, G. A. Rodan, and A. A. Reszka, M-CSF, TNFalpha and RANK ligand promote osteoclast survival by signaling through mTOR/S6 kinase. *Cell Death Differ* **10**, 1165–1177 (2003).

95. D. L. Lacey, H. L. Tan, J. Lu, S. Kaufman, G. Van, W. Qiu, A. Rattan, S. Scully, F. Fletcher, T. Juan, M. Kelley, T. L. Burgess, W. J. Boyle, and A. J. Polverino, Osteoprotegerin ligand modulates murine osteoclast survival *in vitro* and *in vivo*. *Am J Pathol* **157**, 435–448 (2000).

96. S. Roux, P. Lambert-Comeau, C. Saint-Pierre, M. Lepine, B. Sawan, and J. L. Parent, Death receptors, Fas and TRAIL receptors, are involved in human osteoclast apoptosis. *Biochem Biophys Res Commun* **333**, 42–50 (2005).

97. J. Xu, C. Wang, R. Han, N. Pavlos, T. Phan, J. H. Steer, A. J. Bakker, D. A. Joyce, and M. H. Zheng, Evidence of reciprocal regulation between the high extracellular calcium and RANKL signal transduction pathways in RAW cell derived osteoclasts. *J Cell Physiol* **202**, 554–562 (2005).

98. R. D. Chapurlat and P. D. Delmas, Drug insight: Bisphosphonates for postmenopausal osteoporosis. *Nat Clin Pract Endocrinol Metab* **2**, 211–219 (2006).

99. K. Kanaoka, Y. Kobayashi, F. Hashimoto, T. Nakashima, M. Shibata, K. Kobayashi, Y. Kato, and H. Sakai, A common downstream signaling activity of osteoclast survival factors that prevent nitric oxide-promoted osteoclast apoptosis. *Endocrinology* **141**, 2995–3005 (2000).

100. M. J. Oursler, E. W. Bradley, S. L. Elfering, and C. Giulivi, Native, not nitrated, cytochrome c and mitochondria-derived hydrogen peroxide drive osteoclast apoptosis. *Am J Physiol Cell Physiol* **288**, C156–C168 (2005).

101. R. Mentaverri, S. Yano, N. Chattopadhyay, L. Petit, O. Kifor, S. Kamel, E. F. Terwilliger, M. Brazier, and E. M. Brown, The calcium sensing receptor is directly involved in both osteoclast differentiation and apoptosis. *FASEB J* **20**, 2562–2564 (2006).

102. M. Zaidi, O. A. Adebanjo, B. S. Moonga, L. Sun, and C. L. Huang, Emerging insights into the role of calcium ions in osteoclast regulation. *J Bone Miner Res* **14**, 669–674 (1999).

103. V. Garcia-Palacios, L. J. Robinson, C. W. Borysenko, T. Lehmann, S. E. Kalla, and H. C. Blair, Negative regulation of RANKL-induced osteoclastic differentiation in RAW264.7 cells by estrogen and phytoestrogens. *J Biol Chem* **280**, 13720–13727 (2005).

104. A. J. Roelofs, K. Thompson, S. Gordon, and M. J. Rogers, Molecular mechanisms of action of bisphosphonates: Current status. *Clin Cancer Res* **12**, 6222s–6230s (2006).

105. M. R. McClung, E. M. Lewiecki, S. B. Cohen, M. A. Bolognese, G. C. Woodson, A. H. Moffett, M. Peacock, P. D. Miller, S. N. Lederman, C. H. Chesnut, D. Lain, A. J. Kivitz, D. L. Holloway, C. Zhang, M. C. Peterson, and P. J. Bekker, Denosumab in postmenopausal women with low bone mineral density. *N Engl J Med* **354**, 821–831 (2006).

106. B. Seriolo, S. Paolino, A. Sulli, V. Ferretti, and M. Cutolo, Bone metabolism changes during anti-TNF-alpha therapy in patients with active rheumatoid arthritis. *Ann N Y Acad Sci* **1069**, 420–427 (2006).

107. K. L. Winthrop, Risk and prevention of tuberculosis and other serious opportunistic infections associated with the inhibition of tumor necrosis factor. *Nat Clin Pract Rheumatol* **2**, 602–610 (2006).

108. M. Munoz-Torres, G. Alonso, and M. P. Raya, Calcitonin therapy in osteoporosis. *Treat Endocrinol* **3**, 117–132 (2004).

109. H. C. Blair, S. L. Teitelbaum, H. L. Tan, and P. H. Schlesinger, Reversible inhibition of osteoclastic activity by bone-bound gallium (III). *J Cell Biochem* **48**, 401–410 (1992).

110. R. P. Warrell Jr., Gallium nitrate for the treatment of bone metastases. *Cancer* **80**, 1680–1685 (1997).

111. J. P. Mattsson, K. Vaananen, B. Wallmark, and P. Lorentzon, Omeprazole and bafilomycin, two proton pump inhibitors: Differentiation of their effects on gastric, kidney and bone H(+)-translocating ATPases. *Biochim Biophys Acta* **1065**, 261–268 (1991).

112. C. Farina and S. Gagliardi, Selective inhibition of osteoclast vacuolar H(+)-ATPase. *Curr Pharm Des* **8**, 2033–2048 (2002).

113. O. Vasiljeva, T. Reinheckel, C. Peters, D. Turk, V. Turk, and B. Turk, Emerging roles of cysteine cathepsins in disease and their potential as drug targets. *Curr Pharm Des* **13**, 385–401 (2007).

114. S. Kumar, L. Dare, J. A. Vasko-Moser, I. E. James, S. M. Blake, D. J. Rickard, S. M. Hwang, T. Tomaszek, D. S. Yamashita, R. W. Marquis, H. Oh, J. U. Jeong, D. F. Veber, M. Gowen, M. W. Lark, and G. Stroup, A highly potent inhibitor of cathepsin K (relacatib) reduces biomarkers of bone resorption both *in vitro* and in an acute model of elevated bone turnover *in vivo* in monkeys. *Bone* **40**, 122–131 (2007).

115. W. J. Boyle, W. S. Simonet, and D. L. Lacey, Osteoclast differentiation and activation. *Nature* **423**, 337–342 (2003).

116. N. Udagawa, N. Takahashi, H. Yasuda, A. Mizuno, K. Itoh, Y. Ueno, T. Shinki, M. T. Gillespie, T. J. Martin, K. Higashio, and T. Suda, Osteoprotegerin produced by osteoblasts is an important regulator in osteoclast development and function. *Endocrinology* **141**, 3478–3484 (2000).

117. H. C. Blair and N. A. Athanasou, Recent advances in osteoclast biology and pathological bone resorption. *Histol Histopathol* **19**, 189–199 (2004).

118. K. E. Armour, K. J. Armour, M. E. Gallagher, A. Godecke, M. H. Helfrich, D. M. Reid, and S. H. Ralston, Defective bone formation and anabolic response to exogenous estrogen in mice with targeted disruption of endothelial nitric oxide synthase. *Endocrinology* **142**, 760–766 (2001).

119. Q. Lu, D. C. Pallas, H. K. Surks, W. E. Baur, M. E. Mendelsohn, and R. H. Karas, Striatin assembles a membrane signaling complex necessary for rapid, nongenomic activation of endothelial NO synthase by estrogen receptor alpha. *Proc Natl Acad Sci USA* **101**, 17126–17131 (2004).

120. H. Michael, P. L. Harkonen, H. K. Vaananen, and T. A. Hentunen, Estrogen and testosterone use different cellular pathways to inhibit osteoclastogenesis and bone resorption. *J Bone Miner Res* **20**, 2224–2232 (2005).

121. Y. Kobayashi, T. Mizoguchi, I. Take, S. Kurihara, N. Udagawa, and N. Takahashi, Prostaglandin E$_2$ enhances

osteoclastic differentiation of precursor cells through protein kinase A-dependent phosphorylation of TAK1. *J Biol Chem* **280**, 11395–11403 (2005).

122. I. Take, Y. Kobayashi, Y. Yamamoto, H. Tsuboi, T. Ochi, S. Uematsu, N. Okafuji, S. Kurihara, N. Udagawa, and N. Takahashi, Prostaglandin E$_2$ strongly inhibits human osteoclast formation. *Endocrinology* **146**, 5204–5214 (2005).

123. S. Y. Han, N. K. Lee, K. H. Kim, I. W. Jang, M. Yim, J. H. Kim, W. J. Lee, and S. Y. Lee, Transcriptional induction of cyclooxygenase-2 in osteoclast precursors is involved in RANKL-induced osteoclastogenesis. *Blood* **106**, 1240–1245 (2005).

124. E. Canalis, Mechanisms of glucocorticoid-induced osteoporosis. *Curr Opin Rheumatol* **15**, 454–457 (2003).

125. A. W. Eberhardt, A. Yeager-Jones, and H. C. Blair, Regional trabecular bone matrix degeneration and osteocyte death in femora of glucocorticoid-treated rabbits. *Endocrinology* **142**, 1333–1340 (2001).

126. B. S. Moonga, S. Li, J. Iqbal, R. Davidson, V. S. Shankar, P. J. Bevis, A. Inzerillo, E. Abe, C. L. Huang, and M. Zaidi, Ca(2+) influx through the osteoclastic plasma membrane ryanodine receptor. *Am J Physiol Renal Physiol* **282**, F921–F932 (2002).

127. L. Sun, J. Iqbal, S. Dolgilevich, T. Yuen, X. B. Wu, B. S. Moonga, O. A. Adebanjo, P. J. Bevis, F. Lund, C. L. Huang, H. C. Blair, E. Abe, and M. Zaidi, Disordered osteoclast formation and function in a CD38 (ADP-ribosyl cyclase)-deficient mouse establishes an essential role for CD38 in bone resorption. *FASEB J* **17**, 369–375 (2003).

128. I. R. Garrett, B. F. Boyce, R. Oreffo, L. F. Bonewald, J. W. Poser, and G. R. Mundy, Oxygen-derived free radicals stimulate osteoclastic bone resorption in rodent bone *in vitro* and *in vivo*. *J Clin Invest* **85**, 632–639 (1990).

129. X. C. Bai, D. Lu, A. L. Liu, Z. M. Zhang, X. M. Li, Z. P. Zou, W. S. Zeng, B. L. Cheng, and S. Q. Luo, Reactive oxygen species stimulates receptor activator of NF-kappaB ligand expression in osteoblast. *J Biol Chem* **280**, 17497–17506 (2005).

130. R. J. van't Hof and S. H. Ralston, Cytokine-induced nitric oxide inhibits bone resorption by inducing apoptosis of osteoclast progenitors and suppressing osteoclast activity. *J Bone Miner Res* **12**, 1797–1804 (1997).

131. M. A. Taubman, P. Valverde, X. Han, and T. Kawai, Immune response: The key to bone resorption in periodontal disease. *J Periodontol* **76**, 2033–2041 (2005).

132. T. N. Crotti, M. D. Smith, D. M. Findlay, H. Zreiqat, M. J. Ahern, H. Weedon, G. Hatzinikolous, M. Capone, C. Holding, and D. R. Haynes, Factors regulating osteoclast formation in human tissues adjacent to peri-implant bone loss: Expression of receptor activator NFkappaB, RANK ligand and osteoprotegerin. *Biomaterials* **25**, 565–573 (2004).

133. M. T. Manley, K. L. Ong, and S. M. Kurtz, The potential for bone loss in acetabular structures following THA. *Clin Orthop Relat Res* **453**, 246–253 (2006).

134. M. L. Wang, P. F. Sharkey, and R. S. Tuan, Particle bioreactivity and wear-mediated osteolysis. *J Arthroplasty* **19**, 1028–1038 (2004).

135. Y. Bi, J. M. Seabold, S. G. Kaar, A. A. Ragab, V. M. Goldberg, J. M. Anderson, and E. M. Greenfield, Adherent endotoxin on orthopedic wear particles stimulates cytokine production and osteoclast differentiation. *J Bone Miner Res* **16**, 2082–2091 (2001).

136. G. D. Roodman and J. J. Windle, Paget disease of bone. *J Clin Invest* **115**, 200–208 (2005).

137. M. Bendre, D. Gaddy, R. W. Nicholas, and L. J. Suva, Breast cancer metastasis to bone: It is not all about PTHrP. *Clin Orthop Relat Res* **415**, S39–S45 (2003).

138. E. Y. Lin and J. W. Pollard, Macrophages: Modulators of breast cancer progression. *Novartis Found Symp* **256**, 158–168 (2004).

139. M. Abe, K. Hiura, J. Wilde, A. Shioyasono, K. Moriyama, T. Hashimoto, S. Kido, T. Oshima, H. Shibata, S. Ozaki, D. Inoue, and T. Matsumoto, Osteoclasts enhance myeloma cell growth and survival via cell–cell contact: A vicious cycle between bone destruction and myeloma expansion. *Blood* **104**, 2484–2491 (2004).

140. S. J. Choi, J. C. Cruz, F. Craig, H. Chung, R. D. Devlin, G. D. Roodman, and M. Alsina, Macrophage inflammatory protein 1-alpha is a potential osteoclast stimulatory factor in multiple myeloma. *Blood* **96**, 671–675 (2000).

Osteocytes

LYNDA F. BONEWALD

I. INTRODUCTION

Osteocytes are defined as cells embedded in the mineralized bone matrix. Therefore, they are defined by their location, not by their function as is the case for osteoblasts and osteoclasts. This lack of a functional definition implies a lack of knowledge of function. The fact that osteocytes compose over 90–95% of all bone cells [1] yet a clear function has not been ascribed to these cells is disconcerting. Current opinion is that the major function of osteocytes is to translate mechanical strain into biochemical signals between osteocytes and cells on the bone surface to effect (re)modeling, yet this remains to be proven. Osteocytes are thought to respond to mechanical strain to send signals of resorption or formation [2]. They are regularly dispersed throughout the mineralized matrix, connected to each other and cells on the bone surface through slender, cytoplasmic processes radiating in all directions but generally perpendicular to the bone surface. The cell processes or dendrites pass through the bone in thin canals called canaliculi connecting osteocytes with cells on the bone surface. Osteocytes are thought to function as a network of sensor cells mediating the effects of mechanical loading through their extensive communication network referred to as a "syncytium." The term "syncytium" used here is to describe the linked three-dimensional network of cells in bone (not the same definition as syncytium in microbiology, which defines cells with shared cytoplasm). Not only do these cells communicate with each other and with cells on the bone surface, but their dendritic processes are in contact with the bone marrow [3]. Multiple connections through the tips of their dendritic processes imply that osteocytes function as "communicators." (See Figure 6-1.)

II. OSTEOCYTE ONTOGENY

Osteoprogenitor cells residing in the bone marrow give rise to osteoblasts that progress through a series of maturational stages resulting in the mature osteocyte. This review focuses on events occurring during and after the embedding process—specifically on the osteoid cell and the mature osteocyte and on their potential functions. Biomarkers and functional assays have been used to discriminate between these various stages. Whereas numerous markers for osteoblasts are available (cbfa1, osterix, alkaline phosphatase, collagen type I, osteocalcin, etc.; see Chapter 4, by Stein, on osteoblasts), few markers have been available for osteocytes until recently. It would be expected that osteocytes would share some markers with their progenitors, osteoblasts, but would also express unique markers based on their morphology and potential function. Kalajzic and coworkers used promoters for osteocalcin and collagen type I linked to green fluorescent protein (GFP) to examine transgene expression during osteoblast differentiation [4]. Osteocalcin-GFP was expressed in a few osteoblastic cells lining the endosteal bone surface and in scattered osteocytes, whereas GFP driven by the collagen type I promoter was strongly expressed in osteoblasts and osteocytes. Recently, these investigators generated an osteocyte-selective promoter, the 8 kb Dentin Matrix Protein 1 (DMP1), driving GFP that showed exclusive expression in osteocytes [5].

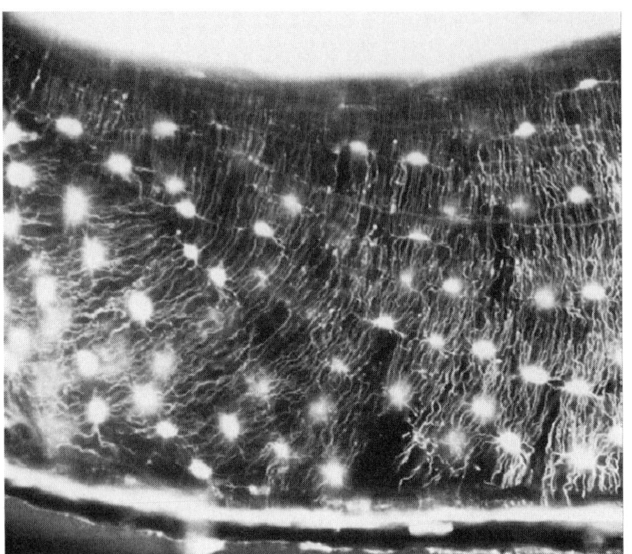

FIGURE 6-1   Procion red staining of the osteocyte lacuno-canalicular system in mouse cortical bone. Note the complexity of the network, yet the orderly alignment of lacunae. (Figure provided by Dr. Jian Feng, UMKC.)

The differentiating osteoblast has one of three fates: It can become embedded in its own osteoid and continue differentiation into an osteocyte; it can quiesce into a lining cell; or more likely, it can undergo apoptosis (for review, see Manolagas [6]). Karsdal and coworkers proposed that matrix metalloproteinase activation of latent transforming growth factor β (TGF-β) blocks osteoblast apoptosis, thereby delaying

differentiation into osteocytes [7]. Identification of mechanisms responsible for osteoblast apoptosis has implications for development of strategies to reduce or inhibit osteoblast apoptosis that could potentially increase bone mass. However, inhibition of osteocyte apoptosis may have beneficial or nonbeneficial effects on bone depending on condition, as addressed later in this chapter. (See Figure 6-2.)

III.   OSTEOID-OSTEOCYTES

Osteoblasts, osteoid cells, and osteocytes may play different roles in the initiation and regulation of mineralization of bone. In 1976 and 1981, Bordier and coworkers [8] and Nijweide and coworkers [9] proposed that osteoid-osteocytes play an important role in the initiation and control of mineralization of the bone matrix. Osteoid-osteocytes were described by Palumbo [10] to be cells actively making matrix and calcifying this matrix. Like osteoblasts, their activity was polarized toward the mineralization front to which their cellular processes were oriented, whereas processes oriented toward blood vessels only began to appear when mineralization began to spread around the cell. The cell body reduces in size in parallel with the formation of cytoplasmic processes with a reduction of about 30% at the osteoid-osteocyte stage and 70% with complete maturation of the osteocyte. During the time an osteoblast becomes an osteocyte, the cell manufactures three times its own volume in matrix [11]. For a review of the osteoblast-to-osteocyte transformation,

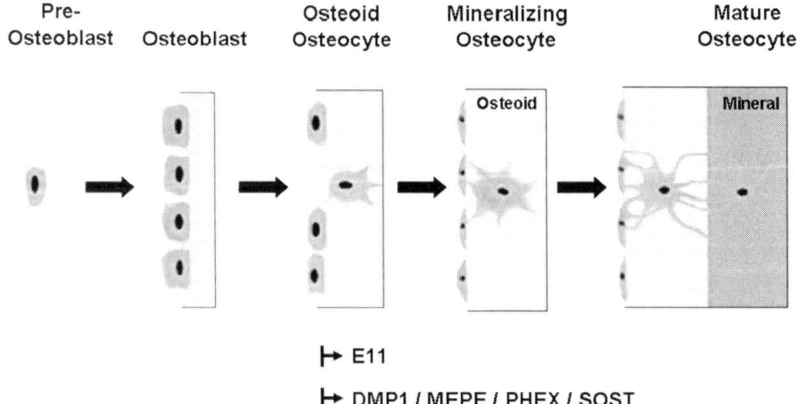

FIGURE 6-2   Osteoblast to osteocyte ontogeny. This diagram represents the process of differentiation from osteoblast precursors to matrix-producing cells, to cells embedded in osteoid, to cells embedded in the mineralized matrix. The markers listed below are relative and overlapping. E11 is the earliest marker specific for the embedding osteocyte [24, 26]. There appears to be some early expression of Dmp1 and PHEX in osteoblasts, but greatest expression is found in osteocytes [32, 33, 36]. The expression of sclerostin or SOST in osteocytes appears delayed compared to other markers for osteocytes [49].

see Franz-Odendaal and colleagues [12]. These authors suggested that, once a cell is surrounded by osteoid, the differentiation process does not end but should be viewed as a continuum of differentiation.

One cell line has been proposed to represent the osteoid-osteocyte. MLO-A5 cells, a postosteoblast/preosteocyte-like cell line established from the long bones of 14-day-old mice expressing the large T-antigen driven by the osteocalcin promoter, differentiate into osteoid-osteocyte-like cells [13]. These cells will mineralize in the absence of beta glycerolphosphate in 6–7 days in sheets, not nodules, but this process is accelerated by the addition of an external source of phosphate. Fourier transform infrared spectra of these cultures is very similar to normal bone [13]. MLO-A5 cells express all of the markers of the late osteoblast such as high alkaline phosphatase, bone sialoprotein, PTH type 1 receptor, and osteocalcin. In culture, these cells begin to express markers of osteocytes as they generate cell processes. Studies show that these cells generate spherical structures that are fully mineralized on their developing cellular processes, and as the cellular process narrows in diameter, these mineralized structures become associated with and initiate collagen-mediated mineralization [14].

Mikuni-Takagaki and colleagues proposed that casein kinase II, produced in high amounts by embedding osteoid-osteocytes and not by osteoblasts, is responsible for phosphorylation of matrix proteins essential for mineralization [15]. Phosphoproteins appear to be essential for bone mineralization as evidenced by *in vitro* crystal nucleation assays [16, 17] and *in vivo* by osteomalacia in animal models with deletion of specific genes such as dentin matrix protein 1 (DMP1) and phosphate-regulating neutral endopeptidase on the chromosome X (PHEX) [18, 19]. Deletion of inhibitors of mineralization such as sclerostin (SOST) and osteoblast/osteocyte factor 45/matrix extracellular phosphoglycoprotein (MEPE) results in osteopetrosis [20, 21]. These phosphoproteins are expressed late in osteoblast differentiation and are all molecules that are highly expressed in osteocytes. Therefore, the embedding osteoid cell and the osteocyte probably play roles in the mineralization process and potentially in phosphate metabolism (see following sections).

IV. OSTEOCYTE SELECTIVE GENES/ PROTEINS AND THEIR POTENTIAL FUNCTIONS

Markers for osteocytes have been minimal, ranging from low alkaline phosphatase to high casein kinase and high osteocalcin protein expression [22]. Antigens such as E11 have been identified that are specific for

osteocytes compared to osteoblasts, and antigens like PHEX, Dmp1, MEPE, and SOST have been found that are more highly expressed in osteocytes compared to osteoblasts. Franz-Odendaal and coauthors provided a list of molecular markers for the preosteoblast to the osteocyte [12].

E11 is the name given to a molecule that is expressed in early osteocytes [23] and found only on the dendritic processes of osteocytes, not osteoblasts *in vivo* [24]. A punctate antibody reaction at the interface between osteoblasts and uncalcified osteoid was described. Less reactivity was observed with osteocytes deeper in the bone matrix. This same antibody also reacted with cementocytes [25]. The major function of E11 may be in the formation of dendritic processes, as reduction in protein expression led to a decrease in dendrite extension in MLO-Y4 osteocyte-like cells [26], and overexpression in an osteoblast-like cell line led to the generation of extended cytoplasmic processes [27]. Ectopic overexpression in keratinocytes induces plasma membrane extensions, a major reorganization of the actin cytoskeleton, and relocalization of ezrin to cell projections [28]. The molecule co-localizes with ezrin, radixin, and moesin (ERMs) [28], proteins that are concentrated in cell-surface projections where they link the actin cytoskeleton to plasma membrane proteins. ERMs play structural roles and are involved in cell motility [29]. E11 was also found to be physically associated with CD44 in tumor vascular endothelial cells [30]. CD44 is highly expressed in osteocytes compared to osteoblasts [31]. Together these data suggest that E11 associates with CD44 and the ERMs to induce and regulate the formation of dendritic processes in osteoid-osteocytes and osteocytes.

Nijweide and coworkers found that their osteocyte specific antibody, Mab OB7.3, recognizes PHEX [32]. This antibody allowed them to purify avian osteocytes from enzymatically isolated bone cells for studies. PHEX was originally described on the plasma membrane of osteoblasts and osteocytes [33], and loss of function mutations in this gene results in X-linked hypophosphatemic rickets [34]. PHEX is a metalloendoproteinase whose substrate is not known. The precise function of PHEX is unclear, but it certainly plays a role in phosphate homeostasis and bone mineralization. These investigators propose that the osteocyte syncytium may be considered a gland that regulates bone phosphate metabolism through expression of PHEX.

Another protein highly expressed in osteocytes is DMP1. Feng and colleagues [35] found the gene expressed in early embryonic bone development in hypertrophic chondrocytes and osteoblasts and later during postnatal bone formation where it is highly expressed in osteocytes, consistent with the observations of Toyosawa, who observed high expression in osteocytes, but not in osteoblasts [36]. DMP1 is specifically expressed along and in the canaliculi of osteocytes within the bone matrix [37]. Potential roles for DMP1 in osteocytes may be related to the post-translational processing and modifications of the protein as a highly phosphorylated protein and regulator of hydroxyapatite formation [38]. Deletion of this gene in mice results in a phenotype similar, if not identical to, the HYP-phenotype [39], suggesting that Dmp1 and PHEX are interactive and essential for phosphate metabolism.

Osteoblast/osteocyte factor 45 (OF45), also known as MEPE (matrix extracellular phosphoglycoprotein), is also highly expressed in osteocytes as compared to osteoblasts. *MEPE* was isolated and cloned from a tumor-induced osteomalacia (TIO) tumor cDNA library [40]. Independently, others isolated and cloned the rat and mouse homologues based on the ability of MEPE to regulate mineralization [41, 42]. The MEPE protein is highly phosphorylated in a region called the ASARM region. Cathepsin D or B can cleave MEPE, releasing the C-terminal phosphoprotein region. This C-terminal ASARM region is a potent inhibitor of mineralization *in vitro* [43–45], and high ASARM peptide production by osteocytes correlates to an osteomalacia-type phenotype in the X-linked rickets mouse model (HYP). Messenger RNA expression for OF45/MEPE begins at E20 in more differentiated osteoblasts that have become encapsulated by bone matrix [42]. These authors placed the sequence of expression of osteoblast-to-osteocyte transition markers as osteocalcin during encapsulation, followed by Dmp1, followed by OF45 as a marker of the mature osteocyte. Deletion of this gene in mice results in increased bone formation and bone mass and resistance to age-associated trabecular bone loss [21]. The authors speculated that, as terminally differentiated osteoblasts become embedded in the bone matrix, OF45 expression is increased and maintained in mature osteocytes and that osteocytes act directly on osteoblasts through OF45 to inhibit their bone-forming activity. Interestingly, Dmp1 and OF45/MEPE belong to the SIBLING (Small, Integrin-Binding LIgand, N-linked Glycoprotein) family that also includes bone sialoprotein, osteopontin, and sialophosphoprotein [46]. This family of proteins may function differently in osteocytes compared to other cell types especially upon phosphorylation by casein kinase.

The *SOST* gene encodes a protein, sclerostin, that is highly expressed in osteocytes and appears to inhibit bone formation [47]. The human condition of sclerostosis is due to a premature termination of the SOST gene [48]. Transgenic mice lacking sclerostin have increased bone mass. Clearly, sclerostin is a negative regulator of bone formation. Controversy exists as to whether sclerostin is a BMP antagonist or functions as a Wnt antagonist [49]. Therefore, sclerostin may be an antagonist of Lrp5, a gene shown to be important as a positive regulator of bone mass [50]. It is suggested that sclerostin may be transported through canaliculi to the bone surface to inhibit bone-forming osteoblasts. It has also been proposed that the anabolic effects of PTH are through inhibition of SOST expression [51].

Another molecule found to be a major component of the osteocyte extracellular matrix and more highly expressed in osteocytes than osteoblasts is CD44 [31]. CD44 is a membrane bound protein and hyaluronic acid receptor that interacts with the ERM (ezrin, radixin, moesin) family of adapter proteins that link to actin in the cytoskeleton. CD44 is a major component of the osteocyte pericellular matrix. CD44 has been shown to be associated with E11 [30] and with osteopontin [52], another member of the SIBLING family, suggesting that other members of this family such as Dmp1 and MEPE may also interact with CD44.

Osteocytes have also been found to be intensively immunoreactive for neurokinin-1, whereas lining cells were found to be positive for neurokinin-2 [53]. Neurokinin-1 and neurokinin-2 are tachykinin receptors for neuropeptides. The presence of these receptors suggests that sensory nerves may regulate the function of bone cells. For additional hypotheses concerning the possible relationship of the neural system to bone, see the review by Turner [54].

V. MORPHOLOGY OF OSTEOCYTES: LACUNOCANALICULAR SYSTEM AND DENDRITE FORMATION

The transformation of a plump polygonal osteoblast to a dendritic osteocyte is striking and dramatic and clearly requires extensive reorganization of the cytoskeleton. The osteocyte loses the typical apical and basolateral plasma membrane polarization characteristic of osteoblasts [55]. Actin filaments were found to be crucial for the maintenance of the osteocyte processes, and two actin-bundling proteins, alpha-actinin and fimbrin, were shown to be useful as markers for osteocytes [56]. Stronger signals of fimbrin were observed at branching points in dendrites. Villin, another actin-bundling protein, is also higher in osteocytes than osteoblasts. Staining patterns were distinct between osteoblasts and osteocytes with filamin along stress fibers in osteoblasts, but only at the base of processes in osteocytes. Staining for spectrin was punctate in osteoblasts, but filamentous in osteocytes [57].

A hydrophobic membrane protein called E11 appears to play a role in dendrite formation. Although known as E11 in osteocytes, it is known by other names (gp38/podoplanin/T1alpha) in other cell types (endothelial cells/podocytes in kidney/type II alveolar lung cells). The earliest description of the gene for E11 was in 1990 as an unknown phorbol ester inducible gene in MC3T3 osteoblast-like cells, called OTS-8 [58]. A common feature of virtually all the cell types that express E11 is their extended cytoplasm or dendritic nature. The fact that E11 is often found in cells that are exposed to an external or internal fluid compartment and is highly negatively charged and resistant to proteases suggests the molecule provides a physical barrier playing a role in protecting cells. Deletion of E11 results in mice that die at birth due to respiratory failure caused by a failure of type II alveolar lung cells to differentiate into type I alveolar lung cells [59]. A potential function in osteocytes was shown by reducing E11 protein expression using an siRNA approach, which prevented dendrite elongation in MLO-Y4 cells in response to shear stress [26].

Dendrite formation is an active process. Osteocytogenesis has been thought to be a passive process whereby some osteoblasts become passively encased in osteoid that passively mineralizes. However, Holmbeck and colleagues [60] showed osteocytogenesis to be an active invasive process requiring cleavage of collagen and potentially other matrix molecules. Osteocytes in mice null for the metalloproteinase MT1-MMP have significantly reduced number and length of dendritic processes. MT1-MMP is a membrane-anchored proteinase that can cleave collagens type I, II, and III; fibrin; fibronectin; and other matrix molecules. In this mouse model, the almost complete lack of dendritic processes did not appear to affect viability or density of osteocytes. This is in contrast to studies by Zhao and coworkers [61] where osteocytes in a mouse model of collagenase resistant type I collagen did show increased apoptosis. However, in the MT1-MMP null mouse, it is difficult to determine the effect of a lack of dendritic processes on either osteocyte function or effects on the skeleton, as this mouse exhibits multiple defects, such as dwarfism due to a lack of MT1-MMP in other skeletal tissues [62]. Interestingly, these investigators and others [63] showed an increase in number of canaliculi between young and adult animals suggesting either that new bone made in the adult or aging animals generates osteocytes with more canaliculi or that embedded osteocytes can generate new dendrites. (See Figure 6-3.)

The osteocyte has been viewed as a quiescent cell type. However, evidence is accumulating that these cells are more active than previously thought. Dallas and colleagues used calvarial explants from transgenic mice with green fluorescent protein (GFP) expression targeted to osteocytes [5] and time lapse dynamic imaging to image living osteocytes within their lacunae [64]. Surprisingly, these studies revealed that, far from being a static cell, the osteocyte may be highly dynamic. Embedded osteocytes expand and contract their cell body within the boundaries of their lacunae and extend and retract their dendrites over a 24-hour

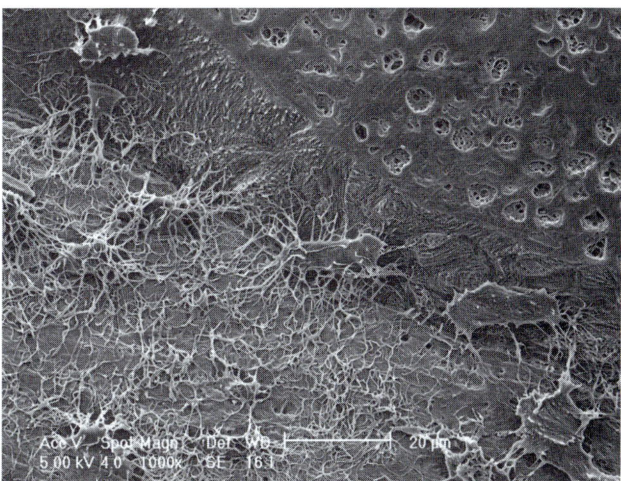

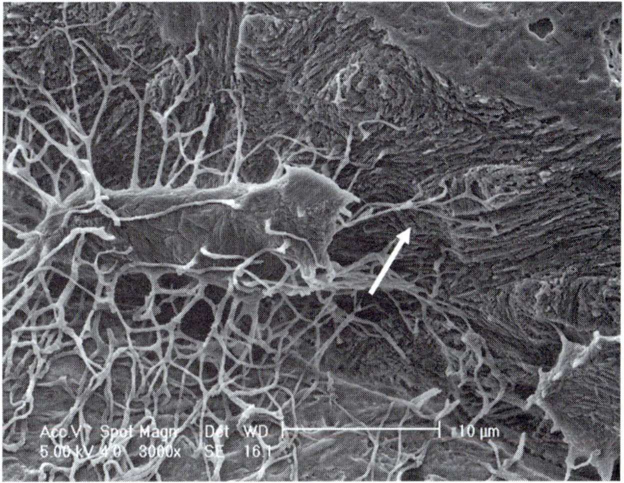

FIGURE 6-3   Osteocytes make contact with cells on the bone surface. The image is an acid-etched resin embedded murine bone visualized by scanning electron microscopy showing the high interconnectivity of the osteocyte lacunocanalicular system. The top panel shows the fully resin-embedded marrow on the top and the complex osteocyte lacunocanalicular network below where the mineral has been removed by acid etching. The bottom panel is a magnification showing canaliculi in contact with the bottom of a cell on the surface of the bone (arrow).

period. These data suggest that dendrites, rather than being permanent connections between osteocytes and between osteocytes and surface cells, may be dynamic structures that can be altered in response to stimuli.

VI. OSTEOCYTE CELL MODELS

There are several reasons why much less is known concerning osteocyte function compared to osteoblasts and osteoclasts. They include the fact that it is difficult to isolate sufficient numbers of osteocytes from the mineralized bone matrix for many types of studies, that it is difficult to maintain their differentiated function *in vitro*, that there is a lack of suitable cell lines, and there is a lack of availability of osteocyte-specific promoters for targeted transgenic approaches. Primary cultures of osteocyte-like cells can be prepared by sequential alternating digestions of fetal rat and chick calvaria with collagenase and EDTA [22, 65]. Cells removed in early digests are fibroblasts/osteoblasts, and the late-released cells represent a population enriched for osteocytes. An osteocyte-specific antibody for avian osteocytes, Mab OB7.3, has also been successfully used in antibody panning techniques to obtain an essentially pure population of avian osteocytes [65]. These primary osteocyte culture systems have been useful in beginning to define the properties of these cells and investigate their biochemistry.

Mice have recently been generated in which the 8 kb DMP1 promoter drives GFP expression [5], and this promoter has been shown to be regulated by mechanical strain [66]. As this promoter is specifically expressed in osteocytes, these mice can be used to study osteocytes especially in conjunction with fluorescence-activated cell sorting to obtain a highly purified population. However, the yields of primary osteocytes are low, thereby making it difficult to obtain large enough numbers of cells for detailed or extensive biochemical studies. To compensate for these difficulties, investigators have attempted to make osteocyte cell lines. To date, only two cell lines with osteocyte-like characteristics have been described. One model of the pre- or early osteocyte is the HOB-01-C1 human bone cell line [67], a temperature-sensitive line that proliferates at 34°C and stops growing at 39°C and has cellular processes, low alkaline phosphatase expression, and high-expressing osteocalcin and CD44. Another model for early osteocytes is the murine MLO-Y4 osteocyte-like cell line [68]. This cell line was derived from a transgenic mouse in which the immortalizing T-antigen was expressed under control of the osteocalcin promoter. MLO-Y4 cells exhibit properties of osteocytes including high expression of osteocalcin, low expression of alkaline phosphatase, high expression of connexin 43 and the antigen E11, a known marker of osteocytes. MLO-Y4 cells retain a dendritic morphology, similar to that observed in primary osteocyte cultures. Numerous laboratories have used this cell line to investigate osteocyte cell function including references [26, 69–93] in addition to others not listed here.

Osteocytic cell lines have been generated from mice lacking the type 1 PTH/PTHrp receptor [75]. These cells have proved useful in determining the effects of PTH on osteoblasts and osteocytes and in the discovery of a receptor that binds to the carboxy terminus of PTH [76]. Expression of CPTH-R is greater on osteocytes than on other bone-derived cell types. The CPTH portion of PTH is cleaved as a normal process; however, its function is not clear. Functional studies of CPTH-Rs in osteocytic cells have suggested the involvement in cell survival and intracellular communication, and in proapoptotic and antiresorptive actions.

As discussed previously, the MLO-A5 cell line has characteristics of a postosteoblast/preosteocyte. These cells are very large, over 100 nm; express all of the markers of the late osteoblast such as extremely high expression of alkaline phosphatase, bone sialoprotein, PTH type 1 receptor, and osteocalcin; and do rapidly mineralize in sheets, not nodules [13]. In culture, these cells begin to express markers of osteocytes such as E11 as they generate cell processes [14]. Bellido and coworkers found that SOST expression is regulated by PTH in these cells [51].

VII. MECHANISMS AND RESPONSE OF OSTEOCYTES TO MECHANICAL FORCES

A known key regulator of osteoblast and osteoclast activity in bone is mechanical strain. Under normal conditions, bone formation and bone resorption are balanced to maintain bone mass. However, by the process of adaptive remodeling, the skeleton is able to continually adapt to mechanical loading by adding new bone to withstand increased amounts of loading and removing bone in response to unloading or disuse (reviewed in [94, 95]). It was actually Galileo in 1638 who first documented this concept suggesting that the shape of bones is related to loading. Julius Wolff in 1892 more eloquently wrote that bone accommodates or responds to strain. The cells of bone with the potential for sensing mechanical strain and translating these forces into biochemical signals include bone lining cells, osteoblasts, and osteocytes. Of these, the osteocytes, with their distribution throughout the bone matrix and their high degree of interconnectivity, are thought to be one of the major cell types responsible for sensing mechanical strain and translating that strain into biochemical signals related to the intensity and distribution of the strain signals [2].

Various studies have demonstrated load-related responses in osteocytes, supporting their proposed role as mechanotransducers in bone. Within a few minutes of loading, glucose 6-phosphate dehydrogenase, a marker of cell metabolism, is increased in osteocytes and lining cells [96–98]. By 2 hours, c-fos mRNA is evident in osteocytes, and by 4 hours, transforming growth factor β (TGF-β) and insulin-like growth factor-1 (IGF-1) mRNAs are increased [99]. The DMP1 gene is activated in response to mechanical loading in osteocytes in the tooth movement model [100] and in the mouse ulna loading model of bone formation [66]. E11 is also increased in response to mechanical load, not only in cells near the bone surface but also in deeply embedded osteocytes [26].

The parameters for inducing bone formation or bone resorption *in vivo* are fairly well known and well characterized. Bone mass is influenced by peak applied strain as shown by Rubin and Lanyon [101, 102]. Bone formation rate is related to loading rate as shown by varying the frequency of applied bending while keeping the magnitude of applied load constant [103]. At bending frequencies of 0.5 to 2.0 Hz, bone formation rate increased as much as 4-fold while no increase was observed at frequencies lower than 0.5 Hz. When rest periods are inserted, the loaded bone shows increased bone formation rates and mechanical properties when compared to bone subjected to a single bout of mechanical loading [104]. Frequency, intensity, and timing of loading are all important parameters. Improved bone structure and strength are greatest if loading is applied in shorter versus longer increments [105]. By studying the effect of frequency and peak strain on mechanically induced bone formation in the rat ulna loading model, Hsieh and Turner [106] built a model that assumed bone cells are activated by fluid shear stress and that stiffness of the cells and the matrix around the cells increases at higher loading frequencies because of viscoelasticity. In this model there is a strain threshold for an osteogenic response that varies with location. For example, in the proximal region of the ulna, the strain required to achieve new bone formation is 1,300 microstrains, whereas different bone formation thresholds exist at the mid-shaft (2,200 microstrains) and the distal region (3,000 microstrains) [107]. The major challenge has been to translate *in vivo* parameters of mechanical loading to *in vitro* cell culture models.

Even though osteocytes are thought to be mechanosensors [108–110], key questions such as how mechanical loading is sensed, how these signals are conveyed to other nonsensing cells, and how these signals are translated into biochemical signals remained to be answered. The application of force to bone results in several potential stimuli for osteocyte function including hydrostatic pressure and fluid flow–induced shear stress. Over the years, various theoretical and experimental studies argued that flow of interstitial fluid driven by extravascular pressure as well as by the applied cyclic mechanical loading is likely the means by which bone cells are informed of mechanical loading [108, 111–113]. It has been found that mechanical forces applied to bone cause fluid flow through the canaliculi surrounding the osteocyte that is probably responsible for the deformation of the cell membrane [111, 114, 115]. Fluid flow imposes a shear stress on osteocytes, thus deforming the cells within their lacunae and the dendrites within their canaliculi. Recently, first real-time attempts to measure solute transport in bone through dye diffusion within the lacunar canalicular system have been conducted *in vivo* [116]. It is hoped that future studies will permit analysis of mechanical loading and blood pressure to this process. (See Figure 6-4.)

A model of strain amplification in osteocyte cell processes was proposed by Weinbaum and coworkers [117]. A recent TEM-based model of the osteocyte process within its canaliculi with a predicted environment was used to build a more detailed theoretical model as to how strain is amplified at the osteocyte dendrite level. One of the requirements of the model is that osteocyte dendritic processes be tethered within canaliculi to the surrounding mineralized matrix through structural components, such as CD44, laminins, and a variety of other unknown proteins and proteoglycans present in the pericellular matrix surrounding the osteocyte. Another major requirement of the model is the formation of hexagonal actin bundles within the cell processes of the osteocyte. A relatively stiff structure can be generated with predominantly fimbrin cross-linked to actin bundles. The actin bundle is then attached to integrin-related proteins through myosin type proteins, ERMs, and others. The model predicts that fluid flow through this structure will deform the shape of these tethering elements, creating a drag force predominantly in this highly viscous, yet sieving pericellular matrix that then imposes a hoop strain on the central actin bundles in the osteocyte cell process. Theoretical modeling predicts osteocyte wall shear stresses resulting from peak physiologic loads *in vivo* in the range of 8 to 30 dynes/cm^2 [111].

Models have been used to predict the effects of canalicular fluid flow on osteocytes. Petrov and Pollack proposed that neither diffusion- nor stress-induced fluid flow is capable of sustaining osteocyte viability,

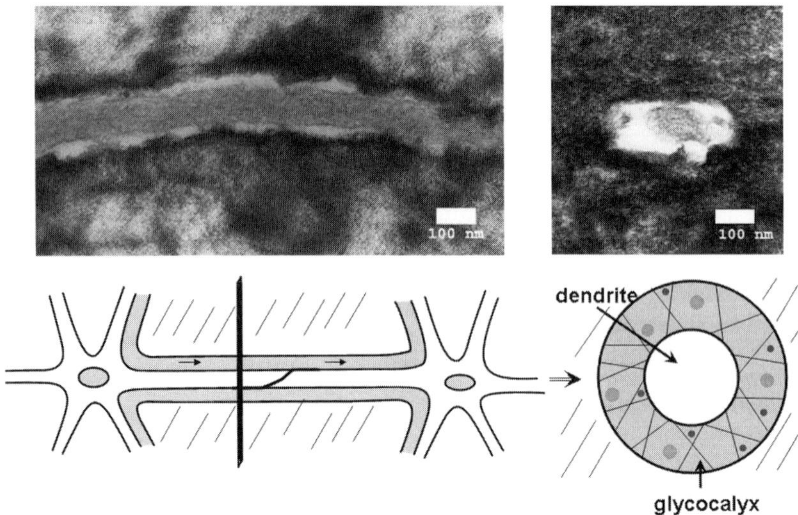

FIGURE 6-4   Canaliculi, dendrites, and fluid flow: It has been proposed that molecules travel in the bone fluid through a glycocalyx that surrounds the dendritic processes within the canaliculi [222]. The dendritic process appears to be anchored to the wall of the canaliculi by integrins [135]. The glycocalyx acts as a sieve or "fishnet" to allow molecules below a size of approximately 7 nm to pass [223]. Studies suggest that molecules as large as albumin can pass through the canaliculi and that the bone fluid serves to provide nutrients to the osteocyte. It has been proposed that immobilization causes a lack of bone fluid flow, which in turn causes hypoxia, followed by osteocyte cell death [224]. A recent report showed that fluid flow–induced PGE_2 release by MLO-Y4 cells is reduced by the degradation of the glycocalyx on the cell surface, a hypothesized mechanosensor in osteocytes [91]. Clearly, the dendritic processes of osteocytes serve numerous functions.

but that cyclic stress stimulates an active nutrient transport system [118]. Tami and coworkers used a model and preliminary *in vivo* data to show that fatigue damage impedes transport from the blood supply, which in turn depletes downstream areas of damage [119]. Smit and coworkers propose that fluid stasis occurs within the bone tissue in front of the cutting cone of the basic multicellular unit where osteoclasts are actively resorbing, while strong fluid flow occurs in the wall of the resting zone where osteoblast precursors are being recruited and in the closing cone where osteoblasts are actively forming bone [120]. Together these models suggest that without sufficient fluid flow, osteocytes support osteoclastic resorption, but with higher magnitudes of flow, osteocytes support osteoblastic bone formation.

A new approach to tackle the question of what magnitude of strain an osteocyte perceives and how magnitude correlates with biological response was instituted by Kotha and coworkers [121] and Harris and coworkers [122]. These investigators determined magnitude of strain (the effector) with mapped gene expression (early biological response) with bone formation (end biological result). This information was used to generate a three-dimensional model correlating magnitude of strain with magnitude and location of gene expression (DMP1 and MEPE, as these are highly expressed in osteocytes) with resulting areas

of new bone formation on the bone surface. The data to date show that osteocytes can respond as a population to increased strain and that the response of each individual osteocyte also correlates with magnitude of strain in its local environment.

It has also been proposed that mechanical information is relayed in part by cell deformation [123–125]. Typical *in vivo* strains in humans are on the order of 1,200 uE (principal compressive strain) to 1,900 uE (maximum shear strain) [126]. These strains were determined using strain gauges that covered an area approximately 1.8 mm by 3.6 mm containing thousands of cells and are therefore averages of osteocyte strain. Variations resulting from microstructural features or discontinuities in the bone matrix will affect the local strain or deformation sensed by individual bone cells. Measured microstructural strains at or near osteocyte lacunae were found to be up to 3 times greater than the average strains measured with an external strain gauge [124, 125]. If bone damage (microcracks) is present, the peri-lacunar strain magnification near a microcrack tip can be as high as 15 times *in vivo* measured bone strain. However, neither fluid flow nor the resulting osteocyte deformations in bone *in vivo* have been directly measured; therefore, theoretical predictions of *in vivo* flow shear stress have not yet been validated.

In *in vitro* cell culture, numerous investigators have used osteoblast cell lines under the assumption that osteocytes will respond in a similar manner. However, primary osteocytes have been shown to be more sensitive than primary osteoblasts in the release of PGE_2 following both hydrostatic compression and pulsatile fluid flow treatment, with pulsatile fluid flow being most effective [127]. Osteoblast-like cells are less responsive to oscillatory flow (applied fluid shear stresses of −20 to +20 dynes/cm^2) than pulsatile fluid flow (applied fluid shear stresses of 0 to 20 dynes/cm^2) and steady fluid flow (applied fluid shear stresses 20 dynes/cm^2) [128]. Correlation and validation of shear stress used in tissue culture with those *in vivo* remain to be performed.

Recently, it was hypothesized that the bending of primary cilia of an osteocyte by extracellular fluid sends signals into cells through gap junctions [129]. However, it is not clear how a single cilium on an osteocyte cell body can mediate this effect. Clearly, PKD1 and PKD2, known mechanosensory proteins in the kidney, do play a role in normal bone structure [130]. It remains to be determined whether the bone defect in these animals is due to a defective mechanosensory function, as has been shown in the kidney [131].

Integrins have been proposed to play a role in mechanotransduction. Integrins, composed of heterodimers of α and β subunits, are major receptors/transducers that connect the cytoskeleton to the extracellular matrix [132] and have been proposed to be candidate mechanosensors in bone cells [133]. Stretch and fluid flow shear stress stimulate pathways that are regulated by integrin binding to the extracellular matrix [134]. Among various isotypes of integrins, $\alpha 5$ and $\beta 1$ integrins are expressed in virtually all cell types in bone [135, 136]. The evidence for the involvement of integrins in gap junction communication and Cx43 expression has been reported [137, 138]. Integrins interact with plasma membrane proteins such as metalloproteases, receptors, transporters, and channels mainly through the extracellular domain of their α subunits [139]. The integrin $\alpha 5$ may act as a tethering protein that responds to shear stress by opening hemichannels in osteocytes [140].

In addition to mechanical loading, both ultrasound and electromagnetic fields have been thought to affect bone cell function. Low intensity pulsed ultrasound is a form of mechanical energy used to accelerate fracture repair and distraction osteogenesis. Osteoblasts respond to ultrasound by increased expression of osteocalcin and insulin-like growth factor 1, while osteocytes do not [141]. Conversely, substrate stretch and PTH increase Ca^{2+} influx in osteocytes, not osteoblasts, suggesting that the anabolic effects of ultra-

sound are through the osteoblast and that osteoblasts and osteocytes can respond distinctly to various forms of mechanical force.

Pulsed electromagnetic fields increase TGF-β and PGE_2 in the osteocyte-like cell line, MLO-Y4, but decrease Cx43 expression in these cells as well as ROS 17/2.8 osteoblast-like cells [86]. As pulsed electromagnetic fields have been used to treat ununited fractures, these healing effects may be partially mediated by the induction of bone anabolic factors such as TGF-β and PGE_2 and by reducing osteocyte communication through Cx43 containing gap junctions. TGF-β produced by osteocytes could be delaying osteoblast differentiation while increasing bone matrix volume [7, 142].

VIII. OSTEOCYTE SIGNALS FOR BONE FORMATION

Nitric oxide (NO) is a short-lived free radical important for the function of many tissues and organs. In bone, NO inhibits resorption and promotes bone formation. Both osteoblasts and osteocytes release NO in response to mechanical strain or fluid flow shear stress [143]. NO can be generated from any of three isoforms of nitric oxide synthase, known as neural (n), endothelial (e), and inducible (i) NOS. Osteoblasts and osteocytes have highest expression of eNOS compared to the other synthases. eNOS positive osteocytes in cases of femoral hip fracture are reduced in the inferior but not the superior region of the femoral neck compared to normal controls [144], suggesting that eNOS positive osteocytes act as sentinels to confine osteoclast activity to stay within single osteons. Even though studies have shown no or little expression of iNOS in osteocytes, mice lacking this enzyme fail to regain bone after immobilization [145]. These mice show no significant bone abnormalities, unlike mice lacking eNOS in which bone growth is retarded. Surprisingly, iNOS has no effect on resorption in the unloading phase but is essential for bone formation in the reloading phase. iNOS expression was found only after unloading and reloading of bone, not in the normal loaded state.

Clearly, prostaglandin is a bone anabolic factor and osteocytes produce prostaglandin in response to load. Prostaglandins are generally thought to be skeletal anabolic agents, as their administration can increase bone mass in humans and animals [146, 147], stimulate bone formation *in vitro* in organ culture [148], and increase nodule formation in rat calvarial osteoblasts [149]. Primary osteocytes and primary calvarial bone cells have been shown to release prostaglandins in response to fluid flow treatment [150]. A number of studies have suggested that osteocytes are the primary source of these load-induced prostaglandins [114, 151]. *In vivo* studies have shown that new bone formation induced by loading can be blocked by the prostaglandin inhibitor, indomethacin [152], and that it is the inducible COX-2 pathway that is primarily involved. Agonists of the prostaglandin receptors have been shown to increase new bone formation [153]. However, others have found that COX-2 null mice are still responsive to mechanotransduction [154]. These authors suggested compensation through COX-1 elevation.

ATP is released within seconds in osteoblasts in response to mechanotransduction [155] and initiates intracellular calcium release. The P2X7 nucleotide receptor is an ATP-gated ion channel expressed in many cell types but appears to play a role in skeletal mechanotransduction [156]. Deletion of this receptor results in mice with an attenuated inflammatory response and reduced bone formation [157]. Macrophages from these animals do not release IL-1 in response to ATP. Skeletal sensitivity to mechanical loading was reduced about 70% in these null mice [156]. Fluid flow shear stress did not induce prostaglandin release in cells isolated from these mice. Blockers of P2X7 receptors suppressed prostaglandin release, whereas agonists enhanced release in MC3T3 osteoblast and MLO-Y4 osteocyte cells. The authors concluded that P2X7 receptor is necessary for release of prostaglandin in response to mechanical load.

It was hypothesized as early as 2002 that Lrp5 is a major factor in the way that bone cells sense and respond to mechanical load [158]. These investigators were responsible for the discovery of the high bone mass (HBM) gene, a mutation in the Lrp5 receptor [159] (see Chapter 15, Johnson, on Lrp5). They reasoned that the HBM mutation results in a skeleton that is overadapted in relation to the actual loads being applied, but yet the skeleton is in homeostatic equilibrium. They found that wild-type bone experienced 40% greater strain than HBM bone with the same load. Based on these observations in humans and mice, the authors hypothesized that the setpoint for load responsiveness was lower in the HBM skeleton. Loss of function mutations in Lrp5 result in low bone mass and osteopenia [160] but, more importantly, do not respond to mechanical load [161], again supporting the notion that Lrp5 is involved in mechanosensation.

Estrogen has been proposed to modulate skeletal response to strain. Ehrlich and coworkers found that about 14% of all osteocytes were positive for estrogen receptor, ERα, under normal locomotion, but this number was decreased to 7.5% after a 2-week loading regimen that resulted in new bone formation in rat ulnae [95]. The distribution of positive cells was uniform

and did not correlate with peak strain magnitude, suggesting that osteocytes respond to strain as a population. The response of mice deficient in the ERα and ERβ is inadequate to mechanical loading [162, 163]. It has been proposed that TGFb III present in MLO-Y4 conditioned media enhances the production of estrogen, which inhibits osteoclastic bone resorption [79]. Conditioned media from osteocyte-like MLO-Y4 cells has also been shown to selectively stimulate the proliferation of mesenchymal stem cells and their differentiation into osteoblasts, but the factors responsible are not known [80]. Estrogen has also been proposed to be an antiapoptotic factor for osteocytes (see the following section).

IX. OSTEOCYTE SIGNALS FOR BONE RESORPTION

Power and coworkers found elevated osteocyte density and lacunar occupancy in resorbing and forming osteons compared to quiescent osteons, leading to their conclusion that osteocytes may contribute to processes initiating or maintaining bone resorption [164]. Osteocytes have been proposed to send signals for bone resorption. Isolated avian osteocytes have been shown to support osteoclast formation and activation [165]. Like isolated chick osteocytes, the osteocyte-like cell line, MLO-Y4, was also found to support osteoclast formation; however, unlike any previously reported stromal cell lines, the cells did so in the absence of any osteotropic factors [93]. These cells express RANK Ligand along their dendritic processes and secrete large amounts of macrophage colony–stimulating factor, both essential for osteoclast formation. Expression of RANK Ligand along osteocyte dendritic processes offers a potential means for osteocytes within bone to interact and stimulate osteoclast precursors at the bone surface. It is interesting that MLO-Y4 cells can support both mesenchymal stem cell and osteoblast differentiation and also support osteoclast formation. It remains to be determined if primary osteocytes can perform all three functions. If so, this supports the hypothesis that osteocytes have the capacity to regulate all phases of bone remodeling.

One of the major means by which osteocytes may support osteoclast activation and formation is through their death. Osteocyte apoptosis can occur at sites of microdamage, and it is proposed that dying osteocytes are targeted for removal by osteoclasts. Verborgt and coworkers mapped the expression of an antiapoptotic molecule called Bcl-2 and a proapoptotic molecule called Bax in osteocytes surrounding microcracks [166] and found that Bax was elevated in osteocytes immediately at the microcrack locus, whereas Bcl-2 was expressed 1–2 mm from the microcrack. The authors proposed that those osteocytes that do not undergo apoptosis are prevented from doing so by active protection mechanisms, suggesting that damaged yet viable osteocytes can send signals.

X. OSTEOCYTE APOPTOSIS

It has been proposed that the purpose and function of osteocytes is to die, thereby releasing signals of remodeling and serving to target particular skeletal sites at selected time points for resorption [167]. Osteocyte apoptosis can occur by aging, immobilization, microdamage, lack of estrogen, elevated cytokines such as TNF-α as occurs in menopause, and treatment with glucocorticoids. Osteocyte cell death can occur in association with pathological conditions, such as osteoporosis and osteoarthritis, leading to increased fragility [168–170]. Such fragility is considered to be due to loss of the ability to sense microdamage and signal to other bone cells for repair [6]. Osteocyte apoptosis has been implicated to play an important role in targeting bone remodeling processes, since it occurs in association with areas of microdamage followed by osteoclastic resorption in mechanically challenged bone [171]. The apoptotic region around microcracks was found to be surrounded by surviving osteocytes expressing Bcl-2, whereas dying osteocytes appeared to be the target of resorbing osteoclasts [166, 172].

In addition to microdamage, other skeletal insults cause osteocyte apoptosis. Oxygen deprivation has been shown to promote osteocyte apoptosis, especially as occurs with immobilization. Hypoxia-inducing factor alpha is elevated, leading to apoptosis and induction of the osteoclastogenic factor, VEGF [77], and osteopontin, a mediator of environmental stress and a potential chemoattractant for osteoclasts [78]. Withdrawal of estrogen results in osteocyte apoptosis [173], as does glucocorticoid treatment [169]. These observations are relevant to disease, as cytokines such as TNF-α and interleukin-1 (IL-1) have been reported to increase with estrogen deficiency [174, 175]. Apoptosis may also play an important role in the third most common cause of osteoporosis: glucocorticoid-induced osteoporosis [6].

Several agents have been found to reduce or inhibit osteoblast and osteocyte apoptosis; they include estrogen and selective estrogen receptor modulators [176], bisphosphonates and calcitonin [87], CD40 Ligand [69], Calbindin-D28k [85], and monocyte chemotactic proteins MCP-1 and -3 [177]. The pathways for some of these antiapoptosis agents have been extensively studied and dissected. For example, the bisphosphonates appear to inhibit apoptosis through interaction with hemichannels and the ERK pathway [88], and Fas/CD95 plays a role in glucocorticoid-induced osteocyte apoptosis [83]. Interestingly, one of the antiapoptotic agents has been shown to be selective for one apoptosis agent over another. MCP-3 will inhibit only glucocorticoid-induced apoptosis of MLO-Y4 osteocyte cells, and not TNF-α induced apoptosis, which is not the case for the other agents [177]. MCP-3 is produced by osteocytes and is regulated by mechanical strain and therefore may selectively protect strained osteocytes.

Hence, osteocyte viability may play a significant role in the maintenance of bone homeostasis and integrity. However, whereas blocking osteocyte apoptosis may improve diseases such as bone loss due to aging or to glucocorticoid therapy, osteocyte apoptosis may be essential for damage repair and normal skeletal replacement. Any agents that block this process may exacerbate conditions in which repair is required.

XI. OSTEOCYTE MODIFICATION OF ITS MICROENVIRONMENT

Over five decades ago, it was proposed that osteocytes may resorb their lacunar wall under certain conditions [178]. The term "osteolytic osteolysis" was initially used to describe the enlarged lacunae in patients with hyperparathyroidism [179] and later in immobilized rats [180]. "Osteolytic osteolysis" has frequently been confused with the resorption mechanisms used by osteoclasts. When primary avian osteocytes were seeded onto dentin slices, no resorption was detected; therefore, these investigators concluded that osteocytes cannot remove mineralized matrix [181]. However, one must keep in mind that removal of mineral by osteocytes (weeks/months) would certainly be slower than osteoclastic resorption (days) and therefore not detectable using this approach. Bonucci and Gherardi [182] suggested that poor mineralization when the osteocyte is being embedded is the reason for enlarged lacunae with renal osteodystrophy. The term "osteocyte halos" was used by Heuck [183] to describe pericanicular demineralization in rickets and later by others to describe periosteocytic lesions in X-linked hypophosphatemic rickets [184], a condition due to an inactivating mutation in PHEX. Such periosteocytic lesions are not present in other chronic hypophosphatemic states. The capacity to deposit or remove mineral from lacunae and canaliculi has important implications with regards to magnitude of fluid shear stress and mechanical properties of bone.

Glucocorticoids, in addition to having effects on apoptosis, may have direct effects on osteocytes, resulting in modification of their microenvironment. It appears that glucocorticoid-treated subjects fracture at higher BMD values than postmenopausal women, but the reason is unclear [185, 186]. Mice injected with pellets releasing prednisolone showed an enlargement of osteocyte lacunae in trabecular bone and the generation of a surrounding sphere of hypomineralized bone [187]. Lacunae act as stress concentrators in bone; therefore, it was proposed that these highly localized changes in bone properties may influence fracture risk in glucocorticoid-treated patients [187]. It was suggested that glucocorticoid may alter or compromise the metabolism and function of the osteocyte, not just induce cell death.

Over four decades ago, it was suggested that the osteocyte has both matrix-forming and matrix-destroying activities [188] and that the osteocyte can remodel its local environment including lacunae and canaliculi [189]. Osteocyte lacunae were shown to take up tetracycline, called "periosteocytic perilacunar tetracycline labeling," indicating the ability to calcify or form bone. In contrast, these early investigators also found acid phosphatase positive osteocytes near endosteal osteoclastic resorbing surfaces, suggesting potential capacity to resorb. Greater solubility of the intralacunar mineral surrounding the normal osteocyte was also found [179]. These observations suggest that the osteocyte can both add and remove mineral from its lacunae and canaliculi.

XII. OSTEOCYTE DENSITY

It is not clear if a relationship exists between osteocyte density and bone volume and remodeling. Jordan and coworkers hypothesized that in cases of osteoarthritis, increased TGF-β may decrease the conversion of osteoblasts to osteocytes, thereby decreasing osteocyte density and increasing bone mass [142] based on studies showing that inactivation of the TGF-β pathway leads to the opposite effects [190]. They examined patients with cox arthrosis known to have elevated TGF-β and found a reduction in osteocyte lacunar density and an increase in wall width in femoral neck biopsies consistent with their prediction. These observations support those of Karsdal [7] showing that osteoblast life span and matrix production before incorporation into matrix as an osteocyte appear to be regulated by TGF-β. In contrast, Vashishth and coworkers found that increasing osteocyte density was associated with increases in bone volume and that osteocyte lacunar density predicts cancellous and cortical bone

volume [191]. Qiu and coworkers found a correlation of increased osteocyte density with less bone remodeling [192, 193]. They found that osteocyte density declines with age but not with menopause, in deep but not superficial bone, and suggest that it is the age of the bone and not the age of the subject that determines osteocyte density. They proposed that one function of remodeling is to maintain osteocyte viability. They also found that fracture patients had fewer osteocytes than healthy controls [194] and concluded that osteocyte deficiency may contribute to bone fragility by impairing osteocyte detection of microdamage or by a reduction in canalicular fluid flow. These authors also found that Black women have higher osteocyte density than White women [195], perhaps playing a role in increased bone strength. In Black women as in White women, more empty lacunae were found in deep than in superficial bone and there was age-related loss of osteocytes.

Robling and Turner did not find a correlation of osteocyte density with mechanosensitivity in three strains of mice. They suggested that genetic components other than osteocyte density regulate mechanosensitivity [196]. Clearly, further study is required to clarify the importance of osteocyte density in osteocyte function and disease in bone.

XIII. ROLE OF GAP JUNCTIONS AND HEMICHANNELS IN OSTEOCYTE COMMUNICATION

Clearly, osteocytes can communicate extracellularly through the production of small molecules such as NO, ATP, prostaglandins, and secretion of larger proteins such as DMP1, MEPE, and SOST. Turner and colleagues suggested that bone cells may communicate in a fashion similar to neural cells [54] through molecules such as glutamate, serotonin, leptin, and neuropeptide Y2 that are responsible for habituation, sensitization, and long-term memory. Osteocytes do not express functional glutamate receptor but do express GLAST, a molecule that sequesters glutamate, suggesting that the osteocyte may signal to responding osteoblasts and osteoclasts that do express the receptor [81]. Serotonin receptors have also been found on osteocytes, the 5-HT(2B) receptor is higher on avian osteocytes than osteoblasts [197], and recently serotonin was shown to increase bone mineral density [198]. Though intriguing to view bone as a neuronal network, further studies are required.

Another means by which osteocytes communicate is intracellularly through gap junctions. The cell processes of osteocytes are connected with each other and lining cells via gap junctions [199, 200], thereby allowing direct cell-to-cell coupling. Gap junctions are transmembrane channels, which connect the cytoplasm of two adjacent cells. These channels permit molecules with molecular weights less than 1 kDa to pass through and have been shown to modulate cell signaling and tissue function in many organs and cells [201, 202]. Gap junction channels are formed by members of a family of proteins known as connexins. Functional gap junctions in osteoblasts were first identified with injection of fluorescent dye into rat calvarial subperiosteal osteoblasts that spread to neighboring osteoblastic cells [203]. Gap junctions and Cx43 are important for osteoblast differentiation, and the functions and expression of gap junctions and Cx43 are regulated by prostaglandins, hormones, and other signaling molecules. Cx43-null mice have delayed ossification, craniofacial abnormalities, and osteoblast dysfunction [204].

It has been proposed that gap junctions function through the propagation of intracellular signals contributing to mechanotransduction in bone, thereby regulating bone cell differentiation [205]. A dominant negative mutant of Cx43 diminishes fluid flow–induced release of PGE_2, but not Ca^{2+} responses [206]. In addition, the fluid flow–induced PGE_2 response of osteoblastic ROS17/2.8 cells is gap junction–mediated and independent of intracellular Ca^i [207]. Fluid flow–induced shear stress stimulates gap junction–mediated intercellular communication and increases Cx43 expression in osteocyte-like MLO-Y4 cells [72]. PGE_2 is released in response to fluid flow functions in an autocrine fashion to activate EP_2 receptor signaling, including increased intracellular cAMP and activated PKA, which in turn stimulates gap junction function and Cx43 expression [71]. Oscillating fluid flow has been shown to upregulate gap junction communication in MLO-Y4 cells by an ERK1/2 MAP kinase–dependent mechanism [70]. Yellowley and coworkers showed that the osteocyte-like MLO-Y4 cells can couple through gap junctions to osteoblast-like MC3T3 cells [92].

Recently, hemichannels were identified in osteocytes in addition to other potential openings or channels to the extracellular bone fluid such as calcium, ion, voltage, stretch-activated channels, and others [208, 209]. Osteocytes and MLO-Y4 osteocyte-like cells [68] express large amounts of Connexin 43, the component of gap junctions, but these cells are in contact only through the tips of their dendritic processes. This raised the question concerning how Cx43 located in the rest of the cell membrane could be functioning. Recently, it was shown that connexins can form and function as unapposed halves of gap junction channels called hemichannels, localized at the cell surface, independent of physical contact with adjacent cells

[210]. Functional hemichannels formed by Cx43 have been reported in neural progenitors and neurons, astrocytes, heart, and osteoblasts and osteocytes. The opening of hemichannels appears to provide a mechanism for ATP and NAD^+ release, which raises intracellular Ca^{2+} activity and promotes Ca^{2+} wave propagation in astrocytes, bone cells, epithelial cells, and outer retina. Hemichannels expressed in bone cells such as MLO-Y4 cells appear to function as essential transducers of the antiapoptotic effects of bisphosphonates [89]. Hemichannels formed by Cx43 directly serve as the pathway for the exit of elevated intracellular PGE_2 in osteocytes induced by fluid flow shear stress [74]. This is the first report of modulation of hemichannel function in response to mechanical stress. Therefore, gap junctions at the tip of dendrites appear to mediate a form of intracellular communication, and hemichannels along the dendrite appear to mediate a form of extracellular communication in osteocytes.

XIV. OSTEOCYTES IN THE EMBRYONIC AND THE ADULT SKELETON

Mechanical strain is required for postnatal, but not for prenatal, skeletal development and maintenance. Mice lacking Dmp1, PHEX, MEPE, SOST, and other proteins that are highly expressed in osteocytes do not show a phenotype until days to weeks or even months after birth [21, 211, 212]. One potential explanation for this is that functional osteocytes are not required in the embryo. Osteocytes may act as "placeholders" in the embryo until they can assume their functions as mechanosensors in the postnatal or adult skeleton. Also, *in utero*, although subjected to some mechanical loading via muscle insertions, the skeleton is not subjected to significant loading from weight-bearing activity. Therefore, responses of load-related bone remodeling are less significant in the developing embryo. Growth and development are the overriding signals prenatally compared to any loading or unloading signals. Their extensive dendrite connections also may not be required because the bone cortices and trabeculae are relatively thin and poorly mineralized, and the cells are near the surface [26]. Thus, nutrients may be able to diffuse readily to the osteocytes without requiring an extensive canalicular system. Therefore, molecules that play a role in the responses of osteocytes to mechanical strain may not reveal their importance for normal skeletal physiology until postnatally or in the adult animal. Osteocyte biology and function may be more relevant to adult disease than to development.

XV. THE IMPLICATIONS OF OSTEOCYTE BIOLOGY FOR BONE DISEASE

Osteocyte viability may play a significant role in the maintenance and integrity of bone. Bone loss due to osteoporosis may be due in part to osteocyte cell death [6, 167]. Manolagas and coworkers have been pioneers in dissecting out the mechanisms and signaling pathways of factors such as estrogens, bisphosphonates, and parathyroid hormone on osteoblast and osteocyte viability and of glucocorticoid on osteoblast and osteocyte apoptosis. In the process, they have identified estrogen receptor ligands called ANGELS for "activators of nongenotropic estrogen-like signaling" that lack transcriptional activity but do have nongenotropic activity on osteoblast and osteocyte viability. It is speculated that ANGELS may be more beneficial than genotropic estrogens in the prevention of osteoporosis.

Osteocyte dendricity may play a role in bone disease. The early formation of dendrites by embedding osteoid-osteocytes is polarized toward the mineralization front to which cellular processes are oriented. Cellular processes toward blood vessels begin to appear only when the mineralization begins to spread around the cell [10]. Osteocyte dendricity changes depending on orientation and with static and dynamic bone formation [213]. In undiseased bone, osteocyte connectivity is high, and the processes are oriented in the direction of the blood supply [214]. In osteoporotic bone there is a marked decrease in connectivity as well as disorientation of the dendrites, which increases in severity. In contrast, in osteoarthritic bone, a decrease in connectivity is observed, but the orientation is intact. In osteomalacic bone, the osteocytes appear viable with high connectivity, but the processes are distorted and the network chaotic [214]. Changes in osteocyte dendricity could have a dramatic effect not only on osteocyte function and viability, but also on the mechanical properties of bone. An equilibrium must be met between number and branching of dendrites to preserve function and viability versus the number that would decrease bone strength. (See Figure 6-5.)

Osteonecrosis is "dead" bone that does not remodel. As the osteocytes are dead or missing in necrotic bone, and as necrotic bone does not remodel, this suggests that viable osteocytes are necessary to send signals of (re)modeling. Osteonecrosis can be due to glucocorticoid treatment, lipid disorders, alcohol abuse, radiation, trauma, sickle cell anemia, and recently to bisphosphonate-induced osteonecrosis of the jaw. Proposed mechanisms responsible for osteonecrosis include a mechanical theory, whereby osteoporosis and the accumulation of unhealed trabecular microcracks result in fatigue fractures; a vascular theory, in which ischemia is caused by microscopic fat emboli and increased intraosseous pressure due to fat accumulation leads to a mechanical impingement on the sinusoidal vascular bed and decreased blood flow; and a new theory involving osteocyte apoptosis, where agents induce osteocyte cell death, which results in dead bone that does not remodel. A number of articles support the mechanism of a lack of vascular supply due to microcracks or fat emboli [215–217]; however, more recent papers suggest that the osteocyte is the

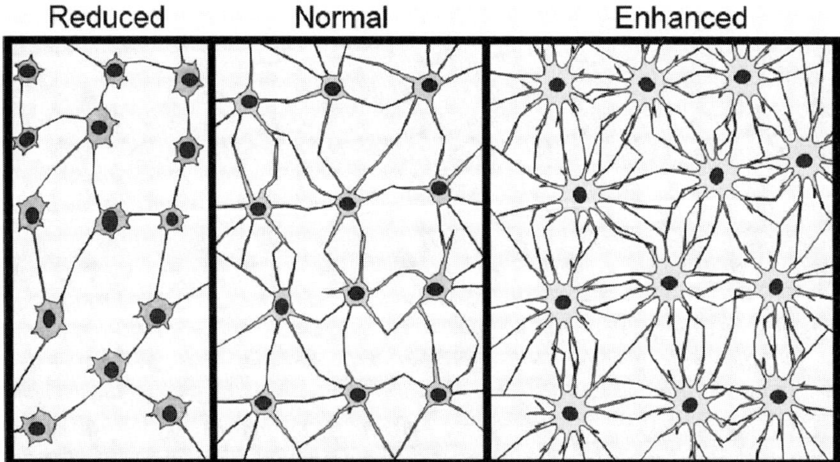

FIGURE 6-5   The effects of lacunocanalicular system complexity are not known. Complexity may increase with age of the animal. Disruptions to this system may occur with disease [214]. Theoretically, changes in osteocyte dendricity would have a dramatic effect on osteocyte function and viability and on the mechanical properties of bone.

target [169, 218–220]. If these conditions are mediated through osteocyte cell death, then new therapeutics to prevent this occurrence are in order.

Osteoid cells or osteocytes may play a role in phosphate homeostasis. Once the osteoblast begins to embed in osteoid, burying itself, molecules such as Dmp1, PHEX, MEPE, and SOST are more highly expressed. MEPE and SOST are thought to be inhibitors of mineralization as null mice have greater bone. Recently, it was found that Dmp1 null mice have a similar phenotype to HYP mice in which PHEX is mutated, and both models are osteomalacic with elevated FGF23 levels [221]. FGF23 has also been found to be highly expressed in osteocytes [221]. Autosomal recessive hypophosphatemic rickets in patients has been found to be due to mutations in Dmp1 [39]. Taken in combination, these molecules may control phosphate metabolism through regulation of this phosphaturic factor. It was also proposed that the osteocyte syncytium could be viewed as an endocrine organ regulating phosphate metabolism. The unraveling of the interactions of these molecules should lead to insight into diseases of hyper- and hypophosphatemia.

XVI. CONCLUSIONS

Bone histomorphologists in the 1940s through the late 1960s were pioneers who generated novel hypotheses regarding the function of osteocytes using only histological observations, their intellect, and their imaginations. Now technology has allowed further observation, has allowed further validation of decades-old hypotheses, has allowed novel extensions of earlier work, and has generated totally unexpected discoveries. Evidence is accumulating that osteocytes are important for bone health; therefore, a revival of interest in this cell is occurring within the bone community. These fascinating cells continue to challenge and stimulate.

ACKNOWLEDGMENT

The author's work in osteocyte biology is supported by the National Institutes of Health AR-46798.

REFERENCES

1. A. M. Parfitt, The cellular basis of bone turnover and bone loss: A rebuttal of the osteocytic resorption-bone flow theory. *Clin Orthop Relat Res,* 236–247 (1977).

2. L. E. Lanyon, Osteocytes, strain detection, bone modeling and remodeling. *Calcif Tissue Int,* **53**, S102–106; discussion S106–107 (1993).

3. H. Kamioka, T. Honjo, and T. Takano-Yamamoto, A three-dimensional distribution of osteocyte processes revealed by the combination of confocal laser scanning microscopy and differential interference contrast microscopy. *Bone,* **28**, 145–149 (2001).

4. I. Kalajzic, *et al.,* Use of type I collagen green fluorescent protein transgenes to identify subpopulations of cells at different stages of the osteoblast lineage. *J Bone Miner Res,* **17**, 15–25 (2002).

5. I. Kalajzic, *et al.,* Dentin matrix protein 1 expression during osteoblastic differentiation, generation of an osteocyte GFP-transgene. *Bone,* **35**, 74–82 (2004).

6. S. C. Manolagas, Birth and death of bone cells: Basic regulatory mechanisms and implications for the pathogenesis and treatment of osteoporosis. *Endocr Rev,* **21**, 115–137 (2000).

7. M. A. Karsdal, *et al.,* Matrix metalloproteinase-dependent activation of latent transforming growth factor-beta controls the conversion of osteoblasts into osteocytes by blocking osteoblast apoptosis. *J Biol Chem,* **277**, 44061–44067 (2002).

8. P. J. Bordier, L. Miravet, A. Ryckerwaert, and H. Rasmussen, Morphological and morphometrical characteristics of the mineralization front. A vitamin D regulated sequence of bone remodeling. In *Bone Histomorphometry* (P. J. Bordier, ed.), pp. 335–354. Armour Montagu, Paris (1976).

9. P. J. Nijweide, A. van der Plas, and J. P. Scherft, Biochemical and histological studies on various bone cell preparations. *Calcif Tissue Int,* **33**, 529–540 (1981).

10. C. Palumbo, A three-dimensional ultrastructural study of osteoid-osteocytes in the tibia of chick embryos. *Cell Tissue Res,* **246**, 125–131 (1986).

11. M. Owen, Cell population kinetics of an osteogenic tissue. I. 1963. *Clin Orthop Relat Res,* 3–7 (1995).

12. T. A. Franz-Odendaal, B. K. Hall, and P. E. Witten, Buried alive: How osteoblasts become osteocytes. *Dev Dyn,* **235**, 176–190 (2006).

13. Y. Kato, *et al.,* Establishment of an osteoid preosteocyte-like cell MLO-A5 that spontaneously mineralizes in culture. *J Bone Miner Res,* **16**, 1622–1633 (2001).

14. C. Barragan-Adjemian, *et al.,* Mechanism by which MLO-A5 late osteoblasts/early osteocytes mineralize in culture: Similarities with mineralization of lamellar bone. *Cal Tiss Int,* **79**, 340–353.

15. Y. Mikuni-Takagaki, *et al.,* Matrix mineralization and the differentiation of osteocyte-like cells in culture. *J Bone Miner Res,* **10**, 231–242 (1995).

16. A. Boskey, Matrix proteins and mineralization: An overview. *Connect Tissue Res,* **35**, 357–363 (1996).

17. G. K. Hunter, P. V. Hauschka, A. R. Poole, L. C. Rosenberg, and H. A. Goldberg, Nucleation and inhibition of hydroxyapatite formation by mineralized tissue proteins. *Biochem J,* **317**(Pt 1), 59–64 (1996).

18. T. M. Strom, *et al.,* Pex gene deletions in Gy and HYP mice provide mouse models for X-linked hypophosphatemia. *Hum Mol Genet,* **6**, 165–171 (1997).

19. H. J. Blair, E. Gormally, I. C. Uwechue, and Y. Boyd, Mouse mutants carrying deletions that remove the genes mutated in Coffin-Lowry syndrome and lactic acidosis. *Hum Mol Genet,* **7**, 549–555 (1998).

20. D. G. Winkler, *et al.,* Osteocyte control of bone formation via sclerostin, a novel BMP antagonist. *EMBO J,* **22**, 6267–6276 (2003).

21. L. C. Gowen, *et al.,* Targeted disruption of the osteoblast/osteocyte factor 45 gene (OF45) results in increased bone formation and bone mass. *J Biol Chem,* **278**, 1998–2007 (2003).

22. Y. Mikuni-Takagaki, Y. Suzuki, T. Kawase, and S. Saito, Distinct responses of different populations of bone cells to mechanical stress. *Endocrinology,* **137**, 2028–2035 (1996).

23. A. Wetterwald, *et al.*, Characterization and cloning of the E11 antigen, a marker expressed by rat osteoblasts and osteocytes. *Bone,* **18**, 125–132 (1996).

24. E. Schulze, M. Witt, M. Kasper, C. W. Lowik, and R. H. Funk, Immunohistochemical investigations on the differentiation marker protein E11 in rat calvaria, calvaria cell culture and the osteoblastic cell line ROS 17/2.8. *Histochem Cell Biol,* **111**, 61–69 (1999).

25. D. Tenorio, A. Cruchley, and F. J. Hughes, Immunocytochemical investigation of the rat cementoblast phenotype. *J Periodontal Res,* **28**, 411–419 (1993).

26. K. Zhang, *et al.*, E11/gp38 selective expression in osteocytes: Regulation by mechanical strain and role in dendrite elongation. *Mol Cell Biol,* **26**, 4539–4552 (2006).

27. L. W. A. Sprague, U. Heinzman, and M. J. Atkinson, Phenotypic changes following over-expression of sense or antisense E11 cDNA in ROS 17/2.8 cells. *J Bone Miner Res,* **11**, S132 (1996).

28. F. G. Scholl, C. Gamallo, S. Vilar, and M. Quintanilla, Identification of PA2.26 antigen as a novel cell-surface mucin-type glycoprotein that induces plasma membrane extensions and increased motility in keratinocytes. *J Cell Sci,* **112**(Pt 24), 4601–4613 (1999).

29. P. Mangeat, C. Roy, and M. Martin, ERM proteins in cell adhesion and membrane dynamics. *Trends Cell Biol,* **9**, 187–192 (1999).

30. I. Ohizumi, *et al.*, Association of CD44 with OTS-8 in tumor vascular endothelial cells. *Biochim Biophys Acta,* **1497**, 197–203 (2000).

31. D. E. Hughes, D. M. Salter, and R. Simpson, CD44 expression in human bone: A novel marker of osteocytic differentiation. *J Bone Miner Res,* **9**, 39–44 (1994).

32. I. Westbroek, K. E. De Rooij, and P. J. Nijweide, Osteocyte-specific monoclonal antibody MAb OB7.3 is directed against PHEX protein. *J Bone Miner Res,* **17**, 845–853 (2002).

33. A. F. Ruchon, *et al.*, Developmental expression and tissue distribution of PHEX protein: Effect of the HYP mutation and relationship to bone markers. *J Bone Miner Res,* **15**, 1440–1450 (2000).

34. F. Francis, *et al.*, A gene (PEX) with homologies to endopeptidases is mutated in patients with X-linked hypophosphatemic rickets. The HYP Consortium. *Nat Genet,* **11**, 130–136 (1995).

35. J. Q. Feng, *et al.*, DMP-1 deficient mice develop dwarfism, chondrodysplasia, and disorganized bone remodeling and mineralization during postnatal development. Vol. Abstract, American Society for Bone and Mineral Research (2002).

36. S. Toyosawa, *et al.*, Dentin matrix protein 1 is predominantly expressed in chicken and rat osteocytes but not in osteoblasts. *J Bone Miner Res,* **16**, 2017–2026 (2001).

37. J. Q. Feng, *et al.*, Loss of Dmp1/DMP1 causes defects in skeletal mineralization and in phosphate handling: Potential role of the osteocyte in mineral metabolism. submitted (2006).

38. G. He and A. George, Dentin matrix protein 1 immobilized on type I collagen fibrils facilitates apatite deposition in vitro. *J Biol Chem,* **279**, 11649–11656 (2004).

39. J. Feng, *et al.*, Loss of DMP1 causes rickets and osteomalacia and identifies a role for osteocytes in mineral metabolism. *Nat Genet,* **38**, 1310–1315 (2006).

40. P. S. Rowe, *et al.*, MEPE, a new gene expressed in bone marrow and tumors causing osteomalacia. *Genomics,* **67**, 54–68 (2000).

41. D. N. Petersen, G. T. Tkalcevic, A. L. Mansolf, R. Rivera-Gonzalez, and T. A. Brown, Identification of osteoblast/osteocyte factor 45 (OF45), a bone-specific cDNA encoding an RGD-containing protein that is highly expressed in osteoblasts and osteocytes. *J Biol Chem,* **275**, 36172–36180 (2000).

42. M. Igarashi, N. Kamiya, K. Ito, and M. Takagi, In situ localization and in vitro expression of osteoblast/osteocyte factor 45 mRNA during bone cell differentiation. *Histochem J,* **34**, 255–263 (2002).

43. D. Bresler, J. Bruder, K. Mohnike, W. D. Fraser, and P. S. Rowe, Serum MEPE-ASARM-peptides are elevated in X-linked rickets (HYP): Implications for phosphaturia and rickets. *J Endocrinol,* **183**, R1–9 (2004).

44. P. S. Rowe, *et al.*, Surface plasmon resonance (SPR) confirms that MEPE binds to PHEX via the MEPE-ASARM motif: A model for impaired mineralization in X-linked rickets (HYP). *Bone,* **36**, 33–46 (2005).

45. P. S. Rowe, *et al.*, MEPE has the properties of an osteoblastic phosphatonin and minhibin. *Bone,* 34, 303–319 (2004).

46. L. W. Fisher and N. S. Fedarko, Six genes expressed in bones and teeth encode the current members of the SIBLING family of proteins. *Connect Tissue Res,* **44**(Suppl 1), 33–40 (2003).

47. K. E. Poole, *et al.*, Sclerostin is a delayed secreted product of osteocytes that inhibits bone formation. *FASEB J,* **19**, 1842–1844 (2005).

48. W. Balemans, *et al.*, Increased bone density in sclerosteosis is due to the deficiency of a novel secreted protein (SOST). *Hum Mol Genet,* **10**, 537–543 (2001).

49. R. L. van Bezooijen, *et al.*, Sclerostin is an osteocyte-expressed negative regulator of bone formation, but not a classical BMP antagonist. *J Exp Med,* **199**, 805–814 (2004).

50. X. Li, *et al.*, Sclerostin binds to LRP5/6 and antagonizes canonical Wnt signaling. *J Biol Chem,* **280**, 19883–19887 (2005).

51. T. Bellido, *et al.*, Chronic elevation of parathyroid hormone in mice reduces expression of sclerostin by osteocytes: A novel mechanism for hormonal control of osteoblastogenesis. *Endocrinology,* **146**, 4577–4583 (2005).

52. G. F. Weber, S. Ashkar, M. J. Glimcher, and H. Cantor, Receptor-ligand interaction between CD44 and osteopontin (Eta-1). *Science,* **271**, 509–512 (1996).

53. I. Fristad, V. Vandevska-Radunovic, K. Fjeld, S. J. Wimalawansa, and I. Hals Kvinnsland. NK1, NK2, NK3 and CGRP1 receptors identified in rat oral soft tissues, and in bone and dental hard tissue cells. *Cell Tissue Res,* **311**, 383–391 (2003).

54. C. H. Turner, A. G. Robling, R. L. Duncan, and D. B. Burr, Do bone cells behave like a neuronal network? *Calcif Tissue Int,* **70**, 435–442 (2002).

55. G. Gu, M. Nars, T. A. Hentunen, K. Metsikko, and H. K. Vaananen, Isolated primary osteocytes express functional gap junctions in vitro. *Cell Tissue Res,* **323**, 263–271 (2006).

56. K. Tanaka-Kamioka, H. Kamioka, H. Ris, and S. S. Lim, Osteocyte shape is dependent on actin filaments and osteocyte processes are unique actin-rich projections. *J Bone Miner Res,* **13**, 1555–1568 (1998).

57. H. Kamioka, Y. Sugawara, T. Honjo, T. Yamashiro, and T. Takano-Yamamoto, Terminal differentiation of osteoblasts to osteocytes is accompanied by dramatic changes in the distribution of actin-binding proteins. *J Bone Miner Res,* **19**, 471–478 (2004).

58. K. Nose, H. Saito, and T. Kuroki, Isolation of a gene sequence induced later by tumor-promoting 12-O-tetradecanoylphorbol-13-acetate in mouse osteoblastic cells (MC3T3–E1) and expressed constitutively in ras-transformed cells. *Cell Growth Differ,* **1**, 511–518 (1990).

59. M. I. Ramirez, *et al.*, T1alpha, a lung type I cell differentiation gene, is required for normal lung cell proliferation and alveolus formation at birth. *Dev Biol,* **256**, 61–72 (2003).

60. K. Holmbeck, *et al.*, The metalloproteinase MT1-MMP is required for normal development and maintenance of osteocyte processes in bone. *J Cell Sci,* **118**, 147–156 (2005).

61. W. Zhao, M. H. Byrne, Y. Wang, and S. M. Krane, Osteocyte and osteoblast apoptosis and excessive bone deposition accompany failure of collagenase cleavage of collagen. *J Clin Invest,* **106**, 941–949 (2000).

62. K. Holmbeck, *et al.*, MT1-MMP-deficient mice develop dwarfism, osteopenia, arthritis, and connective tissue disease due to inadequate collagen turnover. *Cell,* **99**, 81–92 (1999).

63. S. Okada, S. Yoshida, S. H. Ashrafi, and D. E. Schraufnagel, The canalicular structure of compact bone in the rat at different ages. *Microsc Microanal,* **8**, 104–115 (2002).

64. P. Veno, *et al.*, Live imaging of osteocytes within their lacunae reveals cell body and dendrite motions. *J Bone Miner Res,* **21**, S38 (2006).

65. A. van der Plas and P. J. Nijweide, Isolation and purification of osteocytes. *J Bone Miner Res,* **7**, 389–396 (1992).

66. W. Yang, *et al.*, Dentin matrix protein 1 gene cis-regulation: Use in osteocytes to characterize local responses to mechanical loading in vitro and in vivo. *J Biol Chem,* **280**, 20680–20690 (2005).

67. P. V. Bodine, S. K. Vernon, and B. S. Komm, Establishment and hormonal regulation of a conditionally transformed pre-osteocytic cell line from adult human bone. *Endocrinology,* **137**, 4592–4604 (1996).

68. Y. Kato, J. J. Windle, B. A. Koop, G. R. Mundy, and L. F. Bonewald, Establishment of an osteocyte-like cell line, MLO-Y4. *J Bone Miner Res,* **12**, 2014–2023 (1997).

69. S. S. Ahuja, *et al.*, CD40 ligand blocks apoptosis induced by tumor necrosis factor alpha, glucocorticoids, and etoposide in osteoblasts and the osteocyte-like cell line murine long bone osteocyte-Y4. *Endocrinology,* **144**, 1761–1769 (2003).

70. A. I. Alford, C. R. Jacobs, and H. J. Donahue, Oscillating fluid flow regulates gap junction communication in osteocytic MLO-Y4 cells by an ERK1/2 MAP kinase-dependent mechanism small star, filled. *Bone,* **33**, 64–70 (2003).

71. B. Cheng, *et al.*, PGE(2) is essential for gap junction-mediated intercellular communication between osteocyte-like MLO-Y4 cells in response to mechanical strain. *Endocrinology,* **142**, 3464–3473 (2001).

72. B. Cheng, *et al.*, Expression of functional gap junctions and regulation by fluid flow in osteocyte-like MLO-Y4 cells. *J Bone Miner Res,* **16**, 249–259 (2001).

73. P. P. Cherian, *et al.*, Effects of mechanical strain on the function of Gap junctions in osteocytes are mediated through the prostaglandin EP2 receptor. *J Biol Chem,* **278**, 43146–43156 (2003).

74. P. P. Cherian, *et al.*, Mechanical strain opens connexin 43 hemichannels in osteocytes: A novel mechanism for the release of prostaglandin. *Mol Biol Cell,* **16**, 3100–3106 (2005).

75. P. Divieti, *et al.*, Receptors for the carboxyl-terminal region of pth(1–84) are highly expressed in osteocytic cells. *Endocrinology,* **142**, 916–925 (2001).

76. P. Divieti, A. I. Geller, G. Suliman, H. Juppner, and F. R. Bringhurst, Receptors specific for the carboxyl-terminal region of parathyroid hormone on bone-derived cells: Determinants of ligand binding and bioactivity. *Endocrinology,* **146**, 1863–1870 (2005).

77. T. S. Gross, *et al.*, Selected contribution: Osteocytes upregulate HIF-1alpha in response to acute disuse and oxygen deprivation. *J Appl Physiol,* **90**, 2514–2519 (2001).

78. T. S. Gross, K. A. King, N. A. Rabaia, P. Pathare, and S. Srinivasan, Upregulation of osteopontin by osteocytes deprived of mechanical loading or oxygen. *J Bone Miner Res,* **20**, 250–256 (2005).

79. T. J. Heino, T. A. Hentunen, and H. K. Vaananen, Osteocytes inhibit osteoclastic bone resorption through transforming growth factor-beta: Enhancement by estrogen. *J Cell Biochem,* **85**, 185–197 (2002).

80. T. J. Heino, T. A. Hentunen, and H. K. Vaananen, Conditioned medium from osteocytes stimulates the proliferation of bone marrow mesenchymal stem cells and their differentiation into osteoblasts. *Exp Cell Res,* **294**, 458–468 (2004).

81. J. F. Huggett, A. Mustafa, L. O'Neal, and D. J. Mason, The glutamate transporter GLAST-1 (EAAT-1) is expressed in the plasma membrane of osteocytes and is responsive to extracellular glutamate concentration. *Biochem Soc Trans,* **30**, 890–893 (2002).

82. M. A. Karsdal, T. A. Andersen, L. Bonewald, and C. Christiansen, Matrix, metalloproteinases (MMPs) safeguard osteoblasts from apoptosis during transdifferentiation into osteocytes: MT1-MMP maintains osteocyte viability. *DNA Cell Biol,* **23**, 155–165 (2004).

83. G. Kogianni, *et al.*, Fas/CD95 is associated with glucocorticoid-induced osteocyte apoptosis. *Life Sci,* **75**, 2879–2895 (2004).

84. K. Kurata, T. J. Heino, H. Higaki, and H. K. Vaananen, Bone marrow cell differentiation induced by mechanically damaged osteocytes in 3D gel-embedded culture. *J Bone Miner Res,* **21**, 616–625 (2006).

85. Y. Liu, *et al.*, Prevention of glucocorticoid-induced apoptosis in osteocytes and osteoblasts by calbindin-D28k. *J Bone Miner Res,* **19**, 479–490 (2004).

86. C. H. Lohmann, *et al.*, Pulsed electromagnetic fields affect phenotype and connexin 43 protein expression in MLO-Y4 osteocyte-like cells and ROS 17/2.8 osteoblast-like cells. *J Orthop Res,* **21**, 326–334 (2003).

87. L. I. Plotkin, *et al.*, Prevention of osteocyte and osteoblast apoptosis by bisphosphonates and calcitonin. *J Clin Invest,* **104**, 1363–1374 (1999).

88. L. I. Plotkin and T. Bellido, Bisphosphonate-induced, hemichannel-mediated, anti-apoptosis through the Src/ERK pathway: A gap junction-independent action of connexin43. *Cell Commun Adhes,* **8**, 377–382 (2001).

89. L. I. Plotkin, S. C. Manolagas, and T. Bellido, Transduction of cell survival signals by connexin-43 hemichannels. *J Biol Chem,* **277**, 8648–8657 (2002).

90. L. I. Plotkin, S. C. Manolagas, and T. Bellido, Dissociation of the pro-apoptotic effects of bisphosphonates on osteoclasts from their anti-apoptotic effects on osteoblasts/osteocytes with novel analogs. *Bone,* **39**, 443–452 (2006).

91. G. C. Reilly, T. R. Haut, C. E. Yellowley, H. J. Donahue, and C. R. Jacobs, Fluid flow induced PGE2 release by bone cells is reduced by glycocalyx degradation whereas calcium signals are not. *Biorheology,* **40**, 591–603 (2003).

92. C. E. Yellowley, Z. Li, Z. Zhou, C. R. Jacobs, and H. J. Donahue, Functional gap junctions between osteocytic and osteoblastic cells. *J Bone Miner Res,* **15**, 209–217 (2000).

93. S. Zhao, Y. K. Zhang, S. Harris, S. S. Ahuja, and L. F. Bonewald, MLO-Y4 osteocyte-like cells support osteoclast formation and activation. *J Bone Miner Res,* **17**, 2068–2079 (2002).

94. D. B. Burr, A. G. Robling, and C. H. Turner, Effects of biomechanical stress on bones in animals. *Bone,* **30**, 781–786 (2002).

95. P. J. Ehrlich, *et al.*, The effect of in vivo mechanical loading on estrogen receptor alpha expression in rat ulnar osteocytes. *J Bone Miner Res,* **17**, 1646–1655 (2002).

96. T. M. Skerry, L. Bitensky, J. Chayen, and L. E. Lanyon, Early strain-related changes in enzyme activity in osteocytes following bone loading in vivo. *J Bone Miner Res,* **4**, 783–788 (1989).

97. S. L. Dallas, G. Zaman, M. J. Pead, and L. E. Lanyon, Early strain-related changes in cultured embryonic chick tibiotarsi parallel those associated with adaptive modeling in vivo. *J Bone Miner Res,* **8**, 251–259 (1993).

98. R. A. Dodds, N. Ali, M. J. Pead, and L. E. Lanyon, Early loading-related changes in the activity of glucose 6-phosphate dehydrogenase and alkaline phosphatase in osteocytes and periosteal osteoblasts in rat fibulae in vivo. *J Bone Miner Res,* **8**, 261–267 (1993).

99. D. M. Raab-Cullen, M. A. Thiede, D. N. Petersen, D. B. Kimmel, and R. R. Recker, Mechanical loading stimulates rapid changes in periosteal gene expression. *Calcif Tissue Int,* **55**, 473–478 (1994).

100. J. Gluhak-Heinrich, *et al.*, Mechanical loading stimulates dentin matrix protein 1 (DMP1) expression in osteocytes in vivo. *J Bone Miner Res,* **18**, 807–817 (2003).

101. C. T. Rubin and L. E. Lanyon, Regulation of bone formation by applied dynamic loads. *J Bone Joint Surg Am,* **66**, 397–402 (1984).

102. C. T. Rubin and L. E. Lanyon, Regulation of bone mass by mechanical strain magnitude. *Calcif Tissue Int,* **37**, 411–417 (1985).

103. C. H. Turner, M. R. Forwood, and M. W. Otter, Mechanotransduction in bone: Do bone cells act as sensors of fluid flow? *FASEB J,* **8**, 875–878 (1994).

104. A. G. Robling, D. B. Burr, and C. H. Turner, Partitioning a daily mechanical stimulus into discrete loading bouts improves the osteogenic response to loading. *J Bone Miner Res,* **15**, 1596–1602 (2000).

105. A. G. Robling, F. M. Hinant, D. B. Burr, and C. H. Turner, Shorter, more frequent mechanical loading sessions enhance bone mass. *Med Sci Sports Exerc,* **34**, 196–202 (2002).

106. Y. F. Hsieh and C. H. Turner, Effects of loading frequency on mechanically induced bone formation. *J Bone Miner Res,* **16**, 918–924 (2001).

107. Y. F. Hsieh, A. G. Robling, W. T. Ambrosius, D. B. Burr, and C. H. Turner, Mechanical loading of diaphyseal bone in vivo: The strain threshold for an osteogenic response varies with location. *J Bone Miner Res,* **16**, 2291–2297 (2001).

108. S. C. Cowin, L. Moss-Salentijn, and M. L. Moss, Candidates for the mechanosensory system in bone. *J Biomech Eng,* **113**, 191–197 (1991).

109. E. M. Aarden, E. H. Burger, and P. J. Nijweide, Function of osteocytes in bone. *J Cell Biochem,* **55**, 287–299 (1994).

110. E. H. Burger and J. Klein-Nulend, Mechanotransduction in bone—Role of the lacuno-canalicular network. *FASEB J,* **13**(Suppl), S101–112 (1999).

111. S. Weinbaum, S. C. Cowin, and Y. Zeng, A model for the excitation of osteocytes by mechanical loading-induced bone fluid shear stresses. *J Biomech,* **27**, 339–360 (1994).

112. L. Wang, S. P. Fritton, S. Weinbaum, and S. C. Cowin, On bone adaptation due to venous stasis. *J Biomech,* **36**, 1439–1451 (2003).

113. M. L. Knothe Tate, "Whither flows the fluid in bone?" An osteocyte's perspective. *J Biomech,* **36**, 1409–1424 (2003).

114. N. E. Ajubi, *et al.*, Pulsating fluid flow increases prostaglandin production by cultured chicken osteocytes—A cytoskeleton-dependent process. *Biochem Biophys Res Commun,* **225**, 62–68 (1996).

115. L. Wang, S. C. Cowin, S. Weinbaum, and S. P. Fritton, Modeling tracer transport in an osteon under cyclic loading. *Ann Biomed Eng,* **28**, 1200–1209 (2000).

116. L. Wang, *et al.*, In situ measurement of solute transport in the bone lacunar-canalicular system. *Proc Natl Acad Sci USA,* **102**, 11911–11916 (2005).

117. Y. Han, S. C. Cowin, M. B. Schaffler, and S. Weinbaum, Mechanotransduction and strain amplification in osteocyte cell processes. *Proc Natl Acad Sci USA,* **101**, 16689–16694 (2004).

118. N. Petrov and S. R. Pollack, Comparative analysis of diffusive and stress induced nutrient transport efficiency in the lacunar-canalicular system of osteons. *Biorheology,* **40**, 347–353 (2003).

119. A. E. Tami, P. Nasser, O. Verborgt, M. B. Schaffler, and M. L. Knothe Tate, The role of interstitial fluid flow in the remodeling response to fatigue loading. *J Bone Miner Res,* **17**, 2030–2037 (2002).

120. T. H. Smit, E. H. Burger, and J. M. Huyghe, A case for strain-induced fluid flow as a regulator of BMU-coupling and osteonal alignment. *J Bone Miner Res,* **17**, 2021–2029 (2002).

121. S. Kotha, J. Gluhak-Heinrich, M. B. Schaffler, S. E. Harris, and L. Bonewald, A 3D model of osteocyte gene expression in response to mechanical loading—Correlation of Dmp1 gene expression in osteocytes with local strain environment. *J Bone Miner Res,* **20**(Suppl 1), S24 (2005).

122. J. Gluhak-Heinrich, *et al.*, Mechanically induced DMP1 and MEPE expression in osteocytes: Correlation to mechanical strain, osteogenic response and gene expression threshold. *J Bone Miner Res,* **20**(Suppl 1), S73 (2005).

123. F. M. Pavalko, *et al.*, A model for mechanotransduction in bone cells: The load-bearing mechanosomes. *J Cell Biochem,* **88**, 104–112 (2003).

124. D. P. Nicolella, L. F. Bonewald, D. E. Moravits, and J. Lankford, Measurement of microstructural strain in cortical bone. *Eur J Morphol,* **42**, 23–29 (2005).

125. D. P. Nicolella, D. E. Moravits, A. M. Gale, L. F. Bonewald, and J. Lankford, Osteocyte lacunae tissue strain in cortical bone. *J Biomech,* **39**, 1735–1743 (2006).

126. D. B. Burr, *et al.*, In vivo measurement of human tibial strains during vigorous activity. *Bone,* **18**, 405–410 (1996).

127. J. Klein-Nulend, *et al.*, Sensitivity of osteocytes to biomechanical stress in vitro. *FASEB J,* **9**, 441–445 (1995).

128. C. R. Jacobs, *et al.*, Differential effect of steady versus oscillating flow on bone cells. *J Biomech,* **31**, 969–976 (1998).

129. J. F. Whitfield, Primary cilium—Is it an osteocyte's strain-sensing flowmeter? *J Cell Biochem,* **89**, 233–237 (2003).

130. Z. Xiao, *et al.*, Cila-like structures and polycystin 1 in osteoblasts/osteocytes and associated abnormalities in skeletogenesis and Runx 2 expression. *J Biol Chem,* **281**, 884–895.

131. S. M. Nauli, *et al.*, Polycystins 1 and 2 mediate mechanosensation in the primary cilium of kidney cells. *Nat Genet,* **33**, 129–137 (2003).

132. R. O. Hynes, Integrins: Bidirectional, allosteric signaling machines. *Cell,* **110**, 673–687 (2002).

133. D. M. Salter, J. E. Robb, and M. O. Wright, Electrophysiological responses of human bone cells to mechanical stimulation: Evidence for specific integrin function in mechanotransduction. *J Bone Miner Res,* **12**, 1133–1141 (1997).

134. A. Katsumi, A. W. Orr, E. Tzima, and M. A. Schwartz, Integrins in mechanotransduction. *J Biol Chem,* **279**, 12001–12004 (2004).

135. D. E. Hughes, D. M. Salter, S. Dedhar, and R. Simpson, Integrin expression in human bone. *J Bone Miner Res,* **8**, 527–533 (1993).

136. J. H. Bennett, D. H. Carter, A. L. Alavi, J. N. Beresford, and S. Walsh, Patterns of integrin expression in a human mandibular explant model of osteoblast differentiation. *Arch Oral Biol,* **46**, 229–238 (2001).

137. P. D. Lampe, *et al.,* Cellular interaction of integrin alpha-3beta1 with laminin 5 promotes gap junctional communication. *J Cell Biol,* **143**, 1735–1747 (1998).

138. Y. Guo, C. Martinez-Williams, C. E. Yellowley, H. J. Donahue, and D. E. Rannels, Connexin expression by alveolar epithelial cells is regulated by extracellular matrix. *Am J Physiol Lung Cell Mol Physiol,* **280**, L191–202 (2001).

139. F. G. Giancotti, Complexity and specificity of integrin signalling. *Nat Cell Biol,* **2**, E13–14 (2000).

140. A. J. Sillar-Jackson, *et al.,* The role of a5 integrin as a mechanosensor in the regulation of connexin 43 hemichannel release of prostaglandin in response to mechanical stress. *J Bone Miner Res,* **21**, S72 (2006).

141. K. Naruse, A. Miyauchi, M. Itoman, and Y. Mikuni-Takagaki, Distinct anabolic response of osteoblast to low-intensity pulsed ultrasound. *J Bone Miner Res,* **18**, 360–369 (2003).

142. G. R. Jordan, *et al.,* The ratio of osteocytic incorporation to bone matrix formation in femoral neck cancellous bone: An enhanced osteoblast work rate in the vicinity of hip osteoarthritis. *Calcif Tissue Int,* **72**, 190–196 (2003).

143. A. D. Bakker, K. Soejima, J. Klein-Nulend, and E. H. Burger, The production of nitric oxide and prostaglandin E(2) by primary bone cells is shear stress dependent. *J Biomech,* **34**, 671–677 (2001).

144. N. Loveridge, *et al.,* Patterns of osteocytic endothelial nitric oxide synthase expression in the femoral neck cortex: Differences between cases of intracapsular hip fracture and controls. *Bone,* **30**, 866–871 (2002).

145. M. Watanuki, *et al.,* Role of inducible nitric oxide synthase in skeletal adaptation to acute increases in mechanical loading. *J Bone Miner Res,* **17**, 1015–1025 (2002).

146. L. G. Raisz and B. E. Kream, Regulation of bone formation (second of two parts). *N Engl J Med,* **309**, 83–89 (1983).

147. W. S. Jee, K. Ueno, Y. P. Deng, and D. M. Woodbury, The effects of prostaglandin E2 in growing rats: Increased metaphyseal hard tissue and cortico-endosteal bone formation. *Calcif Tissue Int,* **37**, 148–157 (1985).

148. L. G. Raisz, *et al.,* Effects of prostaglandin E2 on bone formation in cultured fetal rat calvariae: Role of insulin-like growth factor-I. *Endocrinology,* **133**, 1504–1510 (1993).

149. T. Nagata, *et al.,* Effect of prostaglandin E2 on mineralization of bone nodules formed by fetal rat calvarial cells. *Calcif Tissue Int,* **55**, 451–457 (1994).

150. J. Klein-Nulend, E. H. Burger, C. M. Semeins, L. G. Raisz, and C. C. Pilbeam, Pulsating fluid flow stimulates prostaglandin release and inducible prostaglandin G/H synthase mRNA expression in primary mouse bone cells. *J Bone Miner Res,* **12**, 45–51 (1997).

151. N. E. Ajubi, J. Klein-Nulend, M. J. Alblas, E. H. Burger, and P. J. Nijweide, Signal transduction pathways involved in fluid flow-induced PGE2 production by cultured osteocytes. *Am J Physiol,* **276**, E171–178 (1999).

152. M. R. Forwood, Inducible cyclo-oxygenase (COX-2) mediates the induction of bone formation by mechanical loading in vivo. *J Bone Miner Res,* **11**, 1688–1693 (1996).

153. H. Hagino, M. Kuraoka, Y. Kameyama, T. Okano, and R. Teshima, Effect of a selective agonist for prostaglandin E receptor subtype EP4 (ONO-4819) on the cortical bone response to mechanical loading. *Bone,* **36**, 444–453 (2005).

154. I. Alam, S. J. Warden, A. G. Robling, and C. H. Turner, Mechanotransduction in bone does not require a functional cyclooxygenase-2 (COX-2) gene. *J Bone Miner Res,* **20**, 438–446 (2005).

155. D. C. Genetos, D. J. Geist, D. Liu, H. J. Donahue, and R. L. Duncan, Fluid shear-induced ATP secretion mediates prostaglandin release in MC3T3–E1 osteoblasts. *J Bone Miner Res,* **20**, 41–49 (2005).

156. J. Li, D. Liu, H. Z. Ke, R. L. Duncan, and C. H. Turner, The P2X7 nucleotide receptor mediates skeletal mechanotransduction. *J Biol Chem,* **280**, 42952–42959 (2005).

157. J. M. Labasi, *et al.,* Absence of the P2X7 receptor alters leukocyte function and attenuates an inflammatory response. *J Immunol,* **168**, 6436–6445 (2002).

158. M. L. Johnson, J. L. Picconi, and R. R. Recker, The gene for high bone mass. *The Endocrinologist,* **12**, 445–453 (2002).

159. R. D. Little, *et al.,* A mutation in the LDL receptor-related protein 5 gene results in the autosomal dominant high-bone-mass trait. *Am J Hum Genet,* **70**, 11–19 (2002).

160. M. P. Akhter, *et al.,* Bone biomechanical properties in LRP5 mutant mice. *Bone,* **35**, 162–169 (2004).

161. K. Sawakami, *et al.,* Site-specific osteopenia and decreased mechanoreactivity in Lrp5 mutant mice. *J Bone Miner Res,* **19**, S1, S38 (2004).

162. K. C. Lee, H. Jessop, R. Suswillo, G. Zaman, and L. E. Lanyon, The adaptive response of bone to mechanical loading in female transgenic mice is deficient in the absence of oestrogen receptor-alpha and -beta. *J Endocrinol,* **182**, 193–201 (2004).

163. K. C. Lee and L. E. Lanyon, Mechanical loading influences bone mass through estrogen receptor alpha. *Exerc Sport Sci Rev,* **32**, 64–68 (2004).

164. J. Power, N. Loveridge, N. Rushton, M. Parker, and J. Reeve, Osteocyte density in aging subjects is enhanced in bone adjacent to remodeling Haversian systems. *Bone,* **30**, 859–865 (2002).

165. K. Tanaka, Y. Yamaguchi, and Y. Hakeda, Isolated chick osteocytes stimulate formation and bone-resorbing activity of osteoclast-like cells. *J Bone Miner Metab,* **13**, 61–70 (1995).

166. O. Verborgt, N. A. Tatton, R. J. Majeska, and M. B. Schaffler, Spatial distribution of Bax and Bcl-2 in osteocytes after bone fatigue: Complementary roles in bone remodeling regulation? *J Bone Miner Res,* **17**, 907–914 (2002).

167. S. C. Manolagas, Choreography from the tomb: An emerging role of dying osteocytes in the purposeful, and perhaps not so purposeful, targeting of bone remodeling. *BoneKey,* **3**, 5–14 (2006).

168. C. R. Dunstan, R. A. Evans, E. Hills, S. Y. Wong, and R. J. Higgs, Bone death in hip fracture in the elderly. *Calcif Tissue Int,* **47**, 270–275 (1990).

169. R. S. Weinstein, R. W. Nicholas, and S. C. Manolagas, Apoptosis of osteocytes in glucocorticoid-induced osteonecrosis of the hip. *J Clin Endocrinol Metab,* **85**, 2907–2912 (2000).

170. S. Y. Wong, *et al.,* The pathogenesis of osteoarthritis of the hip. Evidence for primary osteocyte death. *Clin Orthop Relat Res,* 305–312 (1987).

171. B. S. Noble, *et al.,* Mechanical loading: Biphasic osteocyte survival and targeting of osteoclasts for bone destruction in rat cortical bone. *Am J Physiol Cell Physiol,* **284**, C934–943 (2003).

172. O. Verborgt, G. J. Gibson, and M. B. Schaffler, Loss of osteocyte integrity in association with microdamage and bone remodeling after fatigue in vivo. *J Bone Miner Res,* **15**, 60–67 (2000).

173. A. Tomkinson, J. Reeve, R. W. Shaw, and B. S. Noble, The death of osteocytes via apoptosis accompanies estrogen withdrawal in human bone. *J Clin Endocrinol Metab,* **82,** 3128–3135 (1997).

174. R. Pacifici, *et al.,* Effect of surgical menopause and estrogen replacement on cytokine release from human blood mononuclear cells. *Proc Natl Acad Sci USA,* **88,** 5134–5138 (1991).

175. D. Rickard, G. Russell, and M. Gowen, Oestradiol inhibits the release of tumour necrosis factor but not interleukin 6 from adult human osteoblasts in vitro. *Osteoporos Int,* **2,** 94–102 (1992).

176. S. Kousteni, *et al.,* Nongenotropic, sex-nonspecific signaling through the estrogen or androgen receptors: Dissociation from transcriptional activity. *Cell,* **104,** 719–730 (2001).

177. Y. Kitase, J. X. Jiang, and L. Bonewald, The anti-apoptotic effects of mechanical strain on osteocytes are mediated by PGE2 and monocyte chemotactic protein (MCP-3): Selective protection by MCP3 against glucocorticoid (GC) and not TNF-α induced apoptosis. *J Bone Miner Res,* **21,** S48 (2006).

178. M. Heller-Steinberg, Ground substance, bone salts, and cellular activity in bone formation and destruction. *Am J Anat,* **89,** 347–379 (1951).

179. L. F. Belanger, Osteocytic osteolysis. *Calcif Tissue Res,* **4,** 1–12 (1969).

180. B. Kremlien, C. Manegold, E. Ritz, and J. Bommer, The influence of immobilization on osteocyte morphology: Osteocyte differential count and electron microscopic studies. *Virchows Arch A Pathol Anat Histol,* **370,** 55–68 (1976).

181. A. van der Plas, *et al.,* Characteristics and properties of osteocytes in culture. *J Bone Miner Res,* **9,** 1697–1704 (1994).

182. E. Bonucci and G. Gherardi, Osteocyte ultrastructure in renal osteodystrophy. *Virchows Arch A Pathol Anat Histol,* **373,** 213–231 (1977).

183. F. Heuck, Comparative investigations of the function of osteocytes in bone resorption. *Calcif Tissue Res,* **4**(Suppl) 148–149 (1970).

184. P. J. Marie and F. H. Glorieux, Relation between hypomineralized periosteocytic lesions and bone mineralization in vitamin D-resistant rickets. *Calcif Tissue Int,* **35,** 443–448 (1983).

185. N. E. Lane, An update on glucocorticoid-induced osteoporosis. *Rheum Dis Clin North Am,* **27,** 235–253 (2001).

186. K. G. Saag, Glucocorticoid-induced osteoporosis. *Endocrinol Metab Clin North Am,* **32,** 135–157, vii (2003).

187. N. E. Lane, *et al.,* Glucocorticoid-treated mice have localized changes in trabecular bone material properties and osteocyte lacunar size that are not observed in placebo-treated or estrogen-deficient mice. *J Bone Miner Res,* **21,** 466–476 (2006).

188. C. A. Baud, and D. H. Dupont, *Electron Microscopy,* QQ-10, Academic Press, New York (1962).

189. D. J. Baylink and J. E. Wergedal, Bone formation by osteocytes. *Am J Physiol,* **221,** 669–678 (1971).

190. A. J. Borton, J. P. Frederick, M. B. Datto, X. F. Wang, and R. S. Weinstein, The loss of Smad3 results in a lower rate of bone formation and osteopenia through dysregulation of osteoblast differentiation and apoptosis. *J Bone Miner Res,* **16,** 1754–1764 (2001).

191. D. Vashishth, O. Verborgt, G. Divine, M. B. Schaffler, and D. P. Fyhrie, Decline in osteocyte lacunar density in human cortical bone is associated with accumulation of microcracks with age. *Bone,* **26,** 375–380 (2000).

192. S. Qiu, D. S. Rao, S. Palnitkar, and A. M. Parfitt, Relationships between osteocyte density and bone formation rate in human cancellous bone. *Bone,* **31,** 709–711 (2002).

193. S. Qiu, D. S. Rao, S. Palnitkar, and A. M. Parfitt, Age and distance from the surface but not menopause reduce osteocyte density in human cancellous bone. *Bone,* **31,** 313–318 (2002).

194. S. Qiu, D. S. Rao, S. Palnitkar, and A. M. Parfitt, Reduced iliac cancellous osteocyte density in patients with osteoporotic vertebral fracture. *J Bone Miner Res,* **18,** 1657–1663 (2003).

195. S. Qiu, D. S. Rao, S. Palnitkar, and A. M. Parfitt, Differences in osteocyte and lacunar density between Black and White American women. *Bone,* **38,** 130–135 (2006).

196. A. G. Robling and C. H. Turner, Mechanotransduction in bone: Genetic effects on mechanosensitivity in mice. *Bone,* **31,** 562–569 (2002).

197. I. Westbroek, A. van der Plas, K. E. de Rooij, J. Klein-Nulend, and P. J. Nijweide, Expression of serotonin receptors in bone. *J Biol Chem,* **276,** 28961–28968 (2001).

198. B. I. Gustafsson, *et al.,* Long-term serotonin administration leads to higher bone mineral density, affects bone architecture, and leads to higher femoral bone stiffness in rats. *J Cell Biochem,* **97,** 1283–1291 (2006).

199. S. B. Doty, Morphological evidence of gap junctions between bone cells. *Calc Tissue Int,* **33,** 509–512 (1981).

200. C. Palumbo, S. Palazzini, and G. Marotti, Morphological study of intercellular junctions during osteocyte differentiation. *Bone,* **11,** 401–406 (1990).

201. M. V. Bennett and D. A. Goodenough, Gap junctions, electronic coupling, and intercellular communication. *Neurosci Res Program Bull,* **16,** 1–486 (1978).

202. D. A. Goodenough, J. A. Golinger, and D. L. Paul, Connexins, connexons, and intercellular communication. *Annu Rev Biochem,* **65,** 475–502 (1996).

203. B. G. Jeansonne, F. F. Feagin, R. W. McMinn, R. L. Shoemaker, and W. S. Rehm, Cell-to-cell communication of osteoblasts. *J Dent Res,* **58,** 1415–1423 (1979).

204. F. Lecanda, *et al.,* Connexin43 deficiency causes delayed ossification, craniofacial abnormalities, and osteoblast dysfunction. *J Cell Biol,* **151,** 931–944 (2000).

205. H. J. Donahue, Gap junctions and biophysical regulation of bone cell differentiation. *Bone,* **26,** 417–422 (2000).

206. M. M. Saunders, *et al.,* Gap junctions and fluid flow response in MC3T3-E1 cells. *Am J Physiol Cell Physiol,* **281,** C1917–1925 (2001).

207. M. M. Saunders, *et al.,* Fluid flow-induced prostaglandin E2 response of osteoblastic ROS 17/2.8 cells is gap junction-mediated and independent of cytosolic calcium. *Bone,* **32,** 350–356 (2003).

208. S. C. Rawlinson, A. A. Pitsillides, and L. E. Lanyon, Involvement of different ion channels in osteoblasts' and osteocytes' early responses to mechanical strain. *Bone,* **19,** 609–614 (1996).

209. D. L. Ypey, *et al.,* Voltage, calcium, and stretch activated ionic channels and intracellular calcium in bone cells. *J Bone Miner Res,* **7**(Suppl 2), S377–387 (1992).

210. D. A. Goodenough and D. L. Paul, Beyond the gap: Functions of unpaired connexon channels. *Nat Rev Mol Cell Biol,* **4,** 285–294 (2003).

211. L. Ye, *et al.,* Dmp1-deficient mice display severe defects in cartilage formation responsible for a chondrodysplasia-like phenotype. *J Biol Chem,* **280,** 6197–6203 (2005).

212. N. Sun, Y. Gao, J. Pretoriu, S. Morony, P. J. Kostenuik, S. Simonet, D. L. Lacey, I. Sarosi, C. Kurahara, and C. Patszty, High bone mineral density in SOST knock-out mice demonstrates functional conservation of osteocyte mediated

bone homeostasis in mouse and human. *J Bone Miner Res,* **18**(Suppl 2), S7 (2003).

213. C. Palumbo, M. Ferretti, and G. Marotti, Osteocyte dendrogenesis in static and dynamic bone formation: An ultrastructural study. *Anat Rec A Discov Mol Cell Evol Biol,* **278**, 474–480 (2004).

214. M. L. Knothe Tate, J. R. Adamson, A. E. Tami, and T. W. Bauer, The osteocyte. *Int J Biochem Cell Biol,* **36**, 1–8 (2004).

215. R. L. Cruess, D. Ross, and E. Crawshaw, The etiology of steroid-induced avascular necrosis of bone. A laboratory and clinical study. *Clin Orthop Relat Res,* 178–183 (1975).

216. J. E. Kenzora and M. J. Glimcher, Accumulative cell stress: The multifactorial etiology of idiopathic osteonecrosis. *Orthop Clin North Am,* **16**, 669–679 (1985).

217. C. J. Lavernia, R. J. Sierra, and F. R. Grieco, Osteonecrosis of the femoral head. *J Am Acad Orthop Surg,* **7**, 250–261 (1999).

218. R. S. Weinstein, R. L. Jilka, A. M. Parfitt, and S. C. Manolagas, Inhibition of osteoblastogenesis and promotion of apoptosis of osteoblasts and osteocytes by glucocorticoids. Potential mechanisms of their deleterious effects on bone. *J Clin Invest,* **102**, 274–282 (1998).

219. C. Zalavras, S. Shah, M. J. Birnbaum, and B. Frenkel, Role of apoptosis in glucocorticoid-induced osteoporosis and osteonecrosis. *Crit Rev Eukaryot Gene Expr,* **13**, 221–235 (2003).

220. J. D. Calder, L. Buttery, P. A. Revell, M. Pearse, and J. M. Polak, Apoptosis—A significant cause of bone cell death in osteonecrosis of the femoral head. *J Bone Joint Surg Br,* **86**, 1209–1213 (2004).

221. S. Liu, *et al.,* Elevated levels of FGF23 in dentin matrix protein 1 (DMP1) null mice potentially explain phenotypic similarities to HYP mice. *J Bone Miner Res,* **21**, S51 (2006).

222. S. C. Cowin, S. Weinbaum, and Y. Zeng, A case for bone canaliculi as the anatomical site of strain generated potentials. *J Biomech,* **28**, 1281–1297 (1995).

223. L. Wang, C. Ciani, S. B. Doty, and S. P. Fritton, Delineating bone's interstitial fluid pathway in vivo. *Bone,* **34**, 499–509 (2004).

224. J. S. Dodd, J. A. Raleigh, and T. S. Gross, Osteocyte hypoxia: A novel mechanotransduction pathway. *Am J Physiol,* **277**, C598–602 (1999).

CHAPTER 7

The Regulatory Role of Matrix Proteins in Mineralization of Bone

WEI ZHU, PAMELA GEHRON ROBEY, AND ADELE L. BOSKEY

I. INTRODUCTION

The skeleton is essentially responsible for not only providing structural support and protection to the body's organs but also serving as a reservoir for calcium, magnesium, and phosphate—ions that are of critical importance in physiology. The fabric of bone is a unique composite of living cells embedded in a remarkable three-dimensional structure of extracellular matrix that is stabilized by mineral, which is a carbonate-rich analogue of the geologic mineral hydroxyapatite.

A. Bone Tissue: Composition

During development, mesenchymal cells form the skeleton via two basic pathways [1]. Intramembranous bone is formed by direct differentiation of mesenchymal cells, whereas endochondral bone is formed by an initial condensation of mesenchymal cells that leads to morphogenesis of a cartilaginous structure. Serving as a temporary model, the cartilage becomes calcified and the provisional calcified cartilagenous precursor is subsequently replaced by bone.

Invasion by blood vessels brings in the cells that remove bone (osteoclasts) and, in addition, the osteoblastic precursors that will replace the calcified cartilage with bona fide bone. The initial bone formed, woven bone, is a rather unorganized conglomeration of collagenous and noncollagenous proteins that induce the precipitation of mineral. Through modeling by osteoclasts, this primordial bone is removed and replaced by the formation of lamellar bone, a more highly organized structure with alternating layers of mineralized extracellular matrix, whose plywood-like structure provides bone with its mechanical strength.

Although the mineralized matrices were originally thought to be composed of a unique set of matrix proteins, many of these proteins are also synthesized by nonskeletal cells, with the exception of a few truly bone-specific proteins, which are involved in mineral deposition [2–9]. Bone is composed of 70–90% mineral and only 10–30% represented by protein, with collagenous protein comprising ~90% of the bone matrix and noncollagenous proteins accounting for the remaining ~10%. In addition, virtually all of the known collagenous and noncollagenous proteins in bone studied to date differ from those in other tissues in their chemical nature. These diverse forms are a result of alternative splicing of mRNA and different post-translational modifications, such as glycosylation, phosphorylation, and sulfation. These chemical differences most likely influence the physiological function of these proteins, and the appropriate mixture provides the extracellular matrix with the ability to calcify. Moreover, because the extracellular matrix proteins are the secretory products of cells in the osteoblastic lineage, they represent biochemical markers of maturation stages of cells during the formation process (Figure 7-1) or the resorption process (in their degraded form) of bone. This chapter describes major types of proteins synthesized by osteoblastic cells that are present in bone matrix and discusses their potential roles in the regulation of mineralization.

B. Mineral: Calcification of Bone

The mechanical strength of bone is attributable to the presence of mineral that converts the pliable organic matrix into a more rigid structure [10, 11]. A variety of structural analyses, including x-ray and electron diffraction [12–14], infrared spectroscopy [15], high-voltage electron microscopy [16], nuclear magnetic resonance (NMR), and x-ray absorption fine structure analysis [17–19], have shown that mineral crystals within bone are analogous to the naturally occurring

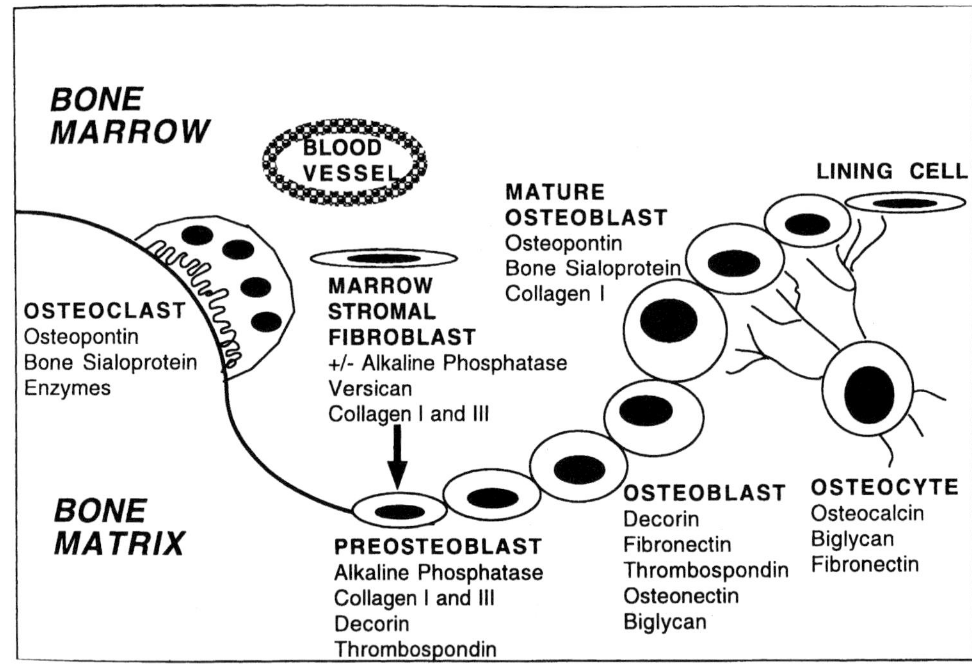

FIGURE 7-1   Maturational stage and bone matrix gene expression. Osteoblastic cells pass through a series of maturational stages, each of which can be partially characterized by the bone matrix proteins that they produce. In addition, osteoclasts also secrete proteins that become incorporated into mineralized matrix.

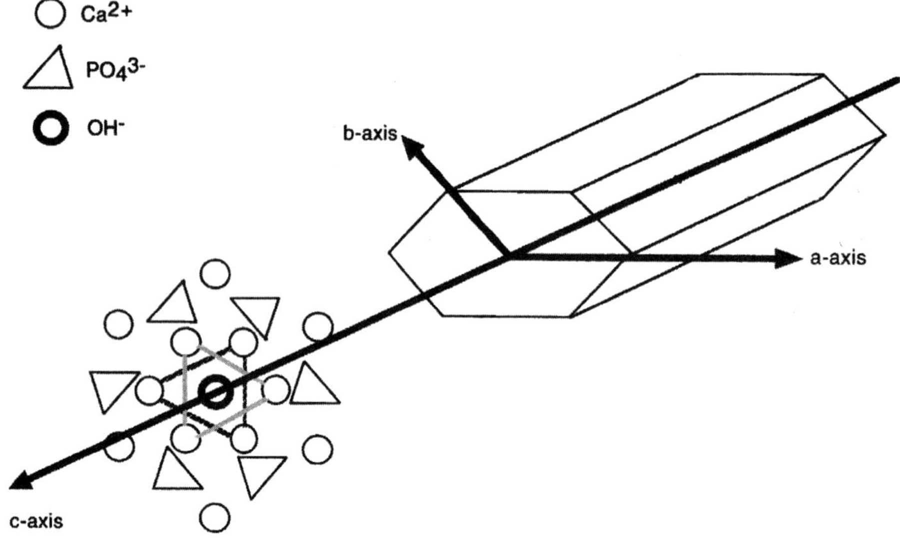

FIGURE 7-2   Crystal lattice structure. A portion of the apatite structure is depicted as it would be viewed along the length (c axis) of the hydroxyapatite crystal, showing the hexagonal arrangement of the Ca^{2+} and PO_4^{3-} ions about the OH^- position.

geologic mineral, hydroxyapatite ($Ca_{10}[PO_4]_6[OH]_2$) (Figure 7-2).

However, in bone, the mineral includes numerous ions not found in pure hydroxyapatite. For example, HPO_4^{2-}, CO_3^{2-}, Mg^{2+}, Na^+, F^-, and citrate are adsorbed onto the crystal surfaces and/or substituted in the lattice for the constituent Ca^{2+}, PO_4^{3-}, and OH^- ions [20–26]. This poorly crystalline apatite in bone, because of its small crystal size and large number of lattice-substituted and surface-adsorbed ion impurities, can be dissolved

more readily than the larger, more perfect crystals of geologic hydroxyapatite. Moreover, this altered solubility allows bone mineral to play an important role in Ca^{2+}, Mg^{2+}, and PO_4^{3-} ion homeostasis [27]. Despite claims of the presence of other mineral phases in bone (e.g., brushite [28], octacalcium phosphate [29], amorphous calcium phosphate [30, 31], and whitlockite [32]), current evidence supports the view that bone mineral is predominantly apatitic, with numerous, and perhaps unique, impurities [23]. This chapter discusses how the initiation of mineral deposition and the growth of mineral crystals are regulated by matrix proteins.

II. COLLAGENOUS PROTEINS

In the skeleton, the major (~90%) structural protein is collagenous in nature. Bone collagen is predominantly composed of type I collagen, which most likely serves a mechanical function providing tensile strength [33]. Collagen may not directly induce mineral deposition in bone matrix; however, it serves as an important "backbone" in support of initial mineral deposition and the organization of crystal growth by providing appropriate scaffolding and orientation of nucleators of mineralization.

A. Structure of the Molecule

Collagen is defined as a trimeric molecule composed of α chain subunits [34, 35]. A significant feature of the component α chains is that their primary sequence is almost entirely made up of a repeating triplet sequence, Gly-X-Y, where X is often proline and Y is often hydroxyproline [34, 35]. Collagenous proteins are either homotrimeric, composed of three identical α chains, or heterotrimeric, with two or three different α chains. Individual α chains of the collagen molecule coil together to form an extended rigid triple helix. The structure is stabilized by hydrogen bonding between OH groups on hydroxyproline and intrachain water [36] and by aldehyde-derived cross-links [37–39].

Type I collagen is a heterotrimer, $(\alpha_1[I]_2, \alpha_2[I])$, and the human gene for $\alpha_1(I)$, COL1A1, located on chromosome 17q21.3–q22, is 18 kb in length and contains 51 exons [40, 41]. The COL1A2 gene, located on chromosome 7q22.1, is 35 kb in length and contains 52 exons [41–43]. The promoters for COL1A1 and COL1A2 have been characterized in detail and contain similar but not identical promoter elements [44–48]. At −29 bp from the transcription start site, the COL1A1 promoter contains a TATA box, whereas it is absent in the COL1A2 promoter. Farther upstream, both contain a CCAAT sequence (−100 bp in COL1A1 and −82 bp in COL1A2), as well as a long stretch of C's and T's, which confer S1 nuclease and DNAse hypersensitivity, implying a relatively open structure. Other elements include a vitamin D response element (VDRE) in COL1A1 [49] and a CAAT-like region that binds to nuclear factor 1 (NF1) in the COL1A2 promoter [50]. It is of interest that the amount of mRNA for COL1A1 is twice the amount of COL1A2, a ratio that is reflected in the final triple-helical molecule. Currently, the collagen family is composed of 23 collagen types and 38 genetically distinct α chains, 4 of which are currently under characterization [51–54].

Based on their structural features, collagens can be generally divided into two groups: fibrillar and non-fibrillar [51, 52, 54]. Collagen fibrils are formed by the mature collagen molecules via head-to-tail associations and different types of fibrillar collagens share a strong structural similarity in that the major part of each molecule is formed by an uninterrupted triple-helical domain. In addition, they are all synthesized as precursors that are proteolytically trimmed of their noncollagenous ends to yield mature molecules [54–57]. Once fibers have been formed in the extracellular environment, they are further stabilized by the formation of inter- and intramolecular cross-links.

Fibrillar collagens (types I, II, III, V, and XI) are by far the most abundant forms and are formed in the interstitial spaces of connective tissues throughout the body [51, 52, 54]. Type I collagen, the predominant collagen of skin, tendon, and bone, forms the major scaffolding of virtually all connective tissues except cartilage, because cartilage contains predominantly type II collagen ($[\alpha_1(II)_3]$) with limited amounts of other collagens. Type III collagen, composed of three identical $\alpha_1(III)$ chains, is found in many tissues rich in type I collagen. Quantitatively minor fibrillar collagens, types V and XI, associated with collagen I and II, respectively, are located on the periphery of the collagen fibrils. Contrary to other fibrillar collagens, their N-terminal extensions are retained and project onto the fibril surface. This feature, together with the correct molar ratios of I/V and II/XI collagens in fibrils, is significant in the regulation of fibril diameter [54, 58]. Structural analysis of fibrillar type I collagen suggests that individual collagen fibrils are aligned in a quarter-staggered array, with a 280-nm periodicity. As a result of the quarter stagger, there are gaps (holes) within the fibrillar structures, and it is these gaps and in the overlapping regions adjacent to them (e band) that bone mineral crystals first appear [35, 59–63].

The nonfibrillar collagens are characterized by triple-helical domains that are either shorter or longer than those of the fibrillar types, and they may contain stretches of non-triple-helical sequences [35, 54]. Several subfamilies can be further distinguished according to similarities in the domain organization, supermolecular structures, and types of extracellular networks they form: (1) collagens that are located on the surfaces of fibrils and are called fibril-associated collagens with interrupted triple helices (FACITs; types IX, XII, XIV, XVI, and XIX), (2) collagens that form hexagonal networks (types VIII and X), (3) type IV collagen found in basement membranes, (4) type VI collagen that forms beaded filaments, (5) type VII collagen that forms anchoring fibrils of basement membranes, (6) collagens with transmembrane domains (types XIII and XVII), and (7) the family of type XV and XVIII collagens [54, 64]. Furthermore, the newly identified collagen types XX and XXIII most likely belong to the nonfibrillar group of collagens [54]. This group of collagens is structurally and functionally very heterogeneous, and only some members are known to be found in bone and cartilage.

Among these nonfibrillar collagens, type X collagen is found in calcified cartilage. The localization of type X to hypertrophic chondrocytes is highly specific, but it does not appear to have a major role in cartilage calcification. Type IV collagen, with the composition $\alpha_1(IV)_2\alpha_2(IV)$, is found in basement membranes, including those that surround vascular endothelial cells that invade bone during osteogenesis. Type VI collagen is significantly shorter than other collagen types and is composed of three distinct α chains that form ropelike microfibrillar structures. Anchoring fibrils are composed of type VII collagen, which is 1.5 times longer than type I collagen. Another short-chain collagen, type VIII, is found in Descemet's membrane of the eye, is synthesized by endothelial cells in culture, and may be related to type X collagen. Type IX is homologous to type V and is a minor constituent in cartilage. Type IX is composed of three different types of α chains, $\alpha_1(IX)$, $\alpha_2(IX)$, and $\alpha_3(IX)$, which form a short and a long triple helix joined by a flexible hinge region. A glycosaminoglycan chain is also attached to one of the α chains at the amino terminus, making this collagen a proteoglycan as well. Type IX has been found as a coating of type II collagen fibrils (the major collagen in cartilage) and covalently attached to it. Type XII is similar to type IX but has three projections extending from the triple helix. This type may also be associated with type I fibrils in tendon. Type XIV (as well as type XII) is structurally related to type IX collagen fibrils, which associate with type II collagen in cartilage.

B. Bone Matrix Collagen(s)

Bone matrix proper contains a rather limited array of collagen types (Table 7-1). Although bone matrix has been reported to contain predominantly type I collagen, other types are certainly present but at lower levels compared to soft connective tissues. Several FACITs (types XII and XIV) have been detected in bone [35, 54, 65, 66], and there are occasional reports of low levels of type III and type V molecules as well [35, 54, 67, 68]. Given the potential role of these low-abundancy collagens in regulating fibril diameter, it is possible that collagen fibrils in bone grow to much larger diameters than in soft tissues due to the reduced proportion of these diameter-regulating collagen types. Moreover, the FACIT collagens seem to have a fundamental role in determining matrix structure, as demonstrated by animals lacking or containing mutated forms of the FACIT collagens [35, 54, 69]. These animals exhibit a spectrum of bone and cartilage disorders, presumably due to abnormal fibril formation.

Whereas there is only one copy of the genes that code for the COLA1 and COLA2 in mammalian genomes, the regulation of type I collagen production in bone is somewhat different from that in soft connective tissues. In bone cell and organ cultures, collagen synthesis is increased by heparin [70], organic phosphate [71], interleukin (IL)-4 [72], and gallium [73]. In contrast, collagen synthesis is decreased by prostaglandin E_2 [74], 1,25-dihydroxyvitamin D_3 [75], cortisol [76], parathyroid hormone (PTH) [77], epidermal growth factor (EGF) [78], basic fibroblast growth factor [79], IL-10 [80], and lead [81]. Although the COL1A1 promoter contains a VDRE, binding of this element by the vitamin D receptor along with its ligand inhibits expression. In addition, removing this element from the promoter does not totally abolish the inhibitory effect of 1,25-dihydroxyvitamin D_3, indicating that other *cis*- and/or *trans*-acting factors are

TABLE 7-1   Collagen Types Found in Bone Matrix

Collagen	Location/function	Molecular structure
Type I: $[\alpha_1(I)_2\alpha(I)]$	Constitutes 90% of matrix in the bone matrix	67-nm banded fibrils
and $[\alpha_1(I)_2]$	Acts as scaffolding and binds to other proteins that initiate hydroxyapatite deposition	
Type III: $[\alpha_1(III)_3]$	Present only in trace amounts and can regulate collagen fiber stickiness	67-nm banded, coats type I fibrils
Type: $[\alpha_1(V)_2\alpha_2(V)]$ and $[\alpha_1(V)\alpha_2(V)]$	Their absence can result in collagen fibrils of large diameter	67-nm banded, coats type I fibrils in some tissues
Type X: $[\alpha_1(X)_3]$	Present in hypertrophic cartilage and can be involved in matrix organization via formation of the template for type I collagen	Probably fishnet-like lattice

involved [75]. Depending on the concentration and the stage of the cell maturation, dexamethasone can either increase or decrease collagen synthesis [82, 83].

In all connective tissues, the collagens serve mechanical functions, providing elasticity and strength for the component tissues [84]. The importance of type I collagen in bone is well demonstrated by various forms of osteogenesis imperfecta (OI; brittle bone disease) in human and animal models, in which bone fragility has been associated with alterations in the type I collagen genes [35, 54, 85–90]. For example, bone fragility and skeletal deformity have been detected in Mov-13 mice, in which a viral insertion within the first intron totally silences the α_1(I) gene [87]. A similar finding was also detected in a knock-in murine model of OI carrying a typical glycine substitution in type I collagen that reproduced a mutation in a type IV OI child [90]. Moreover, brittle bone attributed by reduced strength of bone matrix has been reported in a murine model of skeletal fragility (SAMP6), in which the matrix weakness is caused primarily by poor organization of collagen fibers and reduced collagen content compared to their age-matched controls [91].

The mineral crystals in the bones of patients and transgenic animals with OI tend to be smaller than those in age-matched control bones [92, 93]. In the OI mouse (oim) that lacks the α_2(I) chain [88], tendon [94] and bone [95] mineralization is aberrant. In the oim tendon, the crystals occasionally appear outside of the collagen matrix, a feature never noted when collagen production is normal [96]. Similarly, in the oim bones, the pattern of initial mineral deposition and crystal growth along the collagen differs from normal; the crystals appear both outside of the collagen matrix and within regions of collagen, which are less mineralized than those in the normal controls [93]. In addition, there are thinner fibrils in OI patients that may be insufficient to provide nucleation and scaffolding sites for mineral deposition and can potentially translate into fragile bones [35, 54, 97]. It is not known whether mineral seen away from the collagen fibrils was formed in the absence of a collagen backbone or whether it "broke away" and was later seen in the matrix because the collagen structure was not sufficient to support it. Collagen per se does not initiate mineral deposition; that is, it is not a mineral nucleator since it lacks the appropriate conformation that matches the ion surface of the deposited mineral surface [35, 54, 98]. Nonetheless, data from OI tissues clearly demonstrate the importance of collagen for providing a scaffold to organize the mineral.

As discussed later, other noncollagenous matrix proteins, which are "held" within the collagen matrix, appear to initiate and regulate the mineral deposition in bone [3, 5, 99, 100].

III. INTERMEDIATE CARTILAGE MATRIX

Endochondral bone formation is mediated by a cartilage template, and cartilage macromolecules can be in close proximity to forming bone and may actually be incorporated into the initial boney tissue [1]. The basic scaffolding on which cartilage matrix is built is type II collagen. In addition, a number of proteoglycans have been identified in cartilage matrix, primarily the large proteoglycans, such as aggrecan and versican, and small leucine-rich repeat proteoglycans, such as decorin and biglycan, which are also present in bone matrix [101–105]. Other proteins, including COMP, CD-RAP, chondroadherin, and matrilin-1, are present in cartilage matrix but at much lower levels than type II collagen and aggrecan [106, 107].

Proteoglycans are a class of macromolecules characterized by the covalent attachment of long chains of repeating disaccharides that are often sulfated, termed glycosaminoglycans (GAGs). Based on the sugar composition of the repeating disaccharides, GAGs are divided into subtypes such as chondroitin sulfate (CS), dermatan sulfate (DS), keratan sulfate (KS), heparan sulfate (HS), and hyaluronan (HA; unsulfated) (Figure 7-3). Aggrecan is one of the large CS molecules and has the ability to form aggregates with HA.

A. Large Proteoglycans

I. AGGRECAN

The human aggrecan gene is located on chromosome 15q26 [108]. However, the complete genomic sequence has been reported only in rat and is 63 kb in length, containing 18 exons encoding for structural domains of the molecule [109, 110]. The rat gene promoter lacks a TATA box, and the major transcription start site is located in close proximity to a number of SP1 sites. In addition, there are four AP2 sites located −120 kb upstream in a GC-rich region and two of the SP1 sites overlap [109, 110]. The resulting mRNA species of 8.2 and 8.9 kb predict a 19-residue signal peptide and 2015-residue mature protein, in which a stretch of 1164 residues contains Ser–Gly repeats, the CS attachment site [109, 110]. Intact aggrecan has a molecular weight of approximately 2.5 million Da, with a core protein ranging in apparent molecular weight between 180 and 370 kDa with slightly more than 100 GAG chains (mostly CS, but with some KS) of approximately 25 kDa.

Based on enzymatic cleavage and sequence homology, five domains have been defined in the core protein of aggrecan (Figure 7-4): three globular (G) domains [111], two of which bind to hyaluronic acid (G1 and G2) with

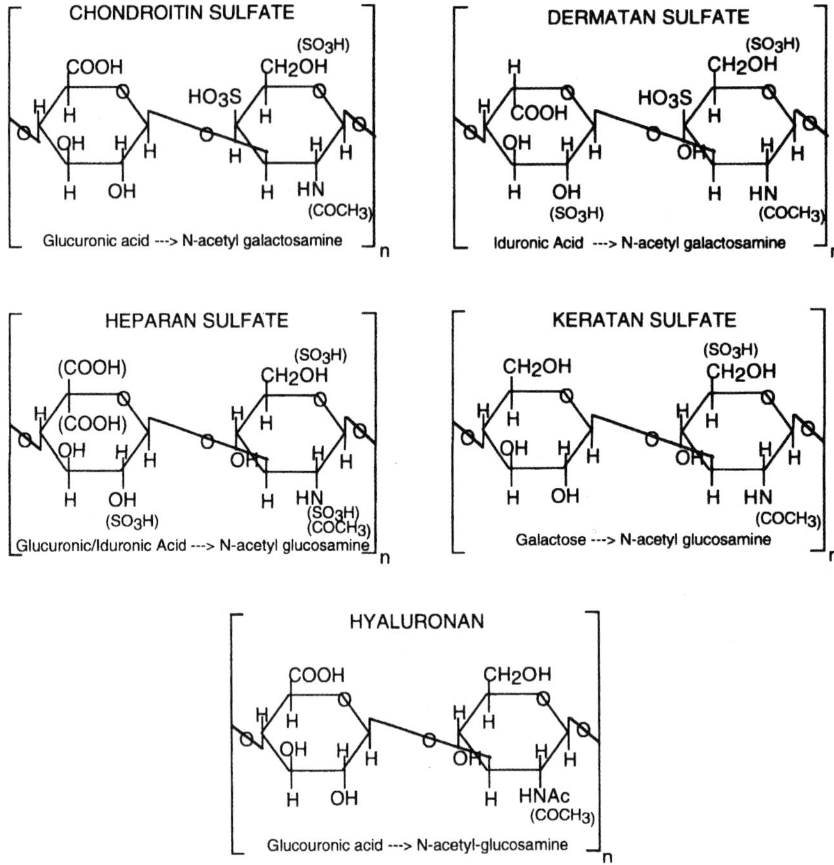

FIGURE 7-3   Disaccharide composition of glycosaminoglycans (GAGs). The GAG side chains that are covalently attached to proteoglycan core proteins are composed of repeating disaccharide units. The composition of the disaccharides, along with modifications by acetylation, results in the formation of chondroitin sulfate, which is epimerized to form dermatan sulfate, heparan sulfate, and keratan sulfate. Hyaluronan is the sole GAG that remains unsulfated and is not covalently linked to core proteins.

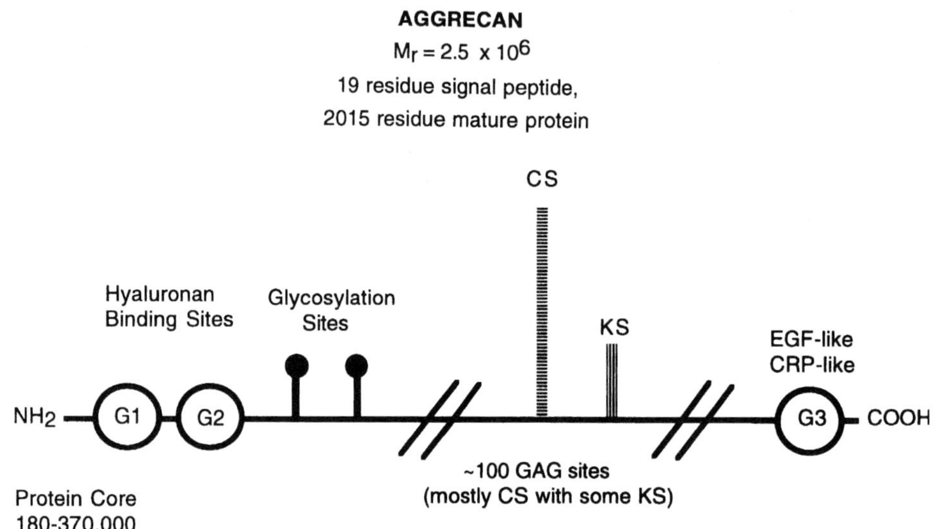

FIGURE 7-4   A representation of the chemical features of the large hyaluronic acid–binding proteoglycan, aggrecan. GAG, glycosaminoglycan; CS, chondroitin sulfate; KS, keratan sulfate; G1, G2, G3, globular domains (see text for description); EGF, epidermal growth factor; CRP, C-reactive protein.

the third one, G3, located at the C terminus; an interglobular domain; and a central domain rich in serine–glycine repeats to which the CS and KS-GAG chains are attached. The G1 domain in the N terminus is structurally homologous to "link protein" [112], a small glycoprotein that stabilizes the interaction between the proteoglycan and hyaluronic acid in cartilage, forming a unique gel-like moiety providing resistance to compression in joints [113]. The adjacent G2 domain provides a flexible hinge. The C-terminal G3 domain contains a set of EGF-like and complement regulatory protein (CRP)-like sequences [114, 115]. The individual GAG chains form extended flexible structures, whereas the serines in the central domain have β-D-xylose attachments with restricted orientation [116]. Electron microscopic analysis showed that the protein core of the aggrecan-like proteoglycans (CS/KS-containing) is fairly homologous in a wide variety of tissues, ranging from tadpole tails to human articular cartilage [117].

Mice with cartilage matrix deficiency (cmd), which is caused by a functional null mutation of aggrecan gene, are characterized by perinatal lethal dwarfism and craniofacial abnormalities, suggesting an important role of this proteoglycan in skeletal development [118]. Moreover, in addition to the hydrodynamic function, aiding in the retention of both water and cations and the exclusion of anions in cartilage [119], proteoglycans are also responsible for matrix maintenance and organization [115], in part through interactions with the GAG chain of type IX collagen that trims the type II collagen fibrils [120]. Furthermore, proteoglycans may also play a role in the regulation of cartilage calcification [121]. The large aggregating cartilage proteoglycans can inhibit hydroxyapatite formation and growth in solution [122–125], and they can also chelate calcium [124, 126] and serve as a source of calcium ions for mineralization if they are degraded into non-Ca^{2+}-binding fragments. Although there is debate as to whether this chelation is involved in the inhibition of mineralization, it is clear that proteoglycans and their component GAGs sterically block hydroxyapatite formation and growth [123].

The amount of aggrecan in bone is much lower than that in cartilage, and whether its presence in bone represents residual calcified cartilage is largely unknown. The presence of elevated amounts of CS proteoglycans in the bones of osteopetrotic animals with defective osteoclasts was linked to the inability of these animals to resorb calcified cartilage [127]. The functions of aggrecan in bone are also unknown. Because of its relatively low concentration, it seems less likely that it has a critical role in preventing osteoid mineralization, similar to its role in preventing cartilage calcification.

2. VERSICAN

Versican is another CS proteoglycan related to aggrecan but found at relatively lower levels in cartilage and bone. Versican has been so termed based on its high variety of forms in a large number of extracellular matrices ("versatility"). The protein core, with a molecular weight of approximately 360 kDa, has a structure similar to that of aggrecan with the exception that it lacks the G2 domain. In addition, versican contains only 12–15 CS side chains (~45 kDa) in contrast to approximately 100 in aggrecan [114].

The versican gene localizes to human chromosome 5q12–q14 with a length of more than 90 kb [128, 129]. The promoter region contains a TATA box, an XRE (xenobiotic responsive element), SP1 binding sites, a CRE (cyclic AMP responsive element), and a CCATT transcription factor binding site [130]. Based on differential splicing and polyadenylation, three mRNA species of 10, 9, and 8 kb are produced [130]. The sequence predicts a 20-residue signal peptide and a 2389-residue mature protein [131]. The gene is composed of 15 exons with a splice variant that utilizes an additional exon [132, 133] (Figure 7-5). The hyaluronic acid (HA) binding region (G1) is in exons 3–6. These exons share homology with the other HA-binding protein, the link protein. This region also contains an Ig-like protein conformation whose function is unknown. Exons 6 and 7 are differentially utilized and contain GAG attachment sites. The carboxy-terminal domain (G3), which contains homology to selectins, EGF, and CRP, is contained within exons 9–14.

A report of rat bone development found that versican was expressed during osteogenesis, where it was more abundant in woven than lamellar bone [134]. Fibroblast growth factor-2 has been found to upregulate versican gene expression in human chondrocytes [135], whereas it appears to be upregulated by transforming growth factor-β (TGF-β) in adult human bone cells and fetal bovine long bone cells (P. Gehron Robey, unpublished data). The function of versican in cartilage and bone is also largely unknown. Potentially, it may serve as a bridge between the extracellular environment and the cell by binding to HA via the amino-terminal binding region and to molecules that have yet to be identified on the cell via the carboxy-terminal domain [114]. In addition, versican stimulates chondrocyte proliferation [136]. The EGF like sequence (G3 domain) may serve to stimulate proliferation of osteoprogenitors because EGF has been reported to stimulate proliferation of osteoblastic cells *in vitro* [137]. There are no studies to date on the role of versican in mineralization of cartilage or bone.

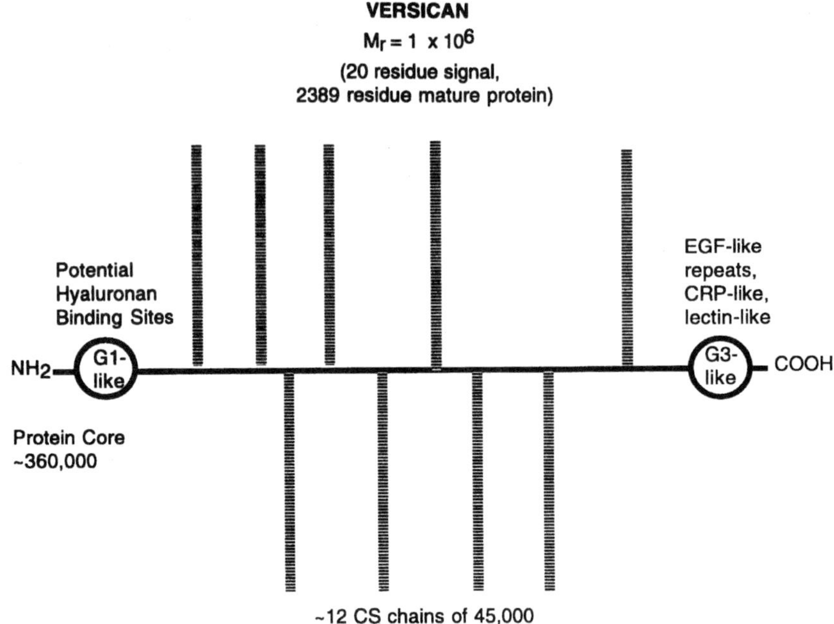

FIGURE 7-5   A representation of the chemical features of the widely distributed proteoglycan that is related to, but not identical to, aggrecan. CS, chondroitin sulfate; G1 and G3, globular domains (see text for description); EGF, epidermal growth factor; CRP, C-reactive protein.

B.   Small Leucine-Rich Repeat Proteoglycans

In addition to aggrecan and versican, another family of proteoglycans is represented by a group whose protein core is characterized by a smaller size and a leucine-rich repeat sequence (SLRP) that is approximately 20–30 amino acids in length [104, 138]. The SLRP family has been subdivided into three classes based on their similarity in gene and amino acid structures [138]. The class I members include decorin, biglycan, and asporin; class II includes fibromodulin, lumican, PRELP, keratocan, and osteoadherin; and the class III members are epiphycan/PG-Lb, mimecan/osteoglycin, and opticin. In cartilage and bone, there are several members of this SLRP family, predominantly including decorin, biglycan, fibromodulin, osteoadherin, and osteoglycin. Although SLRPs are highly homologous, they exhibit distinctly different patterns of expression and tissue localization, indicative of divergent functions within these tissues.

I.   DECORIN

Decorin, so named for its ability to bind to and "decorate" collagen fibrils, has also been called PG-II and PG-40 [117, 138, 139]. The human decorin gene is localized to chromosome 12q23 and is more than 38 kb in length, containing nine exons [140–142]. In mouse, the gene is located on chromosome 10, just proximal to the Steel gene locus, which encodes for stem cell factor, also named mast cell growth factor [143]. Decorin has a core protein of approximately 38 kDa, which includes 10 of the leucine-rich repeat sequences. Although there are three potential GAG attachment sites, only one is utilized for the attachment of a single GAG chain of approximately 40 kDa, resulting in a molecule with an apparent molecular weight of approximately 130 kDa as determined by sodium dodecyl sulfate–polyacrylamide gel electrophoresis (SDS–PAGE) [144] (Figure 7-6). The decorin gene shares 55% homology with, and is organized in a similar fashion to, the biglycan gene (described later) except that the intronic sequences are much longer (two of which are 5.4 kb and >13.2 kb) (141). Gene transcription results in a major mRNA species of 1.6 kb and a minor species of 1.9 kb [145, 146]. The sequence predicts a 359-residue protein that includes a 30-residue prepropeptide.

The synthesis of decorin is downregulated by TGF-β1 and BMP2 in rodent osteoblastic cultures [147, 148]. However, its expression is upregulated by dexamethasone [149] and phytoestrogen ipriflavone metabolite III [150]. Mechanical loading also stimulates the synthesis of decorin [151]. Although it appears that the propeptide is cleaved from the mature decorin in bone, evidence indicates that it is maintained in other tissues such as cartilage [152].

Decorin has been shown to bind to and regulate the fibrillogenesis of type I, II, and VI collagens [153, 154]. In bone, the proposed functions of decorin are the regulation of collagen fibril diameter and fibril orientation, and possibly the prevention of premature

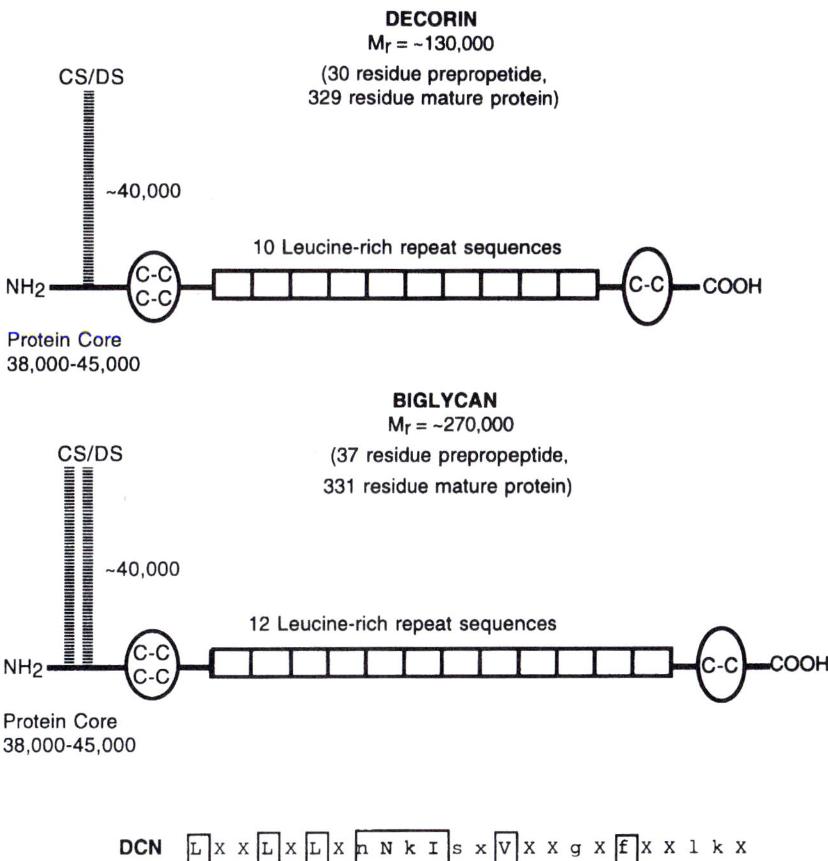

FIGURE 7-6   The two most abundant proteoglycans present in bone matrix are the small chondroitin sulfate/dermatan sulfate proteoglycans, decorin and biglycan. The core protein of each is highly homologous to a number of proteins due to the presence of a leucine-rich repeat sequence. CS, chondroitin sulfate; DS, dermatan sulfate; C–C, disulfide bonding.

osteoid calcification. It is interesting to note that targeted disruption of the decorin gene results primarily in skin laxity and fragility in mice, whereas disruption of the biglycan gene results in reduced skeletal growth and bone mass leading to generalized osteopenia [155]. Moreover, the decorin and biglycan double knock-out mice have additive deficiency in dermis and synergistic effects in bone, and ultrastructural analysis of these mice reveals a complete loss of the basic fibril geometry with the emergence of marked "serrated fibril" morphology [155]. In addition, decorin expression is reduced in certain skin diseases characterized by excessive keratinization [156], stressing the physiologic importance of decorin in regulating fibril formation and collagen–matrix interactions. It is also of interest to note that there is a decreased expression of decorin in some patients with OI [157–159], in which abnormal mineral deposition has been detected outside the collagen matrix.

In cartilage, decorin is present in very low levels and is restricted to the interterritorial matrix [160]. As bone is formed, it is produced by preosteoblasts and osteo-blasts, but its synthesis is not maintained by osteocytes [160], indicating a role of decorin in the regulation of initial mineral deposition. It is not clear if decorin within the tissue is actually inhibitory to matrix mineralization because decorin has a low affinity to hydroxyapatite in contrast to a high affinity to type I collagen in solution [161, 162]. However, studies indicate a role of decorin in matrix mineralization since proteoglycans with low molecular weight are present in the d and e bands of type I collagen fibrils but then disappear when mineralization occurs [57, 138]. The bones of decorin knock-out mice were reported to have no visible bone phenotype, but their teeth do show alterations in matrix properties, and dentin in these mice was found to be hypomineralized [155, 163].

2. BIGLYCAN

Biglycan, also known as PG-I and PG-S, is another small proteoglycan present in both cartilage and bone [164, 165]. Biglycan is highly homologous to decorin. The gene for biglycan is 7 kb in size, containing eight exons, and is localized to Xq27-ter in humans, the only known matrix protein that is not on an autosomal chromosome [166]. The promoter does not contain a TATA box but has a number of *cis*-acting elements, including SP1, AP1, AP2, NF1, and NF-κb binding sites [166, 167]. The gene encodes for a 368-residue proform that is processed to become a mature core with 331 residues [164, 166] with a molecular weight of approximately 37 kDa to which (in most forms) two GAG side chains are attached. The amino-terminal domain contains the GAG attachment sites, followed by 12 of the leucine-rich repeat sequences (Figure 7-6). The first and last repeats contain a characteristic pattern of cysteinyl residues that result in a particular pattern of intramolecular disulfide bonding [164, 168]. The carboxy domain has a sequence that is unique to biglycan and differs from decorin and other leucine-rich repeat sequence–containing proteins.

Biglycan synthesis is regulated in a manner distinct from that of decorin. TGF-β1 and BMP2, which are known to decrease decorin expression, have been reported to either upregulate or have no effects on biglycan expression in rodent osteoblastic cells [147, 148, 169]. TGF-β1 also increases biglycan expression in MC3T3 cells, and IGF-1 and -2 increase its expression in other cell lines [170]. Retinoic acid suppresses biglycan in chondrocytes [171], and dexamethasone and 1,25-dihydroxyvitamin D_3 have been reported to decrease its expression in human bone and marrow cultures [172, 173]. Fluoride, at clinically relevant concentrations, also decreases GAG chain length and composition of biglycan in rat osteoblastic cells [174].

The functions of biglycan in cartilage and bone mineralization remain to be determined. In solution, biglycan at low concentrations can promote apatite formation, whereas at higher concentrations it inhibits the growth and proliferation of mineral crystals [162]. These effects appear to be due to the highly specific high-affinity binding of biglycan for apatite (K_D = 294 μg/μmol). Compared to the decorin knock-out mice, the biglycan knock-out mice have similar structural abnormalities in collagen fibrils but with more serious deficiency in bone than in dermis [155]. In addition, the biglycan knock-out mice have shorter femora, decreased bone density, and failure in achieving peak bone mass compared to controls [175]. The mineral within these bones has increased crystal size relative to wild-type controls [176], also indicating an inhibitory role of this protein. However, the low amount of biglycan present in bone matrix relative to other mineral nucleators and its absence from bone collagen fibrils suggest that its primary function may not be directly related to mineral deposition of bone.

3. Fibromodulin

Another SLRP proteoglycan, fibromodulin, is found predominantly in articular cartilage but also exists in bone [177, 178]. The human gene encoding for fibromodulin is located at chromosome 1q32, is at least 8.5 kb in length, and is only partially characterized [179]. It has an intron–exon organization that differs markedly from that of decorin and biglycan. The intact protein is approximately 59 kDa, and the core protein shares a high homology with decorin and biglycan but bears KS-GAG chains linked to asparaginyl residues rather than CS or DS linked to serinyl/threoninyl residues.

Decorin and fibromodulin are the most active collagen-binding proteins in cartilage and bone, binding to completely different regions on collagen fibrils [180]. Fibromodulin interacts with triple-helical types I and II collagen [180]. In cartilage, the amount of fibromodulin correlates with the size of collagen fibrils [181]. In developing bone induced by demineralized bone matrix, fibromodulin is heavily localized to fibrillar bundles [182]. Observations from the fibromodulin knock-out mouse have indicated that in the absence of functional fibromodulin, collagen fibrils in tail tendon and predentin are abnormal [183–185]. Although no bone phenotype has been reported [183], impaired dentin mineralization and enamel formation have been detected in these mice [185], suggesting a role of fibromodulin in collagen fibrillogenesis and mineralization.

4. Osteoadherin and Osteoglycin

Osteoadherin has been isolated as a minor, leucine- and aspartic acid–rich keratin sulfate proteoglycan found in the mineralized matrix of bone [186, 187] and dentin [188]. It was originally identified from bovine bone and the osteoadherin content of bone extracts has been shown to be 0.4 mg/g tissue wet weight, whereas none was found in extracts of various other bovine tissues [186, 187].

The entire primary sequence has been determined by nucleotide sequencing of a cDNA clone, 4.5 kb in length, from a primary bovine osteoblast expression library [187]. The gene contains four putative sites for tyrosine sulfation, three of which are at the N terminal end of the molecule. The molecular weight of the protein is 49,116 Da, with a calculated isoelectric point for the mature protein of 5.2. The dominating feature is a central region consisting of 11 B-type, leucine-rich repeats ranging in length from 20 to 30 residues. There are six potential sites for N-linked glycosylation. The distribution of cysteine residues resembles that of other leucine-rich repeat proteins except for two centrally located cysteines. Unique to osteoadherin is the presence of a large and very acidic C-terminal domain.

Osteoadherin is synthesized by bovine primary osteoblasts and is exclusively identified in the primary spongiosa by immunohistochemical studies of the bovine fetal rib growth plate, suggesting a role of this protein in bone development [186]. This gene has also been identified in human odontoblastic cells [188]. The detailed functions of osteoadherin remain to be investigated. The primary function of this protein is to bind cells since it has been shown to be as efficient as fibronectin in promoting osteoblast attachment *in vitro* via integrin, $\alpha_V\beta_3$ [186]. In addition, osteoadherin binds well to hydroxyapatite [186], indicating a potential role of this protein in mineralization of bone.

Another leucine-rich repeat protein, osteoglycin, isolated from demineralized bone, was originally named osteoinductive factor [189]. However, it was determined later that copurifying BMPs were the source of its growth stimulatory activity in this preparation, and thus the protein and its gene were renamed osteoglycin [190]. This 12-kDa proteoglycan is a proteolytic product of mimecan, and Western and Northern blotting show that this protein is common in connective tissues but most abundant in eye tissue [190]. Since immunoreactive material was not abundant in extracts of bone [190], its expression is not as bone specific compared to osteoadherin. The functions of osteoglycin are largely unknown.

IV. BONE-ENRICHED MATRIX PROTEINS

In bone, the remaining matrix proteins are mainly composed of two major types: glycoproteins and γ-carboxyglutamic acid (Gla)-containing proteins. The most relevant and abundant glycoproteins are represented by alkaline phosphatase, osteonectin, and the cell attachment proteins, which include, but are not limited to, sialoproteins. Of the Gla-containing proteins, osteocalcin is the major representative. These bone matrix proteins have divergent biochemical properties and play particular roles in the regulation of matrix mineralization.

A. Glycoproteins

This class of proteins is characterized by the covalent linkage of sugar moieties attached via asparaginyl or serinyl residues. Collagen also contains another form of glycosylation (galactosyl and glucosyl-galactosyl-hydroxylysine), which is virtually specific to collagen. These glycoproteins may also be further modified by post-translational sulfation and phosphorylation.

I. ALKALINE PHOSPHATASE

Although the enzymatic activity of alkaline phosphatase is shared by many types of tissues, there is no doubt that induction of alkaline phosphatase activity in

uncommitted progenitors marks the entry of a cell into the osteoblastic lineage and is a hallmark in bone formation. Although alkaline phosphatase is not typically thought of as a matrix protein, studies indicate alkaline phosphatase can be shed from the cell surface of osteogenic cells or in a membrane-bound form (matrix vesicles) [4, 7, 8, 191, 192].

The human gene for alkaline phosphatase is located on chromosome 1 with a length of ~50 kb [193–195]. It contains 12 exons and has a restriction fragment length polymorphism (RFLP) [195, 196]. The rat gene is at least 49 kb with 13 exons and has a similar gene organization [197, 198]. The gene predicts a protein with 524 amino acids that includes a 17–amino acid signal peptide. The enzyme exists as a dimer and the identical monomers have a molecular weight of 50–85 kDa, depending on animal species and degree of post-translational modification, since there are five potential glycosylation sites. Each monomer consists of a central 10-stranded β-sheet surrounded by 15 α-helices of various lengths [195, 198–200]. The active sites are at the carboxyl end of the central β-sheet, and their binding to two zinc and one magnesium ions is thought to be responsible for the dephosphorylation reactions [195, 198, 201]. The C-terminal region is hydrophobic, as would be expected for a protein that is linked to the cell membrane. The glycosylated enzyme is attached to the cytoplasmic membrane on the external surface through a phosphatidyl-inositol-glycan group, which can be cleaved by phospholipase C, thereby releasing it from the cell surface [4, 7, 8, 202–204].

The regulation of the bone/liver/kidney alkaline phosphatase isozyme is controlled by two leader exons, 1A and 1B, with alternative promoters separated by 25 kb [205–207]. The upstream promoter is used preferentially by bone cells and facilitates the high-level expression of alkaline phosphatase in this cell type [206–208]. The downstream promoter is constitutively active, produces low levels of activity, and is used in the kidney [206–208]. Three mRNA species of 2.5, 4.1, and 4.7 kb are produced as the result of differential splicing [206, 207, 209].

The list of factors that regulate alkaline phosphatase in bone cell cultures is quite lengthy and the results are extremely variable. In human and rat osteoblastic cell cultures, 1,25-hydroxyvitamin D_3 upregulates alkaline phosphatase activity [210–212]. In rat and murine osteoblastic cells, alkaline phosphatase activity is upregulated by retinoic acid [213–215]. Dexamethasone, along with ascorbic acid and β-glycerophosphate, promotes alkaline phosphatase activity in human and rat osteoblastic cells [216–218]. BMPs also enhance the expression of alkaline phosphatase in rat and mice osteoblastic cells [219, 220], whereas no significant effects were detected on human bone and marrow cultures [221, 222]. In addition, IL-4 [72] and calcitonin [223] also increase alkaline phosphatase activity in osteosarcoma cells, and IGF-1 [224] is known to increase its activity in deer antler cells. IL-10 [80] and lead [81] have been found to decrease alkaline phosphatase activity.

Histological localization of alkaline phosphatase in developing human subperiosteal bone (Figure 7-7) marked its very specific expression by preosteoblasts and osteoblasts in areas that are destined to become new bone, whereas less expression was found in mineralized matrix [225–228], suggesting this enzyme as a marker for osteoblastic cells at less mature stages. Developmental studies *in vivo* and *in vitro* have also shown that the expression of alkaline phosphatase precedes mineralization and is maintained during early stages of hydroxyapatite deposition [229–231],

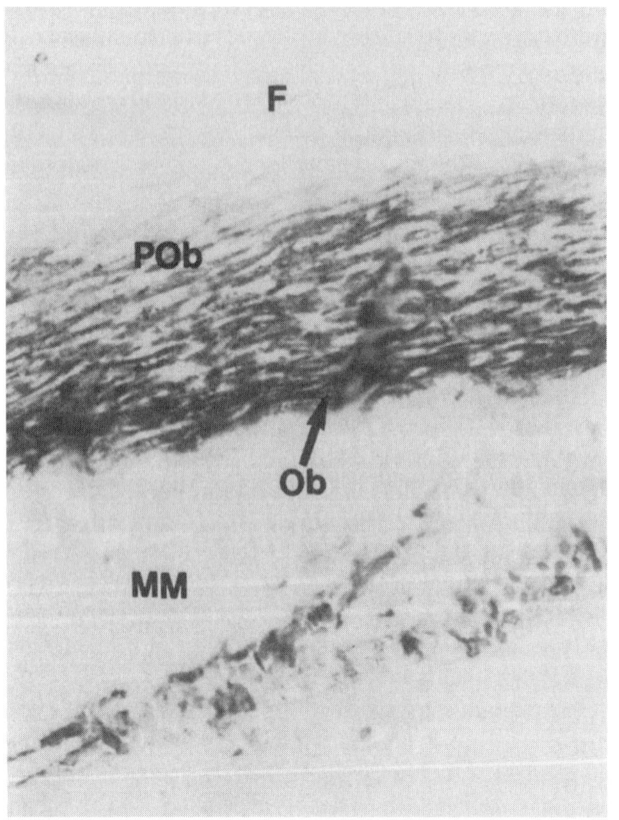

FIGURE 7-7   Alkaline phosphatase in developing bone. By histochemical staining for alkaline phosphatase activity during development, areas that are destined to become bone, as shown here in developing human subperiosteal bone, can be clearly illustrated. The fibrous layer (F) of the periosteum is negative, whereas preosteoblasts (POb) and osteoblasts (Ob) produce high levels of activity. Although a glycoprotein with alkaline phosphatase activity has been isolated from the bone matrix, it is not easily detected in mineralized matrix (MM) by this histochemical assay. Courtesy of Dr. Paolo Bianco.

suggesting a role for this enzyme in mineral deposition. Although the specific mechanisms are largely unknown, its abundance in matrix vesicles is believed to be essential for matrix vesicle–mediated mineralization (*vide infra*) [4, 7, 8]. The crucial role of alkaline phosphatase in mineralization has also been confirmed by the discovery of mutations in this gene in hypophosphatasia, a disease characterized by improper mineral deposition [232, 233], and by the observation that cells that do not normally mineralize will form a mineralized matrix when transfected with the alkaline phosphatase gene [4, 234]. Mice with null mutations for the tissue-nonspecific alkaline phosphatase also provide evidence of the importance of alkaline phosphatase for mineralization [235–237] and show increased osteoid and defective growth plate development. Other functions of alkaline phosphatase associated with mineralization may include its hydrolyzing activity on phosphate esters to provide a source of inorganic phosphate [4, 7, 8, 238] and its activity as a potential phosphate transferase in bone [239].

2. OSTEONECTIN

With the development of novel techniques for the extraction of bone matrix proteins in a nondegraded form [10, 14–20], one of the first noncollagenous bone matrix proteins to be isolated and characterized was osteonectin [240, 241]. Osteonectin, which is also named SPARC (secreted phosphoprotein acidic and rich in cysteine) or BM-40 (basement membrane tumor factor 40), is expressed in a number of tissues during development and by many cell types. In bone, osteonectin can constitute up to 15% of the noncollagenous protein depending on the developmental age and the animal species [241, 242].

There is a single gene (>20,000 kb) encoding for osteonectin located on human chromosome at 5q31–q33 [243] and with one RFLP in the 5′ region [244]. This gene contains 10 exons and the coding sequence predicts a 17-residue signal peptide and a 286-residue mature protein. Domains defined in osteonectin are the EF hand domain (high-affinity Ca^{2+}-binding structure) in the C terminus, a disulfide-rich domain in a cysteine-rich region with homology to an ovomucoid-like (serine protease inhibitor) sequence, and a pentapeptide KKGHK domain [244–249] (Figure 7-8). The promoter does not contain a TATA box or CCAAT sequences but contains a purine-rich region with GA repeats between −55 and −126 [249–253].

Osteonectin has an apparent molecular weight of approximately 35 kDa without reduction of disulfide bonds and appears to increase in size up to approximately 40–46 kDa following reduction, indicative of intrachain disulfide bonds (Figure 7-8). Due to the nature of the amino acid composition and of the post-translational

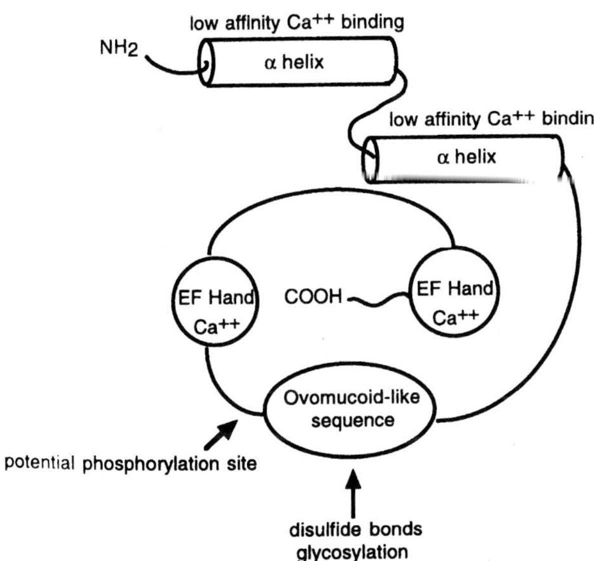

FIGURE 7-8   The chemical characteristics of osteonectin indicate the presence of two α-helical regions at the amino terminus, along with an ovomucoid-like sequence with extensive disulfide bonding and two EF hand structures.

modifications, osteonectin is acidic with a pI of ~5 [250, 254, 255]. Osteonectin may be differentially glycosylated and/or phosphorylated [250, 256] because there are at least two potential N-glycosylation sites that bear diantennary oligosaccharides (an intermediate between high mannose and complex type oligosaccharides that contains variable amounts of sialic acid and fucose) [250, 257].

Factors that regulate the biosynthesis of osteonectin in bone cultures are not well understood. In bovine bone cell cultures that exhibit extensive mineralization, osteonectin appeared at early stages and remained high thereafter [258]. The effect of TGF-β is variable, and a stimulation [259] as well as a lack of effect [260, 261] has been reported. The expression of osteonectin by normal human bone cells is not altered dramatically by any treatment [262], although very modest increases with dexamethasone, retinoic acid, IGF-I, and dibutyryl cAMP have been reported in other systems [249, 250, 263–265].

Osteonectin and its metalloprotease cleavage products bind to type I collagen [240, 266], types III and V collagens [267, 268], and thrombospondin, a known matrix organizer [269], suggesting a role for this protein in determining the organization of the osteoid in bone. Moreover, young osteonectin knock-out mice developed cataracts [270]. More recent studies of older

mice indicate that the mice develop osteopenia with a significant loss of trabecular bone associated with a decreased rate in bone formation [271, 272], also suggesting a role of osteonectin in bone development.

The initial investigations of osteonectin's function demonstrated that when associated with denatured collagen, osteonectin bound calcium and phosphate ions, suggesting that it was promoting mineral deposition [240]. NMR evaluations also showed the presence of a typical EF hand [273] in osteonectin protein structure, which in other systems is involved in calcium chelation and calcium transport (Figure 7-9). However, the tissue distribution of osteonectin within bone suggests that it is not involved in the initiation of mineralization [273, 274]. Expressed by cells in both soft and hard connective tissues, osteonectin accumulates only within mineralized matrix. Whether it has a specific function in further regulating growth and proliferation of mineral crystals or simply accumulates within the mineralized tissue because of its affinity for hydroxyapatite (K_D = 8×10^{-8}, ~11.3 mg osteonectin/g apatite [268]) remains to be determined.

3. TETRANECTIN

Another glycoprotein expressed by osteoblastic cultures undergoing matrix mineralization and immunolocalized in developing woven bone is tetranectin [275]. The gene is 12 kb in length and contains three exons [276]. It has sequence homology with asialoprotein receptor and the G3 domain of aggrecan and versican core proteins (described previously) [277]. The cDNA predicts for a 21-residue signal peptide and a 181-residue mature protein. Tetranectin is a tetrameric protein with a molecular weight of ~21 kDa (subunits with a molecular weight of ~5.8 kDa) that was first isolated from serum and found to bind to the kringle 4 domain of plasminogen [278]. Overexpression of tetranectin by tumor cells caused an increase in matrix mineralization upon implantation into nude mice [275], suggesting a role for tetranectin in mineral deposition. The loss of tetranectin has been correlated to retinoic acid inhibition on mineralization of human osteoblastic cells [279], further identifying a role of this protein in matrix mineralization.

4. RGD-CONTAINING GLYCOPROTEINS

In bone matrix, there are a number of glycoproteins that also have the amino acid sequence Arg-Gly-Asp (RGD). These RGD sequences can be recognized by cell surface receptors as a "cell attachment sequence," which bridges the attachment between extracellular matrix to cells and thus arranges the cells in matrix [280]. Most of these cell surface receptors are integrins formed by one α subunit and one β subunit, each of

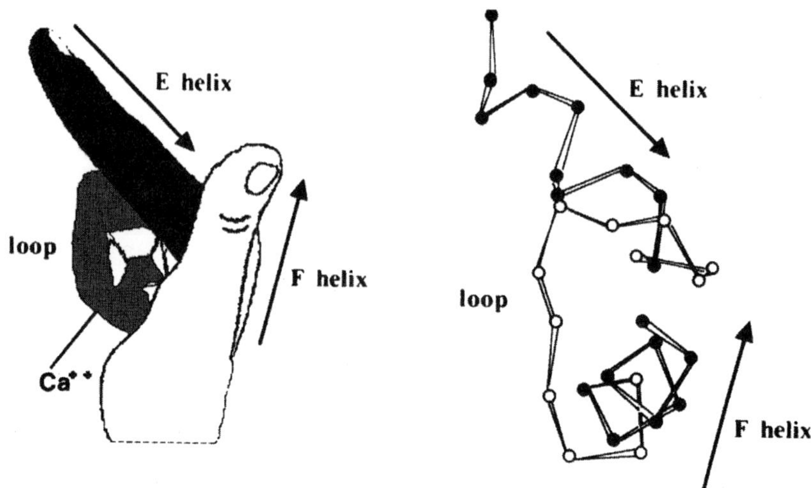

FIGURE 7-9 Structure of an EF hand high-affinity Ca^{2+}-binding site. Depiction of the theoretical structure and the amino acid sequence for the EF hand, which has an extremely high affinity for ionized calcium. Courtesy of Dr. Neal S. Fedarko.

which has a cytoplasmic extension that associates with intracellular signaling pathways, a transmembrane domain, and an extracellular domain [281, 282]. The extracellular domains of the α and β subunits configure a binding pocket that recognizes the RGD sequences in the extracellular matrix proteins and thus mediates the cell–matrix interactions [280–282]. These RGD-containing proteins include collagen (described previously), thrombospondin, fibronectin, vitronectin, and a family of small integrin-binding ligand, N-linked glycoproteins (SIBLINGs) expressed in bones and teeth. The SIBLINGs have been identified by a cluster of genes including osteopontin (OPN) and bone sialo-protein (BSP), dentin matrix protein-1 (DMP-1), dentin sialophosphoprotein (DSPP), matrix extracellular phosphoglycoprotein (MEPE), and enamelin [283].

a. Thrombospondin(s)

These complex modular glycoproteins are relatively less abundant in mineralized matrix of bone relative to other glycoproteins, and they have been found in a large variety of connective tissues, particularly in areas of demarcation [284]. It is now known that in humans there are at least five genes encoding for thrombospond-ins (TSPs). These genes are located on chromosomes 1 (TSP-3), 5 (TSP-4), 6 (TSP-2), 15 (TSP-1), and 19 (TSP-5, which is also known as the cartilage molecule, COMP) [285–290], all with a length of at least 16 kb. Although the coding sequences are all highly homologous and differ only in the number of times that the type I (properdin-like), II (EGF-like), and III (calmodulin-like) sequences are repeated, they utilize distinct promoters [291]. A promoter from the TSP-1 gene has been isolated and characterized [292, 293]. It contains a TATA box and an Egr1 site that is flanked by overlapping GC boxes, followed by a GC-rich region. Binding sites for NFY, AP2, SP1, and an SRE have also been identified. Based on the inhibition of TSP-1 transcription by c-*jun*, an AP1 site may also be present [294]. The resulting mRNA is 6.1 kb [295]. The organization of the TSP-2 and TSP-3 promoters is similar [296–298]. The entire pattern of expression of the different thrombospondin genes is not complete [299], although it is known that TSP-1, TSP-2, and TSP-3 are all expressed in bone [284, 300, 301].

Thrombospondin is a highly complex molecule with a molecular weight of approximately 450 kDa [302] (Figure 7-10), composed of three identical subunits ranging from 150 to 180 kDa that are held together by disulfide bonds. Each monomer has a number of intramolecular disulfide bonds that give rise to a molecule with a roughly dumbbell shape with distinct functional domains. The small amino-terminal globular domain contains a fibrinogen-like sequence along with a region that may have cell binding activities [303, 304] and heparin and platelet binding sites. In addition to homologies to the propeptide of the α(1)I chains of types I and III collagen, von Willebrand factor, and the circumsporozoite protein from *Plasmodium falciparum*, this small globular domain is attached to an extended stalk region that contains three type I and three type II repeat sequences. There is a cluster of cysteine residues in the stalk region that participate in the cross-linking of the monomers and binding sites for types I and V collagens, thrombin, fibrinogen, laminin, plasminogen, and plasminogen activator, indicating a role of thrombospondin in organizing matrix proteins. A large disulfide bonded domain makes up the carboxy-terminal region of the molecule and contains sequence homologies to parvalbumin and fibrinogen, with seven type III repeat sequences, although this sequence does not take on the EF hand structure [305, 306]. This region binds to the histidine-rich glycoprotein of serum, activates platelet aggregation, and has

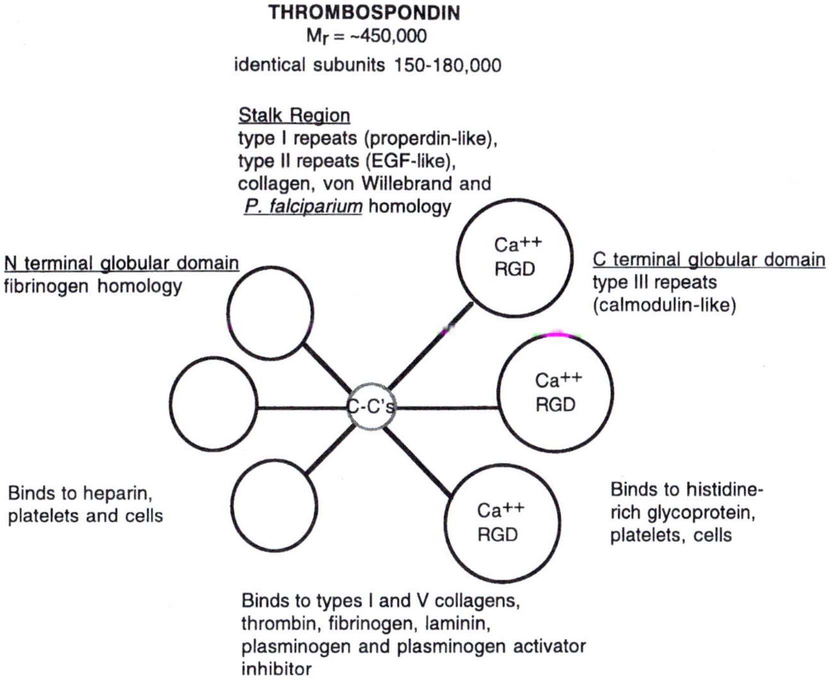

FIGURE 7-10   Thrombospondin is a disulfide-linked trimer that has globular domains at the amino and carboxy terminus interconnected by a stalk region. Each of these domains has a number of binding sites for other proteins, suggesting numerous potential functions in cell–matrix interactions. The cell attachment consensus sequence, RGD, is in the carboxy-terminal domain; however, its availability depends on the calcium ion concentration, which is known to affect the conformation of this region.

multiple Ca^{2+} binding sites. Ca^{2+} binding participates in the conformation of the globular domain. The RGD sequence is also within the Ca^{2+} binding region; however, it is not clear whether under normal physiological conditions the RGD is actually active in mediating cell attachment.

Thrombospondin synthesis has been demonstrated in several cell culture systems, including adult human bone cells [307, 308], rat marrow stromal cells [309], and osteoblastic cells [310, 311]. Its synthesis appears to be inhibited by dexamethasone [311] but increased by TGF-β [307].

Although the precise functions of the thrombospondins in bone are not known, they have been postulated to play a role in bone development and remodeling [312]. Immunohistochemical localization indicated low levels of expression in the periosteum, with primary localization in developing osteoid by osteoblastic cells [307]. There is moderate accumulation of thrombospondin in mineralized matrix [313], and by Western blotting the protein can also be detected in bone matrix extracts [307]. Mice that lack thrombospondin (TSP-2 null) have disordered collagen in their soft tissues (which exhibit fragility), increased cortical bone thickness and density [314–316], and altered fibroblast cell attachment [317]. Bone mineral properties have not been determined in these mice. However, the properties of these mutant animals confirm the importance

of thrombospondin in bone development and collagen fibrillogenesis and possibly in matrix organization [314–316]. It has also been shown *in vitro* that thrombospondins bind to decorin [318], known to regulate collagen fibrillogenesis and to interfere with cell attachment to fibronectin [319], and thrombospondins may bind to growth factors such as TGF-β that later serve as cell signals [309]. Although thrombospondins may be active in the attachment of osteoblastic cells to the $\alpha_v\beta_3$ receptor, which binds to other molecules such as vitronectin, in an RGD-dependent manner [284, 307], thrombospondins do not mediate osteoclast cell attachment as do the other RGD proteins [284, 320, 321].

b.   Fibronectin

Fibronectin is one of the most abundant extracellular matrix proteins in bone and is also a major constituent of serum. It is produced by virtually all connective tissue cells at some stage of development and accumulates in extracellular matrices throughout the body [284, 322].

The chicken gene for fibronectin is 50 kb [323]. In the human gene, six RFLPs have been identified [324], and the gene is located on chromosome 7 and is very complex, with up to 50 exons [325]. The functional domains, composed of type I, II, and III repeat sequences, are each coded for by an exon. The gene is transcribed to form mRNA of ~7.5 kb, but as might be anticipated, there is a great deal of heterogeneity based on differential splicing and up to 20 different mRNA species have been identified [326, 327]. Within the mRNA sequences, there are three major regions, EIIIA, EIIIB, and V, which can be inserted or deleted depending on the tissue. An example is seen in the differences between plasma (void of E but containing V regions) and tissue fibronectins (which contain various combinations of E's), which are the result of exon skipping. Differences between fibronectin produced by different cell types have also been found to be the result of exon subdivision (splicing within an exon). Factors that regulate differential splicing are not well known, nor is the nature of the splice variant produced by bone cells.

The human gene promoter has been identified and it contains TATA and CCAAT boxes, is GC rich, and has an SP1 and a CRE binding site [328, 329]. Promoter analysis indicates that the CCAAT and the CRE located between −164 and −90 are essential for gene activity. However, gel shift analysis indicates that there may be different complexes of proteins that bind to this region depending on the tissue source [330, 331].

Fibronectin is a dimeric protein with a molecular weight of ~400 kDa composed of two subunits of ~250 kDa that are highly homologous but variable depending on the cell source, held together by two disulfide bonds near the carboxy termini (Figure 7-11). Each of the subunits has multiple domains that bind to fibrin, heparin, certain bacteria, gelatin and collagen, DNA, cell surfaces via its RGD site, and another heparin binding site, indicating an important role for this protein in matrix organization. The overall structure of protein consists of 35% antiparallel β-sheets and no α-helices. There are three major types of domains: two unique β-sheets containing type I moieties at the N terminus; 12 type II domains, each with a hydrophobic pocket; and 17–19 type III domains spreading to the C terminus [332, 333]. The RGD sequence, located in a type III domain approximately one-third of the way from the C terminus, is thought to modulate the interactions with cells [334–336], and the C terminus appears to be needed to stimulate fibronectin's own synthesis [337]. The N-terminal domain seems to be required for extracellular matrix deposition [338], and another region is required for binding to chondroitin sulfate [339].

There is not much information on the nature of factors that regulate the synthesis of fibronectin in bone cells. In human and rat bone cell cultures, TGF-β and PTH are known to increase fibronectin synthesis [261]. Estrogen caused a decrease in fibronectin expression of PTH-stimulated levels but had no effect on TGF-β stimulated levels [340]. Gallium nitrate, under investigation as a therapeutic compound for increasing bone mass, also stimulates fibronectin synthesis in rat calvarial cells and ROS 17/2.8 osteosarcoma cells [73].

Fibronectin appears to be important in bone development. Osteoblasts and osteocytes stain intensely for fibronectin, and it is also accumulated in mineralized matrix [313] at an early stage of bone formation during development [313] or during induction by demineral-

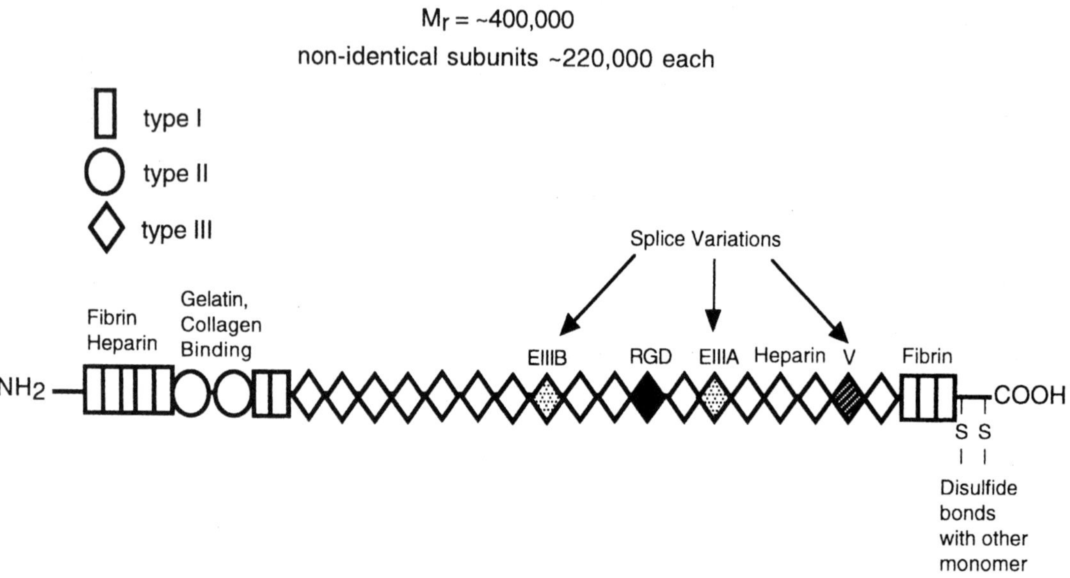

FIGURE 7-11   Fibronectin is composed of nonidentical subunits that are disulfide bonded at their carboxy termini. The molecule is composed of a series of repeating units (types I, II, and III) that give rise to domains with affinities for other proteins. There are several known splice variants (with or without EIIIB, EIIIA, and V; see text for description). The splice variant present in bone is not known. The cell attachment consensus sequence in a type III unit is RGD; however, other sequences that participate in cell attachment have been identified.

ized bone matrix [341, 342]. Western blotting analysis of bone extracts also indicates that it is relatively abundant. Possibly due to its wide expression in connective tissues, the elimination of the fibronectin gene in transgenic animals (and all its variants) is lethal *in utero*; connective tissues do not form, indicating that fibronectin is a component that is essential for development of these tissues [343].

Although the gene and protein properties of fibronectin indicate a role in matrix deposition and organization by interacting with a number of matrix proteins, its actual function is not clear. Fibronectin has been shown to support apatite formation in solution [344]. High-resolution electron microscopy studies have also demonstrated that fibronectin can play a role in early biological crystal nucleation, which may be of significance in ectopic calcification, primary nucleation in calcified tissue and bone in growth on ceramic implants [345]. Like thrombospondin, fibronectin also mediates cell attachments via either RGD-dependent or -independent pathways [284, 313, 322, 346].

c. Vitronectin

Vitronectin, also termed the S-protein of the complement system, is produced predominantly by the liver. It is found in serum at concentrations of 200–400 µg/mL and in bone matrices at low levels [322, 347]. Although it also appears in basement membranes, it is generally found in most matrices containing the fibrillar collagens.

The human gene encoding for vitronectin is located on chromosome 17q [348]. The protein has a molecular weight of ~70 kDa, and the primary structure of human vitronectin was predicted from cDNA analysis by Oldberg *et al.* [349] and Jenne and Stanley [350]. Several homologous domains in the mammalian vitronectin sequences obtained from different sources have been defined [351]. From the amino to the carboxy terminus there is a "somatomedin B" domain which is rich in cysteines, followed by an RGD cell attachment site, a collagen-binding domain, a cross-linking site for transglutaminase, a plasminogen binding site, a heparin binding site, a PAI binding site, and an endogenous cleavage site. Sites for sulfation and cAMP-dependent phosphorylation are also present.

In vitro, vitronectin may be a biosynthetic product of osteoblastic cells [352]. Vitronectin is very active in mediating attachment of all cell types. Bone cells, including osteoclasts, attach very strongly to vitronectin [313, 322, 353, 354], mainly via the receptor integrin, $\alpha_v\beta_3$ [313, 353]. Vitronectin is detectable in developing bone by immunohistochemistry and is found in a very limited number of cells lying on the surface of newly formed bone [352]. However, it is not clear that these cells are in fact osteoblasts. Mice deficient in the vitronectin gene have been shown to have a thrombolytic

phenotype, but there is no report on whether skeletal defects were apparent in these mice [355].

Vitronectin inhibits secondary nucleation of apatite crystals *in vitro* [356], whereas a direct effect on mineral deposition has not been established. Bone matrix is only faintly stained by immunological techniques, indicating accumulation of vitronectin in matrix at very low levels [313]. However, prior to mineral deposition, vitronectin is increased in concentration in the unmineralized osteoid [352], implying that it may be involved in preparing the matrix for mineral deposition.

d. Small Integrin-Binding Ligand, N-Linked Glycoproteins

The SIBLING family of glycoproteins includes OPN, BSP, DMP-1, DSPP, MEPE, and enamelin. These genetically related members are clustered on human chromosome 4, and it is believed to be the result of duplication and subsequent divergent evolution of a single ancient gene. The Human Genome Project has not completed this portion of chromosome 4, so the exact distances between the genes are not known, but currently six members are thought to be within an estimated 372,000-kbp segment and five of those within a single 250-kbp domain [357]. MEPE, the most different member of the family, is located in the center of this cluster of genes [357].

Besides the completely conserved integrin-binding tripeptide, RGD, this family of proteins has a few short sequences that are conserved among members, including the NXS/T motif for N-linked oligosaccharides and a number of casein kinase II–type phosphorylation sites, which together form an acidic serine–aspartate-rich motif (ASARM) that is thought to interact with hydroxyapatite crystals in regulation of the mineralization process [357–360]. The fact that five of the SIBLINGs are very closely spaced causes a significant problem in producing double knock-out mice because cross-breeding single knock-out mice cannot easily be done.

d.1. Osteopontin

This acidic glycoprotein, which was previously termed bone sialoprotein-1 in bone, was also described as a secreted phosphoprotein and pp66, a protein that is dramatically upregulated by cell transformation and in association with tumor progression [358, 361, 362].

The osteopontin gene is localized to 4q21.3 in humans [357]. This gene shares with the other members of the SIBLING family similar intron–exon boundaries and the biochemical similarities of their corresponding exons (Figure 7-12). The gene contains seven exons. Exon 1 is noncoding, exon 2 encodes for the leader sequence plus the first two amino acids of the mature protein, exons 3 and 5 contain sequences for casein kinase II phosphorylation (SSEE), exon 4 is a proline-rich region (PPPP), exon 6 contains the RGD

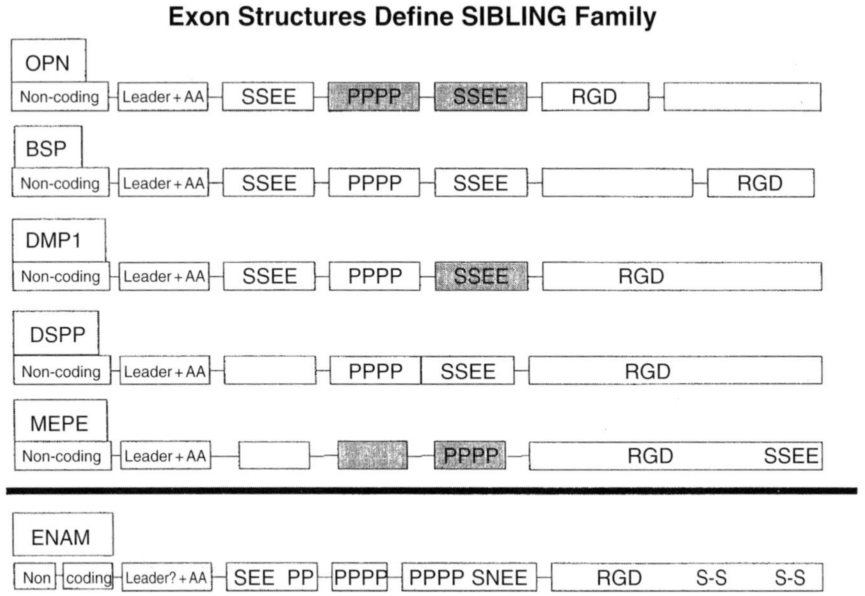

Exon Structures Define SIBLING Family

FIGURE 7-12 Exon structure defines the SIBLING family. The exon structures of the six candidate genes for the SIBLING family are illustrated. Exons are drawn as boxes and introns as connecting lines. Exon 1 is noncoding. For all but ENAM, exon 2 encodes for the leader sequence plus the first two amino acids of the mature protein. Exon 3 often contains the consences for casein kinase II phosphorylation (SSEE), as does exon 5. Exon 4 is usually relatively proline rich (PPPP). The last one or two exons encode the vast majority of the protein (figure not drawn to scale) and always contain the integrin-binding tripeptide ArgGlyAsp (RGD). The shadowing of exons illustrates those exons known to be involved in splice variants. ENAM is a more distantly related gene that has two noncoding 5´ and is also likely to contain disulfide bonds (S–S) that the other SIBLINGs do not.

sequences, and the last exon encodes the vast majority of the protein. Although the amino acid sequence is highly conserved, there are significant differences that appear to be the result of differential splicing of certain exons in different tissues [357, 363–365]. In bone, the mRNA predicts a 301-residue protein that includes a 16-residue signal peptide [364, 366], whereas osteopontin from osteosarcoma appears to have an insertion due to alternative splicing [365].

The osteopontin promoter is highly complex, as would be expected given the range of tissues in which it is synthesized at very precise times and locations. The first kilobase of the mouse osteopontin promoter has been intensely studied. It contains a TATA box, an inverted CCAAT, and a GC box going from 3′ to 5′ upstream from the transcription start site. There is a positive enhancer between −543 and −253 bp and a negative element between −777 and −543 bp [367]. There are five PEA-3 (polyoma enhancer activator) sites, multiple TPA sites, SP1, thyroid hormone response, growth hormone factor, AP4, AP5, AP1, ras activation element sites, and a VDRE site [368]. Transcription in bone gives rise to a 1.6-kb mRNA.

The molecular weight of osteopontin is in the range of 44–75 kDa depending on the method of analysis and the extent of post-translational modification [144, 369]

(Figure 7-13). Due to the nature of post-translational modifications, it does not stain well with Coomassie brilliant blue but becomes blue with Stains All [139, 370], in agreement with its acidic p*I* of 5.0. The structure of osteopontin was originally predicted by Prince from the primary sequence of bovine osteopontin [366, 371], and the structures of osteopontin and BSP have been solved by NMR [372]. There is an RGD cell-binding domain and a single polyaspartyl repeat sequence. This polyaspartyl sequence is highly conserved in all species, implying a functional importance for this domain. Both the RGD cell-binding domain and a non-RGD cell-binding domain in the N terminus have the structures required for integrin interactions needed for cell attachment [373]. The protein in solution has a predominantly random coil structure, but it acquires some β-sheet conformation when bound to hydroxyapatite [374]. Direct analysis of the bone protein indicates that the bone form has an N-linked oligosaccharide, five or six O-linked chains, 12 phosphoserine residues, and one phosphothreonine residue [375]. The chick, rat, mouse, and human proteins show considerable homology, although potential phosphorylation sites vary [376]. In a post-translational modification, osteopontin becomes cross-linked to fibronectin through the action of transglutaminase [377], which may further stabilize its deposition in bone matrix.

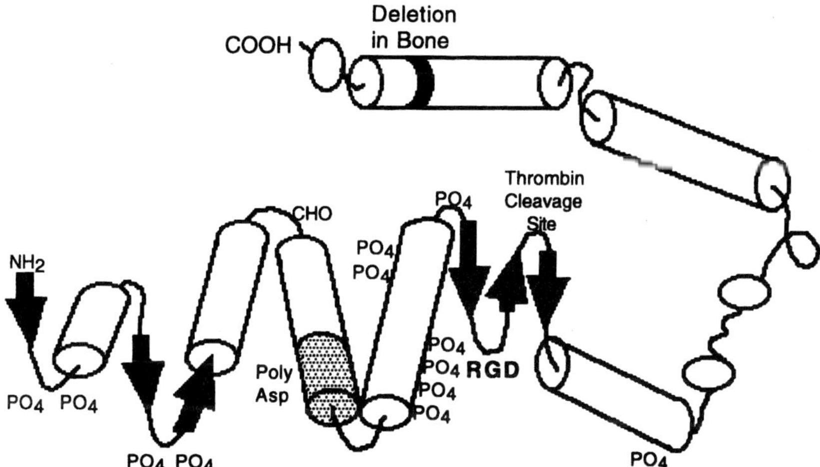

FIGURE 7-13   The osteopontin molecule is composed of numerous stretches of α helix (depicted as cylinders) interconnected in several cases by β-pleated sheets, one of which contains the cell attachment consensus sequence (RGD). A stretch of polyaspartic acid (Poly Asp), along with phosphorylated residues (PO_4), makes osteopontin a highly acidic molecule. Adapted from Denhardt and Guo [368].

Osteopontin promotes osteoblastic and osteoclastic cell attachment *in vitro* [313, 346, 357, 374] and therefore may be important in determining the arrangement of cells in the matrix. In addition, osteopontin is largely accumulated in bone matrix and is highly enriched at cement lines [378, 379]. Inspection of osteopontin production at the cellular level during subperiosteal bone formation indicates that it is produced by osteoblasts and, to a lesser extent, by osteocytes, making it a late marker of osteoblastic differentiation and an early marker of matrix mineralization [357, 379–382].

Due to the correlation of osteopontin production with initial matrix mineralization, there have been many studies on the effect of growth factors and hormones on osteopontin synthesis [383, 384]. In rat osteoblastic cells, osteopontin is stimulated by 1,25-dihydroxyvitamin D_3 [385, 386] and TGF-β [387]. However, long-term treatment with TGF-β caused a decrease in expression of osteopontin, indicating a decrease in osteoblastic phenotype [388]. Osteopontin synthesis is also enhanced by dexamethasone and PTH in culture [389, 390].

Sequence analysis demonstrates that osteopontin and the other phosphorylated sialoproteins have structural features (β-pleated sheets containing anionic and phosphorylated residues) that make them well suited for interactions with hydroxyapatite [372, 374, 391]. In fact, it has been shown that osteopontin binds to hydroxyapatite with both high specificity ($N = 0.026\,\mu\text{mol/m}^2$) and high affinity ($K_D = 1,087\,\mu\text{g}/\mu\text{mol}$)

[392–394]. However, dephosphorylated osteopontin lacks the ability to inhibit hydroxyapatite formation or growth [374, 393], indicating the importance of the phosphate residues (and other post-translational modifications of protein) for interacting with hydroxyapatite and explaining, in part, why osteopontin from different tissues with varying degrees of phosphorylation [395] may have diverse effects on mineral formation and growth.

Based on the EM appearance of apatite crystals grown in the presence of 0–100 μg/mL osteopontin, it appears that this protein blocks crystal elongation [392] rather than secondary nucleation, as is the case for a dentin protein, phosphophoryn [396]. This implies that osteopontin binds with high affinity to one or more apatite crystal faces and further inhibits the growth of crystal. With respect to bone, studies from osteopontin knock-out mice show a distinct bone phenotype with increased mineral crystallinity and increased mineral content [397, 398], also indicating an inhibitory role of this protein in mineralization of bone.

d.2. Bone Sialoprotein

Phosphoproteins in general have long been linked to the mineralization process based on their accumulation at the mineralization front [399, 400] and on the inability of dephosphorylated bone matrices to support mineralization in metastable calcium phosphate solutions [401, 402]. In addition to osteopontin, bone sialoprotein is another major noncollagenous SIBLING that accumulates in cement lines and in spaces between mineralized collagen fibrils [357, 358, 402]. This glycoprotein, somewhat more bone specific than osteopontin, is a heavily sialylated glycoprotein, formerly known as BSP-II [358, 402, 403]. Bone sialoprotein can comprise up to 10% of the noncollagenous protein of bone, depending on the animal species and the type of bone analyzed.

The human gene for BSP is localized to 4q21.3, clustered together with DSPP, DMP-1, MEPE, and OPN [357, 404–406]. It is approximately 15 kb in length, containing a similar seven-exon structure (Figure 7-12) as that of osteopontin, except that the RGD sequence is located in exon 7, whereas exon 6 encodes the vast majority of the protein [357, 405, 406]. The cDNA codes for a 320-residue protein that includes a 16-residue propeptide such that the mature protein (unglycosylated) has a predicted molecular weight of 33.6 kDa [404].

The promoter region of the BSP gene has some unusual characteristics [406, 407]. There is an inverted TATA and CCAAT box in close proximity to an AP1 site (−148 to −142 bp), a CRE (−122 to −116 bp), and a homeobox binding site (−200 to −191 bp). A retinoic acid response element (RARE) is present and overlaps with a glucocorticoid response element (−1,038 to −1,022 bp). A VDRE overlapping the inverted TATA has also been identified [405]. There is a polypurine (CT-rich) stretch that is also found in the osteopontin promoter [408], which can possibly take on a DNA triplex conformation [409]. An AC-rich region is also present that may take on a left-handed helical configuration. This type of structure can either stimulate or inhibit

transcription of the gene [410]. A functional YY-1 site has been identified in intron 1 [406, 410]. However, the elements that convey tissue specificity to the expression of this gene have not been determined. Transcription of the gene results in an mRNA of 2.0 kb, although higher molecular forms have been described [404].

BSP has an apparent molecular weight of approximately 75 kDa as judged by SDS–PAGE and is composed of 50% carbohydrate (12% sialic acid, 7% glucosamine, and 6% galactosamine) (Figure 7-14). It is also rich in aspartic acid, glutamic acid, and glycine, and due to this unique composition, it does not stain well with Coomassie brilliant blue but is stained by Stains All [402, 411]. BSP, distinct from osteopontin, has two or three sets of polyglutamic acid stretches, each starting with a serine/phosphoserine, and tends to be more highly glycosylated and less phosphorylated [383]. Structure analysis [372, 412] places the polyglutamate stretches in an α-helical domain, whereas the proline-rich cell-binding RGD-containing domain would occur at a V-shaped segment, with the arms of the V highly anionic. In addition to glycosylation and phosphorylation, BSP can also be sulfated [413]. The sulfate may be localized to either the carbohydrate side chains or the tyrosine residues [414]. From sequence homologies, the region for such tyrosine sulfation was noted to be between the postulated apatite and the RGD cell binding sites [412]. The RGD cell attachment domain in BSP is located near the C terminus and is recognized by the vitronectin receptor [357, 402, 415], and it facilitates the *in vitro* attachment of fibroblasts

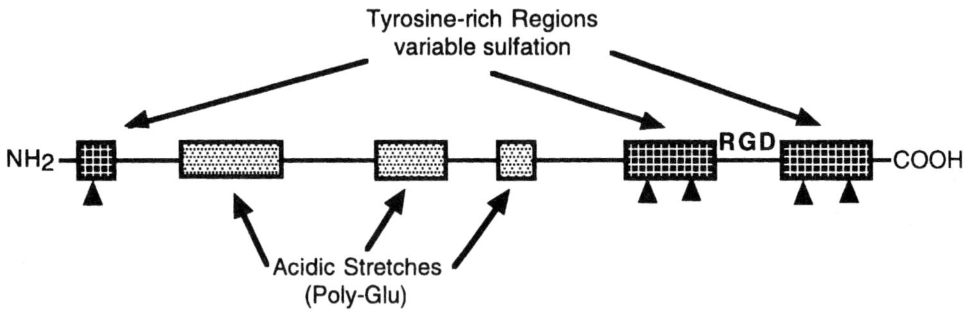

FIGURE 7-14 Sequence analysis predicts the presence of multiple stretches of polyglutamic acid (Poly-Glu) in the first half of the molecule and tyrosine-rich regions in the amino- and carboxy-terminal domains. In the carboxy-terminal region, many of these tyrosines are sulfated. The cell attachment consensus sequence (RGD) is flanked by such regions at the carboxy terminus of the molecule. The molecule is composed of ~50% carbohydrate, including a high concentration of sialic acid residues. Glycosylation is somewhat restricted to the amino-terminal 50% of the molecule. Adapted from Fisher *et al.* [404].

[357, 402, 416], osteoblastic cells [313, 353, 357, 402], and osteoclasts [357, 402, 417].

Biosynthesis of BSP is tightly coordinated with the maturational stage of osteoblastic cells, and it is only produced in cultures that are actively mineralizing. Studies utilizing 1,25-dihydroxyvitamin D_3 have shown that, unlike osteopontin, BSP synthesis is decreased [418]. A derivative of ipriflavone (metabolite III) has been reported to increase the synthesis of BSP [151].

BSP expression is highly enriched in mineralized tissues [404]. It is found in bone, dentin, cementum, and certain regions of hypertrophic chondrocytes [144, 370, 411, 419]. During subperiosteal bone formation, cells in the osteoblastic layer contain BSP, which appears just before or coincident with mineralization [383, 419]. However, after the initial deposition of mineral, the same cells that were previously BSP positive become devoid of BSP, suggesting that the secretion of BSP is not constitutive but, rather, regulated precisely during the initiation of mineralization [225, 383]. However, this is not confirmed by *in vivo* data. The BSP knock-out mice [420], which have a totally nonfunctional BSP gene, were reported to be indistinguishable from wild-type mice at birth, 8½ days, and 1 month, although at 1 year they were 25% smaller than the wild-type mice. X-ray diffraction of homogenized bones of the knock-out animals revealed no differences in mineral crystal relative to controls [420]. Detailed analyses of spatial changes in mineral properties have not been reported.

In vitro, BSP acts as a hydroxyapatite nucleator [383, 394, 404, 421]. When the effect of BSP on mineralization is monitored in an agar gel or at constant pH in solution, it facilitates hydroxyapatite deposition [394, 422], although BSP can also block seed growth [423]. Blocking the carboxylic groups, presumably those in the polyglutamyl domains, destroys BSP's nucleation abilities, whereas dephosphorylating the molecule has less of an effect. This suggests that apatite–BSP interactions occur predominantly through the polyglutamyl repeats; however, other portions of the molecule are also involved [423]. Although the solution data do not prove that BSP has this same function *in situ*, they do demonstrate the nature of the interaction between BSP and hydroxyapatite.

d.3. Bone Acidic Glycoprotein-75 and Dentin Matrix Protein-1

Another sialoprotein originally isolated from rat bone has an apparent molecular weight of ~75kDa and hence is called bone acidic glycoprotein-75 (BAG-75) [424–426]. This protein is heavily glycosylated and contains 7% sialic acid and 8% phosphate. Thirty percent of the residues in this protein are acidic in nature. Whereas in culture, cells from soft connective tissues have been found to synthesize low levels of this protein, BAG-75 is found only in bone, dentin, and growth plate cartilage.

The cDNA and the gene have not been cloned for this molecule. However, there are some data available from direct amino acid sequencing. The amino terminus is approximately 30% homologous with osteopontin. In fact, it does contain polyacid stretches, as do osteopontin and bone sialoprotein [427–429]. In addition, BAG-75 contains both polyaspartate and polyglutamate domains, as well as several phosphorylation sites and an RGD cell binding site [391].

The BAG-75 protein binds with high affinity to both hydroxyapatite and Ca^{2+} ions, as well as to collagen [429]. Immunolocalized next to cells in bone and concentrated in newly formed osteoid, this protein may combine the properties of osteopontin (a mineralization inhibitor) and bone sialoprotein (a nucleator) [425, 426]. BAG-75 also inhibits the resorptive activity of osteoclasts, presumably by blocking its access to bone mineral [430].

Related to BAG-75 is its homologue, DMP-1 [400, 431], another member of SIBLINGs, which is expressed specifically in mineralized tissues by hypertrophic chondrocytes, osteoblasts, and osteocytes [432]. The Human Genome Project has shown that DMP-1 is also located at 4q21.3 in human, closely between DSPP and BSP genes, and contains the similar exon–intron structures [357]. The RGD sequences in DMP-1 are located at the last exon, which also encodes the vast majority of the protein (Figure 7-12). To date, a 2,512-bp upstream segment of the human DMP-1 gene has been isolated and characterized. A CCAAT site was identified in the promoter and a *cis*-regulatory element located between −150 and −63 was found to act as a specific silencer for the gene regulation in some culture systems [433, 434]. Transgenic mice utilizing a mouse DMP-1 promoter *cis*-regulatory system to drive a GFP marker have been generated [435]. In these mice, osteocyte-restricted expression of GFP was observed in histological sections of femur and calvaria and in primary cell cultures, further stressing a role of DMP-1 in mineralization rather than early development of skeleton.

DMP-1 was originally cloned from teeth and expressed as an unphosphorylated 37-kDa fragment, which functioned as a weaker nucleator or inhibitor in solution [400]. A phosphorylated 57-kDa C-terminal peptide of DMP-1 was also identified from teeth and was an effective nucleator of hydroxyapatite formation [400, 436–438]. However, the full-length phosphorylated form of DMP-1, which has been shown to be expressed by bone marrow stromal cells, is an effective mineralization inhibitor [357, 438]. The DMP-1 knock-out mice have hypomineralized bones and teeth [439, 440], also indicating an inhibitory role of this protein. In addition, these mice were shown to overexpress MEPE [441], another potential mineralization inhibitor that was found in rodent bones and teeth in a maturation-dependent manner [442, 443].

d.4.   Dentin Sialophosphoprotein, Matrix Extracellular Phosphoglycoprotein, and Enamelin

The Human Genome Project suggests that DSPP and MEPE are also closely located with other SIBLING members at 4q21.3 in humans, whereas the enamelin gene is located near the centromere at a position of 4q13 [357]. The exon–intron structures of these genes are similar to those of other members of SIBLINGs, such as osteopontin, bone sialoprotein, and DMP-1 (Figure 7-12). However, enamelin is a more distantly related gene that has two noncoding 5′ sequences and is also likely to contain intramolecular disulfide bonds that the other SIBLINGs do not have. In addition, whereas BSP, DMP-1, DSPP, and OPN are all acidic with predicted isoelectric points of 3.4–4.3 (without post-translational modifications), enamelin is neutral and MEPE is strongly basic (pI = 9.2).

DSPP is expressed in a highly regulated fashion during tooth development [433, 434]. As a single gene, an intact protein has not been isolated. However, two DSPP products, DSP and DPP, which are differentially phosphorylated and glycosylated, are coexpressed by odontoblasts and pre-ameloblasts at a time when predentin is being secreted [444]. Only DPP has been reported to regulate type I collagen fibrillogenesis [61, 445] and serve as an effective nucleator for hydroxyapatite formation at lower concentrations and an inhibitor at higher concentrations [446], whereas DSP was not an effective modulator of *in vitro* mineralization [447]. Confirmed by atomic force microscopy, DPP has a distinct pattern of binding to larger (enamel) hydroxyapatite crystals—a pattern not found with DSP [448]. In addition, crystals formed in the presence of DPP were larger than those formed in its absence, suggesting that secondary nucleation is blocked [446]. Furthermore, studies suggest that unphosphorylated DPP has no effects on mineralization, whereas the intact protein is a nucleator [449], but the sites that must be phosphorylated for mineralization to occur, and for proper interaction with fibrillar collagen, are not known. The *in vivo* data have shown that DSPP knock-out mice have decreased mineral content in both their dentin and their bones, stressing the important role of DSPP in mineralization of hard tissues [450].

MEPE, another member of SIBLINGs [357], is 525 residues in length with a short N-terminal signal peptide. This protein was originally identified in oncogenic hypophosphatemic osteomalacia tumors, which are characterized as a bone disease with abnormalities in

mineralization [451]. MEPE appears to be a mineralization inhibitor. Rat and mouse osteoblast cultures lacking MEPE show increased mineralization and human osteoblasts decrease MEPE expression as mineralization progresses [443]. In addition, in cell-free mineralization assays [443], preliminary studies show that the fully phosphorylated recombinant MEPE promotes crystal growth, whereas its C-terminal ASARM peptide (a 23–amino acid peptide from the middle of the molecule containing the RGD) inhibits growth [452, 453]. This suggests that MEPE acts as a nucleator before cleavage and an inhibitor after, which is opposite that of DMP-1 and DSPP, which become nucleators after cleavage. MEPE's interaction with collagen is not reported and the precise role of intact and post-translationally modified MEPE in the mineralization process remains controversial [452, 453]. *In vivo*, the MEPE knock-out mice have increased trabecular bone at 1 year, are more resistant to remodeling, and have increased dentin mineralization, which is the opposite of what is seen in DSPP knock-out mice [454].

Enamelin is the largest protein in the enamel matrix of developing teeth usually expressed by ameloblasts [455, 456]. During the secretory stage of enamel formation, enamelin is found among the crystallites in the rod and interrod enamel and comprises approximately 5% of total matrix protein [456]. The restricted pattern of enamelin expression makes the human enamelin gene a prime candidate in the etiology of amelogenesis imperfecta, a genetic disease in which defects of enamel formation occur in the absence of nondental symptoms [456]. Although the function of enamelin is unknown, it is thought to participate in enamel crystal nucleation and extension and in the regulation of crystal habit [455, 456]. Enamelin is predominantly expressed in developing teeth rather than any other tissues. Thus, the potential role of enamelin in bone mineralization is less likely.

B.   Gla-Containing Proteins

Bone contains a number of proteins that are post-translationally modified by vitamin K–dependent enzymes to form the amino acid, Gla. Due to the sequence requirements of the carboxylating enzymes, the Gla proteins of bone share some sequence homology with certain blood coagulation factors that require γ-carboxylation to maintain their activity. Osteocalcin is the major Gla-containing protein, playing an important role in mineralization of bone, whereas matrix Gla protein is known to be more involved in regulating the calcification of cartilage.

I. OSTEOCALCIN

Osteocalcin was first isolated by the use of nondegradative techniques from acid demineralized bone [457, 458]. It comprises up to 15% of the noncollagenous protein, although the level is variable depending on the animal species [242], and accounts for up to 80% of the total Gla content of mature bone [459]. Extensive screening of protein and RNA extracts [460, 461] and tissue sections by immunohistochemistry [462, 463] from virtually all tissues has failed to detect osteocalcin in any tissue other than dentin and bone, with one exception (in marrow megakaryocytes and platelets) [464]. Thus, osteocalcin was initially reported to be virtually exclusive to bone and was considered the only bone-specific protein.

The human osteocalcin gene is localized on chromosome 1 [465, 466]. The gene is ~1.2 kb in length with four exons that predict a protein of 125 amino acids. The signal peptide contains 26 amino acids in exon 1, a propeptide of 49 amino acids in exon 2 along with the γ-carboxylation recognition sequence, two stretches that become γ-carboxylated in exon 3, and the remainder of the molecule and untranslated region in exon 4 [467, 468]. Interestingly, the mouse genome contains three osteocalcin genes, two of which are activated in bone and one is activated in the kidney [469].

Although some of the basic elements have been determined in the human promoter, most of the extensive characterization has been done primarily in rodent promoters. It contains a TATA box and a CCAAT box. In addition, there is one NF1 binding site and one AP2 binding site, a viral core enhancer, and a CRE. There is also a VDRE at –463 to –437 bp [470] that is flanked by other nuclear binding sites [471–474]. Because of the highly specific nature of osteocalcin expression, the promoter has been intensely scrutinized to determine what properties convey tissue specificity. This has led to the characterization of the "osteocalcin box" [475, 176], located between –99 and –76 bp, which is functionally active [477, 478] and contains a binding site for Msx-1 or Msx-2 (homeodomain proteins). Further characterization of this promoter led to the identification of a binding site, OSE2, located between bp –146 and –132 that binds the transcription factor cbfa1, the so-called osteogenic "master gene" [479].

The protein has a molecular weight of 5.3 kDa but migrates with an apparent molecular weight of ~14 kDa on SDS–PAGE [480, 481]. Depending on the animal species, there is one intramolecular disulfide bond and three to five residues of γ-carboxy glutamic acid [458] (Figure 7-15). The original structural was predicted [482] based on circular dichroism, suggesting that osteocalcin had a structure with extensive (40%) α-helix in the presence of calcium ions. As detailed elsewhere [483], the predicted structure of osteocalcin in the presence of Ca^{2+} consists of two antiparallel α-helical domains, one containing the γ-carboxy glutamic acid residues and one

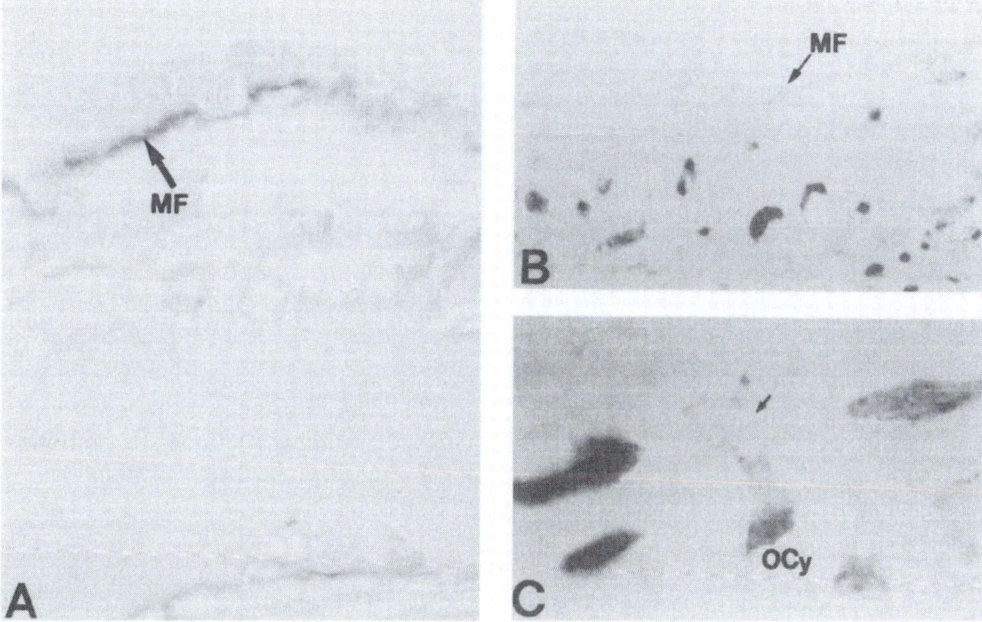

FIGURE 7-15   Osteocalcin immunolocalization in developing bone. Localization of osteocalcin using an antibody against the mature secreted form of the protein sharply demarcates the mineralization front (MF) in developing bone (A). Note, however, the lack of localization within cells that should be synthesizing this molecule. However, when utilizing an antibody raised against the precursor peptide (which is not maintained within mineralized matrix), it can be seen that osteoid osteocytes and osteocytes contain high levels of the proform of the molecule (B and C). Courtesy of Dr. Paolo Bianco.

rich in acidic amino acids. Both of these domains were proposed as sites for calcium chelation. The γ-carboxy glutamic acids were calculated to be 0.5 nm apart, corresponding to the 0.55-nm interatomic spacing of Ca^{2+} ions in the 001 plane of the apatite lattice, suggesting that this domain might be involved in binding to the mineral. A β-pleated sheet in the C terminus was suggested as a cell binding site. Recent insight into the osteocalcin structure comes from comparisons of the NMR data for the Ca^{2+} and Pb^{2+} salts [484] and the Ca^{2+} and Lu^{3+} (lutecium) salts [485]. These NMR studies show that Pb^{2+} and Lu^{3+} compete for the Ca^{2+} binding sites. Since in solution Pb^{2+} blocks the binding of osteocalcin to hydroxyapatite, such data imply that the osteocalcin–apatite interaction occurs through the same domain as Ca^{2+} chelation in solution. Comparison of the Lu^{3+} and Ca^{2+} data for the dog apoprotein demonstrates the presence of two high-affinity binding sites for Ca^{2+} and the conformational changes that occur when Ca^{2+} is present.

The biosynthesis of osteocalcin varies in culture systems and with the length of time in culture. 1,25-Dihydroxyvitamin D_3 [418, 472–474, 476, 486] and 22-oxacalcitriol [487] are known to upregulate osteocalcin expression. BMPs also upregulate osteocalcin in rat and mouse osteoblastic cultures [219, 220]. In general, most factors decrease osteocalcin expression, such as PTH [488], glucocorticoids [489, 490], TGF-β [491, 492], PGE-2 [488], IL-1 [493, 494], tumor necrosis factor-α (TNF-α) [493], IL-10 [80], and lead [81]. Mechanical loading has also been reported to have a negative effect [495].

The proposed functions for osteocalcin in later stages of bone formation and remodeling have been extensively reviewed [496, 497]. During bone development, osteocalcin production is very low and does not reach maximal levels until late stages of mineralization [496–498]. By immunohistochemistry, the mineralization front is intensely stained for osteocalcin, but it has been difficult to demonstrate osteocalcin in osteoid and in cells. However, using an antibody against the precursor form of osteocalcin, the primary cell type that is stained in developing human subperiosteal bone is osteocytes. This antibody stained the cell processes in canaliculi intensely [499], suggesting that perhaps osteocalcin bypasses the osteoid layer by being secreted directly at the mineralization front through the osteocytic cell processes (Figure 7-16). Osteocalcin also appears to be important for induction of the osteoclast phenotype [500]. This concept is supported by the defective osteocalcin production noted in some humans and animals with osteopetrosis, a severely deforming disease characterized by the failure to remodel bone and calcified cartilage [501–503].

Since osteocalcin has a high and relatively specific affinity for apatite, probably due to the binding of the Gla domain to the 100 (*a* axis) face of the apatite crystal,

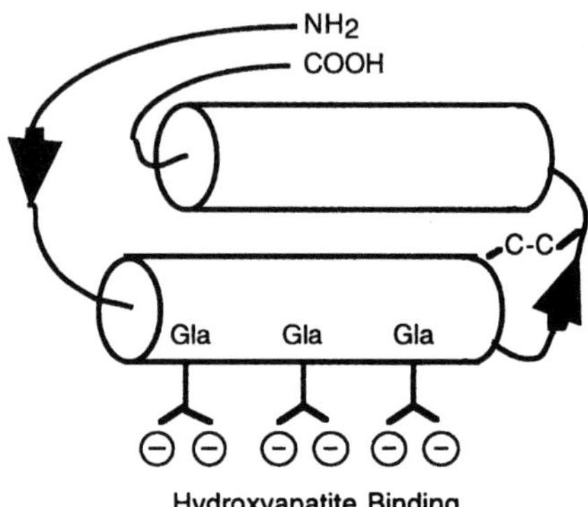

FIGURE 7-16 This small molecule contains two stretches of α-helix (depicted as cylinders) and two regions of β-pleated sheet (arrows). The γ-carboxylated residues of glutamic acid in the amino-terminal helix orient the carboxy groups to the exterior, thereby conferring calcium ion binding with relatively high affinity. There is one intramolecular disulfide bridge (C–C) in the middle region of the molecule. Adapted from Hauschka and Carr. *Biochemistry* **21**, 258–272 (1985).

the protein has been proposed as a specific regulator of the length of the mineral crystals in bone. Osteocalcin is not expressed in culture until mineralization starts [496, 497, 504], which fits the model that it is a regulator of the size and habit of the mineral crystals rather than a promoter of mineral crystal formation. Similarly, during new bone formation, osteocalcin staining and expression occur after mineralization starts [505, 506], and the mineral crystals in the bones of the osteocalcin-null animals fail to mature [507], also demonstrating its role in regulating bone mineral maturation rather than initiation.

2. MATRIX GLA PROTEIN

In addition to osteocalcin, the other major Gla-containing protein in the skeleton is matrix Gla protein (MGP), which was first isolated from bone due to its copurification with BMP [508–510]. MGP has also been found to be expressed in a variety of soft tissues [510, 511].

MGP has a molecular weight of approximately 15 kDa, although it migrates as a substantially larger molecule on SDS–PAGE. The secreted form contains five residues of Gla and one disulfide bridge in a 77- to 79–amino acid residue protein. It also appears that there is a propeptide present at the C terminus that is removed to form the mature protein [512]. MGPs from five different species have phosphorylated serine residues [513, 514]. Thus, the protein is a phosphorylated Gla protein. A distinct physical property of MGP is its insolubility in physiologic solutions (<10 µg/mL) and its tendency to self-associate via hydrophobic interactions.

Due to its insolubility, along with difficulties in isolating a purified protein, the primary structure of MSP was predicted from the cDNA sequences from several species [513, 514]. The MGP gene has been localized to chromosome 12p in human [515]. The gene is approximately 3.9 kb long and contains four exons. The signal peptide is coded for by exon 1 and an α-helical region by exon 2. The recognition sequence for the carboxylating enzymes is found in exon 3 and a sequence that actually becomes α-carboxylated is in exon 4. There are a series of *Alu*I repeats in the 3' untranslated region of the gene. The promoter has been characterized and found to contain a TATA box and a CCAAT box, along with a perfect palindromic sequence that is similar to a RARE [516].

Although there is little information on the developmental expression of MGP throughout the body, it is known that MGP is more abundant in cartilage than in bone [517]. In the skeleton, MGP expression appears early and remains at the same level at all subsequent stages of development [510]. MGP (along with osteocalcin) was initially suggested to be important for the process of endochondral ossification because warfarin-treated rats showed premature epiphyseal closure [518], indicative of impaired remodeling of calcified cartilage.

There is convincing evidence that MGP is an *in vivo* inhibitor of mineralization for cartilage. Mice in which the MGP gene was deleted died prematurely because of massive calcification of their tracheal cartilage and blood vessels [519]. The endochondral cartilage in these animals was also excessively mineralized, but trabecular and cortical bones appeared comparable in mineral properties to age-matched controls [520]. This has further been shown in cell culture studies in which ablation of the MGP in sternal chondrocyte cultures resulted in dystrophic mineralization, whereas addition of exogenous MGP prevented calcification under conditions in which mineralization is normally observed [521]. In light of the data in the knock-out animals and in cell culture, it seems likely that expression of this protein may be a protective action by the cell against unwanted calcification.

C. Other Proteins Involved in Mineralization

I. PROTEOLIPIDS

Proteolipids, as a general class of macromolecules, are membrane components consisting of a hydrophobic protein component and covalently bound lipid [522,

523]. Proteolipids have been isolated from a variety of connective tissues, including bone [524–527] and calcified cartilage [528, 529], in which there are cell and matrix vesicle membrane components [530]. Based on analyses of the apoprotein amino acid compositions, it is clear that there may be more than one type of proteolipid component in bone and cartilage, including but not limited to annexins, lipocortin, calpactin, endonexin, chromobindin, and anchorin.

Structures of the bone and cartilage proteolipids have not been described. However, the structures of several other proteolipids that appear to have common features have been determined by using NMR [531], fluorescent labeling [532, 533], electron microscopy and quasi-elastic light scattering [534], and Fourier transform infrared spectroscopy [535]. In general, these transmembrane proteins have hydrophobic domains throughout the molecule, including the N and C termini. In many cases, the transmembrane domain α-helices are highly ordered, although they have abundant hydrophobic residues (e.g., polyvaline). These hydrophobic domains facilitate interactions with the lipids in the membranes in which the proteolipids are contained. In contrast, the N and C termini are generally flexibly disordered and contain the covalently linked lipids such as palmitoyl-cysteine or acidic phospholipids.

Bone and related cartilage proteolipids have several functions involved in mineralization. They have been demonstrated to act as hydroxyapatite nucleators *in vitro* [524, 526, 529], in a gelatin gel [536], and when implanted in a millipore chamber *in vivo* [537]. They have also been shown to act as ion transporters, such as the annexins [538, 539]. The calcifiable proteolipids are associated with a complex [540, 541] consisting of decreasing molar amounts of calcium, the acidic phospholipids, and inorganic phosphate [542]. These complexes are known to be components of the membranes of extracellular matrix vesicles [543], where they are involved in the initiation of the calcification. Thus, proteolipids in general seem to be important in accumulating ions within the cell and/or extracellular matrix vesicles. As ions accumulate within vesicles, in the presence of the proteolipids, phosphatidylserine, and alkaline phosphatase, mineral crystal formation is initiated and associated with the membranes.

Among these proteolipids, annexins are synthesized by both osteoblasts [544] and chondrocytes [545, 546] and are abundant in matrix vesicle membranes. Lipocortin is a phospholipase A_2 inhibitor [547], and anchorin is a collagen- and cytoskeletal-binding protein [548]. These proteolipids share a 17–amino acid residue homology, which is probably important for the Ca^{2+}-dependent phospholipid binding [548, 530].

2. Serum Proteins

The list of nonstructural proteins that have been identified in bone that originate from serum and become entrapped in bone is quite lengthy [549]. Albumin, α_2-HS glycoprotein (also known as fetuin A), transferrin, α_1-antitrypsin, α_1-antichymotrypsin, IgG, haptoglobulin, hemopexin, serum cholinesterase, and soluble fibronectin are among the plasma proteins that accumulate in bone in detectable amounts [550–554]. Their accumulation is most likely due to their binding to hydroxyapatite. Among these, α_2-HS glycoprotein is thought to have a role in the regulation of matrix mineralization.

Human α_2-HS glycoprotein, which is produced in the liver and circulates in the bloodstream [555], specifically accumulates in mineralized tissues. The human gene sequence for α_2-HS glycoprotein is on chromosome 3, and two RFLPs have been identified [553, 556]. The single mRNA species predicts an 18-residue signal peptide, followed by a sequence that codes for the A and B peptides with an intervening sequence between them. This sequence is presumably lost during cleavage of the precursor to form the mature molecule.

α_2-HS glycoprotein consists of two nonidentical glycosylated peptide chains (chains A and B) that are held together by disulfide bonds [557–560]. These subunits are characterized by repeating Ala-Ala and Pro-Pro sequences. In addition to the single disulfide bond linking the two individual chains by their extreme N and C termini, there are five intradisulfide bonds in the A (heavy) chain. The light B chain has no intrachain S–S bonds. The A chain is composed of three domains consisting of S–S-linked loops. Of the five loops that span 4–19 amino acid residues, two highly homologous loops form one domain, flanked on either side by the other tandem repeats. A mineral binding structure has been proposed for domain 1 of this protein, suggesting that the calcium-binding EF hand motif does not exist [557]. The α_2-HS glycoprotein is also homologous to a nonphosphorylated sialoprotein found in rodent bone [561, 562]; however, the rodent counterpart of α_2-HS glycoprotein consists of one chain rather than two chains.

α_2-HS glycoprotein appears to have a higher affinity for calcium phosphates than other serum proteins since addition of calcium and phosphate to serum led to the removal of all the α_2-HS glycoprotein but removed less than 1% of the albumin [551]. In fact, the ability of serum to inhibit the solution-mediated conversion of amorphous calcium phosphate to hydroxyapatite [557, 563, 564] was attributed to the presence of this high-affinity glycoprotein. α_2-HS glycoprotein is also believed to be involved in preventing unwanted mineralization [550, 557, 565, 566]. The α_2-HS glycoprotein–deficient mouse did not show skeletal abnormalities [566]; however, the serum from these animals did not inhibit apatite formation as efficiently as that from wild-type animals. In addition, the mutant animals developed ectopic calcifications in various organs, confirming the role of this protein as a serum inhibitor of calcification [550, 557, 565].

Whole serum [567] and albumin have been shown to inhibit hydroxyapatite growth in solution [567–569]. The ability of albumin to inhibit apatite growth is attributed to albumin's affinity for apatite [570–573]. Specifically, albumin at 50–250 µg/mL alters the linear rate of growth of apatite seed crystals by binding to the mineral on several faces [568] and blocking the growth of crystal agglomerates [569]. The primary function of albumin in bone is not likely to be one of regulation of mineralization since the extent of inhibition of hydroxyapatite growth in solution indicated that phosphoproteins were more effective inhibitors than albumin.

Although transferrin [574], IgG, IgE, and the other serum proteins also bind to apatite, studies from the Boskey laboratory and from Brigid Heywood's group indicated that IgG had no effect on hydroxyapatite formation, morphology, or growth. However, studies on fractionation of serum from rat bones have revealed that there is a high-molecular-weight fraction (55–150 kDa) composed of one or more proteins that promotes remineralization of decalcified rat tibia, whereas other fractions do not support such recalcification [575, 576]. The nature of the mineralization-inducing proteins in this serum fraction remains to be determined.

3. Matrix Metalloproteinases and Matrix Phosphoprotein Kinases

The degradation of the extracellular matrix to facilitate bone remodeling by osteoclasts has always been considered to be the major function of matrix metalloproteinases (MMPs). These enzymes cleave matrix proteins, which serve as nucleators or inhibitors of mineral formation, to prevent unwanted mineralization from occurring or balance the mineralization rate of the matrix. To date, 24 MMP genes have been identified in humans and 26 well-characterized members have been reported [577–579]. MMPs are classified into six groups based on their structural homology and their substrate specificity (Table 7-2): collagenases (MMP-1, -8, -13, and -18), gelatinases (MMP-2 and -9), stromelysins (MMP-3, -10, and -11), transmembrane MMPs (MT-MMPs, MMP-14, -15, -16, -17, -24, and -25), matrilysins (MMP-7 and -26), and "others" (MMP-12, -19, -20, -21, -22, -23, -27, and -28).

All MMPs share a common domain structure, although not all domains are represented in all family

TABLE 7-2   Currently Known MMPs and Their Substrates[a]

MMP	Alternative names	Substrates
MMP-1	Collagenase-1	Collagens (I, II, III, VII, VIII, X, XI), gelatin, aggrecan, hyaluronidase-treated versican, proteoglycan link protein, large tenascin-C, entactin (nidogen), fibronectin, vitronectin
		Perlecan, ProTNF-α, L-selectin, IL-1β, IGF-BP2, IGF-BP5, IGF-BP3, α_1-P1,[b] α_1-AC, α_2-MG[c]
		MMP-2, MMP-9
MMP-2	Gelatinase A	Collagens (I, III, IV, V, VII, X, XI, XIV), gelatin, elastin, fibronectin, laminin-1, laminin-5, galectin-3, aggrecan, decorin, hyaluronidase-treated versican, proteoglycan link protein, osteonectin, tenascin, vitronectin
		TGF-β, TGF-β_2; IL-1β, TNF-α, α_1-AC, α_1-P1, IGF-BP5, IGF-BP3, FGF R1
		MMP-1, MMP-9, MMP-13
MMP-3	Stromelysin-1	Collagens (III, IV, V, VII, IX, X, XI), elastin, gelatin, aggrecan, versican and hyaluronidose-treated versican, decorin, proteoglycan link protein, large tenascin-C, fibronectin, laminin, entactin, osteonectin, casein, fibrinogen, cross-linked fibrin
		Perlacon, plasminogen, HB-EGF, E-cadherin, α_1-P1, antithrombin-III, substance P, TNF-α, IL-1β, IGF-BP3, α_1-AC, α_2-MG
		MMP-1 "superactivation," MMP-2/TIMP-2 complex, MMP-7, MMP-8, MMP-9, MMP-13
MMP-7	Matrilysin	Collagens (I, IV, X), gelatin, aggrecan, decorin, proteoglycan link protein, fibronectin and laminin, insoluble fibronectin fibrils, entactin, large and small tenascin-C, asteonectin, β_4-integrin, elastin, casein, vitronectin FASL, transferrin, E-cadherin, HB-EGF, α_1-P1, TNF-α, plasminogen
		MMP-1, MMP-2, MMP-9, MMP-9/TIMP-1 complex
MMP-8	Collagenase 2	Collagens (I, II, III, V, VII, VIII, X), gelatin, aggrecan, fibronectin
		α_1-P1, α_2-MG
MMP-9	Gelatinase B	Collagens (IV, V, VII, X, XI, XIV), gelatin, elastin, decorin, laminin, galectin-3, aggrecan, hyaluronidase-treated versican, proteoglycan link protein, fibronectin, entactin, osteonectin, vitronectin
		TGF-β_2, TNF-α, 1L-1β, 1L-2Rα, plasminogen, α_1-AC, α_2-MG, α_1-P1
MMP-10	Stromelysin-2	Collagens (III, IV, V), gelatin, casein, aggrecan, elastin, proteoglycan link protein, laminin, fibronectin MMP-1, MMP-8
MMP-11	Stromelysin-3	Human enzyme, α_1-P1, casein, IGF-BP1, α_2-MG
MMP-12	Metalloelastase	Collagens (I, IV), gelatin, elastin and κ-elastin, casein, fibronectin, aggrecan, vitronectin, decorin, laminin, entactin, proteoglycan monomer, fibrinogen, fibrin
		α_1-P1, α_2-MG, plasminogen
MMP-13	Collagenase-3	Collagens (I, II, III, IV, VI, IX, X, XIV), gelatin, aggrecan, perlecan, large tenascin-C, fibronectin, asteonectin, plasminogen activator inhibitor 2, α_2-MG
		MMP-9
MMP-14	MT1-MMP	Collagens (I, II, III), gelatin, casein, κ-elastin, fibronectin, laminin, vitronectin, proteoglycans, large tenascin-C, entactin, aggrecan
		α_1-P1, α_2-MG, CD44, transglutaminase
		MMP-2, MMP-13
MMP-15	MT2-MMP	Fibronectin, large tenascin-C, entactin, laminin, aggrecan, perlecan
		Transglutaminase
		MMP-2
MMP-16	MT3-MMP	Collagen III, gelatin, casein, fibronectin
		Transglutaminase
		MMP-2
MMP-17	MT4-MMP	Gelatin
		α_2-MG, TNF-α

(Continued)

TABLE 7-2   Currently Known MMPs and Their Substrates[a]—Cont'd

MMP	Alternative names	Substrates
MMP-18	Collagenase-4 (*Xenopus*)	Collagen 1
MMP-19	RASI	Collagens (I, IV), gelatin, fibronectin, laminin, aggrecan, entactin, tenascin, COMP[d]
MMP-20	Enamelysin	Amelogenin, collagen XVIII, aggrecan, COMP
MMP-21	XMMP (*Xenopus*)	ND[e]
MMP-22	CMMP (chicken)	Gelatin
MMP-23	CA-MMP (cysteine array MMP)	Gelatin
MMP-24	MT5-MMP	Collagen I, gelatin, fibronectin, laminin
MMP-25	MT6-MMP	Collagen IV, gelatin, fibronectin
MMP-26	Matrilysin-2	Collagen IV, gelatin, fibronectin
	Endometase	α_1-P1
MMP-27		ND
MMP-28	Epilysin	Casein

[a]From Chaussairn-Miller *et al.* [579].
[b]α_1-P1, α_1-proteinase inhibitor.
[c]α_2-MG, α_2-macroglobulin.
[d]COMP, cartilage oligomeric matrix protein.
[e]ND, not determined.

members [579]. They all have a signal peptide sequence, an amino-terminal catalytic domain containing the highly conserved zinc binding site, and a hemopexin-like carboxy-terminal domain. The latency of the enzymes is maintained by an unpaired cysteine sulfhydryl group in the propeptide domain, which interacts with the active site zinc ion. Activation requires that this cysteine–zinc interaction be perturbed by normal proteolytic removal of the propeptide domain or by ectopically induced conformational change. In addition, the catalytic domain is connected to the hemopexin domain by a hinge region, which is important in determining the substrate specificity of the MMP as well as interactions with tissue inhibitors of metalloproteinases (TIMPs), although this hinge region is lacking in the two matrilysins.

There are four human TIMPs, all of which are low-molecular-weight secreted proteins that bind noncovalently to the active site of MMPs at a 1:1 ratio [578, 579]. However, it is not clear at which point this inhibitory activity is produced. It is apparent that depending on the cell culture system there is a great deal of variability in the ability to produce MMPs and inhibitors. In those cell culture systems producing MMPs, it has been found that MMP activity is stimulated by PTH [580, 581], TNF-α [582], and retinoic acid [583].

Among these MMPS, osteoblastic cells have been found to bear a cell surface receptor for collagenase, and osteoclasts have also been reported to contain collagenases [584, 585]. Collagenase cleaves at a unique site in the collagen triple helix [54, 586] and at a minor site in the nonhelical N-terminal region.

Activated by tyrosine kinase–dependent phosphorylation, collagenase-mediated turnover of the bone matrix seems essential during growth and repair but not during early development. Thus, homozygous transgenic mice whose type I collagen does not contain the unique cleavage site appear normal at birth but develop thickened skin, uteri, and bone during growth and have impaired fracture healing [587].

Although a number of transgenic animals have been generated that are deficient in an MMP, they generally have not displayed a skeletal defect [577–579]. The one exception is the MT1–MMP knock-out mouse, which although normal at birth quickly develops a severe skeletal phenotype characterized by dwarfism, osteopenia, and arthritis [588]. This transmembrane MMP member is known to cleave a number of bone and cartilage matrix proteins, including collagens, aggrecan, fibronectin, and vitronectin.

In bone, osteoclasts produce several cysteine proteases as well as the metalloproteases. It is believed that during osteoclastic resorption the cysteine proteases that have acidic optimal pH act first [589, 590]. Then, as the mineral is dissolved in the acidic environment and the acidity turns to be neutralized, the MMPs function. Specific inhibitors of the cysteine proteases have been used effectively to inhibit osteoclastic resorption [591]. Such inhibitors, while blocking the actions of the cysteine proteases, increase activities of some lysosomal enzymes [592]. One of the most important of the cysteine protease degradative enzymes in bone may be cathepsin K since this enzyme is expressed mainly by osteoclasts, and it appears to initiate the bone degradation process. In this light, cathepsin K knock-out animals have osteopetrosis associated with abnormal matrix degradation but normal mineral resorption [593, 594], as do patients with pycnodysostosis who similarly have abnormal cathepsin K activity [595].

In addition to MMPs, another category of enzymes that appear to be critical for the formation of the mineralized connective tissues are the extracellular matrix phosphoprotein kinases (MPKs). The MPKs isolated from bone and dentin [596–598] are casein II kinases, whose activities can be inhibited by heparin and 2,3-diphosphoglycerate. Analogous to some tyrosine kinases found in the extracellular matrix [599], these phosphoprotein kinases are responsible for the extracellular addition of phosphate to the noncollagenous matrix proteins [600]. Deficient phosphorylation due to altered casein kinase II activity has been reported in the hypophosphatemic mouse, an animal model of human hypophosphatemic rickets, which is resistant to phosphate and vitamin D treatment [601]. Furthermore, since there are protein kinases in the extracellular matrix, it is likely that phosphoprotein phosphatases are also present.

V. THE MINERALIZATION OF BONE MATRIX

Bone mineral (apatite) crystals have a platelike habit [602], are arranged in an oriented fashion on a collagen-based matrix, and have a very limited size of distribution [603]. In general, the mineral crystals in bone (and dentin) are smaller than those in enamel [604] and in dystrophic deposits in severely atherosclerotic plaques [605] or other soft tissue calcifications [124]. Although bone mineral crystals do vary in size with tissue site, age, and disease [92, 606], the range in the lengths of the smallest bone mineral crystals and their orientation imply that their growth must be precisely regulated. Bone mineralization is thus distinct from solution-mediated Ca^{2+} phosphate precipitation, in which similarly sized, nonoriented small crystals are formed and ripen to appreciably larger sizes [607]. This is also distinct from geologic apatite formation, in which high temperatures and pressures yield extremely large single crystals.

A. Requirements for Matrix Mineralization

Analyses of diseased tissues and tissues from transgenic animals indicate that there are a number of cellular and extracellular factors essential for physiologic mineral deposition, which is defined as highly ordered and finely regulated mineral deposition upon a collagenous matrix. For a physiologic mineral deposition, there must be an appropriate collagen-based matrix. This is emphasized by (1) the smaller size of hydroxyapatite crystals in OI bones [90–92] and (2) the relative abundance of mineral that is not associated with collagen in these bones with deficient and/or impaired collagen production [54, 90, 91, 93]. Although in some cases the defective mineralization in the OI bones may also be attributed to altered matrix protein production [608] or retention, collagen is clearly an absolute requirement for physiologic bone mineralization. Similarly, since fibronectin forms the basis on which collagen is deposited, it must also be a requirement.

Equally apparent from requirements for physical mineral deposition is the essential presence of Ca^{2+} and inorganic phosphate (P_i) ions. Calcium ions may be supplied from the cells or from circulating or localized calcium-binding proteins. Phosphate ions may be derived from breakdown of pyrophosphate, an abundant metabolic product; from hydrolysis of phospho-esters or phosphoproteins; or from circulating P_i ions. Studies have shown the requirements of the sodium-dependent P_i transporter, Pit-1 (also named Glvr-1), for calcification in cultures and for mineralization of endochondral bones during embryonic development [609–611]. The exact Ca^{2+} and P_i content of the extracellular fluid of bone is not known, but in cartilage, micropuncture studies showed the pH to be 7.58 and total Ca^{2+} and total P_i to be 1–12 and 3–12 mg/dL, respectively [612].

Subsequently, for the formation of apatite, a basic environment is also essential. Thus, many of the highly anionic matrix proteins, as reviewed in this article, probably contribute to creating this environment by regulating the apatite formation and deposition, the size of mineral growth, or the organization of matrix proteins to be desired for mineral deposition. Which of these matrix proteins is truly crucial in mineralization of bone cannot be determined until the sequence of protein expression is determined precisely and appropriate knock-out and transgenic models are developed. Even in these cases, it may be difficult to prove an essential role for mineralization since it is already apparent that there are compensating mechanisms in the control of this critical process.

Comprehensively, it is certain that the cells are required for the production of a physiologic matrix, synthesizing and exporting necessary enzymes, growth factors, and matrix molecules. In addition, as discussed later, the formation of extracellular matrix vesicles is also apt to prove critical for the initiation of mineralization in some cases.

B. Mineralization Regulated by Matrix Proteins

1. Physical Chemistry of Mineralization

Calcium phosphate precipitation from solution can yield a variety of phases, depending on the pH, $Ca:P_i$ ratio, and solution supersaturation [13, 607, 613]. When the pH is in the physiologic range (7.4–7.8), apatite formation occurs with solution Ca:P molar ratios as high as 2:1 and as low as 1:1 as long as the solution is supersaturated with respect to apatite (i.e., has a $Ca \times P$ product that exceeds the solubility product for apatite). Depending on the supersaturation, intermediate phases such as amorphous calcium phosphate [30, 31, 614], octacalcium phosphate [615–617], or other intermediates may form [13, 618], but in all these cases the final product is apatitic.

Apatite crystals develop in solution when individual ions or ion clusters associate in the same orientation as in the crystal lattice that they are trying to form. When sufficient ion clusters are correctly oriented, they can persist in solution and can serve as a "critical nucleus" for further crystal growth. Homogeneous nucleation, in which crystals form *de novo*, is a rare process [607]. Thus, it is likely that in most instances of solution-mediated apatite deposition, nucleation occurs on foreign materials such as dust, scratches on the container, and buret tips. Such heterogeneous nucleation yields the initial crystals, which then facilitate additional growth by the process of secondary nucleation. In secondary nucleation, growth sites on the preformed apatite crystals serve as branch points for the formation of new crystals, analogous in many ways to the branching of the growing glycogen molecule during glycogenesis. Proliferation by secondary nucleation results in numerous small crystals. Crystal growth in the absence of secondary nucleation would result in fewer but larger crystals. This suggests that most of the crystals in bone form by a secondary nucleation-like process or by growth from individual nuclei. Unfortunately, what regulates crystal size in bone cannot be determined from studies of protein-free solutions.

2. The Role of Matrix Proteins

The mineral in bone, as in the other physiologically calcified tissues, is associated with an organic matrix (Figure 7-17). Protein(s) within such matrices can regulate the nucleation and growth of mineral crystals in several ways. As discussed in this article, these proteins function as

1. Scaffold for mineral deposition (collagens): As the major (~90%) structural proteins in matrix of bone, collagens do not induce the formation of mineral crystals. However, they serve as scaffold to form a highly oriented "backbone" supplying appropriate sites for retention of noncollagenous proteins and initiation of mineral deposition.

2. Nucleators or initiators of mineral crystal formation (bone sialoprotein, osteonectin, bone acidic protein-75, dentin sialophosphoprotein, enamlin, proteolipids, and alkaline phosphatase): The protein(s) may bind Ca^{2+} and/or P_i ions, forming a surface that resembles the apatite surface. In this manner, the protein serves as an epitaxial (similar surface) nucleator, thereby providing a surface for the start of nucleation. Alternatively, the proteins may participate in the formation of membrane matrix vesicle as foci of the initiation of mineralization, such as proteolipids and alkaline phosphatase.

3. Inhibitors of mineral crystal formation (aggrecan, osteopontin, bone acidic protein-75, dentin matrix protein-1, matrix extracellular phosphoglycoprotein, α_2-HS glycoprotein, and albumin): When isolated in an environment relatively free of body fluids, the protein(s) can chelate Ca^{2+} or P_i ions, reducing the fluid supersaturations, which in turn would prevent crystal nucleation and/or growth. Thus, the protein(s) can form a protected environment around the crystal nucleus, sequestering it and thus preventing crystal growth, or stabilizing the nucleus, protecting it from the external environment.

4. Blockade of the growth of mineral (osteocalcin, vitronectin, and matrix Gla protein): The protein(s) may also bind to one or more faces of the growing crystal because its side chains match positions in the lattice, thereby blocking growth in one or more directions or even blocking growth beyond a specific size.

5. Organizers of matrix composition (decorin, thrombospondin, fibronectin, vitronectin, versican, and SIBLINGs): These proteins bind to the collagen backbone of the matrix and other noncollagenous proteins, changing their conformation and their ability to affect crystal nucleation and growth according to the pathways described previously. In addition, these proteins may bind to cell surface via special sequences (RGD) and thus mediate cell–cell and cell–matrix attachments, resulting in a change in the extracellular $Ca \times P$ concentration or the pH of microenvironments.

The ultrastructural studies that combine x-ray crystallographic and electron microscopic techniques provide illustrations for each of these mechanisms for the formation of larger crystals of calcium carbonates, calcium sulfates, brushite, and octacalcium phosphate [98,

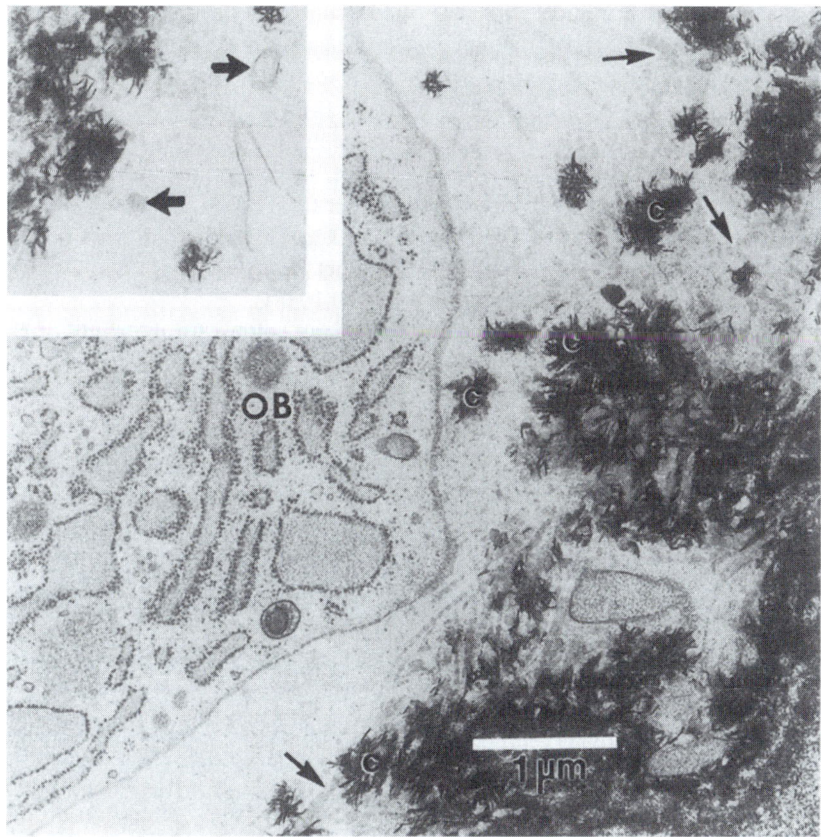

FIGURE 7-17 Cell-mediated matrix mineralization in developing bone. Early mineralization in chick bone. Electron micrograph showing a 17-day-old embryonic tibia, stained with uranyl acetate and lead citrate. Mineral clusters (C) outside the osteoblast (OB) are associated with collagen (thin arrows) and extracellular matrix vesicles (inset). Empty vesicles (thick arrows) as well as vesicles with mineral are seen. Courtesy of Dr. Steven B. Doty.

619, 620]. For example, fibronectin has been shown to bind to the ionic surfaces of calcite that did not include water molecules, but it does not bind at all to brushite whose surfaces all have bound water [621]. The acidic macromolecules from sea animals have been shown to determine the shape of calcite crystals [622]. Cells have been shown to interact with specific faces on such crystals in the presence and absence of RGD-containing macromolecules [280–283, 623]. Scanning electron micrographs have similarly been used to identify the binding sites for polyaspartic acid, mollusk shell proteins, and rat dentin phosphoprotein on the surface of octacalcium phosphate [29, 621].

Although there is no direct evidence of the exact nature of the matrix protein–mineral interaction, there are examples of each of these mechanisms from solution studies of apatite formation. Studies of the effects of bone matrix proteins on apatite formation include those in which preformed seed crystals are added to Ca × P solutions, and the rate of crystal growth is determined at fixed Ca × P and fixed pH [13] or variable Ca × P × OH [618]. Other studies have examined the formation (nucleation and growth) of apatite from

solutions in the presence of insoluble proteins, proteins immobilized on polyanionic beads [624], or proteins in solution [625]. Diffusion studies, in which the protein is held within an agarose [393], silicate [422], or denatured collagen gel [124, 626], have also provided insight into apatite nucleation and growth. From such studies, one can also find examples of the mechanisms listed previously. However, it should be emphasized that because of its affinity for apatite, a protein in low concentrations may act as a nucleator and in higher concentrations may serve to regulate crystal growth. Moreover, promoting or inhibiting mineralization *in situ* is also dependent on the extent of post-translational modification, such as phosphorylation, and on the regulation of collagen fibrillogenesis, which may induce conformational alterations [445, 449, 627].

In addition, the extracellular matrix vesicles and their component lipids may facilitate Ca and P accumulation, while shielding the apatite nucleus as the foci of initial mineralization. Illustrations of this behavior in vesicles have been seen in the iron oxide–forming bacteria and in model liposomes. In the model liposomes, in which Ca^{2+} accumulation is facilitated by an ionophore, initial mineral crystals form inside the liposomes in association with the liposome membrane, where they eventually grow and puncture the liposome membrane and become exposed to the external solution [4, 7, 8, 628–632].

ACKNOWLEDGMENTS

We thank Drs. Paolo Bianco, Larry W. Fisher, Neal S. Fedarko, and Steven B. Doty for providing photographic materials. Dr. Boskey's research as discussed in this article was supported by National Institutes of Health (NIH) grants DE04141, AR037661, and AR41325. Dr. Robey's research discussed in this article was supported by the DIR, National Institute of Dental and Craniofacial Research of the Intramural Research Program, NIH.

REFERENCES

1. P. G. Robey, Normal bone formation–structure. In *Bone Formation and Repair* (C. T. Brighton, G. Friedlander, and J. M. Lan, eds.), pp. 3–12. American Academy of Orthopaedic Surgeons, Rosemont, IL (1994).
2. I. Fernandez-Tresguerres-Hernandez-Gil, M. A. Alobera-Gracia, M. Del-Canto-Pingarron, and L. Blanco-Jerez, Physiological bases of bone regeneration I. Histology and physiology of bone tissue. *Med Oral Patol Oral Cir Bucal* **11**, E47–E51 (2006).
3. H. P. Wiesmann, U. Meyer, U. Plate, and H. J. Hohling, Aspects of collagen mineralization in hard tissue formation. *Int Rev Cytol,* **242**, 121–156 (2005).
4. L. Zhang, M. Balcerzak, J. Radisson, C. Thouverey, S. Pikula, G. Azzar, and R. Buchet, Phosphodiesterase activity of alkaline phosphatase in ATP-initiated Ca(2+) and phosphate deposition in isolated chicken matrix vesicles. *J Biol Chem,* **280**, 37289–37296 (2005).
5. R. J. Waddington, H. C. Roberts, R. V. Sugars, and E. Schonherr, Differential roles for small leucine-rich proteoglycans in bone formation. *Eur Cell Mater,* **6**, 12–21 (2003).
6. E. P. Barboza, R. O. de Souza, A. L. Caula, L. G. Neto, F. C. de Oliveira, and M. E. Duarte, Bone regeneration of localized chronic alveolar defects utilizing cell binding peptide associated with anorganic bovine-derived bone mineral: A clinical and histological study. *J Periodontol,* **73**, 1153–1159 (2002).
7. H. I. Roach, Association of matrix acid and alkaline phosphatases with mineralization of cartilage and endochondral bone. *Histochem J,* **31**, 53–61 (1999).
8. T. Kirsch, H. D. Nah, I. M. Shapiro, and M. Pacifici, Regulated production of mineralization-competent matrix vesicles in hypertrophic chondrocytes. *J Cell Biol,* **137**, 1149–1160 (1997).
9. P. G. Robey and J. D. Termine, Human bone cells *in vitro. Calcif Tissue Int,* **37**, 453–460 (1985).
10. A. L. Boskey, Mineralization, structure, and function of bone. In *Dynamics of Bone and Cartilage Metabolism* (M. J. Seibel, S. P. Robins, and J. P. Bilzekian, eds.), pp. 153–162. Academic Press, New York (1999).
11. S. Coskun, F. Korkusuz, and V. Hasirci, Hydroxyapatite reinforced poly(3-hydroxybutyrate) and poly(3-hydroxybutyrate-co-3-hydroxyvalerate) based degradable composite bone plate. *J Biomater Sci Polym Ed,* **16**, 1485–1502 (2005).
12. E. D. Eanes, Dynamic aspects of apatite phases of mineralized tissues, model studies. In *The Chemistry and Biology of Mineralized Tissues* (W. T. Butler, ed.), pp. 213–221. Ebsco Media, Birmingham, AL (1985).
13. J. Moradian-Oldak, F. Frolow, L. Addadi, and S. Weiner, Interactions between acidic matrix macromolecules and calcium phosphate ester crystals: Relevance to carbonate apatite formation in biomineralization. *Proc R Soc London,* **B247,** 47–55 (1992).
14. O. L. Shtepenko, C. D. Hills, N. J. Coleman, and A. Brough, Characterization and preliminary assessment of a sorbent produced by accelerated mineral carbonation. *Environ Sci Technol,* **39**, 345–354 (2005).
15. N. P. Camacho, S. Rinnerthaler, E. P. Paschalis, R. Mendelsohn, A. L. Boskey, and P. Fratzl, Complementary information on bone ultrastructure from scanning small angle X-ray scattering and Fourier-transform infrared microspectroscopy. *Bone,* **25**, 287–293 (1999).
16. W. J. Landis, K. J. Hodgens, J. Arena, M. J. Song, and B. F. McEwen, Structural relations between collagen and mineral in bone as determined by high voltage electron microscopic tomography. *Microsc Res Tech,* **33**, 192–202 (1996).
17. N. J. Wachter, P. Augat, M. Mentzel, M. R. Sarkar, G. D. Krischak, L. Kinzl, and L. E. Claes, Predictive value of bone mineral density and morphology determined by peripheral quantitative computed tomography for cancellous bone strength of the proximal femur. *Bone,* **28**, 133–139 (2001).
18. B. R. Gomberg, P. K. Saha, and F. W. Wehrli, Method for cortical bone structural analysis from magnetic resonance images. *Acad Radiol,* **12**, 1320–1332 (2005).
19. M. A. Fernandez-Seara, S. L. Wehrli, M. Takahashi, and F. W. Wehrli, Water content measured by proton–deuteron exchange NMR predicts bone mineral density and mechanical properties. *J Bone Miner Res,* **19**, 289–296 (2004).
20. G. Cho, Y. Wu, and J. L. Ackerman, Detection of hydroxyl ions in bone mineral by solid-state NMR spectroscopy. *Science,* **300**, 1123–1127 (2003).
21. M. Kakei, H. Nakahara, N. Tamura, H. Itoh, and M. Kumegawa, Behavior of carbonate and magnesium ions in the initial crystallites at the early developmental stages of the rat calvaria. *Ann Anat,* **179**, 311–316 (1997).
22. M. R. van der Harst, P. A. Brama, C. H. van de Lest, G. H. Kiers, J. DeGroot, and P. R. van Weeren, An integral biochemical analysis of the main constituents of articular cartilage, subchondral and trabecular bone. *Osteoarthritis Cartilage,* **12**, 752–761 (2004).
23. Y. Wu, J. L. Ackerman, E. S. Strawich, C. Rey, H. M. Kim, and M. J. Glimcher, Phosphate ions in bone: Identification of a calcium-organic phosphate complex by 31P solid-state NMR spectroscopy at early stages of mineralization. *Calcif Tissue Int,* **72**, 610–626 (2003).
24. M. Rey, B. Shimizu, M. J. Collins, and M. J. Glimcher, Resolution-enhanced Fourier transform infrared spectroscopy study of the environment of phosphate ion in the early deposits of a solid phase of calcium phosphate in bone and enamel and their evolution with age: 2. Investigations in the $\nu 3PO_4$ domain. *Calcif Tissue Int,* **49**, 383–388 (1991).
25. V. Rey, M. Renugopalakrishnan, B. Shimizu, M. J. Collins, and M. J. Glimcher, A resolution-enhanced Fourier transform infrared spectroscopic study of the environment of the CO_3^2 ion in the mineral phase of enamel during its formation and maturation. *Calcif Tissue Int,* **49**, 2259–2268 (1991).
26. R. Legros, N. Balmain, and G. Bonel, Age-related changes in mineral of rat and bovine cortical bone. *Calcif Tissue Int,* **41**, 137–144 (1987).

27. A. J. Kahn and N. C. Partridge, Bone resorption *in vivo*. In *Bone. Volume 2: The Osteoclast* (B. K. Hall, ed.), pp. 199–240. CRC Press, Boca Raton, FL (1990).

28. A. Kaflak-Hachulska, A. Samoson, and W. Kolodziejski, 1H MAS and 1H → 31P CP/MAS NMR study of human bone mineral. *Calcif Tissue Int,* **73,** 476–486 (2003).

29. M. Jayaraman and M. V. Subramanian, Preparation and characterization of two new composites: Collagen-brushite and collagen octa-calcium phosphate. *Med Sci Monit,* **8,** BR481–BR487 (2002).

30. E. D. Eanes, Amorphous calcium phosphate. *Monogr Oral Sci,* **18,** 130–147 (2001).

31. Y. Politi, T. Arad, E. Klein, S. Weiner, and L. Addadi, Sea urchin spine calcite forms via a transient amorphous calcium carbonate phase. *Science,* **306,** 1161–1164 (2004).

32. R. Lagier and C. A. Baud, Magnesium whitlockite, a calcium phosphate crystal of special interest in pathology. *Pathol Res Pract,* **199,** 329–335 (2003).

33. D. A. D. Parry, The molecular and fibrillar structure of collagen and its relationship to the mechanical properties of connective tissue. *Biophys Chem,* **29,** 195–209 (1988).

34. K. A. Piez, Molecular and aggregate structures of collagens. In *Extracellular Matrix Biochemistry* (K. A. Piez and A. H. Reddi, eds.), pp. 1–40. Elsevier, New York (1984).

35. B. Brodsky and A. V. Persikov, Molecular structure of the collagen triple helix. *Adv Protein Chem,* **70,** 301–339 (2005).

36. C. Hongo, K. Noguchi, K. Okuyama, Y. Tanaka, and N. Nishino, Repetitive interactions observed in the crystal structure of a collagen-model peptide, [(Pro-Pro-Gly)9]3. *J Biochem (Tokyo),* **138,** 135–144 (2005).

37. M. Yamauchi, E. P. Katz, K. Otsubo, K. Teraoka, and G. L. Mechanic, Cross-linking and stereospecific structure of collagen in mineralized skeletal tissues. *Connect Tissue Res,* **21,** 159–169 (1989).

38. D. Jozic, G. Bourenkov, N. H. Lim, R. Visse, H. Nagase, W. Bode, and K. Maskos, X-ray structure of human proMMP-1: New insights into procollagenase activation and collagen binding. *J Biol Chem* **280,** 9578–9585 (2005).

39. J. Bella, M. Eaton, B. Brodsky, and H. M. Berman, Crystal and molecular structure of a collagen-like peptide at 1.9 angstroms. *Science,* **266,** 75–81 (1994).

40. M. L. Chu, W. De Wet, M. Bernard, J. F. Ding, M. Morabito, J. Myers, C. Williams, and F. Ramirez, Human pro alpha 1(I) collagen gene structure reveals evolutionary conservation of a pattern of introns and exons. *Nature,* **310,** 337–340 (1984).

41. F. Ramirez, M. Bernard, M. L. Chu, L. Dickson, F. Sangiorgi, D. Weil, W. De Wet, C. Junien, and M. Sobel, Isolation and characterization of the human fibrillar collagen genes. *Ann N Y Acad Sci,* **460,** 117–129 (1985).

42. W. De Wet, M. Bernard, V. Benson-Chanda, M.-L. Chu, L. Dickson, D. Weil, and F. Ramirez, Organization of the human pro-alpha 2(I) collagen gene. *J Biol Chem,* **262,** 16032–16036 (1987).

43. H. Kuivaniemi, G. Tromp, M. L. Chu, and D. J. Prockop, Structure of a full-length cDNA clone for the preproα2(I) chain of human type I procollagen. Comparison with the chicken gene confirms unusual patterns of gene conservation. *Biochem J,* **252,** 633–640 (1988).

44. M. L. Chu, W. deWet, M. Bernard, and F. Ramirez, Fine structural analysis of the human proα (1) collagen gene: Promoter structure, AluI repeats and polymorphism transcripts. *J Biol Chem* **260,** 2315–2320 (1985).

45. M. H. Finer, S. Aho, and L. Gerstenfeld, Unusual DNA sequences located within the promoter region and the first intron of the chicken proalpha1(I) collagen gene. *J Biol Chem,* **262,** 13323–13332 (1987).

46. A. Lichtler, M. L. Stover, and J. Angilly, Isolation and characterization of the rat α1(I) collagen promoter: Regulation by 1,25-dihydroxyvitamin D_3. *J Biol Chem* **264,** 3072 (1989).

47. A. Schmidt, P. Rossi, and B. de Crombrugghe, Transcriptional control of the mouse α2(I) collagen gene: Functional deletion analysis of the promoter and evidence for cell-specific expression. *Mol Cell Biol* **6,** 347–354 (1986).

48. G. Vogeli, H. Ohkubo, M. E. Sobel, Y. Yamada, I. Pastan, and B. de Crombrugghe, Structure of the promoter for chicken α2 type I collagen gene. *Proc Natl Acad Sci USA,* **78,** 5534–5538 (1981).

49. A. Lichtler, M. L. Stover, J. Angilly, B. Kream, and D. W. Rowe, Isolation and characterization of the rat alpha1(I) collagen promoter. *J Biol Chem,* **269,** 16518 (1994).

50. R. D'Souza, K. Niederreither, and B. de Crombrugghe, Osteoblast-specific expression of the alpha 2(I) collagen promoter in transgenic mice: Correlation with the distribution of TGF-β 1. *J Bone Miner Res,* **8,** 1127–1136 (1993).

51. G. R. Martin, R. Timpl, P. K. Muller, and K. Kuhn, The genetically distinct collagens. *Trends Biochem Sci,* **251,** 285–287 (1985).

52. K. S. E. Cheah, Collagen genes and inherited connective tissue disease. *Biochem J,* **229,** 287–303 (1985).

53. J. C. Brown and R. Timpl, The collagen superfamily. *Int Arch Allergy Immunol,* **107,** 484–490 (1995).

54. J. Myllyharju and K. I. Kivirikko, Collagens and collagen-related diseases. *Ann Med,* **33,** 7–21 (2001).

55. C. H. Wu, C. B. Donovan, and G. Y. Wu, Evidence for pre-translational regulation of collagen synthesis by procollagen propeptides. *J Biol Chem,* **261,** 10482–10484 (1989).

56. R. A. Fleischmajer and J. S. Perlish, The role of the amino propeptide in collagen fibrillogenesis. *Rheumatology,* **10,** 91 (1986).

57. R. A. Fleischmajer, L. W. Fisher, E. D. MacDonald, L. Jacobs Jr., J. S. Perlish, and J. D. Termine, Decorin interacts with fibrillar collagen of embryonic and adult human skin. *J Struct Biol,* **106,** 82–90 (1991).

58. U. K. Blaschke, E. F. Eikenberry, D. J. Hulmes, H. J. Galla, and P. Bruckner, Collagen XI nucleates self-assembly and limits lateral growth of cartilage fibrils. *J Biol Chem,* **275,** 10370–10378 (2000).

59. T. Ono and T. K. Nemoto, Two forms of apatite deposited during mineralization of the hen tendon. *Matrix Biol,* **24,** 239–244 (2005).

60. W. Traub, T. Arad, and S. Weiner, Origin of mineral crystal growth in collagen fibrils. *Matrix,* **12,** 251–255 (1992).

61. W. Traub, A. Jodaikin, T. Arad, A. Veis, and B. Sabsay, Dentin phosphophoryn binding to collagen fibrils. *Matrix,* **12,** 197–201 (1992).

62. A. L. Arsenault, Crystal–collagen relationships in calcified turkey leg tendons visualized by selected-area dark field electron microscopy. *Calcif Tissue Int,* **43,** 202–212 (1988).

63. M. E. Maitland and A. L. Arsenault, A correlation between the distribution of biological apatite and amino acid sequence of type I collagen. *Calcif Tissue Int,* **48,** 341–352 (1991).

64. D. J. Prockop and K. I. Kivirikko, Collagens: Molecular biology, diseases, and potentials for therapy. *Annu Rev Biochem,* **64,** 403–434 (1995).

65. C. Niyibizi and D. R. Eyre, Identification of the cartilage alpha1(IX) chain in type V collagen from bovine bone. *FEBS Lett,* **242,** 314–318 (1989).

66. T. Pihlajaniemi and M. Tamminen, The alpha 1 chain of type XIII collagen consists of three collagenous and four noncollag-

enous domains, and its primary transcript undergoes complex alternative splicing. *J Biol Chem,* **265,** 16922–16928 (1990).

67. M. van der Rest, Collagen of bone. In *Bone Matrix and Bone Specific Products* (B. K. Hall, ed.), pp. 187–239. CRC Press, Boca Raton, FL (1991).

68. C. Niyibizi and D. R. Eyre, Bone type V collagen: Chain composition and location of a trypsin cleavage site. *Connect Tissue Res,* **20,** 247–250 (1989).

69. B. Budde, K. Blumbach, J. Ylostalo, F. Zaucke, H. W. Ehlen, R. Wagener, L. Ala-Kokko, M. Paulsson, P. Bruckner, and S. Grassel, Altered integration of matrilin-3 into cartilage extracellular matrix in the absence of collagen IX. *Mol Cell Biol,* **25,** 10465–10478 (2005).

70. Y. Shibata, Y. Abiko, K. Goto, Y. Moriya, and H. Takiguchi, Heparin stimulates the collagen synthesis in mineralized cultures of the osteoblast-like cell line, MC3T3-E1. *Biochem Int,* **28,** 335–344 (1992).

71. H. C. Tenenbaum, H. Limeback, C. A. McCulloch, H. Mamujee, B. Sukhu, and M. Torontali, Osteogenic phase-specific co-regulation of collagen synthesis and mineralization by beta-glycerophosphate in chick periosteal cultures. *Bone,* **13,** 129–138 (1992).

72. K. Nohtomi, K. Sato, K. Shizume, K. Yamazaki, H. Demura, K. Hosada, Y. Murata, and H. Seo, Stimulation of interleukin-4 of cell proliferation and mRNA expression of alkaline phosphatase and collagen type I in human osteoblast-like cells of trabecular bone. *Bone Miner,* **27,** 69–79 (1994).

73. R. S. Bockman, P. T. Guidon Jr., L. C. Pan, R. Salvatori, and A. Kawaguchi, Gallium nitrate increases type I collagen and fibronectin mRNA and collagen protein levels in bone and fibroblast cells. *J Cell Biochem,* **52,** 396–403 (1993).

74. L. G. Raisz, P. M. Fall, D. N. Petersen, A. Lichtler, and B. E. Kream, Prostaglandin E2 inhibits alpha 1(I)procollagen gene transcription and promoter activity in the immortalized rat osteoblastic clonal cell line. *Mol Endocrinol,* **7,** 17–22 (1993).

75. D. Pavlin, A. Bedalov, M. S. Kronenberg, B. E. Kream, D. W. Rowe, C. L. Smith, J. W. Pike, and A. C. Lichtler, Analysis of regulatory regions in the COL1A1 gene responsible for 1,25-dihydroxyvitamin D3-mediated transcriptional repression in osteoblastic cells. *J Cell Biochem,* **56,** 490–501 (1994).

76. A. M. Delany, B. Y. Gabbitas, and E. Canalis, Cortisol downregulates osteoblast alpha 1 (I) procollagen mRNA by transcriptional and posttranscriptional mechanisms. *J Cell Biochem,* **57,** 488–494 (1995).

77. B. E. Kream, D. LaFrancis, D. N. Peterson, C. Woody, S. Clark, D. W. Rowe, and A. Lichtler, Parathyroid hormone represses α 1(I) collagen promoter activity in cultured calvariae from neonatal transgenic mice. *Mol Endocrinol,* **7,** 399–408 (1993).

78. R. Hata, H. Hori, Y. Nagai, S. Tanaka, M. Kondo, M. Hiramatsu, N. Utsumi, and M. Kumegawa, Selective inhibition of type I collagen synthesis in osteoblastic cells by epidermal growth factor. *Endocrinology,* **115,** 867–876 (1984).

79. M. A. Fang, C. A. Glackin, A. Sadhu, and S. McDougall, Transcriptional regulation of alpha 2(I) collagen gene expression by fibroblast growth factor-2 in MC3T3-E1 osteoblast-like cells. *J Cell Biochem,* **80,** 550–559 (2001).

80. P. Van Vlasselaer, B. Borremans, U. van Gorp, J. R. Dasch, and R. De Waal-Malefyt, Interleukin 10 inhibits transforming growth factor-beta (TGF-beta) synthesis required for osteogenic commitment of mouse bone marrow cells. *J Cell Biol,* **124,** 569–577 (1994).

81. R. F. Klein and K. M. Wiren, Regulation of osteoblastic gene expression by lead. *Endocrinology,* **132,** 2531–2537 (1993).

82. M. V. Hernandez, N. Guanabens, L. Alvarez, A. Monegal, P. Peris, J. Riba, G. Ercilla, M. J. Martinez de Osaba, and J. Munoz-Gomez, Immunocytochemical evidence on the effects of glucocorticoids on type I collagen synthesis in human osteoblastic cells. *Calcif Tissue Int,* **74,** 284–293 (2004).

83. A. Mahonen, A. Jukkola, L. Risteli, J. Risteli, and P. H. Maenpaa, Type I procollagen synthesis is regulated by steroids and related hormones in human osteosarcoma cells. *J Cell Biochem,* **68,** 151–163 (1998).

84. A. L. Boskey, T. L. Wright, and R. D. Blank, Collagen and bone strength. *J Bone Miner Res,* **3,** 330–335 (1999).

85. P. H. Byers, Brittle bones—Fragile molecules: Disorders of collagen gene structure and expression. *Trends Genet,* **6,** 293–300 (1990).

86. H. Kuivaniemi, G. Tromp, and D. J. Prockop, Mutations in collagen genes: Causes of rare and some common diseases in humans. *FASEB J,* **5,** 2052–2060 (1991).

87. K. Harbers, M. Juehan, H. Delius, and R. Jaenisch, Insertion of retrovirus into the first intron of α1(I) collagen gene leads to embryonic fetal mutation in mice. *Proc Natl Acad Sci USA,* **81,** 1504–1508 (1984).

88. S. D. Chipman, H. O. Sweet, D. J. McBride, M. T. Davidson, S. C. Marks Jr., A. R. Schuldiner, D. W. Rowe, R. J. Wenstrup, and J. R. Shapiro, Defective pro α 2 (I) collagen synthesis in a new recessive mutation in mice: A potential model for human osteogenesis imperfecta. *Proc Natl Acad Sci USA,* **90,** 1701–1705 (1993).

89. J. Bonadio and S. A. Goldstein, Our understanding of inherited skeletal fragility and what this has taught us about bone structure and function. In *Cellular and Molecular Biology of Bone* (M. Noda, ed.), pp. 169–189. Academic Press, San Diego (1993).

90. A. Forlino, F. D. Porter, E. J. Lee, H. Westphal, and J. C. Marini, Use of the Cre/lox recombination system to develop a non-lethal knock-in murine model for osteogenesis imperfecta with an alpha1(I) G349C substitution. Variability in phenotype in BrtlIV mice. *J Biol Chem,* **274,** 37923–37931 (1999).

91. M. J. Silva, M. D. Brodt, B. Wopenka, S. Thomopoulos, D. Williams, M. H. Wassen, M. Ko, N. Kusano, and R. A. Bank, Decreased collagen organization and content are associated with reduced strength of demineralized and intact bone in the SAMP6 mouse. *J Bone Miner Res,* **21,** 78–88 (2006).

92. U. Vetter, E. D. Eanes, J. B. Kopp, J. D. Termine, and P. Gehron Robey, Changes in apatite crystal size in bones of patients with osteogenesis imperfecta. *Calcif Tissue Int,* **49,** 248–250 (1991).

93. W. Traub, T. Arad, U. Vetter, and S. Weiner, Ultrastructural studies of bones from patients with osteogenesis imperfecta. *Matrix Biol,* **14,** 337–345 (1994).

94. P. Fratzl, O. Paris, K. Klaushofer, and W. J. Landis, Bone mineralization in an osteogenesis imperfecta mouse model studied by small-angle x-ray scattering. *J Clin Invest,* **97,** 396–402 (1996).

95. N. P. Camacho, L. Hou, T. R. Toledano, W. A. Ilg, C. F. Brayton, C. L. Raggio, L. Root, and A. L. Boskey, The material basis for reduced mechanical properties in oim mice bones. *J Bone Miner Res,* **14,** 264–272 (1999).

96. J. W. Freeman and F. H. Silver, Analysis of mineral deposition in turkey tendons and self-assembled collagen fibers using mechanical techniques. *Connect Tissue Res,* **45,** 131–141 (2004).

97. P. Sarathchandra, F. M. Pope, and S. Y. Ali, Morphometric analysis of type I collagen fibrils in the osteoid of osteogenesis imperfecta. *Calcif Tissue Int,* **65,** 390–395 (1999).

98. T. Saito, M. Yamauchi, Y. Abiko, K. Matsuda, and M. A. Crenshaw, *In vitro* apatite induction by phosphophoryn immobilized on modified collagen fibrils. *J Bone Miner Res,* **15,** 1615–1619 (2000).

99. G. He and A. George, Dentin matrix protein 1 immobilized on type I collagen fibrils facilitates apatite deposition *in vitro*. *J Biol Chem,* **279,** 11649–11656 (2004).

100. A. L. Boskey, Biomineralization: Conflicts, challenges, and opportunities. *J Cell Biochem Suppl,* **30/31,** 83–91 (1998).

101. N. B. Schwartz, E. W. Pirok 3rd, J. R. Mensch Jr., and M. S. Domowicz, Domain organization, genomic structure, evolution, and regulation of expression of the aggrecan gene family. *Prog Nucleic Acid Res Mol Biol,* **62,** 177–225 (1999).

102. K. M. Dyne, M. Valli, A. Forlino, M. Mottes, H. Kress, and G. Cetta, Deficient expression of the small proteoglycan decorin in a case of severe/lethal osteogenesis imperfecta. *Am J Med Genet,* **53,** 161–166 (1996).

103. N. S. Fedarko, Isolation and purification of proteoglycans. *EXS,* **70,** 9–35 (1994).

104. H. Kresse, H. Hausser, and E. Schinherr, Small proteoglycans. In *Proteoglycans* (P. Jollàs, ed.), pp. 73–100. Birkhauser Verlag, Basel, Switzerland (1994).

105. M. Yanagishita, Function of proteoglycans in the extracellular matrix. *Acta Pathol Japonica,* **43,** 283–293 (1993).

106. P. J. Neame, H. Tapp, and A. Azizan, Noncollagenous, nonproteoglycan macromolecules of cartilage. *Cell Mol Life Sci,* **55,** 1327–1340 (1996).

107. F. Deak, D. Piecha, C. Bachrati, M. Paulsson, and I. Kiss, Primary structure and expression of matrilin-2, the closest relative of cartilage matrix protein within the von Willebrand factor type A-like module superfamily. *J Biol Chem,* **272,** 9268–9274 (1997).

108. J. R. Korenberg, X. N. Chen, K. Doege, J. Grover, and P. J. Roughley, Assignment of the human aggrecan gene (AGC1) to 15q26 using fluorescence *in situ* hybridization analysis. *Genomics,* **16,** 546–548 (1993).

109. K. J. Doege, M. Sasaki, E. Horigan, J. R. Hassell, and Y. Yamada, Complete primary structure of the rat cartilage proteoglycan core protein deduced from cDNA clones. *J Biol Chem,* **262,** 17757–17767 (1987).

110. K. J. Doege, K. Garrison, S. N. Coulter, and Y. Yamada, The structure of the rat aggrecan gene and preliminary characterization of its promoter. *J Biol Chem,* **269,** 29232–29240 (1994).

111. J. D. Sandy, C. R. Flannery, R. E. Boynton, and P. J. Neame, Isolation and characterization of disulfide-bonded peptides from the three globular domains of aggregating cartilage proteoglycan. *J Biol Chem,* **265,** 21108–21113 (1990).

112. M. Morgelin, M. Paulsson, T. E. Hardingham, D. Heinegard, and J. Engel, Cartilage proteoglycans. Assembly with hyaluronate and link protein as studied by electron microscopy. *Biochem J,* **253,** 175–185 (1988).

113. T. E. Hardingham, The role of link protein in the structure of cartilage proteoglycan aggregates. *Biochem J,* **177,** 237–247 (1979).

114. L. Kjellän and U. Lindahl, Proteoglycans: Structures and interactions. *Annu Rev Biochem,* **60,** 443–475 (1991).

115. D. Heinegard and A. Oldberg, Structure and biology of cartilage and bone matrix is noncollagenous macromolecules. *FASEB J,* **3,** 2042–3053 (1989).

116. H. Paulssen, R. Busch, V. Sinnwell, and A. Poller-Kruger, Conformation analysis of D-xylose-containing O-glycopeptide sequences. *Carbohydr Res,* **214,** 227–234 (1991).

117. J. E. Scott, Proteoglycan: Collagen interactions in connective tissues. Ultrastructural, biochemical, functional and evolutionary aspects. *Int J Biol Macromol,* **13,** 157–161 (1991).

118. H. Watanabe and Y. Yamada, Chondrodysplasia of gene knockout mice for aggrecan and link protein. *Glycoconjugate J,* **19,** 269–273 (2002).

119. W. D. Comper and T. C. Laurent, Physiological function of connective tissue polysaccharides. *Physiol Rev,* **58,** 255–315 (1978).

120. I. Nishimura, Y. Muragaki, and B. R. Olsen, Tissue-specific forms of type IX collagen–proteoglycan arise from the use of two widely separated promoters. *J Biol Chem,* **264,** 20033–20041 (1989).

121. G. H. Hunter, Role of proteoglycan in the provisional calcification of cartilage. *Clin Orthop Rel Res,* **262,** 256–263 (1991).

122. C. C. Chen, A. L. Boskey, and L. C. Rosenberg, The inhibitory effect of cartilage proteoglycans on hydroxyapatite growth. *Calcif Tissue Int,* **36,** 285–290 (1984).

123. C. C. Chen and A. L. Boskey, Mechanisms of proteoglycan inhibition of hydroxyapatite growth. *Calcif Tissue Int,* **37,** 395–400 (1985).

124. A. L. Boskey, Hydroxyapatite formation in a dynamic gel system: Effects of type I collagen, lipids, and proteoglycans. *J Phys Chem,* **93,** 1628–1633 (1989).

125. D. E. Eanes, Mixed phospholipid liposome calcification. *Bone Miner,* **17,** 269–272 (1992).

126. G. K. Hunter, An ion-exchange mechanism of cartilage calcification. *Connect Tissue Res,* **16,** 111–120 (1987).

127. A. L. Boskey and S. C. Marks, Mineral and matrix alterations in the bones of incisors-absent (ia/ia) osteopetrotic rat. *Calcif Tissue Int,* **37,** 287–292 (1985).

128. M. Yamagata, T. Shinomura, and K. Kimata, Tissue variation of two large chondroitin sulfate proteoglycans (PG-M/versican and PG-H/aggrecan) in chick embryos. *Anat Embryol,* **187,** 433–444 (1993).

129. R. V. Iozzo, M. F. Naso, L. A. Cannizzaro, J. J. Wasmuth, and J. D. McPherson, Mapping of the versican proteoglycan gene (CSPG2) to the long arm of human chromosome 5 (5q12–5q14). *Genomics,* **14,** 845–851 (1992).

130. K. Ito, T. Shinomura, M. Zako, M. Ujita, and K. Kimata, Multiple forms of mouse PG-M, a large chondroitin sulfate proteoglycan generated by alternative splicing. *J Biol Chem,* **270,** 958–965 (1995).

131. D. R. Zimmermann and E. Ruoslahti, Multiple domains of the large fibroblast proteoglycan, versican. *EMBO J,* **8,** 2575–2588 (1990).

132. M. F. Naso, D. R. Zimmermann, and R. V. Iozzo, Characterization of the complete genomic structure of the human versican gene and functional analysis of its promoter. *J Biol Chem,* **269,** 32999–33008 (1994).

133. M. T. Dours-Zimmermann and D. R. Zimmermann, A novel glycosaminoglycan attachment domain identified in two alternative splice variants of human versican. *J Biol Chem,* **269,** 32992–32998 (1994).

134. M. Nakamura, S. Sone, I. Takahashi, I. Mizoguchi, S. Echigo, and Y. Sasano, Expression of versican and ADAMTS1, 4, and 5 during bone development in the rat mandible and hind limb. *J Histochem Cytochem,* **53,** 1553–1562 (2005).

135. E. W. Mandl, H. Jahr, J. L. Koevoet, J. P. van Leeuwen, H. Weinans, J. A. Verhaar, and G. J. van Osch, Fibroblast growth factor-2 in serum-free medium is a potent mitogen and reduces dedifferentiation of human ear chondrocytes in monolayer culture. *Matrix Biol,* **23,** 231–241 (2004)

136. Y. Zhang, L. Cao, C. Kiani, B. L. Yang, W. Hu, and B. B. Yang, Promotion of chondrocyte proliferation by versican mediated by G1 domain and EGF-like motifs. *J Cell Biochem,* **73,** 445–457 (1999).

137. M. E. Antosz, C. G. Bellows, and J. E. Aubin, Biphasic effects of epidermal growth factor on bone nodule formation by isolated rat calvaria cells *in vitro*. *J Bone Miner Res,* **2,** 385–393 (1987).

138. S. P. Henry, M. Takanosu, T. C. Boyd, P. M. Mayne, H. Eberspaecher, W. Zhou, B. de Crombrugghe, M. Hook, and R. Mayne, Expression pattern and gene characterization of asporin. A newly discovered member of the leucine-rich repeat protein family. *J Biol Chem,* **276,** 12212–12221 (2001).

139. G. A. Pringle and C. M. Dodd, Immunoelectron microscopy localization of the core protein of decorin near the d and e bands of tendon collagen fibrils by use of monoclonal antibodies. *J Histochem Cytochem,* **38,** 1405–1411 (1990).

140. M. F. Young, L. W. Fisher, and O. W. McBride, Chromosomal localization of bone matrix PGI (biglycan), PGII (decorin), osteopontin and bone sialoprotein in the human genome. *J Bone Miner Res,* **4,** S380 (1989).

141. K. G. Danielson, A. Fazzio, I. Cohen, L. A. Cannizzaro, I. Eichstetter, and R. V. Iozzo, The human decorin gene: Intron–exon organization, discovery of two alternatively spliced exons in the 5 untranslated region, and mapping of the gene to chromosome 12q23. *Genomics,* **15,** 146–160 (1993).

142. U. Vetter, W. Vogel, W. Just, M. F. Young, and L. W. Fisher, Human decorin gene: Intron–exon junctions and chromosomal localization. *Genomics,* **15,** 161–168 (1993).

143. T. Scholzen, M. Solursh, S. Suzuki, R. Reiter, J. L. Morgan, A. M. Buchberg, L. D. Siracusa, and R. V. Iozzo, The murine decorin, complete cDNA cloning genomic organization, chromosomal assignment, and expression during organogenesis and tissue differentiation. *J Biol Chem,* **269,** 28270–28281 (1994).

144. L. W. Fisher, G. R. Hawkins, N. Tuross, and J. D. Termine, Purification and partial characterization of small proteoglycans I and II, bone sialoproteins I and II, and osteonectin from the mineral compartment of developing human bone. *J Biol Chem,* **262,** 9702–9708 (1987).

145. T. Krusius, K. R. Gehlsen, and E. A. Ruoslahti, Fibroblast chondroitin sulfate proteoglycan core protein contains lectin-like and growth factor-like sequences. *J Biol Chem,* **262,** 13120–13125 (1987).

146. A. A. Day, C. I. McQuillan, J. D. Termine, and M. R. Young, Molecular cloning and sequence analysis of the cDNA for small proteoglycan II of bovine bone. *Biochem J,* **248,** 801–805 (1987).

147. M. Takagi, T. Yamada, N. Kamiya, T. Kumagai, and A. Yamaguchi, Effects of bone morphogenetic protein-2 and transforming growth factor-beta1 on gene expression of decorin and biglycan by cultured osteoblastic cells. *Histochem J,* **31,** 403–409 (1999).

148. T. Yamada, N. Kamiya, D. Harada, and M. Takagi, Effects of transforming growth factor-beta1 on the gene expression of decorin, biglycan, and alkaline phosphatase in osteoblast precursor cells and more differentiated osteoblast cells. *Histochem J,* **31,** 687–694 (1999).

149. S. Kimoto, S. L. Cheng, S. F. Zhang, and L. V. Avioli, The effect of glucocorticoid on the synthesis of biglycan and decorin in human osteoblasts and bone marrow stromal cells. *Endocrinology,* **135,** 2423–2431 (1994).

150. S. L. Cheng, S. F. Zhang, T. L. Nelson, P. M. Warlow, and R. Civitelli, Stimulation of human osteoblast differentiation and function by ipriflavone and its metabolites. *Calcif Tissue Int,* **55,** 356–362 (1994).

151. N. A. Visser, G. P. Vankampen, M. H. Dekoning, and J. K. Vanderkorst, The effects of loading on the synthesis of biglycan and decorin in intact mature articular cartilage in vitro. *Connect Tissue Res,* **30,** 241–250 (1994).

152. R. S. Sawhney, T. M. Hering, and L. J. Sandell, Biosynthesis of small proteoglycan II (decorin) by chondrocytes and evidence for a procore protein. *J Biol Chem,* **266,** 9231–9240 (1991).

153. K. G. Vogel and J. A. Trotter, The effect of proteoglycans on the morphology of collagen fibrils formed in vitro. *Collagen Rel Res,* **7,** 105–114 (1987).

154. D. J. Bidanset, C. Guidry, L. C. Rosenberg, H. Choi, R. Timpl, and M. Hook, Binding of the proteoglycan decorin to collagen type VI. *J Biol Chem,* **267,** 5250–5256 (1992).

155. A. Corsi, T. Xu, X. D. Chen, A. Boyde, J. Liang, M. Mankani, B. Sommer, R. V. Iozzo, I. Eichstetter, P. G. Robey, P. Bianco, and M. F. Young, Phenotypic effects of biglycan deficiency are linked to collagen fibril abnormalities, are synergized by decorin deficiency, and mimic Ehlers–Danlos-like changes in bone and other connective tissues. *J Bone Miner Res,* **17,** 1180–1189 (2002).

156. P. M. Steinlen, E. Maessen, J. Kresse, and I. M. Van Vlijmen, Expression of tenascin, biglycan and decorin in disorders of keratinization. *Br J Dermatol,* **130,** 564–568 (1994).

157. N. S. Fedarko, M. Moerike, R. Brenner, P. Gehron Robey, and U. Vetter, Extracellular matrix formation by osteoblasts from patients with osteogenesis imperfecta. *J Bone Miner Res,* **7,** 921–930 (1992).

158. N. S. Fedarko, P. Gehron Robey, and U. K. Vetter, Extracellular matrix stoichiometry in osteoblasts from patients with osteogenesis imperfecta. *J Bone Miner Res,* **10,** 1122–1129 (1995).

159. K. M. Dyne, M. Valli, M. Forlino, M. Mottes, and G. Cetta, Decreased decorin expression in patients with osteogenesis imperfecta. FECTS meeting, Lyon, France, E6 (1994).

160. P. Bianco, L. W. Fisher, M. F. Young, J. D. Termine, and P. G. Robey, Expression and localization of the two small proteoglycans biglycan and decorin in developing human skeletal and non-skeletal tissues. *J Histochem Cytochem,* **38,** 1549–1563 (1990).

161. D. V. Brown and K. C. Vogel, Characteristics of the in vitro interaction of a small proteoglycan (PGII) of bovine tendon with type I collagen. *Matrix,* **9,** 468–478 (1989).

162. A. L. Boskey, L. Spevak, S. B. Doty, and L. Rosenberg, Effects of bone CS-proteoglycans, DS-decorin, and DS-biglycan on hydroxyapatite formation in a gelatin gel. *Calcif Tissue Int,* **61,** 298–305 (1997).

163. M. Goldberg, D. Septier, O. Rapoport, R. V. Iozzo, M. F. Young, and L. G. Ameye, Targeted disruption of two small leucine-rich proteoglycans, biglycan and decorin, excerpts divergent effects on enamel and dentin formation. *Calcif Tissue Int,* **77,** 297–310 (2005).

164. L. W. Fisher, J. D. Termine, and M. F. Young, Deduced protein sequence of bone small proteoglycan I (biglycan) shows homology with proteoglycan II (decorin) and several non-connective tissue proteins in a variety of species. *J Biol Chem,* **264,** 4571–4576 (1989).

165. T. Krusius and E. Ruoslahti, Primary structure of an extracellular matrix proteoglycan core protein deduced from cloned cDNA. *Proc Natl Acad Sci USA,* **83,** 7683–7687 (1986).

166. L. W. Fisher, A. M. Heegaard, U. Vetter, W. Vogel, W. Just, J. D. Termine, and M. F. Young, Human biglycan gene. *J Biol Chem,* **266,** 14371–14377 (1991).

167. A. M. Heegaard, P. Gehron Robey, W. Vogel, W. Just, R. L. Widom, J. Scholler, J. Fisher, and M. F. Young, Functional characterization of the human biglycan 5′-flanking DNA and binding of the transcription factor c-Krox. *J Bone Miner Res,* **12,** 2050–2060 (1997).

168. P. G. Neame, H. C. Choi, and L. C. Rosenberg, The primary structure of the core protein of the small leucine-rich proteoglycan (PG-I) from bovine articular cartilage. *J Biol Chem,* **264,** 8653–8661 (1989).

169. A. M. Heegaard, Z. Xie, M. F. Young, and K. L. Nielsen, Transforming growth factor beta stimulation of biglycan gene expression is potentially mediated by sp1 binding factors. *J Cell Biochem*, **93,** 463–475 (2004).

170. Y. Takeuchi, T. Matsumoto, E. Ogata, and Y. Shishiba, Effects of transforming growth factor α1 and L-ascorbate on synthesis and distribution of proteoglycans in murine osteoblast-like cells. *J Bone Miner Res*, **8,** 823–829 (1993).

171. D. Pearson and J. Sasse, Differential regulation of biglycan and decorin by retinoic acid in bovine chondrocytes. *J Biol Chem*, **267,** 25364–25370 (1992).

172. P. G. Robey, N. S. Fedarko, and W. Lindner, Biosynthesis of a small proteoglycan (PG-I) by human bone cells. *J Cell Biol*, **105,** 299a (1987).

173. S. Kimoto, S.-L. Cheng, S.-F. Zhang, and L. V. Avioli, The effect of glucocorticoid on the synthesis of biglycan and decorin in human osteoblasts and bone marrow stromal cells. *Endocrinology*, **135,** 2423–2431 (1994).

174. R. J. Waddington and M. S. Langley, Structural analysis of proteoglycans synthesized by mineralizing bone cells *in vitro* in the presence of fluoride. *Matrix Biol*, **17,** 255–268 (1998).

175. L. Ameye and M. F. Young, Mice deficient in small leucine-rich proteoglycans: Novel *in vivo* models for osteoporosis, osteoarthritis, Ehlers–Danlos syndrome, muscular dystrophy, and corneal diseases. *Glycobiology*, **12,** 107R–116R (2002).

176. T. Xu, P. Bianco, L. W. Fisher, G. Longenecker, E. Smith, S. Goldstein, J. Bonadio, A. L. Boskey, A. M. Heegaard, B. Sommer, K. Satomura, P. Dominguez, C. Zhao, A. B. Kulkarni, P. G. Robey, and M. F. Young, Targeted disruption of the biglycan gene leads to an osteoporosis-like phenotype in mice. *Nat Genet*, **20,** 78–82 (1998).

177. H. Greiling, Structure and biological functions of keratan sulfate proteoglycans. *EXS*, **70,** 101–102 (1994).

178. S. Ekman and D. Heinegard, Immunohistochemical localization of matrix proteins in the femoral joint cartilage of growing commercial pigs. *Vet Pathol*, **29,** 514–520 (1992).

179. R. Sztrolovics, X. N. Chen, J. Grover, P. J. Roughley, and J. R. Korenberg, Localization of the human fibromodulin gene (FMOD) to chromosome 1q32 and completion of the cDNA sequence. *Genomics*, **23,** 715–717 (1994).

180. E. Hedbom and D. Heinegård, Binding of fibromodulin and decorin to separate sites on fibrillar collagens. *J Biol Chem*, **268,** 27307–27312 (1993).

181. H. Hedlund, S. Mengarelli-Widholm, D. Heinegard, F. P. Reinhold, and O. Svensson, Fibromodulin distribution and association with collagen. *Matrix Biol*, **14,** 227–232 (1994).

182. A. Hulth, O. Johnell, L. Lindberg, and D. Heinegard, Sequential appearance of macromolecules in bone induction in the rat. *J Orthop Res*, **11,** 367–378 (1993).

183. Y. Asawa, K. Aoki, K. Ohya, H. Ohshima, and Y. Takano, Appearance of electron-dense segments: Indication of possible conformational changes of pre-mineralizing collagen fibrils in the osteoid of rat bones. *J Electron Microsc (Tokyo)*, **53,** 423–433 (2004).

184. L. Svensson, A. Aszodi, F. P. Reinholt, R. Fassler, D. Heinegard, and A. Oldberg, Fibromodulin-null mice have abnormal collagen fibrils, tissue organization, and altered lumican deposition in tendon. *J Biol Chem*, **274,** 9636–9647 (1999).

185. M. Goldberg, D. Septier, A. Oldberg, M. F. Young, and L. G. Ameye, Fibromodulin-deficient mice display impaired collagen fibrillogenesis in predentin as well as altered dentin mineralization and enamel formation. *J Histochem Cytochem*, **65,** 11545–11552 (2005).

186. M. Wendel, Y. Sommarin, and D. Heinegard, Bone matrix proteins: Isolation and characterization of a novel cell-binding keratan sulfate proteoglycan (osteoadherin) from bovine bone. *J Cell Biol*, **141,** 839–847 (1998).

187. Y. Sommarin, M. Wendel, Z. Shen, U. Hellman, and D. Heinegard, Osteoadherin, a cell-binding keratan sulfate proteoglycan in bone, belongs to the family of leucine-rich repeat proteins of the extracellular matrix. *J Biol Chem*, **273,** 16723–16729 (1998).

188. R. Buchaille, M. L. Couble, H. Magloire, and F. Bleicher, Expression of the small leucine-rich proteoglycan osteoadherin/osteomodulin in human dental pulp and developing rat teeth. *Bone*, **27,** 265–270 (2000).

189. L. Madisen, M. Neubauer, G. Plowman, D. Rosen, P. Segarini, J. Dasch, A. Thompson, J. Ziman, H. Bentz, and A. F. Purchio, Molecular cloning of a novel bone-forming compound: Osteoinductive factor. *DNA Cell Biol*, **9,** 303–309 (1990).

190. J. L. Funderburgh, L. M. Corpuz, M. R. Roth, M. L. Funderburgh, E. S. Tasheva, and G. W. Conrad, Mimecan, the 25-kDa corneal keratan sulfate proteoglycan, is a product of the gene producing osteoglycin. *J Biol Chem*, **272,** 28089–28095 (1997).

191. N. S. Fedarko, P. Bianco, U. K. Vetter, and P. G. Robey, Human bone cell enzyme expression and cellular heterogeneity: Correlation of alkaline phosphatase enzyme activity with cell cycle. *J Cell Physiol*, **144,** 115–121 (1990).

192. B. de Bernard, P. Bianco, E. Bonucci, M. Costantini, G. C. Lunazzi, P. Martinuzzi, C. Modricky, L. Moro, E. Panfili, and P. Pollesello, Biochemical and immunohistochemical evidence that in cartilage an alkaline phosphatase is a Ca⁺⁺-binding glycoprotein. *J Cell Biol*, **103,** 1615–1623 (1986).

193. D. M. Swallow, Mapping of the gene coding for the human liver/bone/kidney isozyme of alkaline phosphatase to chromosome 1. *Ann Hum Genet*, **60,** 229–235 (1986).

194. H. H. Hsu, The partial sequencing of human nonspecific (bone/liver/kidney) alkaline phosphatase gene. *Int J Biochem*, **23,** 391–393 (1991).

195. P. Llinas, E. A. Stura, A. Menez, Z. Kiss, T. Stigbrand, J. L. Millan, and M. H. Le Du, Structural studies of human placental alkaline phosphatase in complex with functional ligands. *J Mol Biol*, **350,** 441–451 (2005).

196. J. E. Beever and H. A. Lewin, RFLP at the bovine liver, bone and kidney alkaline phosphatase (ALPL) locus. *Anim Genet*. **23,** 577 (1992).

197. M. Terao, M. Studer, M. Gianni, and E. Garattini, Isolation and characterization of the mouse liver/bone/kidney-type alkaline phosphatase gene. *Biochem J*, **268,** 641–648 (1990).

198. T. Harada, I. Koyama, T. Matsunaga, A. Kikuno, T. Kasahara, M. Hassimoto, D. H. Alpers, and T. Komoda, Characterization of structural and catalytic differences in rat intestinal alkaline phosphatase isozymes. *FEBS J*, **272,** 2477–2486 (2005).

199. E. E. Kim and H. W. Wyckoff, Reaction mechanisms of alkaline phosphatase based on crystal structures. Two-metal ion catalysis. *J Mol Biol (England)* **218,** 449–464 (1991).

200. S. Hansen, L. K. Hansen, and E. Hough, The crystal structure of tris-inhibited phospholipase C from *Bacillus cereus* at 1.9 A resolution: The nature of the metal ion in site 2. *J Mol Biol*, **231,** 870–876 (1993).

201. S. Nagoya, T. Uede, T. Wada, S. Ishii, S. Yamawaki, and K. Kikuchi, Detection of bone-type alkaline phsophatase by monoclonal antibodies reacting with human osteosarcoma-associated antigen. *Jpn J Cancer Res*, **82,** 862–870 (1991).

202. M. G. Low, The glycosyl-phosphatidylinositol anchor of membrane proteins. *Biochim Biophys Acta*, **988,** 427–454 (1989).

203. M. Noda, K. Yoon, G. A. Rodan, and D. E. Koppel, High lateral mobility of endogenous and transfected alkaline phosphatase:

A phosphatidylinositol anchored membrane protein. *J Cell Biol*, **105**, 1671–1677 (1987).

204. V. V. Golovastov, N. I. Mikhaleva, and M. A. Nesmeyanova, Depletion of phosphatidylethanolamine—the major membrane phospholipid of *Escherichia coli*—depresses post-translocational modification of alkaline phosphatase in the periplasm. *Biochemistry (Moscow)*, **67**, 765–769 (2002).

205. M. Studer, M. Terao, M. Gianni, and E. Garattini, Characterization of a second promoter for the mouse liver/bone/kidney-type alkaline phosphatase gene: Cell and tissue specific expression. *Biochem Biophys Res Commun*, **179**, 1352–1360 (1991).

206. N. Yusa, K. Watanabe, S. Yoshida, N. Shirafuji, S. Shimomura, K. Tani, S. Asano, and N. Sato, Transcription factor Sp3 activates the liver/bone/kidney-type alkaline phosphatase promoter in hematopoietic cells. *J Leukocyte Biol*, **68**, 772–777 (2000).

207. Q. Xie and D. H. Alpers, The two isozymes of rat intestinal alkaline phosphatase are products of two distinct genes. *Physiol Genomics*, **3**, 1–8 (2000).

208. J. Zernik, B. Kream, and K. Twarog, Tissue-specific and dexamethasone-inducible expression of an alkaline phosphatase from alternative promoters of the rat bone/liver/kidney/placenta gene. *Biochem Biophys Res Commun*, **176**, 1149–1156 (1991).

209. S. Matsuura, F. Kishi, and T. Kajii, Characterization of a 5′-flanking region of the human liver/bone/kidney alkaline phosphatase gene: Two kinds of mRNA from a single gene. *Biochem Biophys Res Commun*, **168**, 993–1000 (1990).

210. E. Kyeyune-Nyombi, K. H. Lau, D. J. Baylink, and D. D. Strong, 1,25-Dihydroxyvitamin D3 stimulates both alkaline phosphatase gene transcription and mRNA stability in human bone cells. *Arch Biochem Biophys*, **291**, 316–325 (1991).

211. M. J. Municio and M. L. Traba, Effects of 24,25(OH)2D3, 1,25(OH)2D3 and 25(OH)D3 on alkaline and tartrate-resistant acid phosphatase activities in fetal rat calvaria. *J Physiol Biochem*, **60**, 219–224 (2004).

212. B. D. Boyan, R. Batzer, K. Kieswetter, Y. Liu, D. L. Cochran, S. Szmuckler-Moncler, D. D. Dean, and Z. Schwartz, Titanium surface roughness alters responsiveness of MG63 osteoblast-like cells to 1 alpha,25-(OH)2D3. *J Biomed Mater Res*, **39**, 77–85 (1998).

213. J. K. Heath, L. J. Suva, K. Yoon, M. Kiledjian, T. J. Martin, and G. A. Rodan, Retinoic acid stimulates transcriptional activity from the alkaline phosphatase promoter in the immortalized rat calvarial cell line, RCT-1. *Mol Endocrinol*, **6**, 636–646 (1992).

214. H. Nagasawa, S. Takahashi, A. Kobayashi, H. Tazawa, Y. Tashima, and K. Sato, Effect of retinoic acid on murine preosteoblastic MC3T3-E1 cells. *J Nutr Sci Vitaminol (Tokyo)*, **51**, 311–318 (2005).

215. H. M. Song, R. P. Nacamuli, W. Xia, A. S. Bari, Y. Y. Shi, T. D. Fang, and M. T. Longaker, High-dose retinoic acid modulates rat calvarial osteoblast biology. *J Cell Phsiol*, **202**, 255–262 (2005).

216. M. Subramaniam, D. Colvard, P. E. Keeting, K. Rasmussen, B. L. Riggs, and T. C. Spelsberg, Glucocorticoid regulation of alkaline phosphatase, osteocalcin, and proto-oncogenes in normal human osteoblast-like cells. *J Cell Biochem*, **50**, 411–424 (1992).

217. H. Perinpanayagam, T. Martin, V. Mithal, M. Dahman, N. Marzec, J. Lampasso, and R. Dziak, Alveolar bone osteoblast differentiation and Runx2/Cbfa1 expression. *Arch Oral Biol*. **51**, 406–415 (2005).

218. L. Stangenberg, D. J. Schaefer, O. Buettner, J. Ohnolz, D. Mobest, R. E. Horch, G. B. Stark, and U. Kneser, Differentiation of osteoblasts in three-dimensional culture in processed cancellous bone matrix: Quantitative analysis of gene expression based on real-time reverse transcription-polymerase chain reaction. *Tissue Eng*, **11**, 855–864 (2005).

219. M. Takagi, N. Kamiya, T. Takahashi, S. Ito, M. Hasegawa, N. Suzuki, and K. Nakanishi, Effects of bone morphogenetic protein-2 and transforming growth factor beta1 on gene expression of transcription factors, AJ18 and Runx2 in cultured osteoblastic cells. *J Mol Histol*, **35**, 81–90 (2004).

220. W. Zhu, B. A. Rawlins, O. Boachie-Adjei, E. R. Myers, J. Arimizu, E. Choi, J. R. Lieberman, R. G. Crystal, and C. Hidaka, Combined bone morphogenetic protein-2 and -7 gene transfer enhances osteoblastic differentiation and spine fusion in a rodent model. *J Bone Miner Res*, **19**, 2021–2032 (2004).

221. A. M. Osyczka, D. L. Diefenderfer, G. Bhargave, and P. S. Leboy, Different effects of BMP-2 on marrow stromal cells from human and rat bone. *Cell Tissues Organs*, **176**, 109–119 (2004).

222. D. L. Diefenderfer, A. M. Osyczka, G. C. Reilly, and P. S. Leboy, BMP responsiveness in human mesenchymal stem cells. *Connect Tissue Res*, **44**(Suppl. 1), 305–311 (2003).

223. J. R. Farley, S. L. Hall, and S. Herring, Calcitonin acutely increases net 45Ca uptake and alters alkaline phosphatase specific activity in human osteosarcoma cells. *Metabolism*, **42**, 97–104 (1993).

224. P. J. Price, B. O. Oyajobi, R. O. Oreffo, and R. G. G. Russell, Cells cultured from the growing tip of red dear antler express alkaline phosphatase and proliferate in response to insulin-like growth factor-I. *J Endocrinol*, **143**, R9–R16 (1994).

225. P. Bianco, M. Riminucci, G. Silvestrini, E. Bonucci, J. D. Termine, L. W. Fisher, and P. Gehron Robey, Localization of bone sialoprotein (BSP) to Golgi and post-Golgi secretory structures in osteoblasts and to discrete sites in early bone matrix. *J Histochem Cytochem*, **41**, 193–203 (1993).

226. P. Bianco and A. Boyde, Alkaline phosphatase cytochemistry in confocal scanning light microscopy for imaging the bone marrow stroma. *Basic Appl Histochem*, **33**, 17–23 (1989).

227. P. Bianco, N. S. Fedarko, E. Bonucci, J. D. Termine, and P. Gehron Robey, Acid and alkaline phosphatase activities of bone cells revisited. In *Proceedings of the International Society for Bioanaloging Skeletal Implants*, pp. 17–27 (1991).

228. P. Bianco, P. Ballanti, S. Mazzaferro, G. Coen, and E. Bonucci, Combined use of tetracycline fluorescence and alkaline phosphatase cytochemistry in the evaluation of osteoblastic activity in bone biopsies. *Basic Appl Histochem*, **31**(Suppl.), 29 (1987).

229. J. Zernik, K. Twarog, and W. B. Upholt, Regulation of alkaline phosphatase and $\alpha 2(\mathrm{I})$ procollagen synthesis during early intramembranous bone formation in the rat mandible. *Differentiation*, **44**, 207–215 (1990).

230. S. Sakano, Y. Murata, T. Miura, H. Iwata, K. Sato, N. Matsui, and H. Seo, Collagen and alkaline phosphatase gene expression during bone morphogenetic protein (BMP)-induced cartilage and bone differentiation. *Clin Orthop*, **292**, 337–344 (1993).

231. P. Collin, J. R. Nefussi, A. Wetterwald, V. Nicolas, M. L. Boy-Lefever, H. Fleisch, and N. Forest, Expression of collagen, osteocalcin and bone alkaline phosphatase in a mineralizing rat osteoblastic cell culture. *Calcif Tissue Int*, **50**, 175–183 (1992).

232. M. J. Weiss, D. E. C. Cole, K. Ray, M. P. Whyte, M. A. Lafferty, R. A. Mulivor, and H. Harris, A misuse mutation in the human liver/bone/kidney alkaline phosphatase causing a lethal form of hypophosphatasia. *Proc Natl Acad Sci USA,* **85,** 7666–7669 (1989).

233. I. Brun-Heath, A. Taillandier, J. L. Serre, and E. Mornet, Characterization of 11 novel mutations in the tissue non-specific alkaline phosphatase gene responsible for hypophosphatasia and genotype–phenotype correlations. *Mol Genet Metab,* **84,** 273–277 (2005).

234. K. Yoon, E. Golub, and G. A. Rodan, Alkaline phosphatase transfected cells promote calcium and phosphate deposition. *Connect Tissue Res,* **22,** 17–25 (1989).

235. K. N. Fedde, L. Blair, J. Silverstein, S. P. Coburn, L. M. Ryan, R. S. Weinstein, K. Waymire, S. Narisawa, J. L. Millan, G. R. MacGregor, and M. P. Whyte, Alkaline phosphatase knock-out mice recapitulate the metabolic and skeletal defects of infantile hypophosphatasia. *J Bone Miner Res,* **14,** 2015–2026 (1999).

236. A. Kozlenkov, M. H. Le Du, P. Cuniasse, T. Ny, M. F. Hoylaerts, and J. L. Millan, Residues determining the binding specificity of uncompetitive inhibitors to tissue-nonspecific alkaline phosphatase. *J Bone Miner Res,* **19,** 1862–1872 (2004).

237. S. Narisawa, N. Frohlander, and J. L. Millan, Inactivation of two mouse alkaline phosphatase genes and establishment of a model of infantile hypophosphatasia. *Dev Dyn,* **208,** 432–446 (1997).

238. J. E. Puzas and J. S. Brand, Bone cell phosphotyrosine phosphatase: Characterization and regulation by calcitropic hormones. *Endocrinology* **6,** 2463–2468 (1985).

239. W. M. Burch, G. Hammer, and R. E. Wuthier, Phosphotyrosine and phosphoprotein phosphatase activity of alkaline phosphatase in mineralizing cartilage. *Metabolism,* **34,** 169–175 (1985).

240. J. D. Termine, H. K. Kleinman, S. W. Whitson, K. M. Conn, M. L. McGarvey, and G. R. Martin, Osteonectin, a bone-specific protein linking mineral to collagen. *Cell,* **26,** 99–105 (1981).

241. K. Hoshi and H. Ozawa, Control of calcification by extracellular matrix. *Clin Calcium,* **14,** 16–22 (2004).

242. K. M. Conn and J. D. Termine, Matrix protein profiles in calf bone development. *Bone,* **6,** 33–36 (1985).

243. A. Swaroop, B. L. M. Hogan, and U. Francke, Molecular analysis of the cDNA for human SPARC/Osteonectin/BM-40: Sequence, expression and localization of the gene to chromosome 5q31–q33. *Genomics,* **2,** 37 (1988).

244. R. C. Schwartz, M. F. Young, and P. Tsipouras, Two RFLPs in the 5′ end of the human osteonectin (ON) gene. *Nucleic Acids Res,* **16,** 9076 (1988).

245. M. E. Bolander, M. F. Young, L. W. Fisher, Y. Yamada, and J. D. Termine, Osteonectin cDNA sequence reveals potential binding regions for calcium and hydroxyapatite and shows homologies with both a basement membrane protein (SPARC) and a serine–protease inhibitor (ovomucoid). *Proc Natl Acad Sci USA,* **85,** 2919–2923 (1988).

246. D. M. Findlay, L. W. Fisher, C. I. McQuillan, J. D. Termine, and M. F. Young, Isolation of the osteonectin gene: Evidence that a variable region of the osteonectin molecule is encoded within one exon. *Biochemistry,* **27,** 1483–1489 (1988).

247. H. H. McVey, S. Nomura, P. Kelly, I. J. Mason, and B. L. Hogan, Characterization of the mouse SPARC/Osteonectin gene: Intron/exon organization and an unusual promoter region. *J Biol Chem,* **263,** 11111–11116 (1988).

248. J. A. Bassuk, M. L. Iruela-Arispa, T. F. Lane, J. M. Benson, R. A. Berg, and E. H. Sage, Molecular analyses of chicken embryo sparc (osteonectin). *Eur J Biochem,* **218,** 117–127 (1993).

249. V. Laize, A. R. Pombinho, and M. L. Cancela, Characterization of *Sparus aurata* osteonectin cDNA and *in silico* analysis of protein conserved features: Evidence for more than one osteonectin in *Salmonidae. Biochimie,* **87,** 411–420 (2005).

250. R. Thomas, L. D. True, J. A. Bassuk, P. H. Lange, and R. L. Vessella, Differential expression of osteonectin/SPARC during human prostate cancer progression. *Clin Cancer Res,* **6,** 1140–1149 (2000).

251. M. F. Young, D. M. Findlay, and P. Dominguez, Osteonectin promoter: DNA sequence analysis and S1 endonuclease site, potentially associated with transcriptional control in bone cells. *J Biol Chem,* **264,** 450 (1989).

252. P. Dominguez, K. Ibaraki, P. Gehron Robey, T. Hefferan, J. D. Termine, and M. F. Young, The expression of the osteonectin gene is potentially controlled by multiple *cis* and *trans* acting elements. *J Bone Miner Res,* **6,** 1127–1136 (1991).

253. K. Ibaraki, P. Gehron Robey, and M. F. Young, The transcription of the osteonectin gene is potentially controlled by AP1, AP2, AP3 and a novel GAGA binding factor. *Connect Tissue Res,* **27,** 130 (1992).

254. R. W. Romberg, P. G. Werness, P. Lollar, B. L. Riggs, and K. G. Mann, Isolation and characterization of native adult osteonectin. *J Biol Chem,* **260,** 2728–2736 (1985).

255. R. Fujisawa, Y. Wada, Y. Nodasaka, and Y. Kuboki, Acidic amino acid-rich sequences as binding sites of osteonectin to hydroxyapatite crystals. *Biochim Biophys Acta,* **1292,** 53–60 (1996).

256. L. Malaval, B. Darbouret, C. Preaudat, J. P. Jolu, and P. D. Delmas, Intertissular variations in osteonectin: A monoclonal antibody directed to bone osteonectin shows reduced affinity for platelet osteonectin. *J Bone Miner Res,* **6,** 315–323 (1991).

257. R. C. Hughes, A. Taylor, H. Sage, and B. L. M. Hogan, Distinct patterns of glycosylation of colligin, a collagen-binding glycoprotein, and SPARC (osteonectin), a secreted Ca^{++}-binding glycoprotein: Evidence for the localization of colligin in the endoplasmic reticulum. *Eur J Biochem,* **163,** 57–65 (1987).

258. K. Ibaraki, S. W. Whitson, J. D. Termine, and M. F. Young, Bone matrix mRNA expression in differentiating fetal bovine osteoblasts. *J Bone Miner Res,* **7,** 743–754 (1992).

259. M. Noda and G. A. Rodan, Type beta transforming growth factor regulates expression of genes encoding bone matrix proteins. *Connect Tissue Res,* **21,** 71–75 (1989).

260. E. Lieb, T. Vogel, S. Milz, M. Dauner, and M. B. Schulz, Effects of transforming growth factor beta1 on bonelike tissue formation in three-dimensional cell culture. II: Osteoblastic differentiation. *Tissue Eng,* **10,** 1414–1425 (2004).

261. J. L. Wrana, M. Maeno, B. Hawrylyshyn, K. L. Yao, C. Domenicucci, and J. Sodek, Differential effects of transforming growth factor-β on the synthesis of extracellular matrix proteins by normal fetal rat calvarial bone cell populations. *J Cell Biol,* **106,** 915–924 (1988).

262. I. Shur, F. Lokiec, I. Bleiberg, and D. Benayahu, Differential gene expression of cultured human osteoblasts. *J Cell Biochem,* **83,** 547–553 (2001).

263. D. Thiebaud, H. L. Guenther, A. Porret, P. Burckhardt, H. Fleisch, and W. Hofstetter, Regulation of collagen type I and biglycan mRNA levels by hormones and growth factors in normal and immortalized osteoblastic cell lines. *J Bone Miner Res,* **9,** 1347–1354 (1994).

264. I. J. Mason, A. Taylor, J. G. Williams, H. Sage, and B. L. M. Hogan, Evidence from molecular cloning that SPARC, a major product of mouse embryo parietal endoderm, is related to endothelial cell "culture shock" glycoprotein of MR-43000. *EMBO J, 5,* 1465–1472 (1986).

265. H. Zhou, R. G. Hammonds Jr., D. M. Findlay, P. J. Fuller, T. J. Martin, and K. W. Ng, Retinoic acid modulation of mRNA levels in malignant, nontransformed, and immortalized osteoblasts. *J Bone Miner Res, 6,* 767–777 (1991).

266. B. Tyree, The partial degradation of osteonectin by a bone-derived metalloprotease enhances binding to type I collagen. *J Bone Miner Res, 4,* 877–883 (1989).

267. R. J. Kelm Jr. and K. G. Mann, The collagen binding specificity of bone and platelet osteonectin is related to differences in glycosylation. *J Biol Chem, 266,* 9632–9639 (1991).

268. H. Sage, R. B. Vernon, S. E. Funk, E. A. Everitt, and J. Angello, SPARC, a secreted protein associated with cellular proliferation, inhibits cell spreading *in vitro* and exhibits Ca+2 dependent binding to the extracellular matrix. *J Cell Biol, 109,* 341–356 (1989).

269. P. Clezardin, L. Malavel, A.-S. Ehrensperger, P. Delmas, M. Dechavanne, and J. L. McGregor, Complex formation of human thrombospondin with osteonectin. *Eur J Biochem, 175,* 275–284 (1988).

270. D. T. Gilmour, G. J. Lyon, M. B. Carlton, J. R. Sanes, J. M. Cunningham, J. R. Anderson, B. L. Hogan, M. J. Evans, and W. H. Colledge, Mice deficient for the secreted glycoprotein SPARC/osteonectin/BM40 develop normally but show severe age-onset cataract formation and disruption of the lens. *EMBO J, 17,* 1860–1870 (1998).

271. A. M. Delany, M. Amling, M. Priemel, C. Howe, R. Baron, and E. Canalis, Osteopenia and decreased bone formation in osteonectin-deficient mice. *J Clin Invest, 105,* 915–923 (2000).

272. A. L. Boskey, D. J. Moore, M. Amling, E. Canalis, and A. M. Delany, Infrared analysis of the mineral and matrix in bones of osteonectin-null mice and their wild-type controls. *J Bone Miner Res, 18,*1005–1011 (2003).

273. E. Hohenester and J. Engel, Domain structure and organisation in extracellular matrix proteins. *Matrix Biol, 21,* 115–128 (2002).

274. H. J. Roach, Why does bone matrix contain non-collagenous proteins? The possible roles of osteocalcin, osteonectin, osteopontin, and bone sialoprotein in mineralisation and resorption. *Cell Biol Int, 18,* 617–628 (1994).

275. U. M. Wewer, K. Ibaraki, P. Schjorring, M. E. Durkin, M. F. Young, and R. Albrechtsen, A potential role for tetranectin in mineralization during osteogenesis. *J Cell Biol, 127,* 1767–1775 (1994).

276. L. Berglund and T. E. Petersen, The gene structure of tetra nectin, a plasminogen binding protein. *FEBS Lett, 309,* 15–19 (1992).

277. I. Fuhlendorff, S. Clemmensen, and S. Magnusson, Primary structure of tetranectin, a plasminogen kringle 4 binding plasma protein: Homology with a sialoglycoprotein receptor and cartilage proteoglycan core protein. *Biochemistry, 26,* 6757–6764 (1987).

278. I. Clemmensen, L. C. Petersen, and C. Kluft, Purification and characterization of a novel, oligomeric, plasminogen kringle 4 binding protein from human plasma: Tetranectin. *Eur J Biochem, 156,* 327–333 (1986).

279. K. Iba, H. Chiba, T. Yamashita, S. Ishii, and N. Sawada, Phase-independent inhibition by retinoic acid of mineralization correlated with loss of tetranectin expression in a human osteoblastic cell line. *Cell Struct Funct, 26,* 227–233 (2001).

280. J. Takagi, Structural basis for ligand recognition by RGD (Arg-Gly-Asp)-dependent integrins. *Biochem Soc Trans, 32*(Pt. 3), 403–406 (2004).

281. R. O. Hynes, The emergence of integrins: A personal and historical perspective. *Matrix Biol, 23,* 333–340 (2004).

282. R. O. Hynes, Integrins: Bidirectional, allosteric signaling machines. *Cell, 110,* 673–687 (2002).

283. L. W. Fisher and N. S. Fedarko, Six genes expressed in bones and teeth encode the current members of the SIBLING family of proteins. *Connect Tissue Res, 44*(Suppl. 1), 33–40 (2003).

284. J. C. Adams and J. Lawler, The thrombospondins. *Int J Biochem Cell Biol, 36,* 961–968 (2004).

285. F. W. Wolf, R. L. Eddy, T. B. Shows, and V. M. Dixit, Structure and chromosomal localization of the human thrombospondin gene. *Genomics, 6,* 685–691 (1990).

286. E. Jaffe, P. Bornstein, and C. M. Disteche, Mapping of the thrombospondin gene to human chromosome 15 and mouse chromosome 2 by *in situ* hybridization. *Genomics, 7,* 123–126 (1990).

287. T. L. LaBell and P. H. Byers, Sequence and characterization of the complete human thrombospondin 2 cDNA: Potential regulatory role for the 3′ untranslated region. *Genomics, 17,* 225–229 (1993).

288. H. L. Vos, S. Devarayalu, Y. de Vries, and P. Bornstein, Thrombospondin 3 (Thbs3), a new member of the thrombospondin gene family. *J Biol Chem, 267,* 12192–12196 (1992).

289. G. Newton, S. Weremowicz, C. C. Morton, N. A. Jenkins, D. J. Gilbert, N. G. Copeland, and J. Lawler, The thrombospondin-4 gene. *Mamm Genome, 10,* 1010–1016 (1999).

290. J. T. Hecht, M. Deere, E. Putnam, W. Cole, B. Vertel, H. Chen, and J. Lawler, Characterization of cartilage oligomeric matrix protein (COMP) in human normal and pseudoachondroplasia musculoskeletal tissues. *Matrix Biol, 17,* 269–278 (1998).

291. T. Shingu and P. Bornstein, Characterization of the mouse thrombospondin 2 gene. *Genomics, 16,* 78–84 (1993).

292. P. Framson and P. Bornstein, A serum response element and a binding site for NF-Y mediate the serum response of the human thrombospondin 1 gene. *J Biol Chem, 268,* 4989–4996 (1993).

293. T. Shingu and P. Bornstein, Overlapping Egr-1 and Sp1 sites function in the regulation of transcription of the mouse thrombospondin 1 gene. *J Biol Chem, 269,* 32551–32557 (1994).

294. A. Mettouchi, F. Cabon, N. Montreau, P. Vernier, G. Mercier, D. Blangy, H. Tricoire, P. Vigier, and B. Binetruy, SPARC and thrombospondin genes are repressed by the c-jun oncogene in rat embryo fibroblasts. *EMBO J, 13,* 5668–5678 (1994).

295. D. B. Donoviel, P. Framson, C. F. Eldridge, M. Cooke, S. Kobayashi, and P. Bornstein, Structural analysis and expression of the human thrombospondin gene promoter. *J Biol Chem, 263,* 18590–18593 (1988).

296. P. Bornstein, S. Devarayalu, P. Li, C. M. Disteche, and P. Framson, A second thrombospondin gene in the mouse is similar in organization to thrombospondin 1 but does not respond to serum. *Proc Natl Acad Sci USA, 88,* 8636–8640 (1991).

297. K. W. Adolph, G. L. Long, S. Winfield, E. I. Ginns, and P. Bornstein, Structure and organization of the human thrombospondin 3 gene (THBS3). *Genomics, 27,* 329–336 (1995).

298. K. W. Adolph, D. J. Liska, and P. Bornstein, Analysis of the promoter and transcription start sites of the human thrombospondin 2 gene (THBS2). *Gene, 193,* 5–11 (1997).

299. J. Adams and J. Lawler, Extracellular matrix: The thrombospondin family. *Curr Biol, 3,* 188–190 (1993).

300. J. A. Carron, W. B. Bowler, S. C. Wagstaff, and J. A. Gallagher, Expression of members of the thrombospondin

family by human skeletal tissues and cultured cells. *Biochem Biophys Res Commun,* **263,** 389–391 (1999).

301. M. Melnick, H. Chen, Y. Zhou, and T. Jaskoll, Thrombospondin-2 gene expression and protein localization during embryonic mouse palate development. *Arch Oral Biol,* **45,** 19–25 (2000).

302. P. Bornstein, Thrombospondins: Structure and regulation of expression. *FASEB J,* **6,** 3290–3299 (1992).

303. D. D. Roberts, D. M. Haverstick, V. M. Dixit, W. A. Frazier, S. A. Santoro, and V. Ginsburg, The platelet glycoprotein thrombospondin binds specifically to sulfated glycolipids. *J Biol Chem,* **260,** 9405–9411 (1985).

304. C. A. Elzie and J. E. Murphy-Ullrich, The N-terminus of thrombospondin: The domain stands apart. *Int J Biochem Cell Biol,* **36,** 1090–1101 (2004).

305. X. Sun, D. F. Mosher, and A. Rapraeger, Heparan sulfate-mediated binding of epithelial cell surface proteoglycan to thrombospondin. *J Biol Chem,* **264,** 2885–2889 (1989).

306. J. C. Adams, Functions of the conserved thrombospondin carboxy-terminal cassette in cell–extracellular matrix interactions and signaling. *Int J Biochem Cell Biol,* **36,** 1102–1114 (2004).

307. P. Gehron Robey, M. F. Young, L. W. Fisher, and T. D. McClain, Thrombospondin is an osteoblast-derived component of mineralized extracellular matrix. *J Cell Biol,* **108,** 719–727 (1989).

308. W. J. Grzesik, C. R. Frazier, J. R. Shapiro, P. D. Sponseller, P. G. Robey, and N. S. Fedarko, Age-related changes in human bone proteoglycan structure. Impact of osteogenesis imperfecta. *J Biol Chem,* **277,** 43638–43647 (2002).

309. L. Malaval, D. Modrowski, A. K. Gupta. and J. E. Aubin, Cellular expression of bone related proteins during *in vitro* osteogenesis in rat bone marrow stromal cell cultures. *J Cell Physiol,* **158,** 555–572 (1994).

310. N. V. Sherbina and P. Bornstein, Modulation of thrombospondin gene expression during osteoblast differentiation in MC3T3-E1 cells. *Bone,* **13,** 197–201 (1992).

311. Z. Abbadia, J. Amiral, M. C. Trzeciak, P. D. Delmas, and P. Clezardin, The growth-supportive effect of thrombospondin (TSP1) and the expression of TSP1 by human MG-63 osteoblastic cells are both inhibited by dexamethasone. *FEBS Lett,* **335,** 161–166 (1993).

312. J. A. Carron, W. C. Fraser, and J. A. Gallagher, Thrombospondin promotes resorption by osteoclasts *in vitro. Biochem Biophys Res Commun,* **213,** 1017–1025 (1995).

313. W. J. Grzesik and P. Gehron Robey, Bone matrix RGD glycoproteins: Immunolocalization and interaction with human primary osteoblastic bone cells *in vitro. J Bone Miner Res,* **9,** 487–496 (1994).

314. K. D. Hankenson, S. D. Bain, T. R. Kyriakides, E. A. Smith, S. A. Goldstein, and P. Bornstein, Increased marrow-derived osteoprogenitor cells and endosteal bone formation in mice lacking thrombospondin 2. *J Bone Miner Res,* **15,** 851–862 (2000).

315. P. Bornstein, T. R. Kyriakides, Z. Yang, L. C. Armstrong, and D. E. Birk, Thrombospondin 2 modulates collagen fibrillogenesis and angiogenesis. *J Invest Dermatol Symp Proc,* **5,** 61–66 (2000).

316. P. Bornstein, A. Agah, and T. R. Kyriakides, The role of thrombospondins 1 and 2 in the regulation of cell–matrix interactions, collagen fibril formation, and the response to injury. *Int J Biochem Cell Biol,* **36,** 1115–1125 (2004).

317. T. R. Kyriakides, Y. H. Zhu, L. T. Smith, S. D. Bain, Z. Yang, M. T. Lin, K. G. Danielson, R. V. Iozzo, M. LaMarca, C. E. McKinney, E. I. Ginns, and P. Bornstein, Mice that lack

thrombospondin 2 display connective tissue abnormalities that are associated with disordered collagen fibrillogenesis, an increased vascular density, and a bleeding diathesis. *J Cell Biol,* **140,** 419–430 (1998).

318. B. Merle, H. V. Choi, L. Rosenberg, J. Lawler, P. D. Delmas, and P. Clezardin, *In vitro* interaction between decorin and thrombospondin. FECTS XIV meeting, Lyon, France, p. F13 (1994).

319. M. Winnemoller, P. Schîn, P. Vischer, and H. Kresse, Interactions between thrombospondin and the small proteoglycan decorin: Interference with cell attachment. *Eur J Biol,* **59,** 47–55 (1992).

320. M. Grano, P. Zigrino, S. Colucci, and G. Zambonin, Adhesion properties and integrin expression of cultured human osteoclast-like cells. *Exp Cell Res,* **212,** 209–218 (1994).

321. M. A. Horton, E. L. Dorey, S. A. Nesbitt, J. Samanen, F. E. Ali, J. M. Stadel, A. Nichols, R. Greig, and M. H. Helfrich, Modulation of vitronectin receptor-mediated osteoclast adhesion by Arg-Gly-Asp peptide analogs: A structure–function analysis. *J Bone Miner Res,* **8,** 239–247 (1993).

322. S. B. Rodan and G. A. Rodan, Integrin function in osteoclasts. *J Endocrinol,* **154**(Suppl.), S47–S56 (1997).

323. H. Hirano, Y. Yamada, M. Sullivan, B. de Crombrugghe, I. Pastan, and K. M. Yamada, Isolation of genomic DNA clones spanning the entire fibronectin gene. *Proc Natl Acad Sci USA,* **80,** 46–50 (1983).

324. R. Gardella, M. Colombi, and S. Barlati, Human fibronectin gene (FN1) RFLPs: Mapping and linkage disequilibrium analysis. *Hum Genet,* **92,** 639–641 (1993).

325. K. Vibe-Pedersen, S. Magnusson, and F. E. Baralle, Donor and acceptor splice signals within an exon of the human fibronectin gene: A new type of differential splicing. *FEBS Lett,* **207,** 287–291 (1986).

326. A. R. Kornblihtt, K. Umezawa, K. Vibe-Pedersen, and F. E. Baralle, Primary structure of human fibronectin: Differential splicing may generate at lease 10 polypeptides from a single gene. *EMBO J,* **4,** 1755–1759 (1985).

327. R. S. Patel, E. Odermatt, J. E. Schwarzbauer, and R. O. Hynes, Organization of the fibronectin gene provides evidence for exon shuffling during evolution. *EMBO J,* **6,** 2565–2572 (1987).

328. D. C. Dean, C. L. Bowlus, and S. Bourgeois, Cloning and analysis of the promoter region of the human fibronectin gene. *Proc Natl Acad Sci USA,* **84,** 1876–1880 (1987).

329. S. Han, J. D. Ritzenthaler, H. N. Rivera, and J. Roman, Peroxisome proliferator-activated receptor-gamma ligands suppress fibronectin gene expression in human lung carcinoma cells: Involvement of both CRE and Sp1. *Am J Physiol Lung Cell Mol Physiol,* **289,** L419–L428 (2005).

330. S. Miao, P. K. Suri, L. Shu-Ling, A. Abraham, N. Cook, P. Milos, and M. A. Zern, Role of the cyclic AMP response element in rat fibronectin gene expression. *Hepatology,* **17,** 882–890 (1993).

331. A. F. Muro, V. A. Bernath, and A. R. Kornblihtt, Interaction of the −170 cyclic AMP response element with the adjacent CCAAT box in the human fibronectin gene promoter. *J Biol Chem,* **267,** 12767–12774 (1992).

332. A. M. Krezel, G. Wagner, J. Seymour-Ulmer, and R. A. Lazarus, Structure of the RGD protein decorsin: Conserved motif and distinct function in leech proteins that affect blood clotting. *Science,* **264,** 1944–1947 (1994).

333. R. F. Cachau, E. H. Serpersu, A. S. Mildvan, T. August, and L. M. Amzel, Recognition in cell adhesion. A compulsive study of RGD-containing peptides by Monte Carlo and NMR methods. *J Mol Recog,* **2,** 179–186 (1989).

334. R. A. Clark, J. Q. An, D. Greiling, A. Khan, and J. E. Schwarzbauer, Fibroblast migration on fibronectin requires

three distinct functional domains. *J Invest Dermatol,* **121,** 695–705 (2003).

335. J. Sottile, J. Schwarzbauer, J. Seleque, and D. F. Mosher, Five type I modules of fibronectin form a functional unit that binds to fibroblasts and *Staphylococcus aureus. J Biol Chem,* **266,** 12840–12843 (1991).

336. T. Darribere, V. E. Koteliansky, M. A. Chernousov, and S. K. Akiyama, Distinct regions of human fibronectin are essential for fibril assembly in an *in vivo* developing system. *Dev Dyn,* **194,** 63–70 (1992).

337. K. Ichihara-Tanaka, T. Maeda, K. Titani, and K. Sekiguchi, Matrix assembly of recombinant fibronectin polypeptide consisting of amino-terminal 70 kDa and carboxyl-terminal 37 kDa regions. *FEBS Lett,* **299,** 155–158 (1992).

338. J. E. Schwarzbauer, Identification of the fibronectin sequences required for assembly of a fibrillar matrix. *J Cell Biol,* **113,** 463–473 (1991).

339. F. J. Barkalow and J. E. Schwarzbauer, Interactions between fibronectin and chondroitin sulfate are modulated by molecular context. *J Biol Chem,* **269,** 3957–3962 (1994).

340. C. Eielson, D. Kaplan, M. A. Mitnick, I. Paliwal, and K. Insogna, Estrogen modulates parathyroid hormone-induced fibronectin production in human and rat osteoblast-like cells. *Endocrinology,* **135,** 1639–1644 (1994).

341. R. E. Weiss and A. H. Reddi, Appearance of fibronectin during the differentiation of cartilage, bone and bone marrow. *J Cell Biol,* **88,** 630 (1981).

342. M. J. Weiss, P. S. Henthorn, M. A. Lafferty, C. Slaughter, M. Raducha, and H. Harris, Isolation and characterization of a cDNA encoding a human liver/bone/kidney-type alkaline phosphatase. *Proc Natl Acad Sci USA,* **83,** 7182–7186 (1986).

343. E. L. George, E. N. Georges-Labouesse, R. S. Patel-King, H. Rayburn, and R. O. Hynes, Defects in mesoderm, neural tube and vascular development in mouse embryos lacking fibronectin. *Development,* **119,** 1079–1091 (1993).

344. G. Daculsi, P. Pilet, M. Cottrel, and G. Guicheux, Role of fibronectin during biological apatite crystal nucleation: Ultrastructural characterization. *J Biomed Mater Res,* **47,** 228–233 (1997).

345. G. Daculsi, P. Pilet, M. Cottrel, and G. Guicheux, Role of fibronectin during biological apatite crystal nucleation: Ultrastructural characterization. *J Biomed Mater Res,* **47,** 228–233 (1999).

346. R. J. Majeska, M. Port, and T. A. Einhorn, Attachment to extracellular matrix molecules by cells differing in the expression of osteoblastic traits. *J Bone Miner Res,* **8,** 277–289 (1993).

347. D. Seiffert, T. J. Podor, and D. J. Loskutoff, Distribution of vitronectin. In *Biology of Vitronectins and Their Receptors* (K. T. Preissner, S. Rosenblatt, C. Kost, J. Wegerhoff, and D. F. Mosher, eds.), pp. 75–81. Excerpta Medica, Amsterdam (1993).

348. K. T. Preissner and D. Jenne, Vitronectin: A new molecular connection in haemostasis. *Thromb Haemost,* **66,** 189–194 (1991).

349. A. Oldberg, E. G. Hayman, M. D. Pierschbacher, and E. Ruoslahti, Complete amino acid sequence of human vitronectin deduced from cDNA. Similarity of cell attachment sites in vitronectin and fibronectin. *EMBO J,* **4,** 2519–2524 (1985).

350. D. Jenne and K. K. Stanley, Molecular cloning of S-protein, a link between complement coagulation and cell-substrate adhesion. *EMBO J,* **4,** 3153–3157 (1985).

351. H. J. Ehrlich, B. Richter, D. von der Ahe, and K. T. Preissner, Primary structure of vitronectins and homology with other proteins. In *Biology of Vitronectins and Their Receptors* (K. T. Preissner, S. Rosenblatt, C. Kost, J. Wegerhoff, and D. F. Mosher, eds.), pp. 59–66. Excerpta Medica, Amsterdam (1993).

352. T. Kumagai, I. Lee, Y. Ono, M. Maeno, and M. Takagi, Ultrastructural localization and biochemical characterization of vitronectin in developing rat bone. *Histochem J,* **30,** 111–119 (1998).

353. K. P. Mintz, W. J. Grzesik, R. J. Midura, P. G. Robey, J. D. Termine, and L. W. Fisher, Purification and fragmentation of nondenatured bone sialoprotein: Evidence for a cryptic, RGD-resistant cell attachment domain. *J Bone Miner Res,* **8,** 985–995 (1993).

354. J. Davies, J. Warwick, N. Totty, R. Philip, M. Helfrich, and M. Horton, The osteoclast functional antigen, implicated in the regulation of bone resorption, is biochemically related to the vitronectin receptor. *J Cell Biol,* **109,** 1817–1826 (1980).

355. S. Koschnick, S. Konstantinides, K. Schafer, K. Crain, and D. J. Loskutoff, Thrombotic phenotype of mice with a combined deficiency in plasminogen activator inhibitor 1 and vitronectin. *J Thromb Haemost,* **3,** 2290–2295 (2005).

356. R. P. M. Rohanizadeh, J. M. Bouler, D. Couchourel, Y. Fortun, and G. Daculsi, Apatite precipitation after incubation of biphasic calcium-phosphate ceramic in various solutions: Influence of seed species and proteins. *J Biomed Mater Res,* **42,** 530–539 (1998).

357. L. W. Fisher and N. S. Fedarko, Six genes expressed in bones and teeth encode the current members of the SIBLING family of proteins. *Connect Tissue Res,* **44**(Suppl. 1), 33–40 (2003).

358. N. L. Huq, K. J. Cross, M. Ung, and E. C. Reynolds, A review of protein structure and gene organisation for proteins associated with mineralised tissue and calcium phosphate stabilisation encoded on human chromosome 4. *Arch Oral Biol,* **50,** 599–609 (2005).

359. P. S. Rowe, Y. Kumagai, G. Gutierrez, I. R. Garrett, R. Blacher, D. Rosen, J. Cundy, S. Navvab, D. Chen, M. K. Drezner, L. D. Quarles, and G. R. Mundy, MEPE has the properties of an osteoblastic phosphatonin and minhibin. *Bone,* **34,** 303–319 (2004).

360. S. Sekine, K. Kataoka, M. Tanaka, H. Nagata, T. Kawakami, K. Akaji, S. Aimoto, and S. Shizukuishi, Active domains of salivary statherin on apatitic surfaces for binding to *Fusobacterium nucleatum* cells. *Microbiology,* **150,** 2373–2379 (2004).

361. H. Rangaswami, A. Bulbule, and G. C. Kundu, Osteopontin: Role in cell signaling and cancer progression. *Trends Cell Biol,* **16,** 79–87 (2006).

362. D. Chabas, Osteopontin, a multi-faceted molecule. *Med Sci (Paris),* **21,** 832–838 (2005).

363. K. Rafidi, I. Simkina, E. Johnson, M. A. Moore, and L. C. Gerstenfeld, Characterization of the chicken osteopontin-encoding gene. *Gene,* **140,** 163–169 (1994).

364. M. F. Young, J. M. Kerr, J. D. Termine, U. M. Wewer, M. G. Wang, O. W. McBride, and L. W. Fisher, cDNA cloning, mRNA distribution and heterogeneity, chromosomal location, and RFLP analysis of human osteopontin (OPN). *Genomics,* **7,** 491–502 (1990).

365. M. C. Kieffer, D. M. Aauer, and P. J. Barr, The cDNA and derived amino acid sequence for human osteopontin. *Nucleic Acids Res,* **17,** 3306 (1989).

366. J. M. Kerr, L. W. Fisher, J. D. Termine, and M. F. Young, The cDNA cloning and RNA distribution of bovine osteopontin. *Gene,* **108,** 237–243 (1991).

367. A. M. Craig and D. T. Denhardt, The murine gene encoding secreted phosphoprotein 1 (osteopontin): Promoter structure, activity, and induction *in vivo* by estrogen and progesterone. *Gene,* **100,** 163–171 (1991).

368. D. T. Denhardt and X. Guo, Osteopontin: A protein with diverse functions. *FASEB J, 7,* 1475–1482 (1993).

369. K. T. Preissner and D. Jenne, Structure of vitronectin and its biological role in haemostasis. *Thromb Haemost, 66,* 123–132 (1991).

370. A. Franzen and D. Heinegard, Isolation and characterization of two sialoproteins present only in bone calcified matrix. *Biochem J, 232,* 715–724 (1985).

371. C. W. Prince, Secondary structure predictions for rat osteopontin. *Connect Tissue Res, 21,* 15–20 (1989).

372. L. W. Fisher, D. A. Torchia, B. Fohr, M. F. Young, and N. S. Fedarko, Flexible structures of SIBLING proteins, bone sialoprotein, and osteopontin. *Biochem Biophys Res Commun, 280,* 460–465 (2001).

373. S. Van Dijk, J. A. D'Errico, M. J. Somerman, M. C. Farach-Carson, and W. T. Butler, Evidence that a non-RGD domain in rat osteopontin is involved in cell attachment. *J Bone Miner Res, 8,* 1499–1506 (1993).

374. A. Gericke, C. Qin, L. Spevak, Y. Fujimoto, W. T. Butler, E. S. Sorensen, and A. L. Boskey, Importance of phosphorylation for osteopontin regulation of biomineralization. *Calcif Tissue Int, 77,* 45–54 (2005).

375. C. W. Prince, T. Oosawa, W. T. Butler, M. Tomana, A. S. Bhon, and R. E. Schrohenloher, Isolation, characterization and biosynthesis of a phosphorylated glycoprotein formed at bone. *J Biol Chem, 262,* 2900–2907 (1987).

376. W. T. Butler, The nature and significance of osteopontin. *Connect Tissue Res, 23,* 123–126 (1989).

377. S. Beninati, D. R. Senger, E. Cordella-Miele, A. B. Mukherjee, I. Chackalaparampil, V. Shanmugam, K. Singh, and B. B. Mukherjee, Osteopontin: Its transglutaminase-catalyzed posttranslational modifications and cross-linking to fibronectin. *J Biochem (Tokyo), 115,* 675–682 (1994).

378. M. D. McKee and A. Nanci, Osteopontin and the bone remodeling sequence. Colloidal-gold immunocytochemistry of an interfacial extracellular matrix protein. *Ann N Y Acad Sci, 760,* 177–189 (1995).

379. M. P. Mark, C. W. Prince, T. Oosawa, S. Gay, A. L. Bronckers, and W. T. Butler, Immunohistochemical demonstration of a 44-KD phosphoprotein in developing rat bones. *J Histochem Cytochem, 35,* 707–715 (1987).

380. J. Chen, K. Singh, B. B. Mukherjee, and J. Sodek, Developmental expression of osteopontin (OPN) mRNA in rat tissues: Evidence for a role for OPN in bone formation and resorption. *Matrix, 13,* 113–123 (1993).

381. P. G. Strauss, E. I. Closs, J. Schmidt, and V. Erfle, Gene expression during osteogenic differentiation in mandibular condyles *in vitro. J Cell Biol, 110,* 1369–1378 (1990).

382. P. Gehron Robey, P. Bianco, and J. D. Termine, The cellular biology and molecular biochemistry of bone formation. In *Disorders of Mineral Metabolism* (F. L. Coe and M. J. Favus, eds.), pp. 241–263. Raven Press, New York (1992).

383. J. Sodek, J. Chen, S. Kasugai, T. Nagata, Q. Zhang, M. D. McKee, and A. Nanci, Elucidating the functions of bone sialoprotein and osteopontin in bone formation. In *The Chemistry and Biology of Mineralized Tissues* (H. Slavkin and P. Price, eds.), pp. 297–306. Elsevier, Amsterdam (1992).

384. M. Noda and D. T. Denhardt, Regulation of osteopontin gene expression in osteoblasts. *Ann N Y Acad Sci, 760,* 242–248 (1995).

385. C. Zierold, J. A. Mings, J. M. Prahl, G. G. Reinholz, and H. F. DeLuca, Protein synthesis is required for optimal induction of 25-hydroxyvitamin D(3)-24-hydroxylase, osteocalcin, and osteopontin mRNA by 1,25-dihydroxyvitamin D(3). *Arch Biochem Biophys, 404,* 18–24 (2002).

386. C. G. Bellows, S. M. Reimers, and J. N. Heersche, Expression of mRNAs for type-I collagen, bone sialoprotein, osteocalcin, and osteopontin at different stages of osteoblastic differentiation and their regulation by 1,25 dihydroxyvitamin D3. *Cell Tissue Res, 297,* 249–259 (1999).

387. P. J. Fagenholz, S. M. Warren, J. A. Greenwald, P. J. Bouletreau, J. A. Spector, F. E. Crisera, and M. T. Longaker, Osteoblast gene expression is differentially regulated by TGF-beta isoforms. *J Craniofac Surg, 12,* 183–190 (2001).

388. S. E. Harris, L. F. Bonewald, M. A. Harris, and M. Sabatini, Effects of transforming growth factor beta on bone nodule formation and expression of bone morphogenetic protein 2, osteocalcin, osteopontin, alkaline phosphatase, and type I collagen mRNA in long-term cultures of fetal rat calvarial osteoblasts. *J Bone Miner Res, 9,* 855–863 (1994).

389. K. Motomura, A. Ohtsuru, H. Enomoto, T. Tsukazaki, H. Namba, Y. Tsuji, and S. Yamashita, Osteogenic action of parathyroid hormone-related peptide (1–141) in rat ROS cells. *Endocrinol J, 43,* 527–535 (1996).

390. D. Ikedo, K. Ohishi, N. Yamauchi, M. Kataoka, J. Kido, and T. Nagata, Stimulatory effects of phenytoin on osteoblastic differentiation of fetal rat calvaria cells in culture. *Bone, 25,* 653–660 (1999).

391. J. P. Gorski, Acidic phosphoproteins from bone matrix: A structural rationalization of their role in biomineralization. *Calcif Tissue Int, 50,* 391–396 (1992).

392. A. L. Boskey, M. Maresca, W. Ullrich, S. B. Doty, W. T. Butler, and C. W. Prince, Osteopontin–hydroxyapatite interactions *in vitro*: Inhibition of hydroxyapatite formation and growth in a gelatin-gel. *Bone Miner, 22,* 147–159 (1993).

393. G. K. Hunter, C. L. Kyle, and H. A. Goldberg, Modulation of crystal formation by bone phosphoproteins: Structural specificity of the osteopontin-mediated inhibition of hydroxyapatite formation. *Biochem J, 300,* 723–728 (1994).

394. G. K. Hunter and H. A. Goldberg, Nucleation of hydroxyapatite by bone sialoprotein. *Proc Natl Acad Sci USA, 90,* 8562–8565 (1993).

395. K. Singh, A. B. Mukherjee, and M. W. DeVouge, Physiological properties and differential glycosylation of phosphorylated and non-phosphorylated forms of osteopontin secreted by normal rat kidney cells. *J Biol Chem, 265,* 18696–18701 (1992).

396. A. L. Boskey, M. Maresca, S. Doty, B. Sabsay, and A. Veis, Concentration dependent effect of dentin-phosphophoryn in the regulation of *in vitro* hydroxyapatite formation and growth. *Bone Miner, 11,* 55–65 (1990).

397. A. L. Boskey, L. Spevak, E. Paschalis, S. B. Doty, and M. D. McKee, Osteopontin deficiency increases mineral content and mineral crystallinity in mouse bone. *Calcif Tissue Int, 71,* 145–154 (2002).

398. S. R. Rittling, H. N. Matsumoto, M. D. McKee, A. Nanci, X. R. An, K. E. Novick, A. J. Kowalski, M. Noda, and D. T. Denhardt, Mice lacking osteopontin show normal development and bone structure but display altered osteoclast formation *in vitro. J Bone Miner Res, 13,* 1101–1111 (1998).

399. A. Veis, C. Sfeir, and C. B. Wu, Phosphorylation of the proteins of the extracellular matrix of mineralized tissues by casein kinase-like activity. *Crit Rev Oral Biol Med, 8,* 360–379 (1997).

400. C. Qin, O. Baba, and W. T. Butler, Post-translational modifications of sibling proteins and their roles in osteogenesis and dentinogenesis. *Crit Rev Oral Biol Med, 15,* 126–136 (2004).

401. A. Endo, Potential role of phosphoprotein in collagen mineralization—An experimental study *in vitro. J Orthop Assoc, 61,* 563–569 (1987).

402. B. Ganss, R. H. Kim, and J. Sodek, Bone sialoprotein. *Crit Rev Oral Biol Med,* **10,** 79–98 (1999).

403. A. T. Andrews, G. M. Herring, and P. W. Kent, Some studies on the composition of bovine cortical-bone sialoprotein. *Biochem J,* **104,** 705–715 (1967).

404. L. W. Fisher, O. W. McBride, J. D. Termine, and M. F. Young, Human bone sialoprotein. Deduced protein sequence and chromosomal localization. *J Biol Chem,* **265,** 2347–2351 (1990).

405. J. J. Li and J. Sodek, Cloning and characterization of the rat bone sialoprotein gene promoter. *Biochem J,* **289,** 625–629 (1993).

406. J. M. Kerr, L. W. Fisher, J. D. Termine, M. G. Wang, O. W. McBride, and M. F. Young, The human bone sialoprotein gene (IBSP): Genomic localization and characterization. *Genomics,* **17,** 408–415 (1993).

407. R. H. Kim, H. S. Shapiro, J. J. Li, J. L. Wrana, and J. Sodek, Characterization of the human bone sialoprotein (BSP) gene and its promoter sequence. *Matrix Biol,* **14,** 31–40 (1994).

408. Y. Miyazaki, M. Setoguchi, S. Yoshida, Y. Higuchi, S. Akizuki, and S. Yamamoto, The mouse osteopontin gene. Expression in monocytic lineages and complete nucleotide sequence. *J Biol Chem,* **265,** 14432–14438 (1990).

409. A. Bianchi, R. D. Wells, N. H. Heintz, and M. S. Caddle, Sequences near the origin of replication of the DHFR locus of Chinese hamster ovary cells adopt left-handed Z-DNA and triplex structures. *J Biol Chem,* **265,** 21789–21796 (1990).

410. M. F. Young, K. Ibaraki, J. M. Kerr, and A. M. Heegaard, Molecular and cellular biology of the major noncollagenous proteins in bone. In *Cellular and Molecular Biology of Bone* (M. Noda, ed.), pp. 191–234. (1993).

411. L. W. Fisher, S. W. Whitson, L. V. Avioli, and J. D. Termine, Matrix sialoprotein of developing bone. *J Biol Chem,* **258,** 12723–12727 (1983).

412. B. Ecarot-Charrier, F. Bouchard, and C. Delloye, Bone sialoprotein II synthesized by cultured osteoblasts contains tyrosine sulfate. *J Biol Chem,* **264,** 20049–20053 (1989).

413. R. J. Midura, D. J. McQuillan, K. J. Benham, L. W. Fisher, and V. C. Hascall, A rat osteogenic cell line (UMR106-01) synthesizes a highly sulfated form of bone sialoprotein. *J Biol Chem,* **265,** 5285–5291 (1990).

414. H. S. Shapiro, J. Chen, J. L. Wrana, Q. Zhang, M. Blum, and J. Sodek, Characterization of porcine bone sialoprotein: Primary structure and cellular expression. *Matrix,* **13,** 431–440 (1993).

415. A. Oldberg, A. Franzen, D. Heinegard, M. Pierschbacher, and E. Ruoslahti, Identification of a bone sialoprotein receptor in osteosarcoma cells. *J Biol Chem,* **263,** 19433–19436 (1988).

416. M. J. Somerman, L. W. Fisher, R. A. Foster, and J. J. Sauk, Human bone sialoproteins I and II enhance fibroblast attachment *in vitro. Calcif Tissue Int,* **43,** 50–53 (1988).

417. F. P. Ross, J. Chappel, J. I. Alvarez, D. Sander, W. T. Butler, M. C. Farach-Carson, K. A. Mintz, P. G. Robey, S. L. Teitelbaum, and D. A. Cheresh, Interactions between the bone matrix proteins osteopontin and bone sialoprotein and the osteoclast integrin alpha v beta 3 potentiate bone resorption. *J Biol Chem,* **268,** 9901–9907 (1993).

418. Y. Kumei, S. Morita, H. Nakamura, H. Katano, K. Ohya, H. Shimokawa, C. F. Sams, and P. A. Whitson, Osteoblast responsiveness to 1alpha,25-dihydroxyvitamin D3 during spaceflight. *Ann N Y Acad Sci,* **1030,** 121–124 (2004).

419. P. Bianco, L. W. Fisher, M. F. Young, J. D. Termine, and P. Gehron Robey, Expression of bone sialoprotein (BSP) in developing human tissues. *Calcif Tissue Int,* **49,** 421–426 (1991).

420. J. E. Aubin, A. K. Gupta, R. Zirngbl, and J. Rossant, Knockout mice lacking bone sialoprotein expression have bone abnormalities. *J Bone Miner Res,* **11S,** S29 (1996).

421. T. Nagata, H. A. Goldberg, Q. Zhang, C. Domenicucci, and J. Sodek, Biosynthesis of bone proteins by fetal porcine calvariae *in vitro.* Rapid association of sulfated sialoproteins (secreted phosphoprotein-1 and bone sialoprotein) and chondroitin sulfate proteoglycan (CS-PGIII) with bone mineral. *Matrix,* **11,** 86–100 (1991).

422. G. K. Hunter and H. A. Goldberg, Modulation of crystal formation by bone phosphoproteins: Role of glutamic acid-rich sequences in the nucleation of hydroxyapatite by bone sialoprotein. *Biochem J,* **302,** 175–179 (1994).

423. J. T. Stubbs Jr., K. P. Mintz, E. D. Eanes, D. A. Torchia, and L. W. Fisher, Characterization of native and recombinant bone sialoprotein: Delineation of the mineral-binding and cell adhesion domains and structural analysis of the RGD domain. *J Bone Miner Res,* **12,** 1210–1222 (1997).

424. C. Qin, J. C. Brunn, J. Jones, A. George, A. Ramachandran, J. P. Gorski, and W. T. Butler, A comparative study of sialic acid-rich proteins in rat bone and dentin. *Eur J Oral Sci,* **109,** 133–141 (2001).

425. J. P. Gorski, A. Wang, D. Lovitch, D. Law, K. Powell, and R. J. Midura, Extracellular bone acidic glycoprotein-75 defines condensed mesenchyme regions to be mineralized and localizes with bone sialoprotein during intramembranous bone formation. *J Biol Chem,* **279,** 25455–25463 (2004).

426. R. J. Midura, A. Wang, D. Lovitch, D. Law, K. Powell, and J. P. Gorski, Bone acidic glycoprotein-75 delineates the extracellular sites of future bone sialoprotein accumulation and apatite nucleation in osteoblastic cultures. *J Biol Chem,* **279,** 25464–25473 (2004).

427. J. P. Gorski, D. Griffin, G. Dudley, C. Stanford, R. Thomas, C. Huang, E. Lai, B. Karr, and M. Solursh, Bone acidic glycoprotein-75 is a major synthetic product of osteoblastic cells and localized as 75- and/or 50-kDa forms in mineralized phases of bone and growth plate and in serum. *J Biol Chem,* **265,** 14956–14963 (1990).

428. A. George, J. Gui, N. A. Jenkins, D. J. Gilbert, N. G. Copeland, and A. Veis, *In situ* localization and chromosomal mapping of the AG1 (Dmp1) gene. *J Histochem Cytochem,* **42,** 1527–1531 (1994).

429. Y. Chen, B. S. Bal, and J. P. Gorski, Calcium and collagen binding properties of osteopontin, bone sialoprotein, and bone acidic glycoprotein-75 from bone. *J Biol Chem,* **267,** 24871–24878 (1992).

430. M. Sato, W. Grasser, S. Harm, C. Fullenkamp, and J. P. Gorski, Bone acidic glycoprotein 75 inhibits resorption activity of isolated rat and chicken osteoclasts. *FASEB J,* **6,** 2966–2976 (1992).

431. R. Srinivasan, B. Chen, J. P. Gorski, and A. George, Recombinant expression and characterization of dentin matrix protein 1. *Connect Tissue Res,* **40,** 251–258 (1999).

432. J. Q. Feng, H. Huang, Y. Lu, L. Ye, Y. Xie, T. W. Tsutsui, T. Kunieda, T. Castranio, G. Scott, L. F. Bonewald, and Y. Mishina, The Dentin matrix protein 1 (Dmp1) is specifically expressed in mineralized, but not soft, tissues during development. *J Dent Res,* **82,** 776–780 (2003).

433. S. Chen, A. Unterbrink, S. Kadapakkam, J. Dong, T. T. Gu, J. Dickson, H. H. Chuang, and M. MacDougall, Regulation of the cell type-specific dentin sialophosphoprotein gene expression in mouse odontoblasts by a novel transcription repressor and an activator CCAAT-binding factor. *J Biol Chem,* **279,** 42182–42191 (2004).

434. C. Chen, N. Inozentseva-Clayton, J. Dong, T. T. Gu, and M. MacDougall, Binding of two nuclear factors to a novel silencer element in human dentin matrix protein 1 (DMP1) promoter regulates the cell type-specific DMP1 gene expression. *J Cell Biochem,* **92,** 332–349 (2004).

435. I. Kalajzic, A. Braut, D. Guo, X. Jiang, M. S. Kronenberg, M. Mina, M. A. Harris, S. E. Harris, and D. W. Rowe, Dentin matrix protein 1 expression during osteoblastic differentiation, generation of an osteocyte GFP-transgene. *Bone,* **35,** 74–82 (2004).

436. G. He, T. Dahl, A. Veis, and A. George, Nucleation of apatite crystals in vitro by self-assembled dentin matrix protein 1. *Nat Mater,* **2,** 552–558 (2003).

437. G. He and A. George, Dentin matrix protein 1 immobilized on type I collagen fibrils facilitates apatite deposition *in vitro. J Biol Chem,* **279,** 11649–11656 (2004).

438. P. H. Tartaix, M. Doulaverakis, A. George, L. W. Fisher, W. T. Butler, C. Qin, E. Salih, M. Tan, Y. Fujimoto, L. Spevak, and A. L. Boskey, *In vitro* effects of dentin matrix protein-1 on hydroxyapatite formation provide insights into *in vivo* functions. *J Biol Chem,* **279,** 18115–18120 (2004).

439. Y. Ling, H. F. Rios, E. R. Myers, Y. Lu, J. Q. Feng, and A. L. Boskey, DMP1 depletion decreases bone mineralization *in vivo*: An FTIR imaging analysis. *J Bone Miner Res,* **20,** 2169–2177 (2005).

440. L. Ye, Y. Mishina, D. Chen, H. Huang, S. L. Dallas, M. R. Dallas, P. Sivakumar, T. Kunieda, T. W. Tsutsui, A. Boskey, L. F. Bonewald, and J. Q. Feng, Dmp1-deficient mice display severe defects in cartilage formation responsible for a chondrodysplasia-like phenotype. *J Biol Chem,* **280,** 6197–6203 (2005).

441. S. E. Harris, J. Cluhak-Heinrich, M. A. Harris, Y. Yang, P. S. N. Rowe, Y. Xie, H. Rios, S. Zhang, M. D. McKee, L. Bonewald, and J. Q. Feng, MEPE is overexpressed in osteocytes of the DMP1 null mouse, contributing to the hypophosphatemia and osteomalacia. 27th ASBMR, SA462 (2005).

442. C. Lu, S. Huang, T. Miclau, J. A. Helms, and C. Colnot, Mepe is expressed during skeletal development and regeneration. *Histochem Cell Biol,* **121,** 493–499 (2004).

443. H. Siggelkow, E. Schmidt, B. Hennies, and M. Hufner, Evidence of downregulation of matrix extracellular phosphoglycoprotein during terminal differentiation in human osteoblasts. *Bone,* **35,** 570–576 (2004).

444. J. Hao, G. He, K. Narayanan, B. Zou, L. Lin, T. Muni, A. Ramachandran, and A. George, Identification of differentially expressed cDNA transcripts from a rat odontoblast cell line. *Bone,* **37,** 578–588 (2005).

445. T. Dahl and A. Veis, Electrostatic interactions lead to the formation of asymmetric collagen–phosphophoryn aggregates. *Connect Tissue Res,* **44**(Suppl. 1), 206–213 (2003).

446. A. L. Boskey, M. Maresca, S. Doty, B. Sabsay, and A. Veis, Concentration-dependent effects of dentin phosphophoryn in the regulation of *in vitro* hydroxyapatite formation and growth. *Bone Miner,* **11,** 55–65 (1990).

447. A. L. Boskey, L. Spevak, M. Tan, S. B. Doty, and W. T. Butler, Dentin sialoprotein (DSP) has limited effects on *in vitro* apatite formation and growth. *Calcif Tissue Int,* **67,** 472–478 (2000).

448. M. L. Wallwork, J. Kirkham, H. Chen, S. X. Chang, C. Robinson, D. A. Smith, and B. H. Clarkson, Binding of dentin noncollagenous matrix proteins to biological mineral crystals: An atomic force microscopy study. *Calcif Tissue Int,* **71,** 249–255 (2002).

449. G. He, A. Ramachandran, T. Dahl, S. George, D. Schultz, D. Cookson, A. Veis, and A. George, Phosphorylation of phosphophoryn is crucial for its function as a mediator of biomineralization. *J Biol Chem,* **280,** 33109–33114 (2005).

450. T. Sreenath, T. Thyagarajan, B. Hall, G. Longenecker, R. D'Souza, S. Hong, J. T. Wright, M. MacDougall, J. Sauk, and A. B. Kulkarni, Dentin sialophosphoprotein knockout mouse teeth display widened predentin zone and develop defective dentin mineralization similar to human dentinogenesis imperfecta type III. *J Biol Chem,* **278,** 24874–24880 (2003).

451. P. S. Rowe, P. A. de Zoysa, R. Dong, H. R. Wang, K. E. White, M. J. Econs, and C. L. Oudet, MEPE, a new gene expressed in bone marrow and tumors causing osteomalacia. *Genomics,* **67,** 54–68 (2000).

452. L. D. Quarles, FGF23, PHEX, and MEPE regulation of phosphate homeostasis and skeletal mineralization. *Am J Physiol Endocrinol Metab,* **285,** E1–E9 (2003).

453. P. S. Rowe, The wrickkened pathways of FGF23, MEPE and PHEX. *Crit Rev Oral Biol Med,* **15,** 264–281 (2004).

454. L. C. Gowen, D. N. Petersen, A. L. Mansolf, H. Qi, J. L. Stock, G. T. Tkalcevic, H. A. Simmons, D. T. Crawford, K. L. Chidsey-Frink, H. Z. Ke, J. D. McNeish, and T. A. Brown, Targeted disruption of the osteoblast/osteocyte factor 45 gene (OF45) results in increased bone formation and bone mass. *J Biol Chem,* **278,** 1998–2007 (2003).

455. C. C. Hu, J. P. Simmer, J. D. Bartlett, Q. Qian, C. Zhang, O. H. Ryu, J. Xue, M. Fukae, T. Uchida, and M. MacDougall, Murine enamelin: cDNA and derived protein sequences. *Connect Tissue Res,* **39,** 47–61 (1998).

456. C. C. Hu, T. C. Hart, B. R. Dupont, J. J. Chen, X. Sun, Q. Qian, C. H. Zhang, H. Jiang, V. L. Mattern, J. T. Wright, and J. P. Simmer, Cloning human enamelin cDNA, chromosomal localization, and analysis of expression during tooth development. *J Dent Res,* **79,** 912–919 (2000).

457. P. A. Price, A. S. Otsuka, and J. W. Poser, Characterization of gamma-carboxyglutamic acid containing protein from bone. *Proc Natl Acad Sci USA,* **73,** 1447 (1976).

458. P. V. Hauschka, J. B. Lian, and P. M. Gallop, Direct identification of the calcium-binding amino acid, gamma-carboxyglutamic acid in mineralized tissue. *Proc Natl Acad Sci USA,* **72,** 3925–3929 (1975).

459. P. A. Price, A. S. Otsuka, J. W. Poser, J. Kristaponis, and N. Raman, Characterization of gamma-carboxyglutamic-acid containing protein from bone. *Proc Natl Acad Sci USA,* **73,** 1447–1451 (1976).

460. J. D. Fraser and P. A. Price, Lung, heart and kidney express high levels of mRNA for the vitamin K-dependent matrix Gla protein: Implications for the possible functions of matrix Gla protein and for the tissue distribution of the α-carboxylase. *J Biol Chem,* **263,** 11033–11036 (1988).

461. P. A. Price, Gla-containing proteins of mineralized tissues. In *Chemistry and Biology of Mineralized Tissues* (H. Slavkin and P. Price, eds.), pp. 169–176. Elsevier, Amsterdam (1992).

462. A. Bronckers, S. Gay, M. T. DiMuzio, and W. T. Butler, Immunolocalization of α-carboxyglutamic acid containing proteins in developing rat bones. *Coll Relat Res,* **5,** 273–281 (1985).

463. A. J. Camarda, W. T. Butler, R. D. Finkelman, and A. Nanci, Immunocytochemical localization of α-carboxyglutamic acid-containing proteins and osteocalcin in rat bone and dentin. *Calcif Tissue Int,* **40,** 349–355 (1987).

464. J. C. Fleet and J. M. Hock, Identification of osteocalcin mRNA in nonsteoid tissue of rats and humans by reverse transcription-polymerase chain reaction. *J Bone Miner Res,* **9,** 1565–1573 (1994).

465. E. Puchacz, J. G. Lian, G. S. Stein, J. Wozney, K. Heubner, and C. Croce, Chromosomal localization of the human osteocalcin gene. *Endocrinology,* **124,** 2648 (1989).

466. A. J. Celeste, V. Rosen, and J. L. Buecker, Isolation of the human gene for bone Gla protein. *EMBO J,* **5,** 1885–1889 (1986).

467. L. C. Pan and P. A. Price, The propeptide of rat bone gamma-carboxyglutamic acid protein shares homology with other vitamin K-dependent protein precursors. *Proc Natl Acad Sci USA,* **82,** 6109–6113 (1985).

468. L. C. Pan, M. K. Williamson, and P. A. Price, Sequence of the precursor to rat bone gamma-carboxyglutamic acid protein that assimilates in warfarin-treated osteosarcoma cells. *J Biol Chem,* **260,** 13398–13401 (1985).

469. K. Horiuchi, N. Amizuka, S. Takeshita, H. Takamatsu, M. Katsuura, H. Ozawa, Y. Toyama, L. F. Bonewald, and A. Kudo, Identification and characterization of a novel protein, periostin, with restricted expression to periosteum and periodontal ligament and increased expression by transforming growth factor beta. *J Bone Miner Res,* **14,** 1239–1249 (1999).

470. S. A. Kerner, R. A. Scott, and J. W. Pike, Sequence elements in the human osteocalcin gene confer basal activation and inducible response to hormonal vitamin D_3. *Proc Natl Acad Sci USA,* **84,** 4455–4459 (1989).

471. J. P. Bidwell, A. J. van Wijnen, E. G. Fey, S. Dworetzky, S. Penman, J. L. Stein, J. B. Lian, and G. S. Stein, Osteocalcin gene promoter binding factors are tissue-specific nuclear matrix components. *Proc Natl Acad Sci USA,* **90,** 3162–3166 (1993).

472. M. D. Demay, D. A. Roth, and H. M. Kronenberg, Regions of the rat osteocalcin gene which mediate the effect of 1,25-dihydroxyvitamin D_3 on gene transcription. *J Biol Chem,* **264,** 2279–2282 (1989).

473. M. D. Demay, J. M. Gerardi, H. F. DeLuca, and H. M. Kronenberg, DNA sequences in the rat osteocalcin gene that bind the 1,25-dihydroxyvitamin D_3 receptor and confer responsive to 1,25-dihydroxyvitamin D_3. *Proc Natl Acad Sci USA,* **87,** 369–373 (1990).

474. K. Yoon, S. J. C. Rutledge, R. F. Buenaga, and G. A. Rodan, Characterization of the rat osteocalcin gene: Stimulation of promoter activity by 1,25-dihydroxy vitamin D_3. *Biochemistry,* **27,** 8521–8526 (1988).

475. J. Lian, C. Stewart, E. Puchacz, S. Mackowiak, V. Shalhoub, D. Collart, G. Zambetti, and G. S. Stein, Structure of the rat osteocalcin gene and regulation of vitamin D-dependent expression. *Proc Natl Acad Sci USA,* **86,** 1143–1147 (1989).

476. E. R. Markose, J. L. Stein, G. S. Stein, and J. B. Lian, Vitamin D-mediated modifications in protein–DNA interactions at two promoter elements of the osteocalcin gene. *Proc Natl Acad Sci USA,* **87,** 1701–1705 (1990).

477. A. A. J. Heinrichs, C. Banerjee, R. Bortell, T. A. Owen, J. L. Stein, G. S. Stein, and J. B. Lian, Identification and characterization of two proximal elements in the rat osteocalcin gene promoter that may confer species-specific regulation. *J Cell Biochem,* **53,** 240–250 (1993).

478. A. J. J. Heinrichs, R. Bortell, M. Bourke, J. B. Lian, G. S. Stein, and J. L. Stein, Proximal promoter binding protein contributes to developmental, tissue-restricted expression of the rat osteocalcin gene. *J Cell Biochem,* **57,** 90–100 (1995).

479. P. Ducy, R. Zhang, R, V. Geoffroy, A. L. Ridall, and G. Karsenty, Osf2/Cbfa1: A transcriptional activator of osteoblast differentiation. *Cell,* **89,** 747–754 (1997).

480. L. C. Pan and P. A. Price, The propeptide of rat bone gamma-carboxyglutamic acid protein share homology with other vitamin K-dependent protein precursors. *Proc Natl Acad Sci USA,* **82,** 6109–6113 (1985).

481. J. W. Poser, F. S. Esch, and N. C. Ling, Isolation and sequence of the vitamin K-dependent protein from human bone: Undercarboxylation of the first glutamic acid residue. *J Biol Chem,* **255,** 8685 (1980).

482. P. V. Hauschka and S. A. Carr, Calcium-dependent alpha-helical structure in osteocalcin. *Biochemistry,* **21,** 2538–2547 (1982).

483. P. V. Hauschka and F. H. Wians Jr., Osteocalcin–hydroxyapatite interactions in the extracellular organic matrix of bone. *Anat Rec,* **224,** 180–188 (1988).

484. T. L. Dowd, J. F. Rosen, C. M. Gundberg, and R. K. Gupta, The displacement of calcium from osteocalcin at submicrolar concentration of free lead. *Biochem Biophys Acta,* **1226,** 131–137 (1994).

485. D. T. Isbell, S. Du, A. G. Schroering, G. Colombo, and J. G. Shelling, Metal ion binding to dog osteocalcin studied by ^{1}H NMR spectroscopy. *Biochemistry,* **32,** 11352–11362 (1993).

486. R. Paredes, G. Arriagada, F. Cruzat, J. Olate, A. Van Wijnen, J. Lian, G. Stein, J. Stein, and M. Montecino, The Runx2 transcription factor plays a key role in the 1alpha,25-dihydroxy vitamin D3-dependent upregulation of the rat osteocalcin (OC) gene expression in osteoblastic cells. *J Steroid Biochem Mol Biol,* **89/90,** 269–271 (2004).

487. E. F. Barroga, T. Kadosawa, M. Okumura, and T. Fujinaga, Influence of vitamin D and retinoids on the induction of functional differentiation *in vitro* of canine osteosarcoma clonal cells. *Vet J,* **159,** 186–193 (2000).

488. D. Lajeunesse, G. M. Kiebzak, C. Frondoza, and B. Sacktor, Regulation of osteocalcin secretion by human primary bone cells and by the human osteosarcoma cell line MG-63. *Bone Miner,* **14,** 237–250 (1991).

489. A. M. Delany, Y. Dong, and E. Canalis, Mechanisms of glucocorticoid action in bone cells. *J Cell Biochem,* **56,** 295–302 (1994).

490. G. Schepmoes, E. Breen, T. A. Owen, M. A. Aronow, G. S. Stein, and J. B. Lian, Influence of dexamethasone on the vitamin D mediated regulation of the osteocalcin gene expression. *J Cell Biochem,* **47,** 184–196 (1991).

491. A. Pirskanen, T. Jaaskelainen, and P. H. Maenpaa, Effects of transforming growth factor beta 1 on the regulation of osteocalcin synthesis in human MG-63 osteosarcoma cells. *J Bone Miner Res,* **9,** 1635–1642 (1994).

492. R. T. Ingram, S. K. Bonde, B. L. Riggs, and L. A. Fitzpatrick, Effects of transforming growth factor beta (TGF beta) and 1,25 dihydroxyvitamin D3 on the function, cytochemistry and morphology of normal human osteoblast-like cells. *Differentiation,* **55,** 153–163 (1994).

493. R. S. Taichman and P. V. Hauschka, Effects of interleukin-1 beta and tumor necrosis factor-alpha on osteoblastic expression of osteocalcin and mineralized extracellular matrix *in vitro*. *Inflammation,* **16,** 587–601 (1992).

494. D. B. Evans, R. A. Bunning, and R. G. Russell, The effects of recombinant human interleukin-1 beta on cellular proliferation and the production of prostaglandin E2, plasminogen activator, osteocalcin and alkaline phosphatase by osteoblast-like cells derived from human bone. *Biochem Biophys Res Commun,* **166,** 208–216 (1990).

495. D. M. Raab-Cullen, M. A. Thiede, D. N. Petersen, D. B. Kimmel, and R. R. Recker, Mechanical loading stimulates rapid changes in periosteal gene expression. *Calcif Tissue Int,* **55,** 473–478 (1994).

496. C. M. Gundberg, Biochemical markers of bone formation. *Clin Lab Med,* **20,** 489–501 (2000).

497. M. C. de Vernejoul, Markers of bone remodelling in metabolic bone disease. *Drugs Aging,* **12**(Suppl. 1), 9–14 (1998).

498. P. A. Price, J. W. Lothringer, S. A. Baukol, and A. H. Reddi, Developmental appearance of the vitamin K-dependent proteins in bone metabolism. *Annu Rev Nutr,* **8,** 865 (1988).

499. R. Kasai, P. Bianco, P. Gehron Robey, and A. J. Kahn, Production and characterization of an antibody against the human bone GLA protein (BGP/osteocalcin) propeptide and its use in immunocytochemistry of bone cells. *Bone Miner,* **25,** 167–182 (1994).

500. W. H. Liggett, J. B. Lian, J. S. Greenberger, and J. Glowacki, Osteocalcin promotes differentiation of osteoclast progenitors from murine long-term bone marrow cultures. *J Cell Biochem,* **55,** 190–199 (1994).

501. J. B. Lian and S. C. Marks Jr., Osteopetrosis in the rat: Coexistence of reductions in osteocalcin and bone resorption. *Endocrinology,* **126,** 955–962 (1990).

502. V. Shalhoub, B. Betterncourt, M. E. Jackson, C. A. Mackay, J. Lian, and G. S. Stein, Abnormalities of phosphoprotein gene expression in three osteopetrotic rat mutations: Elevated mRNA transcripts, protein synthesis, and accumulation in bone of mutant animals. *J Cell Physiol,* **158,** 110–120 (1994).

503. J. Bollerslev, S. C. Marks Jr., L. Mosekilde, and J. B. Lian, Cortical bone osteocalcin content and matrix composition in autosomal dominant osteopetrosis type I. *Eur J Endocrinol,* **130,** 592–594 (1994).

504. L. C. Gerstenfeld, S. D. Chipman, J. Glowacki, and J. B. Lian, Expression to differentiate function by mineralizing cultures of chicken osteoblasts. *Dev Biol,* **122,** 49–60 (1986).

505. C. G. Groot, J. K. Danes, J. Blok, A. Hoogendijk, and P. V. Hauschka, Light and electron microscopic demonstration of osteocalcin antigenicity in embryonic and adult rat bone. *Bone,* **7,** 379–385 (1986).

506. J. N. M. Heersche, S. Reimers, J. L. Wrana, M. M. Y. Waye, and A. K. Gupta, Changes in expression of alpha I collagen and osteocalcin mRNA in osteoblasts and odontoblasts at different stages of maturity as shown by *in situ* hybridization. *Proc Finn Dent Soc,* **88,** 173–182 (1992).

507. A. L. Boskey, S. Gadaleta, C. Gundberg, S. B. Doty, P. Ducy, and G. Karsenty, Fourier transform infrared microspectroscopic analysis of bones of osteocalcin-deficient mice provides insight into the function of osteocalcin. *Bone,* **23,** 187–196 (1998).

508. P. A. Price, M. R. Urist, and Y. Otawara, Matrix Gla protein, a new g-carboxyglutamic acid-containing protein which is associated with the organic matrix of bone. *Biochem Biophys Res Commun,* **117,** 765–771 (1983).

509. P. A. Price and M. K. Williamson, Primary structure of bovine matrix Gla protein, a new vitamin K-dependent bone protein. *J Biol Chem,* **260,** 14791–14975 (1985).

510. M. Murshed, T. Schinke, M. D. McKee, and G. Karsenty, Extracellular matrix mineralization is regulated locally: Different roles of two Gla-containing proteins. *J Cell Biol,* **165,** 625–630 (2004).

511. S. R. Rannels, M. L. Cancela, E. B. Wolpert, and P. A. Price, Matrix Gla protein mRNA expression in cultured type II pneumocytes. *Am J Physiol,* **265,** L270–L278 (1993).

512. J. E. Hale, M. K. Williamson, and P. A. Price, Carboxyl-terminal proteolytic processing of matrix Gla protein. *J Biol Chem,* **266,** 21145–21149 (1991).

513. P. A. Price, J. S. Rice, and M. K. Williamson, Conserved phosphorylation of serines in the Ser-x-Glu/Ser (P) sequences of the vitamin K-dependent matrix Gla protein from shark, lamb, rat, cow, and human. *Protein Sci,* **5,** 822–830 (1994).

514. M. T. B. Wiedemann and D. Belluoccio, Molecular cloning of avian matrix Gla protein. *Biochim Biophys Acta,* **1395,** 47–49 (1998).

515. M. L. Cancela, C.-L. Hsieh, U. Francke, and P. A. Price, Molecular structure, chromosome assignment and promoter organization of the human matrix Gla protein gene. *J Biol Chem,* **265,** 15040–15048 (1990).

516. M. L. Cancela and P. A. Price, Retinoic acid induces matrix Gla protein gene expression in human cells. *Endocrinology,* **130,** 102–108 (1992).

517. J. S. Rice, M. K. Williamson, and P. A. Price, Isolation and sequence of the vitamin K-dependent matrix Gla protein from the calcified cartilage of the soupfin shark. *J Bone Miner Res,* **9,** 567–576 (1994).

518. P. A. Price, M. K. Williamson, T. Haba, R. B. Dell, and W. S. S. Jee, Excessive mineralization with growth plate closure in rats on chronic warfarin treatment. *Proc Natl Acad Sci USA,* **79,** 7734–7738 (1982).

519. G. Luo, P. Ducy, M. D. McKee, G. J. Pinero, E. Loyer, R. R. Behringer, and G. Karsenty, Spontaneous calcification of arteries and cartilage in mice lacking matrix GLA protein. *Nature,* **386,** 78–81 (1997).

520. A. L. Boskey, M. D. McKee, and G. P. Karsenty, Mineral characterization of bones and soft tissues in matrix Gla protein-deficient mice. In *Chemistry and Biology of Mineralized Tissues* (M. Goldberg, A. Boskey, and C. Robinson, eds.), pp. 63–67 (2000).

521. K. Yagami, J. Y. Suh, M. Enomoto-Iwamoto, E. Koyama, W. R. Abrams, I. M. Shapiro, M. Pacifici, and M. Iwamoto, Matrix GLA protein is a developmental regulator of chondrocyte mineralization and, when constitutively expressed, blocks endochondral and intramembranous ossification in the limb. *J Cell Biol,* **147,** 1097–1108 (1999).

522. J. Folch-Pi and P. Stoffyn, Proteolipids from membrane systems. *Ann N Y Acad Sci,* **195,** 86–107 (1972).

523. J. Folch-Pi and J. Sakura, Preparation of the proteolipid apoprotein from bovine heart, liver and kidney. *Biochim Biophys Acta,* **427,** 401–427 (1977).

524. J. Ennever, B. Boyan-Salyers, and L. J. Riggan, Proteolipid and bone matrix calcification in vitro. *J Dent Res,* **56,** 967–970 (1977).

525. J. Ennever, J. J. Vogel, and L. J. Riggan, Phospholipids of a bone matrix calcification nucleator. *J Dent Res,* **57,** 731–734 (1978).

526. A. L. Boskey, Phospholipids and calcification: An overview. In *Cell Mediated Calcification and Matrix Vesicles* (S. Yousuf Ali, ed.), pp. 175–179. Elsevier, Amsterdam (1986).

527. B. D. Boyan, Proteolipid-dependent calcification. In *The Chemistry and Biology of Mineralized Tissues* (W. F. Butler, ed.), pp. 125–131. EBSCO Media, Birmingham, AL (1985).

528. B. D. Boyan and N. M. Ritter, Proteolipid–lipid relationships in normal and vitamin D-deficient chick cartilage. *Calcif Tissue Int,* **36,** 332–337 (1984).

529. B. D. Boyan, Z. Schwartz, L. D. Swain, and A. Khare, Role of lipids in calcification of cartilage. *Anat Rec,* **224,** 211–219 (1989).

530. L. N. Wu, T. Yoshimori, B. R. Genge, G. R. Sauer, and R. E. Wuthier, Characterization of the nucleational core complex responsible for mineral induction by growth plate cartilage matrix vesicles. *J Biol Chem,* **268,** 25084–25094 (1993).

531. K. Jung, H. Jung, and H. R. Kaback, Dynamics of lactose permease of *Escherichia coli* determined by site-directed fluorescence labeling. *Biochemistry,* **33,** 3980–3985 (1994).

532. J. Stim, A. A. Bernardo, F. T. Kear, Y. Y. Qiu, and J. A. Arruda, Renal cortical basolateral Na+/HC03- co-transporter:

II. Detection of conformational changes with fluorescein isothiocyanate labeling. *J Member Biol,* **140,** 39–46 (1994).

533. M. B. Beest, K. Hoekstra, A. Sein, and D. Hoekstra, Reconstruction of proteolipid protein: Some properties and its role in interlamellar attachment. *Biochem J,* **300,** 545–552 (1994).

534. G. Fernandez-Ballester, J. Castesana, A. M. Fernandez, and J. L. Arrondo, A role for cholesterol as a structural effector of the nicotinic acetylcholine receptor. *Biochemistry,* **33,** 4065–4071 (1994).

535. B. D. Boyan, W. J. Landis, J. Knight, G. Dereszewski, and J. Zeagler, Microbial hydroxyapatite formation as a model of proteolipid-dependent membrane-mediated calcification. *Scan Electron Microsc,* **4,** 1793–1800 (1984).

536. B. D. Boyan-Salyers and A. L. Boskey, Relationship between proteolipids and calcium–phospholipid–phosphate complexes in *Bacterionema matruchotii* calcification. *Calcif Tissue Int,* **30,** 167–174 (1980).

537. C. L. Raggio, B. D. Boyan, and A. L. Boskey, *In vivo* hydroxyapatite formation induced by lipids. *J Bone Miner Res,* **1,** 409–415 (1986).

538. A. Yaari, I. Shapiro, and A. L. Boskey, Ionophoretic activity of a Ca–PL–PO4 complex. *Calcif Tissue Int,* **36,** 317–319 (1984).

539. E. Rojas, H. B. Pollard, H. T. Haigler, C. Parra, and A. L. Burns, Calcium-activated endonexin II forms calcium channels across acidic phospholipids bilayer membranes. *J Biol Chem,* **265,** 21207–21215 (1990).

540. A. L. Boskey and B. D. Boyan-Salyers, Relationship between proteolipids and calcium–phospholipid–phosphate complexes in *Bacterionema matruchotii* calcification. *Calcif Tissue Int,* **30,** 167–174 (1978).

541. A. L. Boskey and A. S. Posner, The role of synthetic and bone extracted calcium–phospholipid–phosphate complexes in hydroxyapatite formation. *Calcif Tissue Res,* **23,** 251–258 (1977).

542. B. D. Boyan and A. L. Boskey, Co-isolation of proteolipids and calcium–phospholipid–phosphate complexes. *Calcif Tissue Int,* **36,** 214–218 (1984).

543. R. E. Wuthier and S. Gore, Partition of inorganic ion and phospholipids in isolated cell, membrane and matrix vesicle fractions. Evidence for Ca:Pi acidic phospholipid complexes. *Calcif Tissue Res,* **24,** 163–171 (1977).

544. F. Suarez, B. Rothhut, C. Comera, and L. Touqui, Expression of annexin I, II, V and VI by rat osteoblasts in primary culture: Stimulation of annexin I expression by dexamethasone. *J Bone Miner Res,* **8,** 1201–1210 (1993).

545. S. van Dijk, D. D. Dean, Y. Liu, Y. Zhao, J. M. Chirgwin, Z. Schwartz, and B. D. Boyan, Purification, amino acid sequence, and cDNA sequence of a novel calcium-precipitating proteolipid involved in calcification of *Corynebacterium matruchotii. Calcif Tissue Int,* **62,** 350–358 (1998).

546. B. R. Genge, L. N. Wu, and R. E. Wuthier, Differential fractionation of matrix vesicle proteins. Further characterization of the acidic phospholipid-dependent Ca^{++}-binding proteins. *J Biol Chem,* **265,** 10678–10685 (1990).

547. G. Weber and E. Ferber, Selective inhibition of phospholipase A2 by different lipocortins. *Biol Chem Hoppe–Seyler,* **371,** 725–731 (1990).

548. M. J. Geisow, U. Fritsche, J. M. Hexham, B. Dash, and T. Johnson, A consensus amino-acid sequence repeat in Torpedo and mammalian Ca^{++}-dependent membrane-binding proteins. *Nature,* **320,** 636–638 (1986).

549. P. D. Delmas, R. P. Tracy, and B. L. Riggs, Identification of the noncollagenous proteins of bovine bone by two-dimensional gel electrophoresis. *Calcif Tissue Int,* **36,** 308 (1984).

550. M. Ketteler, C. Vermeer, C. Wanner, R. Westenfeld, W. Jahnen-Dechent, and J. Floege, Novel insights into uremic vascular calcification: Role of matrix Gla protein and alpha-2-Heremans Schmid glycoprotein/fetuin. *Blood Purif,* **20,** 473–476 (2002).

551. B. A. Ashton, H.-J. Hohling, and J. T. Triffit, Plasma proteins present in human cortical bone: Enrichment of the α_2HS-glycoprotein. *Calcif Tissue Res,* **22,** 27–33 (1976).

552. R. A. Terkeltaub, D. A. Santoro, G. Mandel, and N. Mandel, Serum and plasma inhibit neutrophil stimulation by hydroxyapatite crystals. Evidence that serum alpha 2-HS glycoprotein is a potent and specific crystal-bound inhibitor. *Arthritis Rheum,* **31,** 1081–1089 (1988).

553. D. L. Christie, K. M. Dziegielewska, R. M. Hill, and N. R. Sanders, Fetuin: The bovine homologue of human α_2HS glycoprotein. *FEBS Lett,* **214,** 226–227 (1987).

554. I. R. Dickson, M. K. Bagga, and C. R. Paterson, Variations in the serum concentration and urine excretion of α_2-HS-glycoprotein, a bone-related protein, in normal individuals and in patients with osteogenesis imperfecta. *Calcif Tissue Int,* **35,** 16–20 (1983).

555. H. E. Schultze, K. Heide, and H. Haupt, Charterisierung eines niedermolekularen a_2 mukoids aus human serum. *Natur-wissenschaften,* **49,** 440–453 (1962).

556. C. C. Lee, B. H. Bowman, and F. M. Yang, Human alpha 2-HS-glycoprotein: The A and B chains with a connecting sequence are encoded by a single mRNA transcript. *Proc Natl Acad Sci USA,* **84,** 4403–4407 (1987).

557. W. Jahnen-Dechent, C. Schafer, A. Heiss, and J. Grotzinger, Systemic inhibition of spontaneous calcification by the serum protein alpha 2-HS glycoprotein/fetuin. *Z Kardiol,* **90**(Suppl. 3), 47–56 (2001).

558. R. H. Troxler and K. Schmid, The complete amino acid sequence of the A-chain of human plasma α2-HS glycoprotein. *J Biol Chem,* **261,** 1665–1676 (1986).

559. T. Araki, Y. Yoshioka, and K. Schmid, The position of the disulfide bonds in human plasma α2 HS-glycoprotein and the repeating double disulfide bonds in the domain structure. *Biochim Biophys Acta,* **994,** 195–199 (1989).

560. J. Kellerman, H. Haupt, E.-A. Auerswald, and W. Muller-Esterl, The arrangement of disulfide loops in human α_2-HS glycoprotein. Similarity to the disulfide bridge structures of cystatins and kininogens. *J Biol Chem,* **264,** 14121–14128 (1989).

561. T. Ohnishi, N. Arakaki, O. Nakamura, S. Hirono, and Y. Daikuhara, Purification, characterization, and studies on biosynthesis of a 59-kDa bone sialic acid-containing protein (BSP) from rat mandible using a monoclonal antibody. Evidence from rat mandible using a monoclonal antibody that 59-kDa BSP may be the rat counterpart of human α_2-HS glycoprotein and is synthesized by both hepatocytes and osteoblasts. *J Biol Chem,* **266,** 14636–14645 (1991).

562. T. Ohnishi, O. Nakamura, M. Ozawa, N. Arakaki, T. Muramatsu, and Y. Daikuhara, Molecular cloning and sequence analysis of cDNA for a 59 kD bone sialoprotein of the rat: Demonstration that it is a counterpart of human a_2-HS glycoprotein and bovine fetuin. *J Bone Miner Res,* **8,** 367–377 (1993).

563. N. C. Blumenthal, F. Betts, and A. S. Posner, Effect of carbonate and biological macromolecules on formation and properties of hydroxyapatite. *Calcif Tissue Res,* **18,** 81–90 (1975).

564. J. T. Triffit, U. Gehauer, B. A. Ashton, M. H. Owen, and J. J. Reynolds, Origin of plasma α_2-HS-glycoprotein and its accumulation in bone. *Nature,* **262,** 226–227 (1976).

565. C. Schafer, A. Heiss, A. Schwarz, R. Westenfeld, M. Ketteler, J. Floege, W. Muller-Esterl, T. Schinke, and W. Jahnen-Dechent, The serum protein alpha 2-Heremans–Schmid glycoprotein/fetuin-A is a systemically acting inhibitor of ectopic calcification. *J Clin Invest,* **112,** 357–366 (2003).

566. W. Jahnen-Dechent, T. Schinke, A. Trindl, W. Muller-Esterl, F. Sablitzky, S. Kaiser, and M. Blessing, Cloning and targeted deletion of the mouse fetuin gene. *J Biol Chem,* **272,** 31496–31503 (1997).

567. J. Garnett and P. Dieppe, The effects of serum and human albumin on calcium hydroxyapatite crystal growth. *Biochem J,* **266,** 863–868 (1990).

568. H. Gilman and D.W. Hukins, Seeded growth of hydroxyapatite in the presence of dissolved albumin. *J Inorg Biochem,* **55,** 21–30 (1994).

569. H. Gilman and D. W. Hukins, Seeded growth of hydroxyapatite in the presence of dissolved albumin at constant composition. *J Inorg Biochem,* **55,** 31–39 (1994).

570. H. Iwase, K. Gondo, T. Koike, and I. Ono, Novel precolumn deproteinization method using a hydroxyapatite cartridge for the determination of theophylline and diazepam in human plasma by high-performance liquid chromatography with ultraviolet detection. *J Chromatogr B Biomed Appl,* **655,** 73–81 (1994).

571. H. Furedi-Milhofer, Adsorption of human serum albumin on precipitated hydroxyapatite. *J Colloid Interface Sci,* **69,** 460–480 (1979).

572. M. J. Mura-Galelli, J. C. Voegel, S. Behr, E. F. Bres, and P. Schaaf, Adsorption/desorption of human serum albumin on hydroxyapatite: A critical analysis of the Langmuir model. *Proc Natl Acad Sci USA,* **88,** 5557–5561 (1991).

573. Y. Yoshiuka, F. Gejyo, T. Marti, E. E. Ricki, W. Burgi, G. D. Offner, S. Shimabayashi, Y. Tanizawa, and K. Ishida, Effect of phosphorylated organic compound on the adsorption of bovine serum albumin by hydroxyapatite. *Chem Pharm Bull,* **39,** 2183–2188 (1991).

574. H. Iwase, K. Gondo, T. Koike, and I. Ono, Novel precolumn deproteinization method using a hydroxyapatite cartridge for the determination of theophylline and diazepam in human plasma by high-performance liquid chromatography with ultraviolet detection. *J Chromatogr B Biomed Appl,* **665,** 73–81 (1994).

575. N. J. Hamlin and P. A. Price, Mineralization of decalcified bone occurs under cell culture conditions and requires bovine serum but not cells. *Calcif Tissue Int,* **75,** 231–242 (2004).

576. P. A. Price, H. H. June, N. J. Hamlin, and M. K. Williamson, Evidence for a serum factor that initiates the re-calcification of demineralized bone. *J Biol Chem,* **279,** 19169–19180 (2004).

577. C. E. Brinckerhoff and L. M. Matrisian, Matrix metalloproteinases: A tail of a frog that became a prince. *Nat Rev Mol Cell Biol,* **3,** 207–214 (2002).

578. R. Visse and H. Nagase, Matrix metalloproteinases and tissue inhibitors of metalloproteinases: Structure, function, and biochemistry. *Circ Res,* **92,** 827–839 (2003).

579. C. Chaussain-Miller, F. Fioretti, M. Goldberg, and S. Menashi, The role of matrix metalloproteinases (MMPs) in human caries. *J Dent Res,* **85,** 22–32 (2006).

580. D. K. Scott, K. D. Brinckerhoff, J. C. Clohisy, C. O. Quinn, and N. C. Partridge, Parathyroid hormone induces transcription of collagenase in rat osteoblastic cells by a mechanism using cyclic adenosine $3',5'$-monophosphate and requiring protein synthesis. *Mol Endocrinol,* **6,** 2153–2159 (1992).

581. A. T. Vernillo, N. S. Ramamurthy, H. M. Lee, and B. R. Rifkin, Parathyroid hormone regulation of matrix degrading enzymes in rat osteoblastic osteosarcoma 17/2.8 cells. *Res Commun Chem Pathol Pharmacol,* **75,** 323–339 (1992).

582. F. S. Panagakos and S. Kumar, Modulation of proteases and their inhibitors in immortal human osteoblast-like cells by tumor necrosis factor-alpha *in vitro. Inflammation,* **18,** 243–265 (1994).

583. S. Varghese, S. Rydziel, J. J. Jeffrey, and E. Canalis, Regulation of interstitital collagenase expression and collagen degradation by retinoic acid in bone cells. *Endocrinology,* **134,** 2438–2444 (1994).

584. J. M. Delaise, Y. Eeckout, L. Neff, C. Francois-Gillet, A. Herriet, Y. Su, G. Vaes, and R. Baron, (Pro)collagenase (matrix metalloproteinases-1) is present in rodent osteoclasts and in the underlying bone-resorbing compartment. *J Cell Sci,* **106,** 1071–1082 (1993).

585. G. Leloup, J. M. Delaisse, and G. Vaes, Relationship of the plasminogen activator/plasmin cascade to osteoclast invasion and mineral resorption in explanted fetal metatarsal bones. *J Bone Miner Res,* **9,** 891–902 (1994).

586. M. F. French, A. Bhown, and H. E. Van Wart, Identification of clostridium collagenase hyperactive sites in type I, II, and III collagens: Lack of correlation with local triple helical stability. *J Protein Chem,* **11,** 83–97 (1992).

587. H. Wu, X. Liu, M. Bynne, Y. Harada, S. M. Krane, and R. Jaenisch, Collagenases are critical for skeletal and soft connective tissue remodeling: Abnormalities in mice that express, in the germline, an engineered mutation in the Col 1a1 gene that encodes collagenase resistance. *J Bone Miner Res,* **9,** S141 (1994).

588. K. Holmbeck, P. Bianco, J. Caterina, S. Yamada, M. Kromer, S. A. Kuznetsov, M. Mankani, P. G. Robey, A. R. Poole, I. Pidoux, J. M. Ward, and H. Birkedal-Hansen, MT1-MMP-deficient mice develop dwarfism, osteopenia, arthritis, and connective tissue disease due to inadequate collagen turnover. *Cell,* **99,** 81–92 (1999).

589. V. Everts, J. M. Delaisse, W. Korper, A. Niehof, G. Vaes, and W. Beertsen, Degradation of collagen in the bone-resorbing compartment underlying the osteoclast involves both cysteine-proteinases and matrix metalloproteases. *J Cell Physiol,* **150,** 221–231 (1992).

590. G. Vaes, J. M. Delaisse, and Y. Eeckhout, Relative roles of collagenase and lysosomal cysteine-proteinases in bone resorption. *Matrix* **1**(Suppl. 1), 383–388 (1992).

591. P. A. Hill, D. J. Buttle, S. J. Jones, A. Boyde, M. Murata, J. J. Reynolds, and M. C. Meikle, Inhibition of bone resorption by selective inactivators of cysteine proteinases. *J Cell Biochem,* **45,** 118–130 (1994).

592. J. P. Montenez, J. M. Delaisse, P. M. Tulkens, and B. K. Kishore, Increased activities of cathepsin B and other lysosomal hydrolases in fibroblasts and bone tissue cultured in the presence of cysteine proteinase inhibitors. *Life Sci,* **55,** 1199–1208 (1994).

593. M. Gowen, F. Lazner, R. Dodds, R. Kapadia, J. Feild, M. Tavaria, I. Bertoncello, F. Drake, S. Zavarselk, I. Tellis, P. Hertzogm, C. Debouck, and I. Kola, Cathepsin K knockout mice develop osteopetrosis due to a deficit in matrix degradation but not demineralization. *J Bone Miner Res,* **14,** 1654–1663 (1999).

594. P. Saftig, E. Hunziker, V. Everts, S. Jones, A. Boyde, O. Wehmeyer, A. Suter, and K. von Figura, Functions of cathepsin K in bone resorption. Lessons from cathepsin K deficient mice. *Adv Exp Med Biol,* **477,** 293–303 (2000).

595. N. Fratzl-Zelman, A. Valenta, P. Roschger, A. Nader, B. D. Gelb, P. Fratzl, and K. Klaushofer, Decreased bone turnover

and deterioration of bone structure in two cases of pycnodysostosis. *J Clin Endocrinol Metab,* **89,** 1538–1547 (2004).

596. C. Sfeir and A. Veis, Casein kinase localization in the endoplasmic reticulum of the ROS 17/2.8 cell line. *J Bone Miner Res,* **10,** 607–615 (1995).

597. Y. Suzuki, A. Yamaguchi, T. Ikeda, T. Kawase, S. Saito, and Y. Mikuni-Takagaki, *In situ* phosphorylation of bone and dentin proteins by the casein kinase II-like enzyme. *J Dent Res,* **77,** 1799–1806 (1998).

598. A. Veis, C. Sfeir, and C. B. Wu, Phosphorylation of the proteins of the extracellular matrix of mineralized tissues by casein kinase-like activity. *Crit Rev Oral Biol Med,* **8,** 360–379 (1997).

599. E. H. Fischer, H. Charbonneau, and N. K. Tonks, Protein tyrosine phosphatases: A diverse family of intracellular and transmembrane enzymes. *Science,* **253,** 401–420 (1991).

600. E. Salih, S. Ashkar, L. C. Gerstenfeld, and M. J. Glimcher, Identification of the phosphorylated sites of metabolically 32P-labeled osteopontin from cultured chicken osteoblasts. *J Biol Chem* **272,** 13966–13973 (1997).

601. L. Rifas, S. Cheng, L. R. Halstead, A. Gupta, K. A. Hruska, and L. V. Avioli, Skeletal casein kinase activity defect in the HYP mouse. *Calcif Tissue Int,* **61,** 256–259 (1997).

602. D. D. Lee and M. J. Glimcher, Three-dimensional spatial relationship between the collagen fibrils and the inorganic calcium phosphate crystals of pickeral (*Americanus americanus*) and herring (*Clupea harengus* bone). *J Mol Biol,* **217,** 487–501 (1991).

603. J. Christoffersen and W. J. Landis, A contribution with review to the description of mineralization of bone and other calcified tissues *in vivo. Anat Rec,* **230,** 435–450 (1991).

604. S. Weiner, Organization of extracellularly mineralized tissues: A comparative study of biological crystal growth. *CRC Crit Rev Biochem,* **20,** 365–408 (1986).

605. E. Dmitrovsky and A. L. Boskey, Calcium-acidic phospholipid phosphate complexes in human atherosclerotic aorta. *Calcif Tissue Int,* **37,** 121–125 (1985).

606. J. M. Burnell, D. J. Baylink, C. H. Chestnut 3rd, and E. J. Teubner, The role of skeletal calcium deficiency in postmenopausal osteoporosis. *Calcif Tissue Int,* **38,** 187–192 (1986).

607. A. E. N. Nielsen, *Kinetics of Precipitation.* Pergamon, Oxford (1964).

608. U. Vetter, L. W. Fisher, K. P. Mintz, J. B. Kopp, N. Tuross, J. D. Termine, and P. Gehron Robey, Osteogenesis imperfecta: Changes in noncollagenous proteins in bone. *J Bone Miner Res,* **6,** 501–505 (1991).

609. G. Palmer, J. Zhao, J. Bonjour, W. Hofstetter, and J. Caverzasio, *In vivo* expression of transcripts encoding the Glvr-1 phosphate transporter/retrovirus receptor during bone development. *Bone,* **24,** 1–7 (1999).

610. J. Guicheux, G. Palmer, C. Shukunami, Y. Hiraki, J. P. Bonjour, and J. A. Caverzasio, Novel *in vitro* culture system for analysis of functional role of phosphate transport in endochondral ossification. *Bone,* **27,** 69–74 (2000).

611. C. M. Giachelli, Vascular calcification: *In vitro* evidence for the role of inorganic phosphate. *J Am Soc Nephrol,* **14,** S300–S304 (2003).

612. D. S. Howell, J. C. Pita, J. F. Marquez, and J. E. Madruga, Partition of calcium, phosphate and protein in the fluid phase aspirated at calcifying sites in epiphyseal cartilage. *J Clin Invest,* **47,** 1121–1132 (1968).

613. G. H. Nancollas and B. Tomazic, Growth of calcium phosphate on hydroxyapatite crystals. Effect of supersaturation and ionic medium. *J Phys Chem,* **78,** 2218–2225 (1974).

614. A. L. Boskey and A. S. Posner, Conversion of amorphous calcium phosphate to microcrystalline hydroxyapatite. A pH-dependent, solution-mediated, solid–solid conversion. *J Phys Chem,* **77,** 2313–2317 (1973).

615. W. T. Brown and L. C. Chow, Chemical properties of bone mineral. *Annu Rev Mater Sci,* **6,** 213–262 (1976).

616. E. D. Eanes and J. L. Meyer, The maturation of crystalline calcium phosphates in aqueous suspensions at physiologic pH. *Calcif Tissue Res,* **23,** 259–269 (1977).

617. W. E. Brown, N. Eidelman, and B. Tomazic, Octacalcium phosphate as a precursor in biomineral formation. *Adv Dent Res,* **1,** 306–313 (1987).

618. L. J. Brécević and H. Füredi-Milhofer, Precipitation of calcium phosphates from electrolyte solutions: II. The formation and transformation of precipitates. *Calcif Tissue Res,* **10,** 82–90 (1972).

619. J. Moradian-Oldak, S. Weiner, L. Addadi, W. J. Landis, and W. Traub, Electron imaging and diffraction study of individual crystals of bone, mineralized tendon and synthetic carbonate apatite. *Connect Tissue Res,* **25,** 219–228 (1991).

620. H. Füredi-Milhofer, J. Moradian-Oldak, S. Weiner, A. Veis, K. P. Mintz, and L. Addadi, Interactions of matrix proteins from mineralized tissues with octacalcium phosphate. *Connect Tissue Res,* **30,** 251–264 (1994).

621. D. Hanein, B. Geiger, and L. Addadi, Fibronectin adsorption to surfaces of hydrated crystals. An analysis of the importance of bound water in protein–substrate interactions. *Langmuir,* **9,** 1058–1065 (1993).

622. J. Aizenberg, S. Albeck, S. Weiner, and L. Addadi, Crystal–protein interactions studied by overgrowth of calcite on biogenic skeletal elements. *J Crystal Growth,* **142,** 156–164 (1994).

623. D. Hanein, H. Sabanay, L. Addadi, and B. Geiger, Selective interactions of cells with crystal surfaces. Implications for the mechanism of cell adhesion. *J Cell Sci,* **104,** 275–288 (1993).

624. A. Linde, A. Lussi, and M. A. Crenshaw, A mineral induction by immobilized polyanionic proteins. *Calcif Tissue Int,* **44,** 286–295 (1989).

625. R. W. Romberg, P. G. Werness, B. L. Riggs, and K. G. Mann, Inhibition of hydroxyapatite crystal growth by bone-specific and other calcium-binding proteins. *Biochemistry,* **25,** 1176–1180 (1986).

626. L. Silverman and A. L. Boskey, Diffusion systems for evaluation of biomineralization. *Calcif Tissue Int,* **75,** 494–501 (2004).

627. T. Dahl, B. Sabsay, and A. Veis, Type I collagen–phosphophoryn interactions: Specificity of the monomer–monomer binding. *J Struct Biol,* **123,** 162–168 (1998).

628. G. K. Huneter, *In vitro* studies on matrix-mediated mineralization. In *Bone Metabolism and Mineralization* (B. K. Hall, ed.), pp. 225–247. CRC Press, Boca Raton, FL (1992).

629. B. R. Heywood and E. D. Eanes, An ultrastructural study of the effects of acidic phospholipid substitutions on calcium phosphate precipitation in anionic liposomes. *Calcif Tissue Int,* **50,** 149–156 (1992).

630. E. D. Eanes, A. W. Hailer, R. J. Midura, and V. C. Hascall, Proteoglycan inhibition of calcium phosphate precipitation in liposomal suspensions. *Glycobiology,* **2,** 571–578 (1992).

631. D. Skrtic and E. D. Eanes, Effect of different phospholipid–cholesterol membrane compositions on liposome-mediated formation of calcium phosphates. *Calcif Tissue Int,* **50,** 253–260 (1992).

632. D. Skrtic and E. D. Eanes, Membrane-mediated precipitation of calcium phosphate in model liposomes with matrix vesicle-like lipid composition. *Bone Miner,* **16,** 109–119 (1992).

CHAPTER 8

Development of the Skeleton

SYLVAIN PROVOT, ERNESTINA SCHIPANI, JOY WU, AND HENRY KRONENBERG

I. INTRODUCTION

The skeletal system consists of 206 bones of strikingly varying shapes, sizes, and functions. More than with any other organ, the specific shapes and sizes of these bones are crucial to their functions of providing levers for movement and protection of soft tissues. Despite the striking diversity of the sizes and shapes of individual bones, all bones form through one of two distinct processes: endochondral bone formation, used for the generation of most bones, and intramembranous bone formation, used to form the flat bones of the skull and parts of several other bones. In each of these processes, local paracrine signals and systemic hormonal signals trigger characteristic transcription programs and activation of kinase cascades that orchestrate the generation of the skeleton. In this chapter, we consider the strategies used to pattern the skeleton and then consider the processes of endochondral and intermembranous bone formation during development. Particular attention is paid to the description of the signaling systems and transcription factors that coordinate formation of the skeleton.

II. PATTERNING THE SKELETON

The skeleton is one of the most highly patterned structures in higher organisms. The 206 bones of the adult human vary greatly in size and shape; for instance, the femur may exceed 45 cm in length and its cylindrical shaft is remarkably different from the microscopic shape of ossicles of the inner ear. The anatomical aspect of bones also greatly differs between species; the skeleton of an elephant can be easily distinguished from that of a bat. However, the skeleton is patterned by mechanisms that are similar among all vertebrates. Our knowledge of these mechanisms largely derives from studies of chicken and mouse embryology.

Each tissue present in any organism originates from one of the three embryonic layers that are defined during gastrulation. The ectoderm is the outermost layer and is responsible for the formation of the tegument and associated structures, the epidermis, the nervous system, and some cranial bones. The innermost layer, the endoderm, forms the digestive tract and associated glands, the lungs, the liver, and the pancreas. The mesoderm constitutes the intermediate layer and gives rise to the skeleton (with the exception of several cranial bones), the muscles, and some internal organs such as the kidneys. We commonly distinguish several types of mesoderm, which each form different derivatives, depending on the position of this tissue in the embryo along a radial axis (Figure 8-1). The paraxial mesoderm (PM) corresponds to the tissue immediately adjacent to the neural tube and to the notochord, tissues that form a central anteroposterior (AP) axis in the embryo. The paraxial mesoderm will give rise to the axial skeleton (the vertebral column and associated ribs) and to some bones of the skull. The lateral plate mesoderm (LPM) corresponds to the mesoderm present farthest radially on each side of the neural tube; it will give rise to the appendicular skeleton (the limbs). Lastly, the intermediate mesoderm, located in between the PM and the LPM, does not give rise to any part of the skeleton but is necessary for the formation of internal organs such as the kidneys. Whereas with few exceptions, the embryonic origins of all the bones are known, the molecular signals and tissue interactions required for patterning the skeleton constitute an active field of investigation. We provide a summary of the complex sequence of events that generate the axial, craniofacial, and appendicular skeleton.

A. Axial Skeleton Development

The axial skeleton of vertebrates consists of the vertebrae and the intervertebral discs forming the vertebral column and also the ribs. The vertebral column is an essential element of support and motility of the vertebrate body. It provides attachment for many tendons/muscles in addition to the ribs and some organs, and it protects the spinal cord, which controls most bodily functions. It is also one of the most obviously segmented structures in animals.

Formation of the AP axis of vertebrates occurs through a rostral-to-caudal progression of signals from the embryonic organizer known as the node and Hensen's node in mammals and birds, respectively. These nodal signals result in the production of a variety of tissues including the paraxial mesoderm that gives rise to the axial skeleton. The axial skeleton displays a metameric organization that consists of a series of equivalent units distributed along the AP axis, each comprising a vertebra and its associated muscles, peripheral nerves, and blood vessels. This segmental pattern is established during embryogenesis through the process of somitogenesis, by which the caudal mesenchymal portion of the paraxial mesoderm, called the presomitic mesoderm (PSM), becomes segmented into epithelial somites on each side of the neural tube (Figures 8-1 and 8-2). Fate-mapping studies have defined the progeny of different regions of the somite. Interspecies grafts of portions of the somite between chick and quail embryos allow the distinction of cells derived from the grafted tissue by histological staining (nucleolar quail–chick markers developed by Nicole Le Douarin [1]) or by immunohistochemical staining with species-specific antibodies. These experiments have established that each somite differentiates such that the ventral region becomes mesenchymal and forms the sclerotome, precursor of the vertebrae and the medial portion of the ribs, whereas the dorsal region, termed the dermomyotome, remains epithelial and forms skeletal muscles of the back, body walls, and limbs, the dermis of the back, a portion of the scapula, and perhaps the distal portion of the ribs, although this is controversial [2, 3] (Figure 8-3). Interestingly, the most distal portion of the ribs, called the sternal ribs, derives from the LPM, suggesting that ribs may derive from three different tissues (the sclerotome, dermomyotome, and LPM).

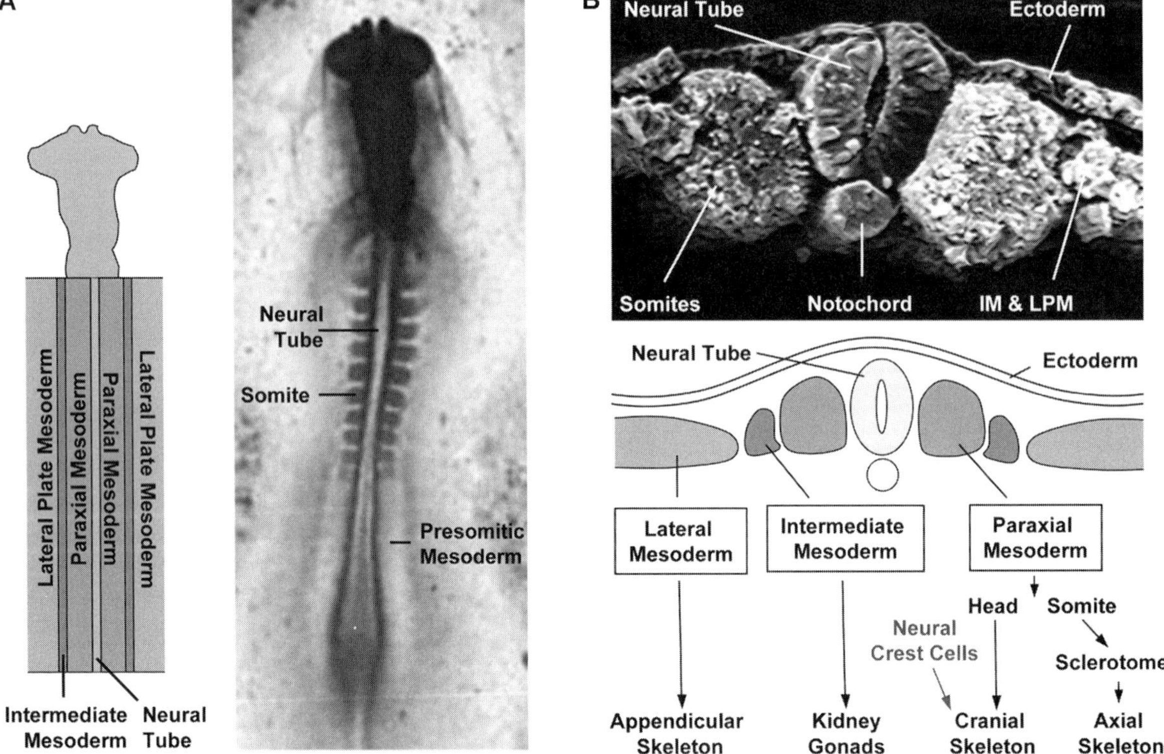

FIGURE 8-1   Overview of the early chicken embryo, the different mesoderms, and their derivatives. Photographs and corresponding schematic representation of dorsal view (A) and a view of a transverse section (B) are shown. The paraxial mesoderm corresponds to the tissue immediately adjacent to the neural tube and to the notochord. It gives rise to the axial skeleton and some bones of the skull (neural crest cells, which do not have a mesodermal but an ectodermal origin, contribute largely to the craniofacial skeleton). The paraxial mesoderm undergoes segmentation of presomitic mesoderm that forms somites thereafter. Somites further mature into sclerotome, which is at the origin of the axial skeleton. The lateral plate mesoderm (LPM) corresponds to the mesoderm present farthest radially on each side of the neural tube. It gives rise to the appendicular skeleton. The intermediate mesoderm (IM) is located in between the paraxial and the lateral plate mesoderm. It gives rise to internal organs, such as kidneys, and to the gonads. The scanning electron microscopy image in B is reprinted with permission from A. H. Monsoro-Burq, Sclerotome development and morphogenesis: When experimental embryology meets genetics. *Int J Dev Biol* **49**, 301–308 (2005).

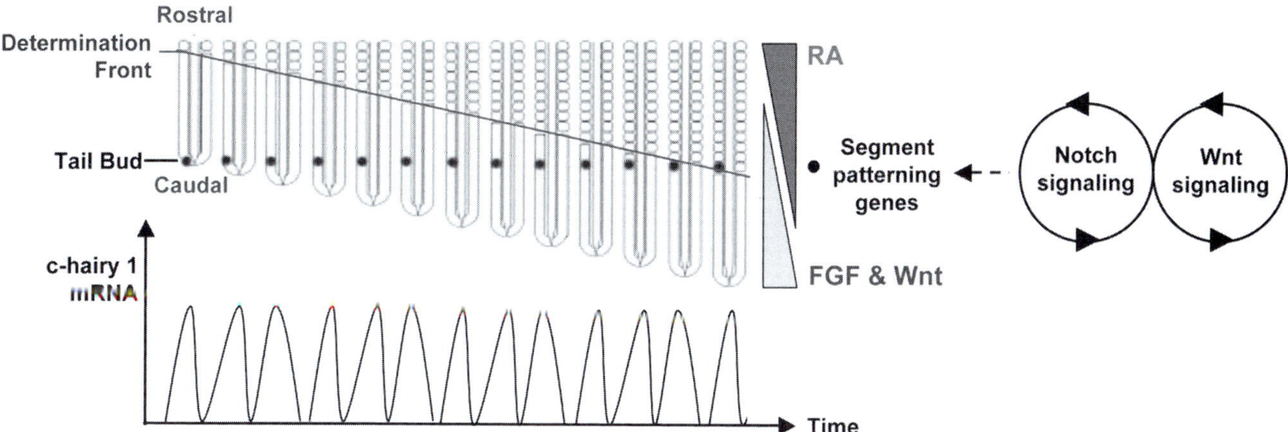

FIGURE 8-2   Schematic representation of somite formation. Somites form by segmentation of the presomitic mesoderm at a regular pace (every 90 minutes in the chick and every 120 minutes in the mouse) in a rostral-to-caudal sequence. This process relies on two distinct processes: (1) the generation of a determination front (solid blue line) that moves posteriorly and (2) the action of an oscillating biological clock that determines the temporal periodicity of the somite formation. Activation of the Notch and Wnt signaling pathway oscillates (c-hairy 1 mRNA expression downstream of Notch signaling is shown as an example). This ensures the temporal periodicity of induction of patterning genes that are responsible for the segmentation. The determination front is generated by two antagonizing gradients of morphogens along the anterior–posterior axis: FGF and Wnt have a higher expression caudally and oppose a gradient of retinoid acid (RA) more highly expressed rostrally. The presomitic cells are thought to be generated by a domain of self-renewing stem cells (tail bud), which become incorporated into a somite, the 12th and last somite formed in chicken embryo. Figure adapted from O. Pourquie, Vertebrate somitogenesis. *Annu Rev Cell Dev Biol* **17**, 311–350 (2001), with permission.

In addition to the dorsoventral regionalization of the somite, the sclerotome is further subdivided into several compartments organized along rostrocaudal and mediolateral axes, increasing the complexity of axial skeleton patterning. For instance, the medial part of the rostral sclerotome (mediorostral quadrant) forms the vertebral body, whereas the mediocaudal quadrant leads to the intervertebral disc, and the laterocaudal

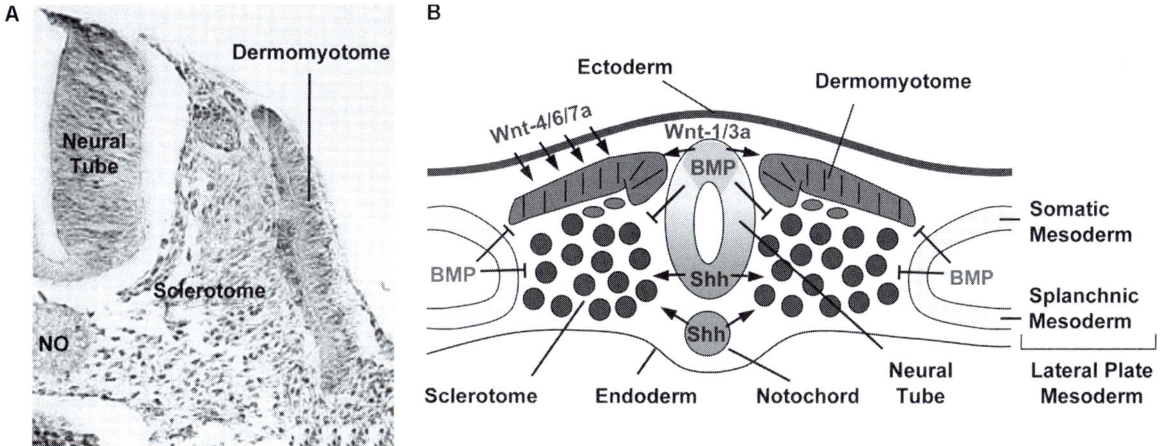

FIGURE 8-3   Compartmentalization of the somite into sclerotome and dermomyotome. Transversal section (A) and schematic representation (B) illustrating the compartmentalized somite. The ventral portion of the epithelial somite generates the sclerotome, whereas the dorsal part remains epithelial and becomes the dermomyotome, which gives rise to dermis and muscle. The myotome (brown staining product depicting desmin immunoreactivity) forms between dermomyotome and sclerotome. The sclerotome undergoes differentiation upon exposure of positive and negative signals released from the surrounding tissues (B). Sonic hedgehog (Shh) secreted by the notochord (NO) and the floor plate of the neural tube is a critical positive inducer of the sclerotome and its differentiation into cartilaginous tissue. Conversely, Wnt signals from the ectoderm and the roof plate of the neural tube promote dermomyotome formation and inhibit chondrogenesis. Bone morphogenetic protein (BMP) signals from the lateral plate mesoderm and the roof plate of the neural tube antagonize Shh signals early during sclerotome formation, but they later cooperate with Shh to promote chondrogenesis. Pax1 is a marker for the early sclerotome, whereas Pax3 expression is restricted to the prospective dermomyotome. (A) Reprinted from C. Kalcheim and R. Ben-Yair, Cell rearrangements during development of the somite and its derivatives. *Curr Opin Genet Dev* **15**, 371–380 (2005), with permission.

quadrant leads to the neural arches, the pedicules of the vertebrae, and the ribs [2]. The rostrocaudal compartmentalization has strong implications for the formation of vertebrae. One vertebra is not produced from one sclerotome but, rather, from the fusion of the caudal half of the sclerotome of one somite with the rostral half of the following somite in a process called resegmentation [4]. In contrast, each axial myotome that gives rise to the muscles of the back derives from a single somite. The consequence of this is that the muscles of the back are attached to two successive vertebrae, allowing the vertebral column to bend.

The last level of regionalization of the somitic mesoderm corresponds to a specification of this tissue according to the AP axis of the embryo. Thus, the somitic mesoderm is patterned into cervical, thoracic, lumbar, sacral, and caudal regions. This regionalization is established early during development, probably before the formation of the PSM, and relies on a changing expression of a family of homeobox-containing transcription factors, the Hox genes, along the AP axis [5]. Hox genes are sequentially activated in time and space in a way that reflects their organization into clusters in the genome. This phenomenon is known as the principle of colinearity. The molecular control ensuring the colinear expression of Hox genes in the body is still not clearly understood [5]. This colinearity of expression of the Hox genes has been conserved during evolution, whereas the number of Hox genes, and therefore the complexity of the specific combinations of Hox genes expressed along the AP axis, has increased in parallel with the increased complexity of the organisms. Primitive fishes as well as more recent vertebrate animals evolved from a common invertebrate ancestor. It is estimated that the first spineless creatures that lived hundreds of million years ago had up to 13 Hox genes to direct their development. When the first mammals appeared much later, they had four sets of 13 Hox genes distributed on different chromosomes. Since then, some genes have been lost during evolution, and recent mammals such as mice and men have a total of 49 Hox genes distributed in four different clusters. Thus, the regionalization of the somitic mesoderm, defined by the specific combination of Hox genes (the Hox code) expressed in specific regions along the AP axis, can explain why some animals, such as snakes, have ribs all the way from the neck to the tail, whereas humans have only 12 pairs of ribs, attached to the thoracic vertebrae. The importance of this regionalization and the Hox code is demonstrated by a very simple experiment: When a piece of PSM dissected from the thoracic domain of paraxial mesoderm (that will form the ribs) of a donor embryo is grafted into the cervical domain of the paraxial mesoderm of a host embryo, this tissue later develops ectopic ribs in the cervical region [4, 6].

The process of somite formation is elaborately regulated (Figure 8-2). Somites form at a regular pace that is species dependent: One somite is formed every 90 minutes in the chick and every 120 minutes in the mouse, in a rostral-to-caudal sequence. The segmentation of the PSM into somites relies on two distinct processes: the generation of a determination front that moves posteriorly during somitogenesis and the actions of an oscillating biological clock that determines the temporal periodicity of somite formation [4, 7]. This oscillator involves the periodic activation of the Notch and the Wnt signaling pathways [8]. As in many biological clocks, such as the circadian rhythm, molecular oscillations are generated through negative transcriptional feedback loops. The precise molecular role of the segmentation clock in somitogenesis remains unclear. Whereas the segmentation clock controls the temporal periodicity of somite formation, the periodicity of the somite distribution in space is mediated by a moving front of cell competence (called the determination front) that travels along the AP axis of the embryo. Two dynamic and antagonizing gradients of morphogens control the progression of this front along the AP axis. Wnt and fibroblast growth factor (FGF) signals are generated rostrally and lead to gradients with decreased expression of these factors anteriorly; this gradient is opposed by a gradient of retinoic acid (RA), which is synthesized maximally anteriorly. The position of the determination front has been proposed to be defined by a threshold level of FGF signaling, which has been suggested to be activated downstream of Wnt signaling. As the embryo extends posteriorly, the position of the determination front recedes along the AP axis. The current model argues that PSM cells become competent for segmentation when FGF signaling drops below a certain threshold, and then the cells adopt a boundary fate if they are juxtaposed to cells in a different phase of the segmentation cycle. This juxtaposition is realized when the wave of expression of cyclic genes sweeps the embryo from the tail to the head. Because of the oscillating nature of the cycling genes, this interface is transient, and a somite is therefore generated once per clock cycle. Thus, somitogenesis, and ultimately anteroposterior patterning of the embryo, requires a precise orchestration of multiple molecular signals responsible for the segmentation clock and the progression on the determination front. As a consequence, any alteration of these signals leads to severe segmentation defects, which often cause axis truncation and early embryonic lethality [8].

A different group of signals specify the compartments of the somite (dermomyotome and sclerotome) (Figure 8-3). After segmentation, positive and negative signals released from the surrounding tissues induce the differentiation of the sclerotome [2]. The notochord and the ventral part (floor plate) of the neural tube secrete the morphogen Sonic hedgehog (Shh), which leads to the formation of the sclerotome, in addition to promoting growth and survival of somitic cells. Conversely, Wnt signals from the ectoderm inhibit chondrogenesis and promote dermomyotome formation dorsally. Lastly, bone morphogenetic protein (BMP) signals from the LPM and the roof plate of the neural tube antagonize Shh ventralizing activity. The BMP signals are antagonized by the presence ventrally of Noggin, a BMP inhibitor secreted by the notochord. The first molecular markers detectable at the time of somitic compartmentalization are the Shh target gene Pax1 in the ventral part of the somite, whereas Pax3 expression is restricted to the prospective dermomyotome. Mice engineered to lack Shh fail to form vertebrae [2], indicating that this factor is absolutely required for the formation of vertebrae *in vivo*. Several observations, however, suggest that both Shh and Pax1 may be required early in the formation of vertebral cartilage but that later steps of somitic chondrogenesis can occur in the absence of these signals. Sclerotomal cells that move dorsally to give rise to the neural arches stop expressing Pax1. Consistent with this observation, Pax1 knockout mice do not develop the ventral part of the vertebra but have normal neural arches [2].

BMP signals are required for formation of the axial skeleton. Misexpression of noggin in the somite, for example, leads to truncation or loss of the vertebrae and the ribs [9]. The requirement for BMP signals to allow sclerotome chondrogenesis stands in contrast to the observation that the same signals earlier inhibit the elaboration of most, if not all, somitic lineages, including sclerotome [10]. Shh expression is the key to explaining these contrasting actions of BMPs. Sclerotomal cells that have been exposed to Shh signals become competent to initiate chondrogenesis upon subsequent exposure to BMP signals, whereas cells directly exposed to BMP signals in the absence of Shh no longer exhibit sclerotomal characteristics but express LPM markers instead [9]. Shh induces the homeobox-containing transcription factor, Nkx3.2/Bapx1, and the high-mobility group (HMG)-box containing transcription factor Sox9; these transcription factors confer a chondrogenic response to BMP signals in the sclerotome [11, 12].

The winged helix transcription factor Foxc2/MFH1 is critical for the formation of the whole vertebra, but other genes control the formation of specific parts of the vertebra. As previously mentioned, the absence of Pax1 leads to only the absence of the ventral part of the vertebra. Lack of Nkx3.2/Bapx1 and Nkx3.1 leads to a total absence of the ventromedial part of the vertebra, as well as to hypoplasia of the neural arches [2]. In contrast, Zic1 mutant mice exhibit defects primarily in the neural arches [2]. These examples indicate that complex and precise vertebral morphogenesis relies on an intricate pattern of expression of multiple genetic activities.

B. Craniofacial Bone Development

Craniofacial bones have both an ectodermal and a mesodermal origin. Most of the cranial skeleton derives from neural crest (NC) cells [13–15], which have an ectodermal origin. Because the NC cells give rise to derivatives generally produced by the mesoderm, the tissue formed by these cells is called mesectoderm, which forms the ectomesenchyme (cranial NC-derived mesenchyme as opposed to the mesodermal mesenchyme). NC cells emerge from the dorsal midline of the neural tube and migrate extensively to form various derivatives, both in the trunk and in the head of the embryo [14, 16]. In the head, NC cells migrate to colonize the pharyngeal arches (also called branchial arches) and other structures more rostral that are surrounded by a layer of ectoderm (including the neuroectoderm and the facial, or surface, ectoderm) in order to form the connective tissue associated with head muscles, tendon, bone, cartilage, and dermis. Modern techniques such as vital dye (DiI) labeling, construction of chimeric embryos (quail–chick or quail–duck chimeras), and the discovery of neural crest–specific markers (Wnt1) have shown that the extensive NC-derived ectomesenchyme primarily occupies the ventral part of the vertebrate head, whereas mesenchyme derived from the mesoderm occupies the dorsal part of the head. Thus, bones of the face and frontal bones derive from NC cells, whereas part of the otic bones and occipital bones at the base of the back of the skull have a mesodermal origin. The tissue origin of the parietal bones that form the skull vault, however, is still a subject of controversy [13, 14]. The construction of quail–chick chimeras predicts that these bones derive from NC cells [17]. By contrast, data based on tracking cells descended from those expressing Wnt1 indicate a mesodermal origin for parietal bones in the mouse [18]. In any event, the large majority of cranial bones, including the parietal bones, are formed through an intramembranous process, with the exception that the occipital bones are formed through endochondral ossification.

Some of the principles underlying the patterning of the complex structure of the cranial bones have emerged. Depending on their precise origins, the NC cells contribute to patterning in a cell-autonomous fashion. On the other hand, signals from the ectoderm and the endoderm to which the NC cells migrate also contribute to the patterning of the skull [15, 19]. The cell-autonomous properties of NC cells are illustrated by the behavior of NC cells that migrate from the neural tube in the trunk. These cells have an extremely low skeletogenetic capacity compared to cranial NC cells [14, 16, 19]. When trunk NC cells were transplanted to the head, these cells did not form cartilage or bone, unless the grafts contained only a small number of such cells [19]. These observations raised the possibility that trunk NC cells may exert inhibitory effects, and that these effects are not strong enough when the number of grafted cells is too low to prevent the positive signals generated by the surrounding environment. Thus, this experiment and others suggest that both the environment and a cell autonomous program are important in the patterning of the cranial bones [15, 19, 20]. The surface ectoderm, the neuroectoderm, and the pharyngeal endoderm are important sources of craniofacial patterning information. In the chick embryo, when the frontonasal zone of the ectoderm is ectopically transplanted to a more dorsal or a more ventral position of the frontonasal prominence, this results in duplications of the upper or lower beak structures, respectively [21]. In this experiment, the transplanted ectoderm expresses two potent morphogens, fibroblast growth factor 8 (FGF8) and Shh, which are able to reprogram the fate of NC cells at the transplant site. Similarly, the neuroectoderm has an important influence on patterning the NC cells into the craniofacial skeleton since blocking of the Shh signal provided by this tissue leads to craniofacial syndromes such as holoprosencephaly [15]. Lastly, the role of pharyngeal endoderm was demonstrated by the fact that removal of this tissue results in reduction or absence of some facial bones, whereas ectopic grafting of this tissue results in supernumerary lower jaws. At least some of the patterning information in the pharyngeal endoderm is mediated via FGF signaling [15, 19, 20].

The role of Shh as a crucial craniofacial morphogen is particularly emphasized by the devastating effects in craniofacial development produced when Shh signals are inhibited [15, 19] in humans, birds, mice, and even fish. Exposing avian embryos to cyclopamine, a potent inhibitor of the Hedgehog signaling pathway, can induce cyclopic defects that are characterized by a single central eye and no discernable nose. Humans with mutations in SHH or downstream effectors also may exhibit cyclopia. Furthermore, retinoic acid

signaling in the rostral head plays an important function since blocking its signaling induces defects that are reminiscent of those induced by Shh inhibition [19]. Another secreted molecule, bone morphogenetic protein 4 (BMP4), has been shown to play a role in the morphological variations of the beak observed in different species of birds [15, 19]. Ectopic overexpression of BMP or, conversely, inhibition of BMP signaling induces an increase or a decrease in the size of the beak, respectively.

In addition to these secreted molecules, several transcription factors have been shown to play critical roles in the genetic control of the head morphogenesis, and thus craniofacial bones. In the craniofacial region, the homeotic genes of the Hox gene family are expressed by NC cells, both before and after their migration to the arches [14, 16]. In the head, two different domains are defined by the presence or absence of Hox gene expression in NC cells: NC cells located in the first branchial arch and anterior structures (anterior hindbrain, midbrain, and forebrain) do not express any Hox gene, whereas NC cells present in the second and more posterior branchial arches express Hoxa2 [14, 15]. Interestingly, quail–chick chimera experiments have shown that Hox-negative NC crest cells are the origin of the entire facial skeleton [14]. Conversely, Hox-positive cranial NC cells are unable to generate any membranous bone. Induced loss of Hoxa2 expression in the second arch results in duplication of maxillary and mandibular structures, which normally arise from the first arch. By contrast, ectopic expression of Hox genes in rostral domains blocks the capacity of cephalic NC cells to differentiate into skeletal structures [14]. It has been shown that NC cells expressing Hoxa2 are more constrained in their ability to respond to local cues from the epithelial environment compared to cells devoid of Hox gene expression [14, 15]. These results further illustrate the fact that both intrinsic genetic program and environmental signals are important, and that they must work together to achieve proper cranial bone patterning. Other homeobox genes have been shown to pattern the cranial skeleton, as well. Mice lacking Distal-less homeobox gene 5 (Dlx5) and Dlx6 exhibit a duplication of the upper jaw and an absence of mandible [19].

Another important determinant of cranial patterning is a series of signals responsible for the generation of and the migration of the NC cells from the neural tube to specific locations. Several craniofacial malformations, called neurocristopathies, can be attributed to defects in the generation and migration of NC cells [22]. The tyrosine receptor ErbB4, for example, is expressed in the neural ectoderm and has been shown to be required in cultured embryos for the proper migration of NC cells [19].

C. Appendicular/Limb Development

Despite the enormous variety of shapes, sizes, and functions of animal limbs, all vertebrate limbs are similarly organized into three fragments distributed in a proximal-to-distal fashion. In the generic limb, the stylopod is the most proximal fragment of the limb (the closest to the body wall) and contains only one bone (humerus or femur). The zeugopod corresponds to the middle fragment and generally contains two bones (ulna and radius, or tibia and fibula). The autopod represents the most distal part of the limb. It contains a variable number of bones (in the wrist, ankle, and digits) and thus corresponds to the most divergent part of the limb in the multiple existing vertebrates. Like the bones of the axial skeleton, but unlike most of the bone in the skull, bones in the limbs are formed through the process of endochondral bone formation.

The limbs originate from the lateral plate mesoderm, which also give rise to parts of the scapula and pelvic bones. In response to complex molecular signals not completely elucidated, the limbs, two forelimbs and two hindlimbs, emerge from the body wall and are initially composed of undifferentiated mesenchymal cells covered by a layer of surface ectoderm. Mesenchymal cells actively proliferate and are exposed to proximal–distal, anterior–posterior, and dorsal–ventral signals responsible for patterning the limb. Later in limb development, a decrease in the mitotic activity of the cells destined to form bones correlates with the aggregation of these cells into mesenchymal condensations. This process precedes the differentiation of the mesenchymal cell population into chondrocytes and into connective tissues such as tendon and muscle sheaths. The different muscles necessary for the limb movements are formed from cells that migrate from the lateral edge of the dermomyotome. Here, we describe the main mechanisms involved in the induction and growth of the limb, its proximal–distal, anterior–posterior, and dorsal–ventral polarity, and we present briefly the processes of formation of mesenchymal condensations and joints, two phenomena that ultimately dictate the number and, very likely, the shape and size of bones.

I. FORMATION OF THE EARLY LIMB

The limbs originate from the LPM as swellings in the body wall called limb buds. Limb buds are present at specific locations along the AP axis of the embryo and are composed of apparently homogeneous mesenchymal cells actively proliferating and covered by a layer of surface ectoderm. As for many other embryonic structures and organs, several lines of evidence suggest that limb buds are induced in the embryonic flank at precise positions that are determined by expression of a specific

combination of Hox genes. For example, the specific combination of Hoxc6, Hoxc8, and Hoxb5 expression directs the formation of forelimbs. In different vertebrates, this combination is present at different levels in the trunk, and in each case this "Hox code" always correlates with the region of forelimb formation and is lost in limbless vertebrates. Interestingly, mice lacking the Hoxb5 gene have the shoulder girdle slightly shifted, an observation that confirms a role of Hox genes in allocating the region that will form the limbs.

The molecular mechanisms responsible for limb bud induction are not completely understood. It is known that signals from several axial tissues medial to the LPM, including the intermediate mesoderm, are important for induction of the limb bud. However, signals from the ectoderm that forms the external layer of the bud also seem to play a role in the initiation of the bud. In both cases, FGF8 and FGF10 are known to control this step, in cooperation with Wnt signaling. The current model is that FGF8 activates Wnt signals, which in turn can restrict the expression of FGF10 in the area of the LPM where the bud will form. Consistent with this model, targeted mutation of the FGF10 gene in mice results in the absence of the limbs, most likely as a consequence of the interruption of limb bud formation.

Despite the fact that the two pairs of limb buds emerging from the embryonic flanks look very similar at the earliest stages of development, major morphological differences appear thereafter. Early experiments performed in the chick embryo demonstrated that the specification between either forelimb or hindlimb is established at the earliest stages of limb development, before formation of the limb bud. When LPM cells that belong to the forelimb field are taken prior to limb budding and transplanted into an ectopic location, the ectopic limb generated always develops as a forelimb. Several genes encoding transcription factors have been found to be exclusively expressed in forelimb or hindlimb in multiple organisms. For instance, the T-box transcription factor Tbx5 and the Hox transcription factors Hoxc4 and Hoxc5 are specifically expressed in the presumptive forelimb area, whereas expression of Tbx4- and the Otx-related homeodomain factor Pitx1 is restricted to the presumptive hindlimb. More important, loss-of-function and gain-of-function studies in mice and chick embryos, respectively, demonstrated the role of Tbx5 in determining the forelimb identity and the role of Tbx4 together with Pitx1 in hindlimb identity. Besides these few genes, most of the key molecules in limb development are similarly expressed in both forelimbs and hindlimbs, and thus the sequences of events responsible for skeletal limb patterning are extremely similar in each limb.

2. PROXIMAL–DISTAL PATTERNING OF THE LIMB

The apical ectodermal ridge (AER) is a specialized structure present at the distal tip of the limb bud and corresponds to a thickening of the ectoderm that runs along the AP axis of the limb bud, separating the dorsal side of the limb from the ventral side (Figure 8-4). Beneath the AER, a zone of undifferentiated cells called the progress zone is responsible for most of the proximal–distal growth of the limb. The fundamental role of the AER in limb growth was demonstrated almost 60 years ago by microsurgical removal of the AER from chick embryos. Interestingly, when the AER was removed early in development, only the proximal (stylopod) part of the limb was formed, whereas when it was removed later, only the autopod was absent. This experiment was particularly important because it demonstrated that the region corresponding to the stylopod differentiates first, followed by that of the zeugopod and then that of the autopod. The AER has been found to maintain proliferation and survival of the cells present in the progress zone by secreting growth factors of the FGF family. When beads coated with FGFs are implanted into chicken limb buds in which the AER has been ablated, limb development occurs relatively normally. Mice in which FGF4 and FGF8 have been conditionally inactivated in the AER do not form limbs. Interestingly, when some FGF signals are inactivated after the early steps of limb patterning have occurred, limbs develop with an abnormal skeletal pattern that suggests that FGFs determine the number of cells that will form the skeletal elements by controlling cell survival.

Currently, one of the most actively debated issues regarding proximal–distal limb patterning is the question of how mesenchymal cells are specified to form the different structures of the limb skeleton (stylopod, zeugopod, and autopod). Two models have been proposed. The first model proposed was the "progress zone model," which postulates that the cells acquire positional information progressively, in a proximal-to-distal sequence. According to this model, cells present in the progress zone are subjected to signals from the AER for different periods of time. As cells leave this zone progressively, their fate is determined by their time in the progress zone: The first cells leaving the progress zone have been exposed for a short period of time and become stylopod progenitors, cells leaving later are exposed longer to AER signals and become zeugopod progenitors, and cells leaving even later become autopod progenitors. Since its description, this first model has been challenged by fate-mapping studies that revealed that different groups of cells present at different depths within the progress zone contribute specifically to the formation of only one skeletal compartment. These observations led to a second model (the "early specification model") that proposes that different groups of mesenchymal cells present in the early limb bud are already specified to

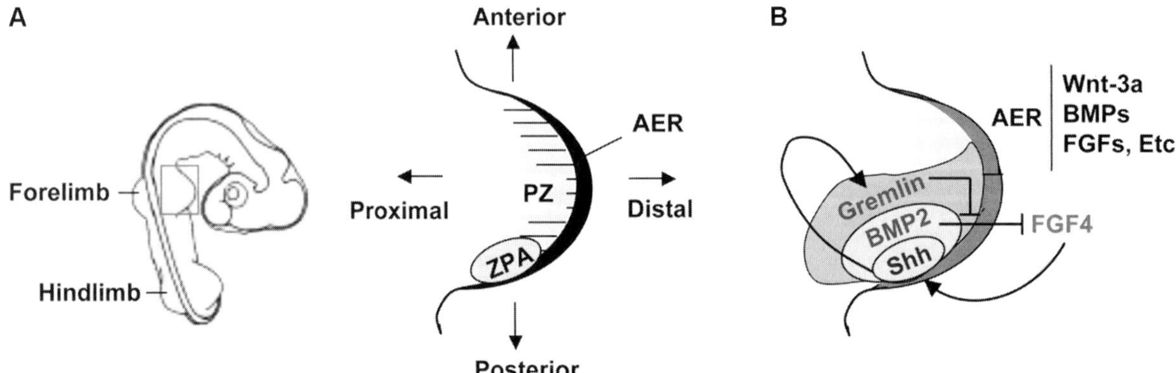

FIGURE 8-4   Schematic representation of early limb formation. (A) Dorsal view of the limb primordium (limb bud), which is composed of mesenchymal cells encased in an ectodermal jacket and contains specific regions that pattern the bud along the anterior–posterior (AP), dorsal–ventral (DV), and proximal–distal (PD) axes. The zone of polarizing activity (ZPA) patterns the AP axis, and the apical ectodermal ridge (AER) maintains outgrowth of the limb bud, keeping underlying mesenchymal cells in the progress zone (PZ) in an undifferentiated state. The dorsal and ventral ectoderms determine the DV polarity of the distal part of the limb (not shown). In fish and amphibians, the region corresponding to the AER is broader and is called apical epidermal cap. The AER is characterized by the expression of several specific genes (B), among which FGFs (particularly FGF4) play a critical role for limb growth and its proximal–distal patterning. The polarizing activity of the ZPA is mediated by sonic hedgehog (Shh), which is required to maintain the AER integrity (B). Shh acts indirectly through the induction of the expression of a BMP inhibitor, Gremlin. Because BMP present in the limb mesoderm suppresses FGF4 expression in the AER, the net action of Shh is to stimulate the production of the FGFs in the AER and thus maintain AER function. FGF4 and other FGFs signal back to the limb bud mesenchyme to maintain the expression of Shh, forming a positive feedback loop. Adapted from J. Capdevila and C. Izpisua Belmonte, Patterning mechanisms controlling vertebrate limb development. *Annu Rev Cell Dev Biol* **17**, 87–132 (2001), with permission.

form the stylopod, zeugopod, and autopod; these cells proliferate to expand longitudinally and become determined to form particular structures in response to local signals. Further experiments are needed to distinguish these and possibly other models.

3. THE ANTEROPOSTERIOR PATTERNING OF THE LIMB

Although the mechanisms determining proximal–distal patterning of the limb are thought to be distinct from those controlling AP limb morphogenesis (thumb to little finger), development along these distinct axes is coordinated. The zone of polarizing activity (ZPA) corresponds to a group of cells located in the posterior mesenchyme of the limb bud that act as an organizer of the AP polarity of the limb (Figure 8-4). In chicken embryos, when the ZPA from one limb bud is grafted into the anterior margin of a host limb, a duplication of the digits is produced such that the ectopic digits form a mirror image of the normal digits present posteriorly. These observations were initially interpreted in terms of a morphogen gradient that diffuses across the limb bud to determine the pattern in a concentration-dependent manner. The polarizing activity of the ZPA is mediated by the secreted factor Shh. Ectopic expression of Shh in the anterior part of the limb bud mimics the effects of the ZPA grafts, suggesting that this factor plays an important function in polarizing the limb. However, Shh null mice have limbs with more than simple AP axis abnormalities [23]. Shh null limbs have preserved proximal structures (stylopod), but intermediate structures (zeugopod) are severely truncated and fused, whereas the autopod is almost completely absent. Thus, Shh is not required to initiate limb development and is not involved in patterning the most proximal limb structures (stylopod). However, the dramatic abnormalities in proximal–distal patterning suggest that Shh acts to maintain the AER. In fact, Shh and FGFs form a regulatory loop between the ZPA and the AER (Figure 8-4). Indeed, removal of the AER leads to the loss of Shh expression, whereas the graft of cells expressing Shh in the limb bud induces ectopic FGF4 expression. Shh acts indirectly on the AER, through induction of the expression of a BMP inhibitor, gremlin, made in the limb mesenchyme. Since BMPs suppress FGF4 expression in the AER, the net action of Shh is to stimulate the production of FGFs in the AER and maintain AER function. Thus, Shh indirectly controls the proximal–distal development of the limb, in addition to its AP polarizing activity. Remarkably, deletion of both Shh and gli3, a transcription factor that mediates many of the actions of Shh, rescues the abnormalities of proximal–distal patterning in the Shh mutant

alone and blocks the actions of Shh on AP patterning as well. These data and others show that much of the action of Shh in the limb results from the suppression of expression of the inhibitory form of gli3.

Several genes are known to control Shh expression in the ZPA. HoxD genes are expressed at specific times during limb morphogenesis in overlapping domains distributed along the AP axis. In turn, Shh controls HoxD gene expression, thus forming a regulatory loop. The basic helix–loop–helix transcription factor dHAND also controls Shh expression in the limb bud. Mice deficient in dHAND expression die on approximately embryonic (E) day 10.5 and present small limbs with no detectable expression of Shh. Lastly, RA has also been shown to control the AP polarization of the limb: Inhibition of RA signaling prevents the establishment of the ZPA, the appearance of Shh expression, and the outgrowth of the limb bud, whereas grafts of beads coated with RA induce Shh and an ectopic ZPA in the limb bud. This effect of RA is thought to depend on the induction of Hox gene expression.

4. DORSOVENTRAL PATTERNING OF THE LIMB

The dorsoventral polarity of the limb is particularly evident when one considers epidermal-associated structures such as hair or feathers. In the case of the human hand, the back of the hand is dorsal and the palm is ventral; muscles and tendons are found in an orderly pattern along this axis. Surgical manipulations involving the rotation of the ectoderm according to the dorsal–ventral axis have not demonstrated that this tissue is responsible for determining dorsal–ventral polarity. Wnt7a is secreted by the dorsal ectoderm and controls the expression of the LIM-homeodomain factor Lmx1, which is specifically expressed in the dorsal mesoderm beneath the ectoderm. Combined data from experiments involving ectopic expression in chicken embryos and targeted gene disruption in mice have demonstrated that Wnt7a and Lim1 are involved in the specification of the dorsal identity of the limb. Conversely, the expression of engrailed-1 (En1), which is restricted to the ventral ectoderm, is required for the specification of the ventral fate since limbs of En1 null mice present structures ventrally that are similar to those normally observed only dorsally. Strikingly, En1 null limbs present ectopic digits ventrally, as a consequence of the formation of a second ectopic AER in the ventral limb, and demonstrate abnormalities of proximal–distal patterning as well. Thus, the molecular signals that control dorsal–ventral patterning also indirectly affect the proximal–distal organization of the limb and, in this way, resemble the determinants of AP patterning.

5. MESENCHYMAL CONDENSATION AND LIMB PATTERNING

An important step in skeletogenesis is the condensation of mesenchymal cells because this step is a prerequisite for chondrogenesis, which precedes bone formation in the limbs. Expression of the transcription factor Sox9 is required for condensation, although its role in condensation is not understood. *In vitro* studies indicate that cell adhesion molecules such as N-cadherin and NCAM are important for the aggregation of mesenchymal cells. Several other molecules are expressed in mesenchymal condensations, including the growth and differentiation factor GDF5 (a member of the BMP family); other secreted factors of the BMP, FGF, and Wnt families; the homeobox transcription factor Barx2; and Hox genes. FGFs may be important for stimulating SOX9 expression in condensations since such stimulation is observed in mesenchymal cells cultured *in vitro* [24]. The precise roles of FGFs and other factors, however, have not been demonstrated.

6. JOINT FORMATION AND LIMB PATTERNING

Joint formation occurs between two adjacent condensations, or within a single condensation, and constitutes another important mechanism for limb skeletal patterning since it determines the number and the size of some skeletal elements. The digits, for instance, initially constitute individual entities called digital rays (one per digit) that are then subdivided into phalanges through the process of joint formation. Typically, the prospective joint region is initially characterized by a group of cells that are denser and flatter than the chondrocytes present on both sides of this zone, called the interzone. Cells of the interzone then undergo cell death, which creates the joint space. Little is known about the molecular and cellular mechanisms that lead to joint formation. GDF5, a member of the BMP family, is strongly expressed in the interzone and is required for the formation of particular joints. Mice lacking GDF5 present several skeletal abnormalities, including the loss of some specific joints in the autopod. GDF6 and GDF7 are also expressed in joints, but their expression is weaker and more restricted than that of GDF5. The loss of GDF6 function results in joint fusions, principally in the ankle and wrist, in sites distinct from those seen in GDF5 mutants. These results demonstrate that the GDF family plays a key role in establishing boundaries between skeletal elements during development, and they suggest that GDF members share the same task, with each of them determining a subset of joints. Deficiency in the BMP antagonist Noggin causes a failure of joint formation in the autopod, both in humans and in mice.

This failure may be due at least in part to an effect on GDF5 expression. In addition to BMP signaling, Wnt–β-catenin signaling plays a fundamental role in inducing joint formation. Ectopic expression of Wnt14 in chicken limb buds and mouse chondrocytes *in vivo* is able to induce GDF5 expression and ectopic joint formation, suggesting that this member of the Wnt family could play a physiological role in normal joint formation. Conversely, mice that lack β-catenin in chondrocytes present a lack of joint formation with the fusion of several bones.

III. ENDOCHONDRAL BONE FORMATION

As noted previously, the craniofacial skeleton, the axial skeleton, and the skeleton of the limbs each begins as mesenchymal condensations. The sizes and shapes of these condensations, as well as the mechanisms controlling the dramatic variation in the sizes and shapes of the condensations, are regulated differently in each body region. Nevertheless, the next steps whereby mesenchymal condensations become bone follow one of only two fairly uniform processes: endochondral bone formation and intramembranous bone formation. We consider endochondral bone formation first (Figure 8-5).

Mesenchymal cells in condensations differentiate into chondrocytes, round cells that secrete a matrix rich in collagen II and aggrecan. These chondrocytes proliferate, enlarging the bone anlage. In response to unknown signals, certain chondrocytes in the center of the anlage then stop proliferating, enlarge (hypertrophy), and change their genetic program to secrete a matrix rich in collagen X. These postmitotic hypertrophic chondrocytes direct the mineralization of the matrix surrounding them and signal to perichondrial cells to influence their differentiation along the osteoblast lineage. These hypertrophic chondrocytes also signal to adjacent blood vessels to invade the bone anlage. Osteoclasts, cells of hematopoietic origin that can digest the extracellular matrix of bone or cartilage, also enter the bone anlage at this time. Hypertrophic chondrocytes then die an apoptotic death, leaving behind a mineralized matrix that serves as a scaffold for formation of a collagen I–rich matrix generated by osteoblasts. While this replacement of hypertrophic chondrocytes by osteoblasts occurs at the center of the bone anlage, chondrocytes farther from the center of the bone continue to proliferate. The chondrocytes closest to the hypertrophic chondrocytes proliferate at a particularly high rate and flatten out, forming columns of flat proliferating chondrocytes that contribute to

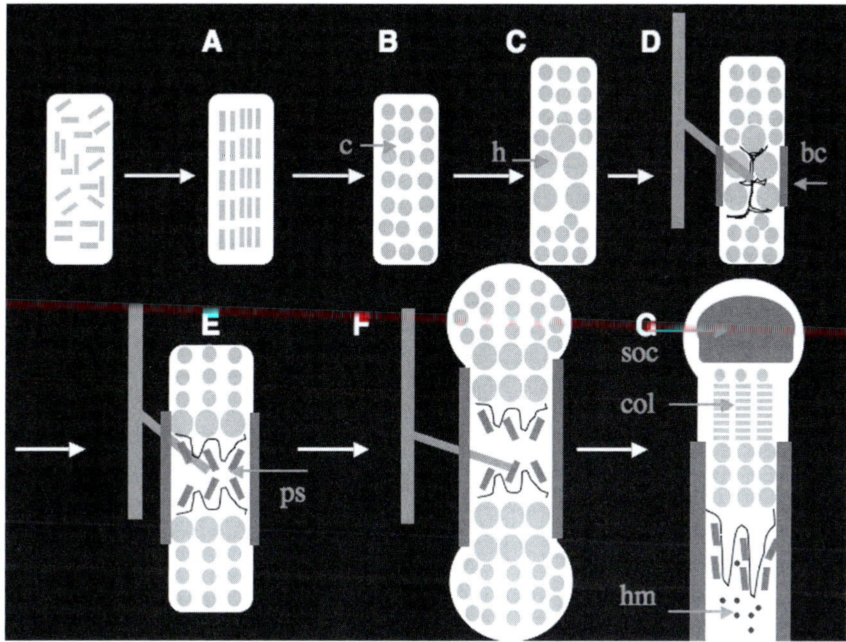

FIGURE 8-5 Endochondral bone formation. (A) Mesenchymal cells condense. (B) Cells of condensations become chondrocytes (c). (C) Chondrocytes at the center of condensation stop proliferating and become hypertrophic (h). (D) Perichondrial cells adjacent to hypertrophic chondrocytes become osteoblasts, forming bone collar (bc). Hypertrophic chondrocytes direct formation of mineralized matrix, attract blood vessels, and undergo apoptosis. (E) Osteoblasts of primary spongiosa accompany vascular invasion forming primary spongiosa (ps). (F) Chondrocytes continue to proliferate, lengthening bone. Osteoblasts of primary spongiosa are precursors of eventual trabecular bone; osteoblasts of bone collar become cortical bone. (G) At the end of bone, a secondary ossification center (soc) forms through the cycle of chondrocyte hypertrophy, vascular invasion, and osteoblast activity. Growth plate below secondary center of ossification forms orderly columns of proliferating chondrocytes (col). Hematopoietic marrow (hm) expands in marrow space along with stromal cells.

the asymmetric expansion of the cartilage mold. As bones enlarge further, secondary sites of ossification form by a mechanism that appears to repeat the process just described. In the center of an enlarging region of round proliferating chondrocytes, certain chondrocytes stop proliferating and become hypertrophic. This process is followed by vascular invasion and deposition of a bone matrix by osteoblasts that replace the hypertrophic chondrocytes that die through apoptosis. In the limbs, the growth cartilage that remains between the primary and secondary ossification centers forms a disc of tissue called a growth plate. This growth plate continues to act as an engine for bone lengthening for characteristic periods postnatally. Depending on the bone and the species, growth plates can persist for life or disappear through poorly understood processes (called growth plate fusion). Here, we first consider the transcriptional mechanisms that regulate the differentiation and activities of chondrocytes, osteoblasts, and osteoclasts during development and then consider the signaling mechanisms responsible for the coordinated events of the endochondral sequence.

A. Chondrocytes

Chondrocytes are cells that produce and maintain a characteristic and abundant extracellular matrix. The cartilaginous matrix is composed of two components, the proteoglycans and the collagens. Proteoglycans are macromolecules containing a core protein with multiple attached polysaccharide chains [25, 26]. Because of their high content of charged polysaccharides, proteoglycans are highly hydrated. The polysaccharide chains in proteoglycans, called glycosaminoglycans (GAGs), are long repeating polymers of specific disaccharides. One of the most important extracellular proteoglycans is aggrecan, the predominant proteoglycan in cartilage. Aggrecan forms large aggregates; a single aggregate can be more than 4 mm long and have a volume larger than that of a bacterial cell. These aggregates give to cartilage its unique gel-like properties and its resistance to deformation. The GAGs covalently attached to aggrecan are keratan sulfate and chondroitin sulfate. The central component of the cartilage proteoglycan aggregate is a long molecule of hyaluronic acid. Hyaluronic acid is a large polysaccharide that forms a highly hydrated gel. Hyaluronic acid is bound to aggrecan in a noncovalent fashion and is the only extracellular oligosaccharide that is not covalently linked to a protein. The binding of hyaluronic acid to aggrecan is facilitated by link proteins that bind to the aggrecan core protein and to hyaluronic acid. The finding of an aggrecan mutation in cmd (cartilage matrix deficiency) mice, which are characterized by cleft palate and short limbs, confirms the critical role of aggrecan in cartilage formation [26]. Notably, no mutations in the human aggrecan gene have been identified. Sulfation is an important translational modification in proteoglycan synthesis [27]. Several human genetic disorders associated with defects in transport of sulfate into the cells also lead to undersulfated proteoglycans and chondrodysplasias [27, 28].

There are three different types of collagens in the growth plate matrix:

1. Fibrillar collagens: collagen type II and collagen type XI. Collagen type II is also found in the vitreous. In cartilage, collagen type II is produced by proliferating chondrocytes and by the upper hypertrophic chondrocytes. Collagen type II is the most abundant collagen of the cartilaginous matrix.
2. Fibril-associated collagen: collagen type IX. Collagen type IX is also found in the vitreous and binds GAGs.
3. Sheet-forming collagen: collagen type X. Collagen type X is exclusively expressed by hypertrophic chondrocytes.

Numerous chondrodysplasias are caused by mutations in each of these collagen genes [27, 29–31].

Mutations in the collagen type X gene, which is expressed exclusively by growth plate chondrocytes, result in the relatively mild Schmid type of metaphyseal chondrodysplasia [32, 33]. The mutations appear to result variably in haploinsufficiency for the protein or dominant negative effects, and they lead to a short growth plate. Structural mutations in mature, type X collagen have not been reported in humans. The precise pathogenesis of Schmid chondrodysplasia is not certain [34].

Chondrocytes originate from condensed mesenchymal cells. Many lines of evidence have shown that Sox proteins are the master transcription factors for chondrogenesis [35–38]. Sox9 as well as L-Sox5 and Sox6 are members of the Sox family of transcription factors that are characterized by the presence of the HMG-box DNA-binding domain. Sox9 is required during sequential steps of the chondrocyte differentiation pathway; it is critical for commitment of mesenchymal cells toward the chondrocyte lineage, and it upregulates expression of critical cartilaginous matrix

genes such as those encoding collagen type II, collagen type IX, collagen type XI, and aggrecan. In the growth plate of long bones, Sox9 is expressed in proliferative chondrocytes and not in hypertrophic cells [39]. Several genetic approaches in the mouse (gain as well as loss of function) have demonstrated that Sox9 positively regulates proliferation and suppresses chondrocyte hypertrophy [37, 40]. In humans, heterozygous missense mutations resulting in haploinsufficiency for expression of the Sox9 gene cause campomelic dysplasia, a rare disorder of skeletal development that results in deformities of most of the bones of the body [41]. Most affected infants die from respiratory failure due to poorly formed tracheal and rib cartilage. Sox9 activity is regulated by phosphorylation in a protein kinase A (PKA)-dependent manner [42]. Phosphorylation of SOX9 by PKA enhances its transcriptional and DNA-binding activity. Parathyroid hormone–related protein (PTHrP)-stimulated activation of PKA is the major regulatory pathway stimulating phosphorylation of SOX9 since in PTHrP null mutants SOX9 phosphorylation is not seen. The increased activity of Sox9 caused by PKA-mediated phosphorylation mediates in part the effect of PTHrP to maintain the chondrocytes as nonhypertrophic chondrocytes [43]. Two other members of the Sox family, L-Sox5 and Sox6, are required for chondrogenesis. Whereas individual L-Sox5 or Sox6 knock-out mice are born with minor cartilage defects, double knock-out animals develop a severe, generalized chondrodysplasia characterized by a virtual absence of mature cartilage, secondary to a defect of cell proliferation and impairment of cartilage matrix production [44, 45]. Similarly to Sox9, L-Sox5 and Sox6 also control sequential steps of growth plate chondrocyte differentiation [44, 45]. The expression of L-Sox5 and Sox6 requires Sox9, and Sox9 may directly cooperate with these transcription factors in controlling expression of cartilage matrix genes. Notably, both L-Sox5 and Sox6 lack transactivation or transrepression domains and may thus act mainly to facilitate organization of transcription complexes.

Little is known about the factors that positively regulate chondrocyte hypertrophic differentiation. Among them is Runx2. Runx2 belongs to the Runt transcription factor family [46] and was initially characterized as a molecule essential for osteoblast differentiation [47–50]. This conclusion was based on the complete absence of osteoblasts in Runx2-deficient mice [49, 50] and the ability of this gene to promote osteoblast-specific gene expression *in vitro* [47]. Inactivating mutations of human Runx2 cause cleidocranial dysplasia 48 in the heterozygous state. More detailed investigations have revealed that Runx2 is also expressed in chondrocytes as they initiate hypertrophy, and loss of this factor in genetically engineered mice severely delays chondrocyte maturation in a number of developing bones [51, 52]. When Runx2 is ectopically expressed in immature chondrocytes, it drives premature maturation of chondrocytes by inducing expression of collagen type X and other hypertrophic markers, both *in vivo* [53–55] and in cultured chondrocytes [56]. The observation that chondrocyte hypertrophy in Runx2-deficient mice is not completely blocked in all bones indicates that additional factors are involved in this process. The transcription factor Runx3 plays a critical role in inducing chondrocyte hypertrophy in cooperation with Runx2 [57]; mice null for both Runx2 and Runx3 have no hypertrophic chondrocytes. In addition, core-binding factor β (CBFβ) has also been identified as a positive regulator of chondrocyte hypertrophy [58, 59]. Indeed, growth plates deficient for CBFβ expression display a phenotype similar to that of Runx2-deficient mice. CBFβ is a transcription factor that forms heterodimers with Runx proteins. In chondrocytes, this factor interacts with Runx2 and is necessary for the efficient DNA binding and transcriptional activity of Runx2. Independently of its role in chondrocyte hypertrophy, Runx2 plays a critical role in vascular invasion of cartilage because there is almost no vascular invasion in most skeletal elements of Runx2-deficient mice [60]. Both the lack of expression of the angiogenic factor, vascular endothelial growth factor (VEGF), which is normally expressed in hypertrophic chondrocytes, and the observation that Runx2 binds to and activates the VEGF promoter *in vitro* suggest that VEGF mediates the Runx2-dependent regulation of blood vessel invasion [60]. The role of Runx2 in osteoblast biology is discussed in Chapter 4 (Stein).

The mammalian fetal growth plate is a virtually avascular tissue, but it requires an angiogenic switch in order to be replaced by bone. It is also a highly hypoxic tissue. The transcription factor hypoxia-inducible factor-1 (HIF-1) is the major mediator of response to hypoxia in mammalian tissues and belongs to the PAS subfamily of bHLH transcription factors [61]. HIF-1 is composed of two subunits, HIF-1α and -β. HIF-1β is constitutively expressed, whereas HIF-1α protein is highly unstable and its accumulation is regulated by the von Hippel–Lindau (VHL) protein, an E3-ubiquitin ligase. Under normoxic condition, this ligase targets HIF-1α to the proteasome, which destroys HIF-1α. Conversely, in hypoxic conditions, HIF-1α is not recognized by VHL, so it accumulates and translocates to the nucleus and activates target genes. One target of HIF-1α is VEGF. Strikingly, each individual component of this VHL–HIF-1α–VEGF pathway is critically involved in chondrocyte survival. In agreement with its angiogenic function, the conditional knock-out of

VEGF in chondrocytes results in delayed blood vessel invasion [62]. More surprisingly, the lack of VEGF generates massive cell death in the epiphyseal regions of the bones, in both the resting and the proliferating zone of the growth plate [62]. A similar phenotype is also observed in growth plates lacking HIF-1α [63]. Taken together, these data demonstrate that VEGF and HIF-1α are key components of a critical pathway that supports chondrocyte survival during endochondral bone formation. In addition to cell death, an increase of chondrocyte proliferation was observed in growth plate chondrocytes lacking HIF-1α. This suggests that in addition to promoting survival of chondrocytes, HIF-1α negatively regulates their rate of proliferation. In agreement with this observation, VHL null growth plate chondrocytes display a significantly reduced proliferation rate [64]. Deletion of VHL in chondrocytes also resulted in accumulation of matrix deposition in the growth plate. The phenotype of growth plate lacking both VHL and HIF-1α is virtually identical to the HIF-1α null growth plate phenotype [64]. This indicates that HIF-1α is likely to be the major target of VHL action in chondrocytes, and it is a key coordinator of chondrocyte survival and proliferation.

Numerous lines of evidence suggest that chondrocytes and osteoblasts originate from a common osteochondroprogenitor [47] and that activation of the transcription factor β-catenin induces osteoblastic and suppresses chondrocytic differentiation in early osteochondroprogenitors [65–67]. β-Catenin is the key downstream signaling molecule of the Wnt canonical signaling pathway [68]. Conditional deletion of β-catenin in limb and head mesenchyme during early embryonic development results in arrest of osteoblastic differentiation and lack of mature osteoblasts in membranous bones. Furthermore, in the absence of β-catenin, osteochondroprogenitors differentiate into chondrocytes instead of osteoblasts. Sox9 can bind β-catenin and block its actions. Thus, the opposing actions of SOX9 and β-catenin help determine the commitment of osteochondroprogenitors and the pace of differentiation of chondrocytes [69]. The Wnt–β-catenin system is described in Chapter 15 (Johnson).

B. Osteoblasts

During intramembranous bone formation, osteoblasts derive from condensed mesenchymal cells [70, 71]. The origin of osteoblasts in endochondral bone formation is more complex. Osteoblasts that form the cortical bone differentiate from the mesenchymal cells of the perichondrium; the osteoblastic cells that will give origin to trabecular bone probably result from migration of osteoblast precursors from the perichondrium as well [72]. Independently of their origin, mesenchymal cells require three critical transcription factors in order to become osteoblasts during embryonic development: β-catenin (as already noted), Runx2, and Osterix [47, 73]. Committed osteoprogenitor cells then proliferate, differentiate into postmitotic osteoblasts that synthesize and mineralize bone matrix, and finally become terminally differentiated osteocytes or quiescent bone lining cells. Osteoblast differentiation is characterized by a loss of proliferative capacity and by a sequential increase in the expression of characteristic proteins, such as alkaline phosphatase, bone sialoprotein, collagen type I, PTH/PTHrP receptor, osteopontin, and finally osteocalcin and matrix metalloproteinase 13 [74]. The mature osteoblast is found adjacent to the bone surface and has morphological and ultrastructural properties that are typical of cells engaged in secretion of a connective tissue matrix. Osteoblasts lay down bone matrix (osteoid) that is composed predominantly (90%) of collagen type I, along with noncollagenous proteins such as osteocalcin, osteopontin, osteonectin, and growth factors. After synthesizing and secreting the bone matrix and directing its mineralization, osteoblasts die or undergo two alternative fates: On quiescent bone surfaces, single layers of flattened, inactive osteoblasts are called bone lining cells. Alternatively, osteoblasts undergo a dramatic change in morphology and become buried in bone matrix as osteocytes. Osteocytes develop by forming numerous dendrite-like cytoplasmic processes that connect with adjacent cells to ensure their viability within the mineralized osteoid and to allow signaling. The cell bodies of osteocytes are found in lacunae, and the numerous processes lie in canaliculi. Osteocytes no longer synthesize collagen and appear to function as mechanosensors; they can reside in healthy bone for long periods of time, but in aging bone, empty lacunae are observed, suggesting that osteocytes undergo apoptosis [75]. Osteocytes secrete modulators of osteoblast activity such as sclerostin [76], suggesting the possibility that osteocytes directly or indirectly regulate the activity of osteoblasts. Osteocytes are discussed in Chapter 6 (Bonewald).

The transcription factor Runx2 is absolutely required during embryonic development for differentiation of mesenchymal cells into osteoblasts throughout the skeleton, during both endochondral and intramembranous ossification. Runx2-deficient mice have a cartilaginous skeleton without any osteoblasts because their differentiation is arrested as early as E12.5 [47, 49, 50]. In mice lacking only one allele of the Runx2, an abnormality in osteoblast differentiation is limited to bones forming through intramembranous ossification [49]. A similar phenotype has been reported in

humans with heterozygous loss-of-function mutations of Runx2 and is called cleidocranial dysplasia (CCD) [48]. CCD is an autosomal dominant condition characterized by hypoplasia of the clavicles, patent fontanelles, supernumerary teeth, short stature, and changes in skeletal patterning and growth. Runx2 activity is controlled by various extracellular signaling pathways; the activity and stability of Runx family members are modified by phosphorylation, acetylation, and ubiquitination [77]. Mitogen-activated protein kinase (MAPK) and the PKA can phosphorylate and thereby activate Runx2 [78]. Schnurri-3, a mammalian homolog of the *Drosophila* zinc finger adapter protein, promotes Runx2 degradation [79]. Lack of Schnurri-3 leads to a dramatic increase in postnatal bone formation, at least in part by increasing the levels of Runx2 protein [79]. Runx2 target genes include genes expressed by mature osteoblasts, such as osteocalcin, bone sialoprotein, osteopontin, and collagen type I [80]. Runx2 expression is directly or indirectly regulated by the homeobox transcription factor Msx2. Msx2 inactivation in mice causes a marked delay of ossification in the bones of the skull [81]. This phenotype is concomitant with downregulation of Runx2 expression. Notably, one human syndrome characterized by increased bone formation at cranial sutures, Boston-type craniosynostosis, is caused by activating mutation in Msx2 [82]. Paradoxically, *in vitro* Msx2 can suppress Runx2 promoter activation [83]; thus, the actions of Msx2 in regulating Runx2 are likely to be multiple and context dependent. In an *in vitro* model of differentiation of early mesenchymal cells, for example, forced Msx2 expression stimulates these cells along the osteoblastic pathway and away from the adipocytic pathway [84]. Thus, Msx2 may act early in the osteoblast pathway to increase the number of osteoblast-specific cells.

Runx2 action is required for the expression of osterix, a zinc finger protein related to Sp1 that is expressed in late chondrocytes and osteoblast progenitors [73]. Like Runx2, osterix is required for differentiation of osteoblasts. Osterix binds the transcription factor NFATc1 and activates the Col Ia1 promoter, thereby stimulating collagen I synthesis [85].

Another major regulator of osteoblast differentiation is the transcription factor ATF4. ATF4 deficiency results in delayed bone formation during embryonic development [86]. Although Col Ia1 gene transcription does not depend on ATF4, the synthesis of collagen I protein is dramatically lowered in ATF4 knockout mice because amino acid availability is limiting. A major action of ATF4 is to regulate transporters that move amino acids into osteoblasts. Strikingly, high levels of amino acids in tissue culture medium correct the defect in collagen synthesis in ATF4 (–/–) osteoblasts, and a high protein diet prevents the osteopenia in ATF4 (−/−) mice [87]. ATF4 also regulates the expression of genes expressed late during osteoblast differentiation, such as that encoding osteocalcin.

C. Osteoclasts

Osteoclasts are multinucleated cells that uniquely degrade mineralized matrix [88]. Studies of osteopetrotic mice and humans have made clear that osteoclasts are essential for resorption of the matrix left behind by dying chondrocytes during endochondral bone development. In the complete absence of osteoclasts, no marrow space for hematopoiesis is formed, although vascular invasion and production of bone adjacent to the growth plate (primary spongiosa) still occurs.

Cells of the osteoblast lineage direct the differentiation and activation of osteoclasts by expressing the key regulators of these processes, macrophage colony-stimulating factor (M-CSF) and RANK ligand (RANKL). Insights into the signaling pathways regulating osteoclast development have come from studying genetically altered mice with osteopetrosis, a condition characterized by the failure of bone resorption due to defective osteoclastogenesis. A characteristic finding in osteopetrotic bones is persistence of cartilage matrix remnants in what should be the marrow space. The spontaneous mutant *op/op* mouse displays an osteopetrotic phenotype with impaired osteoclast differentiation. The mutation occurs in the gene encoding the cytokine M-CSF, and calvarial osteoblasts from these mice cannot support osteoclast development when co-cultured with spleen cells [89, 90]. The addition of recombinant M-CSF can restore bone resorption *in vivo* [91, 92], and supplementation of co-cultures with M-CSF results in the formation of osteoclasts in response to $1,25(OH)_2$-vitamin D [93–95].

RANKL is a type II transmembrane protein of the tumor necrosis factor (TNF) family [96, 97]. It is expressed at highest levels in the bone and bone marrow, but it is also found in lymphoid tissues. A soluble fragment of RANKL, missing its transmembrane domain, along with M-CSF can, by themselves, induce osteoclast formation in the absence of supporting osteoblasts or stromal cells, suggesting that these proteins are the crucial osteoclastogenic factors produced by cells of the osteoblast lineage [97]. RANKL expression is strongly stimulated by known activators of osteoclasts, including PTH, interleukin (IL)-6, IL-11, and $1,25(OH)_2$-vitamin D_3. RANKL null mice develop osteopetrosis with occlusion of the marrow space; they lack differentiated osteoclasts but do have precursors that can differentiate normally when co-cultured with

wild-type cells of the osteoblast lineage. In addition to defects in tooth eruption and mammary gland development, RANKL-deficient mice lack lymph nodes and demonstrate impaired differentiation of B and T lymphocytes [98].

The sole receptor for RANKL, found on osteoclasts and their precursors, is RANK, a member of the TNF receptor family [99]. Polyclonal antibodies against the RANK extracellular domain can induce osteoclast formation in spleen co-cultures when M-CSF is present [100]. Furthermore, an anti-RANKL antibody lacking the Fc domain markedly inhibits RANKL-mediated osteoclastogenesis [100]. As expected, mice with a targeted deletion of RANK have osteopetrosis. As with RANKL-deficient mice, mice lacking RANK have defective B and T cell maturation and lack peripheral lymph nodes, although thymic development proceeds normally [101].

Osteoprotegerin (OPG) is a key modulator of activation of RANK by RANKL. OPG is a soluble member of the TNF receptor family [102, 103]. Production of OPG is strongly upregulated by estrogen, TNF-α, growth hormone, and transforming growth factor-β (TGF-β), whereas it is suppressed by PTH and glucocorticoids. In co-culture experiments, OPG potently inhibits osteoclast induction by vitamin D, PTH, PGE_2, or IL-11 [104]. Hepatic overexpression of OPG in transgenic mice results in osteopetrosis with impaired thymocyte development [102], whereas administration of OPG to rodents results in dramatic increases in bone density with decreased osteoclast number and reduced serum calcium levels [105–107]. Conversely, mice with a targeted deletion of OPG have severe osteoporosis due to increased bone resorption and increased numbers of osteoclasts [108, 109]. This finding demonstrates the important role of OPG in normal bone remodeling. Interestingly, the OPG knock-out mice also have significant calcifications of the aorta and renal arteries [108].

Osteoclasts derive from monocyte/macrophage precursors of the hematopoietic lineage. The Ets family transcription factor, PU.1, is critically important for the earliest events in osteoclastogenesis [110]. PU.1-deficient mice lack not only osteoclasts but also macrophages, while preserving the ability to produce early monocytic cells [110]. PU.1 is thought to regulate the transcription of the M-CSF receptor c-fms [111].

Two main transcription factor complexes, AP-1 and MITF1, are downstream of the M-CSF pathway in osteoclastogenesis. Mice lacking the AP-1 component, c-Fos, lack osteoclasts and are osteopetrotic but have an increased number of bone marrow macrophages [112]. These data suggest that c-Fos has an important role in differentiation of hematopoietic precursors into osteoclasts rather than macrophages. Genetically, c-Fos can be thus placed downstream of PU.1 in the pathway leading to a fully differentiated osteoclast. The microphthalmia gene product, MITF, is mutated in the mi/mi mouse, which is characterized by pigmentation (melanocyte) defects, mast cell defects, and osteopetrosis [113]. MITF is essential for the fusion of mononuclear precursors into multinucleated osteoclasts, and it directly regulates genes important for osteoclast function such as those encoding TRAP, cathepsin K, and osteoclast-associated receptor [113].

Downstream of RANK activation, TRAF6 is required for transduction of the RANK signal (Figure 8-6). TRAF6 interacts with the cytoplasmic tail of RANK, and deletion of this interaction domain abolishes RANK-mediated activation of NF-κB [114]. That NF-κB is critical to osteoclastogenesis has been demonstrated by the finding that mice lacking the p50 and p52 subunits of NF-κB have osteopetrosis, a phenotype that can be rescued by bone marrow transplantation [115, 116]. c-Fos, a member of the AP-1 family mentioned previously as downstream of PU.1, also serves an important function downstream of RANKL signaling [117]. NFATc1 has been identified as a target of both TRAF6 and c-Fos pathways, and it may act as a major regulator of terminal osteoclast differentiation [118]. Notably, a constitutively active form of NFATc1

in c-Fos null cells restores expression of osteoclast-specific genes, demonstrating that NFATc1 is a critical transcriptional regulator downstream of c-Fos during osteoclastogenesis [119]. Activation of NFATc1 requires calcium signals that originate perhaps from immunoreceptor tyrosine-based activation motif signals (Figure 8-6).

A number of clinical disorders have been linked to alterations in the OPG/RANK/RANKL signaling system. Activating mutations in RANK have been associated with familial expansile osteolysis and familial Paget's disease, whereas inactivating mutations in OPG occur in juvenile Paget's disease and idiopathic hyperphosphatasia [120, 121]. The ratio of OPG to RANKL may be important in determining the balance of bone formation and resorption in conditions such as osteoporosis. Consistent with this, glucocorticoids inhibit OPG and stimulate RANKL [122], whereas estrogen increases OPG production by osteoblasts and stromal cells [123]. Continuous PTH exposure, which is associated with increases in activation of osteoclasts and bone resorption, increases the RANKL:OPG ratio. Dysregulation of these factors has also been implicated in various malignancies with a predilection for bone; for example, myeloma cells can augment RANKL production and suppress OPG [124].

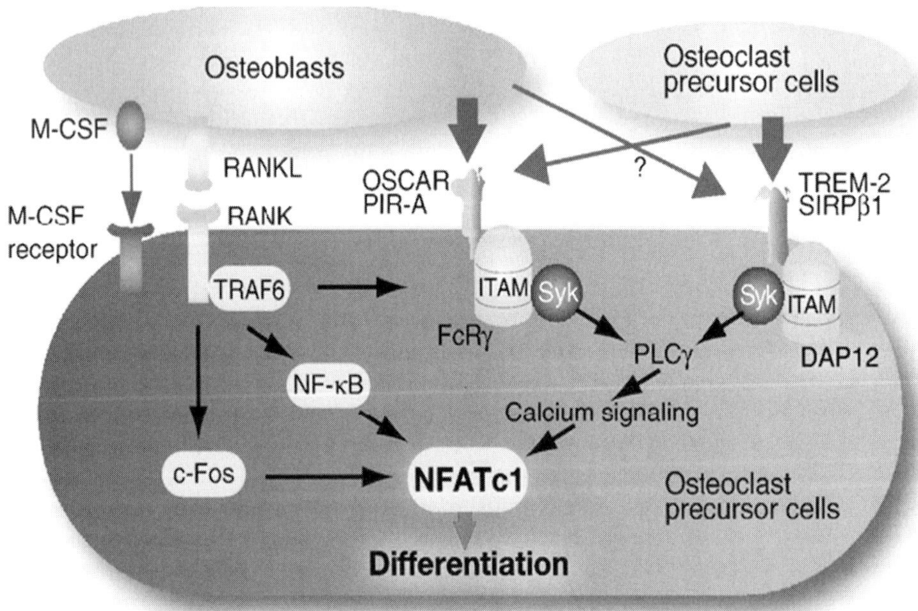

FIGURE 8-6   Cooperation of M-CSF, RANKL, and immunoreceptor tyrosine-based activation motif (ITAM) signals in osteoclastogenesis. Reproduced from M. Asagiri and H. Takayanagi, The molecular understanding of osteoclast differentiation. *Bone* **40**(2), 251–264 (2007), with permission.

D. Signaling Pathways That Regulate Endochondral Bone Formation

Endochondral bone formation requires precise coordination of the proliferation, differentiation, and migration of chondrocytes, osteoblasts, osteoclasts, and cells of the vasculature (Figure 8-5). Many of the same signaling pathways important during early embryogenesis and organogenesis in other systems play important roles in endochondral bone formation. Here, we summarize the roles of signaling by BMPs, wnts, FGFs, Indian hedgehog, PTHrP, C-type natriuretic peptide (CNP), and Delta/Notch during endochondral bone formation.

1. BONE MORPHOGENETIC PROTEIN SIGNALING

The BMP family, originally identified as proteins with the ability to induce ectopic cartilage and bone formation after subcutaneous injection, is the largest within the TGF-β superfamily, consisting of more than 20 members. These ligands bind to and activate the BMP receptors, of which there are two types; each type of receptor can independently bind BMPs and each contains an intracellular serine–threonine protein kinase [125]. The type I receptors include ALK2, ALK3 (BMPRIA), and ALK6 (BMPRIB). Upon ligand binding, the type I receptors heterodimerize with type II receptors (e.g., BMPRII, ActRII, and ActRIIB) that contain constitutively active serine/threonine kinase domains. Phosphorylation of the type I receptors by the type II receptors then leads the type I receptors to phosphorylate members of the Smad family of intracellular proteins. In particular, Smads1, -5, and -8 are activated by BMP type I receptors and are termed R-Smads. R-Smads in turn complex with Smad4, translocate to the nucleus, and there regulate gene transcription.

Negative regulation of BMP signaling occurs at multiple levels. BMP antagonists such as noggin, chordin, and gremlin are secreted and inhibit BMP interaction with the receptors [126]. A novel member of the BMP family, BMP3, can block signaling through the type II BMP receptor ActRII. Another member of the TGF-β family, inhibin, binds type II BMP receptors in the presence of the coreceptor betaglycan and blocks BMP signaling. Intracellularly, the inhibitory Smads (I-Smads) Smad6 and Smad7 can bind to activated type I receptors, competing with R-Smads for activation. Finally, ubiquitin-mediated degradation of R-Smads and receptors is regulated by the E3 ligases, Smurf1 and Smurf2.

Although BMPs can robustly induce ectopic bone formation after injection, their roles in skeletal development have been more challenging to elucidate because of the pleotropic effects of BMP signaling

early in embryogenesis [127]. When BMP signaling is blocked in early chick limbs, mesenchymal condensation is blocked [128]. Furthermore, when BMPR1A and BMPR1B are both ablated from early cartilage elements, the dramatic loss of chondrocyte development suggests that signaling by these receptors is vital for the conversion of prechondrocytes to chondrocytes [129]. The subsequent differentiation of chondrocytes into hypertrophic chondrocytes was disrupted as well. Lack of BMPR1A and BMPR1B signaling led to loss of SOX9, L-SOX5, and SOX6 expression in these cells. Thus, BMP signaling is vital for chondrocyte differentiation. When BMP-2 and -4 were selectively knocked out of early limb buds, the cartilage phenotypes of the resultant mice were less severe than those of the double BMPR knock-out, suggesting roles for other BMPs [130]. Earlier results suggested roles of BMP-5, -6, and -7 in bone development as well [131–134]. Gene targeting of GDF5 and/or GDF6 results in mice with alterations in the lengths and numbers of bones, implicating these factors in joint specification [135, 136].

BMP signaling is also crucial for osteoblast development. The double knockout of BMP-2 and BMP-4 in the limb completely disrupted osteoblast differentiation, demonstrating the crucial roles of these two BMPs in osteoblast differentiation [130]. Effects of BMP signaling in later stages of the osteoblast differentiation are suggested by studies using BMP antagonists. Targeting of noggin overexpression to differentiated osteoblasts by the osteocalcin promoter results in osteopenia by 8 months of age [137]. Likewise, overexpression of gremlin, another BMP antagonist, in differentiated osteoblasts results in reduced bone mineral density and fractures [138].

2. WNT SIGNALING

As noted previously, the downregulation of wnt signaling is essential for the conversion of mesenchymal cells into chondrocytes. This downregulation may be required in part because of the mutual inhibition of the actions of SOX9 and β-catenin [69]. Canonical Wnt signaling also plays a crucial role in osteoblastogenesis. Wnt ligands interact with Frizzled family receptors and coreceptors from the LRP5/6 family. Activation of the canonical signaling pathway, mediated intracellularly by dishevelled (Dsh) proteins, inhibits phosphorylation of β-catenin by a complex containing axin, adenomatous polyposis coli, and glycogen synthase kinase-3β. β-Catenin is therefore allowed to translocate into the nucleus, where it functions as a coactivator in Tcf/Lef1-mediated transcription.

An important role for Wnt signaling in bone first became evident with the identification of clinically important mutations in LRP5, a Wnt coreceptor.

Loss-of-function mutations in LRP5 are associated with osteoporosis–pseudoglioma syndrome, characterized by low bone mass and abnormalities of retinal vasculature [139]. In addition, heterozygote carriers have demonstrably reduced bone mineral density. LRP5-deficient mice are viable and fertile but have decreased bone mass [140, 141]. The bone formation rate is significantly lower than that of wild type, with normal differentiation of osteoblasts. In contrast, kindreds with the G171V mutation in LRP5 were found to have high bone mass [142, 143]. This phenotype is recapitulated in mice engineered to express the same mutation [144]. This mutation maps to one of the YWTD-type β-propellors in the extracellular domain of LRP5, and it appears to inhibit binding of Dkk1, an inhibitor of Wnt signaling [142, 145], and the binding of other wnt antagonists as well, such as sclerostin [146].

A direct role for β-catenin has been documented at multiple stages of osteoblastic differentiation. Deletion of β-catenin in mesenchymal progenitors using a Cre recombinase driven by Dermo1 results in attenuation of both endochondral and intramembranous ossification [67]. Ectopic chondrocytes appear in both calvaria and long bones, suggesting that in the absence of β-catenin a bipotential osteochondroprogenitor will preferentially differentiate toward the cartilage lineage. A similar finding was obtained using a Cre recombinase under the control of the Prx1 promoter [66]. Again, terminal osteoblastogenesis was defective, with the appearance of chondrocytes at the sites of the periosteum and calvarial mesenchyme. Cultured mesenchymal progenitors from both mutant lines demonstrate abnormal differentiation into chondrocytes rather than osteoblasts. Similar effects of removing β-catenin were found when Cre was driven by the osterix promoter, expressed soon after commitment to the osteoblast lineage [147]. In more differentiated osteoblasts, β-catenin plays a critical role in osteoblast-mediated support of osteoclastogenesis, likely via regulation of OPG levels [148]. Finally, deletion of β-catenin in terminally differentiated osteoblasts leads to postnatal growth retardation and early mortality, with reduced trabecular and cortical bone [149]. β-Catenin–deficient osteoblasts demonstrate impaired nodule formation when cultured *in vitro*.

3. FGF, CNP, AND MAPK SIGNALING

We have already discussed some of the roles of FGF signaling in patterning the skeleton. Several of the 22 distinct FGF genes and four FGF receptors (FGFRs) are also expressed at later stages of endochondral bone formation and have been shown to regulate chondrocyte proliferation and maturation [24, 150]. Human craniosynostosis and dwarfism syndromes caused by mutations in multiple FGFRs [24, 151] dramatically illustrate the importance of FGF signaling in bone development.

a. FGF Signaling in Endochondral Bone Development
In later stages of endochondral bone formation, FGFR3 expression becomes restricted to the reserve and proliferating zones of the growth plate, whereas FGFR1 expression is detected in pre- and hypertrophic chondrocytes, and FGFR2 is expressed in osteoblasts in mature bones [24, 152]. These differences in FGFR expression suggest that these receptors may play different functions in endochondral development. Several FGFs, including FGF2, FGF7, FGF8, FGF9, FGF17, and FGF18, are also expressed in developing endochondral bone, principally in the perichondrium and periosteum [152]. Weak expression of FGF8 and FGF12 occurs in proliferating chondrocytes at relatively early stages of growth plate development, and this expression decreases as maturation takes place [152]. The study of the specific roles of individual FGF ligands is complicated by the fact that functional redundancy exists among them and that genetic inactivation of certain FGFs leads to early embryonic lethality. The role of FGF18 is best understood. Mice without FGF18 have an increase in chondrocyte proliferation in later fetal development, an expanded zone of hypertrophic chondrocytes, and delay in ossification [153, 154]. Because the chondrocyte phenotype of these mutant mice closely resembles that observed in FGFR3 knockout animals, FGF18 might be a relevant ligand for FGFR3 in growth plate chondrocytes [153, 154]. Point mutations in FGFR3 leading to a ligand-independent constitutive activation of this receptor are found in humans affected by achondroplasia [151]. Mice overexpressing an activated form of FGFR3 (FGFR3[ach] mice) develop skeletal dwarfism, characterized by a decrease in chondrocyte proliferation and a decrease in the expression of Ihh and type X collagen, markers of pre- and hypertrophic chondrocytes, respectively [151]. The inhibition of chondrocyte proliferation induced by FGFR3 signals occurs at least partly through activation of the Janus kinase-Signal transducer and activator of transcription 1 (JAK-STAT1) pathway [24, 155–157]. When FGFR3 activates Stat1, the expression of the cell cycle inhibitor p21 is stimulated, thus inducing the growth arrest of chondrocytes. Studies *in vitro* indicate that FGFR3 also decreases chondrocyte proliferation indirectly through the suppression of Ihh expression, which has been observed both *in vitro* and *in vivo* [158]. The molecular signals generated by FGFR3 that are responsible for its effect(s) on chondrocyte maturation probably involve activation of the MAPK pathway that activates the extracellular regulated kinase 1 and 2

(ERK1/2) [159, 160]. Further discussion of the FGF system can be found in Chapter 17 (Canalis).

b. MAPK Signaling in Endochondral Bone Development

The role of MAPK pathways in endochondral bone development has been mostly studied *in vitro* with established cell lines, using pharmacological inhibitors. These studies have led to a confusing literature [161], which suggests, nevertheless, that the MAPKs ERK and p38 play a role in regulating chondrocyte differentiation and/or proliferation. Thus far, only a few publications have reported *in vivo* studies of the role of MAPK pathways in cartilage. Misexpression in chondrocytes of a constitutively active form of MKK6, a MAPK kinase that specifically activates the MAPK p38, leads to a dwarf phenotype characterized by reduced chondrocyte proliferation and delayed hypertrophic chondrocyte differentiation [162]. Similarly, a constitutively active form of MEK1 (caMEK), which specifically activates ERKs, has also been misexpressed in chondrocytes and also leads to a dwarf phenotype characterized by a delayed hypertrophic chondrocyte differentiation [159]. Thus, this phenotype resembles that found in achondroplasia caused by activating mutations in FGFR3 (FGFR3[ach]) [24, 163]. Interestingly, caMEK misexpression can rescue the abnormal differentiation of the FGFR3 null mouse phenotype, suggesting that the MAPK ERK acts downstream of this receptor in chondrocytes. However, caMEK expression was not able to reverse the suppression of chondrocyte proliferation seen in the FGFR3[ach] mouse. This suggests that a MAPK–ERK pathway may not be an important regulator of chondrocyte proliferation *in vivo*.

The studies of MAPK signaling discussed so far all involve increasing the activity of the pathway. Studies of CNP signaling have allowed examination of the effects of decreasing MAPK signaling. CNP is related to the natriuretic peptides ANP and BNP, although it does not act as a natriuretic peptide *in vivo*. Instead, in blood vessels, chondrocytes, and some other cell types, CNP has paracrine actions mediated by the natriuretic receptor-B (NP-B), a membrane-bound guanylyl cyclase. CNP null mice are dwarfed, with a phenotype similar to that observed in NP-B knock-out mice. Homozygous inactivating mutations of NP-B in humans result in acromesomelic dysplasia, Maroteaux type, a severe form of disproportionate dwarfism [164]. The CNP null mice also resemble the FGFR3[ach] mice [165]. Conversely, CNP treatment increases longitudinal bone growth by stimulating chondrocyte proliferation and maturation [166, 167]. When ERK activity was repressed through misexpression of CNP in chondrocytes, this misexpression reversed the decreased matrix synthesis seen in the FGFR3[ach] mice

[160]. This repression of ERK activity, however, had no effect on either the rate of proliferation of chondrocytes or their pace of differentiation. Thus, these studies suggest that FGF signaling and CNP signaling have opposite effects on matrix synthesis through regulation of MAPK activity. They also raise the possibility that the major effects of MAPK action in chondrocytes primarily involve regulation of matrix production rather than direct effects on chondrocyte differentiation.

4. INDIAN HEDGEHOG AND PTHrP SIGNALING

Indian hedgehog (Ihh) is a member of the hedgehog family of paracrine factors that regulate development of multiple tissues. Of the three hedgehogs expressed in mammals, sonic hedgehog, Indian hedgehog, and desert hedgehog, only Ihh is expressed in cartilage during endochondral bone development. There, Ihh is synthesized by chondrocytes leaving the proliferative pool (prehypertrophic chondrocytes) and by early hypertrophic chondrocytes. The receptor for Ihh is patched-1 (Ptch-1). Through still poorly understood mechanisms, binding of Ihh to Ptch-1 leads to movement of smoothened (Smo) to the plasma membrane and activation of Smo, a seven-pass transmembrane protein resembling G protein–coupled receptors. Active Smo then triggers a cascade that leads to gene activation. Since Ihh action increases expression of Ptch-1, changes in levels of Ptch-1 mRNA offer a convenient assay for evidence of Ihh action at the cellular level. The phenotype of mice missing Ihh dramatically illustrates the importance of Ihh in endochondral bone formation. Ihh (–/–) mice have normal bones at the condensation stage but subsequently develop dramatic abnormalities of bone development [168]. All cartilage elements are small because of a dramatic decrease in chondrocyte proliferation. The proliferative effect of Ihh is likely to be a direct action on chondrocytes. Cartilage-specific knock-out of Smo leads to decreased proliferation of chondrocytes, and chondrocyte-specific transgenic expression of either Ihh or a constitutively active form of Smo increases chondrocyte proliferation [169].

A second abnormality in Ihh (–/–) mice is an increase in the fraction of chondrocytes that are hypertrophic. This abnormality occurs because chondrocytes leave the pool of proliferating chondrocytes prematurely. The ability of Ihh to keep chondrocytes proliferating is an indirect one, caused by the stimulation of PTHrP synthesis by Ihh [170]. PTHrP is a protein secreted during fetal life by perichondrial cells at the ends of cartilage anlage and by early proliferative chondrocytes. PTHrP acts on the same G protein–coupled receptor used by parathyroid hormone, the calcium-regulating hormone. These PTH/PTHrP receptors (PPRs) are expressed at low levels by proliferating chondrocytes and at high levels by prehypertrophic/early hypertrophic chondrocytes. In PTHrP (−/−) or PPR (−/−) mice, chondrocytes become hypertrophic prematurely, close to the ends of the bones [171, 172]. Hedgehog protein blocked hypertrophy of mouse chondrocytes from wild-type limbs but had no effect on chondrocyte differentiation when added to limbs from PTHrP (−/−) or PPR (−/−) mice [170]. Thus, the stimulation of PTHrP synthesis and secretion mediates the action of Ihh to delay hypertrophy.

The perichondrial cells and chondrocytes competent to synthesize PTHrP are many cell diameters away from the prehypertrophic and hypertrophic cells that synthesize Ihh. Nevertheless, a preliminary report suggests that the effects of Ihh on PTHrP synthesis are actions of Ihh directly on these cells since removal of Smo from groups of chondrocytes that normally synthesize PTHrP leads to the absence of PTHrP synthesis [173]. Deletion of the transcription factor gli3 rescues the expression of PTHrP in some perichondrial cells of Ihh (−/−) mice, suggesting that Ihh stimulates PTHrP synthesis at least partly through suppression of gli3 action [174, 175].

Thus, Ihh, through PTHrP, regulates the transition from flat, proliferating chondrocytes to prehypertrophic chondrocytes. Ihh also has an additional effect on chondrocyte differentiation. Studies in a variety of genetically altered mice demonstrate that Ihh also accelerates the differentiation of round, proliferative chondrocytes in the periarticular regions of growth plates into flat chondrocytes that form columns of proliferating chondrocytes [175, 176]. Both of these actions serve to lengthen the columns of chondrocytes in the growth plate. This lengthening serves to separate the distinct regions of the growth plate that synthesize Ihh and PTHrP. This distancing weakens the ability of Ihh to stimulate the synthesis of PTHrP. Because PTHrP delays the differentiation of chondrocytes into postmitotic cells that synthesize Ihh, PTHrP actions serve to delay the production of Ihh. Thus, PTHrP and Ihh together, through a negative feedback loop, determine the length of the columns of proliferative chondrocytes.

5. VEGF SIGNALING

VEGF is a 45-kDa homodimeric glycoprotein that belongs to the dimeric cysteine–knot growth factor superfamily, and it is also one of the most potent angiogenic factors identified so far. The VEGF gene encodes three isoforms: VEGF120, VEGF164, and VEGF188. All are products of alternative splicing of a single gene [177, 178]. In contrast to the other two isoforms, VEGF120 does not bind the extracellular matrix component heparan sulfate [177, 178].

VEGF is highly expressed by late hypertrophic chondrocytes, and at this stage it has a critical role in blood vessel invasion and replacement of cartilage by bone [179]. Injection of a soluble VEGF receptor in mice leads to impaired angiogenesis, decreased trabecular bone formation, and expansion of the hypertrophic zone in the growth plate [179]. In addition, vessel invasion into the primary ossification center is severely delayed in mice that only express one isoform of VEGF, VEGF120 [180, 181], and in mice in which VEGF expression is abolished [62, 182]. The upregulated expression of VEGF in hypertrophic chondrocytes results in the sprouting of endothelial cells in perichondrial blood vessels. Consistent with these findings, the expression of VEGF receptor 1 (VEGFR1 or Flt-1) and VEGF receptor 2 (VEGFR2 or Flk-1) in the perichondrial endothelium is upregulated as a result of VEGF expression in the hypertrophic cartilage [60, 181, 183]. Invasion of vessels into hypertrophic cartilage thus involves an active cross-talk between hypertrophic chondrocytes and endothelial cells. Expansion of the zone of hypertrophic chondrocytes in the growth plate is also observed following targeted inactivation of the genes encoding matrix metalloprotease-9 [184], the transcription factor Runx2 [60], and connective tissue growth factor [185]. All three genes have been reported to affect VEGF activity in the growth plate.

Studies of chondrocyte-specific knock-out of VEGF and its isoforms have identified VEGF as a critical factor for survival of chondrocytes [62, 186]. Similar massive cell death has been described in epiphyseal cartilage of mice in which the transcription factor HIF-1α, a major regulator of VEGF synthesis, is conditionally inactivated in chondrocytes [63]. The nature of the cellular events regulated by the HIF-1/VEGF pathway in epiphyseal chondrocytes is still largely unknown. Lack of VEGF may affect the number of blood vessels surrounding epiphyseal cartilage, or it may have a direct effect on cartilage. Of note, expression in epiphyseal chondrocytes of neuropilin 1 and neuropilin 2, which are coreceptors for VEGF164 and can potentiate signaling through VEGFR2, makes these receptors potential candidates for mediating actions of VEGF in chondrocytes [62, 186].

Vessel invasion into cartilage is a complex process involving the coordinated activities of both endothelial and osteo(chondro)clastic cells. In addition to controlling endothelial cell activities, VEGF also regulates osteoclastic differentiation, migration, and activity. Cells of the monocyte lineage express VEGFR1, and VEGF can substitute for M-CSF as a costimulator [187].

VEGF has also been reported to be a critical ligand for osteoblast differentiation. Inhibition of VEGF activity results not only in impaired angiogenesis but also in impaired trabecular bone formation [179]. Furthermore, in mice that express only the VEGF120 isoform, the ossification of membranous bones is reduced and osteoblastic differentiation is altered [181]. Finally, VEGFR1, VEGFR2, and neuropilins are all expressed by osteoblasts [188, 189].

6. DELTA/NOTCH SIGNALING

The Notch gene was identified almost a century ago in mutant flies that present notches in their wings, due to the requirement for this gene in limb outgrowth [190]. In many species and many developmental processes, by mediating cell–cell communication, Notch signaling has been shown to be a key regulator of cell fate [191, 192]. Notch signaling is often associated with an inhibitory effect on cell differentiation (e.g., in neurogenesis, myogenesis, and cardiogenesis), but Notch signaling also has an inductive action on cell fate in some settings (the absence of Notch signals results in a lack of wing margin specification in the fly) [191, 192]. Notch is a single-pass transmembrane receptor, which is activated upon binding to the extracellular domain of its ligands. These ligands are transmembrane proteins on adjacent cells. Once activated, Notch is cleaved by γ-secretase, and Notch's intracellular domain translocates into the nucleus and activates, in mammals, the transcription factor C-promoter binding factor-1, also known as recombination signal binding protein Jk. This in turn leads to the activation of transcription of target genes of the HES (Hairy/Enhancer of split) family, which encodes bHLH transcriptional repressors that affect the regulation of downstream target genes. In vertebrates, there are four Notch receptors (Notch 1–4) and five ligands named Delta-1 to -3 (or Delta-like-1 to -3) and Jagged 1 and 2 [192].

Relatively little is known about the role of this pathway in endochondral bone development. The expression patterns of all the Notch receptors and most of the Notch ligands have been observed in the mouse at several stages (ranging from E15.5 to 3 months of age) [193]. This analysis reveals that some of these Notch signaling genes are expressed during articular cartilage formation (Notch1 in particular) and, strikingly, that all of them are easily detected in hypertrophic chondrocytes. The study of the function of Notch signaling in growth plate chondrocytes has long been problematic due to early lethality observed in universal knock-out mice. To circumvent the problem, gain-of-function experiments misexpressed Delta-1 in the chick embryonic limb bud [194]. This study demonstrated that Delta-1 misexpression blocks chondrocyte maturation, resulting in limb shortening. Thus, this suggests that one of the physiological roles of Notch signaling might be the repression of chondrocyte

maturation. Confirmation of this hypothesis requires studies involving loss of function of Notch signaling components. Mice deficient specifically for the signals induced by Jagged 2 die perinatally, allowing the study of their skeleton [195]. These mice present major craniofacial defects, as well as syndactyly, but the endochondral process is not altered. Double conditional knock-out mice for Notch 1 and Notch 2 receptors have been generated using a Prx1-Cre transgenic mouse line. This strategy permits the deletion of floxed genes specifically and early in limb bud mesenchyme [196]. The resultant mice live and do not present any gross skeletal abnormalities in whole mount preparations. Interestingly, however, removal of Notch 1 and 2 in the limb ectoderm recapitulates the phenotype observed in the Jagged 2 null mice, indicating that Notch signaling plays a role in digit septation [196]. Preliminary studies reported in abstract form [197] suggest that the bones of these mice do have an expansion of the hypertrophic region and increased bone mass due to changes in both osteoblasts and osteoclasts. The genetic ablation of presenilins (the catalytic subunit of γ-secretase) is thought to suppress completely the canonical Notch pathway [198]. Mesenchymal-specific removal of presenilins leads to a mild phenotype in the size and shape of bones, restricted to the distal phalanges [196]. As with the Notch 1 and 2 knock-out, however, again histologic analysis reveals an increase in bone mass and an expansion of the hypertrophic region in the growth plate [197]. These studies suggest roles for Notch signaling in several cell types in bone. Further studies should clarify which effects represent direct actions and which represent secondary responses to those effects.

IV. INTRAMEMBRANOUS BONE FORMATION

Both endochondral ossification and intramembranous ossification begin with formation of mesenchymal condensations. During endochondral ossification, these condensations form a cartilage matrix; during intramembranous ossification, mesenchymal condensations differentiate directly into osteoblasts without an intervening chondrogenic phase. In the calvaria, mesenchymal blastemas prefigure sites of future skull bones, and calvarial sutures develop where two opposing bone fronts appose (Figure 8-7). The sutures are the predominant sites of bone growth, which must be carefully coordinated with enlargement of the underlying brain. The most actively proliferating cells are located at the edges of bone fronts, and this is where the differentiation of cells along the osteoblast lineage occurs. Mutations that affect intramembranous ossification are generally manifest as either craniosynostosis, resulting from premature fusion of sutures, or enlarged fontanels, when two skull bones fail to appose correctly. Although the molecular mechanisms underlying intramembranous ossification are not well understood, genetic mutations found in human syndromes have led to the identification of numerous important regulators.

One of the first gene products in which mutations were identified was Msx2 [199]. Mutations in Msx2 result in Boston-type craniosynostosis [199] and lead to enhanced binding of Msx2 to target DNA sequences [82]. Conversely, haploinsufficiency of Msx2 leads to wide-open fontanels in humans [200]. In mice, targeted deletion of Msx2 leads to an ossification defect of the frontal bone, with decreased osteoblast proliferation [81]. The mechanisms of action of Msx2 are unknown, but it may serve to inhibit expression of bone-specific genes such as collagen I [201] and osteocalcin [202] and direct precursors along the osteoblast lineage [84]. How, if at all, these actions contribute to the craniosynostosis phenotype is uncertain.

Although Msx2 was the first mutated gene product linked to craniosynostosis, most craniosynostoses are associated with mutations in FGF receptors. FGFs signal via four tyrosine kinase receptors, and craniosynostosis syndromes have been linked to mutations in FGFR1, FGFR2, and FGFR3. The majority of these syndromes is associated with mutations in FGFR1 and FGFR2, and in fact mice lacking FGFR3 do not have any apparent defects in cranial development [203, 204]. Mutations in FGFR1 and FGFR2 associated with Crouzon, Pfeiffer, and Jackson–Weiss syndromes generally result in gain of function, for example, by causing ligand-independent dimerization by stabilizing intermolecular disulfide bones [205–208]. Two specific missense mutations in FGFR2 lead to increased receptor signaling because the mutant receptors are activated by FGF ligands that do not normally activate the receptor [209].

Mice genetically manipulated to express the P250R mutant form of FGFR1, the ortholog of which causes Crouzon syndrome in humans, demonstrate premature fusion of cranial sutures accompanied by increased expression of the osteoblastic transcription factor Runx2 [210]. Similarly, activating mutations of FGFR2 in mice result in coronal synostosis [211] reminiscent of Apert's syndrome.

The relevant FGF ligands involved in cranial development are being investigated. Multiple FGFs are expressed during intramembranous ossification, including FGF2, FGF4, FGF9, FGF18, and FGF20 [163]. Ectopic expression of FGF2 in mice leads to macrocephaly [212] and coronal synostosis [213]. In addition, retroviral insertion in the region between FGF3 and FGF4 leads to increased expression of both FGF3 and FGF4 in the cranial sutures and Crouzon-like craniosynostosis in mice [214]. In contrast, mice deficient in FGF18 have craniofacial defects and delayed ossification [153, 154].

Twist 1 and Twist 2 are basic helix–loop–helix transcription factors that inhibit the actions of Runx2 in osteoblast development. Twist 1 is coexpressed with Runx2 in calvarial bones, whereas Twist 2 is expressed in the axial skeleton. As in the human craniosynostotic Saethre–Chotzen syndrome, caused by heterozygous inactivating mutations in Twist 1, haploinsufficiency of Twist 1 in mice leads to craniofacial abnormalities [215, 216]. Furthermore, haploinsufficiency of Twist 1 can rescue the delayed fontanels seen with haploinsufficiency of Runx2, demonstrating a role for Twist 1 in inhibiting Runx2, through the interaction of the twist box of Twist with the runt domain of Runx2 [217].

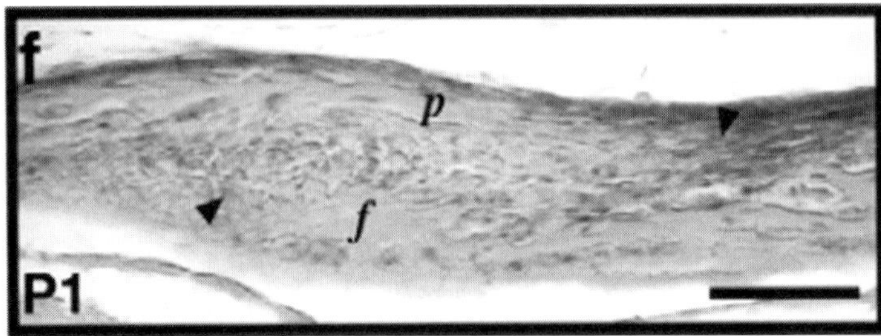

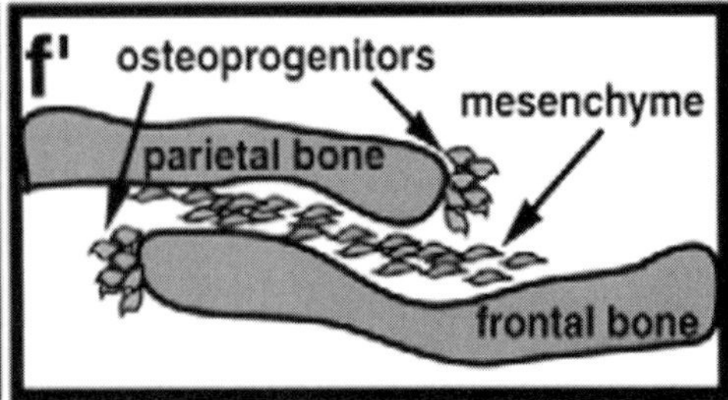

FIGURE 8-7   Coronal suture at P1 in the mouse. This suture occurs at the border of the parietal (p) and frontal (f) bones. Arrows point to expression of the *engrailed 1* gene in osteoprogenitors. Reprinted from R. A. Deckelbaum, A. Majithia, T. Booker, J. E. Henderson, and C. A. Loomis, The homeoprotein engrailed 1 has pleiotropic functions in calvarial intramembranous bone formation and remodeling. *Development* **133**(1), 63–74 (2006), with permission.

REFERENCES

1. N. Le Douarin, Details of the interphase nucleus in Japanese quail (*Coturnix coturnix japonica*). *Bull Biol Fr Belg* **103**, 435–452 (1969).

2. A. H. Monsoro-Burq, Sclerotome development and morphogenesis: When experimental embryology meets genetics. *Int J Dev Biol* **49**, 301–308 (2005).

3. H. Aoyama, S. Mizutani-Koseki, and H. Koseki, Three developmental compartments involved in rib formation. *Int J Dev Biol* **49**(2–3), 325–333 (2005).

4. J. Dubrulle and O. Pourquie, From head to tail: Links between the segmentation clock and antero-posterior patterning of the embryo. *Curr Opin Genet Dev* **12**(5), 519–523 (2002).

5. J. Deschamps and J. van Nes, Developmental regulation of the Hox genes during axial morphogenesis in the mouse. *Development* **132**(13), 2931–2942 (2005).

6. O. Pourquie, Vertebrate somitogenesis: A novel paradigm for animal segmentation? *Int J Dev Biol* **47**(7–8), 597–603 (2003).

7. O. Pourquie, Vertebrate somitogenesis. *Annu Rev Cell Dev Biol* **17**, 311–350 (2001).

8. J. Dubrulle and O. Pourquie, Coupling segmentation to axis formation. *Development* **131**(23), 5783–5793 (2004).

9. L. C. Murtaugh, J. H. Chyung, and A. B. Lassar, Sonic hedgehog promotes somitic chondrogenesis by altering the cellular response to BMP signaling. *Genes Dev* **13**(2), 225–237 (1999).

10. R. Reshef, M. Maroto, and A. B. Lassar, Regulation of dorsal somitic cell fates: BMPs and Noggin control the timing and pattern of myogenic regulator expression. *Genes Dev* **12**(3), 290–303 (1998).

11. L. C. Murtaugh, L. Zeng, J. H. Chyung, and A. B. Lassar, The chick transcriptional repressor Nkx3.2 acts downstream of Shh to promote BMP-dependent axial chondrogenesis. *Dev Cell* **1**, 411–422 (2001).

12. L. Zeng, H. Kempf, L. C. Murtaugh, M. E. Sato, and A. B. Lassar, Shh establishes an Nkx3.2/Sox9 autoregulatory loop that is maintained by BMP signals to induce somitic chondrogenesis. *Genes Dev* **16**(15), 1990–2005 (2002).

13. S. Kuratani, Craniofacial development and the evolution of the vertebrates: The old problems on a new background. *Zool Sci* **22**(1), 1–19 (2005).

14. S. Creuzet, G. Couly, and N. M. Le Douarin, Patterning the neural crest derivatives during development of the vertebrate head: Insights from avian studies. *J Anat* **207**(5), 447–459 (2005).

15. J. A. Helms, D. Cordero, and M. D. Tapadia, New insights into craniofacial morphogenesis. *Development* **132**(5), 851–861 (2005).

16. M. Bronner-Fraser, Neural crest cell formation and migration in the developing embryo. *FASEB J* **8**(10), 699–706 (1994).

17. G. F. Couly, P. M. Coltey, and N. M. Le Douarin, The triple origin of skull in higher vertebrates: A study in quail–chick chimeras. *Development* **117**(2), 409–429 (1993).

18. X. Jiang, S. Iseki, R. E. Maxson, H. M. Sucov, and G. M. Morriss-Kay, Tissue origins and interactions in the mammalian skull vault. *Dev Biol* **241**(1), 106–116 (2002).

19. J. A. Helms and R. A. Schneider, Cranial skeletal biology. *Nature* **423**(6937), 326–331 (2003).

20. M. D. Tapadia, D. R. Cordero, and J. A. Helms, It's all in your head: New insights into craniofacial development and deformation. *J Anat* **207**(5), 461–477 (2005).

21. D. Hu, R. S. Marcucio, and J. A. Helms, A zone of frontonasal ectoderm regulates patterning and growth in the face. *Development* **130**(9), 1749–1758 (2003).

22. R. P. Bolande, Neurocristopathy: Its growth and development in 20 years. *Pediatr Pathol Lab Med* **17**(1), 1–25 (1997).

23. B. Robert and Y. Lallemand, Anteroposterior patterning in the limb and digit specification: Contribution of mouse genetics. *Dev Dyn* **235**(9), 2337–2352 (2006).

24. D. M. Ornitz, FGF signaling in the developing endochondral skeleton. *Cytokine Growth Factor Rev* **16**(2), 205–213 (2005).

25. N. Schwartz and M. Domowicz, Chondrodysplasias due to proteoglycan defects. *Glycobiology* **12**, 57R–68R (2002).

26. H. Watanabe and Y. Yamada, Chondrodysplasia of gene knockout mice for aggrecan and link protein. *Glyconj J* **19**, 269–273 (2002).

27. D. Cohn, Defects in extracellular matrix structural proteins in the osteochondrodysplasias. *Novartis Foundation Symp* **232**, 195–212 (2001).

28. L. Everett and E. Green, A family of mammalian anion transporters and their involvement in human genetic diseases. *Hum Mol Genet* **8**, 1883–1891 (1999).

29. M. Briggs and K. Chapman, Pseudoachondroplasia and multiple epiphyseal dysplasia: Mutation review, molecular interactions, and genotype to phenotype correlation. *Hum Mutat* **19**, 465–478 (2002).

30. S. Mundlos and B. R. Olsen, Diseases of the skeleton: Part I. Molecular insights into skeletal development. Transcription factors and signaling pathways. *FASEB J* **11**, 125–132 (1997).

31. Y. Li and B. Olsen, Murine models of human genetic skeletal disorders. *Matrix Biol* **16**, 49–52 (1997).

32. M. Warman, M. Abbott, S. Apte, T. Hefferon, I. McIntosh, D. Cohn, *et al.*, A type X collagen mutation causes Schmid metaphyseal chondrodysplasia. *Nat Gen* **5**, 79–82 (1993).

33. O. Jacenko, P.A. LuValle, and B. R. Olsen, Spondylometaphyseal dysplasia in mice carrying a dominant negative mutation in a matrix protein specific for cartilage-to-bone transition. *Nature* **365**, 56–61 (1993).

34. K. M. Kwan, M. K. Pang, S. Zhou, S. K. Cowan, R. Y. Kong, T. Pfordte, *et al.*, Abnormal compartmentalization of cartilage matrix components in mice lacking collagen X: Implications for function. *J Cell Biol* **136**(2), 459–471 (1997).

35. W. Bi, J. M. Deng, Z. Zhang, R. R. Behringer, and B. de Crombrugghe, Sox9 is required for cartilage formation. *Nat Genet* **22**(1), 85–89 (1999).

36. V. Lefebvre, P. Li, and B. de Crombrugghe, A new long form of Sox5 (L-Sox5), Sox6 and Sox9 are coexpressed in chondrogenesis and cooperatively activate the type II collagen gene. *EMBO J* **17**(19), 5718–5733 (1998).

37. B. de Crombrugghe, V. Lefebvre, and K. Nakashima, Regulatory mechanisms in the pathways of cartilage and bone formation. *Curr Opin Cell Biol* **13**(6), 721–727 (2001).

38. H. Kronenberg, Developmental regulation of the growth plate. *Nature* **423**, 332–336 (2003).

39. H. Akiyama, M. C. Chaboissier, J. F. Martin, A. Schedl, and B. de Crombrugghe, The transcription factor Sox9 has essential roles in successive steps of the chondrocyte differentiation pathway and is required for expression of Sox5 and Sox6. *Genes Dev* **16**(21), 2813–2828 (2002).

40. V. Lefebvre and P. Smits, Transcriptional control of chondrocyte fate and differentiation. *Birth Defects Res C* **75**, 200–212 (2005).

41. J. Giordano, H. Prior, J. Bamforth, and M. Walter, Genetic study of SOX9 in a case of campomelic dysplasia. *Am J Med Genet* **98**, 176–181 (2001).

42. W. Huang, X. Zhou, V. Lefebvre, and B. de Crombrugghe, Phosphorylation of SOX9 by cyclic AMP-dependent protein kinase A enhances SOX9's ability to transactivate a Col2a1 chondrocyte-specific enhancer. *Mol Cell Biol* **20**(11), 4149–4158 (2000).

43. W. Huang, U. I. Chung, H. M. Kronenberg, and B. de Crombrugghe, The chondrogenic transcription factor Sox9 is a target of signaling by the parathyroid hormone-related peptide in the growth plate of endochondral bones. *Proc Natl Acad Sci USA* **98**(1), 160–165 (2001).

44. P. Smits, P. Li, J. Mandel, Z. Zhang, J. M. Deng, R. R. Behringer, *et al.*, The transcription factors L-Sox5 and Sox6 are essential for cartilage formation. *Dev Cell* **1**(2), 277–290 (2001).

45. P. Smits, P. Dy, S. Mitra, and V. Lefebvre, Sox5 and Sox6 are needed to develop and maintain source, columnar, and hypertrophic chondrocytes in the cartilage growth plate. *J Cell Biol* **164**(5), 747–758 (2004).

46. E. Ogawa, M. Maruyama, H. Kagoshima, M. Inuzuka, J. Lu, M. Satake, *et al.*, PEBP2/PEA2 represents a family of transcription factors homologous to the products of the *Drosophila runt* gene and the human *AML1* gene. *Proc Natl Acad Sci USA* **90**, 6859–6863 (1993).

47. P. Ducy, R. Zhang, V. Geoffroy, A. L. Ridall, and G. Karsenty, Osf2/Cbfa1: A transcriptional activator of osteoblast differentiation. *Cell* **89**(5), 747–754 (1997).

48. S. Mundlos, F. Otto, C. Mundlos, J. B. Mulliken, A. S. Aylsworth, S. Albright, *et al.*, Mutations involving the transcription factor CBFA1 cause cleidocranial dysplasia. *Cell* **89**(5), 773–779 (1997).

49. F. Otto, A. P. Thornell, T. Crompton, A. Denzel, K. C. Gilmour, I. R. Rosewell, *et al.*, Cbfa1, a candidate gene for cleidocranial dysplasia syndrome, is essential for osteoblast differentiation and bone development. *Cell* **89**(5), 765–771 (1997).

50. T. Komori, H. Yagi, S. Nomura, A. Yamaguchi, K. Sasaki, K. Deguchi, *et al.*, Targeted disruption of Cbfa1 results in a complete lack of bone formation owing to maturational arrest of osteoblasts. *Cell* **89**(5), 755–764 (1997).

51. M. Inada, T. Yasui, S. Nomura, S. Miyake, K. Deguchi, M. Himeno, *et al.*, Maturational disturbance of chondrocytes in Cbfa1-deficient mice. *Dev Dyn* **214**(4), 279–290 (1999).

52. I. Kim, F. Otto, B. Zabel, and S. Mundlos, Regulation of chondrocyte differentiation by Cbfa1. *Mech Dev* **80**, 159–170 (1999).

53. C. Ueta, M. Iwamoto, N. Kanatani, C. Yoshida, Y. Liu, M. Enomoto-Iwamoto, *et al.*, Skeletal malformations caused by overexpression of Cbfa1 or its dominant negative form in chondrocytes. *J Cell Biol* **153**(1), 87–100 (2001).

54. S. Takeda, J. P. Bonnamy, M. J. Owen, P. Ducy, and G. Karsenty, Continuous expression of Cbfa1 in nonhypertrophic chondrocytes uncovers its ability to induce hypertrophic chondrocyte differentiation and partially rescues Cbfa1-deficient mice. *Genes Dev* **15**(4), 467–481 (2001).

55. S. Stricker, R. Fundele, A. Vortkamp, and S. Mundlos, Role of Runx genes in chondrocyte differentiation. *Dev Biol* **245**(1), 95–108 (2002).

56. H. Enomoto, M. Enomoto-Iwamoto, M. Iwamoto, S. Nomura, M. Himeno, Y. Kitamura, *et al.*, Cbfa1 is a positive regulatory factor in chondrocyte maturation. *J Biol Chem* **275**, 8695–8702 (2000).

57. C. A. Yoshida, H. Yamamoto, T. Fujita, T. Furuichi, K. Ito, K.-I. Inoue, *et al.*, Runx2 and Runx3 are essential for chondrocyte maturation, and Runx2 regulates limb growth through induction of Indian hedgehog. *Genes Dev* **18**, 952–963 (2004).

58. M. Kundu, A. Javed, J. Jeon, A. Horner, L. Shum, M. Eckhaus, *et al.*, Cbfbeta interacts with Runx2 and has a critical role in bone development. *Nat Genet* **32**, 633–638 (2002).

59. C. Yoshida, T. Furuichi, T. Fujita, R. Fukuyama, N. Kanatani, S. Kobayashi, *et al.*, Core-binding factor beta interacts with Runx2 and is required for skeletal development. *Nat Genet* **32**, 633–638 (2002).

60. E. Zelzer, D. J. Glotzer, C. Hartmann, D. Thomas, N. Fukai, S. Soker, *et al.*, Tissue specific regulation of VEGF expression during bone development requires Cbfa1/Runx2. *Mech Dev* **106**(1–2), 97–106 (2001).

61. E. Schipani, Hypoxia and Hif-1alpha in chondrogenesis. *Sem Cell Dev Biol* **16**, 539–546 (2005).

62. E. Zelzer, R. Mamluk, N. Ferrara, R. Johnson, E. Schipani, and B. Olsen, VEGFA is necessary for chondrocyte survival during bone development. *Development* **131**, 2161–2171 (2004).

63. E. Schipani, H. E. Ryan, S. Didrickson, T. Kobayashi, M. Knight, and R. S. Johnson, Hypoxia in cartilage: HIF-1alpha is essential for chondrocyte growth arrest and survival. *Genes Dev* **15**(21), 2865–2876 (2001).

64. D. Pfander, T. Kobayashi, M. Knight, E. Zelzer, D. Chan, B. Olsen, *et al.*, Deletion of Vhlh in chondrocytes reduces cell proliferation and increases matrix deposition during growth plate development. *Development* **131**, 2497–2508 (2004).

65. H. Hu, M. Hilton, K. Yu, D. Ornitz, and F. Long, Sequential roles of Hedgehog and Wnt signaling in osteoblast development. *Development* **132**, 49–60 (2005).

66. T. P. Hill, D. Spater, M. M. Taketo, W. Birchmeier, and C. Hartmann, Canonical Wnt/beta-catenin signaling prevents osteoblasts from differentiating into chondrocytes. *Dev Cell* **8**(5), 727–738 (2005).

67. T. F. Day, X. Guo, L. Garrett-Beal, and Y. Yang, Wnt/beta-catenin signaling in mesenchymal progenitors controls osteoblast and chondrocyte differentiation during vertebrate skeletogenesis. *Dev Cell* **8**(5), 739–750 (2005).

68. D. A. Glass 2nd and K. Karsenty, Molecular bases of the regulation of bone remodeling by the canonical Wnt signaling pathway. *Curr Topics Dev Biol* **73**, 43–84 (2006).

69. H. Akiyama, J. P. Lyons, Y. Mori-Akiyama, X. Yang, R. Zhang, Z. Zhang, *et al.*, Interactions between Sox9 and beta-catenin control chondrocyte differentiation. *Genes Dev* **18**(9), 1072–1087 (2004).

70. G. Karsenty and E. F. Wagner, Reaching a genetic and molecular understanding of skeletal development. *Dev Cell* **2**(4), 389–406 (2002).

71. E. Zelzer and B. Olsen, The genetic basis for skeletal diseases. *Nature* **423**, 343–348 (2003).

72. C. Colnot, C. Lu, D. Hu, and J. A. Helms, Distinguishing the contributions of the perichondrium, cartilage, and vascular endothelium to skeletal development. *Dev Biol* **269**(1), 55–69 (2004).

73. K. Nakashima, X. Zhou, G. Kunkel, Z. Zhang, J. M. Deng, R. R. Behringer, *et al.*, The novel zinc finger-containing transcription factor osterix is required for osteoblast differentiation and bone formation. *Cell* **108**(1), 17–29 (2002).

74. J. E. Aubin and J. T. Triffitt, Mesenchymal stem cells and osteoblast differentiation. In *Principles of Bone Biology* (J. P. Bilezikian, L. G. Raisz, and G. A. Rodan, eds.), 2nd ed., pp. 59–81. Academic Press, San Diego (2002).

75. J. Iqbal and M. Zaidi, Molecular regulation of mechanotransduction. *Biochem Biophys Res Commun* **18**, 751–755 (2005).

76. R. V. Bezooijen, P. T. Dijke, S. Papapoulos, and C. Lowik, SOST/sclerostin, an osteocyte-derived negative regulator of bone formation. *Cytokine Growth Factor Rev* **16**, 319–327 (2005).

77. S. Bae and Y. Lee, Phosphorylation, acetylation and ubiquitination: The molecular basis of Runx regulation. *Gene* **366**, 58–66 (2006).

78. R. T. Franceschi and G. Xiao, Regulation of the osteoblast-specific transcription factor, Runx2: Responsiveness to multiple signal transduction pathways. *J Cell Biochem* **88**(3), 446–454 (2003).

79. D. Jones, M. Wein, M. Oukka, J. Hofstaetter, M. Glimcher, and L. Glimcher, Regulation of adult bone mass by the zinc finger adapter protein Schnurri-3. *Science* **312**, 1223–1227 (2006).

80. P. Ducy, Cbfa1: A molecular switch in osteoblast biology. *Dev Dyn* **219**(4), 461–471 (2000).
81. I. Satokata, L. Ma, H. Ohshima, M. Bei, I. Woo, K. Nishizawa, *et al.*, Msx2 deficiency in mice causes pleiotropic defects in bone growth and ectodermal organ formation. *Nat Genet* **24**(4), 391–395 (2000).
82. L. Ma, S. Golden, L. Wu, and R. Maxson, The molecular basis of Boston-type craniosynostosis. The Pro148→His mutation in the N-terminal arm of the MSX2 homeodomain stabilizes DNA binding without altering nucleotide sequence preferences. *Hum Mol Genet* **5**(12), 1915–1920 (1996).
83. M. Q. Hassan, R. S. Tare, S. H. Lee, M. Mandeville, M. I. Morasso, A. Javed, *et al.*, BMP2 commitment to the osteogenic lineage involves activation of Runx2 by DLX3 and a homeodomain transcriptional network. *J Biol Chem* **281**(52), 40515–40526 (2006).
84. S. L. Cheng, J. S. Shao, N. Charlton-Kachigian, A. P. Loewy, and D. A. Towler, MSX2 promotes osteogenesis and suppresses adipogenic differentiation of multipotent mesenchymal progenitors. *J Biol Chem* **278**(46), 45969–45977 (2003).
85. T. Koga, Y. Matsui, M. Asagiri, T. Kodama, B. de Crombrugghe, K. Nakashima, *et al.*, NFAT and Osterix cooperatively regulate bone formation. *Nat Med* **11**(8), 880–885 (2005).
86. X. Yang, K. Matsuda, P. Bialek, S. Jacquot, H. C. Masuoka, T. Schinke, *et al.*, ATF4 is a substrate of RSK2 and an essential regulator of osteoblast biology: Implication for Coffin–Lowry syndrome. *Cell* **117**(3), 387–398 (2004).
87. F. Elefteriou, M. D. Benson, H. Sowa, M. Starbuck, X. Liu, D. Ron, *et al.*, ATF4 mediation of NF1 functions in osteoblast reveals a nutritional basis for congenital skeletal dysplasiae. *Cell Metab* **4**(6), 441–451 (2006).
88. W. Boyle, W. Simonet, and D. Lacey, Osteoclast differentiation and activation. *Nature* **423**, 337–342 (2003).
89. W. Wiktor-Jedrzejczak, A. Bartocci, A. W. Ferrante Jr., A. Ahmed-Ansari, K. W. Sell, J. W. Pollard, *et al.*, Total absence of colony-stimulating factor 1 in the macrophage-deficient osteopetrotic (op/op) mouse. *Proc Natl Acad Sci USA* **87**(12), 4828–4832 (1990).
90. H. Yoshida, S. Hayashi, T. Kunisada, M. Ogawa, S. Nishikawa, H. Okamura, *et al.*, The murine mutation osteopetrosis is in the coding region of the macrophage colony stimulating factor gene. *Nature* **345**(6274), 442–444 (1990).
91. R. Felix, M. G. Cecchini, and H. Fleisch, Macrophage colony stimulating factor restores *in vivo* bone resorption in the op/op osteopetrotic mouse. *Endocrinology* **127**(5), 2592–2594 (1990).
92. H. Kodama, A. Yamasaki, M. Nose, S. Niida, Y. Ohgame, M. Abe, *et al.*, Congenital osteoclast deficiency in osteopetrotic (op/op) mice is cured by injections of macrophage colony-stimulating factor. *J Exp Med* **173**(1), 269–272 (1991).
93. G. Hattersley, J. Owens, A. M. Flanagan, and T. J. Chambers, Macrophage colony stimulating factor (M-CSF) is essential for osteoclast formation *in vitro*. *Biochem Biophys Res Commun* **177**(1), 526–531 (1991).
94. H. Kodama, M. Nose, S. Niida, and A. Yamasaki, Essential role of macrophage colony-stimulating factor in the osteoclast differentiation supported by stromal cells. *J Exp Med* **173**(5), 1291–1294 (1991).
95. N. Takahashi, N. Udagawa, T. Akatsu, H. Tanaka, Y. Isogai, and T. Suda, Deficiency of osteoclasts in osteopetrotic mice is due to a defect in the local microenvironment provided by osteoblastic cells. *Endocrinology* **128**(4), 1792–1796 (1991).
96. D. L. Lacey, E. Timms, H. L. Tan, M. J. Kelley, C. R. Dunstan, T. Burgess, *et al.*, Osteoprotegerin ligand is a cytokine that regulates osteoclast differentiation and activation. *Cell* **93**(2), 165–176 (1998).
97. H. Yasuda, N. Shima, N. Nakagawa, K. Yamaguchi, M. Kinosaki, S. Mochizuki, *et al.*, Osteoclast differentiation factor is a ligand for osteoprotegerin/osteoclastogenesis-inhibitory factor and is identical to TRANCE/RANKL. *Proc Natl Acad Sci USA* **95**(7), 3597–3602 (1998).
98. Y. Y. Kong, H. Yoshida, I. Sarosi, H. L. Tan, E. Timms, C. Capparelli, *et al.*, OPGL is a key regulator of osteoclastogenesis, lymphocyte development and lymph-node organogenesis. *Nature* **397**(6717), 315–323 (1999).
99. D. M. Anderson, E. Maraskovsky, W. L. Billingsley, W. C. Dougall, M. E. Tometsko, E. R. Roux, *et al.*, A homologue of the TNF receptor and its ligand enhance T-cell growth and dendritic-cell function. *Nature* **390,** 175–179 (1997).
100. N. Nakagawa, M. Kinosaki, K. Yamaguchi, N. Shima, H. Yasuda, K. Yano, *et al.*, RANK is the essential signaling receptor for osteoclast differentiation factor in osteoclastogenesis. *Biochem Biophys Res Commun* **253**(2), 395–400 (1998).
101. J. Li, I. Sarosi, X. Yan, S. Morony, H. Tan, *et al.*, RANK is the intrinsic hematopoietic cell surface receptor that controls osteoclastogenesis and regulation of bone mass and calcium metabolism. *Proc Natl Acad Sci USA* **97,** 1566–1571 (2000).
102. W. S. Simonet, D. L. Lacey, C. R. Dunstan, M. Kelley, M. S. Chang, R. Luthy, *et al.*, Osteoprotegerin: A novel secreted protein involved in the regulation of bone density. *Cell* **89**(2), 309–319 (1997).
103. E. Tsuda, M. Goto, S. Mochizuki, K. Yano, F. Kobayashi, T. Morinaga, *et al.*, Isolation of a novel cytokine from human fibroblasts that specifically inhibits osteoclastogenesis. *Biochem Biophys Res Commun* **234**(1), 137–142 (1997).
104. T. Suda, N. Takahashi, N. Udagawa, E. Jimi, M. T. Gillespie, and T. J. Martin, Modulation of osteoclast differentiation and function by the new members of the tumor necrosis factor receptor and ligand families. *Endocr Rev* **20**(3), 345–357 (1999).
105. T. Akatsu, T. Murakami, K. Ono, M. Nishikawa, E. Tsuda, S. I. Mochizuki, *et al.*, Osteoclastogenesis inhibitory factor exhibits hypocalcemic effects in normal mice and in hypercalcemic nude mice carrying tumors associated with humoral hypercalcemia of malignancy. *Bone* **23**(6), 495–498 (1998).
106. A. Tomoyasu, M. Goto, N. Fujise, S. Mochizuki, H. Yasuda, T. Morinaga, *et al.*, Characterization of monomeric and homodimeric forms of osteoclastogenesis inhibitory factor. *Biochem Biophys Res Commun* **245**(2), 382–387 (1998).
107. M. Yamamoto, T. Murakami, M. Nishikawa, E. Tsuda, S. Mochizuki, K. Higashio, *et al.*, Hypocalcemic effect of osteoclastogenesis inhibitory factor/osteoprotegerin in the thyroparathyroidectomized rat. *Endocrinology* **139**(9), 4012–4015 (1998).
108. N. Bucay, I. Sarosi, C. R. Dunstan, S. Morony, J. Tarpley, C. Capparelli, *et al.*, Osteoprotegerin-deficient mice develop early onset osteoporosis and arterial calcification. *Genes Dev* **12**(9), 1260–1268 (1998).
109. A. Mizuno, N. Amizuka, K. Irie, A. Murakami, N. Fujise, T. Kanno, *et al.*, Severe osteoporosis in mice lacking osteoclastogenesis inhibitory factor/osteoprotegerin. *Biochem Biophys Res Commun* **247**(3), 610–615 (1998).
110. M. Tondravi, S. McKercher, K. Anderson, J. Erdmann, M. Quiroz, R. Maki, *et al.*, Osteopetrosis in mice lacking haematopoietic transcription factor PU.1. *Nature* **386,** 81–84 (1997).
111. T. Inaba, T. Gotoda, S. Ishibashi, K. Harada, J. Ohsuga, K. Ohashi, *et al.*, Transcription factor PU.1 mediates induction of c-fms in vascular smooth muscle cells: A mechanism

for phenotypic change to phagocytic cells. *Mol Cell Biol* **16**, 2264–2273 (1996).

112. A. E. Grigoriadis, Z. Q. Wang, M. G. Cecchini, W. Hofstetter, R. Felix, H. A. Fleisch, *et al.*, c-Fos: A key regulator of osteoclast–macrophage lineage determination and bone remodeling. *Science* **266**, 443–448 (1994).

113. C. Hershey and D. Fisher, Mitf and Tfe3: Members of a b-HLH-ZIP transcription factor family essential for osteoclast development and function. *Bone* **34**, 689–696 (2004).

114. L. Galibert, M. E. Tometsko, D. M. Anderson, D. Cosman, and W. C. Dougall, The involvement of multiple tumor necrosis factor receptor (TNFR)-associated factors in the signaling mechanisms of receptor activator of NF-kappaB, a member of the TNFR superfamily. *J Biol Chem* **273**(51), 34120–34127 (1998).

115. G. Franzoso, L. Carlson, L. Xing, L. Poljak, E. W. Shores, K. D. Brown, *et al.*, Requirement for NF-kappaB in osteoclast and B-cell development. *Genes Dev* **11**(24), 3482–3496 (1997).

116. V. Iotsova, J. Caamano, J. Loy, Y. Yang, A. Lewin, and R. Bravo, Osteopetrosis in mice lacking NF-kappaB1 and NF-kappaB2. *Nat Med* **3**(11), 1285–1289 (1997).

117. H. Takayanagi, S. Kim, K. Matsuo, H. Suzuki, T. Suzuki, K. Sato, *et al.*, RANKL maintains bone homeostasis through c-Fos-dependent induction of interferon-beta. *Nature* **416**(6882), 744–749 (2002).

118. H. Takayanagi, S. Kim, T. Koga, H. Nishina, M. Isshiki, H. Yoshida, *et al.*, Induction and activation of the transcription factor NFATc1 (NFAT2) integrate RANKL signaling in terminal differentiation of osteoclasts. *Dev Cell* **3**(6), 889–901 (2002).

119. K. Matsuo, D. Galson, C. Zhao, L. Peng, C. Laplace, K. Wang, *et al.*, Nuclear factor of activated T-cells (NFAT) rescues osteoclastogenesis in precursors lacking c-Fos. *J Biol Chem* **279**, 475–480 (2004).

120. L. C. Hofbauer and M. Schoppet, Clinical implications of the osteoprotegerin/RANKL/RANK system for bone and vascular diseases. *JAMA* **292**(4), 490–495 (2004).

121. T. Wada, T. Nakashima, N. Hiroshi, and J. M. Penninger, RANKL-RANK signaling in osteoclastogenesis and bone disease. *Trends Mol Med* **12**(1), 17–25 (2006).

122. L. C. Hofbauer, F. Gori, B. L. Riggs, D. L. Lacey, C. R. Dunstan, T. C. Spelsberg, *et al.*, Stimulation of osteoprotegerin ligand and inhibition of osteoprotegerin production by glucocorticoids in human osteoblastic lineage cells: Potential paracrine mechanisms of glucocorticoid-induced osteoporosis. *Endocrinology* **140**(10), 4382–4389 (1999).

123. L. C. Hofbauer, S. Khosla, C. R. Dunstan, D. L. Lacey, T. C. Spelsberg, and B. L. Riggs, Estrogen stimulates gene expression and protein production of osteoprotegerin in human osteoblastic cells. *Endocrinology* **140**(9), 4367–4370 (1999).

124. N. Giuliani, R. Bataille, C. Mancini, M. Lazzaretti, and S. Barille, Myeloma cells induce imbalance in the osteoprotegerin/osteoprotegerin ligand system in the human bone marrow environment. *Blood* **98**(13), 3527–3533 (2001).

125. V. Rosen, BMP and BMP inhibitors in bone. *Ann N Y Acad Sci* **1068**, 19–25 (2006).

126. M. Yanagita, BMP antagonists: Their roles in development and involvement in pathophysiology. *Cytokine Growth Factor Rev* **16**(3), 309–317 (2005).

127. G. Q. Zhao, Consequences of knocking out BMP signaling in the mouse. *Genesis* **35**(1), 43–56 (2003).

128. S. Pizette and L. Niswander, BMPs are required at two steps of limb chondrogenesis: Formation of prechondrogenic condensations and their differentiation into chondrocytes. *Dev Biol* **219**(2), 237–249 (2000).

129. B. S. Yoon, D. A. Ovchinnikov, I. Yoshii, Y. Mishina, R. R. Behringer, and K. M. Lyons, Bmpr1a and Bmpr1b have overlapping functions and are essential for chondrogenesis *in vivo*. *Proc Natl Acad Sci USA* **102**(14), 5062–5067 (2005).

130. A. Bandyopadhyay, K. Tsuji, K. Cox, B. D. Harfe, V. Rosen, and C. J. Tabin, Genetic analysis of the roles of BMP2, BMP4, and BMP7 in limb patterning and skeletogenesis. *PLoS Genet* **2**(12), e216 (2006).

131. D. M. Kingsley, A. E. Bland, J. M. Grugger, P. C. Marker, L. B. Russell, N. G. Copeland, *et al.*, The mouse short ear skeletal morphogenesis locus is associated with defects in a bone morphogenetic member of the TGFb superfamily. *Cell* **71**, 399–410 (1992).

132. G. Luo, C. Hofmann, A. L. Bronckers, M. Sohocki, A. Bradley, and G. Karsenty, BMP-7 is an inducer of nephrogenesis and is also required for eye development and skeletal patterning. *Genes Dev* **9**(22), 2808–2820 (1995).

133. N. Jena, C. Martin-Seisdedos, P. McCue, and C. M. Croce, BMP7 null mutation in mice: Developmental defects in skeleton, kidney, and eye. *Exp Cell Res* **230**(1), 28–37 (1997).

134. M. J. Solloway, A. T. Dudley, E. K. Bikoff, K. M. Lyons, B. L. Hogan, and E. J. Robertson, Mice lacking Bmp6 function. *Dev Genet* **22**(4), 321–339 (1998).

135. E. E. Storm, T. V. Huynh, N. G. Copeland, N. A. Jenkins, D. M. Kingsley, and S. J. Lee, Limb alterations in brachypodism mice due to mutations in a new member of the TGF beta-superfamily. *Nature* **368**(6472), 639–643 (1994).

136. S. H. Settle Jr., R. B. Rountree, A. Sinha, A. Thacker, K. Higgins, and D. M. Kingsley, Multiple joint and skeletal patterning defects caused by single and double mutations in the mouse Gdf6 and Gdf5 genes. *Dev Biol* **254**(1), 116–130 (2003).

137. X. B. Wu, Y. Li, A. Schneider, W. Yu, G. Rajendren, J. Iqbal, *et al.*, Impaired osteoblastic differentiation, reduced bone formation, and severe osteoporosis in noggin-overexpressing mice. *J Clin Invest* **112**(6), 924–934 (2003).

138. E. Gazzerro, R. C. Pereira, V. Jorgetti, S. Olson, A. N. Economides, and E. Canalis, Skeletal overexpression of gremlin impairs bone formation and causes osteopenia. *Endocrinology* **146**(2), 655–665 (2005).

139. Y. Gong, R. B. Slee, N. Fukai, G. Rawadi, S. Roman-Roman, A. M. Reginato, *et al.*, LDL receptor-related protein 5 (LRP5) affects bone accrual and eye development. *Cell* **107**(4), 513–523 (2001).

140. M. Kato, M. S. Patel, R. Levasseur, I. Lobov, B. H. Chang, D. A. Glass 2nd, *et al.*, Cbfa1-independent decrease in osteoblast proliferation, osteopenia, and persistent embryonic eye vascularization in mice deficient in Lrp5, a Wnt coreceptor. *J Cell Biol* **157**(2), 303–314 (2002).

141. T. Fujino, H. Asaba, M. J. Kang, Y. Ikeda, H. Sone, S. Takada, *et al.*, Low-density lipoprotein receptor-related protein 5 (LRP5) is essential for normal cholesterol metabolism and glucose-induced insulin secretion. *Proc Natl Acad Sci USA* **100**(1), 229–234 (2003).

142. L. M. Boyden, J. Mao, J. Belsky, L. Mitzner, A. Farhi, M. A. Mitnick, *et al.*, High bone density due to a mutation in LDL-receptor-related protein 5. *N Engl J Med* **346**(20), 1513–1521 (2002).

143. R. D. Little, J. P. Carulli, R. G. Del Mastro, J. Dupuis, M. Osborne, C. Folz, *et al.*, A mutation in the LDL receptor-related protein 5 gene results in the autosomal dominant high-bone-mass trait. *Am J Hum Genet* **70**(1), 11–19 (2002).

144. P. Babij, W. Zhao, C. Small, Y. Kharode, P. J. Yaworsky, M. L. Bouxsein, *et al.*, High bone mass in mice expressing a mutant LRP5 gene. *J Bone Miner Res* **18**(6), 960–974 (2003).

145. M. Ai, S. L. Holmen, W. Van Hul, B. O. Williams, and M. L. Warman, Reduced affinity to and inhibition by DKK1 form a common mechanism by which high bone mass-associated missense mutations in LRP5 affect canonical Wnt signaling. *Mol Cell Biol* **25**(12), 4946–4955 (2005).

146. X. Li, Y. Zhang, H. Kang, W. Liu, P. Liu, J. Zhang, *et al.*, Sclerostin binds to LRP5/6 and antagonizes canonical Wnt signaling. *J Biol Chem* **280**(20), 19883–19887 (2005).

147. S. J. Rodda and A. P. McMahon, Distinct roles for Hedgehog and canonical Wnt signaling in specification, differentiation and maintenance of osteoblast progenitors. *Development* **133**(16), 3231–3244 (2006).

148. D. A. Glass 2nd, P. Bialek, J. D. Ahn, M. Starbuck, M. S. Patel, H. Clevers, *et al.*, Canonical Wnt signaling in differentiated osteoblasts controls osteoclast differentiation. *Dev Cell* **8**(5), 751–764 (2005).

149. S. L. Holmen, C. R. Zylstra, A. Mukherjee, R. E. Sigler, M. C. Faugere, M. L. Bouxsein, *et al.*, Essential role of beta-catenin in postnatal bone acquisition. *J Biol Chem* **280**(22), 21162–21168 (2005).

150. H. M. Kronenberg, Developmental regulation of the growth plate. *Nature* **423**(6937), 332–336 (2003).

151. D. M. Ornitz and P. J. Marie, FGF signaling pathways in endochondral and intramembranous bone development and human genetic disease. *Genes Dev* **16**(12), 1446–1465 (2002).

152. E. Minina, S. Schneider, M. Rosowski, R. Lauster, and A. Vortkamp, Expression of Fgf and Tgfbeta signaling related genes during embryonic endochondral ossification. *Gene Expr Patterns* **6**(1), 102–109 (2005).

153. N. Ohbayashi, M. Shibayama, Y. Kurotaki, M. Imanishi, T. Fujimori, N. Itoh, *et al.*, FGF18 is required for normal cell proliferation and differentiation during osteogenesis and chondrogenesis. *Genes Dev* **16**(7), 870–879 (2002).

154. Z. Liu, J. Xu, J. S. Colvin, and D. M. Ornitz, Coordination of chondrogenesis and osteogenesis by fibroblast growth factor 18. *Genes Dev* **16**(7), 859–869 (2002).

155. M. Sahni, R. Raz, J. D. Coffin, D. Levy, and C. Basilico, STAT1 mediates the increased apoptosis and reduced chondrocyte proliferation in mice overexpressing FGF2. *Dev Suppl* **128**(11), 2119–2129 (2001).

156. M. Sahni, D. C. Ambrosetti, A. Mansukhani, R. Gertner, D. Levy, and C. Basilico, FGF signaling inhibits chondrocyte proliferation and regulates bone development through the STAT-1 pathway. *Genes Dev* **13**(11), 1361–1366 (1999).

157. W. C. Su, M. Kitagawa, N. Xue, B. Xie, S. Garofalo, J. Cho, *et al.*, Activation of Stat1 by mutant fibroblast growth-factor receptor in thanatophoric dysplasia type II dwarfism. *Nature* **386**(6622), 288–292 (1997).

158. E. Minina, C. Kreschel, M. C. Naski, D. M. Ornitz, and A. Vortkamp, Interaction of FGF, Ihh/Pthlh, and BMP signaling integrates chondrocyte proliferation and hypertrophic differentiation. *Dev Cell* **3**(3), 439–449 (2002).

159. S. Murakami, G. Balmes, S. McKinney, Z. Zhang, D. Givol, and B. de Crombrugghe, Constitutive activation of MEK1 in chondrocytes causes Stat1-independent achondroplasia-like dwarfism and rescues the Fgfr3-deficient mouse phenotype. *Genes Dev* **18**(3), 290–305 (2004).

160. A. Yasoda, Y. Komatsu, H. Chusho, T. Miyazawa, A. Ozasa, M. Miura, *et al.*, Overexpression of CNP in chondrocytes rescues achondroplasia through a MAPK-dependent pathway. *Nat Med* **10**(1), 80–86 (2004).

161. L. A. Stanton, T. M. Underhill, and F. Beier, MAP kinases in chondrocyte differentiation. *Dev Biol* **263**(2), 165–175 (2003).

162. R. Zhang, S. Murakami, F. Coustry, Y. Wang, and B. de Crombrugghe, Constitutive activation of MKK6 in chondrocytes of transgenic mice inhibits proliferation and delays endochondral bone formation. *Proc Natl Acad Sci USA* **103**(2), 365–370 (2006).

163. D. M. Ornitz and P. J. Marie, FGF signaling pathways in endochondral and intramembranous bone development and human genetic disease. *Genes Dev* **16**(12), 1446–1465 (2002).

164. S. Schulz, C-type natriuretic peptide and guanylyl cyclase B receptor. *Peptides* **26**(6), 1024–1034 (2005).

165. H. Chusho, N. Tamura, Y. Ogawa, A. Yasoda, M. Suda, T. Miyazawa, *et al.*, Dwarfism and early death in mice lacking C-type natriuretic peptide. *Proc Natl Acad Sci USA* **98**(7), 4016–4021 (2001).

166. J. Jaubert, F. Jaubert, N. Martin, L. L. Washburn, B. K. Lee, E. M. Eicher, *et al.*, Three new allelic mouse mutations that cause skeletal overgrowth involve the natriuretic peptide receptor C gene (Npr3). *Proc Natl Acad Sci USA* **96**(18), 10278–10283 (1999).

167. N. Matsukawa, W. Grzesik, N. Takahashi, K. Pandey, S. Pang, M. Yamauchi, *et al.*, The natriuretic peptide clearance receptor locally modulates the physiological effects of the natriuretic peptide system. *Proc Natl Acad Sci USA* **96**, 7403–7408 (1999).

168. B. St.-Jacques, M. Hammerschmidt, and A. P. McMahon, Indian hedgehog signaling regulates proliferation and differentiation of chondrocytes and is essential for bone formation. *Genes Dev* **13**(16), 2072–2086 (1999). [Erratum appears in *Genes Dev* 1999 Oct 1;**13**(19):2617.]

169. F. Long, X. M. Zhang, S. Karp, Y. Yang, and A. P. McMahon, Genetic manipulation of hedgehog signaling in the endochondral skeleton reveals a direct role in the regulation of chondrocyte proliferation. *Development* **128**(24), 5099–5108 (2001).

170. A. Vortkamp, K. Lee, B. Lanske, G. V. Segre, H. M. Kronenberg, and C. J. Tabin, Regulation of rate of cartilage differentiation by Indian hedgehog and PTH-related protein. *Science* **273**(5275), 613–622 (1996).

171. A. C. Karaplis, A. Luz, J. Glowacki, R. T. Bronson, V. L. Tybulewicz, H. M. Kronenberg, *et al.*, Lethal skeletal dysplasia from targeted disruption of the parathyroid hormone-related peptide gene. *Genes Dev* **8**(3), 277–289 (1994).

172. B. Lanske, A. C. Karaplis, K. Lee, A. Luz, A. Vortkamp, A. Pirro, *et al.*, PTH/PTHrP receptor in early development and Indian hedgehog-regulated bone growth. *Science* **273**(5275), 663–666 (1996).

173. M. J. Hilton and F. Long, Temporal control of gene deletion in the cartilage using a tamoxifen-inducible cre transgenic mouse. *J Bone Miner Res* **20**, 539 (2005).

174. M. J. Hilton, X. Tu, J. Cook, H. Hu, and F. Long, Ihh controls cartilage development by antagonizing Gli3, but requires additional effectors to regulate osteoblast and vascular development. *Development* **132**(19), 4339–4351 (2005).

175. L. Koziel, M. Wuelling, S. Schneider, and A. Vortkamp, Gli3 acts as a repressor downstream of Ihh in regulating two distinct steps of chondrocyte differentiation. *Development* **132**(23), 5249–5260 (2005).

176. T. Kobayashi, D. W. Soegiarto, Y. Yang, B. Lanske, E. Schipani, A. P. McMahon, *et al.*, Indian hedgehog stimulates periarticular chondrocyte differentiation to regulate growth plate length independently of PTHrP. *J Clin Invest* **115**(7), 1734–1742 (2005).

177. N. Ferrara, H. Gerber, and J. LeCouter, The biology of VEGF and its receptors. *Nat Med* **9**, 669–676 (2003).

178. E. Zelzer and B. Olsen, Multiple roles of vascular endothelial growth factor (VEGF) in skeletal development, growth, and repair. *Curr Topics Dev Biol* **65**, 169–187 (2005).

179. H. P. Gerber, T. H. Vu, A. M. Ryan, J. Kowalski, Z. Werb, and N. Ferrara, VEGF couples hypertrophic cartilage remodeling, ossification and angiogenesis during endochondral bone formation. *Nat Med* **5**(6), 623–628 (1999).

180. C. Maes, P. Carmeliet, K. Moermans, I. Stockmans, N. Smets, D. Collen, *et al.*, Impaired angiogenesis and endochondral bone formation in mice lacking the vascular endothelial growth factor isoforms VEGF164 and VEGF188. *Mech Dev.* **111**, 61–73 (2002).

181. E. Zelzer, W. McLean, Y. Ng, N. Fukai, A. Reginato, S. Lovejoy, *et al.*, Skeletal defects in VEGF120/120 mice reveal multiple roles for VEGF in skeletogenesis. *Development* **129**, 1893–1904 (2002).

182. J. J. Haigh, H. P. Gerber, N. Ferrara, and E. F. Wagner, Conditional inactivation of VEGF-A in areas of collagen2a1 expression results in embryonic lethality in the heterozygous state. *Development* **127**(7), 1445–1453 (2000).

183. C. Colnot and J. Helms, A molecular analysis of matrix remodeling and angiogenesis during long bone development. *Mech Dev* **100**, 245–250 (2001).

184. T. H. Vu, J. M. Shipley, G. Bergers, J. E. Berger, J. A. Helms, D. Hanahan, *et al.*, MMP-9/gelatinase B is a key regulator of growth plate angiogenesis and apoptosis of hypertrophic chondrocytes. *Cell* **93**(3), 411–422 (1998).

185. S. Ivkovic, B. Yoon, S. Popoff, F. Safadi, D. Libuda, R. Stephenson, *et al.*, Connective tissue growth factor coordinates chondrogenesis and angiogenesis during skeletal development. *Development* **130**, 2779–2791 (2003).

186. C. Maes, I. Stockmans, K. Moermans, R. V. Looveren, N. Smets, P. Carmeliet, *et al.*, Soluble VEGF isoforms are essential for establishing epiphyseal vascularization and regulating chondrocyte development and survival. *J Clin Invest* **113**, 188–199 (2004).

187. S. Niida, T. Kondo, S. Hiratsuka, S. Hayashi, N. Amizuka, T. Noda, *et al.*, VEGF receptor 1 signaling is essential for osteoclast development and bone marrow formation in colony-stimulating factor 1-deficient mice. *Proc Natl Acad Sci USA* **102**, 14016–14021 (1999).

188. J. Harper, L. Gestenfeld, and M. Klagsbrun, Neuropilin-1 expression in osteogenic cells: Downregulation during differentiation of osteoblasts into osteocytes. *J Cell Biochem* **278**, 48745–48753 (2001).

189. J. Street, M. Bao, L. deGuzman, S. Bunting, F. Peale, N. Ferrara, *et al.*, Vascular endothelial growth factor stimulates bone repair by promoting angiogenesis and bone turnover. *Proc Natl Acad Sci USA* **99**, 9656–9661 (2002).

190. T. M. Morgan, The theory of the gene. *Am Nat* **51**, 513–544 (1917).

191. S. Artavanis-Tsakonas, M. D. Rand, and R. J. Lake, Notch signaling: Cell fate control and signal integration in development. *Science* **284**(5415), 770–776 (1999).

192. E. C. Lai, Notch signaling: Control of cell communication and cell fate. *Development* **131**(5), 965–973 (2004).

193. A. J. Hayes, G. P. Dowthwaite, S. V. Webster, and C. W. Archer, The distribution of Notch receptors and their ligands during articular cartilage development. *J Anat* **202**(6), 495–502 (2003).

194. R. Crowe, J. Zikherman, and L. Niswander, Delta-1 negatively regulates the transition from prehypertrophic to hypertrophic chondrocytes during cartilage formation. *Development* **126**(5), 987–998 (1999).

195. R. Jiang, Y. Lan, H. D. Chapman, C. Shawber, C. R. Norton, D. V. Serreze, *et al.*, Defects in limb, craniofacial, and thymic development in Jagged2 mutant mice. *Genes Dev* **12**(7), 1046–1057 (1998).

196. Y. Pan, Z. Liu, J. Shen, and R. Kopan, Notch1 and 2 cooperate in limb ectoderm to receive an early Jagged2 signal regulating interdigital apoptosis. *Dev Biol* **286**(2), 472–482 (2005).

197. J. J. Hilton, S. Bai, R. Kopan, F. P. Ross, S. L. Teitelbaum, and F. Long, Notch signaling represses osteoblast activity but promotes osteoclast function *in vivo*. *J Bone Miner Res* **21**, 566 (2006).

198. Y. Pan, M. H. Lin, X. Tian, H. T. Cheng, T. Gridley, J. Shen, *et al.*, Gamma-secretase functions through Notch signaling to maintain skin appendages but is not required for their patterning or initial morphogenesis. *Dev Cell* **7**(5), 731–743 (2004).

199. E. W. Jabs, U. Muller, X. Li, L. Ma, W. Luo, I. S. Haworth, *et al.*, A mutation in the homeodomain of the human MSX2 gene in a family affected with autosomal dominant craniosynostosis. *Cell* **75**(3), 443–450 (1993).

200. A. O. Wilkie, Z. Tang, N. Elanko, S. Walsh, S. R. Twigg, J. A. Hurst, *et al.*, Functional haploinsufficiency of the human homeobox gene MSX2 causes defects in skull ossification. *Nat Genet* **24**(4), 387–390 (2000).

201. M. Dodig, M. S. Kronenberg, A. Bedalov, B. E. Kream, G. Gronowicz, S. H. Clark, *et al.*, Identification of a TAAT-containing motif required for high level expression of the COL1A1 promoter in differentiated osteoblasts of transgenic mice. *J Biol Chem* **271**(27), 16422–16429 (1996).

202. D. A. Towler, S. J. Rutledge, and G. A. Rodan, MSX-2/Hox 8.1: A transcriptional regulator of the rat osteocalcin promoter. *Mol Endocrinol* **8**, 1484–1493 (1994).

203. J. S. Colvin, B. A. Bohne, G. W. Harding, D. G. McEwen, and D. M. Ornitz, Skeletal overgrowth and deafness in mice lacking fibroblast growth factor receptor 3. *Nat Genet* **12**(4), 390–397 (1996).

204. C. Deng, A. Wynshaw-Boris, F. Zhou, A. Kuo, and P. Leder, Fibroblast growth factor receptor 3 is a negative regulator of bone growth. *Cell* **84**(6), 911–921 (1996).

205. K. M. Neilson and R. E. Friesel, Constitutive activation of fibroblast growth factor receptor-2 by a point mutation associated with Crouzon syndrome. *J Biol Chem* **270**(44), 26037–26040 (1995).

206. A. O. Wilkie, G. M. Morriss-Kay, E. Y. Jones, and J. K. Heath, Functions of fibroblast growth factors and their receptors. *Curr Biol* **5**(5), 500–507 (1995).

207. B. D. Galvin, K. C. Hart, A. N. Meyer, M. K. Webster, and D. J. Donoghue, Constitutive receptor activation by Crouzon syndrome mutations in fibroblast growth factor receptor (FGFR)2 and FGFR2/Neu chimeras. *Proc Natl Acad Sci USA* **93**(15), 7894–7899 (1996).

208. S. C. Robertson, A. N. Meyer, K. C. Hart, B. D. Galvin, M. K. Webster, and D. J. Donoghue, Activating mutations in the extracellular domain of the fibroblast growth factor receptor 2 function by disruption of the disulfide bond in the third immunoglobulin-like domain. *Proc Natl Acad Sci USA* **95**(8), 4567–4572 (1998).

209. K. Yu, A. B. Herr, G. Waksman, and D. M. Ornitz, Loss of fibroblast growth factor receptor 2 ligand-binding specificity in Apert syndrome. *Proc Natl Acad Sci USA* **97**(26), 14536–14541 (2000).

210. Y. X. Zhou, X. Xu, L. Chen, C. Li, S. G. Brodie, and C. X. Deng, A Pro250Arg substitution in mouse Fgfr1 causes increased expression of Cbfa1 and premature fusion of calvarial sutures. *Hum Mol Genet* **9**(13), 2001–2008 (2000).

211. M. K. Hajihosseini, S. Wilson, L. De Moerlooze, and C. Dickson, A splicing switch and gain-of-function mutation in FgfR2-IIIc hemizygotes causes Apert/Pfeiffer-syndrome-like phenotypes. *Proc Natl Acad Sci USA* **98**(7), 3855–3860 (2001).

212. J. D. Coffin, R. Z. Florkiewicz, J. Neumann, T. Mort-Hopkins, G. W. Dorn 2nd, P. Lightfoot, *et al.*, Abnormal bone growth and selective translational regulation in basic fibroblast growth factor (FGF-2) transgenic mice. *Mol Biol Cell* **6**(12), 1861–1873 (1995).

213. J. A. Greenwald, B. J. Mehrara, J. A. Spector, S. M. Warren, P. J. Fagenholz, L. E. Smith, *et al.*, *In vivo* modulation of FGF biological activity alters cranial suture fate. *Am J Pathol* **158**(2), 441–452 (2001).

214. M. B. Carlton, W. H. Colledge, and M. J. Evans, Crouzon-like craniofacial dysmorphology in the mouse is caused by an insertional mutation at the Fgf3/Fgf4 locus. *Dev Dyn* **212**(2), 242–249 (1998).

215. V. el Ghouzzi, M. Le Merrer, F. Perrin-Schmitt, E. Lajeunie, P. Benit, D. Renier, *et al.*, Mutations of the TWIST gene in the Saethre–Chotzen syndrome. *Nat Genet* **15**(1), 42–46 (1999).

216. T. D. Howard, W. A. Paznekas, E. D. Green, L. C. Chiang, N. Ma, R. I. Ortiz de Luna, *et al.*, Mutations in TWIST, a basic helix–loop–helix transcription factor, in Saethre–Chotzen syndrome. *Nat Genet* **15**(1), 36–41 (1997).

217. P. Bialek, B. Kern, X. Yang, M. Schrock, D. Sosic, N. Hong, *et al.*, A twist code determines the onset of osteoblast differentiation. *Dev Cell* **6**(3), 423–435 (2004).

CHAPTER 9

Mouse Genetics as a Tool to Study Bone Development and Physiology

CLIFFORD J. ROSEN AND WESLEY G. BEAMER

I. INTRODUCTION: HISTORICAL PERSPECTIVE AND SIGNIFICANCE

Low bone mineral density (BMD) has become the most established and identifiable risk factor for osteoporotic fractures. The proliferation of newer tools for measuring bone mass has resulted in widespread testing and has also led to the realization that BMD is a complex trait normally distributed across various populations. In addition, the data produced by these tools also provided the first clues that a syndrome once characterized as an age-related disorder associated with back pain and fractures is, in fact, a heritable disease. BMD studies of mother–daughter pairs, twins, and large sib cohorts estimated the heritability of this trait to be between 50% and 70% [1]. This finding led most investigators somewhat hastily to conclude that the genetic influences of BMD were "oligogenic"; that is, the phenotypic variation in BMD was caused by the actions of a limited number of genes with discrete effects. Fueled by this concept, the past 10 years have been characterized by a flood of candidate-gene association studies in both small and large unrelated cohorts [1]. Although data from these papers were conflicted and failed to yield major genes that defined osteoporotic risk, such studies, combined with genome-wide scanning of multigenerational families, served to reinforce the complex and polygenic nature of the genetic influence of bone acquisition. Despite the initial momentum, we are now left with the daunting task of defining the role of multiple genes that individually, or in concert, moderate the acquisition and maintenance of peak bone mass.

In addition to the complex multifactorial nature of genetic influences, three other factors have emerged that further complicate our search for osteoporosis genes. First, it is now clear that individual genetic determinants of BMD are strongly influenced by other genes that do not have direct effects on BMD (i.e., epistasis) [1, 2]. Second, there are numerous environmental factors that may modulate expression of one or more genes [1]. Sorting these genetic and environmental interactions will require complex modeling that must control for nutritional, hormonal, mechanical, and lifestyle factors. Third, some investigators have failed to recognize that the BMD phenotype is only a surrogate for fracture. Defining "bone density genes" provides limited information with respect to prediction of future fractures, in part because there are other contributing factors that affect the eventual status of any given bone. The emergence of measurement tools with superb resolution of bone microstructure has further heightened this awareness. Such technology has resulted in a better understanding of what bone "quality" represents and how this is related to skeletal frailty. Finally, and probably most important, there is a growing realization that BMD represents the sum of several temporally related or stochastic processes beginning with skeletal development and including modeling, remodeling, and consolidation. Deconvolution of bone mass into intermediate phenotypes, such as BMD, cross-sectional area, shape, or a biochemical marker such as insulin-like growth factor (IGF)-I, is likely to yield more mechanistic insights not only into the overall processes of peak acquisition but also into the determinants of skeletal strength. Most certainly, finding genes that predict the risk of fractures in humans is going to require novel strategies and will remain a challenging endeavor for the foreseeable future.

Enter the mouse. Rodent models for testing hypotheses related to skeletal disorders are not new. The ovariectomized rat is a well-established tool suitable for testing new therapies for osteoporosis, as well as for understanding how estrogen deprivation affects the bone remodeling unit. Adding to data from the rat are new models of laboratory mice that carry specific gene deletions (knock-outs), gene additions (transgenic), or spontaneous mutations. These mice are currently on the frontier of basic research, specifically to test how a known gene may regulate diverse skeletal actions. For example, targeted overexpression of IGF-I in transgenic mice using the osteocalcin promoter is characterized by a marked increase in both cortical and trabecular bone density at 6 weeks of age [3]. Similarly, knock-out of the IGF-I receptor in bone leads to a skeletal phenotype characterized by impaired mineralization [4]. Also, mice globally lacking expression of the *Cbfal* gene (i.e., null mutation) are characterized by the absence of osteoblast differentiation, failure to mineralize bone, and lethality at birth [5]. Finally, spontaneous mutants such as the osteopetrotic mouse (*op/op*), which lacks a functional *csf* gene and its product, fail to exhibit differentiated osteoclasts that are required for normal bone resorption [6]. There are thousands of examples of these types of gene mutations, induced or spontaneous, which have already helped to elucidate the potential role of single gene action in bone biology. However, these models represent not only the phenotypes related to gene deletion but also the compensatory mechanisms inherent in the mouse, which are employed to promote survival. Delineating those responses, and separating them from the direct result of a gene deletion, can be extraordinarily difficult.

An entirely different approach utilizes the power of the mouse as a genetic tool to uncover genes whose normal allelic variation regulates BMD. During the past 30 years, inbred strains of mice have helped identify genetic determinants of various disease states with both single and polygenic bases [7]. Although investigators in the bone field have been late in recognizing these models, several factors have hastened their utilization. First, technology was developed to measure BMD accurately, easily, and relatively inexpensively in mice. Use of peripheral quantitative computed tomography (pQCT), peripheral dual x-ray absorptiometry (DXA), and full-body DXA, by both *ex vivo* and *in vivo* methods, now allows investigators to measure BMD and appreciate large differences among knock-outs, transgenics, mutants, and healthy inbred strains [8]. Micro-CT has, for the first time, provided an opportunity to define three-dimensional microstructural aspects of bone, and in conjunction with newer methods of measuring bone strength has opened the door for identifying determinants of bone quality [9]. Finally, the power of breeding strategies to isolate quantitative trait loci (QTL), and to test their effects either singly or in combination with other genetic determinants, has permitted hypothesis testing for individual or clusters of genetic loci [10].

There are several confounding factors that have plagued human genetics studies in the past two decades, making gene identification exceedingly difficult. These include the complex nature of the phenotype regulated by numerous genes, significant gene–gene and environment–gene interaction, and the multifactorial nature of bone quality [1, 11]. Mouse studies have made these much more amenable to resolution. Furthermore, the homology between human and mouse genomes, as well as the intense efforts to map every gene in both species, provides more impetus to use this animal as a tool for defining the heritable determinants of osteoporotic risk. In this article, we describe the role of several mouse model systems for determining the polygenic basis of osteoporosis. In addition, we define their relevance for subsequent human studies. We do not examine mutant, transgenic, or knock-out models, in part because we want to de-emphasize the role of single genes in producing extreme pathology or in defining complex traits. Rather, this review focuses on normal allelic variation in inbred strains of mice, animal models more directly applicable to understanding the BMD trait, and hence osteoporosis, in humans.

II. INTRODUCTION TO MOUSE SKELETAL PHYSIOLOGY

The inbred strains of mouse commonly used in various laboratories, although exhibiting allelic differences in many genes, share some common skeletal characteristics. For example, 95% of the skeleton is cortical bone. The trabecular components are predominantly found in the vertebrae and distal metaphysis of the tibia and femur. Mice, like rats, have an open epiphysis even in old age; hence, there is continued, albeit slow, growth beyond puberty. Weaning generally occurs at 3 weeks of age, and rapid linear growth occurs in most inbred strains between 4 and 8 weeks, a time that parallels the peak of circulating IGF-I. Males tend to be larger in body size than females and thus have greater cortical and trabecular mass than females. Peak cortical bone acquisition occurs by approximately 16 weeks of age, whereas trabecular bone reaches its maximum size between 6 and 10 weeks. Cortical bone is relatively constant across the life span of most inbred strains; in contrast, both femoral and vertebral trabecular bone begins to be lost soon after peak acquisition is attained (M. Bouxsein, personal communication). There is constant modeling occurring in mice even with advanced age, but Haversian remodeling is not present. In part, this is because there is spatial remodeling from periosteum to endosteum throughout murine life. On the other hand, there is trabecular bone turnover, which can be affected by local and systemic hormones. Ovariectomy or orchiectomy, in virtually all inbred strains, results in significant trabecular and cortical bone loss, whether it is performed early in life or in retired breeders. On the other hand, glucocorticoid administration does not perturb the mouse skeleton as much as what has been seen in rats or humans. Overall, the mouse is an excellent model for skeletal phenotyping, particularly because of the ease of imaging animals and their maintenance on a fixed genetic background. Combining these characteristics with the ever-expanding databases for genomic haplotypes provides a sound rationale for using the mouse in skeletal genetic studies.

III. INBRED STRAINS OF MICE

There are many types of mice available for genetic and biologic studies. In general, mice have become the workhorses of biomedical research because of their ease of breeding and reproductive capacity, their relatively short life span, and the availability of large numbers of genetic markers in the mouse genome. Probably more important, however, and unlike the rat, dozens of inbred mouse strains have been available since the early decades of the 20th century. These inbred strains were developed by repeated matings between siblings for at least 20 consecutive generations [7]. This resulted in nearly 100% homozygosity at all alleles across the mouse genome. By continuing the process until the 60th generation, inbred mice eventually became 100% homozygous at all loci (except for any spontaneous mutations that could arise), thereby providing researchers with a plethora of genomically identical mice. More than 70 pure inbred strains are currently available at the Jackson Laboratory alone, including C57BL/6J (B6), the standard strain for many laboratories performing genetic as well as biologic analyses.

The second feature of inbred strains that makes them powerful genetic and physiologic tools is that an individual inbred strain differs from all other strains through alleles in a number of genes. Each strain has its own set of phenotypic characteristics that make it unique and allow innumerable differences in physiologic behavior. One such difference is in the wide variation in BMD among inbred strains. Thus, by choosing two inbred strains that differ in a trait of interest, a cross can be made to enumerate, locate, and define heritability of the genes that contribute to that trait.

Crossing two inbred strains of mice results in hybrid F1 mice that are genetically identical with each other and heterozygous at all loci. As illustrated in Figure 9-1, intercrossing F1 mice results in F2 progeny in which

genetic alleles for BMD have randomly resorted into new combinations, such that at any given locus an F2 mouse will be homozygous for either progenitor strain alleles (i.e., *b6/b6* or *c3/c3*) or heterozygous (i.e., *b6/c3*). BMD regulatory loci that are not genetically linked to each other will independently segregate in these F2. The net effect of alleles at all BMD regulatory loci yields the BMD for each mouse,

Since there are now more than 8000 genotypic markers that are variably polymorphic across inbred strains, investigators can identify QTL by genotyping and phenotyping the F2 progeny [12]. Analysis of the F2 progeny in the extremes of the phenotypic distribution allows rapid identification of major effect loci, whereas analysis of all F2 progeny yields major and minor effect loci, as well as the opportunity to assess trait variance accounted for by each locus and gene–gene interactions. Genetic linkage is established by testing for association of progenitor alleles with high or low expression of a particular phenotype using various computer software programs [12, 13]. The initial QTL may reside in chromosomal regions up to 40 cM (recombination distances between specific markers), areas of the chromosome with hundreds of potential candidate genes. Fine mapping and congenic

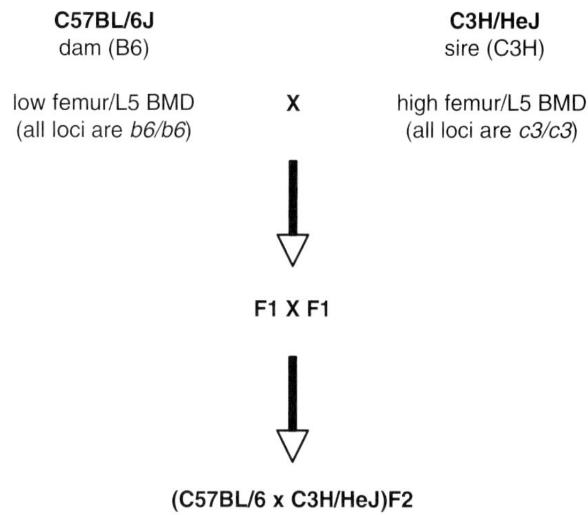

FIGURE 9-1   The use of two inbred mouse strains for analysis of a genetically regulated phenotypic trait. The C57BL/6J and C3H/HeJ strains characterized by low and high volumetric bone mineral density (BMD) are depicted. F2 progeny analyses are equally effective for mapping single gene traits and polygenic traits that are either quantitative or qualitative in nature.

construction define narrower regions of the chromosome and allow for positional cloning and gene sequencing to be undertaken. This type of QTL strategy has been successful in identifying genes associated with obesity, atherosclerosis, epilepsy, and cancer susceptibility in the mouse.

Beamer *et al.* [8] first described large differences in femoral and vertebral BMD, as measured by pQCT, among various inbred strains of mice. Subsequently, other investigators reported similar findings using planar radiography, whole body BMD by DXA, and site-specific regions of interest (i.e., vertebrae, femur, and tibia) by peripheral DXA technology [14–16]. Micro-CT has been utilized to detect differences in trabecular bone structural parameters in vertebrae, femorae, and tibiae among inbred strains [9]. For each pair of strains differing in a bone phenotype, the strategy has been to map QTLs by crossing the progenitor strains and then intercrossing their F1 hybrids to produce large numbers of F2 progeny (range, 250–1000 males and females). Data from several sets of such F2 progeny have been reported from the BMD phenotype and include C3H/HeJ (C3H:) versus *C57BU6J* (B6: low BMD), Castaneus/EiJ (CAST: high BMD) versus B6 (low BMD), SAMR1 (high BMD) versus SAMP6 (low BMD), AKR/J (high) versus SAMP6 (low BMD), and B6 (high BMD) versus DBA/2J (D2 low BMD) [14–18].

As noted previously, bone mass is a complex phenotype that includes mineral content, size (length, width, and cross-sectional area), trabecular connectivity, and shape. Bone strength is determined by these dimensions as well as other variables affecting overall bone quality. Depending on the exact measurement, and the instrument used to define it, a bone density phenotype varies within a given strain as well as between inbred strains. For example, B6 is a low bone density strain compared to C3H when defined by volumetric measurements of the femur such as pQCT, but B6 is a high bone density strain in comparison to D2 when whole body BMD using DXA technology is the bone density phenotype of choice. In part, this can be related to the shape as well as the size of bone and its individual components. The B6 femurs have thinner cortices, a more elliptical shape, and lower volumetric BMD than those of C3H animals. However, in comparison to D2 mice, the periosteal circumference of B6 is greater; hence, areal measurement of the femur by DXA can actually show relatively greater apparent BMD for this strain. Since size, shape, and mineral are all critical components of strength, the phenotype under study becomes critical not only for assigning QTLs but also for attempting to understand the biomechanical mechanisms that ultimately define both bone morphology and strength.

Mapping "bone density genes" can be extremely productive for F2 mice because there is independent segregation of unlinked genes for this polygenic phenotype. Surprisingly, progenitor differences are not mandatory before performing genetic analyses of polygenic traits because the F2 population, with its independent assortment of loci, will reveal genetic determinants whose actions might not be evident in the progenitor inbred strains or may be hidden by actions of epistatic loci [19]. Moreover, QTL in the F2s can display alleles with actions that appear contrary to expectations based on the progenitor strain's phenotype. For example, the high BMD C3H inbred strain has been found in one QTL analysis to be carrying two genes yielding a "low density" in F2 progeny [18]. Similarly, Benes *et al.* [14] found that two AKR Qtls (i.e., chromosomes 7 and 11) were associated with low areal BMD when these alleles were homozygous in F2 mice from an AKR (high density) × SAMP6 (low density) cross. Thus, F2 mice provide an invaluable tool for locating and enumerating QTL, as well as delineating allelic effects. Moreover, the phenomena of transgenesis (gene recombinations in the F2s resulting in phenotypic values that are greater or lesser than the progenitor strain phenotype) can offer critical insight into the complex genetic influence on a specific phenotype [14].

Another exciting aspect of gene mapping in mice is the evolving story surrounding trabecular structure and strength. Newer technology with micro-CT has provided investigators with the opportunity to study trabecular spacing, number, connectivity, and three-dimensional structure in the vertebrae and femorae of mice. These determinants are likely to produce even more phenotypes for QTL analyses, especially with respect to understanding "bone quality." Work by Turner *et al.* [9] suggests that vertebral and femoral neck-breaking strength in one inbred strain can differ considerably from that at the mid-diaphysis of the femur in the same strain. For example, C3H femurs have a high BMD and thick cortex and therefore are stronger than B6 using three-point bending in the mid-diaphysis. However, vertebral strength by compression testing is reduced in C3H compared to B6. Micro-CT analysis has revealed a markedly reduced number of trabeculae in C3H vertebrae [9]. These data are consistent with the fact that although C3H has high apparent BMD (even in the vertebrae) by projectional methodology, trabecular BMD, when quantitated properly, is actually reduced. This suggests that there are distinct genetic determinants within a given strain that define cortical versus trabecular BMD. Moreover, these findings support long-standing clinical observations that different bone components acquire peak bone mass at different times, and that there can be very disparate BMD values between the spine and hip in an individual. Furthermore, response to both antiresorptive and anabolic treatment clearly differs from one skeletal site to another. Hence, there must be a multitude of osteoporosis genes, some of which regulate trabecular and cortical components, some of which must determine distinct skeletal compartments, and some of which must be temporally activated or suppressed during the acquisition and consolidation phase of acquiring bone mass.

Finally, it should be noted that the search for mouse "bone density" genes permits a more in-depth analysis of the biology related to acquisition and maintenance of BMD. For example, C3H mice have high cortical bone density, in part because their bone marrow stromal cells proliferate more rapidly, their osteoblasts are programmed to synthesize more IGF-I and alkaline phosphatase than B6 bone cells, and their rate of bone formation is much higher than that of B6 mice. By histomorphometry, C3H bone shows high rates of bone formation and reduced bone resorption rates compared to that of B6 mice [20, 21]. These data, combined with biochemical markers of bone turnover reflecting similar alterations in bone turnover, demonstrate that the genetic determinants of bone density affect the basic cellular process of bone. Intermediate skeletal phenotypes, such as IGF-I, N-telopeptides, or alkaline phosphatase, can also be examined in F2 mice for QTL analysis, thereby providing additional approaches to understanding the cellular mechanisms responsible for acquisition of peak bone mass. One study illustrates the power of QTL among inbred strains. Klein *et al.* [22] identified a very strong QTL for areal BMD on mouse chromosome 11 in an inbred cross between B6 and DBA. A congenic strain developed from this QTL showed marked changes in areal BMD and bone strength. Expression studies identified *Alox15* as a potential candidate gene because it was markedly up-regulated in comparison to the progenitor. Knock-out of the *Alox15* gene resulted in a high bone mass phenotype, which is thought to result from changes in the PPAR-gamma pathway. Hence, gene identification and discovery of new skeletal pathways can be a direct result of QTL analyses using inbred strains.

IV. RECOMBINANT INBRED STRAINS

Recombinant inbred (RI) strains are generated by outcrossing two progenitor strains and then intercrossing these F1 hybrids to produce F2 progeny, as illustrated in Figure 9-2. Then pairs of sister–brother F2 mice are selected at random to serve as founders for each RI strain. These F2 mice are mated to produce F3-generation mice, and this process of sibling mating within each RI line is repeated for 20 generations. The result is new

inbred strains nearly 100% homozygous at all loci, each of which has a different combination of genes from the same original progenitor strains [7]. The constitution can be maintained indefinitely by continual brother–sister matings. Since the original founders were selected at random, many distinct RI strains can be regenerated from the original progenitors (more than 12 such sets of RI strains exist at the Jackson Laboratory). These RI strain sets are named by the capital letter of each strain separated by "X" (e.g., B6 and D2 gave rise to BXD) plus a number for the particular strain (e.g., BXD-16) [7]. Unlike classic inbred strains, the genotype in an RI strain is somewhat limited by the fact that there are only two possible alleles (e.g., B6 or D2) at each locus. More important, there is limited recombination because

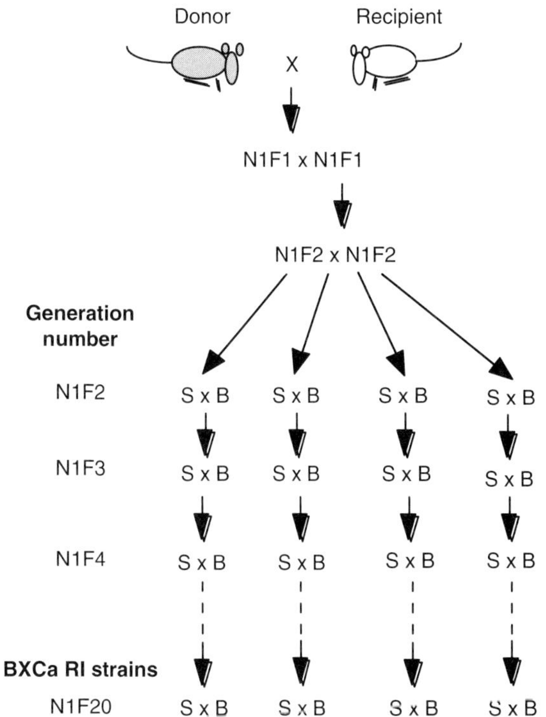

FIGURE 9-2 The method for developing a set of recombinant inbred (RI) strains. The C57BL6J and CAST1EiJ inbred strains are presented as the progenitors. Randomly selected sister (S) and brother (B) sibling N1F2 pairs are mated to produce N1F3 progeny, which in turn are mated to produce N1F4 progeny, etc. At the N1F20 generation, the mice within a strain will be inbred, with better than 99% of all loci homozygous for either B6 or CAST alleles. Each RI strain will have inherited on average one-half of its genes for each of the progenitor strains.

homozygosity sets in relatively quickly during the intercrosses. In addition, because both D2 and B6 inbred strains were originally derived from the same founder animals more than 70 years ago, a significant number of loci are not polymorphic. However, the strength of the RI strategy is that hundreds of genomic markers for each strain are identified, and publicly available databases can be utilized for rapid linkage studies and for determining map distances, with as few as one or two mice phenotyped from each RI strain. To establish genetic linkage, the investigator simply phenotypes each of the strains in the RI set (e.g., 12 BXD strains) to obtain a strain distribution pattern for the phenotype. The strain distribution pattern for the new phenotype is compared with strain distribution patterns of known polymorphic loci previously established in that RI set. When congruence between strain distribution patterns is found, linkage of the phenotype to a specific chromosomal region is established. By using known recombination sites in the mouse RI strains, one can mate RIs to parental strains (e.g., BXD-8 to DBA/2 and B6) to obtain new F1 progeny. Then, by linkage between the genotype and phenotype, the QTL can be placed either above or below the recombination break point. This approach is called the recombinant inbred segregation test strategy (RIST) and has been successfully used by one group to help refine QTL locations for bone density genes [15, 23]. Using published databases containing more than 1500 genetic markers, Klein *et al.* [24] identified QTLs on chromosomes 2, 7, and 11 (Table 9-1) with major effects. Interestingly, some QTLs were present only in males, whereas others were present only in females, and three QTL (chromosomes 1, 18, and 19) appeared to be gender independent [25]. Moreover, several of these QTLs were found to be similar to those identified independently by other groups using different strategies (i.e., F2 matings). Finally, from this same group, Klein *et al.* [24] reported that RIST allowed them to narrow the QTLs on chromosomes 2 and 11 by more than 10 eM. Hence, RI strains have provided a tool for rapidly establishing linkage of whole body areal BMD QTL, as well as for further resolution of large QTL regions into smaller segments.

V. CONGENIC STRAINS

Congenic strains are generated to test the effect of individual or multiple linked loci from a donor strain placed on the genetic background of a recipient strain [26]. As illustrated in Figure 9-3, the strategy is based on repetitive backcrossing to the recipient strain while genotyping for the donor strain's alleles in the subsequent backcross generations. Procedurally, the donor and recipient inbred strains are mated; the F1 progeny are then backcrossed to the recipient successively for 10 generations (NIO). With each generation, the homozygosity of the recipient background increases from 50% at NIFI to 99% at NIOFI. The residual heterozygosity resides at the region of interest carrying the donor alleles. Hence, the genetic background of the chromosomal region of interest has been switched from donor to recipient. Congenic strains are particularly useful for confirming the QTL existence, fine mapping of the QTL, and testing the quantitative effect of individual QTLs. For example, if there is a strong BMD QTL on chromosome 1 from C3H mice, as found by F2 analysis, congenic strain mice in which the chromosome 1 QTL is now placed on a B6 background allow the investigator to test the effect of this single genetic locus on a low bone density background. The congenic strain would be named B6.C3H-I (recipient, donor, and chromosome). Multiple QTLs can be combined to test for gene–gene interactions. Congenic strain construction takes approximately 18 months and may be labor-intensive, but it provides an essential means of evaluating the biology regulated by the QTL, as well as refining the map position of individual loci. However, there are two caveats to this approach: (1) "passenger" loci adjacent to the QTL of interest travel with the donor QTL and may affect the phenotype and mapping precision; (2) the QTL phenotype may be the net result of linked genes with different effects within the QTL region. In each case, fine mapping and generation of nested congenics can overcome these problems [7].

Congenics have become extremely valuable tools for bone biologists, not only for studying the quantitative effect of individual QTLs but also for more completely understanding the phenotype and its underlying physiology. Thus, one moves from breeding strategies and QTL analysis to individual congenic construction in order to define the locus of interest, as well as to test precisely how that locus could affect the phenotype. Two groups have reported generations of congenics using their most promising QTL for either whole body BMD or femoral BMD. These loci include chromosomes 1, 2, 4, 5, 6, 11, 13, 14, and 18 (Table 9-1) [27]. In addition, one group has developed congenic strains for the serum IGF-I phenotype on chromosomes 6 and 15. The power of this strategy is illustrated by several studies [21, 22].

Klein and colleagues [24] constructed congenic mice for the chromosome 2 areal BMD QTL and reported that after the fourth generation, D2.B6-2 homozygous D2 background had a difference of nearly one standard deviation in whole body BMD compared to inbred D2 mice. Moreover, this effect occurred only in female mice. Beamer *et al.* [27] reported that the chromosome 1 QTL from C3H mice had an approxi-

TABLE 9-1   Summary of QTL for BMD Found in F2 and Recombinant Inbred (RI) Progenies in Female Mice from 11 Reports[a]

	Chromosome	Cross-reference	Physical (Mb)	Method and bone	Human homology
Chr 1	BXD RI	65.09	DEXA total body	2q33–q35	Orwoll *et al.* [25]
	B6xC3H	73.16	μCT L5	2q33–q35	Bouxsein *et al.* [28]
	B6x129	110.72	DEXA total body	18q21	Ishimori *et al.* [34]
	B6xC3H	158.61	pQCT femur, L5	1q24–q25	Beamer *et al.* [18]
	B6xC3H	158.61	μCT L5 BV/TV	1q24–q25	Bouxsein *et al.* [28]
	B6xCAST	170.19	pQCT femur	1q21–q24	Beamer *et al.* [17]
	B6x129	175.51	DEXA vertebrae	1q22	Ishimori *et al.* [34]
	B6xDBA/2	186.43	DEXA total body	1q41–q42	Klein *et al.* [24]
	MRLxSJL	186.43	pQCT femur, DEXA total body	1q41–q42	Li *et al.* [35]
	MRLxSJL	190.95	pQCT femur, DEXA total body	1q32	Masinde *et al.* [36]
Chr 2	SAMP6xAKR/J	18.51	DEXA spine	10p15–p11	Benes *et al.* [14]
	BXD RI	71.51	DEXA total body	2q33–q36	Orwoll *et al.* [25]
	B6xDBA/2	79.97	DEXA total body	2q31–q2	Klein *et al.* [24]
	BXD RI	112.43	DEXA total body	18q21; 2q13–31	Orwoll *et al.* [25]
	MRLxSJL	117.80	pQCT total body	11p14/15q15	Masinde *et al.* [36]
	MRLxSJL	162.04	pQCT total body	20q11–q13	Masinde *et al.* [36]
	B6xC3H	168.62	pQCT femur	20q11–q13	Beamer *et al.* [18]
Chr 3	NZBxRF	14.63	pQCT femur	8q13–q22	Wergedal *et al.* [37]
Chr 4	MRLxSJL	45.67	pQCT total body	9q21–q34	Masinde *et al.* [36]
	B6xC3H	100.24	∝CT L5 BV/TV	1p32–p31	Bouxsein *et al.* [28]
	B6xC3H	123.81	pQCT femur, L5	1p34–p33	Beamer *et al.* [18]
	MRLxSJL	132.69	pQCT femur	1p36–p34	Masinde *et al.* [36]
	B6xDBA/2	141.16	DEXA total body	1p36	Klein *et al.* [24]
Chr 5	B6xCAST	77.52	pQCT femur	4q11–q13	Beamer *et al.* [17]
	B6xDBA/2	87.25	∝CT tibial trab BV/TV	4q12–q13	Bower *et al.* [33]
Chr 6	B6x129	75.47	DEXA total body, spine	2p11–p13	Ishimori *et al.* [34]
	B6xC3H	116.12	pQCT femur	3p25–p24	Beamer *et al.* [18]
Chr 7	SAMP6xAKR/J	30.83	DEXA spine	19q12–q13	Benes *et al.* [14]
	BXD RI	27.3–109.5	DEXA total body	19q/11p/15q	Orwoll *et al.* [25]
	NZBxRF	65 cM	pQCT femur	15q11–q13	Wergedal *et al.* [37]
	B6x129	90.68	DEXA vertebrae	11q13–q14	Ishimori *et al.* [34]
	B6xC3H	133.87	pQCT L5	10q26	Beamer *et al.* [18]
Chr 8	B6xC3H	85.25	∝CT L5 BV/TV	19p13	Bouxsein *et al.* [28]
	BXD RI	112.46	DEXA total body	16q22–q23	Orwoll *et al.* [25]
Chr 9	MRLxSJL	32.25	pQCT total body	11q23–q24	Masinde *et al.* [36]
	B6xC3H	33.81	∝CT L5 BV/TV	11q23–q24	Bouxsein *et al.* [28]
	B6xC3H	65.97	∝CT L5 BV/TV	15q21–q23	Bouxsein *et al.* [28]
	MRLxSJL	75.90	pQCT femur	6p12 /15q21–q22	Masinde *et al.* [36]
	B6xC3H	85.69	pQCT L5	6q12–q15	Beamer *et al.* [18]
Chr 10	B6xC3H	113.78	∝CT L5 BV/TV	12q14–q15	Bouxsein *et al.* [28]
	NZBxRF	117.09	pQCT femur	12q14–q15	Wergedal *et al.* [37]
	B6x129	121.60	DEXA total body	12q14–q15	Ishimori *et al.* [34]

(Continued)

TABLE 9-1   Summary of QTL for BMD Found in F2 and Recombinant Inbred (RI) Progenies in Female Mice from 11 Reports[a]—Cont'd

	Chromosome	Cross-reference	Physical (Mb)	Method and bone	Human homology
Chr 11	SAMP6xAKR/J	89.79	DEXA	17q21–q24	Benes et al. [14]
	B6xDBA/2	55.63	DEXA total body	5q31–q35	Klein et al. [24]
	B6xC3H	63.24	pQCT femur, L5	17p13–p11	Beamer et al. [18]
	NZBxRF	70.05	pQCT femur	17p13–p12	Wergedal et al. [37]
	MRLxSJL	83.65	pQCT total body	17q11–q23	Masinde et al. [36]
	SAMP6xSAMP2	88.79	Radiographic CTI	17q21–q23	Shimizu et al. [16]
	BXD RI	119.08	DEXA total body	17q24–q25	Orwoll et al. [25]
Chr 12	B6xC3H	7.63	pQCT femur, L5	2p24–p23	Beamer et al. [18]
	B6xC3H	7.63	μQCT L5 BV/TV	2p24–p23	Bouxsein et al. [28]
	B6xC3H	71.00	μCT L5 BV/TV	14q23–q24	Bouxsein et al. [28]
	B6xC3H	71.00	μCT L5 BV/TV	14q23–q24	Bouxsein et al. [28]
	MRLxSJL	80.37	pQCT femur	14q21–q24	Masinde et al. [36], Li et al. [35]
	B6xC3H	110.41	μCT L5 BV/TV	14q32	Bouxsein et al. [28]
Chr 13	SAMP6xSAMP2	12.49	Radiographic CTI	6p23–p21/7p15–p13	Shimizu et al. [16]
	B6xCAST	20.30	pQCT femur	6p23–p21/7p15–p13	Beamer et al. [17]
	B6xC3H	45.07	μCT L5 BV/TV	6p24–p23	Bouxsein et al. [28]
	B6xC3H	56.49	pQCT femur	5q23–q35	Beamer et al. [18]
Chr 14	MRLxSJL	9.29	pQCT femur	Unknown	Masinde et al. [36]
	B6xC3H	66.69	pQCT femur, L5	8p21–p11	Beamer et al. [18]
	B6xC3H	66.69	μCT L5 BV/TV	8p21–p11	Bouxsein et al. [28]
	MRLxSJL	92.71	pQCT total body	13q14–q21	Masinde et al. [36]
Chr 15	MRLxSJL	13.43	pQCT total body	8q22–q23	Masinde et al. [36]
	B6xCAST	74.07	pQCT femur	8q24	Beamer et al. [17]
Chr 16	B6xC3H	39.03	pQCT femur	3q13–q29	Beamer et al. [18]
	SAMP6xAKR/J	42.16	DEXA spine	3q12–q13	Benes et al. [14]
	BXD RI	45.05	DEXA total body	3q12–13	Orwoll et al. [25]
Chr 17	B6xC3H	3.88	μCT L5 BV/TV	6q25–q27	Bouxsein et al. [28]
	MRLxSJL	31.57	pQCT femur	19p13	Masinde et al. [36]
	MRLxSJL	45.20	pQCT femur	6p21–p12	Li et al. [35]
Chr 18	B6xC3H	44.66	pQCT femur, L5	5q22–q23	Beamer et al. [18]
	MRLxSJL	69.15	pQCT femur	18p11–q21	Masinde et al. [36]
	NZBxRF	80.36	pQCT femur	18q12–q23	Wergedal et al. [37]
Chr 19	MRLxSJL	45.20	pQCT femur	10q11–q23	Masinde et al. [36], Li et al. [35]

[a]Map position in mouse genome given in mega base pairs (Mbp); human homologous regions drawn from ±5 Mbp of published best position.

CTI, cortical thickness index; L5, lumbar 5 vertebra.

mately 8% increase in femoral BMD when placed on a B6 background in N6 congenics. Similar findings were also noted for the chromosome 4 QTL. On the other hand, a "low bone density QTL" from C3H in chromosome 6 was associated with a 3.5% reduction in femoral vBMD and 20% lower BV/TV when donated to a B6 background. Not unlike the studies by Orwoll et al. [25], this effect was noted only in female mice.

Hence, congenic strains can provide insight into the effect of individual QTLs with respect to bone and to the mechanism of such action.

The power of the congenics extends beyond simple tests for QTL effects. For example, Beamer et al. [27] reported that although the congenic on chromosome 1 had a statistically significant effect on femur BMD, micro-CT analysis revealed a much greater

Mating system to develop
a congenic strain

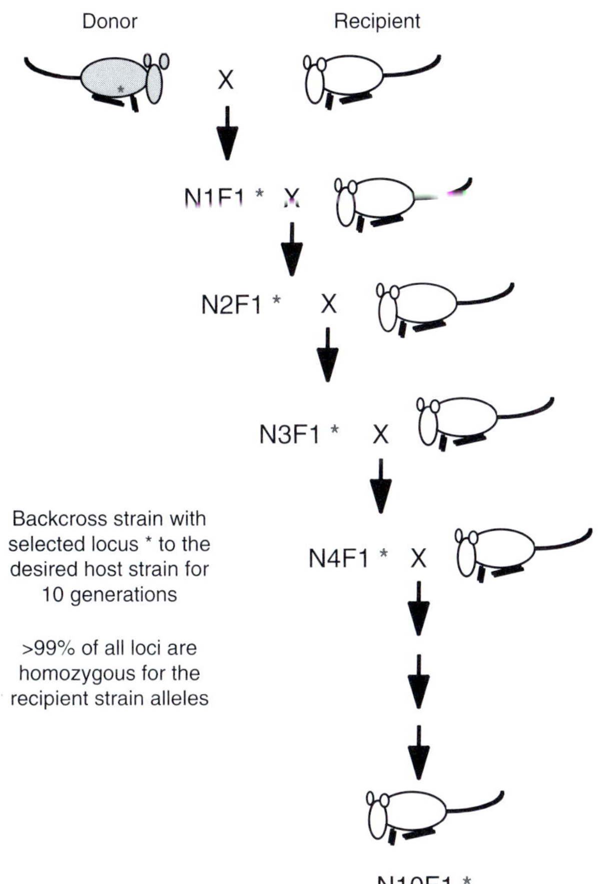

FIGURE 9-3   The method for producing a congenic strain that carries a segment of a chromosome transferred from a donor strain to a recipient strain. The transfer is accomplished by at least 10 cycles of backcrossing and F1 to a recipient strain. The region containing the gene or QTL of interest is found by genotyping each generation of progeny and mating the identified carrier of the donor segment to the recipient strain.

effect on vertebral trabecular bone. In fact, vertebral bone density (measured as BV/TV) was nearly 35% greater in the chromosome 1 congenic compared to progenitor B6 mice at 16 weeks. These findings, also noted for the chromosome 4 QTL congenic, support the thesis that bone microstructure may be altered dramatically while BMD may change only modestly. If confirmed, these data provide more impetus for defining aspects of bone quality and their relationship to skeletal response to long-term antiresorptive therapy. Rosen *et al.* [21] noted that one of the strongest QTLs for serum IGF-I in B6C3F2 mice is likely to be the chromosome 6 QTL noted for BMD. This region encompasses the PPAR-gamma gene and is consistent with findings by Klein *et al.* [22] demonstrating that

Alox15 generates PGJ2, an important ligand for the nuclear receptor, PPAR-gamma. Moreover, the congenic B6.C3H-6 mice at the 10th generation reported by Rosen *et al.* [21] and Bouxsein *et al.* [28] showed not only reduced femoral and vertebral BMD but also markedly lower serum IGF-I concentrations and evidence of increased PPAR-gamma activation compared to B6. Thus, the congenic model provides further proof that these strains can be used not only to map candidate genes but also to provide insight into the mechanisms whereby peak bone mass is acquired.

VI. RECOMBINANT CONGENIC STRAINS

A fourth system available for genetic and biological studies of polygenic traits, such as BMD, is illustrated in Figure 9-4. Recombinant congenic (RC) strains represent a combination of the attributes found in RI strains and congenic strains of mice. As can be discerned from Figure 9-4, two backcrosses are made to a recipient strain to achieve progeny that carry 12.5% of genes from the donor strain. Sibling progeny from the N3F1 cross are then incrossed to inbred status as shown. The intent of this system is to isolate small

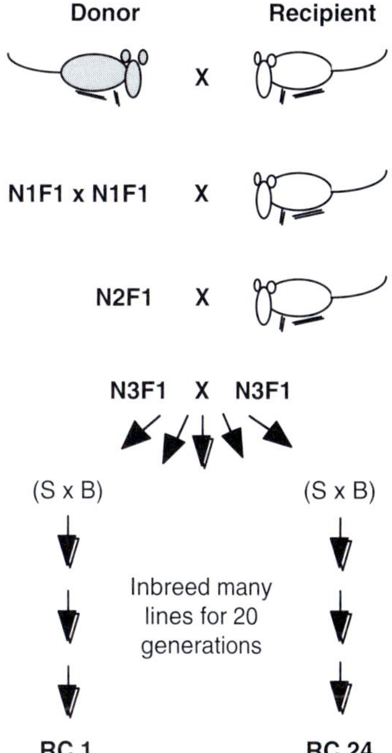

FIGURE 9-4   The method for development of a set of recombinant congenic strains, each of which carries 12.5% of its genes from the donor strain and 87.5% of the genes from the recipient strain. The goal is to capture a subset of genes that regulate a complex trait in a series of related but distinct inbred strains. This method is particularly useful for analyses of phenotypes that depend on modifier genes for expression.

subsets of genes that regulate a complex trait within distinct inbred strains. Demant and colleagues successfully used recombinant congenic (RC) strains to genetically analyze both colon and lung cancer in mice [29, 30]. In addition, preliminary studies [31] reported that genetic linkage to regions on 11 different chromosomes could be demonstrated for bone strength, ash percentage, and morphological parameters using the set of 27 HcB/Dem RC strains. These RC strains were derived from donor C57BL/IOSc-SnA and recipient C3H/DiSnA strains. Thus, RC strains are very suitable for analyses of complex traits and may be most valuable for assessment of genes that interact in subtle ways not easily identified by standard statistical means.

VII. SUMMARY

This review has discussed some of the models that investigators have used to define the genetic determinants of bone density in mice. However, it is quite obvious that despite major efforts by several groups, no mouse bone density gene has yet been cloned. Still, inbred, recombinant inbred, and recombinant congenic strains as model systems offer a wealth of information related to acquisition and maintenance of peak bone mass. With the advent of more rapid genotyping and congenic generation techniques, it seems certain that these putative QTLs will yield numerous genes that contribute to the variance in bone density within a mouse strain. Moving from mouse genes to human genes using published maps, in addition to data from ongoing genome sequencing projects, may actually turn out to be easier than once thought [32]. More of a challenge, however, will be to understand the full effects of a particular gene on bone cell function, the interactions with environmental factors, and perhaps even more important, the regulators of such genes. Notwithstanding those tasks, it has become clear that the power of the mouse for bone biologists lies in several relevant factors:

1. The strong homology (60–70%) between human and mouse genomes (Table 9-1)
2. The successful efforts to map the mouse genome, thereby permitting more rapid identification of putative bone density genes
3. The ease and rapidity of conducting crosses among various strains of mice
4. The relative control that investigators have over environmental factors that modulate genetic determinants of bone density
5. The rapid acceleration in knock-out and transgenic technology permitting functional testing of putative bone density "genes."

Clearly, the last two factors are the most appealing and compelling. In contrast to human studies, strict regulation of dietary factors, physical activity, lifestyle, and environment are relatively straightforward in the mouse. Moreover, except for the sex chromosome, each of the mice in the inbred is an identical twin to the next, carrying the same genome as all others within that strain. This makes it considerably easier to refine the search for various bone density genes and, more important, to be able to test their responsiveness to various perturbations, all within a defined life cycle. Finding "osteoporosis" genes in mice provides an unmatched opportunity to test their role in all aspects of bone biology, and indeed, such findings can then be used to further our understanding of the pathophysiology of this debilitating disease.

REFERENCES

1. J. Eisman, Genetics of osteoporosis. *Endocr Rev,* **20,** 788–804 (1999).
2. W. F. Frankel, Taking stock of complex trait genetics in mice. *Trends Genet,* **11,** 471–477 (1995).
3. B. Zhao, M. Monier-Faugere, M. Langub, Z. Geng, T. Nakayma, J. W. Pike, S. Chernausek, C. J. Rosen, L. R. Donahue, H. Malluche, E. Ha, and T. Clemens, Targeted overexpression of insulin-like growth factor I to osteoblasts of transgenic mice: Increased trabecular bone volume without increased osteoblast proliferation. *Endocrinology,* **141,** 2674–2682 (2000).
4. M. Zhang, S. Xuan, M. L. Bouxsein, D. Von Stechow, N. Akeno, M. C. Faugere, H. Malluche, G. Zhao, C. J. Rosen, A. Efstratiadis, and T. L. Clemens, Osteoblast-specific knockout of the insulin-like growth factor (IGF) receptor gene reveals an essential role of IGF signalling in bone matrix mineralization. *J Biol Chem,* **277,** 44005–44012 (2002).
5. T. Komori, H. Vagi, S. Nomura, Y. H. Yamaguchi, K. Sasaki, K. Deguchi, Y. Shimizu, R. T. Bronson, Y. H. Gao, M. Inada, M. Sato, R. Okamoto, Y. Kitamura, and T. Kishimoto, Targeted disruption of *cbfal* results in a complete lack of bone formation owing to maturation arrest of osteoblasts. *Cell,* **89,** 755–764 (1997).
6. T. Suda, N. Takahashi, N. Udagawa, E. Jimi, M. Gillespie, and T. Martin, Modulation of osteoclast differentiation and function by the new members of the tumor necrosis factor receptor and ligand families. *Endocr Rev,* **20,** 345–357 (1999).
7. L. Silver and J. Nadeau, Encyclopedia of the mouse genome. *Mamm Genome,* **S1,** S388 (1997).
8. W. G. Beamer, L. R. Donahue, C. J. Rosen, and D. J. Baylink, Genetic variability in adult bone density among inbred strains of mice. *Bone,* **8,** 397–403 (1996).
9. C. H. Turner, Y. F. Hsieh, R. Muller, M. B. Bouxsein, D. J. Baylink, C. J. Rosen, M. D. Grynpas, L. R. Donahue, and W. G. Beamer, Genetic regulation of cortical and trabecular bone strength and microstructure in inbred strains of mice. *J Bone Miner Res,* **15,** 1126–1131 (2000).
10. Z. B. Zeng, Precision mapping of quantitative trait loci. *Genetics,* **136,** 1457–1468 (1994).

11. P. Kelley, J. Eisman, and E. Sambrook, Interaction of genetic and environmental influences on peak bone density. *Osteoporosis Int,* **1,** 56–60 (1990).

12. E. S. Lander and D. Bostein, Mapping Mendelian factors underlying quantitative traits using RFLP linkage maps. *Genetics,* **121,** 185–199 (1989).

13. E. Lander and L. Kruglyak, Genetic dissection of complex traits: Guidelines for interpreting and reporting results. *Nat Genet,* **11,** 241–247 (1995).

14. W. Benes, R. S. Weinstein, W. Zheng, J. J. Thaden, R. L. Jilka, S. C. Manolagos, and R. J. Smookler Reis, Chromosomal mapping of osteopenia-associated quantitative trait loci using closely related mouse strains. *J Bone Miner Res,* **15,** 626–633 (2000).

15. R. F. Klein, S. R. Mitchell, T. J. Phillips, J. K. Belknap, and E. S. Orwoll, Quantitative trait loci affecting peak bone mineral density in mice. *J Bone Miner Res,* **13,** 1648–1656 (1998).

16. M. Shimizu, K. Higuchi, B. Bennett, C. Xia, T. Tsuboyama, S. Kasai, T. Chiba, H. Fujisawa, K. Kogishi, H. Kitado, M. Kimoto, N. Takeda, M. Matsuchita, H. Okumura, T. Serikawa, T. Nakamura, T. E. Johnson, and M. Hosokawa, Identification of peak bone mass QTL in a spontaneously osteoporotic mouse strain. *Mamm Genome,* **10,** 81–87 (1999).

17. W. G. Beamer, K. L. Shultz, G. A. Churchill, W. A. Frankel, D. J. Baylink, C. J. Rosen, and L. R. Donahue, Quantitative trait loci for bone density in C57Bu6J and CASTIED inbred mice. *Mamm Genome,* **10,** 1043–1049 (1999).

18. W. Beamer, K. Shultz, L. Donahue, G. Churchill, S. Sen, J. Wergedal, D. Baylink, and C. Rosen, Quantitative trait loci for femoral and lumbar vertebral bone mineral density in C57BL/6J and C3H/HeJ inbred strains of mice. *J Bone Miner Res,* **16,** 1195–1206 (2001).

19. M. Soller, T. Brody, and A. Denizi, On the power of experimental designs for detection of linkage between marker loci and quantitative loci in crosses between inbred lines. *Theor Appl Genet,* **47,** 35–39 (1976).

20. C. J. Rosen, H. P. Damai, D. Vereault, L. R. Donahue, W. G. Beamer, J. Farley, S. Linkhart, T. Linkhart, S. Mohan, and D. J. Baylink, Circulating and skeletal insulin-like growth factor-I (IGF-I) concentrations in two inbred strains of mice with different bone densities. *Bone,* **21,** 217–233 (1997).

21. C. J. Rosen, G. A. Churchill, L. R. Donahue, K. L. Shultz, J. K. Burgess, D. R. Powell, and W. G. Beamer, Mapping quantitative trait loci for serum insulin-like growth factor-I levels in mice. *Bone,* **27,** 521–528 (2000).

22. R. F. Klein, J. Allard, Z. Avnur, T. Nikolcheva, D. Rotstein, A. S. Carlos, M. Shea, R. V. Waters, J. K. Belknap, G. Peltz, and E. S. Orwoll, Regulation of bone mass in mice by the lipoxygenase gene Alox15. *Science,* **303,** 229–232 (2004).

23. A. Darvasi, Experimental strategies for the genetic dissection of complex traits in animal models. *Nat Genet,* **18,** 19–24 (1998).

24. R. F. Klein, A. Carlos, K. Vartanian, V. Chambers, R. Turner, T. Phillips, J. Belknap, and E. Orwoll, Confirmation and fine mapping of chromosomal regions influencing peak bone mass in mice. *J Bone Miner Res,* **16,** 1953–1961 (2001).

25. E. Orwoll, J. Bellknap, and R. Klein, Gender specifically in the genetic determinants of peak bone mass. *J Bone Miner Res,* **16,** 1962–1971 (2001).

26. L. M. Silver, *Mouse Genetics.* Oxford University Press, New York (1995).

27. W. G. Beamer, L. R. Donahue, K. L. Shults, C. J. Rosen, G. A. Churchill, and D. J. Baylink, Genetic regulation of BMD in low density C57BU6J mice carrying donated QTLs from high density C3H1HeJ mice. *J Bone Miner Res,* **15**(Suppl. 1, Abstract 1192), S186 (2000).

28. M. Bouxsein, T. Uchiyama, C. J. Rosen, K. L. Shultz, L. R. Donahue, C. Turner, C. Sen, G. Churchill, R. Muller, and W. G. Beamer, Mapping quantitative trait loci for vertebral bone volume fraction and microarchitecture in mice. *J Bone Miner Res,* **19,** 587–599 (2004).

29. R. J. A. Fijneman, S. S. de Vries, R. C. Jansen, and P. Demant, Complex interactions of new quantitative trait loci, Sluc1, Sluc2, Sluc3, and Sluc4, that influence the susceptibility to lung cancer in the mouse. *Nat Genet,* **14,** 465–467 (1996).

30. C. J. A. Moen, M. A. van der Valk, M. Snock, B. F. M. van Zutphen, O. von Deimling, A. A. M. Hart, and P. Demant, The recombinant congenic strains—A novel genetic tool applied to the study of colon tumor development in the mouse. *Mamm Genome,* **1,** 217–227 (1991).

31. R. Blank, Y. Yershov, T. Baldini, E. Demant, and R. Bockman, Localization of genes contributing to failure load and related phenotypes in HelD Em recombinant congenic mice. *J Bone Miner Res,* **14**(Suppl. 1), 1039 (1999).

32. J. A. Blake, J. E. Richardson, M. T. Davisson, and J. T. Eppig, The Mouse Genome Database (MGD). A comprehensive public resource of genetic, phenotypic and genomic data. *Nucleic Acids Res,* **25,** 85–91 (1997).

33. A. L. Bower, D. H. Lang, G. P. Vogler, D. J. Vandenbergh, D. A. Blizzard, J. T. Stout, G. E. McClearn, and N. A. Sharkey, QTL analysis of trabecular bone in BXD F2 and RI mice. *J Bone Miner Res,* **21,** 1267–1275 (2006).

34. N. Ishimori, R. Li, K. Walsh, R. Korstanje, J. Rollins, P. Petkov, M. Pletcher, T. Wiltshire, L. Donahue, C. Rosen, W. Beamer, G. Churchill, and B. Paigen, Quantitative trait loci that determine BMDE in C57BL/6J and 129S1/S1/SvImJ inbred mice. *J Bone Miner Res,* **21,** 105–112 (2006).

35. X. M. Li, G. Masinde, W. Gu, J. Wergedal, S. Mohan, and D. J. Baylink, Genetic dissection of femur breaking strength in a large population (MRL/MpJ × SJL/J) of F2 mice: Single QTL effects, epistasis, and pleiotropy. *Genomics,* **79,** 421–428 (2002).

36. G. Masinde, X. M. Li, W. Gu, J. Wergedal, S. Mohan, and D. Baylink, Quantitative trait loci for bone density in mice: The genes determining total skeletal density and femur density show little overlap in F2 mice. *Calcif Tissue Int,* **71,** 421–428 (2002).

37. J. Wergedal, C. Ackert-Bicknell, S. W. Tsaih, M. H. C. Sheng, R. Li, S. Mohan, W. G. Beamer, G. Churchill, and D. Baylink, Femur mechanical properties in the F2 progeny of an NZB/B1NJ × RF/J cross are regulated predominantly by genetic loci that regulate bone geometry. *J Bone Miner Res,* **21,** 1256–1266 (2006).

Parathyroid Hormone and Parathyroid Hormone-Related Protein

ROBERT A. NISSENSON

I. INTRODUCTION

Parathyroid hormone (PTH) and PTH-related protein (PTHrP) are major polypeptide factors that regulate skeletal physiology and mineral homeostasis. The appearance of the parathyroid glands during the evolution of terrestrial vertebrates underscores the primary functional role of PTH—the maintenance of adequate levels of plasma ionized calcium in the face of a calcium-deficient terrestrial environment. The secretion of PTH by the parathyroid glands is stimulated when plasma ionized calcium activity falls. Once secreted, PTH acts to restore normal levels of ionized calcium through an integrated series of actions on bone, kidney, and (indirectly) the intestine. For an excellent review of PTH with a historical perspective, see Potts [1]. PTHrP, when present as a circulating factor, produces target cell effects that resemble those of PTH. This is most evident in malignancy-associated hypercalcemia, in which tumors elaborate sufficient quantities of PTHrP to produce biochemical abnormalities overlapping those seen in primary hyperparathyroidism. However, the major physiological function of PTHrP is to act as a local (paracrine) factor that controls the development, morphogenesis, and function of a variety of tissues, including (but not limited to) those involved in skeletal and mineral homeostasis. PTH and PTHrP are tied together historically in that PTHrP was discovered as a result of the quest to understand the pathogenesis of malignancy-associated hypercalcemia. However, they are also related structurally and produce their major physiological effects by activating a common receptor, the PTH/PTHrP receptor. This chapter focuses on the current understanding of the physiology and mechanism of action of these two polypeptides.

II. SECRETION OF PARATHYROID HORMONE

The parathyroid glands first appear during evolution with the movement of animals from an aquatic environment to a terrestrial environment deficient in calcium. Maintenance of adequate levels of plasma ionized calcium (1.0–1.3 mM) is required for normal neuromuscular function, bone mineralization, and many other physiological processes. The parathyroid gland secretes PTH in response to very small decrements in blood ionized calcium in order to maintain the normocalcemic state. As discussed later, PTH accomplishes this task by promoting bone resorption and releasing calcium from the skeletal reservoir, by inducing renal conservation of calcium and excretion of phosphate, and by indirectly enhancing intestinal calcium absorption by increasing the renal production of the active vitamin D metabolite $1,25(OH)_2$ vitamin D. The parathyroid gland functions in essence as a "calciostat," sensing the prevailing blood ionized calcium level and adjusting the secretion of PTH accordingly (Figure 10-1) [2]. The relationship between ionized calcium and PTH secretion is a steep sigmoidal one, allowing significant changes in PTH secretion in response to very small changes in plasma ionized calcium.

In addition to providing acute regulation of PTH secretion, ionized calcium is also a primary factor controlling chronic secretion of the hormone. Thus, sustained hypocalcemia promotes increased expression of the *PTH* gene [3, 4] and results in parathyroid hyperplasia [5]. A common example of the latter is the marked parathyroid hyperplasia (secondary hyperparathyroidism) that frequently accompanies chronic renal failure. $1,25(OH)_2$ vitamin D also serves as a negative regulator of *PTH* gene expression and parathyroid cell hyperplasia. In chronic renal failure, both

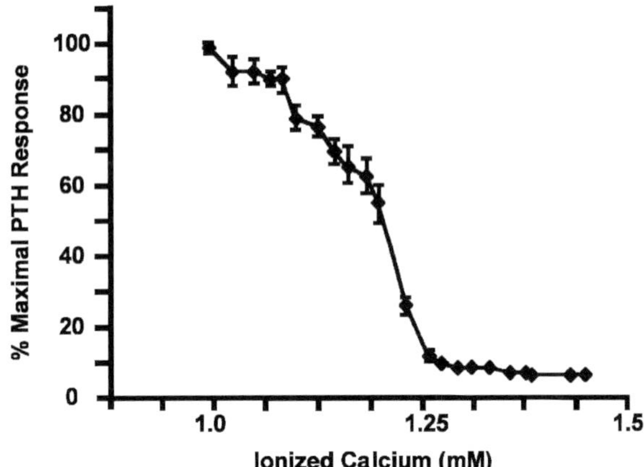

FIGURE 10-1   Relationship between plasma levels of ionized calcium and the release of PTH(1–84) in normal humans. Variations in plasma ionized calcium were achieved by the infusion of calcium or EDTA. Note the sigmoidal relationship, ensuring significant changes in PTH secretion with small variations in ionized calcium. Reproduced with permission from Brown [2].

hypocalcemia and reduced circulating concentrations of $1,25(OH)_2$ vitamin D presumably contribute to the progression of secondary hyperparathyroidism [6].

During the past several years, substantial progress has been made in our understanding of how extracellular calcium controls PTH secretion [7–11]. The plasma membrane of parathyroid cells contains high levels of a calcium-sensing receptor (CaR) [12]. Unlike intracellular calcium-binding proteins that have an affinity for free calcium in the nanomolar range (consistent with intracellular levels of free calcium), the CaR binds calcium in the millimolar range. The receptor is a member of the G protein–coupled receptor superfamily. It contains calcium binding elements in its extracellular domain and signaling determinants in its cytoplasmic regions. Calcium binding to the receptor triggers activation of the G proteins G_q and (to a lesser extent) G_i, resulting in stimulation of phospholipase C and inhibition of adenylyl cyclase, respectively [11, 13]. This results in an increase in intracellular calcium and a decrease in cyclic AMP content of parathyroid cells. By mechanisms that are not yet clear, these signaling pathways serve to suppress the synthesis and secretion of PTH. When blood ionized calcium falls, there is less signaling by the CaRs on the parathyroid cell and PTH secretion consequently increases. The essential role of the CaR can best be seen in humans bearing loss-of-function mutations in the *CaR* gene. In the heterozygous state, such mutations result in familial hypocalciuric hypercalcemia, characterized by an inappropriately high degree of PTH secretion in the face of hypercalcemia [14, 15]. These individuals are quantitatively resistant to the suppressive effect of calcium on PTH secretion due to the reduced number of parathyroid

CaRs. In the homozygous state, patients display a severe increase in PTH secretion with life-threatening hypercalcemia (neonatal severe primary hyperparathyroidism). Mice with homozygous and heterozygous disruption of the *CaR* gene display similar phenotypes [16]. Point mutations in the CaR that produce constitutive signaling have also been described, and these are associated with autosomal dominant hypocalcemia in humans [17].

Pharmacological ligands for the CaR have been developed, and these are effective in altering the ability of the CaR to signal [8]. Calcimimetic drugs bind to transmembrane regions in the CaR and increase the receptor's sensitivity to extracellular calcium. This results in an increase in receptor signaling and thus suppression of PTH secretion. Calcimimetic drugs have clinical utility in the medical management of hyperparathyroidism [18, 19]. Calcilytic drugs act as pharmacological antagonists of the CaR, thereby increasing the secretion of PTH.

III.   METABOLISM OF PARATHYROID HORMONE

Studies carried out more than 30 years ago demonstrated that PTH circulates in multiple forms that can be distinguished by radioimmunoassays specific for different regions of the PTH molecule [20–22]. This heterogeneity has two origins (Figure 10-2). PTH(1–84) is subject to metabolism within the parathyroid gland, resulting in secretion of PTH fragments as well as the intact molecule. In addition, PTH(1–84) is metabolized in peripheral tissues. Midregion and carboxyl-terminal fragments of PTH have a much longer half-life in the circulation than does PTH(1–84) [23–26]. As a result, midregion and carboxyl-terminal fragments of PTH circulate at much higher concentrations than intact PTH(1–84) [27]. Rapid plasma clearance of PTH is due primarily to hepatic metabolism, with a lesser contribution by the kidneys [28–30]. Peripheral metabolism generates mid- and carboxyl-terminal fragments of PTH that resemble those secreted by the parathyroid gland. Mid- and carboxyl-terminal PTH fragments are cleared by renal excretion, and thus circulating levels of these fragments are highly dependent on renal function. Extremely high concentrations of PTH detected with antibodies against the mid- and carboxyl regions of the hormone in many patients with end-stage renal disease thus reflect a combination of secondary hyperparathyroidism and reduced renal clearance of PTH fragments.

Mid- and carboxyl region PTH fragments lack the amino-terminal 1–34 sequence of the hormone required for binding to PTH/PTHrP receptors and producing the classical effects of PTH on kidney and bone. Metabolism of PTH could produce biologically active amino-terminal

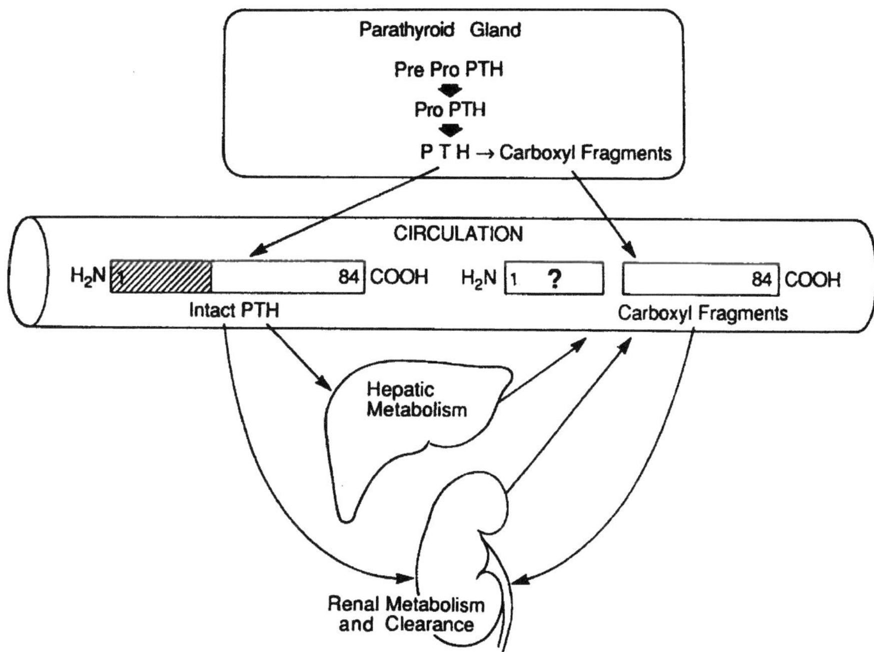

FIGURE 10-2   Metabolism and clearance of PTH. PTH is subject to proteolytic cleavage in the parathyroid gland, as well as in liver and kidney, resulting in the presence of inactive midregion and carboxyl-terminal PTH fragments in the circulation. Amino-terminal PTH fragments are apparently rapidly degraded and do not accumulate in the circulation. Intact PTH has a short half-life in the circulation (2–4 min) due to hepatic and renal metabolism. Midregion and carboxy-terminal PTH fragments are cleared by glomerular filtration. They have a much longer half-life that is dependent on the level of renal function. A large C-terminal fragment, PTH(7–84), that could serve as a PTH/PTHrP receptor antagonist has been identified in the circulation. Reproduced with permission from Endres *et al.* [27].

fragments of PTH, but there is little evidence for the presence of significant levels of amino-terminal PTH fragments in the circulation [31] or for significant secretion of such fragments by the parathyroid gland [32]. Presumably, both the parathyroid gland and the peripheral organs contain enzymes that degrade amino-terminal fragments of PTH. This ensures that circulating levels of biologically active PTH are derived exclusively from glandular secretion of PTH(1–84). There is evidence for potential biological effects of mid- or carboxyl region fragments of PTH [33–36], and there is also evidence for the existence of membrane receptors for these fragments [37–41]. However, the biological role of mid- and carboxyl-region PTH fragments remains unclear.

Calcium-sensitive cathepsins are responsible for cleaving PTH(1–84) within the parathyroid gland. Intraglandular cleavage occurs between residues 34 and 35 or between residues 36 and 37 [42, 43], and a greater proportion of PTH is cleaved under conditions of hypercalcemia [44]. The amino-terminal fragments so produced are rapidly degraded within the parathyroid gland, and thus calcium-sensitive cleavage constitutes a mechanism for inactivation of PTH. Therefore, the level of plasma calcium determines not only the rate of synthesis and secretion of PTH but also the extent to which secreted PTH is biologically active.

A large fragment of PTH identified as PTH(7–84) has been identified in the circulation [45–47]. This fragment is secreted from the parathyroid glands [48] following calcium-dependent intraglandular proteolysis of the amino-terminus of PTH(1–84). It may also arise from peripheral metabolism of PTH(1–84) [49]. PTH(7–84) lacks the amino-terminal residues required for activation of PTH/PTHrP receptors. However, this fragment is detected in some radioimmunoassays for "intact" PTH resulting in overestimation of levels of circulating, biologically active PTH [50]. PTH(7–84) is known to bind with low affinity to PTH/PTHrP receptors, thereby antagonizing the actions of PTH(1–84) [51]. However, it remains uncertain whether endogenous PTH(7–84) circulates at sufficient levels to effectively suppress the target cell actions of PTH under normal physiological conditions.

IV. BONE RESORBING ACTION OF PARATHYROID HORMONE

The major physiological role of PTH is to regulate plasma calcium homeostasis. When dietary calcium intake is inadequate, PTH maintains the level of plasma calcium by mobilizing calcium from the vast reservoir present in bone in the form of the mineral hydroxyapatite. This is accomplished by a direct action of PTH on bone that results in increased osteoclastic bone resorption and increased flux of calcium from bone into blood. Administration of PTH produces rapid movement of calcium out of bone, an effect that is associated with structural changes in cells lining the endosteal surface [52]. It has been suggested that these lining cells form an epithelial-like barrier between the circulation and the bone extracellular fluid [53, 54], and that PTH may act on these cells to promote calcium transport. PTH enhances osteoclastic bone resorption within 15 minutes of its administration [55] and produces a sustained increase in bone resorption that appears to require the recruitment and differentiation of new osteoclasts. PTH-induced bone resorption involves the dissolution of hydroxyapatite bone mineral in the acidic microenvironment created by the osteoclast, as well as the degradation of collagen and other matrix proteins by proteolytic enzymes.

The mechanism by which PTH promotes osteoclastic bone resorption has been a subject of intensive interest. There are conflicting data as to whether functional PTH receptors are present in osteoclasts [56–61], and the bone resorbing actions of PTH are likely to be mediated mainly by activation of PTH/PTHrP receptors present in cells of the osteoblast lineage [56, 57, 62, 63]. The rapid effect of PTH on bone resorption may be due to an action of the hormone on osteoblast lining cells, altering their attachment to the surface of bone or reducing cell–cell interactions, allowing osteoclasts to gain access to the mineralized bone surface. Indeed, PTH has dramatic effects on the morphology of isolated osteoblasts [64] and alters osteoblast expression of connexin 43, a protein involved in cell–cell communication [65–67]. In addition, osteoblasts are known to respond to PTH by secreting proteins such as collagenase [68–71] and plasminogen activators [72–74], which may facilitate osteoclastic bone resorption [75, 76].

The long-term effect of PTH to promote bone resorption involves an action of the hormone to enhance the differentiation of osteoclasts from precursor cells in the monocyte/macrophage lineage. This again results indirectly from the action of PTH on osteoblastic cells. Osteoblasts secrete several cytokines that could potentially influence osteoclastogenesis activity by a paracrine mechanism [77–79]. However, it appears that direct contact between the accessory cells and osteoclasts is required for PTH-induced osteoclast activation [80]. An explanation for this derives from the discovery of the role of rank ligand (RANKL) and its receptor (RANK) in the regulation of osteoclast differentiation and function [81–88]. RANK is a tumor necrosis factor-α (TNF-α) receptor–related protein receptor that is expressed on the surface of osteoclast precursors as well as in differentiated osteoclasts. RANK signaling in osteoclast precursors promotes differentiation to functional osteoclasts, and RANK signaling in differentiated osteoclasts enhances bone resorption and inhibits apoptosis [89–92]. In both cases, RANKL binding to RANK is required for signaling. RANKL is not a secreted protein but, rather, is an intrinsic membrane protein expressed on the surface of cells of the osteoblast lineage. Thus, direct contact between cells of the osteoblast lineage and osteoclasts or their precursors is required for the engagement of RANKL with RANK leading to osteoclast differentiation and activation. RANKL is required for normal osteoclast development and function, and mice lacking RANKL show a loss of functional osteoclasts and osteopetrosis [93]. Cells in the microenvironment of bone also secrete a truncated TNF-α receptor-like molecule termed osteoprotegerin (OPG), which functions as a "decoy receptor" by binding to RANKL and thereby preventing initiation of RANK signaling [94–96]. The importance of OPG as a tonic suppressor of bone turnover is evident from findings in mice lacking functional expression of OPG. These animals display increased bone resorption and osteoporosis [97, 98].

Abundant evidence demonstrates that the RANKL/RANK system plays a major role in PTH-induced bone resorption and calcium mobilization (Figure 10-3). Administration of soluble RANKL to mice elicits severe hypercalcemia within 1 day of administration, and increased osteoclast activity and bone loss are evident within 3 days [89]. Administration of OPG (RANKL antagonist) blocks the calcemic action of exogenous PTH *in vivo* [94]. Addition of OPG also inhibits PTH-induced osteoclast activation and bone resorption *in vitro* and *in vivo* [99–102]. PTH produces an increase in the ratio of RANKL:OPG expressed by osteoblastic cells, an effect that is due to the ability of PTH to increase the expression of RANKL and to inhibit the expression of OPG [91, 100, 103–105]. Similar effects have been observed *in vivo* following exogenous administration of PTH [106]. The effect of PTH on RANKL is exerted at the level of gene transcription. Nonetheless, this action of PTH is very rapid (evident within 1 hour) and thus upregulation of RANKL could contribute not only to osteoclastogenesis but also to the rapid increase in the activity of mature osteoclasts seen in response to PTH.

V. EFFECTS OF PARATHYROID HORMONE ON BONE FORMATION

Administration of PTH intermittently to animals or humans produces a marked anabolic response of the

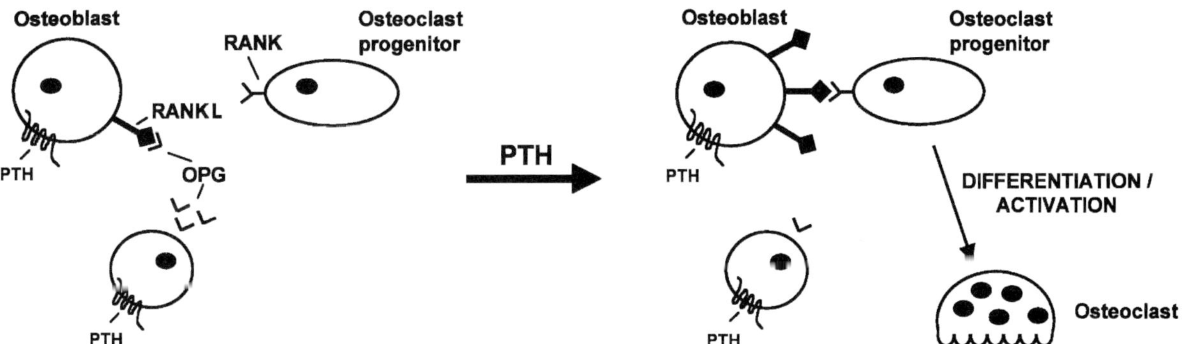

FIGURE 10-3   Regulation of osteoclast differentiation and activation by PTH. Binding of PTH to receptors on osteoblasts results in increased expression of RANKL on the cell surface. Activation of PTH receptors also reduces the secretion of the RANKL inhibitor osteoprotegerin (OPG), which is produced by cells in the bone microenvironment. These effects of PTH promote the action of RANKL on its receptor (RANK) on the surface of osteoclast precursors and mature osteoclasts. RANK signaling, together with the action of macrophage colony–stimulating factor, stimulates the differentiation of osteoclast precursors and promotes the activation of mature osteoclasts.

skeleton [107–118]. This results from a direct effect of PTH on cells of the osteoblast lineage to promote bone formation. PTH promotes bone formation in both trabecular and cortical bone, and these actions are associated with increased trabecular thickness and increased bone strength [113, 119–125]. High levels of PTH are known to produce an increase in the number of osteoblasts, which results in part from the coupling between increased osteoclastic resorption and new bone formation. However, intermittent treatment with low doses of PTH produces an additional direct positive effect on osteoblastic bone formation. The cellular basis for the anabolic action of PTH is not fully understood (Figure 10-4). In principle, PTH could increase the number of mature osteoblasts and/or increase the functional (bone-forming) activity of osteoblasts. PTH receptors are present on osteoblast precursors including bone marrow stromal cells [126–128]. Available evidence indicates that PTH increases the number of active osteoblasts but its direct effect on the replication of osteoblastic cells is variable [129–132]. Model systems for osteoblast differentiation *in vitro* reveal a positive effect of PTH on differentiation, depending on the dose and mode of exposure, with intermittent treatment with low doses being most consistently effective [133–137]. PTH has been shown to down-regulate the expression of two factors, dkk-1 [138] and sclerostin [139, 140], that are negative regulators of canonical wnt signaling. As discussed in detail in Chapter 15 (Johnson), this signaling pathway promotes the differentiation of committed osteoblast precursors [141]. Therefore, it is possible that PTH treatment dampens constitutive inhibition of osteoblast differentiation resulting from expression

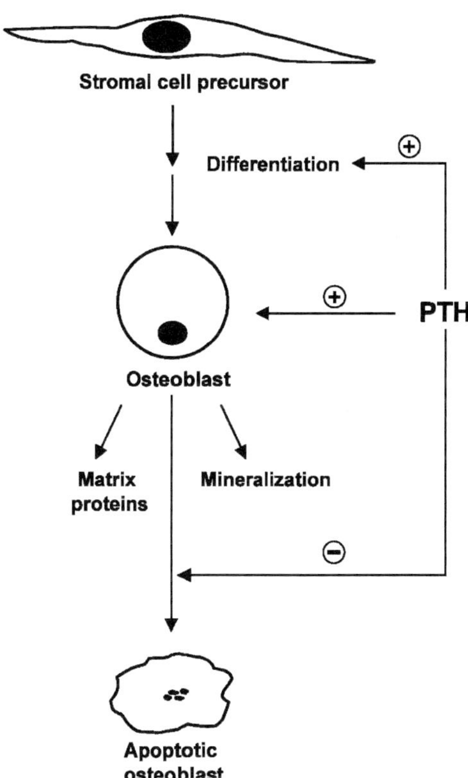

FIGURE 10-4   Possible mechanisms contributing to the anabolic skeletal effect of intermittent PTH administration. PTH may act on bone marrow stromal cell precursors to promote their differentiation to functional osteoblasts. PTH could also act directly on osteoblasts to increase their number or their functional activity. Finally, PTH could increase the life span of mature osteoblasts by inhibiting their death via apoptosis. There is evidence that intermittent treatment with PTH utilizes multiple anabolic mechanisms.

of these wnt pathway inhibitors. PTH also appears to extend the life span of active osteoblasts by inhibiting their apoptosis both *in vivo* [142] and *in vitro* [137]. Taken together, the available data support the notion that intermittent administration of PTH elicits an increase in osteoblastic bone formation via actions to promote osteoblast differentiation and to inhibit osteoblast apoptosis.

Intermittent (e.g., once daily) treatment with PTH elicits skeletal effects in which increased bone formation predominates, whereas continuous treatment with high doses of PTH results in a major increase in bone resorption. Continuous treatment of target cells with high doses of PTH results in a loss of responsiveness (desensitization), and it is possible that the anabolic effects of PTH are particularly sensitive to hormone-induced desensitization. Indeed, *in vivo* inhibition of G protein–coupled receptor kinase (GRK-2), an enzyme known to promote desensitization of the PTH/PTHrP receptor [143–145], enhances the anabolic response to exogenous PTH [146], whereas overexpression of GRK-2 in osteoblasts results in bone loss [147]. Intermittent administration of PTH could allow for resensitization of the anabolic response prior to administration of a subsequent dose of hormone. However, continuous administration of lower doses of PTH also elicits an anabolic skeletal response, suggesting that the balance between bone resorption and anabolism may be related to the dose of PTH rather than to its intermittent administration. The effects of PTH also differ depending on the nature of the skeletal site, with trabecular bone displaying the greatest increase in mass in response to PTH. At doses of PTH that are anabolic in trabecular bone, cortical bone displays increased bone resorption as well as increased bone formation. The net effect of PTH treatment on cortical bone mass is thus variable.

To further complicate matters, it has recently been reported that the anabolic effect of PTH is diminished in patients treated with bisphosphonates to suppress bone resorption [148, 149]. This suggests that some level of ongoing bone resorption is permissive for intermittent PTH to enhance osteoblastic bone formation [150, 151]. Osteoclasts may secrete a permissive factor(s) and/or may liberate such a factor(s) from the bone matrix during the process of bone resorption. Indeed, growth factors such as insulin-like growth factor-1 (IGF-1) and fibroblast growth factor-2 are present in bone matrix, and the ability of exogenous PTH to elicit an anabolic effect in bone is diminished in mice lacking expression of these growth factors [152–155].

VI. RENAL ACTIONS OF PARATHYROID HORMONE

PTH produces a series of renal actions that help to ensure that calcium mobilized from bone contributes optimally to the maintenance of plasma ionized calcium levels. The renal actions of PTH include inhibition of renal phosphate reabsorption, stimulation of renal calcium reabsorption, and increased production of $1,25(OH)_2$ vitamin D. The ability of PTH to inhibit renal phosphate reabsorption has been known for many years, providing the basis for the clinical Ellsworth–Howard test of renal responsiveness to the hormone [156]. Patients with primary hyperparathyroidism display hypophosphatemia and decreased renal tubular reabsorption of phosphate, whereas hypoparathyroid patients are hyperphosphatemic and have increased phosphate reabsorption. Phosphate forms a complex with free calcium in blood. Thus, for a given level of serum calcium, ionized calcium will be reduced as serum phosphate increases. Under conditions of relative hypocalcemia (e.g., during chronic dietary calcium deficiency), PTH secretion is increased, resulting in increased bone resorption. Both calcium and phosphate are released from hydroxyapatite during the process of bone resorption. By promoting renal excretion of phosphate, PTH facilitates a rise in ionized as well as total plasma calcium.

Phosphate reabsorption in the proximal renal tubule is dependent in part on the activity of the type IIa sodium–phosphate cotransporter (NaPi-IIa). The phosphaturic action of PTH derives from the action of the hormone to inhibit the function of this transporter [157, 158]. NaPi-IIa is located in the apical plasma membrane and permits the coupled transport of sodium and phosphate from the tubule into the renal cell. Exposure of proximal tubular cells to PTH results in a reduced V_{max} of the transporter [159, 160], and this is associated with a decrease in the amount of the transporter in the apical plasma membrane [161]. Acute exposure of the proximal tubular cells to PTH enhances the endocytosis and subsequent lysosomal degradation of NaPi-IIa, and this may be the major mechanism responsible for rapid PTH-induced inhibition of renal phosphate reabsorption [162–165]. PTH appears to regulate NaPi-IIa by enhancing its rate of turnover rather than by suppressing its synthesis [166]. Downregulation of NaPi-IIa by PTH involves the action of a Na/H exchange–regulatory molecule (NHERF-1) [165, 167–169], although the precise role for this protein has not been defined. Regulation of renal phosphate handling is further discussed in Chapter 14 (Kumar).

PTH also acts to increase renal calcium reabsorption, thus ensuring that only small amounts of calcium released during PTH-induced bone resorption are lost via renal excretion. The major sites for this effect of PTH are in the distal convoluted tubule and the thick ascending limb of Henle's loop [170, 171]. Available evidence indicates that distal renal tubular calcium reabsorption is an active process that requires calcium influx through dihydropyridine-sensitive calcium channels located in the apical plasma membrane [172]. Drugs that inhibit these channels are effective in blocking PTH-induced renal calcium reabsorption. Unlike voltage-sensitive calcium channels in excitable tissues, PTH-responsive calcium channels in the distal nephron are activated by membrane hyperpolarization [173]. PTH appears to open calcium channels by inducing hyperpolarization of the apical plasma membrane. Calcium entering the distal renal tubular cell in this manner is transported into the extracellular compartment via a sodium–calcium exchanger present on the basolateral plasma membrane [174]. PTH may also act by increasing the expression of transcellular calcium transport proteins in the distal nephron [175].

PTH promotes intestinal calcium reabsorption indirectly, through an action to increase circulating levels of 1,25(OH)$_2$ vitamin D. This vitamin D metabolite acts directly on intestinal epithelial cells to increase the efficiency of calcium (and phosphate) absorption. Primary hyperparathyroidism is commonly associated with increased circulating levels of 1,25(OH)$_2$ vitamin D, whereas reduced levels of this metabolite are present in hypoparathyroidism [176]. PTH produces this effect by increasing the rate of production of 1,25(OH)$_2$ vitamin D through activation of the 25(OH) vitamin D-1-hydroxylase enzyme located in the proximal renal tubule [177–179]. The gene encoding this enzyme has been cloned in multiple laboratories [180–182]. Studies *in vivo* as well as in cultured renal cell lines indicate that PTH increases the expression of the 25(OH) vitamin D-1-hydroxylase gene through a transcriptional mechanism [183–187]. See Chapter 11 (Feldman) for further discussion of this important pathway of PTH action.

VII. PARATHYROID HORMONE–RELATED PROTEIN AS A MEDIATOR OF MALIGNANCY-ASSOCIATED HYPERCALCEMIA

The frequent occurrence of hypercalcemia in individuals with a variety of malignancies has been recognized for many years. An important clue as to the pathogenesis of malignancy-associated hypercalcemia (MAH) came with the recognition that many such individuals display increased excretion of

renal-derived ("nephrogenous") cyclic AMP [188]. Activation of the renal PTH receptor by elevated circulating levels of PTH in hyperparathyroidism was the only known cause of increased nephrogenous cyclic AMP, and thus it was suggested that malignant tumors are capable of producing a factor that activates PTH receptors. Plasma levels of immunoreactive PTH were found to be low in patients with MAH [188], indicating that the relevant circulating factor was not PTH.

Using the activation of PTH receptors as an assay, multiple groups succeeded in isolating and ultimately identifying the PTH-like etiologic factor in MAH [189–192]. This factor was termed PTH-related protein (PTHrP) because of its ability to bind to and activate the PTH receptor and because of its limited sequence similarity to PTH [193–195]. The *PTHrP* gene is subject to alternative splicing, resulting in the production of three protein products ranging from 139 to 173 amino acids differing only in their carboxyl-terminal sequence [196, 197]. PTHrP is capable of reproducing the major target cell actions of PTH and (like PTH) does so via the amino-terminal 34 amino acids or so of the protein. A comparison of the 1–34 sequences of PTH and PTHrP reveals significant amino acid homology, with identity in 8 of the 13 amino-terminal residues. Two of the known contact sites between PTH and the PTH/PTHrP receptor are within this 13–amino acid homologous region [198], indicating that these ligands use very similar mechanisms to activate their common receptor. The molecular mechanisms underlying the overexpression of PTHrP by malignant tumors remain unclear. As the mass of PTHrP-expressing tumor cells expands, systemic levels of PTHrP eventually increase sufficiently to allow the peptide to elicit endocrine effects on PTH/PTHrP receptors in bone and kidney, resulting in MAH.

VIII. PHYSIOLOGICAL ROLES OF PARATHRYOID HORMONE–RELATED PROTEIN

Although PTHrP produces PTH-like target cell effects in patients with MAH, circulating levels of PTHrP are very low to undetectable in normal individuals. This, coupled with the widespread expression of the PTHrP gene in normal tissues, suggested that PTHrP was likely to have physiological functions as a local paracrine factor rather than as a systemic hormone. Subsequent studies have confirmed that PTHrP indeed plays an important role as a paracrine factor in a wide variety of tissues (Table 10-1) [199–203], as summarized here.

TABLE 10-1   Physiological Roles of PTHrP

Target tissues	Actions
Cartilage	Inhibits terminal chondrocyte differentiation; increases chondrocyte proliferation
Bone	Maintains bone mass; promotes bone resorption during lactation
Mammary gland	Facilitates branching morphogenesis of mammary epithelium
Skin	Inhibits terminal differentiation of keratinocytes; promotes normal hair follicle development
Teeth	Promotes normal tooth eruption
Extraembryonic endoderm	Enhances the differentiation of primitive endoderm to parietal endoderm
Smooth muscle	Serves as a general smooth muscle relaxant
Central nervous system	Inhibits neuronal L-type calcium channel activity; protects neurons from excitotoxicity
Placenta	Maintains the positive maternal–fetal transplacental calcium gradient

A. Endochondral Bone Development

The first direct evidence concerning a physiological role for PTHrP appeared in 1994 with the report of the phenotype of mice lacking expression of PTHrP due to targeted gene ablation [204]. These animals died soon after birth and were found to display a form of short-limbed dwarfism with generalized chondrodysplasia. The most striking feature of mice lacking expression of PTHrP is the disruption of normal endochondral ossification. Although the most obvious gross phenotypic abnormality is short-limbed dwarfism, the defect in endochondral bone formation is generalized. The role of PTHrP is best understood in the context of the homeostatic mechanisms regulating the differentiation of cartilage and bone during endochondral bone formation (see Chapter 8, Kronenberg). In the long bones, chondrogenesis is initiated by the differentiation of mesenchymal cell precursors that form nodules and begin to express characteristic genes including those encoding type II collagen and other cartilage matrix proteins [205, 206]. These early chondrocytes are mitotically active, but the cells in the center of the nodule become hypertrophic, cease dividing, and express gene products characteristic of mature chondrocytes (e.g., type X collagen). Hypertrophic chondrocytes undergo programmed cell death (apoptosis), and this is accompanied by vascular invasion. Subsequently, the cartilage scaffold is replaced by bone. In the growing animal, this process is continued in the growth plate, where the differentiation process is subject to tight temporal and spatial control. Mesenchymal cell differentiation and early chondrocyte proliferation occur in a columnar array inward from the articular surface. This spatial profile is extended as the chondrocytes become prehypertrophic and then hypertrophic. After the hypertrophic cells undergo apoptosis, the cartilaginous scaffold is remodeled and subsequently replaced by bone.

The control of endochondral bone formation is maintained by a complex series of extracellular cues and intracellular signaling pathways [207]. One of these factors is Indian hedgehog (Ihh), a member of the ancient hedgehog family of secreted patterning molecules. Ihh functions to promote chondrocyte proliferation and to maintain the pool of proliferating chondrocytes, thus extending the length of the differentiating cartilaginous growth plate prior to terminal differentiation and ossification [208, 209]. Ihh is produced by postmitotic prehypertrophic chondrocytes, suggesting that the factor may serve as a negative feedback signal that slows the rate of transition of chondrocytes from the proliferative to the prehypertrophic pool. Ihh also appears to directly act on cells of the osteoblast lineage to promote their differentiation to mature bone-forming cells [141, 208, 210, 211].

PTHrP appears to mediate some, but not all, of the actions of Ihh on endochondral bone formation [212–215]. PTHrP directly inhibits the differentiation of proliferating chondrocytes to postmitotic prehypertrophic cells. Lack of PTHrP results in accelerated chondrocyte differentiation with shortened growth plates and premature ossification. The cellular composition of the growth plates of *PTHrP–/–* animals is abnormal, with a marked reduction in the number of proliferating chondrocytes. Conversely, overexpression of PTHrP in chondrocytes of mice bearing a collagen II promoter–PTHrP transgene resulted in a distinct form of chondrodysplasia characterized by short-limbed dwarfism and delayed ossification [216]. At birth, these animals displayed a cartilaginous endochondral skeleton, and histological evaluation revealed a marked suppression of the chondrocyte differentiation program. By 7 weeks of age, ossification was evident, but the long bones remained foreshortened and misshapen. Similar abnormalities are seen in humans with hereditary Jansen's metaphyseal chondrodysplasia. The latter disorder has been associated with mutations in the PTH/PTHrP receptor that result in constitutive receptor activation [217, 218].

Ihh acts directly or indirectly on cells in the periarticular perichondrium to increase expression of the *PTHrP* gene [219]. The effect of Ihh (or the related protein Sonic hedgehog) to delay terminal differentiation of chondrocytes in the long bones was not seen in *PTHrP–/–* or in *PTH/PTHrP receptor–/–* mice, indicating an intermediary role of PTHrP in Ihh action in endochondral bone formation [219, 220]. Consistent with this conclusion, a type II collagen promoter-driven constitutively active PTH/PTHrP receptor transgene rescues the abnormally accelerated chondrocyte differentiation program in *Ihh–/–* mice [221]. These animals nonetheless displayed short-limbed dwarfism and decreased chondrocyte proliferation, demonstrating that PTHrP is not the only mediator of the multiple actions of Ihh on endochondral ossification. This conclusion is further supported by the observation that short-limbed dwarfism is much more severe in *Ihh–/–, PTHrP–/–* mice than in *Ihh+/+, PTHrP–/–* mice [221]. It appears that chondrocyte differentiation is regulated in a complex fashion by these two secreted regulatory factors [222, 223].

There is solid evidence that the PTH/PTHrP receptor is responsible for initiating the actions of PTHrP on the differentiation of growth plate chondrocytes. The PTH/PTHrP receptor is expressed in proliferating chondrocytes as well as in cells in the transitional zone between proliferating and hypertrophic chondrocytes, where regulation of terminal differentiation occurs [224]. *PTH/PTHrP–/–* mice display growth plate abnormalities similar to those seen in *PTHrP–/–* mice [220]. Patients with inherited mutations in the PTH/PTHrP receptor that cause constitutive (i.e., ligand-independent) signaling (Jansen's metaphyseal chondrodysplasia) display growth plate abnormalities similar to those seen in mice overexpressing a collagen II promoter–PTHrP transgene [217, 225]. Lack of expression of functional PTH/PTHrP receptors in humans is associated with Blomstrand chondrodysplasia [226–228], a lethal disorder characterized by premature endochondral ossification [229]. Precisely how signaling by the PTH/PTHrP receptor results in the maintenance of proliferating chondrocytes and in the delay of chondrocyte differentiation in the transitional zone is unclear. Genetic evidence suggests that PTHrP may serve to regulate expression of the cyclin-dependent kinase inhibitor p57 in chondrocytes, and this could account for proliferative actions of PTHrP

[230]. In addition, it is known that programmed cell death (apoptosis) occurs during the late terminal differentiation of chondrocytes. This process has been shown to be inhibited by PTHrP, which upregulates anti-apoptotic protein bcl-2 through a cyclic AMP-dependent mechanism [231]. Mice lacking expression of a functional *bcl-2* gene are known to display accelerated differentiation of growth plate chondrocytes, although the severity of the phenotype is much less than that seen in *PTHrP–/–* mice. There is also evidence that PTHrP may act to maintain the expression of Runx2 [232] and Nkx3.2 [233], transcription factors that suppress chondrocyte maturation.

B. Bone

Interestingly, mice with haploinsufficiency of PTHrP were reported to develop trabecular osteopenia after 3 months of age [234], suggesting a role for PTHrP in bone formation in the adult animal. An osteopenic phenotype has also been observed in mice with a targeted deletion in PTHrP expression in osteoblasts [235]. This was associated with decreased recruitment of bone stromal cell osteoblast precursors and increased apoptosis of osteoblasts. These findings suggest that production of PTHrP by cells of the osteoblast lineage plays a role in maintaining the pool of active osteoblasts that participate in bone formation. Expression of PTHrP appears to be required for normal formation of intramembranous as well as endochondral bone [236]. The precise nature of the osteoblastic cells that express PTHrP is not clear because PTHrP promoter activity was detected in a number of cell types in bone but not in mature osteoblasts [237]. PTHrP is also expressed in connective tissue cells in the outer layer of the periosteum and at sites of insertion of tendons and ligaments into cortical bone [237], and it is possible that PTHrP serves as a local regulator of bone formation or turnover in response to mechanical stimulation [238].

C. Mammary Gland

Targeted overexpression of PTHrP in mammary myoepithelial cells of transgenic mice provided direct evidence of a possible role for PTHrP in mammary gland development [239]. The mammary ducts of 18- to 21-day-old transgenic mice were normal in terms of both the size of the ducts and the branching morphogenesis of the developing gland. However, by 6 weeks of age, the transgenic animal displayed a delay in the development of the mammary duct system and a reduction in the degree of ductal branching.

The pregnant transgenic animal displayed similar defects, as well as diminished formation of terminal ductules. Overexpression of PTH in mammary myoepithelial cells of transgenic mice produced identical morphogenetic defects, indicating that this action of PTHrP is mediated by the PTH/PTHrP receptor. The postnatal role of PTHrP in mammary gland development was studied in *PTHrP–/–* mice expressing a PTHrP transgene targeted to cartilage [240], allowing postnatal survival. At 4 months of age, female transgenic mice lack mammary glands. The mammary fat pads appear normal, but mammary epithelial ducts are missing. *PTHrP–/–* mice display arrest of mammary duct development beginning between days 15 and 18 of embryogenesis. At this time, there is degeneration of epithelial elements within the ducts, and the initiation of normal branching morphogenesis of the mammary glands does not occur.

In normal animals, PTHrP is expressed in mammary epithelial cells [240, 241], whereas functional PTH/PTHrP receptors are expressed in the underlying mesenchyme [240, 242]. This pattern of expression suggests that PTHrP is an epithelial signal that acts on PTH/PTHrP receptors in mesenchymal cells to promote mammary epithelial morphogenesis. Consistent with this notion, *PTH/PTHrP receptor–/–* mice display the same defects in embryonic mammary development seen in *PTHrP–/–* mice. Moreover, normal morphogenesis requires PTH/PTHrP receptor expression specifically in mammary mesenchymal cells [242]. Humans lacking functional PTH/PTHrP receptors (Blomstrand chondrodysplasia) fail to develop nipples or breasts [243]. The factors that regulate epithelial production of PTHrP, and the nature of the mesenchymal targets of PTH/PTHrP receptor signaling, are unknown. The mesenchymal genes encoding tenascin C and the androgen receptor are induced by PTHrP [244]. *PTHrP–/–* or *PTH/PTHrP receptor–/–* male mice fail to display the normal androgen-dependent apoptotic destruction of the mammary bud, indicating that induction of the androgen receptor by PTHrP is essential for sexual dimorphism during mammary development. PTHrP production by mammary bud epithelial cells is also essential for the induction of nipple skin differentiation during mammary development [245, 246].

A role for PTHrP during lactation was first suggested by the observation that suckling is a powerful stimulus for increased mammary *PTHrP* gene expression [247]. Subsequently, systemic maternal PTHrP levels have been reported to increase during suckling [248] and to be elevated during lactation [249, 250], although not all studies are in agreement on this [251, 252]. Nonetheless, the findings suggest that systemic PTHrP produced by the mammary gland may be important for mobilizing calcium destined for secretion into breast milk during periods of lactation. In support of this, mammary-specific deletion of the *PTHrP* gene in lactating mice was shown to reduce circulating levels of PTHrP and to attenuate bone loss during the lactation period [253]. Signaling by the CaR in mammary epithelial cells downregulates mammary production of PTHrP [254], perhaps providing a mechanism for negative feedback in response to increased maternal levels of blood calcium. Interestingly, extremely large quantities of PTHrP are secreted into milk during lactation [251]. Suckling animals and humans thus ingest large amounts of PTHrP over an extended time period, yet evidence that milk-derived PTHrP is absorbed in an active form and/or is physiologically important in suckling infants or animals is lacking.

D. Skin and Teeth

Keratinocytes were the first normal cells shown to express PTH-like bioactivity [255] and subsequently the *PTHrP* gene [196]. PTHrP is expressed in the basal layer through the granulosa layer of the skin, with epidermal expression detectable as early as day 14 of embryogenesis in the rat [256, 257], although one report suggests that PTHrP expression in the epidermis is limited to the hair follicles [258]. PTH/PTHrP receptors are present in dermal fibroblasts [258, 259] and keratinocytes [260], and novel binding sites for PTHrP have been detected in keratinocytes [261]. In cultured human keratinocytes, suppression of PTHrP production resulted in increased cell proliferation [262] and decreased differentiation [263]. Thus, PTHrP may have a role in the local regulation of epidermal cell proliferation and differentiation.

Targeted overexpression of PTHrP in basal keratinocytes and outer-root sheath cells of hair follicles in transgenic mice resulted in a failure of ventral hair eruption, which was evident within 6 days after birth [264]. Dorsal hair was evident, but its eruption was delayed and the hairs were shorter and thinner compared to those of normal littermates. Histological evaluation of the transgenic mice revealed thickening of the ventral epidermis and expansion and increased cellularity of the dermis. Hair follicle development was substantially delayed in both ventral and dorsal skin of transgenic mice. These effects are probably due to disruption of the normal epithelial–mesenchymal interactions required for proper hair follicle development and epidermal differentiation. PTHrP appears to promote anagen-to-catagen transition during the hair follicle cycle [258], and this may be mediated in part by an angiogenic action of PTHrP [265].

PTHrP–/– mice that have been rescued by expression of a type II collagen–PTHrP transgene display thinning of the epidermis with hypoplastic sebaceous glands and thinning of hair [266]. These abnormalities could be reversed by targeted expression of PTHrP in skin, indicating that PTHrP expression in basal keratinocytes is necessary for maintaining normal epithelial–mesenchymal interactions during epidermal differentiation. Inhibition of PTHrP action in skin was found to produce an increase in the number of follicles involved in active hair growth [267], and topical application of a PTH/PTHrP receptor antagonist stimulates hair growth in mice [268]. These findings further support a role for PTHrP in promoting hair follicle development. PTHrP apparently maintains the pool of proliferating keratinocytes by suppressing their terminal differentiation, but the underlying mechanisms remain obscure.

PTHrP–/– mice display cranial chondrodystrophy with a failure in normal tooth eruption [269, 270]. In normal animals, PTHrP is expressed in the enamel epithelium, whereas the PTH/PTHrP receptor is expressed in the adjacent dental mesenchyme and in alveolar bone. These findings suggest that PTHrP is a regulator of epithelial–mesenchymal interactions during tooth development as well as a promoter of the resorption of alveolar bone that is required for normal tooth eruption. PTHrP increases the ratio of expression of RANKL: OPG by cementoblasts [271], an effect that presumably promotes the osteoclastic resorption required for tooth eruption [272, 273]. This effect is mediated by the PTH/PTHrP receptor since humans lacking this receptor (Blomstrand chondrodysplasia) display a failure of tooth eruption [243].

E. Other Actions of PTHrP

PTHrP is expressed in a variety of smooth muscles, where it functions as a local muscle relaxing agent. Increased intraluminal pressure (either from muscle contraction or from expanding intraluminal contents) is a known stimulus for *PTHrP* gene expression. Myometrial expression of PTHrP peaks just before the end of pregnancy, and this effect is specific for the pregnant uterine horn in unilaterally pregnant animals [274]. Mechanotransduction is likely to be the primary stimulus since physical stretch induces PTHrP expression in the nonpregnant rat uterus [275]. Human amniotic fluid contains high levels of PTHrP [276, 277], and it is possible that PTHrP produced in the amnion plays a role in suppressing myometrial contractions and/or in regulating chorionic blood flow. PTHrP is also expressed in the smooth muscle of the stomach, bladder, and oviduct, and it promotes muscle relaxation

in these tissues in response to distension [278–280]. Pharmacological doses of PTH can reproduce the relaxing effects of PTHrP, strongly indicating the involvement of the PTH/PTHrP receptor.

PTHrP has effects on both the contractility and the proliferation of vascular smooth muscle. PTHrP is widely expressed in vascular smooth muscle, and administration of PTHrP *in vivo* and *in vitro* elicits vasodilatory responses in a variety of vascular beds [281–284]. Expression of PTHrP in vascular smooth muscle is increased in experimental models of hypertension and in response to vasoconstrictors such as angiotensin II [285, 286]. Targeted overexpression of PTHrP in vascular smooth muscle of transgenic mice results in decreased baseline blood pressure as well as in a diminished hypotensive response to exogenous PTHrP, the latter possibly due to desensitization [287, 288]. The role of endogenous PTHrP is seen in transgenic mice overexpressing the PTH/PTHrP receptor in vascular smooth muscle [289, 290]. These animals are hypotensive and (as expected) are hyperresponsive to exogenous PTHrP with respect to vasodilatation. PTHrP appears to serve as an important physiological regulator of static blood pressure and as a counterregulatory factor secreted in response to vasoconstriction. PTHrP is expressed by endothelial cells [291, 292], and this may contribute to the antiangiogenic effects of the protein. PTHrP is also induced in the blood vessels bathing skeletal muscle after muscle stimulation, perhaps promoting new capillary formation in response to increased muscle contraction [293].

The genes encoding PTHrP and the PTH/PTHrP receptor are widely expressed in the central nervous system, with particularly high levels seen in cerebellar granule cells [294, 295]. These cells also express high levels of L-type calcium channels, and expression of PTHrP appears to be induced by depolarization-induced calcium influx through these channels [296]. Cerebellar granule cells are subject to excitatory cell death in response to agents such as kainic acid that trigger calcium entry through L-type calcium channels. PTHrP blocks this excitatory cell death by inhibiting L-type calcium channel activity through a mechanism that probably involves cyclic AMP signaling via the PTH/PTHrP receptor [297]. This is consistent with previous reports that exogenous PTH inhibits L-type calcium channel activity [298]. These findings suggest that PTHrP functions as a neuronal survival factor produced in response to neuroexcitatory stimuli. Addition of a blocking antibody to PTHrP prevents cerebellar granule cell survival under depolarizing conditions, suggesting that PTHrP is the endogenous factor responsible for neuroprotection [299]. Strong support for this concept is derived from studies of mice lacking expression of PTHrP in the brain. Cortical neurons from these animals display a marked increase in sensitivity to kainic acid–induced excitotoxicity [300]. PTHrP expression increases at sites of ischemic brain injury, where it may play a protective role by enhancing blood flow [301].

As discussed previously, PTHrP is expressed in the myometrium during pregnancy in response to distension produced by the growing fetus. By inducing relaxation of uterine smooth muscle, locally produced PTHrP permits progressive intrauterine growth of the fetus and may also assist in maintaining the uterus in a quiescent state until the onset of parturition. PTHrP also plays an important role in the fetal–placental unit during pregnancy. The protein is expressed in human amniotic tissue and may serve to increase chorionic blood flow [276, 277]. A role for fetal PTHrP in placental calcium transport is indicated by studies demonstrating that *PTHrP–/–* fetuses are hypocalcemic and have a reduced ability to accumulate calcium from the mother's circulation [302]. The relevant site of production of PTHrP in the fetus that drives this effect is not entirely clear. The fetal parathyroid gland is a site of expression of PTHrP [303], suggesting that this might be the source of PTHrP responsible for maintaining the positive maternal–fetal calcium gradient. Indeed, the loss of the positive maternal–fetal placental calcium gradient produced by parathyroidectomy of fetal sheep could be restored by perfusion of the placenta with PTHrP [304]. However, studies indicate that the fetal parathyroid glands are not required to maintain normal placental calcium transport [305].

IX. MECHANISM OF ACTION OF PARATHYROID HORMONE AND PARATHYROID HORMONE–RELATED PROTEIN

A. Signal Transduction

Many of the actions of PTH and PTHrP are initiated by binding of these proteins to the PTH/PTHrP receptor, a G protein–coupled receptor that activates two G proteins and thereby two major signal transduction pathways (Figure 10-5). Soon after the discovery of the cyclic AMP signaling pathway, it was found that PTH is capable of increasing levels of cyclic AMP in target cells through activation of the enzyme adenylyl cyclase [306–309]. Cyclic AMP is a second messenger in the cellular action of a wide variety of hormones and other extracellular regulatory molecules. It activates cyclic AMP-dependent protein kinase (PKA), which in

turn phosphorylates and thereby regulates key proteins that participate in physiological responses. Relatively little is known about the identity of key substrates of PKA that are phosphorylated in response to PTH/PTHrP receptor activation. These presumably include transcription factors, ion channels, transporters, and enzymes involved in cellular metabolism. PTH/PTHrP receptors also activate phospholipase C (PLC), an enzyme that hydrolyzes the plasma membrane phospholipid phosphatidylinositol-4,5-bisphosphate to produce diacylglycerol (DG) and soluble 1,4,5-inositol trisphosphate (IP_3). DG and IP_3 function as second messengers—the former by activating protein kinase C (PKC), and the latter by binding to and opening calcium channels on the membrane of the endoplasmic reticulum, thereby increasing cytosolic free calcium.

The PTH/PTHrP receptor is clearly required for PTH-stimulated bone resorption [310], and a number of studies have been carried out to identify the nature of the relevant signaling pathway(s). Agents that raise cellular cyclic AMP levels (e.g., analogs of cyclic AMP and forskolin) are capable of eliciting bone resorption in organ culture [311–315]. In addition, inhibition of cyclic AMP phosphodiesterase (thus augmenting the cellular cyclic AMP response to PTH) potentiates PTH-induced bone resorption [316]. Activation of PLC-related pathways with calcium ionophores and phorbol esters also promotes bone resorption in organ culture [317–319], and inhibition of PKC is reported to block PTH-stimulated bone resorption [320, 321]. However, at least in mouse calvarial cultures, the effects of calcium ionophores and phorbol esters require the intermediary synthesis of prostaglandins, whereas PTH-induced bone resorption does not [322]. Moreover, in some circumstances, these agents can inhibit bone resorption [323–325]. Thus, available evidence indicates that the cyclic AMP pathway plays a primary second messenger role in the stimulation of bone resorption by PTH.

PTH-induced differentiation of hematopoietic precursors to osteoclast-like cells involves the cyclic AMP pathway [326–328], although the PLC pathway may also contribute [329]. As discussed previously, PTH produces its effects on osteoclast differentiation and function by upregulating expression of RANKL and downregulating expression of OPG in osteoblastic cells. In cell culture models, these effects of PTH are mimicked by agents that raise cellular cyclic AMP levels [330–334] and inhibited by pharmacological agents that disrupt cellular cyclic AMP signaling [331, 333]. These effects are exerted, at least in part, at the level of RANKL and OPG gene transcription [335–338]. Molecular genetic studies *in vivo* further demonstrate an important role for osteoblast cyclic AMP signaling

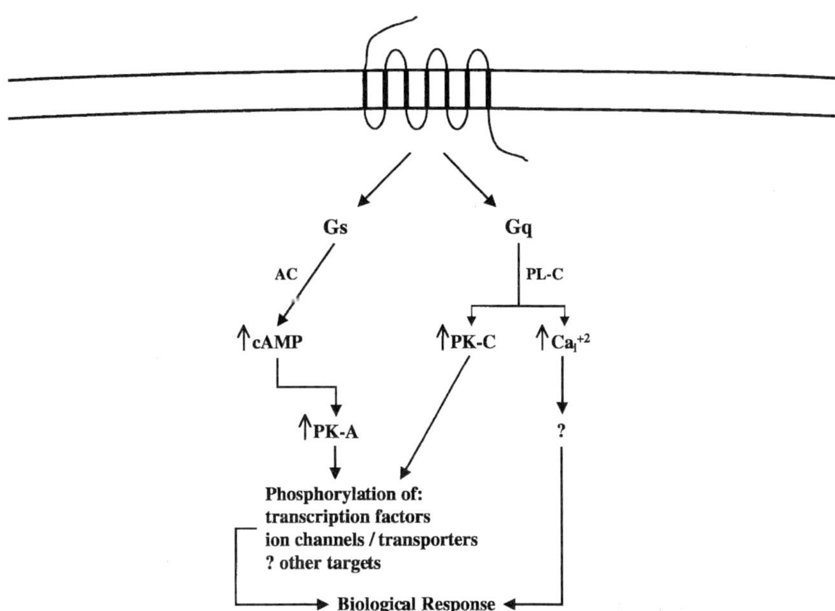

FIGURE 10-5　Signal transduction by the PTH/PTHrP receptor. PTH and PTHrP bind to determinants in the extracellular domain and in the body of the receptor. This leads to conformational changes in the transmembrane helices and consequent structural changes in the cytoplasmic domain. The latter permit productive interaction between the receptor and the G proteins G_s and G_q, activating the adenylyl cyclase (AC) and phospholipase C (PLC) signaling pathways, respectively. These pathways are thought to cooperate in determining the cellular response to the receptor activation. Most available evidence supports a primary role of the cyclic AMP/protein kinase A (PKA) pathway in mediating biological effects of PTH/PTHrP receptor activation, with the PLC pathway playing a modulatory role.

in supporting osteoclastic bone resorption. Thus, targeted deletion in osteoblasts of the alpha subunit of G_s (a protein that couples PTH/PTHrP receptors and other receptors to the production of cyclic AMP) results in mice that display a marked reduction in endosteal osteoclast number and bone resorption [339].

There has been great interest in defining the signaling events that are responsible for the anabolic response of the skeleton to intermittent administration of PTH. Progress in this area has been hampered by the paucity of *in vitro* model systems for investigation of the anabolic response to PTH and the uncertainty about the cellular basis of this effect. In principle, PTH could act to increase the number of mature osteoblasts and/or to increase the functional (bone-forming) activity of mature osteoblasts. PTH generally has been reported to have an antiproliferative effect on cultured osteoblasts, although it is reported to promote proliferation in an osteoblast precursor model [340]. PTH can also promote osteoblast differentiation *in vitro*, depending on the time and duration of treatment [134–136, 341]. *In vivo* studies have demonstrated that amino-terminal fragments of both PTH and PTHrP are anabolic, implicating the PTH/PTHrP receptor as the likely initiator of this skeletal response. Interestingly, PTH(1–30) and PTH(1–31), which activate adenylyl cyclase but have a greatly reduced ability to activate PLC, are effective as anabolic agents in bone [342, 343]. This suggests

that the cyclic AMP pathway is the major mediator of the anabolic actions of PTH. Indeed, genetic deletion of the alpha subunit of G_s in osteoblasts results in marked suppression of trabecular bone formation [339]. Cyclic AMP signaling has been implicated as a mediator of the anti-apoptotic action of PTH in osteoblasts [142] and has been linked to the activation of runx2 and osterix [344–346], transcription factors that are essential for bone formation. However, it should be noted that several studies have demonstrated that activation of cyclic AMP signaling results in inhibition of osteoblast proliferation and differentiation *in vitro* [347–351]. Taken together, these findings indicate that activation of the G_s–cAMP pathway is important for the anabolic response of the skeleton to PTH but that the complex *in vivo* skeletal milieu contributes to this effect in ways that remain to be revealed.

Microdissection studies revealed the presence of PTH-stimulated cyclic AMP generation in the proximal convoluted tubule where sodium-dependent phosphate cotransport occurs [352, 353]. Analogs of cyclic AMP were found to be effective in reproducing the phosphaturic effect of PTH [354–357]. In pseudohypoparathyroidism Ia, genetic deficiency of the alpha subunit of G_s is associated with resistance to the phosphaturic action of PTH [358–361]. With the discovery that an opossum kidney cell line (OK) retains PTH receptors [362] and PTH-inhibited sodium–phosphate cotransport [363], it became possible to carry out studies on the mechanisms of PTH inhibition of phosphate transport. Cyclic AMP clearly has a primary, although not exclusive, role in the negative regulation of sodium–phosphate cotransport by PTH [363–367]. Cyclic AMP (like PTH) promotes rapid downregulation of the type IIa sodium–phosphate cotransporter (NaPi-IIa) in OK cells via enhanced transporter endocytosis and lysosomal degradation [158, 161, 163, 368–370]. Activation of PKC by the PTH/PTHrP receptor may also contribute to inhibition of phosphate transport since treatment of OK cells with PMA or other phorbol esters substantially inhibits sodium–phosphate cotransport and reduces the expression of the type II cotransporter in some [365, 371–374] but not all [368] studies.

The cyclic AMP pathway is known to be important in mediating the effect of PTH to increase the activity of the 25(OH) vitamin D-1-hydroxylase in the proximal renal tubule [179, 375, 376]. PTH has a positive effect on the renal expression of the 1-hydroxylase mRNA *in vivo* [183, 184]. This appears to occur at the level of gene transcription [185, 377], and upstream elements in the 5′ region of the 1-hydroxylase gene confer transcriptional responses to PTH and forskolin in cultured kidney cells [378–380]. The precise elements in the promoter responsible for these effects have not been identified, but putative binding sites for the transcription factors CREB, AP-1, and CCAAT box binding protein are present and represent possible targets [187, 380–382]. PTH-stimulated PLC activation might also contribute to the 1-hydroxylase response since the combination of a calcium ionophore and PMA was shown to promote a sustained increase in 1,25(OH)$_2$ vitamin D production in perifused rat proximal tubule cells [383]. In some circumstances, inhibitors of PKC have been shown to suppress PTH-induced renal production of 1,25(OH)$_2$ vitamin D [384]. In light of these findings, it is possible that PLC has a role in the transcriptional response of the 1-hydroxylase gene to PTH.

The PTH-induced stimulation of renal calcium transport in the distal convoluted tubule appears to require activation of both the PKA and PKC pathways [170]. Inhibition of either of these kinases suppresses PTH-induced calcium uptake by distal tubular cells [385]. Moreover, simultaneous activation of both kinases was shown to be necessary and sufficient to reproduce the effect of PTH on calcium uptake [386]. PTH does not appear to increase the activity of PLC in the distal renal tubule [387], suggesting that an alternative mechanism exists for the PTH-induced generation of diacylglycerol. In this regard, PTH is capable of increasing the activity of phospholipase D, an enzyme that hydrolyzes phosphatidylcholine to produce phosphatidic acid and, indirectly, diacylglycerol [387, 388]. It is possible that

activation of phospholipase D participates in the activation of PKC that is reported to occur in response to PTH as well as amino-terminally truncated PTH fragments [389].

It is likely that the cyclic AMP signaling pathway is of primary importance as a mediator of the developmental and morphogenetic actions of PTHrP. Thus, genetic deficiency of the alpha subunit of G$_s$ in humans produces a constellation of developmental abnormalities (e.g., Abright's hereditary osteodystrophy) that overlap those seen in animals lacking PTHrP or the PTH/PTHrP receptor [390]. Moreover, targeted deletion of this gene in chondrocytes produces neonatal lethality and growth plate defects that closely resemble those seen in the absence of expression of PTHrP or the PTH/PTHrP receptor [391]. However, little is known about the molecular events that link cyclic AMP (or other second messengers) to the developmental and morphogenetic actions of PTHrP.

B. PTH/PTHrP Receptors

I. ACTIVATION OF G PROTEINS

Early studies on the PTH/PTHrP receptor demonstrated a prominent role for GTP and its analogs in regulating ligand–receptor affinity and signaling, suggesting that this receptor couples to GTP-binding (G) proteins [392–397]. The cloning of the cDNA encoding the PTH/PTHrP receptor [398] revealed a predicted protein sequence containing seven putative membrane spanning domains (Figure 10-6), a topology characteristic of members of the G protein–coupled receptor (GPCR) superfamily [399, 400]. In the case of the PTH/PTHrP receptor, the major G proteins that can be activated are G$_s$ and G$_q$. Activation of G$_s$ leads to increased adenylyl cyclase activity, resulting in increased cellular levels of cyclic AMP and activation of PKA. Activation of G$_q$ results in stimulation of PLC, resulting in mobilization of intracellular calcium and activation of PKC. Preference of the PTH/PTHrP receptor for the cyclic AMP signaling pathway is suggested by studies on PTH target cells *in vitro*, in which activation of adenylyl cyclase generally occurs at lower concentrations of added PTH than does activation of PLC [401]. These findings are consistent with the observation that the cyclic AMP pathway is most closely associated with most of the physiological effects of PTH on bone and kidney, with activation of PLC playing a lesser, modulatory role.

2. RECEPTOR ACTIVATION MECHANISMS

When the cDNA sequence of the PTH/PTHrP receptor was first delineated [398], it was apparent that it encoded a protein with a predicted overall structure consistent with those of other known GPCRs. In particular,

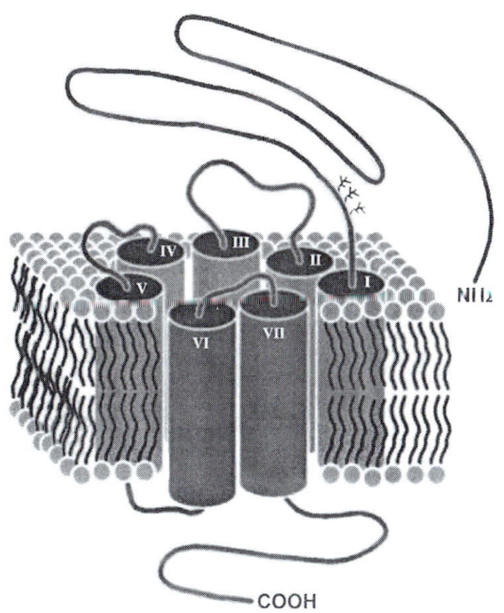

FIGURE 10-6   Structural model of the PTH/PTHrP receptor, indicating the presence of seven membrane spanning helices that surround a central polar cavity. The receptor contains a large, glycosylated N-terminal extracellular domain and a long C-terminal cytoplasmic tail. Agonist binding to the receptor alters the relative orientation of the transmembrane helices, promoting activation of specific G proteins.

the receptor was modeled as containing seven membrane spanning helices, with a large amino-terminal extracellular domain, three extracellular loops, three intracellular loops, and a large carboxy-terminal cytoplasmic tail (Figure 10-6). However, the PTH/PTHrP receptor does not share a number of the specific sequence motifs present in the largest subfamily of GPCRs (the so-called class I family, which includes receptors for a diverse group of ligands ranging from photons to polypeptide hormones). Rather, the PTH/PTHrP receptor is a member of a second GPCR subfamily (class II) that includes receptors for calcitonin, glucagon, and a number of other polypeptide ligands [402]. Members of the class II GPCR subfamily are presumed to share a common basic mechanism of G protein activation but have evolved determinants of specificity that permit binding and activation by only the appropriate peptide ligand.

Mutagenesis studies have been performed to investigate the structural features in the PTH/PTHrP receptor that are important for agonist binding and for maintaining receptor specificity. These studies have demonstrated that the large amino-terminal extracellular domain of the receptor contains critical determinants of agonist binding affinity [403–405]. However, the body of the receptor, which includes the extracellular loops and the transmembrane domains, also plays a role in ligand binding as well as in maintaining ligand specificity

[404, 406–408]. Sites of interaction between amino-terminal PTH fragments and the PTH/PTHrP receptor have been mapped in a series of elegant biochemical studies. There appear to be multiple points of contact between the 1–34 region of PTH/PTHrP and the receptor [408, 409]. Initially, residues in the 23– to 33–amino acid region of PTH(1–34) or PTHrP(1–34) interact with the N-terminal extracellular domain of the PTH/PTHrP receptor [410, 411]. This facilitates an additional interaction between the N-terminus of the ligand and the transmembrane domain of the receptor [412]. This latter interaction is presumably required to initiate the conformation shift in the transmembrane domain of the receptor that is required for signal transduction [413]. This involves the exposure of key amino acids in the second and third cytoplasmic loops of the PTH/PTHrP receptor that are required for activation of G_s and G_q [414, 415]. Additional interactions have been observed between the middle portion of the 1–34 ligands and the extracellular ends of transmembrane helices 1 and 2 [416–418]. These may help to dock the ligand in a position that promotes the association of the N-terminus with the sixth transmembrane domain, the key step in receptor activation.

3. RECEPTOR REGULATION

Signal transduction by GPCRs is generally subject to tight regulatory control. This control can occur in response to agonist binding (homologous regulation) or in response to factors acting through separate pathways (heterologous regulation). Acute control of signaling is accomplished by blocking the ability of agonist-occupied receptors to sustain activation of G proteins (desensitization) or by physically moving the receptors into an intracellular compartment effectively separating them from G proteins (sequestration). Chronic regulation of receptor signaling is accomplished by agonist-induced changes in steady-state levels of expression of receptors due to increased receptor catabolism following receptor internalization (downregulation) and to changes in *de novo* receptor synthesis. Homologous regulation commonly involves all of these mechanisms, whereas heterologous regulation most often occurs through changes in steady-state levels of receptor expression.

Many studies have documented homologous regulation of PTH/PTHrP receptor signaling. Treatment of cultured bone and kidney cells with PTH generally dampens the adenylyl cyclase and PLC responses to a second addition of the hormone [419–428]. In most studies, desensitization of the PTH response occurs rapidly, within minutes of initial exposure to PTH, suggesting that the PTH/PTHrP receptor has become acutely uncoupled from its cognate G proteins. The mechanisms underlying acute desensitization have been well studied for GPCRs such as rhodopsin and β-adrenergic receptors [429–431]. The major mechanism underlying acute desensitization of these receptors is phosphorylation of the cytoplasmic domain of the receptor by a GPCR kinase (GRK). GRKs are serine/threonine kinases that phosphorylate only the agonist-occupied receptor, and phosphorylation facilitates the interaction of the receptor with a member of the arrestin protein family. Arrestin binding to the receptor sterically interferes with the interaction between the receptor and G proteins, thus preventing signal transmission. There is strong evidence that a similar mechanism applies to desensitization of PTH/PTHrP receptor signaling. The PTH/PTHrP receptor is subject to phosphorylation in response to agonist binding [432, 433], and this appears to occur largely if not exclusively on serine residues in the cytoplasmic tail [433–435]. The kinase involved appears to be a member of the GRK family, possibly GRK-2 [434, 436, 437], and a dominant inhibitor of GRK function can suppress PTH/PTHrP receptor desensitization in human osteoblast-like cells [143]. The importance of phosphorylation of the PTH/PTHrP receptor in limiting target cell responsiveness to PTH has been demonstrated *in vivo* [438].

Long-term treatment with PTH results in a loss of cellular PTH/PTHrP receptors (downregulation) and a corresponding reduction in the maximal signaling response to the hormone [427, 439–442]. There is evidence that this process may have pathophysiological relevance. For example, vitamin D deficiency can be associated with target cell resistance to PTH [443–445]. In animal studies, this resistance can be reversed by parathyroidectomy, suggesting that it is the secondary hyperparathyroidism that is responsible for target cell resistance [446]. Infusion of PTH to levels seen in severe secondary hyperparathyroidism produces downregulation of PTH/PTHrP receptors and a reduction in the adenylyl cyclase response to PTH [439]. In chronic renal failure, factors other than hyperparathyroidism may also contribute to reduced target cell expression of PTH/PTHrP receptors [447]. The initial step in downregulation of PTH/PTHrP receptors appears to be agonist-induced accumulation of the receptor in plasma membrane clathrin-coated pits [56, 448]. These pits are endocytic organelles that pinch off from the plasma membrane, thus becoming endocytic vesicles. Once internalized, PTH/PTHrP receptors can be recycled to the plasma membrane or can presumably progress farther down the endocytic pathway to the lysosomes for degradation. The molecular mechanisms underlying the agonist-induced internalization of the PTH/PTHrP receptor are not entirely clear. Agonist-stimulated receptor phosphorylation may facilitate internalization of the PTH/PTHrP receptor [143, 449], although receptor phosphorylation is not required for endocytosis in all cellular settings [435]. Arrestins have been implicated as mediators of GPCR endocytosis, and it is clear that arrestins can become associated with the PTH/PTHrP receptor following agonist binding [450, 451]. In addition, the cytoplasmic tail of the PTH/PTHrP receptor contains a tyrosine-based sequence that has been implicated in promoting internalization of other membrane receptors. Mutation of this sequence markedly inhibits agonist-induced PTH/PTHrP receptor endocytosis [448]. Interestingly, there is evidence that arrestin binding to the PTH/PTHrP receptor can also contribute to activation of the MAP kinase pathway by PTH [452–454] and to the anabolic effect of PTH *in vivo* [455].

Another mechanism for regulation of PTH/PTHrP receptor levels is through changes in expression of the receptor gene. In osteoblastic cells, PTH is reported to decrease levels of PTH/PTHrP receptor mRNA by a mechanism involving the cyclic AMP pathway [456, 457]. This may be due to direct transcriptional activation of the *PTH/PTHrP receptor* gene by PKA-activated transcription factors [458], but the details of this pathway have yet to be elucidated. Homologous control of PTH/PTHrP receptor expression appears to be target cell specific in that PTH reportedly does not reduce expression of the *PTH/PTHrP receptor* gene in the kidneys of rats with secondary hyperparathyroidism [447, 459]. Heterologous factors are also reported to regulate levels of PTH/PTHrP receptor expression in bone and kidney. The cytokine TGF-β upregulates the expression of the PTH/PTHrP receptor in osteoblastic osteosarcoma cells [460], although the opposite effect is reported in primary cultures of fetal rat osteoblasts [461] and in OK cells [462]. IGF-1 downregulates the expression of the PTH/PTHrP receptor by a transcriptional mechanism [463]. Dexamethasone treatment produces an increase in expression of the PTH/PTHrP receptor in osteoblastic cells but not in kidney cells [464, 465], whereas $1,25(OH)_2$ vitamin D downregulates expression of the *PTH/PTHrP receptor* gene [466]. Thyroid hormone upregulates expression of the PTH/PTHrP receptor [467]. It should be noted that most of these studies have been carried out in cultured bone and kidney cells *in vitro*, and much more needs to be done to establish the physiological relevance of these effects.

C. Nontraditional Mechanisms of Action of PTHrP

The discovery of PTHrP was based on the PTH-like endocrine actions of this peptide in patients with malignancy-associated hypercalcemia. The classical mechanism of action of PTHrP is thus to bind to and activate the widely expressed PTH/PTHrP receptor.

The amino-terminal 1–34 domain of PTHrP is responsible for binding to the PTH/PTHrP receptor, thus initiating signal transduction. However, it appears that the PTH/PTHrP receptor does not mediate all of the physiological actions of PTHrP. Two additional mechanisms have been identified by which PTHrP can potentially influence cellular function (Figure 10-7). One involves the notion of PTHrP as a polyhormone that yields mid- and carboxyl-region fragments with distinct biological activities that are presumably mediated by novel cell surface receptors. The second mechanism relates to the ability of PTHrP to translocate to the nucleus of cells in which it is expressed, thereby altering cell proliferation and/or gene expression.

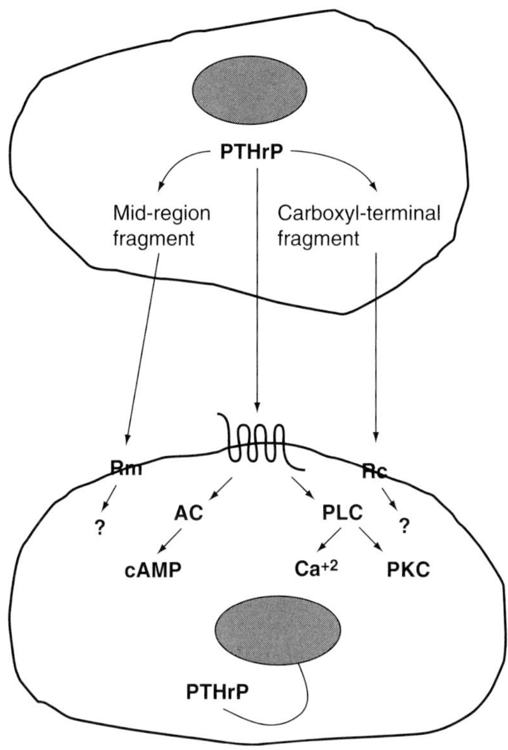

FIGURE 10-7   Mechanisms of action of PTHrP. The majority of the actions of PTHrP result from the binding of the amino-terminal portion of the protein to the PTH/PTHrP receptor, leading to the activation of adenylyl cyclase (AC) and phospholipase C (PLC). Activation of these effector enzymes results in increased cellular levels of cyclic AMP (cAMP), intracellular calcium, and protein kinase C (PKC). PTHrP is also processed post-translationally, producing midregion and C-terminal fragments of the protein. These fragments have cellular effects that are presumably mediated by novel membrane receptors (Rm and Rc), acting through unknown signaling pathways. PTHrP has also been localized to the nucleus of cells (intracrine action), where it may regulate nuclear functions such as mitosis, apoptosis, and RNA processing.

1. PTHrP as a Polyhormone

The PTHrP gene is subject to alternative splicing, resulting in multiple protein products (ranging from 139 to 173 amino acids) that differ only in the extent of their C-termini [202]. Only the N-terminal 34 amino acids are needed to produce all of the PTH-like actions of PTHrP on the PTH/PTHrP receptor, and several groups have been interested in assessing a possible biological role for the remainder of the molecule. Indeed, PTHrP is subject to post-translational proteolytic processing [468], and the cleavage products include a midregion fragment (amino acids 38–94) and a C-terminal fragment (amino acids 107–139) as well as PTHrP(1–36) [200]. Fragments of PTHrP are secreted by some cells, at least *in vitro*, and thus have the potential to elicit biological responses in a paracrine or endocrine fashion. Synthetic PTHrP(107–139) has been reported to elicit biological effects such as inhibition of bone resorption [469], stimulation of osteoblast proliferation [470], and stimulation of interleukin-6 expression in osteoblasts [471]. The nature of the receptor and signaling pathway responsible for these actions of PTHrP is unclear, although the latter effect appeared to involve activation of PKC. This peptide is also reported to activate voltage-sensitive calcium channels in osteoblastic cells [472].

A physiological role for PTHrP fragments is suggested by studies of placental calcium transport. The normal positive maternal–fetal calcium gradient can be restored in parathyroidectomized sheep fetuses by administration of midregion fragments of PTHrP but not by PTH or by N-terminal PTHrP fragments [302, 473]. This effect must therefore be initiated by a receptor distinct from the classical PTH/PTHrP receptor.

2. Intracrine Actions of PTHrP

Several studies have demonstrated that, once synthesized, PTHrP can localize to the nucleolus as well as be secreted [474, 475]. Nucleolar localization requires the presence of a targeting signal in the carboxyl region of the molecule [476] and occurs through an interaction with the targeting protein importin-β. Secreted PTHrP can also be taken up by cells and translocated to the nucleus, and this appears to involve a receptor distinct from the PTH/PTHrP receptor [477]. Although the functional significance of nuclear PTHrP has yet to be definitely established, a number of intriguing findings have been reported. Intracellular expression of PTHrP has been shown to protect chondrocytes from apoptosis induced by serum deprivation, and this effect was dependent on the presence of an intact nucleolar localization signal [478]. Targeting of PTHrP to the nucleus may involve synthesis of PTHrP from alternative translational start sites [479, 480]. Nuclear localization of PTHrP is associated with mitogenesis in cultured vascular smooth muscle cells [481, 482]. This effect requires serine and threonine residues in the mid- to C-terminal region of PTHrP and involves phosphorylation of the cell cycle checkpoint retinoblastoma protein [483]. Proliferative effects of PTHrP are associated with downregulation of the cyclin-dependent kinase inhibitor p57 in chondrocytes and in vascular smooth muscle cells [230, 484]. By contrast, secreted PTHrP inhibits proliferation of vascular smooth muscle cells via activation of cyclic AMP signaling by the PTH/PTHrP receptor [286, 485]. In cultured keratinocytes, PTHrP is present in the nucleolus during the G_1 phase of the cell cycle but redistributes to the cytoplasm during cell division [486]. Interestingly, PTHrP is phosphorylated by the cell cycle regulatory kinase CDC2–CDK2, and this appears to promote translocation of the PTHrP from the nucleus to the cytoplasm [487]. Nuclear export of PTHrP is also regulated by a leucine-rich region in the C-terminal domain of PTHrP [488]. It is possible that PTHrP acts, at least in part, through direct interaction with ribonucleoprotein complexes since PTHrP is capable of binding directly to RNA via a polybasic region within the nuclear localization signal [477]. Further work is needed to more clearly define the physiological significance of intracrine signaling by PTHrP.

ACKNOWLEDGMENTS

Portions of this work were supported by National Institutes of Health grants DK35323 and DK072071 and by the Medical Research Service of the Department of Veterans' Affairs.

REFERENCES

1. J. T. Potts, Parathyroid hormone: Past and present. *J Endocrinol* **187,** 311–325 (2005).
2. E. M. Brown, Extracellular Ca^{2+} sensing, regulation of parathyroid cell function, and role of Ca^{2+} and other ions as extracellular (first) messengers. *Physiol Rev* **71,** 371–411 (1991).
3. J. Russell, D. Lettieri, and L. M. Sherwood, Direct regulation by calcium of cytoplasmic messenger ribonucleic acid coding for pre-proparathyroid hormone in isolated bovine parathyroid cells. *J Clin Invest* **72,** 1851–1855 (1983).
4. T. Naveh-Many and J. Silver, Regulation of parathyroid hormone gene expression by hypocalcemia, hypercalcemia, and vitamin D in the rat. *J Clin Invest* **86,** 1313–1319 (1990).
5. Y. C. Li, M. Amling, A. E. Pirro, M. Priemel, J. Meuse, R. Baron, G. Delling, and M. B. Demay, Normalization of mineral ion homeostasis by dietary means prevents hyperparathyroidism, rickets, and osteomalacia, but not alopecia in vitamin D receptor-ablated mice. *Endocrinology* **139,** 4391–4396 (1998).
6. J. Silver and R. Levi, Cellular and molecular mechanisms of secondary hyperparathyroidism. *Clin Nephrol* **63,** 119–126 (2005).
7. N. Chattopadhyay and E. M. Brown, Role of calcium-sensing receptor in mineral ion metabolism and inherited disorders of calcium-sensing. *Mol Genet Metab* **89,** 189–202 (2006).
8. A. M. Hofer and E. M. Brown, Extracellular calcium sensing and signalling. *Nat Rev Mol Cell Biol* **4,** 530–538 (2003).
9. E. M. Brown, Calcium receptor and regulation of parathyroid hormone secretion. *Rev Endocr Metab Disord* **1,** 307–315 (2000).
10. W. Chang and D. Shoback, Extracellular Ca^{2+}-sensing receptors—An overview. *Cell Calcium* **35,** 183–196 (2004).
11. E. M. Brown and R. J. MacLeod, Extracellular calcium sensing and extracellular calcium signaling. *Physiol Rev* **81,** 239–297 (2001).
12. E. M. Brown, G. Gamba, D. Riccardi, M. Lombardi, R. Butters, O. Kifor, A. Sun, M. A. Hediger, J. Lytton, and S. C. Hebert, Cloning and characterization of an extracellular Ca(2+)-sensing receptor from bovine parathyroid. *Nature* **366,** 575–580 (1993).
13. W. Chang, T. H. Chen, S. Pratt, and D. Shoback, Amino acids in the second and third intracellular loops of the parathyroid Ca^{2+}-sensing receptor mediate efficient coupling to phospholipase C. *J Biol Chem* **275,** 19955–19963 (2000).
14. S. H. Pearce, C. Williamson, O. Kifor, M. Bai, M. G. Coulthard, M. Davies, N. Lewis-Barned, D. McCredie, H. Powell, P. Kendall-Taylor, E. M. Brown, and R. V. Thakker, A familial syndrome of hypocalcemia with hypercalciuria due to mutations in the calcium-sensing receptor. *N Engl J Med* **335,** 1115–1122 (1996).
15. M. R. Pollak, C. E. Seidman, and E. M. Brown, Three inherited disorders of calcium sensing. *Medicine (Baltimore)* **75,** 115–123 (1996).
16. C. Ho, D. A. Conner, M. R. Pollak, D. J. Ladd, O. Kifor, H. B. Warren, E. M. Brown, J. G. Seidman, and C. E. Seidman, A mouse model of human familial hypocalciuric hypercalcemia and neonatal severe hyperparathyroidism. *Nat Genet* **11,** 389–394 (1995).
17. F. Raue, C. Haag, E. Schulze, and K. Frank-Raue, The role of the extracellular calcium-sensing receptor in health and disease. *Exp Clin Endocrinol Diabetes* **114,** 397–405 (2006).
18. S. C. Hebert, Therapeutic use of calcimimetics. *Annu Rev Med* **57,** 349–364 (2006).
19. S. J. Steddon and J. Cunningham, Calcimimetics and calcilytics—Fooling the calcium receptor. *Lancet* **365,** 2237–2239 (2005).
20. S. A. Berson and R. S. Yalow, Immunochemical heterogeneity of parathyroid hormone in plasma. *J Clin Endocrinol Metab* **28,** 1037–1047 (1968).
21. J. M. Canterbury and E. Reiss, Fractionation of circulating parathyroid hormone (PTH) in man. *J Lab Clin Med* **78,** 814 (1971).
22. C. D. Arnaud, Immunochemical heterogeneity of circulating parathyroid hormone in man: Sequel to an original observation by Berson and Yalow. *Mt Sinai J Med* **40,** 422–432 (1973).
23. S. B. Oldham, E. J. Finck, and F. R. Singer, Parathyroid hormone clearance in man. *Metabolism* **27,** 993–1001 (1978).
24. G. V. Segre, H. D. Niall, J. F. Habener, and J. T. Potts, Jr., Metabolism of parathyroid hormone: Physiologic and clinical significance. *Am J Med* **56,** 774–784 (1974).
25. J. Fox, M. Scott, R. A. Nissenson, and H. Heath, 3rd, Effect of plasma calcium concentration on the metabolic clearance rate of parathyroid hormone in the dog. *J Lab Clin Med* **102,** 70–77 (1983).

26. S. K. Libutti, H. R. Alexander, D. L. Bartlett, M. L. Sampson, M. E. Ruddel, M. Skarulis, S. J. Marx, A. M. Spiegel, W. Simmonds, and A. T. Remaley, Kinetic analysis of the rapid intraoperative parathyroid hormone assay in patients during operation for hyperparathyroidism. *Surgery* **126,** 1145–1151 (1999).

27. D. B. Endres, R. Villanueva, C. F. Sharp, Jr., and F. R. Singer, Measurement of parathyroid hormone. *Endocrinol Metab Clin North Am* **18,** 611–629 (1989).

28. K. J. Martin, K. A. Hruska, J. J. Freitag, S. Klahr, and E. Slatopolsky, The peripheral metabolism of parathyroid hormone. *N Engl J Med* **301,** 1092–1098 (1979).

29. K. Martin, K. Hruska, A. Greenwalt, S. Klahr, and E. Slatopolsky, Selective uptake of intact parathyroid hormone by the liver: Differences between hepatic and renal uptake. *J Clin Invest* **58,** 781–788 (1976).

30. K. A. Hruska, A. Korkor, K. Martin, and E. Slatopolsky, Peripheral metabolism of intact parathyroid hormone. Role of liver and kidney and the effect of chronic renal failure. *J Clin Invest* **67,** 885–892 (1981).

31. F. R. Bringhurst, A. M. Stern, M. Yotts, N. Mizrahi, G. V. Segre, and J. T. Potts, Jr., Peripheral metabolism of PTH: Fate of biologically active amino terminus *in vivo. Am J Physiol* **255,** E886–E893 (1988).

32. J. A. Flueck, F. P. Di Bella, A. J. Edis, J. M. Kehrwald, and C. D. Arnaud, Immunoheterogeneity of parathyroid hormone in venous effluent serum from hyperfunctioning parathyroid glands. *J Clin Invest* **60,** 1367–1375 (1977).

33. S. Erdmann, W. Muller, S. Bahrami, S. I. Vornehm, H. Mayer, P. Bruckner, K. von der Mark, and H. Burkhardt, Differential effects of parathyroid hormone fragments on collagen gene expression in chondrocytes. *J Cell Biol* **135,** 1179–1191 (1996).

34. M. Nasu, T. Sugimoto, H. Kaji, J. Kano, and K. Chihara, Carboxyl-terminal parathyroid hormone fragments stimulate type-1 procollagen and insulin-like growth factor-binding protein-5 mRNA expression in osteoblastic UMR-106 cells. *Endocr J* **45,** 229–234 (1998).

35. T. Tsuboi and A. Togari, Comparison of the effects of carboxyl-terminal parathyroid hormone peptide[53–84] and aminoterminal peptide[1–34] on mouse tooth germ *in vitro. Arch Oral Biol* **43,** 335–339 (1998).

36. C. Duvos, A. Scutt, and H. Mayer, hPTH-fragments (53–84) and (28–48) antagonize the stimulation of calcium release and repression of alkaline phosphatase activity by hPTH-(1–34) *in vitro. FEBS Lett* **580,** 1509–1514 (2006).

37. N. Inomata, M. Akiyama, N. Kubota, and H. Juppner, Characterization of a novel parathyroid hormone (PTH) receptor with specificity for the carboxyl-terminal region of PTH-(1–84). *Endocrinology* **136,** 4732–4740 (1995).

38. H. Takasu, H. Baba, N. Inomata, Y. Uchiyama, N. Kubota, K. Kumaki, A. Matsumoto, K. Nakajima, T. Kimura, S. Sakakibara, T. Fujita, K. Chihara, and I. Nagai, The 69–84 amino acid region of the parathyroid hormone molecule is essential for the interaction of the hormone with the binding sites with carboxyl-terminal specificity. *Endocrinology* **137,** 5537–5543 (1996).

39. P. Divieti, N. Inomata, K. Chapin, R. Singh, H. Juppner, and F. R. Bringhurst, Receptors for the carboxyl-terminal region of pth(1–84) are highly expressed in osteocytic cells. *Endocrinology* **142,** 916–925 (2001).

40. P. Divieti, A. I. Geller, G. Suliman, H. Juppner, and F. R. Bringhurst, Receptors specific for the carboxyl-terminal region of parathyroid hormone on bone-derived cells: Determinants of

41. T. M. Murray, L. G. Rao, P. Divieti, and F. R. Bringhurst, Parathyroid hormone secretion and action: Evidence for discrete receptors for the carboxyl-terminal region and related biological actions of carboxyl-terminal ligands. *Endocr Rev* **26,** 78–113 (2005).

42. R. R. MacGregor, J. W. Hamilton, G. N. Kent, R. E. Shofstall, and D. V. Cohn, The degradation of proparathormone and parathormone by parathyroid and liver cathepsin B. *J Biol Chem* **254,** 4428–4433 (1979).

43. J. W. Hamilton, R. L. Jilka, and R. R. MacGregor, Cleavage of parathyroid hormone to the 1–34 and 35–84 fragments by cathepsin D-like activity in bovine parathyroid gland extracts. *Endocrinology* **113,** 285–292 (1983).

44. G. P. Mayer, J. A. Keaton, J. G. Hurst, and J. F. Habener, Effects of plasma calcium concentration on the relative proportion of hormone and carboxyl fragments in parathyroid venous blood. *Endocrinology* **104,** 1778–1784 (1979).

45. P. D'Amour, J. H. Brossard, L. Rousseau, L. Nguyen-Yamamoto, E. Nassif, C. Lazure, D. Gauthier, J. R. Lavigne, and R. J. Zahradnik, Structure of non-(1–84) PTH fragments secreted by parathyroid glands in primary and secondary hyperparathyroidism. *Kidney Int* **68,** 998–1007 (2005).

46. P. D'Amour, Circulating PTH molecular forms: What we know and what we don't. *Kidney Int Suppl* S29–S33 (2006).

47. P. A. Friedman and W. G. Goodman, PTH(1–84)/PTH(7–84): A balance of power. *Am J Physiol Renal Physiol* **290,** F975–F984 (2006).

48. H. Yamashita, P. Gao, T. Cantor, T. Futata, T. Murakami, S. Uchino, S. Watanabe, H. Kawamoto, M. Fukagawa, and S. Noguchi, Large carboxy-terminal parathyroid hormone (PTH) fragment with a relatively longer half-life than 1–84 PTH is secreted directly from the parathyroid gland in humans. *Eur J Endocrinol* **149,** 301–306 (2003).

49. L. Nguyen-Yamamoto, L. Rousseau, J. H. Brossard, R. Lepage, P. Gao, T. Cantor, and P. D'Amour, Origin of parathyroid hormone (PTH) fragments detected by intact-PTH assays. *Eur J Endocrinol* **147,** 123–131 (2002).

50. H. Juppner and J. T. Potts, Jr., Immunoassays for the detection of parathyroid hormone. *J Bone Miner Res* **17**(Suppl. 2), N81–N86 (2002).

51. N. Horiuchi, M. F. Holick, J. T. Potts, Jr., and M. Rosenblatt, A parathyroid hormone inhibitor *in vivo*: Design and biological evaluation of a hormone analog. *Science* **220,** 1053–1055 (1983).

52. R. Talmage, The demand for bone calcium in maintenance of plasma calcium concentration. In *Mechanisms of Localized Bone Loss* (R. Horton, T. M. Tarplay, and W. F. Davis, eds.), pp. 73–92. Info Retrieval, Washington, DC (1967).

53. R. A. M. Talmage, Physiological role of parathyroid hormone. In *Handbook of Physiology* (R. A. A. Grup, ed.), pp. 343–351. American Physiological Society, Washington, DC (1967).

54. W. F. Neuman, M. W. Neuman, and C. R. Myers, Blood:bone disequilibrium: III. Linkage between cell energetics and Ca fluxes. *Am J Physiol* **236,** C244–C248 (1979).

55. M. E. Holtrop and G. J. King, The ultrastructure of the osteoclast and its functional implications. *Clin Orthop Relat Res* **123,** 177–196 (1977).

56. C. M. Silve, G. T. Hradek, A. L. Jones, and C. D. Arnaud, Parathyroid hormone receptor in intact embryonic chicken bone: Characterization and cellular localization. *J Cell Biol* **94,** 379–386 (1982).

57. M. F. Rouleau, H. Warshawsky, and D. Goltzman, Parathyroid hormone binding *in vivo* to renal, hepatic, and skeletal tissues

of the rat using a radioautographic approach. *Endocrinology* **118**, 919–931 (1986).

58. A. Teti, R. Rizzoli, and A. Zambonin Zallone, Parathyroid hormone binding to cultured avian osteoclasts. *Biochem Biophys Res Commun* **174**, 1217–1222 (1991).

59. N. Agarwala and C. V. Gay, Specific binding of parathyroid hormone to living osteoclasts. *J Bone Miner Res* **7**, 531–539 (1992).

60. L. G. Rao, T. M. Murray, and J. N. Heersche, Immunohistochemical demonstration of parathyroid hormone binding to specific cell types in fixed rat bone tissue. *Endocrinology* **113**, 805–810 (1983).

61. D. W. Dempster, C. E. Hughes-Begos, K. Plavetic-Chee, A. Brandao-Burch, F. Cosman, J. Nieves, S. Neubort, S. S. Lu, A. Iida-Klein, T. Arnett, and R. Lindsay, Normal human osteoclasts formed from peripheral blood monocytes express PTH type 1 receptors and are stimulated by PTH in the absence of osteoblasts. *J Cell Biochem* **95**, 139–148 (2005).

62. T. J. Chambers, P. M. McSheehy, B. M. Thomson, and K. Fuller, The effect of calcium-regulating hormones and prostaglandins on bone resorption by osteoclasts disaggregated from neonatal rabbit bones. *Endocrinology* **116**, 234–239 (1985).

63. R. L. Jilka, Are osteoblastic cells required for the control of osteoclast activity by parathyroid hormone? *Bone Miner* **1**, 261–266 (1986).

64. S. S. Miller, A. M. Wolf, and C. D. Arnaud, Bone cells in culture: Morphologic transformation by hormones. *Science* **192**, 1340–1343 (1976).

65. R. Civitelli, K. Ziambaras, P. M. Warlow, F. Lecanda, T. Nelson, J. Harley, N. Atal, E. C. Beyer, and T. H. Steinberg, Regulation of connexin 43 expression and function by prostaglandin E2 (PGE2) and parathyroid hormone (PTH) in osteoblastic cells. *J Cell Biochem* **68**, 8–21 (1998).

66. P. C. Schiller, B. A. Roos, and G. A. Howard, Parathyroid hormone upregulation of connexin 43 gene expression in osteoblasts depends on cell phenotype. *J Bone Miner Res* **12**, 2005–2013 (1997).

67. H. J. Donahue, K. J. McLeod, C. T. Rubin, J. Andersen, E. A. Grine, E. L. Hertzberg, and P. R. Brink, Cell-to-cell communication in osteoblastic networks: Cell line-dependent hormonal regulation of gap junction function. *J Bone Miner Res* **10**, 881–889 (1995).

68. J. E. Puzas and J. S. Brand, Parathyroid hormone stimulation of collagenase secretion by isolated bone cells. *Endocrinology* **104**, 559–562 (1979).

69. J. K. Heath, S. J. Atkinson, M. C. Meikle, and J. J. Reynolds, Mouse osteoblasts synthesize collagenase in response to bone resorbing agents. *Biochim Biophys Acta* **802**, 151–154 (1984).

70. Y. Eeckhout, J. M. Delaisse, and G. Vaes, Direct extraction and assay of bone tissue collagenase and its relation to parathyroid-hormone-induced bone resorption. *Biochem J* **239**, 793–796 (1986).

71. N. C. Partridge, J. J. Jeffrey, L. S. Ehlich, S. L. Teitelbaum, C. Fliszar, H. G. Welgus, and A. J. Kahn, Hormonal regulation of the production of collagenase and a collagenase inhibitor activity by rat osteogenic sarcoma cells. *Endocrinology* **120**, 1956–1962 (1987).

72. J. A. Hamilton, S. Lingelbach, N. C. Partridge, and T. J. Martin, Regulation of plasminogen activator production by bone-resorbing hormones in normal and malignant osteoblasts. *Endocrinology* **116**, 2186–2191 (1985).

73. G. Leloup, C. Peeters-Joris, J. M. Delaisse, G. Opdenakker, and G. Vaes, Tissue and urokinase plasminogen activators in bone tissue and their regulation by parathyroid hormone. *J Bone Miner Res* **6**, 1081–1090 (1991).

74. T. Nonaka, H. Matsumoto, W. Shimada, K. Okada, H. Fukao, S. Ueshima, H. Kikuchi, S. Tanaka, and O. Matsuo, Effect of bone resorbing factors on μ-PA and its specific receptor in osteosarcoma cell line. *Clin Chim Acta* **223**, 129–142 (1993).

75. W. Zhao, M. H. Byrne, B. F. Boyce, and S. M. Krane, Bone resorption induced by parathyroid hormone is strikingly diminished in collagenase-resistant mutant mice. *J Clin Invest* **103**, 517–524 (1999).

76. R. Chiusaroli, A. Maier, M. C. Knight, M. Byrne, L. M. Calvi, R. Baron, S. M. Krane, and E. Schipani, Collagenase cleavage of type I collagen is essential for both basal and parathyroid hormone (PTH)/PTH-related peptide receptor-induced osteoclast activation and has differential effects on discrete bone compartments. *Endocrinology* **144**, 4106–4116 (2003).

77. S. Amano, S. Hanazawa, K. Hirose, Y. Ohmori, and S. Kitano, Stimulatory effect on bone resorption of interleukin-1-like cytokine produced by an osteoblast-rich population of mouse calvarial cells. *Calcif Tissue Int* **43**, 88–91 (1988).

78. M. C. Horowitz, D. L. Coleman, P. M. Flood, T. S. Kupper, and R. L. Jilka, Parathyroid hormone and lipopolysaccharide induce murine osteoblast-like cells to secrete a cytokine indistinguishable from granulocyte–macrophage colony-stimulating factor. *J Clin Invest* **83**, 149–157 (1989).

79. D. B. Evans, R. A. Bunning, J. Van Damme, and R. G. Russell, Natural human IL-1 beta exhibits regulatory actions on human bone-derived cells *in vitro*. *Biochem Biophys Res Commun* **159**, 1242–1248 (1989).

80. E. Jimi, I. Nakamura, H. Amano, Y. Taguchi, T. Tsurukai, M. Tamura, N. Takahashi, and T. Suda, Osteoclast function is activated by osteoblastic cells through a mechanism involving cell-to-cell contact. *Endocrinology* **137**, 2187–2190 (1996).

81. T. Wada, T. Nakashima, N. Hiroshi, and J. M. Penninger, RANKL-RANK signaling in osteoclastogenesis and bone disease. *Trends Mol Med* **12**, 17–25 (2006).

82. T. Suda, N. Takahashi, N. Udagawa, E. Jimi, M. T. Gillespie, and T. J. Martin, Modulation of osteoclast differentiation and function by the new members of the tumor necrosis factor receptor and ligand families. *Endocr Rev* **20**, 345–357 (1999).

83. L. C. Hofbauer, S. Khosla, C. R. Dunstan, D. L. Lacey, W. J. Boyle, and B. L. Riggs, The roles of osteoprotegerin and osteoprotegerin ligand in the paracrine regulation of bone resorption. *J Bone Miner Res* **15**, 2–12 (2000).

84. L. C. Hofbauer, C. A. Kuhne, and V. Viereck, The OPG/RANKL/RANK system in metabolic bone diseases. *J Musculoskelet Neuronal Interact* **4**, 268–275 (2004).

85. G. D. Roodman, Regulation of osteoclast differentiation. *Ann N Y Acad Sci* **1068**, 100–109 (2006).

86. S. Tanaka, K. Nakamura, N. Takahasi, and T. Suda, Role of RANKL in physiological and pathological bone resorption and therapeutics targeting the RANKL-RANK signaling system. *Immunol Rev* **208**, 30–49 (2005).

87. W. J. Boyle, W. S. Simonet, and D. L. Lacey, Osteoclast differentiation and activation. *Nature* **423**, 337–342 (2003).

88. J. E. Aubin and E. Bonnelye, Osteoprotegerin and its ligand: A new paradigm for regulation of osteoclastogenesis and bone resorption. *Osteoporos Int* **11**, 905–913 (2000).

89. D. L. Lacey, E. Timms, H. L. Tan, M. J. Kelley, C. R. Dunstan, T. Burgess, R. Elliott, A. Colombero, G. Elliott, S. Scully, H. Hsu, J. Sullivan, N. Hawkins, E. Davy, C. Capparelli, A. Eli, Y. X. Qian, S. Kaufman, I. Sarosi, V. Shalhoub, G. Senaldi, J. Guo, J. Delaney, and W. J. Boyle, Osteoprotegerin ligand is a cytokine that regulates osteoclast differentiation and activation. *Cell* **93**, 165–176 (1998).

90. T. L. Burgess, Y. Qian, S. Kaufman, B. D. Ring, G. Van, C. Capparelli, M. Kelley, H. Hsu, W. J. Boyle, C. R. Dunstan, S. Hu, and D. L. Lacey, The ligand for osteoprotegerin (OPGL) directly activates mature osteoclasts. *J Cell Biol* **145**, 527–538 (1999).

91. H. Yasuda, N. Shima, N. Nakagawa, K. Yamaguchi, M. Kinosaki, S. Mochizuki, A. Tomoyasu, K. Yano, M. Goto, A. Murakami, E. Tsuda, T. Morinaga, K. Higashio, N. Udagawa, N. Takahashi, and T. Suda, Osteoclast differentiation factor is a ligand for osteoprotegerin/osteoclastogenesis-inhibitory factor and is identical to TRANCE/RANKL. *Proc Natl Acad Sci USA* **95**, 3597–3602 (1998).

92. J. M. Quinn, J. Elliott, M. T. Gillespie, and T. J. Martin, A combination of osteoclast differentiation factor and macrophage-colony stimulating factor is sufficient for both human and mouse osteoclast formation *in vitro*. *Endocrinology* **139**, 4424–4427 (1998).

93. Y. Y. Kong, H. Yoshida, I. Sarosi, H. L. Tan, E. Timms, C. Capparelli, S. Morony, A. J. Oliveira-dos-Santos, G. Van, A. Itie, W. Khoo, A. Wakeham, C. R. Dunstan, D. L. Lacey, T. W. Mak, W. J. Boyle, and J. M. Penninger, OPGL is a key regulator of osteoclastogenesis, lymphocyte development and lymph-node organogenesis. *Nature* **397**, 315–323 (1999).

94. W. S. Simonet, D. L. Lacey, C. R. Dunstan, M. Kelley, M. S. Chang, R. Luthy, H. Q. Nguyen, S. Wooden, L. Bennett, T. Boone, G. Shimamoto, M. DeRose, R. Elliott, A. Colombero, H. L. Tan, G. Trail, J. Sullivan, E. Davy, N. Bucay, L. Renshaw-Gegg, T. M. Hughes, D. Hill, W. Pattison, P. Campbell, S. Sander, G. Van, J. Tarpley, P. Derby, R. Lee, and W. J. Boyle, Osteoprotegerin: A novel secreted protein involved in the regulation of bone density. *Cell* **89**, 309–319 (1997).

95. H. Yasuda, N. Shima, N. Nakagawa, S. I. Mochizuki, K. Yano, N. Fujise, Y. Sato, M. Goto, K. Yamaguchi, M. Kuriyama, T. Kanno, A. Murakami, E. Tsuda, T. Morinaga, and K. Higashio, Identity of osteoclastogenesis inhibitory factor (OCIF) and osteoprotegerin (OPG): A mechanism by which OPG/OCIF inhibits osteoclastogenesis *in vitro*. *Endocrinology* **139**, 1329–1337 (1998).

96. E. Tsuda, M. Goto, S. Mochizuki, K. Yano, F. Kobayashi, T. Morinaga, and K. Higashio, Isolation of a novel cytokine from human fibroblasts that specifically inhibits osteoclastogenesis. *Biochem Biophys Res Commun* **234**, 137–142 (1997).

97. N. Bucay, I. Sarosi, C. R. Dunstan, S. Morony, J. Tarpley, C. Capparelli, S. Scully, H. L. Tan, W. Xu, D. L. Lacey, W. J. Boyle, and W. S. Simonet, Osteoprotegerin-deficient mice develop early onset osteoporosis and arterial calcification. *Genes Dev* **12**, 1260–1268 (1998).

98. A. Mizuno, N. Amizuka, K. Irie, A. Murakami, N. Fujise, T. Kanno, Y. Sato, N. Nakagawa, H. Yasuda, S. Mochizuki, T. Gomibuchi, K. Yano, N. Shima, N. Washida, E. Tsuda, T. Morinaga, K. Higashio, and H. Ozawa, Severe osteoporosis in mice lacking osteoclastogenesis inhibitory factor/osteoprotegerin. *Biochem Biophys Res Commun* **247**, 610–615 (1998).

99. J. Padagas, M. Colloton, V. Shalhoub, P. Kostenuik, S. Morony, L. Munyakazi, M. Guo, D. Gianneschi, E. Shatzen, Z. Geng, H. L. Tan, C. Dunstan, D. Lacey, and D. Martin, The receptor activator of nuclear factor-kappaB ligand inhibitor osteoprotegerin is a bone-protective agent in a rat model of chronic renal insufficiency and hyperparathyroidism. *Calcif Tissue Int* **78**, 35–44 (2006).

100. J. C. Huang, T. Sakata, L. L. Pfleger, M. Bencsik, B. P. Halloran, D. D. Bikle, and R. A. Nissenson, PTH differentially regulates expression of RANKL and OPG. *J Bone Miner Res* **19**, 235–244 (2004).

101. B. S. Kwon, S. Wang, N. Udagawa, V. Haridas, Z. H. Lee, K. K. Kim, K. O. Oh, J. Greene, Y. Li, J. Su, R. Gentz, B. B. Aggarwal, and J. Ni, TR1, a new member of the tumor necrosis factor receptor superfamily, induces fibroblast proliferation and inhibits osteoclastogenesis and bone resorption. *FASEB J* **12**, 845–854 (1998).

102. K. Tsukii, N. Shima, S. Mochizuki, K. Yamaguchi, M. Kinosaki, K. Yano, O. Shibata, N. Udagawa, H. Yasuda, T. Suda, and K. Higashio, Osteoclast differentiation factor mediates an essential signal for bone resorption induced by 1 alpha,25-dihydroxyvitamin D3, prostaglandin E2, or parathyroid hormone in the microenvironment of bone. *Biochem Biophys Res Commun* **246**, 337–341 (1998).

103. T. Murakami, M. Yamamoto, K. Ono, M. Nishikawa, N. Nagata, K. Motoyoshi, and T. Akatsu, Transforming growth factor-beta1 increases mRNA levels of osteoclastogenesis inhibitory factor in osteoblastic/stromal cells and inhibits the survival of murine osteoclast-like cells. *Biochem Biophys Res Commun* **252**, 747–752 (1998).

104. N. J. Horwood, J. Elliott, T. J. Martin, and M. T. Gillespie, Osteotropic agents regulate the expression of osteoclast differentiation factor and osteoprotegerin in osteoblastic stromal cells. *Endocrinology* **139**, 4743–4746 (1998).

105. S. K. Lee and J. A. Lorenzo, Parathyroid hormone stimulates TRANCE and inhibits osteoprotegerin messenger ribonucleic acid expression in murine bone marrow cultures: Correlation with osteoclast-like cell formation. *Endocrinology* **140**, 3552–3561 (1999).

106. Y. L. Ma, R. L. Cain, D. L. Halladay, X. Yang, Q. Zeng, R. R. Miles, S. Chandrasekhar, T. J. Martin, and J. E. Onyia, Catabolic effects of continuous human PTH (1–38) *in vivo* is associated with sustained stimulation of RANKL and inhibition of osteoprotegerin and gene-associated bone formation. *Endocrinology* **142**, 4047–4054 (2001).

107. W. B. High, H. E. Black, and C. C. Capen, Histomorphometric evaluation of the effects of low dose parathyroid hormone administration on cortical bone remodeling in adult dogs. *Lab Invest* **44**, 449–454 (1981).

108. T. Yonaga, Action of parathyroid hormone, with special reference to its anabolic effect on different kinds of tissues in rats (I). *Bull Tokyo Med Dent Univ* **25**, 237–248 (1978).

109. J. Reeve, P. J. Meunier, J. A. Parsons, M. Bernat, O. L. Bijvoet, P. Courpron, C. Edouard, L. Klenerman, R. M. Neer, J. C. Renier, D. Slovik, F. J. Vismans, and J. T. Potts, Jr., Anabolic effect of human parathyroid hormone fragment on trabecular bone in involutional osteoporosis: A multicentre trial. *Br Med J* **280**, 1340–1344 (1980).

110. K. E. Poole and J. Reeve, Parathyroid hormone—A bone anabolic and catabolic agent. *Curr Opin Pharmacol* **5**, 612–617 (2005).

111. K. T. Brixen, P. M. Christensen, C. Ejersted, and B. L. Langdahl, Teriparatide (biosynthetic human parathyroid hormone 1–34): A new paradigm in the treatment of osteoporosis. *Basic Clin Pharmacol Toxicol* **94**, 260–270 (2004).

112. C. J. Rosen, What's new with PTH in osteoporosis: Where are we and where are we headed? *Trends Endocrinol Metab* **15**, 229–233 (2004).

113. M. Girotra, M. R. Rubin, and J. P. Bilezikian, The use of parathyroid hormone in the treatment of osteoporosis. *Rev Endocr Metab Disord* **7**, 113–121 (2006).

114. M. Gunness-Hey and J. M. Hock, Increased trabecular bone mass in rats treated with human synthetic parathyroid hormone. *Metab Bone Dis Relat Res* **5**, 177–181 (1984).

115. T. J. Wronski, C. F. Yen, H. Qi, and L. M. Dann, Parathyroid hormone is more effective than estrogen or bisphosphonates

for restoration of lost bone mass in ovariectomized rats. *Endocrinology* **132,** 823–831 (1993).

116. C. C. Liu and D. N. Kalu, Human parathyroid hormone-(1–34) prevents bone loss and augments bone formation in sexually mature ovariectomized rats. *J Bone Miner Res* **5,** 973–982 (1990).

117. J. F. Whitfield, P. Morley, and G. E. Willick, The bone-building action of the parathyroid hormone: Implications for the treatment of osteoporosis. *Drugs Aging* **15,** 117–129 (1999).

118. D. W. Dempster, F. Cosman, M. Parisien, V. Shen, and R. Lindsay, Anabolic actions of parathyroid hormone on bone. *Endocr Rev* **14,** 690–709 (1993).

119. T. Hirano, D. B. Burr, C. H. Turner, M. Sato, R. L. Cain, and J. M. Hock, Anabolic effects of human biosynthetic parathyroid hormone fragment (1–34), LY333334, on remodeling and mechanical properties of cortical bone in rabbits. *J Bone Miner Res* **14,** 536–545 (1999).

120. L. Zhang, N. Endo, N. Yamamoto, T. Tanizawa, and H. E. Takahashi, Effects of single and concurrent intermittent administration of human PTH (1–34) and incadronate on cancellous and cortical bone of femoral neck in ovariectomized rats. *Tohoku J Exp Med* **186,** 131–141 (1998).

121. M. Sato, G. Q. Zeng, and C. H. Turner, Biosynthetic human parathyroid hormone (1–34) effects on bone quality in aged ovariectomized rats. *Endocrinology* **138,** 4330–4337 (1997).

122. B. D. Baumann and T. J. Wronski, Response of cortical bone to antiresorptive agents and parathyroid hormone in aged ovariectomized rats. *Bone* **16,** 247–253 (1995).

123. N. E. Lane, D. B. Kimmel, M. H. Nilsson, F. E. Cohen, S. Newton, R. A. Nissenson, and G. J. Strewler, Bone-selective analogs of human PTH(1–34) increase bone formation in an ovariectomized rat model. *J Bone Miner Res* **11,** 614–625 (1996).

124. N. E. Lane, J. M. Thompson, G. J. Strewler, and J. H. Kinney, Intermittent treatment with human parathyroid hormone (hPTH[1–34]) increased trabecular bone volume but not connectivity in osteopenic rats. *J Bone Miner Res* **10,** 1470–1477 (1995).

125. L. Mosekilde, C. H. Sogaard, C. C. Danielsen, and O. Torring, The anabolic effects of human parathyroid hormone (hPTH) on rat vertebral body mass are also reflected in the quality of bone, assessed by biomechanical testing: A comparison study between hPTH-(1–34) and hPTH-(1–84). *Endocrinology* **129,** 421–428 (1991).

126. R. W. Zhang, S. C. Supowit, X. Xu, H. Li, M. D. Christensen, R. Lozano, and D. J. Simmons, Expression of selected osteogenic markers in the fibroblast-like cells of rat marrow stroma. *Calcif Tissue Int* **56,** 283–291 (1995).

127. K. C. Hicok, T. Thomas, F. Gori, D. J. Rickard, T. C. Spelsberg, and B. L. Riggs, Development and characterization of conditionally immortalized osteoblast precursor cell lines from human bone marrow stroma. *J Bone Miner Res* **13,** 205–217 (1998).

128. M. F. Rouleau, J. Mitchell, and D. Goltzman, Characterization of the major parathyroid hormone target cell in the endosteal metaphysis of rat long bones. *J Bone Miner Res* **5,** 1043–1053 (1990).

129. X. Y. Zang, Y. B. Tan, Z. L. Pang, W. Z. Zhang, and J. Zhao, Effects of parathyroid hormone and estradiol on proliferation and function of human osteoblasts from fetal long bone. An *in vitro* study. *Chin Med J (Engl)* **107,** 600–603 (1994).

130. M. Sabatini, C. Lesur, M. Pacherie, P. Pastoureau, N. Kucharczyk, J. L. Fauchere, and J. Bonnet, Effects of parathyroid hormone and agonists of the adenylyl cyclase and protein kinase C pathways on bone cell proliferation. *Bone* **18,** 59–65 (1996).

131. B. R. MacDonald, J. A. Gallagher, and R. G. Russell, Parathyroid hormone stimulates the proliferation of cells derived from human bone. *Endocrinology* **118,** 2445–2449 (1986).

132. N. C. Partridge, A. L. Opie, R. T. Opie, and T. J. Martin, Inhibitory effects of parathyroid hormone on growth of osteogenic sarcoma cells. *Calcif Tissue Int* **37,** 519–525 (1985).

133. S. Nishida, A. Yamaguchi, T. Tanizawa, N. Endo, T. Mashiba, Y. Uchiyama, T. Suda, S. Yoshiki, and H. E. Takahashi, Increased bone formation by intermittent parathyroid hormone administration is due to the stimulation of proliferation and differentiation of osteoprogenitor cells in bone marrow. *Bone* **15,** 717–723 (1994).

134. P. C. Schiller, G. D'Ippolito, B. A. Roos, and G. A. Howard, Anabolic or catabolic responses of MC3T3-E1 osteoblastic cells to parathyroid hormone depend on time and duration of treatment. *J Bone Miner Res* **14,** 1504–1512 (1999).

135. T. Ishizuya, S. Yokose, M. Hori, T. Noda, T. Suda, S. Yoshiki, and A. Yamaguchi, Parathyroid hormone exerts disparate effects on osteoblast differentiation depending on exposure time in rat osteoblastic cells. *J Clin Invest* **99,** 2961–2970 (1997).

136. Y. H. Wang, Y. Liu, K. Buhl, and D. W. Rowe, Comparison of the action of transient and continuous PTH on primary osteoblast cultures expressing differentiation stage-specific GFP. *J Bone Miner Res* **20,** 5–14 (2005).

137. Y. H. Wang, Y. Liu, and D. W. Rowe, Effects of transient PTH on early proliferation, apoptosis, and subsequent differentiation of osteoblast in calvarial osteoblast cultures. *Am J Physiol Endocrinol Metab* **292,** E594–E603 (2007).

138. N. H. Kulkarni, D. L. Halladay, R. R. Miles, L. M. Gilbert, C. A. Frolik, R. J. Galvin, T. J. Martin, M. T. Gillespie, and J. E. Onyia, Effects of parathyroid hormone on Wnt signaling pathway in bone. *J Cell Biochem* **95,** 1178–1190 (2005).

139. H. Keller and M. Kneissel, SOST is a target gene for PTH in bone. *Bone* **37,** 148–158 (2005).

140. T. Bellido, A. A. Ali, I. Gubrij, L. I. Plotkin, Q. Fu, C. A. O'Brien, S. C. Manolagas, and R. L. Jilka, Chronic elevation of parathyroid hormone in mice reduces expression of sclerostin by osteocytes: A novel mechanism for hormonal control of osteoblastogenesis. *Endocrinology* **146,** 4577–4583 (2005).

141. S. J. Rodda and A. P. McMahon, Distinct roles for Hedgehog and canonical Wnt signaling in specification, differentiation and maintenance of osteoblast progenitors. *Development* **133,** 3231–3244 (2006).

142. R. L. Jilka, R. S. Weinstein, T. Bellido, P. Roberson, A. M. Parfitt, and S. C. Manolagas, Increased bone formation by prevention of osteoblast apoptosis with parathyroid hormone. *J Clin Invest* **104,** 439–446 (1999).

143. S. Fukayama, G. Kong, J. L. Benovic, E. Meurer, and A. H. Tashjian, Jr., Beta-adrenergic receptor kinase-1 acutely regulates PTH/PTHrP receptor signalling in human osteoblastlike cells. *Cell Signal* **9,** 469–474 (1997).

144. J. P. Vilardaga, M. Frank, C. Krasel, C. Dees, R. A. Nissenson, and M. J. Lohse, Differential conformational requirements for activation of G proteins and the regulatory proteins arrestin and G protein-coupled receptor kinase in the G protein-coupled receptor for parathyroid hormone (PTH)/PTH-related protein. *J Biol Chem* **276,** 33435–33443 (2001).

145. P. J. Flannery and R. F. Spurney, Domains of the parathyroid hormone (PTH) receptor required for regulation by G protein-coupled receptor kinases (GRKs). *Biochem Pharmacol* **62,** 1047–1058 (2001).

146. L. Wang, L. D. Quarles, and R. F. Spurney, Unmasking the osteoinductive effects of a G-protein-coupled recep-

147. L. Wang, S. Liu, L. D. Quarles, and R. F. Spurney, Targeted overexpression of G protein-coupled receptor kinase-2 in osteoblasts promotes bone loss. *Am J Physiol Endocrinol Metab* **288,** E826–E834 (2005).

148. D. M. Black, S. L. Greenspan, K. E. Ensrud, L. Palermo, J. A. McGowan, T. F. Lang, P. Garnero, M. L. Bouxsein, J. P. Bilezikian, and C. J. Rosen, The effects of parathyroid hormone and alendronate alone or in combination in postmenopausal osteoporosis. *N Engl J Med* **349,** 1207–1215 (2003).

149. J. S. Finkelstein, A. Hayes, J. L. Hunzelman, J. J. Wyland, H. Lee, and R. M. Neer, The effects of parathyroid hormone, alendronate, or both in men with osteoporosis. *N Engl J Med* **349,** 1216–1226 (2003).

150. T. J. Martin, Does bone resorption inhibition affect the anabolic response to parathyroid hormone? *Trends Endocrinol Metab* **15,** 49–50 (2004).

151. A. J. Koh, B. Demiralp, K. G. Neiva, J. Hooten, R. M. Nohutcu, H. Shim, N. S. Datta, R. S. Taichman, and L. K. McCauley, Cells of the osteoclast lineage as mediators of the anabolic actions of parathyroid hormone in bone. *Endocrinology* **146,** 4584–4596 (2005).

152. D. D. Bikle, T. Sakata, C. Leary, H. Elalieh, D. Ginzinger, C. J. Rosen, W. Beamer, S. Majumdar, and B. P. Halloran, Insulin-like growth factor I is required for the anabolic actions of parathyroid hormone on mouse bone. *J Bone Miner Res* **17,** 1570–1578 (2002).

153. N. Miyakoshi, Y. Kasukawa, T. A. Linkhart, D. J. Baylink, and S. Mohan, Evidence that anabolic effects of PTH on bone require IGF-I in growing mice. *Endocrinology* **142,** 4349–4356 (2001).

154. R. M. Locklin, S. Khosla, R. T. Turner, and B. L. Riggs, Mediators of the biphasic responses of bone to intermittent and continuously administered parathyroid hormone. *J Cell Biochem* **89,** 180–190 (2003).

155. M. M. Hurley, Y. Okada, L. Xiao, Y. Tanaka, M. Ito, N. Okimoto, T. Nakamura, C. J. Rosen, T. Doetschman, and J. D. Coffin, Impaired bone anabolic response to parathyroid hormone in Fgf2−/− and Fgf2+/− mice. *Biochem Biophys Res Commun* **341,** 989–994 (2006).

156. J. H. R. Ellsworth, Studies on physiology of parathyroid glands: Some responses of normal human kidneys and blood to intravenous parathyroid extract. *Bull Johns Hopkins Hosp* **55,** 296 (1934).

157. N. Zhao and H. S. Tenenhouse, Npt2 gene disruption confers resistance to the inhibitory action of parathyroid hormone on renal sodium-phosphate cotransport. *Endocrinology* **141,** 2159–2165 (2000).

158. H. Murer, I. Forster, H. Hilfiker, M. Pfister, B. Kaissling, M. Lotscher, and J. Biber, Cellular/molecular control of renal Na/Pi-cotransport. *Kidney Int Suppl* **65,** S2–S10 (1998).

159. T. A. K. Berndt, Renal regulation of phosphate excretion. In *The Kidney: Physiology and Pathophysiology* (D. A. G. Seldin, ed.), pp. 2511–2532. Raven Press, New York (1992).

160. H. Murer, Homer Smith Award. Cellular mechanisms in proximal tubular Pi reabsorption: Some answers and more questions. *J Am Soc Nephrol* **2,** 1649–1665 (1992).

161. H. Murer, I. Forster, N. Hernando, G. Lambert, M. Traebert, and J. Biber, Posttranscriptional regulation of the proximal tubule NaPi-II transporter in response to PTH and dietary P(i). *Am J Physiol* **277,** F676–F684 (1999).

162. I. Keusch, M. Traebert, M. Lotscher, B. Kaissling, H. Murer, and J. Biber, Parathyroid hormone and dietary phosphate provoke a lysosomal routing of the proximal tubular Na/Pi-cotransporter type II. *Kidney Int* **54,** 1224–1232 (1998).

163. Y. Zhang, J. M. Norian, C. E. Magyar, N. H. Holstein-Rathlou, A. K. Mircheff, and A. A. McDonough, *In vivo* PTH provokes apical NHE3 and NaPi2 redistribution and Na-K-ATPase inhibition. *Am J Physiol* **276,** F711–F719 (1999).

164. M. Traebert, J. Roth, J. Biber, H. Murer, and B. Kaissling, Internalization of proximal tubular type II Na-P(i) cotransporter by PTH: Immunogold electron microscopy. *Am J Physiol Renal Physiol* **278,** F148–F154 (2000).

165. D. Bacic, C. A. Wagner, N. Hernando, B. Kaissling, J. Biber, and H. Murer, Novel aspects in regulated expression of the renal type IIa Na/Pi-cotransporter. *Kidney Int Suppl,* **91,** S5–S12 (2004).

166. S. A. Kempson, M. Lotscher, B. Kaissling, J. Biber, H. Murer, and M. Levi, Parathyroid hormone action on phosphate transporter mRNA and protein in rat renal proximal tubules. *Am J Physiol* **268,** F784–F791 (1995).

167. M. J. Mahon, J. A. Cole, E. D. Lederer, and G. V. Segre, Na+/H+ exchanger-regulatory factor 1 mediates inhibition of phosphate transport by parathyroid hormone and second messengers by acting at multiple sites in opossum kidney cells. *Mol Endocrinol* **17,** 2355–2364 (2003).

168. R. Cunningham, D. Steplock, S. Shenolikar, and E. J. Weinman, Defective PTH regulation of sodium-dependent phosphate transport in NHERF-1−/− renal proximal tubule cells and wild-type cells adapted to low-phosphate media. *Am J Physiol Renal Physiol* **289,** F933–F938 (2005).

169. R. Cunningham, D. Steplock, R. S. Biswas, F. Wang, S. Shenolikar, and E. J. Weinman, Adenoviral expression of NHERF-1 in NHERF-1 null mouse renal proximal tubule cells restores Npt2a regulation by low phosphate media and parathyroid hormone. *Am J Physiol Renal Physiol* **291,** F896–F901 (2006).

170. P. A. Friedman and F. A. Gesek, Calcium transport in renal epithelial cells. *Am J Physiol* **264,** F181–F198 (1993).

171. D. W. Seldin, Renal handling of calcium. *Nephron* **81**(Suppl. 1), 2–7 (1999).

172. B. J. Bacskai and P. A. Friedman, Activation of latent Ca^{2+} channels in renal epithelial cells by parathyroid hormone. *Nature* **347,** 388–391 (1990).

173. F. A. Gesek and P. A. Friedman, On the mechanism of parathyroid hormone stimulation of calcium uptake by mouse distal convoluted tubule cells. *J Clin Invest* **90,** 749–758 (1992).

174. P. A. Friedman, Codependence of renal calcium and sodium transport. *Annu Rev Physiol* **60,** 179–197 (1998).

175. M. van Abel, J. G. Hoenderop, A. W. van der Kemp, M. M. Friedlaender, J. P. van Leeuwen, and R. J. Bindels, Coordinated control of renal Ca(2+) transport proteins by parathyroid hormone. *Kidney Int* **68,** 1708–1721 (2005).

176. N. A. Breslau, Normal and abnormal regulation of 1,25-(OH)2D synthesis. *Am J Med Sci* **296,** 417–425 (1988).

177. D. R. Fraser and E. Kodicek, Regulation of 25-hydroxycholecalciferol-1-hydroxylase activity in kidney by parathyroid hormone. *Nat New Biol* **241,** 163–166 (1973).

178. R. Kremer and D. Goltzman, Parathyroid hormone stimulates mammalian renal 25-hydroxyvitamin D3-1 alpha-hydroxylase *in vitro. Endocrinology* **110,** 294–296 (1982).

179. C. R. Rost, D. D. Bikle, and R. A. Kaplan, *In vitro* stimulation of 25-hydroxycholecalciferol 1 alpha-hydroxylation by parathyroid hormone in chick kidney slices. Evidence for a role for adenosine 3′,5′-monophosphate. *Endocrinology* **108,** 1002–1006 (1981).

180. T. Monkawa, T. Yoshida, S. Wakino, T. Shinki, H. Anazawa, H. F. Deluca, T. Suda, M. Hayashi, and T. Saruta, Molecular

cloning of cDNA and genomic DNA for human 25-hydroxyvitamin D3 1 alpha-hydroxylase. *Biochem Biophys Res Commun* **239**, 527–533 (1997).

181. G. K. Fu, D. Lin, M. Y. Zhang, D. D. Bikle, C. H. Shackleton, W. L. Miller, and A. A. Portale, Cloning of human 25-hydroxyvitamin D-1 alpha-hydroxylase and mutations causing vitamin D-dependent rickets type 1. *Mol Endocrinol* **11**, 1961–1970 (1997).

182. F. H. Glorieux and R. St.-Arnaud, Molecular cloning of (25-OH D)-1 alpha-hydroxylase: An approach to the understanding of vitamin D pseudo-deficiency. *Recent Prog Horm Res* **53**, 341–350 (1998).

183. A. Murayama, K. Takeyama, S. Kitanaka, Y. Kodera, Y. Kawaguchi, T. Hosoya, and S. Kato, Positive and negative regulations of the renal 25-hydroxyvitamin D3 1alpha-hydroxylase gene by parathyroid hormone, calcitonin, and 1alpha,25(OH)2D3 in intact animals. *Endocrinology* **140**, 2224–2231 (1999).

184. T. Shinki, Y. Ueno, H. F. DeLuca, and T. Suda, Calcitonin is a major regulator for the expression of renal 25-hydroxyvitamin D3-1alpha-hydroxylase gene in normocalcemic rats. *Proc Natl Acad Sci USA* **96**, 8253–8258 (1999).

185. H. L. Brenza and H. F. DeLuca, Regulation of 25-hydroxyvitamin D3 1alpha-hydroxylase gene expression by parathyroid hormone and 1,25-dihydroxyvitamin D3. *Arch Biochem Biophys* **381**, 143–152 (2000).

186. J. L. Omdahl, H. A. Morris, and B. K. May, Hydroxylase enzymes of the vitamin D pathway: Expression, function, and regulation. *Annu Rev Nutr* **22**, 139–166 (2002).

187. H. J. Armbrecht, T. L. Hodam, and M. A. Boltz, Hormonal regulation of 25-hydroxyvitamin D3-1alpha-hydroxylase and 24-hydroxylase gene transcription in opossum kidney cells. *Arch Biochem Biophys* **409**, 298–304 (2003).

188. A. F. Stewart, R. Horst, L. J. Deftos, E. C. Cadman, R. Lang, and A. E. Broadus, Biochemical evaluation of patients with cancer-associated hypercalcemia: Evidence for humoral and nonhumoral groups. *N Engl J Med* **303**, 1377–1383 (1980).

189. S. B. Rodan, K. L. Insogna, A. M. Vignery, A. F. Stewart, A. E. Broadus, S. M. D'Souza, D. R. Bertolini, G. R. Mundy, and G. A. Rodan, Factors associated with humoral hypercalcemia of malignancy stimulate adenylate cyclase in osteoblastic cells. *J Clin Invest* **72**, 1511–1515 (1983).

190. G. J. Strewler, R. D. Williams, and R. A. Nissenson, Human renal carcinoma cells produce hypercalcemia in the nude mouse and a novel protein recognized by parathyroid hormone receptors. *J Clin Invest* **71**, 769–774 (1983).

191. S. A. Rabbani, J. Mitchell, D. R. Roy, R. Kremer, H. P. Bennett, and D. Goltzman, Purification of peptides with parathyroid hormone-like bioactivity from human and rat malignancies associated with hypercalcemia. *Endocrinology* **118**, 1200–1210 (1986).

192. J. M. Moseley, M. Kubota, H. Diefenbach-Jagger, R. E. Wettenhall, B. E. Kemp, L. J. Suva, C. P. Rodda, P. R. Ebeling, P. J. Hudson, J. D. Zajac, *et al.*, Parathyroid hormone-related protein purified from a human lung cancer cell line. *Proc Natl Acad Sci USA* **84**, 5048–5052 (1987).

193. L. J. Suva, G. A. Winslow, R. E. Wettenhall, R. G. Hammonds, J. M. Moseley, H. Diefenbach-Jagger, C. P. Rodda, B. E. Kemp, H. Rodriguez, E. Y. Chen, *et al.*, A parathyroid hormone-related protein implicated in malignant hypercalcemia: Cloning and expression. *Science* **237**, 893–896 (1987).

194. G. J. Strewler, P. H. Stern, J. W. Jacobs, J. Eveloff, R. F. Klein, S. C. Leung, M. Rosenblatt, and R. A. Nissenson, Parathyroid hormone-like protein from human renal carcinoma cells. Structural and functional homology with parathyroid hormone. *J Clin Invest* **80**, 1803–1807 (1987).

195. M. Mangin, A. C. Webb, B. E. Dreyer, J. T. Posillico, K. Ikeda, E. C. Weir, A. F. Stewart, N. H. Bander, L. Milstone, D. E. Barton, *et al.*, Identification of a cDNA encoding a parathyroid hormone-like peptide from a human tumor associated with humoral hypercalcemia of malignancy. *Proc Natl Acad Sci USA* **85**, 597–601 (1988).

196. M. Mangin, K. Ikeda, B. E. Dreyer, L. Milstone, and A. E. Broadus, Two distinct tumor-derived, parathyroid hormone-like peptides result from alternative ribonucleic acid splicing. *Mol Endocrinol* **2**, 1049–1055 (1988).

197. M. A. Thiede, G. J. Strewler, R. A. Nissenson, M. Rosenblatt, and G. A. Rodan, Human renal carcinoma expresses two messages encoding a parathyroid hormone-like peptide: Evidence for the alternative splicing of a single-copy gene. *Proc Natl Acad Sci USA* **85**, 4605–4609 (1988).

198. M. Mannstadt, H. Juppner, and T. J. Gardella, Receptors for PTH and PTHrP: Their biological importance and functional properties. *Am J Physiol* **277**, F665–F675 (1999).

199. G. J. Strewler, The physiology of parathyroid hormone-related protein. *N Engl J Med* **342**, 177–185 (2000).

200. J. J. Wysolmerski and A. F. Stewart, The physiology of parathyroid hormone-related protein: An emerging role as a developmental factor. *Annu Rev Physiol* **60**, 431–460 (1998).

201. B. Lanske and H. M. Kronenberg, Parathyroid hormone-related peptide (PTHrP) and parathyroid hormone (PTH)/ PTHrP receptor. *Crit Rev Eukaryot Gene Expr* **8**, 297–320 (1998).

202. W. M. Philbrick, J. J. Wysolmerski, S. Galbraith, E. Holt, J. J. Orloff, K. H. Yang, R. C. Vasavada, E. C. Weir, A. E. Broadus, and A. F. Stewart, Defining the roles of parathyroid hormone-related protein in normal physiology. *Physiol Rev* **76**, 127–173 (1996).

203. M. E. Dunbar, J. J. Wysolmerski, and A. E. Broadus, Parathyroid hormone-related protein: From hypercalcemia of malignancy to developmental regulatory molecule. *Am J Med Sci* **312**, 287–294 (1996).

204. A. C. Karaplis, A. Luz, J. Glowacki, R. T. Bronson, V. L. Tybulewicz, H. M. Kronenberg, and R. C. Mulligan, Lethal skeletal dysplasia from targeted disruption of the parathyroid hormone-related peptide gene. *Genes Dev* **8**, 277–289 (1994).

205. A. Erlebacher, E. H. Filvaroff, S. E. Gitelman, and R. Derynck, Toward a molecular understanding of skeletal development. *Cell* **80**, 371–378 (1995).

206. S. Provot and E. Schipani, Molecular mechanisms of endochondral bone development. *Biochem Biophys Res Commun* **328**, 658–665 (2005).

207. H. M. Kronenberg, Developmental regulation of the growth plate. *Nature* **423**, 332–336 (2003).

208. B. St.-Jacques, M. Hammerschmidt, and A. P. McMahon, Indian hedgehog signaling regulates proliferation and differentiation of chondrocytes and is essential for bone formation. *Genes Dev* **13**, 2072–2086 (1999).

209. L. P. Lai and J. Mitchell, Indian hedgehog: Its roles and regulation in endochondral bone development. *J Cell Biochem* **96**, 1163–1173 (2005).

210. F. Long, U. I. Chung, S. Ohba, J. McMahon, H. M. Kronenberg, and A. P. McMahon, Ihh signaling is directly required for the osteoblast lineage in the endochondral skeleton. *Development* **131**, 1309–1318 (2004).

211. H. Hu, M. J. Hilton, X. Tu, K. Yu, D. M. Ornitz, and F. Long, Sequential roles of Hedgehog and Wnt signaling in osteoblast development. *Development* **132**, 49–60 (2005).

212. U. Chung and H. M. Kronenberg, Parathyroid hormone-related peptide and Indian hedgehog. *Curr Opin Nephrol Hypertens* **9,** 357–362 (2000).

213. A. C. Karaplis, PTHrP: Novel roles in skeletal biology. *Curr Pharm Des* **7,** 655–670 (2001).

214. H. M. Kronenberg, PTHrP and skeletal development. *Ann N Y Acad Sci* **1068,** 1–13 (2006).

215. H. Juppner, Role of parathyroid hormone-related peptide and Indian hedgehog in skeletal development. *Pediatr Nephrol* **14,** 606–611 (2000).

216. E. C. Weir, W. M. Philbrick, M. Amling, I. A. Neff, R. Baron, and A. E. Broadus, Targeted overexpression of parathyroid hormone-related peptide in chondrocytes causes chondrodysplasia and delayed endochondral bone formation. *Proc Natl Acad Sci USA* **93,** 10240–10245 (1996).

217. E. Schipani, K. Kruse, and H. Juppner, A constitutively active mutant PTH-PTHrP receptor in Jansen-type metaphyseal chondrodysplasia. *Science* **268,** 98–100 (1995).

218. E. Schipani and S. Provot, PTHrP, PTH, and the PTH/PTHrP receptor in endochondral bone development. *Birth Defects Res C Embryo Today* **69,** 352–362 (2003).

219. A. Vortkamp, K. Lee, B. Lanske, G. V. Segre, H. M. Kronenberg, and C. J. Tabin, Regulation of rate of cartilage differentiation by Indian hedgehog and PTH-related protein. *Science* **273,** 613–622 (1996).

220. B. Lanske, A. C. Karaplis, K. Lee, A. Luz, A. Vortkamp, A. Pirro, M. Karperien, L. H. Defize, C. Ho, R. C. Mulligan, A. B. Abou-Samra, H. Juppner, G. V. Segre, and H. M. Kronenberg, PTH/PTHrP receptor in early development and Indian hedgehog-regulated bone growth. *Science* **273,** 663–666 (1996).

221. S. J. Karp, E. Schipani, B. St.-Jacques, J. Hunzelman, H. Kronenberg, and A. P. McMahon, Indian hedgehog coordinates endochondral bone growth and morphogenesis via parathyroid hormone related-protein-dependent and -independent pathways. *Development* **127,** 543–548 (2000).

222. T. Kobayashi, U. I. Chung, E. Schipani, M. Starbuck, G. Karsenty, T. Katagiri, D. L. Goad, B. Lanske, and H. M. Kronenberg, PTHrP and Indian hedgehog control differentiation of growth plate chondrocytes at multiple steps. *Development* **129,** 2977–2986 (2002).

223. T. Kobayashi, D. W. Soegiarto, Y. Yang, B. Lanske, E. Schipani, A. P. McMahon, and H. M. Kronenberg, Indian hedgehog stimulates periarticular chondrocyte differentiation to regulate growth plate length independently of PTHrP. *J Clin Invest* **115,** 1734–1742 (2005).

224. K. Lee, J. D. Deeds, and G. V. Segre, Expression of parathyroid hormone-related peptide and its receptor messenger ribonucleic acids during fetal development of rats. *Endocrinology* **136,** 453–463 (1995).

225. M. Jansen, Uber atypische chondrodystrophie (achondroplasie) und uber eine noch night beschriebene angeborene wachstumsstarung des knochensystems: metaphysare dysostosis. *Z Orthop Chir* **61,** 253–286 (1934).

226. A. S. Jobert, P. Zhang, A. Couvineau, J. Bonaventure, J. Roume, M. Le Merrer, and C. Silve, Absence of functional receptors for parathyroid hormone and parathyroid hormone-related peptide in Blomstrand chondrodysplasia. *J Clin Invest* **102,** 34–40 (1998).

227. A. C. Karaplis, B. He, M. T. Nguyen, I. D. Young, D. Semeraro, H. Ozawa, and N. Amizuka, Inactivating mutation in the human parathyroid hormone receptor type 1 gene in Blomstrand chondrodysplasia. *Endocrinology* **139,** 5255–5258 (1998).

228. R. A. Nissenson, Parathyroid hormone (PTH)/PTHrP receptor mutations in human chondrodysplasia. *Endocrinology* **139,** 4753–4755 (1998).

229. S. Blomstrand, I. Claesson, and J. Save-Soderbergh, A case of lethal congenital dwarfism with accelerated skeletal maturation. *Pediatr Radiol* **15,** 141–143 (1985).

230. H. E. MacLean, J. Guo, M. C. Knight, P. Zhang, D. Cobrinik, and H. M. Kronenberg, The cyclin-dependent kinase inhibitor p57(Kip2) mediates proliferative actions of PTHrP in chondrocytes. *J Clin Invest* **113,** 1334–1343 (2004).

231. M. Amling, L. Neff, S. Tanaka, D. Inoue, K. Kuida, E. Weir, W. M. Philbrick, A. E. Broadus, and R. Baron, Bcl-2 lies downstream of parathyroid hormone-related peptide in a signaling pathway that regulates chondrocyte maturation during skeletal development. *J Cell Biol* **136,** 205–213 (1997).

232. J. Guo, U. I. Chung, D. Yang, G. Karsenty, F. R. Bringhurst, and H. M. Kronenberg, PTH/PTHrP receptor delays chondrocyte hypertrophy via both Runx2-dependent and -independent pathways. *Dev Biol* **292,** 116–128 (2006).

233. S. Provot, H. Kempf, L. C. Murtaugh, U. I. Chung, D. W. Kim, J. Chyung, H. M. Kronenberg, and A. B. Lassar, Nkx3.2/Bapx1 acts as a negative regulator of chondrocyte maturation. *Development* **133,** 651–662 (2006).

234. N. Amizuka, A. C. Karaplis, J. E. Henderson, H. Warshawsky, M. L. Lipman, Y. Matsuki, S. Ejiri, M. Tanaka, N. Izumi, H. Ozawa, and D. Goltzman, Haploinsufficiency of parathyroid hormone-related peptide (PTHrP) results in abnormal postnatal bone development. *Dev Biol* **175,** 166–176 (1996).

235. D. Miao, B. He, Y. Jiang, T. Kobayashi, M. A. Soroceanu, J. Zhao, H. Su, X. Tong, N. Amizuka, A. Gupta, H. K. Genant, H. M. Kronenberg, D. Goltzman, and A. C. Karaplis, Osteoblast-derived PTHrP is a potent endogenous bone anabolic agent that modifies the therapeutic efficacy of administered PTH 1–34. *J Clin Invest* **115,** 2402–2411 (2005).

236. N. Suda, O. Baba, N. Udagawa, T. Terashima, Y. Kitahara, Y. Takano, T. Kuroda, P. V. Senior, F. Beck, and V. E. Hammond, Parathyroid hormone-related protein is required for normal intramembranous bone development. *J Bone Miner Res* **16,** 2182–2191 (2001).

237. X. Chen, C. M. Macica, B. E. Dreyer, V. E. Hammond, J. R. Hens, W. M. Philbrick, and A. E. Broadus, Initial characterization of PTH-related protein gene-driven lacZ expression in the mouse. *J Bone Miner Res* **21,** 113–123 (2006).

238. X. Chen, C. M. Macica, K. W. Ng, and A. E. Broadus, Stretch-induced PTH-related protein gene expression in osteoblasts. *J Bone Miner Res* **20,** 1454–1461 (2005).

239. J. J. Wysolmerski, J. F. McCaughern-Carucci, A. G. Daifotis, A. E. Broadus, and W. M. Philbrick, Overexpression of parathyroid hormone-related protein or parathyroid hormone in transgenic mice impairs branching morphogenesis during mammary gland development. *Development* **121,** 3539–3547 (1995).

240. J. J. Wysolmerski, W. M. Philbrick, M. E. Dunbar, B. Lanske, H. Kronenberg, and A. E. Broadus, Rescue of the parathyroid hormone-related protein knockout mouse demonstrates that parathyroid hormone-related protein is essential for mammary gland development. *Development* **125,** 1285–1294 (1998).

241. S. L. Ferrari, R. Rizzoli, and J. P. Bonjour, Parathyroid hormone-related protein production by primary cultures of mammary epithelial cells. *J Cell Physiol* **150,** 304–311 (1992).

242. M. E. Dunbar, P. Young, J. P. Zhang, J. McCaughern-Carucci, B. Lanske, J. J. Orloff, A. Karaplis, G. Cunha, and J. J. Wysolmerski, Stromal cells are critical targets in the regulation of mammary ductal morphogenesis by parathyroid hormone-related protein. *Dev Biol* **203,** 75–89 (1998).

243. J. J. Wysolmerski, S. Cormier, W. M. Philbrick, P. Dann, J. P. Zhang, J. Roume, A. L. Delezoide, and C. Silve, Absence of functional type 1 parathyroid hormone (PTH)/PTH-related protein receptors in humans is associated with abnormal breast development and tooth impaction. *J Clin Endocrinol Metab* **86,** 1788–1794 (2001).

244. M. E. Dunbar, P. R. Dann, G. W. Robinson, L. Hennighausen, J. P. Zhang, and J. J. Wysolmerski, Parathyroid hormone-related protein signaling is necessary for sexual dimorphism during embryonic mammary development. *Development* **126,** 3485–3493 (1999).

245. J. Foley, P. Dann, J. Hong, J. Cosgrove, B. Dreyer, D. Rimm, M. Dunbar, W. Philbrick, and J. Wysolmerski, Parathyroid hormone-related protein maintains mammary epithelial fate and triggers nipple skin differentiation during embryonic breast development. *Development* **128,** 513–525 (2001).

246. T. Kobayashi, H. M. Kronenberg, and J. Foley, Reduced expression of the PTH/PTHrP receptor during development of the mammary gland influences the function of the nipple during lactation. *Dev Dyn* **233,** 794–803 (2005).

247. M. A. Thiede and G. A. Rodan, Expression of a calcium-mobilizing parathyroid hormone-like peptide in lactating mammary tissue. *Science* **242,** 278–280 (1988).

248. M. Yamamoto, L. T. Duong, J. E. Fisher, M. A. Thiede, M. P. Caulfield, and M. Rosenblatt, Suckling-mediated increases in urinary phosphate and 3′,5′-cyclic adenosine monophosphate excretion in lactating rats: Possible systemic effects of parathyroid hormone-related protein. *Endocrinology* **129,** 2614–2622 (1991).

249. K. J. Mather, C. L. Chik, and B. Corenblum, Maintenance of serum calcium by parathyroid hormone-related peptide during lactation in a hypoparathyroid patient. *J Clin Endocrinol Metab* **84,** 424–427 (1999).

250. J. N. VanHouten and J. J. Wysolmerski, Low estrogen and high parathyroid hormone-related peptide levels contribute to accelerated bone resorption and bone loss in lactating mice. *Endocrinology* **144,** 5521–5529 (2003).

251. A. A. Budayr, B. P. Halloran, J. C. King, D. Diep, R. A. Nissenson, and G. J. Strewler, High levels of a parathyroid hormone-like protein in milk. *Proc Natl Acad Sci USA* **86,** 7183–7185 (1989).

252. M. Sowers, D. Zhang, B. W. Hollis, B. Shapiro, C. A. Janney, M. Crutchfield, M. A. Schork, F. Stanczyk, and J. Randolph, Role of calciotrophic hormones in calcium mobilization of lactation. *Am J Clin Nutr* **67,** 284–291 (1998).

253. J. N. VanHouten, P. Dann, A. F. Stewart, C. J. Watson, M. Pollak, A. C. Karaplis, and J. J. Wysolmerski, Mammary-specific deletion of parathyroid hormone-related protein preserves bone mass during lactation. *J Clin Invest* **112,** 1429–1436 (2003).

254. L. Ardeshirpour, P. Dann, M. Pollak, J. Wysolmerski, and J. VanHouten, The calcium-sensing receptor regulates PTHrP production and calcium transport in the lactating mammary gland. *Bone* **38,** 787–793 (2006).

255. J. J. Merendino, Jr., K. L. Insogna, L. M. Milstone, A. E. Broadus, and A. F. Stewart, A parathyroid hormone-like protein from cultured human keratinocytes. *Science* **231,** 388–390 (1986).

256. R. V. Campos, S. L. Asa, and D. J. Drucker, Immunocytochemical localization of parathyroid hormone-like peptide in the rat fetus. *Cancer Res* **51,** 6351–6357 (1991).

257. E. J. Atillasoy, W. J. Burtis, and L. M. Milstone, Immunohistochemical localization of parathyroid hormone-related protein (PTHRP) in normal human skin. *J Invest Dermatol* **96,** 277–280 (1991).

258. Y. M. Cho, G. L. Woodard, M. Dunbar, T. Gocken, J. A. Jimenez, and J. Foley, Hair-cycle-dependent expression of parathyroid hormone-related protein and its type I receptor: Evidence for regulation at the anagen to catagen transition. *J Invest Dermatol* **120,** 715–727 (2003).

259. K. K. Pun, C. D. Arnaud, and R. A. Nissenson, Parathyroid hormone receptors in human dermal fibroblasts: Structural and functional characterization. *J Bone Miner Res* **3,** 453–460 (1988).

260. A. Errazahi, Z. Bouizar, M. Lieberherr, E. Souil, and M. Rizk-Rabin, Functional type I PTH/PTHrP receptor in freshly isolated newborn rat keratinocytes: Identification by RT-PCR and immunohistochemistry. *J Bone Miner Res* **18,** 737–750 (2003).

261. J. J. Orloff, M. B. Ganz, A. E. Ribaudo, W. J. Burtis, M. Reiss, L. M. Milstone, and A. F. Stewart, Analysis of PTHRP binding and signal transduction mechanisms in benign and malignant squamous cells. *Am J Physiol* **262,** E599–E607 (1992).

262. S. M. Kaiser, P. Laneuville, S. M. Bernier, J. S. Rhim, R. Kremer, and D. Goltzman, Enhanced growth of a human keratinocyte cell line induced by antisense RNA for parathyroid hormone-related peptide. *J Biol Chem* **267,** 13623–13628 (1992).

263. S. M. Kaiser, M. Sebag, J. S. Rhim, R. Kremer, and D. Goltzman, Antisense-mediated inhibition of parathyroid hormone-related peptide production in a keratinocyte cell line impedes differentiation. *Mol Endocrinol* **8,** 139–147 (1994).

264. J. J. Wysolmerski, A. E. Broadus, J. Zhou, E. Fuchs, L. M. Milstone, and W. M. Philbrick, Overexpression of parathyroid hormone-related protein in the skin of transgenic mice interferes with hair follicle development. *Proc Natl Acad Sci USA* **91,** 1133–1137 (1994).

265. A. G. Diamond, R. M. Gonterman, A. L. Anderson, K. Menon, C. D. Offutt, C. H. Weaver, W. M. Philbrick, and J. Foley, Parathyroid hormone-related protein and the PTH receptor regulate angiogenesis of the skin. *J Invest Dermatol* **126,** 2127–2134 (2006).

266. J. Foley, B. J. Longely, J. J. Wysolmerski, B. E. Dreyer, A. E. Broadus, and W. M. Philbrick, PTHrP regulates epidermal differentiation in adult mice. *J Invest Dermatol* **111,** 1122–1128 (1998).

267. M. F. Holick, S. Ray, T. C. Chen, X. Tian, and K. S. Persons, A parathyroid hormone antagonist stimulates epidermal proliferation and hair growth in mice. *Proc Natl Acad Sci USA* **91,** 8014–8016 (1994).

268. J. D. Safer, S. Ray, and M. F. Holick, A topical PTH/PTHrP receptor antagonist stimulates hair growth in mice. *Endocrinology* **148,** 1167–1170 (2007).

269. W. M. Philbrick, B. E. Dreyer, I. A. Nakchbandi, and A. C. Karaplis, Parathyroid hormone-related protein is required for tooth eruption. *Proc Natl Acad Sci USA* **95,** 11846–11851 (1998).

270. Y. Kitahara, N. Suda, T. Kuroda, F. Beck, V. E. Hammond, and Y. Takano, Disturbed tooth development in parathyroid hormone-related protein (PTHrP)-gene knockout mice. *Bone* **30,** 48–56 (2002).

271. F. Boabaid, J. E. Berry, A. J. Koh, M. J. Somerman, and L. K. McCauley, The role of parathyroid hormone-related protein in the regulation of osteoclastogenesis by cementoblasts. *J Periodontol* **75,** 1247–1254 (2004).

272. J. G. Liu, M. J. Tabata, T. Fujii, T. Ohmori, M. Abe, Y. Ohsaki, J. Kato, S. Wakisaka, M. Iwamoto, and K. Kurisu, Parathyroid hormone-related peptide is involved in protection against invasion of tooth germs by bone via promoting the differentiation

of osteoclasts during tooth development. *Mech Dev* **95,** 189–200 (2000).

273. I. A. Nakchbandi, E. E. Weir, K. L. Insogna, W. M. Philbrick, and A. E. Broadus, Parathyroid hormone-related protein induces spontaneous osteoclast formation via a paracrine cascade. *Proc Natl Acad Sci USA* **97,** 7296–7300 (2000).

274. M. A. Thiede, A. G. Daifotis, E. C. Weir, M. L. Brines, W. J. Burtis, K. Ikeda, B. E. Dreyer, R. E. Garfield, and A. E. Broadus, Intrauterine occupancy controls expression of the parathyroid hormone-related peptide gene in preterm rat myometrium. *Proc Natl Acad Sci USA* **87,** 6969–6973 (1990).

275. A. G. Daifotis, E. C. Weir, B. E. Dreyer, and A. E. Broadus, Stretch-induced parathyroid hormone-related peptide gene expression in the rat uterus. *J Biol Chem* **267,** 23455–23458 (1992).

276. J. E. Ferguson, 2nd, J. V. Gorman, D. E. Bruns, E. C. Weir, W. J. Burtis, T. J. Martin, and M. E. Bruns, Abundant expression of parathyroid hormone-related protein in human amnion and its association with labor. *Proc Natl Acad Sci USA* **89,** 8384–8388 (1992).

277. A. M. Germain, H. Attaroglu, P. C. MacDonald, and M. L. Casey, Parathyroid hormone-related protein mRNA in avascular human amnion. *J Clin Endocrinol Metab* **75,** 1173–1175 (1992).

278. M. Yamamoto, S. C. Harm, W. A. Grasser, and M. A. Thiede, Parathyroid hormone-related protein in the rat urinary bladder: A smooth muscle relaxant produced locally in response to mechanical stretch. *Proc Natl Acad Sci USA* **89,** 5326–5330 (1992).

279. M. Ito, A. Ohtsuru, H. Enomoto, S. Ozeki, M. Nakashima, T. Nakayama, K. Shichijo, I. Sekine, and S. Yamashita, Expression of parathyroid hormone-related peptide in relation to perturbations of gastric motility in the rat. *Endocrinology* **134,** 1936–1942 (1994).

280. M. Francis, M. Arkle, L. Martin, T. M. Butler, M. C. Cruz, G. Opare-Aryee, C. G. Dacke, and J. F. Brown, Relaxant effects of parathyroid hormone and parathyroid hormone-related peptides on oviduct motility in birds and mammals: Possible role of nitric oxide. *Gen Comp Endocrinol* **133,** 243–251 (2003).

281. G. A. Nickols, A. D. Nana, M. A. Nickols, D. J. DiPette, and G. K. Asimakis, Hypotension and cardiac stimulation due to the parathyroid hormone-related protein, humoral hypercalcemia of malignancy factor. *Endocrinology* **125,** 834–841 (1989).

282. T. Massfelder, N. Fiaschi-Taesch, A. F. Stewart, and J. J. Helwig, Parathyroid hormone-related peptide—A smooth muscle tone and proliferation regulatory protein. *Curr Opin Nephrol Hypertens* **7,** 27–32 (1998).

283. F. Meziani, B. Van Overloop, F. Schneider, and A. Gairard, Parathyroid hormone-related protein-induced relaxation of rat uterine arteries: Influence of the endothelium during gestation. *J Soc Gynecol Investig* **12,** 14–19 (2005).

284. Y. Gao and J. U. Raj, Parathyroid hormone-related protein-mediated responses in pulmonary arteries and veins of newborn lambs. *Am J Physiol Lung Cell Mol Physiol* **289,** L60–L66 (2005).

285. K. Takahashi, D. Inoue, K. Ando, T. Matsumoto, K. Ikeda, and T. Fujita, Parathyroid hormone-related peptide as a locally produced vasorelaxant: Regulation of its mRNA by hypertension in rats. *Biochem Biophys Res Commun* **208,** 447–455 (1995).

286. C. J. Pirola, H. M. Wang, A. Kamyar, S. Wu, H. Enomoto, B. Sharifi, J. S. Forrester, T. L. Clemens, and J. A. Fagin, Angiotensin II regulates parathyroid hormone-related protein expression in cultured rat aortic smooth muscle cells through transcriptional and post-transcriptional mechanisms. *J Biol Chem* **268,** 1987–1994 (1993).

287. S. Maeda, R. L. Sutliff, J. Qian, J. N. Lorenz, J. Wang, H. Tang, T. Nakayama, C. Weber, D. Witte, A. R. Strauch, R. J. Paul, J. A. Fagin, and T. L. Clemens, Targeted overexpression of parathyroid hormone-related protein (PTHrP) to vascular smooth muscle in transgenic mice lowers blood pressure and alters vascular contractility. *Endocrinology* **140,** 1815–1825 (1999).

288. M. S. Landa, S. I. Garcia, L. Liberjen, M. L. Schuman, S. Finkielman, and C. J. Pirola, Parathyroid hormone-related protein overexpression decreases blood pressure in spontaneously hypertensive rats. *Clin Exp Hypertens* **27,** 343–354 (2005).

289. J. Qian, J. N. Lorenz, S. Maeda, R. L. Sutliff, C. Weber, T. Nakayama, M. C. Colbert, R. J. Paul, J. A. Fagin, and T. L. Clemens, Reduced blood pressure and increased sensitivity of the vasculature to parathyroid hormone-related protein (PTHrP) in transgenic mice overexpressing the PTH/PTHrP receptor in vascular smooth muscle. *Endocrinology* **140,** 1826–1833 (1999).

290. W. T. Noonan, J. Qian, W. D. Stuart, T. L. Clemens, and J. N. Lorenz, Altered renal hemodynamics in mice overexpressing the parathyroid hormone (PTH)/PTH-related peptide type 1 receptor in smooth muscle. *Endocrinology* **144,** 4931–4938 (2003).

291. T. Massfelder, J. J. Helwig, and A. F. Stewart, Parathyroid hormone-related protein as a cardiovascular regulatory peptide. *Endocrinology* **137,** 3151–3153 (1996).

292. M. M. Bakre, Y. Zhu, H. Yin, D. W. Burton, R. Terkeltaub, L. J. Deftos, and J. A. Varner, Parathyroid hormone-related peptide is a naturally occurring, protein kinase A-dependent angiogenesis inhibitor. *Nat Med* **8,** 995–1003 (2002).

293. A. G. Schneider, K. Leuthauser, and D. Pette, Parathyroid hormone-related protein is rapidly upregulated in blood vessels of rat skeletal muscle by low-frequency stimulation. *Pflugers Arch* **439,** 167–173 (1999).

294. E. C. Weir, M. L. Brines, K. Ikeda, W. J. Burtis, A. E. Broadus, and R. J. Robbins, Parathyroid hormone-related peptide gene is expressed in the mammalian central nervous system. *Proc Natl Acad Sci USA* **87,** 108–112 (1990).

295. D. R. Weaver, J. D. Deeds, K. Lee, and G. V. Segre, Localization of parathyroid hormone-related peptide (PTHrP) and PTH/PTHrP receptor mRNAs in rat brain. *Brain Res Mol Brain Res* **28,** 296–310 (1995).

296. E. H. Holt, A. E. Broadus, and M. L. Brines, Parathyroid hormone-related peptide is produced by cultured cerebellar granule cells in response to L-type voltage-sensitive Ca^{2+} channel flux via a Ca^{2+}/calmodulin-dependent kinase pathway. *J Biol Chem* **271,** 28105–28111 (1996).

297. M. L. Brines, Z. Ling, and A. E. Broadus, Parathyroid hormone-related protein protects against kainic acid excitotoxicity in rat cerebellar granule cells by regulating L-type channel calcium flux. *Neurosci Lett* **274,** 13–16 (1999).

298. P. K. Pang, R. Wang, J. Shan, E. Karpinski, and C. G. Benishin, Specific inhibition of long-lasting, L-type calcium channels by synthetic parathyroid hormone. *Proc Natl Acad Sci USA* **87,** 623–627 (1990).

299. T. Ono, K. Inokuchi, A. Ogura, Y. Ikawa, Y. Kudo, and S. Kawashima, Activity-dependent expression of parathyroid hormone-related protein (PTHrP) in rat cerebellar granule neurons. Requirement of PTHrP for the activity-dependent survival of granule neurons. *J Biol Chem* **272,** 14404–14411 (1997).

300. O. Chatterjee, I. A. Nakchbandi, W. M. Philbrick, B. E. Dreyer, J. P. Zhang, L. K. Kaczmarek, M. L. Brines, and A. E. Broadus, Endogenous parathyroid hormone-related protein functions as a neuroprotective agent. *Brain Res* **930**, 58–66 (2002).

301. J. L. Funk, E. Migliati, G. Chen, H. Wei, J. Wilson, K. J. Downey, P. J. Mullarky, B. M. Coull, P. F. McDonagh, and L. S. Ritter, Parathyroid hormone-related protein induction in focal stroke: A neuroprotective vascular peptide. *Am J Physiol Regul Integr Comp Physiol* **284**, R1021–R1030 (2003).

302. C. S. Kovacs, B. Lanske, J. L. Hunzelman, J. Guo, A. C. Karaplis, and H. M. Kronenberg, Parathyroid hormone-related peptide (PTHrP) regulates fetal–placental calcium transport through a receptor distinct from the PTH/PTHrP receptor. *Proc Natl Acad Sci USA* **93**, 15233–15238 (1996).

303. R. J. MacIsaac, I. W. Caple, J. A. Danks, H. Diefenbach-Jagger, V. Grill, J. M. Moseley, J. Southby, and T. J. Martin, Ontogeny of parathyroid hormone-related protein in the ovine parathyroid gland. *Endocrinology* **129**, 757–764 (1991).

304. C. P. Rodda, M. Kubota, J. A. Heath, P. R. Ebeling, J. M. Moseley, A. D. Care, I. W. Caple, and T. J. Martin, Evidence for a novel parathyroid hormone-related protein in fetal lamb parathyroid glands and sheep placenta: Comparisons with a similar protein implicated in humoral hypercalcaemia of malignancy. *J Endocrinol* **117**, 261–271 (1988).

305. C. S. Kovacs, N. R. Manley, J. M. Moseley, T. J. Martin, and H. M. Kronenberg, Fetal parathyroids are not required to maintain placental calcium transport. *J Clin Invest* **107**, 1007–1015 (2001).

306. L. R. Chase and G. D. Aurbach, Renal adenyl cyclase: Anatomically separate sites for parathyroid hormone and vasopressin. *Science* **159**, 545–547 (1968).

307. L. R. Chase, S. A. Fedak, and G. D. Aurbach, Activation of skeletal adenyl cyclase by parathyroid hormone *in vitro*. *Endocrinology* **84**, 761–768 (1969).

308. R. Marcus and G. D. Aurbach, Bioassay of parathyroid hormone *in vitro* with a stable preparation of adenyl cyclase from rat kidney. *Endocrinology* **85**, 801–810 (1969).

309. T. Dousa and I. Rychlik, The effect of parathyroid hormone on adenyl cyclase in rat kidney. *Biochim Biophys Acta* **158**, 484–486 (1968).

310. B. Lanske, P. Divieti, C. S. Kovacs, A. Pirro, W. J. Landis, S. M. Krane, F. R. Bringhurst, and H. M. Kronenberg, The parathyroid hormone (PTH)/PTH-related peptide receptor mediates actions of both ligands in murine bone. *Endocrinology* **139**, 5194–5204 (1998).

311. G. Vaes, Parathyroid hormone-like action of N62-O-dibutyryladenosine-3′5′ (cyclic)-monophosphate on bone explants in tissue culture. *Nature* **219**, 939–940 (1968).

312. L. G. Raisz, Physiologic and pharmacologic regulation of bone resorption. *N Engl J Med* **282**, 909–916 (1970).

313. M. P. Herrmann-Erlee, Studies on the role of cyclic AMP in parathyroid hormone-induced embryonic bone resorption. *J Endocrinol* **48**, lix–lxi (1970).

314. W. A. Peck and S. Klahr, Cyclic nucleotides in bone and mineral metabolism. *Adv Cyclic Nucleotide Res* **11**, 89–130 (1979).

315. J. A. Lorenzo, S. Sousa, and J. Quinton, Forskolin has both stimulatory and inhibitory effects on bone resorption in fetal rat long bone cultures. *J Bone Miner Res* **1**, 313–317 (1986).

316. M. P. Herrmann-Erlee and J. M. Van Der Meer, The effects of dibutyryl cyclic AMP, aminophylline and propranolol on PTE-induced bone resorption *in vitro*. *Endocrinology* **94**, 424–434 (1974).

317. R. Dziak and P. Stern, Parathyromimetic effects of the ionophore, A23187, on bone cells and organ cultures. *Biochem Biophys Res Commun* **65**, 1343–1349 (1975).

318. J. A. Lorenzo and S. Sousa, Phorbol esters stimulate bone resorption in fetal rat long-bone cultures by mechanisms independent of prostaglandin synthesis. *J Bone Miner Res* **3**, 63–67 (1988).

319. D. C. Abraham, C. L. Wadkins, and H. H. Conaway, Enhancement of fetal rat limb bone resorption by phorbol ester (PMA) and ionophore A-23187. *Calcif Tissue Int* **42**, 191–195 (1988).

320. M. P. Bos, W. Most, J. P. van Leeuwen, and M. P. Herrmann-Erlee, Role of protein kinase C (PKC) in bone resorption: Effect of the specific PKC inhibitor 1-alkyl-2-methylglycerol. *Biochem Biophys Res Commun* **184**, 1317–1323 (1992).

321. S. M. Sprague, M. M. Popovtzer, M. Dranitzki-Elhalel, and H. Wald, Parathyroid hormone-induced calcium efflux from cultured bone is mediated by protein kinase C translocation. *Am J Physiol* **271**, F1139–F1146 (1996).

322. E. Bornefalk, S. Ljunghall, and O. Ljunggren, Bone resorption induced by A23187 is abolished by indomethacin: Implications for second messenger utilised by parathyroid hormone. *Eur J Pharmacol* **345**, 333–338 (1998).

323. J. L. Ivey, D. R. Wright, and A. H. Tashjian, Jr., Bone resorption in organ culture: Inhibition by the divalent cation ionophores A23187 and X-537A. *J Clin Invest* **58**, 1327–1338 (1976).

324. R. Dziak and P. H. Stern, Responses of fetal rat bone cells and bone organ cultures to the ionophore, A23187. *Calcif Tissue Res* **22**, 137–147 (1976).

325. M. Ransjo and U. H. Lerner, 12-O-tetradecanoylphorbol-13-acetate, a phorbol ester stimulating protein kinase C, inhibits bone resorption in vitro induced by parathyroid hormone and parathyroid hormone-related peptide of malignancy. *Acta Physiol Scand* **139**, 249–250 (1990).

326. H. Kaji, T. Sugimoto, M. Kanatani, M. Nasu, and K. Chihara, Estrogen blocks parathyroid hormone (PTH)-stimulated osteoclast-like cell formation by selectively affecting PTH-responsive cyclic adenosine monophosphate pathway. *Endocrinology* **137**, 2217–2224 (1996).

327. M. Ransjo, A. Lie, and E. J. Mackie, Cholera toxin and forskolin stimulate formation of osteoclast-like cells in mouse marrow cultures and cultured mouse calvarial bones. *Eur J Oral Sci* **107**, 45–54 (1999).

328. Y. H. Gao and M. Yamaguchi, Inhibitory effect of genistein on osteoclast-like cell formation in mouse marrow cultures. *Biochem Pharmacol* **58**, 767–772 (1999).

329. T. Sugimoto, M. Kanatani, H. Kaji, T. Yamaguchi, M. Fukase, and K. Chihara, Second messenger signaling of PTH- and PTHRP-stimulated osteoclast-like cell formation from hemopoietic blast cells. *Am J Physiol* **265**, E367–E373 (1993).

330. I. Villa, E. Mrak, A. Rubinacci, F. Ravasi, and F. Guidobono, CGRP inhibits osteoprotegerin production in human osteoblast-like cells via cAMP/PKA-dependent pathway. *Am J Physiol Cell Physiol* **291**, C529–C537 (2006).

331. H. Kondo, J. Guo, and F. R. Bringhurst, Cyclic adenosine monophosphate/protein kinase A mediates parathyroid hormone/parathyroid hormone-related protein receptor regulation of osteoclastogenesis and expression of RANKL and osteoprotegerin mRNAs by marrow stromal cells. *J Bone Miner Res* **17**, 1667–1679 (2002).

332. M. Kanzawa, T. Sugimoto, M. Kanatani, and K. Chihara, Involvement of osteoprotegerin/osteoclastogenesis inhibitory factor in the stimulation of osteoclast formation by parathyroid hormone in mouse bone cells. *Eur J Endocrinol* **142**, 661–664 (2000).

333. S. K. Lee and J. A. Lorenzo, Regulation of receptor activator of nuclear factor-kappa B ligand and osteoprotegerin mRNA

expression by parathyroid hormone is predominantly mediated by the protein kinase a pathway in murine bone marrow cultures. *Bone* **31,** 252–259 (2002).

334. H. Brandstrom, K. B. Jonsson, C. Ohlsson, O. Vidal, S. Ljunghall, and O. Ljunggren, Regulation of osteoprotegerin mRNA levels by prostaglandin E2 in human bone marrow stroma cells. *Biochem Biophys Res Commun* **247,** 338–341 (1998).

335. D. L. Halladay, R. R. Miles, K. Thirunavukkarasu, S. Chandrasekhar, T. J. Martin, and J. E. Onyia, Identification of signal transduction pathways and promoter sequences that mediate parathyroid hormone 1–38 inhibition of osteoprotegerin gene expression. *J Cell Biochem* **84,** 1–11 (2001).

336. Q. Fu, R. L. Jilka, S. C. Manolagas, and C. A. O'Brien, Parathyroid hormone stimulates receptor activator of NFkappa B ligand and inhibits osteoprotegerin expression via protein kinase A activation of cAMP-response element-binding protein. *J Biol Chem* **277,** 48868–48875 (2002).

337. Q. Fu, S. C. Manolagas, and C. A. O'Brien, Parathyroid hormone controls receptor activator of NF-kappaB ligand gene expression via a distant transcriptional enhancer. *Mol Cell Biol* **26,** 6453–6468 (2006).

338. S. Kim, M. Yamazaki, N. K. Shevde, and J. W. Pike, Transcriptional control of receptor activator of nuclear factor-{kappa}B ligand by the protein kinase A activator forskolin and the transmembrane glycoprotein 130-activating cytokine, oncostatin M, is exerted through multiple distal enhancers. *Mol Endocrinol* **21,** 197–214 (2007).

339. A. Sakamoto, M. Chen, T. Nakamura, T. Xie, G. Karsenty, and L. S. Weinstein, Deficiency of the G-protein alpha-subunit G(s)alpha in osteoblasts leads to differential effects on trabecular and cortical bone. *J Biol Chem* **280,** 21369–21375 (2005).

340. T. Onishi, W. Zhang, X. Cao, and K. Hruska, The mitogenic effect of parathyroid hormone is associated with E2F-dependent activation of cyclin-dependent kinase 1 (cdc2) in osteoblast precursors. *J Bone Miner Res* **12,** 1596–1605 (1997).

341. Y. Isogai, T. Akatsu, T. Ishizuya, A. Yamaguchi, M. Hori, N. Takahashi, and T. Suda, Parathyroid hormone regulates osteoblast differentiation positively or negatively depending on the differentiation stages. *J Bone Miner Res* **11,** 1384–1393 (1996).

342. R. H. Rixon, J. F. Whitfield, L. Gagnon, R. J. Isaacs, S. Maclean, B. Chakravarthy, J. P. Durkin, W. Neugebauer, V. Ross, W. Sung, *et al.*, Parathyroid hormone fragments may stimulate bone growth in ovariectomized rats by activating adenylyl cyclase. *J Bone Miner Res* **9,** 1179–1189 (1994).

343. J. F. Whitfield, P. Morley, G. E. Willick, S. MacLean, V. Ross, J. Barbier, and R. J. Isaacs, Stimulation of femoral trabecular bone growth in ovariectomized rats by human parathyroid hormone (hPTH)-(1–30)NH(2). *Calcif Tissue Int* **65,** 143–147 (1999).

344. R. T. Franceschi, G. Xiao, D. Jiang, R. Gopalakrishnan, S. Yang, and E. Reith, Multiple signaling pathways converge on the Cbfa1/Runx2 transcription factor to regulate osteoblast differentiation. *Connect Tissue Res* **44**(Suppl. 1), 109–116 (2003).

345. V. Krishnan, T. L. Moore, Y. L. Ma, L. M. Helvering, C. A. Frolik, K. M. Valasek, P. Ducy, and A. G. Geiser, Parathyroid hormone bone anabolic action requires Cbfa1/Runx2-dependent signaling. *Mol Endocrinol* **17,** 423–435 (2003).

346. B. L. Wang, C. L. Dai, J. X. Quan, Z. F. Zhu, F. Zheng, H. X. Zhang, S. Y. Guo, G. Guo, J. Y. Zhang, and M. C. Qiu, Parathyroid hormone regulates osterix and Runx2 mRNA expression predominantly through protein kinase A signal-ing in osteoblast-like cells. *J Endocrinol Invest* **29,** 101–108 (2006).

347. A. Inoue, Y. Hiruma, S. Hirose, A. Yamaguchi, and H. Hagiwara, Reciprocal regulation by cyclic nucleotides of the differentiation of rat osteoblast-like cells and mineralization of nodules. *Biochem Biophys Res Commun* **215,** 1104–1110 (1995).

348. H. Kaneki, I. Takasugi, M. Fujieda, M. Kiriu, S. Mizuochi, and H. Ide, Prostaglandin E2 stimulates the formation of mineralized bone nodules by a cAMP-independent mechanism in the culture of adult rat calvarial osteoblasts. *J Cell Biochem* **73,** 36–48 (1999).

349. A. J. Koh, C. A. Beecher, T. J. Rosol, and L. K. McCauley, 3′,5′-Cyclic adenosine monophosphate activation in osteoblastic cells: Effects on parathyroid hormone-1 receptors and osteoblastic differentiation *in vitro. Endocrinology* **140,** 3154–3162 (1999).

350. J. Kano, T. Sugimoto, M. Fukase, and T. Fujita, The activation of cAMP-dependent protein kinase is directly linked to the inhibition of osteoblast proliferation (UMR-106) by parathyroid hormone-related protein. *Biochem Biophys Res Commun* **179,** 97–101 (1991).

351. A. Wang, J. A. Martin, L. A. Lembke, and R. J. Midura, Reversible suppression of *in vitro* biomineralization by activation of protein kinase A. *J Biol Chem* **275,** 11082–11091 (2000).

352. D. Chabardes, M. Imbert, A. Clique, M. Montegut, and F. Morel, PTH sensitive adenyl cyclase activity in different segments of the rabbit nephron. *Pflugers Arch* **354,** 229–239 (1975).

353. S. Torikai and M. Imai, A simple method to determine adenylate cyclase activity in isolated single nephron segments by radioimmunoassay for succinyl adenosine 3′,5′-cyclic monophosphate. *Tohoku J Exp Med* **129,** 91–99 (1979).

354. C. A. Harris and J. F. Seely, Resistance to the phosphaturic effect of parathyroid hormone in the hamster. *Am J Physiol* **237,** F175–F181 (1979).

355. T. J. Berndt, M. J. Onsgard, and F. G. Knox, Effect of cAMP analogue infusion on phosphate reabsorption in phosphate-deprived rats. *Am J Physiol* **255,** F96–F99 (1988).

356. N. Beck and B. B. Davis, Impaired renal response to parathyroid hormone in potassium depletion. *Am J Physiol* **228,** 179–183 (1975).

357. T. Mano, K. Uchimura, R. Hayashi, T. Kobahashi, K. Fujiwara, M. Makino, H. Kakizawa, M. Nagata, A. Nakai, M. Wada, A. Nagasaka, and M. Itoh, Increased urinary phosphate excretion in pseudohypoparathyroidism type II with long-term treatment with phosphodiesterase inhibitor. *Horm Metab Res* **31,** 602–605 (1999).

358. A. M. Moses, N. Breslau, and R. Coulson, Renal responses to PTH in patients with hormone-resistant (pseudo) hypoparathyroidism. *Am J Med* **61,** 184–189 (1976).

359. R. M. Neer, G. W. Tregear, and J. T. Potts, Jr., Renal effects of native parathyroid hormone and synthetic biologically active fragments in pseudohypoparathyroidism and hypoparathyroidism. *J Clin Endocrinol Metab* **44,** 420–423 (1977).

360. F. Albright, C. H. Burnett, and P. H. Smith, Pseudohypoparathyroidism: An example of "Seabright–Bantam" syndrome. *Endocrinology* **30,** 922–932 (1942).

361. M. A. Levine, Pseudohypoparathyroidism: From bedside to bench and back. *J Bone Miner Res* **14,** 1255–1260 (1999).

362. A. P. Teitelbaum and G. J. Strewler, Parathyroid hormone receptors coupled to cyclic adenosine monophosphate formation in an established renal cell line. *Endocrinology* **114,** 980–985 (1984).

363. K. Malmstrom and H. Murer, Parathyroid hormone inhibits phosphate transport in OK cells but not in LLC-PK1 and JTC-12.P3 cells. *Am J Physiol* **251,** C23–C31 (1986).

364. J. A. Cole, L. R. Forte, S. Eber, P. K. Thorne, and R. E. Poelling, Regulation of sodium-dependent phosphate transport by parathyroid hormone in opossum kidney cells: Adenosine 3′,5′-monophosphate-dependent and -independent mechanisms. *Endocrinology* **122,** 2981–2989 (1988).

365. G. Quamme, J. Pfeilschifter, and H. Murer, Parathyroid hormone inhibition of Na$^+$/phosphate cotransport in OK cells: Intracellular [Ca^{2+}] as a second messenger. *Biochim Biophys Acta* **1013,** 166–172 (1989).

366. K. J. Martin, C. L. McConkey, A. K. Jacob, E. A. Gonzalez, M. Khan, and J. J. Baldassare, Effect of U-73,122, an inhibitor of phospholipase C, on actions of parathyroid hormone in opossum kidney cells. *Am J Physiol* **266,** F254–F258 (1994).

367. J. H. Segal and A. S. Pollock, Transfection-mediated expression of a dominant cAMP-resistant phenotype in the opossum kidney (OK) cell line prevents parathyroid hormone-induced inhibition of Na-phosphate cotransport. A protein kinase-A-mediated event. *J Clin Invest* **86,** 1442–1450 (1990).

368. E. D. Lederer, S. S. Sohi, J. M. Mathiesen, and J. B. Klein, Regulation of expression of type II sodium-phosphate cotransporters by protein kinases A and C. *Am J Physiol* **275,** F270–F277 (1998).

369. S. J. Khundmiri, M. J. Rane, and E. D. Lederer, Parathyroid hormone regulation of type II sodium–phosphate cotransporters is dependent on an A kinase anchoring protein. *J Biol Chem* **278,** 10134–10141 (2003).

370. M. Traebert, H. Volkl, J. Biber, H. Murer, and B. Kaissling, Luminal and contraluminal action of 1–34 and 3–34 PTH peptides on renal type IIa Na–P(i) cotransporter. *Am J Physiol Renal Physiol* **278,** F792–F798 (2000).

371. M. F. Pfister, J. Forgo, U. Ziegler, J. Biber, and H. Murer, cAMP-dependent and -independent downregulation of type II Na–Pi cotransporters by PTH. *Am J Physiol* **276,** F720–F725 (1999).

372. M. Nakai, Y. Kinoshita, M. Fukase, and T. Fujita, Phorbol esters inhibit phosphate uptake in opossum kidney cells: A model of proximal renal tubular cells. *Biochem Biophys Res Commun* **145,** 303–308 (1987).

373. J. A. Cole, S. L. Eber, R. E. Poelling, P. K. Thorne, and L. R. Forte, A dual mechanism for regulation of kidney phosphate transport by parathyroid hormone. *Am J Physiol* **253,** E221–E227 (1987).

374. K. Malmstrom, G. Stange, and H. Murer, Intracellular cascades in the parathyroid-hormone-dependent regulation of Na$^+$/phosphate cotransport in OK cells. *Biochem J* **251,** 207–213 (1988).

375. J. Welsh, V. Weaver, and M. Simboli-Campbell, Regulation of renal 25(OH)D3 1 alpha-hydroxylase: Signal transduction pathways. *Biochem Cell Biol* **69,** 768–770 (1991).

376. H. L. Henry, Parathyroid hormone modulation of 25-hydroxyvitamin D3 metabolism by cultured chick kidney cells is mimicked and enhanced by forskolin. *Endocrinology* **116,** 503–510 (1985).

377. A. A. Portale and W. L. Miller, Human 25-hydroxyvitamin D-1alpha-hydroxylase: Cloning, mutations, and gene expression. *Pediatr Nephrol* **14,** 620–625 (2000).

378. H. L. Brenza, C. Kimmel-Jehan, F. Jehan, T. Shinki, S. Wakino, H. Anazawa, T. Suda, and H. F. DeLuca, Parathyroid hormone activation of the 25-hydroxyvitamin D3-1alpha-hydroxylase gene promoter. *Proc Natl Acad Sci USA* **95,** 1387–13891 (1998).

379. A. Murayama, K. Takeyama, S. Kitanaka, Y. Kodera, T. Hosoya, and S. Kato, The promoter of the human 25-hydroxyvitamin D3 1 alpha-hydroxylase gene confers positive and negative responsiveness to PTH, calcitonin, and 1 alpha,25(OH)2D3. *Biochem Biophys Res Commun* **249,** 11–16 (1998).

380. X. F. Kong, X. H. Zhu, Y. L. Pei, D. M. Jackson, and M. F. Holick, Molecular cloning, characterization, and promoter analysis of the human 25-hydroxyvitamin D3-1alpha-hydroxylase gene. *Proc Natl Acad Sci USA* **96,** 6988–6993 (1999).

381. X. H. Gao, P. P. Dwivedi, S. Choe, F. Alba, H. A. Morris, J. L. Omdahl, and B. K. May, Basal and parathyroid hormone induced expression of the human 25-hydroxyvitamin D 1alpha-hydroxylase gene promoter in kidney AOK-B50 cells: Role of Sp1, Ets and CCAAT box protein binding sites. *Int J Biochem Cell Biol* **34,** 921–930 (2002).

382. J. N. Flanagan, L. Wang, V. Tangpricha, J. Reichrath, T. C. Chen, and M. F. Holick, Regulation of the 25-hydroxyvitamin D-1alpha-hydroxylase gene and its splice variant. *Recent Results Cancer Res* **164,** 157–167 (2003).

383. H. K. Ro, V. Tembe, and M. J. Favus, Evidence that activation of protein kinase-C can stimulate 1,25-dihydroxyvitamin D3 secretion by rat proximal tubules. *Endocrinology* **131,** 1424–1428 (1992).

384. M. Janulis, V. Tembe, and M. J. Favus, Role of protein kinase C in parathyroid hormone stimulation of renal 1,25-dihydroxyvitamin D3 secretion. *J Clin Invest* **90,** 2278–2283 (1992).

385. J. G. Hoenderop, J. J. De Pont, R. J. Bindels, and P. H. Willems, Hormone-stimulated Ca^{2+} reabsorption in rabbit kidney cortical collecting system is cAMP-independent and involves a phorbol ester-insensitive PKC isotype. *Kidney Int* **55,** 225–233 (1999).

386. G. Hilal, D. Claveau, M. Leclerc, and M. G. Brunette, Ca^{2+} transport by the luminal membrane of the distal nephron: Action and interaction of protein kinases A and C. *Biochem J* **328**(Pt. 2), 371–375 (1997).

387. P. A. Friedman, F. A. Gesek, P. Morley, J. F. Whitfield, and G. E. Willick, Cell-specific signaling and structure–activity relations of parathyroid hormone analogs in mouse kidney cells. *Endocrinology* **140,** 301–309 (1999).

388. A. T. Singh, J. G. Kunnel, P. J. Strieleman, and P. H. Stern, Parathyroid hormone (PTH)-(1–34), [Nle(8,18),Tyr34]PTH-(3–34) amide, PTH-(1–31) amide, and PTH-related peptide-(1–34) stimulate phosphatidylcholine hydrolysis in UMR-106 osteoblastic cells: Comparison with effects of phorbol 12,13-dibutyrate. *Endocrinology* **140,** 131–137 (1999).

389. P. Morley, J. F. Whitfield, and G. E. Willick, Design and applications of parathyroid hormone analogues. *Curr Med Chem* **6,** 1095–1106 (1999).

390. A. M. Spiegel and L. S. Weinstein, Inherited diseases involving g proteins and g protein-coupled receptors. *Annu Rev Med* **55,** 27–39 (2004).

391. A. Sakamoto, M. Chen, T. Kobayashi, H. M. Kronenberg, and L. S. Weinstein, Chondrocyte-specific knockout of the G protein G(s)alpha leads to epiphyseal and growth plate abnormalities and ectopic chondrocyte formation. *J Bone Miner Res* **20,** 663–671 (2005).

392. H. Kather, W. Tschope, and B. Simon, Human fat cell adenylate cyclase. Modulation of parathyroid hormone action by guanine nucleotides. *Res Exp Med (Berlin)* **171,** 201–204 (1977).

393. V. P. Michalangeli, N. H. Hunt, and T. J. Martin, States of activation of chick kidney adenylate cyclase induced by

parathyroid hormone and guanyl nucleotides. *J Endocrinol* **72**, 69–79 (1977).

394. E. Bellorin-Font and K. J. Martin, Regulation of the PTH-receptor-cyclase system of canine kidney: Effects of calcium, magnesium, and guanine nucleotides. *Am J Physiol* **241**, F364–F373 (1981).

395. R. E. Rizzoli, T. M. Murray, S. J. Marx, and G. D. Aurbach, Binding of radioiodinated bovine parathyroid hormone-(1–84) to canine renal cortical membranes. *Endocrinology* **112**, 1303–1312 (1983).

396. A. P. Teitelbaum, R. A. Nissenson, and C. D. Arnaud, Coupling of the canine renal parathyroid hormone receptor to adenylate cyclase: Modulation by guanyl nucleotides and N-ethylmaleimide. *Endocrinology* **111**, 1524–1533 (1982).

397. R. A. Nissenson, S. R. Abbott, A. P. Teitelbaum, O. H. Clark, and C. D. Arnaud, Endogenous biologically active human parathyroid hormone: Measurement by a guanyl nucleotide-amplified renal adenylate cyclase assay. *J Clin Endocrinol Metab* **52**, 840–846 (1981).

398. H. Juppner, A. B. Abou-Samra, M. Freeman, X. F. Kong, E. Schipani, J. Richards, L. F. Kolakowski, Jr., J. Hock, J. T. Potts, Jr., H. M. Kronenberg, *et al.*, A G protein-linked receptor for parathyroid hormone and parathyroid hormone-related peptide. *Science* **254**, 1024–1026 (1991).

399. T. Schoneberg, G. Schultz, and T. Gudermann, Structural basis of G protein-coupled receptor function. *Mol Cell Endocrinol* **151**, 181–193 (1999).

400. C. D. Strader, T. M. Fong, M. P. Graziano, and M. R. Tota, The family of G-protein-coupled receptors. *FASEB J* **9**, 745–754 (1995).

401. M. Babich, H. Choi, R. M. Johnson, K. L. King, G. E. Alford, and R. A. Nissenson, Thrombin and parathyroid hormone mobilize intracellular calcium in rat osteosarcoma cells by distinct pathways. *Endocrinology* **129**, 1463–1470 (1991).

402. R. C. Gensure, T. J. Gardella, and H. Juppner, Parathyroid hormone and parathyroid hormone-related peptide, and their receptors. *Biochem Biophys Res Commun* **328**, 666–678 (2005).

403. H. Juppner, E. Schipani, F. R. Bringhurst, I. McClure, H. T. Keutmann, J. T. Potts, Jr., H. M. Kronenberg, A. B. Abou-Samra, G. V. Segre, and T. J. Gardella, The extracellular amino-terminal region of the parathyroid hormone (PTH)/PTH-related peptide receptor determines the binding affinity for carboxyl-terminal fragments of PTH-(1–34). *Endocrinology* **134**, 879–884 (1994).

404. P. R. Turner, T. Bambino, and R. A. Nissenson, A putative selectivity filter in the G-protein-coupled receptors for parathyroid hormone and secretion. *J Biol Chem* **271**, 9205–9208 (1996).

405. C. Bergwitz, T. J. Gardella, M. R. Flannery, J. T. Potts, Jr., H. M. Kronenberg, S. R. Goldring, and H. Juppner, Full activation of chimeric receptors by hybrids between parathyroid hormone and calcitonin. Evidence for a common pattern of ligand–receptor interaction. *J Biol Chem* **271**, 26469–26472 (1996).

406. T. J. Gardella, H. Juppner, A. K. Wilson, H. T. Keutmann, A. B. Abou-Samra, G. V. Segre, F. R. Bringhurst, J. T. Potts, Jr., S. R. Nussbaum, and H. M. Kronenberg, Determinants of [Arg2]PTH-(1–34) binding and signaling in the transmembrane region of the parathyroid hormone receptor. *Endocrinology* **135**, 1186–1194 (1994).

407. P. R. Turner, T. Bambino, and R. A. Nissenson, Mutations of neighboring polar residues on the second transmembrane helix disrupt signaling by the parathyroid hormone receptor. *Mol Endocrinol* **10**, 132–139 (1996).

408. T. J. Gardella, M. D. Luck, M. H. Fan, and C. Lee, Transmembrane residues of the parathyroid hormone (PTH)/PTH-related peptide receptor that specifically affect binding and signaling by agonist ligands. *J Biol Chem* **271**, 12820–12825 (1996).

409. S. R. Hoare, Mechanisms of peptide and nonpeptide ligand binding to class B G-protein-coupled receptors. *Drug Discov Today* **10**, 417–427 (2005).

410. M. Mannstadt, M. D. Luck, T. J. Gardella, and H. Juppner, Evidence for a ligand interaction site at the amino-terminus of the parathyroid hormone (PTH)/PTH related protein receptor from cross-linking and mutational studies. *J Biol Chem* **273**, 16890–16896 (1998).

411. R. C. Gensure, T. J. Gardella, and H. Juppner, Multiple sites of contact between the carboxyl-terminal binding domain of PTHrP-(1–36) analogs and the amino-terminal extracellular domain of the PTH/PTHrP receptor identified by photoaffinity cross-linking. *J Biol Chem* **276**, 28650–28658 (2001).

412. A. Bisello, A. E. Adams, D. F. Mierke, M. Pellegrini, M. Rosenblatt, L. J. Suva, and M. Chorev, Parathyroid hormone–receptor interactions identified directly by photocross-linking and molecular modeling studies. *J Biol Chem* **273**, 22498–22505 (1998).

413. S. P. Sheikh, J. P. Vilardarga, T. J. Baranski, O. Lichtarge, T. Iiri, E. C. Meng, R. A. Nissenson, and H. R. Bourne, Similar structures and shared switch mechanisms of the beta2-adrenoceptor and the parathyroid hormone receptor. Zn(II) bridges between helices III and VI block activation. *J Biol Chem* **274**, 17033–17041 (1999).

414. Z. Huang, Y. Chen, S. Pratt, T. H. Chen, T. Bambino, R. A. Nissenson, and D. M. Shoback, The N-terminal region of the third intracellular loop of the parathyroid hormone (PTH)/PTH-related peptide receptor is critical for coupling to cAMP and inositol phosphate/Ca^{2+} signal transduction pathways. *J Biol Chem* **271**, 33382–33389 (1996).

415. A. Iida-Klein, J. Guo, M. Takemura, M. T. Drake, J. T. Potts, Jr., A. Abou-Samra, F. R. Bringhurst, and G. V. Segre, Mutations in the second cytoplasmic loop of the rat parathyroid hormone (PTH)/PTH-related protein receptor result in selective loss of PTH-stimulated phospholipase C activity. *J Biol Chem* **272**, 6882–6889 (1997).

416. A. T. Zhou, R. Bessalle, A. Bisello, C. Nakamoto, M. Rosenblatt, L. J. Suva, and M. Chorev, Direct mapping of an agonist-binding domain within the parathyroid hormone/parathyroid hormone-related protein receptor by photoaffinity crosslinking. *Proc Natl Acad Sci USA* **94**, 3644–3649 (1997).

417. R. C. Gensure, N. Shimizu, J. Tsang, and T. J. Gardella, Identification of a contact site for residue 19 of parathyroid hormone (PTH) and PTH-related protein analogs in transmembrane domain two of the type 1 PTH receptor. *Mol Endocrinol* **17**, 2647–2658 (2003).

418. A. Wittelsberger, M. Corich, B. E. Thomas, B. K. Lee, A. Barazza, P. Czodrowski, D. F. Mierke, M. Chorev, and M. Rosenblatt, The mid-region of parathyroid hormone (1–34) serves as a functional docking domain in receptor activation. *Biochemistry* **45**, 2027–2034 (2006).

419. G. L. Wong, Induction of metabolic changes and downregulation of bovine parathyroid hormone-responsive adenylate cyclase are dissociable in isolated osteoclastic and osteoblastic bone cells. *J Biol Chem* **254**, 34–37 (1979).

420. W. A. Peck and G. Kohler, Hormonal and nonhormonal desensitization in isolated bone cells. *Calcif Tissue Int* **32**, 95–103 (1980).

421. S. R. Goldring, J. M. Dayer, and M. Rosenblatt, Factors regulating the response of cells cultured from human giant cell

tumors of bone to parathyroid hormone. *J Clin Endocrinol Metab* **53**, 295–300 (1981).

422. W. I. Chao and L. R. Forte, Rat kidney cells in primary culture: Hormone-mediated desensitization of the adenosine 3′,5′-monophosphate response to parathyroid hormone and calcitonin. *Endocrinology* **111**, 252–259 (1982).

423. H. L. Henry, N. S. Cunningham, and T. A. Noland, Jr., Homologous desensitization of cultured chick kidney cells to parathyroid hormone. *Endocrinology* **113**, 1942–1949 (1983).

424. C. Bergwitz, A. B. Abou-Samra, R. D. Hesch, and H. Juppner, Rapid desensitization of parathyroid hormone dependent adenylate cyclase in perfused human osteosarcoma cells (SaOS-2). *Biochim Biophys Acta* **1222**, 447–456 (1994).

425. S. K. Lee and P. H. Stern, Studies on the mechanism of desensitization of the parathyroid hormone-stimulated calcium signal in UMR-106 cells: Reversal of desensitization by alkaline phosphatase but not by protein kinase C downregulation. *J Bone Miner Res* **9**, 781–789 (1994).

426. A. Fujimori, A. Miyauchi, K. A. Hruska, K. J. Martin, L. V. Avioli, and R. Civitelli, Desensitization of calcium messenger system in parathyroid hormone-stimulated opossum kidney cells. *Am J Physiol* **264**, E918–E924 (1993).

427. S. Fukayama, A. H. Tashjian, Jr., and F. R. Bringhurst, Mechanisms of desensitization to parathyroid hormone in human osteoblast-like SaOS-2 cells. *Endocrinology* **131**, 1757–1769 (1992).

428. K. K. Pun, P. W. Ho, R. A. Nissenson, and C. D. Arnaud, Desensitization of parathyroid hormone receptors on cultured bone cells. *J Bone Miner Res* **5**, 1193–1200 (1990).

429. M. Bunemann and M. M. Hosey, G-protein coupled receptor kinases as modulators of G-protein signalling. *J Physiol* **517**(Pt. 1), 5–23 (1999).

430. R. J. Lefkowitz, G protein-coupled receptors: III. New roles for receptor kinases and beta-arrestins in receptor signaling and desensitization. *J Biol Chem* **273**, 18677–18680 (1998).

431. J. G. Krupnick and J. L. Benovic, The role of receptor kinases and arrestins in G protein-coupled receptor regulation. *Annu Rev Pharmacol Toxicol* **38**, 289–319 (1998).

432. E. Blind, T. Bambino, and R. A. Nissenson, Agonist-stimulated phosphorylation of the G protein-coupled receptor for parathyroid hormone (PTH) and PTH-related protein. *Endocrinology* **136**, 4271–4277 (1995).

433. F. Qian, A. Leung, and A. Abou-Samra, Agonist-dependent phosphorylation of the parathyroid hormone/parathyroid hormone-related peptide receptor. *Biochemistry* **37**, 6240–6246 (1998).

434. E. Blind, T. Bambino, Z. Huang, M. Bliziotes, and R. A. Nissenson, Phosphorylation of the cytoplasmic tail of the PTH/PTHrP receptor. *J Bone Miner Res* **11**, 578–586 (1996).

435. N. Malecz, T. Bambino, M. Bencsik, and R. A. Nissenson, Identification of phosphorylation sites in the G protein-coupled receptor for parathyroid hormone. Receptor phosphorylation is not required for agonist-induced internalization. *Mol Endocrinol* **12**, 1846–1856 (1998).

436. M. Bliziotes, J. Murtagh, and K. Wiren, Beta-adrenergic receptor kinase-like activity and beta-arrestin are expressed in osteoblastic cells. *J Bone Miner Res* **11**, 820–826 (1996).

437. F. Dicker, U. Quitterer, R. Winstel, K. Honold, and M. J. Lohse, Phosphorylation-independent inhibition of parathyroid hormone receptor signaling by G protein-coupled receptor kinases. *Proc Natl Acad Sci USA* **96**, 5476–5481 (1999).

438. G. S. Bounoutas, H. Tawfeek, L. F. Frohlich, U. I. Chung, and A. B. Abou-Samra, Impact of impaired receptor internalization on calcium homeostasis in knock-in mice expressing a phosphorylation-deficient parathyroid hormone (PTH)/PTH-related peptide receptor. *Endocrinology* **147**, 4674–4679 (2006).

439. C. A. Mahoney and R. A. Nissenson, Canine renal receptors for parathyroid hormone. Downregulation *in vivo* by exogenous parathyroid hormone. *J Clin Invest* **72**, 411–421 (1983).

440. A. P. Teitelbaum, C. M. Silve, K. O. Nyiredy, and C. D. Arnaud, Downregulation of parathyroid hormone (PTH) receptors in cultured bone cells is associated with agonist-specific intracellular processing of PTH–receptor complexes. *Endocrinology* **118**, 595–602 (1986).

441. I. Yamamoto, C. Shigeno, J. T. Potts, Jr., and G. V. Segre, Characterization and agonist-induced downregulation of parathyroid hormone receptors in clonal rat osteosarcoma cells. *Endocrinology* **122**, 1208–1217 (1988).

442. J. Mitchell and D. Goltzman, Mechanisms of homologous and heterologous regulation of parathyroid hormone receptors in the rat osteosarcoma cell line UMR-106. *Endocrinology* **126**, 2650–2660 (1990).

443. L. R. Forte, G. A. Nickols, and C. S. Anast, Renal adenylate cyclase and the interrelationship between parathyroid hormone and vitamin D in the regulation of urinary phosphate and adenosine cyclic 3,5-monophosphate excretion. *J Clin Invest* **57**, 559–568 (1976).

444. S. G. Massry, R. Stein, J. Garty, A. I. Arieff, J. W. Coburn, A. W. Norman, and R. M. Friedler, Skeletal resistance to the calcemic action of parathyroid hormone in uremia: Role of 1,25 (OH)2 D3. *Kidney Int* **9**, 467–474 (1976).

445. I. G. Lewin, S. E. Papapoulos, G. N. Hendy, S. Tomlinson, and J. L. O'Riordan, Reversible resistance to the renal action of parathyroid hormone in human vitamin D deficiency. *Clin Sci (London)* **62**, 381–387 (1982).

446. G. A. Nickols, D. L. Carnes, C. S. Anast, and L. R. Forte, Parathyroid hormone-mediated refractoriness of rat kidney cyclic AMP system. *Am J Physiol* **236**, E401–E409 (1979).

447. P. Urena, M. Mannstadt, M. Hruby, A. Ferreira, F. Schmitt, C. Silve, R. Ardaillou, B. Lacour, A. B. Abou-Samra, G. V. Segre, *et al.*, Parathyroidectomy does not prevent the renal PTH/PTHrP receptor downregulation in uremic rats. *Kidney Int* **47**, 1797–1805 (1995).

448. Z. Huang, Y. Chen, and R. A. Nissenson, The cytoplasmic tail of the G-protein-coupled receptor for parathyroid hormone and parathyroid hormone-related protein contains positive and negative signals for endocytosis. *J Biol Chem* **270**, 151–156 (1995).

449. H. A. Tawfeek, F. Qian, and A. B. Abou-Samra, Phosphorylation of the receptor for PTH and PTHrP is required for internalization and regulates receptor signaling. *Mol Endocrinol* **16**, 1–13 (2002).

450. S. L. Ferrari, V. Behar, M. Chorev, M. Rosenblatt, and A. Bisello, Endocytosis of ligand–human parathyroid hormone receptor 1 complexes is protein kinase C-dependent and involves beta-arrestin2. Real-time monitoring by fluorescence microscopy. *J Biol Chem* **274**, 29968–29975 (1999).

451. J. P. Vilardaga, C. Krasel, S. Chauvin, T. Bambino, M. J. Lohse, and R. A. Nissenson, Internalization determinants of the parathyroid hormone receptor differentially regulate beta-arrestin/receptor association. *J Biol Chem* **277**, 8121–8129 (2002).

452. C. A. Syme, P. A. Friedman, and A. Bisello, Parathyroid hormone receptor trafficking contributes to the activation of extracellular signal-regulated kinases but is not required for regulation of cAMP signaling. *J Biol Chem* **280**, 11281–11288 (2005).

453. D. Gesty-Palmer, M. Chen, E. Reiter, S. Ahn, C. D. Nelson, S. Wang, A. E. Eckhardt, C. L. Cowan, R. F. Spurney, L. M. Luttrell, and R. J. Lefkowitz, Distinct beta-arrestin- and G protein-dependent pathways for parathyroid hormone receptor-stimulated ERK1/2 activation. *J Biol Chem* **281,** 10856–10864 (2006).

454. A. Rey, D. Manen, R. Rizzoli, J. Caverzasio, and S. L. Ferrari, Proline-rich motifs in the parathyroid hormone (PTH)/PTH-related protein receptor C terminus mediate scaffolding of c-Src with beta-arrestin2 for ERK1/2 activation. *J Biol Chem* **281,** 38181–38188 (2006).

455. S. L. Ferrari, D. D. Pierroz, V. Glatt, D. S. Goddard, E. N. Bianchi, F. T. Lin, D. Manen, and M. L. Bouxsein, Bone response to intermittent parathyroid hormone is altered in mice null for β-arrestin2. *Endocrinology* **146,** 1854–1862 (2005).

456. E. A. Gonzalez and K. J. Martin, Coordinate regulation of PTH/PTHrP receptors by PTH and calcitriol in UMR 106-01 osteoblast-like cells. *Kidney Int* **50,** 63–70 (1996).

457. J. W. Jongen, E. C. Willemstein-van Hove, J. M. van der Meer, M. P. Bos, H. Juppner, G. V. Segre, A. B. Abou-Samra, J. H. Feyen, and M. P. Herrmann-Erlee, Downregulation of the receptor for parathyroid hormone (PTH) and PTH-related peptide by PTH in primary fetal rat osteoblasts. *J Bone Miner Res* **11,** 1218–1225 (1996).

458. T. Kawane, J. Mimura, T. Yanagawa, Y. Fujii-Kuriyama, and N. Horiuchi, Parathyroid hormone (PTH) downregulates PTH/PTH-related protein receptor gene expression in UMR-106 osteoblast-like cells via a 3,5-cyclic adenosine monophosphate-dependent, protein kinase A-independent pathway. *J Endocrinol* **178,** 247–256 (2003).

459. G. Turner, C. Coureau, M. R. Rabin, B. Escoubet, M. Hruby, O. Walrant, and C. Silve, Parathyroid hormone (PTH)/PTH-related protein receptor messenger ribonucleic acid expression and PTH response in a rat model of secondary hyperparathyroidism associated with vitamin D deficiency. *Endocrinology* **136,** 3751–3758 (1995).

460. L. K. McCauley, C. A. Beecher, M. E. Melton, J. R. Werkmeister, H. Juppner, A. B. Abou-Samra, G. V. Segre, and T. J. Rosol, Transforming growth factor-beta1 regulates steady-state PTH/PTHrP receptor mRNA levels and PTHrP binding in ROS 17/2.8 osteosarcoma cells. *Mol Cell Endocrinol* **101,** 331–336 (1994).

461. J. W. Jongen, E. C. Willemstein-Van Hove, J. M. Van der Meer, M. P. Bos, H. Juppner, G. V. Segre, A. B. Abou-Samra, J. H. Feyen, and M. P. Herrmann-Erlee, Downregulation of the receptor for parathyroid hormone (PTH) and PTH-related peptide by transforming growth factor-beta in primary fetal rat osteoblasts. *Endocrinology* **136,** 3260–3266 (1995).

462. F. Law, J. P. Bonjour, and R. Rizzoli, Transforming growth factor-beta: A downregulator of the parathyroid hormone-related protein receptor in renal epithelial cells. *Endocrinology* **134,** 2037–2043 (1994).

463. T. Kawane, J. Mimura, Y. Fujii-Kuriyama, and N. Horiuchi, Identification of the promoter region of the parathyroid hormone receptor gene responsible for transcriptional suppression by insulin-like growth factor-I. *Arch Biochem Biophys* **439,** 61–69 (2005).

464. P. Urena, A. Iida-Klein, X. F. Kong, H. Juppner, H. M. Kronenberg, A. B. Abou-Samra, and G. V. Segre, Regulation of parathyroid hormone (PTH)/PTH-related peptide receptor messenger ribonucleic acid by glucocorticoids and PTH in ROS 17/2.8 and OK cells. *Endocrinology* **134,** 451–456 (1994).

465. J. Yaghoobian and T. B. Drueke, Regulation of the transcription of parathyroid-hormone/parathyroid-hormone-related peptide receptor mRNA by dexamethasone in ROS 17/2.8 osteosarcoma cells. *Nephrol Dial Transplant* **13,** 580–586 (1998).

466. H. Wald, M. Dranitzki-Elhalel, R. Backenroth, and M. M. Popovtzer, Evidence for interference of vitamin D with PTH/PTHrP receptor expression in opossum kidney cells. *Pflugers Arch* **436,** 289–294 (1998).

467. W. X. Gu, P. H. Stern, L. D. Madison, and G. G. Du, Mutual upregulation of thyroid hormone and parathyroid hormone receptors in rat osteoblastic osteosarcoma 17/2.8 cells. *Endocrinology* **142,** 157–164 (2001).

468. V. Y. Hook, D. Burton, S. Yasothornsrikul, R. H. Hastings, and L. J. Deftos, Proteolysis of ProPTHrP(1–141) by "prohormone thiol protease" at multibasic residues generates PTHrP-related peptides: Implications for PTHrP peptide production in lung cancer cells. *Biochem Biophys Res Commun* **285,** 932–938 (2001).

469. J. Cornish, K. E. Callon, G. C. Nicholson, and I. R. Reid, Parathyroid hormone-related protein-(107–139) inhibits bone resorption *in vivo*. *Endocrinology* **138,** 1299–1304 (1997).

470. J. Cornish, K. E. Callon, C. Lin, C. Xiao, J. M. Moseley, and I. R. Reid, Stimulation of osteoblast proliferation by C-terminal fragments of parathyroid hormone-related protein. *J Bone Miner Res* **14,** 915–922 (1999).

471. F. De Miguel, P. Martinez-Fernandez, C. Guillen, A. Valin, A. Rodrigo, M. E. Martinez, and P. Esbrit, Parathyroid hormone-related protein (107–139) stimulates interleukin-6 expression in human osteoblastic cells. *J Am Soc Nephrol* **10,** 796–803 (1999).

472. A. Valin, C. Guillen, and P. Esbrit, C-terminal parathyroid hormone-related protein (PTHrP) (107–139) stimulates intracellular Ca(2+) through a receptor different from the type 1 PTH/PTHrP receptor in osteoblastic osteosarcoma UMR 106 cells. *Endocrinology* **142,** 2752–2759 (2001).

473. A. D. Care, S. K. Abbas, D. W. Pickard, M. Barri, M. Drinkhill, J. B. Findlay, I. R. White, and I. W. Caple, Stimulation of ovine placental transport of calcium and magnesium by mid-molecule fragments of human parathyroid hormone-related protein. *Exp Physiol* **75,** 605–608 (1990).

474. E. Maioli and V. Fortino, The complexity of parathyroid hormone-related protein signalling. *Cell Mol Life Sci* **61,** 257–262 (2004).

475. N. M. Fiaschi-Taesch and A. F. Stewart, Minireview: Parathyroid hormone-related protein as an intracrine factor—Trafficking mechanisms and functional consequences. *Endocrinology* **144,** 407–411 (2003).

476. M. T. Nguyen and A. C. Karaplis, The nucleus: A target site for parathyroid hormone-related peptide (PTHrP) action. *J Cell Biochem* **70,** 193–199 (1998).

477. M. M. Aarts, D. Levy, B. He, S. Stregger, T. Chen, S. Richard, and J. E. Henderson, Parathyroid hormone-related protein interacts with RNA. *J Biol Chem* **274,** 4832–4838 (1999).

478. J. E. Henderson, N. Amizuka, H. Warshawsky, D. Biasotto, B. M. Lanske, D. Goltzman, and A. C. Karaplis, Nucleolar localization of parathyroid hormone-related peptide enhances survival of chondrocytes under conditions that promote apoptotic cell death. *Mol Cell Biol* **15,** 4064–4075 (1995).

479. N. Amizuka, K. Oda, J. Shimomura, and T. Maeda, Biological action of parathyroid hormone (PTH)-related peptide (PTHrP) mediated either by the PTH/PTHrP receptor or the nucleolar translocation in chondrocytes. *Anat Sci Int* **77,** 225–236 (2002).

480. M. Nguyen, B. He, and A. Karaplis, Nuclear forms of parathyroid hormone-related peptide are translated from non-AUG start sites downstream from the initiator methionine. *Endocrinology* **142,** 694–703 (2001).

481. T. Massfelder, P. Dann, T. L. Wu, R. Vasavada, J. J. Helwig, and A. F. Stewart, Opposing mitogenic and anti-mitogenic actions of parathyroid hormone-related protein in vascular smooth muscle cells: A critical role for nuclear targeting. *Proc Natl Acad Sci USA* **94,** 13630–13635 (1997).

482. E. Schordan, S. Welsch, S. Rothhut, A. Lambert, M. Barthelmebs, J. J. Helwig, and T. Massfelder, Role of parathyroid hormone-related protein in the regulation of stretch-induced renal vascular smooth muscle cell proliferation. *J Am Soc Nephrol* **15,** 3016–3025 (2004).

483. N. Fiaschi-Taesch, K. K. Takane, S. Masters, J. C. Lopez-Talavera, and A. F. Stewart, Parathyroid-hormone-related protein as a regulator of pRb and the cell cycle in arterial smooth muscle. *Circulation* **110,** 177–185 (2004).

484. N. Fiaschi-Taesch, B. M. Sicari, K. Ubriani, T. Bigatel, K. K. Takane, I. Cozar-Castellano, A. Bisello, B. Law, and A. F. Stewart, Cellular mechanism through which parathyroid hormone-related protein induces proliferation in arterial smooth muscle cells: Definition of an arterial smooth muscle PTHrP/p27kip1 pathway. *Circ Res* **99,** 933–942 (2006).

485. W. D. Stuart, S. Maeda, P. Khera, J. A. Fagin, and T. L. Clemens, Parathyroid hormone-related protein induces G1 phase growth arrest of vascular smooth muscle cells. *Am J Physiol Endocrinol Metab* **279,** E60–E67 (2000).

486. M. H. Lam, S. L. Olsen, W. A. Rankin, P. W. Ho, T. J. Martin, M. T. Gillespie, and J. M. Moseley, PTHrP and cell division: Expression and localization of PTHrP in a keratinocyte cell line (HaCaT) during the cell cycle. *J Cell Physiol* **173,** 433–446 (1997).

487. M. H. Lam, C. M. House, T. Tiganis, K. I. Mitchelhill, B. Sarcevic, A. Cures, R. Ramsay, B. E. Kemp, T. J. Martin, and M. T. Gillespie, Phosphorylation at the cyclin-dependent kinases site (Thr85) of parathyroid hormone-related protein negatively regulates its nuclear localization. *J Biol Chem* **274,** 18559–18566 (1999).

488. J. C. Pache, D. W. Burton, L. J. Deftos, and R. H. Hastings, A carboxyl leucine-rich region of parathyroid hormone-related protein is critical for nuclear export. *Endocrinology* **147,** 990–998 (2006).

CHAPTER 11

Vitamin D: Biology, Action, and Clinical Implications

DAVID FELDMAN, PETER J. MALLOY, ARUNA V. KRISHNAN, AND EVA BALINT

I. INTRODUCTION

Vitamin D is the major regulator of calcium homeostasis in the body and is critically important for normal mineralization of bone. The active hormone, 1α,25-dihydroxyvitamin D [1,25(OH)$_2$D], is produced by sequential hydroxylations of vitamin D in the liver (25-hydroxylation) and the kidney (1α-hydroxylation). 1,25(OH)$_2$D, working through the vitamin D receptor (VDR), functions by a genomic mechanism similar to the classical steroid hormones to regulate target gene transcription. The traditional actions of 1,25(OH)$_2$D are to enhance calcium and phosphate absorption from the intestine in order to maintain normal concentrations in the circulation and to provide adequate amounts of these minerals to the bone-forming site to allow mineralization of bone to proceed normally. However, in the past two decades, it has become increasingly clear that vitamin D has many additional functions that implicate the hormone in a wide array of actions relating to bone formation as well as to other areas unrelated to bone or mineral metabolism, including antiproliferative, prodifferentiating, and immunosuppressive activities. In this chapter we describe the basic biology of vitamin D including its metabolism, physiology, mechanism of action, and its diverse functions in the body, including those actions that relate to mineral metabolism as well as the newer actions. Several recent reviews of vitamin D mechanism of action and function have been published [1–7] as well as a comprehensive book addressing all areas of vitamin D [8].

A. Chemistry, Structure, and Terminology

Vitamin D exists in two forms: vitamin D$_3$ (cholecalciferol) and vitamin D$_2$ (ergocalciferol) When written without a subscript, the designation vitamin D denotes either D$_2$ or D$_3$. Sunlight, in the form of UV-B rays, cleaves the B ring between carbon-9 and -10 to open the ring and create a secosteroid structure (Figure 11-1). By this process, the precursor (provitamin) molecules, 7-dehydrocholesterol in animals and ergosterol in plants, are converted to the secosteroids, vitamin D$_3$ and vitamin D$_2$, respectively [9]. The two secosteroids differ only in the presence of a methyl group at carbon 28 and a double bond between carbon 22 and 23 on the side chain of vitamin D$_2$. Vitamin D$_2$ and vitamin D$_3$ are handled identically in the body and converted, via two hydroxylation steps, first in the liver and then in the kidney to the active hormones, 1,25(OH)$_2$D$_2$ or 1,25(OH)$_2$D$_3$ (calcitriol) (see Figure 11-2). The complex conversion of vitamin D to the active hormone by cytochrome P450 enzymes is detailed in Section III of this chapter. 1,25(OH)$_2$D then acts in multiple target tissues throughout the body by binding to its nuclear receptor, the vitamin D receptor (VDR), to regulate gene expression. The mechanism of vitamin D action is discussed in Section IV.

B. History

The unfolding of the story of vitamin D from its discovery as an antirachitic factor and designation as a vitamin to its transition from being considered a vitamin to its recognition as a hormone has all occurred within the past 75 years. Yet the substance appears to be evolutionarily very ancient, produced by phytoplankton exposed to sunlight approximately 750 million years ago [10]. The history of the identification of vitamin D, the beneficial effects of sunlight on rickets, the elucidation of the pathway of conversion of vitamin D to 1,25(OH)$_2$D, and the realization that vitamin D is a steroid hormone have been detailed in multiple reviews [10–14].

FIGURE 11-1   1,25(OH)$_2$D metabolic pathways. UV-B indicates ultraviolet radiation (wavelength 290–320 nm) emitted from the sun. Liver 25 refers to hepatic 25-hydroxylase and kidney 24R and 1α are renal 24-hydroxylase and 1α-hydroxylase, respectively. Reproduced with permission from M. F. Holick, in *Endocrinology* (L. J. DeGroot *et al.*, eds.). Saunders, Philadelphia, 1995.

II. VITAMIN D SYNTHESIS AND METABOLISM

A. Vitamin D Metabolism

Vitamin D is fat-soluble and dietary sources are absorbed via the lymphatics in the proximal small bowel. Factors that are important for absorption include: (1) gastric, pancreatic, and biliary secretions; (2) formation of micelles; (3) diffusion through the unstirred layer adjacent to the intestinal mucosa; (4) brush border membrane uptake; (5) incorporation into chylomicrons; and (6) absorption into the lymphatics. The mechanism of intestinal calcium absorption and

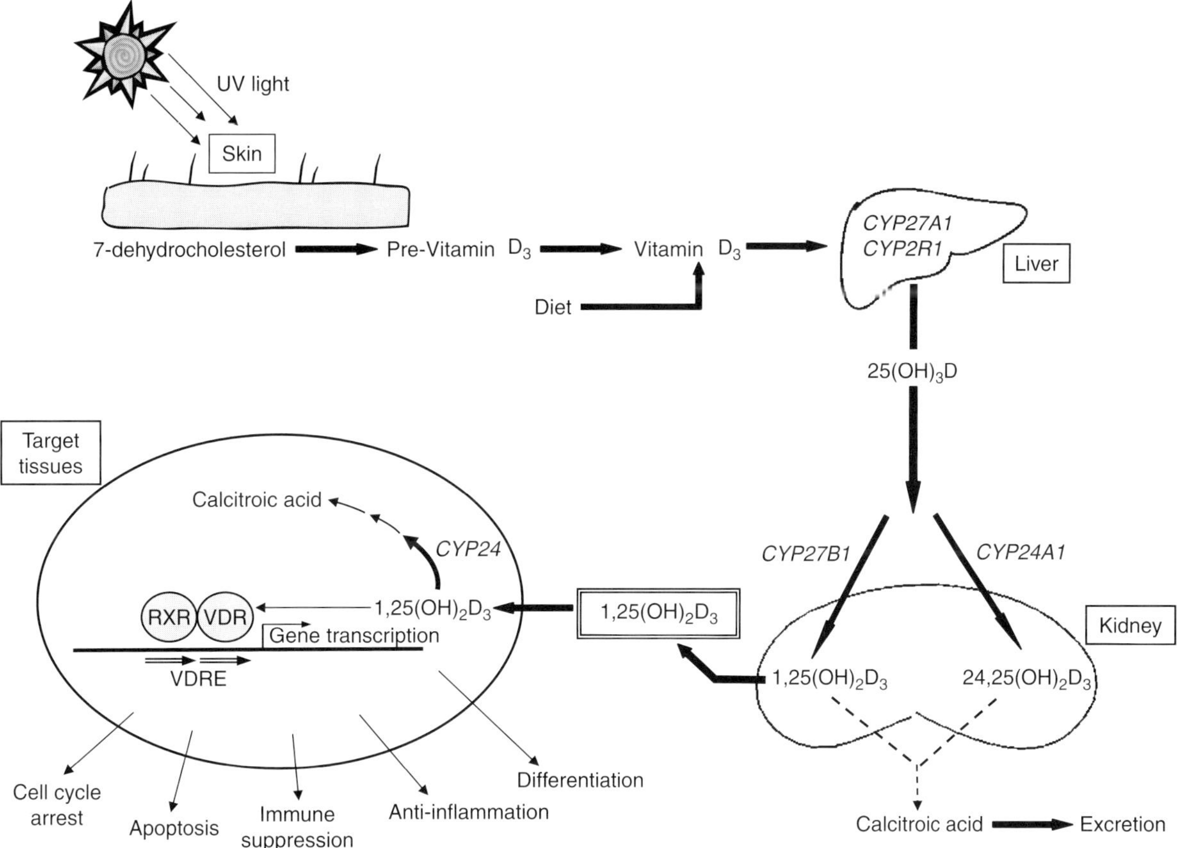

FIGURE 11-2   Overview of the vitamin D metabolic pathway.

its regulation by vitamin D was recently reviewed [15] and is discussed later in this chapter (Section VI).

Disorders that interfere with the preceding processes or that disrupt the small bowel mucosa can interfere with vitamin D absorption and include cystic fibrosis, chronic pancreatitis with pancreatic insufficiency, biliary obstruction, sprue (gluten enteropathy), inflammatory bowel disease involving the small bowel, short bowel syndrome, and gastrointestinal surgery [16]. Assessing vitamin D status is clinically important in patients with these or related conditions. After an oral dose of vitamin D, blood levels begin to rise at 4 hours, peak by 12 hours, and return to close to baseline by 72 hours. This pharmacokinetic profile provides a useful clinical test for assessing adequacy of vitamin D absorption. Serum vitamin D level can be measured 12 hours after an oral dose of 50,000 IU of vitamin D; a value of ≥50 ng/mL is indicative of normal vitamin D absorption, whereas malabsorption is indicated when values are ≤10 ng/mL [17]. Although most cases of

rickets are due to vitamin D deficiency, calcium and iron deficiency might also play a role [18, 19]. Studies of rachitic children in Nigeria, Turkey, and South Africa suggest that calcium deficiency also contributes to this condition [20–22]. The children responded better to treatment with calcium alone or calcium and vitamin D than treatment with vitamin D alone [20, 22]. Iron deficiency might also influence vitamin D metabolism by affecting vitamin D handling in the skin or intestine [19]. A third of children with anemia are also vitamin D deficient, half of vitamin D deficient children are anemic, and iron treatment results in rising vitamin D levels [23].

B.   Photobiology of Vitamin D: Endogenous Production

There are two sources of vitamin D: dietary intake and endogenous production (Figure 11-1). Endogenous vitamin D production occurs in the skin as a result of ultraviolet (UV) radiation from exposure to sunlight, and this synthetic process distinguishes vitamin D from the true vitamins. The subject of the photobiology of vitamin D_3 has recently been reviewed [24]. The UV radiation emitted from the sun and transmitted to the surface of the earth can be broadly divided into two spectra: UV-A (wavelength 320–400 nm) and UV-B (wavelength 290–320 nm). Light energy is transmitted to the epidermis and dermis, where stores of 7-dehydrocholesterol (provitamin D_3) are located. UV-B radiation causes scission of the C9–C10 bond in the steroid, yielding the "split" or secosteroid previtamin D_3. Thermal equilibration within the skin occurs over a day converting previtamin D_3 to vitamin D_3. Vitamin D_3 binds to the circulating vitamin D binding protein (DBP) and thus leaves the skin and enters the circulation (Figure 11-2). During prolonged exposure to UV-B radiation, previtamin D_3 synthesis plateaus at about 15% of the 7-dehydrocholesterol skin content and leads to the increasing production of the biologically inert compounds lumisterol and a small amount of tachysterol from previtamin D_3. This restriction on previtamin D_3 formation may serve as a mechanism to prevent overproduction of vitamin D_3. Several factors have been found to affect the cutaneous synthesis of vitamin D_3, including latitude and seasonal variation, skin pigmentation, the use of topical sunscreens, and age. In addition, $1,25(OH)_2D$ may feed back on the skin to add to the regulation, since it acts on epidermal constituents [25]. In addition, UV-B radiation inhibits levels of VDR, suggesting the existence of a feedback mechanism in that UV-B initiates vitamin D synthesis in keratinocytes and at the same time it limits VDR abundance [26].

i. Latitude and Season

Since the conversion of 7-dehydrocholesterol to previtamin D_3 in the skin requires UV-B radiation, the amount of previtamin D_3 synthesized is related to the amount of UV-B radiation absorbed by the skin. The amount of solar radiation reaching the surface of the earth is limited by the changing zenith angle of the sun and decreases with increasing global latitude. Similarly, the incident radiation on the surface of the earth is diminished during the fall and winter months when the sun is lower in the sky. Therefore, the variation in cutaneous UV-B radiation exposure due to seasonal variation or geographical location can influence the amount of vitamin D_3 synthesized in the skin. As a result, no previtamin D_3 is synthesized in Boston (42° N latitude) from November to February, and 10 degrees farther north, in Edmonton, this period

is extended from October to March. In more southerly locations, such as Los Angeles and Puerto Rico, previtamin D_3 synthesis occurs year round [10]. An interesting commentary on the relative importance of sunlight was described by Holick in a study of naval personnel onboard submarines [10]. Submariners who were not exposed to sunlight for 3 months failed to maintain adequate vitamin D levels even while ingesting 600 IU/day of vitamin D, supporting the concept that 800 IU/day or more may be necessary to maintain normal vitamin D levels in the absence of adequate sunlight.

2. Skin Pigmentation

The degree of skin pigmentation (i.e., melanin content) also affects vitamin D_3 production in the skin. Melanin protects the body from excess sunlight by acting as a sink to absorb UV-B rays, and acts as a competitor of 7-dehydrocholesterol for UV-B radiation. Therefore, the more melanin that is present in the skin, the less UV-B radiation is available for previtamin D synthesis. Melanin thus reduces the efficiency of previtamin D_3 production in response to sunlight. However, individuals with high melanin levels compensate by increasing the conversion of 25(OH)D to $1,25(OH)_2D$ [27]. Loomis raised the hypothesis that melanin pigmentation evolved in people living near the equator to prevent the excessive production of vitamin D due to constant exposure to sunlight [28]. As people migrated away from the equatorial regions, their sunlight exposure was shortened and, in order to allow adequate production of vitamin D and prevent rickets, the melanin levels in their skin diminished. Critics of Loomis's hypothesis point out that there are no reported cases of hypercalcemia secondary to vitamin D toxicity as a sole consequence of prolonged sun exposure.

When individuals of different skin pigmentation were exposed to the same suberythemic dose of UV radiation (27 mJ/cm^2), Whites showed the largest incremental rise in serum vitamin D levels, while Asians showed an intermediate increase and Black individuals the smallest rise [29]. Basal levels of 25(OH)D are lower in young healthy Blacks as compared to young healthy Whites; however, their $1,25(OH)_2D$ levels are higher than Whites, possibly due to relative secondary hyperparathyroidism [30]. Increased skin pigmentation doesn't limit the absolute amount of previtamin D_3 made, but rather it extends the period of sunlight exposure necessary to reach maximum production of previtamin D_3 [31]. This time interval for maximum previtamin D_3 production ranges from 0.5 hour in lightly pigmented individuals to 3 hours or more in darker pigmented people.

3. SUNSCREENS, SUN EXPOSURE, AND AGE

Interestingly, similar to melanin, topical sunscreens act as a competitor of the photochemical production of vitamin D_3 by absorbing UV radiation. Para-amino benzoic acid–based preparations with an SPF 8 rating can significantly block the cutaneous production of vitamin D_3. Age is also a variable that can influence the production of vitamin D_3, since the amount of 7-dehydrocholesterol in the skin and the efficiency of previtamin D_3 photoproduction decreases as a consequence of advancing age [24].

The geographic distribution of various cancers (breast, colon, prostate, bladder, rectal, stomach, uterine, and non-Hodgkin's lymphoma) [32–34], heart disease [35], and multiple sclerosis [36] suggests a correlation of lack of sun exposure and low vitamin D status with morbidity from these conditions. The role of vitamin D in preventing cancer and autoimmune diseases will be discussed in Section X. Grant *et al.* speculated that 50,000 to 63,000 individuals die yearly in the United States secondary to the hypothesis that there is an increased incidence of cancer related to vitamin D insufficiency [37]. To maintain adequate vitamin D levels and prevent vitamin D deficiency–related morbidity and mortality, moderate sun exposure (4–10 minutes/day for fair-skinned and 60–80 minutes for dark-skinned individuals) has been advocated [38]. In contrast, other investigators are concerned about the risk of skin cancer when recommendations of sun exposure are advanced, and they prefer fortification or supplement strategies [39]. While maintaining appropriate vitamin D levels, potential side effects of excessive sun exposure need to be considered. UV radiation is among the known environmental carcinogens [40]. The World Health Organization (WHO) estimated that 1.5 million disability adjusted life years and about 60,000 deaths yearly worldwide are related to malignant skin cancers including malignant melanomas, and about 90% of these cancers are linked to excessive ultraviolet radiation from the sun [41]. Other diseases associated with excessive UV-B radiation include sunburn, skin aging, cataracts, and pterygium, most of which are preventable by proper sun protection measures. Although excessive use of sunscreen could theoretically lead to vitamin D deficiency, this notion is not supported by clinical trials. Sunscreen use at a level sufficient to prevent actinic keratosis did not induce vitamin D deficiency or hyperparathyroidism [42–44], likely related to the fact that sunscreens do not completely block UV-B radiation and emphasizing that even minimal sun exposure can lead to some vitamin D synthesis.

4. BALANCE BETWEEN SUN SAFETY AND ADEQUATE VITAMIN D SYNTHESIS

Neither of the extremes of excessive sunbathing or zero UV exposure is recommended [39, 41]. Sun exposure is the most powerful stimulus for cutaneous previtamin D synthesis, and even casual sun exposure will produce some vitamin D. Excessive sun exposure is not necessary for vitamin D synthesis, since in fair-skinned individuals, maximal vitamin D synthesis occurs rapidly (within 5 minutes) [39]. Regular short sun exposure was shown to have a protective effect against skin cancers, possibly through vitamin D production [45]. Even though sunscreen use does not lead to clinical vitamin D deficiency, complete sun protection can lead to decreased vitamin D photosynthesis. Individuals with very limited sun exposure (institutionalized patients or veiled women) are at risk of developing vitamin D insufficiency. Thus, vitamin D supplementation should be encouraged while promoting UV-B protection. While vitamin D photosynthesis via sun exposure should not be the only vitamin D source, sun exposure in moderation seems to be safe and very efficient in preventing vitamin D deficiency. Clearly, more work is necessary to clarify the optimal amount of sun exposure. Sun safety by judicious use of sunscreen and avoidance of excessive sun exposure to prevent serious skin complications are essential. In addition to moderate sun exposure, obtaining vitamin D from the much safer and readily available supplements seems prudent.

C. Dietary Sources and Food Fortification

The main source of vitamin D in humans is sunlight-dependent synthesis by the skin, in the form of vitamin D_3. A well-balanced, nutritious diet does not necessarily provide sufficient amounts of vitamin D, because the vitamin is present in only a limited number of items of the human diet, either in the form of vitamin D_2 from plant sources or vitamin D_3 from animal sources. Foods naturally containing substantial amounts of vitamin D are relatively few: egg yolks, liver, fatty fish, and fish liver oils (cod liver oil) (Table 11-1) [46, 47].

While the fortification of some staple foods (milk or margarine) is mandatory in the United States, Canada, and Australia, manufacturers voluntarily fortify a large number of foods with vitamin D in the United States and Europe [48]. In the United States the primary dietary source of vitamin D is fortified milk, which nominally contains 400 IU/quart. Vitamin D content is generally expressed as either micrograms (mcg) or international units (IU). The biological activity of 1 mcg vitamin D is equivalent to 40 IU. Vitamin D from fortified food products effectively increases serum vitamin D levels, similar to taking vitamin D–containing supplements [49]. Dairy products made from milk (cheese, yogurt, ice cream) are not always fortified, and if not, they do not contain substantial amounts of vitamin D. Other commonly supplemented sources may include orange juice, cereals, breads, and fortified margarine (Table 11-1).

While food fortification is inexpensive, this means to increase vitamin D intake of the general population has some limitations. Vitamin D content of fortified foods has been found to vary considerably [50]. There are a limited number of fortified food choices, and individuals with restricted diets (elderly, children, vegetarians) may not benefit significantly [51]. Those who consume fortified milk or margarine have higher 25(OH)D levels, but fortification is not always enough to correct or prevent vitamin D deficiency [48]. Considering that fortified staple foods are consumed by a wide age spectrum of the population, age-specific recommendations might not be easy to accomplish by simply increasing the vitamin D content of selected foods. Advocates of supplementation contend that the benefits of fortifying foods with vitamin D outweigh the minimal risk of overdosing. The risk may be great, especially for small children, while aiming to supply sufficient amounts for the elderly [52]. Fortification is prevalent in processed foods. According to a report from the United Kingdom, however, three-quarters of fortified foods are high in fat, sugar, or salt, and manufacturers often use fortification as a marketing tool to promote unhealthy foods [53, 54].

D. Transport in Circulation: Vitamin D Binding Protein (DBP)

Group-specific component (Gc), a 58-kD plasma alpha globulin, was originally described immunologically in 1959, and approximately 16 years later Gc was identified as a vitamin D binding protein (DBP) [55]. DBP is very polymorphic, with over 120 variants being described [56], making it useful in forensic medicine and as a population marker [57]. DBP belongs to the same protein family as human serum albumin, α-fetoprotein, and afamin, exhibiting an all α-helical structure, sequence homology, similar overall folding, and similar free fatty acid binding capacity [58].

TABLE 11-1   Vitamin D Content of Various Foods.

Food	Serving size	Vitamin D content (IU)	% Daily value
Cod liver oil	1 Tbs (15 mL)	1,360	340
Salmon, cooked	3.5 oz	360	90
Mackerel, cooked	3.5 oz	345	90
Sardines canned in oil, drained	1.75 oz	250	70
Tuna, canned in oil	3 oz	200	50
Eel, cooked	15 oz	200	50
Egg	One whole	20	5
Milk, vitamin D fortified	1 cup	98	25
Orange juice, fortified	1 cup	98	25
Margarine, fortified	1 Tbs	60	15
Pudding prepared with fortified milk	0.5 cup	50	10
Ready-to-eat cereals, fortified	0.75–1 cup (serving sizes vary)	40	10
Liver, beef, cooked	3.5 oz	15	4
Cheese, Swiss	1 oz	12	4
Milk, not fortified	1 cup	10	2.5
Human breast milk	1 cup (250 mL)	3.7	1

Percent daily value based on 400 IU recommended daily intake. Adapted from the Dietary Supplements Fact Sheet: Vitamin D, National Institute of Health, retrieved on July 17, 2006, http://dietary-supplements.info.nih.gov/factsheets/vitaminD.asp#h3 and USDA Nutrient Database website: http://www.nal.usda.gov/fnic/cgi-bin/nut_search.pl

Only about 5% of the binding sites are normally occupied, probably due to the high concentration of DBP in the circulation [59]. The binding affinity of DBP for the vitamin D metabolites is as follows: $25(OH)_3D = 24,25(OH)_2D_3 > 1,25(OH)_2D_3 >$ Vitamin $D > 1,24,25(OH)_3D_3$. The affinity of D_2 metabolites is lower than the D_3 metabolites. Vitamin D_3 synthesized in the skin travels in plasma almost entirely bound to DBP, whereas vitamin D_2 obtained in the diet is associated with both lipoproteins (chylomicrons) and DBP [60]. Like other steroid hormones in the circulation, the free or unbound $1,25(OH)_2D$ is in equilibrium with the bound form. It is the free fraction of the $1,25(OH)_2D$ that is hormonally active, and binding to DBP inhibits accessibility of the steroid to the cell and prolongs $1,25(OH)_2D$ half-life [61]. In serum, approximately 0.04% of $25(OH)D$ and 0.4% of $1,25(OH)_2D$ are found in the free form. DBP functions as a reservoir of $25(OH)D$ and serves as a buffer to prevent the too rapid tissue delivery of the steroids to target cells. DBP thereby prevents vitamin D deficiency and presents $25(OH)D$ for renal activation to $1,25(OH)_2D$ [62].

Several findings suggest that DBP may have other critical roles in the body in addition to being the vitamin D transport protein. It circulates at micromolar concentrations, 100-fold in excess of its main ligand $25(OH)D$, and is only 5% occupied with calciferols [59]. DBP binds monomeric G-actin molecules and is part of the extracellular actin scavenger system, and plays a role in the immune response against neoplasia. Additionally, DBP has been shown to be membrane-associated on a number of cell types, either acquired from serum or synthesized by the cell [63]. The function of membrane-associated DBP is unclear, and no specific DBP receptor has been described [59]. Membrane-associated DBP may aid in sterol transport into the cell, or it may play a role in modulating the function of $1,25(OH)_2D$ by limiting its interaction with the cell and the VDR [59].

DBP is primarily synthesized in the liver [59], and serum levels of DBP are increased in pregnancy and in patients treated with estrogens, whereas levels are decreased in liver disease, malnutrition, and nephrotic syndrome. Circulating levels of DBP correlate with survival in patients with hepatic failure [64], sepsis, and multiple organ dysfunction after trauma [65]. Calcitropic hormones do not appear to regulate the synthesis of DBP.

Although there are no reports of patients with DBP deficiency suggesting an essential role of DBP in humans [58], a DBP knockout mouse has been described [66]. The DBP null (–/–) mice are phenotypically normal and fertile. However, they have lower circulating levels of $25(OH)D$ and $1,25(OH)_2D$ when fed a normal diet and exhibit secondary hyperparathyroidism and bone changes when fed a vitamin D–deficient diet. These findings were not seen in the control normal mice and support the concept that DBP acts as a storehouse for vitamin D metabolites, thus protecting the animal in times of vitamin D deficiency. DBP markedly prolonged the serum half-life of $25(OH)D$ and less dramatically prolonged the half-life of vitamin D by slowing its hepatic uptake and increasing the efficiency of its conversion to $25(OH)D$ in the liver. On the other hand, after an overload of vitamin D, DBP–/– mice were less susceptible to hypercalcemia and its toxic effects. The DBP knockout mice show an increase in clearance of vitamin D protecting them from excess circulating hormone levels. Thus, the role of DBP is to maintain stable serum stores of vitamin D metabolites and modulate the rates of its bioavailability, activation, and end-organ responsiveness. These properties may have evolved to stabilize and maintain serum levels of vitamin D in environments with variable vitamin D availability [59, 66].

E. Megalin and Cubilin

Megalin is a large multifunctional endocytic clearance receptor for circulating proteins that has been implicated in vitamin D uptake and delivery to the kidney for activation to $1,25(OH)_2D$ [67]. Knockout of the megalin gene in mice usually is lethal, but the few survivors were characterized as having severe rickets [68]. The findings suggested that DBP may be a ligand for megalin and that megalin is critical for $25(OH)D$ uptake by the kidney. In addition to the classical hypothesis of free vitamin D uptake by diffusion in the proximal convoluted tubules at the basolateral site of the epithelium, recent studies identified an alternative uptake route involving endocytosis of $25(OH)D$-DBP complexes at the luminal surface of the proximal convoluted tubule [67]. Cubilin directly binds to megalin and forms a coreceptor complex. DBP-carrying vitamin D is filtered by the glomerulus and reabsorbed by the "cargo" receptor megalin or the megalin/cubulin complex in tubular cells. The two-receptor model proposes that $25(OH)D$-DBP complexes bind either to megalin followed by endocytosis, or first binds to cubilin and then to megalin followed by endocytosis [69]. In addition to megalin and cubulin, the complex process of endocytosis involves the cellular adaptor disabled 2, the endocytic machinery including voltage-gated chloride channel-5 and vitamin D binding proteins. The internalized

25(OH)D-DBP complexes are degraded in lysosomes, and free 25(OH)D is carried to the mitochondria for hydroxylation via a currently unknown mechanism, likely involving an interaction between megalin and intracellular 25(OH)D binding proteins [67, 70]. The knock-out mice with null (−/−) megalin genotype develop proteinuria [71] and lose their vitamin D-DBP complex into the urine, leading to vitamin D deficiency and rickets [68]. While lack of DBP or megalin results in a total loss of 25(OH)D reabsorption, cubilin deficiency causes only a partial 25(OH)D reabsortion defect [69]. It has been suggested that the expression of megalin in intestine, breast, and prostate indicates the involvement of the endocytic pathway in conjunction with extrarenal 1α-hydroxylase activity [67].

F. Intracellular Vitamin D Response Element Binding Proteins

Adams and his colleagues [70] described intracellular vitamin D binding proteins (IDBPs) that they speculate play a role in the intracellular movement of vitamin D metabolites, interacting with megalin and promoting delivery of 25(OH)D substrate to the inner mitochondrial membrane for 1-hydroxylation [70]. The IDBPs are related to the heat shock 70 (HSP 70) proteins and, as chaperones, contain intracellular organelle targeting sequences to direct bound molecules to various intracellular destinations.

A novel cause of vitamin D–resistant rickets has been described recently, involving the overexpression of a vitamin D response element binding protein (REBiP) [72]. This form of rickets was found to be responsive to high dose 1,25(OH)$_2$D treatment [73]. REBiP directly binds to single- or double-strand nucleic acids and competes with VDR-RXR for vitamin D response element (VDRE) binding in a dominant-negative fashion (see Section IV for details of VDR and VDRE interaction). This mechanism is similar to the previously described vitamin D resistance in New World primates that require very high levels of 1,25(OH)$_2$D to avert rickets [74]. As a compensatory mechanism, IDBPs exhibit high affinity and capacity for 25(OH)D. IDBP-1 was shown to promote 25(OH)D ligand delivery to the VDR, improving its DNA binding ability and antagonizing the dominant-negative effect of REBiP [75].

G. Assays of Vitamin D Metabolites

Assays of 25(OH)D and 1,25(OH)$_2$D provide valuable tools to assess vitamin D status of patients [76]. The best indicator of the overall vitamin D status of an individual, 25(OH)D, was originally measured by competitive binding assay (CBPA), first introduced in 1971, using a reliable but relatively cumbersome procedure [77]. The available methods today include CBPA-based assays, radioimmunoassay (RIA), high performance liquid chromatography (HPLC), and chemiluminescent immunoassay (CLIA) methods, recently reviewed by Hollis [76] and Zerwekh [78].

In a recent study, Binkley *et al.* reported that three methods to measure 25(OH)D, performed by eight different laboratories, showed an unacceptable level of variation between methods and laboratories [79]. Using the same samples, the mean 25(OH)D concentration differed 2-fold between laboratories. Vitamin D insufficiency (25[OH]D) below 32 ng/mL, 80 nmol/L) varied between 17% and 90%, depending solely on the laboratory and test used. The problem is further complicated by the fact that some RIA antibodies recognize both 25(OH)$_2$D and 25(OH)$_3$D, while others grossly underestimate 25(OH)$_2$D levels [80]. HPLC is the gold standard, allowing individual quantitation of 25(OH)$_2$D and 25(OH)$_3$D, but this method is slow and expensive and not widely available [81]. HPLC-tandem mass spectrometry is a recently developed promising approach to accurately quantitate 25(OH)$_2$D and 25(OH)$_3$D, with shorter assay times more suitable for the routine clinical laboratory [82]. In the context of the current epidemic of vitamin D deficiency, international assay standardization is essential and will, one hopes, occur in the near future.

Although measurement of 1,25(OH)$_2$D is more difficult than 25(OH)D because it circulates at approximately 1,000-fold lower concentration than 25(OH)D, i.e., pg/mL instead of ng/mL, [^{125}I]-based radioimmunoassays are now available for determining 1,25(OH)$_2$D concentrations. In the clinical setting, measurement of 25(OH)D is generally more useful for assessing vitamin D status. However, in cases of genetic disease, such as 1α-hydroxylase deficiency (see Section VII. A) or hereditary vitamin D resistant rickets (HVDRR) (see Section VII.B), or in some cases of hypercalcemia, measurement of 1,25(OH)$_2$D is critical to fully understand the pathophysiology.

H. Optimal 25(OH)D Serum Levels

The optimal serum 25(OH)D levels are currently under strenuous debate. Although many authors consider the current normal range (approximately 24.9–169.5 nmol/L; 10–68 ng/mL, depending on the lab) to be too low, there is not yet consensus on what it should be raised to. Several different criteria have been applied, including the 25(OH)D level necessary for maximal suppression of PTH, maximal intestinal calcium absorption, reduced fracture rates, reduced falls, and highest bone mineral density. The necessary 25(OH)D concentration for maximal PTH suppression has been estimated to be between 30 and 99 nmol/L (13.2–39.6 ng/mL), with most estimates clustering at 75–80 nmol/L (30–32 ng/mL) [83]. In respect to calcium absorption, reduction of bone loss, risk of falling, and reduction of fractures, 25(OH)D levels at the 65–100 nmol/L (26–40 ng/mL) range seem to provide the most benefit [84–88]. Based on an evolutionary perspective and data from individuals with high sun exposure (lifeguards, field workers, sunbathers), the "normal" 25(OH)D concentration in humans was suggested by some authors to be in the 150 nmol/L (60 ng/mL) range [89, 90]. According to this view, nutritional vitamin D deficiency could be considered to be present at circulating 25(OH)D values below 80 nmol/L (32 ng/mL), which is much higher than the current low normal value of 37.5 nmol/L (15 ng/mL) [90], and higher than many authors previously considered normal.

The average increment of serum 25(OH)D is 1.2 nmol/L (0.48 ng/mL) for every 1 mcg (40 IU) of vitamin D_3 ingested at low serum 25(OH)D levels, and 0.7 nmol/L (0.28 ng/mL) or less at serum levels above 70 nmol/L (32 ng/mL) [91]. Based on these data, the daily vitamin D requirement is estimated to be at least 15 mcg (600 IU) of vitamin D_3 to reach a serum level of 50 nmol/L (20 ng/mL), and at least 20–25 mcg (800–1,000 IU) to maintain a level of 75 nmol/L (30 ng/mL) [92]. Vitamin D_2 is less effective and gives a smaller increment of only 0.3 nmol/L for every microgram ingested, with an estimated relative potency of D_3:D_2 of 9.5:1 [93].

Currently, several different vitamin D intake recommendations exist. In the United States, the current guidelines suggest 200 IU/day (5 mcg) for children, 200–400 IU/day (5–10 mcg) for most adults, and 600–800 IU/day (15–20 mcg) for the elderly [94, 95]. Daily values on food labels are based on the Food and Drug Administration's reference value of 400 IU/day. However, on average, adult intake is estimated to be less than 100 IU/day, suggesting that dietary sources of vitamin D play only a minor role in vitamin D homeostasis (see Section XI for consequences on bone). Studies suggest that daily intake of 200–400 IU might not be sufficient

to prevent or treat vitamin D insufficiency, especially in those not receiving adequate sunlight exposure [96]. The current tolerable upper intake level is 25 mcg/day (1,000 IU) for infants and 50 mcg/d (2,000 IU) for adults in the United States [94] and 25 mcg/d for the entire population in the United Kingdom [97]. To maintain a serum 25(OH)D level above 80 nmol/L (32 ng/mL) in adults, estimated daily doses as high as 800–2,600 IU might be necessary [90, 98], which are substantially above the currently recommended intake and the current tolerable upper intake levels. Several investigators believe that the current tolerable upper limit has became a barrier for adequate vitamin D supplementation of high-risk populations [90, 99–101]. The current tolerable upper limit is designed to be safe and effective for the population at large, and it succeeds for the majority of individuals who receive sufficient casual sun exposure. It is not designed to accommodate specific individual needs, especially of those with zero sun exposure. Recommending higher vitamin D intake to large populations also carries the potential risk of overdosing certain individuals. While some believe that actual toxicity will not occur below 25(OH)D values of 250 nmol/L (100 ng/mL), which would require a continuing oral intake in excess of 10,000 IU/day (250 µg/d) [98], there is an increased risk of developing renal stones, as evidenced by the Women's Health Initiative (WHI) trial, using relatively small doses of vitamin D_3 (400–1,000 IU/day) [102]. Thus, some investigators subscribe to a more cautious view and urge a more modest and potentially safer increase in recommended doses. While many of the controversies need further clarification, it appears that the current official guidelines [94] are safe and effective in preventing skeletal complications including rickets and osteomalacia. However, for high-risk populations with limited sun exposure, much higher intakes of vitamin D seem necessary to prevent fractures. Moreover, higher than currently recommended doses may be necessary for achieving nonskeletal effects, including inhibition of cancer progression or prevention of autoimmune diseases (see Section X). This upward trend in recommended 25(OH)D levels is reflected by the most recent Dietary Guidelines for Americans 2005, recommending 25 mcg/d (1,000 IU/day) vitamin D intake for high-risk groups (elderly/housebound, dark skin) in order to maintain 25(OH)D values at 80 nmol/L (32 ng/mL) with the aim of reducing bone loss [103]. Based on accumulating evidence, it is likely that normal values of vitamin D levels as well as dietary recommendations will undergo upward changes in the near future [99, 104].

25(OH)D. 1 ng/mL is equivalent to 2,496 nmol/L (conversion factor 2.496).

1,25(OH)$_2$D: 1 pg/mL is equivalent to 2.4 pmol/L (conversion factor 2.40).

The biological activity of 1 mcg vitamin D is equivalent to 40 IU.

I.   Vitamin D Deficiency

Vitamin D deficiency is prevalent worldwide secondary to limited sun exposure and inadequate dietary sources. It is estimated that vitamin D inadequacy is present in 36% of healthy young adults and 57% of general medicine inpatients in the United States [105]. Populations at risk include limited sun exposure, especially those living in countries above 40 degrees latitude north or south of the equator [106], the elderly, the homebound, dark-skinned individuals, submariners, astronauts, veiled and pregnant women (prevalence up to 80% in this group [107]) and mothers of infants treated for rickets (80% prevalence [108]). Exclusively breast-fed infants are particularly vulnerable [52]. Other risk factors include limited intake of vitamin D–rich foods, fortified food products or dietary supplements, fat malabsorption, renal failure, alcoholism, and drug interaction, which may decrease vitamin D levels (corticosteroids, rifampin, antacids, calcium channel blockers, cholestyramine, anticonvulsants).

Even in those people taking supplements, especially the elderly or subjects who are ill and hospitalized, hypovitaminosis D may be common [109] and may contribute to osteoporotic fracture [110]. Evidence that vitamin D supplementation at doses of 17.5–20 mcg/day to maintain serum 25(OH)D levels above 80 nmol/L (32 ng/mL) reduces fractures has been accumulating [86, 111–113]. It is unwise to assume that vitamin D status is normal, even if subjects are taking 400 IU supplementation. Many authors have concluded that 800 IU/day or more would be an effective intake yet still safe. In the pediatric population, vitamin D deficiency is seen secondary to limited sun exposure, inadequate vitamin D supplementation, dietary restrictions, with a higher incidence in dark-skinned individuals and exclusively breast-fed children. The optimal amount of vitamin D supplementation during pregnancy and breast-feeding is unknown. A recent study suggested that even larger than recommended daily doses (800–1,600 IU/day; 20–40 mcg/day) of vitamin D were not sufficient to normalize 25(OH)D levels throughout the pregnancy of vitamin D–insufficient minority women [114]. Using 1,000 IU/day (25 mcg/day) vitamin D supplementation to healthy breast-feeding mothers was not sufficient to maintain adequate vitamin D levels in the infants [115]. Most investigators agree that sun exposure in moderation is safe and effective for vitamin D photosynthesis. In exclusively breast-fed infants, an estimated 2 hours of sun exposure weekly to the hands and face would maintain a serum 25(OH)D concentration above the lower limit of normal (11 ng/mL, 27.5 nmol/L) [116]. Avoidance of UV-B radiation for skin cancer safety should be accompanied by encouragement of vitamin D supplementation (see following sections). The American Academy of Pediatrics guidelines recommend 200 IU of vitamin D daily for breast-fed infants [117].

III.   PATHWAYS OF ACTIVATION AND INACTIVATION OF VITAMIN D

A.   25-Hydroxylation

The pathways of vitamin D activation are diagrammed in Figure 11-2. The first step in the activation of vitamin D to the biologically active hormone $1,25(OH)_2D$ is hydroxylation at the carbon-25 position in the liver [118]. Although the liver parenchymal cells are the primary site for 25-hydroxylation, extrahepatic 25-hydroxylation is seen in many other tissues as well. In the liver 25-hydroxylation is probably carried out by more than one enzyme localized either in the mitochondria (CYP27A1/sterol 27-hydroxylase) or in the microsomes (CYP2D25 and CYP2R1) [118]. The gene-encoding human CYP27, a cytochrome P450 enzyme, has been cloned [119–121], and localized to chromosome 2q33-qter [120]. The CYP27 gene encodes a protein with both sterol 27-hydroxylase as well as vitamin D 25-hydroxylase activities. The former step is important in the biosynthetic pathway of bile acids, catalyzing the 26- or 27-hydroxylation of cholesterol and bile acid precursors [122]. The capacity of CYP27 for hydroxylation of cholesterol or bile acid intermediates is much greater than the 25-hydroxylation of vitamin D. Among the vitamin D molecules, CYP27 prefers 1α-hydroxylated derivatives of D_3 over the nonhydroxylated derivatives, including the natural substrate D_3, which is hydroxylated at the C-25 position less efficiently than $1\alpha(OH)_3D$ [118]. CYP27A1 hydroxylates vitamin D_3 compounds at C-25 as well as many other positions on the side chain of the molecules [121, 123]. The rare genetic disease cerebrotendinous xanthomatosis is due to a deficiency of CYP27 activity [120]. The deficiency in sterol 27-hydroxylase activity results in the accumulation of bile acid precursors and cholestanol, which deposit in the brain and peripheral nerves forming tuberous xanthomata [124]. The patients with this disease also exhibit low bone mineral density associated with low 25(OH)D levels and increased fracture risk [125]. A deficiency in the enzymatic activity is not clinically apparent unless severe hepatic failure develops. The disturbance in vitamin D metabolism in this disease as well as in CYP27 knock-out mice [126] is quite mild, suggesting that the 25-hydroxylation of vitamin D is not solely dependent on CYP27 activity.

Vitamin D status is an important modulator of the 25-hydroxylation of vitamin D. In patients with hypervitaminosis D, 25(OH)D levels are markedly elevated (as much as 15-fold), while 1,25(OH)$_2$D levels are relatively normal [127]. Production of 25(OH)D is dependent primarily on the concentration of vitamin D; however, higher basal vitamin D and 25(OH)D levels may diminish the production of 25(OH)D *in vivo*. 1,25(OH)$_2$D has been shown to limit the production of 25(OH)D. Treatment with 1,25(OH)$_2$D prevented the increase seen in 25(OH)D levels after oral vitamin D given to volunteers [128]. This effect may be explained by increased metabolism of 25(OH)D to 24R,25-dihydroxyvitamin D [24,25(OH)$_2$D] due to induction of 24-hydroxylase by 1,25(OH)$_2$D (see Section III.C.2) and therefore increased the metabolic clearance rate of 25(OH)D. Intestinal CYP27A1 expression is regulated by the vitamin D metabolites, and the mechanisms include both transcriptional repression and a decrease in CYP27A1 mRNA half-life [118]. Interestingly, calcium may also have a direct modulatory role on the 25-hydroxylase activity. However, *in vivo*, the role of calcium to modulate 25-hydroxylase activity is likely mediated via changes in PTH, which influence the production of 1,25(OH)$_2$D, which in turn increases the metabolism of 25(OH)D through 24-hydroxylation.

The 25-hydroxylation of vitamin D in the microsomes may be catalyzed by more than one cytochrome P450 enzyme. While microsomal CYP2D25 has been shown to be involved in 25-hydroxylation in pig liver and kidney, its physiological contribution in human liver is not yet clear [118]. A recent study [129] in CYP27A1 null mouse liver has identified an evolutionarily conserved orphan cytochrome P450 named CYP2R1, which is demonstrated to exhibit vitamin D 25-hydroxylase enzyme activity. The mouse CYP2R1 sequence is 89% identical to the human enzyme [130]. CYP2R1 is present in high abundance in liver and testis and hydroxylates both D$_2$ and D$_3$ compounds including 1α(OH)$_3$D and is likely to be the high-affinity microsomal vitamin D 25-hydroxylase enzyme [118]. Cheng *et al.* [131] recently elucidated the molecular defect in a patient with the rare autosomal recessive disorder of selective 25(OH)D deficiency. The patient exhibited very low circulating levels of 25-hydroxyvitamin D and classic symptoms of vitamin D deficiency. The patient was found to be homozygous for a transition mutation in exon 2 of the CYP2R1 gene, which eliminated vitamin D 25-hydroxylase enzyme activity. These observations establish CYP2R1 as a biologically important human vitamin D 25-hydroxylase.

B. 25-Hydroxyvitamin D-1α-Hydroxylase

1. THE 25-HYDROXYVITAMIN D-1α-HYDROXYLASE ENZYME

Following hydroxylation in the liver, 25(OH)D is transported in the circulation bound to DBP, and the kidney accomplishes the final step of vitamin D activation, namely 1α-hydroxylation (Figure 11-2). This step is apparently megalin-dependent (see Section II.D). The 25-hydroxyvitamin D-1α-hydroxylase (1α-hydroxylase) is a mitochondrial P450 enzyme present in low abundance and localized to the proximal tubule of the nephron [132]. As a mixed function oxidase the enzyme requires NADPH$^+$, molecular oxygen, ferredoxin, and ferredoxin reductase for activity. The cDNAs for the 1α-hydroxylase from the mouse, rat, and human have been cloned [133–135]. The predicted amino acid sequence confirms that the 1α-hydroxylase gene (CYP1α or CYP27B1) is a member of the cytochrome P450 enzyme superfamily. The 1α-hydroxylase exhibits significant homologies to the vitamin D-25-hydroxylase (CYP27) and the 25-hydroxyvitamin D-24-hydroxylase (CYP24) enzymes. The human 1α-hydroxylase gene is approximately 5 kb in length and is composed of 9 exons. Fluorescent *in situ* hybridization (FISH) analysis localized the gene to chromosome 12q13.3, confirming earlier reports that the gene defect causing 1α-hydroxylase deficiency was linked to chromosome 12q14, close to the gene coding for the vitamin D receptor [136, 137]. The gene is expressed in kidney epithelial cells in both the proximal and distal tubules as well as selected other sites [138].

The kidney is the major source of circulating 1,25(OH)$_2$D. However, humans and animals devoid of functioning renal tissue exhibit low but detectable 1,25(OH)$_2$D concentrations in the circulation [139]. Several extrarenal tissues including skin [134], bone [140], macrophages [141, 142], colon [143], placenta [144], and prostate [145] have now been shown to exhibit 1α-hydroxylase activity. It is clear that the 1α-hydroxylase enzyme expressed in renal and nonrenal tissues is encoded by the same gene since mutations causing 1α-hydroxylase deficiency have been found in both renal [135] and nonrenal tissues including keratinocytes [134] and blood cells [142]. See Section VI.C for discussion of extrarenal 1α-hydroxylase and hypercalcemia.

2. REGULATION OF RENAL 1α-HYDROXYLASE

In contradistinction to the 25-hydroxylase, the renal 1α-hydroxylase is a tightly regulated enzyme and is the critical determinant of 1,25(OH)$_2$D synthesis (Figure 11-3). The overall regulation of the 1α-hydroxylase is determined by the calcium and phosphorus requirements of the organism and is mediated by several bioactive

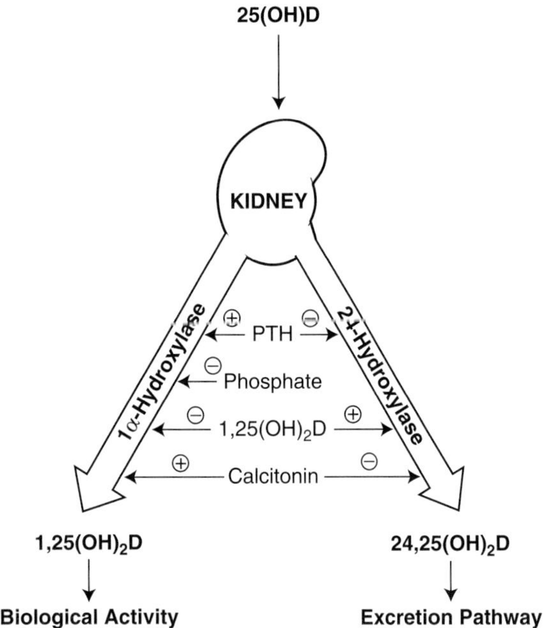

FIGURE 11-3   Regulation of 1α-hydroxylase and 24-hydroxylase activities in kidney.

substances. The principal regulator of renal 1α-hydroxylase is PTH [146]. However, other important regulators include phosphate, 1,25(OH)₂D itself, calcium, and calcitonin. The production of 1,25(OH)₂D also may be modulated by other hormones such as estrogen, prolactin, and growth hormone, but these effects in mammalian systems appear to be small. Analysis of the human 1α-hydroxylase promoter has identified positive response elements for PTH and calcitonin and a negative response element for 1,25(OH)₂D [147, 148]. In normocalcemic rats, the expression of 1α-hydroxylase is determined by the levels of calcitonin and 1,25(OH)₂D [149]. In hypocalcemic states, the expression of 1α-hydroxylase is determined by the levels of PTH and 1,25(OH)₂D [147, 148].

a. PTH

Evidence that PTH is the primary regulator of the 1α-hydroxylase is substantial [132, 150]. 1,25(OH)₂D levels are increased in hyperparathyroidism and reduced in hypoparathyroidism. After parathyroidectomy, 1,25(OH)₂D levels fall and are increased after administration of PTH to normal subjects and to patients with hypoparathyroidism. Moreover, substantial *in vitro* data indicate that PTH markedly stimulates 1α-hydroxylase activity in mammalian renal slices, isolated renal tubules, and cultured renal cells. The stimulatory effect of PTH on 1α-hydroxylase is mediated in part by the cAMP signaling pathway. However, protein kinase C has also been shown to be involved in PTH regulation of 1α-hydroxylase [132].

b. Phosphate

Phosphate is the second most important physiological regulator of the 1α-hydroxylase with high phosphate levels suppressing and low levels stimulating enzyme activity [150, 151]. In mice, dietary phosphate restriction leads to increases in the mRNA and protein levels of 1α-hydroxylase in the proximal renal tubule. In humans, phosphorus restriction increases 1,25(OH)₂D levels to 180% of control, and phosphorus supplementation decreases 1,25(OH)₂D levels by 29%. These changes are due to alterations in the synthetic rate rather than changes in the half-life of the enzyme, demonstrating the important role played by phosphate on the 1α-hydroxylase. The effect of elevated phosphate to inhibit 1α-hydroxylation is a contributing factor in the development of renal osteodystrophy during chronic renal failure and is part of the rationale for using phosphate binder therapy to delay the onset of bone disease in these patients [150, 151].

c. 1,25(OH)₂D

Interestingly, 1,25(OH)₂D regulates its own production. This activity is mediated directly at the level of the 1α-hydroxylase in the kidney and indirectly by inhibition of PTH (as described previously). Low 1,25(OH)₂D levels lead to increased 1α-hydroxylase activity and 1,25(OH)₂D synthesis, whereas high 1,25(OH)₂D levels inhibit the enzyme activity [132, 150]. The ability of 1,25(OH)₂D to inhibit 1α-hydroxylase activity has been demonstrated *in vitro* as well as *in vivo*. This effect involves both PTH-dependent and PTH-independent mechanisms; 1,25(OH)₂D directly (PTH-independent) decreases 1α-hydroxylase activity as well as decreases PTH secretion (PTH-dependent). *In vivo*, however, it is difficult to separate the contribution of changes in calcium or PTH from direct 1,25(OH)₂D actions because of the tight linkage of these systems. In VDR null (−/−) mice, the 1α-hydroxylase gene expression is increased, a phenomenon used to help in the cloning of this elusive gene [133], and the upregulation of 1α-hydroxylase by PTH was evident. However, a down-regulation of 1α-hydroxylase gene expression by 1,25(OH)₂D₃ was not observed, implying that the VDR is essential for the negative regulation of this gene by 1,25(OH)₂D₃ probably via an effect on PTH transcription [133, 147]. *In vivo* another complexity is the finding that administration of 1,25(OH)₂D chronically can regulate its serum concentration by increasing its metabolic clearance rate by induction of the 24-hydroxylase enzyme (see Section III.C). Several *in vitro* studies have examined the regulatory effects of 1,25(OH)₂D on the 1α-hydroxylase promoter. The results are mixed, and possibly several mechanisms are involved in the regulation of 1α-hydroxylase by 1,25(OH)₂D, including decreases in transcription and modulation of post-transcriptional and/or post-translational processes [132].

d. Calcium

Although regulation of 1α-hydroxylase in response to changes in serum calcium levels is mainly due to changes in PTH, calcium may act independently as well. The effect of calcium in the regulation of 1α-hydroxylase may explain why some patients with severe hyperparathyroidism and very high serum calcium levels exhibit low 1,25(OH)$_2$D values [152]. Although the underlying mechanism for this finding is obscure, one might speculate that the calcium-sensing receptor (CaR) originally described in parathyroid glands [153] and also found in the kidney [154] may mediate this effect [155]. However, studies in VDR null (−/−) mice indicate that calcium is likely an indirect modulator of 1α-hydroxylase, since in the absence of a 1,25(OH)$_2$D action, changes in calcium did not alter the levels of 1α-hydroxylase activity [147].

e. Calcitonin

Calcitonin can also stimulate 1,25(OH)$_2$D synthesis in thyroparathyroidectomized rats [156]. Similarly, 1,25(OH)$_2$D levels increase after calcitonin administration to patients with X-linked hypophosphatemic rickets [157] as well as in the HYP mouse [158], where the 1α-hydroxylase response to PTH is abnormal. In normocalcemic rats where PTH levels are relatively low, calcitonin has been shown to be a major regulator of the renal enzyme [149]. Analysis of the human 1α-hydroxylase gene promoter has demonstrated a positive regulatory region for calcitonin [147].

f. Chronic Renal Failure

In the 5/6ths nephrectomized rat model of renal failure, the renal 1α-hydroxylase gene expression decreased, and the positive effects of PTH and calcitonin were diminished [147]. This study, and others like it, also showed that PTH and calcitonin positively regulate renal 1α-hydroxylase gene expression via PKA-dependent and -independent pathways, respectively, and that 1,25(OH)$_2$D$_3$ is a negative regulator. Furthermore, in a moderate state of chronic renal failure, renal cells expressing the 1α-hydroxylase gene appear to have diminished potential to respond to the positive regulators, PTH and calcitonin [139, 151, 159].

g. The Klotho Gene Product

The klotho gene encodes a membrane-bound glycosidase expressed in kidney tubular cells, and a homozygous mutation in this gene in mice displays disorders similar to those seen in human aging [160]. These mice and others null for the klotho gene exhibit greatly elevated plasma 1,25(OH)$_2$D$_3$ and 1α-hydroxylase mRNA, suggesting that klotho is a negative regulator of 1α-hydroxylase [160].

3. REGULATION AND SIGNIFICANCE OF EXTRARENAL 1α-HYDROXYLASE

In recent years the presence of extrarenal 1α-hydroxylase has been demonstrated in several tissues, which contributes to the local production of 1,25(OH)$_2$D$_3$ within the tissue. The extrarenal synthesis of 1,25(OH)$_2$D$_3$ does not significantly affect serum levels of 1,25(OH)$_2$D$_3$ likely because of autocrine induction of vitamin D-24-hydroxylase (see Section III.C) in these tissues [161]. However, in cases of an increased macrophage pool in the body, 1,25(OH)$_2$D production by these cells can lead to hypercalcemia with suppressed PTH [138, 139]. The usual regulators of renal 1α-hydroxylase—PTH, calcitonin, and 1,25(OH)$_2$D—apparently do not play a primary role in controlling extrarenal 1α-hydroxylase activity [162]. However, other hormones and factors are known to regulate extrarenal 1α-hydroxylase. In prostate cells, epidermal growth factor (EGF) has been shown to upregulate 1α-hydroxylase promoter activity, and the MAPK pathway may be involved in this regulation [163]. Potential regulators of 1α-hydroxylase in macrophages include cytokines and the nitric oxide system [138, 139, 164].

Extrarenal 1α-hydroxylase might be of significance in several settings. For example, toxicity due to excessive vitamin D intake is characterized by hypercalcemia and elevated plasma levels of vitamin D$_3$ and 25(OH)$_3$D but not 1,25(OH)$_2$D$_3$ [165]. We speculate that the increases in serum calcium levels in the face of normal or very slightly elevated concentrations of 1,25(OH)$_2$D$_3$ could be explained by the presence of 1α-hydroxylase and local conversion of 25(OH)$_3$D to 1,25(OH)$_2$D$_3$ in intestinal and bone cells, causing the enhancement of intestinal calcium absorption and calcium release from the bone and the resultant hypercalcemia.

In many normal and malignant cells, 1,25(OH)$_2$D$_3$ has been shown to exhibit antiproliferative and prodifferentiation effects [166], raising the possibility of its use as an anticancer agent (see Section X.B). The presence of 1α-hydroxylase in some of these cells has led to speculation that 25(OH)$_3$D can be used in cancer therapy, since it can be converted locally within the cancer tissue to the active hormone 1,25(OH)$_2$D$_3$. This strategy could potentially inhibit cell proliferation without causing the systemic effect of hypercalcemia [145]. The significance of the extrarenal 1α-hydroxylase activity in the anticancer actions and antituberculosis activity of 1,25(OH)$_2$D$_3$ is further discussed in detail in Section X.B.

C. 25-Hydroxyvitamin D-24-Hydroxylation in Kidney and Other Sites

1. THE 25-HYDROXYVITAMIN D-24-HYDROXYLASE ENZYME (CYP24)

25-Hydroxyvitamin D-24-hydroxylase (24-hydroxylase, CYP24) is a mitochondrial P450 enzyme, which, in general, is expressed in all the cells that are responsive to 1,25(OH)$_2$D [167]. The enzyme catalyzes the hydroxylation on carbon 24 of both 25(OH)D and 1,25(OH)$_2$D. 24 Hydroxylase converts 25(OH)D to 24,25(OH)$_2$D, which may have some biological activity (see Section III.C.3).

However, the formation of $24,25(OH)_2D$ is generally considered to represent the first step in the degradative and excretory pathway of vitamin D (Figures 11-1 and 11-2). The enzyme hydroxylates $1,25(OH)_2D$ to form $1,24,25(OH)_3D$, initiating the inactivation pathway of the active hormone. Thus, 24-hydroxylase acts to protect the body from the overproduction of $1,25(OH)_2D$ [168]. In addition to initiating the catabolic pathway of $25(OH)_3D$ and $1,25(OH)_2D_3$ by 24-hydroxylation, the enzyme also catalyzes the dehydrogenation of the 24-OH group and performs 23-hydroxylation, resulting in $24\text{-}oxo\text{-}1,23,25(OH)_3D_3$ [167]. This C_{24} oxidation pathway leads to the formation of calcitroic acid, the major end product of $1,25(OH)_2D_3$ catabolism. The intestine is a major site of hormonal inactivation by virtue of its abundant 24-hydroxylase activity. In the nephron, the enzyme is distributed in the proximal and distal tubules, the glomerulus, and the mesangium. The human 24-hydroxylase gene has been cloned and shown to be present on chromosome 20q13 [169], and its promoter region has been characterized [167, 170].

2. REGULATION OF 24-HYDROXYLASE ACTIVITY

The regulation of the 24-hydroxylase activity (see Figure 11-3) has been reviewed recently [167]. $1,25(OH)_2D$ is the primary regulator of the 24-hydroxylase, causing a marked induction of enzymatic activity and mRNA levels via a VDR-mediated genomic pathway (see Section IV.H). Recently, two vitamin D response elements (VDREs, see Section IV.F) were identified in the promoter of the 24-hydroxylase gene [170, 171]. Since 24-hydroxylase can be induced by $1,25(OH)_2D$ in many VDR containing cells, induction of 24-hydroxylase has proven to be an excellent marker of $1,25(OH)_2D$ biological activity. Therefore, the levels of 24-hydroxylase mRNA become undetectable in VDR-*null* mice [172]. Measurement of 24-hydroxylase enzyme activity and induction of mRNA by $1,25(OH)_2D$ has been extensively employed in studies of cultured dermal fibroblasts from hereditary vitamin D resistant rickets (HVDRR) patients harboring mutations in the VDR gene [2, 173] (see Section VII.B).

In the kidney, PTH stimulates 1α-hydroxylase and inhibits 24-hydroxylase [174], effects that are opposite to those of $1,25(OH)_2D$. Calcitonin has been shown to down-regulate 24-hydroxylase mRNA and enzyme activity in rat intestine *in vivo* [175], suggesting the presence of an intestinal calcitonin receptor and an unanticipated function for this hormone. Other factors that influence the stimulation of 24-hydroxylase expression by $1,25(OH)_2D_3$ include activators of protein kinase C and glucocorticoids in kidney, intestinal, and bone cells and interferon-gamma in monocytes/macrophages [167].

3. CONTROVERSY OVER WHETHER $24,25(OH)_2D$ EXHIBITS DISTINCT BIOLOGICAL ACTIVITY

24-Hydroxylation of the substrate $25(OH)D$ results in the formation of $24,25(OH)_2D$. Controversy has existed over whether $24,25(OH)_2D$ has biological activity [176]. $24,25(OH)_2D$ can bind to the VDR and exhibit some biological activity at high concentration [177]. A 24-hydroxylase knock-out mouse model has been generated to address the physiological role of $24,25(OH)_2D$ [168]. However, since 24-hydroxylase initiates $1,25(OH)_2D_3$ inactivation, the 24-hydroxylase null mice have high $1,25(OH)_2D_3$ levels. To rule out the contribution of high $1,25(OH)_2D$ to the bone phenotype found in this study of 24-hydroxylase null mice, a subsequent study examined a double knock-out mouse generated by crossing the 24-hydroxylase (–/–) mice with VDR (–/–) mice. The animals were fed a high calcium diet to maintain normal calcium concentrations in the serum [178]. While the 24-hydroxylase (–/–), VDR (–/+) mice showed reduced amounts of mineralized tissue in the mandible and cranial bones, the 24-hydroxylase (–/–), VDR (–/–) double knock-out mice showed normal bone formation at all sites. The data indicate that the impaired mineralization phenotype seen in the 24-hydroxylase (–/–) mice was due to the increase in $1,25(OH)_2D_3$ action on the bone because of loss of the 24-hydroxylase inactivation pathway. The authors concluded that $24,25(OH)_2D_3$ is not an essential hormone for bone formation [178].

4. OTHER METABOLITES

The 24-hydroxylation of the active hormone $1,25(OH)_2D$ initiates its inactivation and production of more polar metabolites, eventually leading to calcitroic acid [179, 180]. The affinity of the 24-hydroxylase enzyme is 5–10 times greater for $1,25(OH)_2D$ than $25(OH)D$, making $1,25(OH)_2D$ the preferred substrate. The resulting product $1,24,25(OH)_3D$ binds to the VDR, but with lower affinity, and exhibits diminished potency when compared to $1,25(OH)_2D$ in biological effects such as stimulation of intestinal calcium absorption, mobilization of calcium from bone, and antirachitic activity in rats [181]. 24-Hydroxylase also catalyzes the dehydrogenation of the 24-OH group and performs 23-hydroxylation, resulting in $24\text{-}oxo\text{-}1,23,25(OH)_3D_3$ and the catabolic pathway initiated by 23-hydroxylation eventually leads to the formation of $1,25(OH)_2D_3\text{-}26,23$ lactone [182].

In recent years a new pathway of C-3 epimerization of vitamin D metabolites has been discovered that seems to occur in selective target cells in addition to the C-24 and C-23 oxidation pathways [183]. The C-3 epimerization appears to be a common pathway for all the major vitamin D_3 metabolites.

IV. MECHANISM OF 1,25(OH)$_2$D ACTION

The classical actions of 1,25(OH)$_2$D include the regulation of calcium and phosphate metabolism, actions that determine the quality of bone mineralization. These classical 1,25(OH)$_2$D actions prevent rickets in children and osteomalacia in adults as well as play a role in the prevention of osteoporosis. The biological actions of 1,25(OH)$_2$D are mediated by the VDR, a member of the steroid-thyroid-retinoid receptor superfamily of ligand-activated transcription factors. The VDR belongs to the subfamily of nuclear receptors that form heterodimers with the retinoid X receptor (RXR) and includes the thyroid hormone receptor (TR), retinoic acid receptor (RAR), peroxisome-proliferator activated receptor (PPAR), farnesoid X receptor (FXR), and a number of orphan receptors for which ligands have not been identified. 1,25(OH)$_2$D binds to the VDR, dimerizes with RXR, and the complex binds to VDREs in the promoter regions of target genes to regulate the expression, either up or down, of multiple vitamin D responsive genes (Figure 11-4). Several reviews of the 1,25(OH)$_2$D-VDR system have been published [2–7], and the subject is extensively covered in the book *Vitamin D* [8].

A. The Vitamin D Receptor (VDR)

In 1987 the cloning of chick VDR cDNA was reported by McDonnell *et al.* [184]. This milestone in research subsequently led to the cloning of the human VDR cDNA [185]. The human VDR cDNA contained ~4,800 nucleotides and encoded a protein of 427 amino acids with a predicted molecular mass of 48,000 Da [185]. The VDR exhibits a modular domain structure

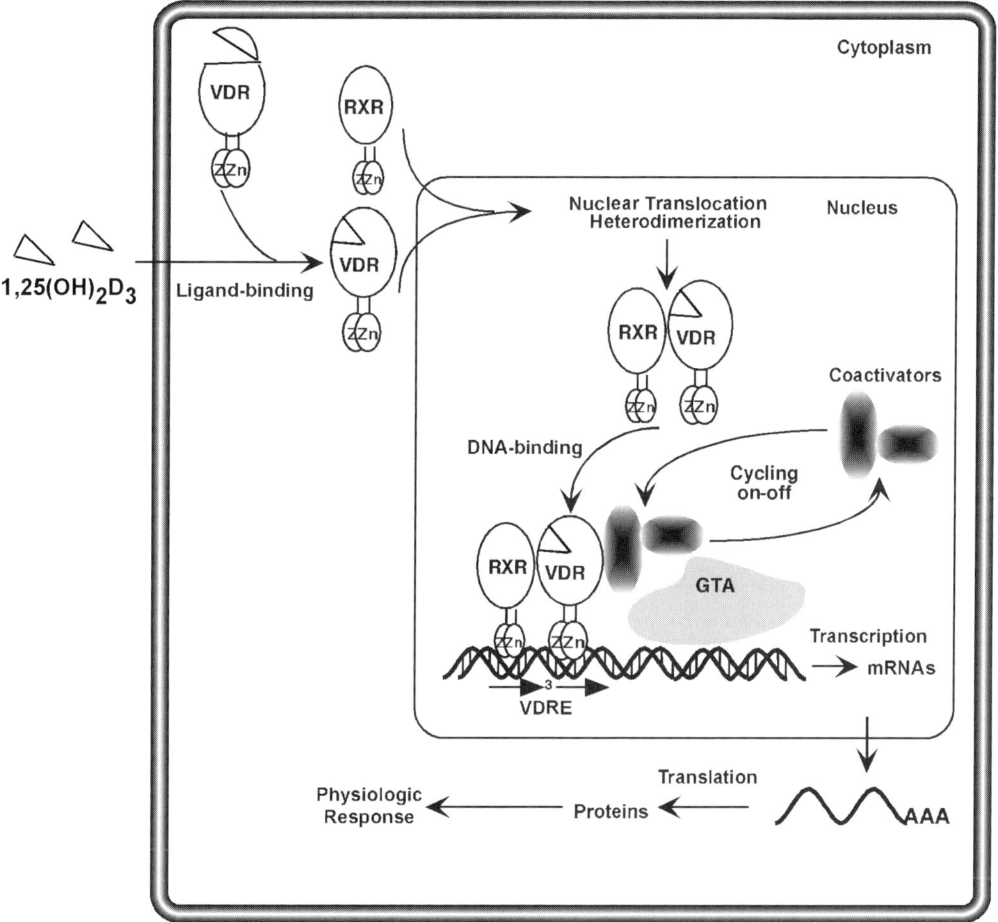

FIGURE 11-4   Overview of 1,25(OH)$_2$D-mediated gene transactivation by VDR. In this model circulating 1,25(OH)$_2$D enters the cell and binds to the VDR in the cytoplasm or the nucleus. The liganded VDR then heterodimerizes with RXR and translocates to the nucleus if it was originally in the cytoplasm. In the nucleus the VDR-RXR heterodimer binds to VDREs in promoters of target genes. Coactivators such as SRC-1 are recruited to the complex to modify the chromatin. These coactivators are then released, allowing interaction with the DRIP coactivator complex and the general transcription machinery to promote gene transcription.

designated A–F similar to that of other members of the nuclear receptor gene superfamily (Figure 11-5). At the N-terminus of the VDR is the A/B domain that is approximately 24 amino acids long although it can extend up to 74 amino acids due to alternative splicing and differential promoter usage [186, 187]. The A/B domain is the most variable region of the nuclear receptors. In some receptors the A/B domain contains an activation function referred to as activation function 1 (AF-1) that mediates ligand-independent transcriptional enhancement. The VDR A/B domain is relatively short compared to the other members of the superfamily and does not exhibit AF-1 activity. The C domain contains the highly conserved DNA-binding domain (DBD). At the carboxy-terminus the E region binds ligand and comprises the ligand-binding domain (LBD). The D domain or "hinge," the least conserved domain among the nuclear receptors, connects the DBD and LBD. The VDR has no F domain.

I. The DNA-Binding Domain (DBD)

The DBD of the VDR contains nine highly conserved cysteine residues that comprise a two zinc finger structure (Figure 11-5). Four of these highly conserved cysteine residues tetrahedrally coordinate the binding of a single zinc atom in each zinc finger module. The two zinc modules of the VDR are not topologically equivalent and serve different functions within the protein. The first zinc finger module contains an α-helix known as the P-box (aa residues 42–46) that functions to direct specific DNA-binding in the major groove of the DNA binding site. The second zinc finger module contains an α-helix known as the D-box (aa residue 61–65) that serves as a dimerization interface for interaction with retinoid X receptor (RXR). An α-helix immediately downstream of the second zinc finger (aa residues 90–101) termed the T-box may also provide an interaction surface for partner proteins. The DBD also contains sites for serine phosphorylation and nuclear localization [188].

2. The Hinge

The hinge region links the DBD to the LBD and encompasses amino acid residues 88–120. Two stretches of basic amino acid residues (aa residues 102–104 and 109–111) are required for transactivation and binding to VDREs. These basic amino acids are important, since replacing them with alanines failed to

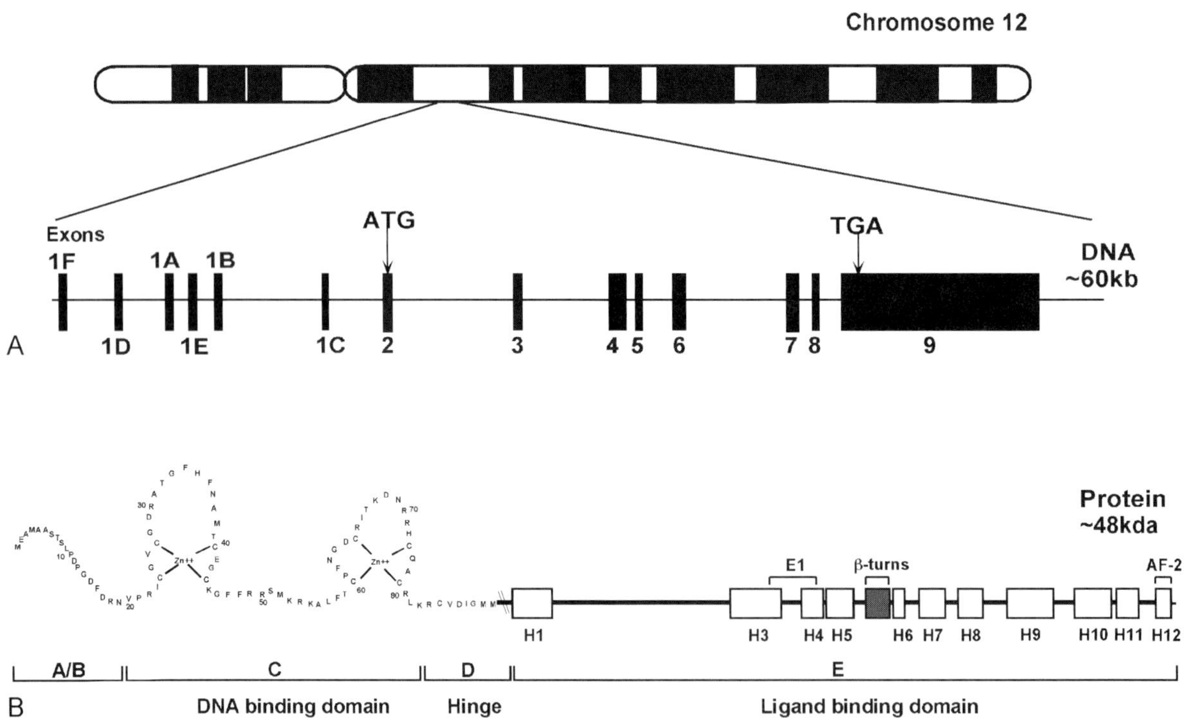

FIGURE 11-5   The VDR. (A) Organization of the VDR chromosomal gene. The human VDR gene is located on chromosome 12q13-14 and spans approximately 60 kilobases of DNA. The gene is composed of at least 5 noncoding exons and 8 coding exons. Alternative splicing results in at least 14 types of transcripts. The translation start site (ATG) and termination (TGA) signals are shown. (B) Domains A–E are shown below the protein model. The DNA-binding domain consists of two zinc finger modules located at the amino terminal portion of the receptor. The ligand-binding domain contains 12 α-helices shown as open boxes and 3 β-turns shown as a filled box. The E1 and AF-2 subregions of the receptor are important in transactivation.

restore transactivation [188]. Crystallographic analysis of the VDR DBD homodimer bound to VDREs showed that amino acid residues 97–121 form a long continuous α-helix [189]. C-terminal deletion of five to nine amino acids of the hinge (Δ114–120 and Δ112–120) reduced transactivation by more than 50%, while deletion of 13 amino acids (Δ108–120) abolished transactivation. On the other hand, replacing amino acid residues 114–120 with alanines did not alter transactivation, suggesting that this section of the hinge acts as a sequence-independent spacer [190].

3. THE LIGAND-BINDING DOMAIN (LBD)

a. 1,25(OH)$_2$D$_3$ Binding Pocket

Binding of 1,25(OH)$_2$D to the VDR LBD leads to conformational changes that increase its capacity to dimerize with RXR and stimulate DNA binding. Ligand binding also exposes surfaces of the VDR that act to recruit proteins active in modifying chromatin such as SRC-1 and the DRIP complex or proteins such as TFIIB or the TAFs that are associated with the core transcriptional machinery (see Section IV.F).

The crystal structure of the VDR LBD bound to 1,25(OH)$_2$D$_3$ was determined by Rochel *et al.* [191]. As shown in Figure 11-6, the VDR LBD is composed of 13 α-helices and 3 β-sheets. The ligand-binding pocket forms a large cavity of 693 Å and is lined with hydrophobic amino acid residues. When bound to the VDR, the A ring of 1,25(OH)$_2$D$_3$ embraces helix H3 and orients toward the C-terminus of helix H5. The 1α-OH group forms hydrogen bonds with Ser237 (H3) and Arg274 (H5) and the 3β-OH group forms bonds with Ser278 (H5) and Tyr143. The conjugated triene connecting the A and C rings fits into a hydrophobic channel formed between Ser275 (loop H5–β) and Trp286 (β1) on one side and Leu233 (H3) on the other side. The C ring contacts Trp286, and the C18 methyl group is aimed at Val234 in helix H3. The 25-OH group forms hydrogen bonds with His305 (loop H6–H7) and His397 (H11).

The AF-2 domain is contained within helix H12. From crystallographic studies of other receptors [192, 193], the H12 α-helix is repositioned following ligand binding such that the repositioning locks the ligand in the cavity of the ligand-binding pocket. The repositioning of H12 also leads to the formation of a complex high-affinity protein surface that allows interactions with specific comodulators such as SRC-1 and DRIP205 that are critical for transcriptional activation (see Section IV.F).

Upon ligand binding, the position of helix H12 is stabilized by hydrophobic interactions involving helix H12, helix H3, helix H5, and helix H11. In addition, a salt bridge contributes to the repositioning of helix H12, all of which are controlled by 1,25(OH)$_2$D$_3$ binding [191].

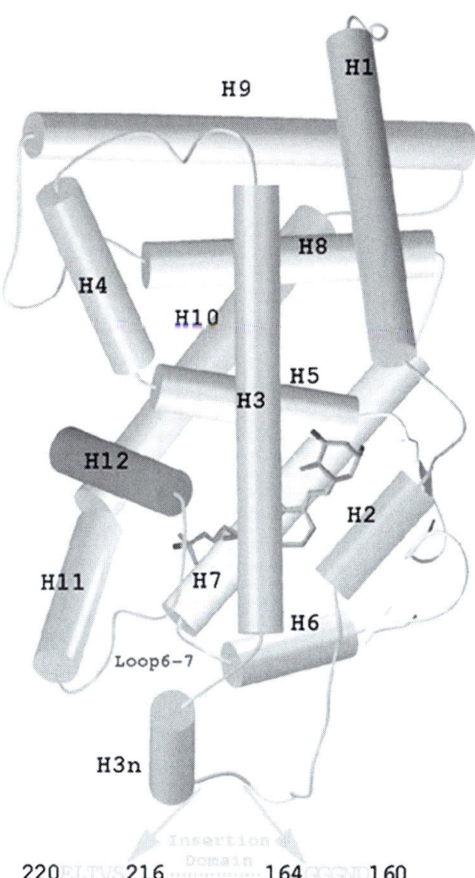

FIGURE 11-6   Three-dimensional structure of the holo-VDR LBD. The α-helices are shown as cylinders, and the three β sheets located between helix 5 and helix 6 as arrows. Helix 12 is shown in purple, and the ligand 1,25(OH)$_2$D$_3$ is in yellow. The location of the insertion domain deleted from the LBD is shown in green. Reproduced with permission from [191].

Several natural mutations that occur at amino acids that contact 1,25(OH)$_2$D$_3$ in the VDR LBD have been identified as the molecular basis of vitamin D resistance in patients with hereditary vitamin D–resistant rickets (HVDRR) (see Section VII) [2, 173]. Mutations have also been created in several amino acids predicted to be important in ligand binding. The naturally occurring mutations as well as the artificially created mutations demonstrate the importance of each of these amino acids in binding 1,25(OH)$_2$D$_3$. Ligand binding modeling has also been extended to docking vitamin D analogs [191]. The large volume of the binding pocket accommodates structural differences in ligand but does not as yet explain the differential activity of various vitamin D analogs (see Section VIII).

b. Alternative Binding Pocket

The genomic responses by 1,25(OH)$_2$D are mediated by the 6-s-*trans* form. On the other hand, the 6-s-*cis*-locked analog 1α,25(OH)$_2$lumisterol$_3$ is a weak activator of genomic responses but a full agonist for 1,25(OH)$_2$D-mediated rapid responses (see Section V). In computer modeling studies when 1α,25(OH)$_2$lumisterol$_3$ was docked to the VDR LBD, a potential alternative ligand-binding pocket was discovered. The putative alternative pocket (A pocket) partially overlaps the 1,25(OH)$_2$D binding pocket or genomic pocket (G pocket) [194]. Both the 6-s-*cis* and 6-s-*trans* forms of 1,25(OH)$_2$D can bind to the VDR A pocket. Whether ligand binding to the A pocket is the mechanism whereby the VDR mediates the rapid responses is still hypothetical at this time.

B. The VDR Gene

In humans, the VDR gene is located on chromosome 12q13-14, in close proximity to the 25-hydroxyvitamin D-1α-hydroxylase gene (CYP27B1) [137]. The VDR gene is composed of at least 11 exons that span 60 kb of DNA (Figure 11-5) [186, 187]. The VDR protein is encoded by exons 2–8. Exon 2 contains the translation initiation site and encodes the first zinc finger module, and exon 3 encodes the second zinc finger module. The 13 α-helices and 3 β-sheets of the ligand-binding domain [191] are encoded by exons 4 and 6–9. Exon 5 encodes a unique loop in the VDR that lacks structure and is unconserved. Exon 9 also contains approximately 3,200 nucleotides of 3′ noncoding sequence as well [186]. Exons 1A–1F are located at the 5′ end of the VDR gene. The expression of the VDR gene is directed by multiple promoters upstream of exon 1A, 1D, and 1F. Differential promoter usage and alternative splicing generate up to 14 mRNA transcripts [186, 187]. Two of the transcripts originating from a promoter upstream of exon 1D encode N-terminal variants that are 23 or 50 amino acids longer and encode VDRs of 450 and 477 amino acids, respectively [187]. Transcripts originating from exon 1F, the most distal exon, were expressed only in the parathyroids, kidney, and intestine, tissues involved in calcium regulation [187].

A putative promoter sequence was identified upstream of exon 1A in the human VDR gene. The GC-rich sequence contains potential binding sites for the transcription factors SP-1, AP1, AP2, C/EBP, and the nuclear factor (NF)-κB but lacked a TATA box [186]. The VDR is also a downstream target of the Wilm's tumor suppressor protein WT1. A WT1 responsive element was located in the upstream region of exon 1A at −308 to −300 [195]. Also a sequence located between exon 1C and exon 2 was shown to be capable of responding to retinoic acid [186]. Enhancer elements for VDR-RXR have also been identified in the VDR gene and are responsible for 1,25(OH)$_2$D$_3$-mediated upregulation (homologous upregulation) of

the VDR [196]. The VDR promoter is also induced by p63, a member of the p53 family of transcription factors [197]. The p53 protein also binds to conserved intronic sequences of the VDR gene *in vivo* [198].

In the intestine, the caudal-related homeodomain protein Cdx-2 contributes to the transcriptional regulation of the VDR gene. Cdx-2 binds to the sequence 5′-ATAAAAACTTAT-3′ at −3,731 to −3,720 bp relative to the transcription start site in the VDR promoter [199]. A polymorphism was identified in the core sequence 5′-A/GTAAAAACTTAT-3′ in the Cdx-2 binding site in the VDR gene promoter [200]. The G allele exhibited 70% lower transcriptional activity than that of the A allele, suggesting that the polymorphism may affect the expression of VDR in the small intestine.

C. Heterodimerization

Early studies in yeast examining the interaction of the VDR with the osteocalcin VDRE demonstrated that a protein from a nuclear extract from mammalian cells was required for DNA binding. The protein was later identified as a mixture of the retinoid X receptors (RXRα, RXRβ, and RXRγ) [201, 202]. RXR is a 55-kDa protein that binds 9-*cis*-retinoic acid as its ligand [203, 204] and is found widely distributed in cells and tissues, including those that do not express the VDR. RXR has now been shown to be the heterodimerization partner of a number of receptors in the steroid-thyroid-retinoid gene superfamily including VDR, TR, RAR, PPAR FXR, and a number of orphan receptors [205]. Utilizing an extensive series of internal deletions of the VDR, two regions located within the LBD, the E1 region (overlapping helixes H3 and H4) and helix H10, were shown to be essential for dimerization with RXR (Figure 11-5). Other regions of the receptor may also contribute to the RXR interface [1, 191]. In the presence of 1,25(OH)$_2$D, the RXR is allosterically modified by the VDR. In the absence of the RXR ligand, the unliganded RXR assumes the liganded conformation and acquires the capability to recruit coactivators and therefore acts as a major contributor to 1,25(OH)$_2$D-dependent transcription [206].

D. Post-Translational Modification of the VDR

The VDR is phosphorylated in a ligand-dependent manner in intact cells. Phosphorylation of the VDR occurs prior to the initiation of calcium uptake and induction of calcium-binding protein. The VDR is phosphorylated on serine residues by several different protein kinases. Ser208 is the major site phosphorylated by casein kinase II following addition of $1,25(OH)_2D_3$. A variety of data suggest that VDR phosphorylation may be linked to transactivation [188]. On the other hand, phosphorylation of Ser51 by protein kinase C (PKC) diminished DNA binding and nuclear localization of the VDR, while phosphorylation of VDR at Ser182 by protein kinase A (PKA) reduced RXR heterodimerization and transactivation in response to $1,25(OH)_2D_3$ [188]. Post-translational modification of RXR is also important, as phosphorylation of RXR by mitogen-activated protein kinase (MAPK) was shown to inhibit $1,25(OH)_2D$ signaling [207]. These findings suggest that differential phosphorylation of the VDR or RXR plays a role in determining the functional activity of the VDR.

Regulation of VDR content is an important element that contributes to the magnitude of $1,25(OH)_2D$ responsiveness. The VDR undergoes homologous (autoregulation) or heterologous (regulation by other factors) regulation, which is discussed in Section IV.J.

A number of nuclear receptors, including the VDR, are degraded by the ubiquitin (Ub)-proteasome pathway, and this pathway is thought to provide a means of preventing overstimulation by hormones. Proteolytic degradation by the Ub-proteasome system involves the covalent attachment of Ub molecules to the target protein, followed by degradation through the 26S proteasome. SUG1, a component of the 26S proteosome, binds to the VDR AF-2 domain in a $1,25(OH)_2D$-dependent manner, and overexpression of SUG1 inhibits $1,25(OH)_2D$-induced transactivation by the VDR. Furthermore, the proteosome inhibitor MG132 protected the VDR from degradation and increased $1,25(OH)_2D$ responses [208].

E.   Vitamin D Response Elements and Target Genes

Transcriptional activation of target genes by $1,25(OH)_2D$ is complex and involves a sequence of events centered around the VDR (Figure 11-7). The VDR acts as a *trans*-acting factor that interacts with specific VDREs located in the promoter regions of $1,25(OH)_2D$-responsive genes. $1,25(OH)_2D$ induces a wide array of biological responses, some resulting in an upregulation of specific mRNAs and others that down-regulate protein expression. Stimulatory or inhibitory actions may be tissue specific or depend on the state of cellular differentiation. The first vitamin D response element (VDRE) was identified in the promoter region of the human osteocalcin (OC) gene. The OC VDRE sequence GGGTGAacgGGGGCA is an imperfect hexanucleotide direct repeat that is separated by a 3 nucleotide spacer,

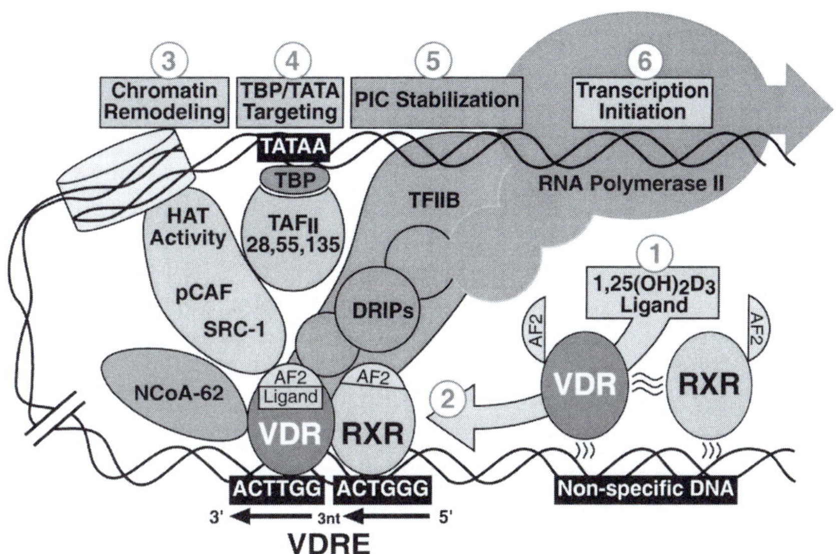

FIGURE 11-7   Model of $1,25(OH)_2D$ gene transactivation. Upon entering the cell, $1,25(OH)_2D_3$ binds to the VDR leading to the formation of a VDR:RXR heterodimer (1). The heterodimeric complex subsequently binds to vitamin D response elements (VDREs) in promoter regions of target genes through their cognate DNA-binding domains (2). Conformational changes in the VDR:RXR heterodimer initiate recruitment of coactivating proteins, including SRC-1 and NCoA-62 to the oligomeric complex. The histone deacetylase activity of SRC-1 modifies the chromatin structure and facilitates essential contact with the general transcription apparatus (3). Additional proteins are recruited to the complex, such as TBP and TAFs for targeting promoter elements (4). Binding of TFIIB and DRIPs to the complex stabilizes the preinitiation complex (5). Once the proteins have been assembled, transcription is initiated by RNA polymerase II (6). Reproduced with permission from [3].

a so-called direct repeat 3 (DR3) motif. Since the initial characterization of the OC VDRE, many other VDREs have been identified. On most VDREs, a polarity exists where RXR binds to the upstream hexanucleotide and the VDR binds to the downstream hexanucleotide. The diversity of VDRE sequences is becoming most apparent as more and more VDREs are characterized [209].

With the advent of microarray technology, a number of analyses have been performed in a variety of cells and tissues that have identified many novel as well as previously characterized genes that are up- or down-regulated by $1,25(OH)_2D$. Analyses of many of the genes regulated by $1,25(OH)_2D$ have identified one or more VDREs in their promoters. Chromatin immunoprecipitation (ChIP) has also aided in the identification of functional VDREs *in vivo*. In most cells and tissues, the most highly induced gene is 24-hydroxylase (CYP24A1) that contains two VDREs in its promoter. In the intestine, a classical target tissue, $1,25(OH)_2D$ induces the transport proteins for transepithelial absorption of Ca^{2+} including TRPV5 (ECAC1/CAT2) and TRPV6 (ECAC2/CAT1), and the calbindins (calbindin-D9k, calbindin-D28k). In the kidney, $1,25(OH)_2D$ induces the sodium-dependent phosphate cotransporter 2 (NPT2) and down-regulates 1α-hydroxylase (CYP27B1). In bone cells $1,25(OH)_2D$ induces OC, osteopontin, MN1, and RANKL. In keratinocytes, $1,25(OH)_2D$ induces involucrin, transglutaminase, and the corepressor *Hairless*, among many others. As $1,25(OH)_2D$ causes cell cycle arrest in many cells, a number of proteins that regulate cell cycle progression have been shown to be induced by $1,25(OH)_2D$, including cyclin C, p21, p27, IGFBP-3, and MKP5. Some of the many proteins down-regulated by $1,25(OH)_2D$ include collagen, PTH, PTHrP, calcitonin, IL-2, atrial natriuretic peptide, and c-*myc*.

F. VDR Interacting Proteins: Coregulators

A large number of proteins have been found to interact with the VDR as well as other nuclear receptors [1, 4, 210]. The VDR interacts with proteins that are required for or augment transcriptional activation, termed coactivators, and proteins that repress gene transactivation, termed corepressors. The particular coactivator protein recruited to the VDR may contribute to the tissue-specific function of VDR [211]. The list of VDR-interacting proteins continues to grow. A few of the coactivators and corepressors that interact with the VDR are briefly discussed in the following subsections. Figure 11-7 depicts the interaction of VDR with some of the critical interacting proteins.

1. Coactivators

The p160 class of coactivators that includes SRC-1, GRIP-1/TIF2/SRC-2, and ACTR/AIB1/SCR-3 binds to the VDR when ligand is present. SRC-1 exhibits histone acetyltransferase (HAT) activity that is thought to relax the chromatin structure and allow for transcription to begin. SRC-1, like many other coactivators, contains one or more nuclear receptor–interacting domains composed of conserved LxxLL interaction motifs. Upon $1,25(OH)_2D_3$-induced repositioning of helix H12, a hydrophobic cleft is formed on the VDR surface that functions as a docking site for the coactivator LxxLL motif interaction [212]. This interaction can be inhibited by synthetic LxxLL antagonists that prevent VDR transactivation [213]. The p160 coactivators bound to the liganded-nuclear receptors use at least three different activation domains to recruit additional coactivators. The histone acetyltransferases such as p300, CBP (CREB-binding protein), and pCAF (p300/CBP-associated factor) bind to the p160 coactivators and the histone arginine methyltransferases and modify chromatin through histone acetylation and methylation, further remodeling the nucleosomal structure.

The VDR interacts with the DRIP/TRAP (vitamin D receptor–interacting proteins/thyroid hormone receptor–associated proteins) complex in a ligand-dependent manner [214, 215]. At least 13 proteins constitute the DRIP/TRAP complex although only DRIP205/TRAP220 binds directly to the VDR. Other DRIPs/TRAPs are recruited to the growing complex of proteins subsequent to DRIP205/TRAP220 binding. DRIP205/TRAP220 binding to VDR is enhanced by the phosphatase inhibitor okadaic acid, suggesting that phosphorylation of the VDR may be an important mechanism in creating an active surface with DRIP205/TRAP220 [216]. In cell-free transcription assays, DRIPs/TRAPs mediated the ligand-dependent gene transcription by the VDR [214]. Recent findings suggest that the VDR interacts with a number of previously characterized or yet to be discovered complexes that may allow for the tissue-specific regulation of the VDR [217]. ChIP assays indicate that p160 coactivators and CBP and p300 are recruited to the VDR target genes CYP24A1 and osteopontin rapidly (15 minutes to 1 hour) after ligand binding [218]. DRIP205/TRAP220 and RNA polymerase II are subsequently recruited to the receptor-bound promoter. ChIP assays show periodic cycling of p160 coactivators and DRIP205/TRAP220. Also, $1,25(OH)_2D_3$ treatment strongly enhanced histone 4 acetylation on the CYP24A1 promoter. These findings suggest that p160 coactivators and CBP/p300 modify chromatin and allow for the subsequent recruitment of the DRIP/TRAP complex, which may target the RNA polymerase II apparatus [218].

The VDR interacts with the basal transcription factor TFIIB and TATA-binding protein (TBP)-associated factors (TAFs), proteins associated with the basal transcriptional machinery. TAFs bind to a region from helix H3 to helix H5 and to helix H8 of the VDR, and these interactions appear to enhance transcription through direct stabilization of the transcriptional machinery.

NCoA-62 (nuclear receptor coactivator; 62,000 Da) also known as Ski-interacting protein (SKIP) interacts with the VDR in a ligand-dependent manner [219]. The p160 coactivator GRIP1 and NCoA62/SKIP synergistically enhance ligand-dependent VDR transcriptional activity by forming a ternary complex with VDR [220]. NCoA62/SKIP has been identified as a component of the spliceosome machinery and may have a role in coupling transcriptional regulation by VDR to RNA splicing [221].

WINAC (Williams syndrome transcription factor [WSTF] including nucleosome assembly complex) recruits the unliganded VDR to promoters of VDR target genes. WINAC can stimulate $1,25(OH)_2D_3$-induced gene activation or repression by VDR [222]. WINAC is required for the $1,25(OH)_2D_3$-mediated repression of the 1α-hydroxylase (CYP27A1) gene [223]. WINAC exhibits an ATP-dependent chromatin-remodeling activity, and the loss of this activity may contribute to Williams syndrome [224].

Many other proteins have been shown to interact with the VDR such as Smad3, BCL2-associated athanogene (BAG1), retinoblastoma binding protein 2, cJun, STAT1, thymine-DNA glycosylase, transcription intermediary factor 1 (TIF-1), promyelocytic leukemia zinc finger (PLZF), and nuclear receptor coregulator (NRC).

2. COREPRESSORS

Several corepressor proteins have been shown to interact with the VDR and inhibit basal transcription. Corepressors recruit histone deacetylases (HDACs) that repress transcription by stabilizing chromatin. VDR-RXR heterodimers can bind to a wide range of hormone response elements. In the absence of $1,25(OH)_2D$, the VDR repressed basal transcription on thyroid hormone response elements and blocked triiodothyronine (T3)-mediated gene transactivation. VDR repression of T3-mediated transcription may be significant in tissues where VDR and TR are coexpressed and compete for RXR [225]. A direct interaction of the VDR with the corepressors NCoR, SMRT, and Allen was shown to be independent of the VDR AF-2 domain but sensitive to $1,25(OH)_2D_3$ [226]. NCoR is increased in some breast cancer cells and suppresses VDR target genes causing $1,25(OH)_2D_3$

resistance [227]. In prostate cancer cells, elevated SMRT levels suppress the target genes associated with the antiproliferative actions of $1,25(OH)_2D_3$ [228].

The *Hairless* gene product HR binds directly to VDR both *in vitro* and *in vivo*. HR binds to the central portion of the VDR LBD and is independent of the AF-2 domain [229]. VDR-mediated transactivation is strongly inhibited by HR. In mouse skin VDR and HR are found in cells of the hair follicle [229]. HR and VDR are also coexpressed in keratinocytes [230]. When HR is overexpressed in human keratinocytes, induction of $1,25(OH)_2D_3$-responsive genes by VDR is suppressed. When associated with VDREs *in vivo*, HR-VDR interactions were disrupted by $1,25(OH)_2D_3$, allowing recruitment of coactivators [230]. The role of HR and the unliganded VDR in regulating the hair cycle is discussed in Section IV.I.

G. Nuclear Translocation

Translocation of the VDR between the cytoplasm and the nucleus is a complex event. Deletion studies have shown that there are both ligand-dependent and -independent pathways underlying the nuclear transport of VDR. The VDR contains at least three nuclear localization signals (NLS). One NLS resides in the DBD between the first and second zinc finger modules (aa 49–55) [231]. A second NLS is represented by residues 76–102 immediately C-terminal to the second zinc finger [232]. A third NLS is a bipartite sequence located in the hinge region (aa 154–173) [233]. The transport of VDR from cytoplasm to nucleus was examined using fluorescent protein-tagged chimeras of full-length or truncated constructs of the VDR [234]. In the presence of $1,25(OH)_2D$ the cytoplasmic VDR was translocated to the nucleus and colocalized with RXR [235]. Truncation of either the LBD or the AF-2 region of VDR abolished ligand-dependent translocation and transactivation. The findings support a model of ligand-dependent VDR translocation and indicate that translocation from cytoplasm to nucleus is part of the receptor activation process [234]. Photobleaching experiments have demonstrated that the VDR shuttles back and forth between the cytoplasm and the nucleus and that $1,25(OH)_2D$ increases the nuclear accumulation of VDR [235]. A putative nuclear export signal is located at amino acids 320–325 in the VDR LBD. Export of the unliganded VDR is mediated by the CRM-1 export receptor [236]. Calreticulin binding is also critical to VDR and RXR export from the nucleus, as mutations of the calreticulin binding sites in VDR and RXR DBDs inhibit nuclear export [236].

H. Transactivation of Target Genes by 1,25(OH)₂D

An overview model of 1,25(OH)₂D-regulated gene transactivation is shown in Figure 11-4 and the detailed model in Figure 11-7. In the absence of 1,25(OH)₂D, both VDR and RXR can be detected in the cytoplasm and in the nucleus. Upon 1,25(OH)₂D binding, the VDR-RXR heterodimer in the cytoplasm translocates to the nucleus and forms a high-affinity complex that acquires the ability to recognize and bind with high affinity to VDREs through their cognate DBDs. During 1,25(OH)₂D binding, helix H12 is repositioned, forming a high-affinity protein surface capable of interacting with specific coactivator proteins required for transactivation. The liganded VDR-RXR heterodimer attracts p160 coactivator proteins such as SRC-1 and with its intrinsic histone acetyltransferase activity derepresses the chromatin so that nucleosomes are rearranged and naked DNA becomes accessible. SRC-1 is cycled off, and other coactivators such as the DRIP complex cycle on, allowing for the assembly of the transcriptional apparatus [218]. TATA binding protein-associated factors (TAFs) are also recruited to target TATA/TBP binding sites. Other proteins including TFIIB serve to stabilize the complex. Transcription is then initiated by RNA polymerase II. The specifically induced mRNA transcripts are translated into proteins, eliciting the downstream actions of the hormone.

I. Unliganded Actions of the VDR

Recently, there has been evidence accumulating from research on several different nuclear receptors for a gene-silencing role of the unliganded nuclear receptors. The role of the unliganded VDR and its associated proteins in gene silencing is just emerging. From studies of patients with HVDRR and VDR KO mice, it is now becoming clear that the unliganded VDR plays a major role in the regulation of the hair cycle.

Some patients with HVDRR, but not all, exhibit total body alopecia and skin lesions [2]. Histological examination of the skin of HVDRR patients revealed the absence of hair follicles and the presence of dermal cysts. Miller *et al.* [237] noted that the hair loss and skin lesions in their HVDRR patient were a phenocopy of the disorder atrichia and papular lesions (APL) that is caused by mutations in the *Hairless* gene [238, 239]. The *Hairless* gene product HR is thought to coordinate the balance between cell proliferation, differentiation, and/or apoptosis in the epidermis and hair follicle [240]. Since the mutations in VDR and HR result in the same phenotype in respect to the hair loss and skin deformities,

it has been hypothesized that VDR and HR regulate a common pathway that controls postnatal cycling of the hair follicle [237]. The discovery that HR functions as a corepressor of VDR provided a possible connection between APL and the alopecia in HVDRR [229]. During the hair cycle, hair follicles undergo a cyclical process of rest (telogen), active growth and hair shaft generation (anagen), and apoptosis-driven regression (catagen) [241]. The VDR KO mice also exhibit alopecia and skin wrinkling due to the presence of dermal cysts. In VDR KO mice, the transition from telogen to anagen is inhibited [242, 243]. Suppression or neutralization of a potential tonic inhibitor(s) of hair growth that is thought to exist in telogen skin is thought to trigger the telogen-to-anagen transition [229]. Parathyroid hormone-related peptide (PTHrP) or inhibitors of the Wnt signaling pathway are attractive candidates as potential tonic inhibitors, since overexpression of PTHrP or disruption of Wnt signaling interferes with hair follicle development [244, 245].

It has been hypothesized that the unliganded VDR has a role in gene silencing during the hair cycle [229, 246]. From studies of VDR mutations that cause HVDRR with alopecia, it is apparent that DNA binding and RXR heterodimerization are essential functions of the VDR that are required to prevent alopecia, since defects in these critical regions of the VDR lead to alopecia [2]. Also, mice with defective RXRs in the skin have alopecia, demonstrating that RXR is a critical factor in regulating hair growth [247]. From studies of the VDR mutations that cause HVDRR without alopecia [2, 246, 248, 249], ligand binding, coactivator binding, and 1,25-(OH)₂D-mediated gene transactivation are dispensable functions of the VDR in regulating the hair cycle, since defects that disrupt these activities do not cause alopecia. In further support of this hypothesis, Skorija and colleagues showed that targeted expression to keratinocytes of either a Leu233Ser mutant VDR that does not bind ligand or a Leu417Ser mutant VDR that exhibits defective coactivator binding can restore hair growth to VDR knock-out mice that have alopecia [250]. Thus, it appears that the unliganded VDR-RXR heterodimer together with HR silence the expression of a specific gene or set of genes at critical times during the hair cycle and that dysregulation of these genes due to VDR or HR mutations can cause alopecia. While these conclusions are drawn from observations in regard to hair growth and alopecia, gene silencing by the VDR may not be limited to genes involved in hair growth but may occur in other tissues where VDR and HR or other corepressors are coexpressed. Microarray analyses of gene expression in kidney from normal mice with WT VDR and VDR knock-out mice have revealed that a number of genes are derepressed in the absence of the VDR, suggesting that they are regulated by the unliganded VDR [251].

J. Regulation of VDR Abundance

Within each target tissue, the amount of VDR protein expressed in a cell is not fixed but rather is dynamically regulated by a variety of physiological and developmental signals. This is important, since the level of VDR expressed in a target cell determines the amplitude of the response evoked by 1,25(OH)$_2$D. Upregulation of VDR enhances the response to 1,25(OH)$_2$D, whereas down-regulation of the VDR diminishes the response [252–256]. Of the many factors that regulate VDR expression, the ligand 1,25(OH)$_2$D itself is an important modulator that increases the receptor abundance (homologous upregulation). Other regulators that may up- or down-regulate VDR levels (heterologous regulation) include steroid and peptide hormones, growth factors, activators of second messenger pathways, and intracellular calcium [256]. In some cases the VDR levels are dependent on the proliferation/differentiation status of the target cell. Changes in VDR levels are also observed during neonatal development in different tissues [256].

1. HOMOLOGOUS REGULATION

The VDR is upregulated by 1,25(OH)$_2$D and other vitamin D metabolites that bind to the VDR (homologous regulation), and this has been observed both *in vitro* [257] and *in vivo* [258–260]. The magnitude of homologous upregulation varies from 2- to 10-fold depending on the target cell. In pig kidney cells, human skin fibroblasts, and human mammary cancer cells (MCF-7), the VDR level increases when the cells are treated with 1,25(OH)$_2$D$_3$, 1,24,25(OH)$_3$D$_3$, 24,25(OH)$_2$D$_3$, and 25(OH)D$_3$, and the concentrations required for maximal upregulation closely reflect the affinities of the various metabolites for the VDR [257]. Several studies have shown that the upregulation of the VDR is due to an increase in the transcription of the VDR gene [184, 260, 261]. Zella *et al.* [196] demonstrated that 1,25(OH)$_2$D$_3$ induced VDR gene expression in mouse bone *in vivo* and in mouse osteoblastic cells. Using chromatin immunoprecipitation-DNA microarray (ChIP-chip) analysis, they identified a conserved region 27 kb downstream of the transcription start site that was able to confer 1,25(OH)$_2$D$_3$ regulation to downstream promoters. These studies in mice subsequently led to the identification of a highly conserved region within the human VDR gene that was capable of mediating 1,25(OH)$_2$D$_3$ induction [196].

Other studies have shown that upregulation of the VDR by 1,25(OH)$_2$D$_3$ is mainly due to the stabilization of the ligand-occupied VDR [262–264]. Either one or both of these phenomena (increased synthesis vs. stabilization) may be operative depending on the target cell [257, 263]. In pig kidney cells, about two-thirds of the upregulation appeared to be due to the stabilization of the VDR and one-third due to the increased synthesis of the VDR protein [257]. Homologous upregulation of VDR may have an important role in the treatment of psoriasis, a hyperproliferative skin disorder. Chen *et al.* [265] showed that the therapeutic response to 1,25(OH)$_2$D treatment in patients with psoriasis correlated with the upregulation of VDR in psoriatic skin. In patients who showed clinical improvement with treatment, significant upregulation of VDR mRNA was observed in the psoriatic lesions, while there was no upregulation in patients who did not respond to 1,25(OH)$_2$D.

2. HETEROLOGOUS REGULATION

Various hormones including steroid and peptide hormones and growth factors regulate VDR expression (heterologous regulation) in a cell- and tissue-specific manner. In cultured cells, VDR expression has been shown to be closely related to the rate of cell proliferation, with VDR levels being higher in proliferating cells than in quiescent cells [266, 267]. The human VDR promoter sequence upstream of exon 1A contains several potential binding sites for the SP1 transcription factor and other transcriptional activators including cAMP response elements [186]. In NIH-3T3 mouse fibroblasts forskolin or dibutyl-cAMP increased VDR mRNA expression and VDR protein levels (8- to 12-fold), possibly by a mechanism involving protein kinase A [254, 268]. In the osteoblast cell line, UMR 106 activation of the cAMP signal pathway by PTH increased VDR mRNA levels [255, 269]. Prostaglandin E2 also upregulated VDR abundance, possibly by a mechanism involving cAMP [270]. The caudal-related homeodomain transcription factor Cdx-2 was identified as a regulator of VDR transcription in the intestine [199]. Cdx-2 is able to activate VDR gene transcription in the intestine by binding to a *cis*-element in the human VDR gene promoter.

In contrast, mitogens such as basic fibroblast growth factor and phorbol esters that activate protein kinase-C lead to a significant decrease in VDR abundance in spite of stimulating cell proliferation [271]. Elevating intracellular Ca^{2+} levels by calcium ionophores also decreases VDR abundance. The down-regulation of VDR is the result of a decrease in VDR gene transcription and/or destabilization of the VDR mRNA [271]. Also, in some cell systems, induction of cell differentiation leads to a decrease in VDR abundance [261, 271–273].

Glucocorticoids [274–277], estrogens [278, 279], retinoids [280, 281], and PTH [255, 269, 282] also regulate VDR expression. Changes in VDR abundance elicited by these hormones are reflected in the magnitude of $1,25(OH)_2D$ responsiveness. However, there are species differences between various rodent models so that extrapolation to humans from animal experiments is not always possible. Even within a species, there may be tissue-specific differences. The intron 3′ of exon 1C of the human VDR gene responds to retinoic acid, suggesting a direct effect on the VDR gene as the molecular mechanism for the regulation of VDR by retinoids [186]. Excess glucocorticoids down-regulate the VDR and cause $1,25(OH)_2D$ resistance, whereas PTH upregulates the VDR and enhances $1,25(OH)_2D$ responsiveness [255, 269]. Thus, these hormones modulate target cell sensitivity to $1,25(OH)_2D$ in part through regulation of VDR levels.

Analysis of the VDR in parathyroid glands has been extensively studied [159]. It has been postulated that reduced levels of VDR in parathyroid glands may be related to lack of $1,25(OH)_2D$ suppression of parathyroid hormone secretion and parathyroid cell hyperplasia and may contribute to the pathogenesis of secondary hyperparathyroidism in chronic renal failure [283–285]. The low serum levels of $1,25(OH)_2D$ in chronic renal failure may further accentuate this effect. Similarly, vitamin D status may alter the pattern of signs and symptoms in primary hyperparathyroidism [286–288].

K. VDR Knockout Mice

Several groups have generated VDR KO mice. In the original VDR KO mouse models, exon 2 in the VDR gene was disrupted in one model [172] and exon 3 was deleted in the other [289, 290]. The mice containing the disruption in exon 2 eliminated the first zinc finger but expressed a truncated VDR that retained $1,25(OH)_2D_3$ binding [291]. In both VDR KO models, the mice were phenotypically normal at birth, suggesting that $1,25(OH)_2D_3$ actions are not necessary for normal embryogenesis. After weaning, the mice became hypocalcemic and developed rickets similar to patients with HVDRR. Alopecia also appeared progressively as the mice aged. Most of the VDR KO mice generated by disruption of exon 2 were infertile and died by 15 weeks after birth [172]. These mice were also noted to have uterine hypoplasia and impaired folliculogenesis. The VDR KO mice generated by deleting exon 3 survived at least 6 months [289]. In both VDR KO mouse models, the survival of the mice was enhanced by a high calcium diet supplemented with lactose [290].

Many but not all of the abnormalities in the reproductive organs were eliminated by maintenance of normal calcium levels with the "rescue diet" [292]. Estrogen levels were only partially corrected by calcium repletion [292], suggesting a role for $1,25(OH)_2D_3$ in regulating aromatase gene expression [293]. A VDR KO mouse model was also generated where a lacZ reporter gene was expressed from the endogenous VDR promoter [294]. These mice expressed lacZ and a truncated VDR due to initiation from exon 3. The lacZ homozygous mice showed growth retardation, rickets, secondary hyperparathyroidism, and alopecia. LacZ expression was strongly expressed in bones, cartilage, intestine, kidney, skin, brain, heart, and parathyroid glands. When fed the rescue diet, the serum calcium and PTH levels were normalized. However, in the kidney a profound calcium "leak" was noted in homozygous mutant mice [294]. In the duodenum, expression of TRPV6/CAT1 and TRPV5/CAT2 was considerably reduced in VDR KO mice fed a normal calcium diet [295]. As in the human disease HVDRR, normalization of calcium did not resolve the alopecia.

Studies employing 1α-hydroxylase, VDR, and the double KO mice showed that the calcium ion and the $1,25(OH)_2D_3$-VDR system exert discrete effects on skeletal and calcium homeostasis. Both calcium and $1,25(OH)_2D_3$ regulated parathyroid gland size and the development of the cartilaginous growth plate independently of the VDR. Calcium levels were associated with PTH secretion and mineralization of bone, while increased calcium absorption and optimal osteoclastic bone resorption and osteoblastic bone formation were modulated by the VDR and $1,25(OH)_2D_3$ [296–299].

Analyses of the VDR KO mice have revealed many new and unknown aspects of VDR actions as well as supported earlier findings in cultured cells. For example, VDR KO mice exhibit enhanced thrombogenicity, suggesting that the VDR has a role in maintaining antithrombotic homeostasis [300, 301]. Also, VDR KO mice were hypertensive and had increased renin expression and plasma angiotensin II production, suggesting that the VDR is a negative regulator of the renin-angiotensin system [302, 303]. A role for the VDR in the generation of Th-2–driven inflammation was demonstrated by the failure of the VDR KO mice to develop experimental allergic asthma [304]. VDR knockout mice also exhibited severe inflammation of the gastrointestinal tract in two different experimental models of inflammatory bowel disease (IBD) [305]. In the CD45RB transfer model of IBD, T cells from VDR KO mice induced more severe colitis than wild-type T cells. In the second model of IBD, VDR/IL-10 double KO mice developed accelerated IBD and rectal bleeding. By 8 weeks of age, all of the double KO mice had died, whereas all of the VDR and IL-10 single KO mice were healthy. The data suggest that the VDR has an important role in regulating inflammation in the gastrointestinal tract [305]. VDR KO mice also exhibit dysregulation of myoregulatory transcription factors myf5, myogenin, E2A, and early myosin heavy chain isoforms in muscle and increased numbers of dendritic epidermal T cells (DETC) [306]. In chemical carcinogen-induced tumorigenesis models, VDR KO mice that were fed the rescue diet to normalize calcium exhibited an increased incidence of mammary gland hyperplasia and tumor development in epidermis and lymphoid tissues [307]. VDR ablation did not affect tumor development in ovary, uterus, lung, or liver. These data suggest that VDR signaling may act to suppress tumorigenesis [307].

V. NONGENOMIC ACTIONS OF VITAMIN D

In addition to the classical VDR-mediated genomic pathway, $1,25(OH)_2D$ also has been shown to elicit rapid responses [308]. The term "rapid response" is used to describe the biological effects of $1,25(OH)_2D$ that occur within a few minutes after hormone treatment and are considered too rapid to be explained by a VDR-mediated genomic pathway. Rather, the rapid responses are thought to be mediated by a direct action of $1,25(OH)_2D$ on the plasma membrane of target cells stimulating a signal transduction pathway involving the rapid opening of voltage-sensitive Ca^{2+} channels and activation of protein kinases [309]. Some of the $1,25(OH)_2D$-induced rapid responses include changes in intracellular calcium flux, alteration in phospholipid metabolism and phosphate transport, and changes in alkaline phosphatase and adenylate cyclase activities. Also, "transcaltachia," a process of transluminal transport of Ca^{2+} across the intestine, has been shown to occur rapidly when vitamin D–replete animals are treated with $1,25(OH)_2D_3$. The rapid Ca^{2+} transport is thought to be facilitated by endocytic and lysosomal vesicles that deliver the Ca^{2+} to the basolateral membrane where it is released by exocytosis into the lamina propria. However, because the transcaltachia response requires vitamin D–replete animals, a pre-existing condition induced by $1,25(OH)_2D$ may be operative, and thus, transcaltachia may be dependent upon a $1,25(OH)_2D$-VDR–mediated genomic pathway.

Several lines of evidence support the existence of a nongenomic $1,25(OH)_2D$-mediated signal transduction pathway. For instance, the antagonist $1\beta,25(OH)_2D_3$, which has minimal effect on $1,25\alpha(OH)_2D$-induced genomic actions, blocks the effect of $1\alpha,25(OH)_2D_3$ on transcaltachia [310]. Similarly, some vitamin D analogs such as the 6-s-cis blocked conformer that binds poorly to the VDR are able to generate the transcaltachia response in perfused chick intestine and Ca^{2+} influx in ROS 17/2.8 cells [311]. In NB4 cells, an acute promyelocytic leukemia cell line, the 6-s-cis blocked conformer was 20 times more effective at priming the cells for monocytic differentiation than the natural hormone. This response was attenuated by the $1\beta,25(OH)_2D_3$, a specific antagonist of the nongenomic response [312]. The 6-s-cis locked analog, $1\alpha,25(OH)_2lumisterol_3$, also induces transcaltachia and stimulates Ca^{2+} uptake in the osteosarcoma cell line [311]. $1\alpha,25(OH)_2lumisterol_3$ was shown to augment glucose-induced insulin secretion in rat pancreatic islet cells while also increasing intracellular Ca^{2+} concentrations [313]. $1\alpha,25(OH)_2lumisterol_3$ also protected skin cells from UV-induced cell loss and cyclobutane pyrimidine dimer damage to an extent comparable with that of $1,25(OH)_2D$, suggesting that the photoprotective effects of $1,25(OH)_2D$ are mediated via the rapid response pathway(s) [314].

In osteoblasts the plasma membrane VDR is localized in plasma membrane caveolae and is thought to mediate the rapid effects of $1,25(OH)_2D$. The presence of saturable and specific $[^3H]$-$1,25(OH)_2D$ binding sites in caveolae supports membrane VDR as the $1,25(OH)_2D$-binding protein in the membrane-enriched fraction [315]. In osteoblasts isolated from WT and VDR KO mice, $1,25(OH)_2D$ modulated ion channel activities only in WT cells, demonstrating that a functional VDR is required for the rapid modulation of electric currents by $1,25(OH)_2D$ [316]. Also, rapid responses to $1,25(OH)_2D_3$ in osteoblasts were abrogated in homozygous mice expressing a mutant VDR with a deletion of the DBD, supporting the conclusion that the nuclear VDR mediates the nongenomic actions of $1,25(OH)_2D$ [294]. On the other hand, $1,25(OH)_2D$ induced a rapid increase in Ca^{2+} and PKC activity in osteoblasts from both WT and VDR KO mice, arguing that the VDR is not essential for these rapid actions [317]. Thus, the nature of the receptor that mediates rapid, nongenomic actions is still unclear.

An alternate potential membrane receptor for $1,25(OH)_2D$ was isolated from chick intestinal basolateral membranes. The 65-kDa membrane receptor termed 1,25D(3)-MARRS (membrane-associated, rapid-response steroid-binding) was subsequently cloned from a chicken cDNA library [318]. The 1,25D(3)-MARRS protein is identical to the multifunctional protein ERp57. Ribozyme inactivation of 1,25D(3)-MARRS decreased specific membrane-associated $1,25(OH)_2D_3$ binding, while nuclear receptor binding remained unaffected. $1,25(OH)_2D_3$ dependent stimulation of protein kinase C activity was also reduced in the presence of the ribozyme [319].

Knockout mouse models for either CYP27B1 (1α-hydroxylase) or the VDR or the double knockout have demonstrated that calcium absorption and bone and cartilage remodeling require both 1,25(OH)₂D and the VDR. On the other hand, 1,25(OH)₂D actions independent of VDR were speculated to play a role in the development of the growth plate as well as parathyroid gland function [297]. In rat costochondral growth plate, chondrocytes 1,25(OH)₂D and 24R,25(OH)₂D cause a rapid increase in PKC activity that resulted in the activation of the ERK1/2 family of MAP kinases [320]. In ROS 17/2.8 cells and mouse primary osteoblasts, 1,25(OH)₂D₃ promotes the rapid potentiation of outward Cl(−) currents. The rapid actions of 1,25(OH)₂D₃ on Cl(−) and Ca(2+) channels seem to couple to secretory activities, thus contributing to bone mass formation [321].

VI. PHYSIOLOGY: REGULATION OF SERUM CALCIUM

A. Interaction of PTH and Vitamin D to Regulate Serum Calcium

The concentration of Ca²⁺ in plasma and extracellular fluid is maintained within a narrow range, variations up or down being associated with untoward effects [285, 322, 323]. In the balanced state, the dietary intake of approximately 1,000 mg of calcium is equal to the combined excretion in feces ($\cong$720 mg) and urine ($\cong$280 mg). Coordinated interaction of 1,25(OH)₂D and PTH to regulate 1α-hydroxylase activity plays a major role in the maintenance of calcium balance (Figure 11-8). Small decreases in serum calcium result in increases in PTH secretion, which stimulates upregulation of 1α-hydroxylase activity, and increased renal phosphate excretion. The combination of increased PTH and decreased phosphate leads to enhanced 1α-hydroxylase activity. The regulation of phosphate homeostasis is discussed in Chapter 14 (Tebben). The augmented synthesis of 1,25(OH)₂D enhances intestinal calcium absorption to restore the calcium concentration toward normal levels, which in turn feeds back to diminish PTH secretion, thereby limiting the further production of 1,25(OH)₂D. In addition, 1,25(OH)₂D feeds back on the kidney to inhibit further production of 1,25(OH)₂D by down-regulating 1α-hydroxylase gene expression while stimulating 24-hydroxylase gene expression. Furthermore, serum calcium is maintained by the combined actions of PTH and 1,25(OH)₂D on the bone to increase bone resorption and by the action of PTH on the kidney to increase calcium reabsorption. In hypercalcemic states, PTH is suppressed by a signal transmitted via the

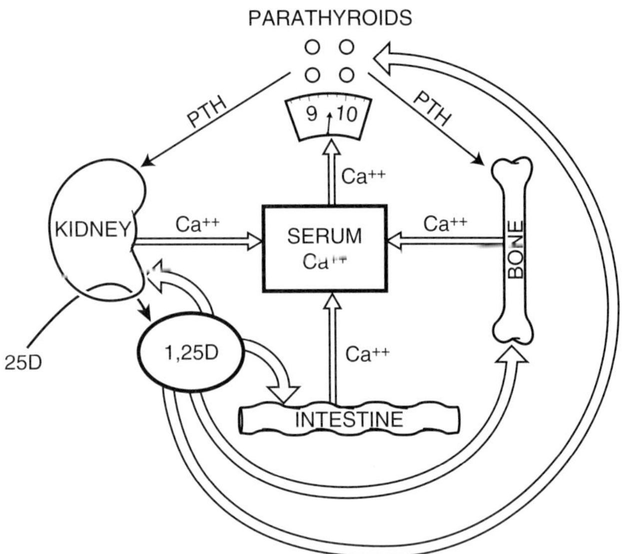

FIGURE 11-8   Regulation of Ca²⁺ levels in the blood by 1,25(OH)₂D and PTH.

parathyroid calcium-sensing receptor (CaR) [146], and the entire process is reversed. In rat parathyroid glands and kidney, the expression of the CaR gene is increased by 1,25(OH)₂D but not by Ca²⁺. Upregulation of the CaR is thought to be involved in the suppressive effects of vitamin D compounds on PTH secretion. The selective action of less calcemic vitamin D analogs that have a greater suppressive effect on PTH expression may allow for their potential use in therapeutic situations with elevated PTH concentrations [285] (see Section VIII on Analogs).

B. Vitamin D Toxicity and Hypercalcemia

Hypervitaminosis D occurs when large doses of vitamin D are administered, or in the context of large topical doses, or via increased endogenous production in several disease states (see Section VI.C). Vitamin D toxicity is characterized by various clinical manifestations of hypercalcemia, including hypercalciuria, ectopic calcifications, hyperphosphatemia, renal stones, polyuria and polydipsia, hypertension, anorexia, nausea, vomiting, and constipation. Excessive sun exposure does not cause symptomatic hypercalcemia, due to photodegradation of previtamin D₃ to inactive sterols (tachysterol and lumisterol) in the skin, as well as the protective effect of melanin production in the irradiated skin, which acts as a natural sunscreen. Toxic doses of vitamin D have not been established for all age groups. Although vitamin D toxicity generally occurs with the administration of daily doses greater than 10,000 IU/day (250 mcg/day), the increased incidence of nephrocalcinosis was found with much lower doses of vitamin D_3 (400–1,000 IU/day, 10–25 mcg/day) in the Women's Health Initiative (WHI) clinical trial [102]. Thus, toxicity monitoring is recommended when administering vitamin D doses above 50 mcg/day (2,000 IU/day) or lower doses chronically. Vitamins D_2 and D_3 cause more prolonged toxicity than 25(OH)D or 1,25(OH)D because of increased lipid solubility, resulting in the potential for prolonged toxicity up to 18 months [165]. Treatment of vitamin D–mediated hypercalcemia includes dietary vitamin D and calcium restriction, avoidance of sunlight, and pharmacological therapy. Reduced oxalate intake is also recommended, since hyperoxaluria increases the risk of kidney stone formation. Symptomatic patients should be treated with normal saline for hydration, with or without a loop diuretic [165]. Thiazide diuretics should be avoided because they can worsen hypercalcemia. Bisphosphonates or calcitonin could be considered to inhibit bone resorption [165]. Treatment with glucocorticoids is effective in hypercalcemia associated with vitamin D intoxication due to lymphoma and granulomatous disease. Glucocorticoids act primarily on the lymphomatous or macrophage lesions to cause regression but also may have some benefit by acting in the intestines (reduce calcium absorption), the kidneys (increase excretion), and bone (inhibit resorption). In sarcoidosis, aminoquinolines help to correct hypercalcemia by reducing 1,25(OH)$_2$D levels, but due to side effects, this therapy is reserved for those who fail glucocorticoids [138]. The use of the antifungal drug ketoconazole as a diagnostic test or as therapy for hypercalcemic states has been suggested [324, 325]. Ketoconazole inhibits fungal growth by blocking the P450 enzyme 14-demethylase in the pathway to ergosterol synthesis [326]. The drug has been shown to inhibit mammalian P450 enzymes including 24-hydroxylase [327] and 1α-hydroxylase [328].

C. Extrarenal 1,25(OH)$_2$D Synthesis and Hypercalcemic States

Under normal physiological conditions, the kidney is the primary site of 1,25(OH)$_2$D formation. However, small amounts of 1,25(OH)$_2$D are produced locally in various other tissues, and in selected pathological conditions the extrarenal production of 1,25(OH)$_2$D may significantly contribute to alterations in calcium homeostasis [138]. Tissues shown to synthesize 1,25(OH)$_2$D from 25(OH)D include human decidua and placenta, bone cells, keratinocytes, colon, breast, prostate, spleen, melanoma cells, hepatoma cells, and synovial and pulmonary monocytes and macrophages. Although renal and extrarenal 1α-hydroxylase are

identical [329], there are major differences in their regulation. While renal 1α-hydroxylase is under the feedback control of calcium, parathyroid hormone, and phosphate but not steroids, extrarenal macrophage 1α-hydroxylase shows increased susceptibility to corticosteroids but is unresponsive to the regulatory effect of PTH or phosphate [138]. In a model proposed by Hewison and Adams, the macrophage that lacks 24-hydroxylase would escape another negative feedback mechanism that shunts 1,25(OH)$_2$D toward inactive metabolites [138]. Accumulating evidence suggests that macrophage 1α-hydroxylase is regulated by cytokines, lipopolysaccharide, nitric oxide, and intracellular vitamin D–binding proteins. Hypercalcemia can be expected to occur in 7–24% of patients with sarcoidosis [330]. Proof of the clinical significance of extrarenal production of 1,25(OH)$_2$D was first provided from studies on an anephric patient with sarcoidosis who developed hypercalcemia [331]. Cultured pulmonary alveolar macrophages from patients with diffuse pulmonary sarcoidosis have been shown to be capable of producing 1,25(OH)$_2$D in excess, compared to macrophages from patients with less severe disease [138]. In addition to sarcoidosis, other granulomatous disorders have been associated with hypercalcemia and elevated 1,25(OH)$_2$D levels, including tuberculosis, leprosy, silicone-induced granulomatosis, and disseminated candidiasis [138].

Hypercalcemia in lymphoma patients is often due to elevations in 1,25(OH)$_2$D. Both Hodgkin's and non-Hodgkin's lymphoma have been associated with elevated 1,25(OH)$_2$D levels [332]. Hypercalcemia in these disorders is estimated to occur in 5% of patients with Hodgkin's disease and in 15% of patients with non-Hodgkin's lymphoma. In one report, 1,25(OH)$_2$D levels were elevated in 55% of a group of 22 hypercalcemic patients with non-Hodgkin's lymphoma, and many of the normocalcemic patients with non-Hodgkin's lymphoma had evidence of dysregulated 1,25(OH)$_2$D synthesis [333]. Lymphocytes transformed with HTLV-1 have been shown to convert 25(OH)D to 1,25(OH)$_2$D *in vitro*, indicating that these lymphoma-like cells have 1α-hydroxylase activity, and there is evidence that lymphomatous tissue *in vitro* can convert 25(OH)D to 1,25(OH)$_2$D. However, recent studies suggest that the lymphoma cell itself may not be responsible for the 1α-hydroxylase activity found in lymphoma patients, but rather it is the associated macrophages that produce 1,25(OH)$_2$D [334].

Elevated 1,25(OH)$_2$D levels are observed in pregnancy and appear to increase as gestation progresses [335]. DBP is stimulated by estrogens, and both the total and free 1,25(OH)$_2$D levels are elevated during pregnancy and estrogen therapy [336, 337]. Only the free hormone is thought to be active [338]. The increased 1,25(OH)$_2$D may augment the intestinal absorption of calcium that occurs during pregnancy, which is necessary to supply calcium to the developing fetal skeleton. The metabolism of vitamin D during pregnancy has been recently reviewed [339].

VII. GENETIC DISORDERS

Examples of both over- and underproduction of the 1α-hydroxylated vitamin D sterols are not uncommon. Disorders associated with increased renal production of 1,25(OH)$_2$D include hyperparathyroidism and tumoral calcinosis. Conditions that have decreased production of 1,25(OH)$_2$D as part of their clinical picture include hypoparathyroidism and pseudohypoparathyroidism, renal failure, X-linked hypophosphatemic rickets, tumor-induced osteomalacia (TIO) or oncogenic osteomalacia, and hereditary 1α-hydroxylase deficiency [340–344].

A. 1α-Hydroxylase Deficiency (VDDR-I, PDDR)

The clinical findings of hereditary complete deficiency of renal 1α-hydroxylase were first described in 1961 by Prader *et al.* [345]. 1α-Hydroxylase deficiency is caused by mutations in the cytochrome P450 1α-hydroxylase gene (referred to as either CYP27B1 or CYP1α). This disease has been previously referred to as vitamin D dependent rickets type I (VDDR-I), pseudo vitamin D deficiency type I, and pseudo vitamin D deficiency rickets (PDDR). 1α-Hydroxylase deficiency is a rare autosomal recessive disease that is manifested at an early age [134, 340, 344]. Hypocalcemia, elevated PTH levels, increased alkaline phosphatase, and low urine calcium are found. Affected children present with hypotonia, muscle weakness, growth failure, and rickets. Tetany and convulsions may occur with severe hypocalcemia. Patients with 1α-hydroxylase deficiency have normal serum 25(OH)D concentrations and low levels of 1,25(OH)$_2$D. Circulating 1,25(OH)$_2$D does not increase after PTH infusion, consistent with defective 1α-hydroxylase activity. Very large doses of vitamin D or 25(OH)D are required for adequate treatment of 1α-hydroxylase deficiency; often 20,000 to over 100,000 IU of vitamin D daily is needed. On the other hand, modest doses of 1,25(OH)$_2$D (0.25–2 μg/day), which bypass the deficient enzyme, tend to be sufficient to restore calcium to normal and heal the rickets [340].

A number of mutations scattered throughout the entire region of the CYP27B1 gene have been identified that disrupt the enzyme activity [133–135, 344].

An R389G mutation totally abolished enzyme activity, while L343F and E189G mutations retained 2.3% and 22% of wild-type activity, respectively [346]. The two mutations that confer partial enzyme activity *in vitro* were found in patients with mild laboratory abnormalities, suggesting that such mutations contribute to the phenotypic variation observed in patients with 1α-hydroxylase deficiency.

A 1α-hydroxylase KO mouse model has also been generated [347]. These mice develop hypocalcemia, secondary hyperparathyroidism, and rickets similar to the patients with 1α-hydroxylase deficiency. The 1α-hydroxylase KO mice exhibited altered noncollagenous matrix protein expression and reduced numbers of osteoclasts in bone. The female mutant knockout mice exhibited uterine hypoplasia with absent corpora lutea and were infertile. The knockout mice also had reduced levels of CD4- and CD8-positive peripheral T lymphocytes. The 1α-hydroxylase enzyme, presumably by synthesis of 1,25(OH)$_2$D, appears to play a critical role in mineral and skeletal homeostasis as well as in female reproduction and immune function [347]. When 1α-hydroxylase KO mice were fed the rescue diet, the hypocalcemia and secondary hyperparathyroidism were corrected, and the rickets and osteomalacia were cured. The diet did not entirely correct bone growth, as femur size in the 1α-hydroxylase KO mice remained significantly smaller than that of control mice [348].

B. Hereditary 1,25-Dihydroxyvitamin D–Resistant Rickets (HVDRR)

Hereditary 1,25-dihydroxyvitamin D–resistant rickets (HVDRR), also known as vitamin D dependent rickets type II (VDDR-II) or pseudo vitamin D deficiency type II, is a rare genetic disease that arises as a result of mutations in the gene encoding the VDR [2, 173]. The clinical manifestations include early onset rickets, hypocalcemia, secondary hyperparathyroidism, and elevated 1,25(OH)$_2$D levels. The parents who are heterozygotic carriers of the mutations have no evidence of bone disease. Consanguinity is present in most cases. In many patients, total body alopecia, including eyebrows and eyelashes, accompanies the disease and provides initial evidence of the HVDRR syndrome [2, 173]. The patients with alopecia may also have skin lesions or dermal cysts [237].

The molecular basis of HVDRR is due to heterogeneous mutations (see Figure 11-9) in the VDR gene that lead to changes in critical amino acids that interfere with an essential step in the hormone action pathway [2, 173]. Mutations have been described that introduce premature stops that truncate the VDR. These mutations are the result of nonsense mutations, deletions, or caused by mutations that introduce splicing errors [237, 349–359]. Mutations have also been identified in the DBD that interfere with DNA binding [360–366]. A number of mutations have been identified in the VDR LBD (Figure 11-9) [237, 246, 248, 249, 351, 354, 367–371]. In one HVDRR case, Arg274, the contact point for the 1α-OH group of 1,25(OH)$_2$D$_3$, was mutated to leucine. The Arg274Leu mutation reduced the binding affinity for [^{3}H]1,25(OH)$_2$D$_3$ by about 1,000-fold [351]. In a second HVDRR case, His305, the contact site for the 25-OH group of 1,25(OH)$_2$D$_3$, was mutated to glutamine. The His305Gln mutation lowered the affinity for

1,25(OH)$_2$D$_3$ by 5–10-fold and caused a similar reduction in gene transactivation [368]. In a third HVDRR case, Trp286 that contacts the C-ring of 1,25(OH)$_2$D$_3$ was mutated to arginine. The Trp286Arg mutation severely reduced ligand binding and caused complete loss of transactivation [370]. Several patients were found to have mutations in the VDR LBD (Phe251Cys, Gln259Pro, or Arg391Cys) that disrupted heterodimerization with RXR [354, 367, 369]. One patient was shown to have a Glu420Lys mutation in the VDR AF-2 domain in helix H12 that eliminated coactivator binding [246]. The mutations either reduce or abolish 1,25(OH)$_2$D-mediated transactivation. Analyses of these mutations

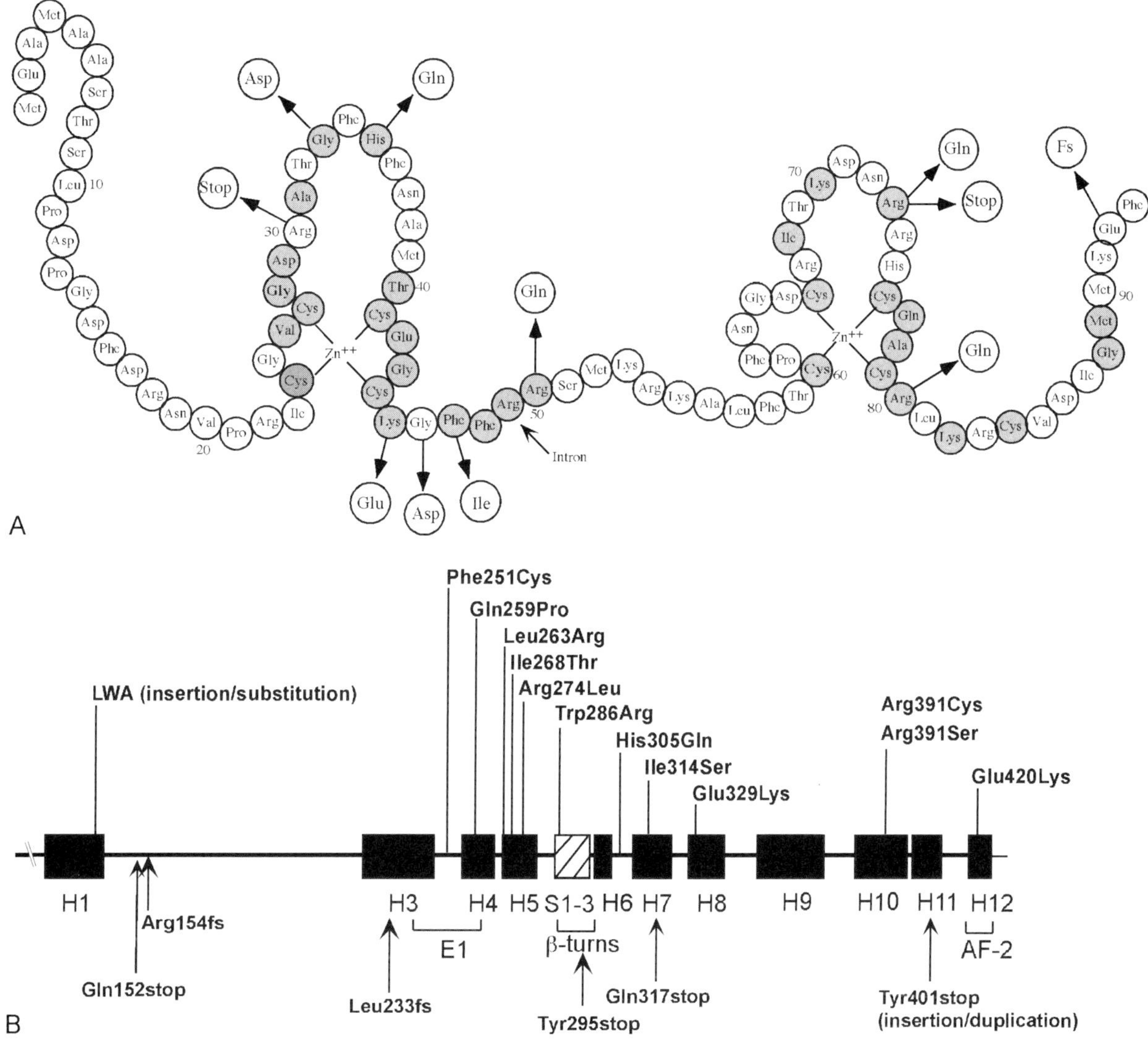

A

B

FIGURE 11-9 Mutations in the VDR causing hereditary vitamin D–resistant rickets (HVDRR). Panel A depicts the two zinc finger modules and the amino acid composition of the DBD. Conserved amino acids are depicted as shaded circles. Natural mutations are indicated by large arrows. The location of the intron separating exon 2 and exon 3, which encode the separate zinc finger modules, is indicated by an arrow labeled intron. Numbers specify amino acid number. Panel B depicts the location of the α-helices (H1–H12) of the VDR LBD. The α-helices are depicted as filled boxes, and the region containing the β-turns is drawn as a cross-hatched box. The E1 and AF-2 regions are shown above the α-helices. The location of the mutations is indicated by arrows. The Fs refers to a frameshift mutation.

have provided valuable insight into the many aspects of VDR function in gene transactivation. They also have provided essential clues as to which functions of the VDR are required for normal hair growth, as not all mutations result in alopecia. To date, all patients with DBD mutations and RXR heterodimerization mutations had alopecia, while patients with mutations that affect ligand binding or coactivator interactions did not have alopecia. Most mutations that truncate the VDR also cause alopecia. Recently, two HVDRR patients with sparse hair but without total alopecia were described with mutations that truncate the VDR [358, 359]. In one case the truncated protein was predicted to have 154 amino acids of the wild-type receptor and 23 additional amino acids and eliminated the LBD [358]. However, no studies on the truncated VDR were presented. In the second case the truncated protein was 400 amino acids in length and eliminated helix H12. The truncated VDR failed to bind ligand and coactivators but interacted with RXR and HR [359]. Cumulatively, these data suggest that the unliganded VDR with the ability to bind to DNA and heterodimerize with RXR functions to prevent alopecia. Ligand binding and coactivator interactions as well as gene transactivation appear to be dispensable functions of the VDR in regulating hair regrowth.

The successful treatment of children with HVDRR, who are unresponsive to large doses of vitamin D derivatives or oral calcium supplements, has been achieved by the chronic intravenous administration of calcium [372–374]. The intravenous calcium infusions were given nightly over a period of many months. By bypassing the intestinal defect in calcium absorption, over time they were able to correct the hypocalcemia. The treatment eventually resulted in normalization of serum calcium levels, correction of secondary hyperparathyroidism, and healing of rickets on x-ray and with apparent normal mineralization of bone. The clinical improvement can be sustained if adequate serum calcium and phosphorus concentrations are maintained. Despite healing of the rickets, the alopecia does not improve as a consequence of the treatment. Amniotic fluid cells or chorionic villus samples have been used in prenatal testing for HVDRR [375]. HVDRR has also been confirmed by assaying [³H]1,25(OH)$_2$D$_3$ binding and 1,25(OH)$_2$D$_3$-induced 24-hydroxylase activity as well as by examining restriction fragment length polymorphisms (RFLP) generated by the mutations [376].

C. VDR Polymorphisms

Osteoporosis has strong polygenic influences, and variance in bone mineral density (BMD) is estimated to be 50–80% heritable [377–379]. Several reviews [378–380] extensively discuss the role of VDR polymorphisms and the risk of osteoporosis.

VIII. 1,25(OH)$_2$D$_3$ ANALOGS WITH DECREASED CALCEMIC ACTIVITY

A. Agonists

In addition to being a major regulator of calcium metabolism, 1,25(OH)$_2$D exhibits many nonclassical actions in the body, including inhibiting cell growth, promoting cell differentiation, and suppressing the immune response (see Section X). These properties make 1,25(OH)$_2$D$_3$ an attractive candidate for treating a number of serious diseases. However, to effectively treat these diseases as well as osteoporosis, the dose of 1,25(OH)$_2$D$_3$ might well be in the range that would induce hypercalciuria, hypercalcemia, and renal stones, and therefore these unfavorable side effects limit its clinical utility. On the other hand, structural analogs of 1,25(OH)$_2$D$_3$ have been developed that exhibit a reduced calcemic response compared to 1,25(OH)$_2$D$_3$ yet retain many of the other therapeutically useful properties of the hormone, thus increasing their therapeutic potential [381–387].

Multiple analogs have been developed by the Roche company, the BioXell company, the Leo Company, the Chugai Company, Cytochroma, and others as well as by various investigators (see Figure 11-10) [8, 385, 386]. Changes that have been made in the 1,25(OH)$_2$D$_3$ molecule to create these analogs include insertion of extra carbons, oxygen, or unsaturation in the carbon side chain, 16-ene derivatives, 19-nor derivatives, 20-epi derivatives, 3-epi derivatives, and 1-hydroxymethyl derivatives. Scientists at the Roche and then BioXell company (Nutley, NJ) have synthesized the so-called Gemini analogs with two side chains emanating at carbon 20 and in collaboration with academic scientists are investigating their potential as drug candidates for the treatment of bone diseases, hypertension, acute allograft rejection, and colon cancer [387]. Both side chains of the Gemini analog are accommodated in the ligand binding pocket of the VDR and contribute to the transcriptional activity of the molecule [387]. Several novel nonsecosteroids have also been identified that exhibit activity by binding to the VDR [384]. The structures of a few of the more clinically available analogs are depicted in Figure 11-10. Many analogs have been shown to have a reduced calcemic response and/or a greater growth inhibitory potency and therefore a wider therapeutic index when compared to 1,25(OH)$_2$D$_3$. The mechanism for the differential activity displayed by the analogs is not totally clear but may be related to a number of properties: (a) decreased binding

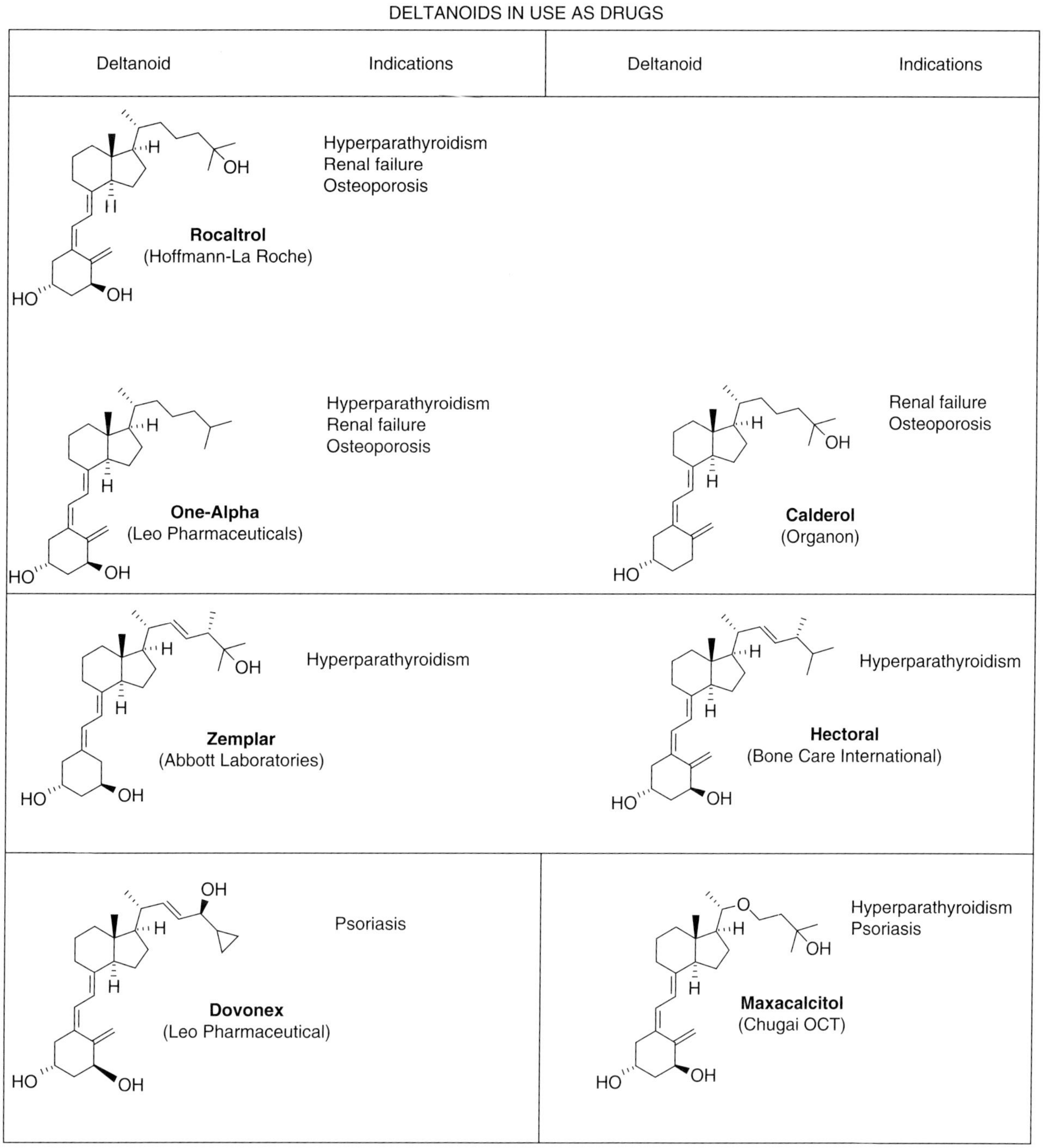

FIGURE 11-10   Structure of $1,25(OH)_2D_3$ (Rocaltrol) and six analogs in clinical use. Reproduced with permission from [386].

to DBP [388], (b) altered metabolic clearance and/or production of metabolites that retain significant biological activity [389–392], (c) increased ability to induce dimerization with RXR [393] or recruit coregulatory proteins [394], (d) increased ability to act preferentially to maintain an active conformation of the VDR within selected target tissues or upon a limited number of target genes [395, 396], and (e) ability to prevent degradation of the VDR [397].

Vitamin D analogs in general exhibit increased antiproliferative activity and decreased calcemic effects. Based on a number of *in vitro* and *in vivo* studies, these analogs are currently in use or being evaluated for use in many diseases, including osteoporosis, secondary hyperparathyroidism, psoriasis, autoimmune disorders, a variety of cancers (also see Section X), benign prostatic hyperplasia (BPH), transplant rejection, and other conditions requiring immunosuppression [5, 384, 386,

398]. We briefly discuss the analogs currently in use for osteoporosis and secondary hyperparathyroidism. The vitamin D analog 22-oxa-1,25(OH)$_2$D$_3$ (OCT or maxacalcitol), developed by Chugai Pharmaceuticals [399], has a lower affinity for VDR than 1,25(OH)$_2$D$_3$ but is 10 times more potent than 1,25(OH)$_2$D$_3$ in differentiating the myeloid leukemia cell line HL-60 and 100-fold less active in bone mobilization. OCT, like 1,25(OH)$_2$D$_3$, also suppresses PTH production and is a potent inhibitor of the renal 1α-hydroxylase activity. OCT is used in Japan in chronic renal failure patients to inhibit excessive PTH secretion [400, 401]. Studies using another analog developed by Chugai, ED-71 (2β-[3-hydroxypropyl] calcitriol), indicate that it is a potent inhibitor of bone resorption as well as a stimulator of bone formation, and this analog is being used for the treatment of osteoporosis in Japan [399].

Chronic renal failure is frequently associated with the development of secondary hyperparathyroidism due to low serum 1,25(OH)$_2$D$_3$ levels and phosphate retention. Treatment with 1,25(OH)$_2$D$_3$ must be carefully monitored, since too high a dose can result in hypercalcemia and an exacerbation of hyperphosphatemia. The consequent elevation in serum calcium and phosphate might lead to an increased risk of vascular calcification and coronary artery disease in the patients [402]. Several vitamin D analogs have been developed that appear to exhibit reduced calcemic effects while retaining the suppressive effect on parathyroid glands and therefore may represent a safer and more effective way of controlling secondary hyperparathyroidism. These analogs include OCT (maxacalcitol) and 1,25(OH)$_2$-26,27-F$_6$-D$_3$ (falecalcitriol), which are available in Japan, and 19-nor-1,25(OH)$_2$D$_2$ (paricalcitol, Zemplar) and 1α(OH)$_2$D (doxercalciferol, Hecterol), available in the United States. In a randomized, double-blind, placebo-controlled study in patients with stage 3 and 4 chronic kidney disease, doxercalciferol was shown to significantly suppress serum iPTH levels with reduced hypercalcemia, hypercalciuria, and hyperphosphatemia [403]. A multicenter, double-blind, randomized study comparing the efficacy of paricalcitol and 1,25(OH)$_2$D$_3$ in renal disease patients undergoing hemodialysis has demonstrated that paricalcitol is more effective in reducing serum PTH with fewer instances of hypercalcemia as compared to 1,25(OH)$_2$D$_3$ [404]. Teng *et al.* [405] assessed a large clinical database of about 67,000 patients undergoing hemodialysis receiving either paricalcitol or 1,25(OH)$_2$D$_3$ and demonstrated that paricalcitol was associated with a significantly lower mortality rate, especially those caused by cardiovascular events, over the 36-month follow-up, when compared to 1,25(OH)$_2$D$_3$. Vitamin D analogs exhibit significant differences in hypercalcemic properties, potentially via differential effects on intestinal and/or renal calcium handling. The low-calcemic vitamin D analogs 22-oxacalcitrol and paricalcitol were shown to be less potent in inducing intestinal calcium absorption and in stimulating the expression of TRPV6, calbindin-D$_{9K}$, and PMCA1, as compared to 1,25(OH)$_2$D$_3$ [406]. While 1,25(OH)$_2$D$_3$ upregulates VDR in the intestine, paricalcitol was shown to suppress intestinal VDR expression [407]. The novel 1,25(OH)$_2$D$_3$ analog, ZK191784, was recently shown to selectively decrease intestinal calcium absorption in both wild-type and TRPV5 knockout mice, and is speculated to be less calcemic in humans as well, secondary to exhibiting selective 1,25(OH)$_2$D$_3$ antagonist effect in the intestine and acting as an agonist in the kidneys [408].

B. Antagonists

Novel analogs with antagonistic activity, 1α,25(OH)$_2$D$_3$-26,23-lactams, have been designed based on the principle of regulation of the folding of helix 12 in the VDR, and these analogs have been shown to inhibit the differentiation of HL-60 cells induced by 1,25(OH)$_2$D$_3$ [409]. The analog (23S)-25-dehydro-1α-hydroxyvitamin D$_3$-26,23-lactone (TEI-9647) has been shown to exhibit antagonist activity by binding to the VDR and preventing the dimerization with RXR and subsequent recruitment of the co-activator SRC-1 [410]. TEI-9647 has a small amount of agonist activity, suggesting it is a partial agonist/antagonist [411]. However, its major antagonistic action may be clinically useful in selected states of hypercalcemia. Recently, hybrid analogs, which act as potent antagonists, have been designed based on the hybridization of structural motifs in the A-ring and in the side chain of the molecule [412].

IX. ACTIONS OF VITAMIN D IN CLASSICAL TARGET ORGANS TO REGULATE MINERAL HOMEOSTASIS

The classical actions of 1,25(OH)$_2$D on intestine, bone, and kidney include improved efficiency of intestinal calcium absorption, increased calcium mobilization from bone, and maintenance of adequate concentrations of calcium and phosphate in the extracellular fluid to promote normal mineralization of bone. Calcium enters the body via the intestine, and its loss is regulated by the kidneys. Calcium transport across the renal and intestinal epithelial surface is almost identical. In recent years our understanding of the molecular mechanism of calcium entry across epithelial surfaces has undergone major changes, directed by the discovery of the epithelial calcium channels in the intestine and kidney and by generating several knockout models (VDR, 1α-hydroxylase, and double knockout). These knockout models also helped to elucidate vitamin D–dependent and –independent regulatory mechanisms in maintaining calcium homeostasis. The overview of our current understanding of calcium transport is detailed in the first part of this section. Additional mechanisms by which 1,25(OH)$_2$D modulates calcium homeostasis including autoregulation of 1,25(OH)$_2$D synthesis as well as regulation of the calciotropic peptides PTH and calcitonin are discussed later in this section. The nonclassical, newly recognized actions of 1,25(OH)$_2$D on many additional target cells, apparently unrelated to maintenance of systemic mineral homeostasis, are discussed in Section X.

A.   Overview of Calcium Absorption across Renal and Intestinal Epithelia and the Role of Vitamin D

The process of calcium transport across renal and intestinal epithelia has been clarified in recent years and reviewed in detail [15, 413]. Calcium is transported across epithelia via paracellular and transcellular pathways. The paracellular transport of calcium is a passive process, regulated by tight junctions. Transcellular calcium transport, a process similar in renal and enterocyte epithelial cells, is carried out in three steps: Following entry through the calcium channels at the luminal surface, calcium translocates to the basolateral membrane via calbindins and is extruded to the interstitial space at the basolateral membrane via plasma membrane calcium pumps (see Figure 11-11). To date, two epithelial calcium channels have been described: TRPV5 (ECAC1/CAT2) and TRPV6 (ECAC2/CAT1). Their name reflects that these receptors belong to the vanilloid (V) receptor subfamily of transient receptor potential (TRP) channels. TRPV5 and TRPV6 on the luminal membrane play distinctive roles in the kidney and small intestine, respectively [414]. The plasma membrane calcium pumps consist of an ATP-dependent Ca^{2+}-ATP-ase (PMCA1b) and a Na$^+$/Ca^{2+} (NCX1) exchanger mechanism. 1,25(OH)$_2$D stimulates several steps of epithelial calcium transport by upregulating calcium channels TRPV5 and TRPV6, calcium transport proteins (calbindins), and the plasma membrane calcium pump (see Figure 11-11, thin arrows inside the cell).

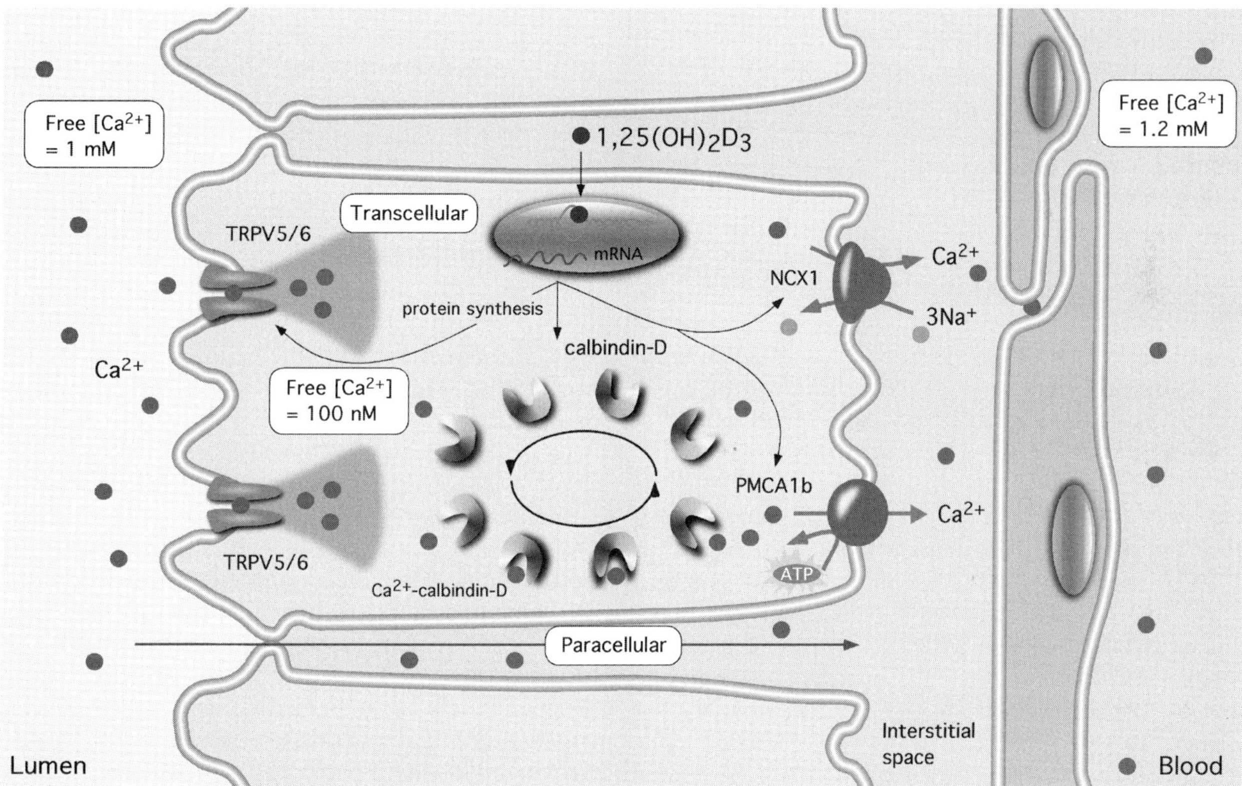

FIGURE 11-11   Mechanism of epithelial calcium transport. Paracellular calcium transport through tight junctions is represented by the paracellular arrow. Transcellular calcium transport is carried out in three steps: (1) following entry through the calcium channels TRPV5 and TRPV6, (2) calcium will diffuse across the cell bound to calbindin, and (3) be extruded at the basolateral membrane via an ATP-dependent Ca^{2+}-ATP-ase (PMCA1b) and Na$^+$/Ca^{2+} (NCX1) exchanger mechanism. 1,25(OH)$_2$D increases the expression of calcium channels, calbindins, and the extrusion systems (arrows). Reproduced with permission from [15].

Calcium absorption from the lumen is mediated by TRPV5 and 6, which share 80% sequence homology [415]. They are constitutively active calcium channels, and the most calcium-selective members of the TRP superfamily [416]. TRPV channels contain six transmembrane domains with a pore-forming region between domains 5 and 6 and large cytosolic C- and N-tails. The transmembrane domains surround a central pore in a tetrameric configuration [15]. Intracellular calcium exhibits feedback inhibition of TRPV5 and 6. Expression of TRPV5 and 6 is tightly controlled by $1,25(OH)_2D_3$, and vitamin D upregulates the expression of TRPV5 in renal cells and TRPV6 in duodenum [295, 417–419]. TRPV5 knockout mice, characterized by renal calcium wasting, show compensatory increase in intestinal calcium absorption and increased intestinal TRPV6 levels [414]. TRPV5 knockout mice exhibit a bone phenotype with reduced bone thickness [414], possibly as a result of renal calcium wasting and a direct effect of TRPV5 on bone.

Calcium translocation across the epithelial cell is mediated by the vitamin D–dependent calcium-binding proteins calbindin-D_{9K} and calbindin-D_{28K} (see Figure 11-11). Calbindin-D_{9K} expression is stimulated by $1,25(OH)_2D_3$ [420], and functional VDREs are present in calbindin promoters [421]. Calbindin-D_{28K}-knockout mice develop normally but have increased urinary calcium excretion compared to wild-type littermates, but normal serum calcium and PTH levels, suggesting compensatory mechanisms such as increased intestinal calcium absorption [422, 423].

Calcium extrusion against an electrochemical gradient at the interstitial surface is mediated by two calcium transporters: the Ca^{2+}-ATP-ase (PMCA) and Na^+/Ca^{2+} exchanger (NCX). PMCAs are calcium pumps present in all eukaryotic cells. All four known isoforms are present in the kidneys, with the highest activity in the distal convoluted tubule. PTH stimulates calcium reabsorption in the nephron via a cAMP-dependent increase of NCX1 activity; however, the exact mechanism is not clear. The effect of $1,25(OH)_2D_3$ on basolateral extrusion systems remains controversial but appears to be organ-specific. In small intestine, most studies found PMCA1b to be the $1,25(OH)_2D_3$-regulated element [295, 420]. On the other hand, in renal models, $1,25(OH)_2D_3$ seems to mediate upregulation of NCX1 but failed to show a consistent effect on PMCA1b expression [424].

B. $1,25(OH)_2D$ Actions in Intestine

I.   INTESTINAL CALCIUM ABSORPTION

Three mechanisms for intestinal calcium absorption have been described [413]: The first is the trans-cellular, vitamin D–regulated process involving the calcium channel TRPV6, calbindins, and the plasma membrane calcium pump, with a mechanism similar in the intestine and kidney (described in Section IX.A; see Figure 11-11). The second is the paracellular passive route; the rate of absorption is driven by mass action and is a function of the calcium concentration. The third is transcaltachia, the process of very rapid change in calcium flux that occurs within minutes in isolated perfused duodenum [425] (further described in Section IV on nongenomic actions of vitamin D).

Calcium and phosphate are absorbed along the length of the small intestine. Using rate of absorption and transit time in that segment, it has been estimated earlier and confirmed more recently that calcium is mostly absorbed at the lower segments of the small intestine in rats and dogs, 0–2% in the stomach, 8–15% in the duodenum, 4–25% in the jejunum, and 62–88% in the ileum [413, 426]. VDRs are present along the entire course of the small intestine, with the highest concentration proximally and the levels decreasing distally [427]. The abundance of VDRs in the duodenum is the highest of all organs reported, and at any cross-sectional level along the intestine VDR content is highest in crypts and decreases as the cells progress up the villus [428]. VDRs are also present throughout the colon [429] and are expressed in colon cancer cell lines as well as in cancer specimens removed at surgery [261, 430, 431]. Epithelial calcium channels are expressed mainly in the duodenum, less in the stomach and jejunum, but no expression was found in the ileum [432]. Even though the duodenum possesses the most active known transcellular calcium transport system including epithelial calcium channels, calbindins, and plasma membrane calcium pump, it is surprising that it is in the ileum where the most calcium uptake takes place. The reason for this seemingly inconsistent finding is currently under debate [413]. Some have suggested that calcium absorption in the ileum might be passive and vitamin D independent [433]. According to others, calcium is actively transported along all segments of the intestine in a vitamin D–dependent manner [413]. Considering that calbindin-D_{9K} [434] and plasma membrane calcium pump [435] are described in the ileum, but epithelial calcium channels are not present, it was suggested that calcium might enter these cells passively, with the transcellular calcium transport being similar to the duodenum [413]. Although the ileal calcium transport is probably slower compared to the duodenum, the larger amount of calcium absorbed in the ileum could be secondary to the significantly longer transit time in that segment of the intestine [413].

2. ACTION OF VITAMIN D METABOLITES ON CALCIUM ABSORPTION

Heaney and colleagues [322, 436, 437] have investigated the calcium absorptive response to graded doses of vitamin D_3, 25(OH)D, and 1,25(OH)$_2$D in healthy adult men. While no relationship was found between baseline absorption and serum vitamin D metabolite levels, all three vitamin D compounds significantly elevated ^{45}Ca absorption from a 300-mg calcium load given as part of a standard test meal. 1,25(OH)$_2$D was active even at the lowest dose (0.5 μg/day), and the slope was such that doubling of absorption would occur at an oral dose of approximately 3 μg/day. 25(OH)D was also active in elevating absorption and did so without raising total circulating 1,25(OH)$_2$D$_3$ levels. On the basis of the dose-response curves for 1,25(OH)$_2$D and 25(OH)D, the two compounds exhibited a molar ratio for physiological potency of approximately 100:1. The absorptive effect of vitamin D_3 was seen only at the highest dose level (1,250 μg, or 50,000 IU/day) and was apparently mediated by conversion to 25(OH)D. Analysis of the pooled 25(OH)D data from both the 25(OH)D and vitamin D_3–treated groups suggests that approximately one-eighth of circulating vitamin D–like absorptive activity under untreated conditions in winter may reside in 25(OH)D. This is a substantially larger share than has been predicted from studies of *in vitro* receptor binding [322, 436, 437]. We hypothesize that local conversion of 25(OH)D to 1,25(OH)$_2$D accounts for the activity of 25(OH)D.

3. CHANGES IN CALCIUM ABSORPTION WITH AGE

Whether intestinal calcium absorption changes with age in healthy individuals is currently under debate [438]. A recent study did not show age-related changes in intestinal calcium absorption using a stable isotope approach in rats [439]. On the other hand, osteoporosis is often associated with decreased intestinal calcium absorption with increasing age, and this phenomenon is speculated to contribute to its pathogenesis [440, 441]. Duodenal calcium channel TRPV6 expression was found to decline with age in rats, and the changes correlated with duodenal calcium transport and calbindin D levels [442]. Dietary vitamin D or 1,25(OH)$_2$D had a reversal effect upregulating TRPV6 levels [442]. In rats there is an age-related decrease in the induction of calbindin protein in response to 1,25(OH)$_2$D in the duodenum, but not in the ileum or kidney [443]. This decline in protein expression may be due to decreased translation of calbindin-D$_{9k}$ mRNA in the duodenum with age. Several earlier studies suggested that intestinal VDR declines with age in the rat [444, 445]. Duodenal biopsies of human subjects showed a slight trend toward a decrease of VDR abundance in the intestine with age [446]. However, the change in VDR abundance did not correlate with calcium absorption efficiency [447]. Estradiol may be an additional regulator of calcium absorption, since a direct effect of estradiol on intestinal calcium absorption independent of 1,25(OH)$_2$D has been demonstrated [448].

4. HYPERCALCIURIA

Idiopathic hypercalciuria, the most common form of renal stone disease, is characterized by hyperabsorption of calcium, hypercalciuria, and normal or elevated 1,25(OH)$_2$D levels [449]. Hypercalciuria in genetic hypercalciuric stone-forming (GHS) rats has been studied as a model for human intestinal calcium hyperabsorptive conditions [449, 450]. The GHS rats with normal serum 1,25(OH)$_2$D levels are hyperabsorptive and have a greater number of VDRs than normal in intestine, kidney, and bone. Post-transcriptional dysregulation of VDR with increased VDR half-life and calbindin-D$_{9K}$ accumulation in rat duodenum was described after 1,25(OH)$_2$D administration in these animals [450]. Administration of 1,25(OH)$_2$D$_3$ increases VDR gene expression significantly in GHS but not normocalciuric animals. The results suggest that GHS rats hyper-respond to modest doses of 1,25(OH)$_2$D$_3$ by upregulating VDR gene expression. This unique characteristic suggests that GHS rats may be susceptible to small fluctuations in serum 1,25(OH)$_2$D$_3$, which may pathologically amplify the actions of 1,25(OH)$_2$D$_3$ on calcium metabolism that thus contributes to the hypercalciuria and stone formation [450]. Whether this mechanism also causes some forms of human hypercalciuria and renal stones remains to be proven.

C. 1,25(OH)$_2$D Actions in Bone

Bone undergoes constant remodeling involving osteoclast-mediated bone resorption and osteoblast-mediated bone formation (see Chapter 3, Parfitt). 1,25(OH)$_2$D is a major regulator of both formation and resorption. The detailed actions of 1,25(OH)$_2$D on bone are more completely discussed in Chapter 4 (Lian) on osteoblasts; and Chapter 5 (Blair) on osteoclasts. Vitamin D is necessary for normal mineralization of the skeleton, and when it is deficient, a mineralization defect develops, causing rickets in growing children and osteomalacia in adults [451]. 1,25(OH)$_2$D actions on bone are complex, and both direct and indirect effects have been described. Direct actions on the bone are further complicated because 1,25(OH)$_2$D appears to affect several cell types, including osteoblasts, bone stromal cells, and osteoclasts. In addition, the nature of the response to $1,25(OH)_2D$ is dependent on the differentiation state of the bone cell [452].

VDRs are expressed in osteoblasts, and direct actions of $1,25(OH)_2D_3$ on these cells include modulation of cell growth and stimulation of differentiation [253, 453, 454]. $1,25(OH)_2D_3$ induces osteoblasts to progress from immature, proliferating cells to differentiated, non-dividing cells that synthesize matrix proteins and mineralize bone. Many $1,25(OH)_2D_3$-regulated gene products have been identified, including osteocalcin, Eta-1 (osteopontin), alkaline phosphatase, collagen, transforming growth factor-β (TGF-β), vascular endothelial growth factor (VEGF), matrix metalloproteinase-9 (MMP-9), integral membrane calcium-binding protein (IMCAL), receptor activator of NF-κB ligand (RANKL), Runx-2/Cbfa1 transcription factor, as well as a wide range of extracellular matrix, attachment, and signaling proteins identified by microarray approaches [455, 456].

Although $1,25(OH)_2D$ has been well known to promote bone mineralization since its discovery as an antirachitic agent many years ago [10–14], there is no definitive evidence that direct actions of $1,25(OH)_2D$ on bone are required for normal bone mineralization. The effects to promote mineralization appear to be due mainly to $1,25(OH)_2D$ actions on the intestine to enhance calcium and phosphate absorption to ensure optimal delivery of these ions to the bone-forming cells. This concept of permissive action is supported by studies showing restoration of normal bone mineralization in the absence of vitamin D action when adequate calcium and phosphorus are provided by rescue diets or intravenous infusion to vitamin D–deficient rats, VDR knockout mice, and children with HVDRR [2, 290]. In the latter situation, chronically administered IV calcium infusions, which bypass the intestinal site of $1,25(OH)_2D$ action, can achieve normalization of serum calcium levels, reverse secondary hyperparathyroidism, and promote healing of the mineralization defect of rickets despite the fact that $1,25(OH)_2D$ action at the bone is prevented because of defective VDR [2]. These studies highlight the essential role of $1,25(OH)_2D$ action on the intestine and indicate that the hormone's actions on bone are indirect in regard to the process of mineralization. In a recent study, knockouts for either VDR or 1α-hydroxylase and double knockouts for both were compared [299]. Despite normalizing serum calcium levels with rescue diet, these knockouts exhibited a subtle bone phenotype, as well as abnormalities in intestinal calcium absorption and parathyroid size. These findings indicate that calcium acts either independently or in concert with the $1,25(OH)_2D$/VDR system in the context of bone metabolism and calcium homeostasis, but normalization is not complete without vitamin D action.

There are nonetheless many consequential effects of $1,25(OH)_2D$ on bone, often in conjunction with PTH. It has been known for many years that $1,25(OH)_2D$ stimulates bone resorption [457]. This effect appears to be due to $1,25(OH)_2D$ actions to directly stimulate the differentiation of precursor cells, mononuclear phagocytes of the macrophage lineage, to fuse into mature multinucleated osteoclasts [456]. This process, osteoclastogenesis, involves a complex interaction of osteoclast precursor cells, osteoblasts, and bone stromal cells. Together with other factors, $1,25(OH)_2D$ promotes the early stages of osteoclastogenesis by direct actions on the osteoclast precursor cells. During the later stages of this differentiation process, the developing osteoclasts seem to lose their VDR, and $1,25(OH)_2D$ stimulation of differentiation becomes indirect by acting on cells in the osteoblast lineage, possibly osteoblast stromal cells, to induce osteoclast differentiating–inducing factor(s).

Osteoclastogenesis is regulated by receptor activator of NF-κB ligand (RANKL), an osteoclastogenic factor of osteoblastic origin, and its interaction with the osteoclast precursor receptor RANK. RANKL induces osteoclastogenesis from the circulating osteoclast precursor cell and promotes osteoclast activity (see Chapter 5, Blair). Osteoprotegerin (OPG) is the decoy receptor for RANKL, and osteoclastogenesis is regulated by the relative OPG/RANKL abundance. $1,25(OH)_2D$ appears to regulate both RANKL and OPG. It has been suggested that $1,25(OH)_2D$ directly stimulates osteoclastogenesis via VDR/RXR binding to a VDRE on the RANKL promoter in osteoblasts [458]. Recent studies suggested that vitamin D is able to regulate both bone formation and resorption by a location- and differentiation-specific action on osteoblasts [459].

D. $1,25(OH)_2D$ Actions in Kidney

The most important renal actions of $1,25(OH)_2D$ are probably the regulation of the 1α- and 24-hydroxylases (see Figure 11-3). $1,25(OH)_2D$ has a short and a long loop feedback to regulate its own production (see Figure 11-8). In the presence of adequate $1,25(OH)_2D$ levels, the short loop feedback is a direct renal action of $1,25(OH)_2D$ to inhibit 1α-hydroxylase and to induce 24-hydroxylase gene expression. In the presence of sufficient $1,25(OH)_2D$, the two actions coordinately drive $25(OH)D$ into $24,25(OH)_2D$, an inactivation pathway, and inhibit further $1,25(OH)_2D$ synthesis. The long loop feedback is via $1,25(OH)_2D$ inhibition of PTH gene expression, since PTH is the major stimulator of 1α-hydroxylase activity. The $1,25(OH)_2D$ action on PTH is also mediated indirectly via $1,25(OH)_2D$ regulation of serum Ca^{2+} concentration, which will rise subsequent to the calcemic actions of 1,25(OH)$_2$D on intestine and bone (see Figure 11-8).

Calcium transport across epithelia is very similar in the kidneys and intestine, regulated by vitamin D and described in Section IX.A (see Figure 11-11).

In chronic kidney disease (CKD), as the mass of functional renal tissue declines, the production of 1,25(OH)$_2$D diminishes with resultant vitamin D insufficiency, secondary hyperparathyroidism, and with coexisting hyperphosphatemia further suppressing 1,25(OH)$_2$D synthesis, all leading to the development of renal osteodystrophy [146, 285]. In addition to prevention of hyperphosphatemia with phosphate binders, vitamin D replacement has become a cornerstone of managing patients with CKD. Initially, oral 1,25(OH)$_2$D$_3$ and then intravenous 1,25(OH)$_2$D$_3$ were used. Hypercalcemia, a frequent side effect of 1,25(OH)$_2$D$_3$ treatment, raised concerns about accelerated vascular calcification and cardiovascular complications, which was a major factor in the development of less calcemic vitamin D analogs. Currently, several new, less calcemic analogs have become available worldwide (see Section VIII for details). These analogs show promising results in both animal models of uremia and in clinical trials. The mechanism by which these analogs exert their more selective actions to suppress the parathyroid glands while inducing less intestinal epithelial calcium transport, to effectively suppress secondary hyperparathyroidism without causing hypercalcemia, is still under investigation [285, 396, 460]. Less calcemic activity of these analogs was suggested to be secondary to shorter half-life, altered binding for DBP and the vitamin D receptor, and lack of intestinal VDR upregulation.

A recent large, retrospective, uncontrolled study involved the chart analysis of over 60,000 patients on hemodialysis [405]. The data showed a survival advantage of those patients receiving paricalcitol versus calcitriol. This result raised speculation about whether this finding represents a true survival advantage of paricalcitol or a disadvantage of calcitriol therapy owing to hypercalcemia with accelerated vascular calcification and death from cardiovascular complications. Although this study had limitations (retrospective analysis; lack of controls; nonrandom assignment to therapy; a difference in calcium intake between the treatment groups; better predictors of outcome, nutritional status, vascular access in the paricalcitol group), the survival advantage of the paricalcitol-treated patients is too important and warrants follow-up investigation. Clearly, more research is necessary to confirm these results in a randomized, controlled, prospective fashion, which is currently under way.

E. 1,25(OH)$_2$D Action on the Parathyroid Glands and Regulation of PTH

The parathyroid glands possess VDR and are an important component of the systemic regulation of calcium homeostasis by 1,25(OH)$_2$D [146, 159, 285]. The major effect of 1,25(OH)$_2$D in this site is to suppress PTH secretion by inhibiting mRNA and protein synthesis. The other major regulator of PTH secretion is serum Ca^{2+}, which acts via the calcium-sensing receptor (CaR) in the parathyroid glands [461]. 1,25(OH)$_2$D also regulates the expression of the CaR. VDREs have been identified in the CaR promoter, and transcription was increased by 1,25(OH)$_2$D administration in parathyroid, thyroid C-cells, and kidney in rats [462]. It has been suggested that the weight of parathyroid adenomas is related to vitamin D nutrition, indicating the importance of the feedback of vitamin D to inhibit parathyroid growth [288].

Patients with chronic renal failure develop secondary hyperparathyroidism in part due to decreased renal production of 1,25(OH)$_2$D by the diseased kidneys. In addition, inappropriately elevated PTH secretion may result from decreased levels of VDR in the parathyroid glands of uremic patients, resulting in less efficient suppression of PTH synthesis by 1,25(OH)$_2$D [285]. Studies indicate that the decrease in VDR is not uniformly distributed in parathyroid glands from chronic renal failure patients and that selected areas of low VDR content exhibit the most severe hyperplasia [463]. Suppression of elevated PTH in secondary hyperparathyroidism of chronic renal failure may be accomplished by administration of 1,25(OH)$_2$D$_3$ or its analogs as described previously.

Vitamin D deficiency is increasingly common worldwide and is often seen in patients with primary hyperparathyroidism [464]. In addition to other causes of vitamin D deficiency, accelerated catabolism of 25(OH)D because of elevated 1,25(OH)$_2$D induction of 24-hydroxylase may also contribute. Uncertainty exists about whether to treat the vitamin D deficiency associated with hyperparathyroidism or whether correction of vitamin D deficiency will worsen hypercalcemia [287]. In a recent report, vitamin D replacement of patients with primary hyperparathyroidism and vitamin D insufficiency did not worsen hypercalcemia but improved PTH levels and bone turnover [465].

F. Regulation of PTHrP and Calcitonin

1,25(OH)$_2$D$_3$ inhibits PTHrP expression in many normal tissues as well as malignant cells [466] but not all tissues (e.g., prostate) [467]. This may add to the beneficial effects of $1,25(OH)_2D_3$ in the treatment of cancer with metastases to bone and/or in humoral hypercalcemia of malignancy. The less calcemic analog of $1,25(OH)_2D_3$, EB1089, was shown to adequately suppress PTHrP production by a squamous cell cancer xenografted into mice and reverse the hypercalcemic state caused by excess PTHrP [468]. Although use of a vitamin D preparation in a hypercalcemic state might at first appear counterintuitive, the less calcemic analogs may have a role in suppressing pathologic levels of PTHrP in humoral hypercalcemia of malignancy. Calcitonin is another calciotropic peptide hormone regulated by $1,25(OH)_2D_3$ [469]. Inhibition of mRNA and protein expression has been demonstrated *in vivo* in rat and *in vitro* in medullary thyroid cancer cells. These issues are discussed in more detail in Chapter 10 (Nissenson).

X. ACTIONS OF $1,25(OH)_2D$ IN NONCLASSICAL TARGET ORGANS

In recent years a number of additional actions of $1,25(OH)_2D$ beyond regulating mineral homeostasis have been discovered in numerous nonclassical target organs. Many of these actions involve promotion of cell differentiation and inhibition of cell proliferation and appear to be unrelated to the regulation of total body calcium metabolism by $1,25(OH)_2D_3$. VDR expression and $1,25(OH)_2D_3$ effects have been demonstrated in a variety of tissues and cells including hematopoietic, immunologic, and epidermal cells, and many cancer cells. These diverse actions of $1,25(OH)_2D_3$ and its analogs have been the subject of several recent reviews [5, 166, 385, 470] and an entire book [8].

A. Vitamin D and Cancer

1. EPIDEMIOLOGY

A number of studies have found a protective relationship between vitamin D status and decreased risk of cancer. Most of these epidemiological studies have not directly measured the circulating vitamin D concentrations or dietary intake on cancer incidence or mortality. Nevertheless, higher rates of cancer mortality have been observed in regions with less UV-B radiation, among African Americans, and among overweight people, each associated with lower levels of circulating 25(OH)D, suggesting the beneficial effect of vitamin D on cancer mortality [471]. Garland *et al.* [472] analyzed 63 observational studies of vitamin D status in relation to the risk of colon, breast, prostate, and ovarian cancers and found that the preponderance of data indicates that

vitamin D insufficiency raises cancer risk. Several studies have demonstrated an inverse correlation between sunlight exposure and the incidence of colon and prostate cancers [473–476]. Epidemiological studies assessing the inverse association between dietary vitamin D intake and incidence of colon or prostate cancer are still considered inconclusive [473]. Studies correlating the measured plasma levels of vitamin D metabolites with cancer incidence have shown an inverse relationship between plasma 25(OH)D levels and colorectal cancer, whereas in the case of prostate cancer, the results have been variable [472, 473]. Several studies have also examined the association between polymorphisms in the VDR gene and the risk for colon and prostate cancers, and the results have also been variable [472, 473, 477].

2. $1,25(OH)_2D_3$ EFFECTS ON CELL GROWTH AND DIFFERENTIATION

VDRs are expressed in many normal and malignant cell types, indicating a wide array of previously unrecognized potential targets for $1,25(OH)_2D$ action [5]. In many of these normal and malignant cells, $1,25(OH)_2D$ and its analogs exert pleiotropic actions to inhibit cell proliferation and promote differentiation [166, 385, 431, 477–482]. A number of important mechanisms have been implicated in $1,25(OH)_2D_3$-mediated growth inhibition. A primary mechanism appears to be the induction of cell cycle arrest in the G_1/G_0 phase, due to an increase in the expression of cyclin-dependent kinase inhibitors such as $p21^{Waf/Cip1}$ and $p27^{Kip}$, inhibition of cyclin-dependent kinase activity, and regulation of the phosphorylation status of the retinoblastoma protein (pRb) [483–487]. As the loss of the expression of cell cycle regulators has been associated with a more aggressive cancer phenotype and decreased prognosis and poorer survival, these observations suggest that $1,25(OH)_2D_3$ may be a suitable therapy to inhibit cancer progression. In addition, $1,25(OH)_2D_3$ induces apoptosis in some cancer cells and down-regulates antiapoptotic genes like *bcl-2* [478, 488, 489]. Other mechanisms include the stimulation of differentiation, modulation of growth factor actions, and regulation of the expression and function of oncogenes and tumor suppressor genes [431, 483, 490]. The inhibition of invasion and metastasis of tumor cells as well as the suppression of angiogenesis have also been shown to contribute to the antitumor effects of $1,25(OH)_2D_3$ [477, 478]. Recent studies in prostate cancer have revealed anti-inflammatory effects of $1,25(OH)_2D_3$ through the inhibition of prostaglandin synthesis and actions as well as the inactivation of stress-induced kinase signaling and downstream production of inflammatory cytokines, suggesting a role for vitamin D in cancer chemoprevention [491, 492], since inflammation has clearly been associated with carcinogenesis and cancer progression [493, 494].

3. VITAMIN D METABOLIZING ENZYMES AS REGULATORS OF THE ANTICANCER EFFECTS OF $1,25(OH)_2D_3$

a. Role of 1-Hydroxylase

The presence of extrarenal 1α-hydroxylase has been demonstrated in several tissues, which contributes to the local production of $1,25(OH)_2D_3$ within the tissue. In some cancers such as prostate and parathyroid carcinomas, the expression levels and activity of 1α-hydroxylase in the cancer cells are lower than in the normal cells [495–497]. However, in other malignant cells such as colon cancer cells, 1α-hydroxylase levels are elevated at least before the cancers progress to an advanced stage [498]. In prostate cancer cells the decrease in 1α-hydroxylase appears to be due to decreased 1α-hydroxylase promoter activity in these cells [495]. The reduction in 1α-hydroxylase may endow the malignant cells with an intrinsic growth advantage because of the resultant decrease in the local production of the growth inhibitory agent $1,25(OH)_2D_3$. In addition, local deficiency of $1,25(OH)_2D_3$ may allow cellular de-differentiation and invasion, hallmarks of malignancy that may represent an important mechanism that contributes to cancer development and/or progression. These observations also suggest that the administration of the precursor $25(OH)_3D$ might be an effective cancer chemopreventive strategy while 1α-hydroxylase is initially still high within the tissue [145].

b. Role of 24-Hydroxylase

$1,25(OH)_2D_3$ induces the expression of 24-hydroxylase in many target cells, including cancer cells, which catalyzes the initial step in the conversion of the active molecule $1,25(OH)_2D_3$ or the precursor molecule $25(OH)D$ into less active metabolites. Therefore, the degree of growth inhibitory response elicited by $1,25(OH)_2D_3$ is inversely proportional to the 24-hydroxylase activity in these cells. For example, among human PCa cell lines, the magnitude of $1,25(OH)_2D_3$-mediated growth inhibition is inversely proportional to 24-hydroxylase activity in these cells [499, 500]. Prostate cells that have high 24-hydroxylase expression exhibit decreased sensitivity to $1,25(OH)_2D_3$, resulting in negligible to a very low degree of growth inhibition following $1,25(OH)_2D_3$ treatment. However, co-addition of inhibitors of P450 hydroxylases including 24-hydroxylase, such as liarozole and ketoconazole or genistein (a soy isoflavone that directly inhibits vitamin D–24-hydroxylase enzyme activity), renders the cells more responsive to $1,25(OH)_2D_3$ [501–503]. These observations suggest that co-administration of $1,25(OH)_2D_3$ with inhibitors of 24-hydroxylase such as ketoconazole, liarozole, or genistein may enhance

its antitumor effects, and combination therapy will be a useful strategy in cancer treatment. The combination approach may also allow the use of $1,25(OH)_2D_3$ at lower concentrations and thereby reduce its hypercalcemic side effects. Alternatively $1,25(OH)_2D_3$ analogs that resist 24-hydroxylation may well be more biologically active in these settings (see Sections III.C.4 and VIII).

4. ROLE OF VITAMIN D IN CANCER PREVENTION OR THERAPY

Because of its actions to inhibit cell proliferation and promote differentiation, $1,25(OH)_2D$ has been considered a good candidate for possible "chemoprevention" or "differentiation" therapy in a number of malignant cell types that possess VDR [166, 385].

a. Colon Cancer

VDRs are present in the colon [429], in colon cancer cell lines, as well as in surgically removed colon cancers [430, 431]. The possibility that calcium and/or vitamin D may be active in decreasing colon cancer has been examined by several groups, and an adequate intake of calcium (in the range of 1,800 mg/day) and vitamin D (800–1,000 IU/day) has been found in some studies to have a protective effect against the development of colon cancer [504, 505]. Eisman and coworkers in an earlier study showed that $1,25(OH)_2D_3$ administration could inhibit the growth of colon cancer xenografts in nude mice [506]. Studies in a number of colon cancer models have demonstrated the tumor inhibitory and prodifferentiation effects of $1,25(OH)_2D$ or its analogs both *in vitro* and *in vivo* [431, 507–509]. A recent study in the APC(min) mouse model demonstrated that both vitamin D and calcium individually exert inhibitory effects on the development of polyps and exhibit a synergistic effect when used together [510]. VDR expression correlates with colon cancer prognosis: High VDR levels are associated with favorable prognosis, and VDR expression is down-regulated in high-grade tumors. An inverse correlation was recently described between the expression of VDR and SNAIL, a zinc finger transcription factor, in human colon cancer [511]. SNAIL down-regulates VDR expression transcriptionally and post-transcriptionally, resulting in a loss of vitamin D–mediated antiproliferating effect *in vitro* and *in vivo*.

b. Breast Cancer

VDRs are present in normal breast and breast cancer cell lines and in many human cancer specimens [478, 486]. Adequate calcium and vitamin D intake has been shown to enhance survival rates among breast cancer patients in some studies [472]. $1,25(OH)_2D_3$ suppresses the growth of human breast cancer cell lines in culture and also *in vivo* in xenografts of human breast cancer cells in nude mice and nitrosomethylurea (NMU), induced breast cancer in rats [478]. A number of investigators have shown that

1,25(OH)$_2$D or its analogs exhibit antiproliferative effects in breast cancer cells through a number of different mechanisms [478, 512]. 1,25(OH)$_2$D$_3$ has also been shown to decrease estrogen receptor–alpha levels in breast cancer cells and inhibit estrogen stimulation of breast cancer cell growth [513–515]. In addition to its antiproliferative effects, 1,25(OH)$_2$D stimulates apoptosis in some breast cancer cells [516] and may enhance the responsiveness of breast cancer cells to conventional cytotoxic agents [517]. Studies in VDR-null mice reveal that vitamin D participates in the negative growth control of normal mammary gland and that the disruption of VDR signaling results in abnormal morphology of the mammary ducts, an increase in preneoplastic lesions, and accelerated mammary tumor development, suggesting that vitamin D compounds may play a beneficial role in the chemoprevention of breast cancer [518]. The analog EB1089 inhibits proliferation of MCF-7 human breast cancer cells *in vitro* and exhibits more potency than 1,25(OH)$_2$D$_3$ in inhibiting tumor growth induced by the carcinogen NMU and therefore has therapeutic potential [478]. However, in a small phase I study of EB1089 in breast cancer and colon cancer patients, no clear antitumor effects were seen, although some patients exhibited disease stabilization over a few months [519].

c. Prostate Cancer

In a prediagnostic study with stored sera, low 1,25(OH)$_2$D blood levels were found to be an important predictor for palpable and anaplastic tumors in men over 57 years of age but not for incidentally discovered or well-differentiated tumors [520]. VDRs are present in prostate cancer cell lines [500, 521] and in normal prostate [522] and 1,25(OH)$_2$D$_3$ inhibits the growth of all these cell types in culture [477]. 1,25(OH)$_2$D$_3$ and vitamin D analogs exert antiproliferative effects in multiple prostate cancer models, and several mechanisms mediate these effects [477, 479, 481–483, 523]. The induction of apoptosis may also play some role in the growth-inhibitory activity of 1,25(OH)$_2$D$_3$ in some prostate cancer cells [488]. One of the recently discovered molecular mechanisms mediating 1,25(OH)$_2$D$_3$ effects in prostate cells is the inhibition of the synthesis and actions of growth-stimulatory prostaglandins, through multiple 1,25(OH)$_2$D$_3$ actions, including a decrease in the expression of the proinflammatory molecule, cyclooxygenase-2 (COX-2) [491]. Moreover, 1,25(OH)$_2$D$_3$ has been shown to cause synergistic inhibition of prostate cell growth when combined with nonsteroidal anti-inflammatory drugs (NSAIDs), suggesting that a combination of vitamin D or its analogs with NSAIDs may be useful in prostate cancer therapy [491]. 1,25(OH)$_2$D$_3$ also induces the expression of MAP kinase phosphatase-5 in primary prostate cells, leading to the inactivation of the stress kinase p38 and

inhibition of interleukin-6 production [492]. These new mechanisms of action support an anti-inflammatory role for 1,25(OH)$_2$D$_3$ in prostate cancer and suggest that it may have beneficial prostate cancer chemopreventive effects. The efficacy of 1,25(OH)$_2$D$_3$ as a chemopreventive agent was recently evaluated using Nkx3.1; Pten mutant mice, which recapitulate stages of prostate carcinogenesis from prostate intraepithelial neoplasia (PIN) lesions to high-grade PIN to adenocarcinoma [524]. The findings revealed that 1,25(OH)$_2$D$_3$ is beneficial in early-stage preventing the development of high-grade PIN rather than advanced disease, providing support for its use in the chemoprevention of prostate cancer. Several vitamin D analogs exhibit greater antiproliferative potency than 1,25(OH)$_2$D$_3$, raising the possibility of the therapeutic potential of these drugs in the treatment of prostate cancer [525]. Clinical trials have begun to address the utility of 1,25(OH)$_2$D$_3$ in treating prostate cancer patients [526, 527]. Studies by Beer, Trump, and coworkers demonstrated that intermittent administration of very high doses of 1,25(OH)$_2$D$_3$ are well tolerated by prostate cancer patients without significant toxicity or renal calculi [528, 529]. In combination with the chemotherapy drug docetaxel, 1,25(OH)$_2$D$_3$ given weekly at 45 mcg produced favorable effects on the time to disease progression and survival [530, 531]. An unanticipated benefit of the combination was decreased side effects of docetaxel [530]. A phase III placebo-controlled randomized trial is currently under way testing the safety and efficacy of this combination in prostate cancer patients.

d. Hematological Malignancies

In addition to promoting osteoclastogenesis from macrophage precursors as described previously in the section on bone (Section IX.B), 1,25(OH)$_2$D$_3$ has been shown to stimulate a variety of immature hematopoietic myeloid cells to differentiate into mature cells, including M-1 mouse myeloid leukemic cells, HL-60 human promyelocytic leukemia cells, U-937 human monocytic cells, and peripheral human monocytes [480]. Expression of VDR is found in various normal hematopoietic cells as well as leukemic cells. O'Kelly *et al.* [532] investigated the possible role of VDR in hematopoiesis using VDR knockout mice and found that although there was normal myelopoiesis in these mice, the T lymphocyte responses were abnormal. 1,25(OH)$_2$D$_3$ and its analogs induce differentiation and inhibit the proliferation of several acute myeloid leukemia cell lines [480]. In HL-60 cells, the 1,25(OH)$_2$D$_3$-induced response is the stimulation of terminal differentiation into cells with characteristics of macrophages, and the response appears to be mediated by inhibition of the expression of the c-*myc* oncogene [533]. Liu *et al.* [484] showed that 1,25(OH)$_2$D stimulates myeloid leukemic cell lines to terminally differentiate into monocytes/macrophages. Using the myelomonocytic U937 cell line, they showed that 1,25(OH)$_2$D induces the expression of the Cdk inhibitor p21WAF1 (CIP1), which caused the cells to terminally differentiate. Other molecular mechanisms mediating the antiproliferative and differentiation-inducing effects of vitamin D compounds in myeloid leukemic cells include the upregulation of the homeobox genes such as HoxA 10 and HoxB 4, down-regulation of bcl$_2$, and the modulation of the intracellular kinase pathways p38, MAPK, ERK, and PI3–K [480]. Several *in vitro* studies have also reported the differentiation-promoting effects of several vitamin D analogs in leukemic cells. The effects on leukemic cells *in vitro* as well as the prolongation of survival time in mice inoculated with myeloid leukemia cells in an early study [534] have led to the consideration of using 1,25(OH)$_2$D$_3$ or its analogs therapeutically in human leukemia as a "differentiation" therapy [480]. Even though 1,25(OH)$_2$D$_3$ and its analogs have shown promise in laboratory studies, clinical trials of vitamin D compounds in leukemia and myelodysplastic syndrome have so far yielded only mediocre results [535]. *In vitro* and animal studies suggest that a number of agents including dexamethasone, retinoids, histone deacetylase inhibitors, and chemotherapy drugs may synergize with vitamin D analogs, and clinical trials testing these combinations in leukemia patients appear to be warranted.

e. Other Malignancies

Vitamin D compounds have been shown to demonstrate anticancer effects in several other malignancies as well. The growth inhibitory effect of 1,25(OH)$_2$D$_3$ on tumor cells was first demonstrated in human melanoma cells [536]. Since then a large body of evidence has accumulated, indicating the antiproliferative and pro-differentiation effects of 1,25(OH)$_2$D in melanocytes as well as malignant melanoma cells and melanoma xenografts [537, 538]. Genetic variants in VDR may alter the risk of cutaneous melanoma [539]. In a phase II trial of EB1089 (Seocalcitol) in patients with inoperable hepatic cancer, some reduction of bulky tumor mass was achieved. Of 33 evaluable patients, 2 had a complete response; 12, stable disease; and 19, progressive disease [540]. Recent cell culture or animal model research on 1,25(OH)$_2$D$_3$ and its analogs also provides evidence for a potential beneficial role of these compounds in ovarian [541], pancreatic [542, 543], and lung cancers [541, 543–545].

B. Immune System: 1,25(OH)$_2$D Actions on Immunosuppression and Cytokine Production

In addition to 1,25(OH)$_2$D$_3$ effects on myeloid cells described previously and on monocytic/macrophage precursors that are differentiated into osteoclasts (described in Section IX.B), 1,25(OH)$_2$D$_3$ has many important immunomodulatory effects [5, 381, 382, 470, 546–553]. VDR is present in most cell types of the immune system, particularly in antigen-presenting cells (APCs) such as monocyte/macrophages and dendritic cells. Circulating resting T and B cells do not express VDR, but when blast-transformed or mitogen-activated, these cells do express VDR and respond to 1,25(OH)$_2$D$_3$ [554]. Many studies report the beneficial effects of 1,25(OH)$_2$D$_3$ and its analogs in autoimmune diseases [5, 470, 549]. 1,25(OH)$_2$D$_3$ inhibits antigen-induced T-cell proliferation and cytokine production as well as selectively suppresses the development of helper T cell subset type 1 (Th1) by inhibiting the production of cytokines such as IL-2, IL-12, and interferon-gamma (IFN-γ) [470, 555]. 1,25(OH)$_2$D$_3$ has also been shown to enhance the development of Th2 cells [556], and this action might contribute to its beneficial effect in the treatment of autoimmune diseases and allograft rejection [470]. In addition, the modulation of APC function by vitamin D may also play a role in the development of T cell responses. 1,25(OH)$_2$D$_3$ and its analogs have been shown to inhibit the differentiation and maturation of dendritic cells, APCs that play a key role in the induction of T-cell–mediated immune responses. *In vivo* studies in allograft rejection models demonstrate that 1,25(OH)$_2$D$_3$ and its analogs induce dendritic cells with tolerogenic properties as well as CD4$^+$CD25$^+$ regulatory T cells that are able to mediate transplantation tolerance [557]. Activated macrophages synthesize 1,25(OH)$_2$D$_3$ as they express 1α-hydroxylase (described in Section VI.C). The regulation of macrophage 1α-hydroxylase differs from that of the renal enzyme and is mediated by immune signals such as IFN-γ that stimulate the enzyme [558]. The macrophage enzyme is also not suppressed by the end-product 1,25(OH)$_2$D$_3$, which might explain the hypercalcemia associated with conditions of macrophage overactivation such as tuberculosis and sarcoidosis [138]. 1,25(OH)$_2$D$_3$ also regulates the secretion of PGE$_2$ and granulocyte-macrophage colony-stimulating factor by monocyte-derived cells [470].

In various animal models, 1,25(OH)$_2$D$_3$ reduces immune responses when administered prior to induction or early in the disease process [549]. 1,25(OH)$_2$D$_3$ and its analogs inhibit the development of several autoimmune diseases such as experimental allergic encephalomyelitis, multiple sclerosis, systemic lupus erythematosis, thyroiditis, collagen-induced arthritis, inflammatory bowel disease, and type I diabetes [5, 470]. 1,25(OH)$_2$D$_3$ and its analogs have also been tested alone or in combination with other immunosuppressive agents such as cyclosporine in many experimental models for their ability to suppress transplant rejection, and the results suggest that they potentially can be used for the prevention of transplant rejection [546].

1,25(OH)$_2$D immunosuppressive activity has been well studied in the autoimmune model of diabetes that spontaneously develops in nonobese diabetic (NOD) mice [547, 550]. Type I diabetes can be prevented without generalized immunosuppression by using 1,25(OH)$_2$D$_3$ and less calcemic analogs of 1,25(OH)$_2$D when treatment is started early, i.e., before the autoimmune attack, reflected by insulitis. In fact, administration of 1,25(OH)$_2$D$_3$ before the onset of insulitis has been shown to prevent the progression of diabetes in NOD mice [547, 550]. Even if the autoimmune disease is already active, treatment with 1,25(OH)$_2$D analogs can prevent clinical diabetes when this therapy is combined with a short induction course of an immunosuppressant such as cyclosporin A.

Vitamin D deficiency is known to be associated with tuberculosis [548], and sunlight exposure plays a beneficial role in the treatment of this disease [553]. A recent study by Liu *et al.* [559] provided a mechanism for this beneficial effect by demonstrating that 1,25(OH)$_2$D$_3$ production and action in human macrophages contributes to innate immunity and microbicidal effects in tuberculosis. The study showed that the activation of the Toll-like receptors of human macrophages increases the expression of VDR and 1α-hydroxylase genes in these cells, leading to the induction of the antimicrobial peptide cathelicidin and killing of intracellular *Mycobacterium tuberculosis*. Recent studies also suggest an association between vitamin D status and disease progression following human immunodeficiency virus (HIV) infection and demonstrate a positive correlation between 1,25(OH)$_2$D levels and CD4$^+$ cell counts [552]. Additional studies are needed to evaluate the potential beneficial role of vitamin D supplementation to HIV-infected patients.

C. 1,25(OH)$_2$D Effects on Skin: Use in the Treatment of Psoriasis

Skin, in addition to being the site of initiation of vitamin D synthesis, is also a 1,25(OH)$_2$D target organ [25]. Human dermal fibroblasts and keratinocytes possess VDR and are 1,25(OH)$_2$D$_3$-responsive [560]. For this reason, cultured dermal fibroblasts are frequently used to study HVDRR [2, 561, 562]. Keratinocytes are capable of the synthesis of vitamin D from endogenous sources of 7-dehydrocholesterol as well as the metabolic activation of vitamin D via the 25-hydroxylase and 1α-hydroxylase steps to 1,25(OH)$_2$D$_3$ and thus are capable

of the entire vitamin D synthetic pathway [25]. However, when the renal production of 1,25(OH)$_2$D$_3$ is normal, circulating levels of 1,25(OH)$_2$D$_3$ limit the contribution from epidermal production through the induction of 24-hydroxylase within the keratinocytes, which catabolizes the endogenously produced 1,25(OH)$_2$D$_3$ [563].

1,25(OH)$_2$D$_3$ inhibits the proliferation and promotes the terminal differentiation of keratinocytes, including the stimulation of involucrin, cornified envelope development, and transglutaminase I, the enzyme that cross-links the components of the cornified envelope [25]. Other cells within the skin also contain VDR and appear to be 1,25(OH)$_2$D$_3$ targets as well. Melanoma cells express VDR, and 1,25(OH)$_2$D$_3$ induces differentiation and inhibits cell proliferation [536]. 1,25(OH)$_2$D$_3$ is likely to be an autocrine or paracrine regulator of epidermal differentiation, since it is produced by the keratinocytes. 1,25(OH)$_2$D$_3$ and its analogs have also been shown to protect human skin from UV radiation–induced damage and apoptosis of skin cells via response pathways involving nitric oxide as well as increased p53 expression favoring DNA repair over apoptosis [314, 564]. The protective effect of vitamin D compounds against DNA photo damage has also been demonstrated *in vivo* in hairless SKh:HR1 mice [564].

Psoriasis is a hyperproliferative disorder of the epidermis, which is also characterized by abnormal keratinocyte differentiation and infiltration of immune cells into the epidermis and dermis. Psoriasis responds to treatment with vitamin D preparations applied topically or administered systemically [565]. The antipsoriatic effect may be due to the antiproliferative and prodifferentiation actions of 1,25(OH)$_2$D$_3$ but may also involve immunosuppressive and anti-inflammatory properties of the hormone [546, 565, 566]. Interestingly, in keratinocytes, the VDR levels are down-regulated within a few hours after UV-B irradiation [26]. These results strongly suggest the existence of a feedback mechanism in that UV-B initiates vitamin D synthesis in keratinocytes and at the same time limits VDR abundance. The findings provide a potential explanation for the reported lack of any additive effect between 1,25(OH)$_2$D and UV-B phototherapy in the treatment of psoriasis. Newer vitamin D analogs with reduced calcemic activity are being developed to improve the therapeutic potential of treating psoriasis (see Figure 11-10). Clinical trials using oral 1α-hydroxyvitamin D$_3$ and topical calcipotriol (marketed as Dovonex) in psoriasis patients have yielded promising results [546, 565, 566]. Clinical studies using combinations of topical calcitriol or calcipotriol with potent topical steroids such as betamethasone demonstrate an increased efficacy, a more rapid onset of action, and better tolerance of the combination regimen as compared to the individual treatments [567].

D. 1,25(OH)$_2$D Actions in the Nervous System: NGF, Alzheimer's Disease, and Aging

The first evidence for the presence of VDR in brain came from autoradiographic studies using [^{3}H]1,25(OH)$_2$D$_3$ to localize the receptor [568]. In rodents [^{3}H]1,25(OH)$_2$D$_3$-binding sites were located throughout the brain from basal forebrain to midbrain and hindbrain [569]. Calbindin-D$_{28k}$ in the brain is not vitamin D dependent; however, 1,25(OH)$_2$D$_3$ was found to stimulate choline acetyltransferase activity in the bed nucleus of the stria terminalis [570]. Furthermore, nerve growth factor (NGF) mRNA levels were stimulated by 1,25(OH)$_2$D$_3$ in mouse L929 fibroblasts, an *in vitro* model of nerve cell function [571, 572], and other studies demonstrated that 1,25(OH)$_2$D$_3$ induced NGF mRNA levels in hippocampus and cortex [573]. In the intact organism, 1,25(OH)$_2$D$_3$ treatment results in improved memory performance of young adult rats in the Morris water-maze test [574]. Interestingly, VDR mRNA expression is decreased in the hippocampus of patients with Alzheimer's disease (AD) [575]. A possible role of decreased 1,25(OH)$_2$D or VDR with aging leading to decreased NGF production in the brain has raised conjecture about a possible role of decreased vitamin D action in the neurodegeneration found with aging or AD [569]. VDR levels have been thought to possibly decrease with aging in intestine [446], and although a connection to the brain is highly speculative at this time, some role for 1,25(OH)$_2$D in the central nervous system seems clear. Long-term experiments in aging rats suggest that chronic treatment with 1,25(OH)$_2$D$_3$ increases neuronal density in the middle regions of hippocampus in these animals, suggesting that 1,25(OH)$_2$D$_3$ reduces biomarkers of aging [576].

AD patients are susceptible to hypovitaminosis D due to their age and being confined to a hospital or a nursing facility. A study of 46 ambulatory elderly women with AD showed that 26% had decreased 25(OH)D (5–10 ng/mL) and 54% had osteomalacic levels (<5 ng/mL) [577]. Those with decreased vitamin D had increased PTH and decreased BMD. Many AD patients were sunlight-deprived and consumed less than 100 IU of vitamin D per day. Vitamin D deficiency due to sunlight deprivation and malnutrition, together with compensatory hyperparathyroidism, contributes significantly to decreased BMD and increased risk of falls and hip fractures in patients with AD [577]. In a recent randomized, prospective study Sato *et al.* evaluated the effect of sunlight exposure in elderly women with AD and concluded that sunlight exposure increased the BMD in these patients by increasing serum 25(OH)D concentrations, leading to a reduction in the incidence of nonvertebral fractures [578].

E. 1,25(OH)$_2$D Actions on the Reproductive System

The role of 1,25(OH)$_2$D in reproduction has been examined, and the hormone appears to play a role in normal ovulation, fetal and neonatal bone development, milk production, and maintenance of normocalcemia and mineral homeostasis in the neonate [339, 579]. Extrarenal synthesis of 1,25(OH)$_2$D takes place in the placenta, which also expresses VDR. In addition, 1,25(OH)$_2$D stimulates human placental lactogen (hPL) expression from trophoblast cells, and a VDRE has been demonstrated in the 5′ upstream region of the hPL gene supporting a role for 1,25(OH)D in placental function [580]. VDR and vitamin D–dependent Ca^{2+} binding protein are found in a number of additional tissues including testis, uterus, pancreas, pituitary, thyroid, gonads, and muscle including the heart [581–583], but the functional role of 1,25(OH)$_2$D in these sites is unclear and will require further investigation.

In the VDR-ablated mice, uterine hypoplasia and ovarian abnormalities were detected in females and testicular defects and sperm abnormalities in males [172]. However, many of these defects were improved after calcium nutrition was normalized by the rescue diet [292]. However, in females the estradiol levels were still somewhat reduced and gonadotropin levels somewhat elevated, suggesting a residual defect in estrogen synthesis unrelated to calcium. These parameters were normalized by estradiol administration. Female mice that developed 1,25(OH)$_2$D$_3$ deficiency due to targeted ablation of the 1α-hydroxylase gene also exhibit infertility, uterine hypoplasia, ovarian abnormalities, and absent corpora lutea [347]. Since the expression of the aromatase gene is regulated by vitamin D [293], an effect on estradiol synthesis may affect fertility in VDR-ablated mice and HVDRR subjects [292]. Both calcium and 1,25(OH)$_2$D$_3$ play a role in oocyte maturation and follicular development. Abnormalities in calcium homeostasis may in part be responsible for the arrested follicular development in women with polycystic ovarian syndrome (PCOS), and vitamin D and calcium therapy may help to normalize the menstrual abnormalities seen in women with PCOS [584]. Studies in VDR-null mice also reveal that 1,25(OH)$_2$D$_3$ and VDR play a role in ductal elongation and branching morphogenesis during pubertal development of the mammary gland [512]. 1,25(OH)$_2$D$_3$ appears to play a role in the differentiation of endometrial cells to decidual cells, a crucial step in pregnancy, and is necessary to maintain calcium and bone metabolism in the fetus and the neonates in humans [579].

Müllerian inhibiting substance (MIS), also known as anti-müllerian hormone (AMH), plays an important role during fetal sexual development in males, causing the regression of the müllerian ducts by inducing apoptosis [585]. MIS also has postnatal actions to inhibit steroidogenesis [586] and may be involved in the pathogenesis of PCOS [587]. MIS has been shown also to inhibit ovarian and cervical cancers, both of which are tumors of müllerian origin [588]. Recently, MIS was demonstrated to have antiproliferative and proapoptotic effects in prostate, breast, and uterine cancers [589–592]. Recent studies from our lab reveal that the expression of the MIS gene is increased by $1,25(OH)_2D_3$ in prostate cancer cells [593]. The promoter sequence of the human MIS gene contains a positive VDRE and is responsive to $1,25(OH)_2D_3$. Because of its potential therapeutic benefits, MIS is under active development as a cancer therapy. Induction of MIS expression may play an important role in the anticancer actions of $1,25(OH)_2D_3$.

Recently, data were reported demonstrating the benefit of a $1,25(OH)_2D_3$ analog in benign prostatic hyperplasia (BPH) and indicating that the human urinary bladder may also be a vitamin D target [398, 594, 595].

F. $1,25(OH)_2D$ Effects on Blood Pressure

Hypertension contributes significantly to the morbidity and mortality associated with cardiovascular disease, stroke, and end-stage renal disease. The renin-angiotensin system and the adrenal steroid hormone aldosterone regulate mammalian blood pressure and salt/water homeostasis. Renin, a protease secreted by juxtaglomerular cells in the kidney, cleaves liver-derived angiotensinogen to angiotensin I, which is further acted upon by the angiotensin converting enzyme (ACE) to produce angiotensin II. Angiotensin II modulates blood pressure by being a potent vasoconstrictor as well as by stimulating the adrenal synthesis of aldosterone, which increases renal sodium retention. These two actions effectively increase blood volume leading to hypertension. Recent research indicates that $1,25(OH)_2D_3$ is a negative regulator of the renin-angiotensin system and may play an important role in the control of blood pressure. Many studies have shown an inverse association between serum $1,25(OH)_2D_3$ levels and blood pressure in normotensive and hypertensive subjects [596]. Circulating $1,25(OH)_2D$ levels have been shown to be inversely related to plasma renin activity in patients with essential hypertension [597], suggesting that $1,25(OH)_2D_3$ may be a negative regulator of renin. Recent studies in VDR-null mice revealed substantial elevations in renin mRNA and protein levels in the kidney [302]. The plasma levels of angiotensin II, which is a downstream product of renin, and the plasma and urinary levels of aldosterone are also markedly increased in these mice [302, 303]. As a consequence of the overstimulation of

the renin-angiotensin system, the VDR-null mice develop high blood pressure, increased water intake, and cardiac hypertrophy with accompanying compensatory increase in the expression of atrial natriuretic peptide [302]. These abnormalities can be corrected by treating the VDR-null mice with captopril, an ACE inhibitor, or an angiotensin II AT1 receptor antagonist confirming that the underlying defect is the overstimulation of the renin-angiotensin system [303]. *In vitro* studies reveal that the negative regulatory effect of $1,25(OH)_2D_3$ on renin expression is due to a direct repression of renin gene transcription possibly through nVDREs in the renin promoter [596]. The renin-angiotensin system is an important drug target for therapeutic intervention of hypertension. ACE inhibitors and angiotensin II receptor antagonists are among the most popular antihypertensive drugs that are currently used. The finding that $1,25(OH)_2D_3$ suppresses renin expression suggests that $1,25(OH)_2D_3$ has the potential to be used as an antihypertensive agent. Intravenous $1,25(OH)_2D_3$ treatment of hemodialysis patients with secondary hyperparathyroidism results in significant decreases in plasma renin and angiotensin II levels and a concomitant regression of myocardial hypertrophy seen in these patients [598]. Thus, $1,25(OH)_2D_3$ may exert beneficial effects on the regulation of renocardiovascular functions and blood pressure, and the less calcemic analogs of $1,25(OH)_2D_3$ have the potential to be developed into antihypertensive drugs.

G. Antithrombotic Effects of $1,25(OH)_2D_3$

$1,25(OH)_2D_3$ has been shown to exhibit anticoagulant effects by upregulating the expression of thrombomodulin and down-regulating that of tissue factor in monocytes and several monocytic leukemia cells [300, 599]. Studies in VDR knockout mice demonstrate the antithrombotic effects of $1,25(OH)_2D_3$ *in vivo* [301]. A clinical observation supporting the antithrombotic effect of $1,25(OH)_2D_3$ in humans was reported in a recent study by Beer *et al.* in a placebo-controlled, randomized trial of high-dose $1,25(OH)_2D_3$ with the chemotherapy drug docetaxel in prostate cancer patients [600]. Addition of high-dose $1,25(OH)_2D_3$ to docetaxel caused a statistically significant reduction in the incidence of thrombotic events as compared to the placebo, suggesting that $1,25(OH)_2D_3$ may act as an antithrombotic agent.

XI. VITAMIN D AND OSTEOPOROSIS

The importance of vitamin D in the etiology and treatment of osteoporosis will be discussed in detail in a number of subsequent chapters in this book. The use of vitamin D and its analogs to prevent and treat osteoporosis was recently reviewed [601–609].

In brief, several potential mechanisms have been put forward to implicate vitamin D in the development of osteoporosis. (1) The possibility that polymorphisms within the gene encoding the VDR contribute substantially to genetic differences in osteoporosis risk was raised by Morrison *et al.* [610]. The basis for this genetic effect on osteoporosis risk is presumably as a hereditary factor affecting "peak bone mass," but the mechanism is unknown. At this time the VDR genotype hypothesis remains controversial, since a recent meta-analysis of a large number of studies has found minimal if any correlation of VDR polymorphism to osteoporotic risk [380]. (2) An age-related decline in renal $1,25(OH)_2D$ production, due in part to a diminished renal response to PTH and reduced intestinal calcium absorption [438]. There appears to be a defect in renal response to PTH so that older women with osteoporosis require greater amounts of PTH to stimulate $1,25(OH)_2D$ production. (3) A relative decrease in $1,25(OH)_2D$ levels has been considered a contributing factor in the development of senile osteoporosis [611]. A low vitamin D state from inadequate diet and decreased exposure to sunlight as people age, especially in the house-bound elderly, contributes to malabsorption of calcium and vitamin D "insufficiency" in the elderly [612]. Other recent studies concur that there is a high prevalence of vitamin D insufficiency in the elderly, even in the active community. These individuals may have established vertebral osteoporosis with increased bone turnover, decreased BMD at the hip, and thus enhanced risk of further osteoporotic fractures in comparison with vitamin D–sufficient subjects [613]. (4) An age-related decline in intestinal VDR creating a relative $1,25(OH)_2D$ resistant state and impairing intestinal calcium absorption [446]. All of these factors coordinately contribute to age-related bone loss that, according to some studies, can be ameliorated by vitamin D and calcium supplements [87, 614, 615]. It is interesting to note that following resolution of vitamin D insufficiency, there is a rapid recovery of BMD [616]. (5) Failure to reach peak bone mass in the absence of adequate vitamin D supplementation. A recent study found that lower 25(OH)D levels correlate with a lower BMD accumulation rate of young women living in northern latitudes, suggesting that these women, commonly suffering from vitamin D insufficiency, might not reach peak bone mass [617]. Some studies have found that vitamin D supplementation modestly reduces fracture rate, prevents bone loss, and improves BMD [614, 618, 619].

Vitamin D has direct actions to affect estrogen synthesis by regulating the activity of aromatase in osteoblasts [293] and estrogen half-life by regulating 17β-hydroxysteroid dehydrogenase in keratinocytes [620]. The impact of these effects that may take place in multiple organs and their potential role in modulating vitamin D actions remain to be fully clarified.

Vitamin D has effects on muscle, and recent findings suggest that vitamin D insufficiency may be associated with decreased muscle strength [85, 621, 622] and therefore increased rates of falling [623, 624]. Vitamin D supplementation improves mobility and reduces falls in the elderly [625]. In ambulant nursing home and hostel residents and community-dwelling elderly, recent studies found that a low 25(OH)D level was an independent risk factor for recurrent falls [621, 626]. Vitamin D replacement reduces the incidence of falls in nursing home residents, even in non-vitamin D–deficient individuals [627]. Increased rate of falling can contribute to increased fractures in vitamin D–insufficient individuals.

Glucocorticoid-induced osteoporosis is the most prevalent form of secondary osteoporosis, due to the widespread use of steroids for autoimmune, gastrointestinal, and rheumatologic conditions; organ transplantation; and antineoplastic treatment regimens. Glucocorticoid-induced osteoporosis involves an initial phase of rapid bone loss secondary to increased bone resorption followed by a second phase of progressive impairment of bone formation [628]. The ultimate goal of therapy is prevention of bone loss and fractures using reduction of corticosteroid doses to a minimum, appropriate calcium and vitamin D doses, and other antiresorptive or anabolic agents when indicated. Vitamin D doses of 800 IU/day (20 mcg/day) with appropriate calcium supplementation have been recommended for osteoporosis prevention for those receiving glucocorticoid therapy [629]. For short courses of low-dose glucocorticoid therapy (below 7.5 mg/day prednisone or equivalent), depending on BMD and other risk factors, some investigators consider that vitamin D and calcium supplementation may be sufficient for bone protection [630, 631]. Treatment with vitamin D plus calcium, as a first-line therapy, should be recommended to patients receiving long-term corticosteroids [632]. No difference between the various vitamin D analogs on bone loss or fracture prevention in glucocorticoid-induced osteoporosis has been detected thus far [633, 634].

REFERENCES

1. M. R. Haussler, G. K. Whitfield, C. A. Haussler, J. Hsieh, P. D. Thompson, S. H. Selznick, C. E. Dominguez, and P. W. Jurutka, The nuclear vitamin D receptor: Biological and molecular regulatory properties revealed. *J Bone Miner Res,* **13**, 325–349 (1998).

2. P. J. Malloy, J. W. Pike, and D. Feldman, The vitamin D receptor and the syndrome of hereditary 1,25-dihydroxyvitamin D-resistant rickets. *Endocr Rev, 20*, 156–188 (1999).

3. G. K. Whitfield, P. W. Jurutka, C. A. Haussler, and M. R. Haussler, Steroid hormone receptors: Evolution, ligands, and molecular basis of biologic function. *J Cell Biochem,* Suppl 32–33, 110–122 (1999).

4. C. Rachez, and L. P. Freedman, Mechanisms of gene regulation by vitamin D(3) receptor: A network of coactivator interactions. *Gene, 246*, 9–21 (2000).

5. S. Nagpal, S. Na, and R. Rathnachalam, Noncalcemic actions of vitamin D receptor ligands. *Endocr Rev, 26*, 662–687 (2005).

6. S. Christakos, P. Dhawan, Q. Shen, X. Peng, B. Benn, and Y. Zhong, New insights into the mechanisms involved in the pleiotropic actions of 1,25 dihydroxyvitamin D_3. *Ann NY Acad Sci, 1068*, 194–203 (2006).

7. M. B. Demay, Mechanism of vitamin D receptor action. *Ann NY Acad Sci, 1068*, 204–213 (2006).

8. D. Feldman, J. W. Pike, and F. H. Glorieux, *Vitamin D,* 2nd ed. Elsevier Academic Press, San Diego (2005).

9. R. L. Horst, and T. A. Reinhardt, Vitamin D metabolism. In *Vitamin D* (D. Feldman, J. W. Pike, and F. Glorieux, eds.), pp. 13–31. Academic Press, San Diego (1997).

10. M. F. Holick, McCollum Award Lecture, 1994, Vitamin D—New horizons for the 21st century. *Am J Clin Nutr, 60*, 619–630 (1994).

11. E. Kodicek, The story of vitamin D, from vitamin to hormone. *Lancet, 1*, 325–329 (1974).

12. H. F. DeLuca, The vitamin D story: A collaborative effort of basic science and clinical medicine. *FASEB J, 2*, 224–236 (1988).

13. H. F. DeLuca, Historical perspective. In *Vitamin D,* 2nd ed. (D. Feldman, J. W. Pike, F. Glorieux, eds.), pp. 3–12. Elsevier Academic Press, San Diego (2005).

14. J. L. O'Riordan, Rickets in the 17th century. *J Bone Miner Res, 21*, 1506–1510 (2006).

15. J. G. Hoenderop, B. Nilius, and R. J. Bindels, Calcium absorption across epithelia. *Physiol Rev, 85*, 373–422 (2005).

16. M. Davies, J. L. Berry, and A. P. Mee, Bone disorders associated with gastrointestinal and hepatobiliary diseases. In *Vitamin D,* 2nd ed. (D. Feldman, J. W. Pike, and F. Glorieux, eds.), pp. 1293–1312. Elsevier Academic Press, San Diego (2005).

17. C. W. Lo, P. W. Paris, T. L. Clemens, J. Nolan, and M. F. Holick, Vitamin D absorption in healthy subjects and in patients with intestinal malabsorption syndromes. *Am J Clin Nutr, 42*, 644–649 (1985).

18. T. D. Thacher, P. R. Fischer, M. A. Strand, and J. M. Pettifor, Nutritional rickets around the world: Causes and future directions. *Ann Trop Paediatr, 26*, 1–16 (2006).

19. B. Wharton, and N. Bishop, Rickets. *Lancet, 362*, 1389–1400 (2003).

20. T. D. Thacher, P. R. Fischer, J. M. Pettifor, J. O. Lawson, C. O. Isichei, J. C. Reading, and G. M. Chan, A comparison of calcium, vitamin D, or both for nutritional rickets in Nigerian children. *N Engl J Med, 341*, 563–568 (1999).

21. L. M. Oginni, C. A. Sharp, C. Worsfold, O. S. Badru, and M. W. Davie, Healing of rickets after calcium supplementation. *Lancet, 353*, 296–297 (1999).

22. G. Kutluk, F. Cetinkaya, and M. Basak, Comparisons of oral calcium, high dose vitamin D and a combination of these in the treatment of nutritional rickets in children. *J Trop Pediatr, 48*, 351–353 (2002).

23. D. Heldenberg, G. Tenenbaum, and Y. Weisman, Effect of iron on serum 25-hydroxyvitamin D and 24,25-dihydroxyvitamin D concentrations. *Am J Clin Nutr, 56*, 533–536 (1992).

24. M. Holick, Photobiology of vitamin D. In *Vitamin D,* 2nd ed. (D. Feldman, J. W. Pike, and F. Glorieux, eds.), pp. 37–45. Elsevier Academic Press, San Diego (2005).

25. D. D. Bikle, Vitamin D: Role in skin and hair. In *Vitamin D,* 2nd ed. (D. Feldman, J. W. Pike, and F. Glorieux, eds.), pp. 609–630. Elsevier Academic Press, San Diego (2005).

26. S. J. Courtois, S. Segaert, H. Degreef, R. Bouillon, and M. Garmyn, Ultraviolet B suppresses vitamin D receptor gene expression in keratinocytes. *Biochem Biophys Res Commun, 246*, 64–69 (1998).

27. L. Y. Matsuoka, J. Wortsman, T. C. Chen, and M. F. Holick, Compensation for the interracial variance in the cutaneous synthesis of vitamin D. *J Lab Clin Med, 126*, 452–457 (1995).

28. W. F. Loomis, Skin-pigment regulation of vitamin-D biosynthesis in man. *Science, 157*, 501–506 (1967).

29. L. Y. Matsuoka, J. Wortsman, J. G. Haddad, P. Kolm, and B. W. Hollis, Racial pigmentation and the cutaneous synthesis of vitamin D. *Arch Dermatol, 127*, 536–538 (1991).

30. S. S. Harris, Vitamin D and African Americans. *J Nutr, 136*, 1126–1129 (2006).

31. M. F. Holick, J. A. MacLaughlin, and S. H. Doppelt, Regulation of cutaneous previtamin D3 photosynthesis in man: Skin pigment is not an essential regulator. *Science, 211*, 590–593 (1981).

32. G. G. Schwartz, and T. C. Chen, Vitamin D, sunlight, and the natural history of prostate cancer. In *Vitamin D,* 2nd ed. (D. Feldman, J. W. Pike, and F. Glorieux, eds.), pp. 1599–1615. Elsevier Academic Press, San Diego (2005).

33. F. C. Garland, C. F. Garland, E. D. Gorham, and J. F. Young, Geographic variation in breast cancer mortality in the United States: A hypothesis involving exposure to solar radiation. *Prev Med, 19*, 614–622 (1990).

34. W. B. Grant, An estimate of premature cancer mortality in the U.S. due to inadequate doses of solar ultraviolet-B radiation. *Cancer, 94*, 1867–1875 (2002).

35. E. A. Mortimer, Jr., R. R. Monson, and B. MacMahon, Reduction in mortality from coronary heart disease in men residing at high altitude. *N Engl J Med, 296*, 581–585 (1977).

36. J. F. Kurtzke, Multiple sclerosis in time and space—Geographic clues to cause. *J Neurovirol, 6*(Suppl 2), S134–140 (2000).

37. W. B. Grant, C. F. Garland, and M. F. Holick, Comparisons of estimated economic burdens due to insufficient solar ultraviolet irradiance and vitamin D and excess solar UV irradiance for the United States. *Photochem Photobiol, 81*, 1276–1286 (2005).

38. M. F. Holick, Sunlight and vitamin D for bone health and prevention of autoimmune diseases, cancers, and cardiovascular disease. *Am J Clin Nutr, 80*, 1678S–1688S (2004).

39. D. Wolpowitz, and B. A. Gilchrest, The vitamin D questions: How much do you need and how should you get it? *J Am Acad Dermatol, 54*, 301–317 (2006).

40. U.S. Department of Health and Human Services, *Report on Carcinogens*, Eleventh Edition. Public Health Service, National Toxicology Program.

41. M. T. R. Lucas, W. Smith, and B. Armstrong, Solar ultraviolet radiation: Global burden of disease from solar ultraviolet radiation. World Health Organization, Geneva (2006).

42. R. Marks, P. A. Foley, D. Jolley, K. R. Knight, J. Harrison, and S. C. Thompson, The effect of regular sunscreen use on vitamin D levels in an Australian population. Results of a randomized controlled trial. *Arch Dermatol, 131*, 415–421 (1995).

43. J. Farrerons, M. Barnadas, J. Rodriguez, A. Renau, B. Yoldi, A. Lopez-Navidad, and J. Moragas, Clinically prescribed sunscreen (sun protection factor 15) does not decrease serum vitamin D concentration sufficiently either to induce changes in

parathyroid function or in metabolic markers. *Br J Dermatol,* **139**, 422–427 (1998).

44. R. B. Sollitto, K. H. Kraemer, and J. J. DiGiovanna, Normal vitamin D levels can be maintained despite rigorous photoprotection: Six years' experience with xeroderma pigmentosum. *J Am Acad Dermatol,* **37**, 942–947 (1997).

45. S. Gandini, F. Sera, M. S. Cattaruzza, P. Pasquini, O. Picconi, P. Boyle, and C. F. Melchi, Meta-analysis of risk factors for cutaneous melanoma. II. Sun exposure. *Eur J Cancer,* **41**, 45–60 (2005).

46. U.S. Department of Agriculture ARS, USDA Nutrient Database for Standard Reference, Release 16 (2003).

47. National Institutes of Health, Office of Dietary Supplements: Dietary Supplements Fact Sheet: Vitamin D (2006).

48. M. S. Calvo, S. J. Whiting, and C. N. Barton, Vitamin D fortification in the United States and Canada: Current status and data needs. *Am J Clin Nutr,* **80**, 1710S–1716S (2004).

49. A. M. Natri, P. Salo, T. Vikstedt, A. Palssa, M. Huttunen, M. U. Karkkainen, H. Salovaara, V. Piironen, J. Jakobsen, and C. J. Lamberg-Allardt, Bread fortified with cholecalciferol increases the serum 25-hydroxyvitamin D concentration in women as effectively as a cholecalciferol supplement. *J Nutr,* **136**, 123–127 (2006).

50. M. F. Holick, Q. Shao, W. W. Liu, and T. C. Chen, The vitamin D content of fortified milk and infant formula. *N Engl J Med,* **326**, 1178–1181 (1992).

51. M. S. Calvo, S. J. Whiting, and C. N. Barton, Vitamin D intake: A global perspective of current status. *J Nutr,* **135**, 310–316 (2005).

52. M. S. Calvo, and S. J. Whiting, Public health strategies to overcome barriers to optimal vitamin D status in populations with special needs. *J Nutr,* **136**, 1135–1139 (2006).

53. G. Bonner, H. Warwick, M. Barnardo, and T. Lobstein, *Fortification Examined: How Added Nutrients Can Undermine Good Nutrition.* The Food Commission (UK) Ltd, London (1999).

54. H. M. Meltzer, A. Aro, N. L. Andersen, B. Koch, and J. Alexander, Risk analysis applied to food fortification. *Public Health Nutr,* **6**, 281–291 (2003).

55. S. P. Daiger, M. S. Schanfield, and L. L. Cavalli-Sforza, Group-specific component (Gc) proteins bind vitamin D and 25-hydroxyvitamin D. *Proc Natl Acad Sci USA,* **72**, 2076–2080 (1975).

56. H. Cleve, and J. Constans, The mutants of the vitamin-D-binding protein: More than 120 variants of the GC/DBP system. *Vox Sang,* **54**, 215–225 (1988).

57. L. L. Cavalli-Sforza, P. Menozzi, and A. Piazza, *The History and Geography of Human Genes.* Princeton University Press, Princeton, NJ (1994).

58. I. Bogaerts, C. Verboven, H. Van Baelen, and R. Bouillon, New aspects of DBP. In *Vitamin D,* 2nd ed. (D. Feldman, J. W. Pike, and F. Glorieux, eds.), pp. 135–152. Elsevier Academic Press, San Diego (2005).

59. C. J. Laing, and N. E. Cooke, Vitamin D binding protein. In *Vitamin D,* 2nd ed. (D. Feldman, J. W. Pike, and F. Glorieux, eds.), pp. 117–134. Elsevier, San Diego (2005).

60. J. G. Haddad, L. Y. Matsuoka, B. W. Hollis, Y. Z. Hu, and J. Wortsman, Human plasma transport of vitamin D after its endogenous synthesis. *J Clin Invest,* **91**, 2552–2555 (1993).

61. R. Bouillon, K. Allewaert, D. Z. Xiang, B. K. Tan, and H. van Baelen, Vitamin D analogs with low affinity for the vitamin D binding protein: Enhanced in vitro and decreased in vivo activity. *J Bone Miner Res,* **6**, 1051–1057 (1991).

62. J. P. Berg, Vitamin D-binding protein prevents vitamin D deficiency and presents vitamin D for its renal activation. *Eur J Endocrinol,* **141**, 321–322 (1999).

63. M. Guoth, A. Murgia, R. M. Smith, M. B. Prystowsky, N. E. Cooke, and J. G. Haddad, Cell surface vitamin D-binding protein (GC-globulin) is acquired from plasma. *Endocrinology,* **127**, 2313–2321 (1990).

64. F. V. Schiodt, L. Rossaro, R. T. Stravitz, A. O. Shakil, R. T. Chung, and W. M. Lee, Gc-globulin and prognosis in acute liver failure. *Liver Transpl,* **11**, 1223–1227 (2005).

65. B. Dahl, F. V. Schiodt, P. Ott, F. Wians, W. M. Lee, J. Balko, and G. E. O'Keefe, Plasma concentration of Gc-globulin is associated with organ dysfunction and sepsis after injury. *Crit Care Med,* **31**, 152–156 (2003).

66. F. F. Safadi, P. Thornton, H. Magiera, B. W. Hollis, M. Gentile, J. G. Haddad, S. A. Liebhaber, and N. E. Cooke, Osteopathy and resistance to vitamin D toxicity in mice null for vitamin D binding protein. *J Clin Invest,* **103**, 239–251 (1999).

67. T. E. Willnow, and A. Nykjaer, Endocytic pathways for 25-(OH) vitamin D3. In *Vitamin D,* 2nd ed. (D. Feldman, J. W. Pike, and F. Glorieux, eds.), pp. 153–163. Elsevier Academic Press, San Diego (2005).

68. A. Nykjaer, D. Dragun, D. Walther, H. Vorum, C. Jacobsen, J. Herz, F. Melsen, E. I. Christensen, and T. E. Willnow, An endocytic pathway essential for renal uptake and activation of the steroid 25-(OH) vitamin D_3. *Cell,* **96**, 507–515 (1999).

69. A. Nykjaer, J. C. Fyfe, R. Kozyraki, J. R. Leheste, C. Jacobsen, M. S. Nielsen, P. J. Verroust, M. Aminoff, A. de la Chapelle, S. K. Moestrup, R. Ray, J. Gliemann, T. E. Willnow, and E. I. Christensen, Cubilin dysfunction causes abnormal metabolism of the steroid hormone 25(OH) vitamin D(3). *Proc Natl Acad Sci USA,* **98**, 13895–13900 (2001).

70. J. S. Adams, "Bound" to work: The free hormone hypothesis revisited. *Cell,* **122**, 647–649 (2005).

71. J. R. Leheste, B. Rolinski, H. Vorum, J. Hilpert, A. Nykjaer, C. Jacobsen, P. Aucouturier, J. O. Moskaug, A. Otto, E. I. Christensen, and T. E. Willnow, Megalin knockout mice as an animal model of low molecular weight proteinuria. *Am J Pathol,* **155**, 1361–1370 (1999).

72. H. Chen, M. Hewison, B. Hu, and J. S. Adams, Heterogeneous nuclear ribonucleoprotein (hnRNP) binding to hormone response elements: A cause of vitamin D resistance. *Proc Natl Acad Sci USA,* **100**, 6109–6114 (2003).

73. M. Hewison, A. R. Rut, K. Kristjansson, R. E. Walker, M. J. Dillon, M. R. Hughes, and J. L. O'Riordan, Tissue resistance to 1,25-dihydroxyvitamin D without a mutation of the vitamin D receptor gene. *Clin Endocrinol,* **39**, 663–670 (1993).

74. H. Chen, B. Hu, E. A. Allegretto, and J. S. Adams, The vitamin D response element-binding protein. A novel dominant-negative regulator of vitamin D-directed transactivation. *J Biol Chem,* **275**, 35557–35564 (2000).

75. S. Wu, S. Ren, H. Chen, R. F. Chun, M. A. Gacad, and J. S. Adams, Intracellular vitamin D binding proteins: Novel facilitators of vitamin D-directed transactivation. *Mol Endocrinol,* **14**, 1387–1397 (2000).

76. B. Hollis, Detection of vitamin D and its major metabolites. In *Vitamin D,* 2nd ed. (D. Feldman, J. W. Pike, and F. Glorieux, eds.), Elsevier Academic Press, San Diego (2005).

77. J. G. Haddad, and K. J. Chyu, Competitive protein-binding radioassay for 25-hydroxycholecalciferol *J Clin Endocrinol Metab,* **33**, 992–995 (1971).

78. J. E. Zerwekh, The measurement of vitamin D: Analytical aspects. *Ann Clin Biochem,* **41**, 272–281 (2004).

79. N. Binkley, D. Krueger, C. S. Cowgill, L. Plum, E. Lake, K. E. Hanson, H. F. DeLuca, and M. K. Drezner, Assay variation confounds the diagnosis of hypovitaminosis D: A call for standardization. *J Clin Endocrinol Metab,* **89**, 3152–3157 (2004).

80. B. W. Hollis, Editorial: The determination of circulating 25-hydroxyvitamin D: No easy task. *J Clin Endocrinol Metab,* **89**, 3149–3151 (2004).

81. G. L. Lensmeyer, D. A. Wiebe, N. Binkley, and M. K. Drezner, HPLC method for 25-hydroxyvitamin D measurement: Comparison with contemporary assays. *Clin Chem,* **52**, 1120–1126 (2006).

82. A. K. Saenger, T. J. Laha, D. E. Bremner, and S. M. Sadrzadeh, Quantification of serum 25-hydroxyvitamin D(2) and D(3) using HPLC-tandem mass spectrometry and examination of reference intervals for diagnosis of vitamin D deficiency. *Am J Clin Pathol,* **125**, 914–920 (2006).

83. M. C. Chapuy, and P. J. Meunier, Vitamin D insufficiency in adults and the elderly. In *Vitamin D* (D. Feldman, J. W. Pike, and F. Glorieux, eds.), pp. 679–693. Academic Press, San Diego (1997).

84. H. A. Bischoff, H. B. Stahelin, W. Dick, R. Akos, M. Knecht, C. Salis, M. Nebiker, R. Theiler, M. Pfeifer, B. Begerow, R. A. Lew, and M. Conzelmann, Effects of vitamin D and calcium supplementation on falls: A randomized controlled trial. *J Bone Miner Res,* **18**, 343–351 (2003).

85. H. A. Bischoff-Ferrari, T. Dietrich, E. J. Orav, F. B. Hu, Y. Zhang, E. W. Karlson, and B. Dawson-Hughes, Higher 25-hydroxyvitamin D concentrations are associated with better lower-extremity function in both active and inactive persons aged > or =60 y. *Am J Clin Nutr,* **80**, 752–758 (2004).

86. M. C. Chapuy, M. E. Arlot, P. D. Delmas, and P. J. Meunier, Effect of calcium and cholecalciferol treatment for three years on hip fractures in elderly women. *BMJ,* **308**, 1081–1082 (1994).

87. M. C. Chapuy, M. E. Arlot, F. Duboeuf, J. Brun, B. Crouzet, S. Arnaud, P. D. Delmas, and P. J. Meunier, Vitamin D3 and calcium to prevent hip fractures in the elderly women. *N Engl J Med,* **327**, 1637–1642 (1992).

88. R. P. Heaney, Vitamin D, nutritional deficiency, and the medical paradigm. *J Clin Endocrinol Metab,* **88**, 5107–5108 (2003).

89. R. Vieth, What is the optimal vitamin D status for health? *Prog Biophys Mol Biol,* **92**, 26–32 (2006).

90. B. W. Hollis, Circulating 25-hydroxyvitamin D levels indicative of vitamin D sufficiency: Implications for establishing a new effective dietary intake recommendation for vitamin D. *J Nutr,* **135**, 317–322 (2005).

91. R. P. Heaney, K. M. Davies, T. C. Chen, M. F. Holick, and M. J. Barger-Lux, Human serum 25-hydroxycholecalciferol response to extended oral dosing with cholecalciferol. *Am J Clin Nutr,* **77**, 204–210 (2003).

92. B. Dawson-Hughes, R. P. Heaney, M. F. Holick, P. Lips, P. J. Meunier, and R. Vieth, Estimates of optimal vitamin D status. *Osteoporos Int,* **16**, 713–716 (2005).

93. L. A. Armas, B. W. Hollis, and R. P. Heaney, Vitamin D$_2$ is much less effective than vitamin D$_3$ in humans. *J Clin Endocrinol Metab,* **89**, 5387–5391 (2004).

94. Institute of Medicine, Food and Nutrition Board. *Dietary Reference Intakes for Calcium, Phosphorus, Magnesium, Vitamin D and Fluoride.* National Academy Press, Washington, D.C. (1999).

95. RDA Standing Committee on the Scientific Evaluation of Dietary Reference Intakes. *Dietary Reference Intakes for Calcium, Phosphorus, Magnesium, Vitamin D and Fluoride.* National Academy Press, Washington, D.C. (1997).

96. M. Lehtonen-Veromaa, T. Mottonen, I. Nuotio, K. Irjala, and J. Viikari, The effect of conventional vitamin D(2) supplementation on serum 25(OH)D concentration is weak among peripubertal Finnish girls: A 3-y prospective study. *Eur J Clin Nutr,* **56**, 431–437 (2002).

97. Safer Upper Levels for Vitamins and Minerals. Food Standards Agency, Great Britain (2003).

98. R. P. Heaney, The vitamin D requirement in health and disease. *J Steroid Biochem Mol Biol,* **97**, 13–19 (2005).

99. R. Vieth, Critique of the considerations for establishing the tolerable upper intake level for vitamin D: Critical need for revision upwards. *J Nutr,* **136**, 1117–1122 (2006).

100. R. P. Heaney, Barriers to optimizing vitamin D3 intake for the elderly. *J Nutr,* **136**, 1123–1125 (2006).

101. S. J. Whiting, and M. S. Calvo, Overview of the proceedings from Experimental Biology 2005 symposium: Optimizing vitamin D intake for populations with special needs: Barriers to effective food fortification and supplementation. *J Nutr,* **136**, 1114–1116 (2006).

102. R. D. Jackson, A. Z. LaCroix, M. Gass, R. B. Wallace, J. Robbins, C. E. Lewis, T. Bassford, S. A. Beresford, H. R. Black, P. Blanchette, D. E. Bonds, R. L. Brunner, R. G. Brzyski, B. Caan, J. A. Cauley, R. T. Chlebowski, S. R. Cummings, I. Granek, J. Hays, G. Heiss, S. L. Hendrix, B. V. Howard, J. Hsia, F. A. Hubbell, K. C. Johnson, H. Judd, J. M. Kotchen, L. H. Kuller, R. D. Langer, N. L. Lasser, M. C. Limacher, S. Ludlam, J. E. Manson, K. L. Margolis, J. McGowan, J. K. Ockene, M. J. O'Sullivan, L. Phillips, R. L. Prentice, G. E. Sarto, M. L. Stefanick, L. Van Horn, J. Wactawski-Wende, E. Whitlock, G. L. Anderson, A. R. Assaf, and D. Barad, Calcium plus vitamin D supplementation and the risk of fractures. *N Engl J Med,* **354**, 669–683 (2006).

103. U.S. Department of Health and Human Services and U.S. Department of Agriculture. *Dietary Guidelines for Americans,* pp. 19–20 (2005).

104. S. J. Whiting, and M. S. Calvo, Dietary recommendations for vitamin D: A critical need for functional end points to establish an estimated average requirement. *J Nutr,* **135**, 304–309 (2005).

105. M. F. Holick, High prevalence of vitamin D inadequacy and implications for health. *Mayo Clin Proc,* **81**, 353–373 (2006).

106. A. C. Looker, B. Dawson-Hughes, M. S. Calvo, E. W. Gunter, and N. R. Sahyoun, Serum 25-hydroxyvitamin D status of adolescents and adults in two seasonal subpopulations from NHANES III. *Bone,* **30**, 771–777 (2002).

107. S. R. Grover, and R. Morley, Vitamin D deficiency in veiled or dark-skinned pregnant women. *Med J Aust,* **175**, 251–252 (2001).

108. J. M. Nozza, and C. P. Rodda, Vitamin D deficiency in mothers of infants with rickets. *Med J Aust,* **175**, 253–255 (2001).

109. M. K. Thomas, D. M. Lloyd-Jones, R. I. Thadhani, A. C. Shaw, D. J. Deraska, B. T. Kitch, E. C. Vamvakas, I. M. Dick, R. L. Prince, and J. S. Finkelstein, Hypovitaminosis D in medical inpatients. *N Engl J Med,* **338**, 777–783 (1998).

110. M. S. LeBoff, L. Kohlmeier, S. Hurwitz, J. Franklin, J. Wright, and J. Glowacki, Occult vitamin D deficiency in postmenopausal US women with acute hip fracture. *JAMA,* **281**, 1505–1511 (1999).

111. M. C. Chapuy, R. Pamphile, E. Paris, C. Kempf, M. Schlichting, S. Arnaud, P. Garnero, and P. J. Meunier, Combined calcium and vitamin D3 supplementation in elderly women: Confirmation of reversal of secondary hyperparathyroidism and hip fracture risk: The Decalyos II study. *Osteoporos Int,* **13**, 257–264 (2002).

112. B. Dawson-Hughes, S. S. Harris, E. A. Krall, and G. E. Dallal, Effect of calcium and vitamin D supplementation on bone density in men and women 65 years of age or older. *N Engl J Med,* **337**, 670–676 (1997).

113. D. P. Trivedi, R. Doll, and K. T. Khaw, Effect of four monthly oral vitamin D3 (cholecalciferol) supplementation on fractures and mortality in men and women living in the community: Randomised double blind controlled trial. *BMJ,* **326,** 469 (2003).

114. S. Datta, M. Alfaham, D. P. Davies, F. Dunstan, S. Woodhead, J. Evans, and B. Richards, Vitamin D deficiency in pregnant women from a non-European ethnic minority population— An interventional study. *BJOG,* **109,** 905–908 (2002).

115. M. Ala-Houhala, 25-hydroxyvitamin D levels during breastfeeding with or without maternal or infantile supplementation of vitamin D. *J Pediatr, Gastroenterol Nutr,* **4,** 220–226 (1985).

116. B. L. Specker, B. Valanis, V. Hertzberg, N. Edwards, and R. C. Tsang, Sunshine exposure and serum 25-hydroxyvitamin D concentrations in exclusively breast-fed infants. *J Pediatr,* **107,** 372–376 (1985).

117. L. M. Gartner, J. Morton, R. A. Lawrence, A. J. Naylor, D. O'Hare, R. J. Schanler, and A. I. Eidelman, Breastfeeding and the use of human milk. *Pediatrics,* **115,** 496–506 (2005).

118. M. Gascon-Barre, The vitamin D 25-hydroxylase. In *Vitamin D,* 2nd ed. (D. Feldman, J. W. Pike, and F. Glorieux, eds.), pp. 46–67. Elsevier Academic Press, San Diego (2005).

119. E. Usui, M. Noshiro, and K. Okuda, Molecular cloning of cDNA for vitamin D3 25-hydroxylase from rat liver mitochondria. *FEBS Lett,* **262,** 135–138 (1990).

120. J. J. Cali, C. L. Hsieh, U. Francke, and D. W. Russell, Mutations in the bile acid biosynthetic enzyme sterol 27-hydroxylase underlie cerebrotendinous xanthomatosis. *J Biol Chem,* **266,** 7779–7783 (1991).

121. Y. D. Guo, S. Strugnell, D. W. Back, and G. Jones, Transfected human liver cytochrome P-450 hydroxylates vitamin D analogs at different side-chain positions. *Proc Natl Acad Sci USA,* **90,** 8668–8872 (1993).

122. K. I. Okuda, Liver mitochondrial P450 involved in cholesterol catabolism and vitamin D activation. *J Lipid Res,* **35,** 361–372 (1994).

123. F. J. Dilworth, I. Scott, A. Green, S. Strugnell, Y. D. Guo, E. A. Roberts, R. Kremer, M. J. Calverley, H. L. Makin, and G. Jones, Different mechanisms of hydroxylation site selection by liver and kidney cytochrome P450 species (CYP27 and CYP24) involved in vitamin D metabolism. *J Biol Chem,* **270,** 16766–16774 (1995).

124. I. Bjorkhem, and S. Skrede, Familial diseases with storage of sterols other than cholesterol: Cerebrotendinous xanthomatosis and phytosterolemia. In *The Metabolic Basis of Inherited Disease,* 6th ed. (C. R. Scriver, A. L. Beaudet, W. S. Sly, and D. Valle, eds.), pp. 1283–1302. McGraw Hill, New York (1989).

125. V. M. Berginer, S. Shany, D. Alkalay, J. Berginer, S. Dekel, G. Salen, G. S. Tint, and D. Gazit, Osteoporosis and increased bone fractures in cerebrotendinous xanthomatosis. *Metabolism,* **42,** 69–74 (1993).

126. H. Rosen, A. Reshef, N. Maeda, A. Lippoldt, S. Shpizen, L. Triger, G. Eggertsen, I. Bjorkhem, and E. Leitersdorf, Markedly reduced bile acid synthesis but maintained levels of cholesterol and vitamin D metabolites in mice with disrupted sterol 27-hydroxylase gene. *J Biol Chem,* **273,** 14805–14812 (1998).

127. M. R. Hughes, D. J. Baylink, P. G. Jones, and M. R. Haussler, Radioligand receptor assay for 25-hydroxyvitamin D2/D3 and 1 alpha, 25 dihydroxyvitamin D2/D3. *J Clin Invest,* **58,** 61–70 (1976).

128. N. H. Bell, S. Shaw, and R. T. Turner, Evidence that 1,25-dihydroxyvitamin D3 inhibits the hepatic production of

25-hydroxyvitamin D in man. *J Clinical Invest,* **74,** 1540–1544 (1984).

129. J. B. Cheng, D. L. Motola, D. J. Mangelsdorf, and D. W. Russell, De-orphanization of cytochrome P450 2R1, a microsomal vitamin D 25-hydroxylase. *J Biol Chem,* **278,** 38084–38093 (2003).

130. D. R. Nelson, Comparison of P450s from human and fugu: 420 million years of vertebrate P450 evolution. *Arch Biochem Biophys,* **409,** 18–24 (2003).

131. J. B. Cheng, M. A. Levine, N. H. Bell, D. J. Mangelsdorf, and D. W Russell, Genetic evidence that the human CYP2R1 enzyme is a key vitamin D 25-hydroxylase. *Proc Natl Acad Sci USA,* **101,** 7711–7715 (2004).

132. H. L. Henry, The 25-hydroxyvitamin D 1α-hydroxylase. In *Vitamin D,* 2nd ed. (D. Feldman, J. W. Pike, and F. Glorieux, eds.), pp. 68–83. Elsevier Academic Press, San Diego (2005).

133. K. Takeyama, S. Kitanaka, T. Sato, M. Kobori, J. Yanagisawa, and S. Kato, 25-hydroxyvitamin D_3 1alpha-hydroxylase and vitamin D synthesis. *Science,* **277,** 1827–1830 (1997).

134. G. K. Fu, D. Lin, M. Y. Zhang, D. D. Bikle, C. H. Shackleton, W. L. Miller, and A. A. Portale, Cloning of human 25-hydroxyvitamin D-1 alpha-hydroxylase and mutations causing vitamin D-dependent rickets type 1. *Mol Endocrinol,* **11,** 1961–1970 (1997).

135. S. Kitanaka, K. Takeyama, A. Murayama, T. Sato, K. Okumura, M. Nogami, Y. Hasegawa, H. Niimi, J. Yanagisawa, T. Tanaka, and S. Kato, Inactivating mutations in the 25-hydroxyvitamin D_3 1(alpha)-hydroxylase gene in patients with pseudovitamin D-deficiency rickets. *N Engl J Med,* **338,** 653–661 (1998).

136. M. Labuda, K. Morgan, and F. H. Glorieux, Mapping autosomal recessive vitamin D dependency type I to chromosome 12q14 by linkage analysis. *Am J Hum Genet,* **47,** 28–36 (1990).

137. M. Labuda, T. M. Fujiwara, M. V. Ross, K. Morgan, J. Garcia-Heras, D. H. Ledbetter, M. R. Hughes, and F. H. Glorieux, Two hereditary defects related to vitamin D metabolism map to the same region of human chromosome 12q13–14. *J Bone Miner Res,* **7,** 1447–1453 (1992).

138. M. Hewison, and J. S. Adams, Extra-renal 1α-hydroxylase activity and human disease. In *Vitamin D,* 2nd ed. (D. Feldman, J. W. Pike, and F. Glorieux, eds.), pp. 1378–1400. Elsevier Academic Press, San Diego (2005).

139. N. H. Bell, Renal and nonrenal 25-hydroxyvitamin D-1alpha-hydroxylases and their clinical significance. *J Bone Miner Res,* **13,** 350–353 (1998).

140. G. A. Howard, R. T. Turner, D. J. Sherrard, and D. J. Baylink, Human bone cells in culture metabolize 25-hydroxyvitamin D3 to 1,25-dihydroxyvitamin D3 and 24,25-dihydroxyvitamin D3. *J Biol Chem,* **256,** 7738–7740 (1981).

141. J. S. Adams, O. P. Sharma, M. A. Gacad, and F. R. Singer, Metabolism of 25-hydroxyvitamin D3 by cultured pulmonary alveolar macrophages in sarcoidosis. *J Clinical Invest,* **72,** 1856–1860 (1983).

142. S. J. Smith, A. K. Rucka, J. L. Berry, M. Davies, S. Mylchreest, C. R. Paterson, D. A. Heath, M. Tassabehji, A. P. Read, A. P. Mee, and E. B. Mawer, Novel mutations in the 1alpha-hydroxylase (P450c1) gene in three families with pseudovitamin D-deficiency rickets resulting in loss of functional enzyme activity in blood-derived macrophages. *J Bone Miner Res,* **14,** 730–739 (1999).

143. H. S. Cross, M. Peterlik, G. S. Reddy, and I. Schuster, Vitamin D metabolism in human colon adenocarcinoma-derived Caco-2 cells: Expression of 25-hydroxyvitamin D3-1alpha-

hydroxylase activity and regulation of side-chain metabolism. *J Steroid Biochem Mol Biol,* **62**, 21–28 (1997).

144. Y. Weisman, A. Harell, S. Edelstein, M. David, Z. Spirer, and A. Golander, 1 alpha, 25-dihydroxyvitamin D3 and 24,25-dihydroxyvitamin D3 in vitro synthesis by human decidua and placenta. *Nature,* **281**, 317–319 (1979).

145. G. G. Schwartz, L. W. Whitlatch, T. C. Chen, B. L. Lokeshwar, and M. F. Holick, Human prostate cells synthesize 1,25-dihydroxyvitamin D3 from 25-hydroxyvitamin D3. *Cancer Epidemiol Biomarkers Prev,* **7**, 391–395 (1998).

146. S. Dusso, A. J. Brown, and E. A. Slatopolsky, Vitamin D and renal failure. In *Vitamin D,* 2nd ed. (D. Feldman, F. Glorieux, and J. W. Pike, eds.), pp. 1313–1338. Elsevier Academic Press, San Diego (2005).

147. A. Murayama, K. Takeyama, S. Kitanaka, Y. Kodera, Y. Kawaguchi, T. Hosoya, and S. Kato, Positive and negative regulations of the renal 25-hydroxyvitamin D3 1alpha-hydroxylase gene by parathyroid hormone, calcitonin, and 1alpha,25(OH)2D3 in intact animals. *Endocrinology,* **140**, 2224–2231 (1999).

148. X. F. Kong, X. H. Zhu, Y. L. Pei, D. M. Jackson, and M. F. Holick, Molecular cloning, characterization, and promoter analysis of the human 25-hydroxyvitamin D3-1alpha-hydroxylase gene. *Proc Natl Acad Sci USA,* **96**, 6988–6993 (1999).

149. T. Shinki, Y. Ueno, H. F. DeLuca, and T. Suda, Calcitonin is a major regulator for the expression of renal 25-hydroxyvitamin D3-1alpha-hydroxylase gene in normocalcemic rats. *Proc Natl Acad Sci USA,* **96**, 8253–8258 (1999).

150. P. Beckerman, and J. Silver, Vitamin D and the parathyroid. *Am J Med Sci,* **317**, 363–369 (1999).

151. E. Slatopolsky, A. Dusso, and A. J. Brown, The role of phosphorus in the development of secondary hyperparathyroidism and parathyroid cell proliferation in chronic renal failure. *Am J Med Sci,* **317**, 370–376 (1999).

152. J. Wortsman, J. G. Haddad, J. T. Posillico, and E. M. Brown, Primary hyperparathyroidism with low serum 1,25-dihydroxyvitamin D levels. *J Clin Endocrinol Metab,* **62**, 1305–1308 (1986).

153. E. M. Brown, G. Gamba, D. Riccardi, M. Lombardi, R. Butters, O. Kifor, A. Sun, M. Hediger, J. Lytton, and S. C. Hebert, Cloning, expression, and characterization of an extracellular Ca²⁺ sensing receptor from bovine parathyroid. *Nature,* **366**, 575–580 (1993).

154. D. Riccardi, J. Park, W. Lee, G. Gamba, E. M. Brown, and S. Hebert, Cloning and functional expression of a rat kidney extracellular calcium/polyvalent cation-sensing receptor. *Proc Natl Acad Sci USA,* **92**, 131–135 (1995).

155. N. Chattopadhyay, A. Mithal, and E. M. Brown, The calcium-sensing receptor: A window into the physiology and pathophysiology of mineral ion metabolism. *Endocr Rev,* 17, 289 (1996).

156. P. Jaeger, W. Jones, T. L. Clemens, and J. P. Hayslett, Evidence that calcitonin stimulates 1,25-dihydroxyvitamin D production and intestinal absorption of calcium in vivo. *J Clin Invest,* **78**, 456–461 (1986).

157. M. J. Econs, B. Lobaugh, and M. K. Drezner, Normal calcitonin stimulation of serum calcitriol in patients with X-linked hypophosphatemic rickets. *J Clin Endocrinol Metab,* **75**, 408–411 (1992).

158. T. Nesbitt, B. Lobaugh, and M. K. Drezner, Calcitonin stimulation of renal 25-hydroxyvitamin D-1 alpha-hydroxylase activity in hypophosphatemic mice. Evidence that the regulation of calcitriol production is not universally abnormal in X-linked hypophosphatemia. *J Clin Invest,* **79**, 15–19 (1987).

159. J. Silver, and T. Naveh-Many, Vitamin D and the parathyroid glands. In *Vitamin D,* 2nd ed. (D. Feldman, J. W. Pike, F. Glorieux, eds.), pp. 537–550. Elsevier Academic Press, San Diego (2005).

160. H. Tsujikawa, Y. Kurotaki, T. Fujimori, K. Fukuda, and Y. Nabeshima, Klotho, a gene related to a syndrome resembling human premature aging, functions in a negative regulatory circuit of vitamin D endocrine system. *Mol Endocrinol,* **17**, 2393–2403 (2003).

161. M. Hewison, D. Zehnder, R. Chakraverty, and J. S. Adams, Vitamin D and barrier function: A novel role for extra-renal 1 alpha-hydroxylase. *Mol Cell Endocrinol,* **215**, 31–38 (2004).

162. M. V. Young, G. G. Schwartz, L. Wang, D. P. Jamieson, L. W. Whitlatch, J. N. Flanagan, B. L. Lokeshwar, M. F. Holick, and T. C. Chen, The prostate 25-hydroxyvitamin D-1 alpha-hydroxylase is not influenced by parathyroid hormone and calcium: Implications for prostate cancer chemoprevention by vitamin D. *Carcinogenesis,* **25**, 967–971.

163. L. Wang, J. N. Flanagan, L. W. Whitlatch, D. P. Jamieson, M. F. Holick, and T. C. Chen, Regulation of 25-hydroxyvitamin D-1alpha-hydroxylase by epidermal growth factor in prostate cells. *J Steroid Biochem Mol Biol,* **89–90**, 127–130 (2004).

164. J. S. Adams, and S. Y. Ren, Autoregulation of 1,25-dihydroxyvitamin D synthesis in macrophage mitochondria by nitric oxide. *Endocrinology,* **137**, 4514–4517 (1996).

165. M. R. Rubin, M. Thys-Jacobs, F. K. W. Chan, L. M. C. Koberle, and J. P. Bilezikian, Hypercalcemia due to vitamin D toxicity. In *Vitamin D,* 2nd ed. (D. Feldman, J. W. Pike, and F. Glorieux, eds.), pp. 1355–1377. Elsevier Academic Press, San Diego (2005).

166. J. P. van Leeuwen, and H. A. P. Pols, Vitamin D: Cancer and differentiation. In *Vitamin D,* 2nd ed. (D. Feldman, J. W. Pike, and F. Glorieux, eds.), pp. 1571–1597. Elsevier Academic Press, San Diego (2005).

167. J. Omdhal, and B. May, The 25-hydroxyvitamin D 24-hydroxylase. In *Vitamin D,* 2nd ed. (D. Feldman, J. W. Pike, and F. Glorieux, eds.), pp. 84–104. Elsevier Academic Press, San Diego (2005).

168. R. St. Arnaud, Targeted inactivation of vitamin D hydroxylases in mice. *Bone,* **25**, 127–129 (1999).

169. C. N. Hahn, E. Baker, P. Laslo, B. K. May, J. L. Omdahl, and G. R. Sutherland, Localization of the human vitamin D 24-hydroxylase gene (CYP24) to chromosome 20q13. 2->q13. 3. *Cytogenet Cell Genet,* **62**, 192–193 (1993).

170. K. S. Chen, and H. F. DeLuca, Cloning of the human 1 alpha,25-dihydroxyvitamin D₃ 24-hydroxylase gene promoter and identification of two vitamin D-responsive elements. *Biochim Biophys Acta,* **1263**, 1–9 (1995).

171. Y. Ohyama, K. Ozono, M. Uchida, T. Shinki, S. Kato, T. Suda, O. Yamamoto, M. Noshiro, and Y. Kato, Identification of a vitamin D-responsive element in the 5′-flanking region of the rat 25-hydroxyvitamin D₃ 24-hydroxylase gene. *J Biol Chem,* **269**, 10545–10550 (1994).

172. T. Yoshizawa, Y. Handa, Y. Uematsu, S. Takeda, K. Sekine, Y. Yoshihara, T. Kawakami, K. Arioka, H. Sato, Y. Uchiyama, S. Masushige, A. Fukamizu, T. Matsumoto, and S. Kato, Mice lacking the vitamin D receptor exhibit impaired bone formation, uterine hypoplasia and growth retardation after weaning. *Nat Genet,* **16**, 391–396 (1997).

173. P. J. Malloy, J. W. Pike, and D. Feldman, Hereditary 1,25-dihydroxyvitamin D resistant rickets. In *Vitamin D,* 2nd ed. (D. Feldman, J. W. Pike, and F. Glorieux, eds.), pp. 1207–1238. Elsevier Academic Press, San Diego (2005).

174. T. Shigematsu, N. Horiuchi, Y. Ogura, T. Miyahara, and T. Suda, Human parathyroid hormone inhibits renal 24-hydrox-

ylase activity of 25-hydroxyvitamin D3 by a mechanism involving adenosine 3′,5′-monophosphate in rats. *Endocrinology,* **118**, 1583–1589 (1986).

175. M. J. Beckman, J. P. Goff, T. A. Reinhardt, D. C. Beitz, and R. L. Horst, *In vivo* regulation of rat intestinal 24-hydroxylase: Potential new role of calcitonin. *J Clin Endocrinol Metab,* **135**, 1951–1955 (1994).

176. R. St. Arnaud, and F. H. Glorieux, 24,25-dihydroxyvitamin D—Active metabolite or inactive catabolite? *Endocrinology,* **139**, 3371–3374 (1998).

177. M. Uchida, K. Ozono, and J. Pike, Activation of the human osteocalcin gene by 24R,25-dihydroxyvitamin D3 occurs through the vitamin D receptor and the vitamin D-responsive element. *J Bone Miner Res,* **9**, 1981–1987 (1994).

178. R. St. Arnaud, A. Arabian, R. Travers, F. Barletta, M. Raval-Pandya, K. Chapin, J. Depovere, C. Mathieu, S. Christakos, M. B. Demay, and F. H. Glorieux, Deficient mineralization of intramembranous bone in vitamin D-24-hydroxylase-ablated mice is due to elevated 1,25-dihydroxyvitamin D and not to the absence of 24,25-dihydroxyvitamin D. *Endocrinology,* **141**, 2658–2666 (2000).

179. R. L. Horst, T. A. Reinhardt, and G. S. Reddy, Vitamin D metabolism. In *Vitamin D,* 2nd ed. (D. Feldman, F. Glorieux, and J. W. Pike, eds.), pp. 15–36. Elsevier Academic Press, San Diego (2005).

180. G. Jones, Analog metabolism. In *Vitamin D,* 2nd ed. (D. Feldman, F. Glorieux, and J. W. Pike, eds.), pp. 1423–1448. Elsevier Academic Press, San Diego (2005).

181. L. Castillo, Y. Tanaka, H. F. DeLuca, and N. Ikekawa, On the physiological role of 1,24,25–trihydroxyvitamin D3. *Miner Electrolyte Metab,* **1**, 198–207 (1978).

182. M. L. Siu-Caldera, L. Zou, M. G. Ehrlich, E. R. Schwartz, S. Ishizuka, and G. S. Reddy, Human osteoblasts in culture metabolize both 1 alpha, 25-dihydroxyvitamin D3 and its precursor 25-hydroxyvitamin D3 into their respective lactones. *Endocrinology,* **136**, 4195–4203 (1995).

183. M. G. Bischof, M. L. Siu-Caldera, A. Weiskopf, P. Vouros, H. S. Cross, M. Peterlik, and G. S. Reddy, Differentiation-related pathways of 1 alpha,25-dihydroxycholecalciferol metabolism in human colon adenocarcinoma-derived Caco-2 cells: Production of 1 alpha,25-dihydroxy-3epi-cholecalciferol. *Exp Cell Res,* **241**, 194–201 (1998).

184. D. P. McDonnell, D. J. Mangelsdorf, J. W. Pike, M. R. Haussler, and B. W. O'Malley, Molecular cloning of complementary DNA encoding the avian receptor for vitamin D. *Science,* **235**, 1214–1217 (1987).

185. A. R. Baker, D. P. McDonnell, M. Hughes, T. M. Crisp, D. J. Mangelsdorf, M. R. Haussler, J. W. Pike, J. Shine, and B. W. O'Malley, Cloning and expression of full-length cDNA encoding human vitamin D receptor. *Proc Natl Acad Sci USA,* **85**, 3294–3298 (1988).

186. K. Miyamoto, R. A. Kesterson, H. Yamamoto, Y. Taketani, E. Nishiwaki, S. Tatsumi, Y. Inoue, K. Morita, E. Takeda, and J. W. Pike, Structural organization of the human vitamin D receptor chromosomal gene and its promoter. *Mol Endocrinol,* **11**, 1165–1179 (1997).

187. L. A. Crofts, M. S. Hancock, N. A. Morrison, and J. A. Eisman, Multiple promoters direct the tissue-specific expression of novel N-terminal variant human vitamin D receptor gene transcripts. *Proc Natl Acad Sci USA,* **95**, 10529–10534 (1998).

188. G. K. Whitfield, P. W. Jurutka, C. A. Haussler, J. C. Hsieh, T. K. Barthel, E. T. Jacobs, C. E. Dominguez, M. L. Thatcher, and M. R. Haussler, Nuclear vitamin D receptor: Structure-function, molecular control of gene transcription, and novel bioreactions.

In *Vitamin D,* 2nd ed. (D. Feldman, F. Glorieux, and J. W. Pike, eds.), pp. 219–261. Elsevier Academic Press, San Diego (2005).

189. P. L. Shaffer, and D. T. Gewirth, Structural basis of VDR-DNA interactions on direct repeat response elements. *EMBO J,* **21**, 2242–2252 (2002).

190. P. L. Shaffer, D. P. McDonnell, and D. T. Gewirth, Characterization of transcriptional activation and DNA-binding functions in the hinge region of the vitamin D receptor. *Biochemistry,* **44**, 2678–2685 (2005).

191. N. Rochel, J. M. Wurtz, A. Mitschler, B. Klaholz, and D. Moras, The crystal structure of the nuclear receptor for vitamin D bound to its natural ligand. *Mol Cell,* **5**, 173–179 (2000).

192. K. Paech, P. Webb, G. G. Kuiper, S. Nilsson, J. Gustafsson, P. J. Kushner, and T. S. Scanlan, Differential ligand activation of receptors ERalpha and ERbeta at AP1 sites. *Science,* **277**, 1508–1510 (1997).

193. A. K. Shiau, D. Barstad, P. M. Loria, L. Cheng, P. J. Kushner, D. A. Agard, and G. L. Greene, The structural basis of estrogen receptor/coactivator recognition and the antagonism of this interaction by tamoxifen. *Cell,* **95**, 927–937 (1998).

194. M. T. Mizwicki, D. Keidel, C. M. Bula, J. E. Bishop, L. P. Zanello, J. M. Wurtz, D. Moras, and A. W. Norman, Identification of an alternative ligand-binding pocket in the nuclear vitamin D receptor and its functional importance in 1alpha,25(OH)2-vitamin D3 signaling. *Proc Natl Acad Sci USA,* **101**, 12876–12881 (2004).

195. T. H. Lee, and J. Pelletier, Functional characterization of WT1 binding sites within the human vitamin D receptor gene promoter. *Physiol Genomics,* **7**, 187–200 (2001).

196. L. A. Zella, S. Kim, N. K. Shevde, and J. W. Pike, Enhancers located within two introns of the vitamin D receptor gene mediate transcriptional autoregulation by 1,25-dihydroxyvitamin D3. *Mol Endocrinol,* **20**, 1231–1247 (2006).

197. R. Kommagani, T. M. Caserta, and M. P. Kadakia, Identification of vitamin D receptor as a target of p63. *Oncogene,* **25**, 3745–3751 (2006).

198. R. Maruyama, F. Aoki, M. Toyota, Y. Sasaki, H. Akashi, H. Mita, H. Suzuki, K. Akino, M. Ohe-Toyota, Y. Maruyama, H. Tatsumi, K. Imai, Y. Shinomura, and T. Tokino, Comparative genome analysis identifies the vitamin D receptor gene as a direct target of p53-mediated transcriptional activation. *Cancer Res,* **66**, 4574–4583 (2006).

199. H. Yamamoto, K. Miyamoto, B. Li, Y. Taketani, M. Kitano, Y. Inoue, K. Morita, J. W. Pike, and E. Takeda, The caudal-related homeodomain protein Cdx-2 regulates vitamin D receptor gene expression in the small intestine. *J Bone Miner Res,* **14**, 240–247 (1999).

200. H. Arai, K. I. Miyamoto, M. Yoshida, H. Yamamoto, Y. Taketani, K. Morita, M. Kubota, S. Yoshida, M. Ikeda, F. Watabe, Y. Kanemasa, and E. Takeda, The polymorphism in the caudal-related homeodomain protein Cdx-2 binding element in the human vitamin D receptor gene. *J Bone Miner Res,* **16**, 1256–1264 (2001).

201. S. A. Kliewer, K. Umesono, D. J. Mangelsdorf, and R. M. Evans, Retinoid X receptor interacts with nuclear receptors in retinoic acid, thyroid hormone and vitamin D3 signalling. *Nature,* **355**, 446–449 (1992).

202. X. K. Zhang, B. Hoffmann, P. B. Tran, G. Graupner, and M. Pfahl, Retinoid X receptor is an auxiliary protein for thyroid hormone and retinoic acid receptors. *Nature,* **355**, 441–446 (1992).

203. A. A. Levin, L. J. Sturzenbecker, S. Kazmer, T. Bosakowski, C. Huselton, G. Allenby, J. Speck, C. Kratzeisen, M. Rosenberger, A. Lovey, and J. F. Grippo, 9-cis retinoic acid stereoisomer binds

and activates the nuclear receptor RXR alpha. *Nature,* **355**, 359–361 (1992).

204. R. A. Heyman, D. J. Mangelsdorf, J. A. Dyck, R. B. Stein, G. Eichele, R. M. Evans, and C. Thaller, 9-cis retinoic acid is a high affinity ligand for the retinoid X receptor. *Cell,* **68**, 397–406 (1992).

205. D. J. Mangelsdorf, and R. M. Evans, The RXR heterodimers and orphan receptors. *Cell,* **83**, 841–850 (1995).

206. D. J. Bettoun, T. P. Burris, K. A. Houck, D. W. Buck, 2nd, K. R. Stayrook, B. Khalifa, J. Lu, W. W. Chin, and S. Nagpal, Retinoid X receptor is a nonsilent major contributor to vitamin D receptor-mediated transcriptional activation. *Mol Endocrinol,* **17**, 2320–2328 (2003).

207. C. Solomon, J. H. White, and R. Kremer, Mitogen-activated protein kinase inhibits 1,25-dihydroxyvitamin D3-dependent signal transduction by phosphorylating human retinoid X receptor alpha. *J Clin Invest,* **103**, 1729–1735 (1999).

208. H. Masuyama, and P. N. MacDonald, Proteasome-mediated degradation of the vitamin D receptor (VDR) and a putative role for SUG1 interaction with the AF-2 domain of VDR. *J Cell Biochem,* **71**, 429–440 (1998).

209. T. T. Wang, L. E. Tavera-Mendoza, D. Laperriere, E. Libby, N. B. MacLeod, Y. Nagai, V. Bourdeau, A. Konstorum, B. Lallemant, R. Zhang, S. Mader, and J. H. White, Large-scale in silico and microarray-based identification of direct 1,25-dihydroxyvitamin D3 target genes. *Mol Endocrinol,* **19**, 2685–2695 (2005).

210. N. J. McKenna, R. B. Lanz, and B. W. O'Malley, Nuclear receptor coregulators: Cellular and molecular biology. *Endocr Rev,* **20**, 321–344 (1999).

211. D. Robyr, A. P. Wolffe, and W. Wahli, Nuclear hormone receptor coregulators in action: Diversity for shared tasks. *Mol Endocrinol,* **14**, 329–347 (2000).

212. W. Feng, R. C. Ribeiro, R. L. Wagner, H. Nguyen, J. W. Apriletti, R. J. Fletterick, J. D. Baxter, P. J. Kushner, and B. L. West, Hormone-dependent coactivator binding to a hydrophobic cleft on nuclear receptors. *Science,* **280**, 1747–1749 (1998).

213. P. Pathrose, O. Barmina, C. Y. Chang, D. P. McDonnell, N. K. Shevde, and J. W. Pike, Inhibition of 1,25-dihydroxyvitamin D3-dependent transcription by synthetic LXXLL peptide antagonists that target the activation domains of the vitamin D and retinoid X receptors. *J Bone Miner Res,* **17**, 2196–2205 (2002).

214. C. Rachez, Z. Suldan, J. Ward, C. P. Chang, D. Burakov, H. Erdjument-Bromage, P. Tempst, and L. P. Freedman, A novel protein complex that interacts with the vitamin D3 receptor in a ligand-dependent manner and enhances VDR transactivation in a cell-free system. *Genes Dev,* **12**, 1787–1800 (1998).

215. C. Rachez, B. D. Lemon, Z. Suldan, V. Bromleigh, M. Gamble, A. M. Naar, H. Erdjument-Bromage, P. Tempst, and L. P. Freedman, Ligand-dependent transcription activation by nuclear receptors requires the DRIP complex. *Nature,* **398**, 824–828 (1999).

216. F. Barletta, L. P. Freedman, and S. Christakos, Enhancement of VDR-mediated transcription by phosphorylation: Correlation with increased interaction between the VDR and DRIP205, a subunit of the VDR-interacting protein coactivator complex. *Mol Endocrinol,* **16**, 301–314 (2002).

217. K. Yamaoka, M. Shindo, K. Iwasaki, I. Yamaoka, Y. Yamamoto, H. Kitagawa, and S. Kato, Multiple co-activator complexes support ligand-induced transactivation function of VDR. *Arch Biochem Biophys,* **460**, 166–171 (2006).

218. S. Kim, N. K. Shevde, and J. W. Pike, 1,25-dihydroxyvitamin D3 stimulates cyclic vitamin D receptor/retinoid X receptor DNA-binding, co-activator recruitment, and histone acetylation in intact osteoblasts. *J Bone Miner Res,* **20**, 305–317 (2005).

219. T. A. Baudino, D. M. Kraichely, S. C. Jefcoat, Jr., S. K. Winchester, N. C. Partridge, and P. N. MacDonald, Isolation and characterization of a novel coactivator protein, NCoA-62, involved in vitamin D-mediated transcription. *J Biol Chem,* **273**, 16434–16441 (1998).

220. J. B. Barry, G. M. Leong, W. B. Church, L. L. Issa, J. A. Eisman, and E. M. Gardiner, Interactions of SKIP/NCoA-62, TFIIB, and retinoid X receptor with vitamin D receptor helix H10 residues. *J Biol Chem,* **278**, 8224–8228 (2003).

221. P. N. MacDonald, D. R. Dowd, C. Zhang, and C. Gu, Emerging insights into the coactivator role of NCoA62/SKIP in vitamin D-mediated transcription. *J Steroid Biochem Mol Biol,* **89–90**, 179–186 (2004).

222. S. Kato, R. Fujiki, and H. Kitagawa, Vitamin D receptor (VDR) promoter targeting through a novel chromatin remodeling complex. *J Steroid Biochem Mol Biol,* **89–90**, 173–178 (2004).

223. R. Fujiki, M. S. Kim, Y. Sasaki, K. Yoshimura, H. Kitagawa, and S. Kato, Ligand-induced transrepression by VDR through association of WSTF with acetylated histones. *EMBO J,* **24**, 3881–3894 (2005).

224. H. Kitagawa, R. Fujiki, K. Yoshimura, Y. Mezaki, Y. Uematsu, D. Matsui, S. Ogawa, K. Unno, M. Okubo, A. Tokita, T. Nakagawa, T. Ito, Y. Ishimi, H. Nagasawa, T. Matsumoto, J. Yanagisawa, and S. Kato, The chromatin-remodeling complex WINAC targets a nuclear receptor to promoters and is impaired in Williams syndrome. *Cell,* **113**, 905–917 (2003).

225. P. M. Yen, Y. Liu, A. Sugawara, and W. W. Chin, Vitamin D receptors repress basal transcription and exert dominant negative activity on triiodothyronine-mediated transcriptional activity. *J Biol Chem,* **271**, 10910–10916 (1996).

226. P. Polly, M. Herdick, U. Moehren, A. Baniahmad, T. Heinzel, and C. Carlberg, VDR-alien: A novel, DNA-selective vitamin D(3) receptor-corepressor partnership. *FASEB J,* **14**, 1455–1463 (2000).

227. C. M. Banwell, D. P. MacCartney, M. Guy, A. E. Miles, M. R. Uskokovic, J. Mansi, P. M. Stewart, L. P. O'Neill, B. M. Turner, K. W. Colston, and M. J. Campbell, Altered nuclear receptor corepressor expression attenuates vitamin D receptor signaling in breast cancer cells. *Clin Cancer Res,* **12**, 2004–2013 (2006).

228. F. L. Khanim, L. M. Gommersall, V. H. Wood, K. L. Smith, L. Montalvo, L. P. O'Neill, Y. Xu, D. M. Peehl, P. M. Stewart, B. M. Turner, and M. J. Campbell, Altered SMRT levels disrupt vitamin D3 receptor signalling in prostate cancer cells. *Oncogene,* **23**, 6712–6725 (2004).

229. J. C. Hsieh, J. M. Sisk, P. W. Jurutka, C. A. Haussler, S. A. Slater, M. R. Haussler, and C. C. Thompson, Physical and functional interaction between the vitamin D receptor and hairless corepressor, two proteins required for hair cycling. *J Biol Chem,* **278**, 38665–38674 (2003).

230. Z. Xie, S. Chang, Y. Oda, and D. D. Bikle, Hairless suppresses vitamin D receptor transactivation in human keratinocytes. *Endocrinology,* **147**, 314–323 (2006).

231. J. C. Hsieh, Y. Shimizu, S. Minoshima, N. Shimizu, C. A. Haussler, P. W. Jurutka, and M. R. Haussler, Novel nuclear localization signal between the two DNA-binding zinc fingers in the human vitamin D receptor. *J Cell Biochem,* **70**, 94–109 (1998).

232. Z. Luo, J. Rouvinen, and P. H. Maenpaa, A peptide C-terminal to the second Zn finger of human vitamin D receptor is able to specify nuclear localization. *Eur J Biochem,* **223**, 381–387 (1994).

233. T. Michigami, A. Suga, M. Yamazaki, C. Shimizu, G. Cai, S. Okada, and K. Ozono, Identification of amino acid sequence in the hinge region of human vitamin D receptor that transfers a cytosolic protein to the nucleus. *J Biol Chem,* **274**, 33531–33538 (1999).

234. A. Racz, and J. Barsony, Hormone-dependent translocation of vitamin D receptors is linked to transactivation. *J Biol Chem,* **274**, 19352–19360 (1999).

235. J. Barsony, and K. Prufer, Vitamin D receptor and retinoid X receptor interactions in motion. *Vitam Horm,* **65**, 345–376 (2002).

236. K. Prufer, and J. Barsony, Retinoid x receptor dominates the nuclear import and export of the unliganded vitamin D receptor. *Mol Endocrinol,* **16**, 1738–1751 (2002).

237. J. Miller, K. Djabali, T. Chen, Y. Liu, M. Ioffreda, S. Lyle, A. M. Christiano, M. Holick, and G. Cotsarelis, Atrichia caused by mutations in the vitamin D receptor gene is a phenocopy of generalized atrichia caused by mutations in the hairless gene. *J Invest Dermatol,* **117**, 612–617 (2001).

238. M. Paradisi, G. S. Chuang, C. Angelo, C. Pedicelli, A. Martinez-Mir, and A. M. Christiano, Atrichia with papular lesions resulting from a novel homozygous missense mutation in the hairless gene. *Clin Exp Dermatol,* **28**, 535–538 (2003).

239. A. Zlotogorski, Z. Hochberg, P. Mirmirani, A. Metzker, D. Ben-Amitai, A. Martinez-Mir, A. A. Panteleyev, and A. M. Christiano, Clinical and pathologic correlations in genetically distinct forms of atrichia. *Arch Dermatol,* **139**, 1591–1596 (2003).

240. A. A. Panteleyev, N. V. Botchkareva, J. P. Sundberg, A. M. Christiano, and R. Paus, The role of the hairless (hr) gene in the regulation of hair follicle catagen transformation. *Am J Pathol,* **155**, 159–171 (1999).

241. M. H. Hardy, The secret life of the hair follicle. *Trends Genet,* **8**, 55–61 (1992).

242. Y. Sakai, and M. B. Demay, Evaluation of keratinocyte proliferation and differentiation in vitamin D receptor knockout mice. *Endocrinology,* **141**, 2043–2049 (2000).

243. C. H. Chen, Y. Sakai, and M. B. Demay, Targeting expression of the human vitamin D receptor to the keratinocytes of vitamin D receptor null mice prevents alopecia. *Endocrinology,* **142**, 5386–5389 (2001).

244. J. J. Wysolmerski, A. E. Broadus, J. Zhou, E. Fuchs, L. M. Milstone, and W. M. Philbrick, Overexpression of parathyroid hormone-related protein in the skin of transgenic mice interferes with hair follicle development. *Proc Natl Acad Sci USA,* **91**, 1133–1137 (1994).

245. G. M. Beaudoin, 3rd, J. M. Sisk, P. A. Coulombe, and C. C. Thompson, Hairless triggers reactivation of hair growth by promoting Wnt signaling. *Proc Natl Acad Sci USA,* **102**, 14653–14658 (2005).

246. P. J. Malloy, R. Xu, L. Peng, P. A. Clark, and D. Feldman, A novel mutation in helix 12 of the vitamin D receptor impairs coactivator interaction and causes hereditary 1,25-dihydroxyvitamin D-resistant rickets without alopecia. *Mol Endocrinol,* **16**, 2538–2546 (2002).

247. M. Li, H. Chiba, X. Warot, N. Messaddeq, C. Gerard, P. Chambon, and D. Metzger, RXR-alpha ablation in skin keratinocytes results in alopecia and epidermal alterations. *Development,* **128**, 675–688 (2001).

248. P. J. Malloy, R. Xu, L. Peng, S. Peleg, A. Al-Ashwal, and D. Feldman, Hereditary 1,25-dihydroxyvitamin D resistant rickets due to a mutation causing multiple defects in vitamin D receptor function. *Endocrinology,* **145**, 5106–5114 (2004).

249. P. J. Malloy, R. Xu, A. Cattani, L. Reyes, and D. Feldman, A unique insertion/substitution in helix H1 of the vitamin D receptor ligand binding domain in a patient with hereditary 1,25-dihydroxyvitamin D-resistant rickets. *J Bone Miner Res,* **19**, 1018–1024 (2004).

250. K. Skorija, M. Cox, J. M. Sisk, D. R. Dowd, P. N. MacDonald, C. C. Thompson, and M. B. Demay, Ligand-independent actions of the vitamin D receptor maintain hair follicle homeostasis. *Mol Endocrinol,* **19**, 855–862 (2005).

251. X. Li, W. Zheng, and Y. C. Li, Altered gene expression profile in the kidney of vitamin D receptor knockout mice. *J Cell Biochem,* **89**, 709–719 (2003).

252. M. Hirst, and D. Feldman, Regulation of 1,25(OH)2 vitamin D3 receptor content in cultured LLC-PK1 kidney cells limits hormonal responsiveness. *Biochem Biophys Res Commun,* **116**, 121–127 (1983).

253. T. L. Chen, J. M. Li, T. VanYe, C. M. Cone, and D. Feldman, Hormonal responses to 1,25-dihydroxyvitamin D3 in cultured mouse osteoblast-like cells—Modulation by changes in receptor level. *J Cell Physiol,* **126**, 21–28 (1986).

254. A. V. Krishnan, and D. Feldman, Cyclic adenosine 3′,5′-monophosphate up-regulates 1,25-dihydroxyvitamin D3 receptor gene expression and enhances hormone action. *Mol Endocrinol,* **6**, 198–206 (1992).

255. A. V. Krishnan, S. D. Cramer, F. R. Bringhurst, and D. Feldman, Regulation of 1,25-dihydroxyvitamin D3 receptors by parathyroid hormone in osteoblastic cells: Role of second messenger pathways. *Endocrinology,* **136**, 705–712 (1995).

256. A. V. Krishnan, and D. Feldman, Regulation of vitamin D receptor abundance. In *Vitamin D* (D. Feldman, F. Glorieux, and J. W. Pike, eds.), pp. 179–200. Academic Press, San Diego (1997).

257. E. M. Costa, M. A. Hirst, and D. Feldman, Regulation of 1,25-dihydroxyvitamin D3 receptors by vitamin D analogs in cultured mammalian cells. *Endocrinology,* **117**, 2203–2210 (1985).

258. E. M. Costa, and D. Feldman, Homologous up-regulation of the 1,25 (OH)2 vitamin D3 receptor in rats. *Biochem Biophys Res Commun,* **137**, 742–747.

259. M. J. Favus, D. J. Mangelsdorf, V. Tembe, B. J. Coe, and M. R. Haussler, Evidence for in vivo upregulation of the intestinal vitamin D receptor during dietary calcium restriction in the rat. *J Clin Invest,* **82**, 218–224 (1988).

260. M. Strom, M. E. Sandgren, T. A. Brown, and H. F. DeLuca, 1,25-dihydroxyvitamin D3 up-regulates the 1,25-dihydroxyvitamin D3 receptor in vivo. *Proc Natl Acad Sci USA,* **86**, 9770–9773 (1989).

261. X. Zhao, and D. Feldman, Regulation of vitamin D receptor abundance and responsiveness during differentiation of HT-29 human colon cancer cells. *Endocrinology,* **132**, 1808–1814 (1993).

262. R. J. Wiese, A. Uhland-Smith, T. K. Ross, J. M. Prahl, and H. F. DeLuca, Up-regulation of the vitamin D receptor in response to 1,25-dihydroxyvitamin D3 results from ligand-induced stabilization. *J Biol Chem,* **267**, 20082–20086 (1992).

263. D. Santiso-Mere, T. Sone, G. T. Hilliard, J. W. Pike, and D. P. McDonnell, Positive regulation of the vitamin D receptor by its cognate ligand in heterologous expression systems. *Mol Endocrinol,* **7**, 833–839 (1993).

264. N. C. Arbour, J. M. Prahl, and H. F. DeLuca, Stabilization of the vitamin D receptor in rat osteosarcoma cells through the action of 1,25-dihydroxyvitamin D3. *Mol Endocrinol, 7*, 1307–1312 (1993).

265. M. L. Chen, A. Perez, D. K. Sanan, G. Heinrich, T. C. Chen, and M. F. Holick, Induction of vitamin D receptor mRNA expression in psoriatic plaques correlates with clinical response to 1,25-dihydroxyvitamin D3. *J Invest Dermatol, 106*, 637–641 (1996).

266. T. L. Chen, and D. Feldman, Regulation of 1,25-dihydroxyvitamin D3 receptors in cultured mouse bone cells. Correlation of receptor concentration with the rate of cell division. *J Biol Chem, 256*, 5561–5566 (1981).

267. A. Krishnan, and D. Feldman, Stimulation of 1,25-dihydroxyvitamin D3 receptor gene expression in cultured cells by serum and growth factors. *J Bone Miner Res, 6*, 1099–1107 (1991).

268. F. Jehan, and H. F. DeLuca, Cloning and characterization of the mouse vitamin D receptor promoter. *Proc Natl Acad Sci USA, 94*, 10138–10143 (1997).

269. J. P. van Leeuwen, J. C. Birkenhager, T. Vink-van Wijngaarden, G. J. van den Bemd, and H. A. Pols, Regulation of 1,25-dihydroxyvitamin D3 receptor gene expression by parathyroid hormone and cAMP-agonists. *Biochem Biophys Res Commun, 185*, 881–886 (1992).

270. S. J. Smith, L. M. Green, M. E. Hayes, and E. B. Mawer, Prostaglandin E2 regulates vitamin D receptor expression, vitamin D-24-hydroxylase activity and cell proliferation in an adherent human myeloid leukemia cell line (Ad-HL60). *Prostaglandins Other Lipid Mediat, 57*, 73–85 (1999).

271. A. V. Krishnan, and D. Feldman, Activation of protein kinase-C inhibits vitamin D receptor gene expression. *Mol Endocrinol, 5*, 605–612 (1991).

272. S. Pillai, D. D. Bikle, and P. M. Elias, 1,25-dihydroxyvitamin D production and receptor binding in human keratinocytes varies with differentiation. *J Biol Chem, 263*, 5390–5395 (1988).

273. X. P. Yu, H. Mocharla, F. G. Hustmyer, and S. C. Manolagas, Vitamin D receptor expression in human lymphocytes. Signal requirements and characterization by Western blots and DNA sequencing. *J Biol Chem, 266*, 7588–7595 (1991).

274. T. L. Chen, C. M. Cone, E. Morey-Holton, and D. Feldman, Glucocorticoid regulation of 1,25(OH)2-vitamin D3 receptors in cultured mouse bone cells. *J Biol Chem, 257*, 13564–13569 (1982).

275. M. Hirst, and D. Feldman, Glucocorticoid regulation of 1,25(OH)2vitamin D3 receptors: Divergent effects on mouse and rat intestine. *Endocrinology, 111*, 1400–1402 (1982).

276. M. Hirst, and D. Feldman, Glucocorticoids down-regulate the number of 1,25-dihydroxyvitamin D3 receptors in mouse intestine. *Biochem Biophys Res Commun, 105*, 1590–1596 (1982).

277. H. K. Nielsen, E. F. Eriksen, T. Storm, and L. Mosekilde, The effects of short-term, high-dose treatment with prednisone on the nuclear uptake of 1,25-dihydroxyvitamin D3 in monocytes from normal human subjects. *Metabolism, 37*, 109–114 (1988).

278. M. R. Walters, An estrogen-stimulated 1,25-dihydroxyvitamin D3 receptor in rat uterus. *Biochem Biophys Res Commun, 103*, 721–726 (1981).

279. Y. Liel, S. Kraus, J. Levy, and S. Shany, Evidence that estrogens modulate activity and increase the number of 1,25-dihydroxyvitamin D receptors in osteoblast-like cells (ROS 17/2. 8). *Endocrinology, 130*, 2597–2601 (1992).

280. P. M. Petkovich, J. N. Heersche, D. O. Tinker, and G. Jones, Retinoic acid stimulates 1,25-dihydroxyvitamin D3 binding in rat osteosarcoma cells. *J Biol Chem, 259*, 8274–8280 (1984).

281. T. L. Chen, and D. Feldman, Retinoic acid modulation of 1,25(OH)2 vitamin D3 receptors and bioresponse in bone cells: Species differences between rat and mouse. *Biochem Biophys Res Comm, 132*, 74–80 (1985).

282. T. A. Reinhardt, and R. L. Horst, Parathyroid hormone down-regulates 1,25-dihydroxyvitamin D receptors (VDR) and VDR messenger ribonucleic acid in vitro and blocks homologous up-regulation of VDR in vivo. *Endocrinology, 127*, 942–948 (1990).

283. A. B. Korkor, Reduced binding of [^{3}H]1,25-dihydroxyvitamin D$_3$ in the parathyroid glands of patients with renal failure. *N Engl J Med, 316*, 1573–1577 (1987).

284. S. B. Brown, T. T. Brierley, N. Palanisamy, I. B. Salusky, W. Goodman, M. L. Brandi, T. B. Drueke, E. Sarfati, P. Urena, R. S. Chaganti, J. W. Pike, and A. Arnold, Vitamin D receptor as a candidate tumor-suppressor gene in severe hyperparathyroidism of uremia. *J Clin Endocrinol Metab, 85*, 868–872 (2000).

285. A. S. Dusso, A. J. Brown, and E. Slatopolsky, Vitamin D. *Am J Physiol Renal Physiol, 289*, F8–28 (2005).

286. S. J. Silverberg, E. Shane, D. W. Dempster, and J. P. Bilezikian, The effects of vitamin D insufficiency in patients with primary hyperparathyroidism. *Am J Med, 107*, 561–567 (1999).

287. D. Feldman, Vitamin D, parathyroid hormone, and calcium: A complex regulatory network. *Am J Med, 107*, 637–639 (1999).

288. D. S. Rao, M. Honasoge, G. W. Divine, E. R. Phillips, M. W. Lee, M. R. Ansari, G. B. Talpos, and A. M. Parfitt, Effect of vitamin D nutrition on parathyroid adenoma weight: Pathogenetic and clinical implications. *J Clin Endocrinol Metab, 85*, 1054–1058 (2000).

289. Y. C. Li, A. E. Pirro, M. Amling, G. Delling, R. Baron, R. Bronson, and M. B. Demay, Targeted ablation of the vitamin D receptor: An animal model of vitamin D-dependent rickets type II with alopecia. *Proc Natl Acad Sci USA, 94*, 9831–9835 (1997).

290. M. Amling, M. Priemel, T. Holzmann, K. Chapin, J. M. Rueger, R. Baron, and M. B. Demay, Rescue of the skeletal phenotype of vitamin D receptor-ablated mice in the setting of normal mineral ion homeostasis: Formal histomorphometric and biomechanical analyses. *Endocrinology, 140*, 4982–4987 (1999).

291. C. M. Bula, J. Huhtakangas, C. Olivera, J. E. Bishop, A. W. Norman, and H. L. Henry, Presence of a truncated form of the vitamin D receptor (VDR) in a strain of VDR-knockout mice. *Endocrinology, 146*, 5581–5586 (2005).

292. K. Kinuta, H. Tanaka, T. Moriwake, K. Aya, S. Kato, and Y. Seino, Vitamin D is an important factor in estrogen biosynthesis of both female and male gonads. *Endocrinology, 141*, 1317–1324 (2000).

293. S. Tanaka, M. Haji, R. Takayanagi, Y. Sugioka, and H. Nawata, 1,25-dihydroxyvitamin D3 enhances the enzymatic activity and expression of the messenger ribonucleic acid for aromatase cytochrome P450 synergistically with dexamethasone depending on the vitamin D receptor level in cultured human osteoblasts. *Endocrinology, 137*, 1860–1869 (1996).

294. R. G. Erben, D. W. Soegiarto, K. Weber, U. Zeitz, M. Lieberherr, R. Gniadecki, G. Moller, J. Adamski, and R. Balling, Deletion of deoxyribonucleic acid binding domain of the vitamin D

receptor abrogates genomic and nongenomic functions of vitamin D. *Mol Endocrinol,* **16**, 1524–1537 (2002).

295. S. J. Van Cromphaut, M. Dewerchin, J. G. Hoenderop, I. Stockmans, E. Van Herck, S. Kato, R. J. Bindels, D. Collen, P. Carmeliet, R. Bouillon, and G. Carmeliet, Duodenal calcium absorption in vitamin D receptor-knockout mice: Functional and molecular aspects. *Proc Natl Acad Sci USA,* **98**, 13324–13329 (2001).

296. D. Goltzman, D. Miao, D. K. Panda, and G. N. Hendy, Effects of calcium and of the vitamin D system on skeletal and calcium homeostasis: Lessons from genetic models. *J Steroid Biochem Mol Biol,* **89–90**, 485–489 (2004).

297. G. N. Hendy, and D. Goltzman, Does calcitriol have actions independent from the vitamin D receptor in maintaining skeletal and mineral homeostasis? *Curr Opin Nephrol Hypertens,* **14**, 350–354 (2005).

298. G. N. Hendy, K. A. Hruska, S. Mathew, and D. Goltzman, New insights into mineral and skeletal regulation by active forms of vitamin D. *Kidney Int,* **69**, 218–223 (2006).

299. D. K. Panda, D. Miao, I. Bolivar, J. Li, R. Huo, G. N. Hendy, and D. Goltzman, Inactivation of the 25-hydroxyvitamin D 1alpha-hydroxylase and vitamin D receptor demonstrates independent and interdependent effects of calcium and vitamin D on skeletal and mineral homeostasis. *J Biol Chem,* **279**, 16754–16766 (2004).

300. T. Koyama, M. Shibakura, M. Ohsawa, R. Kamiyama, and S. Hirosawa, Anticoagulant effects of 1alpha,25-dihydroxyvitamin D3 on human myelogenous leukemia cells and monocytes. *Blood,* **92**, 160–167 (1998).

301. K. Aihara, H. Azuma, M. Akaike, Y. Ikeda, M. Yamashita, T. Sudo, H. Hayashi, Y. Yamada, F. Endoh, M. Fujimura, T. Yoshida, H. Yamaguchi, S. Hashizume, M. Kato, K. Yoshimura, Y. Yamamoto, S. Kato, and T. Matsumoto, Disruption of nuclear vitamin D receptor gene causes enhanced thrombogenicity in mice. *J Biol Chem,* **279**, 35798–35802 (2004).

302. Y. C. Li, J. Kong, M. Wei, Z. F. Chen, S. Q. Liu, and L. P. Cao, 1,25-dihydroxyvitamin D(3) is a negative endocrine regulator of the renin-angiotensin system. *J Clin Invest,* **110**, 229–238 (2002).

303. J. Kong, and Y. C. Li, Effect of ANG II type I receptor antagonist and ACE inhibitor on vitamin D receptor-null mice. *Am J Physiol Regul Integr Comp Physiol,* **285**, R255–261 (2003).

304. A. Wittke, V. Weaver, B. D. Mahon, A. August, and M. T. Cantorna, Vitamin D receptor-deficient mice fail to develop experimental allergic asthma. *J Immunol,* **173**, 3432–3436 (2004).

305. M. Froicu, V. Weaver, T. A. Wynn, M. A. McDowell, J. E. Welsh, and M. T. Cantorna, A crucial role for the vitamin D receptor in experimental inflammatory bowel diseases. *Mol Endocrinol,* **17**, 2386–2392 (2003).

306. I. Endo, D. Inoue, T. Mitsui, Y. Umaki, M. Akaike, T. Yoshizawa, S. Kato, and T. Matsumoto, Deletion of vitamin D receptor gene in mice results in abnormal skeletal muscle development with deregulated expression of myoregulatory transcription factors. *Endocrinology,* **144**, 5138–5144 (2003).

307. G. M. Zinser, M. Suckow, and J. Welsh, Vitamin D receptor (VDR) ablation alters carcinogen-induced tumorigenesis in mammary gland, epidermis and lymphoid tissues. *J Steroid Biochem Mol Biol,* **97**, 153–164 (2005).

308. A. W. Norman, 1a,25(OH)$_2$-vitamin D$_3$ mediated rapid and genomic responses are dependent upon critical structure-function relationships for both the ligand and receptor(s). In *Vitamin D* (D. Feldman, F. Glorieux, and J. W. Pike, eds.), pp. 381–407. Elsevier Academic Press, San Diego (2005).

309. M. C. Farach-Carson, and J. J. Bergh, Effects of 1,25-dihydroxyvitamin D$_3$ on voltage-sensitive calcium channels in the vitamin D endocrine system. In *Vitamin D*, 2nd ed. (D. Feldman, F. Glorieux, and J. W. Pike, eds.), pp. 751–760. Elsevier Academic Press, San Diego (2005).

310. A. W. Norman, R. Bouillon, M. C. Farach-Carson, J. E. Bishop, L. X. Zhou, I. Nemere, J. Zhao, K. R. Muralidharan, and W. H. Okamura, Demonstration that 1 beta,25-dihydroxyvitamin D3 is an antagonist of the nongenomic but not genomic biological responses and biological profile of the three A-ring diastereomers of 1 alpha,25-dihydroxyvitamin D3. *J Biol Chem,* **268**, 20022–20030 (1993).

311. A. W. Norman, W. H. Okamura, M. W. Hammond, J. E. Bishop, M. C. Dormanen, R. Bouillon, H. van Baelen, A. L. Ridall, E. Daane, R. Khoury, and M. C. Farach-Carson, Comparison of 6-s-cis- and 6-s-trans-locked analogs of 1alpha,25-dihydroxyvitamin D3 indicates that the 6-s-cis conformation is preferred for rapid nongenomic biological responses and that neither 6-s-cis- nor 6-s-trans-locked analogs are preferred for genomic biological responses. *Mol Endocrinol,* **11**, 1518–1531 (1997).

312. M. Bhatia, J. B. Kirkland, and K. A. Meckling-Gill, Monocytic differentiation of acute promyelocytic leukemia cells in response to 1,25-dihydroxyvitamin D3 is independent of nuclear receptor binding. *J Biol Chem,* **270**, 15962–15965.

313. M. Kajikawa, H. Ishida, S. Fujimoto, E. Mukai, M. Nishimura, J. Fujita, Y. Tsuura, Y. Okamoto, A. W. Norman, and Y. Seino, An insulinotropic effect of vitamin D analog with increasing intracellular Ca2+ concentration in pancreatic beta-cells through nongenomic signal transduction. *Endocrinology,* **140**, 4706–4712 (1999).

314. G. Wong, R. Gupta, K. M. Dixon, S. S. Deo, S. M. Choong, G. M. Halliday, J. E. Bishop, S. Ishizuka, A. W. Norman, G. H. Posner, and R. S. Mason, 1,25-dihydroxyvitamin D and three low-calcemic analogs decrease UV-induced DNA damage via the rapid response pathway. *J Steroid Biochem Mol Biol,* **89–90**, 567–570 (2004).

315. J. A. Huhtakangas, C. J. Olivera, J. E. Bishop, L. P. Zanello, and A. W. Norman, The vitamin D receptor is present in caveolae-enriched plasma membranes and binds 1 alpha,25(OH)2-vitamin D3 in vivo and in vitro. *Mol Endocrinol,* **18**, 2660–2671 (2004).

316. L. P. Zanello, and A. W. Norman, Rapid modulation of osteoblast ion channel responses by 1alpha,25(OH)2-vitamin D3 requires the presence of a functional vitamin D nuclear receptor. *Proc Natl Acad Sci USA,* **101**, 1589–1594 (2004).

317. R. K. Wali, J. Kong, M. D. Sitrin, M. Bissonnette, and Y. C. Li, Vitamin D receptor is not required for the rapid actions of 1,25-dihydroxyvitamin D3 to increase intracellular calcium and activate protein kinase C in mouse osteoblasts. *J Cell Biochem,* **88**, 794–801 (2003).

318. I. Nemere, S. E. Safford, B. Rohe, M. M. DeSouza, and M. C. Farach-Carson, Identification and characterization of 1,25D3-membrane-associated rapid response, steroid (1,25D3-mARRS) binding protein. *J Steroid Biochem Mol Biol,* **89–90**, 281–285 (2004).

319. I. Nemere, M. C. Farach-Carson, B. Rohe, T. M. Sterling, A. W. Norman, B. D. Boyan, and S. E. Safford, Ribozyme knockdown functionally links a 1,25(OH)2D3 membrane binding protein (1,25D3-mARRS) and phosphate uptake in intestinal cells. *Proc Natl Acad Sci USA,* **101**, 7392–7397 (2004).

320. B. D. Boyan, and Z. Schwartz, Rapid vitamin D-dependent PKC signaling shares features with estrogen-dependent

PKC signaling in cartilage and bone. *Steroids*, **69**, 591–597 (2004).

321. L. P. Zanello, and A. Norman, 1alpha,25(OH)2 vitamin D3 actions on ion channels in osteoblasts. *Steroids*, **71**, 291–297 (2006).

322. R. P. Heaney, Vitamin D: Role in the calcium economy. In *Vitamin D*, 2nd ed. (D. Feldman, F. Glorieux, and J. W. Pike, eds.), pp. 773–787. Elsevier Academic Press, San Diego (2005).

323. G. Carmeliet, S. Van Cromphaut, E. Daci, C. Maes, and R. Bouillon, Disorders of calcium homeostasis. *Best Pract Res Clin Endocrinol Metab*, **17**, 529–546 (2003).

324. A. R. Glass, and C. Eil, Ketoconazole-induced reduction in serum 1,25-dihydroxyvitamin D and total serum calcium in hypercalcemic patients. *J Clin Endocrinol Metab*, **66**, 934–938 (1988).

325. J. S. Adams, O. P. Sharma, M. M. Diz, and D. B. Endres, Ketoconazole decreases the serum 1,25-dihydroxyvitamin D and calcium concentration in sarcoidosis-associated hypercalcemia. *J Clin Endocrinol Metab*, **70**, 1090–1095 (1990).

326. D. Feldman, Ketoconazole and other imidazole derivatives as inhibitors of steroidogenesis. *Endocrine Rev*, **7**, 409–420 (1986).

327. D. S. Loose, P. B. Kan, M. A. Hirst, R. A. Marcus, and D. F. Feldman, Ketoconazole blocks adrenal steroidogenesis by inhibiting cytochrome P450-dependent enzymes. *J Clin Invest*, **71**, 1495–1499 (1983).

328. H. L. Henry, Effect of ketoconazole and miconazole on 25-hydroxyvitamin D3 metabolism by cultured chick kidney cells. *J Steroid Biochem*, **23**, 991–994 (1985).

329. D. Zehnder, R. Bland, M. C. Williams, R. W. McNinch, A. J. Howie, P. M. Stewart, and M. Hewison, Extrarenal expression of 25-hydroxyvitamin d(3)-1 alpha-hydroxylase. *J Clin Endocrinol Metab*, **86**, 888–894 (2001).

330. L. E. Siltzbach, D. G. James, E. Neville, J. Turiaf, J. P. Battesti, O. P. Sharma, Y. Hosoda, R. Mikami, and M. Odaka, Course and prognosis of sarcoidosis around the world. *Am J Med*, **57**, 847–852 (1974).

331. G. L. Barbour, J. W. Coburn, E. Slatopolsky, A. W. Norman, and R. L. Horst, Hypercalcemia in an anephric patient with sarcoidosis: Evidence for extrarenal generation of 1,25-dihydroxyvitamin D. *N Engl J Med*, **305**, 440–443 (1981).

332. J. F. Seymour, and R. F. Gagel, Calcitriol: The major humoral mediator of hypercalcemia in Hodgkin's disease and non-Hodgkin's lymphomas. *Blood*, **82**, 1383–1394 (1993).

333. J. F. Seymour, R. F. Gagel, F. B. Hagemeister, M. A. Dimopoulos, and F. Cabanillas, Calcitriol production in hypercalcemic and normocalcemic patients with non-Hodgkin lymphoma. *Ann Intern Med*, **121**, 633–640 (1994).

334. M. Hewison, V. Kantorovich, H. R. Liker, A. J. Van Herle, P. Cohan, D. Zehnder, and J. S. Adams, Vitamin D-mediated hypercalcemia in lymphoma: Evidence for hormone production by tumor-adjacent macrophages. *J Bone Miner Res*, **18**, 579–582 (2003).

335. R. Kumar, W. R. Cohen, P. Silva, and F. H. Epstein, Elevated 1,25-dihydroxyvitamin D plasma levels in normal human pregnancy and lactation. *J Clin Invest*, **63**, 342–344 (1979).

336. D. D. Bikle, E. Gee, B. Halloran, and J. G. Haddad, Free 1,25-dihydroxyvitamin D levels in serum from normal subjects, pregnant subjects, and subjects with liver disease. *J Clin Invest*, **74**, 1966–1971 (1984).

337. C. Cheema, B. F. Grant, and R. Marcus, Effects of estrogen on circulating "free" and total 1,25-dihydroxyvitamin D and on the parathyroid-vitamin D axis in postmenopausal women. *J Clin Invest*, **83**, 537–542 (1989).

338. S. Vargas, R. Bouillon, H. van Baelen, and L. G. Raisz, Effects of vitamin D-binding protein on bone resorption stimulated by 1,25 dihydroxyvitamin D3. *Calcif Tissue Int*, **47**, 164–168 (1990).

339. H. J. Kalkwarf, and B. L. Specker, Vitamin D metabolism in pregnancy and lactation. In *Vitamin D*, 2nd ed. (D. Feldman, J. W. Pike, and F. Glorieux, eds.), pp. 839–850. Elsevier Academic Press, San Diego (2005).

340. F. H. Glorieux, and R. St. Arnaud, Vitamin D pseudodeficiency. In *Vitamin D*, 2nd ed. (D. Feldman, J. W. Pike, and F. Glorieux, eds.), pp. 1197–1205. Elsevier Academic Press, San Diego (2005).

341. M. K. Drezner, Clinical disorders of phosphate homeostasis. In *Vitamin D*, 2nd ed. (D. Feldman, J. W. Pike, and F. Glorieux, eds.), pp. 1159–1187. Elsevier Academic Press, San Diego (2005).

342. T. O. Carpenter, and K. L. Insogna, The hypocalcemic disorders: Differential diagnosis and therapeutic use of vitamin D. In *Vitamin D*, 2nd ed. (D. Feldman, J. W. Pike, and F. Glorieux, eds.), pp. 1049–1063. Elsevier Academic Press, San Diego (2005).

343. R. Bouillon, The many faces of rickets. *N Engl J Med*, **338**, 681–682 (1998).

344. W. L. Miller, and A. A. Portale, Genetic disorders of vitamin D biosynthesis. *Endocrinol Metab Clin North Am*, **28**, 825–840 (1999).

345. V. A. Prader, R. Illig, and E. Heierli, Eine besondere form der primaren vitamin-D-resistenten rachitis mit hypocalcamie und autosomal-dominantem erbgang: Die hereditare pseudo-mangelrachitis. *Helvetica Paediatrica Acta*, **16**, 452–468 (1961).

346. X. Wang, M. Y. Zhang, W. L. Miller, and A. A. Portale, Novel gene mutations in patients with 1alpha-hydroxylase deficiency that confer partial enzyme activity in vitro. *J Clin Endocrinol Metab*, **87**, 2424–2430 (2002).

347. D. K. Panda, D. Miao, M. L. Tremblay, J. Sirois, R. Farookhi, G. N. Hendy, and D. Goltzman, Targeted ablation of the 25-hydroxyvitamin D 1alpha-hydroxylase enzyme: Evidence for skeletal, reproductive, and immune dysfunction. *Proc Natl Acad Sci USA*, **98**, 7498–7503 (2001).

348. O. Dardenne, J. Prudhomme, S. A. Hacking, F. H. Glorieux, and R. St. Arnaud, Rescue of the pseudo-vitamin D deficiency rickets phenotype of CYP27B1-deficient mice by treatment with 1,25-dihydroxyvitamin D3, biochemical, histomorphometric, and biomechanical analyses. *J Bone Miner Res*, **18**, 637–643 (2003).

349. H. H. Ritchie, M. R. Hughes, E. T. Thompson, P. J. Malloy, Z. Hochberg, D. Feldman, J. W. Pike, and B. W. O'Malley, An ochre mutation in the vitamin D receptor gene causes hereditary 1,25-dihydroxyvitamin D_3-resistant rickets in three families. *Proc Natl Acad Sci USA*, **86**, 9783–9787 (1989).

350. P. J. Malloy, Z. Hochberg, D. Tiosano, J. W. Pike, M. R. Hughes, and D. Feldman, The molecular basis of hereditary 1,25-dihydroxyvitamin D_3 resistant rickets in seven related families. *J Clin Invest*, **86**, 2071–2079 (1990).

351. K. Kristjansson, A. R. Rut, M. Hewison, J. L. O'Riordan, and M. R. Hughes, Two mutations in the hormone binding domain of the vitamin D receptor cause tissue resistance to 1,25 dihydroxyvitamin D_3. *J Clin Invest*, **92**, 12–16 (1993).

352. R. J. Wiese, H. Goto, J. M. Prahl, S. J. Marx, M. Thomas, A. al-Aqeel, and H. F. DeLuca, Vitamin D-dependency rickets type II: Truncated vitamin D receptor in three kindreds. *Mol Cell Endocrinol*, **90**, 197–201 (1993).

353. N. S. Hawa, F. J. Cockerill, S. Vadher, M. Hewison, A. R. Rut, J. W. Pike, J. L. O'Riordan, and S. M. Farrow,

Identification of a novel mutation in hereditary vitamin D resistant rickets causing exon skipping. *Clin Endocrinol,* **45**, 85–92 (1996).

354. F. J. Cockerill, N. S. Hawa, N. Yousaf, M. Hewison, J. L. O'Riordan, and S. M. Farrow, Mutations in the vitamin D receptor gene in three kindreds associated with hereditary vitamin D resistant rickets. *J Clin Endocrinol Metab,* **82**, 3156–3160 (1997).

355. J. B. Mechica, M. O. Leite, B. B. Mendonca, E. S. Frazzatto, A. Borelli, and A. C. Latronico, A novel nonsense mutation in the first zinc finger of the vitamin D receptor causing hereditary 1,25-dihydroxyvitamin D_3–resistant rickets. *J Clin Endocrinol Metab,* **82**, 3892–3894 (1997).

356. W. Zhu, P. J. Malloy, E. Delvin, G. Chabot, and D. Feldman, Hereditary 1,25-dihydroxyvitamin D-resistant rickets due to an opal mutation causing premature termination of the vitamin D receptor. *J Bone Miner Res,* **13**, 259–264 (1998).

357. P. J. Malloy, W. Zhu, R. Bouillon, and D. Feldman, A novel nonsense mutation in the ligand binding domain of the vitamin D receptor causes hereditary 1,25-dihydroxyvitamin D-resistant rickets. *Mol Genet Metab,* **77**, 314–318 (2002).

358. P. Katavetin, S. Wacharasindhu, and V. Shotelersuk, A girl with a novel splice site mutation in VDR supports the role of a ligand-independent VDR function on hair cycling. *Horm Res,* **66**, 273–276 (2006).

359. P. J. Malloy, J. Wang, L. Peng, S. Nayak, J. M. Sisk, C. C. Thompson, and D. Feldman, A unique insertion/duplication in the VDR gene that truncates the VDR causing hereditary 1,25-dihydroxyvitamin D-resistant rickets without alopecia. *Arch Biochem Biophys,* **460**, 285–292 (2007).

360. M. R. Hughes, P. J. Malloy, D. G. Kieback, R. A. Kesterson, J. W. Pike, D. Feldman, and B. W. O'Malley, Point mutations in the human vitamin D receptor gene associated with hypocalcemic rickets. *Science,* **242**, 1702–1705 (1988).

361. T. Sone, S. J. Marx, U. A. Liberman, and J. W. Pike, A unique point mutation in the human vitamin D receptor chromosomal gene confers hereditary resistance to 1,25-dihydroxyvitamin D_3. *Mol Endocrinol,* **4**, 623–631 (1990).

362. T. Saijo, M. Ito, E. Takeda, A. H. Huq, E. Naito, I. Yokota, T. Sone, J. W. Pike, and Y. Kuroda, A unique mutation in the vitamin D receptor gene in three Japanese patients with vitamin D-dependent rickets type II: Utility of single-strand conformation polymorphism analysis for heterozygous carrier detection. *Am J Hum Genet,* **49**, 668–673 (1991).

363. H. Yagi, K. Ozono, H. Miyake, K. Nagashima, T. Kuroume, and J. W. Pike, A new point mutation in the deoxyribonucleic acid-binding domain of the vitamin D receptor in a kindred with hereditary 1,25-dihydroxyvitamin D-resistant rickets. *J Clin Endocrinol Metab,* **76**, 509–512 (1993).

364. P. J. Malloy, Y. Weisman, and D. Feldman, Hereditary 1 alpha,25-dihydroxyvitamin D-resistant rickets resulting from a mutation in the vitamin D receptor deoxyribonucleic acid-binding domain. *J Clin Endocrinol Metab,* **78**, 313–316 (1994).

365. A. R. Rut, M. Hewison, K. Kristjansson, B. Luisi, M. R. Hughes, and J. L. O'Riordan, Two mutations causing vitamin D resistant rickets: Modelling on the basis of steroid hormone receptor DNA-binding domain crystal structures. *Clin Endocrinol,* **41**, 581–590 (1994).

366. N. U. Lin, P. J. Malloy, N. Sakati, A. Al-Ashwal, and D. Feldman, A novel mutation in the deoxyribonucleic acid-binding domain of the vitamin D receptor causes hereditary 1,25-dihydroxyvitamin D-resistant rickets. *J Clin Endocrinol Metab,* **81**, 2564–2569 (1996).

367. G. K. Whitfield, S. H. Selznick, C. A. Haussler, J. C. Hsieh, M. A. Galligan, P. W. Jurutka, P. D. Thompson, S. M. Lee, J. E. Zerwekh, and M. R. Haussler, Vitamin D receptors from patients with resistance to 1,25-dihydroxyvitamin D_3, point mutations confer reduced transactivation in response to ligand and impaired interaction with the retinoid X receptor heterodimeric partner. *Mol Endocrinol,* **10**, 1617–1631 (1996).

368. P. J. Malloy, T. R. Eccleshall, C. Gross, L. Van Maldergem, R. Bouillon, and D. Feldman, Hereditary vitamin D resistant rickets caused by a novel mutation in the vitamin D receptor that results in decreased affinity for hormone and cellular hyporesponsiveness. *J Clin Invest,* **99**, 297–304 (1997).

369. P. J. Malloy, W. Zhu, X. Y. Zhao, G. B. Pehling, and D. Feldman, A novel inborn error in the ligand-binding domain of the vitamin D receptor causes hereditary vitamin D-resistant rickets. *Mol Genet Metab,* **73**, 138–148 (2001).

370. T. M. Nguyen, P. Adiceam, M. L. Kottler, H. Guillozo, M. Rizk-Rabin, F. Brouillard, P. Lagier, C. Palix, J. M. Garnier, and M. Garabedian, Tryptophan missense mutation in the ligand-binding domain of the vitamin D receptor causes severe resistance to 1,25-dihydroxyvitamin D. *J Bone Miner Res,* **17**, 1728–1737 (2002).

371. M. Nguyen, A. d'Alesio, J. M. Pascussi, R. Kumar, M. D. Griffin, X. Dong, H. Guillozo, M. Rizk-Rabin, C. Sinding, P. Bougneres, F. Jehan, and M. Garabedian, Vitamin D-resistant rickets and type 1 diabetes in a child with compound heterozygous mutations of the vitamin D receptor (L263R and R391S): Dissociated responses of the CYP-24 and rel-B promoters to 1,25-dihydroxyvitamin D3. *J Bone Miner Res,* **21**, 886–894 (2006).

372. S. Balsan, M. Garabedian, M. Larchet, A. M. Gorski, G. Cournot, C. Tau, A. Bourdeau, C. Silve, and C. Ricour, Long-term nocturnal calcium infusions can cure rickets and promote normal mineralization in hereditary resistance to 1,25-dihydroxyvitamin D. *J Clin Invest,* **77**, 1661–1667 (1986).

373. Z. Hochberg, D. Tiosano, and L. Even, Calcium therapy for calcitriol-resistant rickets. *J Ped,* **121**, 803–808 (1992).

374. A. al-Aqeel, P. Ozand, S. Sobki, W. Sewairi, and S. Marx, The combined use of intravenous and oral calcium for the treatment of vitamin D dependent rickets type II (VDDRII). *Clin Endocrinol,* **39**, 229–237 (1993).

375. Y. Weisman, N. Jaccard, C. Legum, Z. Spirer, G. Yedwab, L. Even, S. Edelstein, A. M. Kaye, and Z. Hochberg, Prenatal diagnosis of vitamin D-dependent rickets, type II: Response to 1,25-dihydroxyvitamin D in amniotic fluid cells and fetal tissues. *J Clin Endocrinol Metab,* **71**, 937–943 (1990).

376. Y. Weisman, P. J. Malloy, A. V. Krishnan, N. Jaccard, D. Feldman, and Z. Hochberg, Prenatal diagnosis of calcitriol resistant rickets (CRR) by 1,25(OH)$_2$D$_3$ binding, 24-hydroxylase induction and RFLP analysis. Vitamin D: A pluripotent steroid hormone: Structural studies, molecular endocrinology, and clinical applications: Proceedings of the ninth workshop on vitamin D, Orlando, 106 (1994).

377. A. G. Uitterlinden, H. Burger, Q. Huang, E. Odding, C. M. van Duijn, A. Hofman, J. C. Birkenhager, J. P. van Leeuwen, and H. A. Pols, Vitamin D receptor genotype is associated with radiographic osteoarthritis at the knee. *J Clin Invest,* **100**, 259–263 (1997).

378. J. A. Eisman, Genetics of osteoporosis. *Endocr Rev,* **20**, 788–804 (1999).

379. J. M. Zmuda, J. A. Cauley, and R. E. Ferrell, Recent progress in understanding the genetic susceptibility to osteoporosis. *Genet Epidemiol,* **16**, 356–367 (1999).

380. A. G. Uitterlinden, S. H. Ralston, M. L. Brandi, A. H. Carey, D. Grinberg, B. L. Langdahl, P. Lips, R. Lorenc, B. Obermayer-Pietsch, J. Reeve, D. M. Reid, A. Amidei, A. Bassiti, M. Bustamante, L. B. Husted, A. Diez-Perez, H. Dobnig, A. M. Dunning, A. Enjuanes, A. Fahrleitner-Pammer, Y. Fang, E. Karczmarewicz, M. Kruk, J. P. van Leeuwen, C. Mavilia, J. B. van Meurs, J. Mangion, F. E. McGuigan, H. A. Pols, W. Renner, F. Rivadeneira, N. M. van Schoor, S. Scollen, R. E. Sherlock, and J. P. Ioannidis, The association between common vitamin D receptor gene variations and osteoporosis: A participant-level meta-analysis. *Ann Intern Med*, **145**, 255–264 (2006).

381. L. Adorini, 1,25-dihydroxyvitamin D3 analogs as potential therapies in transplantation. *Curr Opin Investig Drugs*, **3**, 1458–1463 (2002).

382. L. Adorini, Intervention in autoimmunity: The potential of vitamin D receptor agonists. *Cell Immunol*, **233**, 115–124 (2005).

383. R. Bouillon, L. Verlinden, G. Eelen, P. De Clercq, M. Vandewalle, C. Mathieu, and A. Verstuyf, Mechanisms for the selective action of vitamin D analogs. *J Steroid Biochem Mol Biol*, **97**, 21–30 (2005).

384. Y. Ma, B. Khalifa, Y. K. Yee, J. Lu, A. Memezawa, R. S. Savkur, Y. Yamamoto, S. R. Chintalacharuvu, K. Yamaoka, K. R. Stayrook, K. S. Bramlett, Q. Q. Zeng, S. Chandrasekhar, X. P. Yu, J. H. Linebarger, S. J. Iturria, T. P. Burris, S. Kato, W. W. Chin, and S. Nagpal, Identification and characterization of noncalcemic, tissue-selective, nonsecosteroidal vitamin D receptor modulators. *J Clin Invest*, **116**, 892–904 (2006).

385. S. Masuda, and G. Jones, Promise of vitamin D analogues in the treatment of hyperproliferative conditions. *Mol Cancer Ther*, **5**, 797–808 (2006).

386. G. H. Posner, and M. Kahraman, Overview: Rational design of 1a-25-dihydroxyvitamin D3 analogs (deltanoids). In *Vitamin D*, 2nd ed. (D. Feldman, J. W. Pike, and F. Glorieux, eds.), pp. 1405–1422. Elsevier Academic Press, San Diego (2005).

387. M. R. Uskokovic, H. Maehr, G. S. Reddy, Y. C. Li, L. Adorini, and M. F. Holick, Gemini: The 1,25-dihydroxy vitamin D analogs with two side-chains. In *Vitamin D*, 2nd ed. (D. Feldman, F. Glorieux, and J. W. Pike, eds.), pp. 1511–1524. Elsevier Academic Press, San Diego (2005).

388. A. J. Brown, and E. Slatopolsky, Mechanisms for the selective actions of vitamin D analogs. In *Vitamin D*, 2nd ed. (D. Feldman, J. W. Pike, and F. Glorieux, eds.), pp. 1449–1470. Elsevier Academic Press, San Diego (2005).

389. A. J. Brown, C. Ritter, E. Slatopolsky, K. R. Muralidharan, W. H. Okamura, and G. S. Reddy, 1alpha,25-dihydroxy-3-epi-vitamin D3, a natural metabolite of 1alpha,25-dihydroxyvitamin D3, is a potent suppressor of parathyroid hormone secretion. *J Cell Biochem*, **73**, 106–113 (1999).

390. F. J. Dilworth, G. R. Williams, A. M. Kissmeyer, J. L. Nielsen, E. Binderup, M. J. Calverley, H. L. Makin, and G. Jones, The vitamin D analog, KH1060, is rapidly degraded both in vivo and in vitro via several pathways: Principal metabolites generated retain significant biological activity. *Endocrinology*, **138**, 5485–5496 (1997).

391. G. Jones, Analog metabolism. In *Vitamin D* (D. Feldman, F. Glorieux, and J. W. Pike, eds.), pp. 973–994. Academic Press, San Diego (1997).

392. S. Swami, X. Y. Zhao, S. Sarabia, M. L. Siu-Caldera, M. Uskokovic, S. G. Reddy, and D. Feldman, A low-calcemic vitamin D analog (Ro 25–4020) inhibits the growth of LNCaP human prostate cancer cells with increased potency by producing an active 24-oxo metabolite (Ro 29–9970). *Recent Results Cancer Res*, **164**, 349–352 (2003).

393. X. Y. Zhao, T. R. Eccleshall, A. V. Krishnan, C. Gross, and D. Feldman, Analysis of vitamin D analog-induced heterodimerization of vitamin D receptor with retinoid X receptor using the yeast two-hybrid system. *Mol Endocrinol*, **11**, 366–378 (1997).

394. W. Yang, and L. P. Freedman, 20-Epi analogues of 1,25-dihydroxyvitamin D3 are highly potent inducers of DRIP coactivator complex binding to the vitamin D3 receptor. *J Biol Chem*, **274**, 16838–16845 (1999).

395. S. Peleg, M. Sastry, E. D. Collins, J. E. Bishop, and A. W. Norman, Distinct conformational changes induced by 20-epi analogues of 1 alpha,25-dihydroxyvitamin D_3 are associated with enhanced activation of the vitamin D receptor. *J Biol Chem*, **270**, 10551–10558 (1995).

396. S. Peleg, Molecular basis for differential action of vitamin D analogs. In *Vitamin D*, 2nd ed. (D. Feldman, J. W. Pike, and F. Glorieux, eds.), pp. 1471–1488. Elsevier Academic Press, San Diego (2005).

397. T. Jaaskelainen, S. Ryhanen, A. Mahonen, H. F. DeLuca, and P. H. Maenpaa, Mechanism of action of superactive vitamin D analogs through regulated receptor degradation. *J Cell Biochem*, **76**, 548–558 (2000).

398. M. Maggi, C. Crescioli, A. Morelli, E. Colli, and L. Adorini, Pre-clinical evidence and clinical translation of benign prostatic hyperplasia treatment by the vitamin D receptor agonist BXL-628 (Elocalcitol). *J Endocrinol Invest*, **29**, 665–674 (2006).

399. N. Kubodera, Development of OCT and ED-71. In *Vitamin D*, 2nd ed. (D. Feldman, F. Glorieux, and J. W. Pike, eds.), pp. 1525–1541. Elsevier Academic Press, San Diego (2005).

400. K. Murakami, H. Miyachi, A. Watanabe, N. Kawamura, M. Fujii, S. Koide, M. Murase, H. Kushimoto, M. Hasegawa, M. Tomita, Y. Hiki, and S. Sugiyama, Suppression of parathyroid hormone secretion in CAPD patients by intraperitoneal administration of maxacalcitol. *Clin Exp Nephrol*, **8**, 134–138 (2004).

401. Y. Nishii, and T. Okano, History of the development of new vitamin D analogs: Studies on 22-oxacalcitriol (OCT) and 2beta-(3-hydroxypropoxy)calcitriol (ED-71). *Steroids*, **66**, 137–146 (2001).

402. A. J. Brown, and D. W. Coyne, Vitamin D analogs: New therapeutic agents for secondary hyperparathyroidism. *Treat Endocrinol*, **1**, 313–327 (2002).

403. J. W. Coburn, H. M. Maung, L. Elangovan, M. J. Germain, J. S. Lindberg, S. M. Sprague, M. E. Williams, and C. W. Bishop, Doxercalciferol safely suppresses PTH levels in patients with secondary hyperparathyroidism associated with chronic kidney disease stages 3 and 4. *Am J Kidney Dis*, **43**, 877–890 (2004).

404. S. M. Sprague, F. Llach, M. Amdahl, C. Taccetta, and D. Batlle, Paricalcitol versus calcitriol in the treatment of secondary hyperparathyroidism. *Kidney Int*, **63**, 1483–1490 (2003).

405. M. Teng, M. Wolf, E. Lowrie, N. Ofsthun, J. M. Lazarus, and R. Thadhani, Survival of patients undergoing hemodialysis with paricalcitol or calcitriol therapy. *N Engl J Med*, **349**, 446–456 (2003).

406. A. J. Brown, J. Finch, and E. Slatopolsky, Differential effects of 19-nor-1,25-dihydroxyvitamin D(2) and 1,25-dihydroxyvitamin D(3) on intestinal calcium and phosphate transport. *J Lab Clin Med*, **139**, 279–284 (2002).

407. F. Takahashi, J. L. Finch, M. Denda, A. S. Dusso, A. J. Brown, and E. Slatopolsky, A new analog of 1,25-(OH)2D3,

19-NOR-1,25-(OH)2D2, suppresses serum PTH and parathyroid gland growth in uremic rats without elevation of intestinal vitamin D receptor content. *Am J Kidney Dis,* **30**, 105–112 (1997).

408. T. Nijenhuis, B. C. van der Eerden, U. Zugel, A. Steinmeyer, H. Weinans, J. G. Hoenderop, J. P. van Leeuwen, and R. J. Bindels, The novel vitamin D analog ZK191784 as an intestine-specific vitamin D antagonist. *FASEB J,* **20**, 2171–2173 (2006).

409. Y. Kato, Y. Nakano, H. Sano, A. Tanatani, H. Kobayashi, R. Shimazawa, H. Koshino, Y. Hashimoto, and K. Nagasawa, Synthesis of 1alpha,25-dihydroxyvitamin D3-26,23-lactams (DLAMs), a novel series of 1 alpha,25-dihydroxyvitamin D3 antagonist. *Bioorg Med Chem Lett,* **14**, 2579–2583 (2004).

410. K. Ozono, M. Saito, D. Miura, T. Michigami, S. Nakajima, and S. Ishizuka, Analysis of the molecular mechanism for the antagonistic action of a novel 1alpha,25-dihydroxyvitamin D(3) analogue toward vitamin D receptor function. *J Biol Chem,* **274**, 32376–32381 (1999).

411. M. Perakyla, F. Molnar, and C. Carlberg, A structural basis for the species-specific antagonism of 26,23-lactones on vitamin D signaling. *Chem Biol,* **11**, 1147–1156 (2004).

412. T. Fujishima, Y. Kojima, I. Azumaya, A. Kittaka, and H. Takayama, Design and synthesis of potent vitamin D receptor antagonists with A-ring modifications: Remarkable effects of 2alpha-methyl introduction on antagonistic activity. *Bioorg Med Chem,* **11**, 3621–3631 (2003).

413. R. H. Wasserman, Vitamin D and the intestinal absorption of calcium: A view and overview. In *Vitamin D,* 2nd ed. (D. Feldman, J. W. Pike, and F. Glorieux, eds.), pp. 411–428. Elsevier Academic Press, San Diego (2005).

414. J. G. Hoenderop, J. P. van Leeuwen, B. C. van der Eerden, F. F. Kersten, A. W. van der Kemp, A. M. Merillat, J. H. Waarsing, B. C. Rossier, V. Vallon, E. Hummler, and R. J. Bindels, Renal Ca2+ wasting, hyperabsorption, and reduced bone thickness in mice lacking TRPV5. *J Clin Invest,* **112**, 1906–1914 (2003).

415. S. F. van de Graaf, J. G. Hoenderop, and R. J. Bindels, Regulation of TRPV5 and TRPV6 by associated proteins. *Am J Physiol Renal Physiol,* **290**, F1295–1302 (2006).

416. R. Vennekens, J. G. Hoenderop, J. Prenen, M. Stuiver, P. H. Willems, G. Droogmans, B. Nilius, and R. J. Bindels, Permeation and gating properties of the novel epithelial Ca(2+) channel. *J Biol Chem,* **275**, 3963–3969 (2000).

417. J. G. Hoenderop, D. Muller, A. W. van der Kemp, A. Hartog, M. Suzuki, K. Ishibashi, M. Imai, F. Sweep, P. H. Willems, C. H. Van Os, and R. J. Bindels, Calcitriol controls the epithelial calcium channel in kidney. *J Am Soc Nephrol,* **12**, 1342–1349 (2001).

418. T. Nijenhuis, J. G. Hoenderop, A. W. van der Kemp, and R. J. Bindels, Localization and regulation of the epithelial Ca2+ channel TRPV6 in the kidney. *J Am Soc Nephrol,* **14**, 2731–2740 (2003).

419. M. B. Meyer, M. Watanuki, S. Kim, N. K. Shevde, and J. W. Pike, The human transient receptor potential vanilloid type 6 distal promoter contains multiple vitamin D receptor binding sites that mediate activation by 1,25-dihydroxyvitamin D3 in intestinal cells. *Mol Endocrinol,* **20**, 1447–1461 (2006).

420. M. van Abel, J. G. Hoenderop, A. W. van der Kemp, J. P. van Leeuwen, and R. J. Bindels, Regulation of the epithelial Ca2+ channels in small intestine as studied by quantitative mRNA detection. *Am J Physiol Gastrointest Liver Physiol,* **285**, G78–85 (2003).

421. R. K. Gill, and S. Christakos, Identification of sequence elements in mouse calbindin-D28k gene that confer 1,25-dihydroxyvitamin D₃– and butyrate-inducible responses. *Proc Natl Acad Sci USA,* **90**, 2984–2988 (1993).

422. K. Sooy, J. Kohut, and S. Christakos, The role of calbindin and 1,25dihydroxyvitamin D3 in the kidney. *Curr Opin Nephrol Hypertens,* **9**, 341–347 (2000).

423. W. Zheng, Y. Xie, G. Li, J. Kong, J. Q. Feng, and Y. C. Li, Critical role of calbindin-D28k in calcium homeostasis revealed by mice lacking both vitamin D receptor and calbindin-D28k. *J Biol Chem,* **279**, 52406–52413 (2004).

424. J. G. Hoenderop, O. Dardenne, M. van Abel, A. W. van der Kemp, C. H. Van Os, R. St. Arnaud, and R. J. Bindels, Modulation of renal Ca2+ transport protein genes by dietary Ca2+ and 1,25-dihydroxyvitamin D3 in 25-hydroxyvitamin D3-1alpha-hydroxylase knockout mice. *FASEB J,* **16**, 1398–1406 (2002).

425. A. W. Norman, Vitamin D receptor (VDR): New assignments for an already busy receptor. *Endocrinology,* **147**, 5542–5548 (2006).

426. C. Duflos, C. Bellaton, D. Pansu, and F. Bronner, Calcium solubility, intestinal sojourn time and paracellular permeability codetermine passive calcium absorption in rats. *J Nutr,* **125**, 2348–2355 (1995).

427. D. Feldman, T. A. McCain, M. A. Hirst, T. L. Chen, and K. W. Colston, Characterization of a cytoplasmic receptor-like binder for 1 alpha, 25-dihydroxycholecalciferol in rat intestinal mucosa. *J Biol Chem,* **254**, 10378–10384 (1979).

428. S. D. Chan, and D. Atkins, The temporal distribution of the 1 alpha,25-dihydroxycholecalciferol receptor in the rat jejunal villus. *Clin Sci,* **67**, 285–290 (1984).

429. M. A. Hirst, and D. Feldman, 1,25-dihydroxyvitamin D3 receptors in mouse colon. *J Steroid Biochem,* **14**, 315–319 (1981).

430. P. Lointier, F. Meggouh, P. Dechelotte, D. Pezet, C. Ferrier, J. Chipponi, and S. Saez, 1,25-dihydroxyvitamin D3 receptors and human colon adenocarcinoma. *Br J Surg,* **78**, 435–439 (1991).

431. H. S. Cross, Vitamin D and colon cancer. In *Vitamin D,* 2nd ed. (D. Feldman, J. W. Pike, and F. Glorieux, eds.), pp. 1709–1725. Elsevier Academic Press, San Diego (2005).

432. J. B. Peng, X. Z. Chen, U. V. Berger, S. Weremowicz, C. C. Morton, P. M. Vassilev, E. M. Brown, and M. A. Hediger, Human calcium transport protein CaT1. *Biochem Biophys Res Commun,* **278**, 326–332 (2000).

433. F. Bronner, B. Slepchenko, R. J. Wood, and D. Pansu, The role of passive transport in calcium absorption. *J Nutr,* **133**, 1426; author reply 1427 (2003).

434. H. J. Armbrecht, M. Boltz, R. Strong, A. Richardson, M. E. Bruns, and S. Christakos, Expression of calbindin-D decreases with age in intestine and kidney. *Endocrinology,* **125**, 2950–2956 (1989).

435. H. J. Armbrecht, M. A. Boltz, and V. B. Kumar, Intestinal plasma membrane calcium pump protein and its induction by 1,25(OH)(2)D(3) decrease with age. *Am J Physiol,* **277**, G41–47 (1999).

436. R. P. Heaney, Lessons for nutritional science from vitamin D. *Am J Clin Nutr,* **69**, 825–826 (1999).

437. R. P. Heaney, M. J. Barger-Lux, M. S. Dowell, T. C. Chen, and M. F. Holick, Calcium absorptive effects of vitamin D and its major metabolites. *J Clin Endocrinol Metab,* **82**, 4111–4116 (1997).

438. B. P. Halloran, and A. A. Portale, Vitamin D metabolism and aging. In *Vitamin D,* 2nd ed. (D. Feldman, J. W. Pike, and F. Glorieux, eds.), pp. 823–838. Elsevier Academic Press, San Diego (2005).

439. C. Coudray, C. Feillet-Coudray, M. Rambeau, J. C. Tressol, E. Gueux, A. Mazur, and Y. Rayssiguier, The effect of aging

on intestinal absorption and status of calcium, magnesium, zinc, and copper in rats: A stable isotope study. *J Trace Elem Med Biol,* **20,** 73–81 (2006).

440. S. A. Abrams, Calcium turnover and nutrition through the life cycle. *Proc Nutr Soc,* **60,** 283–289 (2001).

441. B. L. Riggs, Role of the vitamin D-endocrine system in the pathophysiology of postmenopausal osteoporosis. *J Cell Biochem,* **88,** 209–215 (2003).

442. A. J. Brown, I. Krits, and H. J. Armbrecht, Effect of age, vitamin D, and calcium on the regulation of rat intestinal epithelial calcium channels. *Arch Biochem Biophys,* **437,** 51–58 (2005).

443. H. J. Armbrecht, M. A. Boltz, S. Christakos, and M. E. Bruns, Capacity of 1,25-dihydroxyvitamin D to stimulate expression of calbindin D changes with age in the rat. *Arch Biochem Biophys,* **352,** 159–164 (1998).

444. R. L. Horst, J. P. Goff, and T. A. Reinhardt, Advancing age results in reduction of intestinal and bone 1,25-dihydroxyvitamin D receptor. *Endocrinology,* **126,** 1053–1057 (1990).

445. C. T. Liang, J. Barnes, S. Imanaka, and H. F. DeLuca, Alterations in mRNA expression of duodenal 1,25-dihydroxyvitamin D3 receptor and vitamin D-dependent calcium binding protein in aged Wistar rats. *Exp Gerontol,* **29,** 179–186 (1994).

446. P. R. Ebeling, M. E. Sandgren, E. P. DiMagno, A. W. Lane, H. F. DeLuca, and B. L. Riggs, Evidence of an age-related decrease in intestinal responsiveness to vitamin D: Relationship between serum 1,25-dihydroxyvitamin D3 and intestinal vitamin D receptor concentrations in normal women. *J Clin Endocrinol Metab,* **75,** 176–182 (1992).

447. M. J. Barger-Lux, R. P. Heaney, S. J. Lanspa, J. C. Healy, and H. F. DeLuca, An investigation of sources of variation in calcium absorption efficiency [published erratum appears in *J Clin Endocrinol Metab,* **80**(7), 2068 (1995)]. *J Clin Endocrinol Metab,* **80,** 406–411 (1995).

448. E. M. Colin, G. J. van den Bemd, M. Van Aken, S. Christakos, H. R. De Jonge, H. F. Deluca, J. M. Prahl, J. C. Birkenhager, C. J. Buurman, H. A. Pols, and J. P. Van Leeuwen, Evidence for involvement of 17beta-estradiol in intestinal calcium absorption independent of 1,25-dihydroxyvitamin D3 level in the rat. *J Bone Miner Res,* **14,** 57–64 (1999).

449. M. J. Favus, and F. L. Coe, Idiopathic hypercalciuria and nephrolithiasis. In *Vitamin D,* 2nd ed. (D. Feldman, J. W. Pike, and F. Glorieux, eds.), pp. 1339–1354. Elsevier Academic Press, San Diego (2005).

450. J. Yao, P. Kathpalia, D. A. Bushinsky, and M. J. Favus, Hyperresponsiveness of vitamin D receptor gene expression to 1,25-dihydroxyvitamin D3. A new characteristic of genetic hypercalciuric stone-forming rats. *J Clin Invest,* **101,** 2223–2232 (1998).

451. M. A. Parfitt, Vitamin D and the pathogenesis of rickets and osteomalacia. In *Vitamin D,* 2nd ed. (D. Feldman, F. Glorieux, and J. W. Pike, eds.), pp. 1029–1048. Elsevier Academic Press, San Diego (2005).

452. G. S. Stein, J. B. Lian, M. Montecino, J. L. Stein, A. J. van Wijnen, A. Javed, J. Y. Choi, K. S. Zaidi, S. Gutierrez, J. Shen, S. Pockwinse, and D. Young, Intranuclear organization of the regulatory machinery for vitamin D-mediated control of skeletal gene expression. In *Vitamin D,* 2nd ed. (D. Feldman, F. Glorieux, and J. W. Pike, eds.), pp. 327–340. Elsevier Academic Press, San Diego (2005).

453. B. E. Kream, M. Jose, S. Yamada, and H. F. DeLuca, A specific high-affinity binding macromolecule for 1,25-dihydroxyvitamin D_3 in fetal bone. *Science,* **197,** 1086–1088 (1977).

454. T. L. Chen, M. A. Hirst, and D. Feldman, A receptor-like binding macromolecule for 1 alpha, 25-dihydroxycholecalciferol in cultured mouse bone cells. *J Biol Chem,* **254,** 7491–7494 (1979).

455. M. C. Farach-Carson, and Y. Xu, Microarray detection of gene expression changes induced by 1,25(OH)(2)D(3) and a Ca(2+) influx-activating analog in osteoblastic ROS 17/2.8 cells. *Steroids,* **67,** 467–470 (2002).

456. P. H. Stern, Bone. In *Vitamin D,* 2nd ed. (D. Feldman, J. W. Pike, and F. Glorieux, eds.), pp. 565–573. Elsevier Academic Press, San Diego (2005).

457. L. G. Raisz, C. L. Trummel, M. F. Holick, and H. F. DeLuca, 1,25-dihydroxycholecalciferol: A potent stimulator of bone resorption in tissue culture. *Science,* **175,** 768–769 (1972).

458. R. Kitazawa, and S. Kitazawa, Vitamin D(3) augments osteoclastogenesis via vitamin D-responsive element of mouse RANKL gene promoter. *Biochem Biophys Res Commun,* **290,** 650–655 (2002).

459. P. A. Baldock, G. P. Thomas, J. M. Hodge, S. U. Baker, U. Dressel, D. P. O'Loughlin, G. C. Nicholson, K. H. Briffa, J. A. Eisman, and E. M. Gardiner, Vitamin D action and regulation of bone remodeling: Suppression of osteoclastogenesis by the mature osteoblast. *J Bone Miner Res,* **21,** 1618–1626 (2006).

460. A. J. Brown, and E. A. Slatopolsky, Mechanisms for the selective actions of vitamin D analogs. In *Vitamin D,* 2nd ed. (D. Feldman, J. W. Pike, and F. Glorieux, eds.), pp. 1449–1469. Elsevier Academic Press, San Diego (2005).

461. E. M. Brown, Calcium-sensing receptor. In *Vitamin D,* 2nd ed. (D. Feldman, F. Glorieux, and J. W. Pike, eds.), pp. 551–562. Elsevier Academic Press, San Diego (2005).

462. L. Canaff, and G. N. Hendy, Human calcium-sensing receptor gene. Vitamin D response elements in promoters P1 and P2 confer transcriptional responsiveness to 1,25-dihydroxyvitamin D. *J Biol Chem,* **277,** 30337–30350 (2002).

463. N. Fokuda, H. Tanaka, Y. Tominaga, M. Fukagawa, K. Kurokawa, and Y. Seino, Decreased 1,25-dihydroxyvitamin D3 receptor density is associated with a more severe form of parathyroid hyperplasia in chronic uremic patients. *J Clin Invest,* **92,** 1436–1443 (1993).

464. S. J. Silverberg, and J. P. Bilezikian, The diagnosis and management of asymptomatic primary hyperparathyroidism. *Nat Clin Pract Endocrinol Metab,* **2,** 494–503 (2006).

465. A. Grey, J. Lucas, A. Horne, G. Gamble, J. S. Davidson, and I. R. Reid, Vitamin D repletion in patients with primary hyperparathyroidism and coexistent vitamin D insufficiency. *J Clin Endocrinol Metab,* **90,** 2122–2126 (2005).

466. J. P. van Leeuwen, and H. A. Pols, Vitamin D: Cancer and differentiation. In *Vitamin D,* 2nd ed. (D. Feldman, F. Glorieux, and J. W. Pike, eds.), pp. 1571–1597. Elsevier Academic Press, San Diego (2005).

467. S. D. Cramer, D. M. Peehl, M. G. Edgar, S. T. Wong, L. J. Deftos, and D. Feldman, Parathyroid hormone-related protein (PTHrP) is an epidermal growth factor-regulated secretory product of human prostatic epithelial cells. *Prostate,* **29,** 20–29 (1996).

468. El K. Abdaimi, V. Papavasiliou, S. A. Rabbani, J. S. Rhim, D. Goltzman, and R. Kremer, Reversal of hypercalcemia with the vitamin D analogue EB1089 in a human model of squamous cancer. *Cancer Res,* **59,** 3325–3328 (1999).

469. A. F. Russo, and R. F. Gagel, Vitamin D control of the calcitonin gene in thyroid C cells. In *Vitamin D,* 2nd ed. (D. Feldman, J. W. Pike, and F. Glorieux, eds.), pp. 687–702. Elsevier Academic Press, San Diego (2005).

470. C. Mathieu, and L. Adorini, The coming of age of 1,25-dihydroxyvitamin D(3) analogs as immunomodulatory agents. *Trends Mol Med,* **8,** 174–179 (2002).

471. E. Giovannucci, The epidemiology of vitamin D and cancer incidence and mortality: A review (United States). *Cancer Causes Control,* **16,** 83–95 (2005).

472. C. F. Garland, F. C. Garland, E. D. Gorham, M. Lipkin, H. Newmark, S. B. Mohr, and M. F. Holick, The role of vitamin D in cancer prevention. *Am J Public Health,* **96,** 252–261 (2006).

473. E. Giovannucci, and E. A. Platz, Epidemiology of cancer risk: Vitamin D and calcium. In *Vitamin D,* 2nd ed. (D. Feldman, J. W. Pike, and F. Glorieux, eds.), pp. 1617–1634. Elsevier Academic Press, San Diego (2005).

474. J. Moan, A. C. Porojnicu, T. E. Robsahm, A. Dahlback, A. Juzeniene, S. Tretli, and W. Grant, Solar radiation, vitamin D and survival rate of colon cancer in Norway. *J Photochem Photobiol B,* **78,** 189–193 (2005).

475. G. G. Schwartz, Vitamin D and the epidemiology of prostate cancer. *Semin Dial,* **18,** 276–289 (2005).

476. G. G. Schwartz, and B. S. Hulka, Is vitamin D deficiency a risk factor for prostate cancer? (Hypothesis). *AntiCancer Res,* **10,** 1307–1311 (1990).

477. A. V. Krishnan, D. M. Peehl, and D. Feldman, Vitamin D and prostate cancer. In *Vitamin D,* 2nd ed. (D. Feldman, J. W. Pike, and F. Glorieux, eds.), pp. 1679–1707. Elsevier Academic Press, San Diego (2005).

478. K. Colston, and J. Welsh, Vitamin D and breast cancer. In *Vitamin D,* 2nd ed. (D. Feldman, J. W. Pike, and F. Glorieux, eds.), pp. 1663–1677. Elsevier Academic Press, San Diego (2005).

479. C. S. Johnson, P. A. Hershberger, and D. L. Trump, Vitamin D-related therapies in prostate cancer. *Cancer Metastasis Rev,* **21,** 147–158 (2002).

480. J. O'Kelly, R. Morosetti, and H. P. Koeffler, Vitamin D and hematological malignancy. In *Vitamin D,* 2nd ed. (D. Feldman, J. W. Pike, and F. Glorieux, eds.), pp. 1727–1740. Elsevier Academic Press, San Diego (2005).

481. L. V. Stewart, and N. L. Weigel, Vitamin D and prostate cancer. *Exp Biol Med (Maywood),* **229,** 277–284 (2004).

482. G. P. Studzinski, and M. Danilenko, Differentiation and the cell cycle. In *Vitamin D,* 2nd ed. (D. Feldman, J. W. Pike, and F. Glorieux, eds.), pp. 1635–1661. Elsevier Academic Press, San Diego (2005).

483. A. V. Krishnan, D. M. Peehl, and D. Feldman, Inhibition of prostate cancer growth by vitamin D: Regulation of target gene expression. *J Cell Biochem,* **88,** 363–371 (2003).

484. M. Liu, M. H. Lee, M. Cohen, M. Bommakanti, and L. P. Freedman, Transcriptional activation of the Cdk inhibitor p21 by vitamin D_3 leads to the induced differentiation of the myelomonocytic cell line U937. *Genes Dev,* **10,** 142–153 (1996).

485. W. H. Park, J. G. Seol, E. S. Kim, C. W. Jung, C. C. Lee, L. Binderup, H. P. Koeffler, B. K. Kim, and Y. Y. Lee, Cell cycle arrest induced by the vitamin D(3) analog EB1089 in NCI-H929 myeloma cells is associated with induction of the cyclin-dependent kinase inhibitor p27. *Exp Cell Res,* **254,** 279–286 (2000).

486. J. Welsh, J. A. Wietzke, G. M. Zinser, B. Byrne, K. Smith, and C. J. Narvaez, Vitamin D-3 receptor as a target for breast cancer prevention. *J Nutr,* **133,** 2425S–2433S (2003).

487. S. H. Zhuang, and K. L. Burnstein, Antiproliferative effect of 1alpha,25-dihydroxyvitamin D3 in human prostate cancer cell line LNCaP involves reduction of cyclin-dependent kinase 2 activity and persistent G1 accumulation. *Endocrinology,* **139,** 1197–1207 (1998).

488. S. E. Blutt, T. J. McDonnell, T. C. Polek, and N. L. Weigel, Calcitriol-induced apoptosis in LNCaP cells is blocked by overexpression of Bcl-2. *Endocrinology,* **141,** 10–17 (2000).

489. A. J. Watson, An overview of apoptosis and the prevention of colorectal cancer. *Crit Rev Oncol Hematol,* **57,** 107–121 (2006).

490. S. P. Xie, S. Y. James, and K. W. Colston, Vitamin D derivatives inhibit the mitogenic effects of IGF-I on MCF-7 human breast cancer cells. *J Endocrinol,* **154,** 495–504 (1997).

491. J. Moreno, A. V. Krishnan, S. Swami, L. Nonn, D. M. Peehl, and D. Feldman, Regulation of prostaglandin metabolism by calcitriol attenuates growth stimulation in prostate cancer cells. *Cancer Res,* **65,** 7917–7925 (2005).

492. L. Nonn, L. Peng, D. Feldman, and D. M. Peehl, Inhibition of p38 by vitamin D reduces interleukin-6 production in normal prostate cells via mitogen-activated protein kinase phosphatase 5, implications for prostate cancer prevention by vitamin D. *Cancer Res,* **66,** 4516–4524 (2006).

493. M. S. Lucia, and K. C. Torkko, Inflammation as a target for prostate cancer chemoprevention: Pathological and laboratory rationale. *J Urol,* **171,** S30–34; discussion S35 (2004).

494. J. Vakkila, and M. T. Lotze, Inflammation and necrosis promote tumour growth. *Nat Rev Immunol,* **4,** 641–648 (2004).

495. T. C. Chen, L. Wang, L. W. Whitlatch, J. N. Flanagan, and M. F. Holick, Prostatic 25-hydroxyvitamin D-1alpha-hydroxylase and its implication in prostate cancer. *J Cell Biochem,* **88,** 315–322 (2003).

496. J. Y. Hsu, D. Feldman, J. E. McNeal, and D. M. Peehl, Reduced 1alpha-hydroxylase activity in human prostate cancer cells correlates with decreased susceptibility to 25-hydroxyvitamin D3-induced growth inhibition. *Cancer Res,* **61,** 2852–2856 (2001).

497. U. Segersten, P. Correa, M. Hewison, P. Hellman, H. Dralle, T. Carling, G. Akerstrom, and G. Westin, 25-hydroxyvitamin D(3)-1alpha-hydroxylase expression in normal and pathological parathyroid glands. *J Clin Endocrinol Metab,* **87,** 2967–2972 (2002).

498. P. Bareis, G. Bises, M. G. Bischof, H. S. Cross, and M. Peterlik, 25-hydroxy-vitamin D metabolism in human colon cancer cells during tumor progression. *Biochem Biophys Res Commun,* **285,** 1012–1017 (2001).

499. G. J. Miller, G. E. Stapleton, T. E. Hedlund, and K. A. Moffat, Vitamin D receptor expression, 24-hydroxylase activity, and inhibition of growth by 1alpha,25-dihydroxyvitamin D3 in seven human prostatic carcinoma cell lines. *Clin Cancer Res,* **1,** 997–1003 (1995).

500. R. Skowronski, D. Peehl, and D. Feldman, Vitamin D and prostate cancer: 1,25-dihydroxyvitamin D_3 receptors and actions in prostate cancer cell lines. *Endocrinology,* **132,** 1952–1960 (1993).

501. L. H. Ly, X. Y. Zhao, L. Holloway, and D. Feldman, Liarozole acts synergistically with 1alpha,25-dihydroxyvitamin D3 to inhibit growth of DU 145 human prostate cancer cells by blocking 24-hydroxylase activity. *Endocrinology,* **140,** 2071–2076 (1999).

502. D. M. Peehl, E. Seto, J. Y. Hsu, and D. Feldman, Preclinical activity of ketoconazole in combination with calcitriol or the vitamin D analogue EB 1089 in prostate cancer cells. *J Urol,* **168,** 1583–1588 (2002).

503. S. Swami, A. V. Krishnan, D. M. Peehl, and D. Feldman, Genistein potentiates the growth inhibitory effects of 1,25-dihydroxyvitamin D(3) in DU145 human prostate cancer cells: Role of the direct inhibition of CYP24 enzyme activity. *Mol Cell Endocrinol,* **12,** 12 (2005).

504. C. F. Garland, F. C. Garland, and E. D. Gorham, Calcium and vitamin D. Their potential roles in colon and breast cancer prevention. *Ann NY Acad Sci,* **889,** 107–119 (1999).

505. E. D. Gorham, C. F. Garland, F. C. Garland, W. B. Grant, S. B. Mohr, M. Lipkin, H. L. Newmark, E. Giovannucci, M. Wei, and M. F. Holick, Vitamin D and prevention of colorectal cancer. *J Steroid Biochem Mol Biol,* **97,** 179–194 (2005).

506. J. A. Eisman, D. H. Barkla, and P. J. Tutton, Suppression of *in vivo* growth of human cancer solid tumor xenografts by 1,25-dihydroxyvitamin D3. *Cancer Res,* **47,** 21–25 (1987).

507. G. D. Diaz, C. Paraskeva, M. G. Thomas, L. Binderup, and A. Hague, Apoptosis is induced by the active metabolite of vitamin D3 and its analogue EB1089 in colorectal adenoma and carcinoma cells: Possible implications for prevention and therapy. *Cancer Res,* **60,** 2304–2312.

508. S. R. Evans, A. M. Schwartz, E. I. Shchepotin, M. Uskokovic, and I. B. Shchepotin, Growth inhibitory effects of 1,25-dihydroxyvitamin D3 and its synthetic analogue, 1alpha,25-dihydroxy-16-ene-23yne-26,27-hexafluoro-19-nor-cholecalciferol (Ro 25–6760), on a human colon cancer xenograft. *Clin Cancer Res,* **4,** 2869–2876 (1998).

509. S. Huerta, R. W. Irwin, D. Heber, V. L. Go, H. P. Koeffler, M. R. Uskokovic, and D. M. Harris, 1alpha,25-(OH)(2)-D(3) and its synthetic analogue decrease tumor load in the Apc(min) mouse. *Cancer Res,* **62,** 741–746 (2002).

510. D. M. Harris, and V. L. Go, Vitamin D and colon carcinogenesis. *J Nutr,* **134,** 3463S–3471S (2004).

511. H. G. Palmer, M. J. Larriba, J. M. Garcia, P. Ordonez-Moran, C. Pena, S. Peiro, I. Puig, R. Rodriguez, R. de la Fuente, A. Bernad, M. Pollan, F. Bonilla, C. Gamallo, A. G. de Herreros, and A. Munoz, The transcription factor SNAIL represses vitamin D receptor expression and responsiveness in human colon cancer. *Nat Med,* **10,** 917–919 (2004).

512. J. Welsh, J. A. Wietzke, G. M. Zinser, S. Smyczek, S. Romu, E. Tribble, J. C. Welsh, B. Byrne, and C. J. Narvaez, Impact of the vitamin D$_3$ receptor on growth-regulatory pathways in mammary gland and breast cancer. *J Steroid Biochem Mol Biol,* **83,** 85–92 (2002).

513. F. Davoodi, R. V. Brenner, S. R. Evans, L. M. Schumaker, M. Shabahang, R. J. Nauta, and R. R. Buras, Modulation of vitamin D receptor and estrogen receptor by 1,25(OH)2-vitamin D3 in T-47D human breast cancer cells. *J Steroid Biochem Mol Biol,* **54,** 147–153 (1995).

514. S. Y. James, A. G. Mackay, L. Binderup, and K. W. Colston, Effects of a new synthetic vitamin D analogue, EB1089, on the oestrogen-responsive growth of human breast cancer cells. *J Endocrinol,* **141,** 555–563 (1994).

515. S. Swami, A. V. Krishnan, and D. Feldman, 1alpha,25-dihydroxyvitamin D3 down-regulates estrogen receptor abundance and suppresses estrogen actions in MCF-7 human breast cancer cells. *Clin Cancer Res,* **6,** 3371–3379 (2000).

516. J. Welsh, K. VanWeelden, L. Flanagan, I. Byrne, E. Nolan, and C. J. Narvaez, The role of vitamin D3 and antiestrogens in modulating apoptosis of breast cancer cells and tumors. *Subcell Biochem,* **30,** 245–270 (1998).

517. L. Lowe, C. M. Hansen, S. Senaratne, and K. W. Colston, Mechanisms implicated in the growth regulatory effects of vitamin D compounds in breast cancer cells. *Recent Results Cancer Res,* **164,** 99–110 (2003).

518. J. Welsh, Vitamin D and breast cancer: Insights from animal models. *Am J Clin Nutr,* **80,** 1721S–1724S (2004).

519. T. Gulliford, J. English, K. W. Colston, P. Menday, S. Moller, and R. C. Coombes, A phase I study of the vitamin D analogue EB 1089 in patients with advanced breast and colorectal cancer. *Br J Cancer,* **78,** 6–13 (1998).

520. E. H. Corder, H. A. Guess, B. S. Hulka, G. D. Friedman, M. Sadler, R. T. Vollmer, B. Lobaugh, M. K. Drezner, J. H. Vogelman, and N. Orentreich, Vitamin D and prostate cancer: A prediagnostic study with stored sera. *Cancer Epi Biomak Prevent,* **2,** 467–472 (1993).

521. G. J. Miller, G. E. Stapleton, J. A. Ferrara, M. S. Lucia, S. Pfister, T. E. Hedlund, and P. Upadhya, The human prostatic carcinoma cell line LNCaP expresses biologically active, specific receptors for 1 alpha,25-dihydroxyvitamin D3. *Cancer Res,* **52,** 515–520 (1992).

522. D. M. Peehl, R. J. Skowronski, G. K. Leung, S. T. Wong, T. A. Stamey, and D. Feldman, Antiproliferative effects of 1,25-dihydroxyvitamin D3 on primary cultures of human prostatic cells. *Cancer Res,* **54,** 805–810 (1994).

523. B. R. Konety, and R. H. Getzenberg, Vitamin D and prostate cancer. *Urol Clin North Am,* **29,** 95–106, ix (2002).

524. W. Banach-Petrosky, X. Ouyang, H. Gao, K. Nader, Y. Ji, N. Suh, R. S. DiPaola, and C. Abate-Shen, Vitamin D inhibits the formation of prostatic intraepithelial neoplasia in Nkx3.1;Pten mutant mice. *Clin Cancer Res,* **12,** 5895–5901 (2006).

525. T. C. Chen, M. F. Holick, B. L. Lokeshwar, K. L. Burnstein, and G. G. Schwartz, Evaluation of vitamin D analogs as therapeutic agents for prostate cancer. *Recent Results Cancer Res,* **164,** 273–288 (2003).

526. C. Gross, T. Stamey, S. Hancock, and D. Feldman, Treatment of early recurrent prostate cancer with 1,25-dihydroxyvitamin D3 (calcitriol) [published erratum appears in *J Urol,* **160**(3 Pt 1), 840 (1998)]. *J Urol,* **159,** 2035–2039; discussion 2039–2040 (1998).

527. D. L. Trump, J. Muindi, M. Fakih, W. D. Yu, and C. S. Johnson, Vitamin D compounds: Clinical development as cancer therapy and prevention agents. *AntiCancer Res,* **26,** 2551–2556 (2006).

528. T. M. Beer, M. Javle, G. N. Lam, W. D. Henner, A. Wong, and D. L. Trump, Pharmacokinetics and tolerability of a single dose of DN-101, a new formulation of calcitriol, in patients with cancer. *Clin Cancer Res,* **11,** 7794–7799 (2005).

529. T. M. Beer, D. Lemmon, B. A. Lowe, and W. D. Henner, High-dose weekly oral calcitriol in patients with a rising PSA after prostatectomy or radiation for prostate carcinoma. *Cancer,* **97,** 1217–1224 (2003).

530. T. M. Beer, C. W. Ryan, P. M. Venner, D. P. Petrylak, G. S. Chatta, J. D. Ruether, C. H. Redfern, L. Fehrenbacher, M. N. Saleh, D. M. Waterhouse, M. A. Carducci, D. Vicario, R. Dreicer, C. S. Higano, F. R. Ahmann, K. N. Chi, W. D. Henner, A. Arroyo, and F. W. Clow, Double-blinded randomized study of high-dose calcitriol plus docetaxel compared with placebo plus docetaxel in androgen-independent prostate cancer: A report from the ASCENT investigators. *J Clin Oncol,* **25,** 669–674 (2007).

531. T. M. Beer, A. Myrthue, and K. M. Eilers, Rationale for the development and current status of calcitriol in androgen-independent prostate cancer. *World J Urol,* **23,** 28–32 (2005).

532. J. O'Kelly, J. Hisatake, Y. Hisatake, J. Bishop, A. Norman, and H. P. Koeffler, Normal myelopoiesis but abnormal T lymphocyte responses in vitamin D receptor knockout mice. *J Clin Invest,* **109,** 1091–1099 (2002).

533. P. H. Reitsma, P. G. Rothberg, S. M. Astrin, J. Trial, Z. Bar-Shavit, A. Hall, S. L. Teitelbaum, and A. J. Kahn, Regulation of myc gene expression in HL-60 leukaemia cells by a vitamin D metabolite. *Nature,* **306,** 492–494 (1983).

534. Y. Honma, M. Hozumi, E. Abe, K. Konno, M. Fukushima, S. Hata, Y. Nishii, H. F. DeLuca, and T. Suda, 1 alpha,25-dihydroxyvitamin D3 and 1 alpha-hydroxyvitamin D3 prolong survival time of mice inoculated with myeloid leukemia cells. *Proc Nat Acad Sci USA,* **80,** 201–204 (1983).

535. Q. T. Luong, and H. P. Koeffler, Vitamin D compounds in leukemia. *J Steroid Biochem Mol Biol,* **97**, 195–202 (2005).

536. K. Colston, J. M. Colston, and D. Feldman, 1,25-dihydroxyvitamin D3 and malignant melanoma: The presence of receptors and inhibition of cell growth in culture. *Endocrinology,* **108**, 1083–1086 (1981).

537. J. E. Osborne, and P. E. Hutchinson, Vitamin D and systemic cancer: Is this relevant to malignant melanoma? *Br J Dermatol,* **147**, 197–213 (2002).

538. M. Seifert, M. Rech, V. Meineke, W. Tilgen, and J. Reichrath, Differential biological effects of 1,25-dihydroxyvitamin D3 on melanoma cell lines in vitro. *J Steroid Biochem Mol Biol,* **89–90**, 375–379 (2004).

539. C. Li, Z. Liu, Z. Zhang, S. S. Strom, J. E. Gershenwald, V. G. Prieto, J. E. Lee, M. I. Ross, P. F. Mansfield, J. N. Cormier, M. Duvic, E. A. Grimm, and Q. Wei, Genetic variants of the vitamin D receptor gene alter risk of cutaneous melanoma. *J Invest Dermatol* **127**, 276–280 (2007).

540. K. Dalhoff, J. Dancey, L. Astrup, T. Skovsgaard, K. J. Hamberg, F. J. Lofts, O. Rosmorduc, S. Erlinger, J. Bach Hansen, W. P. Steward, T. Skov, F. Burcharth, and T. R. Evans, A phase II study of the vitamin D analogue seocalcitol in patients with inoperable hepatocellular carcinoma. *Br J Cancer,* **89**, 252–257 (2003).

541. X. Zhang, S. V. Nicosia, and W. Bai, Vitamin D receptor is a novel drug target for ovarian cancer treatment. *Curr Cancer Drug Targets,* **6**, 229–244 (2006).

542. S. Kawa, K. Yoshizawa, T. Nikaido, and K. Kiyosawa, Inhibitory effect of 22-oxa-1,25-dihydroxyvitamin D3, maxacalcitol, on the proliferation of pancreatic cancer cell lines. *J Steroid Biochem Mol Biol,* **97**, 173–177 (2005).

543. G. G. Schwartz, D. Eads, A. Rao, S. D. Cramer, M. C. Willingham, T. C. Chen, D. P. Jamieson, L. Wang, K. L. Burnstein, M. F. Holick, and C. Koumenis, Pancreatic cancer cells express 25-hydroxyvitamin D-1 alpha-hydroxylase and their proliferation is inhibited by the prohormone 25-hydroxyvitamin D3. *Carcinogenesis,* **25**, 1015–1026 (2004).

544. K. Nakagawa, Y. Sasaki, S. Kato, N. Kubodera, and T. Okano, 22-oxa-1alpha,25-dihydroxyvitamin D3 inhibits metastasis and angiogenesis in lung cancer. *Carcinogenesis,* **26**, 1044–1054 (2005).

545. W. Zhou, R. Suk, G. Liu, S. Park, D. S. Neuberg, J. C. Wain, T. J. Lynch, E. Giovannucci, and D. C. Christiani, Vitamin D is associated with improved survival in early-stage non-small cell lung cancer patients. *Cancer Epidemiol Biomarkers Prev,* **14**, 2303–2309 (2005).

546. L. Adorini, Regulation of immune responses by vitamin D receptor ligands. In *Vitamin D,* 2nd ed. (D. Feldman, F. Glorieux, and J. W. Pike, eds.), pp. 631–648. Elsevier Academic Press, San Diego (2005).

547. K. M. Casteels, C. Mathieu, M. Waer, D. Valckx, L. Overbergh, J. M. Laureys, and R. Bouillon, Prevention of type I diabetes in nonobese diabetic mice by late intervention with nonhypercalcemic analogs of 1,25-dihydroxyvitamin D3 in combination with a short induction course of cyclosporin A. *Endocrinology,* **139**, 95–102 (1998).

548. T. Y. Chan, Vitamin D deficiency and susceptibility to tuberculosis. *Calcif Tissue Int,* **66**, 476–478 (2000).

549. J. Lemire, Vitamin D3, autoimmunity and immunosuppression. In *Vitamin D,* 2nd ed. (D. Feldman, J. W. Pike, and F. Glorieux, eds.), pp. 1753–1762. Elsevier Academic Press, San Diego (2005).

550. C. Mathieu, C. Gysemans, and R. Bouillon, Vitamin D and diabetes. In *Vitamin D,* 2nd ed. (D. Feldman, J. W. Pike, and F. Glorieux, eds.), pp. 1763–1778. Elsevier Academic Press, San Diego (2005).

551. E. Seibert, N. W. Levin, and M. K. Kuhlmann, Immunomodulating effects of vitamin D analogs in hemodialysis patients. *Hemodial Int,* **9**(Suppl 1), S25–29 (2005).

552. E. Villamor, A potential role for vitamin D on HIV infection? *Nutr Rev,* **64**, 226–233 (2006).

553. M. Zasloff, Fighting infections with vitamin D. *Nat Med,* **12**, 388–390 (2006).

554. S. C. Manolagas, X. P. Yu, G. Girasole, and T. Bellido, Vitamin D and the hematolymphopoietic tissue: A 1994 update. *Sem Nephrol,* **14**, 129–143 (1994).

555. J. M. Lemire, D. C. Archer, L. Beck, and H. L. Spiegelberg, Immunosuppressive actions of 1,25-dihydroxyvitamin D3, preferential inhibition of Th1 functions. *J Nutr,* **125**, 1704S–1708S (1995).

556. A. Boonstra, F. J. Barrat, C. Crain, V. L. Heath, H. F. Savelkoul, and A. O'Garra, 1alpha,25-dihydroxyvitamin D3 has a direct effect on naive CD4(+) T cells to enhance the development of Th2 cells. *J Immunol,* **167**, 4974–4980 (2001).

557. M. D. Griffin, W. Lutz, V. A. Phan, L. A. Bachman, D. J. McKean, and R. Kumar, Dendritic cell modulation by 1alpha,25 dihydroxyvitamin D3 and its analogs: A vitamin D receptor-dependent pathway that promotes a persistent state of immaturity in vitro and in vivo. *Proc Natl Acad Sci USA,* **98**, 6800–6805 (2001).

558. L. Overbergh, B. Decallonne, D. Valckx, A. Verstuyf, J. Depovere, J. Laureys, O. Rutgeerts, R. St. Arnaud, R. Bouillon, and C. Mathieu, Identification and immune regulation of 25-hydroxyvitamin D-1-alpha-hydroxylase in murine macrophages. *Clin Exp Immunol,* **120**, 139–146 (2000).

559. P. T. Liu, S. Stenger, H. Li, L. Wenzel, B. H. Tan, S. R. Krutzik, M. T. Ochoa, J. Schauber, K. Wu, C. Meinken, D. L. Kamen, M. Wagner, R. Bals, A. Steinmeyer, U. Zugel, R. L. Gallo, D. Eisenberg, M. Hewison, B. W. Hollis, J. S. Adams, B. R. Bloom, and R. L. Modlin, Toll-like receptor triggering of a vitamin D-mediated human antimicrobial response. *Science,* **311**, 1770–1773 (2006).

560. D. Feldman, T. Chen, M. Hirst, K. Colston, M. Karasek, and C. Cone, Demonstration of 1,25-dihydroxyvitamin D_3 receptors in human skin biopsies. *J Clin Endocrinol Metab,* **51**, 1463–1465 (1980).

561. D. Feldman, and P. J. Malloy, Hereditary 1,25-dihydroxyvitamin D resistant rickets: Molecular basis and implications for the role of 1,25$(OH)_2D_3$ in normal physiology. *Mol Cell Endocrinol,* **72**, C57–62 (1990).

562. M. R. Hughes, P. J. Malloy, B. W. O'Malley, J. W. Pike, and D. Feldman, Genetic defects of the 1,25-dihydroxyvitamin D_3 receptor. *J Receptor Res,* **11**, 699–716 (1991).

563. Z. Xie, S. J. Munson, N. Huang, A. A. Portale, W. L. Miller, and D. D. Bikle, The mechanism of 1,25-dihydroxyvitamin D(3) autoregulation in keratinocytes. *J Biol Chem,* **277**, 36987–36990 (2002).

564. K. M. Dixon, S. S. Deo, G. Wong, M. Slater, A. W. Norman, J. E. Bishop, G. H. Posner, S. Ishizuka, G. M. Halliday, V. E. Reeve, and R. S. Mason, Skin cancer prevention: A possible role of 1,25dihydroxyvitamin D3 and its analogs. *J Steroid Biochem Mol Biol,* **97**, 137–143 (2005).

565. J. Reichrath, and M. F. Holick, Psoriasis and other skin diseases. In *Vitamin D,* 2nd ed. (D. Feldman, J. W. Pike, and F. Glorieux, eds.), pp. 1793–1804. Elsevier Academic Press, San Diego (2005).

566. J. Reichrath, S. M. Muller, A. Kerber, H. P. Baum, and F. A. Bahmer, Biologic effects of topical calcipotriol (MC 903) treatment in psoriatic skin. *J Am Acad Dermatol,* **36**, 19–28 (1997).

567. A. Charakida, O. Dadzie, F. Teixeira, M. Charakida, G. Evangelou, and A. C. Chu, Calcipotriol/betamethasone dipropionate for the treatment of psoriasis. *Expert Opin Pharmacother,* **7**, 597–606 (2006).

568. W. E. Stumpf, H. J. Bidmon, L. Li, C. Pilgrim, A. Bartke, A. Mayerhofer, and C. Heiss, Nuclear receptor sites for vitamin D-soltriol in midbrain and hindbrain of Siberian hamster (*Phodopus sungorus*) assessed by autoradiography. *Histochemistry*, **98**, 155–164 (1992).

569. P. Brachet, I. Neveu, P. Naveilhan, E. Garcion, and D. Wion, Vitamin D, a neuroactive hormone: From brain development to pathological disorders. In *Vitamin D*, 2nd ed. (D. Feldman, J. W. Pike, and F. Glorieux, eds.), pp. 1779–1789. Elsevier Academic Press, San Diego (2005).

570. J. Sonnenberg, V. N. Luine, L. C. Krey, and S. Christakos, 1,25-dihydroxyvitamin D3 treatment results in increased choline acetyltransferase activity in specific brain nuclei. *Endocrinology*, **118**, 1433–1439 (1986).

571. D. Wion, D. MacGrogan, I. Neveu, F. Jehan, R. Houlgatte, and P. Brachet, 1,25-dihydroxyvitamin D3 is a potent inducer of nerve growth factor synthesis. *J Neurosci Res*, **28**, 110–114 (1991).

572. I. M. Musiol, and D. Feldman, 1,25-dihydroxyvitamin D3 induction of nerve growth factor in L929 mouse fibroblasts: Effect of vitamin D receptor regulation and potency of vitamin D3 analogs. *Endocrinology*, **138**, 12–18 (1997).

573. M. S. Saporito, E. R. Brown, K. C. Hartpence, H. M. Wilcox, J. L. Vaught, and S. Carswell, Chronic 1,25-dihydroxyvitamin D3-mediated induction of nerve growth factor mRNA and protein in L929 fibroblasts and in adult rat brain. *Brain Res*, **633**, 189–196 (1994).

574. P. A. De Viragh, D. Wolfer, H. A. Lipp, and M. R. Celio (eds.), *Behavioral Changes in Chronically D-Hyper-Vitaminotic Animals*. Walter de Gruyter, New York (1988).

575. M. K. Sutherland, M. J. Somerville, L. K. Yoong, C. Bergeron, M. R. Haussler, and D. R. McLachlan, Reduction of vitamin D hormone receptor mRNA levels in Alzheimer as compared to Huntington hippocampus: Correlation with calbindin-28k mRNA levels. *Brain Res, Mol Brain Res*, **13**, 239–250 (1992).

576. P. W. Landfield, and L. Cadwallader-Neal, Long-term treatment with calcitriol (1,25(OH)2 vit D3) retards a biomarker of hippocampal aging in rats. *Neurobiol Aging*, **19**, 469–477 (1998).

577. Y. Sato, T. Asoh, and K. Oizumi, High prevalence of vitamin D deficiency and reduced bone mass in elderly women with Alzheimer's disease. *Bone*, **23**, 555–557 (1998).

578. Y. Sato, J. Iwamoto, T. Kanoko, and K. Satoh, Amelioration of osteoporosis and hypovitaminosis D by sunlight exposure in hospitalized, elderly women with Alzheimer's disease: A randomized controlled trial. *J Bone Miner Res*, **20**, 1327–1333 (2005).

579. K. Ozono, S. Nakajima, and T. Michigami, Vitamin D and reproductive organs. In *Vitamin D*, 2nd ed. (D. Feldman, F. Glorieux, and J. W. Pike, eds.), pp. 851–861. Elsevier Academic Press, San Diego (2005).

580. A. Stephanou, R. Ross, and S. Handwerger, Regulation of human placental lactogen expression by 1,25-dihydroxyvitamin D3. *Endocrinology*, **135**, 2651–2656 (1994).

581. H. Reichel, H. P. Koeffler, and A. W. Norman, The role of the vitamin D endocrine system in health and disease. *N Engl J Med*, **320**, 980–991 (1989).

582. M. R. Walters, Newly identified actions of the vitamin D endocrine system. *Endocrine Rev*, **13**, 719–764 (1992).

583. D. D. Bikle, Clinical counterpoint: Vitamin D: New actions, new analogs, new therapeutic potential. *Endocrine Rev*, **13**, 765–784 (1992).

584. S. Thys-Jacobs, D. Donovan, A. Papadopoulos, P. Sarrel, and J. P. Bilezikian, Vitamin D and calcium dysregulation in the polycystic ovarian syndrome. *Steroids*, **64**, 430–435 (1999).

585. J. Teixeira, S. Maheswaran, and P. K. Donahoe, Mullerian inhibiting substance: An instructive developmental hormone with diagnostic and possible therapeutic applications. *Endocr Rev*, **22**, 657–674 (2001).

586. A. M. Trbovich, P. M. Sluss, V. M. Laurich, F. H. O'Neill, D. T. MacLaughlin, P. K. Donahoe, and J. Teixeira, Mullerian inhibiting substance lowers testosterone in luteinizing hormone-stimulated rodents. *Proc Natl Acad Sci USA*, **98**, 3393–3397 (2001).

587. C. L. Cook, Y. Siow, A. G. Brenner, and M. E. Fallat, Relationship between serum mullerian-inhibiting substance and other reproductive hormones in untreated women with polycystic ovary syndrome and normal women. *Fertil Steril*, **77**, 141–146 (2002).

588. T. U. Barbie, D. A. Barbie, D. T. MacLaughlin, S. Maheswaran, and P. K. Donahoe, Mullerian inhibiting substance inhibits cervical cancer cell growth via a pathway involving p130 and p107. *Proc Natl Acad Sci USA*, **100**, 15601–15606 (2003).

589. Y. Hoshiya, V. Gupta, D. L. Segev, M. Hoshiya, J. L. Carey, L. M. Sasur, T. T. Tran, T. U. Ha, and S. Maheswaran, Mullerian inhibiting substance induces NFkB signaling in breast and prostate cancer cells. *Mol Cell Endocrinol*, **211**, 43–49 (2003).

590. D. L. Segev, T. U. Ha, T. T. Tran, M. Kenneally, P. Harkin, M. Jung, D. T. MacLaughlin, P. K. Donahoe, and S. Maheswaran, Mullerian inhibiting substance inhibits breast cancer cell growth through an NFkappa B-mediated pathway. *J Biol Chem*, **275**, 28371–28379 (2000).

591. D. L. Segev, Y. Hoshiya, M. Hoshiya, T. T. Tran, J. L. Carey, A. E. Stephen, D. T. MacLaughlin, P. K. Donahoe, and S. Maheswaran, Mullerian-inhibiting substance regulates NF-kappa B signaling in the prostate in vitro and in vivo. *Proc Natl Acad Sci USA*, **99**, 239–244 (2002).

592. T. T. Tran, D. L. Segev, V. Gupta, H. Kawakubo, G. Yeo, P. K. Donahoe, and S. Maheswaran, Mullerian inhibiting substance regulates androgen-induced gene expression and growth in prostate cancer cells through a nuclear factor-kappaB-dependent Smad-independent mechanism. *Mol Endocrinol*, **20**, 2382–2391 (2006).

593. P. J. Malloy, L. Peng, and D. Feldman, Mullerian inhibiting substance (MIS) is up-regulated by 1,25-dihydroxyvitamin D3 in LNCaP prostate cancer cells via a direct interaction of the vitamin D receptor with a vitamin D response element in the MIS promoter. The Endocrine Society's 88th Annual Meeting, Boston, MA, pp. P3–55, p. 650 (2006).

594. C. Crescioli, A. Morelli, L. Adorini, P. Ferruzzi, M. Luconi, G. B. Vannelli, M. Marini, S. Gelmini, B. Fibbi, S. Donati, D. Villari, G. Forti, E. Colli, K. E. Andersson, and M. Maggi, Human bladder as a novel target for vitamin D receptor ligands. *J Clin Endocrinol Metab*, **90**, 962–972 (2005).

595. E. Colli, P. Rigatti, F. Montorsi, W. Artibani, S. Petta, N. Mondaini, R. Scarpa, P. Usai, L. Olivieri, and M. Maggi, BXL628, a novel vitamin D3 analog arrests prostate growth in patients with benign prostatic hyperplasia: A randomized clinical trial. *Eur Urol*, **49**, 82–86 (2006).

596. Y. C. Li, Vitamin D and the renin-angiotensin system. In *Vitamin D*, 2nd ed. (D. Feldman, J. W. Pike, and F. Glorieux, eds.), pp. 871–881. Elsevier Academic Press, San Diego (2005).

597. L. M. Resnick, F. B. Muller, and J. H. Laragh, Calcium-regulating hormones in essential hypertension. Relation to plasma renin activity and sodium metabolism. *Ann Intern Med*, **105**, 649–654 (1986).

598. C. W. Park, Y. S. Oh, Y. S. Shin, C. M. Kim, Y. S. Kim, S. Y. Kim, E. J. Choi, Y. S. Chang, and B. K. Bang, Intravenous calcitriol regresses myocardial hypertrophy in

hemodialysis patients with secondary hyperparathyroidism. *Am J Kidney Dis,* **33**, 73–81 (1999).

599. M. Ohsawa, T. Koyama, K. Yamamoto, S. Hirosawa, S. Kamei, and R. Kamiyama, 1alpha,25-dihydroxyvitamin D(3) and its potent synthetic analogs downregulate tissue factor and upregulate thrombomodulin expression in monocytic cells, counteracting the effects of tumor necrosis factor and oxidized LDL. *Circulation,* **102**, 2867–2872 (2000).

600. T. M. Beer, P. M. Venner, C. W. Ryan, D. P. Petrylak, G. Chatta, Dean J. Ruether, K. N. Chi, J. G. Curd, and T. G. Deloughery, High dose calcitriol may reduce thrombosis in cancer patients. *Br J Haematol,* **135**, 392–394 (2006).

601. R. Eastell, and L. B. Riggs, Vitamin D and osteoporosis. In *Vitamin D,* 2nd ed. (D. Feldman, J. W. Pike, and F. Glorieux, eds.), pp. 1101–1120. Elsevier Academic Press, San Diego (2005).

602. S. Boonen, D. Vanderschueren, P. Haentjens, and P. Lips, Calcium and vitamin D in the prevention and treatment of osteoporosis—A clinical update. *J Intern Med,* **259**, 539–552 (2006).

603. K. F. Mauck, and B. L. Clarke, Diagnosis, screening, prevention, and treatment of osteoporosis. *Mayo Clin Proc,* **81**, 662–672 (2006).

604. S. Khosla, and B. L. Riggs, Pathophysiology of age-related bone loss and osteoporosis. *Endocrinol Metab Clin North Am,* **34**, 1015–1030, xi (2005).

605. N. E. Lane, Epidemiology, etiology, and diagnosis of osteoporosis. *Am J Obstet Gynecol,* **194**, S3–11 (2006).

606. L. Mosekilde, Vitamin D and the elderly. *Clin Endocrinol (Oxf),* **62**, 265–281 (2005).

607. L. G. Raisz, Pathogenesis of osteoporosis: Concepts, conflicts, and prospects. *J Clin Invest,* **115**, 3318–3325 (2005).

608. P. Sambrook, and C. Cooper, Osteoporosis. *Lancet,* **367**, 2010–2018 (2006).

609. J. M. Zmuda, Y. T. Sheu, and S. P. Moffett, Genetic epidemiology of osteoporosis: Past, present, and future. *Curr Osteoporos Rep,* **3**, 111–115 (2005).

610. N. A. Morrison, J. C. Qi, A. Tokita, P. J. Kelly, L. Crofts, T. V. Nguyen, P. N. Sambrook, and J. A. Eisman, Prediction of bone density from vitamin D receptor alleles. *Nature,* **367**, 284–287 (1994).

611. J. C. Gallagher, S. E. Fowler, J. R. Detter, and S. S. Sherman, Combination treatment with estrogen and calcitriol in the prevention of age-related bone loss. *J Clin Endocrinol Metab,* **86**, 3618–3628 (2001).

612. M. C. Chapuy, P. Preziosi, M. Maamer, S. Arnaud, P. Galan, S. Hercberg, and P. J. Meunier, Prevalence of vitamin D insufficiency in an adult normal population. *Osteoporos Int,* **7**, 439–443 (1997).

613. T. Dixon, P. Mitchell, T. Beringer, S. Gallacher, C. Moniz, S. Patel, G. Pearson, and P. Ryan, An overview of the prevalence of 25-hydroxy-vitamin D inadequacy amongst elderly patients with or without fragility fracture in the United Kingdom. *Curr Med Res Opin,* **22**, 405–415 (2006).

614. M. W. Tilyard, G. F. Spears, J. Thomson, and S. Dovey, Treatment of postmenopausal osteoporosis with calcitriol or calcium. *N Engl J Med,* **326**, 357–362 (1992).

615. J. C. Gallagher, Prevention of bone loss in postmenopausal and senile osteoporosis with vitamin D analogues. *Osteoporos Int,* **1**, 172–175 (1993).

616. J. S. Adams, V. Kantorovich, C. Wu, M. Javanbakht, and B. W. Hollis, Resolution of vitamin D insufficiency in osteopenic patients results in rapid recovery of bone mineral density. *J Clin Endocrinol Metab,* **84**, 2729–2730 (1999).

617. M. K. Lehtonen-Veromaa, T. T. Mottonen, I. O. Nuotio, K. M. Irjala, A. E. Leino, and J. S. Viikari, Vitamin D and attainment of peak bone mass among peripubertal Finnish girls: A 3-y prospective study. *Am J Clin Nutr,* **76**, 1446–1453 (2002).

618. E. Papadimitropoulos, G. Wells, B. Shea, W. Gillespie, B. Weaver, N. Zytaruk, A. Cranney, J. Adachi, P. Tugwell, R. Josse, C. Greenwood, and G. Guyatt, Meta-analyses of therapies for postmenopausal osteoporosis. VIII: Meta-analysis of the efficacy of vitamin D treatment in preventing osteoporosis in postmenopausal women. *Endocr Rev,* **23**, 560–569 (2002).

619. H. A. Bischoff-Ferrari, W. C. Willett, J. B. Wong, E. Giovannucci, T. Dietrich, and B. Dawson-Hughes, Fracture prevention with vitamin D supplementation: A meta-analysis of randomized controlled trials. *JAMA,* **293**, 2257–2264 (2005).

620. S. V. Hughes, E. Robinson, R. Bland, H. M. Lewis, P. M. Stewart, and M. Hewison, 1,25-dihydroxyvitamin D3 regulates estrogen metabolism in cultured keratinocytes. *Endocrinology,* **138**, 3711–3718.

621. M. B. Snijder, N. M. van Schoor, S. M. Pluijm, R. M. van Dam, M. Visser, and P. Lips, Vitamin D status in relation to one-year risk of recurrent falling in older men and women. *J Clin Endocrinol Metab,* **91**, 2980–2985 (2006).

622. M. Montero-Odasso, and G. Duque, Vitamin D in the aging musculoskeletal system: An authentic strength preserving hormone. *Mol Aspects Med,* **26**, 203–219 (2005).

623. L. Dukas, H. B. Staehelin, E. Schacht, and H. A. Bischoff, Better functional mobility in community-dwelling elderly is related to D-hormone serum levels and to daily calcium intake. *J Nutr, Health Aging,* **9**, 347–351 (2005).

624. N. K. Latham, C. S. Anderson, and I. R. Reid, Effects of vitamin D supplementation on strength, physical performance, and falls in older persons: A systematic review. *J Am Geriatr Soc,* **51**, 1219–1226 (2003).

625. H. A. Bischoff-Ferrari, B. Dawson-Hughes, W. C. Willett, H. B. Staehelin, M. G. Bazemore, R. Y. Zee, and J. B. Wong, Effect of vitamin D on falls: A meta-analysis. *JAMA,* **291**, 1999–2006 (2004).

626. L. Flicker, K. Mead, R. J. MacInnis, C. Nowson, S. Scherer, M. S. Stein, J. Thomas, J. L. Hopper, and J. D. Wark, Serum vitamin D and falls in older women in residential care in Australia. *J Am Geriatr Soc,* **51**, 1533–1538 (2003).

627. L. Flicker, R. J. MacInnis, M. S. Stein, S. C. Scherer, K. E. Mead, C. A. Nowson, J. Thomas, C. Lowndes, J. L. Hopper, and J. D. Wark, Should older people in residential care receive vitamin D to prevent falls? Results of a randomized trial. *J Am Geriatr Soc,* **53**, 1881–1888 (2005).

628. E. Canalis, J. P. Bilezikian, A. Angeli, and A. Giustina, Perspectives on glucocorticoid-induced osteoporosis. *Bone,* **34**, 593–598 (2004).

629. J. Compston, US and UK guidelines for glucocorticoid-induced osteoporosis: Similarities and differences. *Curr Rheumatol Rep,* **6**, 66–69 (2004).

630. J. L. Shaker, and B. P. Lukert, Osteoporosis associated with excess glucocorticoids. *Endocrinol Metab Clin North Am,* **34**, 341–356, viii–ix (2005).

631. J. P. Devogelaer, S. Goemaere, S. Boonen, J. J. Body, J. M. Kaufman, J. Y. Reginster, S. Rozenberg, and Y. Boutsen, Evidence-based guidelines for the prevention and treatment of glucocorticoid-induced osteoporosis: A consensus document of the Belgian Bone Club. *Osteoporos Int,* **17**, 8–19 (2006).

632. 3. Amin, M. P. LaValley, R. W. Simms, and D. T. Felson, The role of vitamin D in corticosteroid induced osteoporosis: A meta-analytic approach. *Arthritis Rheum,* **42**, 1740–1751 (1999).

633. P. N. Sambrook, M. Kotowicz, P. Nash, C. B. Styles, V. Naganathan, K. N. Henderson-Briffa, J. A. Eisman, and G. C. Nicholson, Prevention and treatment of glucocorticoid-induced osteoporosis: A comparison of calcitriol, vitamin D plus calcium, and alendronate plus calcium. *J Bone Miner Res,* **18**, 919–924 (2003).

634. F. Richy, E. Schacht, O. Bruyere, O. Ethgen, M. Gourlay, and J. Y. Reginster, Vitamin D analogs versus native vitamin D in preventing bone loss and osteoporosis-related fractures: A comparative meta-analysis. *Calcif Tissue Int,* **76**, 176–186 (2005).

Regulation of Bone Cell Function by Estrogens

BARRY S. KOMM, BORIS CHESKIS, AND PETER V. N. BODINE

I. INTRODUCTION

Estrogens and their diverse effects on bone remodeling are perhaps less well characterized than one would expect. The positive impact of estrogens on the skeleton has been well known and documented since the early 1940s, and it continues to be a common treatment modality for osteoporosis [1–6]. However, the mechanism(s) by which estrogens regulate the bone remodeling process and thereby protect the skeleton continues to undergo intense evaluation. New insights into alternative pathways impacted by estrogens in bone and further characterization of genetically modified animals have led to considerable modifications about how we view estrogenic influence on the skeleton.

II. WHAT IS AN ESTROGEN?

Before discussing the role that estrogens play in bone, it is important to define what an estrogen is and the abundance of basic science that describes the multiple facets of estrogenic activity. Estrogens are represented by a large number of molecules, both steroidal and nonsteroidal in nature. The endogenous vertebrate estrogens are 18-carbon, four-ringed structures [7] (Figure 12-1) derived from cholesterol. The most common estrogens in humans include the following steroids: estrone (E1), 17β-estradiol (E2), and estriol (E3). There is an array of estrogenic metabolites that display variable estrogenic activity in addition to several well-characterized B-ring saturated estrogens [8]. In addition to these classic estrogens, several estrogenic substances obtained from plant sources (phytoestrogens), synthetic estrogens (i.e., diethylstilbestrol), and a relatively large group of xenobiotics (e.g., DDT and biphenols) have also been classified as estrogens. Finally, there is a growing number of molecules, originally classified as antiestrogens but currently undergoing reclassification (based on their biological activity), that are represented by a diverse set of chemical structures (Figure 12-1) and are collectively referred to as selective estrogen receptor modulators (SERMs) [9, 10]. Several new molecules that display remarkable specificity for either estradiol receptor α (ERα) or estradiol receptor β (ERβ) provide important tools to aid in the characterization of these receptors' roles without having to genetically manipulate an animal or cell to remove one or both of the receptors. The combination of selective ligands and genetically modified animals provides powerful tools to more thoroughly understand the functional role of the estrogen receptors and how ligands influence their activity.

III. ESTROGEN RECEPTORS

A. Members of the Nuclear Receptor Superfamily

What this assortment of compounds has in common is that they exert their function via a single class of nuclear localized proteins—estrogen receptors. There are currently two members of the estrogen receptor family referred to as ERα [11–13] and ERβ [14, 15]. The estrogen receptors are members of a large superfamily (Table 12-1) of nuclear localized receptors represented by members that bind the classical group of steroid hormones that includes the following: glucocorticoids, progestins, androgens, and mineralocorticoids. In addition to these, other members include the receptors for vitamin D_3, retinoids, thyroid hormones, oxysterols, farnesol, prostanoids, and ecdysone. There are also more than 50 members of this superfamily for which a ligand has not been identified, and they are referred to as orphan nuclear receptors [16–19].

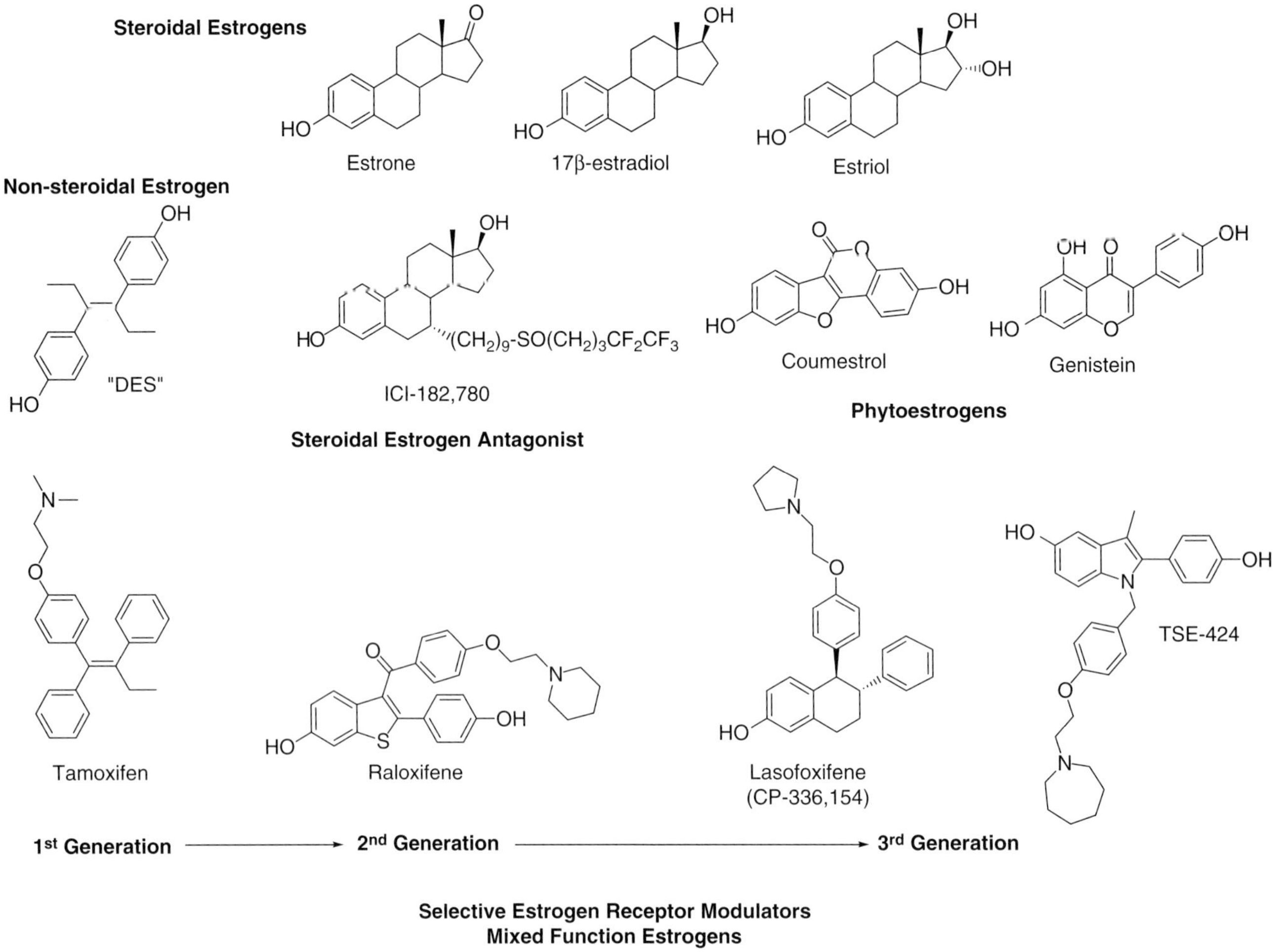

Selective Estrogen Receptor Modulators
Mixed Function Estrogens

FIGURE 12-1   Structures of a variety of compounds that can be classified as members of the family of estrogens. In red are classical steroidal estrogens represented by the three predominant circulating estrogens detected in mammals. In pink is the nonsteroidal and potent estrogen diethylstilbestrol. In green are two phytoestrogens, both nonsteroidal but functionally characterized as estrogens. In black is the potent steroidal antiestrogen ICI-182780. This compound has been described as a pure estrogen receptor antagonist; however, its characterization is still under examination. At the bottom of the figure in blue are three generations of selective estrogen receptor modulators. Originally referred to as antiestrogens, this group of compounds exhibit mixed functional activity, all seemingly transduced by estrogen receptors. What all these compounds (and there are hundreds more) have in common is that they bind to the estrogen receptors and functionally affect estrogen receptor activity. In some cases, the effects are only as agonists, or as relatively potent antagonists, but most commonly they are as mixed function ligands with their effects related to the cellular target and the specific genes that are being monitored.

The receptors in this group share many common features. Structurally, this group of proteins can be dissected into discrete regions with different functions [11, 20]. The regions are designated simply as A–F (Figure 12-2). The unifying characteristic of each nuclear receptor family member is a zinc finger domain (region C) associated with DNA binding (DNA-binding domain [DBD]). The receptors are DNA-binding proteins that interact with specific DNA sequences (e.g., estrogen response element and androgen receptor element) [21, 22] via two cysteine-rich domains that intercalate zinc to form binding "fingers." The homology between members of this family in this domain is relatively high, and although there is amino acid disparity in the DBD, the cysteine residues can be aligned for all of the receptors supporting their derivation from a common ancestral protein. The other domains are a ligand-binding domain (region D, E, and F), a nuclear localization domain (D), and a hinge domain (D). In addition, two transactivation domains, AF-1 and AF-2, are located in the N-terminal (A/B) and C-terminal (E) portions of the protein, respectively [23].

The mechanism through which information is transduced from the ligand by the receptor has been the

TABLE 12-1 Members of the Steroid/Thyroid/Retinoid Nuclear Receptor Superfamily

Androgen	Estrogen (α, β)
Glucocorticoid	Mineralocorticoid
Progesterone (A, B)	Thyroid hormone (α, β)
Vitamin D	Retinoic acid (α, β, γ)
Retinoid X receptor (α, β, γ)	Peroxisome proliferator activating receptor (α, β, γ)
Pregnane receptor	Ecdysone
Orphan receptors (>50)	

subject of intense research for more than 40 years. It has become clear that ligand binding to the estrogen receptor initiates a number of processes. Ligand binding produces a change in conformation that for several members of the family, including the ER, appears to begin with the displacement of heat shock proteins [24, 25]. Subsequently, two liganded estrogen receptors dimerize [26], are biochemically modified (e.g., acetylation and phosphorylation) [27], and then bind to specific DNA sequences. In this simple model, the "activated" ER complex can act as an enhancer or repressor of gene transcriptional activity [28, 29].

B. Coactivators and Corepressors

The model for ER regulation of gene transcription has become more complex with the discovery of several proteins that interact with the ER as well as other members of the steroid hormone receptor superfamily. These proteins are referred to as coregulators and are represented by both coactivators [30, 31] and corepressors [32, 33]. Several coregulators have been identified, represented by a diverse group of proteins and RNA [34]. Not unlike the nuclear receptors, several of these proteins contain specific regions associated with independent function [35], including histone acetylation, CREB-binding protein interaction domains, and a nuclear receptor interaction domain (NRID) [36, 37]. The corepressors contain histone acetylase domains [33]. Within the NRID domain, one or more LXXLL motifs interact with the ER and other members of the superfamily [38, 39]. This binding has been verified by cocrystallization of the ERα ligand-binding domain (LBD) with a small peptide containing an LXXLL domain from the coactivator protein GRIP 1 (SRC-2) [40] and has been shown to interact specifically with a region of the receptor represented by helices 3, 4, 5, and 12 [40, 41]. Interaction of these coactivators via the NRID has also been demonstrated to be associated with increased transcriptional activity of the ER [42]. The transcriptional complex is composed of an array of proteins that include several coactivators whose roles may vary; however, some definitely serve to bridge the enhancer region of ER binding on DNA with the basal transcriptional machinery. The DRIP/TRAP complex of proteins (>10 proteins) has been shown to play the dual role of transcriptional activation and bridging the transcriptional enhancer complex with the

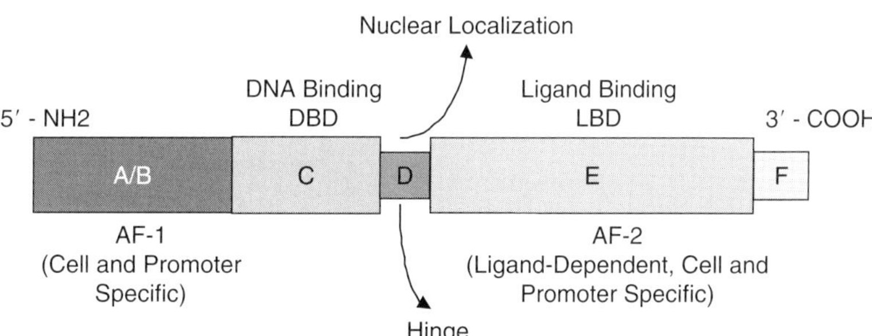

FIGURE 12-2 Schematic structure of nuclear hormone receptors. This family of receptors, which includes the estrogen receptors, can be represented as cassettes with interchangeable units. The A/B domain at the N-terminus contains at least one transactivation domain (AF-1) that is ligand independent. The A/B domain is adjacent to the C domain, which represents the DNA-binding domain containing two cysteine loops that each intercalate one zinc molecule to form DNA-binding fingers. This domain is highly conserved among the family members. The D domain is much less well defined but has been described as the hinge domain and contains a nuclear localization signal; however, other sites in the ER have been linked to nuclear localization outside of the D domain. The E domain represents the ligand-binding domain and is not as conserved as the C domain. Additionally, embedded within the ligand-binding domain is another transactivation domain, AF-2, which is ligand dependent, unlike AF-1. The F domain of the receptor does not display any clear function; however, removal of small parts of this domain can affect receptor function (both ligand binding and transactivation), and it had to be removed to efficiently crystallize the ligand-binding domain of the estrogen receptors.

basal transcriptional complex. Not all proteins in the DRIP complex have been shown to interact with the ER, and this complex does not play a functional role in transcriptional enhancement with nuclear steroid hormone receptors [43–45]. This is not to say that the ER cannot interact directly with proteins associated with the basal transcriptional machinery, as has been suggested for the vitamin D receptor.

C. Alternate Pathways for Estrogenic Activity

In addition to direct regulation of gene expression (genomic action), steroid hormones regulate cell signaling phosphorylation cascades. This process is insensitive to inhibitors of RNA and protein biosynthesis and, in some cases, can take place in the absence of a nucleus, with isolated cell membranes, or enucleated cytoplasts [46]. All members of the steroid hormones, from the corticosteroids (glucocorticoids and mineralocorticoids) to the gonadal hormones (estrogens, progestins, and androgens), vitamin D_3, and thyroid hormone, can exhibit nongenomic effects. These effects range from activation of adenylyl cyclase, mitogen-activated protein kinases (MAPKs), and phosphatidylinositol 3-kinase (PI3K) to increases in intracellular calcium concentrations [47–55].

In recent years, significant progress has been made in understanding the molecular mechanisms of the nongenomic action of the steroid/nuclear receptors. Major issues that remain to be addressed are the nature of receptors that are responsible for hormone-mediated activation of cell signaling pathways, molecular mechanisms that integrate hormonal action in regulation of signaling pathways, and the physiological role of rapid nongenomic actions of steroid hormones.

1. RECEPTORS THAT MEDIATE THE NONGENOMIC ACTION OF STEROID HORMONES

There is substantial evidence that a subpopulation of conventional steroid/nuclear hormone and vitamin D receptors mediate rapid effects of steroid hormones and vitamin D on regulation of cell signal transduction pathways. In experiments conducted in cell lines, rapid activation of various signaling pathways by all classes of steroid hormones and vitamin D has been shown to be dependent on conventional steroid/nuclear receptors by reconstitution experiments with receptor negative cell lines, by knock-down of receptors with siRNA or antisense RNAs, by use of highly specific steroid receptor antagonists, or by studies with receptor knockout mice. Furthermore, the onset of rapid electrical responses to vitamin D_3 was shown to be

lost in primary osteoblast cells derived from vitamin D receptor (VDR) knockout mice [56]. Similarly, it was demonstrated that the nongenomic enhancement by corticosterone of glutamate transmission in the CA1 hippocampal area was missing in a forebrain-specific mineralocorticoid receptor knockout mouse model [57].

However, novel membrane receptors unrelated to conventional steroid receptors have also been implicated. An orphan member of the G protein–coupled receptor (GPCR) superfamily, termed GPR30, has been reported to act independently of classical ERs to trigger rapid signaling by estrogens [58, 59]. E2 treatment of GPR30 transfected cells that apparently lack ER expression caused activation of a stimulatory G protein (Gs) that is directly coupled to this receptor and increased adenylyl cyclase activity [60]. GPR30 is localized to endoplasmic reticulum and binds E2 with nanomolar affinity [59]. A novel GPCR-like membrane progesterone receptor (mPR) in spotted sea trout oocytes has also been identified [61]. mPR binds progesterone with high nanomolar affinity and is involved in mediating progesterone induction of meiosis. The receptor contains seven putative hydrophobic transmembrane domains, and based on sequence, a family of mPR-related proteins has been identified in a number of different species, including frog, human, and mouse [61]. These novel putative membrane steroid receptors and conventional receptors as mediators of rapid steroid signaling are not mutually exclusive; both classical and membrane steroid receptors may be utilized in mediation of the nongenomic action. However, the biological relationship between GPR30 and mPR with conventional ER and PR is not known. It has been suggested that a complex network of proteins that consists of conventional steroid receptors and other steroid-binding proteins, such as GPCR30 and mPR, could mediate rapid steroid signaling [51, 52]. Finally, some rapid actions of steroids mediated at the cell membrane may involve allosteric effects of steroids on other known membrane receptors, enzymes, or ion channels. For example, progesterone can alter oxytocin activation of its receptor and the analgesic effect of progesterone metabolites is mediated by allosteric effects on $GABA_A$ receptors [62, 63].

2. MEMBRANE LOCALIZATION OF "NUCLEAR" RECEPTORS

Although the majority of steroid receptors are localized in the nucleus in the presence of hormone, there is evidence that a small fraction of receptors are localized at or near the cell membrane in either the presence or the absence of ligand. Immunocytochemical staining has demonstrated positive staining of ERα at the plasma membrane of different cells [64, 65]. Confocal microscopy showed that E2 treatment of MCF7 cells rapidly induced membrane ruffles, pseudopodia, and translocation of ERα to the cell membrane. Also, endogenous ERα was biochemically isolated from plasma membranes and caveolae fractions of endothelial cells, and estrogen was able to stimulate signaling in these isolated membrane fractions. How steroid receptors traffic to the cell membrane and what controls the traffic and the precise sites in the cell where steroid receptors mediate their rapid signaling activities are important questions that remain to be addressed. Steroid receptors do not contain transmembrane domains that could mediate their membrane localization; therefore, interactions with other proteins and post-translational receptor modifications have been proposed to be involved. Candidate interacting proteins include caveolin-1 and -2 and the 110-kDa caveolin-binding protein striatin. Caveolae are specialized regions of the plasma membrane that assemble and organize signaling protein complexes [66]. Endogenous ERα has been reported to interact with caveolin-1 and -2 in an E2-dependent manner in MCF-7 and in vascular smooth muscle cells, and over-expression of caveolin-1 in MCF-7 cells increased E2-dependent ERα translocation to the plasma membrane [67]. Striatin is a calmodulin-binding member of the WD-repeat family of proteins that contains several protein–protein interaction domains and is required for estrogen-induced activation of endothelial nitric oxide synthase (eNOS). Striatin binds to amino acids 183–253 of ERα and can bridge it with the G protein–coupled receptor Gαl complex [68, 69]. It has been also proposed that ERα can be targeted to membrane by adaptor protein Shc [70]. In MCF7 cells, the Src homology domain 2 (SH2) of Shc has been shown to directly interact with the N-terminal part of ERα [71]. ERα, Shc, and insulin-like growth factor type 1 receptor (IGF-1R) interact on the cell membrane of MCF7 cells through Shc binding to phosphorylation sites of the intracellular domain of the IGF-1 receptor. Further supporting the importance of this interaction for ERα translocation to cell membrane, treatment of the cells with siRNA for Shc, or IGF-1R, attenuated E2-induced ERα translocation to cell membrane and E2 stimulation of MAPK phosphorylation [72]. Another membrane adaptor protein, p130Cas (Crk-associated substrate), has also been reported to interact with the ERα–cSrc complex in T47D breast cancer cells and to potentiate estrogen activation of Src [73]. p130Cas could potentially be important for membrane localization of the ERα–Src complex and for its integration into the network of membrane signaling molecules. Palmitoyl acyl transferase–dependent S-palmitoylation of ERα has been reported and shown to promote ERα association with the plasma membrane and interaction

with caveolin-1 [68, 74]. Mutation of the palmitoylation site in ERα (cystine 447) amino acid or inhibition of palmitoylation with 2-bromo-palmitate resulted in a significant decrease in receptor localization at the plasma membrane. Furthermore, cystine 447–mutated ERα did not stimulate an activation of MAP and PI3 kinases [74]. A terminally truncated 46-kDa variant of ERα has been found to be preferentially palmitoylated and enriched in plasma membrane of several cell types (endothelial, osteoblasts, and MCF 7 cells) [75–77]. This modification was shown by use of palmitoylation inhibitors to contribute to membrane localization of 46-kDa ERα. It has been suggested that truncated ERα through altered protein folding may expose sites for fatty acid acylation that are not accessible in full-length 66-kDa ERα [78].

3. ACTIVATION OF CELL SIGNALING BY STEROID RECEPTORS

One of the most intriguing questions that remain to be answered is how the conformational changes in receptor molecules induced by the binding of steroid hormones are converted into activation of some kinases. The nature of the upstream receptor targets also remains to be better established. A physical association of ERα with IGF-R, cSrc, and PI3 kinase has been previously reported [78–81]. Multiple lines of evidence suggest that activation of the tyrosine kinase cSrc represents one of the initial steps in ERα-mediated cell signaling, at least in some cells [82]. The Src kinases share common structural organization, differing in the N-terminal 60–80 amino acids [83]. There are several functional motifs common to all Src family members. The N-terminal region, Src homology 4 domain (SH4), contains consensus sequences for myristoylation and palmitoylation [84]. The SH3 domain binds polyproline motifs [85], and the SH2 domain binds to phosphotyrosine-containing sequences [86]. The C-terminal SH1 domain contains the catalytic region and a short regulatory domain with major regulatory tyrosine Y527 [83]. Under basal conditions, the catalytic domain of Src is constrained in an inactive state through intramolecular interactions. Binding of the SH2 domain to the C-terminal phosphorylated tyrosine and the SH3 domain to the proline-rich region in the Src linker domain locks the molecule in an inhibited conformation [87]. Full catalytic activation requires the release of these constraints. cSrc can be activated either by dephosphorylation of the C-terminal inhibitory phosphotyrosine site (or, in oncogenic variants, by loss of the C-terminal tail) or by binding of high-affinity ligands to the SH2 or SH3 domains. These domains are modular polypeptide units that mediate protein–protein interactions and are found together on many proteins, suggesting that their activities can be coordinated and that they can cooperate in Src regulation [85, 88]. The essential role of Src kinase in the nongenomic action of steroid receptors was demonstrated in experiments with embryonic fibroblasts derived from Src$^{-/-}$ mice. These cells did not show rapid activation of the MAPK pathway in response to AR and ERα activation, whereas wild-type Src+/+ cells did show rapid activation [89].

Direct and hormone-dependent interaction of PR$_B$ and the cSrc SH3 domain is necessary and sufficient for activation of cSrc and its downstream targets, leading to phosphorylation/activation of Erk 1/2 [90]. Endogenous as well as overexpressed and purified PR$_B$ and cSrc interact, and this interaction is mediated by a polyproline region encoded by amino acids 421–428 of PR$_B$ with the Src SH3 domain. Mutational analysis of PR and competition experiments with peptides corresponding to the PXXP sequence demonstrate that this motif in the N-terminal part of PR is responsible for PR interaction with the SH3 domain of Src. In addition to cSrc, PR also interacts with SH3 domains of hematopoietic cell kinase (Hck), the regulatory subunit of PI3K (p85), Grb2, and the tyrosine kinases Fyn and Crk [90]. Because the activation constant of HcK by PR is in the low nanomolar range, PR is a potent activator of Src by an SH3 domain displacement mechanism [90]. However, the ability to directly interact with SH3 domains appears to be a unique property of PR. Other steroid receptors, including receptors of the thyroid hormones, either do not have PXXP motifs or, like androgen receptor (AR), contain a short polyproline sequence in the N-terminal part of their molecules but do not directly interact with the SH3 domain of Src [90, 91].

An alternative model of cSrc activation by the PR has also been proposed in which PR-B indirectly interacts with Src through formation of a complex with unliganded ER. ERα in turn is proposed to activate Src by a direct interaction with the Src SH2 domain [78, 82]. Indeed, ERα is able to interact with the SH2 domain of cSrc [90–93] and phosphotyrosine 537 of ERα is required for this interaction [52, 94]. PR–ERα interactions were detected in cells by yeast two-hybrid assay and by co-immunoprecipitation approaches [94], and sites of interaction were mapped to two broad regions of the N-terminal domain of PR flanking the PXXP motif and to the LBD of ER [82, 94]. The polyproline motif of PR was dispensable for progestin activation of Src by this mechanism [94]. Based on these results, it has been proposed that progesterone activation of Src in cells expressing ER may be mediated indirectly through unliganded ERα, and that direct PR interaction with Src through its intrinsic polyproline motif occurs mainly in the absence of ERα. How progestin can transmit a signal to Src through unliganded

ER has not been explored, and whether this indirect mechanism of PR activation of Src is mediated through direct ER–PR contacts or involves another protein is not known. There is no evidence of a direct protein–protein interaction between ER and PR.

Although ERα is capable of interacting directly with the SH2 domain of Src in an estrogen-dependent manner, this interaction does not appear to be sufficient for estrogen-induced activation of Src [90, 93]. An adaptor protein, MNAR (modulator of nongenomic action of estrogen receptor), has been identified that is required for estrogen-induced ERα activation of Src and the downstream MAPK pathway [93]. MNAR is homologous to a protein that was previously isolated by pulldown with the SH2 domain of p56^{lck} (Lck) [95]. The protein, referred to as proline and glutamic acid–rich protein (p160) [95], was later designated PELP1 (proline-, glutamic acid-, leucine-rich protein) [96]. MNAR is an ~120-kDa scaffold protein that contains multiple protein–protein interaction domains. The N-terminal portion of the MNAR molecule contains 10 LXXLL motifs, similar to those in the p160 family of coactivators, that mediate hormone agonist–dependent interaction with AF-2 of nuclear receptors [97], and 3 PXXP motifs that are similar to SH3 domain interaction sequences. Purified MNAR alone simulates cSrc enzymatic activity; however, purified ERα and MNAR together synergize to produce strong estrogen-dependent activation of cSrc [93]. Interaction between endogenous ERα, MNAR, and Src was demonstrated using co-immunoprecipitation from the cell extracts of MCF7 cells. As evidence that MNAR and ERα cooperate to activate Src in intact cells, overexpression of MNAR enhanced estrogen stimulation of Src enzymatic activity and phosphorylation of MAPK in MCF-7 cells, whereas expression of antisense oligonucleotides to MNAR attenuated estrogen activation of the Src/MAPK pathway [93]. Mutational analysis and functional evaluation of MNAR and the use of ERα and cSrc mutants revealed that MNAR interacts with the Src SH3 domain via its N-terminal PXXP motif (designated PXXPP motif 1). Mutation of this motif abolished the MNAR-induced activation of the Src/MAPK pathway. ER interacts with the Src SH2 domain using phosphotyrosine 537, and this complex is further stabilized by MNAR–ER interaction. The region responsible for MNAR interaction with ERα maps to two N-terminal LXXLL motifs of MNAR (designated LXXLL motifs 4 and 5). Mutation of these motifs prevented ERα–MNAR complex formation and eliminated activation of the Src/MAPK pathway [52]. The presence of multiple LXXLL motifs suggests that MNAR can potentially interact with multiple nuclear receptors. Indeed, MNAR also interacts in a hormone agonist–dependent manner with several other steroid receptors, including AR, GR, PR, and VDR [52, 93]. However, it is not clear whether all receptors would require MNAR for activation of cell signaling pathways. Existing data indicate that MNAR is a scaffold that is promoting receptor binding to Src and stabilizing the ERα–Src complex. Therefore, it is reasonable to postulate that the affinity of ERα binding to the Src–MNAR complex is higher than that of ERα binding to Src alone. Thus, formation of this complex can take place at lower concentrations of ERα, Src, and E2. Some receptors, however, may not require an adaptor molecule (e.g., PR) because they may interact with Src with high affinity or their expression level is high.

MNAR has also been implicated in mediating rapid androgen-induced signaling [98]. AR is involved in the development, growth, and progression of prostate cancer (CaP). CaP, however, often progresses from an androgen-dependent to an androgen-independent tumor, making androgen ablation therapy ineffective. The mechanisms that are responsible for the development of androgen-independent CaP are unknown. Unni and coauthors [98] demonstrated that treatment of LNCaP cells with DHT leads to AR–MNAR–Src complex formation and activation of the cSrc/MAPK/CREB pathway. Activation of this pathway correlates well with an increase in DNA biosynthesis and inhibition of apoptosis. In contrast, in LNCaP-HP cells, which are androgen independent, Src is constitutively activated, which is associated with DHT-independent, constitutive interaction between Src, AR, and MNAR. These data suggest that MNAR is involved in AR-mediated activation of the Src/MAPK/CREB pathway [98]. It has been demonstrated that *Xenopus* oocyte maturation is regulated via a "release of inhibition" mechanism whereby constitutive G protein–mediated signals, including Gβγ and Gα, hold cells in meiotic arrest. Steroid-triggered signaling overcomes these inhibitory signals, resulting in meiotic progression. Evidence suggests that androgens play a critical role in regulating oocyte maturation [99]. MNAR is expressed in oocytes, and reduction of its expression by RNA interference markedly enhanced testosterone-triggered maturation and activation of the MAPK pathway. Endogenous MNAR, AR, and Gβ interact, and this interaction requires the N-terminal part of the MNAR molecule, which contains multiple LXXLL motifs [100]. These data suggest that MNAR plays an important role in steroid hormone–induced *Xenopus* oocyte maturation.

4. FUNCTIONAL CONSEQUENCES OF STEROID ACTIVATION OF CELL SIGNALING PATHWAYS

One of the best characterized extranuclear actions of steroids is the rapid activation of the Ras/Raf/MAPK pathway. In nerve cells, E2 rapidly triggers Erk 1/2 activation, leading to c-Fos gene expression [101]. Rapid activation of this pathway was also found in osteoblasts [102] and in white adipocytes [103]. Estrogen-activated growth of the human colon carcinoma–derived Caco-2 cell is mediated through rapid and reversible stimulation of the cSrc and cYes and subsequent activation of Erk1 and Erk2 kinases [104]. In the MCF-7 human breast cancer cell line, E2 triggered a rapid increase in the active form of p21ras, rapid tyrosine phosphorylation of Shc and p190, and association of p190 with the guanosine triphosphatase (GTPase) activating protein. Both Shc and p190 are substrates of activated Src, and once phosphorylated, they can interact with other proteins and stimulate p21ras. Estrogen-mediated stimulation of the Ras/Raf/ERK pathway promotes MCF7 cell proliferation [105]. Rapid progesterone-induced activation of Src and downstream MAPK cascade in a manner dependent on conventional PR has also been observed in different mammalian cells, including breast cancer cell lines. As with estrogens, the proliferative effects of progesterone in breast cancer cells were shown to be dependent on progesterone activation of the cSrc/Raf/MAPK pathway.

The MAPK pathway is involved in the control of many fundamental cellular functions, including cell proliferation, survival, differentiation, apoptosis, motility, and metabolism. Some of these functions are mutually exclusive, such as estradiol proliferation in MCF7 cells [105] versus cell cycle arrest and differentiation in osteoblasts [106]. Activation of the MAPK pathway by sex steroids exerts antiapoptotic effects on osteoblasts/osteocytes but proapoptotic effects on osteoclasts. Apparently, the kinetics of ERK phosphorylation and the length of time that phospho-Erks are retained in the nucleus are responsible for the pro- versus antiapoptotic effects of estrogen on different cell types of bone and perhaps their many other target tissues [106]. It has long been recognized that transient and sustained signaling from the Ras/ERK pathway can lead to the different biological outcomes of proliferation and differentiation, respectively [107].

A well-characterized and biologically important action of estrogen is the acute effect on blood vessels to stimulate vasodilation and protect against vascular injury. This action has been shown to be mediated by a subpopulation of ERα in plasma membrane of endothelial cells through activation of eNOS and stimulation of NO production via the PI3K/Akt signaling pathway. Src, which is upstream of PI3K, also appears to be important. As evidence of the biological importance of this action of estrogen, mice treated with estrogen show increased eNOS activity and decreased vascular leukocyte accumulation after ischemia and reperfusion injury in a manner dependent on PI3K and eNOS. ERα knockout mice lost the acute protective effect of estrogen on the vascular injury response, which indicates that a conventional receptor mediated this rapid effect of estrogen [108]. One of the important downstream targets of PI3K is the threonine–serine kinase Akt/protein kinase B. Activation of PI3K/Akt by estrogens has also been shown to be important in breast cancer cells in mediating estrogen [92], stimulation of cell cycle progression [92], and inhibition of apoptosis [109]. Other SRs, such as AR, PR [90], and GR, also interact with the regulatory subunit of the PI3K, p85 [81].

Many cell signaling pathways converge upon and regulate the phosphorylation status and hence activity of multiple transcription factors, which affects gene expression. Several examples of this mode of regulation have been reported, including ERα-dependent estrogen regulation of the c-fos gene mediated by Src/MAP and Src/PI3K pathways converging on Elk-1 and SRF, respectively; estrogen regulation of cyclin D1 mediated by the PI3K/Akt pathway; and estrogen regulation of the Egr-1 gene mediated by MAPK activation of SRF [110, 111].

Protein phosphorylation cascades rapidly stimulated by steroids also play an important role in gene regulation by affecting receptors' stability and transcriptional activity. PR and retinoic acid receptor-γ2 (RARγ2) undergo ligand-dependent degradation mediated by the ERK and p38 pathways, respectively [112, 113]. ERa is phosphorylated on multiple serine/threonine residues in the N-terminus by MAPK and other kinases, and these phosphorylations are important for intrinsic transcriptional activity of the receptor [114, 115]. SRC-1 and GRIP-1, members of the p160 family of steroid receptor coactivators, are direct targets of MAPKs. In both SRC-1 and GRIP-1, ERK pathway activation leads to enhanced coactivation function [116, 117].

5. Summary

The action of steroid hormones is mediated by a complex interface of direct control of gene expression and by the regulation of cellular phosphorylation cascades. Although the genomic action of nuclear receptors is relatively well understood, the mechanisms that integrate receptors' action in regulation of cell signaling as well as the precise physiological role of the nongenomic action remain poorly defined. Significant progress has been made in our understanding of the molecular mechanisms of receptor-mediated activation of important signaling molecules. Several membrane proteins have been identified that interact with classical receptors and influence the nongenomic action. However, the precise role of these proteins in receptor regulation

of cell signaling remains to be further investigated. It is possible that the composition of the receptor complexes at the plasma membrane is cell type dependent, which may potentially explain cell type selectivity of the nongenomic action. Significant progress has also been made in understanding how interactions between conventional receptors and kinases lead to activation of cell phosphorylation cascades. Direct PR binding and activation of Src suggest that some receptors may directly regulate important signaling molecules. Others, such as ER and AR, may require an adaptor or scaffold protein to facilitate their interaction. A novel adaptor protein termed MNAR, which contains multiple interaction domains and stimulates receptor binding to some kinases, has been identified. It has been demonstrated that MNAR interacts with ERα and -β, PR, AR, GR, and VDR ([95] and Greger and Cheskis, unpublished data). Interaction with MNAR is essential for ER- and AR-mediated activation of Src/MAP [94, 98] and for ER activation of the PI3/Akt pathway (Greger and Cheskis, unpublished data). MNAR also controls AR activation of the G protein–coupled receptors [100]. Data also suggest that in cells treated with growth factors, the MNAR–AR–Src complex also interacts with epithelial growth factor (EGF) receptor [118]. Therefore, interaction with MNAR converts binding of specific ligand and conformational changes in receptor molecules into regulation of signaling molecules that control important cellular functions. Future studies may find that, in addition to MNAR, some other proteins may also play a similar role by integrating receptor actions in the regulation of important cellular processes.

Considering that nongenomic and genomic functions of steroid receptors may potentially regulate different cellular processes, ligands that can differentiate between direct transcriptional and nongenomic mechanisms may represent a new generation of functionally selective regulators of nuclear receptors' actions. These compounds may allow tissue-selective regulation of important physiological processes and may potentially be pharmacologically superior to currently marketed drugs, ligands of steroid receptors.

D. ERβ

Estrogens can elicit a variety of physiological responses, and until 1996, it was believed that transduction of information occurred through one nuclear receptor protein (ER). However, as mentioned previously, a second protein has been identified that also exhibits high-affinity binding for estrogens, which has been called ERβ [14, 15, 119]. Its chromosome location is different from that of the human ERα (14 vs. 6, respectively) [120]. The two transcripts are of different length, with ERβ coding for a protein of 530 amino acids [121] and ERα coding for a protein of 595 amino acids [122]. Additionally, their tissue distribution varies, especially in the central nervous system, ovary, uterus, and prostate [123]. The functional role of ERβ remains controversial; however, data demonstrate a role for ERβ in the skeleton. *In vitro* transcription assays have shown that ERβ, like ERα, dimerizes and binds to DNA (specifically estrogen response elements [EREs]). Yet, it has been shown that under appropriate conditions, ERβ heterodimerizes with ERα, and the resulting complex binds to DNA more avidly than the ERβ homodimer [124]. However, the transcriptional activity of the heterodimer is similar to that of the ERβ homodimer, but it differs from that of the ERα homodimer. The affinity of 17β-estradiol for the two receptors is essentially identical, but clearly under *in vitro* conditions ER is a more effective activator of transcription [121].

Another characteristic difference between these two receptors is their apparent variation in ligand affinity. Whereas 17β-estradiol binding affinity is the same, another estrogen, the phytoestrogen genestein, shows a remarkable preference for ERβ (~30-fold) [125]. The interaction of coactivators with these two proteins is also different. The design of new molecules demonstrates their role in affecting coactivator and corepressor interaction in ERα and ERβ [126]. This information, coupled with the different tissue distribution and apparent differences in ligand preference, suggests that specific ligands may exist that activate one receptor preferentially over the other [127]. If this is the case, then it also seems quite possible that these compounds could be synthesized and specifically activate only one of the receptors. Considerable work in this area of ERα and ERβ selective ligands has provided tools to elucidate the roles of the two receptor proteins [128–130]. The pharmaceutical implications are obvious.

E. Crystallization of ERα and ERβ

Both ERα and ERβ LBDs have been crystallized (without region F, which apparently inhibits efficient crystallization) [40, 131, 132]. ERα cocrystallized with diethylstilbestrol (DES), 17β-estradiol, and 4-OH tamoxifen demonstrates that these ligands generate two different conformations of the ERα LBD. With a natural agonist (17-estradiol) or a synthetic agonist (DES), the ligand fits snuggly into a pocket, and helix 12 (12 of 12 helices in LDB crystallized) appears to cover the binding domain [40]. With the SERM 4-OH tamoxifen, helix 12 no longer covers the binding pocket, and it shifts in position to a region that masks amino acids in

helices 3–5. The hydrophobic surface created by those amino acids is critical for the interaction of members of the p160 coactivator family (SRC1, -2, and -3) [40]. Indeed, transcriptional activation studies performed with these coactivators in the presence of various antiestrogens reveal little to no activity, thereby supporting the structural data and the importance of the AF-2 domain in estrogen receptor transactivation.

F. Tissue Selective Estrogens

It has become clear that estrogen receptors are rather accommodating partners for a wide variety (chemically diverse) of ligands. This is unlike the other members of the steroid receptor superfamily, which demonstrate more stringent binding parameters. Compounds with rather diverse structures have been demonstrated to bind with high affinity to the ER and exhibit various potencies depending on the endpoints evaluated. Classically, the targets of estrogen action were the uterus, breast, and liver. In the past two decades, it has been shown that estrogens directly impact the skeleton, central nervous system, immune system, cardiovascular system, and the gastrointestinal tract. The discovery of ERβ has led to the inclusion of the prostate as an estrogen target tissue in males, along with some tissues common to both sexes (i.e., bone, cardiovascular, and immune). Obviously, depending on the tissue, the genetic response to estrogens varies. There may be a group of genes that respond similarly in all tissues to a particular agonist, but the key end responses are most likely tissue selective as a result of a specific set of genes' responsiveness. Thus, in the uterus a collection of genetic endpoints can be quantitated that are distinct from those of the mammary gland. This is a critical premise defining the role of tissue-selective estrogens (or SERMs) and their clinical applications [8, 10]. Perhaps all estrogens are selective and a change in nomenclature is in order. Nevertheless, one example of a tissue-selective estrogen is a compound that behaves as an estrogen receptor agonist in the skeleton but as an antagonist (actually, no activity but would antagonize estrogens) in the uterus. Tamoxifen, which was originally targeted for contraception, turned out to be a better antiestrogen on breast tissue and was developed as a treatment for hormone (estrogen) responsive breast cancer. As more data were generated, it was seen to affect several other tissues besides the breast [133]. Some of the effects were positive (estrogen agonist activity), such as on the skeleton and lipid profiles, whereas others were considered negative, such as the antagonist effect in the central nervous system and the agonist effect on the uterus [134–137].

How could this be? Clearly, all SERMs do not behave identically. The difference between tamoxifen and the next SERM to follow, raloxifene, is primarily on the uterine endometrium. A number of SERMs have followed, including lasofoxifene [138], bazedoxifene [139], ospemifene [140], SCH 57068 [141], SP500263 [142], and HMR-3339 [143]. Interestingly, all of these SERMs have a surprisingly similar effect on the skeleton despite differences in bioavailability and chemical structure; however, their uterine profile appears to be the major distinguishing characteristic. Because of structural diversity, their impact on estrogen receptor function due to different receptor conformation varies [144], and, conceptually, this must account for the differences in responses that are seen when comparing these compounds.

IV. ERα AND ERβ KNOCKOUT MICE (ERKO AND βERKO)

In an effort to more clearly define the physiologic role(s) of both ERα and ERβ, knockout (KO) mice have been generated [145, 146]. Neither KO is lethal and the phenotype exhibited by mice was not as predictable as anticipated. The ERKO and βERKO (ERβ knockout) animals do not demonstrate a striking skeletal phenotype, suggesting that the presence of either one of the receptors is sufficient to maintain skeletal responsiveness to estrogens. There is a small, but significant, decrease in bone length in both sexes of the ERKO animals. This is not seen in the βERKO animals. Bone mineral density is minimally affected in both KO strains [147]. Ovariectomy of either knockout results in osteopenia, which is typical of wild-type mice and rats, supporting the fact that either receptor is capable of maintaining "normal" modeling in the mouse. Investigations have shown that only ERα regulates bone remodeling in males, whereas in females both receptors play a role and under basal conditions compensate for one another [148]. Yet the data do support the fact that despite the sex of the animal, ERα is the primary effector of 17β-estradiol on the skeleton [149]. Further support for the key role of ERα from knockout experimentation was the demonstration that ERα was required for a full osteogenic response to loading and, in fact, ERβ appeared to depress ERα-mediated strain-related increase in osteoblast number and function [150].

There is one report of a human who suffers from an ERα inactivation (point mutation resulting in a premature stop codon) [151]. This man exhibits an overt phenotype in which longitudinal bone growth has not terminated (no epiphyseal closure) and bone mineral density has been compromised. Although not published, it appears that this man expresses normal ERβ and normal androgen receptors. The skeletal phenotype of this man is opposite of that seen in mice lacking ERα, which should warn us (once again) about extrapolation of results from rodents to man. The human data, at least in this man, also suggest that ERβ and androgen receptors are not sufficient to overcome the inactivation of ERα in all aspects of skeletal function in which estrogens are required.

The ERKO mice are characterized by atrophic uteri, ovarian malfunction, and tremendously increased circulating estrogens. The testes are abnormal in appearance, wet weight, and function. Successful production of ERKO animals requires heterozygote crossing due to the reproductive impairment in both sexes when both ERα alleles are inactivated. The βERKO animals, like their ERKO counterparts, exhibit ovarian changes; however, unlike the ERKO animals, which have hemorrhagic ovaries, the βERKOs demonstrate some mature follicles but reduced numbers compared to normal, wild-type mice, resulting in reduced fecundity. The uteri of these mice are normal and circulating estrogens are normal. Testicular histology and function is normal, as is male reproductive behavior; however, with age, prostate and bladder hyperplasia has been reported. ERβ receptor distribution is clearly distinct from ERα; there is some overlap, but there is absolutely no ERα in specific central nervous system regions, the ovarian granulosa cells, and, in males, the prostate. The animal data indicate that ERα plays a dominant role in the uterus and the ovary, which raises questions as to the absolute necessity of ERβ in the granulosa cells. It is hoped that the double knockout animals that are becoming available will aid in the elucidation of ER function more clearly than the individually knocked-out animal examples. Early data on males revealed that the bone phenotype is like that of the ERKO animals, again bringing into question the role of ERβ in the normal developing and remodeling skeleton [152]. In addition, it has been shown that a functional androgen receptor (AR) is not sufficient to allow 17β-estradiol to prevent loss of bone mass in double knockout animals [153].

V. ESTROGENS AND BONE

Estrogens are important regulators of skeletal development and homeostasis [154]. This is demonstrated by the dramatic loss of bone that occurs after menopause [155, 156]. Moreover, estrogens were considered to be a first-line therapy for the treatment of postmenopausal osteoporosis [5, 157]. The reason for this is that these steroid hormones not only suppress bone resorption and turnover but also relieve additional menopausal symptoms such as hot flashes [5, 180]. However, the impact of estrogens on bone goes beyond the female skeleton. It is becoming increasingly recognized that these hormones not only play a major role in the cause and prevention of postmenopausal or type I osteoporosis but also are contributing factors to the development of type II or senile osteoporosis, which affects both aging women and men [158].

Estrogens have both direct and indirect effects on the skeleton [154, 158, 159]. The extraskeletal actions of these steroids on calcium homeostasis include the regulation of intestinal calcium absorption [160, 161] or secretion [162]. They also include the modulation of serum 1,25-dihydroxy-vitamin D_3 levels, renal calcium excretion, and the secretion of parathyroid hormone (PTH) [158, 159].

The direct action of estrogens on bone cells is the subject of this chapter. Although some of this work has been reviewed previously [154, 163–166], our goal is to provide a comprehensive review of the literature and some insights into the complexities and mechanisms of estrogen action in the skeleton.

VI. ESTROGEN RECEPTORS IN BONE CELLS

Many cell types in the skeleton have been shown to express ERs. These include cells of both osteoblast and osteoclast lineages, as well as chondrocytes and endothelial cells. For historic reasons, our discussion of this work begins with the cells of the osteoblast lineage since these were the first bone-derived cells reported to express the ER.

A. Estrogen Receptors in Osteoblasts

Prior to 1987, bone cells were not generally considered to be direct targets for estrogens [167]. However, this view began to change in 1987 when Gray *et al.* [168] reported that 17β-estradiol decreased proliferation and increased alkaline phosphatase activity in rat UMR-106 osteosarcoma cells, which are an *in vitro* model for the osteoblast or bone-forming cell [169]. This report was followed the subsequent year by four publications that demonstrated that rat and human osteoblastic cells expressed ERs and/or exhibited estrogenic responses. Komm *et al* [170] showed specific binding sites for [^{125}I]-17β-estradiol in nuclear extracts from rat ROS 17/2.8 and human HOS-TE85 osteosarcoma cells, as well as ER mRNA expression by these cells. These authors also reported that 17β-estradiol

upregulated type I procollagen and transforming growth factor (TGF)-β1 mRNA levels in HOS-TE85 cells. On the other hand, Eriksen *et al.* [171] described specific nuclear binding sites for [^{3}H]-17β-estradiol in explant cultures of normal human osteoblasts (hOBs), in addition to ER mRNA expression by these cells. This group also demonstrated that 17β-estradiol upregulated nuclear PR levels in hOB cells. Kaplan *et al.* [172] showed by both immunocytochemistry and ligand-binding assays that osteoblasts in cystic bone lesions from a female patient with McCune–Albright syndrome (fibrous dysplasia) expressed ERs. Finally, Ernst *et al.* [173] reported that 17β-estradiol increased the proliferation of primary rat osteoblasts (ROBs) and upregulated α_1 type I procollagen mRNA levels in these cells. Since these initial observations more than a decade ago, ER expression has been reported to occur in a dozen different *in vitro* osteoblast models as well as in osteoblasts from *in situ* studies of bone (Table 12-1). These models represent a variety of mammalian and avian species. Moreover, ER expression has been determined using Northern blot or reverse-transcriptase polymerase chain reaction (RT-PCR) analysis for mRNA and Western blot or immunocytochemistry for protein. In addition, ER function has been determined by ligand-binding, DNA-binding, and ERE reporter gene assays as well as endogenous responses. Analysis of the ligand-binding data indicates that osteoblasts express relatively low numbers (60–4,500/cell) of high-affinity ERs (K_D = 0.05–1.1 nM for 17β-estradiol) [170–172, 174–180]. Although these levels are much lower than those for uterine and breast cells, which express high amounts of ER, they are consistent with the degree of expression seen in other "nonclassical" estrogen-responsive tissues [181]. Together, these results provide unequivocal evidence that osteoblasts express functional ERs and are one of the direct targets of estrogen action in the skeleton.

In 1996, the discovery of a second ER termed ERβ was reported [182]. This discovery resulted in renaming the original ER as ERα. Since each of these had a distinct, albeit overlapping, tissue distribution, investigators began to reexamine ER expression in osteoblasts in light of these new findings. As outlined in Table 12-1, *in situ* studies of rat and human bone have demonstrated that osteoblasts express both ER isoforms [183–189]. Moreover, several *in vitro* osteoblast models, including primary rat and human osteoblasts, have been shown to express both ERα and ERβ [180, 187, 190–194]. However, after reexamining the early literature, it is unclear in some instances if a specific osteoblastic cell line expresses either one or the other, or both, ER isoforms. This is particularly true for the human osteosarcoma cell lines HOS-TE85, SaOS-2, and MG-63 (Table 12-2). Unpublished results from our laboratory using RT-PCR

TABLE 12-2   Estrogen Receptors in Osteoblasts

Isoform	System	Observations	References
ERα and ERβ	Rat ROS 17/2.8 osteosarcoma cells	mRNA	170
		Ligand binding	178
		Protein	179
		ERE-tk-CAT	190
ERα (?)	Human HOS-TE85 osteosarcoma cells	mRNA	170
		Ligand binding	174
		Protein	360
ERα and ERβ	Primary human OB (hOB) cells	mRNA	171
		Ligand binding	175
		Protein	361
		ERE-tk-Luc/Cat	187, 192, 289, 362
ERα and ERβ	Human bone	Protein	172
		mRNA	183, 184, 187, 189
ERα (?) and ERβ	Human SaOS-2 osteosarcoma cells	Ligand binding	174
		mRNA	187
		Protein	
ERα and ERβ	Rat bone	mRNA	186, 188, 363
ERα and ERβ	Primary rat OB (ROB) cells	mRNA	302
		ERE-tk-CAT	190, 193
ERα (?)	Japanese quail bone	Protein	202
ERα (?)	Immortalized human HOBIT cells	mRNA	176
ERα and ERβ	Immortalized mouse MC-3T3-E1 cells	mRNA	177
		Protein	360, 364
ERα (?)	Primary mouse OB cells	mRNA	360
		Protein	
ERα (?) and ERβ	Human MG-63 osteosarcoma cells	mRNA	365
		Protein	187
ERα and ERβ	Rat UMR-106 osteosarcoma cells	Ligand binding	179
		Protein	364
		mRNA	
		ERE-tk-CAT	
ERα and ERβ	Immortalized human HOB-03-CE6 cells	mRNA	180, 194
		Ligand binding	
		DNA binding	
		ERE-tk-Luc	
ERα	Rabbit bone	mRNA	184
		Protein	
ERα and ERβ	Transformed human SV-HFO cells	mRNA	191
ERβ	Mouse bone	mRNA	187
		Protein	

analysis indicate that these human osteosarcoma cell lines express only ERβ mRNA. Although osteoblasts appear to express both ERα and ERβ, it is not known if the isoforms heterodimerize in these cells and what impact this may have on estrogenic responses. Moreover, the ER isoforms appear to be differentially regulated during osteoblast differentiation, which may contribute to the differential effects of estrogens on these cells. In ROBs [190, 193] and in SV-HFO transformed human fetal osteoblastic cells [191], ERα mRNA expression increases with increasing stage of differentiation. On the other hand, ERβ message levels either remain constant [189] or increase [191] with advancing cellular development. Thus, the ratios of ERα to ERβ in osteoblasts may vary as the cells progress from the preosteoblast to the mature osteocyte. Moreover, this variation might contribute to the differential estrogenic responses that have been observed in these cells [193]. Support for this idea comes from work by Hall and McDonnell [195]. Using transient transfection assays, these authors showed the following: (1) ERβ functions as a transdominant inhibitor of ERα transcriptional activity at subsaturating steroid levels, (2) ERα and ERβ can heterodimerize in cells, and (3) ERβ can interact with target gene promoters in the absence of ligand. Thus, Hall and McDonnell concluded that the relative levels of expression of these two receptor isoforms would determine how a cell responds to either estrogens or antiestrogens.

B. Estrogen Receptors in Osteocytes and Lining Cells

Osteoblasts, which arise from mesenchymal stem cells in the bone marrow, undergo further differentiation to either lining cells or osteocytes [169]. Lining cells are thought to be quiescent osteoblasts that line the mineralized bone matrix and regulate access of the osteoclasts to this tissue [196]. On the other hand, osteocytes are osteoblasts that become embedded within the mineralized matrix and assume a stellate or dendritic morphology [197, 198]. The primary function of osteocytes, which are the most abundant cell type in mature bone, is to serve as mechanosensory cells [197, 198]. As such, these cells are involved in strain perception and the adaptive mediation of physical forces on bone modeling and remodeling [198, 199]. Osteocytes and lining cells may also be targets for estrogens [198].

As outlined in Table 12-3, evidence from *in situ* studies of bone indicates that mammalian and avian osteocytes express ERs. Receptor expression in these cells has been shown to occur using *in situ* hybridization for mRNA and immunocytochemistry for protein. Moreover, as with osteoblasts, human osteocytes have been reported to express both ERα and ERβ [183, 187, 189, 200, 201]. Unpublished observations from our laboratory with a conditionally immortalized human osteocyte cell line (HOB-05-T1) indicate that these cells express both ERα and ERβ mRNA (as measured by RT-PCR), and that these receptors are functional based on the transactivation of an ERE reporter gene by 17β-estradiol. Estrogenic responses in osteocytes are discussed later.

At least two publications document ER expression in bone lining cells. Ohashi *et al.* [202] reported that lining cells in Japanese quail bone contained ERs, whereas Kusec *et al.* [184] showed ERα mRNA and protein expression in human lining cells. Although these studies suggest that estrogens may play a role in the physiology of these cells, there are no identified estrogenic responses in lining cells. One of the limitations to these types of investigations is that there are no *in vitro* models to study lining cell biology.

C. Estrogen Receptors in Bone Marrow Stromal Cells

Pluripotent mesenchymal stem cells of bone marrow have the capacity to become osteoblasts, as well as chondrocytes, adipocytes, myoblasts, and fibroblasts [203, 204]. Like other cells of the osteoblast lineage, these bone marrow stromal cells (BMSCs) express ERs and are estrogen responsive. As summarized in Table 12-4, primary BMSCs from rodents and humans, as well as some immortalized bone marrow stromal cell lines, have been reported to express ERα and ERβ. In these studies,

TABLE 12-3   Estrogen Receptors in Osteocytes

Isoform	System	Observations	References
ERα (?)	Japanese quail bone	Protein	202
ERα and ERβ	Human bone	Protein	200
		mRNA	183, 184, 187, 189, 201
ERα (?)	Pig bone	Protein	200
ERα (?)	Guinea pig bone	Protein	184
ERα	Rabbit bone	mRNA	184
		Protein	
ERβ	Mouse bone	mRNA	187
		Protein	
ERα and ERβ	Immortalized human HOB-05-T1 cells	mRNA	Bodine and Komm, unpublished data

TABLE 12-4   Estrogen Receptors in Bone Marrow Stromal Cells

Isoform	System	Observations	References
ERα	Mouse +/+ LDA11 cells	Ligand binding mRNA	366
ERα	Mouse MBA 13.2 cells	Ligand binding mRNA	366
ERα	Mouse BMSCs	mRNA	364, 366, 367
ERα and ERβ	Rat BMSCs	mRNA	186
ERα and ERβ	Mouse ST2 cells	mRNA Protein	364
ERα and ERβ	Human BMSCs	mRNA	205, 206

ER expression was demonstrated using RT-PCR and Northern hybridization for mRNA, immunocytochemistry for protein, and cytosolic ligand-binding assays for receptor function. Oreffo *et al.* [205] reported that human BMSCs express ERα mRNA based on Northern blot analysis, and that its expression increases as the cells undergo differentiation to osteoblasts. Likewise, Dieudonne *et al.* [206] stated that immortalized human bone marrow stromal fibroblasts (BMSFs) isolated from a patient with a mutated ERα gene, as well as nonimmortalized control BMSFs from normal patients, expressed ERβ mRNA as determined by RT-PCR. Moreover, the nonimmortalized control BMSFs were acknowledged to express the wild-type ERα message. Estrogenic responses in BMSCs are discussed later.

D. Estrogen Receptors in Cells of the Osteoclast Lineage

Osteoclasts are multinucleated giant cells that are responsible for bone resorption [207, 208]. These cells arise from hemopoietic stem cells of the monocyte/macrophage lineages, which, like BMSCs, are found in the bone marrow [208]. Since the primary therapeutic effect of estrogens on the postmenopausal skeleton is to suppress bone resorption [155, 156], it seems logical that cells of the osteoclastic lineage would express ERs. However, the direct action of estrogens on these cells is less accepted by the field than is an indirect effect through the cells of the osteoblast lineage.

Table 12-4 summarizes the evidence for ER expression by osteoclastic cells. In 1990, Pensler *et al.* [209] reported that human osteoclasts isolated from membranous bone (pediatric craniotomies) expressed ERs based on immunocytochemistry of fixed cells and radioimmunoassay of cell lysates. Subsequently, Oursler and colleagues described the presence of ERs in osteoclasts purified from either chicken long bones [210] or human giant cell tumors (hGCTs) of bone (i.e., osteoclastomas) [211]. For these studies, the authors used a monoclonal antibody (121F) generated to chicken osteoclasts to purify mature osteoclasts (≥90% pure) from these tissues. ER expression was then demonstrated using either Northern blot analysis [210] or RT-PCR [211] for ERα mRNA, Western blot analysis for receptor protein [210], and a nuclear ligand-binding assay that indicated that the chicken osteoclasts contained 5,000–6,000 ERs/nucleus [210]. Two groups confirmed that human osteoclasts express ERα mRNA. Hoyland *et al.* [183] used *in situ* RT-PCR to demonstrate the presence of ERα message in normal human bone samples, whereas Sunyer *et al.* [212] used RT-PCR to reveal the expression of this message in purified normal human osteoclasts (hOCLs). ER mRNA has also been reported to be expressed by isolated mature rabbit osteoclasts [213]. Thus, at least five separate laboratories have found evidence for ER expression in osteoclasts. However, in contrast to these observations, Collier *et al.* [214] failed to detect either ERα or ERβ mRNA in pure preparations of microisolated osteoclasts from hGCTs. Moreover, the authors confirmed their results using fluorescence *in situ* hybridization, which showed that the tumor mononuclear cells expressed ERα message, whereas the multinuclear osteoclasts did not express this gene. The reason for this discrepancy is not clear. However, Oursler [207] postulated that prior *in vivo* exposure to estrogens may have downregulated ER levels in the osteoclasts examined by Collier and coworkers [214]. This conclusion is based on the work of Pederson *et al.* [215], who reported that *in vivo* treatment of 5-week-old chickens with 17β-estradiol dramatically suppressed ER protein levels in the purified osteoclasts.

Preosteoclasts also appear to express ERs (Table 12-5). For example, Fiorelli *et al.* [216] used RT-PCR (for ERα), Western blot analysis, a nuclear extract ligand-binding assay, and an ERE reporter gene assay to demonstrate the presence of functional ERs in

TABLE 12-5 Estrogen Receptors in Cells of the Osteoclast Lineage

Isoform	System	Observations	References
ERα and ERβ (?)	Human bone	Protein	209
		mRNA	183, 187
ERα (?)	Chicken osteoclasts	Ligand binding	210
		mRNA	
		Protein	
ERα	Human giant cell tumors	mRNA	211
		Protein	
ERα	Human FLG-29.1 preosteoclastic cells	Ligand binding	216
		mRNA	
		Protein	
		ERE-tk-Cat	
ERα (?)	Rabbit osteoclasts	mRNA	213
ERα	Mouse hemopoietic blast cells	mRNA	217
ERα	Rat preosteoclasts	mRNA	368
ERα	Primary human osteoclasts	mRNA	212
ERα	Human TCG 51 preosteoclastic cells	Protein	369
ERβ (?)	Mouse bone	Protein	187

human leukemic FLG 29.1 cells. The ligand-binding assay showed that this cell line, which can be induced to express an osteoclast-like phenotype, contained approximately 400 ERs/nucleus. Moreover, Kanatani *et al.* [240] demonstrated that mouse hemopoietic blast cells, which contain osteoclast progenitors, express ERα mRNA based on RT-PCR. Estrogenic responses in osteoclastic cells are discussed later.

E. Estrogen Receptors in Chondrocytes and Other Bone-Associated Cells

Estrogens play an important role in the regulation of human longitudinal bone growth and skeletal maturation [154]. These steroid hormones accelerate endochondral bone formation in early adolescence but also initiate epiphyseal growth plate fusion in late adolescence. Consistent with these observations, chondrocytes express both ERα and ERβ. As outlined in Table 12-6, rabbit, mouse, rat, human, and pig chondrocytes have all been reported to possess ERs. These observations are based on *in situ* hybridization for ERα mRNA [184], immunocytochemistry for ERα and ERβ proteins [184, 218–221], and cytosolic ligand-binding assays [222–225]. Scatchard analysis of the ligand-binding data indicates that chondrocytes express relatively low amounts (3.9–11.2 fmol/mg protein) [225] of high-affinity ERs (K_D = 0.12–0.87 nM for 17β-estradiol) [222, 225]. Thus, these receptor kinetics and levels are comparable to

those found in osteoblasts [180]. In human growth plate chondrocytes, ERα was reported to be expressed by resting, proliferative, and hypertrophic cells [184], while ERβ expression was shown to be restricted to the hypertrophic cells [221]. Thus, these ER isoforms may have distinct roles in the regulation of endochondral bone growth and maturation. Estrogenic responses in chondrocytes are discussed later.

At least one report describes the expression of ERs in bone-derived endothelial cells [226]. Using bovine bone endothelial (BBE) cells, the authors showed that these cells expressed ER mRNA by Northern hybridization and contained specific binding sites for [³H]-17β-estradiol (K_D = 17.2 nM, B_{max} = 32,000 sites/cell). Treatment of the cells with 17β-estradiol enhanced proliferation and suppressed PTH-stimulated cyclic-adenosine monophosphate (cAMP) accumulation. As described in more detail later, both of these estrogenic responses have also been observed in osteoblasts. Thus, this study suggests that estrogens may regulate bone angiogenesis as well as bone formation and resorption.

F. Summary

It is clear from the numerous studies reviewed in this section that many cell types in the skeleton express ERs. These estrogen-responsive cell types include bone marrow progenitor cells as well as mature osteoblasts, osteoclasts, and chondrocytes. In the osteoblast lineage, each cell type—from the BMSC to the osteocyte or lining

TABLE 12-6   Estrogen Receptors in Chondrocytes

Isoform	System	Observations	References
ERα	Rabbit chondrocytes	Ligand binding	222
		mRNA	184
		Protein	220
ERα and ERβ	Human chondrocytes	Protein	218, 219
		mRNA	184, 221, 369
ERα (?)	Rat chondrocytes	Protein	220, 225
ERα (?)	Pig bone	Protein	200
ERα (?)	Guinea pig	Protein	200

cell—has been shown to be a potential estrogen target. Thus, the totality of estrogen's effects on the skeleton may, to a large extent, be equivalent to the sum of its action on all of these cell types. In the following section, we review the estrogenic responses of skeletal cells and place them in the context of *in vivo* knowledge of estrogen action.

VII. ESTROGENIC RESPONSES IN BONE CELLS

Consistent with the expression of ERs by many bone cell types, there are also many estrogenic responses in these cells. Our review of these responses, which are sometimes contradictory, will attempt to place them in the context of estrogen's known physiologic and therapeutic function in the skeleton.

A. Estrogenic Responses in Cells of the Osteoblast Lineage

Due to the profusion of *in vitro* models, much of what we know about estrogen action on bone cells is in relationship to the osteoblast. As summarized in Table 12-7, 43 estrogenic responses have been identified in 15 different *in vitro* osteoblast models. In order to make sense of these observations, we have separated them into six different categories: regulation of osteoblast number, regulation of matrix production and mineralization, regulation of growth factor expression and responsiveness, regulation of factors that modulate bone resorption, regulation of receptor expression and signal transduction, and miscellaneous responses. Moreover, we have indicated which *in vitro* models were reported to exhibit each of the responses. The reason for doing this is to determine if a given response is a general estrogenic effect in an osteoblast or whether it might be specific to a particular cell line (e.g., immortalized MC-

3T3-E1 mouse cells) or cell type (e.g., osteosarcoma-derived cells). From our viewpoint, the most pertinent osteoblast models to attempt to translate *in vitro* observations of estrogens into *in vivo* relevance are primary cultures. On the other hand, caution should be applied to observations that are only made in osteosarcoma cells since these are generally considered to be unreliable models of osteoblast biology [227, 228]. When available, we have also noted when an *in vitro* estrogenic response has been observed *in vivo* and therefore may be physiologically or pharmacologically relevant.

I. REGULATION OF OSTEOBLAST NUMBER

Using UMR-106 rat osteosarcoma cells, Gray *et al.* [168] reported that 17β-estradiol decreases osteoblastic cell proliferation. In the same study, 17β-estradiol also increased alkaline phosphatase activity. Given the limitations of osteosarcoma cells as models of osteoblast biology [227, 228], these results suggested that estrogens might potentiate cellular differentiation since the mature rat osteoblast no longer divides and expresses high levels of alkaline phosphatase [169, 228]. Subsequent to this publication, other research has described similar results using four additional *in vitro* osteoblast models (Table 12-7). These models include primary osteoblasts isolated from the tibias of 17β-estradiol-treated ovariectomized (OVX) rats [229, 230]. Moreover, Westerlind *et al.* [231] confirmed these observations *in vivo* by showing that the potent nonsteroidal estrogen DES reduces the [3H]thymidine-labeling index of tibial osteoblasts in OVX rats. Thus, a suppressive effect of estrogens on osteoblast proliferation is consistent with an inhibitory action of the steroid on bone turnover [154–156].

In contrast to these findings, other laboratories using additional *in vitro* models, as well as ROBs, have reported that estrogens increase osteoblast proliferation and DNA synthesis (Table 12-6). There are several possible explanations for these discrepancies. First, with the exception of the studies using UMR-106 and ROBs, the other

TABLE 12-7   Estrogenic Responses in Cells of the Osteoblast Lineage

Response	Systems	References
Regulation of cell number		
Decreases proliferation and decreases DNA synthesis	Rat UMR-106 osteosarcoma cells	168, 319
	Human HTB-96 cells overexpressing ERα	231, 370
	Primary rat OB (ROB) cells	229, 232
	Rat ROS.SMER-14 cells overexpressing ERα	178
	Human hFOB/ER9 cells overexpressing ERα	233, 243, 268, 277
	Rat ROS 17/2.8 osteosarcoma cells	371
	Rat bone	230
Increases proliferation and increases DNA synthesis	Primary rat OB (ROB) cells	173, 232, 246
	Transformed rat RCT-1 and -3 cells	246
	Immortalized mouse MC-3T3-E1 cells	177, 242
	Primary human OB (hOB) cells	241
	Human HOS-TE85 osteosarcoma cells	372
	Primary mouse bone marrow stromal cells	235
	Rat bone	236
	Mouse bone	234
Inhibits glucocorticoid-induced apoptosis	Primary rat OB (ROB) cells	239
	Primary mouse OB cells	239
	Mouse bone	239
Regulation of matrix production and mineralization		
Increases alkaline phosphatase	Rat UMR-106 osteosarcoma cells	168, 319
	Rat ROS.SMER-14 cells overexpressing ERα	178
	Primary human OB (hOB) cells	241
	Immortalized mouse MC-3T3-E1 cells	242
	Human hFOB/ER9 cells overexpressing ERα	233, 243
	Immortalized human HOB-03-CE6 cells	180
	Primary rat OB (ROB) cells	193
Decreases alkaline phosphatase	Primary rat OB (ROB) cells	193
	Rat bone	245
Increases osteocalcin	Primary rat OB (ROB) cells	193
Decreases osteocalcin	Rat ROS 17/2.8 osteosarcoma cells	244
	Human hFOB/ER9 cells overexpressing ERα	233, 243
	Primary rat OB (ROB) cells	193
	Rat bone	244, 245, 247
Increases osteonectin	Primary rat OB (ROB) cells	193
Decreases osteonectin	Primary rat OB (ROB) cells	193
	Rat bone	245
Increases type I collagen	Human HOS-TE85 osteosarcoma cells	170
	Primary rat OB (ROB) cells	173, 174, 193
	Transformed rat RCT-1 and -3 cells	246
	Primary human OB (hOB) cells	175, 192, 362
	Immortalized mouse MC-3T3-E1 cells	242
Decreases type I collagen	Primary human OB (hOB) cells	362
	Rat bone	245, 247
Increases mineralization	Human HOS-TE85 osteosarcoma cells	250
	Primary human OB (SaM-1) cells	250

(Continued)

TABLE 12-7   Estrogenic Responses in Cells of the Osteoblast Lineage—Cont'd

Response	Systems	References
Regulation of growth factor expression and responsiveness		
Increases TGF-β1	Human HOS-TE85 osteosarcoma cells	170
	Rat UMR-106 osteosarcoma cells	262
	Primary human OB (hOB) cells	263
	Primary mouse OB cells	264
	Primary rat OB (ROB) cells	193
	Rat ROS 17/2.8 osteosarcoma cells	244
	Rat bone	244, 264
Increases TGF-β3	Human MG-63 osteosarcoma cells	267
	Rat bone	265
Increases TIEG	Human hFOB/ER9 cells overexpressing ERα	268
Increases BMP-6	Human hFOB/ER9 cells overexpressing ERα	270
Increases IGF-1	Rat UMR-106 osteosarcoma cells	272
	Primary rat OB (ROB) cells	246, 373
	Transformed rat RCT-1 and -3 cells	246
	Human hFOB/ER9 cells overexpressing ERα	374
Increases growth hormone receptor	Rat UMR-106 osteosarcoma cells	273
	Primary human OB (hOB) cells	273
Increases IGF-BPs	Primary rat OB (ROB) cells	276
	Human hFOB/ER9 cells overexpressing ERα	277
	Human SaOS-2 osteosarcoma cells	278
Decreases IGF-BP3	Primary human bone marrow stromal cells	280
Blocks PGE$_2$-induced IGF-1	ROB cells overexpressing ERα	281
Regulation of factors that modulate bone resorption		
Decreases IL-6	Mouse +/+ LDA11 marrow stromal cells	282
	Primary human OB (hOB) cells	282
	Primary rat OB (ROB) cells	282
	Primary mouse OB cells	282, 375
	Immortalized mouse MC-3T3-E1 cells	282
	Human SaOS-2 cells overexpressing ERα	376
	Human hFOB/ER9 cells overexpressing ERα	290
	Immortalized human HOB-03-CE6 cells	180
	Human MG-63 osteosarcoma cells	377
	Primary human bone marrow stromal cells	284
	In vivo (mice)	283, 378
Decreases TNF-β	Primary human OB (hOB) cells	288
Decreases gp80 and gp130	Mouse +/+ LDA11 marrow stromal cells	292
	Immortalized mouse MC-3T3-E1 cells	292
Increases OPG	Human hFOB/ER9 cells overexpressing ERα	294
	Primary human OB (hOB) cells	294
Suppresses PTH action	Human SaOS-2 osteosarcoma cells	278, 297, 298, 300
	Transformed rat RCT-1 and -3 cells	246, 302
	Primary rat OB (ROB) cells	246
	Primary mouse OB cells	299
	Primary human OB (hOB) cells	263
	Immortalized mouse MC-3T3-E1 cells	242

(Continued)

TABLE 12-7   Estrogenic Responses in Cells of the Osteoblast Lineage—Cont'd

Response	Systems	References
	Immortalized human HOB-03-CE6 cells	180
	In vivo (humans)	301
Enhances PTH action	Human SaOS-2 osteosarcoma cells	303, 304
	Primary rat OB (ROB) cells	304
	Primary human OB (hOB) cells	304
Increases IL-1β	Immortalized human HOBIT cells	306
Regulation of receptor expression and signal transduction		
Increases PR	Primary human OB (hOB) cells	171
	Human hFOB/ER9 cells overexpressing ERα	290, 307
Antagonizes VD$_3$ responsiveness	Rat UMR-106 osteosarcoma cells	272
Increases VDR and VD$_3$ responsiveness	Rat ROS 17/2.8 osteosarcoma cells	308
	Human OGA osteosarcoma cells	308
Increases ERα	Primary human OB (hOB) cells	192, 362
	Primary rat OB (ROB) cells	193
	In vivo (human bone)	189, 201
Decreases ERα	Primary rat OB (ROB) cells	193
	In vivo (human bone)	201
Decreases IP$_3$ receptor I	Rat UMR-106 osteosarcoma cells	310, 312
	Human SaOS-2 osteosarcoma cells	310, 312
	Primary rat OB (ROB) cells	310, 312
	Immortalized mouse MC-3T3-E1 cells	310, 312
	G-292 human osteosarcoma cells	310, 312
Increases basal NOS	Human HOS-TE85 osteosarcoma cells	313
	In vivo (rats)	315
Decreases cytokine-induced NO	Immortalized mouse MC-3T3-E1 cells	316
Enhances bradykinin action	Primary human OB (hOB) cells	317
	Miscellaneous responses	
Increases CK	Primary rat OB (ROB) cells	236
	Immortalized mouse MC-3T3-E1 cells	236
	Rat ROS 17/2.8 osteosarcoma cells	236
	Rat bone	236
Increases HSP-27	Immortalized mouse MC-3T3-E1 cells	318
Increases AST, GGT, LDH, and transferrin	Rat UMR-106 osteosarcoma cells	319

publications that showed that 17β-estradiol suppresses proliferation utilized cell lines that overexpressed ERα. Thus, as was concluded by Watts and King [231], overexpression of the ER may inhibit cell proliferation by artifactually interfering with transcription. If this is true, then a transfected ER may not necessarily function the same as the endogenous ER. On the other hand, the studies that reported that 17β-estradiol stimulated osteoblast proliferation all used *in vitro* models that naturally expressed ERs. Second, at least two groups have reported that *in vitro* treatment of ROBs with 17β-estradiol enhances cell proliferation or DNA synthesis [173, 232]. In contrast, Modrowski *et al.* [229] used isolated osteoblasts from *in vivo*–treated OVX rats to show that the steroid inhibits proliferation. Consequently, these two experimental paradigms may generate cells that are in different stages of differentiation (e.g., preosteoblastic vs. mature osteoblasts), and these stages may respond differently to estrogens [193, 233]. Whereas a suppressive effect of estrogens on osteoblast proliferation is consistent with a potentiation of differentiation or a suppression of bone turnover, a stimulatory effect might relate to an expansion of the preosteoblast pool [234]. For example, Qu *et al.* [235] presented evidence that treatment of primary mouse BMSC cultures with 17β-estradiol stimulates cellular proliferation and differentiation into osteoblastic cells. This *in vitro* observation is consistent with an *in vivo* study of Somjen *et al.* [236], which reported that 17β-estradiol stimulates DNA synthesis in rat bone. Moreover, in OVX Swiss–Webster mice, high doses of 17β-estradiol (50–100 μg/mouse/week, s.c., for 4 weeks) increased both endosteal and cancellous bone formation, as well as inhibited bone resorption [237]. Thus, in some circumstances, estrogens may stimulate bone formation [238] as well as inhibit resorption and turnover. However, the stimulatory action of the steroid may represent a pharmacological or toxicological effect rather than a physiological or therapeutic response [239].

In addition to regulating cell division, estrogens have also been shown to control osteoblast apoptosis. Gohel *et al.* [240] reported that 17β-estradiol blocks the induction of apoptosis by cortisol in primary rat and mouse osteoblasts. These *in vitro* observations were confirmed by an *in vivo* experiment that showed that 17β-estradiol decreased the number of apoptotic osteoblasts in the calvaria of dexamethasone-treated mice. Consequently, estrogens may modulate osteoblast number by regulating both proliferation and viability. As reviewed later, estrogens may also suppress osteocyte apoptosis but induce the programmed cell death of osteoclasts.

2. REGULATION OF MATRIX PRODUCTION AND MINERALIZATION

One of the most commonly observed estrogenic responses in osteoblasts is the upregulation of alkaline phosphatase expression, which is an important phenotypic marker of the osteoblast lineage [169]. Estrogens have been reported to increase alkaline phosphatase mRNA levels and/or activity in seven different *in vitro* osteoblast models (Table 12-6). These models include rat osteosarcoma cell lines [168, 178], primary cultures of ROB or hOB cells [193, 241], immortalized mouse MC-3T3-E1 cells [242], and the conditionally immortalized human osteoblast cell lines hFOB/ER9 and HOB-03-CE6 [180, 233, 243]. However, in the case of ROB cells, 17β-estradiol has also been reported to downregulate alkaline phosphatase expression [193]. The explanation for this discrepancy is that 17β-estradiol regulates the steady-state mRNA levels of this enzyme in a differentiation selective manner [193]. In postproliferative/nodule-forming stage ROB cells (i.e., mature osteoblasts), 17β-estradiol suppresses alkaline phosphatase expression, whereas in postmineralization stage cells (i.e., osteocytes) the steroid hormone increases enzyme message levels. This same pattern of regula-

tion also holds true for the noncollagenous bone matrix proteins osteocalcin and osteonectin [193].

Estrogens also regulate the expression of osteocalcin (Table 12-6), which is the most selective phenotypic marker of the osteoblast lineage [169]. As noted previously, 17β-estradiol downregulates steady-state osteocalcin mRNA levels in postproliferative/nodule-forming stage ROB cells but upregulates it in postmineralization stage cells [193]. Moreover, estrogens have been reported to decrease osteocalcin expression in ROS 17/2.8 osteosarcoma cells [244] and in hFOB/ER9 cells, which overexpress human ERα [233, 243]. Confirmation that estrogen downregulates alkaline phosphatase, osteocalcin, and osteonectin mRNA levels *in vivo* comes from the study by Turner *et al.* [245]. These authors reported that DES treatment of OVX rats decreased the expression of these messages in periosteal osteoblasts isolated from lone bones. Again, a suppression of osteoblastic activity as measured by the expression of bone matrix proteins would be consistent with a reduction in bone turnover.

The most abundant bone matrix protein, of course, is type I collagen [169], and it is perhaps not surprising that estrogens have been shown to regulate its expression (Table 12-6). Komm *et al.* [170] and Ernst *et al.* [173] were the first to report that 17β-estradiol upregulated α₁ type I procollagen mRNA levels in HOS-TE85 human osteosarcoma cells and in ROB cells, respectively. Subsequent studies confirmed these observations in hOBs [175], MC-3T3-E1 cells [242], and transformed rat RCT-1 and RCT-3 cell lines [246]. In contrast to these *in vitro* studies, type I collagen expression does not appear to be upregulated by estrogens *in vivo*. In fact, mRNA levels for this bone matrix protein have been reported to increase in OVX rat bones [247, 248], and estrogens have been observed to either suppress this increase [245, 247] or have no effect [249]. Again, these *in vivo* observations are consistent with the concept that estrogen deficiency increases bone resorption and bone turnover, and that estrogens reduce these effects [154–156].

Finally, at least one report describes the effects of estrogens on mineralization. Takeuchi *et al.* [250] showed that 17β-estradiol at concentrations of 1–100 nM increased the calcium content of extracellular matrix that was laid down *in vitro* by either HOS-TE85 human osteosarcoma cells or primary human osteoblasts (referred to as SaM-1 cells).

3. REGULATION OF GROWTH FACTOR EXPRESSION AND RESPONSIVENESS

Another aspect of osteoblast biology that estrogens have been shown to regulate is growth factor expression or growth factor responsiveness. Bone is an abundant reservoir for several growth factors, including isoforms of TGF-β, the bone morphogenetic proteins (BMPs), and the IGFs [251–256]. These peptides are synthesized and secreted by cells of the osteoblast and/or osteoclast lineages, and they regulate the proliferation, differentiation, and activities of these cell types [251, 252, 254–258]. In fact, growth factors, together with other cytokines, provide the elaborate communication network that couples osteoclastic bone resorption to osteoblastic bone formation [154, 164]. Moreover, it is the disruption of this network that, to a large extent, leads to accelerated bone resorption and increased bone turnover after menopause [164, 259–261].

The first bone cell–derived growth factor whose expression was shown to be regulated by estrogens was TGF-β1. Komm *et al.* [170] reported in 1988 that 17β-estradiol treatment of HOS-TE85 human osteosarcoma cells upregulated the steady-state levels of TGF-β1 mRNA. As outlined in Table 12-6, estrogens have also been shown to increase TGF-β1 mRNA expression and/or TGF-β protein secretion in rodent osteosarcoma cell lines [244, 262], as well as primary cultures of human, mouse, and rat osteoblasts [193, 263, 264]. Moreover, estrogens have been observed to increase TGF-β expression in bone *in vivo*. Finkelman *et al.* [264] reported that treatment of OVX rats with 17β-estradiol upregulated TGF-β protein levels in long bones. In another study, Ikeda *et al.* [244] demonstrated that TGF-β1 mRNA levels decreased in the tibia of OVX rats. However, neither Westerlind *et al.* [248] nor Yang *et al.* [265] were able to confirm these findings. Although TGF-β regulates osteoblast proliferation, differentiation, and activity *in vitro* and promotes bone formation *in vivo* [251, 252, 255], it has also been reported to inhibit osteoclast differentiation and activity *in vitro* [154, 164]. Thus, an increase in osteoblastic TGF-β production would be consistent with an antiresorptive effect of estrogens therapeutically [155, 156].

Estrogens, as well as tissue-selective estrogens (TSEs) [157] or SERMs [266], have also been reported by at least one group to increase TGF-β3 expression by osteoblastic cells (Table 12-6). Yang *et al.* [265] observed an increase in TGF-β3 mRNA levels in the femurs of OVX rats that were treated with either 17β-estradiol or the SERM raloxifene; in contrast, the message levels for either TGF-β1 or TGF-β2 were unaffected by these treatments. Although *in situ* studies to identify the cell type(s) that was responsible for this expression were not reported, this same group subsequently demonstrated that 17β-estradiol or raloxifene upregulated TGF-β3 mRNA levels in MG-63 human osteosarcoma cells [267]. These observations were extended by cotransfection studies in MG-63 cells

using human TGF-β3 promoter–reporter gene constructs and human ERβ expression vectors [265, 267]. These experiments indicated that a variety of estrogens and TSEs/SERMs upregulated TGF-β3 promoter activity in an ERβ-dependent manner. Although these results were intriguing, an apparent disconnection occurred between the *in vitro* and *in vivo* pharmacology since the potency and efficacy of compounds in this *in vitro* assay did not correlate with their bone-sparing activities *in vivo*. Moreover, 17β-estradiol was also an antagonist of raloxifene in this *in vitro* system [267]. In any event, as with TGF-β1, an upregulation of TGF-β3 expression in bone by either estrogens or a TSE/SERM would be consistent with an antiresorptive effect since this isoform also inhibits *in vitro* osteoclastic differentiation and activity [265].

In addition to upregulating TGF-β expression in osteoblasts, estrogens may act like these peptides in terms of their downstream effects. For instance, 17β-estradiol has been reported by Tau *et al.* [268] to increase expression of TIEG (TGF-β inducible early gene) in conditionally immortalized hFOB/ER9 human fetal osteoblasts. The expression of this gene is also increased by TGF-β in human osteoblastic cells [269]. Treatment of this cell line with 17β-estradiol, or overexpression of TIEG, causes a reduction in DNA synthesis. These results suggest that at least part of the mechanism by which estrogens inhibit osteoblast proliferation may involve upregulation of TIEG.

Estrogens appear to regulate the expression of additional members of the TGF-β superfamily. In 1998, Rickard *et al.* [270] reported that treatment of hFOB/ER9 cells with 17β-estradiol increased both the steady-state mRNA levels and the protein levels of BMP-6 (Table 12-6). In contrast, the steroid hormone had no effect on TGF-β1, TGF-β2, BMP-2, BMP-4, or BMP-5 expression. Like the TGF-βs, the BMPs also have autocrine and paracrine effects on a variety of skeletal cells [251, 254]. van den Wijngaard *et al.* [271] reported that antiestrogens or TSEs/SERMs such as tamoxifen, raloxifene, and ICI-164,384 upregulated human BMP-4 promoter–luciferase expression in U2-OS human osteosarcoma cells that were cotransfected with hERα but not hERβ. However, this response required expression of relatively high receptor levels and was blocked by cotreatment with 17β-estradiol. Since there is no evidence that endogenous BMP-4 expression is increased in osteoblasts without ER overexpression, it is unclear whether or not this observation has any bearing on the pharmacological actions of TSEs/SERMs in the skeleton.

In addition to members of the TGF-β/BMP family, estrogens have been observed to regulate the expression of components of the osteoblastic IGF/growth hormone (GH) system as well. Gray *et al.* [272] were the first to report that 17β-estradiol treatment upregulated the secretion of IGF-1 and IGF-2 from UMR-106 rat osteosarcoma cells. These results were confirmed, at least for IGF-1, in three additional osteoblast models including ROBs (Table 12-7). Likewise, 17β-estradiol has been reported to increase GH receptor expression and GH action in UMR-106 cells and normal human osteoblast cultures [273]. In contrast, *in vivo* studies by Turner and coworkers [249, 274] in OVX rats failed to verify these *in vitro* observations. In fact, these authors demonstrated that estrogen loss resulted in an increase IGF-1 mRNA expression in calvarial periosteum and that DES treatment suppressed this increase. Since IGFs increase bone formation, resorption, and turnover [252, 253], an upregulation of osteoblastic IGF expression following 17β-estradiol treatment *in vitro* is inconsistent with a suppressive effect of the steroid hormone on resorption and turnover *in vivo* [154–156]. On the other hand, the *in vitro* studies were confirmed by Erdmann *et al.* [275], who showed that supraphysiological doses of 17β-estradiol increased IGF-1 protein levels in femoral shaft bone matrix of OVX rats. However, these authors cautioned that this stimulatory effect of estrogens only occurred at relatively high concentrations of steroid, and that this may not be relevant to the normal physiological actions of the hormone. Since high doses of estrogens stimulate bone formation in OVX mice [234, 237], upregulation of IGF-1 levels in bone may be part of the mechanism by which this pharmacological effect occurs.

Estrogens have also been reported to increase IGF-binding protein (IGF-BP) secretion and expression by ROBs [276], hFOB/ER9 cells [277], and SaOS-2 human osteosarcoma cells [278] (Table 12-6). IGF-BPs are secreted proteins that bind IGF-1 and IGF-2 and regulate their bioavailability and activity [257, 279]. Consequently, the IGF-BPs can either enhance or inhibit IGF action. Moreover, in some instances, these BPs may also act independently of the IGFs. Of the six IGF-BPs, all of which are expressed by human osteoblasts [279], IGF-BP4 is considered to be the most inhibitory to IGF activity [257]. In 1996, Kassem *et al.* [300] demonstrated that 17β-estradiol increased IGF-BP4 mRNA expression and secretion in hFOB/ER9 conditionally immortalized fetal human osteoblasts that overexpress hERα. In contrast, the steroid had no effect on either IGF-2 or IGF-BP3 expression. In addition, 17β-estradiol decreased IGF-BP4 proteolysis. Since 17β-estradiol also inhibited DNA synthesis by these cells, the authors proposed that upregulation of IGF-BP4 levels in the bone microenvironment might contribute to the suppressive action of estrogens on bone formation observed *in vivo* [154]. On the other

hand, Rosen *et al.* [280] reported that 17β-estradiol suppressed IGF-BP3 secretion from a primary culture of human BMSCs.

Another potential mechanism by which estrogens may suppress IGF-dependent bone turnover is through antagonism of induced IGF-1 expression. Using ROBs that were cotransfected with a human ERα expression vector, McCarthy *et al.* [281] reported that 17β-estradiol suppressed PGE$_2$-induced rat IGF-1 promoter-luciferase activity. However, basal promoter function was unaffected by the hormone.

4. Regulation of Factors That Modulate Bone Resorption

As noted previously, the therapeutic actions of estrogens preclinically and clinically primarily involve the suppression of bone resorption and bone turnover [5, 154]. One of the chief estrogenic targets for these antiresorptive effects is the cells of the osteoblast lineage [259], [164, 260, 261]. As outlined in Table 12-6, at least five different effects of estrogens on osteoblasts and their progenitors involve the suppression of cytokine production, cytokine action, or bone resorbing hormone activity.

One of the most commonly reported estrogenic effects in cells of the osteoblast lineage is the downregulation of interleukin (IL)-6 synthesis, which is a cytokine that stimulates the differentiation of osteoclast progenitors to mature bone resorbing cells [203, 259–261]. In 1992, Girasole *et al.* [282] reported that 17β-estradiol suppressed the induction of IL-6 secretion by tumor necrosis factor (TNF)-α or IL-1β in mouse +/+ LDA11 stromal cells, MC-3T3-E1 immortalized mouse osteoblastic cells, or primary cultures of rat and human osteoblasts. Moreover, in neonatal mouse calvarial–derived bone cell cultures that contain osteoblasts as well as osteoclast progenitors, 17β-estradiol inhibited both TNF-α–stimulated IL-6 production and osteoclast development. In addition, a similar suppression was also observed with an anti–IL-6 antibody, indicating that IL-6 was involved in this process. These *in vitro* observations were confirmed later that year by an *in vivo* study in mice that was reported by the same group [283]. These findings were also corroborated by Cheleuitte *et al.* [284], who used cultured BMSCs isolated from postmenopausal women. These authors showed that basal and IL-1β–stimulated IL-6 secretion from the BMSCs *in vitro* was significantly reduced (relative to age-matched controls) when the cells were isolated from women using estrogen replacement therapy (ERT). The mechanism for the inhibition of IL-6 expression by 17β-estradiol was determined by Pottratz *et al.* [285], who showed that it was through an ER-mediated indirect effect on IL-6 promoter activity. Subsequent studies have demonstrated that the ER interferes with nuclear factor (NF)-κB activity, although the precise molecular events involved in this suppression remain to be elucidated [286].

Although several other research groups have corroborated these findings using a variety of *in vitro* osteoblast models (Table 12-7), others have been unable to verify IL-6 as a target for estrogen action [287–290]. These reports used primary cultures of hOBs or human BMSCs, which are known to express relatively low and variable amounts of ER [171, 289]. Our laboratory offered a possible explanation for this discrepancy. Using conditionally immortalized human HOB-03-CE6 cells that naturally express functional ERs [180], we showed that the bone-resorbing cytokines TNF-α and IL-1α/β are potent suppressors of ligand-dependent receptor activity [194]. In this cell line, 17β-estradiol downregulates basal IL-6 mRNA levels [180] but does not block the induction of IL-6 secretion by either TNF-β or IL-1β [194]. Thus, we postulated that in osteoblasts that normally express low ER levels, TNF-β and IL-1α/β may inactivate the receptor before it can blunt IL-6 production. Although Rickard *et al.* [288] were unable to demonstrate that 17β-estradiol suppressed IL-1α–induced IL-6 secretion from hOB cells, they did show that the steroid downregulated the release of TNF-β from these cells in response to IL-1α stimulation.

Estrogens have also been shown to blunt IL-6 responsiveness in osteoblastic and BMSCs cells. The IL-6 receptor is a bipartite complex composed of two transmembrane glycoproteins. One is an 80-kDa protein (gp80) that binds the cytokine, whereas the other is a dimer of a 130-kDa protein (gp130) that is involved in signal transduction to the JAK/STAT (Janus kinase/ signal transducer and activator of transcription) pathway [291]. Lin *et al.* [292] reported that 17β-estradiol downregulated gp80 and gp130 mRNA levels, as well as gp130 protein levels, in +/+ LDA11 stromal cells. Likewise, the steroid hormone also suppressed the induction of gp130 mRNA by PTH, IL-11, or leukemia inhibitory factor in MC-3T3-E1 osteoblastic cells.

Although cells of the osteoblast lineage produce many proteins that potentiate osteoclastogenesis and osteoclastic activity, one termed RANKL (receptor activator of NF-κB ligand) appears to be critical for this process [208, 293]. RANKL is a membrane protein found on the surface of osteoblasts and BMSCs. Moreover, it is the ligand for RANK (receptor activator of NF-κB), a transmembrane protein that is expressed by osteoclast progenitors and mature bone resorbing cells. The binding of RANKL to RANK stimulates the differentiation of osteoclast progenitors to mature osteoclasts. Additionally, it activates the mature cells. However, RANKL is also a ligand for a secreted decoy receptor called osteoprotegerin (OPG). Osteoblasts and BMSCs synthesize OPG as well as RANKL [164, 293], and OPG suppresses bone resorption by sequestering RANKL [164, 208, 293]. Consequently, given the antiresorptive nature of estrogens, it is not surprising that these hormones have been observed to increase OPG expression by osteoblasts. Using both conditionally immortalized hFOB/ER9 fetal human osteoblastic cells and hOBs, Hofbauer *et al.* [294] demonstrated that 17β-estradiol upregulated OPG mRNA levels and increased OPG secretion.

One potential mechanism by which estrogens suppress cytokine expression in BMSCs was elucidated by Srivastava *et al.* [295]. Using primary cultures of BMSCs isolated from mice, these authors showed that ovariectomy results in increased nuclear levels of phosphorylated Egr-1, which is a transcription factor that modulates expression of the cytokine macrophage colony-stimulating factor (M-CSF). M-CSF, in turn, is an important inducer (together with RANKL) of osteoclast differentiation [208]. Compared to nonphosphorylated Egr-1, the phosphorylated protein binds less well to another transcription factor, Sp-1; this results in increased nuclear levels of free Sp-1, which leads to increased transactivation of the M-CSF gene in BMSCs. Conversely, treatment of wild-type OVX mice with 17β-estradiol decreases the levels of phosphorylated Egr-1 in the nucleus of BMSCs and therefore downregulates M-CSF expression. Protein antagonists of IL-1 and TNF-α mimic this downregulation. In contrast, 17β-estradiol has no effect on M-CSF expression in OVX mice that lack Egr-1.

Another commonly reported osteoblastic response to estrogens is the suppression of PTH action. Like estrogens, PTH is an important hormonal regulator of bone metabolism [296]. Osteoblasts are the primary targets for PTH action in bone and mediate both anabolic and catabolic activities of this hormone. In fact, one of the bone resorbing effects of PTH on osteoblastic cells is the upregulation of RANKL expression [208]. As summarized in Table 12-6, treatment of seven different *in vitro* osteoblast models with 17β-estradiol has been shown to block the ability of those cells to respond to PTH. Typically, 17β-estradiol has been observed to inhibit the PTH-stimulated increase in intracellular cAMP levels [180, 242, 246, 297, 298]. However, the steroid has also been reported to interfere with some of the downstream effects of the peptide as well [263, 278, 292, 299, 300]. In at least one instance, PTH has also been shown to block an estrogenic effect in an osteoblast [268]. Furthermore, the suppressive effect of estrogens on PTH activity has also been observed clinically. Using urinary biochemical markers of bone resorption, Cosman *et al.* [301] reported that postmenopausal women treated with estrogens exhibited a markedly blunted response to a continuous intravenous infusion of PTH(1–34).

The mechanism by which estrogens interfere with PTH signaling is not clear. Using SaOS-2 human osteosarcoma cells, Monroe and Tashjian [298] proposed that this suppression was due to a decrease in membrane-associated adenylyl cyclase activity. However, this mechanism does not appear to be applicable to HOB-03-CE6 conditionally immortalized human osteoblasts since the inhibitory actions of 17β-estradiol are selective for PTH over PGE_2- and forskolin-stimulated cAMP production [180]. Ernst et al. [302] suggested that the ability of 17β-estradiol to reduce PTH-stimulated cAMP production in RCT-3 transformed rat osteoblasts was due to a nongenomic action of the steroid because it was observed within 4 hours of treatment and was not enhanced by overexpression of ERα. Although these data are suggestive of a nongenomic effect, they are by no means conclusive.

Although most studies have demonstrated an antagonistic effect of estrogens on PTH activity or cytokine expression, a few reports have shown the opposite to occur (Table 12-7). For example, 17β-estradiol has been observed to enhance PTH responsiveness. In dexamethasone-conditioned SaOS-2 cells, 17β-estradiol and PTH potentiate each other's stimulatory effect on alkaline phosphatase activity [303], whereas in SaOS-2 cells as well as in primary rat and human osteoblasts, the steroid enhances the ability of PTH to stimulate fibronectin production [304]. Although these reports appear to contradict the antagonistic effects of estrogens on PTH activity in osteoblasts, PTH receptors are coupled to at least two signal transduction pathways [305], and estrogens may have different actions on these second messenger systems. Likewise, using a T-antigen transformed human osteoblast cell line (HOBIT), Pivirotto et al. [306] presented evidence that 17β-estradiol upregulates IL-1β mRNA levels. However, since this effect has only been reported to occur in HOBIT cells, its biological significance is questionable.

5. REGULATION OF RECEPTOR EXPRESSION AND SIGNAL TRANSDUCTION

Estrogens have been reported to modulate the expression of several receptors in osteoblasts. At least three members of the nuclear receptor superfamily are known to be regulated by these steroids. As occurs in uterine and breast cells, treatment of either hOBs or conditionally immortalized hFOB/ER9 cells with 17β-estradiol upregulates PR expression [171, 290, 307]. The steroid has also been observed to increase VDR levels and vitamin D_3

responsiveness in two osteosarcoma cell lines [208, 209]. In addition, it either increases [192, 193] or decreases [193] ERα mRNA levels in primary cultures of human and rat osteoblasts, respectively. In the case of ROB cells, our laboratory demonstrated that 17β-estradiol downregulates ERα expression in day 14 nodule-forming cultures (osteoblastic cells), whereas it upregulates receptor expression in day 30 late mineralization–stage cultures (osteocytic cells) [193]. Consistent with these observations, Hoyland et al. [201] reported that ERT or hormone replacement therapy (HRT) decreases the number of ERα mRNA-positive osteoblasts in human bone biopsies. On the other hand, ERT/HRT increases the number of ERα protein-positive osteocytes in these biopsies. Thus, estrogens play a role in both directly regulating osteoblastic activity and modulating the hormonal responsiveness of the cells.

Estrogens have also been reported to regulate additional signal transduction pathways in osteoblasts (Table 12-7). One interesting finding is that 17β-estradiol downregulates mRNA expression of the type I inositol trisphosphate (IP_3) receptor in several in vitro osteoblast models [310]. This receptor is a transmembrane calcium channel found on the "calciosome," which is a specialized component of the endoplasmic reticulum that is involved in the storage and release of IP_3-sensitive intracellular calcium [311]. This receptor is therefore essential for the phosphoinositide signaling pathway. Since bone resorbing agents such as PTH, prostaglandins, and bradykinin utilize this pathway, suppression of type I IP_3 receptor expression by estrogens in osteoblasts may lead to decreased bone resorption and turnover. Although the human type I IP_3 receptor promoter does not contain a consensus ERE, 17β-estradiol nevertheless downregulates promoter–reporter gene constructs when transiently transfected into G-292 human osteosarcoma cells [312].

Another interesting observation is the upregulation of eNOS or NOS-1 mRNA expression and enzyme activity in HOS TE-85 human osteosarcoma cells [313]. Since high NO levels have been reported to inhibit in vitro osteoclastic bone resorption [314], this estrogenic effect is also consistent with an antiresorptive role for the steroid. Moreover, an in vivo study with OVX rats confirmed these results. Wimalawansa et al. [315] reported that treatment of OVX rats with either 17β-estradiol or nitroglycerine (an NO donor) reversed lumbar spine bone loss as measured by dual-energy x-ray absorptiometry. In contrast, cotreatment with 17β-estradiol and N^G-nitro-L-arginine methyl ester (an NOS inhibitor) blocked the bone-sparing effects of the steroid hormone. In contrast to these observations regarding basal NO production, Van Bezooijen et al. [316]

reported that 17β-estradiol treatment of mouse immortalized MC-3T3-E1 osteoblasts suppressed cytokine-induced (NOS-2-mediated) NO synthesis. This finding may reflect the generally antagonistic nature of estrogens toward cytokine action (i.e., IL-1β and TNF-α) in the skeleton.

Finally, pretreatment of hOBs with 17β-estradiol has been reported to increase bradykinin responsiveness as measured by the release of arachidonic acid from the cells [317]. However, since bradykinin stimulates bone resorption, the physiological significance of this observation is unclear.

6. MISCELLANEOUS RESPONSES

As outlined in Table 12-6, treatment of several rodent osteoblastic cell models with 17β-estradiol has been reported to have the following effects: It increases creatine kinase (CK) [236]; increases heat shock protein (HSP)-27 [230]; and increases aspartate aminotransferase (AST), γ-glutamyl transferase (GGT), lactate dehydrogenase (LDH), and transferrin [318]. However, the physiological or therapeutic significance of these responses is unclear. The upregulation of CK activity by 17β-estradiol was also observed in rat bone *in vivo*, and this may represent another anabolic effect of the steroid [236].

7. SUMMARY

As described in the preceding sections, approximately one-third (15/43) of the estrogenic responses observed in a broad range of *in vitro* osteoblast and BMSC models are consistent with the suppressive effects of estrogens on bone resorption and bone turnover *in vivo*. However, in other instances such as the anabolic effects, a disconnection occurs between the *in vitro* responses and the *in vivo* physiology of these steroids. *In vivo*, increased bone turnover upon estrogen depletion is primarily driven by increased osteoclastic bone resorption and the subsequent inadequate ability of osteoblastic bone formation to keep pace with this accelerated bone loss [154–156]. On the other hand, *in vitro* studies with osteoblasts are almost always performed with pure cultures of cells (i.e., cloned osteoblastic cell lines) and in the absence of osteoclasts. Consequently, the opportunity for coupling between the two cell types is lost [320]. Thus, in isolation, estrogens appear to have both stimulatory and inhibitory effects on osteoblastic function. In some *in vitro* models, such as hFOB/ER9 cells [233] or ROBs [193], these differential effects seem to occur as a result of changes that arise during cellular differentiation. However, it is not known if estrogens have divergent actions on osteoblasts as they undergo maturation *in vivo*. Another possible explanation for the apparent

anabolic effects of estrogens on osteoblasts *in vitro* is that these may represent a pharmacological response to the steroid and not a physiological one [239].

B. Estrogenic Responses in Osteocytes

Only a few estrogenic responses have been observed in osteocytes, and all of these reports come from *in situ* studies. In what may well be the first publication on this subject, Whitson [321] described the results of an electron microscopic analysis of metatarsal bones isolated from vehicle and 17β-estradiol-*n*-valerate–treated female rabbits. Although not quantitative, the author noted that the number of tight junctions (possibly gap junctions) formed between osteocytes was greater in bones from the estrogen-treated animals. Moreover, he suggested that this increased tight junction formation might be related to an accelerated osteogenesis.

Twenty-five years later, Tomkinson *et al.* [322] reported the findings of a clinical study of premenopausal women who were treated with a gonadotropin-releasing hormone (GnRH) analogue for endometriosis. Transiliac biopsies were taken from the women before and after GnRH analogue therapy, which resulted in a dramatic decrease in serum 17β-estradiol levels. Although osteocyte lacunae density was not affected by the treatment, the percentage of lacunae containing viable osteocytes (as determined by cell-associated lactate dehydrogenase activity) was reduced in all but one of the six patients. These results suggested that estrogen deficiency is associated with increased osteocyte apoptosis [198]. Since one of the functions of osteocytes is to serve as mechanosensors [197, 199, 323], these observations also implied that estrogen deficiency could lead to increased bone fragility (and therefore increased fracture) at weight-bearing skeletal sites with or without an accompanying net bone loss. The same group confirmed this clinical study the following year using OVX rats [324]. In this preclinical model of estrogen deficiency, OVX increased the number of apoptotic osteocytes (as determined by DNA strand fragmentation) in both trabecular and cortical bone of the tibia. In addition, repletion with 17β-estradiol reversed this increase and returned the apoptotic index to the sham values.

In another *in situ* study of OVX rats, Ikeda *et al.* [325] observed that osteopontin mRNA expression increased after OVX in osteocytes that were located in metaphyseal trabecular bone of the femur but not in those found in the epiphysis. Since osteopontin is one of the bone matrix proteins to which osteoclasts are known to bind [169], these data suggested a possible role for the osteocyte in regulating bone resorption. Our laboratory has also presented evidence that osteocytic cells may play a role in modulating osteoclastic activity [326]. Using a conditionally immortalized human preosteocytic (i.e., osteoid–osteocyte) cell line (HOB-01-C1), we showed that these cells secrete high amounts of IL-6 and monocyte chemoattractant protein (MCP)-1 in response to treatment with the bone resorbing cytokines IL-1β and TNF-α. Together, IL-6 and MCP-1, in addition to other factors, might stimulate osteoclast differentiation and recruitment to a specific bone-remodeling site.

Another potential regulatory target for estrogens in osteocytes is ERα. Using immunofluorescence to study ERα protein expression in human bone biopsies, Braidman and colleagues [189, 201] reported that ERT/HRT increases the number of ERα protein-positive osteocytes and osteoblasts. Curiously, the number of ERα mRNA-positive osteoblasts was observed to decline with ERT/HRT [201].

As noted previously, osteocytes are postulated to serve as mechanosensors [197, 199, 323]. As such, they are thought to translate the effects of weight bearing or weightlessness into either increases or decreases in bone-mineral density, respectively. Several studies suggest that estrogens regulate the process of mechanosensory stimulation, and that mechanical strain and estrogen action may share common signaling pathways. Using organ cultures of rat ulnae isolated from female rats, Cheng *et al.* [327, 328] reported that both 17β-estradiol and mechanical loading stimulated [³H]thymidine and [³H]proline incorporation into the bones. Moreover, when the treatments were combined, a synergistic effect was observed. Thus, estrogens appeared to enhance the osteogenic response of the bones to mechanical strain. A subsequent study by the same group using primary cultures of rat long bone–derived osteoblasts demonstrated that both 17β-estradiol and mechanical strain increase cellular DNA synthesis [232]. Furthermore, these increases were suppressed by cotreatment with the antiestrogen ICI-182,780. Although osteoblasts are probably not the targets for mechanical loading *in vivo* [323], these results nevertheless suggest that mechanical strain can activate the ER.

The observation that mechanical strain and estrogens appear to share common signal transduction pathways is supported by an *in vivo* study by Westerlind *et al.* [329]. Using OVX rats, these authors showed that estrogen deficiency resulted in a preferential loss of cancellous bone from a site that experiences low mechanical strain (distal femur metaphysis), whereas one that experiences high strain energies (distal femur epiphysis) did not lose bone (even though bone turnover was increased at both sites). In addition, increased mechanical loading (treadmill exercise) suppressed OVX-induced cancellous bone loss from the proximal tibial metaphysis. Conversely, treatment of OVX animals with 17β-estradiol suppressed tibial cancellous bone loss that resulted from decreased mechanical loading (unilateral sciatic neurotomy).

Finally, there is also evidence that these preclinical findings may translate to humans. For example, in a small clinical study of postmenopausal women, Kohrt *et al.* [330] reported that HRT and weight-bearing exercise had an additive effect on total body bone mineral accretion. Thus, the efficacies of HRT and weight-bearing exercise on the skeleton seem to be enhanced by concurrent use. Although the previously mentioned studies do not specifically address the role of estrogens in osteocyte biology per se, the implication of this work is that osteocytes—as the major mechanosensory cell in bone—are at least one of the targets for these effects.

C. Estrogenic Responses in Cells of the Osteoclast Lineage

In addition to indirectly inhibiting bone resorption through cells of the osteoblast lineage, estrogens have also been reported to have direct suppressive effects on cells of the osteoclast lineage [207]. The most extensive evidence for a direct inhibitory effect of estrogens on mature osteoclasts comes from the work of Oursler and colleagues [207]. Using both avian and hGCT-derived osteoclasts that were highly purified (≧90% homogeneous) with an osteoclast-specific monoclonal antibody (121F), this group reported that 17β-estradiol inhibits *in vitro* bone resorption by these preparations [210, 211, 215, 331–333]. Estrogenic responses in these studies include the following: the upregulation of c-fos, c-jun, TGF-β2, TGF-β3, and TGF-β4 mRNA levels; the downregulation of tartrate-resistant acid phosphatase (TRAP), cathepsin B, cathepsin D, LEP-100, and lysozyme message levels; the induction of total TGF-β protein secretion (due mostly to an increase in TGF-β3); and the suppression of TRAP, cathepsin B, cathepsin L, and β-glucuronidase activity as well as lysozyme protein production. The majority of these effects are consistent with an estrogen-mediated decrease in osteoclast activity and subsequent bone resorption. For example, TGF-β is an inhibitor of bone resorption, whereas lysosomal proteases such as the cathepsins are involved in digesting the bone matrix [207]. Confirmation that estrogens suppress osteoclastic gene expression *in vivo* comes from the studies of Zheng *et al.* [334], who demonstrated that treatment of OVX rats with 17β-estradiol decreased the expression of TRAP mRNA in bone. Additional support for a direct effect of estrogens on osteoclasts comes from the work of Sunyer *et al.* [212]. Employing normal hOCLs that were also purified to 90% homogeneity with the 121F monoclonal antibody, these authors reported that 17β-estradiol decreased the mRNA levels of the signaling receptor for IL-1 (IL-1RI), and increased the message levels of the IL-1 decoy receptor (IL-1RII). This change in receptor expression correlated with a suppression of IL-1 β-mediated IL-8 expression by the steroid hormone. Moreover, 17β-estradiol pretreatment abrogated the reduction of hOCL apoptosis by IL-1β. Finally, Mano *et al.* [213] demonstrated that 17β-estradiol also inhibits the *in vitro* bone resorption of purified rabbit osteoclasts and reduces the expression of cathepsin K mRNA by these cells.

However, some studies have failed to detect a direct inhibitory effect of estrogens on mature osteoclasts. For example, Williams *et al.* [335] were unable to suppress bone resorption of purified avian osteoclasts with either 17β-estradiol or DES. On the other hand, high (micromolar) levels of the TSE/SERM tamoxifen decreased osteoclast activity. Likewise, calmodulin antagonists had a similar effect. Additional experiments led the authors to conclude that tamoxifen acted through a membrane-associated target to suppress osteoclastic bone resorption independently of the ER. This target appeared to be similar or related to the target for the calmodulin inhibitors.

As indicated previously, estrogens have been observed to increase the expression of TGF-β by both osteoblasts and osteoclasts. In addition, these steroids suppress osteoblast apoptosis but enhance programmed cell death of osteoclasts [336, 337]. Hughes *et al.* [338] elegantly demonstrated a connection between estrogens, TGF-β, and osteoclast apoptosis. These authors showed that treatment of marrow culture–derived murine osteoclasts with 17β-estradiol increased the percentage of cells undergoing apoptosis. Likewise, treatment of the cultures with TGF-β1 also increased osteoclast apoptosis. Moreover, the induction of osteoclast programmed cell death by 17β-estradiol could be blocked by coincubation with a pan-specific TGF-β antibody. Consistent with its bone-sparing effects [266], treatment of the osteoclast-containing cultures with tamoxifen also increased apoptosis of these cells. These *in vitro* observations were confirmed with an *in vivo* study in which OVX mice were treated with 17β-estradiol. Since the marrow culture system used by Hughes *et al.* was a heterogeneous cell population, the promotion of osteoclast apoptosis by 17β-estradiol could have resulted from either a direct action of the steroid on osteoclasts or an indirect effect on another cell type, such as the osteoblasts or BMSCs.

In addition to inducing apoptosis of mature osteoclasts, estrogens may also have similar effects on osteoclast progenitors. Zecchi-Orlandini *et al.* [339] reported that 17β-estradiol induced apoptosis of the human monoblastic leukemia cell line FLG 29.1, which has characteristics resembling preosteoclasts. Moreover, treatment of this cell line with the TSE/SERM raloxifene [266] also induced apoptosis [340]. The FLG 29.1 cells can be stimulated to form osteoclast-like cells *in vitro* by treatment with phorbol ester, vitamin D_3, or osteoblast–derived factors [341]. These agents also induce the expression of a novel superoxide dismutase–related membrane glycoprotein, which is the osteoclast-specific antigen that is recognized by the 121F monoclonal antibody. Incubation of the cells with 17β-estradiol suppresses the induction of this antigen by phorbol ester [341]. Thus, these results suggest that estrogens may also suppress osteoclast differentiation by acting directly on their progenitors.

Additional reports also indicate that estrogens can suppress osteoclast differentiation. Schiller *et al.* [342] demonstrated that 17β-estradiol antagonizes the induction of osteoclast-like cell formation by vitamin D_3 in primary cultures of mouse bone marrow cells. In addition, these authors showed that the ability of vitamin D_3 to stimulate osteoclast differentiation is at least partially mediated by an upregulation of IL-6 secretion, and that 17β-estradiol blocks this effect as well. Estrogens also suppress PTH-stimulated osteoclast formation. Using primary mouse hemopoietic blast cell cultures, which were reportedly free of stromal cells and osteoblasts, Kanatani *et al.* [217] presented evidence that these osteoclast precursors contain PTH receptor mRNA based on RT-PCR. These cells also express ERα message. Treatment of the mouse hemopoietic blast cell cultures with either vitamin D_3 or PTH(1–34) induces the formation of osteoclast-like cells (i.e., TRAP-positive multinucleated cells). On the other hand, cotreatment of the cultures with 17β-estradiol blunts the stimulation of osteoclast differentiation by PTH but not by vitamin D_3. These authors also demonstrated that 17β-estradiol blocks osteoclast-like cell formation induced by agents that activate adenylyl cyclase or mimic cAMP but not ones that activate protein kinase C or increase intracellular calcium. Although an earlier report from the same group suggested that estrogens suppress PTH-induced osteoclast differentiation indirectly through an effect on osteoblasts [300], the study by Kanatani *et al.* concluded that this inhibitory effect might also be due to a direct action on osteoclast progenitor cells.

In summary, there is substantial evidence to conclude that estrogens inhibit osteoclast differentiation and activity in two ways: (1) indirectly via the osteoblast and stromal cell and (2) directly through interaction with the ER in osteoclast progenitors and mature osteoclasts. However, as with other aspects of estrogen action on bone cells, this area of research is controversial.

D.　Estrogenic Responses in Chondrocytes

Another important target cell in the skeleton for estrogens is the chondrocyte. As noted previously, these cells have been shown to express both ERα and ERβ. Moreover, chondrocytes have also been reported to exhibit estrogenic responses. *In vivo*, it is well known that estrogens accelerate endochondral growth during puberty and potentiate epiphyseal closure at the end of the growth spurt [154]. Consistent with these physiological responses, 17β-estradiol has been observed to decrease the *in vitro* proliferation and/or DNA synthesis of embryonic duck [343] and rat chondrocytes [344]. In duck chondrocytes, 17β-estradiol also suppressed sulfated proteoglycan synthesis [343], whereas in fetal rabbit [345] and human chondrocytes [346], the steroid had the opposite effect. Additional *in vitro* estrogenic effects in rat chondrocytes include the upregulation of alkaline phosphatase activity and collagen production, which are consistent with a potentiation of cellular differentiation by the steroid [344].

VIII.　ESTROGEN-RELATED RECEPTOR-α AND OSTEOPONTIN GENE EXPRESSION

In addition to expressing ERα and ERβ, osteoblasts also express a related member of the nuclear receptor superfamily known as estrogen-related receptor (ERR)-1 or -α [347–349]. ERR-α is an orphan receptor that shares 68% amino acid identity with ERα and ERβ in the DNA-binding domain but only 36% identity in the ligand-binding domain [349]. Consequently, it does not bind 17β-estradiol but instead is constitutively active in serum-containing medium [349]. However, this constitutive activity is diminished upon charcoal treatment of the serum [349]. ERR-α, as well as the related ERR-β, transactivates promoters containing either an ERE or an SF-1-response element (SFRE) [349]. ERα also binds to both of these DNA response elements, whereas ERβ does not bind to the SFRE [349]. ERR-α mRNA is highly expressed in the ossification zones of the developing mouse skeleton (long bones, vertebrae, ribs, and skull), as well as in some human osteosarcoma cell lines (HOS-TE85 and SaOS-2)

and hOBs [347]. Given this expression pattern, as well as the knowledge that the osteopontin promoter contains an SFRE, it is perhaps not surprising that cotransfection of rat ROS 17/2.8 osteosarcoma cells with ERR-α and an osteopontin promoter–reporter gene construct resulted in the transactivation of this promoter [347–349]. Moreover, transient transfection of ROS 17/2.8 cells and immortalized mouse MC-3T3-E1 cells with ERR-α produced an upregulation of endogenous osteopontin mRNA levels [348]. Taken together, these data demonstrate that osteopontin gene expression in the osteoblast is regulated not only by ERα in an estrogen-dependent manner but also by ERR-α in an estrogen-independent manner [349]. In contrast, ERβ does not appear to regulate this gene [349]. Thus, these observations also point to a potential functional difference between the biological roles of ERα and ERβ in the osteoblast. However, since osteopontin is an apparent binding site for osteoclasts to the bone matrix [169], the physiological significance of its upregulation by estrogens via either ERα or ERR-α in a ligand-independent manner is unclear.

IX.　NONGENOMIC ACTIONS OF ESTROGENS IN BONE CELLS

Although the majority of estrogenic effects are believed to be mediated by one of the nuclear ERs, some responses may also originate at the plasma membrane [251, 350]. Estrogens have been reported to produce rapid effects (within seconds or minutes) on a variety of cell types, including bone cells [350, 351]. These nongenomic actions are thought to be mediated via a membrane receptor. However, it is unclear whether or not this receptor is a membrane-localized form of a nuclear ER or if it is a distinct transmembrane protein such as a GPCR [350, 351]. In a series of papers on primary female rat osteoblasts, Lieberherr and coworkers presented convincing evidence for rapid, membrane-derived effects of 17β-estradiol [352–354]. Treatment of ROB cells with low concentrations (1 pM to 1 nM) of 17β-estradiol increased intracellular calcium levels within 10–30 seconds [352]. Through the use of various inhibitors, the source of this calcium was shown to be both extracellular via plasma membrane channels and intracellular from the endoplasmic reticulum or calciosome. The cells within the same time frame also produced IP_3 and diacylglycerol (DAG) after treatment with the steroid. Since inhibitors of both phospholipase C (PLC) and G_i proteins blocked the release of IP_3 and DAG, the authors concluded that 17β-estradiol acted through a GPCR [352]. Consistent

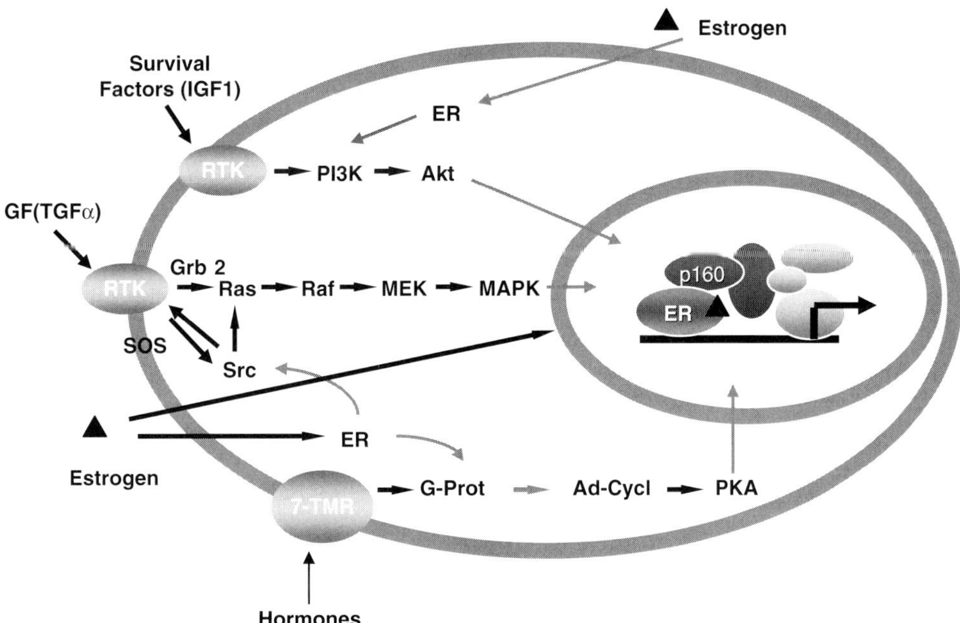

FIGURE 12-3   Estrogens can affect cell function through several pathways. Classically, an estrogen diffuses through the plasma membrane to interact with a nuclear localized receptor (ERα or ERβ or both). The binding of ligand results in a rapid conformational change in the receptor and other biochemical modifications, such as phosphorylation and acetylation. Associated with the changes in conformation are interactions with coactivators (e.g., p160). These proteins form a transcriptional complex linking the receptor DNA complex to the basal transcriptional machinery resulting in changes in transcriptional activity. Alternatively, estrogens have now been shown to activate rapid signaling pathways via PI3 kinase, Src-kinase, and PKA.

with estrogens working through a distinct membrane receptor and not simply a membrane-localized ER, tamoxifen was neither an agonist nor an antagonist of 17β-estradiol. Subsequent studies by this group refined the model to include activation of PLC-β2 by βγ subunits [353, 354]. In contrast, vitamin D_3, which also has rapid effects on female ROB cells, was shown to act via modulation of PLC-β1 by Gα (q/11) [353, 354]. A potential downstream target for the rapid generation of a membrane-derived signal by 17β-estradiol was reported by Endoh *et al.* [355]. These authors showed that treatment of ROS 17/2.8 cells with 17β-estradiol activated the MAPK within 5 minutes.

Estrogens may also produce rapid nongenomic effects in cells of the osteoclast lineage [113, 356–359]. For example, using the human preosteoclastic cell line FLG 29.1, Fiorelli *et al.* [113] demonstrated that 17β-estradiol stimulated an increase in intracellular pH within 50 seconds, as well as an increase in intracellular cAMP and cGMP after 30 minutes. In addition, Brubaker and Gay [359] reported that treatment of isolated avian osteoclasts with 17β-estradiol caused a depolarization of the plasma membrane potential within seconds of adding the steroid to the cells. The mechanism for the depolarization appeared to be due to

regulation of potassium channel activity. The net effect of this rapid, nongenomic estrogenic response could be an inhibition of osteoclastic acidification.

X. CONCLUSION

Estrogens clearly play a critical role in bone biology. The increase in research aimed at elucidating the functional role of estrogens in bone remodeling that has occurred in the past 25 years has led to the discovery of a multitude of potential pathways that are impacted by estrogens in the skeleton. The sheer abundance of estrogenic-related regulated events in bone cells supports the contention that estrogens, working through their receptors, play key roles in the development and maintenance of a normal skeleton. Questions that remain to be answered relate to the differences in the skeletal response to the various types of estrogens (estradiol vs. phytoestrogens vs. SERMS). All estrogens do not evoke the same response in bone, whether considering a specific gene's regulation in isolated osteoblasts or a global skeletal response *in vivo*. Why this occurs is not known. The complexity of the bone remodeling process coupled with the multiple sites where an estrogen could elicit an effect will make it difficult to fully answer the question, but as technology advances, so will the possibility of answering tough questions.

REFERENCES

1. F. Albright, P. H. Smith, and A. M. Richardson, Postmenopausal osteoporosis: Its clinical features. *JAMA* **116**, 2465–2474 (1941).
2. A. W. Diddle and I. Q. Smith, Postmenopausal osteoporosis: The role of estrogens. *Southern Med J* **77**, 868–874 (1984).
3. M. C. Chapuy and P. J. Meunier, Prevention and treatment of osteoporosis. *Aging Clin Exp Res* **7**, 164–173 (1995).
4. J. P. Bilezikian, Estrogens and postmenopausal osteoporosis: Was Albright right after all? *J Bone Miner Res* **13**, 774–776 (1998).
5. R. Eastell, Treatment of postmenopausal osteoporosis. *N Engl J Med* **338**, 736–746 (1998).
6. A. Horsman, M. Jones, R. Francis, and C. Nordin, The effect of estrogen dose on postmenopausal bone loss. *N Engl J Med* **309**, 1404–1407 (1983).
7. T. M. Willson, Chemistry: Structure and function relationships. *J Estrogen Antiestrogens Basic Clin Aspects*, 21–22 (1997).
8. M.B. O'Connell, Pharmacokinetic and pharmacologic variation between different estrogen products. *J Clin Pharmacol* **35**, 18s–24s (1995).
9. H. U. Bryant and W. H. Dere, Selective estrogen receptor modulators: An alternative to hormone replacement therapy (44204). *Proc Soc Exp Biol Med* **217**, 45–52 (1998).
10. T. A. Grese and J. A. Dodge, Selective estrogen receptor modulators (SERMs). *Curr Pharm Design* **4**, 71–92 (1998).
11. V. Kumar, S. Green, G. Stack, M. Berry, J.-R. Jin, and P. Chambon, Functional domains for the human estrogen receptor. *Cell* **51**, 941–951 (1987).
12. S. Green, P. Walter, G. Greene, A. Krust, C. Goffin, E. Jensen, G. Scrace, M. Waterfield, and P. Chambon, Cloning of the human oestrogen receptor cDNA. *J Steroid Biochem* **24**, 77–83 (1986).
13. P. Walter, S. Green, G. Greene, A. Krust, J.-M. Bornert, J.-M. Jeltsch, A. Staub, E. Jensen, G. Scrace, M. Waterfield, and P. Chambon, Cloning of the human estrogen receptor cDNA. *Proc Natl Acad Sci USA* **82**, 7889–7893 (1985).
14. G. J. M. Kuiper, E. Enmark, M. Pelto-Huikko, S. Nilsson, and J. Gustafsson, Cloning of a novel estrogen receptor expressed in rat prostate and ovary. *Proc Natl Acad Sci USA* **93**, 5925–5930 (1996).
15. S. Mosselman, J. Polman, and R. Dijkema, ER-beta: Identification and characterization of a novel human estrogen receptor. *FEBS Lett* **392**, 49–53 (1996).
16. B. O'Malley, The steroid receptor superfamily: More excitement predicted for the future. *Mol Endocrinol* **4**, 363–369 (1990).
17. C. Weinberger, V. Giguere, S. Hollenberg, M. G. Rosenfeld, and R. M. Evans, Human steroid receptors and *erbA* proto-oncogene products: Members of a new superfamily of enhancer binding proteins. *Cold Spring Harb Symp Quant Biol*, **51**, 759–772 (1986).
18. Q. Zhao, S. Khorasanizadeh, Y. Miyoshi, M. A. Lazar, and F. Rastinejad, Structural elements of an orphan nuclear receptor–DNA complex. *Mol Cell* **1**, 849–861 (1998).
19. B. Blumberg and R. M. Evans, Orphan nuclear receptors—New ligands and new possibilities. *Genes Dev* **12**, 3149–3155 (1998).
20. R. Kumar and E. Thompson, The structure of the nuclear hormone receptors. *Steroids* **64**, 310–319 (1999).
21. L. Klein-Hitpass, S. Y. Tsai, G. L. Greene, J. H. Clark, M.-J. Tsai, and B. W. O'Malley, Specific binding of estrogen receptor to the estrogen response element. *Mol Cell Biol* **9**, 43–49 (1989).
22. G. Klock, U. Strahle, and G. Schutz, Oestrogen and glucocorticoid responsive elements are closely related by distinct. *Nature* **329**, 734–736 (1987).
23. M. T. Tzukerman, A. Esty, D. Santiso-Mere, P. Danielian, M. G. Parker, R. B. Stein, J. W. Pike, and D. P. McDonnell, Human estrogen receptor transactivational capacity is determined by both cellular and promoter context and mediated by two functionally distinct intramolecular regions. *Mol Endocrinol* **8**, 21–30 (1994).
24. K. Inano, M. Haino, M. Iwasaki, N. Ono, T. Horigome, and H. Sugano, Reconstruction of the 9 S estrogen receptor with heat shock protein 90. *FEBS Lett* **267**, 157–159 (1990).
25. J. Devin-Leclerc, X. Meng, F. Delahaye, P. Leclerc, E.-E. Baulieu, and M.-G. Catelli, Interaction and dissociation by ligands of estrogen receptor and Hsp90: The antiestrogen RU 58668 induces a protein synthesis-dependent clustering of the receptor in the cytoplasm. *Mol Endocrinol* **12**, 842–854 (1998).
26. V. Kumar and P. Chambon, The estrogen receptor binds tightly to its responsive element as a ligand-induced homodimer. *Cell* **55**, 145–156 (1988).
27. F. Auricchio, Phosphorylation of steroid receptors. *J Steroid Biochem* **32**, 613–622 (1989).
28. W. W. Grody, W. T. Schrader, and B. W. O'Malley, Activation, transformation, and subunit structure of steroid hormone receptors. *Endocr Rev* **3**, 141–163 (1982).
29. B. Katzenellenbogen, B. Bhardwaj, H. Fang, B. Ince, F. Pakdel, J. Reese, D. Schodin, and C. Wrenn, Hormone binding and transcription activation by estrogen receptors: Analyses using mammalian and yeast systems. *J Steroid Biochem Mol Biol* **47**, 39–48 (1993).
30. N. J. McKenna, R. B. Lanz, and B. W. O'Malley, Nuclear receptor coregulators: Cellular and molecular biology. *Endocr Rev* **20**, 321–344 (1999).
31. J. D. Chen and H. Li, Coactivation and corepression in transcriptional regulation by steroid/nuclear hormone receptors. *Crit Rev Euk Gene Expr* **8**, 169–190 (1998).
32. P. Clement-Lacroix, C. Ormandy, L. Lepescheux, P. Ammann, D. Damotte, V. Goffin, B. Bouchard, M. Amling, M. Gaillard-Kelly, N. Binart, R. Baron, and P. A. Kelly, Osteoblasts are a new target for prolactin: Analysis of bone formation in prolactin receptor knockout mice. *Endocrinology* **140**, 96–105 (1999).
33. N. Huang, E. vom Baur, J.-M. Garnier, T. Lerouge, J.-L. Vonesch, Y. Lutz, P. Chambon, and R. Losson, Two distinct nuclear receptor interaction domains in NSD1, a novel SET protein that exhibits characteristics of both corepressors and coactivators. *EMBO J* **17**, 3398–3412 (1998).
34. R. M. Lavinski, K. Jepsen, T. Heinzel, J. Torchia, T.-M. Mullen, R. Schiff, A. L. Del-Rio, M. Ricote, S. Ngo, J. Gemsch, S. G. Hilsenbeck, C. K. Osborne, C. K. Glass, M. G. Rosenfeld, and D. W. Rose, Diverse signaling pathways modulate nuclear receptor recruitment of N-CoR and SMRT complexes. *Proc Natl Acad Sci USA* **95**, 2920–2925 (1998).
35. C. M. Klinge, Estrogen receptor interaction with co-activators and co-repressors. *Steroids* **65**, 227–251 (2000).
36. J. T. Kadonaga, Eukaryotic transcription: An interlaced network of transcription factors and chromatin-modifying machines. *Cell* **92**, 307–313 (1998).
37. E. Korzus, J. Torchia, D. W. Rose, L. Xu, R. Kurokawa, E. M. McInerney, T. Mullen, C. K. Glass, and M. G. Rosenfeld, Transcription factor-specific requirements for coactivators and their acetyltransferase functions. *Science* **279**, 703–707 (1998).

38. H. Y. Mak, S. Hoare, P. M. A. Henttu, and M. G. Parker, Molecular determinants of the estrogen receptor–coactivator interface. *Mol Cell Biol* **19,** 3895–3903 (1999).
39. P. M. A. Henttu, E. Kalkhoven, and M. G. Parker, AF-2 activity and recruitment of steroid receptor coactivator 1 to the estrogen receptor depend on a lysine residue conserved in nuclear receptors. *Mol Cell Biol* **17,** 1832–1839 (1997).
40. A. K. Shiau, D. Barstad, P. M. Loria, L. Cheng, P. J. Kushner, D. A. Agard, and G. L. Greene, The structural basis of estrogen receptor/coactivator recognition and the antagonism of this interaction by tamoxifen. *Cell* **95,** 927–937 (1998).
41. W. Feng, R. C. J. Ribeiro, R. L. Wagner, H. Nguyen, J. W. Apriletti, R. J. Fletterick, J. D. Baxter, P. J. Kushner, and B. L. West, Hormone-dependent coactivator binding to a hydrophobic cleft on nuclear receptors. *Science* **280,** 1747–1749 (1998).
42. W. L. Kraus and J. T. Kadonaga, p300 and estrogen receptor cooperatively activate transcription via differential enhancement of initiation and reinitiation. *Genes Dev* **12,** 331–342 (1998).
43. C. Rachez, Z. Suldan, J. Ward, C.-P. B. Chang, D. Burakov, H. Erdjument-Bromage, P. Tempst, and L. P. Freedman, A novel protein complex that interacts with the vitamin D3 receptor in a ligand-dependent manner and enhances VDR transactivation in a cell-free system. *Genes Dev* **12,** 1787–1800 (1998).
44. L. P. Freedman, Increasing the complexity of coactivation in nuclear receptor signaling. *Cell* **97,** 5–8 (1999).
45. M. Ito, C.-X. Yaun, S. Malik, W. Gu, J. D. Fondell, S. Yamamura, Z.-Y. Fu, X. Zhang, J. Qin, and R. G. Roeder, Identity between TRAP and SMCC complexes indicates novel pathways for the function of nuclear receptors and diverse mammalian activators. *Mol Cell* **3,** 361–370 (1999).
46. W. Welshons, E. Cormier, M. Wolf, P. J. Williams, and V. Jordan, Estrogen receptor distribution in enucleated breast cancer cell lines. *Endocrinology* **122,** 2379–2386 (1988).
47. V. Boonyaratanakornkit and D. P. Edwards, Receptor mechanisms of rapid extranuclear signalling initiated by steroid hormones. *Essays Biochem* **40,** 105–120 (2004).
48. E. R. Levin, Integration of the extranuclear and nuclear actions of estrogen. *Mol Endocrinol* **19,** 1951–1959 (2005).
49. C. S. Watson and B. Gametchu, Proteins of multiple classes may participate in nongenomic steroid actions. *Exp Biol Med* **228,** 1272–1281 (2003).
50. A. W. Norman, M. T. Mizwicki, and D. P. G. Norman, Steroid-hormone rapid actions, membrane receptors, and a conformational ensemble model. *Nat Rev Drug Discov* **3,** 27–41 (2004).
51. B. J. Cheskis, Regulation of cell signalling cascades by steroid hormones. *J Cell Biochem* **93,** 20–27 (2004).
52. F. Barletta, C.-W. Wong, C. McNally, B. S. Komm, B. Katzenellenbogen, and B. J. Cheskis, Characterization of the interactions of estrogen receptor and MNAR in the activation of cSrc. *Mol Endocrinol* **18,** 1096–1108 (2004).
53. E. Falkenstein and M. Wehling, Nongenomically initiated steroid actions. *Eur J Clin Invest* **30,** 51–54 (2000).
54. A. C. Cato, A. Nestl, and S. Mink, Rapid actions of steroid receptors in cellular signaling pathways. *Science* **2002,** RE9 (2002).
55. M. A. Shupnik, Crosstalk between steroid receptors and the c-Src-receptor tyrosine kinase pathways: Implications for cell proliferation. *Oncogene* **23,** 7979–7989 (2004).
56. L. P. Zanello and A. W. Norman, Rapid modulation of osteoblast ion channel responses by 1{α},25(OH)2-vitamin D3 requires the presence of a functional vitamin D nuclear receptor. *Proc Natl Acad Sci USA* **101,** 1589–1594 (2004).
57. H. Karst, S. Berger, M. Turiault, F. Tronche, G. Schutz, and M. Joels, Mineralocorticoid receptors are indispensable for nongenomic modulation of hippocampal glutamate transmission by corticosterone. *Proc Natl Acad Sci USA* **102,** 19204–19207 (2005).
58. E. J. Filardo, J. A. Quinn, K. I. Bland, and A. R. Frackelton Jr., Estrogen-induced activation of Erk-1 and Erk-2 requires the G protein-coupled receptor homolog GPR30 and occurs via trans-activation of the epidermal growth factor receptor through release of HB-EGF. *Mol Endocrinol* **14,** 1649–1660 (2000).
59. C. M. Revankar, D. F. Cimino, L. A. Sklar, J. B. Arterburn, and E. R. Prossnitz, A transmembrane intracellular estrogen receptor mediates rapid cell signaling. *Science* **307,** 1625–1630 (2005).
60. P. Thomas, Y. Pang, E. J. Filardo, and J. Dong, Identity of an estrogen membrane receptor coupled to a G-protein in human breast cancer cells. *Endocrinology*, **146,** 624–632 (2004).
61. Y. Zhu, C. D. Rice, Y. Pang, M. Pace, and P. Thomas, From the cover: Cloning expression and characterization of a membrane progestin receptor and evidence it is an intermediary in meiotic maturation of fish oocytes. *Proc Natl Acad Sci USA* **100,** 2231–2236 (2003).
62. E. Grazzini, G. Guillon, B. Mouillac, and H. H. Zingg, Inhibition of oxytocin receptor function by direct binding of progesterone. *Nature* **392,** 437–438 (1998).
63. J. J. Lambert, D. Belelli, D. R. Peden, A. W. Vardy, and J. A. Peters, Neurosteroid modulation of GABAA receptors. *Prog Neurobiol* **71,** 67–80 (2003).
64. C. S. Watson, C. H. Campbell, and B. Gametchu, Membrane oestrogen receptors on rat pituitary tumour cells: Immuno-identification and responses to oestradiol and xenoestrogens. *Exp Physiol* **84,** 1013–1022 (1999).
65. D. Zivadinovic and C. Watson, Membrane estrogen receptor-alpha levels predict estrogen-induced ERK1/2 activation in MCF-7 cells. *Breast Cancer Res* **7,** R130–R144 (2005).
66. R. G. W. Anderson, The caveolae membrane system. *Annu Rev Biochem* **67,** 199–225 (1998).
67. M. Razandi, P. Oh, A. Pedram, J. Schnitzer, and E. R. Levin, ERs associate with and regulate the production of caveolin: Implications for signaling and cellular actions. *Mol Endocrinol* **16,** 100–115 (2002).
68. F. Acconcia, P. Ascenzi, G. Fabozzi, P. Visca, and M. Marino, S-palmitoylation modulates human estrogen receptor-[alpha] functions. *Biochem Biophys Res Commun* **316,** 878–883 (2004).
69. Q. Lu, D. C. Pallas, H. K. Surks, W. E. Baur, M. E. Mendelsohn, and R. H. Karas, Striatin assembles a membrane signaling complex necessary for rapid nongenomic activation of endothelial NO synthase by estrogen receptor {α}. *Proc Natl Acad Sci USA* **101,** 17126–17131 (2004).
70. G. Pelicci, L. Dente, A. De Giuseppe, B. Verducci-Galletti, S. Giuli, S. Mele, C. Vetriani, M. Giorgio, P. P. Pandolfi, G. Cesareni, and P. G. Pelicci, A family of Shc related proteins with conserved PTB, CH1 and SH2 regions. *Oncogene* **13,** 633–641 (1996).
71. R. X.-D. Song, R. A. McPherson, L. Adam, Y. Bao, M. Shupnik, R. Kumar, and R. J. Santen, Linkage of rapid estrogen action to MAPK activation by ER{α}-Shc association and Shc pathway activation. *Mol Endocrinol* **16,** 116–127 (2002).
72. R. X. Song, C. J. Barnes, Z. Zhang, Y. Bao, R. Kumar, and R. J. Santen, The role of Shc and insulin-like growth factor 1 receptor in mediating the translocation of estrogen receptor {alpha} to the plasma membrane. *Proc Natl Acad Sci USA* **101,** 2076–2081 (2004).
73. S. Cabodi, L. Moro, G. Baj, M. Smeriglio, P. Di Stefano, S. Gippone, N. Surico, L. Silengo, E. Turco, G. Tarone, and P.

Defilippi, p130Cas interacts with estrogen receptor {α} and modulates non-genomic estrogen signaling in breast cancer cells. *J Cell Sci* **117**, 1603–1611 (2004).

74. F. Acconcia, P. Ascenzi, A. Bocedi, E. Spisni, V. Tomasi, A. Trentalance, P. Visca, and M. Marino, Palmitoylation-dependent estrogen receptor {alpha} membrane localization: Regulation by 17{beta}-estradiol. *Mol Biol Cell* **16**, 231–237 (2005).

75. M. Longo, M. Brama, M. Marino, S. Bernardini, K. S. Korach, W. C. Wetsel, R. Scandurra, T. Faraggiana, G. Spera, and R. Baron, Interaction of estrogen receptor [alpha] with protein kinase C [alpha] and c-Src in osteoblasts during differentiation. *Bone* **34**, 100–111 (2004).

76. S. Denger, G. Reid, M. Kos, G. Flouriot, D. Parsch, H. Brand, K. S. Korach, V. Sonntag-Buck, and F. Gannon, ER{alpha} gene expression in human primary osteoblasts: Evidence for the expression of two receptor proteins. *Mol Endocrinol* **15**, 2064–2077 (2001).

77. D. Marquez and R. J. Pietras, Membrane-associated binding sites for estrogen contribute to growth regulation in human breast cancer cells. *Oncogene* **20**, 5420–5430 (2001).

78. L. Li, M. P. Haynes, and J. R. Bender, Plasma membrane localization and function of the estrogen receptor alpha variant (ER46) in human endothelial cells. *Proc Natl Acad Sci USA* **100**, 4807–4812 (2003).

79. A. Schlegel, C. Wang, B. S. Katzenellenbogen, R. G. Pestell, and M. P. Lisanti, Caveolin-1 potentiates estrogen receptor alpha (ERalpha) signaling. Caveolin-1 drives ligand-independent nuclear translocation and activation of ERalpha. *J Biol Chem* **274**, 33551–33556 (1999).

80. S. Kahlert, S. Nuedling, M. van Eickels, H. Vetter, R. Meyer, and C. Grohe, Estrogen receptor alpha rapidly activates the IGF-1 receptor pathway. *J Biol Chem* **275**, 18447–18453 (2000).

81. T. Simoncini, A. Hafezl-Moghadam, D. P. Brazil, K. Ley, W. W. Chin, and J. K. Liao, Interaction of oestrogen receptor with the regulatory subunit of phosphatidylinositol-3-OH kinase. *Nature* **407**, 538–541 (2000).

82. A. Migliaccio, D. Piccolo, G. Castoria, M. Di Domenico, A. Bilancio, M. Lombardi, W. Gong, M. Beato, and F. Auricchio, Activation of the Src/p21ras/Erk pathway by progesterone receptor via cross-talk with estrogen receptor. *EMBO J* **17**, 2008–2018 (1998).

83. A. Migliaccio, G. Castoria, M. Di Domenico, A. De Falco, A. Bilancio, and F. Auricchio, Src is an initial target of sex steroid hormone action. *Ann N Y Acad Sci* **963**, 185–190 (2002).

84. S. M. Thomas and J. S. Brugge, Cellular functions regulated by Src family kinases. *Annu Rev Cell Dev Biol* **13**, 513–609 (1997).

85. M. Resh, Interaction of tyrosine kinase oncoproteins with cellular membranes. *Biochim Biophys Acta* **1155**, 307–322 (1993).

86. G. Cohen, R. Ren, and D. Baltimore, Modular binding domains in signal transduction proteins. *Cell* **80**, 237–248 (1995).

87. M. Matsuda, B. J. Mayer, Y. Fukui, and H. Hanafusa, Binding of transforming protein P47gag-crk to a broad range of phosphotyrosine-containing proteins. *Science* **248**, 1537–1539 (1990).

88. S. R. Hubbard, M. Mohammadi, and J. Schlessinger, Autoregulatory mechanisms in protein-tyrosine kinases. *J Biol Chem* **273**, 11987–11990 (1998).

89. S. Kousteni, T. Bellido, L. I. Plotkin, C. A. O'Brien, D. L. Bodenner, L. Han, K. Han, G. B. DiGregorio, J. A. Katzenellenbogen, B. S. Katzenellenbogen, P. K. Roberson, R. S. Weinstein, R. L. Jilka, and S. C. Manolagas, Nongenotropic sex-nonspecific signaling through the estrogen or androgen receptors: Dissociation from transcriptional activity. *Cell* **104**, 719–730 (2001).

90. V. Boonyaratanakornkit, M. P. Scott, V. Ribon, L. Sherman, S. M. Anderson, J. L. Maller, W. T. Miller, and D. P. Edwards, Progesterone receptor contains a proline-rich motif that directly interacts with SH3 domains and activates c-Src family tyrosine kinases. *Mol Cell* **8**, 269–280 (2001).

91. A. Migliaccio, G. Castoria, M. Di Domenico, A. de Falco, A. Bilancio, M. Lombardi, M. V. Barone, D. Ametrano, M. S. Zannini, C. Abbondanza, and F. Auricchio, Steroid-induced androgen receptor–oestradiol receptor {beta}–Src complex triggers prostate cancer cell proliferation. *EMBO J* **19**, 5406–5417 (2000).

92. G. Castoria, A. Migliaccio, A. Bilancio, M. Di Domenico, A. de Falco, M. Lombardi, R. Fiorentino, L. Varricchio, M. V. Barone, and F. Auricchio, PI3-kinase in concert with Src promotes the S-phase entry of oestradiol-stimulated MCF-7 cells. *EMBO J* **20**, 6050–6059 (2001).

93. C. W. Wong, C. McNally, E. Nickbarg, B. S. Komm, and B. J. Cheskis, Estrogen receptor-interacting protein that modulates its nongenomic activity-crosstalk with Src/Erk phosphorylation cascade. *Proc Natl Acad Sci USA* **99**, 14783–14788 (2002).

94. C. Ballare, M. Uhrig, T. Bechtold, E. Sancho, M. Di Domenico, A. Migliaccio, F. Auricchio, and M. Beato, Two domains of the progesterone receptor interact with the estrogen receptor and are required for progesterone activation of the c-Src/Erk pathway in mammalian cells. *Mol Cell Biol* **23**, 1994–2008 (2003).

95. I. Joung, J. L. Strominger, and J. Shin, Molecular cloning of a phosphotyrosine-independent ligand of the p56lck SH2 domain. *Proc Natl Acad Sci USA* **93**, 5991–5995 (1996).

96. R. K. Vadlamudi, R. A. Wang, A. Mazumdar, Y. Kim, J. Shin, A. Sahin, and R. Kumar, Molecular cloning and characterization of PELP1, a novel human coregulator of estrogen receptor alpha. *J Biol Chem* **276**, 38272–38279 (2001).

97. D. M Heery, E. Kalkhoven, S. Hoare, and M. G. Parker, A signature motif in transcriptional co-activators mediates binding to nuclear receptors. *Nature* **387**, 733–736 (2001).

98. E. Unni, S. Sun, B. Nan, M. J. McPhaul, B. Cheskis, M. A. Mancini, and M. Marcelli, Changes in androgen receptor nongenotropic signaling correlate with transition of LNCaP cells to androgen independence. *Cancer Res* **64**, 7156–7168 (2004).

99. L. B. Lutz, L. M. Cole, M. K. Gupta, K. W. Kwist, R. J. Auchus, and S. R. Hammes, Evidence that androgens are the primary steroids produced by *Xenopus laevis* ovaries and may signal through the classical androgen receptor to promote oocyte maturation. *Proc Natl Acad Sci USA* **98**, 13728–13733 (2001).

100. D. Haas, S. N. White, L. B. Lutz, M. Rasar, and S. R. Hammes, The modulator of nongenomic actions of the estrogen receptor (MNAR) regulates transcription-independent androgen receptor-mediated signaling: Evidence that MNAR participates in G protein-regulated meiosis in *Xenopus laevis* oocytes. *Mol Endocrinol* **19**, 2035–2046 (2005).

101. J. J. Watters, J. S. Campbell, M. J. Cunningham, E. G. Krebs, and D. M. Dorsa, Rapid membrane effects of steroids in neuroblastoma cells: Effects of estrogen on mitogen activated protein kinase signalling cascade and c-fos immediate early gene transcription. *Endocrinology* **138**, 4030–4033 (1997).

102. H. Endoh, H. Sasaki, K. Maruyama, K. Takeyama, I. Waga, T. Shimizu, S. Kato, and H. Kawashima, Rapid activation of MAP kinase by estrogen in the bone cell line. *Biochem Biophys Res Commun* **235**, 99–102 (1997).

103. E. Garcia Dos Santos, M. N. Dieudonne, R. Pecquery, V. Le Moal, Y. Giudicelli, and D. Lacasa, Rapid nongenomic

E2 effects on p42/p44 MAPK activator protein-1 and cAMP response element binding protein in rat white adipocytes. *Endocrinology* **143,** 930–940 (2002).

104. M. Di Domenico, G. Castoria, A. Bilancio, A. Migliaccio, and F. Auricchio, Estradiol activation of human colon carcinoma-derived Caco-2 cell growth. *Cancer Res* **56,** 4516–4521 (1996).

105. A. Migliaccio, M. Di Domenico, G. Castoria, A. de Falco, P. Bontempo, E. Nola, and F. Auricchio, Tyrosine kinase/p21ras/MAP-kinase pathway activation by estradiol receptor complex in MCF-7 cells. *EMBO J* **15,** 1292–1300 (1996).

106. J.-R. Chen, L. I. Plotkin, J. I. Aguirre, L. Han, R. L. Jilka, S. Kousteni, T. Bellido, and S. C. Manolagas, Transient versus sustained phosphorylation and nuclear accumulation of ERKs underlie anti- versus pro-apoptotic effects of estrogens. *J Biol Chem* **280,** 4632–4638 (2005).

107. C. J. Marshall, Specificity of receptor tyrosine kinase signaling: Transient versus sustained extracellular signal-regulated kinase activation. *Cell* **80,** 179–185 (1995).

108. R. H. Karas, H. Schulten, G. Pare, M. J. Aronovitz, C. Ohlsson, J. A. Gustafsson, and M. E. Mendelsohn, Effects of estrogen on the vascular injury response in estrogen receptor alpha beta (double) knockout mice. *Circ Res* **89,** 534–539 (2001).

109. R. A. Campbell, P. Bhat-Nakshatri, N. M. Patel, D. Constantinidou, S. Ali, and H. Nakshatri, PI3 kinase/AKT-mediated activation of estrogen receptor alpha: A new model for antiestrogen resistance. *J Biol Chem* **27,** 9817–9824 (2001).

110. R. Duan, W. Xie, R. C. Burghardt, and S. Safe, Estrogen receptor-mediated activation of the serum response element in MCF-7 cells through MAPK-dependent phosphorylation of Elk-1. *J Biol Chem* **276,** 11590–11598 (2001).

111. R. Duan, W. Xie, X. Li, A. McDougal, and S. Safe, Estrogen regulation of c-fos gene expression through phosphatidylinositol-3-kinase-dependent activation of serum response factor in MCF-7 breast cancer cells. *Biochem Biophys Res Commun* **294,** 384–394 (2002).

112. C. A. Lange, T. Shen, and K. B. Horwitz, Phosphorylation of human progesterone receptors at serine-294 by mitogen-activated protein kinase signals their degradation by the 26S proteasome. *Proc Natl Acad Sci USA* **97,** 1032–1037 (2000).

113. M. Gianni, A. Bauer, E. Garattini, P. Chambon, and C. Rochette-Egly, Phosphorylation by p38MAPK and recruitment of SUG-1 are required for RA-induced RAR gamma degradation and transactivation. *EMBO J* **21,** 3760–3769 (2002).

114. W. Feng, P. Webb, P. Nguyen, X. Liu, J. Li, M. Karin, and P. J. Kushner, Potentiation of estrogen receptor activation function 1 (AF-1) by Src/JNK through a serine 118-independent pathway. *Mol Endocrinol* **15,** 32–45 (2001).

115. C. Rochette-Egly, Nuclear receptors: Integration of multiple signalling pathways through phosphorylation. *Cell Signal* **15,** 355–366 (2003).

116. B. G. Rowan, N. L. Weigel, and B. W. O'Malley, Phosphorylation of steroid receptor coactivator-1. Identification of the phosphorylation sites and phosphorylation through the mitogen-activated protein kinase pathway. *J Biol Chem* **275,** 4475–4483 (2000).

117. G. N. Lopez, C. W. Turck, F. Schaufele, M. R. Stallcup, and P. J. Kushner, Growth factors signal to steroid receptors through mitogen-activated protein kinase regulation of p160 coactivator activity. *J Biol Chem* **276,** 22177–22182 (2001).

118. A. Migliaccio, M. Di Domenico, G. Castoria, M. Nanayakkara, M. Lombardi, A. de Falco, A. Bilancio, L. Varriccchio, A. Ciociola, and F. Auricchio, Steroid receptor regulation of epidermal growth factor signaling through Src in breast and prostate cancer cells: Steroid antagonist action. *Cancer Res* **65,** 10585–10593 (2005).

119. G. Tremblay, A. Tremblay, N. Copeland, D. Gilbert, N. Jenkins, F. Labrie, and V. Giguerc, Cloning chromosomal localization and functional analysis of the murine estrogen receptor β. *Mol Endocrinol* **11,** 353–365 (1997).

120. E. Enmark, P. Pelto-Huikko, K. Grandien, S. Lagercrantz, J. Lagercrantz, G. Fried, M. Nordenskjold, and J.-A. Gustafsson, Human estrogen receptor β-gene structure, chromosomal localization, and expression pattern. *J Clin Endocrinol Metab* **82,** 4258–4265 (1997).

121. R. Bhat, D. Harnish, P. Stevis, C. Lyttle, and B. Komm, A novel human estrogen receptor β: Identification and functional analysis of additional N-terminal amino acids. *J Steroid Biochem Mol Biol* **67,** 233–240 (1998).

122. D. Chen, P. Pace, R. Coombes, and S. Ali, Phosphorylation of human estrogen receptor α by protein kinase A regulates dimerization. *Mol Cell Biol* **19,** 1002–1015 (1999).

123. P. Shughrue, M. Lane, P. Scrimo, and I. Mechenthaler, Comparative distribution of estrogen receptor-α (ER-α) and -β (ER-β) mRNA in the rat pituitary gonad and reproductive tract. *Steroids* **63,** 498–504 (1998).

124. S. Cowley, S. Hoare, S. Mosselman, and M. Parker, Estrogen receptors α and β form heterodimers on DNA. *J Biol Chem* **272,** 19858–19862 (1997).

125. G. Kuiper, S. Nilsson, and J.-A. Gustafsson, Characteristics and function of the novel estrogen receptor β. *Horm Signal* **1,** 89–112 (1998).

126. A. L. Rodriguez, A. Tamrazi, M. L. Collins, and J. A. Katzenellenbogen, Design synthesis and *in vitro* biological evaluation of small molecule inhibitors of estrogen receptor alpha coactivator binding. *J Med Chem* **47,** 600–611 (2004).

127. J. Sun, M. Meyers, B. Fink, R. Rajendran, J. Katzenbellenbogen, and B. Katzenbellenbogen, Novel ligands that function as selective estrogens or antiestrogens for estrogen receptor-α or estrogen receptor-β. *Endocrinology* **140,** 800–804 (1999).

128. M. DeAngelis, F. Stossi, K. A. Carlson, B. S. Katzenellenbogen, and J. A. Katzenellenbogen, Indazole estrogens: Highly selective ligands for the estrogen receptor beta. *J Med Chem* **48,** 1132–1144 (2005).

129. H. A. Harris, J. A. Katzenellenbogen, and B. S. Katzenellenbogen, Characterization of the biological roles of the estrogen receptors ER alpha and ER beta in estrogen target tissues *in vivo* through the use of an ER alpha selective ligand. *Endocrinology* **143,** 4172–4177 (2002).

130. W. R. Harrington, S. Sheng, D. H. Barnett, L. N. Petz, J. A. Katzenellenbogen, and B. S. Katzenellenbogen, Activities of estrogen receptor α- and β-selective ligands at diverse estrogen responsive gene sites mediating transactivation or transrepression. *Mol Cell Endocrinol* **206,** 13–22 (2003).

131. A. Brzozowski, A. Pike, Z. Dauter, R. Hubbard, T. Bonn, O. Engstrom, L. Ohman, G. Greene, J.-A. Gustafsson, and M. Carlquist, Molecular basis of agonism and antagonism in the oestrogen receptor. *Nature* **389,** 753–758 (1997).

132. A. Pike, A. Brzozowski, R. Hubbard, T. Bonn, A.-G. Thorsell, O. Engstrom, J. Ljunggren, J.-A. Gustafsson, and M. Carlquist, Structure of the ligand-binding domain of oestrogen receptor beta in the presence of a partial agonist and a full antagonist. *EMBO J* **18,** 4608–4618 (1999).

133. W. Gradishar and V. Jordan, Clinical potential of new antiestrogens. *J Clin Oncol* **15,** 840–852 (1997).

134. R. Love, H. Barden, R. Mazess, S. Epstein, and R. Chappell, Effect of tamoxifen on lumbar spine bone mineral density in

postmenopausal women after 5 years. *Arch Intern Med* **154,** 2585–2588 (1994).

135. C. A. Frolik, H. U. Bryant, E. C. Black, D. E. Magee, and S. Chandrasekhar, Time-dependent changes in biochemical bone markers and serum cholesterol in ovariectomized rats: Effects of raloxifene HC1, tamoxifen, estrogen, and alendronate. *Bone* **18,** 621–627 (1996).

136. P. Carthew, R. Edwards, and B. Nolan, Uterotrophic effects of tamoxifen, toremifene, and raloxifene do not predict endometrial cell proliferation in the ovariectomized CD1 mouse. *Toxicol Appl Pharmacol* **158,** 24–32 (1999).

137. T. Nickelsen, E. G. Lufkin, B. L. Riggs, D. A. Cox, and T. H. Crook, Raloxifene hydrochloride, a selective estrogen receptor modulator: Safety assessment of effects on cognitive function and mood in postmenopausal women. *Psychoneuroendocrinology* **24,** 115–128 (1999).

138. H. Z. Ke, H. Qi, D. T. Crawford, K. L. Chidsey-Frink, H. A. Simmons, and D. D. Thompson, Lasofoxifene (CP-336,156), a selective estrogen receptor modulator, prevents bone loss induced by aging and orchidectomy in the adult rat. *Endocrinology* **141,** 1338–1344 (2000).

139. B. S. Komm, Y. P. Kharode, P. V. Bodine, H. A. Harris, C. P. Miller, and C. R. Lyttle, Bazedoxifene acetate: A selective estrogen receptor modulator with improved selectivity. *Endocrinology* **146**(9), 3999–4008 (2005).

140. S. K. Voipro, J. Komi, L. Kangas, K. Halonen, M. W. DeGregorio, and R. V. Erkkola, Effects of ospemifene (FC-1271a) on uterine endometrium, vaginal maturation index, and hormonal status in healthy postmenopausal women. *Maturitas* **43,** 207–214 (2002).

141. P. E. Goss, S. Qi, A. M. Cheung, H. Hu, M. Mendes, and K. P. A. Pritzker, The selective estrogen receptor modulator SCH57068 prevents bone loss, reduces serum cholesterol, and blocks estrogen-induced uterine hypertrophy in ovariectomized rats. *J Steroid Biochem Mol Biol* **92,** 79–87 (2004).

142. M. S. K. Sutherland, S. G. Lipps, N. Patnnik, L. M. Guyo-Fung, S. Khammungkune, W. Xie, H. A. Brady, M. S. Barbosa, D. W. Anderson, and B. Stein, SP500263, a novel SERM, blocks osteoclastogenesis in a human bone cell model: Role of IL-6 and GM-CSF. *Cytokine* **23,** 1–14 (2003).

143. P. Amman, S. Bourrin, F. Brunner, J.-M. Meyer, P. Clement-Lacroix, R. Baron, M. Gaillard, and R. Rizzoli, A new selective estrogen receptor modulator HMR-3339 fully corrects bone alterations induced by ovariectomy in adult rats. *Bone* **55,** 151–161 (2004).

144. J. Norris, L. Paige, D. Christensen, C.-Y. Chang, M. Huacani, D. Fan, P. Hamilton, D. Fowlkes, and D. McDonnell, Peptide antagonists of the human estrogen receptor. *Science* **285,** 744–746 (1999).

145. D. Lubahn, J. Moyer, T. Golding, J. Couse, K. Korach, and O. Smithies, Alteration of reproductive function but not prenatal sexual development after insertional disruption of the mouse estrogen receptor gene. *Proc Natl Acad Sci USA* **90,** 11162–11166 (1993).

146. J. Krege, J. Hodgin, J. Couse, E. Enmark, M. Warner, J. Mahler, M. Sar, K. Korach, J.-A. Gustafsson, and O. Smithies, Generation and reproductive phenotypes of mice lacking estrogen receptor β. *Proc Natl Acad Sci USA* **95,** 15677–15682 (1998).

147. J. Couse and K. Korach, Estrogen receptor null mice: What have we learned and where will they lead us? *Endocr Rev* **20,** 358–417 (1999).

148. N. A. Sims, S. Dupont, A. Krust, P. Clement-Lacroix, D. Minet, M. Resche-Rigon, M. Gaillard-Kelly, and R. Barron, Deletion of estrogen receptors reveals a regulatory role for estrogen receptors-β in bone remodeling in females but not in males. *Bone* **30,** 18–25 (2002).

149. N. A. Sims, P. Clement-Lacroix, D. Minet, C. Fraslon-Venhulle, M. Gaillard-Kelly, M. Resche-Rigon, and R. Barron, A functional androgen receptor is not sufficient to allow estradiol to protect bone after gonadectomy in estradiol receptor-deficient mice. *J Clin Invest* **111,** 1319–1327 (2003).

150. K. C. L. Lee, H. Jessop, R. Suswillo, G. Zaman, and L. E. Lanyon, The adaptive response of bone to mechanical loading in female transgenic mice is deficient in the absence of estrogen receptor-α and -β. *J Endocrinol* **182,** 193–201 (2004).

151. E. Smith, J. Boyd, G. Frank, H. Takahashi, R. Cohen, B. Specker, T. Williams, D. Lubahn, and K. Korach, Estrogen resistance caused by a mutation in the estrogen-receptor gene in a man. *N Engl J Med* **331,** 1056–1061 (1994).

152. O. Vidal, M. K. Lindberg, K. Hollberg, D. J. Baylink, G. Andersson, D. B. Lubahn, S. Mohan, J.-A. Gustafsson, and C. Ohlsson, Estrogen receptor specificity in the regulation of skeletal growth and maturation in male mice. *Proc Natl Acad Sci USA* **97,** 5474–5479 (2000).

153. N. A. Sims, P. Clement-Lacroix, D. Minet, C. Fraslon-Venhulle, M. Gaillard-Kelly, M. Resche-Rigon, and R. Barron, A functional androgen receptor is not sufficient to allow estradiol to protect bone after gonadectomy in estradiol receptor-deficient mice. *J Clin Invest* **111,** 1319–1327 (2003).

154. R. T. Turner, B. L. Riggs, and T. C. Spelsberg, Skeletal effects of estrogen. *Endocr Rev* **15,** 275–300 (1994).

155. B. L. Riggs and L. J. I. Melton, The prevention and treatment of osteoporosis. *N Engl J Med* **327,** 620–627 (1992).

156. D. W. Dempster and R. Lindsay, Pathogenesis of osteoporosis. *Lancet* **341,** 797–805 (1993).

157. B. S. Komm and P. V. Bodine, The ongoing saga of osteoporosis treatment. *J Cell Biochem Suppl* **30/31,** 277–283 (1998).

158. B. L. Riggs, S. Khosla, and L. J. Melton 3rd, A unitary model for involutional osteoporosis: Estrogen deficiency causes both type I and type II osteoporosis in postmenopausal women and contributes to bone loss in aging men. *J Bone Miner Res* **13,** 763–773 (1998).

159. R. L. Prince, Counterpoint: Estrogen effects on calciotropic hormones and calcium homeostasis. *Endocr Rev* **15,** 301–309 (1994).

160. Y. Liel, S. Shany, P. Smirnoff, and B. Schwartz, Estrogen increases 1,25-dihydroxyvitamin D receptors expression and bioresponse in the rat duodenal mucosa. *Endocrinology* **140,** 280–285 (1999).

161. M. ten Bolscher, J. C. Netelenbos, R. Barto, L. M. Van Buuren, and W. J. F. Van der Vijgh, Estrogen regulation of intestinal calcium absorption in the intact and ovariectomized adult rat. *J Bone Miner Res* **14,** 1197–1202 (1999).

162. C. R. Draper, I. M. Dick, and R. L. Prince, The effect of estrogen deficiency on calcium balance in mature rats. *Calcif Tissue Int* **64,** 325–328 (1999).

163. M. J. Oursler, M. Kassem, R. Turner, B. L. Riggs, and T. C. Spelsberg, Regulation of bone cell function by gonadal steroids. In *Osteoporosis* (R. Marcus, D. Feldman, and J. Kelsey, eds.), pp. 237–260. Academic Press, San Diego (1996).

164. T. C. Spelsberg, M. Subramaniam, B. L. Riggs, and S. Khosla, The actions and interactions of sex steroids and growth factors cytokines on the skeleton. *Mol Endocrinol* **13,** 819–828 (1999).

165. D. J. Rickard, M. Subramaniam, and T. C. Spelsberg, Molecular and cellular mechanisms of estrogen action on the skeleton. *J Cell Biochem Suppl* **32/33,** 123–132 (1999).

166. R. Bland, Steroid hormone receptor expression and action in bone. *Clin Sci* **98,** 217–240 (2000).

167. B. L. Riggs and L. J. I. Melton, Involutional osteoporosis. *N Engl J Med* **314,** 1676–1686 (1986).

168. T. K. Gray, T. C. Flynn, K. M. Gray, and L. M. Nabell, 17beta-estradiol acts directly on the clonal osteoblastic cell line UMR106. *Proc Natl Acad Sci USA* **84,** 6267–6271 (1987).

169. J. B. Lian, G. S. Stein, E. Canalis, P. Gehron Robey, and A. L. Boskey, Bone formation: Osteoblast lineage cells, growth factors, matrix proteins, and the mineralization process. In *Primer on the Metabolic Bone Diseases and Disorders of Mineral Metabolism* (M. J. Favus, ed.), pp. 14–29. Lippincott Williams & Wilkins, Philadelphia (1999).

170. B. S. Komm, C. M. Terpening, D. J. Benz, K. A. Graeme, A. Gallegos, M. Korc, G. L. Greene, B. W. O'Malley, and M. R. Haussler, Estrogen binding receptor mRNA and biologic response in osteoblast-like osteosarcoma cells. *Science* **241,** 81–84 (1988).

171. E. F. Eriksen, D. S. Colvard, N. J. Berg, M. L. Graham, K. G. Mann, T. C. Spelsberg, and B. L. Riggs, Evidence of estrogen receptors in normal human osteoblast-like cells. *Science* **241,** 84–86 (1988).

172. F. S. Kaplan, M. D. Fallon, S. D. Boden, R. Schmidt, M. Senior, and J. G. Haddad, Estrogen receptors in bone in a patient with polyostotic fibrous dysplasia (McCune–Albright syndrome). *N Engl J Med* **319,** 421–425 (1988).

173. M. Ernst, C. Schmid, and E. R. Froesch, Enhanced osteoblast proliferation and collagen gene expression by estradiol. *Proc Natl Acad Sci USA* **85,** 2307–2310 (1988).

174. M. C. Etienne, J. L. Fischel, G. Milano, P. Formento, J. L. Formento, M. Francoual, M. Frenay, and M. Namer, Steroid receptors in human osteoblast-like cells. *Eur J Cancer* **26,** 807–810 (1990).

175. D. J. Benz, M. R. Haussler, and B. S. Komm, Estrogen binding and estrogenic responses in normal human osteoblast-like cells. *J Bone Miner Res* **6,** 531–541 (1991).

176. P. E. Keeting, R. E. Scott, D. S. Colvard, M. A. Anderson, M. J. Oursler, T. C. Spelsberg, and B. L. Riggs, Development and characterization of a rapidly proliferating well-differentiated cell line derived from normal adult human osteoblast-like cells transfected with SV40 large T antigen. *J Bone Miner Res* **7,** 127–136 (1992).

177. A. Masuyama, Y. Ouchi, T. Sato, T. Hosoi, T. Nakamura, and H. Orimo, Characteristics of steroid receptors in cultured MC3T3-E1 osteoblastic cells and effect of steroid hormones on cell proliferation. *Calcif Tissue Int* **51,** 376–381 (1992).

178. S. Migliaccio, V. L. Davis, M. K. Gibson, T. K. Gray, and K. S. Korach, Estrogens modulate the responsiveness of osteoblast-like cells stably transfected with estrogen receptor. *Endocrinology* **130,** 2617–2624 (1992).

179. V. L. Davis, J. F. Couse, T. K. Gray, and K. S. Korach, Correlation between low levels of estrogen receptors and estrogen responsiveness in two rat osteoblast-like cell lines. *J Bone Miner Res* **9,** 983–991 (1994).

180. P. V. Bodine, J. Green, H. A. Harris, R. A. Bhat, G. S. Stein, J. B. Lian, and B. S. Komm, Functional properties of a conditionally phenotypic estrogen-responsive human osteoblast cell line. *J Cell Biochem* **65,** 368–387 (1997).

181. D. R. Ciocca and L. M. Vargas Roig, Estrogen receptors in human nontarget tissues: Biological and clinical implications. *Endocr Rev* **16,** 35–62 (1995).

182. S. Nilsson, G. Kuiper, and J.-A. Gustafsson, ER-beta: A novel estrogen receptor offers the potential for new drug development. *Trends Endocrinol Metab* **9,** 387–395 (1998).

183. J. A. Hoyland, A. P. Mee, P. Baird, I. P. Braidman, E. B. Mawer, and A. J. Freemont, Demonstration of estrogen receptor mRNA in bone using *in situ* reverse-transcriptase polymerase chain reaction. *Bone* **20,** 87–92 (1997).

184. V. Kusec, A. S. Virdi, R. Prince, and J. T. Triffitt, Localization of estrogen receptor-alpha in human and rabbit skeletal tissues. *J Clin Endocrinol Metab* **83,** 2421–2428 (1998).

185. D. N. Petersen, G. T. Tkalcevic, P. H. Koza-Taylor, T. G. Turi, and T. A. Brown, Identification of estrogen receptor beta2, a functional variant of estrogen receptor beta expressed in normal rat tissues. *Endocrinology* **139,** 1082–1092 (1998).

186. S. K. Lim, Y. J. Won, H. C. Lee, K. B. Huh, and Y. S. Park, A PCR analysis of ER alpha and ER beta mRNA abundance in rats and the effect of ovariectomy. *J Bone Miner Res* **14,** 1189–1196 (1999).

187. O. Vidal, L. G. Kindblom, and C. Ohlsson, Expression and localization of estrogen receptor-beta in murine and human bone. *J Bone Miner Res* **14,** 923–929 (1999).

188. S. H. Windahl, M. Norgard, G. Kuiper, J. A. Gustafsson, and G. Andersson Cellular distribution of estrogen receptor beta in neonatal rat bone. *Bone* **26,** 117–121 (2000).

189. I. Braidman, C. Baris, L. Wood, P. Selby, J. Adams, A. Freemont, and J. Hoyland, Preliminary evidence for impaired estrogen receptor-alpha protein expression in osteoblasts and osteocytes from men with idiopathic osteoporosis. *Bone* **26,** 423–427 (2000).

190. Y. Onoe, C. Miyaura, H. Ohta, S. Nozawa, and T. Suda, Expression of estrogen receptor beta in rat bone. *Endocrinology* **138,** 4509–4512 (1997).

191. J. Arts, G. G. Kuiper, J. M. Janssen, J. A. Gustafsson, C. W. Lowik, H. A. Pols, and J. P. van Leeuwen, Differential expression of estrogen receptors alpha and beta mRNA during differentiation of human osteoblast SV-HFO cells. *Endocrinology* **138,** 5067–5070 (1997).

192. M. A. Ankrom, J. A. Patterson, P. Y. d'Avis, U. K. Vetter, M. R. Blackman, P. D. Sponseller, M. Tayback, P. G. Robey, J. R. Shapiro, and N. S. Fedarko, Age-related changes in human oestrogen receptor alpha function and levels in osteoblasts. *Biochem J* **333,** 787–794 (1998).

193. P. V. Bodine, R. A. Henderson, J. Green, M. Aronow, T. Owen, G. S. Stein, J. B. Lian, and B. S. Komm, Estrogen receptor-alpha is developmentally regulated during osteoblast differentiation and contributes to selective responsiveness of gene expression. *Endocrinology* **139,** 2048–2057 (1998).

194. P. V. N. Bodine, H. A. Harris, and B. S. Komm, Suppression of ligand-dependent estrogen receptor activity by bone-resorbing cytokines in human osteoblasts. *Endocrinology* **140,** 2439–2451 (1999).

195. J. M. Hall and D. P. McDonnell, The estrogen receptor beta-isoform (ERbeta) of the human estrogen receptor modulates ERalpha transcriptional activity and is a key regulator of the cellular response to estrogens and antiestrogens. *Endocrinology* **140,** 5566–5578 (1999).

196. R. R. Recker, Embryology, anatomy, and microstructure of bone. In *Disorders of Bone and Mineral Metabolism* (F. L. Coe and M. J. Favus, eds.), pp. 219–240. Raven Press, New York (1992).

197. E. M. Aarden, E. H. Burger, and P. J. Nijweide, Functions of osteocytes in bone. *J Cell Biochem* **55,** 287–299 (1994).

198. B. S. Noble and J. Reeve, Osteocyte function, osteocyte death, and bone fracture resistance. *Mol Cell Endocrinol* **159,** 7–13 (2000).

199. L. E. Lanyon, Osteocytes, strain detection, bone modeling and remodeling. *Calcif Tissue Int* **53**(Suppl. 1), S102–S107 (1993).

200. I. P. Braidman, L. K. Davenport, D. H. Carter, P. L. Selby, E. B. Mawer, and A. J. Freemont, Preliminary *in situ* identification

of estrogen target cells in bone. *J Bone Miner Res* **10**, 74–80 (1995).

201. J. A. Hoyland, C. Baris, L. Wood, P. Baird, P. L. Selby, A. J. Freemont, and I. P. Braidman, Effect of ovarian steroid deficiency on oestrogen receptor alpha expression in bone. *J Pathol* **188**, 294–303 (1999).

202. T. Ohashi, S. Kusuhara, and K. Ishida, Immunoelectron microscopic demonstration of estrogen receptors in osteogenic cells of Japanese quail. *Histochemistry* **96**, 41–44 (1991).

203. S. C. Manolagas and R. L. Jilka, Bone marrow cytokines and bone remodeling: Emerging insights into the pathophysiology of osteoporosis. *N Engl J Med* **332**, 305–311 (1995).

204. J. E. Aubin, Bone stem cells. *J Cell Biochem Suppl* **30/31**, 73–82 (1998).

205. R. O. C. Oreffo, V. Kusec, S. Romberg, and J. T. Triffitt, Human bone marrow osteoprogenitors express estrogen receptor-alpha and bone morphogenetic proteins 2 and 4 mRNA during osteoblastic differentiation. *J Cell Biochem* **75**, 382–392 (1999).

206. S. C. Dieudonne, T. Xu, J. Y. Chou, S. A. Kuznetsov, K. Satomura, M. Mankani, N. S. Fedarko, E. P. Smith, P. G. Robey, and M. F. Young, Immortalization and characterization of bone marrow stromal fibroblasts from a patient with a loss of function mutation in the estrogen receptor-alpha gene. *J Bone Miner Res* **13**, 598–608 (1998).

207. M. J. Oursler, Estrogen regulation of gene expression in osteoblasts and osteoclasts. *Crit Rev Euk Gene Expr* **8**, 125–140 (1998).

208. T. Suda, N. Takahashi, N. Udagawa, E. Jimi, M. T. Gillespie, and T. J. Martin, Modulation of osteclast differentiation and function by new members of the tumor necrosis factor receptor and ligand families. *Endocr Rev* **20**, 345–357 (1999).

209. J. M. Pensler, C. B. Langman, J. A. Radosevich, M. L. Maminta, M. Mangkornkanok, R. Higbee, and A. Molteni, Sex steroid hormone receptors in normal and dysplastic bone disorders in children. *J Bone Miner Res* **5**, 493–498 (1990).

210. M. J. Oursler, P. Osdoby, J. Pyfferoen, B. L. Riggs, and T. C. Spelsberg, Avian osteoclasts as estrogen target cells. *Proc Natl Acad Sci USA* **88**, 6613–6617 (1991).

211. M. J. Oursler, L. Pederson, L. Fitzpatrick, B. L. Riggs, and T. Spelsberg, Human giant cell tumors of the bone (osteoclastomas) are estrogen target cells. *Proc Natl Acad Sci USA* **91**, 5227–5231 (1994).

212. T. Sunyer, J. Lewis, P. Collin-Osdoby, and P. Osdoby, Estrogen's bone-protective effects may involve differential IL-1 receptor regulation in human osteoclast-like cells. *J Clin Invest* **103**, 1409–1418 (1999).

213. H. Mano, T. Yuasa, T. Kameda, K. Miyazawa, Y. Nakamaru, M. Shiokawa, Y. Mori, T. Yamada, K. Miyata, H. Shindo, H. Azuma, Y. Hakeda, and M. Kumegawa, Mammalian mature osteoclasts as estrogen target cells. *Biochem Biophys Res Commun* **223**, 637–642 (1996).

214. F. M. Collier, W. H. Huang, W. R. Holloway, J. M. Hodge, M. T. Gillespie, L. L. Daniels, M. H. Zheng, and G. C. Nicholson, Osteoclasts from human giant cell tumors of bone lack estrogen receptors. *Endocrinology* **139**, 1258–1267 (1998).

215. L. Pederson, M. Kremer, N. T. Foged, B. Winding, C. Ritchie, L. A. Fitzpatrick, and M. J. Oursler, Evidence of a correlation of estrogen receptor level and avian osteoclast estrogen responsiveness. *J Bone Miner Res* **12**, 742–752 (1997).

216. G. Fiorelli, F. Gori, M. Petilli, A. Tanini, S. Benvenuti, M. Serio, P. Bernabei, and M. L. Brandi, Functional estrogen receptors in a human preosteoclastic cell line. *Proc Natl Acad Sci USA* **92**, 2672–2676 (1995).

217. M. Kanatani, T. Sugimoto, Y. Takahashi, H. Kaji, R. Kitazawa, and K. Chihara, Estrogen via the estrogen receptor blocks

218. cAMP-mediated parathyroid hormone (PTH)-stimulated osteoclast formation. *J Bone Miner Res* **13**, 854–862 (1998).

218. H. Ben-Hur, G. Mor, I. Blickstein, I. Likhman, F. Kohen, R. Dgani, V. Insler, P. Yaffe, and A. Ornoy, Localization of estrogen receptors in long bones and vertebrae of human fetuses. *Calcif Tissue Int* **53**, 91–96 (1993).

219. H. Ben-Hur, H. H. Thole, A. Mashiah, V. Insler, V. Berman, E. Shezen, D. Elias, A. Zuckerman, and A. Ornoy, Estrogen, progesterone and testosterone receptors in human fetal cartilaginous tissue: Immunohistochemical studies. *Calcif Tissue Int* **60**, 520–526 (1997).

220. J. Kennedy, C. Baris, J. A. Hoyland, P. L. Selby, A. J. Freemont, and I. P. Braidman, Immunofluorescent localization of estrogen receptor-alpha in growth plates of rabbits but not in rats at sexual maturity. *Bone* **24**, 9–16 (1999).

221. L. O. Nilsson, A. Boman, L. Savendahl, G. Grigelioniene, C. Ohlsson, E. M. Ritzen, and J. Wroblewski, Demonstration of estrogen receptor-beta immunoreactivity in human growth plate cartilage. *J Clin Endocrinol Metab* **84**, 370–373 (1999).

222. N. Dayani, M. T. Corvol, P. Robel, B. Eychenne, B. Moncharmont, L. Tsagris, and R. Rappaport, Estrogen receptors in cultured rabbit articular chondrocytes: Influence of age. *J Steroid Biochem* **31**, 351–356 (1988).

223. B. A. Monaghan, F. S. Kaplan, C. R. Lyttle, M. D. Fallon, S. D. Boden, and J. G. Haddad, Estrogen receptors in fracture healing. *Clin Orthop Rel Res* **280**, 277–280 (1992).

224. H. Pinus, A. Ornoy, N. Patlas, P. Yaffe, and Z. Schwartz, Specific beta-estradiol binding in cartilage and serum from young mice and rats is age dependent. *Connective Tissue Res* **30**, 85–98 (1993).

225. E. Nasatzky, Z. Schwartz, W. A. Soskolne, B. P. Brooks, D. D. Dean, B. D. Boyan, and A. Ornoy, Evidence for receptors specific for 17beta-estradiol and testosterone in chondrocyte cultures. *Connective Tissue Res* **30**, 277–294 (1994).

226. M. L. Brandi, C. Crescioli, A. Tanini, U. Frediani, D. Agnusdei, and C. Gennari, Bone endothelial cells as estrogen targets. *Calcif Tissue Int* **53**, 312–317 (1993).

227. G. S. Stein and J. B. Lian, Molecular mechanisms mediating proliferation/differentiation interrelationships during progressive development of the osteoblastic phenotype. *Endocr Rev* **14**, 424–442 (1993).

228. J. Clover and M. Gowen, Are MG-63 and HOS TE85 human osteosarcoma cell lines representative models of the osteoblastic phenotype? *Bone* **15**, 585–591 (1994).

229. D. Modrowski, L. Miravet, M. Feuga, and P. J. Marie, Increased proliferation of osteoblast precursor cells in estrogen-deficient rats. *Am J Physiol* **264**, E190–E196 (1993).

230. K. C. Westerlind, G. K. Wakley, G. L. Evans, and R. T. Turner, Estrogen does not increase bone formation in growing rats. *Endocrinology* **133**, 2924–2934 (1993).

231. C. K. Watts and R. J. King, Overexpression of estrogen receptor in HTB 96 human osteosarcoma cells results in estrogen-induced growth inhibition and receptor cross talk. *J Bone Miner Res* **9**, 1251–1258 (1994).

232. E. Damien, J. S. Price, and L. E. Lanyon, The estrogen receptor's involvement in osteoblasts' adaptive response to mechanical strain. *J Bone Miner Res* **13**, 1275–1282 (1998).

233. J. A. Robinson, S. A. Harris, B. L. Riggs, and T. C. Spelsberg, Estrogen regulation of human osteoblastic cell proliferation and differentiation. *Endocrinology* **138**, 2919–2927 (1997).

234. A. Samuels, M. J. Perry, and J. H. Tobias, High-dose estrogen-induced osteogenesis in the mouse is partially suppressed by indomethacin. *Bone* **25**, 675–680 (1999).

235. Q. Qu, P. L. Harkonen, and H. K. Vaananen, Comparative effects of estrogen and antiestrogens on differentiation of osteoblasts in mouse bone marrow culture. *J Cell Biochem* **73,** 500–507.

236. D. Somjen, Y. Weisman, A. Harell, E. Berger, and A. M. Kaye, Direct and sex-specific stimulation by sex steroids of creatine kinase activity and DNA synthesis in rat bone. *Proc Natl Acad Sci USA* **86,** 3361–3365 (1989).

237. S. D. Bain, M. C. Bailey, D. L. Celino, M. M. Lantry, and M. W. Edwards, High-dose estrogen inhibits bone resorption and stimulates bone formation in the ovariectomized mouse. *J Bone Miner Res* **8,** 435–442 (1993).

238. C. H. Turner, Editorial: Do estrogens increase bone formation? *Bone* **12,** 305–306 (1991).

239. R. T. Turner, Mice estrogen and postmenopausal osteoporosis. *J Bone Miner Res* **14,** 187–191 (1999).

240. A. Gohel, M. B. McCarthy, and G. Gronowicz, Estrogen prevents glucocorticoid-induced apoptosis in osteoblasts *in vivo* and *in vitro*. *Endocrinology* **140,** 5339–5347 (1999).

241. H. J. Verhaar, C. A. Damen, S. A. Duursma, and B. A. Scheven, A comparison of the action of progestins and estrogen on the growth and differentiation of normal adult human osteoblast-like cells *in vitro*. *Bone* **15,** 307–311 (1994).

242. R. J. Majeska, J. T. Ryaby, and T. A. Einhorn, Direct modulation of osteoblastic activity with estrogen. *J Bone Joint Surg* **76,** 713–721 (1994).

243. J. A. Robinson, K. M. Waters, R. T. Turner, and T. C. Spelsberg, Direct action of naturally occurring estrogen metabolites on human osteoblastic cells. *J Bone Miner Res* **15,** 499–506 (2000).

244. T. Ikeda, C. Shigeno, R. Kasai, H. Kohno, S. Ohta, H. Okumura, J. Konishi, and T. Yamamuro, Ovariectomy decreases the mRNA levels of transforming growth factor-beta1 and increases the mRNA levels of osteocalcin in rat bone *in vivo*. *Biochem Biophys Res Commun* **194,** 1228–1233 (1993).

245. R. T. Turner, D. S. Colvard, and T. C. Spelsberg, Estrogen inhibition of periosteal bone formation in rat long bones: Down-regulation of gene expression for bone matrix proteins. *Endocrinology* **127,** 1346–1351 (1990).

246. M. Ernst, J. K. Heath, and G. A. Rodan, Estradiol effects on proliferation, messenger ribonucleic acid for collagen and insulin-like growth factor-I, and parathyroid hormone-stimulated adenylate cyclase activity in osteoblastic cells from calvariae and long bones. *Endocrinology* **125,** 825–833 (1989).

247. M. A. Salih, C. C. Liu, B. H. Arjmandi, and D. N. Kalu, Estrogen modulates the mRNA levels for cancellous bone protein of ovariectomized rats. *Bone Miner* **23,** 285–299 (1993).

248. K. C. Westerlind, T. J. Wronski, G. L. Evans, and R. T. Turner, The effect of long-term ovarian hormone deficiency on transforming growth factor-beta and bone matrix protein mRNA expression in rat femora. *Biochem Biophys Res Commun* **200,** 283–289 (1994).

249. R. T. Turner, P. Backup, P. J. Sherman, E. Hill, G. L. Evans, and T. C. Spelsberg, Mechanism of action of estrogen on intramembranous bone formation: Regulation of osteoblast differentiation and activity. *Endocrinology* **131,** 883–889 (1992).

250. M. Takeuchi, M. Tokin, and K. Nagata, Tamoxifen directly stimulates the mineralization of human osteoblast-like osteosarcoma cells through a pathway independent of estrogen response element. *Biochem Biophys Res Commun* **210,** 295–301 (1995).

251. M. Centrella, M. C. Horowitz, J. M. Wozney, and T. L. McCarthy, Transforming growth factor-B gene family members and bone. *Endocr Rev* **15,** 27–39 (1994).

252. T. A. Linkhart, S. Mohan, and D. J. Baylink, Growth factors for bone growth and repair: IGF, TGF-beta and BMP. *Bone* **19,** 1S–12S (1996).

253. R. N. Margolis, E. Canalis, and N. C. Partridge, Invited review of a workshop: Anabolic hormones in bone: Basic research and therapeutic potential. *J Clin Endocrinol Metab* **81,** 872–877 (1996).

254. J. M. Wozney and V. Rosen, Bone morphogenetic protein and bone morphogenetic protein gene family in bone formation and repair. *Clin Orthop Rel Res* **346,** 26–37 (1998).

255. L. F. Bonewald, Regulation and regulatory activities of transforming growth factor beta. *Crit Rev Euk Gene Expr* **9,** 33–44 (1999).

256. C. J. Rosen, Serum insulin-like growth factors and insulin-like growth factor-binding proteins: Clinical implications. *Clin Chem* **45,** 1384–1390 (1999).

257. J. L. Jones and D. R. Clemmons, Insulin-like growth factors and their binding proteins: Biological actions. *Endocr Rev* **16,** 3–34 (1995).

258. P. P.-C. Hu, M. B. Datto, and X.-F. Wang, Molecular mechanisms of transforming growth factor-beta signaling. *Endocr Rev* **19,** 349–363 (1998).

259. M. C. Horowitz, Cytokines and estrogen in bone: Anti-osteoporotic effects. *Science* **260,** 626–627 (1993).

260. R. Pacifici, Estrogen cytokines and pathogenesis of postmenopausal osteoporosis. *J Bone Miner Res* **11,** 1043–1051 (1996).

261. R. L. Jilka, Cytokines bone remodeling and estrogen deficiency: A 1998 update. *Bone* **23,** 75–81 (1998).

262. T. K. Gray, B. Lipes, T. Linkhart, S. Mohan, and D. J. Baylink, Transforming growth factor beta mediates the estrogen induced inhibition of UMR106 cell growth. *Connect Tissue Res* **20,** 23–32 (1989).

263. M. J. Oursler, C. Cortese, P. Keeting, M. A. Anderson, S. K. Bonde, B. L. Riggs, and T. C. Spelsberg, Modulation of transforming growth factor-beta production in normal human osteoblast-like cells by 17 beta-estradiol and parathyroid hormone. *Endocrinology* **129,** 3313–3320 (1991).

264. R. D. Finkelman, N. H. Bell, D. D. Strong, L. M. Demers, and D. J. Baylink, Ovariectomy selectively reduces the concentration of transforming growth factor β in rat bone: Implications for estrogen deficiency-associated bone loss. *Proc Natl Acad Sci USA* **89,** 12190–12193 (1992).

265. N. N. Yang, H. U. Bryant, S. Hardikar, M. Sato, R. J. Galvin, A. L. Glasebrook, and J. D. Termine, Estrogen and raloxifene stimulate transforming growth factor-beta 3 gene expression in rat bone: A potential mechanism for estrogen- or raloxifene-mediated bone maintenance. *Endocrinology* **137,** 2075–2084 (1996).

266. D. P. McDonnell, The molecular pharmacology of SERMs. *Trends Endocrinol Metab* **10,** 301–311 (1999).

267. N. N. Yang, M. Venugopalan, S. Hardikar, and A. Glasebrook, Identification of an estrogen response element activated by metabolites of 17beta-estradiol and raloxifene. *Science* **273,** 1222–1225 (1996).

268. K. R. Tau, T. E. Hefferan, K. M. Waters, J. A. Robinson, M. Subramaniam, B. L. Riggs, and T. C. Spelsberg, Estrogen regulation of a transforming growth factor-beta inducible early gene that inhibits deoxyribonucleic acid synthesis in human osteoblasts. *Endocrinology* **139,** 1346–1353 (1998).

269. M. Subramaniam, S. A. Harris, M. J. Oursler, K. Rasmussen, B. L. Riggs, and T. C. Spelsberg, Identification of a novel TGF-beta-regulated gene encoding a putative zinc finger protein in human osteoblasts. *Nucleic Acids Res* **23,** 4907–4912 (1995).

270. D. J. Rickard, L. C. Hofbauer, S. K. Bonde, F. Gori, T. C. Spelsberg, and B. L. Riggs, Bone morphogenetic protein-6 production in human osteoblastic cell lines. Selective regulation by estrogen. *J Clin Invest* **101**, 413–422 (1998).

271. A. van den Wijngaard, W. R. Mulder, R. Dijkema, C. J. C. Boersma, S. Mosselman, E. J. J. van Zoelen, and W. Olijve, Antiestrogens specifically up-regulate bone morphogenic protein-4 promoter activity in human osteoblastic cells. *Mol Endocrinol* **14**, 623–633 (2000).

272. T. K. Gray, S. Mohan, T. A. Linkhart, and D. J. Baylink, Estradiol stimulates *in vitro* the secretion of insulin-like growth factors by the clonal osteoblastic cell line UMR106. *Biochem Biophys Res Commun* **158**, 407–412 (1989).

273. M. C. Slootweg, D. Swolin, J. C. Netelenbos, O. G. Isaksson, and C. Ohlsson, Estrogen enhances growth hormone receptor expression and growth hormone action in rat osteosarcoma cells and human osteoblast-like cells. *J Endocrinol* **155**, 159–164 (1997).

274. R. T. Turner, L. S. Kidder, M. Zhang, S. A. Harris, K. C. Westerlind, A. Maran, and T. J. Wronski, Estrogen has rapid tissue-specific effects on rat bone. *J Appl Physiol* **86**, 1950–1958 (1999).

275. J. Erdmann, S. Storch, J. Pfeilschifter, P. Ochlich, R. Ziegler, and F. Bauss, Effects of estrogen on the concentration of insulin-like growth factor-1 in rat bone. *Bone* **22**, 503–507 (1998).

276. C. Schmid, M. Ernst, J. Zapf, and E. R. Froesch, Release of insulin-like growth factor carrier proteins by osteoblasts: Stimulation by estradiol and growth hormone. *Biochem Biophys Res Commun* **160**, 788–794 (1989).

277. M. Kassem, R. Okazaki, D. De Leon, S. A. Harris, J. A. Robinson, T. C. Spelsberg, C. A. Conover, and B. L. Riggs, Potential mechanism of estrogen-mediated decrease in bone formation: Estrogen increases production of inhibitory insulin-like growth factor-binding protein-4. *Proc Assoc Am Phys* **108**, 155–164 (1996).

278. Y. Kudo, M. Iwashita, T. Iguchi, Y. Takeda, N. Hizuka, K. Takano, and T. Muraki, Estrogen and parathyroid hormone regulate insulin-like growth factor binding protein-4 in SaOS-2 cells. *Life Sci* **61**, 165–170 (1997).

279. V. Hwa, Y. Oh, and R. G. Rosenfeld, The insulin-like growth factor binding protein (IGFBP) superfamily. *Endocr Rev* **20**, 761–787 (1999).

280. C. J. Rosen, D. Verault, C. Steffens, D. Cheleuitte, and J. Glowacki, Effects of age and estrogen status on the skeletal IGF regulatory system. *Endocrine* **7**, 77–80 (1997).

281. T. L. McCarthy, C. Ji, H. Shu, S. Casinghino, K. Crothers, P. Rotwein, and M. Centrella, 17beta-estradiol potently suppresses cAMP-induced insulin-like growth factor-I gene activation in primary rat osteoblast cultures. *J Biol Chem* **272**, 18132–18139 (1997).

282. G. Girasole, R. L. Jilka, G. Passeri, S. Boswell, G. Boder, D. C. Williams, and S. C. Manolagas, 17b-estradiol inhibits interleukin-6 production by bone marrow-derived stromal cells and osteoblastic cells *in vitro*: A potential mechanism for the antiosteoporotic effect of estrogens. *J Clin Invest* **89**, 883–891 (1992).

283. R. L. Jilka, G. Hangoc, G. Girasole, G. Passeri, D. C. Williams, J. S. Abrams, B. Boyce, H. Broxmeyer, and S. C. Manolagas, Increased osteoclast development after estrogen loss: Mediation by interleukin-6. *Science* **257**, 88–91 (1992).

284. D. Cheleuitte, S. Mizuno, and J. Glowacki, *In vitro* secretion of cytokines by human bone marrow: Effects of age and estrogen status. *J Clin Endocrinol Metab* **83**, 2043–2051 (1998).

285. S. T. Pottratz, T. Bellido, H. Mocharla, D. Crabb, and S. C. Manolagas, 17beta-estradiol inhibits expression of human interleukin-6 promoter–reporter constructs by a receptor-dependent mechanism. *J Clin Invest* **93**, 944–950 (1994).

286. L. I. McKay and J. A. Cidlowski, Molecular control of immune/inflammatory responses: Interactions between nuclear factor-kappaB and steroid receptor-signaling pathways. *Endocr Rev* **20**, 435–459 (1999).

287. L. R. Chaudhary, T. C. Spelsberg, and B. L. Riggs, Production of various cytokines by normal human osteoblast-like cells in response to interleukin-1 beta and tumor necrosis factor-alpha: Lack of regulation by 17 beta-estradiol. *Endocrinology* **130**, 2528–2534 (1992).

288. D. Rickard, G. Russell, and M. Gowen, Oestradiol inhibits the release of tumor necrosis factor but not interleukin-6 from adult human osteoblasts *in vitro*. *Osteoporosis Int* **2**, 94–102 (1992).

289. L. Rifas, J. S. Kenney, M. Marcelli, R. Pacifici, S.-L. Cheng, L. L. Dawson, and L. V. Avioli, Production of interleukin-6 in human osteoblasts and human bone marrow stromal cells: Evidence that induction by interleukin-1β and tumor necrosis factor-α is not regulated by ovarian steroids. *Endocrinology* **136**, 4056–4067 (1995).

290. M. Kassem, S. A. Harris, T. C. Spelsberg, and B. L. Riggs, Estrogen inhibits interleukin-6 production and gene expression in a human osteoblastic cell line with high levels of estrogen receptors. *J Bone Miner Res* **11**, 193–199 (1996).

291. T. Taga and T. Kishimoto, GP130 and the interleukin-6 family of cytokines. *Annu Rev Immunol* **15**, 797–819 (1997).

292. S. C. Lin, T. Yamate, Y. Taguchi, V. Z. Borba, G. Girasole, C. A. O'Brien, T. Bellido, E. Abe, and S. C. Manolagas, Regulation of the gp80 and gp130 subunits of the IL-6 receptor by sex steroids in the murine bone marrow. *J Clin Invest* **100**, 1980–1990 (1997).

293. L. C. Hofbauer, S. Khosla, C. R. Dunstan, D. L. Lacey, W. J. Boyle, and B. L. Riggs, The roles of osteoprotegerin and osteoprotegerin ligand in the paracrine regulation of bone resorption. *J Bone Miner Res* **15**, 2–12 (2000).

294. L. C. Hofbauer, S. Khosla, C. R. Dunstan, D. L. Lacey, T. C. Spelsberg, and B. L. Riggs, Estrogen stimulates gene expression and protein production of osteoprotegerin in human osteoblastic cells. *Endocrinology* **140**, 4367–4370 (1999).

295. S. Srivastava, M. N. Weitzmann, R. B. Kimble, M. Rizzo, M. Zahner, J. Milbrandt, F. P. Ross, and R. Pacifici, Estrogen blocks M-CSF gene expression and osteoclast formation by regulating phosphorylation of Egr-1 and its interaction with Sp-1. *J Clin Invest* **102**, 1850–1859 (1998).

296. T. Chevalley and R. Rizzoli, Influence de l'hormone parathyroidienne sur l'os. *Press Medicale* **28**, 547–553 (1999).

297. S. Fukayama and A. H. Tashjian, Direct modulation by estradiol of the response of human bone cells (SaOS-2) to human parathyroid hormone (PTH) and PTH-related protein. *Endocrinology* **124**, 397–401 (1989).

298. J. J. Monroe and A. H. J. Tashjian, Pretreatment with 17beta-estradiol attenuates basal- and PTH-stimulated membrane adenylyl cyclase activity in human osteoblast-like SAOS-2 cells. *Bichem Biophys Res Commun* **225**, 320–325 (1996).

299. C. C. Pilbeam, J. Klein-Nulend, and L. G. Raisz, Inhibition by 17β-estradiol of PTH-stimulated resorption and prostaglandin production in cultured neonatal mouse calvariae. *Biochem Biophys Res Commun* **163**, 1319–1324 (1989).

300. H. Kaji, T. Sugimoto, M. Kanatani, M. Nasu, and K. Chihara, Estrogen blocks parathyroid hormone (PTH)-stimulated osteoclast-like cell formation by selectively affecting PTH-responsive cyclic adenosine monophosphate pathway. *Endocrinology* **137**, 2217–2224 (1996).

301. F. Cosman, V. Shen, F. Xie, M. Seibel, A. Ratcliffe, and R. Lindsay, Estrogen protects against bone resorbing effects of parathyroid hormone infusion: Assessment by use of biochemical markers. *Annal Internal Med* **118,** 337–343 (1993).

302. M. Ernst, M. G. Parker, and G. A. Rodan, Functional estrogen receptors in osteoblastic cells demonstrated by transfection with a reporter gene containing an estrogen response element. *Mol Endocrinol* **5,** 1597–1606 (1991).

303. L. G. Rao, J. N. Wylie, M. S. K. Sutherland, and T. M. Murray, 17β-Estradiol and parathyroid hormone potentiate each other's stimulatory effects on alkaline phosphatase activity in SaOS-2 cells in a differentiation-dependent manner. *Endocrinology* **134,** 614–620 (1994).

304. C. Eielson, D. Kaplan, M. A. Mitnick, I. Paliwal, and K. Insogna, Estrogen modulates parathyroid hormone-induced fibronectin production in human and rat osteoblast-like cells. *Endocrinology* **135,** 1639–1644 (1994).

305. V. Behar, C. Nakamoto, Z. Greenberg, A. Bisello, L. J. Suva, M. Rosenblatt, and M. Chorev, Histidine at position 5 is the specificity "switch" between two parathyroid hormone receptor subtypes. *Endocrinology* **137,** 4217–4224 (1996).

306. L. A. Pivirotto, D. S. Cissel, and P. E. Keeting, Sex hormones mediate interleukin-1β production by human osteoblastic HOBIT cells. *Mol Cell Endocrinol* **111,** 67–74 (1995).

307. S. A. Harris, K. R. Tau, R. J. Enger, D. O. Toft, B. L. Riggs, and T. C. Spelsberg, Estrogen response in the hFOB 1.19 human fetal osteoblastic cell line stably transfected with the human estrogen receptor gene. *J Cell Biochem* **59,** 193–201 (1995).

308. Y. Liel, S. Kraus, J. Levy, and S. Shany, Evidence that estrogens modulate activity and increase the number of 1,25-dihydroxyvitamin D receptors in osteoblast-like cells (ROS 17/2.8). *Endocrinology* **130,** 2597–2601 (1992).

309. M. Ishibe, T. Nojima, T. Ishibashi, T. Koda, K. Kaneda, R. N. Rosier, and J. E. Puzas, 17β-Estradiol increases the receptor number and modulates the action of 1,25-dihydroxyvitamin D3 in human osteosarcoma-derived osteoblast-like cells. *Calcif Tissue Int* **57,** 430–435 (1995).

310. K. L. Kirkwood, R. Dziak, and P. G. Bradford, Inositol trisphosphate receptor gene expression and hormonal regulation in osteoblast-like cell lines and primary osteoblast cell cultures. *J Bone Miner Res* **11,** 1889–1896 (1996).

311. J. W. J. Putney and G. S. J. Bird, The inositol phosphate-calcium signaling system in nonexcitable cells. *Endocr Rev* **14,** 610–631 (1993).

312. K. L. Kirkwood, K. Homick, M. B. Dragon, and P. G. Bradford, Cloning and characterization of the type I inositol 1,4,5-trisphosphate receptor gene promoter: Regulation by 17beta-estradiol in osteoblasts. *J Biol Chem* **272,** 22425–22431 (1997).

313. K. E. Armour and S. H. Ralston, Estrogen upregulates endothelial constitutive nitric oxide synthase expression in human osteoblast-like cells. *Endocrinology* **139,** 799–802 (1998).

314. S. H. Ralston, L.-P. Ho, M. H. Helfrich, P. S. Grabowski, P. W. Johnston, and N. Benjamin, Nitric oxide: A cytokine-induced regulator of bone resorption. *J Bone Miner Res* **10,** 1040–1049 (1995).

315. S. J. Wimalawansa, G. DeMarco, P. Gangula, and C. Yallampalli, Nitric oxide donor alleviates ovariectomy-induced bone loss. *Bone* **18,** 301–304 (1996).

316. R. L. Van Bezooijen, C. Van der Bent, S. E. Papapoulos, and C. W. G. M. Lowik, Oestrogenic compounds modulate cytokine-induced nitric oxide production in mouse osteoblast-like cells. *J Pharm Pharmacol* **51,** 1409–1414 (1999).

317. D. S. Cissel, M. Murty, D. L. Whipkey, J. D. Blaha, G. M. Graeber, and P. E. Keeting, Estrogen pretreatment increases arachidonic acid release by bradykinin stimulated normal human osteoblast-like cells. *J Cell Biochem* **60,** 260–270 (1996).

318. L. F. Cooper and K. Uoshima, Differential estrogenic regulation of small Mr heat shock protein expression in osteoblasts. *J Biol Chem* **269,** 7869–7873 (1994).

319. D. D. Bankson, N. Rifai, M. E. Williams, L. M. Silverman, and T. K. Gray, Biochemical effects of 17 beta-estradiol on UMR106 cells. *Bone Miner* **6,** 55–63 (1989).

320. E. F. Eriksen, Normal and pathological remodeling of human trabecular bone: Three dimensional reconstruction of the remodeling sequence in normals and in metabolic bone disease. *Endocr Rev* **7,** 379–408 (1984).

321. S. W. Whitson, Tight junction formation in the osteon. *Clin Orthop Rel Res* **86,** 206–213 (1972).

322. A. Tomkinson, J. Reeve, R. W. Shaw, and B. S. Noble, The death of osteocytes via apoptosis accompanies estrogen withdrawal in human bone. *J Clin Endocrinol Metab* **82,** 3128–3135 (1997).

323. R. L. Duncan and C. H. Turner, Mechanotransduction and the functional response of bone to mechanical strain. *Calcif Tissue Int* **57,** 344–358 (1995).

324. A. Tomkinson, E. F. Gevers, J. M. Wit, J. Reeve, and B. S. Noble, The role of estrogen in the control of rat osteocyte apoptosis. *J Bone Miner Res* **13,** 1243–1250 (1998).

325. T. Ikeda, A. Yamaguchi, S. Yokose, Y. Nagal, H. Yamato, T. Nakamura, H. Tsurukami, T. Tanizawa, and S. Yoshiki, Changes in biological activity of bone cells in ovariectomized rats revealed by *in situ* hybridization. *J Bone Miner Res* **11,** 780–788 (1996).

326. P. V. Bodine, S. K. Vernon, and B. S. Komm, Establishment and hormonal regulation of a conditionally transformed pre-osteocytic cell line from adult human bone. *Endocrinology* **137,** 4592–4604 (1996).

327. M. Z. Cheng, G. Zaman, S. C. F. Rawlinson, R. F. L. Suswillo, and L. E. Lanyon, Mechanical loading and sex hormone interactions in organ cultures of rat ulna. *J Bone Miner Res* **11,** 502–511 (1996).

328. M. Z. Cheng, G. Zaman, S. C. F. Rawlinson, A. A. Pitsillides, R. F. L. Suswillo, and L. E. Lanyon, Enhancement by sex hormones of the osteoregulatory effects of mechanical loading and prostaglandins in explants of rat ulna. *J Bone Miner Res* **12,** 1424–1430 (1997).

329. K. C. Westerlind, T. J. Wronski, E. L. Ritman, Z.-P. Luo, K.-N. An, N. H. Bell, and R. T. Turner, Estrogen regulates the rate of bone turnover but bone balance in ovariectomized rats is modulated by prevailing mechanical strain. *Proc Natl Acad Sci USA* **94,** 4199–4204 (1997).

330. W. M. Kohrt, D. B. Snead, E. Slatopolsky, and S. J. J. Birge, Additive effects of weight-bearing exercise and estrogen on bone mineral density in older women. *J Bone Miner Res* **10,** 1303–1311 (1995).

331. M. J. Oursler, L. Pederson, J. Pyfferoen, P. Osdoby, L. Fitzpatrick, and T. C. Spelsberg, Estrogen modulation of avian osteoclast lysosomal gene expression. *Endocrinology* **132,** 1373–1380 (1993).

332. M. Kremer, J. Judd, B. Rifkin, J. Auszmann, and M. J. Oursler, Estrogen modulation of osteoclast lysosomal enzyme secretion. *J Cell Biochem* **57,** 271–279 (1995).

333. J. A. Robinson, B. L. Riggs, T. C. Spelsberg, and M. J. Oursler, Osteoclasts and transforming growth factor-beta: Estrogen-mediated isoform-specific regulation of production. *Endocrinology* **137,** 615–621 (1996).

334. M. H. Zheng, T.-T. A. Lau, R. Prince, A. Criddle, S. Wysocki, M. Beilharz, J. M. Papadimitriou, and D. J. Wood, 17beta-

Estradiol suppresses gene expression of tartrate-resistant acid phosphatase and carbonic anhydrase II in ovariectomized rats. *Calcif Tissue Int* **56**, 166–169 (1995).

335. J. P. Williams, H. C. Blair, M. A. McKenna, S. E. Jordan, and J. M. McDonald, Regulation of avian osteoclastic H+-ATPase and bone resorption by tamoxifen and calmodulin antagonists. Effects independent of steroid receptors. *J Biol Chem* **271**, 12488–12495 (1996).

336. B. F. Boyce, D. E. Hughes, K. R. Wright, L. Xing, and A. Dai, Recent advances in bone biology provide insight into the pathogenesis of bone diseases. *Lab Invest* **79**, 83–94 (1999).

337. S. C. Manolagas, Birth and death of bone cells: Basic regulatory mechanisms for the pathogenesis and treatment of osteoporosis. *Endocr Rev* **21**, 115–137 (2000).

338. D. E. Hughes, A. Dai, J. C. Tiffee, H. H. Li, G. R. Mundy, and B. F. Boyce, Estrogen promotes apoptosis of murine osteoclasts mediated by TGF-beta. *Nat Med* **2**, 1132–1136 (1996).

339. S. Zecchi-Orlandini, L. Formigli, A. Tani, S. Benvenuti, G. Fiorelli, L. Papucci, S. Capaccioli, G. E. Orlandini, and M. L. Brandi, 17 Beta-estradiol induces apoptosis in the preosteoclastic FLG 29.1 cell line. *Biochem Biophys Res Commun* **255**, 680–685 (1999).

340. G. Fiorelli, V. Martineti, F. Gori, S. Benvenuti, U. Frediani, L. Formigli, S. Zecchi, and M. L. Brandi, Heterogeneity of binding sites and bioeffects of raloxifene on the human leukemic cell line FLG 29.1. *Biochem Biophys Res Commun* **240**, 573–579 (1997).

341. Z. Khalkhali-Ellis, P. Collin-Osdoby, L. Li, M. L. Brandi, and P. Osdoby, A human homolog of the 150 kD avian osteoclast membrane antigen related to superoxide dismutase and essential for bone resorption is induced by developmental agents and opposed by estrogen in FLG 29.1 cells. *Calcif Tissue Int* **60**, 187–193 (1997).

342. C. Schiller, R. Gruber, K. Redlich, G. M. Ho, F. Katzgraber, M. Willheim, P. Pietschmann, and M. Peterlik, 17Beta-estradiol antagonizes effects of 1alpha,25-dihydroxyvitamin D3 on interleukin-6 production and osteoclast-like cell formation in mouse bone marrow primary cultures. *Endocrinology* **138**, 4567–4571 (1997).

343. M. M. Takahashi and T. Noumura, Sexually dimorphic and laterally asymmetric development of the embryonic duck syrinx: Effect of estrogen on *in vitro* cell proliferation and chondrogenesis. *Dev Biol* **121**, 417–422 (1987).

344. E. Nasatzky, Z. Schwartz, B. D. Boyan, W. A. Soskolne, and A. Ornoy, Sex-dependent effects of 17beta-estradiol on chondrocyte differentiation in culture. *J Cell Physiol* **154**, 359–367 (1993).

345. M.-T. Corvol, A. Carrascosa, L. Tsagris, O. Blanchard, and R. Rappaport, Evidence for a direct *in vitro* action of sex steroids on rabbit cartilage cells during skeletal growth: Influence of age and sex. *Endocrinology* **120**, 1422–1429 (1987).

346. O. Blanchard, L. Tsagris, R. Rappaport, G. Duval-Beaupere, and M.-T. Corvol, Age-dependent responsiveness of rabbit and human cartilage cells to sex steroids *in vitro*. *J Steroid Biochem Mol Biol* **40**, 711–716 (1991).

347. E. Bonnelye, J. M. Vanacker, T. Dittmar, A. Begue, X. Desbiens, D. T. Denhardt, J. E. Aubin, V. Laudet, and B. Fournier, The ERR-1 orphan receptor is a transcriptional activator expressed during bone development. *Mol Endocrinol* **11**, 905–916 (1997).

348. J. M. Vanacker, C. Delmarre, X. J. Guo, and V. Laudet, Activation of the osteopontin promoter by the orphan nuclear receptor estrogen receptor related alpha. *Cell Growth Differ* **9**, 1007–1014 (1998).

349. J. M. Vanacker, K. Pettersson, J. A. Gustafsson, and V. Laudet, Transcriptional targets shared by estrogen receptor-related receptors (ERRs) and estrogen receptor (ER) alpha but not by ERbeta. *EMBO J* **18**, 4270–4279 (1999).

350. E. R. Levin, Cellular functions of the plasma membrane estrogen receptor. *Trends Endocrinol Metab* **10**, 374–377 (1999).

351. M. J. Kelly and E. J. Wagner, Estrogen modulation of G-protein coupled receptors. *Trends Endocrinol Metab* **10**, 369–374 (1999).

352. M. Lieberherr, B. Grosse, M. Kachkache, and S. Balsan, Cell signaling and estrogens in female rat osteoblasts: A possible involvement of unconventional nonnuclear receptors. *J Bone Miner Res* **8**, 1365–1376 (1993).

353. V. Le Mellay, B. Grosse, and M. Lieberherr, Phospholipase Cβ and membrane action of calcitriol and estradiol. *J Biol Chem* **272**, 11902–11907 (1997).

354. V. Le Mellay, F. Lasmoles, and M. Lieberherr, Gα q/11 and Gβ gamma proteins and membrane signaling of calcitriol and estradiol. *J Cell Biochem* **75**, 138–146 (1999).

355. H. Endoh, H. Sasaki, K. Maruyama, K.-I. Takeyama, I. Waga, T. Shimizu, S. Kato, and H. Kawashima, Rapid activation of MAP kinase by estrogen in the bone cell line. *Biochem Biophys Res Commun* **235**, 99–102 (1997).

356. G. Fiorelli, F. Gori, U. Frediani, F. Franceschelli, A. Tanini, C. Tosti-Guerra, S. Benvenuti, L. Gennari, L. Becherini, and M. L. Brandi, Membrane binding sites and non-genomic effects of estrogen in cultured human pre-osteoclastic cells. *J Steroid Biochem Mol Biol* **59**, 233–240 (1996).

357. C. V. Gay, N. L. Kief, and P. J. Bekker, Effect of estrogen on acidification in osteoclasts. *Biochem Biophys Res Commun* **192**, 1251–1259 (1993).

358. K. D. Brubaker and C. V. Gay, Specific binding of estrogen to osteoclast surfaces. *Biochem Biophys Res Commun* **200**, 899–907 (1994).

359. K. D. Brubaker and C. V. Gay, Depolarization of osteoclast plasma membrane potential by 17 beta-estradiol. *J Bone Miner Res* **14**, 1861–1866 (1999).

360. A. Ikegami, S. Inoue, T. Hosoi, Y. Mizuno, T. Nakamura, Y. Ouchi, and H. Orimo, Immunohistochemical detection and Northern blot analysis of estrogen receptors in osteoblastic cells. *J Bone Miner Res* **8**, 1103–1109 (1993).

361. K. Grandien, M. Backdahl, O. Ljunggren, J.-A. Gustafsson, and A. Berkenstam, Estrogen target tissue determines alternative promoter utilization of the human estrogen receptor gene in osteoblasts and tumor cell lines. *Endocrinology* **136**, 2223–2229 (1995).

362. R. Delaveyne-Bitbol and M. Garabedian, *In vitro* responses to 17 beta-estradiol throughout pubertal maturation in female human bone cells. *J Bone Miner Res* **14**, 376–385 (1999).

363. M. E. Bolander, M. E. Joyce, S. E. Boden, B. Oliver, and A. Heydemann, Estrogen receptor mRNA expression during fracture healing in the rat detected by polymerase chain reaction amplification. In *Calcium Regulation and Bone Metabolism* (D. V. Cohn, F. H. Glorieux, and T. J. Martin, eds.). Elsevier, New York, pp. 382–387 (1990).

364. R. Gruber, K. Czerwenka, F. Wolf, G. M. Ho, M. Willheim, and M. Peterlik, Expression of the vitamin D receptor of estrogen and thyroid hormone receptor alpha- and beta-isoforms and of the androgen receptor in cultures of native mouse bone marrow and of stromal/osteoblastic cells. *Bone* **24**, 465–473 (1999).

365. A. Mahonen and P. H. Maenpaa, Steroid hormone modulation of vitamin D receptor levels in human MG-63 osteosarcoma cells. *Biochem Biophys Res Commun* **205**, 1179–1186 (1994).

366. T. Bellido, G. Girasole, G. Passeri, X. P. Yu, H. Mocharla, R. L. Jilka, A. Notides, and S. C. Manolagas, Demonstration of estrogen and vitamin D receptors in bone marrow-derived stromal cells: Up-regulation of the estrogen receptor by 1,25-dihydroxyvitamin-D3. *Endocrinology* **133,** 553–562 (1993).

367. Q. Qu, M. Perala-Heape, A. Kapanen, J. Dahllund, J. Salo, H. K. Vaananen, and P. Harkonen, Estrogen enhances differentiation of osteoblasts in mouse bone marrow culture. *Bone* **22,** 201–209 (1998).

368. W. H. Huang, A. T. Lau, L. L. Daniels, H. Fujii, U. Seydel, D. J. Wood, J. M. Papadimitriou, and M. H. Zheng, Detection of estrogen receptor alpha carbonic anhydrase II and tartrate-resistant acid phosphatase mRNAs in putative mononuclear osteoclast precursor cells of neonatal rats by fluorescence *in situ* hybridization. *J Mol Endocrinol* **20,** 211–219 (1998).

369. R. O. C. Oreffo, V. Kusec, A. S. Virdi, A. M. Flanagan, M. Grano, A. Zambonin-Zallone, and J. T. Triffitt, Expression of estrogen receptor-alpha in cells of the osteoclastic lineage. *Histochem Cell Biol* **111,** 125–133 (1999).

370. C. K. Watts, M. G. Parker, and R. J. King, Stable transfection of the oestrogen receptor gene into a human osteosarcoma cell line. *J Steroid Biochem* **34,** 483–490 (1989).

371. M. Z. Cheng, G. Zaman, S. C. F. Rawlinson, S. Mohan, D. J. Baylink, and L. E. Lanyon, Mechanical strain stimulates ROS cell proliferation through IGF-II and estrogen through IGF-I. *J Bone Miner Res* **14,** 1742–1750 (1999).

372. A. Ikegami, S. Inoue, T. Hosoi, M. Kaneki, Y. Mizuno, Y. Akedo, Y. Ouchi, and H. Orimo, Cell cycle-dependent expression of estrogen receptor and effect of estrogen on proliferation of synchronized human osteoblast-like osteosarcoma cells. *Endocrinology,* **135,** 782–789 (1994).

373. M. Ernst and G. A. Rodan, Estradiol regulation of insulin-like growth factor-I expression in osteoblastic cells: Evidence for transcriptional control. *Mol Endocrinol* **5,** 1081–1089 (1991).

374. M. Kassem, R. Okazaki, S. A. Harris, T. C. Spelsberg, C. A. Conover, and B. L. Riggs, Estrogen effects on insulin-like growth factor gene expression in a human osteoblastic cell line with high levels of estrogen receptor. *Calcif Tissue Int* **62,** 60–66 (1998).

375. Q. Qu, P. L. Harkonen, J. Monkkonen, and H. K. Vaananen, Conditioned medium of estrogen-treated osteoblasts inhibits osteoclast maturation and function *in vitro. Bone* **25,** 211–215 (1999).

376. B. Huo, D. A. Dossing, and M. T. Dimuzio, Generation and characterization of a human osteosarcoma cell line stably transfected with the human estrogen receptor gene. *J Bone Miner Res* **10,** 769–781 (1995).

377. S. Koka, T. M. Petro, and R. A. Reinhardt, Estrogen inhibits interleukin-1beta-induced interleukin-6 production by human osteoblast-like cells. *J Interferon Cytokine Res* **18,** 479–483 (1998).

378. G. Passeri, G. Girasole, R. L. Jilka, and S. C. Manolagas, Increased interleukin-6 production by murine bone marrow and bone cells after estrogen withdrawal. *Endocrinology* **133,** 822–828 (1993).

Androgens and Skeletal Biology: Basic Mechanisms

KRISTINE M. WIREN

I. INTRODUCTION

The obvious impact of menopause on skeletal health has focused much of the research describing the general action of gonadal steroids on the specific effects of estrogen in bone (see Chapter 12, Komm). However, androgens clearly have important beneficial effects, in both men and women, on skeletal development and on the maintenance of bone mass. Thus, it has been demonstrated that androgens (1) influence growth plate maturation and closure, helping to determine longitudinal bone growth during development; (2) mediate regulation of trabecular (cancellous) and cortical bone mass in a fashion distinct from estrogen, leading to a sexually dimorphic skeleton; (3) modulate peak bone mass acquisition; and (4) inhibit bone loss [1]. In castrate animals, replacement with non-aromatizable androgens (e.g., 5α-dihydrotestosterone [DHT]) yields beneficial effects that are clearly distinct from those observed with estrogen replacement [2, 3]. In intact females, blockade of the androgen receptor (AR) with the specific AR antagonist hydroxyflutamide results in osteopenia [4]. Furthermore, treatment with nonaromatizable androgen alone in females results in improvement in bone mineral density [5]. Finally, combination therapy with estrogen and androgen in postmenopausal women is more beneficial than either steroid alone [6–8], indicating nonparallel and distinct pathways of action. Combined, these reports illustrate the distinct actions of androgens and estrogens on the skeleton. Thus, in both men and women it is probable that androgens and estrogens each have important yet distinct functions during bone development and in the subsequent maintenance of skeletal homeostasis in the adult. With the awakening awareness of the importance of the effects of androgen on skeletal homeostasis, and the potential to make use of this information for the treatment of bone disorders, much remains to be learned.

II. ANDROGENS AND THE ROLE OF ANDROGEN METABOLISM

A. Metabolism of Androgens in Bone: 5α-Reductase, Aromatase, and 17β-Hydroxysteroid Dehydrogenase Activities

Sex steroids, ultimately derived from cholesterol, are synthesized predominantly in gonadal tissue, the adrenal gland, and placenta as a consequence of enzymatic conversions. After peripheral metabolism, androgenic activity is represented in a variety of steroid molecules that include testosterone (Figure 13-1). There is evidence in a range of tissues that the eventual cellular effects of testosterone may not be the result (or not only the result) of direct action of testosterone but may also reflect the effects of sex steroid metabolites formed as a consequence of local enzyme activities. The most important testosterone metabolites in bone are 5α-DHT (the result of 5α reduction of testosterone) and estradiol (formed by the aromatization of testosterone). Testosterone and DHT are the major and most potent androgens, with androstenedione (the major circulating androgen in women) and dehydroepiandrosterone (DHEA) as immediate androgen precursors that exhibit weak androgen activity [9]. In men, the most abundant circulating androgen metabolite is testosterone, whereas concentrations of other weaker androgens such as androstenedione and DHEA-sulfate are similar between males and females. Downstream metabolites of DHT and androstenedione are inactive at the AR and include 5α-androstane-3α or 3β,17β-diol (3α/β-androstanediol) and 5α-androstenedione. Data suggest that aromatase cytochrome P450 (the product of the CYP19 gene), 17β-hydroxysteroid dehydrogenase (17β-HSD), and 5α-reductase activities are all present

Peripheral androgen metabolism

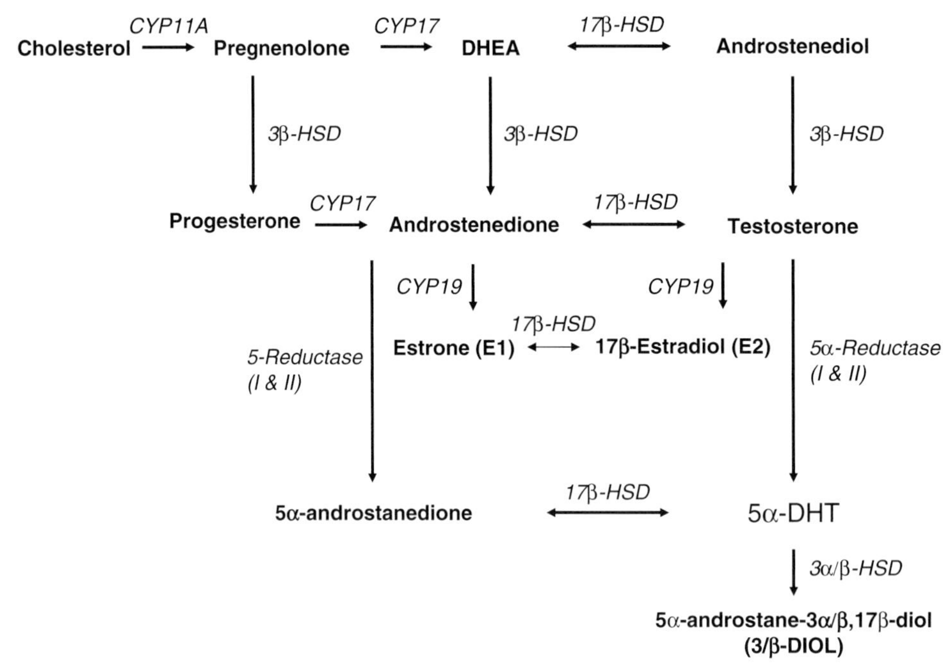

FIGURE 13-1 Principal conversions and major enzyme activities involved in androgen synthesis and metabolism. Steroid hormone synthesis involves metabolism of cholesterol, with dehydrogenation of pregnenolone producing progesterone that can serve as a precursor for the other gonadal steroid hormones. DHEA, dehydroepiandrosterone; CYP11A, cytochrome P450 cholesterol side chain cleavage enzyme; CYP17, cytochrome P450 17α-hydroxylase/17,20 lyase; 17β-HSD, 17β-hydroxysteroid dehydrogenase; CYP19, aromatase cytochrome P450.

in bone tissue, at least to some measurable extent in some compartments, but the biologic relevance of each remains somewhat controversial.

5α-Reductase is an important activity with regard to androgen metabolism in general since testosterone is converted to the more potent androgen metabolite DHT via 5α-reductase action [10]. 5α-Reductase activity was first described in crushed rat mandibular bone [11], with similar findings reported in crushed human spongiosa [12]. Two different 5α-reductase genes encode type 1 and type 2 isozymes in many mammalian species [13]; human osteoblastic cells express the type 1 isozyme [14]. Essentially the same metabolic activities were reported in experiments with human epiphyseal cartilage and chondrocytes [15]. In general, the K_m values for bone 5α-reductase activity are similar to those in other androgen responsive tissues [12, 16]. However, the cellular populations in many of these studies were mixed; hence, the specific cell type responsible for the activity is unknown. Interestingly, Turner *et al.* [17] found that periosteal cells do not have detectable 5α-reductase activity, raising the possibilities that the enzyme may be functional in only selected skeletal compartments and that testosterone may be the active androgen metabolite at this clinically important site.

From a clinical perspective, the general importance of this enzymatic pathway is uncertain because patients with 5α-reductase type 2 deficiency have normal bone mineral density [18], and Bruch *et al.* [10] found no significant correlation between enzyme activities and bone volume. In mutant null mice lacking 5α-reductase type 1 (mice express very little type 2 isozyme), the effect on the skeleton has not been analyzed due to midgestational fetal death as a consequence of estrogen excess [19]. Analysis of the importance of 5α-reductase activity has been approached with the use of finasteride (an inhibitor of 5α-reductase activity); treatment of male animals does not recapitulate the effects of castration [20], strongly suggesting that reduction of testosterone to DHT by 5α-reductase is not the major determinant in the effects of gonadal hormones on bone. Consistent with this finding, testosterone therapy in hypogonadal older men, either when administered alone or when combined with finasteride, increases bone mineral density, again suggesting that DHT is not essential for the beneficial effects of testosterone on bone [21]. Thus, the available clinical data remain uncertain, and the impact of this enzyme, which isozyme may be involved, whether it is uniformly present in all cell types involved in bone modeling/remodeling, or whether local activity is important at all remain unresolved issues.

Another important enzymatic arm of testosterone metabolism involves the biosynthesis of estrogens from androgen precursors, catalyzed by aromatase. Of note, this enzyme is well known to be both expressed and regulated in a very pronounced tissue-specific manner [22], and it also demonstrates species differences, given the low levels in mice. Modest levels of aromatase activity have been reported in bone from mixed cell populations derived from both sexes [23–25] and from osteoblastic cell lines [16, 26, 27]. Aromatase expression in intact bone has also been documented by *in situ* hybridization and immunohistochemical analysis [25]. Aromatase mRNA is expressed predominantly in lining cells, chondrocytes, and some adipocytes; however, there is no detectable expression in osteoclasts or in cortical bone in mice [28]. At least in vertebral bone, the mesenchymal distal promoter I.4 is predominantly utilized [29]. The enzyme kinetics in bone cells seem to be similar to those in other tissues, although the V_{max} may be increased by glucocorticoids [27]. Whether the level of aromatase activity in bone is sufficiently high to produce physiologically relevant concentrations of steroids remains an open question; nevertheless, in the male only 15% of circulating estrogen is produced in the testes, with the remaining 85% produced by peripheral metabolism that could include bone as one site of conversion [30].

Aromatase catalyzes the metabolism of adrenal and testicular C19 androgens (androstenedione and testosterone) to C18 estrogens (estrone and estradiol), thus producing the potent estrogen estradiol (E2) from testosterone and the weaker estrogen estrone (E1) from its adrenal precursors androstenedione and DHEA [23]. Typically, in the circulation, E2 will comprise up to 40% of total estrogen, E1 will comprise up to an additional 40%, with estriol (E3) comprising the remaining 20% of total estrogen [31]. In addition to aromatase, osteoblasts contain enzymes that are able to interconvert estradiol and estrone (17β-HSD) and to hydrolyze estrone sulfate, the most abundant estrogen in the circulation, to estrone (steroid sulfatase) [26, 32]. Nawata *et al.* [23] reported that dexamethasone and $1\alpha,25(OH)_2D_3$ synergistically enhance aromatase activity and aromatase mRNA expression in human osteoblast-like cells. In addition, both leptin and $1\alpha,25(OH)_2D_3$ treatment increased aromatase activity in human mesenchymal stem cells during osteogenesis but not during adipogenesis [33]. Additional studies are needed to better define aromatase expression, given the potential importance of the enzyme, and its regulation by a variety of mechanisms (including androgens and estrogens) in other tissues [22, 34].

The clinical impact of aromatase activity and an indication of the importance of conversion of circulating androgen into estrogen are demonstrated in reports of women and men with aromatase deficiencies who present with a skeletal phenotype [35]. Interestingly, natural mutation is remarkably rare, with only seven males and six females reported to date. The presentation of men with aromatase deficiency is very similar to that of a man with estrogen receptor-α (ERα) deficiency [36]—namely, an obvious delay in bone age, lack of epiphyseal closure, and tall stature with high bone turnover and osteopenia [30]—suggesting that aromatase (and likely estrogen action) has a substantial role to play during skeletal development in the male. In addition, estrogen therapy of males with aromatase deficiency has been associated with an increase in bone mass [30], particularly in the growing skeleton [37]. Inhibition of aromatization pharmacologically with nonsteroidal inhibitors (e.g., vorozole or letrozole) results in modest decreases in bone mineral density and changes in skeletal modeling in young growing orchidectomized males [38], and less dramatically so in boys with constitutional delay of puberty treated for 1 year [39], suggesting that short-term treatment during growth has limited negative consequences in males. Inhibition of aromatization in older orchidectomized males resembles castration, with similar increases in bone resorption and bone loss, suggesting that aromatase activity likely plays a role in skeletal maintenance in males [40]. These studies herald the importance of aromatase activity (and estrogen) in the mediation of some androgen action in bone in both males and females. The finding of these enzymes in bone clearly raises the difficult issue of the origin of androgenic effects in the skeleton: Do they arise solely from direct androgen effects (as is suggested by the actions of nonaromatizable androgens such as DHT) or also from the local or other site production of estrogenic intermediates? The results described previously seem to indicate that both steroids appear to be important to both male and female skeletal health.

The 17β-HSDs (most of which are dehydrogenase reductases, except type 5, which is an aldoketo reductase) have been shown to catalyze either the last step of sex steroid synthesis or the first step of their degradation (to produce weak or potent sex steroids via oxidation or reduction, respectively) and can thus also play a critical role in peripheral steroid metabolism. The oxidative pathway forms 17 ketosteroids, whereas the reductive pathway forms 17β-hydroxysteroids. The enzyme reversibly catalyzes the formation of androstenediol (an androgen) from DHEA, in addition to the biosynthesis of estradiol from estrone, the synthesis of testosterone from androstenedione, and the production of DHT from 5α-androstenedione all via the reductive activity of 17β-HSD. Of the 13 enzyme isotypes of 17β-HSD activity [31], types 1–4 have been demonstrated in human osteoblastic cells [41].

The administration of testosterone can stimulate bone formation and inhibit bone resorption, likely through multiple mechanisms that involve both androgen receptor (AR)- and estrogen receptor (ER)-mediated processes. However, there is substantial evidence that some, if not most, of the biologic actions of androgens in the skeleton are mediated by AR. Both *in vivo* and *in vitro* systems reveal the effects of the nonaromatizable androgen DHT to be essentially the same as those of testosterone (*vida infra*). In addition, blockade of the AR with the receptor antagonist flutamide results in osteopenia as a result of reduced bone formation [4]. In addition, complete androgen insensitivity results in a significant decrease in bone mineral density in spine and hip sites [18] even in the setting of strong compliance with estrogen treatment [42]. These reports clearly indicate that androgens, independent of estrogenic metabolites, have primary effects on osteoblast function. However, the clinical reports of subjects with aromatase deficiency also highlight the relevance of metabolism of androgen to biopotent estrogens, at least in the circulation, to influence bone development and/or maintenance. It thus seems likely that further elucidation of the regulation of steroid metabolism, and the potential mechanisms by which androgenic and estrogenic effects are coordinated, will have physiological, pathophysiological, and therapeutic implications.

B. Synthetic Androgens

In addition to the endogenous steroid metabolites highlighted in Figure 13-1, there are also a variety of drugs with androgenic activity. These include anabolic steroids, such as nonaromatizable oxandrolone, that bind and activate AR (albeit with lower affinity than testosterone [43]) and a class of drugs under extensive development referred to as selective AR modulators (SARMs), which demonstrate tissue-specific agonist or antagonist activities with respect to AR transactivation [44]. These orally active nonsteroidal, nonaromatizable SARMs are being developed to target androgen action in bone, muscle, and fat and to influence libido but to not exacerbate prostate growth, hirsutism, and acne. Several have been identified with beneficial effects on bone mass [45–47] and provide a new alternative to androgen replacement therapy.

III. CELLULAR BIOLOGY OF THE ANDROGEN RECEPTOR IN THE SKELETON

Because there remains confusion interpreting the skeletal actions of sex steroids as previously noted, the specific mechanisms by which androgens affect skeletal homeostasis are becoming the focus of intensified research [1, 48]. As a classic steroid hormone, the biological cellular signaling responses to androgen are mediated through the AR, a ligand-inducible transcription factor. ARs have been identified in a variety of cells found in bone [49]. Characterization of AR expression in these cells thus clearly identifies bone as a target tissue for androgen action. The direct effects of androgen that influence the complex processes of proliferation, differentiation, mineralization, and gene expression in the osteoblast are being characterized, but much remains to be established. Androgen effects on bone may also be indirectly modulated and/or mediated by other autocrine and paracrine factors in the bone microenvironment. The remainder of this chapter reviews progress on the characterization of androgen action in bone through AR signaling.

A. Molecular Mechanisms of Androgen Action in Bone Cells: The Androgen Receptor

Direct characterization of AR expression in a variety of tissues, including bone, was made possible by the cloning of the AR cDNA [50, 51]. The AR is a member of the class I (so-called classical or steroid) nuclear receptor superfamily, as are the ERα and ERβ isoforms, the progesterone receptor, and the mineralocorticoid and glucocorticoid receptors [52]. Steroid receptors are transcription factors with a highly conserved modular design characterized by three functional domains: the transactivation, DNA-binding, and ligand-binding domains. In the absence of ligand, the AR protein is generally localized in the cytoplasmic compartment of target cells in a large complex of molecular chaperones consisting of loosely bound heat shock, cyclophilin, and other accessory proteins [53]. Interestingly, in the unliganded form, AR conformation is unique with a relatively unstructured N-terminal transactivation domain [54]. As lipids, androgens can freely diffuse through the plasma membrane to bind the AR to induce a conformational change. Once bound by ligand, the AR dissociates from the multiprotein complex, translocates to the nucleus, and recruits coactivators or corepressors that demonstrate expression that can be cell-type specific [55], allowing the formation of homodimers (or potentially heterodimers) that activate a cascade of events in the nucleus [56]. Bound to DNA, the AR influences transcription and/or translation of a specific network of genes, leading to the specific cellular response to the steroid.

A steroid hormone target tissue is frequently defined as one that possesses the steroid receptor, at a functional level, with a measurable response in the presence of hormone. Bone tissue clearly meets this standard with respect to androgen. Colvard *et al.* [57] first reported the presence of AR mRNA and specific androgen binding sites in normal human osteoblastic cells. The abundance of both AR and ER proteins was similar, suggesting that androgens and estrogens each play important roles in skeletal physiology (Figure 13-2). Subsequent reports have confirmed AR mRNA expression and/or the presence of androgen binding sites in both normal and clonal transformed osteoblastic cells derived from a variety of species [16, 58–62]. The size of the AR mRNA transcript in osteoblasts (~10 kb) is similar to that described in prostate and other tissues [50], as is the size of the AR protein analyzed by Western blotting (~110 kDa) [16]. There are reports of two isoforms of AR protein in human osteoblast-like cells (~110 and ~97 kDa) [63] as first described in human prostatic tissue [64]. It appears that these isoforms do not possess similar functional activities in bone, particularly with respect to effects on proliferation [65]. The number of specific androgen binding sites in osteoblasts varies, depending on methodology and the cell source, from 1000 to 14,000 sites/cell [16, 61, 63, 66] but is in a range seen in other androgen target tissues. Furthermore, the binding affinity of the AR

found in osteoblastic cells (K_D = 0.5–2 × 10^{-9}) is typical of that found in other tissues. Androgen binding is specific, without significant competition by estrogen, progesterone, or dexamethasone [16, 57, 63]. Finally, testosterone and DHT appear to have relatively similar binding affinities [16, 58]. All these data are consistent with the notion that the direct biologic effects of androgenic steroids in osteoblasts are mediated at least in part via classic mechanisms associated with the AR as a member of the steroid hormone receptor superfamily described previously.

In addition to the classical AR present in bone cells, several other androgen-dependent signaling pathways have been described. Specific binding sites for weaker adrenal androgens (e.g., DHEA) have been described [67]; DHEA can also transactivate AR [9], thus raising the possibility that DHEA or similar androgenic compounds may also have direct effects in bone. DHEA and its metabolites may also bind and activate additional receptors, including ER, peroxisome proliferator activated receptor-α, and pregnane X receptor [68]. Bodine *et al.* [69] showed that DHEA caused a rapid inhibition of c-*fos* expression in human osteoblastic cells that was more robust than that seen with the classical androgens (DHT, testosterone, and androstenedione). In addition, DHEA may inhibit bone resorption by osteoclasts when in the presence of osteoblasts, likely through changes in osteoprotegerin (OPG) and receptor activator of NF-κB ligand (RANKL) concentrations [70]. Alternatively, androgens may be specifically bound in osteoblastic cells by a novel 63-kDa cytosolic protein [71]. In addition, there are reports of distinct AR polymorphisms identified in different races

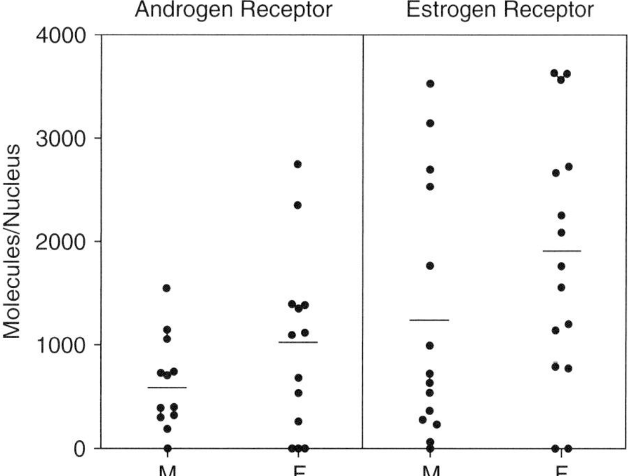

FIGURE 13-2   Nuclear androgen and estrogen receptor binding in normal human osteoblast-like cells. Solid circles represent the mean calculated number of molecules per cell nucleus for each cell strain. (Left) Specific nuclear binding of [^{3}H]R1881 (methyltrienolone, an androgen analog) in 12 strains from normal men and 13 strains from normal women. (Right) Specific nuclear [^{3}H]estradiol binding in 15 strains from men and 15 strains from women. The horizontal lines indicate the mean receptor concentrations. Adapted from Colvard *et al.* [57] with permission.

that may have biological impact on androgen responses [72], but to date none has been shown to have an effect with respect to bone tissue [73]. These different isoforms have the potential to interact in distinct fashions with other signaling molecules, such as c-Jun [74]. Finally, androgens may regulate osteoblast activity via rapid nongenomic mechanisms [75, 76] through membrane receptors displayed at the bone cell surface [77]. The role and biologic significance of these nonclassical signaling pathways in androgen-mediated responses in bone remain controversial, and most data suggest that genomic signaling may be the more significant regulator in bone and other tissues [78–81].

B. Localization of Androgen Receptor Expression in Osteoblastic Populations

Ultimately, bone mass is determined by two biological processes: formation and resorption. Distinct cell types mediate these processes. The bone-forming cell, the osteoblast, synthesizes bone matrix, regulates mineralization, and is responsive to most calciotropic hormones. The osteoclast is responsible for bone resorption. Clues about the potential sequelae of AR signaling might be derived from a better understanding of the cell types in which expression is documented. *In vivo* analysis has demonstrated significant expression of ARs in all cells of the osteoblast lineage, including osteoblasts, osteocytes, and in osteoclasts [82]. Interestingly, ARs are also expressed in bone marrow stromal [83] and mesenchymal precursor cells [84]—pluripotent cells that can differentiate into muscle, bone, and fat. Androgen action may modulate precursor differentiation toward the osteoblast and/or myoblast lineage while inhibiting differentiation toward the adipocyte lineage [85]. These effects on stromal differentiation could underlie some of the well-described consequences of androgen administration on body composition, including increased muscle mass [86]. However, the relevance of the increased muscle mass associated with androgen administration to positively influence bone quality remains unsolved.

In the bone microenvironment, the localization of AR expression has been described in intact human bone by Abu *et al.* [49] using immunocytochemical methods. In developing bone from young adults, ARs were predominantly expressed in active osteoblasts at sites of bone formation (Figure 13-3). ARs were also observed in osteocytes embedded in the bone matrix. Importantly, both the pattern of AR distribution and the level of expression were similar in males and in females. Furthermore, AR was observed in bone marrow and stromal/osteoblast precursor cells [83]. In addition,

expression of the AR has been characterized in cultured osteoblastic cell populations isolated from bone biopsy specimens, determined at the mRNA level and by binding analysis [63]. Expression varied according to the skeletal site of origin and age of the donor of the cultured osteoblastic cells: AR expression was higher at cortical and intramembranous bone sites, and it was lower in trabecular bone. This distribution pattern may correlate with androgen responsiveness in the bone compartment. AR expression was highest in osteoblastic cultures generated from young adults and somewhat lower in samples from either prepubertal or senescent bone. Data indicate preferential nuclear staining of AR in males at sexual maturity, suggesting activation and translocation of the receptor in bone when androgenic steroid levels are elevated, consistent with androgen regulation of AR levels [87, 88]. Again, no differences were found between male and female samples, suggesting that differences in receptor number per se do not underlie development of a sexually dimorphic skeleton. Since androgens are so important in bone development at the time of puberty, it is not surprising that ARs are also present in epiphyseal chondrocytes [49, 89]. The expression of ARs in such a wide variety of cell types known to be important for bone modeling during development, and remodeling in the adult, provides evidence for direct actions of androgens in bone and cartilage tissue. These results illustrate the complexity of androgen effects on bone. Although bone is a target tissue with respect to androgen action, the mechanisms and cell types by which androgens exert their effects on bone biology remain incompletely characterized. An additional complexity in terms of mechanism is that androgens may influence bone directly by activation of the AR or indirectly after aromatization of androgens into estrogens with subsequent activation of the ER, as described previously.

C. Regulation of Androgen Receptor Expression

The regulation of AR expression in osteoblasts is incompletely understood. Homologous regulation of AR mRNA by androgen has been described that is tissue specific; upregulation by androgen exposure is seen in a variety of mesenchymal cells including osteoblasts [60, 62, 87, 88], whereas in prostate and smooth muscle tissue, downregulation is observed after androgen exposure [87, 90] (Figure 13-4). The androgen-mediated upregulation observed in osteoblasts occurs, at least in part, through changes in AR gene transcription [87, 88]. No effect, or even inhibition, of AR mRNA by androgen exposure in other osteoblastic

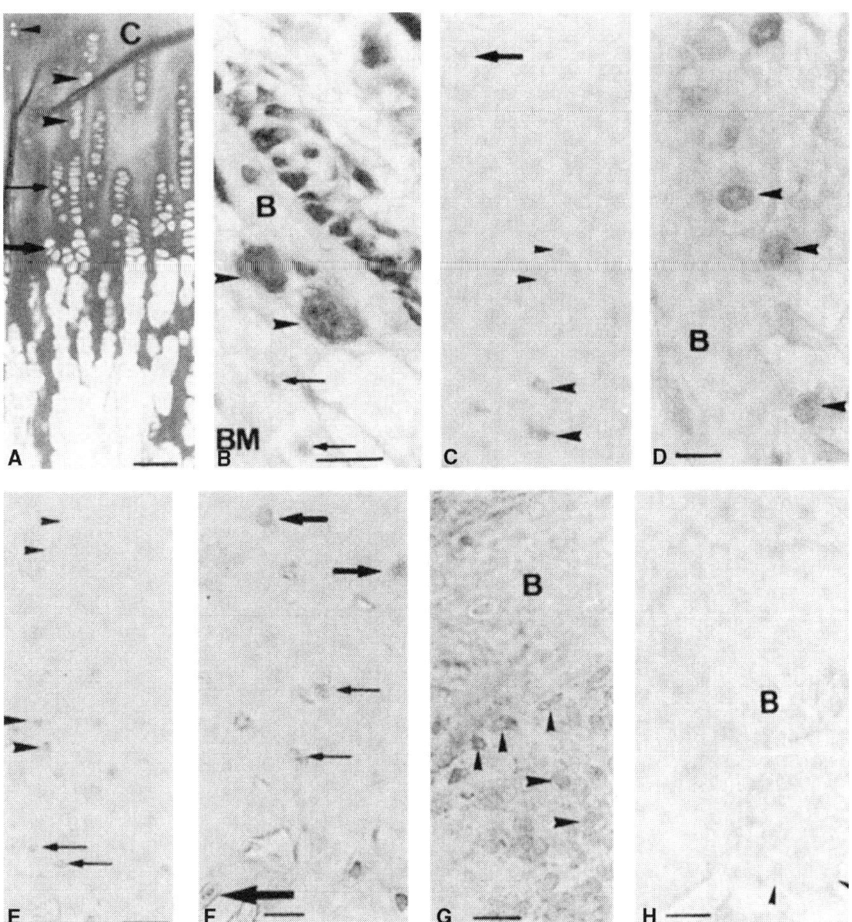

FIGURE 13-3   The localization of AR in normal tibial growth plate and adult osteophytic human bone. (A) Morphologically, sections of the growth plate consist of areas of endochondral ossification with undifferentiated (small arrowhead), proliferating (large arrowheads), mature (small arrow), and hypertrophic (large arrow) chondrocytes. Scale bar = 80 μm. An inset of an area of the primary spongiosa is shown in B. (B) Numerous osteoblasts (small arrowheads) and multinucleated osteoclasts (large arrowheads) on the bone surface. Mononuclear cells within the bone marrow are also present (arrows). Scale bar = 60 μm. (C) In the growth plate, AR is predominantly expressed by hypertrophic chondrocytes (large arrowheads). Minimal expression is observed in the mature chondrocytes (small arrowheads). The receptors are rarely observed in the proliferating chondrocytes (arrow). (D) In the primary spongiosa, the AR is predominantly and highly expressed by osteoblasts at modeling sites (arrowheads). Scale bar = 20 μm. (E) In the osteophytes, AR is also observed at sites of endochondral ossification in undifferentiated (small arrowheads), proliferating (large arrowheads), mature (small arrows), and hypertrophic-like (large arrow) chondrocytes. Scale bar = 80 μm. (F) A higher magnification of E showing proliferating (medium-sized arrows), mature (small arrows), and hypertrophic-like chondrocytes (large arrow). Scale bar = 40 μm. (G) At sites of bone remodeling, the receptors are highly expressed in the osteoblasts (small arrowheads) and also in mononuclear cells in the bone marrow (large arrowheads). Scale bar = 40 μm. (H) AR is not detected in osteoclasts (small arrowheads). Scale bar = 40 μm. B, bone; C, cartilage; BM, bone marrow. Reproduced with permission from E. Abu, A. Horner, J. Triffit, and J. Compston, *J Clin Endocrinol Metab* **82**, 3493–3497. Copyright 1997, The Endocrine Society.

models has also been described [63, 91]. Interestingly, a novel property of the AR is that binding of androgen increases AR protein levels, which has been shown in osteoblastic cells as well [88]. This property distinguishes AR from most other steroid receptor molecules that are downregulated by ligand binding. The elevated AR protein levels may be a consequence of increased stability mediated by androgen binding, resulting from N-terminal and C-terminal interactions [92], but the stability of AR protein in osteoblastic cells has not been determined. The mechanism(s) that underlies

tissue specificity in autologous AR regulation, and the possible biological significance of distinct autologous regulation of AR, is not understood. It is possible that AR upregulation by androgen in bone may result in an enhancement of androgen responsiveness at times when androgen levels are rising or elevated.

Quantitative determination of the level of receptor expression during osteoblast differentiation is difficult to achieve in bone slices. However, analysis of AR, ERα, and ERβ mRNA and protein expression during osteoblast differentiation *in vitro* demonstrates that

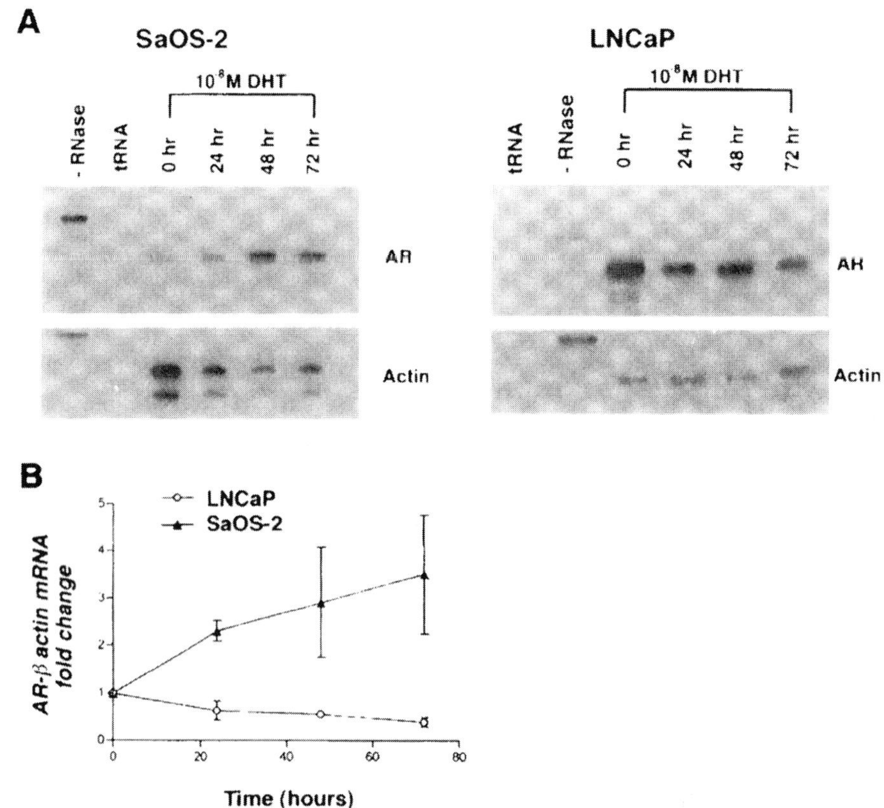

FIGURE 13-4　(A) Dichotomous regulation of AR mRNA levels in osteoblast-like and prostatic carcinoma cell lines after exposure to androgen. Time course of changes in AR mRNA abundance after DHT exposure in human SaOS-2 osteoblastic cells and human LNCaP prostatic carcinoma cells. To determine the effect of androgen exposure on hAR mRNA abundance, confluent cultures of either osteoblast-like cells (SaOS-S) or prostatic carcinoma cells (LNCaP) were treated with 10^{-8} M DHT for 0, 24, 48, or 72 hours. Total RNA was then isolated and subjected to RNase protection analysis with 50 μg total cellular RNA from SaOS-2 osteoblastic cells and 10 μg total RNA from LNCaP cultures. (B) Densitometric analysis of AR mRNA steady-state levels. The AR mRNA to β-actin ratio is expressed as the mean ± SEM compared to the control value from three to five independent assessments. K. Wiren, X. Zhang, C. Chang, E. Keenan, and E. Orwoll, Transcriptional up-regulation of the human androgen receptor by androgen in bone cells, *Endocrinology* **138**, 2291–2300. Copyright 1997, The Endocrine Society.

each receptor displays differentiation stage-distinct patterns in osteoblasts (Figure 13-5) [93]. The levels of AR expression increase throughout osteoblast differentiation, with the highest AR levels seen in mature osteoblast/osteocytic cultures. These results suggest that an important compartment for androgen action may be mature, mineralizing osteoblasts, and they indicate that osteoblast differentiation and steroid receptor regulation are intimately associated. Given that the osteocyte is the most abundant cell type in bone, and a likely mediator of focal bone deposition and response to mechanical strain [94], it is not surprising that androgens may also augment the osteoanabolic effects of mechanical strain in osteoblasts [95].

AR expression in osteoblasts can be upregulated by exposure to other steroid hormones, including glucocorticoids, estrogen, or 1,25-dihydroxyvitamin D_3 [63]. Whether additional hormones, growth factors, or agents influence AR expression in bone is not known. Furthermore, whether the AR in osteoblasts undergoes post-translational processing that might influence receptor signaling (stabilization, phosphorylation, etc.) as described in other tissues [96, 97] and the potential functional implications [98, 99] are also unknown. Ligand-independent activation of AR has also been described in other tissues [100] but has not been explored in bone.

Steroid receptor transcriptional activity, including that of the AR, is strongly influenced by transcriptional regulators such as coactivators or corepressors [101, 102]. These coactivators/corepressors can influence the downstream signaling of nuclear receptors; their levels are influenced by the cellular context, and these coregulators can differentially affect specific promoters. AR-specific coactivators have been identified [103], many of which interact with the ligand-binding domain of the receptor [104]. Expression and regulation of these modulators may thus influence the ability of steroid receptors to regulate gene expression in bone [105], but this remains underexplored with respect to

A

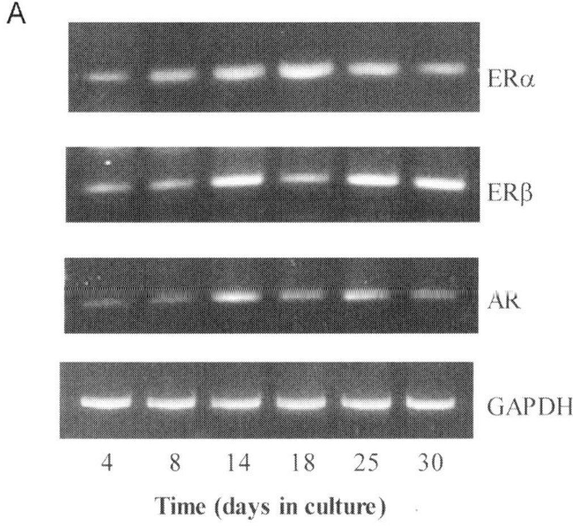

B

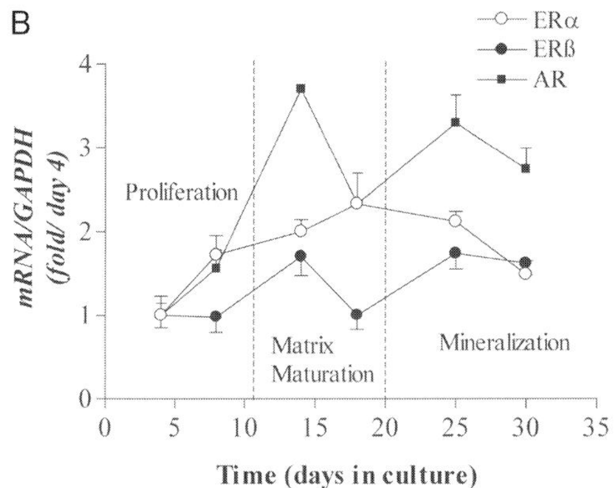

FIGURE 13-5   Expression analyses of ERα, ERβ, and AR during *in vitro* differentiation in normal rat osteoblastic (rOB) cultures. (A) Normal rOB cells were cultured for the indicated number of days during proliferation, matrix maturation, mineralization, and postmineralization stages. Total RNA was isolated and subjected to relative RT-PCR analysis using primers specific for rat ERα, ERβ, and AR or rat GAPDH. Reverse transcription was conducted with PCR carried out for 40 cycles for the steroid receptors, with parallel reactions performed using GAPDH primers for 25 cycles (all in the linear range). Bands for rat ERα at the predicted 240 bp, rat ERβ at 262 bp, rat AR at 276 bp, and GAPDH at 609 bp are shown. (B) Analyses of ERα, ERβ, and AR mRNA relative abundance. Semiquantitative analysis of mRNA steady-state expression by relative RT-PCR was performed after scanning the negative image of the photographed gels. Data are expressed in arbitrary units as the ratio of receptor abundance to GAPDH expression, then normalized to expression values at day 4 in preconfluent cultures. Data represent mean ± SEM. From K. Wiren, A. Chapman Evans, and X. Zhang, Osteoblast differentiation influences androgen and estrogen receptor-alpha and -beta expression. *J Endocrinol* **175**, 683–694 (2002). © Society for Endocrinology (2002). Reproduced by permission.

androgen action. The specific coactivator/corepressor profile present in cells representing different bone compartments (i.e., periosteal cells and proliferating or mineralizing cells) may help determine the activity of the selective receptor modulators such as SARMS.

IV. THE CONSEQUENCES OF ANDROGEN ACTION IN BONE CELLS

A. Effects of Androgens on Proliferation and Apoptosis

Evidence suggests that androgens act directly on the osteoblast and there are reports, some in clonal osteoblastic cell lines, of modulatory effects of gonadal androgen treatment on proliferation, differentiation, matrix production, and mineral accumulation [106]. Not surprisingly, androgen has been shown to influence bone cells in a complex fashion. For example, the effect of androgen on osteoblast proliferation has been shown to be biphasic in nature, with enhancement following short or transient treatment but significant inhibition following longer treatment. As a case in point, Kasperk *et al.* [107, 108] demonstrated in osteoblast-like cells in primary culture (murine, passaged human) that a variety of androgens in serum-free medium increase DNA synthesis ([³H]thymidine incorporation) and cell counts. Testosterone and non-aromatizable androgens (DHT and fluoxymesterone) were nearly equally effective regulators. Yet the same group [109] reported that prolonged DHT treatment inhibited normal human osteoblastic cell proliferation (cell counts) in cultures pretreated with DHT. In addition, Benz *et al.* [58] showed that prolonged androgen exposure in the presence of serum inhibited proliferation (cell counts) by 15–25% in a transformed human osteoblastic line (TE-85). Testosterone and DHT again were nearly equally effective regulators. Hofbauer *et al.* [110] examined the effect of DHT exposure on proliferation in hFOB/AR-6, an immortalized human osteoblastic cell line stably transfected with an AR expression construct (with ~4000 receptors/cell). In this line, DHT treatment inhibited cell proliferation by 20–35%. Consistent with stimulation, Somjen *et al.* [111] demonstrated increased creatine kinase–specific activity in male osteoblastic cells after exposure to DHT for 24 hours. Although these various studies employed different model systems (transformed osteoblastic cells vs. second to fourth passage normal human cells) and culture conditions (including differences in the state of osteoblast differentiation, receptor number, phenol red–containing vs. phenol red–free, or serum containing vs. serum free), it appears that exposure time is an important variable. Clear time dependence for the response to androgen has been shown by Wiren *et al.* [112], where osteoblast proliferation was stimulated at early treatment times, but with more prolonged DHT treatment osteoblast viability decreased (Figure 13-6). This result was AR dependent (inhibitable

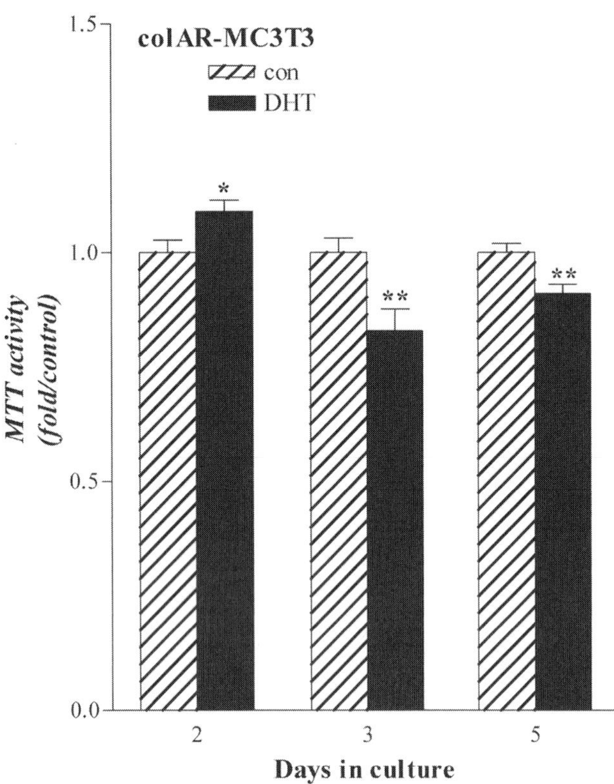

FIGURE 13-6 Complex effect of androgen on DNA accumulation in osteoblastic cultures. Kinetics of DHT response in proliferating colAR-MC3T3 cultures measured with colorimetric [3-(4,5-dimethylthiazol-2-yl)-2,5-diphenyltetrazolium bromide] (MTT) assay. Cultures of stably transfected colAR-MC3T3 continuously with 10^{-8} M DHT for 2 days led to increased MTT accumulation, but longer treatment for 3 or 5 days resulted in inhibition. Data are mean ± SEM of six to eight dishes with six wells/dish. $^{*}p < 0.05$; $^{**}p < 0.01$ (vs. control). From K. Wiren, A. Toombs, and X.-W. Zhang, Androgen inhibition of MAP kinase pathway and Elk-1 activation in proliferating osteoblasts. *J Mol Endocrinol* **32**, 209–226 (2004). © Society for Endocrinology (2004). Reproduced by permission.

by co-incubation with flutamide) and was observed in both normal rat calvarial osteoblasts and AR stably transfected MC3T3 cells. In mechanistic terms, reduced viability was associated with overall reduction in mitogen-activated (MAP) kinase signaling and with inhibition of *elk-1* gene expression, protein abundance, and extent of phosphorylation. The inhibition of MAP kinase activity after chronic androgen treatment again contrasts with stimulation of MAP kinase signaling and AP-1 transactivation observed with brief androgen exposure [112], which may be mediated through nongenomic mechanisms [75, 113, 114].

As a component of control of osteoblast survival, it is also important to consider the process of programmed cell death, or apoptosis [115]. In particular, as the osteoblast population differentiates *in vitro*, the mature bone cell phenotype undergoes apoptosis [116]. With

respect to the effects of androgen exposure, chronic DHT treatment has been shown to result in enhanced osteoblast apoptosis in both proliferating osteoblastic (day 5) and mature osteocytic cultures (day 29) [117]. In this report, stimulation observed with DHT treatment was opposite to the inhibitory effects on apoptosis seen with E_2 treatment (Figure 13-7). An androgen-mediated increase in the Bax/Bcl-2 ratio was also observed, predominantly through inhibition of Bcl-2, and was dependent on functional AR. Overexpression of *bcl-2* or RNAi knock down of *bax* abrogated the effects of DHT, indicating that increased Bax/Bcl-2 was necessary and sufficient for androgen-enhanced apoptosis. The increase in the Bax/Bcl-2 ratio was at least in part a consequence of reductions in Bcl-2 phosphorylation and protein stability, consistent with inhibition of MAP kinase pathway activation after DHT treatment as noted previously. *In vivo* analysis of calvaria in AR-transgenic male mice demonstrated enhanced TUNEL staining in both osteoblasts and osteocytes, and it was observed even in areas of new bone growth [117]. This may not be surprising, given an association between new bone growth and apoptosis [118], as has been observed in other remodeling tissues and/or associated with development and tissue homeostasis [119]. Apoptotic cell death could thus be important in making room for new bone formation and matrix deposition, which may have clinical significance by influencing bone homeostasis and bone mineral density [120]. Thus, mounting evidence suggests that chronic androgen treatment does not increase osteoblast number or viability in the mature bone compartment. It is interesting to speculate that the inhibitory action of androgens in osteoblasts, especially in the endosteal compartment, is important for the relative maintenance of cortical width (which is similar between males and females) given the strong stimulation at the periosteal surface, such that the skeleton does not become excessively large and heavy during development.

B. Effects of Androgens on Differentiation of Osteoblastic Cells

Osteoblast differentiation can be characterized by changes in alkaline phosphatase activity and/or alterations in the expression of important extracellular matrix proteins, such as type I collagen, osteocalcin, and osteonectin. Effects of androgens on expression of these marker activities/proteins are poorly described and inconsistent. For example, enhanced osteoblast differentiation, as measured by increased matrix production, has been shown to result from androgen exposure in both normal osteoblasts and transformed

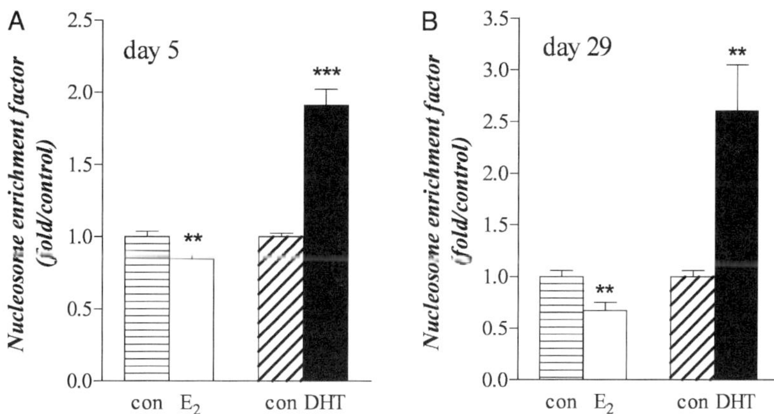

FIGURE 13-7   Characterization of osteoblast apoptosis: results of androgen and estrogen treatment during proliferation (day 5) and during differentiation into mature osteoblast/osteocyte cultures (day 29). Apoptosis was assessed at day 5 or day 29 after continuous DHT and E_2 treatment (both at 10^{-8} M). Apoptosis was induced by etoposide treatment in proliferating cultures and by serum starvation for 48 hours in confluent cultures before isolation, replaced with 0.1% BSA. (A) Analysis of apoptosis after evaluating DNA fragmentation by cytoplasmic nucleosome enrichment at day 5. The data are expressed as mean ± SEM ($n = 6$) from two independent experiments. $^{**}p < 0.01$, $^{***}p < 0.001$ (vs. control). (B) Analysis of apoptosis by cytoplasmic nucleosome enrichment analysis at day 29. The data are expressed as mean ± SEM ($n = 6$) from two independent experiments. $^{**}p < 0.01$ versus control. Reprinted from K. Wiren, A. Toombs, A. Semirale, and X. W. Xhang, Osteoblast and osteocyte apoptosis associated with androgen action in bone: Requirement of increased Bax/Bcl-2 ratio. *Bone* **38**, 637–651. Copyright 2006 with permission from Elsevier.

clonal human osteoblastic cells (TE-89). Androgen treatment appeared to increase the proportion of cells expressing alkaline phosphatase activity, thus representing a shift toward a more differentiated phenotype [107]. Kasperk *et al.* [121] subsequently reported dose-dependent increases in alkaline phosphatase activity in both high- and low-alkaline phosphatase subclones of SaOS2 cells and human osteoblastic cells [109]. However, there are also reports, in a variety of model systems, of androgens either inhibiting [110] or having no effect on alkaline phosphatase activity [62, 122], which may reflect both the complexity and the dynamics of osteoblastic differentiation. Androgen-mediated increases in type I α_1 collagen protein and mRNA levels [58, 121, 122], and increased osteocalcin secretion [109], have also been described. Consistent with increased collagen production, androgen treatment has also been shown to stimulate mineral accumulation in a time- and dose-dependent manner [62, 109, 123]. However, transgenic mice with targeted overexpression of AR in the osteoblast lineage showed decreased levels of most bone markers *in vivo* in total RNA extracts derived from long bone samples, including decreased collagen, osterix, and osteocalcin gene expression [28]. These results suggest that under certain conditions, androgens may enhance osteoblast differentiation and could thus play an important role in the regulation of bone matrix production and/or organization. On the other hand, many positive anabolic effects of androgen may be limited to distinct osteoblastic populations, for example, in the periosteal compartment [1, 28].

C. Direct Effects of Androgens on Other Cell Types in the Skeleton

Potential modulation of osteoclast action by androgen is suggested by reports of AR expression in the osteoclast [82]. Androgen treatment reduces bone resorption of isolated osteoclasts [124], inhibits osteoclast formation [125] and that stimulated by parathyroid hormone (PTH) [126], and may play a direct role in regulating aspects of osteoclast activity in AR null mice [127]. Indirect effects of androgen to modulate osteoclasts via osteoblasts are indicated by the increase in OPG levels following testosterone treatment in osteoblasts [128] and in skeletally targeted AR-transgenic male mice in serum and bone [28]. In addition, DHEA treatment has been shown to increase the OPG/RANKL ratio in osteoblastic cells and inhibit osteoclast activity in co-culture [70]. Androgen may be a less significant determinant of bone resorption *in vivo* than estrogen [129, 130], although this remains controversial [131].

As with effects noted in osteoblastic populations, androgens also regulate chondrocyte proliferation and expression. Although some of the consequences of androgen action are mediated after metabolic conversion to estrogen, which limits long bone growth, non-aromatizable androgen stimulates longitudinal bone growth [132]. AR expression has been demonstrated in cartilage [133], and androgen exposure promotes chondrogenesis as shown with increased creatine kinase and DNA synthesis after androgen exposure in cultured epiphyseal chondrocytes [89, 134]. Increased

[^{35}S]sulfate incorporation into newly synthesized cartilage [135] and increased alkaline phosphatase activity [136] are androgen mediated. Regulation of these effects is obviously complex because they were influenced by the age of the animals and the site from which chondrocytes were derived. Thus, in addition to effects on osteoblasts, multiple cell types in the skeletal milieu are regulated by androgen exposure.

D. Interaction with Other Factors to Modulate Bone Activity

The effects of androgens on osteoblast activity must certainly also be considered in the context of the very complex endocrine, paracrine, and autocrine milieu in the bone microenvironment. Systemic and/or local factors can act in concert, or can antagonize, to influence bone cell function. This has been well described with regard to modulation of the effects of estrogen on bone [137–139]. Androgens have also been shown to regulate well-known modulators of osteoblast proliferation or function. The most extensively characterized growth factor influenced by androgen exposure is transforming growth factor-β (TGF-β). TGF-β is stored in bone (the largest reservoir for TGF-β) in a latent form, and it has been show to be a mitogen for osteoblasts [140, 141]. Androgen treatment has been shown to increase TGF-β activity in human osteoblast primary cultures. The expression of some TGF-β mRNA transcripts (apparently TGF-β2) was increased, but no effect on TGF-β1 mRNA abundance was observed [69, 108, 142]. At the protein level, specific immunoprecipitation analysis reveals DHT-mediated increases in TGF-β activity to be predominantly TGF-β2 [69, 109]. DHT has also been shown to inhibit both TGF-β gene expression and TGF-β-induced early gene expression that correlates with growth inhibition in this cell line [110]. The TGF-β-induced early gene has been shown to be a transcription factor that may mediate some TGF-β effects [143]. These results are consistent with the notion that TGF-β may mediate androgen effects on osteoblast proliferation. On the other hand, TGF-β1 mRNA levels are increased by androgen treatment in human clonal osteoblastic cells (TE-89) under conditions in which osteoblast proliferation is slowed [58]. Thus, the specific TGF-β isoform may determine osteoblast responses. It is interesting to note that *in vivo*, orchiectomy (ORX) drastically reduces bone content of TGF-β levels, and testosterone replacement prevents this reduction [144]. These data support the finding that androgens influence cellular expression of TGF-β and suggest that the bone loss associated with castration is related to a reduction in growth factor abundance induced by androgen deficiency.

Other growth factor systems may also be influenced by androgens. Conditioned media from DHT-treated normal osteoblast cultures are mitogenic, and DHT pretreatment increases the mitogenic response to fibroblast growth factor and to insulin-like growth factor-2 (IGF-2) [108]. In part, this may be due to slight increases in IGF-2 binding in DHT-treated cells [108] since IGF-1 and IGF-2 levels in osteoblast-conditioned media are not affected by androgen [108, 145]. Although most studies have not found regulation of IGF-1 or IGF-2 abundance by androgen exposure [16, 108, 145], there is a report that IGF-1 mRNA levels are significantly upregulated by DHT [146]. Androgens may also modulate expression of components of the AP-1 transcription factor [69] or AP-1 transcriptional activation [112]. Thus, androgens may modulate osteoblast differentiation via a mechanism whereby growth factors or other mediators of differentiation are regulated by androgen exposure.

Androgens may modulate responses to other important osteotropic hormones/regulators. Testosterone and DHT specifically inhibit the cAMP response elicited by PTH or parathyroid hormone-related protein (PTHrP) in the human clonal osteoblast-like cell line SaOS-2, whereas the inactive or weakly active androgen 17α-epitestosterone had no effect. This inhibition may be mediated via an effect on the PTH receptor–G$_s$-adenylyl cyclase [147–149]. The production of prostaglandin E$_2$ (PGE$_2$), another important regulator of bone metabolism, is also affected by androgens. Pilbeam and Raisz [150] showed that androgens (both DHT and testosterone) were potent inhibitors of both parathyroid hormone and interleukin-1-stimulated PGE$_2$ production in cultured neonatal mouse calvaria. The effects of androgens on PTH action and PGE$_2$ production suggest that androgens could act to modulate (reduce) bone turnover in response to these agents.

Finally, both androgen [151] and estrogen [138, 152, 153] inhibit production of interleukin-6 (IL-6) by osteoblastic cells. In stromal cells of the bone marrow, androgens have been shown to have potent inhibitory effects on the production of IL-6 and the subsequent stimulation of osteoclastogenesis by marrow osteoclast precursors [154]. Interestingly, adrenal androgens (androstenediol, androstenedione, and DHEA) have similar inhibitory activities on IL-6 gene expression and protein production by stromal cells [154]. The loss of inhibition of interleukin-6 production by androgen may also contribute to the marked increase in bone remodeling and resorption that follows ORX, in addition to modulation of osteoclast activity through changes in the OPG/RANKL ratio as noted previously. Moreover, androgens inhibit the expression of the genes encoding the two subunits of the IL-6 receptor (gp80 and gp130) in the murine bone marrow, another mechanism that may blunt the effects of this osteoclastogenic cytokine in intact animals [155]. In these aspects, the effects of androgens seem to be very similar to those of estrogen, which may also inhibit osteoclastogenesis via mechanisms that involve IL-6 inhibition and/or OPG/RANKL ratio changes.

V. THE SKELETAL EFFECTS OF ANDROGEN: ANIMAL STUDIES

The effects of androgens on bone remodeling have been examined fairly extensively in animal models. Much of this work has been performed on species not perfectly suited to reflect human bone metabolism (rodents), and certainly the field remains incompletely explored. Nevertheless, animal models do provide valuable insights into the effects of androgens at organ and cellular levels. Many of the studies of androgen action have been performed in male rats, in which rapid skeletal growth occurs until approximately 4 months of age, at which time epiphyseal growth slows markedly (although never completely ceases at some sites). Many studies have also employed mice as genetic models. Because the effects of androgen deficiency may be different in growing and more mature animals [156], it is appropriate to consider the two situations independently.

A. Effects on Epiphyseal Function and Bone Growth during Skeletal Development and Puberty

In most mammals, there is a marked gender difference in bone morphology. The mechanisms responsible for these differences are complex and presumably involve both androgenic and estrogenic actions. Estrogens are particularly important for the regulation of epiphyseal function and act to reduce the rate of longitudinal growth via influences on chondrocyte proliferation and action, as well as on the timing of epiphyseal closure [157]. Androgens appear to have opposite effects and tend to promote long bone growth, chondrocyte maturation, and metaphyseal ossification. Androgen deficiency retards these processes [158]. Nevertheless, excess concentrations of androgen will accelerate aging of the growth plate and reduce growth potential [159], possibly via conversion to estrogens.

Although the specific roles of sex steroids in the regulation of epiphyseal growth and maturation remain somewhat unresolved, there is evidence that androgens do have direct effects independent of those of estrogen. For instance, testosterone injected directly into the growth plates of rats increases plate width [160]. In a model of endochondral bone development based on the subcutaneous implantation of demineralized bone matrix in castrate rats, both testosterone and DHT increased the incorporation of calcium during osteoid formation [123]. Interestingly, in this model androgens reduced the incorporation of [^{35}S]sulfate into glycosaminoglycans early in the developing cartilage. In summary, these data support the contention that androgens play a direct role in chondrocyte physiology, but how these actions are integrated with those of other regulators is unclear.

During childhood and adolescence, skeletal development is characterized by marked expansion of cortical proportions and increasing trabecular density. During this process, the skeleton develops distinctly in males and females, most significantly at the periosteal surface. Thus, sex differences in skeletal morphology and physiology occur during or near puberty. For this reason, it is hypothesized that gender differences, particularly with respect to "bone quality" and architecture (i.e., predominantly bone width), are modulated by the sex steroids estrogen and androgen. Consistent with this, a distinct response to estrogen and androgen has been described *in vivo*, especially in cortical bone. At the periosteum, estrogen suppresses while androgen stimulates new bone formation, yet conversely at the endosteal surface estrogen stimulates but androgen strongly suppresses formation [28]. Again, these two sex steroids may act in opposition in some situations at distinct bone compartments. Thus, estrogen decreases but androgen increases radial growth in cortical bone through periosteal apposition. These distinct responses to estrogen and androgen during growth likely play an important role in determining sexual dimorphism of the skeleton—that is, that male bones are wider but not thicker than those of females [161]. Young men do have larger bone areas than women with increased whole bone cross-sectional area, particularly at peripheral sites [162]. Interestingly, low levels of estrogen (in the obligate presence of androgen) may also be important for stimulation of periosteal bone formation during development [37]. Androgens are also essential for the production of peak total body bone mass in males [163]. Finally, androgens are known to interact with the growth hormone–IGF system in the coordination of skeletal growth. Growth hormone deficiency in males has no net effect on endosteal growth but reduces by half expansion at the periosteal surface [164], underscoring the codependence of these two hormonal systems in the control of pubertal skeletal change.

B. Mature Male Animals

Results from animal studies also support an effect of androgen on bone formation in the mature animal. Experimental strategies such as surgical or pharmacological intervention and examination of genetic models have all been employed to characterize androgen signaling in the adult. In mature rats, castration eventually results in osteopenia and both cortical and trabecular compartments are affected. At a time when longitudinal growth has slowed markedly, pronounced differences as a consequence of castration appear in cortical bone ash weight per unit length, cross-sectional area, cortical thickness, and bone mineral density (Figure 13-8) [165–168]. Castration results in changes in both trabecular and cortical bone compartments, and dramatic bone loss in trabecular bone is noted in both males and females, but sex-specific responses are

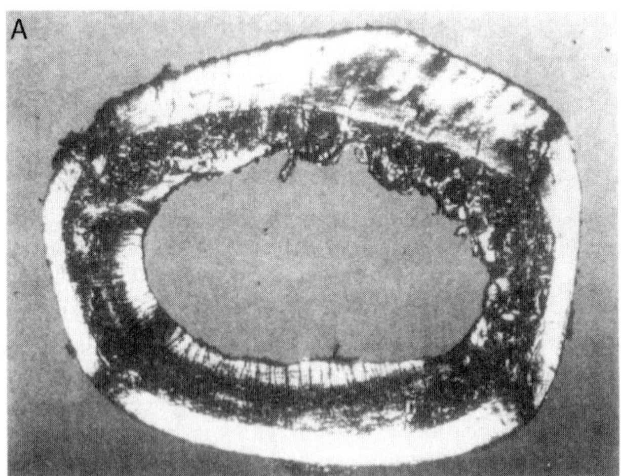

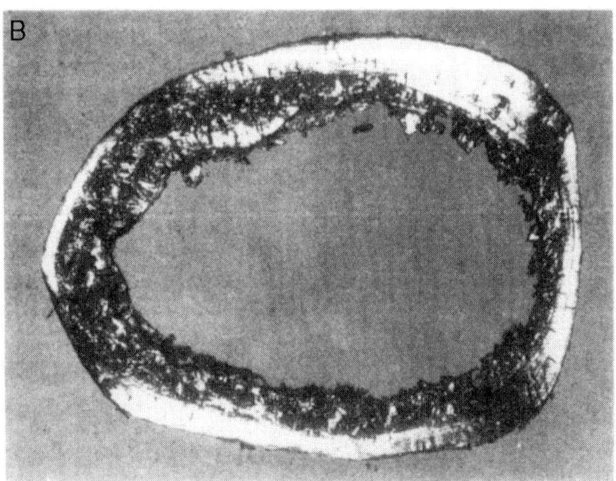

FIGURE 13-8 Microphotographs of 200-μm-thick mid-diaphyseal cross-sections from 24-month-old (A) intact and (B) ORX rats taken in a polarization microscope. Magnification ×14. From Danielsen [166], with kind permission from Springer Science and Business Media.

most dimorphic in cortical bone. For example, distinct effects of androgen are seen with gonadectomy when comparing the effects of ORX in male versus ovariectomy (OVX) in female rats. In Turner *et al.*'s [3] classic study, OVX and the associated loss of sex steroids in the female generally resulted in decreased trabecular area with increased osteoclast number. In cortical bone in OVX females, an increase at the periosteal surface was seen with circumferential enlargement (Figure 13-9A), but a decrease in endosteal labeling was seen. In summary, these results demonstrate that estrogen protects trabecular bone predominantly through inhibition of osteoclast activity/recruitment but has an inhibitory action at the periosteal surface, as noted previously [169]. In the male, ORX with the attendant loss of sex steroids also results in decreased trabecular area with increased osteoclast number. However, in contrast to the female, periosteal formation in cortical bone is reduced with the loss of androgen (Figure 13-9B). Androgen treatment is effective in suppressing the acceleration of bone remodeling normally seen after ORX [170]. This divergent trend in the periosteal response to castration in male and female animals abolishes the sexual dimorphism usually present in radial bone growth. In the intact animal, the stimulation of endosteal formation by estrogen compensates for the lack of periosteal formation, thus leading to no difference in cortical width between the sexes. Nevertheless, factors that influence periosteal apposition may constitute an important therapeutic class since periosteal bone formation is often a neglected determinant of bone strength [161]. ORX shows either little net effect [164] or slight reductions on the endosteal surface in males, likely due to increased resorption. Consistent with this, increased intracortical resorption cavities are reported to result from ORX [165, 171]. As might be expected in light of these changes, breaking strength can be decreased in cortical bone [164]. In addition, it appears that ORX affects cranial development more than OVX [172], suggesting that androgen action is particularly important in intramembranous bone.

In addition to changes in bone size at the periosteal surface, trabecular bone volume is reduced rapidly after castration as well [165, 173], and osteopenia becomes pronounced with time [48]. It is likely that this bone loss results from increased bone resorption because it is associated with increased resorption cavities, osteoclasts, and blood flow [165, 166]. Dynamic histomorphometric and biochemical measures of bone remodeling increase quickly after ORX [173, 174], with evidence of increased osteoclast numbers only 1 week after castration [173]. These changes include an increase in osteoblastic activity as well as increased bone resorption, reflecting an initial high turnover state that is followed

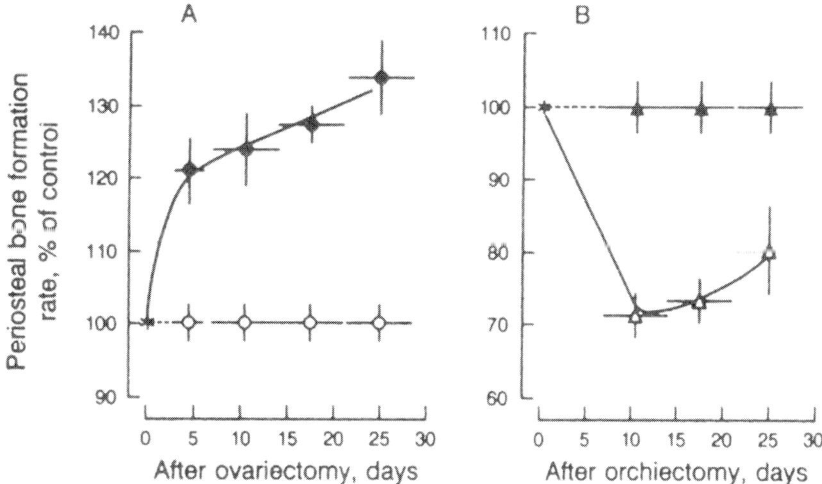

FIGURE 13-9   (A) The effect of ovariectomy (OVX) on periosteal bone formation rate. The mean ± SEM (vertical bar) and tetracycline labeling period (horizontal line) for intact controls (open circles) and OVX (solid circles) rats are shown as a function of time after OVX. $p < 0.01$ for all OVX time points compared to intact controls. (B) The effect of ORX on periosteal bone formation rate. The mean ± SEM and tetracycline labeling periods for intact controls (solid triangles) and ORX (open triangles) are shown as a function of time after ORX. $p < 0.01$ for all ORX time points compared to the same labeling period in intact controls. From R. T. Turner, G. K. Wakley, and K. S. Hannon, Differential effects of androgens on cortical bone histomorphometry in gonadectomized male and female rats. *J Orthop Res* **814**, 612–617. Copyright © 1990. Reprinted with permission of Wiley-Liss, Inc., a subsidiary of John Wiley & Sons, Inc.

by a reduction in remodeling rates and osteopenia. In the SAMP6 mouse, which is a model of accelerated senescence in which osteoblastic function is impaired, the rise in remodeling following ORX is blunted, which has been interpreted as evidence that the early changes after gonadectomy are dependent on osteoblast-derived signals [175]. As noted previously, androgens reduce osteoclast formation and activity [125], which may be partially mediated by increased OPG levels [28, 128]. The initial phase of increased bone remodeling activity subsides with time [166, 174], and by 4 months there is evidence of a depression in bone turnover rates in some skeletal areas (Figure 13-10) [166]. As in younger animals, indices of mineral metabolism are not altered by these changes in skeletal metabolism [168]. Careful histomorphometric analysis of androgen action in ORX male mice by Ohlsson and workers [175] has shown that the bone-sparing effect of AR activation in trabecular bone is distinct from the bone-sparing effect of ERα at that site. The analysis demonstrated that AR activation does preserve the number of trabeculae but does not preserve thickness or volumetric density, nor mechanical strength in cortical bone.

As a potential model for the effects of hypogonadism in humans [48], animal models therefore suggest an early phase of high bone turnover and bone loss after ORX, followed by a reduction in remodeling rates and osteopenia. The remodeling imbalance responsible for loss of bone mass appears complex because there are changes in rates of both bone formation and

resorption and also patterns that vary from one skeletal compartment to another. These overall changes may be similar to those noted in female animals after castration, in which a loss of estrogen signaling has been associated with an early stimulation of osteoblast progenitor differentiation, an even greater increase in osteoclast numbers, with bone resorption and bone loss [177].

C. Androgens in the Female Animal

Of course, androgens are present in females as well as males and may affect bone metabolism. In castrated female rats, DHT administration suppresses elevated concentrations of bone resorption markers as well the increases in osteocalcin levels [178]. However, alkaline phosphatase activity increases further. Additional evidence to support the contention that androgens play a role in females includes the fact that antiandrogens are capable of evoking osteopenia in intact (i.e., fully estrogenized) female rats [4, 179]. This result suggests that androgens can provide crucial support to bone mass independent of estrogens in females. Of interest, the character of the bone loss induced by flutamide suggested that estrogen prevents bone resorption, whereas androgens stimulate bone formation. In periosteal bone, DHT and testosterone appear to stimulate bone formation after ORX in young male rats, whereas in castrated females they suppress bone formation [3], perhaps

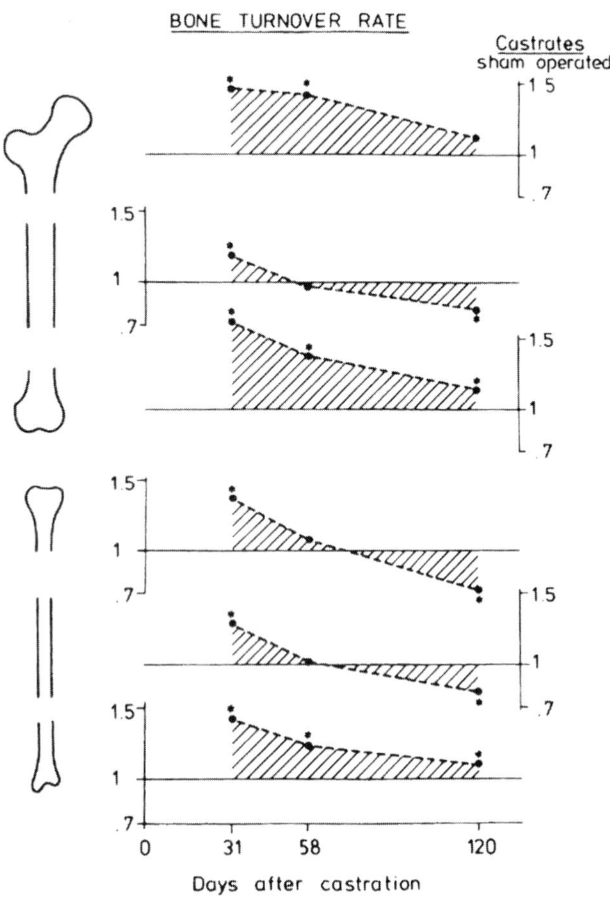

FIGURE 13-10   Evolution of the bone calcium turnover rate after castration (ratio of castrated/sham-operated animals). $^*p <$ 0.05. From M. Verhas, A. Schoutens, M. L'hermite-Baleriaux, N. Dourov, A. Nerschaeren, M. Mone, and A. Heilporn, The effect of orchidectomy on bone metabolism in aging rats. *Calcif Tissue Int* **39**, 74–77 (1986), with kind permission from Springer Science and Business Media.

reflecting an interaction or synergism between sex steroids and their effects on bone. There is also some information concerning androgen action in females in additional animal models, including primates. For instance, in adult female cynomolgus monkeys, testosterone treatment increased cortical and trabecular bone density as well as biomechanical strength [180]. As noted previously, although postmenopausal women can be effectively treated with androgens, combination therapy with estrogen and androgen is more beneficial than either steroid alone [6–8]. This result has been confirmed in an animal model [181].

D. Gender Specificity

In most mammals, there is a marked gender difference in morphology that results in a sexually dimorphic skeleton. The mechanisms responsible for these differences are necessarily complex and presumably involve both androgenic and estrogenic actions on the skeleton. It is becoming increasingly clear that estrogens are particularly important for the regulation of epiphyseal function and act to reduce the rate of longitudinal growth via influences on chondrocyte proliferation and function, as well as on the timing of epiphyseal closure [157]. Androgens, on the other hand, appear to have many opposite effects to estrogen on the skeleton. Androgens tend to promote long bone growth, chondrocyte maturation, and metaphyseal ossification, as noted previously. Furthermore, the most dramatic effect of androgens is on bone size, particularly cortical thickness [180], because androgens appear to have gender-specific effects on periosteal bone formation to inhibit or stimulate growth [3]. Of course, this difference has important biomechanical implications, with thicker bones being stronger bones [161]. Furthermore, the response of the adult skeleton (to the same intervention) results in distinct responses in males and females. For example, in a model of disuse osteopenia, antiorthostatic suspension results in significant reduction in bone formation rate at the endosteal perimeter in males. In females, however, a decrease in bone formation rate occurs along the periosteal perimeter [182]. Gender-specific responses *in vivo* and *in vitro* [111], and the mechanism(s) that underlies such responses in bone cells, may thus have significant implications in treatment options for metabolic bone disease.

VI. ANIMAL MODELS OF ALTERED ANDROGEN RESPONSIVENESS

The specific contribution of AR signaling *in vivo* has also been approached using genetic animal models with global AR modulation, including the testicular feminization (Tfm) model of androgen insufficiency syndrome [169, 183], and with (nontargeted) global AR knockout mice [127, 184]. The Tfm (AR-deficient) male rat provides an interesting model for the study of the unique effects of androgens in bone. In these Tfm rats, androgens are presumed to be incapable of action, but estrogen and androstenedione concentrations are considerably higher than those in normal males [185, 186]. Clear increases also exist in Tfm male rats in serum concentrations of calcium, phosphorus, and osteocalcin, whereas IGF-1 concentrations are decreased. Estimates of bone mass suggest that Tfm rats have reduced longitudinal and radial growth rates, but that trabecular volume and density are similar to those of normal rats. In selected sites, measures of bone mass and remodeling were intermediate between normal male and female values. However, castration reduced bone volume markedly in Tfm male rats, suggesting a major role for estrogens as well in skeletal homeostasis (Figure 13-11). This model again indicates that androgens have an independent role to play in normal bone growth and metabolism, but the model is complex and not easily dissected. Meticulous analysis in Tfm mice by Vanderschueren *et al.* [169] has also shown that the positive effects of testosterone on cortical bone are generally mediated by stimulation of periosteal bone formation, which was absent in Tfm mice. Histomorphometric analysis shows that AR mediated testosterone action is essential for periosteal bone formation (in male mice) and also contributes to trabecular bone maintenance. This is very similar to the study of humans with the androgen insensitivity syndrome. Marcus *et al.* [42] reported that there is a deficit in bone mineral density in women with androgen insensitivity even when compliance with estrogen replacement is excellent. However, inadequate estrogen replacement appeared to worsen the deficit, and other environmental factors are difficult to quantitate. Thus, in Tfm models, ORX demonstrates the importance of AR in mediating the positive effects of androgen to contribute to trabecular bone maintenance, and in cortical bone particularly at the periosteal surface [169, 183].

The bone phenotype that develops in a global AR null (ARKO) male mouse model is a high-turnover osteopenia, with reduced trabecular bone volume and a significant stimulatory effect on osteoclast function [127, 184, 187]. As expected, bone loss with ORX in male ARKO mice was only partially prevented by treatment with aromatizable testosterone due to the lack of AR.

A final model for AR modulation is represented by overexpression of AR in AR-transgenic mice [28], constructed with full-length AR under the control of the 3.6-kb type I collagen promoter, with AR overexpression in osteoblast stromal precursors and throughout the osteoblast lineage. AR-transgenic mice are the only model with skeletally targeted manipulation of AR expression, and they demonstrate enhanced sensitivity to androgen without changes in circulating steroids or androgen administration [28]. AR overexpression in this model results in a complex phenotype predominantly in males, with increased trabecular bone mass (with increased trabecular number but not thickness) in the setting of inhibition of resorption due to reduced osteoclast activity. In addition, cortical formation is altered with periosteal expansion but inhibition of inner endosteal deposition (Figure 13-12), consistent with the known effects of androgen to stimulate periosteal apposition and opposite to the effects of estrogen on these compartments. Inhibition of osteoclastic resorption may be responsible for altered trabecular morphology, consistent with reduced osteoclast activity and increased trabecular bone volume observed with androgen therapy in rodents and humans. The dramatic inhibition of bone formation at the endosteal envelope may underlie the modest decrease in cortical bone area and subsequent reductions in biomechanical properties that are observed. Notably, the bone phenotype observed in AR-transgenic mice is consistent with many of the known effects of androgen treatment on the skeleton. Combined, studies employing genetic models indicate that AR expressed in bone can be a direct mediator of androgen action to influence skeletal development and homeostasis.

VII. EFFECTS ON THE PERIOSTEUM: THE ROLE OF ANDROGEN RECEPTOR VERSUS AROMATIZATION OF TESTOSTERONE

As noted previously, androgen-mediated AR transactivation is likely a key determinant of the sexually dimorphic pattern of periosteal apposition that is most clearly demonstrated in male AR-transgenic mice in the absence of hormone administration [28]. Furthermore, essentially all of the alterations induced by ORX (in both growing and mature animals) can be prevented at least in part by replacement with either testosterone or nonaromatizable androgens [3, 171, 188–192]. These results strongly suggest that aromatization of androgens to estrogens cannot fully explain the actions of androgens on bone metabolism.

However, estrogens also seem to play a role in the effects of androgen on periosteal apposition. Although AR activity is essential, low levels of estrogens are likely required for optimal stimulation of periosteal growth [193], as observed in aromatase deficiency

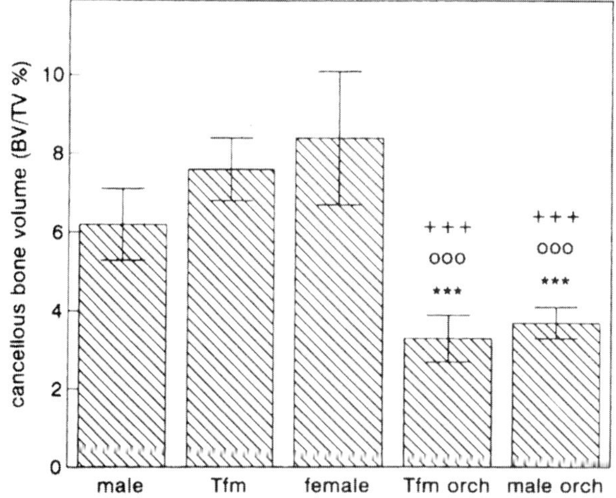

FIGURE 13-11 Cancellous bone volume of the proximal metaphysis of the tibia in male, female, Tfm, and orchiectomized male rats. Adapted from Vanderschueren *et al.* [185], with kind permission from Springer Science and Business Media.

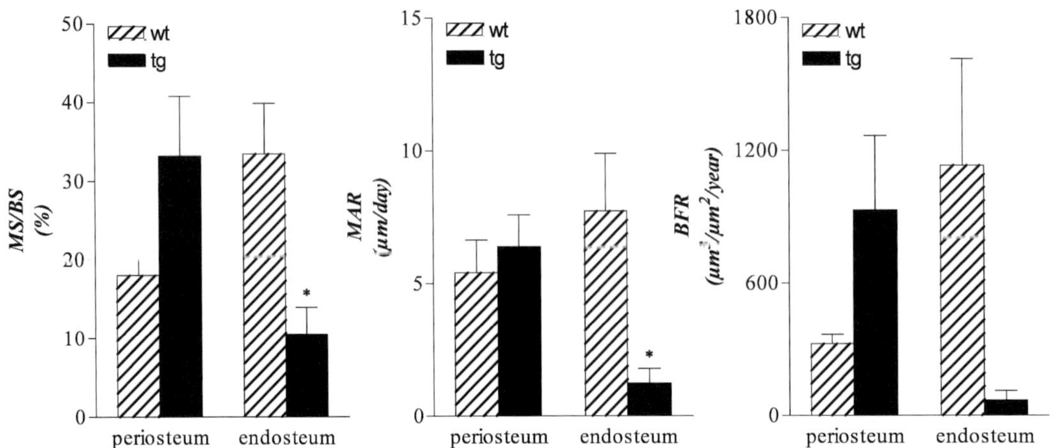

FIGURE 13-12 Characterization of cortical bone formation in AR-transgenic (AR-tg) mice. Dynamic histomorphometric analysis was performed in cortical bone after fluorescent imaging microscopy in AR-tg males ($n = 6$–8). Mineralizing surface as a percentage of bone surface (MS/BS), mineral apposition rate (MAR), and bone formation rate (BFR) at both the endosteal and the periosteal surfaces were determined in wild-type (wt) and AR-tg mice. $^*p < 0.05$. Reproduced with permission from K. Wiren, X.-W. Zhang, A. Toombs, M. Gentile, V. Kasparcova, S.-I. Harada, and K. Jepsen, Targeted overexpression of androgen receptor in osteoblasts: Unexpected complex bone phenotype in growing animals. *Endocrinology* **145**, 3507–3522 (2004). Copyright 2004, The Endocrine Society.

even in males [37]. Estrogens may also help prevent bone loss following castration in male animals. Vanderschueren *et al.* [168] reported that estradiol (and nandrolone) was capable of not only preventing the increase in biochemical indices stimulated by ORX but also preventing cortical and trabecular bone loss. In fact, estradiol resulted in an absolute increase in trabecular bone volume not achieved with androgen replacement. Similarly, estrogen was reported to antagonize the increase in blood flow resulting from castration and to increase bone ash weight more consistently than testosterone. Although the available data are far from complete, these studies raise obvious questions about the overlap between the actions of androgens and estrogens in bone and/or the consequences of skeletal adaptation to changes in bone morphology.

The gender reversal of estrogen replacement in male animals is also instructive. Nonaromatizable androgens are capable of preventing or reversing osteopenia and abnormalities in bone remodeling in OVX females [3, 194]. These actions apparently result from the suppression of trabecular bone resorption as well as stimulation of periosteal bone formation [194]. Very similar results have been reported following the treatment of OVX animals with DHEA [3]. Moreover, blockage of androgen action with an AR antagonist in female rats already treated with an estrogen antagonist increases bone loss and indices of osteoclast activity more than treatment with an estrogen antagonist alone [195], again indicating that ovarian androgens (apart from estrogens) exert a protective effect on bone in females. Analogously,

androstenedione reduces (although does not abrogate) trabecular bone loss and remodeling alterations in OVX animals treated with an aromatase inhibitor [196, 197]. This protective effect was blocked by the addition of an AR antagonist [196]. Finally, whereas aromatase inhibition in male rats reduces bone mass, the large increase in remodeling induced by ORX does not occur in these animals [38]. Also, ORX in ERKO mice further reduces bone mass [80]. The latter observation implicates a role for androgens in the maintenance of bone mass in ERKO mice.

VIII. SUMMARY

The effects of androgens on bone health are obviously both pervasive and complex. Androgens are important in the maintenance of a healthy skeleton and have been shown to stimulate bone formation in the periosteum. Androgens influence skeletal modeling and remodeling by multiple mechanisms through effects on osteoblasts and osteoclasts, and they even perhaps influence the differentiation of pluripotent stem cells toward distinct lineages. The specific effects of androgen on bone cells are mediated directly through an AR signaling pathway, but there are also indirect contributions to overall skeletal health through aromatization and ER signaling. The effects of androgens are particularly dramatic during growth in boys but almost certainly play an important role during this period in girls as well. Throughout the rest of life, androgens affect skeletal function in both sexes. Still poorly characterized, more needs to be done to unravel the mechanisms by which androgens influence the physiology and pathophysiology of bone, and there remains much to be learned about the roles of androgens at all levels. The interaction of androgens and estrogens and how their respective actions can be utilized for specific diagnostic and therapeutic benefit are important but unanswered issues. With an increase in the understanding of the nature of androgen effects will come greater opportunities to use their positive actions in the prevention and treatment of a wide variety of skeletal disorders.

REFERENCES

1. K. Wiren, Androgens and bone growth: It's location, location, location. *Curr Opin Pharmacol* **5,** 626–632 (2005).
2. R. Turner, K. Hannon, L. Demers, J. Buchanan, and N. Bell, Differential effects of gonadal function on bone histomorphometry in male and female rats. *J Bone Miner Res* **4,** 557–563 (1989).
3. R. Turner, G. Wakley, and K. Hannon, Differential effects of androgens on cortical bone histomorphometry in gonadectomized male and female rats. *J Orthop Res* **8,** 612–617 (1990).
4. A. Goulding and E. Gold, Flutamide-mediated androgen blockade evokes osteopenia in the female rat. *J Bone Miner Res* **8,** 763–769 (1993).
5. V. Coxam, B. Bowman, M. Mecham, C. Roth, M. Miller, and S. Miller, Effects of dihydrotestosterone alone and combined with estrogen on bone mineral density, bone growth, and formation rates in ovariectomized rats. *Bone* **19,** 107–114 (1996).
6. B. Miller, M. De Souza, K. Slade, and A. Luciano, Sublingual administration of micronized estradiol and progesterone, with and without micronized testosterone: Effect on biochemical markers of bone metabolism and bone mineral density. *Menopause* **7,** 318–326 (2000).
7. C. Castelo-Branco, J. Vicente, F. Figueras, A. Sanjuan, M. Martinez de Osaba, E. Casals, F. Pons, J. Balasch, and J. Vanrell, Comparative effects of estrogens plus androgens and tibolone on bone, lipid pattern and sexuality in postmenopausal women. *Maturitas* **34,** 161–168 (2000).
8. L. Raisz, B. Wiita, A. Artis, A. Bowen, S. Schwartz, M. Trahiotis, K. Shoukri, and J. Smith, Comparison of the effects of estrogen alone and estrogen plus androgen on biochemical markers of bone formation and resorption in postmenopausal women. *J Clin Endocrinol Metab* **81,** 37–43 (1996).
9. Q. Mo, S. Lu, and N. Simon, Dehydroepiandrosterone and its metabolites: Differential effects on androgen receptor trafficking and transcriptional activity. *J Steroid Biochem Mol Biol* **99,** 50–58 (2006).
10. H. Bruch, L. Wolf, R. Budde, G. Romalo, and H. Scheikert, Androstenedione metabolism in cultured human osteoblast-like cells. *J Clin Endocrinol Metab* **75,** 101–105 (1992).
11. J. Vittek, K. Altman, G. Gordon, and A. Southren, The metabolism of 7alpha-3H-testosterone by rat mandibular bone. *Endocrinology* **94,** 325–329 (1974).
12. H. Schweikert, W. Rulf, N. Niederle, H. Schafer, E. Keck, and F. Kruck, Testosterone metabolism in human bone. *Acta Endocrinol* **95,** 258–264 (1980).
13. D. Russell and J. Wilson, Steroid 5α-reductase: Two genes/two enzymes. *Annu Rev Biochem* **63,** 25–61 (1994).
14. S. Issa, D. Schnabel, M. Feix, L. Wolf, H. Schaefer, D. Russell, and H. Schweikert, Human osteoblast-like cells express predominantly steroid 5α-reductase type 1. *J Clin Endocrinol Metab* **87,** 5401–5407 (2002).
15. L. Audi, A. Carrascosa, and A. Ballabriga, Androgen metabolism by human fetal epiphyseal cartilage and its chondrocytes in primary culture. *J Clin Endocrinol Metab* **58,** 819–825 (1984).
16. Y. Nakano, I. Morimoto, O. Ishida, T. Fujihira, A. Mizokami, A. Tanimoto, N. Yanagihara, F. Izumi, and S. Eto, The receptor, metabolism and effects of androgen in osteoblastic MC3T3-E1 cells. *Bone Miner* **26,** 245–259 (1994).
17. R. Turner, B. Bleiberg, D. Colvard, P. Keeting, G. Evans, and T. Spelsberg, Failure of isolated rat tibial periosteal cells to 5α reduce testosterone to 5α-dihydroxytestosterone. *J Bone Miner Res* **5,** 775–779 (1990).
18. V. Sobel, B. Schwartz, Y. Zhu, J. Cordero, and J. Imperato-McGinley, Bone mineral density in the complete androgen insensitivity and 5α-reductase-2 deficiency syndromes. *J Clin Endocrinol Metab* **91,** 3017–3023 (2006).
19. M. Mahendroo, K. Cala, C. Landrum, and D. Russell, Fetal death in mice lacking 5α-reductase type 1 caused by estrogen excess. *Mol Endocrinol* **11,** 917–927 (1997).
20. H. Rosen, S. Tollin, R. Balena, V. Middlebrooks, A. Moses, M. Yamamoto, A. Zeind, and S. Greenspan, Bone density is normal in male rats treated with finasteride. *Endocrinology* **136,** 1381–1387 (1995).
21. J. Amory, N. Watts, K. Easley, P. Sutton, B. Anawalt, A. Matsumoto, W. Bremner, and J. Tenover, Exogenous testosterone or testosterone with finasteride increases bone mineral density in older men with low serum testosterone. *J Clin Endocrinol Metab* **89,** 503–510 (2004).
22. E. Simpson, M. Mahendroo, G. Means, M. Kilgore, M. Hinshelwood, S. Graham-Lorence, B. Amarneh, Y. Ito, C. Fisher, M. Michael, C. Mendelson, and S. Bulun, Aromatase cytochrome P450, the enzyme responsible for estrogen biosynthesis. *Endocr Rev* **15,** 342–355 (1994).
23. H. Nawata, S. Tanaka, S. Tanaka, R. Takayanagi, Y. Sakai, T. Yanase, S. Ikuyama, and M. Haji, Aromatase in bone cell: Association with osteoporosis in postmenopausal women. *J Steroid Biochem Mol Biol* **53,** 165–174 (1995).
24. H. Schweikert, L. Wolf, and G. Romalo, Oestrogen formation from androstenedione in human bone. *Clin Endocrinol* **43,** 37–42 (1995).
25. H. Sasano, M. Uzuki, T. Sawai, H. Nagura, G. Matsunaga, O. Kashimoto, and N. Harada, Aromatase in human bone tissue. *J Bone Miner Res* **12,** 1416–1423 (1997).
26. A. Purohit, A. Flanagan, and M. Reed, Estrogen synthesis by osteoblast cell lines. *Endocrinology* **131,** 2027–2029 (1992).
27. S. Tanaka, Y. Haji, T. Yanase, R. Takayanagi, and H. Nawata, Aromatase activity in human osteoblast-like osteosarcoma cell. *Calcif Tissue Int* **52,** 107–109 (1993).
28. K. Wiren, X.-W. Zhang, A. Toombs, M. Gentile, V. Kasparcova, S.-I. Harada, and K. Jepsen, Targeted overexpression of androgen receptor in osteoblasts: Unexpected complex bone phenotype in growing animals. *Endocrinology* **145,** 3507–3522 (2004).
29. M. Shozu and E. Simpson, Aromatase expression of human osteoblast-like cells. *Mol Cell Endocrinol* **139,** 117–129 (1998).
30. L. Gennari, R. Nuti, and J. Bilezikian, Aromatase activity and bone homeostasis in men. *J Clin Endocrinol Metab* **89,** 5898–5907 (2004).

31. S. Lin, R. Shi, W. Qiu, A. Azzi, D. Zhu, H. Dabbagh, and M. Zhou, Structural basis of the multispecificity demonstrated by 17beta-hydroxysteroid dehydrogenase types 1 and 5. *Mol Cell Endocrinol* **248,** 38–46 (2006).

32. M. Muir, G. Romalo, L. Wolf, W. Elger, and H. Schweikert, Estrone sulfate is a major source of local estrogen formation in human bone. *J Clin Endocrinol Metab* **89,** 4685–4692 (2004).

33. A. Pino, J. Rodriguez, S. Rios, P. Astudillo, L. Leiva, G. Seitz, M. Fernandez, and J. Rodriguez, Aromatase activity of human mesenchymal stem cells is stimulated by early differentiation, vitamin D and leptin. *J Endocrinol* **191,** 715–725 (2006).

34. S. Abdelgadir, J. Resko, S. Ojeda, E. Lephart, M. McPhaul, and C. Roselli, Androgens regulate aromatase cytochrome P450 messenger ribonucleic acid in rat brain. *Endocrinology* **135,** 395–401 (1994).

35. M. Jones, W. Boon, J. Proietto, and E. Simpson, Of mice and men: The evolving phenotype of aromatase deficiency. *Trends Endocrinol Metab* **17,** 55–64 (2006).

36. E. Smith, J. Boyd, G. Frank, H. Takahashi, R. Cohen, B. Specker, T. Williams, D. Lubahn, and K. Korach, Estrogen resistance caused by a mutation in the estrogen-receptor gene in a man. *N Engl J Med* **331,** 1056–1061 (1994).

37. R. Bouillon, M. Bex, D. Vanderschueren, and S. Boonen, Estrogens are essential for male pubertal periosteal bone expansion. *J Clin Endocrinol Metab* **89,** 6025–6029 (2004).

38. D. Vanderschueren, E. Van Herck, J. Nijs, and A. Ederveen, Aromatase inhibition impairs skeletal modeling and decreases bone mineral density in growing male rats. *Endocrinology* **138,** 2301–2307 (1997).

39. S. Wickman, E. Kajantie, and L. Dunkel, Effects of suppression of estrogen action by the p450 aromatase inhibitor letrozole on bone mineral density and bone turnover in pubertal boys. *J Clin Endocrinol Metab* **88,** 3785–3789 (2003).

40. D. Vanderschueren, E. Van Herck, R. De Coster, and R. Bouillon, Aromatization of androgens is important for skeletal maintenance of aged male rats. *Calcif Tissue Int* **59,** 179–183 (1996).

41. M. Feix, L. Wolf, and H. Schweikert, Distribution of 17beta-hydroxysteroid dehydrogenases in human osteoblast-like cells. *Mol Cell Endocrinol* **171,** 163–164 (2001).

42. R. Marcus, D. Leary, D. Schneider, E. Shane, M. Favus, and C. Quigley, The contribution of testosterone to skeletal development and maintenance: Lessons from the androgen insensitivity syndrome. *J Clin Endocrinol Metab* **85,** 1032–1037 (2000).

43. J. Kemppainen, E. Langley, C. Wong, K. Bobseine, W. Kelce, and E. Wilson, Distinguishing androgen receptor agonists and antagonists: Distinct mechanisms of activation by medroxyprogesterone acetate and dihydrotestosterone. *Mol Endocrinol* **13,** 440–454 (1999).

44. J. Omwancha and T. Brown, Selective androgen receptor modulators: In pursuit of tissue-selective androgens. *Curr Opin Investig Drugs* **7,** 873–881 (2006).

45. J. Miner, W. Chang, M. Chapman, P. Finn, M. Hong, F. Lopez, K. Marschke, J. Rosen, W. Schrader, R. Turner, A. van Oeveren, H. Viveros, L. Zhi, and A. Negro-Vilar, An orally active selective androgen receptor modulator is efficacious on bone, muscle, and sex function with reduced impact on prostate. *Endocrinology* **148,** 363–373 (2007).

46. G. Allan, M. Lai, T. Sbriscia, O. Linton, D. Haynes-Johnson, S. Bhattacharjee, R. Dodds, J. Fiordeliso, J. Lanter, Z. Sui, and S. Lundeen, A selective androgen receptor modulator that reduces prostate tumor size and prevents orchidectomy-induced bone loss in rats. *J Steroid Biochem Mol Biol* **103,** 76–83 (2007).

47. J. Kearbey, W. Gao, R. Narayanan, S. Fisher, D. Wu, D. Miller, and J. Dalton, Selective androgen receptor modulator (SARM) treatment prevents bone loss and reduces body fat in ovariectomized rats. *Pharm Res* **24,** 328–335 (2007).

48. D. Vanderschueren, L. Vandenput, S. Boonen, M. Lindberg, R. Bouillon, and C. Ohlsson, Androgens and bone. *Endocr Rev* **25,** 389–425 (2004).

49. E. Abu, A. Horner, V. Kusec, J. Triffitt, and J. Compston, The localization of androgen receptors in human bone. *J Clin Endocrinol Metab* **82,** 3493–3497 (1997).

50. C. Chang, J. Kokontis, and S. Liao, Structural analysis of complementary DNA and amino acid sequences of human and rat androgen receptors. *Proc Natl Acad Sci USA* **85,** 7211–7215 (1988).

51. D. Lubahn, D. Joseph, P. Sullivan, H. Willard, F. French, and E. Wilson, Cloning of human androgen receptor complementary DNA and localization to the X chromosome. *Science* **240,** 324–326 (1988).

52. D. Mangelsdorf, C. Thummel, M. Beato, P. Herrlich, G. Schutz, K. Umesono, B. Blumberg, P. Kastner, M. Mark, P. Chambon, and R. Evans, The nuclear receptor superfamily: The second decade. *Cell* **83,** 835–839 (1995).

53. D. Picard, Chaperoning steroid hormone action. *Trends Endocrinol Metab* **17,** 229–235 (2006).

54. H. Shen and G. Coetzee, The androgen receptor: Unlocking the secrets of its unique transactivation domain. *Vitam Horm* **71,** 301–319 (2005).

55. S. Kumar, M. Saradhi, N. Chaturvedi, and R. Tyagi, Intracellular localization and nucleocytoplasmic trafficking of steroid receptors: An overview. *Mol Cell Endocrinol* **246,** 147–156 (2006).

56. C. Chang, N. Saltzman, S. Yeh, W. Young, E. Keller, H.-J. Lee, C. Wang, and A. Mizokami, Androgen receptor: An overview. *Crit Rev Eukaryot Gene Expr* **5,** 97–125 (1995).

57. D. Colvard, E. Eriksen, P. Keeting, E. Wilson, D. Lubahn, F. French, B. Riggs, and T. Spelsberg, Identification of androgen receptors in normal human osteoblast-like cells. *Proc Natl Acad Sci USA* **86,** 854–857 (1989).

58. D. Benz, M. Haussler, M. Thomas, B. Speelman, and B. Komm, High-affinity androgen binding and androgenic regulation of a1(I)-procollagen and transforming growth factor-β steady state messenger ribonucleic acid levels in human osteoblast-like osteosarcoma cells. *Endocrinology* **128,** 2723–2730 (1991).

59. E. Orwoll, L. Stribska, E. Ramsey, and E. Keenan, Androgen receptors in osteoblast-like cell lines. *Calcif Tissue Int* **49,** 183–187 (1991).

60. Y. Zhuang, M. Blauer, A. Pekki, and P. Tuohimaa, Subcellular location of androgen receptor in rat prostate, seminal vesicle and human osteosarcoma MG-63 cells. *J Steroid Biochem Mol Biol* **41,** 693–696 (1992).

61. P. Liesegang, G. Romalo, M. Sudmann, L. Wolf, and H. Schweikert, Human osteoblast-like cells contain specific, saturable, high-affinity glucocorticoid, androgen, estrogen, and 1a,25-dihydroxycholecalciferol receptors. *J Androl* **15,** 194–199 (1994).

62. M. Takeuchi, H. Kakushi, and M. Tohkin, Androgens directly stimulate mineralization and increase androgen receptors in human osteoblast-like osteosarcoma cells. *Biochem Biophys Res Commun* **204,** 905–911 (1994).

63. C. Kasperk, A. Helmboldt, I. Borcsok, S. Heuthe, O. Cloos, F. Niethard, and R. Ziegler, Skeletal site-dependent expression of the androgen receptor in human osteoblastic cell populations. *Calcif Tissue Int* **61,** 464–473 (1997).

64. C. Wilson and M. McPhaul, A and B forms of the androgen receptor are present in human genital skin fibroblasts. *Proc Natl Acad Sci USA* **91,** 1234–1238 (1994).

65. U. Liegibel, U. Sommer, I. Boercsoek, U. Hilscher, A. Bierhaus, H. Schweikert, P. Nawroth, and C. Kasperk, Androgen receptor isoforms AR-A and AR-B display functional differences in cultured human bone cells and genital skin fibroblasts. *Steroids* **68,** 1179–1187 (2003).

66. A. Masuyama, Y. Ouchi, F. Sato, T. Hosoi, T. Nakamura, and H. Orimo, Characteristics of steroid hormone receptors in cultured MC3T3-E1 osteoblastic cells and effect of steroid hormones on cell proliferation. *Calcif Tissue Int* **51,** 376–381 (1992).

67. A. Meikle, R. Dorchuck, B. Araneo, J. Stringham, T. Evans, S. Spruance, and R. Daynes, The presence of a dehydroepiandrosterone-specific receptor binding complex in murine T cells. *J Steroid Biochem Mol Biol* **42,** 293–304 (1992).

68. S. Webb, T. Geoghegan, R. Prough, and K. M. Miller, The biological actions of dehydroepiandrosterone involves multiple receptors. *Drug Metab Rev* **38,** 89–116 (2006).

69. P. Bodine, B. Riggs, and T. Spelsberg, Regulation of c-fos expression and TGF-β production by gonadal and adrenal androgens in normal human osteoblastic cells. *J Steroid Biochem Mol Biol* **52,** 149–158 (1995).

70. Y. Wang, L. Wang, D. Li, and W. Wang, Dehydroepiandrosterone inhibited the bone resorption through the upregulation of OPG/RANKL. *Cell Mol Immunol* **3,** 41–45 (2006).

71. K. Wrogemann, G. Podolsky, J. Gu, and E. Rosenmann, A 63-kDa protein with androgen-binding activity is not from the androgen receptor. *Biochem Cell Biol* **69,** 695–701 (1991).

72. C. Pettaway, Racial differences in the androgen/androgen receptor pathway in prostate cancer. *J Natl Med Assoc* **91,** 653–660 (1999).

73. I. Van Pottelbergh, S. Lumbroso, S. Goemaere, C. Sultan, and J. Kaufman, Lack of influence of the androgen receptor gene CAG-repeat polymorphism on sex steroid status and bone metabolism in elderly men. *Clin Endocrinol (Oxford)* **55,** 659–666 (2001).

74. A. Grierson, R. Mootoosamy, and C. Miller, Polyglutamine repeat length influences human androgen receptor/c-Jun mediated transcription. *Neurosci Lett* **277,** 9–12 (1999).

75. H. Kang, C. Cho, K. Huang, J. Wang, Y. Hu, H. Lin, C. Chang, and K. Huang, Nongenomic androgen activation of phosphatidylinositol 3-kinase/Akt signaling pathway in MC3T3-E1 osteoblasts. *J Bone Miner Res* **19,** 1181–1190 (2004).

76. S. Kousteni, J. Chen, T. Bellido, L. Han, A. Ali, C. O'Brien, L. Plotkin, Q. Fu, A. Mancino, Y. Wen, A. Vertino, C. Powers, S. Stewart, R. Ebert, A. Parfitt, R. Weinstein, R. Jilka, and S. Manolagas, Reversal of bone loss in mice by nongenotropic signaling of sex steroids. *Science* **298,** 843–846 (2002).

77. M. Lieberherr and B. Grosse, Androgens increase intracellular calcium concentration and inositol 1,4,5-triphosphate and diacylglycerol formation via a pertussis toxin-sensitive G-protein. *J Biol Chem* **269,** 7217–7223 (1994).

78. M. Centrella, T. McCarthy, W. Chang, D. Labaree, and R. Hochberg, Estren (4-estren-3alpha,17beta-diol) is a prohormone that regulates both androgenic and estrogenic transcriptional effects through the androgen receptor. *Mol Endocrinol* **18,** 1120–1130 (2004).

79. B. van der Eerden, J. Emons, S. Ahmed, H. van Essen, C. Lowik, J. Wit, and M. Karperien, Evidence for genomic and nongenomic actions of estrogen in growth plate regulation in female and male rats at the onset of sexual maturation. *J Endocrinol* **175,** 277–288 (2002).

80. N. Sims, P. Clement-Lacroix, D. Minet, C. Fraslon-Vanhulle, M. Gaillard-Kelly, M. Resche-Rigon, and R. Baron, A functional androgen receptor is not sufficient to allow estradiol to protect bone after gonadectomy in estradiol receptor-deficient mice. *J Clin Invest* **111,** 1319–1327 (2003).

81. S. Hewitt, J. Collins, S. Grissom, K. Hamilton, and K. Korach, Estren behaves as a weak estrogen rather than a nongenomic selective activator in the mouse uterus. *Endocrinology* **147,** 2203–2214 (2006).

82. B. van der Eerden, N. van Til, A. Brinkmann, C. Lowik, J. Wit, and M. Karperien, Gender differences in expression of androgen receptor in tibial growth plate and metaphyseal bone of the rat. *Bone* **30,** 891–896 (2002).

83. R. Gruber, K. Czerwenka, F. Wolf, G. Ho, M. Willheim, and M. Peterlik, Expression of the vitamin D receptor, of estrogen and thyroid hormone receptor alpha- and beta-isoforms, and of the androgen receptor in cultures of native mouse bone marrow and of stromal/osteoblastic cells. *Bone* **24,** 465–473 (1999).

84. I. Sinha-Hikim, W. Taylor, N. Gonzalez-Cadavid, W. Zheng, and S. Bhasin, Androgen receptor in human skeletal muscle and cultured muscle satellite cells: Up-regulation by androgen treatment. *J Clin Endocrinol Metab* **89,** 5245–5255 (2004).

85. R. Singh, J. Artaza, W. Taylor, N. Gonzalez-Cadavid, and S. Bhasin, Androgens stimulate myogenic differentiation and inhibit adipogenesis in C3H 10T1/2 pluripotent cells through an androgen receptor-mediated pathway. *Endocrinology* **144,** 5081–5088 (2003).

86. K. Herbst and S. Bhasin, Testosterone action on skeletal muscle. *Curr Opin Clin Nutr Metab Care* **7,** 271–277 (2004).

87. K. Wiren, X.-W. Zhang, C. Chang, E. Keenan, and E. Orwoll, Transcriptional up-regulation of the human androgen receptor by androgen in bone cells. *Endocrinology* **138,** 2291–2300 (1997).

88. K. Wiren, E. Keenan, X. Zhang, B. Ramsey, and E. Orwoll, Homologous androgen receptor up-regulation in osteoblastic cells may be associated with enhanced functional androgen responsiveness. *Endocrinology* **140,** 3114–3124 (1999).

89. A. Carrascosa, L. Audi, M. Ferrandez, and A. Ballabriga, Biological effects of androgens and identification of specific dihydrotestosterone-binding sites in cultured human fetal epiphyseal chondrocytes. *J Clin Endocrinol Metab* **70,** 134–140 (1990).

90. M. Lin, J. Rajfer, R. Swerdloff, and N. Gonzalez-Cadavid, Testosterone down-regulates the levels of androgen receptor mRNA in smooth muscle cells from the rat corpora cavernosa via aromatization to estrogens. *J Steroid Biochem Mol Biol* **45,** 333–343 (1993).

91. L. Hofbauer, K. Hicok, M. Schroeder, S. Harris, J. Robinson, and S. Khosla, Development and characterization of a conditionally immortalized human osteoblastic cell line stably transfected with the human androgen receptor gene. *J Cell Biochem* **66,** 542–551 (1997).

92. E. Langley, J. Kemppainen, and E. Wilson, Intermolecular NH2-/carboxyl-terminal interactions in androgen receptor dimerization revealed by mutations that cause androgen insensitivity. *J Biol Chem* **273,** 92–101 (1998).

93. K. Wiren, A. Chapman Evans, and X. Zhang, Osteoblast differentiation influences androgen and estrogen receptor-alpha and -beta expression. *J Endocrinol* **175,** 683–694 (2002).

94. E. Seeman, Osteocytes—Martyrs for integrity of bone strength. *Osteoporos Int* **17,** 1443–1448 (2006).

95. U. Liegibel, U. Sommer, P. Tomakidi, U. Hilscher, L. Van Den Heuvel, R. Pirzer, J. Hillmeier, P. Nawroth, and C. Kasperk, Concerted action of androgens and mechanical strain shifts

bone metabolism from high turnover into an osteoanabolic mode. *J Exp Med* **196,** 1387–1392 (2002).

96. J. Kemppainen, M. Lane, M. Sar, and E. Wilson, Androgen receptor phosphorylation, turnover, nuclear transport, and transcriptional activities. *J Biol Chem* **267,** 968–974 (1992).

97. T. Ikonen, J. Palvimo, P. Kallio, P. Reinikainen, and O. Janne, Stimulation of androgen-regulated transactivation by modulators of protein phosphorylation. *Endocrinology* **135,** 1359–1366 (1994).

98. L. Blok, P. de Ruiter, and A. Brinkmann, Androgen receptor phosphorylation. *Endocr Res* **22,** 197–219 (1996).

99. L. Wang, X. Liu, W. Kreis, and D. Budman, Phosphorylation/dephosphorylation of androgen receptor as a determinant of androgen agonistic or antagonistic activity. *Biochem Biophys Res Commun* **259,** 21–28 (1999).

100. S. Dehm and D. Tindall, Ligand-independent androgen receptor activity is activation function-2 independent and resistant to antiandrogens in androgen refractory prostate cancer cells. *J Biol Chem* **281,** 27882–27893 (2006).

101. B. He, R. Gampe Jr., A. Hnat, J. Faggart, J. Minges, F. French, and E. Wilson, Probing the functional link between androgen receptor coactivator and ligand-binding sites in prostate cancer and androgen insensitivity. *J Biol Chem* **281,** 6648–6663 (2006).

102. H. Yoon and J. Wong, The corepressors silencing mediator of retinoid and thyroid hormone receptor and nuclear receptor corepressor are involved in agonist- and antagonist-regulated transcription by androgen receptor. *Mol Endocrinol* **20,** 1048–1060 (2006).

103. H. MacLean, G. Warne, and J. Zajac, Localization of functional domains in the androgen receptor. *J Steroid Biochem Mol Biol* **62,** 233–242 (1997).

104. S. Yeh and C. Chang, Cloning and characterization of a specific coactivator, ARA$_{70}$, for the androgen receptor in human prostate cells. *Proc Natl Acad Sci USA* **93,** 5517–5521 (1996).

105. M. Haussler, C. Haussler, P. Jurutka, P. Thompson, J. Hsieh, L. Remus, S. Selznick, and G. Whitfield, The vitamin D hormone and its nuclear receptor: Molecular actions and disease states. *J Endocrinol* **154**(Suppl.), S57–S73 (1997).

106. M. Notelovitz, Androgen effects on bone and muscle. *Fertil Steril* **77**(Suppl. 4), S34–S41 (2002).

107. C. Kasperk, J. Wergedal, J. Farley, T. Linkart, R. Turner, and D. Baylink, Androgens directly stimulate proliferation of bone cells *in vitro. Endocrinology* **124,** 1576–1578 (1989).

108. C. Kasperk, R. Fitzsimmons, D. Strong, S. Mohan, J. Jennings, J. Wergedal, and D. Baylink, Studies of the mechanism by which androgens enhance mitogenesis and differentiation in bone cells. *J Clin Endocrinol Metab* **71,** 1322–1329 (1990).

109. C. Kasperk, G. Wakley, T. Hierl, and R. Ziegler, Gonadal and adrenal androgens are potent regulators of human bone cell metabolism *in vitro. J Bone Miner Res* **12,** 464–471 (1997).

110. L. Hofbauer, K. Hicok, and S. Khosla, Effects of gonadal and adrenal androgens in a novel androgen-responsive human osteoblastic cell line. *J Cell Biochem* **71,** 96–108 (1998).

111. D. Somjen, S. Katzburg, F. Kohen, B. Gayer, O. Sharon, D. Hendel, and A. Kaye, Responsiveness to estradiol-17beta and to phytoestrogens in primary human osteoblasts is modulated differentially by high glucose concentration. *J Steroid Biochem Mol Biol* **99,** 139–146 (2006).

112. K. Wiren, A. Toombs, and X.-W. Zhang, Androgen inhibition of MAP kinase pathway and Elk-1 activation in proliferating osteoblasts. *J Mol Endocrinol* **32,** 209–226 (2004).

113. S. Kousteni, L. Han, J. Chen, M. Almeida, L. Plotkin, T. Bellido, and S. Manolagas, Kinase-mediated regulation of common transcription factors accounts for the bone-protective effects of sex steroids. *J Clin Invest* **111,** 1651–1664 (2003).

114. Y. Zagar, G. Chaumaz, and M. Lieberherr, Signaling crosstalk from Gbeta4 subunit to Elk-1 in the rapid action of androgens. *J Biol Chem* **279,** 2403–2413 (2004).

115. A. Wyllie, J. Kerr, and A. Currie, Cell death: The significance of apoptosis. *Int Rev Cytol* **68,** 251–307 (1980).

116. M. Lynch, C. Capparelli, J. Stein, G. Stein, and J. Lian, Apoptosis during bone-like tissue development *in vitro. J Cell Biochem* **68,** 31–49 (1998).

117. K. Wiren, A. Toombs, A. Semirale, and X.-W. Zhang, Apoptosis associated with androgen action in bone: Requirement of increased Bax/Bcl-2 ratio. *Bone,* **38,** 637–651 (2006).

118. C. Palumbo, M. Ferretti, and A. De Pol, Apoptosis during intramembranous ossification. *J Anat* **203,** 589–598 (2003).

119. R. Lanz, S. Chua, N. Barron, B. Soder, F. DeMayo, and B. O'Malley, Steroid receptor RNA activator stimulates proliferation as well as apoptosis *in vivo. Mol Cell Biol* **23,** 7163–7176 (2003).

120. M. Miura, X. Chen, M. Allen, Y. Bi, S. Gronthos, B. Seo, S. Lakhani, R. Flavell, X. Feng, P. Robey, M. Young, and S. Shi, A crucial role of caspase-3 in osteogenic differentiation of bone marrow stromal stem cells. *J Clin Invest* **114,** 704–713 (2004).

121. C. Kasperk, K. Faehling, I. Borcsok, and R. Ziegler, Effects of androgens on subpopulations of the human osteosarcoma cell line SaOS2. *Calcif Tissue Int* **58,** 376–382 (1996).

122. C. Gray, K. Colston, A. Mackay, M. Taylor, and T. Arnett, Interaction of androgen and 1,25-dihydroxyvitamin D3: Effects on normal rat bone cells. *J Bone Miner Res* **7,** 41–46 (1992).

123. S. Kapur and A. Reddi, Influence of testosterone and dihydrotestosterone on bone-matrix induced endochondral bone formation. *Calcif Tissue Int* **44,** 108–113 (1989).

124. L. Pederson, M. Kremer, J. Judd, D. Pascoe, T. Spelsburg, B. Riggs, and M. Oursler, Androgens regulate bone resorption acitivity of isolated osteoclasts *in vitro. Proc Natl Acad Sci USA* **96,** 505–510 (1999).

125. D. Huber, A. Bendixen, P. Pathrose, S. Srivastava, K. Dienger, N. Shevde, and J. Pike, Androgens suppress osteoclast formation induced by RANKL and macrophage-colony stimulating factor. *Endocrinology* **142,** 3800–3808 (2001).

126. Q. Chen, H. Kaji, T. Sugimoto, and K. Chihara, Testosterone inhibits osteoclast formation stimulated by parathyroid hormone through androgen receptor. *FEBS Lett* **491,** 91–93 (2001).

127. H. Kawano, T. Sato, T. Yamada, T. Matsumoto, K. Sekine, T. Watanabe, T. Nakamura, T. Fukuda, K. Yoshimura, T. Yoshizawa, K. Aihara, Y. Yamamoto, Y. Nakamichi, D. Metzger, P. Chambon, K. Nakamura, H. Kawaguchi, and S. Kato, Suppressive function of androgen receptor in bone resorption. *Proc Natl Acad Sci USA* **100,** 9416–9421 (2003).

128. Q. Chen, H. Kaji, M. Kanatani, T. Sugimoto, and K. Chihara, Testosterone increases osteoprotegerin mRNA expression in mouse osteoblast cells. *Horm Metab Res* **36,** 674–678 (2004).

129. A. Falahati-Nini, B. Riggs, E. Atkinson, W. O'Fallon, R. Eastell, and S. Khosla, Relative contributions of testosterone and estrogen in regulating bone resorption and formation in normal elderly men. *J Clin Invest* **106,** 1553–1560 (2000).

130. K. Oh, E. Rhee, W. Lee, S. Kim, K. Baek, M. Kang, E. Yun, C. Park, S. Ihm, M. Choi, H. Yoo, and S. Park, Circulating osteoprotegerin and receptor activator of NF-kappaB ligand system are associated with bone metabolism in middle-aged males. *Clin Endocrinol (Oxford)* **62,** 92–98 (2005).

131. B. Leder, K. LeBlanc, D. Schoenfeld, R. Eastell, and J. Finkelstein, Differential effects of androgens and estrogens on bone turnover in normal men. *J Clin Endocrinol Metab* **88**, 204–210 (2003).

132. O. Nilsson, R. Marino, F. De Luca, M. Phillip, and J. Baron, Endocrine regulation of the growth plate. *Horm Res* **64**, 157–165 (2005).

133. O. Nilsson, D. Chrysis, O. Pajulo, A. Boman, M. Holst, J. Rubinstein, E. Martin Ritzen, and L. Savendahl, Localization of estrogen receptors-alpha and -beta and androgen receptor in the human growth plate at different pubertal stages. *J Endocrinol* **177**, 319–326 (2003).

134. D. Somjen, Y. Weisman, Z. Mor, A. Harell, and A. Kaye, Regulation of proliferation of rat cartilage and bone by sex steroid hormones. *J Steroid Biochem Mol Biol* **40**, 717–723 (1991).

135. M. Corvol, A. Carrascosa, L. Tsagris, O. Blanchard, and R. Rappaport, Evidence for a direct *in vitro* action of sex steroids on rabbit cartilage cells during skeletal growth: Influence of age and sex. *Endocrinology* **120**, 1422–1429 (1987).

136. Z. Schwartz, E. Nasatzky, A. Ornoy, B. Brooks, W. Soskolne, and B. Boyan, Gender-specific, maturation-dependent effects of testosterone on chondrocytes in culture. *Endocrinology* **134**, 1640–1647 (1994).

137. M. Horowitz, Cytokines and estrogen in bone: Anti-osteoporotic effects. *Science* **260**, 626–627 (1993).

138. M. Kassem, S. Harris, T. Spelsberg, and B. Riggs, Estrogen inhibits interleukin-6 production and gene expression in a human osteoblastic cell line with high levels of estrogen receptors. *J Bone Miner Res* **11**, 193–199 (1996).

139. H. Kawaguchi, C. Pilbeam, S. Vargas, E. Morse, J. Lorenzo, and L. Raisz, Ovariectomy enhances and estrogen replacement inhibits the activity of bone marrow factors that stimulate prostaglandin production in cultured mouse calvariae. *J Clin Invest* **96**, 539–548 (1995).

140. M. Centrella, M. Horowitz, J. Wozney, and T. McCarthy, Transforming growth factor-β gene family members and bone. *Endocr Rev* **15**, 27–39 (1994).

141. S. Harris, L. Bonewald, M. Harris, M. Sabatini, S. Dallas, J. Feng, N. Ghosh-Choudhury, J. Wozney, and G. Mundy, Effects of transforming growth factor β on bone nodule formation and expression of bone morphogenetic protein 2, osteocalcin, osteopontin, alkaline phosphatase, and type I collagen mRNA in long-term cultures of fetal rat calvarial osteoblasts. *J Bone Miner Res* **9**, 855–863 (1994).

142. X. Wang, Z. Schwartz, P. Yaffe, and A. Ornoy, The expression of transforming growth factor-beta and interleukin-1beta mRNA and the response to 1,25(OH)2D3' 17 beta-estradiol and testosterone is age dependent in primary cultures of mouse-derived osteoblasts *in vitro*. *Endocrine* **11**, 13–22 (1999).

143. M. Subramaniam, S. Harris, M. Oursler, K. Rasmussen, B. Riggs, and T. Spelsberg, Identification of a novel TGF-beta-regulated gene encoding a putative zinc finger protein in human osteoblasts. *Nucleic Acids Res* **23**, 4907–4912 (1995).

144. R. Gill, R. Turner, T. Wronski, and N. Bell, Orchiectomy markedly reduces the concentration of the three isoforms of transforming growth factor beta in rat bone, and reduction is prevented by testosterone. *Endocrinology* **139**, 546–550 (1998).

145. E. Canalis, M. Centrella, and T. McCarthy, Regulation of insulin-like growth factor-II production in bone cultures. *Endocrinology* **129**, 2457–2462 (1991).

146. F. Gori, L. Hofbauer, C. Conover, and S. Khosla, Effects of androgens on the insulin-like growth factor system in an androgen-responsive human osteoblastic cell line. *Endocrinology* **140**, 5579–5586 (1999).

147. S. Fukayama and H. Tashjian, Direct modulation by androgens of the response of human bone cells (SaOS-2) to human parathyroid hormone (PTH) and PTH-related protein. *Endocrinology* **125**, 1789–1794 (1989).

148. A. Gray, H. Feldman, J. McKinlay, and C. Longcope, Age, disease, and changing sex hormone levels in middle-aged men: Results of the Massachusetts male aging study. *J Clin Endocrinol Metab* **73**, 1016–1025 (1991).

149. A. Vermeulen, Clinical review 24: Androgens in the aging male. *J Clin Endocrinol Metab* **73**, 221–224 (1991).

150. C. Pilbeam and L. Raisz, Effects of androgens on parathyroid hormone and interleukin-1-stimulated prostaglandin production in cultured neonatal mouse calvariae. *J Bone Miner Res* **5**, 1183–1188 (1990).

151. L. Hofbauer and S. Khosla, Androgen effects on bone metabolism: Recent progress and controversies. *Eur J Endocrinol* **140**, 271–286 (1999).

152. G. Passeri, G. Girasole, R. Jilka, and S. Manolagas, Increased interleukin-6 production by murine bone marrow and bone cells after estrogen withdrawal. *Endocrinology* **133**, 822–828 (1993).

153. L. Rifas, J. Kenney, M. Marcelli, R. Pacifici, S. Cheng, L. Dawson, and L. Avioli, Production of interleukin-6 in human osteoblasts and human bone marrow stromal cells: Evidence that induction by interleukin-1 and tumor necrosis factor-alpha is not regulated by ovarian steroids. *Endocrinology* **136**, 4056–4067 (1995).

154. T. Bellido, R. Jilka, B. Boyce, G. Girasole, H. Broxmeyer, S. Dalrymple, R. Murray, and S. Manolagas, Regulation of interleukin-6, osteoclastogenesis, and bone mass by androgens. *J Clin Invest* **95**, 2886–2895 (1995).

155. S. Lin, T. Yamate, Y. Taguchi, V. Borba, G. Girasole, C. O'Brien, T. Bellido, E. Abe, and S. Manolagas, Regulation of the gp80 and gp130 subunits of the IL-6 receptor by sex steroids in the murine bone marrow. *J Clin Invest* **100**, 1980–1990 (1997).

156. D. Vanderschueren and R. Bouillon, Androgens and bone. *Calcif Tissue Int* **56**, 341–346 (1995).

157. R. Turner, B. Riggs, and T. Spelsberg, Skeletal effects of estrogen. *Endocr Rev* **15**, 275–300 (1994).

158. H. Lebovitz and G. Eisenbarth, Hormonal regulation of cartilage growth and metabolism. *Vitam Horm* **33**, 575–648 (1975).

159. J. Iannotti, Growth plate physiology and pathology. *Orthop Clin North Am* **21**, 1–17 (1990).

160. S. Ren, S. Malozowski, P. Sanchez, D. Sweet, D. Loriaux, and F. Cassorla, Direct administration of testosterone increases rat tibial epiphyseal growth plate width. *Acta Endocr (Copenhagen)* **21**, 401–405 (1989).

161. E. Seeman, Periosteal bone formation—A neglected determinant of bone strength. *N Engl J Med* **349**, 320–323 (2003).

162. B. Riggs, L. Melton 3rd, R. Robb, J. Camp, E. Atkinson, J. Peterson, P. Rouleau, C. McCollough, M. Bouxsein, and S. Khosla, Population-based study of age and sex differences in bone volumetric density, size, geometry, and structure at different skeletal sites. *J Bone Miner Res* **19**, 1945–1954 (2004).

163. D. Vanderschueren, L. Vandenput, and S. Boonen, Reversing sex steroid deficiency and optimizing skeletal development in the adolescent with gonadal failure. *Endocr Dev* **8**, 150–165 (2005).

164. B. Kim, L. Mosekilde, Y. Duan, X. Zhang, L. Tornvig, J. Thomsen, and E. Seeman, The structural and hormonal basis of sex differences in peak appendicular bone strength in rats. *J Bone Miner Res* **18**, 150–155 (2003).

165. C. Wink and W. L. Felts, Effects of castration on the bone structure of male rats: A model of osteoporosis. *Calcif Tissue Int* **32**, 77–82 (1980).

166. M. Verhas, A. Schoutens, M. L'Hermite-Baleriaux, N. Dourov, A. Verschaeren, M. Mone, and A. Heilporn, The effect of orchidectomy on bone metabolism in aging rats. *Calcif Tissue Int* **39**, 74–77 (1986).

167. C. Danielsen, Long-term effect of orchidectomy on cortical bone from rat femur: Bone mass and mechanical properties. *Calcif Tissue Int* **50**, 169–174 (1992).

168. D. Vanderschueren, E. Van Herck, A. Suiker, W. Visser, L. Schot, and R. Bouillon, Bone and mineral metabolism in aged male rats: Short and long term effects of androgen deficiency. *Endocrinology* **130**, 2906–2916 (1992).

169. L. Vandenput, J. Swinnen, S. Boonen, E. Van Herck, R. Erben, R. Bouillon, and D. Vanderschueren, Role of the androgen receptor in skeletal homeostasis: The androgen-resistant testicular feminized male mouse model. *J Bone Miner Res* **19**, 1462–1470 (2004).

170. K. Venken, S. Boonen, E. Van Herck, L. Vandenput, N. Kumar, R. Sitruk-Ware, K. Sundaram, R. Bouillon, and D. Vanderschueren, Bone and muscle protective potential of the prostate-sparing synthetic androgen 7alpha-methyl-19-nortestosterone: Evidence from the aged orchidectomized male rat model. *Bone* **36**, 663–670 (2005).

171. G. Prakasam, J. Yeh, M. Chen, M. Castro-Magana, C. Liang, and J. Aloia, Effects of growth hormone and testosterone on cortical bone formation and bone density in aged orchiectomized rats. *Bone* **24**, 491–497 (1999).

172. T. Fujita, J. Ohtani, M. Shigekawa, T. Kawata, M. Kaku, S. Kohno, K. Tsutsui, K. Tenjo, M. Motokawa, Y. Tohma, and K. Tanne, Effects of sex hormone disturbances on craniofacial growth in newborn mice. *J Dent Res* **83**, 250–254 (2004).

173. M. Gunness and E. Orwoll, Early induction of alterations in cancellous and cortical bone histology after orchiectomy in mature rats. *J Bone Miner Res* **10**, 1735–1744 (1995).

174. D. Vanderschueren, I. Jans, E. van Herck, K. Moermans, J. Verhaeghe, and R. Bouillon, Time-related increase of biochemical markers of bone turnover in androgen-deficient male rats. *J Bone Miner Res* **26**, 123–131 (1994).

175. R. Weinstein, R. Jilka, A. Parfitt, and S. Manolagas, The effects of androgen deficiency on murine bone remodeling and bone mineral density are mediated via cells of the osteoblastic lineage. *Endocrinology* **138**, 4013–4021 (1997).

176. S. Moverare, K. Venken, A. Eriksson, N. Andersson, S. Skrtic, J. Wergedal, S. Mohan, P. Salmon, R. Bouillon, J. Gustafsson, D. Vanderschueren, and C. Ohlsson, Differential effects on bone of estrogen receptor alpha and androgen receptor activation in orchidectomized adult male mice. *Proc Natl Acad Sci USA* **100**, 13573–13578 (2003).

177. R. Jilka, K. Takahashi, M. Munshi, D. Williams, P. Roberson, and S. Manolagas, Loss of estrogen upregulates osteoblastogenesis in the murine bone marrow. Evidence for autonomy from factors released during bone resorption. *J Clin Invest* **101**, 1942–1950 (1998).

178. R. Mason and H. Morris, Effects of dihydrotestosterone on bone biochemical markers in sham and oophorectomized rats. *J Bone Miner Res* **12**, 1431–1437 (1997).

179. C. Lea, N. Kendall, and A. Flanagan, Casodex (a nonsteroidal antiandrogen) reduces cancellous, endosteal, and periosteal bone formation in estrogen-replete female rats. *Calcif Tissue Int* **58**, 268–272 (1996).

180. M. Kasra and M. Grynpas, The effects of androgens on the mechanical properties of primate bone. *Bone* **17**, 265–270 (1995).

181. A. Tivesten, S. Moverare-Skrtic, A. Chagin, K. Venken, P. Salmon, D. Vanderschueren, L. Savendahl, A. Holmang, and C. Ohlsson, Additive protective effects of estrogen and androgen treatment on trabecular bone in ovariectomized rats. *J Bone Miner Res* **19**, 1833–1839 (2004).

182. T. Bateman, J. Broz, M. Fleet, and S. Simske, Differing effects of two-week suspension on male and female mouse bone metabolism. *Biomed Sci Instrum* **34**, 374–379 (1997).

183. T. Tozum, M. Oppenlander, A. Koh-Paige, D. Robins, and L. McCauley, Effects of sex steroid receptor specificity in the regulation of skeletal metabolism. *Calcif Tissue Int* **75**, 60–70 (2004).

184. S. Yeh, M. Tsai, Q. Xu, X. Mu, H. Lardy, K. Huang, H. Lin, S. Yeh, S. Altuwaijri, X. Zhou, L. Xing, B. Boyce, M. Hung, S. Zhang, L. Gan, C. Chang, and M. Hung, Generation and characterization of androgen receptor knockout (ARKO) mice: An *in vivo* model for the study of androgen functions in selective tissues. *Proc Natl Acad Sci USA* **99**, 13498–13503 (2002).

185. D. Vanderschueren, E. Van Herck, A. Suiker, W. Visser, L. Schot, K. Chung, R. Lucas, T. Einhorn, and R. Bouillon, Bone and mineral metabolism in the androgen-resistant (testicular feminized) male rat. *J Bone Miner Res* **8**, 801–809 (1993).

186. D. Vanderschueren, E. Van Herck, P. Geusens, A. Suiker, W. Visser, K. Chung, and R. Bouillon, Androgen resistance and deficiency have difference effects on the growing skeleton of the rat. *Calcif Tissue Int* **55**, 198–203 (1994).

187. S. Kato, T. Matsumoto, H. Kawano, T. Sato, and K. Takeyama, Function of androgen receptor in gene regulations. *Steroid Biochem Mol Biol* **89/90**(1–5), 627–633 (2004).

188. A. Schoutens, M. Verhas, M. L'Hermite-Baleriaux, M. L'Hermite, A. Verschaeren, N. Dourov, M. Mone, A. Heilporn, and A. Tricot, Growth and bone haemodynamic responses to castration in male rats. Reversibility by testosterone. *Acta Endocrinol* **107**, 428–432 (1984).

189. J. Kapitola, J. Kubickova, and J. Andrle, Blood flow and mineral content of the tibia of female and male rats: Changes following castration and/or administration of estradiol or testosterone. *Bone* **16**, 69–72 (1995).

190. D. Somjen, Z. Mor, and A. Kaye, Age dependence and modulation by gonadectomy of the sex-specific response of rat diaphyseal bone to gonadal steroids. *Endocrinology* **134**, 809–814 (1994).

191. G. Wakley, H. Schutte, K. Hannon, and R. Turner, Androgen treatment prevents loss of cancellous bone in the orchidectomized rat. *J Bone Miner Res* **6**, 325–330 (1991).

192. D. Vanderschueren, E. Van Herck, P. Schot, E. Rush, T. Einhorn, P. Geusens, and R. Bouillon, The aged male rat as a model for human osteoporosis: Evaluation by nondestructive measurements and biomechanical testing. *Calcif Tissue Int* **53**, 342–347 (1993).

193. K. Venken, K. De Gendt, S. Boonen, J. Ophoff, R. Bouillon, J. Swinnen, G. Verhoeven, and D. Vanderschueren, Relative impact of androgen and estrogen receptor activation in the effects of androgens on trabecular and cortical bone in growing male mice: A study in the androgen receptor knockout mouse model. *J Bone Miner Res* **21**, 576–585 (2006).

194. J. Tobias, A. Gallagher, and T. Chambers, 5α-Dihydrotestosterone partially restores cancellous bone volume in osteopenic ovariectomized rats. *Am J Physiol* **267**, E853–E859 (1994).

195. C. Lea and A. Flanagan, Ovarian androgens protect against bone loss in rats made oestrogen deficient by treatment with ICI 182,780. *J Endocrinol* **160**, 111–117 (1999).

196. C. Lea and A. Flanagan, Physiological plasma levels of androgens reduce bone loss in the ovariectomized rat. *Am J Physiol* **274,** E328–E335 (1998).

197. C. Lea, V. Moxham, M. Reed, and A. Flanagan, Androstenedione treatment reduces loss of cancellous bone volume in ovariectomised rats in a dose–responsive manner and the effect is not mediated by oestrogen. *J Endocrinol* **156,** 331–339 (1998).

Phosphatonins

PETER J. TEBBEN, THERESA J. BERNDT, AND RAJIV KUMAR

I. INTRODUCTION

The role of phosphorus in human physiology is diverse and essential for a multitude of systems to function properly. In addition to its critical role in skeletal mineralization, phosphorus is an essential factor in all other tissues. Phosphorus plays an integral part in energy homeostasis, enzyme function, and cell membrane integrity [1–5]. Significant hypophosphatemia can result in skeletal, hematopoietic, muscle, or cardiac dysfunction.

Phosphorus is a key substrate in bone, and appropriate concentrations are required for normal mineralization to occur [6]. Calcium and phosphorus are incorporated into the skeleton primarily in the form of hydroxyapatite. Conditions resulting in chronic hypophosphatemia are associated with abnormal mineralization that manifest as rickets in children and osteomalacia in adults [7, 8]. Mineralization defects can occur in spite of normal concentrations of calcium, $1\alpha,25$-dihydroxyvitamin D_3 [$1\alpha,25(OH)_2D_3$], and parathyroid hormone (PTH) when phosphorus concentrations are insufficient. Diseases such as X-linked hypophosphatemic rickets (XLH), autosomal dominant hypophosphatemic rickets (ADHR), and tumor-induced osteomalacia (TIO) are disorders characterized histologically by widened osteoid seams as a result of defective mineralization due to hypophosphatemia [9, 10]. The majority of patients with these disorders have normal or near normal concentrations of calcium, $1\alpha,25(OH)_2D_3$, and PTH [8, 11]. Although additional factors may be involved, this points to the essential role phosphorus plays in normal skeletal biology.

Many factors affect the absorption/reabsorption of phosphorus in the intestine and kidney that ultimately influence concentrations in the blood (Table 14-1). Classically, the major hormones involved are considered to be $1\alpha,25(OH)_2D_3$ and PTH. However, more recently, it has become clear that newly described phosphaturic peptides play an important role in disorders of phosphate homeostasis and skeletal mineraliza-

tion [12–26]. In this chapter, we discuss the role of $1\alpha,25(OH)_2D_3$ and PTH as well as the potential role of the phosphaturic peptides FGF23, sFRP4, MEPE, and FGF7 in abnormal, and possibly normal, phosphate homeostasis.

II. PHOSPHORUS HOMEOSTASIS

The majority of phosphate in humans is found in the skeleton with the remainder distributed in other tissues and the extracellular space. Phosphorus balance is primarily determined by intestinal absorption and renal reabsorption regulated by vitamin D and PTH, respectively. Serum phosphorus concentrations reflect overall balance and the movement of phosphorus between plasma and bone or soft tissue. The physiological range of circulating phosphorus concentrations in adults is approximately 2.5–4.5 mg/dL. This value is slightly higher in children whose normal range may be as high as 4.3–5.4 mg/dL. Plasma phosphate concentrations are decreased by ingestion of a low-phosphate diet and increased by a high-phosphate intake. These changes are associated with a concomitant reciprocal change in plasma calcium concentration. This in turn causes changes in PTH and vitamin D synthesis to restore phosphate balance. It is important to recognize that the renal and intestinal adaptations in phosphate absorption/reabsorption can also occur independent of PTH and vitamin D. However, our understanding of phosphorus homeostasis is best understood in the context of its interactions with vitamin D and PTH.

A. Role of the Kidney in Phosphate Homeostasis

In states of neutral phosphate balance, the amount of phosphate excreted in the urine is equal to the net amount of intestinal phosphate absorption. Virtually all inorganic phosphate in the serum is filtered by the glomerulus [27–29]. About 80% to 90% of filtered phosphorus is reabsorbed in the kidney, primarily by the proximal tubule. The amount of phosphorus reabsorbed is greatest in the first half of the proximal tubule and exceeds that of sodium [27]. There is evidence for further phosphorus reabsorption by the pars recta portion of the proximal tubule, particularly in the absence of PTH [27]. Little or no phosphorus reabsorption occurs in the loop of Henle or the distal tubule. The reabsorption of phosphorus is sodium-dependent and is mediated by a sodium-phosphate cotransporter (NaPi IIa) [30]. NaPi IIa transporter activity is increased by ingestion of a low-phosphate diet and decreased by ingestion of a high-phosphate diet. The renal adaptation to changes in dietary phosphate intake occurs very rapidly, and changes in phosphate reabsorption can occur independent of PTH. This intrinsic renal adaptation, which is demonstrable *in vivo* and *in vitro*, is mediated by unknown mechanisms.

PTH is recognized as the principal hormonal regulator of renal phosphate reabsorption by the proximal tubule. However, it is important to recognize additional factors modulate the inhibition of phosphate reabsorption by PTH, such as respiratory acidosis or alkalosis, volume status, catecholamines, and growth hormone [27, 31, 32] (Table 14-1).

It is well known that PTH concentrations are exquisitely sensitive to changes in serum calcium concentrations [33, 34]. PTH secretion is also stimulated by high-phosphate ingestion, and this effect is mediated indirectly by decreases in calcium as well as through direct mechanisms [35–38]. Under normal conditions, a phosphate load (intravenous or oral) will stimulate PTH release from parathyroid gland cells, thus increasing renal excretion of phosphorus and maintaining normal serum phosphate concentrations. Likewise, a diet low in phosphorus will result in renal conservation of phosphorus at least partially due to a decrease in PTH secretion. The phosphaturic effect of PTH administration is the result of removal of NaPi IIa transporters from the apical brush border of renal proximal tubule cells both *in vitro* and *in vivo* [39] (Figures 14-1 and 14-2). Chronic exposure to elevated concentrations of PTH in normal animals results in an increased fractional excretion of phosphorus and hypophosphatemia.

Although PTH appears to be primarily responsible for renal phosphate regulation, vitamin D also alters renal phosphate reabsorption. *In vivo* and *in vitro* studies performed by Taketani *et al.* demonstrated that NaPi IIa expression in renal tissues is increased by the administration of $1\alpha,25(OH)_2D_3$ [40]. Others have shown that the effect of $1\alpha,25(OH)_2D_3$ on phosphate reabsorption in the kidney requires the presence of PTH [41–43].

The recently described phosphatonins, fibroblast growth factor 23 (FGF23), secreted frizzled-related protein 4 (sFRP4), matrix extracellular phosphoglycoprotein (MEPE), and fibroblast growth factor 7 (FGF7) also inhibit renal phosphate reabsorption *in vitro* and *in vivo* [44–47]. The effects of phosphatonins on renal phosphate reabsorption will be discussed in detail in the following sections.

B. Role of the Intestine in Phosphate Homeostasis

Phosphorus absorption in the intestine primarily takes place in the proximal small bowel. The intestinal absorption of phosphorus is largely dependent on the amount of phosphorus consumed. Nonhormonal factors such as the availability of phosphorus in the gastrointestinal tract can influence serum phosphorus concentrations. Dietary calcium and other phosphate-binding substances (such as sevelamer hydrochloride) will effectively reduce the amount of intestinal phosphate available for absorption.

The intestinal epithelial apical brush border contains a sodium-dependent phosphate-cotransporter, NaPi IIb. The amount of intestinal apical membrane NaPi IIb is increased in animals fed a low-phosphate diet or after the administration of $1\alpha,25(OH)_2D_3$ [48].

TABLE 14-1   Factors That Alter Renal Phosphate Excretion

Increase	Decrease
• High-phosphate diet	• Low-phosphate diet
• Parathyroid hormone	• Thyroparathyroidectomy
• Increased pCO_2	• Growth hormone
• Calcitonin	• Thyroxine
• Chronic vitamin D	• Acute vitamin D
• Glucagon	• Insulin
• Glucocorticoids	• Volume contraction
• Volume expansion	• Decreased pCO_2
• Chronic acidosis	• Stimulation of α/β adrenoreceptors
• Dopamine	
• Starvation	
• Diuretics	
• "Phosphatonins"	
∘ FGF23	
∘ sFRP4	
∘ MEPE	
∘ FGF7	

Modified from [130].

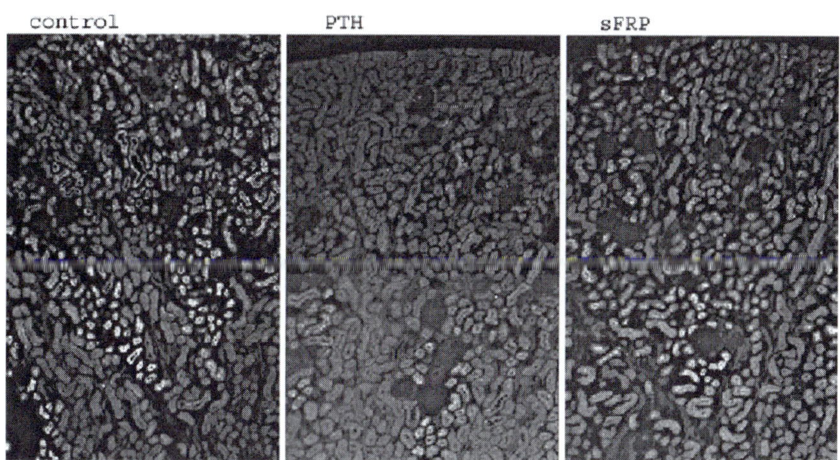

FIGURE 14-1   Immunohistochemical detection of NaPi IIa in rat renal tissue. NaPi IIa protein is present in the apical brush border membranes of proximal tubule cells. Reduced NaPi IIa staining is apparent in the renal slices taken from rats infused with PTH or sFRP4 compared to control rats (original magnification, 40×). Reprinted from [39].

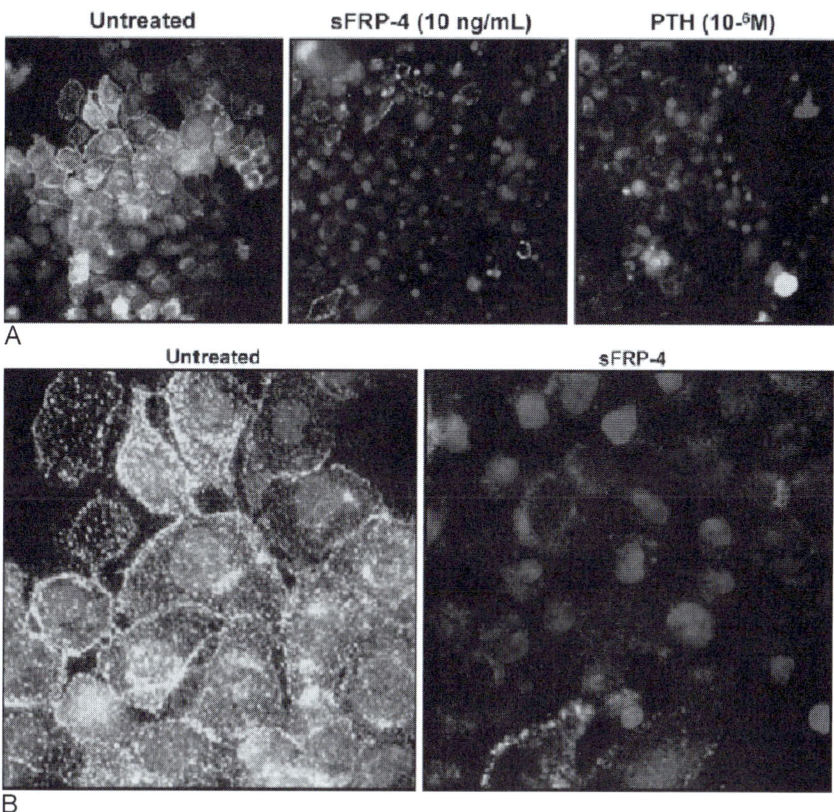

FIGURE 14-2   Opossum kidney (OK) cells expressing chimeric NaPi IIa-V5 were exposed to sFRP4 or PTH. The presence of NaPi IIa-V5 was detected using an antibody directed against the V5 epitope. After exposure of the OK cells expressing NaPi IIa-V5 to sFRP4 or PTH for 3 hours, reduced NaPi IIa-V5 protein was detected compared to untreated cells. (A, original magnification, 200×; B, original magnification, 400×.) Reprinted from [39].

The upregulation of NaPi IIb in the intestine while on a low-phosphate diet is mediated by vitamin D–dependent mechanisms and is independent of PTH [49–51].

Conversely, a high phosphorus diet or elevated serum phosphorus concentrations act to decrease the expression of 25-hydroxyvitamin D-1α-hydroxylase in the renal proximal tubule cells. Decreased conversion of 25-hydroxyvitamin D_3 to $1\alpha,25(OH)_2D_3$ will lead to a decrease in the intestinal absorption of phosphorus returning serum concentrations to the physiological range.

Despite our seemingly robust understanding regarding the various factors involved in normal phosphate physiology, our knowledge is incomplete. The recently described phosphatonins have added significantly to our knowledge of phosphorus and vitamin D metabolism and bone mineralization.

III. PHOSPHATONINS

Several diseases characterized by abnormal phosphorus, vitamin D, and bone metabolism have led to the discovery of factors that may regulate phosphate homeostasis in physiologic and pathophysiologic conditions. Studies of inherited forms of rickets (XLH and ADHR) and TIO have identified proteins that conform to the proposed definition of a phosphatonin. A phosphatonin is considered to be a circulating factor that induces phosphaturia through PTH-independent mechanisms leading to hypophosphatemia. The peptides that fulfill this definition include fibroblast growth factor 23 (FGF23), matrix extracellular phosphoglycoprotein (MEPE), secreted frizzled-related protein 4 (sFRP4), and fibroblast growth factor 7 (FGF7). Several of these peptides also inhibit the formation of $1\alpha,25(OH)_2D_3$ by decreasing the expression of 25-hydroxyvitamin D-1α-hydroxylase.

Prior to the identification of these phosphaturic peptides, it had long been recognized that a circulating factor was likely responsible for the hyperphosphaturia, hypophosphatemia, and rickets/osteomalacia associated with TIO and XLH. Adults with TIO present with classic symptoms of osteomalacia including pain, weakness, and fractures or pseudofractures. Children with TIO have been described with rickets. This form of hypophosphatemic rickets/osteomalacia can be differentiated from the inherited forms of rickets in that it is acquired and can be cured if the offending tumor is removed. The observation that the hypophosphatemia and bone disease completely resolved with removal of the tumor suggested that a circulating factor, presumably arising from the tumor, caused the phosphate abnormalities. Cai *et al.* performed studies in which cells derived from a tumor from a patient with TIO expressed a factor that inhibited phosphate transport in opossum kidney (OK) cells [52]. This factor was present in the supernatant fraction of cultured tumor cells, specifically inhibited sodium-dependent phosphate transport, and did not affect amino acid or glucose transport. Furthermore, when these cells were implanted into nude mice, hypophosphatemia and osteomalacia occurred.

Additional evidence that a circulating factor other than PTH could induce phosphaturia has come from studies of the mouse model of XLH. The *Hyp* mouse has a 3′ deletion of the gene encoding the phosphate-regulating gene with homologies to endopeptidases on the X chromosome (PHEX). These mice display a phenotype consisting of hyperphosphaturia, hypophosphatemia, and osteomalacia. Studies of *Hyp* mice parabiosed with normal mice showed that phosphaturia could be induced in the wild-type mouse, suggesting that a circulating phosphaturic factor was present in the blood of *Hyp* mice [53, 54]. Further evidence for a circulating phosphaturic factor was offered by Nesbitt *et al.*, who performed renal cross-transplantation studies between normal and *Hyp* mice [55]. In these experiments, normal mice receiving a kidney from a *Hyp* mouse had normal phosphate excretion. In contrast, a *Hyp* mouse receiving a kidney from a normal mouse showed no change in its hyperphosphaturia. These studies were consistent with the concept of the existence of a humoral factor being responsible for the phosphaturia in *Hyp* mice and not an intrinsic renal defect. It has been suggested that alterations in PHEX in the *Hyp* mice may be responsible for impaired degradation of a hypothetical phosphaturic factor.

IV. FIBROBLAST GROWTH FACTOR 23

A. Hypophosphatemic Disorders with Defective Mineralization

I. AUTOSOMAL DOMINANT HYPOPHOSPHATEMIC RICKETS

FGF23 is a 251–amino acid peptide encoded on the short arm of chromosome 12 in humans. FGF23 was initially believed to play a role in the function of the ventrolateral thalamic nucleus of the brain based on *in situ* hybridization studies performed in mice [56]. Shortly after this initial report of a novel fibroblast growth factor, the ADHR Consortium identified missense mutations in the gene encoding FGF23 in patients with ADHR [57]. It was speculated that the missense mutations lead to a gain of function in FGF23 and that FGF23 may be a circulating factor capable of inducing hypophosphatemia. Substitution of the arginine residues at amino acid positions 176 or 179 results in the ADHR phenotype (hypophosphatemia, hyperphosphaturia, rickets/osteomalacia, short stature, and dental abscesses) [57]. Shimada *et al.* demonstrated that this mutant FGF23 was resistant to proteolytic cleavage between residues 176 and 180 and was phosphaturic when administered intraperitoneally to mice [58]. Furthermore, when cells expressing mutant or wild-type FGF23 were implanted into athymic nude mice, the animals became hypophosphatemic and had impaired bone mineralization (Figure 14-3) [58]. Bai *et al.* generated transgenic mice overexpressing mutant FGF23 to assess the role of mutant FGF23 in phosphate homeostasis [59]. One- to two-month-old FGF23 transgenic mice exhibited hypophosphatemia, increased renal phosphate excretion, elevated alkaline phosphatase concentrations, and inappropriately low serum $1\alpha,25(OH)_2D_3$ concentrations relative to the degree of hypophosphatemia. Femoral shortening and mineralization defects were also seen in the mutant FGF23 transgenic mice compared to wild-type mice. These biochemical and histological characteristics are reminiscent of the findings in patients with ADHR. In addition to hypophosphatemia and inappropriately low $1\alpha,25(OH)_2D_3$ concentrations, Saito *et al.* also demonstrated that mutant FGF23 reduced sodium-dependent phosphate transport in renal as well as intestinal brush border membrane vesicles [60]. The changes in intestinal phosphate absorption were vitamin D dependent. Several disorders associated with abnormal serum FGF23 and phosphate levels are outlined in Table 14-2.

2. TUMOR-INDUCED OSTEOMALACIA

TIO is an acquired disorder with many clinical and biochemical similarities to patients with ADHR and XLH. Patients exhibit signs and symptoms of rickets/osteomalacia including bone pain, fractures, and weakness. Hemangiopericytoma is the most common histological type of tumor associated with this syndrome although various other types have been described and have been found in virtually all regions of the body [61–68]. Distinguishing TIO from inherited forms of hypophosphatemic rickets/osteomalacia can be difficult if it develops at a young age, since these tumors are notoriously difficult to locate. As previously mentioned, studies by Cai *et al.* demonstrated that a factor (or factors) secreted from cells taken from a tumor in a patient with TIO was able to inhibit sodium-dependent phosphate transport in renal tubular cells. This factor was distinct from PTH and did not alter glucose or amino acid transport in OK cells [52, 69]. The observation that extracts from tumors taken from patients with

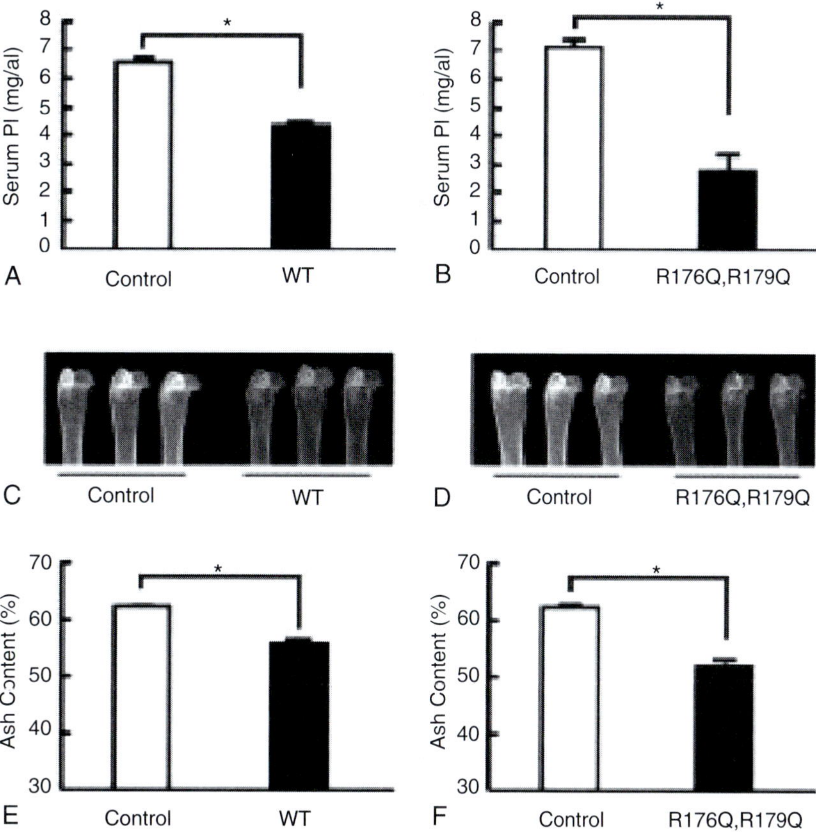

FIGURE 14-3   Chinese hamster ovary (CHO) cells expressing wild-type FGF23 or mutant FGF23 (R176Q or R179Q) were implanted into mice. Serum phosphorus is reduced in the mice exposed to CHO cells expressing wild-type or mutant FGF23 (A, B); Radiographs of femurs (C, D) and ash content of femurs (E, F) demonstrate reduced mineral content in mice exposed to FGF23 (wild-type or mutant) compared to control animals. Reprinted from [58].

TABLE 14-2   Disorders Associated with Abnormalities in FGF23

Disorder	Pi Concentration	FGF23 Concentration		Reference
		Intact	C-terminal	
XLH	Decreased	?	Increased	[11]
ADHR	Decreased	?	Increased	[57]
TIO	Decreased	Increased	Increased	[11, 61, 63–66, 71, 80]
HHM	Decreased	?	Increased	[131]
HLNSS	Decreased	?	Increased	[132]
Fibrous dysplasia	Decreased	?	Increased	[82, 83]
Tumoral calcinosis	Increased	Decreased	Increased	[84–86, 133]
Renal failure	Increased	Increased	Increased	[80, 90–92, 98]
Graves' disease*	Variable	Variable	?	[134]
Ovarian cancer**	Normal	Increased	Increased	[135]

XLH, X-linked hypophosphatemic rickets; ADHR, autosomal dominant hypophosphatemic rickets; TIO, tumor-induced osteomalacia; HHM, humoral hypercalcemia of malignancy; HLNSS, hypophosphatemic linear nevus sebaceous syndrome.
*FGF23 concentrations declined with antithyroid therapy.
**Stage 3 and 4 ovarian cancer.

TIO inhibit phosphate transport in OK cells has been replicated by others [65, 70].

Studies using serial analysis of gene expression (SAGE) demonstrated that in addition to FGF23, other phosphaturic factors including MEPE, sFRP4, and FGF7 are also highly expressed in tumors taken from patients with TIO [68]. The effects of FGF23 on phosphate and vitamin D metabolism have been the most characterized, and the development of an assay for FGF23 in serum has allowed the assessment of the role of FGF23 in disease conditions associated with phosphate wasting. This may explain why not all patients with TIO have elevated serum concentrations of FGF23 and implies that other phosphaturic proteins may also be important in the clinical expression of this disease [11].

Serum concentrations of FGF23 have been measured in patients with known or presumed TIO, and most but not all patients have elevated FGF23 concentrations [11, 61, 64, 71–73]. Furthermore, serum FGF23 concentrations decline into the normal range shortly after removal of the offending tumor [11, 64, 71]. Serum phosphate and $1\alpha,25(OH)_2D_3$ concentrations normalize within hours to days after removal of the offending tumor [11, 62, 64, 71, 74, 75]. However, the histological changes in bone require significantly more time to normalize [72].

Selective venous sampling for determination of FGF23 concentrations has been employed to confirm the location of a TIO tumor prior to surgical excision. Significantly higher FGF23 concentrations were found just proximal to the offending tumor compared to other sampling locations [66]. RT-PCR, *in situ* hybridization,

and immunohistochemical techniques have demonstrated FGF23 mRNA and protein expression in TIO tumors [63, 65]. These observations suggest that the tumor is the source of elevated circulating concentrations of FGF23.

These data provide compelling evidence that FGF23 is a causative factor inducing the biochemical and histological changes seen in TIO.

3. X-LINKED HYPOPHOSPHATEMIC RICKETS

XLH is caused by mutations in the gene encoding PHEX, an endopeptidase on the X chromosome, and is characterized by hypophosphatemia, increased renal fractional excretion of phosphorus, and rickets [76, 77]. As previously mentioned, the *Hyp* mouse is the animal homologue of XLH. Studies of these mice have shown that renal phosphate wasting and the bone phenotype are due to a circulating factor and not an intrinsic renal defect [53–55]. Many patients with XLH have elevated blood concentrations of FGF23 compared to normal controls [11]. Several investigators have suggested that FGF23 is a substrate for PHEX [78, 79]. This provides a possible explanation for the elevated FGF23 concentrations observed in XLH as well as a mechanism for renal phosphate loss. However, not all patients with XLH have elevated concentrations of FGF23, implying that other factors may also be important in the development of hypophosphatemia and rickets [80].

4. FIBROUS DYSPLASIA/McCUNE-ALBRIGHT

Fibrous dysplasia is caused by post-zygotic activating mutations in the *GNAS1* gene. Fibrous dysplasia of one or more bones may be an isolated finding or associated with McCune-Albright syndrome with characteristic café-au-lait macules, precocious puberty, and other related endocrine abnormalities. Some patients with these disorders also display renal phosphate wasting [81]. The degree of phosphate wasting has been correlated with the extent of bone involvement. Phosphate wasting was not related to elevated cAMP, suggesting that an activating *GNAS1* mutation in the kidney was not responsible for the phosphaturia. Others have reported that FGF23 is expressed in the abnormal bone of many patients with isolated fibrous dysplasia [82]. In that study, the intensity of FGF23 staining in bone tissue negatively correlated with serum phosphorus concentrations. Serum FGF23 concentrations are higher in patients with McCune-Albright syndrome or fibrous dysplasia than in age-matched controls [83]. Furthermore, patients with renal phosphate wasting associated with fibrous dysplasia or McCune-Albright have higher serum FGF23 concentrations than those without.

B. Hyperphosphatemic Disorders

1. Tumoral Calcinosis

Tumoral calcinosis is an interesting condition caused in some cases by mutations in the *GALNT3* gene or the *FGF23* gene [84–87]. The phenotype is similar despite the different genetic etiology. Biochemical findings include hyperphosphatemia, increased renal reabsorption of phosphorus, and normal or elevated $1\alpha,25(OH)_2D_3$ [84–88]. These findings are opposite to those found in patients with disorders associated with increased FGF23 activity such as ADHR and TIO. It is interesting to note that when FGF23 concentrations are measured by a technique that identifies carboxy-terminal fragments and intact FGF23, the concentrations are elevated. An explanation for this finding is offered by Benet-Pages *et al.*, who demonstrated altered processing of the mutant form of FGF23 (S71G) [85]. Expression of mutant FGF23 in HEK 293 cells resulted in the secretion of carboxy-terminal fragments of FGF23 but not intact FGF23. The intact protein was retained within the Golgi complex. Araya *et al.* reported similar *in vitro* data [86]. In this report, expression of mutant FGF23 (S129F) resulted in reduced detection of intact and N-terminal FGF23 by Western blotting. Serum FGF23 levels are also elevated in their patients with tumoral calcinosis when measured with an assay that detects carboxy-terminal fragments as well as the intact molecule. However, when measured using an assay that detects only intact FGF23, the concentrations were low. This suggests that biological activity of

FGF23 requires an intact molecule that is not secreted in patients with tumoral calcinosis.

Instead of mineralization defects resulting in rickets or osteomalacia, patients with tumoral calcinosis may have dramatic extraskeletal mineral deposits. A similar clinical and biochemical phenotype is apparent in FGF23 null mice, confirming that FGF23 mutations in patients with tumoral calcinosis represent a loss of function [89].

2. Chronic Kidney Disease

Patients with chronic kidney disease have abnormal phosphate and vitamin D metabolism. As renal function declines, serum phosphorus concentrations increase and $1\alpha,25(OH)_2D_3$ concentrations decrease. PTH levels are frequently elevated but insufficient to correct the hyperphosphatemia and impaired vitamin D production. Several investigators have documented increased serum concentrations of FGF23 in patients with chronic kidney disease. Initial studies were performed with an ELISA utilizing a capture-and-detection antibody that recognizes epitopes within the carboxy-terminal portion of the protein [80, 90, 91]. It was unclear whether the elevation in FGF23 was the result of increased production, decreased clearance, or the accumulation of inactive FGF23 fragments. Subsequent reports have clearly documented that intact FGF23 concentrations in serum are also elevated in patients with renal insufficiency [92, 93].

It has been suggested that increased FGF23 concentrations in renal disease may represent a compensatory mechanism for hyperphosphatemia. Serum FGF23 concentrations correlate positively with serum phosphorus and with the fractional excretion of phosphorus in some patients with CKD [91, 94]. Also, as FGF23 concentrations increase, $1\alpha,25(OH)_2D_3$ concentrations decline. This is not surprising since it has been shown that FGF23 acts in the renal proximal tubule to diminish 25-hydroxyvitamin D_3 1α-hydroxylase expression [95]. A potential feedback loop may exist between FGF23 and vitamin D, since $1\alpha,25(OH)_2D_3$ therapy in patients with CKD decreased serum FGF23 concentrations.

Decreased $1\alpha,25(OH)_2D_3$ may lead to increased PTH production and contribute to secondary hyperparathyroidism in these patients. Kazama *et al.* reported that serum FGF23 concentrations were highly predictive of the development of advanced secondary hyperparathyroidism in patients receiving chronic dialysis [96]. These investigators also found serum FGF23 levels to be predictive of their response to calcitriol therapy [97]. Patients treated with calcitriol had significantly higher serum phosphorus and FGF23 concentrations after 24 weeks of therapy. It is not clear whether the calcitriol therapy or increased serum phosphorus was directly responsible for increased serum FGF23. Another study performed in patients receiving maintenance hemodialysis found that serum phosphorus was positively associated with serum FGF23 concentrations. In this study, subjects were treated with sevelamer hydrochloride and calcium or calcium alone. The subjects receiving combined treatments had a significant reduction in serum phosphorus and FGF23, whereas subjects treated with calcium alone had no changes in either analyte [98].

C. Physiological Effects of FGF23

I. EFFECTS OF FGF23 IN THE KIDNEY AND INTESTINE

Phosphate homeostasis is affected directly by FGF23 as a result of its inhibition of NaPi IIa cotransporter activity and indirectly by inhibition of 25-hydroxyvitamin D_3 1α-hydroxylase expression. Experiments using OK cells (a proximal tubule epithelial cell) have demonstrated that phosphorus uptake is inhibited by FGF23 [58]. As previously mentioned, phosphate transport in the kidney is primarily regulated by the activity of NaPi IIa cotransporters in the apical membrane. FGF23 causes internalization of NaPi IIa cotransporters and degradation in the lysosome resulting in decreased phosphate transport.

Hypophosphatemia and impaired conversion of 25-hydroxyvitamin D_3 to $1\alpha,25(OH)_2D_3$ offer explanations for the impaired mineralization seen in the previously described disorders associated with elevated serum FGF23 concentrations.

$1\alpha,25(OH)_2D_3$ plays an important role in phosphate regulation primarily in the intestine. 25-Hydroxyvitamin D_3 1α-hydroxylase converts the inactive form of vitamin D to its active metabolite $1\alpha,25(OH)_2D_3$, which increases phosphorus transport in the small bowel. XLH and TIO are both examples of hypophosphatemic disorders characterized by inappropriately low or normal $1\alpha,25(OH)_2D_3$ concentrations relative to the degree of hypophosphatemia. This is in contrast to the marked elevation in serum $1\alpha,25(OH)_2D_3$ concentrations that is associated with hypophosphatemia induced by dietary phosphate restriction. In fact, serum $1\alpha,25(OH)_2D_3$ concentrations and renal 25-hydroxyvitamin D_3 1α-hydroxylase expression are decreased in animals exposed to FGF23 [95].

Miyamoto *et al.* performed a set of experiments in wild-type and vitamin D receptor (VDR) null mice [99]. The investigators injected mutant FGF23 (R179Q), which lowered serum phosphorus and $1\alpha,25(OH)_2D_3$ concentrations. Intestinal brush border membrane vesicles of the wild-type mice showed decreased sodium-dependent phosphate transport and reduced amounts of NaPi IIb protein. In contrast, intestinal sodium-dependent phosphate transport was not affected by FGF23 (R179Q) in the VDR null mice. These data suggest that FGF23 indirectly decreases phosphate transport in the intestine by reducing serum $1\alpha,25(OH)_2D_3$ concentrations.

2. EFFECTS OF FGF23 IN BONE

FGF23 has been shown to be expressed in a number of tissues including bone. Perwad *et al.* measured FGF23 mRNA in the calvaria of mice fed a diet containing 0.02% or 1% phosphate. FGF23 mRNA abundance was reduced by 85% in mice fed the low-phosphate diet [100]. In addition, FGF23 mRNA abundance was 30-fold higher in *Hyp* mouse calvaria, a condition known to be associated with elevated serum FGF23 concentrations. These data suggest that expression of FGF23 in bone is responsible for the changes in serum levels in patients or animals with XLH or after dietary phosphate manipulation.

It is clear that humans and mice with altered serum FGF23 levels display distinct bone phenotypes. Hypophosphatemic disorders such as XLH, ADHR, and TIO are characterized by rickets or osteomalacia. Bone histomorphometry reveals a mineralization defect with widened osteoid seams. However, it is not clear whether these changes are due to altered phosphorus and vitamin D metabolism or if there is a direct effect of FGF23 on bone.

Several investigators have determined that FGF23 binds to various fibroblast growth factor receptors (FGFR) [45, 101–103]. Yu *et al.* demonstrated that FGF23 binds to and activates the c-splice isoforms of FGFR1–3 and FGFR4 [102]. Others have also shown that the binding of FGF23 to various FGFRs does so with higher affinity in the presence of the protein klotho [103]. FGFRs are known to play an important role in limb development, including those that appear to interact with FGF23 [104]. Mutations in FGFR3 result in achondroplasia, hypochondroplasia, or thanatophoric dysplasia, which are characterized by various degrees of limb deformity including shortening and bowing. Limb shortening has also been reported in FGF23 null mice [105]. The authors also described narrowed growth plates with decreased numbers of hypertrophic chondrocytes. The ribs and vertebrae of the FGF23 null mice demonstrate a marked increase in woven bone and osteoid. FGF23 null mice have similar biochemical and clinical characteristics to patients with tumoral calcinosis due to mutations in the gene encoding FGF23, including hyperphosphatemia, elevated $1\alpha,25(OH)_2D_3$ concentrations, and extraskeletal mineralization [84–86, 105]. Chefetz *et al.* described a child with tumoral calcinosis due to a homozygous mutation in FGF23 (M96T). Radiographic investigation showed obvious bony abnormalities with areas of sclerosis, bowing of the distal radius, shortening of the ulna, and modeling defects in the distal femur and proximal tibia.

Although these observations suggest that FGF23 plays an important role in skeletogenesis, conclusive data are lacking. It will be difficult to interpret *in vivo* evidence for direct skeletal effects of FGF23 because of its concomitant effects on phosphorus and vitamin D metabolism. Perhaps future studies employing osteoblast cell culture in which vitamin D and mineral concentrations can be held constant may shed some light in this area.

3. FGF23 IN NORMAL PHOSPHATE HOMEOSTASIS

Significant evidence exists supporting the role of changes in serum FGF23 levels on phosphorus and vitamin D metabolism in disease states by mechanisms outlined in Figure 14-4. However, it is not entirely clear whether FGF23 plays a role in normal phosphate and vitamin D physiology. Several investigators have measured serum FGF23 concentrations after dietary manipulation of phosphorus, calcium, and/or vitamin D. Conflicting results have been reported in humans. Larsson *et al.* studied 6 healthy males for 6 days. A normal diet for 1 day was followed by 2 days of low-phosphate intake and subsequently a high-phosphate diet [91]. However, no changes in serum FGF23 levels were noted. A larger study by Ferrari *et al.* evaluated 29 healthy males given a low-phosphate diet for 5 days followed by a high-phosphate diet for 5 days separated by 2 days of a normal diet. These investigators found significantly lower serum FGF23 concentrations during phosphate restriction compared to supplementation

[106]. Experiments in wild-type and VDR null mice have given additional insight into the role of FGF23 during changes in dietary phosphate intake. Wild-type mice fed a low-phosphate diet have significantly lower serum FGF23 concentrations [107]. In the same set of experiments, it was noted that VDR null mice have very low basal FGF23 concentrations. However, when fed a rescue diet designed to normalize calcium and phosphorus, serum FGF23 levels increase dramatically, suggesting the effect of phosphate (and/or calcium) on serum FGF23 does not require vitamin D. Others have also documented an increase in serum FGF23 concentrations in mice with dose-dependent increases in phosphate ingestion [100, 108].

Administration of vitamin D to mice also results in elevations in FGF23 concentrations within 24 hours [109]. Basal concentrations of FGF23 in VDR null mice are low compared to wild-type mice and do not increase after administration of $1\alpha,25(OH)_2D_3$ [110]. Similar to phosphate, $1\alpha,25(OH)_2D_3$ administration to mice results in a significant upregulation of FGF23 mRNA in bone tissue [109]. Using this *in vivo* model, it is difficult to determine whether the changes were directly related to $1\alpha,25(OH)_2D_3$ or if changes in phosphorus, calcium, and/or PTH may have contributed. *In vitro* data would support a direct effect of $1\alpha,25(OH)_2D_3$ on FGF23 expression. Using UMR-106 osteoblast-like cells, Kolek *et al.* reported significant increases in FGF23 mRNA within 4 hours after exposure to $1\alpha,25(OH)_2D_3$. Taken together, these data support a physiological role of FGF23 in vitamin D and phosphorus homeostasis.

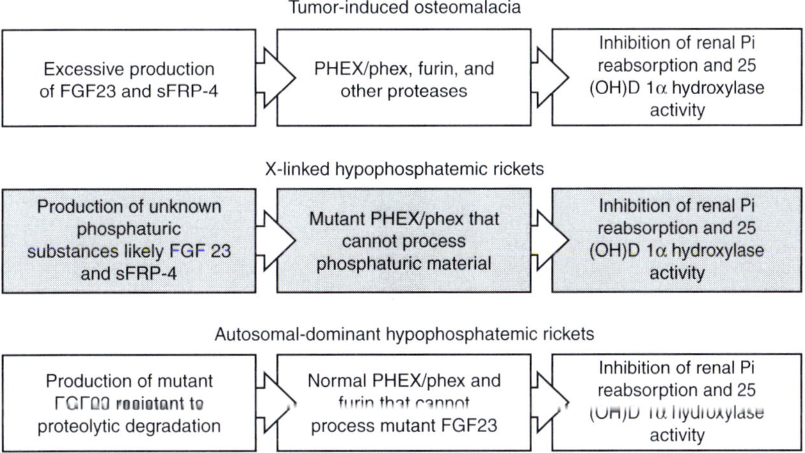

FIGURE 14-4   FGF23, fibroblast growth factor 23; sFRP4, secreted frizzled-related protein 4; PHEX, phosphate-regulating gene with homologies to endopeptidases on the X chromosome.

V. SECRETED FRIZZLED-RELATED PROTEIN 4

We have previously shown that sFRP4 is highly expressed in TIO tumors [68]. The relative expression of this protein was higher than FGF23. sFRP4 inhibits sodium-dependent phosphate transport in OK cells [44]. *In vivo* experiments in which recombinant sFRP4 was infused into normal rats resulted in reduced renal phosphorus reabsorption after 2 hours [44] (Figure 14-5). This effect was noted in intact as well as thyroparathyroidectomized animals, indicating PTH is not necessary for sFRP4 to induce phosphaturia. After 8 hours of sFRP4 infusion, serum phosphorus concentrations declined. However, the expected increase in $1\alpha,25(OH)_2D_3$ did not occur. Thus, sFRP4 may impair 25-hydroxyvitamin D_3 1α-hydroxylase activity similar to the effect of FGF23.

Infusion of sFRP4-reduced sodium-dependent phosphate transport in brush border membrane vesicles was compared to vehicle-infused animals [39]. This effect appears to be due to a reduction in the amount of NaPi IIa protein in renal tubule cells, which is easily appreciated in Figure 14-1A. Figure 14-1B demonstrates the loss of NaPi IIa expression in OK cells exposed to sFRP4, suggesting this is a direct effect and not due to other factors such as changes in PTH.

Secreted frizzled-related proteins including sFRP4 contain cysteine-rich domains similar to Frizzled receptors and act as Wnt antagonists. Subsequent experiments in our laboratory have demonstrated that sFRP4 is able to antagonize the Wnt pathway as demonstrated by reduced β-catenin and increased phosphorylated β-catenin expression [44]. The Wnt signaling pathway is complex and involves several other factors including the low-density lipoprotein receptor–related protein 5/6 (LRP 5/6), Frizzled receptors, and intracellular signaling through β-catenin. When secreted, Wnt proteins bind to Frizzled and the coreceptor LRP 5/6, resulting in the inhibition of intracellular phosphorylation of β-catenin. Nonphosphorylated β-catenin is then able to enter the nucleus and affect gene expression [111].

Wnt signaling plays an important role in normal development and likely is involved in bone and mineral metabolism [112]. Disruptions in this pathway have been described to affect bone biology. Mutations in LRP5 can lead to a high or low bone mass phenotype depending on whether the change leads to a gain or loss of function [113–115]. (See also Chapter 15, Johnson.)

VI. MATRIX EXTRACELLULAR PHOSPHOGLYCOPROTEIN

MEPE is a highly expressed protein in tumors causing TIO [68, 116]. It is also expressed in osteoblasts and osteocytes of mice during skeletogenesis and during fracture repair within fibroblast-like cells, chondrocytes, and osteocytes [117]. Immunohistochemical techniques and *in vitro* data have shown that MEPE is also expressed in osteocytes and osteoblasts of humans [118, 119]. It has been implicated to play a role in XLH since *Hyp* mice exhibit a 3-fold increase in mRNA levels of this protein in bone compared to normal mice [120].

In healthy subjects, serum concentrations of MEPE and phosphate are positively correlated [121]. Further-

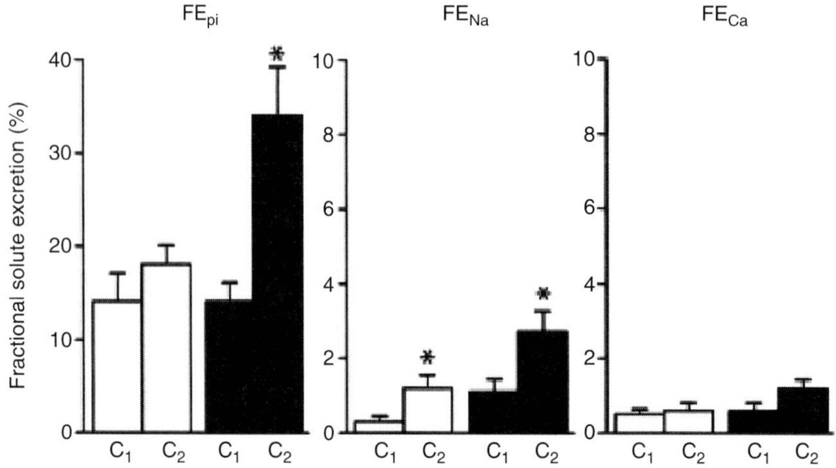

FIGURE 14-5   Effect of infusion of sFRP4 on solute excretion in intact rats. Intact rats were administered sFRP4 (black bars) at a dose of 0.3 μg/kg/h or vehicle (white bars) by intravenous infusion over a period of 2 hours. C_1 indicates equilibrium period prior to the infusion of sFRP4 or vehicle. C_2 indicates the experimental period during which sFRP4 or vehicle was infused. Fractional excretion of inorganic phosphate increased significantly in the rats after infusion of sFRP4. Reprinted from [44].

more, MEPE is positively correlated with hip bone mineral density in subjects over 60 years old [121]. Exposure of mouse osteoblast culture to $1\alpha,25(OH)_2D_3$ significantly reduced MEPE mRNA levels [120]. MEPE expression appears to be controlled at least in part by $1\alpha,25(OH)_2D_3$. These data imply MEPE may be an important factor in phosphorus and bone metabolism.

Recombinant MEPE administered into the peritoneum of mice reduces serum phosphorus and induces phosphaturia (Figure 14-6) [47]. Phosphate uptake in renal proximal tubule cell cultures is dose dependently inhibited by MEPE [47]. However, $1\alpha,25(OH)_2D_3$ concentrations in serum did not decline as is seen with exposure to FGF23 and sFRP4. Elevated MEPE expression in TIO tumors may contribute to the hypophosphatemia in these patients but cannot explain the defect in vitamin D metabolism. Other factors such as FGF23 or sFRP4 must be present to fully explain the biochemical phenotype.

In addition to the effects of MEPE on renal tubular phosphate handling, it may also play a role in mineralization. MEPE is normally cleaved and releases a peptide containing an ASARM sequence that is capable of inhibiting mineralization [122]. PHEX has been shown to interact with MEPE, which prevents proteolysis and release of the ASARM peptide [122]. Patients with XLH and *Hyp* mice have mutant PHEX, which is therefore unable to interact with MEPE and ASARM, resulting in release of ASARM. This can then lead to impairment of mineralization. Elevated concentrations of the ASARM peptide have been measured in the serum of patients with XLH and in *Hyp* mice [123].

VII. FIBROBLAST GROWTH FACTOR 7

FGF7 is a secreted protein also known as keratinocyte growth factor (KGF). FGF7 appears to be involved in the repair of skin injury and has also been implicated to play a role in other diseases such as breast cancer [124–128]. FGFR-2 IIIb is the receptor for FGF7 and is a distinct isoform from that proposed for FGF23 [129]. Carpenter *et al.* recently reported two patients with TIO tumors that abundantly expressed FGF7 [46]. These investigators demonstrated that conditioned media from TIO tumor cell cultures inhibited phosphate transport

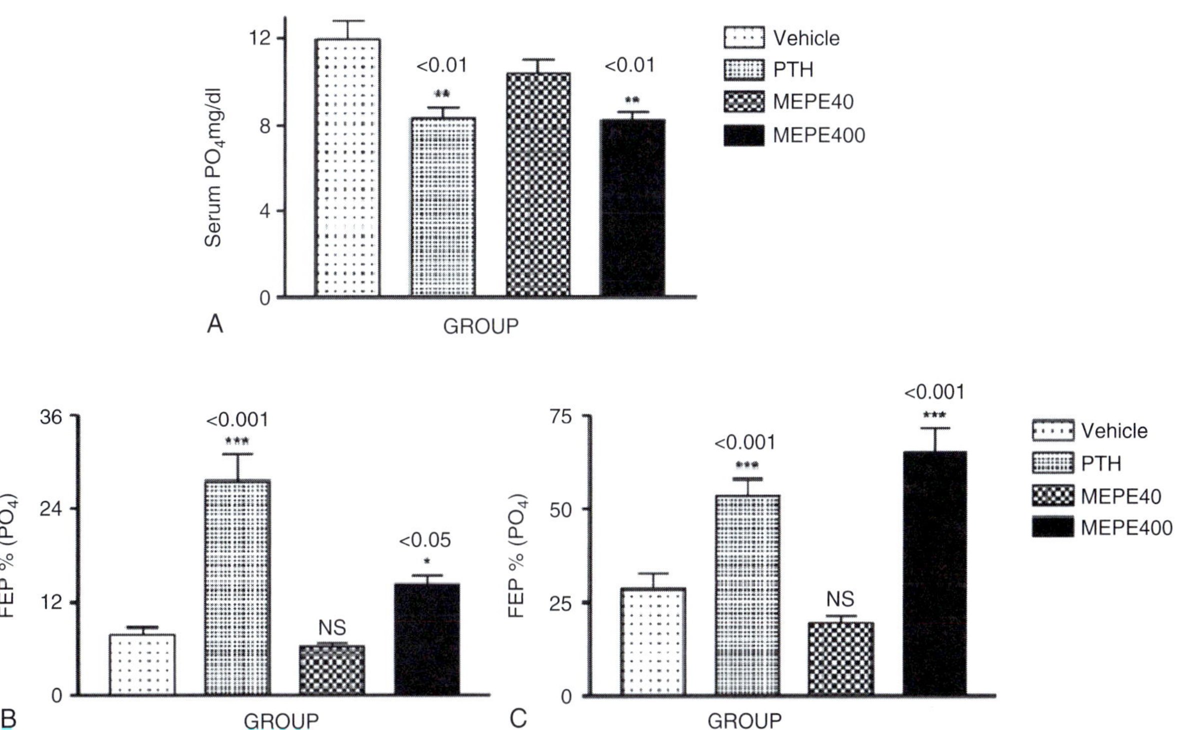

FIGURE 14-6   Exposure of mice to intraperitoneal injections of MEPE or PTH causes a reduction in serum phosphorus concentrations after 31 hours compared to vehicle-treated mice (A). Fractional excretion of phosphorus is increased in mice exposed to intraperitoneal injections of MEPE or PTH after 6 hours (B) or 31 hours (C). MEPE40, 40 µg/kg/30 hours; MEPE400, 400 µg/kg/30 hours; PTH, 80 µg/kg/30 hours. Reprinted from [47].

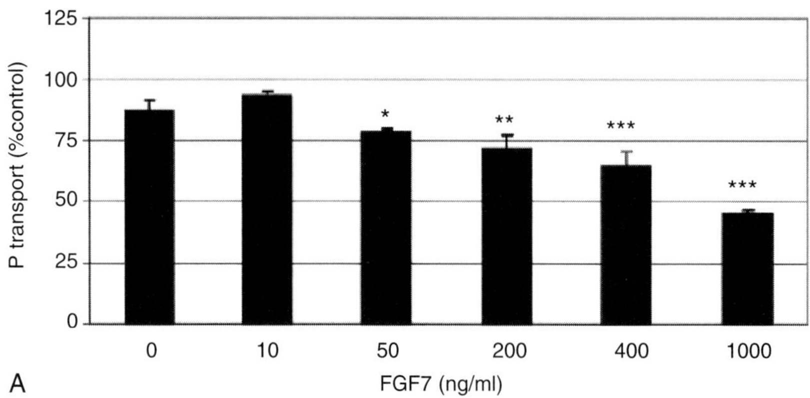

FIGURE 14-7 Phosphate transport was measured in opossum kidney (OK) cells exposed to increasing concentrations of FGF7. A significant and dose-dependent decrease in OK cell phosphate transport was detected at FGF7 concentrations of 50 ng/mL or greater. Reprinted from [46].

in OK cells. In this study, FGF23 concentrations in the conditioned media were not elevated, whereas FGF7 concentrations were higher in conditioned media that inhibited phosphate transport. Exposure of OK cells to FGF7 reduced phosphate transport in a dose-dependent manner (Figure 14-7). This effect could be diminished by the addition of neutralizing FGF7 antibodies.

Lyakhovich *et al.* performed a study using breast cancer cell lines expressing FGF7 [125]. Exposure of this cell line to $1\alpha,25(OH)_2D_3$ increased FGF7 expression in a time-dependent manner. It is intriguing to speculate that FGF7 may be involved in normal phosphate homeostasis and regulated by vitamin D. However, more data will be required to elucidate a possible role of FGF7 in bone and mineral metabolism.

Taken together, these results suggest that several tumor-derived factors most likely contribute to the phosphate wasting seen in patients with TIO. It may also explain why not all patients with TIO have been documented to have elevated serum concentrations of FGF23.

VIII. SUMMARY

The regulation of phosphate metabolism and skeletal mineralization is complex and is mediated by mechanisms that are incompletely understood. Although a variety of newly described phosphaturic factors alter normal phosphate homeostasis, PTH remains the key hormone that regulates phosphate homeostasis. Phosphatonins, which were initially identified as the result of the study of rare disorders characterized by hyperphosphaturia, have expanded our knowledge regarding normal and abnormal bone and mineral homeostasis. FGF23 and sFRP4 are both capable of inhibiting renal phosphate reabsorption and decreasing

the formation of active vitamin D metabolites leading to changes in intestinal phosphate absorption and bone mineralization. The synthesis of these peptides may be influenced by changes in dietary phosphorus intake. MEPE and FGF7 have also been shown to induce hyperphosphaturia, and MEPE may play a significant role in skeletal mineralization. A better understanding of the phosphatonins will provide useful insight into normal and abnormal skeletal biology.

REFERENCES

1. E. G. Krebs, and J. A. Beavo, Phosphorylation-dephosphorylation of enzymes. *Ann Rev Biochem,* **48,** 923–959 (1979).
2. E. G. Krebs, and J. T. Stull, Protein phosphorylation and metabolic control. *Ciba Foundation Symposium,* **31,** 355–367 (1975).
3. S. R. Hubbard, and J. H. Till, Protein tyrosine kinase structure and function. *Ann Rev Biochem,* **69,** 373–398 (2000).
4. P. Cohen, The structure and regulation of protein phosphatases. *Ann Rev Biochem,* **58,** 453–508 (1989).
5. P. Cohen, The discovery of protein phosphatases: From chaos and confusion to an understanding of their role in cell regulation and human disease. *Bioessays,* **16**(8), 583–588 (1994).
6. W. F. Neuman, Bone material and calcification mechanisms. In *Fundamental and Clinical Bone Physiology* (M. R. Urist, ed.), pp. 83–107. Lippincott, Philadelphia (1980).
7. J. L. Berry, M. Davies, and A. P. Mee, Vitamin D metabolism, rickets, and osteomalacia. *Sem Musculoskel Radiol,* **6**(3), 173–182 (2002).
8. M. J. Econs, and P. T. McEnery, Autosomal dominant hypophosphatemic rickets/osteomalacia: Clinical characterization of a novel renal phosphate-wasting disorder. *J Clin Endocrinol Metab,* **82**(2), 674–681 (1997).
9. P. J. Marie, and F. H. Glorieux, Histomorphometric study of bone remodeling in hypophosphatemic vitamin D-resistant rickets. *Metabol Bone Dis Related Res,* **3**(1), 31–38 (1981).
10. S. M. Jan de Beur, Tumor-induced osteomalacia. *JAMA,* **294**(10), 1260–1267 (2005).
11. K. B. Jonsson, R. Zahradnik, T. Larsson, *et al.,* Fibroblast growth factor 23 in oncogenic osteomalacia and X-linked

hypophosphatemia [see comment]. *N Engl J Med,* **348**(17), 1656–1663 (2003).

12. E. A. Imel, and M. J. Econs, Fibroblast growth factor 23, roles in health and disease. *J Am Soc Nephrol,* **16**(9), 2565–2575 (2005).

13. H. S. Tenenhouse, and Y. Sabbagh. Novel phosphate-regulating genes in the pathogenesis of renal phosphate wasting disorders. *Pflugers Archiv, Eur J Physiol,* **444**(3), 317–326 (2002).

14. R. Kumar, Phosphatonin—A new phosphaturetic hormone? (Lessons from tumour-induced osteomalacia and X-linked hypophosphataemia). *Nephrol Dial Transplantation,* **12**(1), 11–13 (1997).

15. P. S. Rowe, The wrickened pathways of FGF23, MEPE and PHEX. *Crit Rev Oral Biol Med,* **15**(5), 264–281 (2004).

16. R. Kumar, Tumor-induced osteomalacia and the regulation of phosphate homeostasis. *Bone,* **27**(3), 333–338 (2000).

17. T. O. Carpenter, Oncogenic osteomalacia—A complex dance of factors. *N Engl J Med,* **348**(17), 1705–1708 (2003).

18. X. Yu, and K. E. White, FGF23 and disorders of phosphate homeostasis. *Cytokine & Growth Factor Reviews,* **16**(2), 221–232 (2005).

19. R. Kumar, New insights into phosphate homeostasis: Fibroblast growth factor 23 and frizzled-related protein-4 are phosphaturic factors derived from tumors associated with osteomalacia. *Cur Opin Nephrol Hypertens,* **11**(5), 547–553 (2002).

20. S. C. Schiavi, and R. Kumar, The phosphatonin pathway: New insights in phosphate homeostasis. *Kidney Int,* **65**(1), 1–14 (2004).

21. G. J. Strewler, FGF23, hypophosphatemia, and rickets: Has phosphatonin been found? *Proc Natl Acad Sci USA,* **98**(11), 5945–5946 (2001).

22. S. Fukumoto, and T. Yamashita, Fibroblast growth factor-23 is the phosphaturic factor in tumor-induced osteomalacia and may be phosphatonin. *Cur Opin Nephrol Hypertens,* **11**(4), 385–389 (2002).

23. C. R. Dunstan, H. Zhou, and M. J. Seibel, Fibroblast growth factor 23, a phosphatonin regulating phosphate homeostasis? *Endocrinology,* **145**(7), 3084–3086 (2004).

24. C. Silve, and L. Beck, Is FGF23 the long sought after phosphaturic factor phosphatonin? *Nephrol Dial Transplantation,* **17**(6), 958–961 (2002).

25. L. D. Quarles, FGF23, PHEX, and MEPE regulation of phosphate homeostasis and skeletal mineralization. *Am J Physiol, Endocrinol Metab,* **285**(1), E1–9 (2003).

26. T. J. Berndt, S. Schiavi, and R. Kumar, "Phosphatonins" and the regulation of phosphorus homeostasis. *Am J Physiol, Renal Physiol,* **289**(6), F1170–1182 (2005).

27. T. J. Berndt, and F. G. Knox, Renal regulation of phosphate excretion. In *The Kidney: Physiology and Pathophysiology,* 2nd ed. (D. W. Seldin and G. H. Giebisch, eds.), pp. 2511–2532. Raven Press, New York (1992).

28. F. G. Knox, and A. Haramati, Renal regulation of phosphate excretion. In *The Kidney: Physiology and Pathophysiology* (D. W. Seldin and G. H. Giebisch, eds.), pp. 1351–1396. Raven Press, New York (1985).

29. F. G. Knox, J. A. Haas, T. Berndt, G. R. Marchand, and S. P. Youngberg, Phosphate transport in superficial and deep nephrons in phosphate-loaded rats. *Am J Physiol,* **233**(2), F150–153 (1977).

30. H. Murer, N. Hernando, I. Forster, and J. Biber, Molecular aspects in the regulation of renal inorganic phosphate reabsorption: The type IIa sodium/inorganic phosphate co-transporter as the key player. *Curr Opin Nephrol Hypertens,* **10**(5), 555–561 (2001).

31. C. B. Woda, N. Halaihel, P. V. Wilson, A. Haramati, M. Levi, and S. E. Mulroney, Regulation of renal NaPi-2 expression and tubular phosphate reabsorption by growth hormone in the juvenile rat. *Am J Physiol, Renal Physiol,* **287**(1), F117–123 (2004).

32. J. B. Puschett, and M. Goldberg, The relationship between the renal handling of phosphate and bicarbonate in man. *J Lab Clin Med,* **73**(6), 956–969 (1969).

33. S. Khosla, P. R. Ebeling, A. F. Firek, M. M. Burritt, P. C. Kao, and H. Heath, 3rd. Calcium infusion suggests a "set-point" abnormality of parathyroid gland function in familial benign hypercalcemia and more complex disturbances in primary hyperparathyroidism. *J Clin Endocrinol Metab,* **76**(3), 715–720 (1993).

34. F. D. Grant, P. R. Conlin, and E. M. Brown, Rate and concentration dependence of parathyroid hormone dynamics during stepwise changes in serum ionized calcium in normal humans. *J Clin Endocrinol Metab,* **71**(2), 370–378 (1990).

35. E. Reiss, J. M. Canterbury, M. A. Bercovitz, and E. L. Kaplan, The role of phosphate in the secretion of parathyroid hormone in man. *J Clin Invest,* **49**(11), 2146–2149 (1970).

36. J. C. Estepa, E. Aguilera-Tejero, I. Lopez, Y. Almaden, M. Rodriguez, and A. J. Felsenfeld, Effect of phosphate on parathyroid hormone secretion in vivo. *J Bone Miner Res,* **14**(11), 1848–1854 (1999).

37. Y. Almaden, A. Hernandez, V. Torregrosa, *et al.,* High phosphate level directly stimulates parathyroid hormone secretion and synthesis by human parathyroid tissue in vitro. *J Am Soc Nephrol,* **9**(10), 1845–1852 (1998).

38. Y. Almaden, A. Canalejo, A. Hernandez, *et al.,* Direct effect of phosphorus on PTH secretion from whole rat parathyroid glands in vitro. *J Bone Miner Res,* **11**(7), 970–976 (1996).

39. T. Berndt, B. Bielesz, T. A. Craig, *et al.,* Secreted frizzled-related protein-4 reduces sodium-phosphate co-transporter abundance and activity in proximal tubule cells. *Pflugers Archiv, Eur J Physiol,* **451**(4), 579–587 (2006).

40. Y. Taketani, H. Segawa, M. Chikamori, *et al.,* Regulation of type II renal Na+-dependent inorganic phosphate transporters by 1,25-dihydroxyvitamin D3. Identification of a vitamin D-responsive element in the human NAPi-3 gene. *J Biol Chem,* **273**(23), 14575–14581 (1998).

41. H. Georgaki, and J. B. Puschett, Acute effects of a "physiological" dose of 1,25-dihydroxy vitamin D3 on renal phosphate transport. *Endocr Res Commun,* **9**(2), 135–143 (1982).

42. S. Madsen, and K. Olgaard, Has vitamin D a direct renal effect on the tubular reabsorption of phosphate? A study in parathyroidectomized (PTX) and non-PTX man. *Adv Exp Med Biol,* **103**, 111–123 (1978).

43. M. A. Burnatowska, C. A. Harris, R. A. Sutton, and J. F. Seely, Effects of vitamin D on renal handling of calcium, magnesium and phosphate in the hamster. *Kidney Int,* **27**(6), 864–870 (1985).

44. T. Berndt, T. A. Craig, A. E. Bowe, *et al.,* Secreted frizzled-related protein 4 is a potent tumor-derived phosphaturic agent. *J Clin Invest, J Clin Invest,* **112**(5), 785–794 (2003).

45. T. Yamashita, M. Konishi, A. Miyake, K. Inui, and N. Itoh, Fibroblast growth factor (FGF)-23 inhibits renal phosphate reabsorption by activation of the mitogen-activated protein kinase pathway. *J Biol Chem,* **277**(31), 28265–28270 (2002).

46. T. O. Carpenter, B. K. Ellis, K. L. Insogna, W. M. Philbrick, J. Sterpka, and R. Shimkets, Fibroblast growth factor 7, an inhibitor of phosphate transport derived from oncogenic osteomalacia causing tumors. *J Clin Endocrinol Metab,* **90**(2), 1012–1020 (2005).

47. P. S. Rowe, Y. Kumagai, G. Gutierrez, *et al.,* MEPE has the properties of an osteoblastic phosphatonin and minhibin. *Bone,* **34**(2), 303–319 (2004).

48. O. Hattenhauer, M. Traebert, H. Murer, and J. Biber, Regulation of small intestinal Na-P(i) type IIb cotransporter by dietary phosphate intake. *Am J Physiol,* **277**(4 Pt 1), G756–762 (1999).

49. H. S. Tenenhouse, Recent advances in epithelial sodium-coupled phosphate transport. *Cur Opin Nephrol Hypertens,* **8**(4), 407–414 (1999).

50. R. W. Gray, and J. L. Napoli, Dietary phosphate deprivation increases 1,25-dihydroxyvitamin D3 synthesis in rat kidney in vitro. *J Biol Chem,* **258**(2), 1152–1155 (1983).

51. H. S. Cross, H. Debiec, and M. Peterlik, Mechanism and regulation of intestinal phosphate absorption. *Miner Electrolyte Metab,* **16**(2–3), 115–124 (1990).

52. Q. Cai, S. F. Hodgson, P. C. Kao, *et al.,* Brief report: Inhibition of renal phosphate transport by a tumor product in a patient with oncogenic osteomalacia [see comment]. *N Engl J Med,* **330**(23), 1645–1649 (1994).

53. R. A. Meyer, Jr, H. S. Tenenhouse, M. H. Meyer, and A. H. Klugerman, The renal phosphate transport defect in normal mice parabiosed to X-linked hypophosphatemic mice persists after parathyroidectomy. *J Bone Miner Res,* **4**(4), 523–532 (1989).

54. R. A. Meyer, Jr., M. H. Meyer, and R. W. Gray, Parabiosis suggests a humoral factor is involved in X-linked hypophosphatemia in mice. *J Bone Miner Res,* **4**(4), 493–500 (1989).

55. T. Nesbitt, T. M. Coffman, R. Griffiths, and M. K. Drezner, Crosstransplantation of kidneys in normal and Hyp mice. Evidence that the Hyp mouse phenotype is unrelated to an intrinsic renal defect. *J Clin Invest,* **89**(5), 1453–1459 (1992).

56. T. Yamashita, M. Yoshioka, and N. Itoh, Identification of a novel fibroblast growth factor, FGF-23, preferentially expressed in the ventrolateral thalamic nucleus of the brain. *Biochem Biophys Res Commun,* **277**(2), 494–498 (2000).

57. K. E. White, W. E. Evans, J. L. H. O'Riordan, M. C. Speer, M. J. Econs, B. Lorenz-Depiereux, M. Grabowski, T. Meitinger, and T. M. Strom, Autosomal dominant hypophosphataemic rickets is associated with mutations in FGF23. *Nat Genetics,* **26**(3), 345–348 (2000).

58. T. Shimada, T. Muto, I. Urakawa, *et al.,* Mutant FGF-23 responsible for autosomal dominant hypophosphatemic rickets is resistant to proteolytic cleavage and causes hypophosphatemia in vivo. *Endocrinology,* **143**(8), 3179–3182 (2002).

59. X. Bai, D. Miao, J. Li, D. Goltzman, and A. C. Karaplis, Transgenic mice overexpressing human fibroblast growth factor 23 (R176Q) delineate a putative role for parathyroid hormone in renal phosphate wasting disorders. *Endocrinology,* **145**(11), 5269–5279 (2004).

60. H. Saito, K. Kusano, M. Kinosaki, *et al.,* Human fibroblast growth factor-23 mutants suppress Na+-dependent phosphate co-transport activity and 1alpha,25-dihydroxyvitamin D3 production. *J Biol Chem,* **278**(4), 2206–2211 (2003).

61. M. B. Zimering, F. A. Caldarella, K. E. White, and M. J. Econs, Persistent tumor-induced osteomalacia confirmed by elevated postoperative levels of serum fibroblast growth factor-23 and 5-year follow-up of bone density changes. *Endocr Pract,* **11**(2), 108–114 (2005).

62. R. A. Sweet, J. L. Males, A. J. Hamstra, and H. F. DeLuca, Vitamin D metabolite levels in oncogenic osteomalacia. *Ann Intern Med,* **93**(2), 279–280 (1980).

63. T. Larsson, R. Zahradnik, J. Lavigne, O. Ljunggren, H. Juppner, and K. B. Jonsson, Immunohistochemical detection of FGF-23 protein in tumors that cause oncogenic osteomalacia. *Eur J Endocrinol,* **148**(2), 269–276 (2003).

64. J. L. Dupond, H. Mahammedi, D. Prie, *et al.,* Oncogenic osteomalacia: Diagnostic importance of fibroblast growth factor 23 and F-18 fluorodeoxyglucose PET/CT scan for the diagnosis and follow-up in one case. *Bone,* **36**(3), 375–378 (2005).

65. A. E. Nelson, R. C. Bligh, M. Mirams, *et al.,* Clinical case seminar: Fibroblast growth factor 23, a new clinical marker for oncogenic osteomalacia. *J Clin Endocrinol Metab,* **88**(9), 4088–4094 (2003).

66. Y. Takeuchi, H. Suzuki, S. Ogura, *et al.,* Venous sampling for fibroblast growth factor-23 confirms preoperative diagnosis of tumor-induced osteomalacia. *J Clin Endocrinol Metab,* **89**(8), 3979–3982 (2004).

67. M. R. John, H. Wickert, K. Zaar, *et al.,* A case of neuroendocrine oncogenic osteomalacia associated with a PHEX and fibroblast growth factor-23 expressing sinusoidal malignant schwannoma. *Bone,* **29**(4), 393–402 (2001).

68. S. M. De Beur, R. B. Finnegan, J. Vassiliadis, *et al.,* Tumors associated with oncogenic osteomalacia express genes important in bone and mineral metabolism. *J Bone Miner Res,* **17**(6), 1102–1110 (2002).

69. R. Kumar, J. D. Haugen, E. D. Wieben, J. M. Londowski, and Q. Cai, Inhibitors of renal epithelial phosphate transport in tumor-induced osteomalacia and uremia. *Proc Assoc Am Physicians,* **107**(3), 296–305 (1995).

70. K. B. Jonsson, M. Mannstadt, A. Miyauchi, *et al.,* Extracts from tumors causing oncogenic osteomalacia inhibit phosphate uptake in opossum kidney cells. *J Endocrinol,* **169**(3), 613–620 (2001).

71. Y. Yamazaki, R. Okazaki, M. Shibata, *et al.,* Increased circulatory level of biologically active full-length FGF-23 in patients with hypophosphatemic rickets/osteomalacia. *J Clin Endocrinol Metab,* **87**(11), 4957–4960 (2002).

72. L. M. Ward, F. Rauch, K. E. White, *et al.,* Resolution of severe, adolescent-onset hypophosphatemic rickets following resection of an FGF-23–producing tumour of the distal ulna. *Bone,* **34**(5), 905–911 (2004).

73. E. A. Imel, M. Peacock, P. Pitukcheewanont, *et al.,* Sensitivity of fibroblast growth factor 23 measurements in tumor induced osteomalacia. *J Clin Endocrinol Metab,* **91**(6), 2055–2061 (2006).

74. E. S. Siris, T. L. Clemens, D. W. Dempster, *et al.,* Tumor-induced osteomalacia. Kinetics of calcium, phosphorus, and vitamin D metabolism and characteristics of bone histomorphometry. *Am J Med,* **82**(2), 307–312 (1987).

75. E. Leicht, G. Biro, and H. J. Langer. Tumor-induced osteomalacia: Pre- and postoperative biochemical findings. *Horm Metab Res,* **22**(12), 640–643 (1990).

76. C. F. Verge, A. Lam, J. M. Simpson, C. T. Cowell, N. J. Howard, and M. Silink, Effects of therapy in X-linked hypophosphatemic rickets. *N Engl J Med,* **325**(26), 1843–1848 (1991).

77. F. Francis, S. Hennig, B. Korn, R. Reinhardt, P. de Jong, A. Poustka, H. Lehrach, P. S. N Rowe, J. N. Goulding, T. Summerfield, R. Mountford, A. P. Read, E. Popowska, E. Pronicka, K. E. Davies, J. L. H. O'Riordan, M. J. Econs, T. Nesbitt, M. K. Drezner, C. Oudet, S. Pannetier, A. Hanauer, T. M. Strom, A. Meindl, B. Lorenz, B. Cagnoli, K. L. Mohnike, J. Murken, and T. Meitinger, A gene (PEX) with homologies to endopeptidases is mutated in patients with X-linked hypophosphatemic rickets. The HYP Consortium. *Nat Genetics,* **11**(2), 130–136 (1995).

78. M. Campos, C. Couture, I. Y. Hirata, *et al.,* Human recombinant endopeptidase PHEX has a strict S1′ specificity for acidic residues and cleaves peptides derived from fibroblast growth factor-23 and matrix extracellular phosphoglycoprotein. *Biochem J,* **373**(Pt 1), 271–279 (2003).

79. A. E. Bowe, R. Finnegan, S. M. Jan De Beur, *et al.*, FGF-23 inhibits renal tubular phosphate transport and is a PHEX substrate. *Biochem Biophys Res Commun,* **284**(4), 977–981 (2001).

80. T. J. Weber, S. Liu, O. S. Indridason, and L. D. Quarles, Serum FGF23 levels in normal and disordered phosphorus homeostasis. *J Bone Miner Res,* **18**(7), 1227–1234 (2003).

81. M. T. Collins, C. Chebli, J. Jones, *et al.*, Renal phosphate wasting in fibrous dysplasia of bone is part of a generalized renal tubular dysfunction similar to that seen in tumor-induced osteomalacia. *J Bone Miner Res,* **16**(5), 806–813 (2001).

82. K. Kobayashi, Y. Imanishi, H. Koshiyama, *et al.*, Expression of FGF23 is correlated with serum phosphate level in isolated fibrous dysplasia. *Life Sci,* **78**(20), 2295–2301 (2006).

83. M. Riminucci, M. T. Collins, N. S. Fedarko, *et al.*, FGF-23 in fibrous dysplasia of bone and its relationship to renal phosphate wasting. *J Clin Invest,* **112**(5), 683–692 (2003).

84. I. Chefetz, R. Heller, A. Galli-Tsinopoulou, *et al.*, A novel homozygous missense mutation in FGF23 causes familial tumoral calcinosis associated with disseminated visceral calcification. *Hum Genetics,* **118**(2), 261–266 (2005).

85. A. Benet-Pages, P. Orlik, T. M. Strom, and B. Lorenz-Depiereux, An FGF23 missense mutation causes familial tumoral calcinosis with hyperphosphatemia. *Hum Mol Genetics,* **14**(3), 385–390 (2005).

86. K. Araya, S. Fukumoto, R. Backenroth, *et al.*, A novel mutation in fibroblast growth factor 23 gene as a cause of tumoral calcinosis. *J Clin Endocrinol Metab,* **90**(10), 5523–5527 (2005).

87. O. Topaz, D. L. Shurman, R. Bergman, *et al.*, Mutations in GALNT3, encoding a protein involved in O-linked glycosylation, cause familial tumoral calcinosis. *Nat Genetics,* **36**(6), 579–581 (2004).

88. R. Steinherz, R. W. Chesney, B. Eisenstein, A. Metzker, H. F. DeLuca, and M. Phelps, Elevated serum calcitriol concentrations do not fall in response to hyperphosphatemia in familial tumoral calcinosis. *Am J Dis Child,* **139**(8), 816–819 (1985).

89. T. Shimada, M. Kakitani, Y. Yamazaki, *et al.*, Targeted ablation of FGF23 demonstrates an essential physiological role of FGF23 in phosphate and vitamin D metabolism. *J Clin Invest,* **113**(4), 561–568 (2004).

90. Y. Imanishi, M. Inaba, K. Nakatsuka, *et al.*, FGF-23 in patients with end-stage renal disease on hemodialysis. *Kidney Int,* **65**(5), 1943–1946 (2004).

91. T. Larsson, U. Nisbeth, O. Ljunggren, H. Juppner, K. B. Jonsson. Circulating concentration of FGF-23 increases as renal function declines in patients with chronic kidney disease, but does not change in response to variation in phosphate intake in healthy volunteers. *Kidney Int,* **64**(6), 2272–2279 (2003).

92. T. Shigematsu, J. J. Kazama, T. Yamashita, *et al.*, Possible involvement of circulating fibroblast growth factor 23 in the development of secondary hyperparathyroidism associated with renal insufficiency. *Am J Kidney Dis,* **44**(2), 250–256 (2004).

93. S. Nakanishi, J. J. Kazama, T. Nii-Kono, *et al.*, Serum fibroblast growth factor-23 levels predict the future refractory hyperparathyroidism in dialysis patients [see comment]. *Kidney Int,* **67**(3), 1171–1178 (2005).

94. O. Gutierrez, T. Isakova, E. Rhee, *et al.*, Fibroblast growth factor-23 mitigates hyperphosphatemia but accentuates calcitriol deficiency in chronic kidney disease. *J Am Soc Nephrol,* **16**(7), 2205–2215 (2005).

95. T. Shimada, Y. Yamazaki, M. Takahashi, *et al.*, Vitamin D receptor-independent FGF23 actions in regulating phosphate and vitamin D metabolism. *Am J Physiol, Renal Physiol,* **289**(5), F1088–1095 (2005).

96. J. J. Kazama, F. Gejyo, T. Shigematsu, and M. Fukagawa, Role of circulating fibroblast growth factor 23 in the development of secondary hyperparathyroidism. *Therapeutic Apheresis & Dialysis,* **9**(4), 328–330 (2005).

97. J. J. Kazama, F. Sato, K. Omori, *et al.*, Pretreatment serum FGF-23 levels predict the efficacy of calcitriol therapy in dialysis patients. *Kidney Int,* **67**(3), 1120–1125 (2005).

98. F. Koiwa, J. J. Kazama, A. Tokumoto, *et al.*, Sevelamer hydrochloride and calcium bicarbonate reduce serum fibroblast growth factor 23 levels in dialysis patients. *Therapeutic Apheresis & Dialysis,* **9**(4), 336–339 (2005).

99. K. Miyamoto, M. Ito, M. Kuwahata, S. Kato, and H. Segawa, Inhibition of intestinal sodium-dependent inorganic phosphate transport by fibroblast growth factor 23. *Therapeutic Apheresis & Dialysis,* **9**(4), 331–335 (2005).

100. F. Perwad, N. Azam, M. Y. Zhang, T. Yamashita, H. S. Tenenhouse, and A. A. Portale, Dietary and serum phosphorus regulate fibroblast growth factor 23 expression and 1,25-dihydroxyvitamin D metabolism in mice. *Endocrinology,* **146**(12), 5358–5364 (2005).

101. X. Yan, H. Yokote, X. Jing, *et al.*, Fibroblast growth factor 23 reduces expression of type IIa Na+/Pi co-transporter by signaling through a receptor functionally distinct from the known FGFRs in opossum kidney cells. *Genes to Cells,* **10**(5), 489–502 (2005).

102. X. Yu, O. A. Ibrahimi, R. Goetz, *et al.*, Analysis of the biochemical mechanisms for the endocrine actions of fibroblast growth factor-23. *Endocrinology,* **146**(11), 4647–4656 (2005).

103. H. Kurosu, Y. Ogawa, M. Miyoshi, *et al.*, Regulation of fibroblast growth factor-23 signaling by klotho. *J Biol Chem,* **281**(10), 6120–6123 (2006).

104. A. O. Wilkie, S. J. Patey, S. H. Kan, A. M. van den Ouweland, and B. C. Hamel, FGFs, their receptors, and human limb malformations: Clinical and molecular correlations. *Am J Med Genetics,* **112**(3), 266–278 (2002).

105. D. Sitara, M. S. Razzaque, M. Hesse, *et al.*, Homozygous ablation of fibroblast growth factor-23 results in hyperphosphatemia and impaired skeletogenesis, and reverses hypophosphatemia in Phex-deficient mice. *Matrix Biology,* **23**(7), 421–432 (2004).

106. S. L. Ferrari, J. P. Bonjour, and R. Rizzoli, Fibroblast growth factor-23 relationship to dietary phosphate and renal phosphate handling in healthy young men. *J Clin Endocrinol Metab,* **90**(3), 1519–1524 (2005).

107. X. Yu, Y. Sabbagh, S. I. Davis, M. B. Demay, and K. E. White, Genetic dissection of phosphate- and vitamin D-mediated regulation of circulating FGF23 concentrations. *Bone,* **36**(6), 971–977 (2005).

108. M. Ito, Y. Sakai, M. Furumoto, *et al.*, Vitamin D and phosphate regulate fibroblast growth factor-23 in K-562 cells. *Am J Physiol, Endocrinol Metab,* **288**(6), E1101–1109 (2005).

109. O. I. Kolek, E. R. Hines, M. D. Jones, *et al.*, 1alpha,25-dihydroxyvitamin D3 upregulates FGF23 gene expression in bone: The final link in a renal-gastrointestinal-skeletal axis that controls phosphate transport. *Am J Physiol, Gastrointest Liver Physiol,* **289**(6), G1036–1042 (2005).

110. H. Saito, A. Maeda, S. Ohtomo, *et al.*, Circulating FGF-23 is regulated by 1alpha,25-dihydroxyvitamin D3 and phosphorus in vivo. *J Biol Chem,* **280**(4), 2543–2549 (2005).

111. R. Nusse, Wnt signaling in disease and in development. *Cell Research,* **15**(1), 28–32 (2005).

112. M. L. Johnson, K. Harnish, R. Nusse, and W. Van Hul, LRP5 and Wnt signaling: A union made for bone. *J Bone Miner Res,* **19**(11), 1749–1757 (2004).

113. Y. Gong, R. B. Slee, N. Fukai, *et al.*, LDL receptor-related protein 5 (LRP5) affects bone accrual and eye development. *Cell,* **107**(4), 513–523 (2001).

114. M. L. Kwee, W. Balemans, E. Cleiren, *et al.*, An autosomal dominant high bone mass phenotype in association with craniosynostosis in an extended family is caused by an LRP5 missense mutation. *J Bone Miner Res,* **20**(7), 1254–1260 (2005).

115. L. Van Wesenbeeck, E. Cleiren, J. Gram, *et al.*, Six novel missense mutations in the LDL receptor-related protein 5 (LRP5) gene in different conditions with an increased bone density. *Am J Hum Genetics,* **72**(3), 763–771 (2003).

116. P. S. Rowe, P. A. de Zoysa, R. Dong, *et al.*, MEPE, a new gene expressed in bone marrow and tumors causing osteomalacia. *Genomics,* **67**(1), 54–68 (2000).

117. C. Lu, S. Huang, T. Miclau, J. A. Helms, and C. Colnot, MEPE is expressed during skeletal development and regeneration. *Histochem Cell Biol,* **121**(6), 493–499 (2004).

118. A. Nampei, J. Hashimoto, K. Hayashida, *et al.*, Matrix extracellular phosphoglycoprotein (MEPE) is highly expressed in osteocytes in human bone. *J Bone Miner Metab,* **22**(3), 176–184 (2004).

119. H. Siggelkow, E. Schmidt, B. Hennies, and M. Hufner, Evidence of downregulation of matrix extracellular phosphoglycoprotein during terminal differentiation in human osteoblasts. *Bone,* **35**(2), 570–576 (2004).

120. L. Argiro, M. Desbarats, F. H. Glorieux, and B. Ecarot, MEPE, the gene encoding a tumor-secreted protein in oncogenic hypophosphatemic osteomalacia, is expressed in bone. *Genomics,* **74**(3), 342–351 (2001).

121. A. Jain, N. S. Fedarko, M. T. Collins, *et al.*, Serum levels of matrix extracellular phosphoglycoprotein (MEPE) in normal humans correlate with serum phosphorus, parathyroid hormone and bone mineral density. *J Clin Endocrinol Metab,* **89**(8), 4158–4161 (2004).

122. P. S. Rowe, I. R. Garrett, P. M. Schwarz, *et al.*, Surface plasmon resonance (SPR) confirms that MEPE binds to PHEX via the MEPE-ASARM motif: A model for impaired mineralization in X-linked rickets (HYP). *Bone,* **36**(1), 33–46 (2005).

123. D. Bresler, J. Bruder, K. Mohnike, W. D. Fraser, and P. S. Rowe, Serum MEPE-ASARM-peptides are elevated in X-linked rickets (HYP), implications for phosphaturia and rickets. *J Endocrinol,* **183**(3), R1–9 (2004).

124. H. D. Beer, M. G. Gassmann, B. Munz, *et al.*, Expression and function of keratinocyte growth factor and activin in skin morphogenesis and cutaneous wound repair. *J Invest Dermatol, Symp Proc,* **5**(1), 34–39 (2000).

125. A. Lyakhovich, N. Aksenov, P. Pennanen, *et al.*, Vitamin D induced up-regulation of keratinocyte growth factor (FGF-7/KGF) in MCF-7 human breast cancer cells. *Biochem Biophys Res Commun,* **273**(2), 675–680 (2000).

126. J. S. Rubin, H. Osada, P. W. Finch, W. G. Taylor, S. Rudikoff, and S. A. Aaronson, Purification and characterization of a newly identified growth factor specific for epithelial cells. *Proc Natl Acad Sci USA,* **86**(3), 802–806 (1989).

127. J. Jacquemier, Z. Z. Sun, F. Penault- Llorca, *et al.*, FGF7 protein expression in human breast carcinomas. *J Pathol,* **186**(3), 269–274 (1998).

128. G. S. Bansal, H. C. Cox, S. Marsh, *et al.*, Expression of keratinocyte growth factor and its receptor in human breast cancer. *Br J Cancer,* **75**(11), 1567–1574 (1997).

129. X. Yu, and K. E. White, Fibroblast growth factor 23 and its receptors. *Therapeutic Apheresis & Dialysis,* **9**(4), 308–312 (2005).

130. P. J. Tebben, and R. Kumar, Vitamin D and the kidney. In *Vitamin D,* vol 1. 2nd ed. (D. Feldman, ed.), pp. 515–536. Elsevier, Burlington, MA (2005).

131. R. J. Singh, and R. Kumar, Fibroblast growth factor 23 concentrations in humoral hypercalcemia of malignancy and hyperparathyroidism. *Mayo Clinic Proc,* **78**(7), 826–829 (2003).

132. W. H. Hoffman, H. W. Jueppner, B. R. Deyoung, O'Dorisio S. M, and K. S. Given, Elevated fibroblast growth factor-23 in hypophosphatemic linear nevus sebaceous syndrome. *Am J Med Genetics, Part A,* **134**(3), 233–236 (2005).

133. T. Larsson, S. I. Davis, H. J. Garringer, *et al.*, Fibroblast growth factor-23 mutants causing familial tumoral calcinosis are differentially processed. *Endocrinology,* **146**(9), 3883–3891 (2005).

134. H. Yamashita, Y. Yamazaki, H. Hasegawa, *et al.*, Fibroblast growth factor-23 in patients with Graves' disease before and after antithyroid therapy: Its important role in serum phosphate regulation. *J Clin Endocrinol Metab,* **90**(7), 4211–4215 (2005).

135. P. J. Tebben, K. R. Kalli, W. A. Cliby, *et al.*, Elevated fibroblast growth factor 23 in women with malignant ovarian tumors. *Mayo Clinic Proc,* **80**(6), 745–751 (2005).

CHAPTER 15

Wnt Signaling in Bone

MARK L. JOHNSON AND ROBERT R. RECKER

Galileo in 1638 published his famous *Discourse on Two New Sciences*, as it has come to be known, and in this text he recognized that the skeleton must be able to adjust its size in order to adapt to changes in load bearing. He wrote:

> [I]t would be impossible to build up the bony structures of men, horses, or other animals so as to hold together and perform their normal functions if these animals were to be increased enormously in height; for this increase in height can be accomplished only by employing a material which is harder and stronger than usual, or by enlarging the size of the bones, thus changing their shape.... Clearly then if one wishes to maintain in a great giant the same proportion of limb as that found in an ordinary man he must either find a harder and stronger material for making the bones or he must admit a diminution of strength in comparison with men of medium stature.... And further Galileo proposed the principle that [i]n the case of two cylinders, one hollow the other solid but having equal volumes and equal lengths, their resistances [bending strengths] are to each other in the ratio of their diameters. [1]

While the importance of Galileo's comments are best viewed in hindsight and are perhaps a mere historical curiosity, two and one-half centuries later Julius Wolff developed a mathematical formula that described bone's ability to resist the load applied to it [2]. Wolff more succinctly stated the proposal that the skeleton must respond to changes in loading. Today, we accept these principles as dogma, yet our understanding of the precise molecular events that drive the adaptive ability of the skeleton is still largely a mystery. Harold Frost gave us a theoretical construct he called the "mechanostat" in which to frame our exploration of how the skeleton responds to mechanical loads [3–5]. If one considers a sequence of events involving first the *perception* of changes in load followed by *transduction* into a biochemical event and then a *response* to those biochemical signals, clearly we have a considerable understanding about the bone formation/resorption response that occurs. We have some understanding of the biochemical signals that are required to elicit those responses, but we know very little about cellular events/mechanisms that are involved in the "perception" of mechanical load and the communication that occurs between bone cells that leads to the ultimate response of bone to changes in load.

Within the past 5 years, we have gained new insights into the regulation of bone mass through a number of genetic studies that have revealed fundamental aspects of bone cell biology and have provided new tools to understanding mechanosensation in bone. This chapter focuses on the role of Wnts and Wnt signaling in bone biology and its emerging role in bone responsiveness to mechanical loading and bone mass acquisition and maintenance. The role of Wnts in the embryonic development and limb patterning of the skeleton has been appreciated since the 1990s [6–8]. However, it wasn't until 2001–2002 with the description of mutations in *LRP5* that gave rise to conditions of low [9] and high bone mass [10] that we began to appreciate the central role Wnt signaling plays in the maintenance of adult bone mass [11, 12]. Thus, of all the chapters in this book, this chapter may reflect a summary of information that is in the greatest state of flux. We have attempted to be as current as possible; suffice it to say that the fourth edition of this book, whenever it is published in the future, will undoubtedly represent a more complete understanding of this emerging field as it relates to the biology of bone.

I. WNTS

The Wnt genes encode a large family of secreted, highly post-translationally modified proteins that play key roles in cell differentiation, proliferation and apoptosis, and function and regulate developmental and homeostatic mechanisms in species found throughout the Animal Kingdom. Wnt was first identified as a mouse proto-oncogene integration site/locus (*int-1*) for the murine mammary tumor virus [13]. Later the *int-1* gene was shown to be homologous to the *Drosophila* segmentation gene *Wingless* (*Wg*), and the name *Wnt* was coined [14, 15]. There are currently 19 known *WNT* genes in humans. They share general homology in the range of 35%. Members within subgroups can have much higher homology, however, and all Wnts contain 23–24 conserved cysteine residues that have similar spacing within the various proteins, suggesting an important conservation of function needed for Wnt activity [16].

Until recently, the Wnt proteins have been extremely difficult to purify, and molecular/genetic approaches have driven the study of these genes. Critical to the function of Wnt proteins is a lipid modification in which a palmitate is added to the first of the conserved cysteine residues, cysteine-77, near the amino terminal end [17]. Removal of the palmitate results in loss of Wnt activity. In *Drosophila* the protein (an acyltransferase enzyme) encoded by the gene *porcupine* and in *C. elegans* by the gene *mom-1* appears to be responsible for the palmitoylation of Wnts, although direct proof is lacking [18, 19]. These are membrane-bound proteins found in the endoplasmic reticulum, and thus Wnts are secreted as lipid modified proteins, which is somewhat unusual, as this modification is more common for anchoring proteins to the cytoplasmic face of the plasma membrane [20]. The hedgehog family of proteins is also palmitoylated and secreted in this manner [19]. The role of the palmitate in Wnt activity/function is not understood. It may be that this targets Wnt to the cell surface and tethers it to the membrane, but at the same time Wnts are known to act as morphogens and at long range [21–23]. This raises the question of whether specific Wnt binding and/or transport proteins exist, but there is no direct evidence to demonstrate the existence of such a protein [19]. Another possibility is that the palmitate participates in the binding of Wnts to the Lrp5/6 and/or frizzled coreceptors [12]. Again, direct evidence for this is lacking. Even less understood is the role of the glycosylation of Wnts, which has been shown not to be essential for activity [24].

II. THE WNT/β-CATENIN SIGNALING PATHWAY

Several reviews of the Wnt signaling pathways [16, 19, 25–38], their role in development [39–44], cancer [45–58] and various other various diseases [59–61], and more recently in the field of bone biology [11, 12, 62–65], can be found in the literature. Also, the potential for manipulating the Wnt pathway in the treatment of disease has been extensively discussed in the literature [66–68].

The best studied of the Wnt signaling pathways is the Wnt/β-catenin signaling pathway [40], which is often referred to as the canonical Wnt pathway (although the origins of this descriptor are vague). The details of this pathway are known to a far greater extent than the other pathways, and all of the currently known proteins involved in canonical Wnt signaling can be found on the website maintained by Dr. Roel Nusse (http://www.stanford.edu/~rnusse/). However, new components in all of the currently known pathways through which Wnts exert their biological action are constantly being identified, and it is safe to say that there are still more protein components of these pathways than are currently known. A simplified view of the Wnt/β-catenin pathway and its core components is shown in Figure 15-1. At the level of the cell membrane, two proteins form a coreceptor complex and bind Wnt in their extracellular domains, and then through a mechanism(s) that is not fully understood, activate an intracellular cascade. The first of these two proteins in the coreceptor complex is Frizzled (Fz), which was the first coreceptor identified and shown to be a receptor for Wnts in various species [69–72]. The second of the coreceptors is a small family of proteins known as the low-density lipoprotein receptor-related proteins (Lrp), which are composed of Lrp5 [73–75] and Lrp6 [76] (in vertebrates) and Arrow [77] (in *Drosophila*). Arrow and Lrp6 were first identified as Wnt coreceptors in 2000 [77–79]; however, the models of Wnt/β-catenin signaling that exist in the pre-2000 literature required only slight modification to incorporate this coreceptor into the pathway. Evidence suggests that the formation of the Wnt-Lrp5/6–Frizzled complex is required to activate the Wnt/β-catenin signaling pathway [78, 80], whereas the other Wnt pathways (discussed later) are activated when only Wnt and Frizzled interact [81].

There are currently 10 known members of the Frizzled family in vertebrates. All of these proteins share basic structural features including a cysteine-rich domain (CRD) at the amino terminal end, seven transmembrane spanning domains, and a cytoplasmic tail [44]. The CRD is thought to be responsible for binding of Wnt. A family of proteins, the secreted frizzled-related proteins (sFRPs) in vertebrates [82, 83] or the Frzb family in *Xenopus* [84, 85], also contains a CRD that can compete for the binding of Wnt and thereby function to modulate its biological activity. In addition to the various members of the Frizzled family, there is also another protein, Smoothened, that is a seven-pass transmembrane protein with homology to the Frizzleds

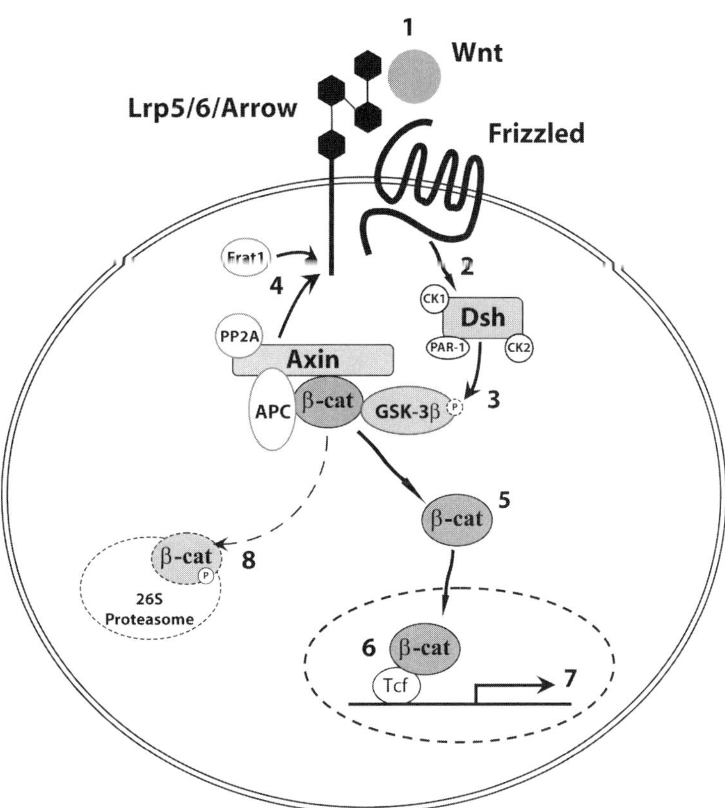

FIGURE 15-1   The Wnt/β-catenin signaling pathway. (1) Wnt binds to the coreceptor complex consisting of either Lrp5 or Lrp6 or Arrow and one of the Frizzleds. (2) Dsh is activated through a mechanism that is not fully understood but likely involves phosphorylation by kinases such as CK1, CK2, and/or Par-1. (3) Activation of Dsh leads to the phosphorylation and inactivation of GSK-3β. (4) Wnt binding also induces the phosphorylation of the cytoplasmic tail of the Lrp coreceptor and binding of Frat-1 and the recruitment of the Axin/APC/GSK-3β degradation complex to the cytoplasmic tail through binding of Axin. (5) β-catenin dissociates from the degradation complex and accumulates in the cytoplasm. (6) β-catenin translocates into the nucleus and forms a complex with the Tcf/LEF-1 family of transcription factors. (7) This binding of β-catenin initiates a conformational change in the chromatin and the initiation of gene transcription. (8) In the absence of Wnt, β-catenin is normally phosphorylated by GSK-3β, which marks it for ubiquitination and subsequent targeting to the 26S proteosome degradation pathway.

and functions as a receptor in the Hedgehog signaling pathway [86, 87]. The precise mechanism through which Frizzled activates the intracellular cascade of events that leads to the stabilization of β-catenin in the cell cytoplasm is still largely unknown. As will be discussed later, Frizzled coreceptors also play an important role in the other Wnt signaling pathways.

The other Wnt coreceptors, Lrp5/6/Arrow, are members of a larger family of low-density lipoprotein (LDL) receptors [88, 89], of which the LDL receptor that mediates the cellular endocytosis of cholesterol is the prototypical member [90]. These proteins are single-pass transmembrane proteins with a large extracellular domain that is responsible for ligand binding and an intracellular domain that is required for signaling. Lrp5 and Lrp6 were long considered orphan members of the LDL receptor family. The earliest studies of Lrp5 focused on it as being involved in the binding of apolipoprotein E and a possible role in Type I diabe-

tes [74, 91]. However, the identification of mutations in *LRP5* that give rise to conditions of low bone mass [9] and high bone mass [10, 92] and shortly thereafter other mutations that give rise to a variety of altered bone mass phenotypes [93–99] have positioned Lrp5 and Lrp6 at center stage for an explosion of Wnt signaling research in the bone field. In addition, mutations in *LRP5* have also been shown to be causal for the loss of vision associated with familial exudative retinopathy (FEVR) [100, 101] and have been associated with some degree of variation in normal bone mass [102–110]. The potential role of LRP5 in other disease processes is also being investigated [111, 112].

The structure/function of Lrp5 has been recently reviewed [33]. The organization of the extracellular domain of Lrp5/6/Arrow is essentially reversed compared to other members of the LDL receptor (LDLR) family [88]. There are four domains formed from six Tyr-Trp-Thr-Asp (YWTD) repeat motifs; however, there is a high degree of degeneracy in these motifs [113, 114]. Each β-propeller domain is separated by an EGF-like repeat element. Between the fourth β-propeller domain and the transmembrane spanning domain is a region containing three LDLR Type A (LA) repeat elements. Jeon *et al.* [115] provided a 1.5 Å resolution crystal structure of a fragment of the LDLR molecule spanning the YWTD repeats and its two flanking EGF modules and compared this model to the second YWTD-EGF domain in Lrp6. These repeats form a 6-bladed β-propeller structure, which in LDLR functions to control lipoprotein release at low pH and receptor recycling. However, Lrp5/6/Arrow function as signaling molecules and participate with Frizzled in the binding of Wnt. The interaction of Lrp6 with Wnt seems to reside in the first two β-propeller domains [116], and these domains have also been shown to be involved in the interaction with Frizzled [117]. Additional evidence for an interaction of Lrp5/6, Frizzled, and Wnt comes from the studies of Tamai *et al.* [78] and Semenov *et al.* [118]. Furthermore, Tolwinski *et al.* [119] provided evidence to suggest that in *Drosophila*, Wg facilitates the formation of a complex between Arrow and Frizzled 2 (Dfz2). Collectively, these data support a model for the formation of a complex between Lrp/5/6/Arrow and Frizzled in the presence of Wnt. Given the number of possible Wnts and Frizzled, it remains a challenge to understand if there is a hierarchical organization in which specific Wnts bind to specific Lrps and Frizzleds and how this relates to downstream cellular events.

A number of mutations have been identified in the human *LRP5* gene, and these, along with other molecular/genetic manipulations, have provided a first-pass functional map of Lrp5/6/Arrow in terms of where various other regulatory proteins may interact with the molecule. There are currently four known members of the Dickkopf (Dkk) family of proteins designated Dkk1–4 [120–122]. The high bone mass (HBM) G171V mutation [10, 92] has been shown to alter Dkk1 inhibition of the pathway, thereby implicating the first β-propeller domain in binding of this negative modulator [92, 123, 124]. However, previously published deletion construct studies on Lrp6 demonstrated a requisite role of the third β-propeller domain in Dkk1 binding [117], and more detailed studies have subsequently mapped the critical amino acids in the third domain for this interaction [125]. Perhaps the G171V mutation in Lrp5 alters the tertiary structure of the protein such that the third propeller domain is no longer able to bind Dkk1. Until crystal structure data are produced, the explanation for this paradox will remain unknown. Interestingly, Dkk2 had been shown to be capable of both inhibition and activation of the Wnt/β-catenin signaling pathway in that it interferes with the binding of Wnt8, but is a

weak activator of Lrp6 [126]. A model to explain this suggests that Dkk2 can interfere with Wnt binding, but in the absence of Wnt or high levels of Lrp5/6 can act as a weak activator [126, 127]. The inhibitory effects of the Dkks are mediated by another single pass transmembrane class of proteins, the Kremens (Kremen 1 or Kremen 2) [128]. The Kremens can form a ternary complex with Dkk and Lpr5/6, and this induces the rapid internalization/endocytosis of the complex and removal of Lrp5/6 from the cell surface. Lrp5/6/Arrow do not contain the NPXY consensus internalization sequence [76, 113] in their cytoplasmic domain and thus do not participate in the endocytic cycle that other members of the Lrp family can undergo upon binding of ligand [129]. The duality of function observed with Dkk2 appears to lie in the presence or absence of Kremen 2. In the absence of Kremen 2, Dkk has weak activating ability when it binds to Lrp6 (as discussed previously), but when Kremen 2 is present, then Dkk2 has inhibitory effects on the pathway [130].

Additional studies have suggested that the major effect of the G171V mutation is by interfering with receptor trafficking to the cell surface [125], which is mediated by a chaperone protein Mesd in mouse [131] or Boca in *Drosophila* [132]. As noted previously, Dkk1 also inhibits Wnt/β-catenin signaling by binding to Lrp5/6 and Kremen, which results in internalization of Lrp5/6 and its subsequent degradation by the 26S proteosome complex [117, 118, 128, 130]. Recent evidence suggests that the interference with receptor trafficking by the G171V mutation may not play the major role that had been proposed [133]. Also, Mesd has been shown to also have the ability to bind mature Lrp6 and antagonize ligand binding at the cell surface [134]. This finding suggests a much more complex role for Mesd than just functioning as a trafficking/protein-folding chaperone.

In addition to Wnt and Dkk binding, two other proteins that play important roles in the regulation of the Wnt/β-catenin signaling pathway through interaction with Lrp5/6 have been described. The protein Wise [116] has been shown to both inhibit and activate the pathway in a context-dependent fashion similar to the duality of regulation shown for Dkk2. It shares the same binding domain in Lrp6 with Wnt8, and a model adapted from Dkk2 has been proposed to produce weak activation by Wise when binding Lrp6 in the absence of Wnt and to compete with Wnt when it is present and thereby inhibit the strong activation by Wnt [116]. Thus, it appears that a similar context-dependent inhibition/activation can occur for both Dkk2 and Wise in the regulation of Lrp5-mediated signaling. Sclerostin [135], the SOST gene product, has been shown to inhibit the activity of the Wnt/β-catenin signaling pathway [136, 137]. Data have been published demonstrating that both SOST and Wise proteins interact with the first two β-propeller domains of Lrp5/6 [116, 136]. The list of interacting proteins with the extracellular domains of Lrp5/6/Arrow and the Frizzleds will undoubtedly continue to grow. Recently, Ai *et al.* [138] showed that FEVR-causing mutations in *LRP5* can affect both Wnt and Norrin signaling, raising the possibility that within the developing eye, LRP5 may be transducing a Norrin signal rather than a Wnt signal.

The intracellular domains of Lrp5/6/Arrow and Frizzled are critical for the stabilization of β-catenin in response to Wnt binding. As mentioned previously, Frizzled is also involved in signaling through other pathways besides the Wnt/β-catenin pathway. The main components of the intracellular compartment— Dishevelled, Axin, GSK-3β, and β-catenin, along with several other ancillary proteins that play important roles in the Wnt/β-catenin signaling pathway—will be considered next. All of these main components were identified prior to the discovery of Lrp5/6/Arrow as coreceptors with Frizzled.

Genetic studies in *Drosophila* first positioned Dishevelled (Dsh) in the Wg signaling pathway at a point downstream from Frizzled and upstream from zesty-white 3 (glycogen synthase kinase-3β [GSK-3β] in vertebrates) [139–142]. Dsh is a branch point between the Wnt/β-catenin pathway and the other pathways (so-called noncanonical pathways) that some Wnts regulate (discussed in following paragraphs). Dsh contains four domains that appear to be critical for its function, although the exact nature of how Dsh works remains a major unsolved mystery. These are the DIX domain, which is also found in Axin; a conserved stretch of basic amino acids; a PDZ domain; and the DEP domain that is found in other vertebrate proteins known to interact with G-proteins [44, 143]. Dsh phosphorylation appears to be a common feature of its participation in all of the signaling pathways through which it is known to act. Several kinases have been implicated as playing a role, including casein kinase 1 [144] and 2 [145] and PAR-1 [146]. Phosphorylation of Dsh can occur through both Lrp5/6-dependent (leading to the β-catenin signaling pathway) and -independent mechanisms (Wnt binding to Frizzled in the absence of Lrp5/6) [147]. Three models of how Frizzled and Dsh may interact in conjunction with Wnt and Lrp5/6/Arrow have been proposed [143], but it is not clear which one is correct. The first model involves a recruitment of Dsh bound to Frizzled and Axin bound to the cytoplasmic tail of Lrp5/6/Arrow to form a complex that ultimately results in β-catenin stabilization (discussed in following paragraphs). The second model involves a Dsh-mediated "vesicular-type" transport of Axin to the cell membrane where it associates with Lrp5/6/Arrow. The third model is a parallel signaling model in which Frizzled-Dsh-Axin and Lrp5/6/Arrow-Axin associations occur in parallel, and both are required for activation of β-catenin signaling. Gonzalez-Sancho *et al.* [147] recently provided evidence supporting a parallel pathway model. Clearly, the role of Dsh is not fully understood, and as will be discussed further in relationship to other Wnt signaling pathways, this protein is multifunctional with regard to its mechanism of action.

Axin is another critical player in the Wnt/β-catenin signaling pathway. Axin was identified as a novel inhibitor of this pathway and cloned in mice as the product of the *Fused* locus [148]. In order to avoid confusion with the *Drosophila fused* gene, it was renamed Axin for *ax*is *in*hibition. In humans, a second *AXIN* gene, *AXIN2*, which bears homology to mouse Conductin, has been identified [149]. Axin functions as a docking or scaffolding protein and was initially shown to bind to Dsh, the adenomatous polyposis coli (APC) protein and GSK-3β [150–158]. Nakamura *et al.* [150] examined the interacting domains of Axin and determined that amino acids 581–616 were responsible for binding β-catenin, while the RGS domain interacts with APC, and GSK-3β appears to bind to amino acids 444–543. The Arm repeats in β-catenin (see following text) mediate the interaction with Axin. Interestingly, in these studies, binding of GSK-3β was observed only in the presence of β-catenin and not by GSK-3β alone.

When Lrp5/6/Arrow were identified as coreceptors in the Wnt signaling pathway, several studies quickly followed that demonstrated the binding of the cytoplasmic tail of Lrp5/6 to Axin [119, 159–161]. LRP5/6 contain consensus amino acid sequences (PPP[T/S]P) that are phosphorylation site motifs in the cytoplasmic tail. Deletion constructs of these motifs showed that decreasing ability to stimulate the LEF-1 promoter and Axin binding were not abolished until all three repeats were removed [159]. Site-specific mutations in LRP6 indicate that the PPP(T/S)P motifs are required for Axin binding [162]. Recent evidence suggests that phosphorylation of the cytoplasmic tail may be mediated by casein kinase 1 (CK1) gamma [163]. Simultaneously, Zeng *et al.* [164] demonstrated that both CK1 and GSK-3 were involved in a dual phosphorylation of Lrp6. A membrane-associated form of GSK-3 (in their model a distinct pool of GSK-3 separate from the cytoplasmic GSK-3 that inhibits β-catenin) was responsible for first phosphorylating followed by a CKI-mediated phosphorylation. This dual phosphorylation is required for Axin binding to Lrp6. The identification of these kinases involved in the phosphorylation of Lrp6 implies the presence of a phosphatase (unknown at present) that coordinately regulates the ability of Lrp6 to bind Axin [143]. Additionally, Hay *et al.* [165] demonstrated that Frat-1 binds to the cytoplasmic tail of LRP5 and is involved in the subsequent inhibition of GSK-3β. Li *et al.* suggested that Frat-1 mediates the dissociation of the GSK-3β from Axin [166]. The region where Frat-1 binds to LRP5 contains one of the PP(S/T)P repeats, but this repeat is not required for Frat-1 binding [165]. Thus, the role of Lrp5/6/Arrow in the regulation of intracellular events involved in Wnt/β-catenin signaling requires the interaction of a number of proteins and undoubtedly many more that have yet to be identified.

Familial adenomatous polyposis (FAP) is a disease that predisposes to colorectal cancer (see review, [167]). Numerous studies had identified the FAP locus in humans on chromosome 5q21, and three different groups identified mutations in the adenomatous polyposis coli (APC) gene that established this gene as causal in FAP [168–170]. Initial studies linked APC to a role in cell adhesion [171, 172]. Subsequent experiments in *Xenopus* demonstrated that APC acts as part of the Wnt/β-catenin signaling pathway [173]. APC appears to function as a docking protein for β-catenin, an association that appears to involve phosphorylation of APC by GSK-3β [174, 175]. The structure of APC includes seven Armadillo (Arm) repeats like those found in Armadillo/β-catenin (see following text), three binding sites for β-catenin, a series of 20 amino acid repeats, a basic domain, and a PDZ domain at the C-terminus [44]. Phosphorylation of APC by GSK-3β in the central region where β-catenin binds to APC is required for that binding to occur [174]. Binding of β-catenin to APC maintains the pool of free β-catenin in the cytoplasm at very low levels. Wnt binding to the coreceptor complex triggers a cascade of events that result in the intracellular accumulation of free β-catenin.

Glycogen synthase kinase-3β is the key regulator of intracellular β-catenin and, as is already evident, is involved in the phosphorylation of a number of the components of this pathway and is regulated through the opposing action of a number of proteins. GSK-3β was first described as an inhibitor of glycogen synthesis by phosphorylation of glycogen synthase (see review [176]). The phosphorylating ability of GSK-3β is enhanced by a priming phosphorylation event, and when multiple GSK-3β phosphorylation sites are present, as in the cytoplasmic tail of Lrp5/6, the enzyme can act in a self-priming fashion. Liu *et al.* [177] showed that an initial "priming" phosphorylation of β-catenin by CK1 is required for GSK-3β phosphorylation. Jho *et al.* [178] showed that GSK-3β also phosphorylates two sites within Axin (T609 and S614) and that this phosphorylation is prerequisite for binding of β-catenin. They proposed a model in which Wnt signaling leads to a dephosphorylation of these sites within GSK-3β and the release of β-catenin from the complex. GSK-3 is itself regulated by a number of upstream kinases, and phosphorylation of Ser9 has been shown to inactivate the enzyme [179]. Protein kinase B/Akt, which is downstream of phosphatidylinositol 3-kinase [180], integrin-linked kinase (ILK) [181], protein kinase C [182], and protein kinase A [183], has been shown to inhibit GSK-3β, although in the case of ILK the phosphorylation is not on Ser9 [181]. There are many other regulators of GSK-3β (#534), but those mentioned here all play roles in bone cell biology and represent possible intersections between their pathways and the Wnt/β-catenin signaling pathway.

Given the large number of protein kinases and phosphorylations that are involved in the regulation of the assembly of the degradation complex, it is not surprising that a protein phosphatase(s) should also play a role in the regulation of β-catenin. Protein phosphatase 2A (PP2A) has been shown to bind axin, APC, and Dsh [184–186] and that CK1 phosphorylation of the degradation complex results in the dissociation of PP2A [187]. The loss of the counterbalancing phosphatase activity results in further CK1-mediated phosphorylations and favors the release of β-catenin from the complex.

The intricate series of interactions initiated by the binding of Wnt to the coreceptor complex are ultimately targeted at regulating the intracellular concentration of β-catenin. β-catenin was originally shown to be associated with E-cadherin and its sequence homologous to the *Drosophila* segment polarity gene *armadillo* [188]. *Armadillo* encodes a protein that was subsequently shown to be the downstream component of the pathway activated by Wingless (Wg) [139, 140]. Both proteins contain a cassette of 12 *Armadillo* (Arm) repeats in the middle of the protein that are interaction sites with APC, E-cadherin, and the nuclear transcription factors of the Tcf/LEF family (lymphoid-specific transcription factors that are members of the high-mobility-group box transcription factors) [27, 44]. Yost *et al.* [189] first demonstrated that mutations in the GSK-3β phosphorylation site near the amino terminal end of β-catenin resulted in activation of β-catenin. Subsequently, Aberle *et al.* [190] demonstrated that the proteosome pathway degrades β-catenin, and mutations in the GSK-3β phosphorylation site in β-catenin result in stabilization of the protein. Therefore, inhibition of GSK-3β allows β-catenin to accumulate within the cytoplasm and then translocate into the nucleus, where it binds to the Tcf/LEF-1 proteins (see reviews [41, 44]) to regulate the transcription of several target genes (for a catalog of known target genes, see the Wnt homepage website at http://www.stanford.edu/~rnusse/). How β-catenin is translocated into the nucleus is not understood, and once in the nucleus, its mechanism of action is only partially understood. The Tcf proteins have no intrinsic ability to activate gene expression and appear to serve mainly a docking function for other coregulatory proteins. In *Drosophila*, Tcf normally acts as a repressor of Wnt/Wg target genes by forming a complex with another nuclear protein, Groucho [191]. The repressing activity of Groucho appears to be regulated by the histone deacetylase enzyme, Rpd3 [192]. Studies in *Xenopus* have shown that the acetyltransferases p300 and CBP cooperate with β-catenin to activate the *siamois* gene promoter [193]. These data suggest that β-catenin induces the formation of larger complexes of proteins that alter chromatin structure and thereby regulate target gene transcription [194].

In summary, the Wnt/β-catenin signaling pathway (see Figure 15-1) involves the initial binding of Wnt to the coreceptor complex consisting of either Lrp5 or Lrp6 (in vertebrates) or Arrow (in *Drosophila*) and a member of the Frizzled family of proteins. Through a mechanism that is not totally understood, Frizzled then activates Dsh, probably through a specific phosphorylation. At the same time, phosphorylation of the cytoplasmic tail of Lrp5/6/Arrow and the binding of FRAT-1 results in the binding of Axin to the tail and the collapse of the APC/β-catenin/GSK-3β degradation complex. Activation of Dsh also results in the phosphorylation and consequent inactivation of GSK-3β. Normally, it appears that β-catenin is bound to APC and APC to Axin as a consequence of specific phosphorylations mediated by GSK-3β (and other kinases), and this results in the phosphorylation of β-catenin and its degradation by the 26S proteosome complex. When Wnt binding to the coreceptor complex occurs, the downstream inhibition of GSK-3β likely results in its exclusion from the degradation complex and further facilitates the collapse of the Axin/APC/GSK-3β/β-catenin degradation complex (perhaps coincident with the binding of Axin to Lrp5/6/Arrow). When β-catenin is no longer degraded, this leads to its accumulation in the cytoplasm. Once cytoplasmic β-catenin levels accumulate, a portion can then translocate into the nucleus where it binds to the Tcf/LEF family of transcription factors and, through the recruitment of additional proteins, regulates the transcription of specific target genes.

III. OTHER WNT PATHWAYS

Three other Wnt signaling pathways have been described in the literature that do not involve β-catenin—namely, the planar cell polarity (PCP) pathway, the Wnt/Ca^{+2} pathway, and a protein kinase A pathway.

A role for Dishevelled has been implicated in the first two of these pathways. These other pathways are shown diagrammatically in Figures 15-2, 15-3, and 15-4.

The planar cell polarity (PCP) pathway (Figure 15-2) is responsible for the proper orientation of wing hairs and thoracic bristles in *Drosophila*, and mutants in which these structures are misoriented were the origin of the *Frizzled* and *Dishevelled* genes (see review [195]). Downstream of Frizzled in this pathway are Dsh and the RhoA G-protein, its associated kinase ROCK/Drok, and the c-Jun N-terminal kinase (JNK) [196]. The seven-transmembrane spanning domain classes of receptors are most commonly associated with trimeric G-protein–coupled signaling. However, until recently, hard evidence that Frizzled, a seven-transmembrane spanning receptor and a known component of the Wnt signaling pathway, was associated with trimeric G-proteins was lacking, although studies in which agents that blocked G-protein signaling could block the rat Frizzled2-induced translocation of protein kinase C (PKC) [197], implying a role for G-proteins in some aspect of Wnt signaling (see following text). Katanaev *et al.* [198] recently provided strong evidence that Frizzled signals through the Gαo subunit in both the Wnt/β-catenin pathway and the PCP pathway. What determines the decision as to whether Dsh signals to one pathway versus another is not fully understood. It may be that Frizzled localization, along with the distribution of downstream components, plays a critical role and/or that the specific Frizzled and Wnt ligands participating in the initial binding events are key [32, 56]. Also, evidence for a two-compartment model involving Dsh has been proposed in which Dsh associated with cytoplasmic vesicles is targeted for signaling through the Wnt/β-catenin pathway and Dsh associated with actin and the plasma membrane signals via the PCP pathway [34, 199].

The Wnt/Ca^{+2} pathway (Figure 15-3) involves Wnt-stimulated intracellular release of Ca^{+2} and the activation of the kinases, the Ca^{+2}–calmodulin-dependent protein kinase II (CamKII) and PKC [30, 37, 163]. Dsh also has been shown to be involved in the early activation of this pathway [200]. The question of how signaling through this pathway is regulated remains open [201]. Currently, the models that have been proposed suggest that some level of discrimination lies in the specific Wnt ligand and Frizzled receptor that trigger the pathway [30, 36, 37].

Recently, a fourth Wnt signaling pathway (Figure 15-4) has been described that is involved in myogenesis [202, 203]. Chen *et al.* [202] showed that the cyclic AMP signaling pathway through protein kinase A (PKA) is downstream of select Wnts [202]. Activation of PKA in embryonic muscle

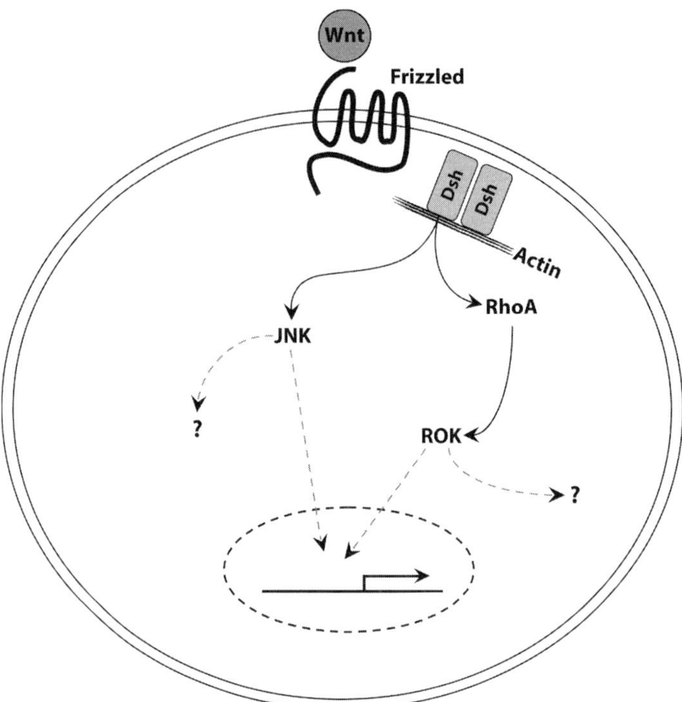

FIGURE 15-2   The planar cell polarity (PCP) pathway of Wnt signaling. Binding of Wnt to the Frizzled receptor leads to activation of a distinct compartment of Dsh associated with the actin cytoskeleton. This Dsh compartment is separate from the Dsh that activates the Wnt/β-catenin pathway. Activation of this actin-associated Dsh leads to the ROK (through RhoA) and Jnk pathways that can then affect either gene expression of have cytoplasmic consequences.

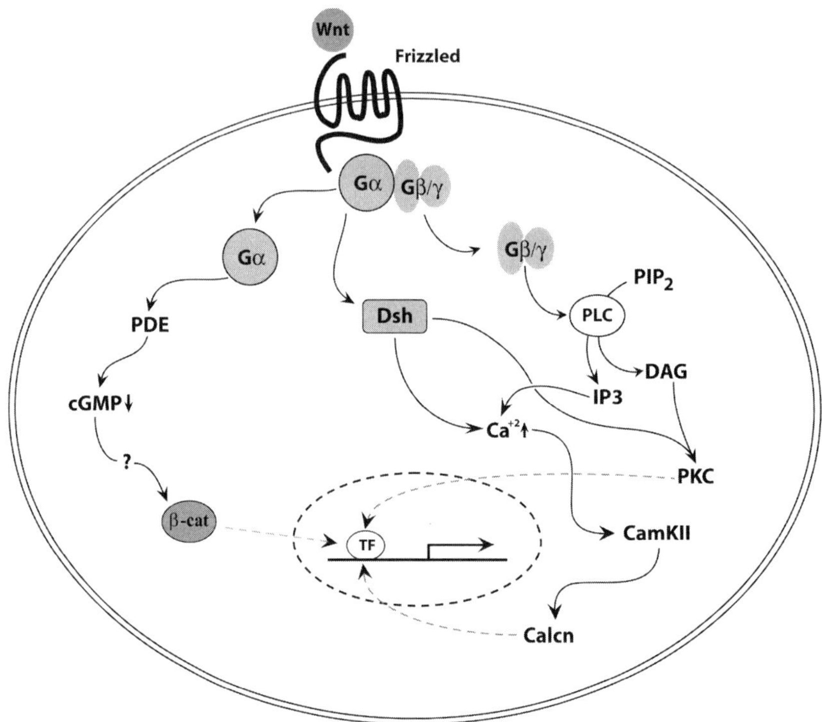

FIGURE 15-3   The Wnt/Ca^{+2} signaling pathway. Binding of Wnt to the Frizzled receptor results in the intracellular release of Ca^{+2} and the activation of phospholipase C (PLC) via a trimeric G-protein and activation of Dsh. Both of these pathways can lead to protein kinase C (PKC) and calmodulin-dependent protein kinase II activation. PKC and calcineurin are able to alter the activity of transcription factors (TF) in the nucleus of various target genes. It is also speculated that the Gα subunit can alter cGMP levels through modulation of phosphodiesterase activity (PDE) and that this may have downstream effects on the intracellular level of β-catenin.

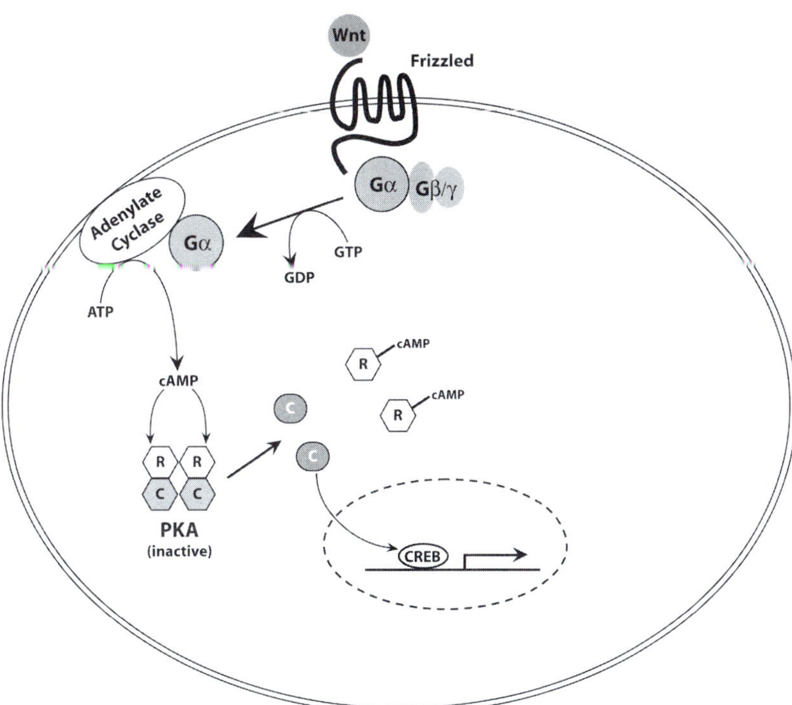

FIGURE 15-4   The Wnt/PKA signaling pathway. This pathway involves the binding of Wnt to Frizzled and the activation of adenylate cyclase, resulting in the production of cAMP. cAMP binds to the inactive form of protein kinase A, releasing the regulatory subunits (R) and enabling the catalytic subunits (C) to enter the nucleus and phosphorylated CREB and regulate gene transcription.

development results in the nuclear phosphorylation of the transcription factor CREB and the subsequent induction of genes such as *Pax3* and *Myf5* [202]. Whether this pathway is unique to muscle development or is functional in other developmental pathways is unknown. Clearly, this discovery further expands both the number of Wnt signaling pathways and the cellular events they orchestrate and suggests that we still have much more to learn about these pathways.

IV. MUTATIONS IN WNT PATHWAY COMPONENTS AND ALTERED BONE MASS

Our initial awareness of the critical role played by Wnt/β-catenin signaling in the regulation of bone mass is a consequence of human genetic studies that identified mutations in *LRP5* as causal for conditions of either extremely low [9] or extremely high bone mass [10, 92]. Several subsequent studies have identified a collection of mutations in *LRP5* that result in altered bone mass phenotypes [93–99], and several polymorphisms in *LRP5* have been associated with normal variation in bone phenotypes [102–110, 204]. The low bone mass mutations generally result in a condition known as osteoporosis pseudoglioma syndrome (OPPG) (see review [138]) although other phenotypes have been attributed to these inactivating types of mutations, such as familial exudative retinopathy (FEVR) [100, 101]. Mutations in *LRP5* that result in gain-of-function and increased bone mass phenotypes have been given any number of clinical descriptors such as high bone mass (HBM), autosomal dominant osteopetrosis, endosteal hyperostosis, Worth disease, Van Buchem disease, and autosomal dominant osteosclerosis [12]. A clinical descriptor to collectively catalog these diseases as "craniotubular hyperostoses" has been proposed [12]. The clinical manifestation of these conditions varies, but all result from mutations in *LRP5*.

Table 15-1 lists cases of LRP5 mutations resulting in gain of function and summarizes the phenotypic features. We are indebted to Van Wesenbecck *et al.* [93] for their review in 2003. Phenotypes of several cases included in their report were described [205–210] prior to genetic analysis identifying the LRP5 mutation. Several features of the phenotypes are worth noting: (1) All have generalized increase in bone mass. (2) All have reduced susceptibility to fracture. (3) For the most part the shapes of the long bones are normal. (4) Some are completely asymptomatic. (5) The majority have more or less severe abnormalities visible in the face and/or skull. (6) The

TABLE 15-1   Reported Cases of High Bone Mass Due to Mutations in Lrp5.

Mutation	Reference	Presenting diagnosis	Z-score spine/hip	Clinical features of the phenotypes
G171V	[10]	High bone mass	5.22/3.35	Normally shaped bones, thickened long bones, resistance to fracture, no accompanying syndrome (see text regarding the proband)
G171V	[92]	High bone mass	6.83/4.42	Normally shaped bones, facial lesions—square jaw, torus lesions of palate, mandible, maxilla; minimal additional pathology
R154M	[97]	Worth disease	8.7/8.5	Thickened cranium, dense base of calvarium, thickened cortices of long bones, sclerosis of vertebrae, torus lesions, and dental pathology
D111Y	[93]	Osteopetrosis	Not reported	Normally shaped bones, thickened cranium, enlarged mandible, thickened cortices of long bones
G171R	[93]	Osteopetrosis	Not reported	Thickened cortices of long bones, normally shaped bones, dense cranial base
A214T	[205]	Endosteal hyperostosis	Not reported	Normally shaped bones, thickened cortices of long bones, large elongated mandible, torus lesions, dense calvarium
A214T	[212]	Autosomal dominant osteopetrosis type II	Not reported	Thickened cranium, craniosynostosis, optic nerve atrophy, hearing loss, developmental delay, prominent jaw, absence of torus lesions, thickened cortices, enlarged mandible
A214V	[208]	Autosomal dominant osteosclerosis	Not reported	Thickened cortices of long bones, large mandible, increased gonial angle
A242T	[206]	Endosteal hyperostosis	Not reported	Cortical thickening, normal shape, elongated mandible, decreased gonial angle, torus lesions, dense calvarium
A242T	[207]	Van Buchem disease	Not reported	Large mandible, thickened calvaria, thickened cortices of long bones
A242T	[205]	Endosteal hyperostosis	Not reported	Cortical thickening, normally shaped bones, elongated mandible, decreased gonial angle, torus lesions, dense calvarium
A242T	[93]	Osteopetrosis	Not reported	Thickened cortices, normal shape, elongated mandible, decreased gonial angle, torus palatinus, dense calvarium
T253I	[209]	Autosomal dominant osteosclerosis	Not reported	Generalized osteosclerosis, pronounced in cranial vault
T253I	[210]	Autosomal dominant osteosclerosis	Hip 3.6	Generalized osteosclerosis, pronounced in cranial vault

The causal LRP5 protein amino acid sequence change is shown in the left column. The clinical diagnosis is shown as described in the literature, and details of changes in bone mineral density as a sex- and age-adjusted Z-score for the spine and hip are given, if possible. References refer to the clinical description summarized in the right columns.

orofacial abnormalities include torus lesions of the maxilla, palate, and mandible; enlarged jaw with increased gonial angle; and large forehead. (7) Very few have neurologic abnormalities in spite of the marked increase in bone mass. (8) In all cases reported to date, the gain of function mutations in the gene encode changes in the amino terminal part of the protein, before the first epidermal growth factor–like domain.

The spectrum of presenting diagnoses reflects the fact that variations occur in the phenotypes, even in kindreds with the same mutation. The differences in phenotypes may be the product of the differences in locations of the mutations in *LRP5*, but differences are occurring in kindreds with the same mutation. Other explanations may be the presence of gene-by-environment or gene-by-gene interactions. One insight into the latter occurred in the kindred in which *LRP5* mutations were first identified as associated with very high bone mass [10, 211]. The proband in this report did not have noticeable orofacial abnormalities when first examined at the age of 18, nor did the other 16 affected members of this kindred of 37. However, when examined 10 years later, she had obvious torus lesions of the mandible, maxilla, and palate. At that time her mother was re-examined and confirmed to have no orofacial lesions though she had the high bone mass phenotype. However, her father had ordinary torus palatinus. Thus, we hypothesize that the proband inherited the G171V mutation in LRP5 from her mother and the torus mutation from her father, and the combined inheritance caused exaggerated expression of the torus trait through gene-by-gene interaction. Surely, other kindreds with increased bone mass due to *LRP5* mutations will exhibit other mechanisms for variation in the phenotypes. With regard to the absence of an obvious orofacial phenotype in the proband of Johnson's HBM kindred [211], others have described the onset of the facial abnormalities in persons with *LRP5* high bone mass mutations as absent in childhood and appearing late in adolescence [93, 205].

Insights into the function of LRP5 come from examination of the phenotypes. For example, it is important to note that, except for the orofacial abnormalities, for the most part the bones are normally shaped. Further, the striking increase in bone mass is usually not accompanied by symptoms and signs of spinal nerve root compression, spinal cord compression, cranial nerve compression, marrow deficiency, or other pathology as occurs, for example, in patients with impaired osteoclast function due to carbonic anhydrase II deficiency. Indeed, while some have suffered significant morbidity [97, 212], many are completely symptom-free and have no important morbidity. In the Johnson HBM kindred, none of the affected individuals had prior diagnoses of any skeletal abnormalities except for the proband, who developed the orofacial phenotype well after adolescence. Many in that kindred reached advanced age into the ninth decade without skeletal morbidity, and none of the affected members of this kindred has ever suffered a fracture. Further, skeletal biomarkers were normal, and a transilial biopsy in one affected member of the kindred showed normal surface-based bone remodeling rates as measured by tetracycline fluorochrome markers (unpublished). Finally, expression of the phenotype seems to plateau when an affected individual reaches adulthood. The Johnson HBM kindred did not show evidence of continued gain in bone mass during adult life, nor did the other reports mention continued gain in bone mass after reaching adulthood.

An important question regarding the LRP5/Wnt signaling pathway is whether polymorphisms contained therein influence variation in bone mass in the general population. Several investigators have studied this question, and the results strongly suggest that "normal" polymorphisms in *LRP5* have a small influence on population variation in bone mass, and in many reports the effect is more significant in males [102–110, 204, 213–215].

Loss of function mutations in *LRP5* result in early childhood abnormalities consisting in low bone mass, frequent low-trauma fractures, and blindness [9, 94, 108]. The reduction in bone mass with this recessive trait is severe, consisting in reduced cortical thickness, reduced trabecular bone volume, and retarded growth. The eye pathology consists of severe disruption of ocular development, with persistent hyperplasia of the primary vitreous. The eye defect can involve all of the eye structures including the cornea. There are no abnormalities in collagen synthesis, and no apparent endocrine abnormalities. The syndrome is called osteoporosis pseudoglioma. Obligate carriers have reduced bone mass, and the heterozygous mutations are associated with primary osteoporosis in children [108]. Mutations in *LRP5* have been shown to underlie some of the cases of common familial exudative vitreoretinopathy (FEVR) [100]. Initially, the locus for FEVR-designated *EVR1* was mapped to 11q [216, 217], and mutations in the gene encoding the Wnt receptor Frizzled-4 were identified as causal [218]. It is notable that mutations in either *LRP5* or *FZD4* can result in FEVR, underlining the significance of Wnt signaling in eye development.

V. WNT SIGNALING AND BONE CELL FUNCTION

The identification of mutations in *LRP5* that result in conditions of altered bone mass has launched a literal explosion of studies aimed at understanding the role of Wnt signaling in the function of bone cells. These studies have implicated Wnts and Wnt signaling in the entire gamut of possible functions from differentiation and proliferation to apoptosis, from cell–cell communication to the functional activity of bone cells, and from embryonic skeletal patterning to the maintenance of adult bone mass and the ability of bone to respond to mechanical load. Some aspects of these studies are mentioned in other chapters within this book. In this section we examine our current understanding of Wnt signaling and its role in the differentiation and function of bone cells.

Our understanding of Wnt/β-catenin signaling in bone cell function has been rapidly advanced through the use cell culture model systems and the development of transgenic and knockout mouse models in which specific genes encoding proteins in the pathway or involved in its regulation have been specifically overexpressed or deleted. One of the first clues that

Wnt signaling might be important in the skeleton came from studies of mice carrying mutations in Wnt genes that failed to develop somites [6–8]. A few years later Pinson *et al.* [79] demonstrated that LRP6−/− mice had limb patterning defects and other severe developmental abnormalities that mimicked those observed in mice carrying Wnt gene mutations. This work, along with the studies by Wehrli *et al.* [77] and Tamai [78], positioned LRP5/6/Arrow in the Wnt/β-catenin signaling pathway. However, it wasn't until the discovery of human mutations in *LRP5* that gave rise to low bone mass in OPPG [9] or high bone mass in two separate kindreds [10, 92] that Wnt/β-catenin signaling in bone became a major focus of research in the bone field.

In their original identification of inactivating mutations in *LRP5* in OPPG, Gong *et al.* [9] provided evidence that LRP5 regulates osteoblast differentiation and proliferation. Kato *et al.* [219] subsequently observed in their studies of the *Lrp5−/−* mouse abnormal osteoblast proliferation despite normal *Cbfa1* expression. These knockout mice develop a low bone mass and eye phenotype indistinguishable from human OPPG [9]. They also reported decreased bone mineral deposition in these knockout mice, which may have been a consequence of decreased osteoblast numbers or could be due to poorly functioning osteoblasts. They further suggested that Lrp5/Wnt represents an independent pathway from the classical Cbfa1/Runx2 pathway [220] regulating osteoblast proliferation, since they observed no decrease in Cbfa1/Runx2 mRNA in the *Lrp5−/−* bones. However, Gaur *et al.* [221] recently showed that Wnts promote osteogenesis by directly stimulating *Runx2* gene expression, suggesting that Wnt signaling is upstream of Runx2. It is not clear what might explain these two seemingly contradictory observations; however, the *Lrp5−/−* mice do have normal expression of *Lrp6*, and perhaps in embryonic development or early differentiation steps of osteoblastogenesis, the Lrp6 coreceptor plays a more important role or can compensate for the loss of Lrp5. Also, Hill *et al.* [222] and Day *et al.* [223] both showed that control of the Wnt/β-catenin signaling dictates skeletal lineage commitment to osteoblastogenesis versus chondrogenesis. Collectively, these and other data have resulted in a working paradigm for how Wnt signaling regulates osteoblastogenesis, but much remains to be understood [65].

As mentioned previously, Boyden *et al.* [92] demonstrated that in NIH3T3 cells transfected with either LRP5 or LRP5[G171V] there was no effect of the G171V mutation on the activity of the pathway in the presence of added Wnt-1, but the inhibition of the pathway by Dkk-1 was almost completely eliminated by the mutation. Zhang *et al.* [125] suggested that the G171V mutation alters Mesd binding and transport of Lrp5 to the cell surface. They proposed a paracrine-autocrine model to explain the effect of the mutation on bone cell function. However, Ai *et al.* [133] provided counterevidence to this model and suggested that reduced affinity for Dkk1 is the main result of the mutation. Regardless of the explanation, the effect of the G171V mutation appears to be consistent with a gain of function and supports an important caveat that the default position for the pathway is in an "off" position due to the presence of these inhibitory regulatory proteins.

Two major unanswered questions are which of the known negative regulators is critical and in which bone cell type does this regulation need to occur? Li *et al.* [224] showed that the *Dkk2−/−* mouse develops osteopenia with decreased trabecular and cortical bone mineral content and decreased mineral apposition rates. Their results suggest a defect in these mice in the terminal differentiation of osteoblasts and mineralization of osteoid. While the Dkks are certainly one candidate, sclerostin, another inhibitor of the Wnt/β-catenin pathway [136, 137], is abundantly expressed by osteocytes [225, 226] and is another candidate. Recently, sclerostin produced by the osteocytes has been proposed to regulate bone formation by osteoblasts and the induction or maintenance of the lining cells on the bone surface [226]. This implies a paracrine type of system in which the osteocyte produces a factor, sclerostin, which regulates Wnt pathway activity in nearby osteoblasts and possibly induces their progression toward a lining cell. Further studies are needed to confirm this hypothesis, but it is an exciting observation regarding how lining cells are formed and regulated. Another class of inhibitory molecules is the secreted frizzled-related proteins [82, 83]. Bodine and colleagues showed that sFRP-1 is an important regulator of apoptosis of osteoblastic and osteocytic cells in culture [227] and that *sFRP-1* knockout mice have increased trabecular bone volume [228], supporting an *in vivo* action for this negative regulator of Wnt/β-catenin signaling.

In addition to the previously mentioned results, several other studies support a role for Wnt/β-catenin signaling in osteoblastogenesis. For example, Rawadi *et al.* [229] showed that in mesenchymal cell lines that are capable of differentiating into osteoblastic cells, BMP-2 and Sonic Hedgehog (Shh) have the capacity to induce alkaline phosphatase (ALP) (as a marker of osteoblasts). In their model BMPs induce the expression of Wnts, which subsequently activate the Wnt/β-catenin signaling pathway and drive the differentiation of these cell lines toward an osteoblast phenotype. However, Winkler *et al.* [230] suggested that the Wnt/β-catenin signaling pathway induces BMP expression and BMP pathway activation induces the C3H10T1/2 cell to differentiate into the osteoblastic cell as evidenced by ALP expression. Regardless of which model is correct and whether Wnts directly or indirectly promote osteoblastogenesis, clearly this pathway is important for osteoblast differentiation.

Other questions that remain open to debate are which of the Wnts function as the endogenous ligand for regulating bone formation and/or do different Wnts serve different roles? Hu *et al.* [231] implicated Wnt signaling at an early event in osteoblastogenesis, but downstream of Indian Hedgehog signaling. They also suggested that Wnt7b is the endogenous ligand regulating osteoblastogenesis. In contrast, Bennett *et al.* [232] presented evidence that Wnt10b is an important endogenous regulator of bone formation, as it inhibits adipogenesis and stimulates osteoblastogenesis of mesenchymal precursors. It seems likely that more than one of the known Wnts are functioning in bone. In support of this, Kennell *et al.* [233] recently suggested that Wnt signaling inhibits adipogenesis through β-catenin and β-catenin-independent pathways.

It appears that Wnt signaling has a potential number of intersections/interactions with other pathways in the regulation of bone mass. For example, the interaction between PTH and the Wnt/β-catenin pathway has been studied by several groups. Kulkarni *et al.* [234] showed that, with continuous PTH treatment of rats *in vivo* or UMR 106 cells in culture, there is a downregulation of *Lrp5* and *Dkk1* and an upregulation of *Lrp6* and *Frz-1*. They suggest that the effects of PTH on the Wnt/β-catenin signaling pathway are in part mediated by a cAMP-PKA pathway. Bodine *et al.* [235] concluded from their studies of PTH action in the *sFrp*-1 knockout mice that PTH and Wnt signaling may share some common components, but PTH action appears to extend beyond the Wnt pathway. Iwaniec *et al.* [236] examined the effects of PTH on the *Lrp5* knockout mouse and concluded that Lrp5 is not required for the stimulatory effect of PTH. Kharode *et al.* [237] studied the effects of PTH on the HBM transgenic mouse skeleton and concluded that PTH and Lrp5 work through complementary pathways. Recently, two groups showed that PTH inhibits *SOST* gene expression, suggesting that PTH reduces the level of this negative regulator of the Wnt pathway and thereby stimulates osteoblastogenesis [238, 239]. Thus, the picture that emerges with PTH is complex. Some of the evidence suggests that PTH works through a complementary pathway, and some of the evidence suggests that PTH alters the regulation of the Wnt pathway. However, at this time it is not possible to exclude the possibility that one of the other pathways through which Wnts can signal may play a role in the action of PTH. Another pathway that may be potentially interacting with Wnt signaling in bone is prostaglandin E_2

(PGE$_2$). Recent work from the cancer field has shown that PGE$_2$ is capable of stimulating the Wnt pathway through PI-3 kinase activation of Akt leading to inhibition of GSK-3β and through the G-protein, Gαs binding to axin and possibly inducing the displacement of APC and the intracellular accumulation of β-catenin [240]. Given the importance of PGE$_2$ in regulating bone formation, this potential interaction creates a new pathway for PGE$_2$ action in bone. However, studies need to be performed in order to determine if this interaction of PGE$_2$ and Wnt/β-catenin signaling functions in bone.

The Wnt/β-catenin signaling pathway is known to play a role in cellular apoptosis [48, 241–247]. The data of Kato *et al.* [219] also implicated a role for Lrp5 in regulating the regression of vascular capillary cells in the hyaloid vessels of the eye due to apoptosis induced by ocular macrophages. Loss of Lrp5 resulted in the retention of these vessels and contributed to the eventual development of blindness. Importantly, these results suggested a role for the Wnt/β-catenin pathway in macrophages and macrophage-mediated cell killing, but how they are connected remains unknown at this time [219]. Babij *et al.* [248] created two transgenic mouse lines using the col1a1 promoter to drive either the normal human LRP5 cDNA (LRP5tg) or the human LRP5[G171V] cDNA (HBMtg). The LRP5tg line had only a modest increase in bone mass, while the HBMtg line had increased bone mass and a phenotype that recapitulated the human HBM phenotype. This suggested that it was the G171V mutation and not overexpression of the gene that largely contributed to the increased bone mass in these mice. The HBMtg mouse had reduced osteoblast and osteocyte apoptosis. This suggests that the increase in bone mass in the affected members of the HBM kindred [10, 211] and in the HBMtg mouse [248] is due to osteoblasts that live longer and thereby make more bone versus a model in which the osteoblasts produce more bone per unit of time. Increased osteocyte survival would be consistent with the concept that in order to maintain bone one needs viable osteocytes (see Chapter 6, Bonewald).

The skeletal effects of mutations in other components of the Wnt/β-catenin pathway have been reported. Loss of Lrp6 is an embryonically lethal condition, but the heterozygous *Lrp6*+/− mouse is viable. Kharode *et al.* [249] and subsequently Holmen *et al.* [250] showed an effect of Lrp6 on bone mineral density in these heterozygous knockout mice. This raises the question of how Lrp5 and Lrp6 function in the regulation of bone mass. Clearly, the loss of one is not compensated completely by the other, suggesting distinct (or perhaps only partially overlapping) roles for each of these coreceptors in bone. As might be expected, the complete loss of β-catenin [251, 252] or APC [253, 254] in a traditional knockout mouse model results in early embryonic lethality. Conditional deletions created postnatally result in osteopenia as evidenced by decreased cortical and trabecular bone volume [255]. Loss of Axin2 has been shown to result in craniosynostosis, implicating the Wnt/β-catenin signaling pathway in skull morphogenesis [256]. Interestingly, the defects in calvarial morphogenesis caused by loss of Axin2 were not manifest during embryonic development, but in the early postnatal period.

Thus, there is compelling evidence supporting a role for regulation of the Wnt/β-catenin signaling pathway at several levels in both osteoblasts and osteocytes. What about osteoclast function and/or differentiation? The osteoclast is the cell responsible for resorbing bone (see Chapter 5, Blair). It is derived from the monocyte/macrophage cell lineage. These precursors are induced to differentiate into osteoclasts by RANKL (the ligand for receptor for activation of nuclear factor kappa B [RANK]) produced by stromal cells and osteoblasts, and by macrophage colony-stimulating factor (M-CSF). Work from Gillespie's laboratory has shown that secreted frizzled-related protein 1 (sFRP-1) inhibits osteoclastogenesis by binding to RANKL [257]. Thus, there appears to be a connection between the Wnt signaling pathway and osteoclastogenesis. Studies from Chatterjee-Kishore and colleagues [258, 259] have shown that the osteoprotegerin (OPG) to RANKL ratio increased in response to *in vivo* mechanical loading of the HBM transgenic mice suggesting that, in response to load, the osteoblast shuts down osteoclastogenesis through these negative modulators of RANKL. We also observed increased *sFrp-1* gene expression in response to loading, which could have a similar effect on osteoclastogenesis [259]. Recently, Glass *et al.* [260] studied the β-catenin over-expressing mouse and showed that the increased bone mass observed in these mice is due to inhibition of osteoclastogenesis by osteoblasts. Similarly, Spencer *et al.* [261] showed that Wnt signaling in osteoblasts regulates *RANKL* gene expression. These data clearly support the concept that activation of the Wnt/β-catenin signaling pathway in osteoblasts can have downstream consequences on osteoclasts and perhaps other cell types through the regulation of specific modulator proteins that are important in orchestrating the activity of cells. However, to date, there is no evidence for a direct role of the Wnt pathway in osteoclasts.

VI. WNT SIGNALING AND THE BONE RESPONSE TO MECHANICAL LOADING

We began this chapter with a brief historical vignette regarding the adaptive response of the skeleton to changes in mechanical loading. The precise mechanism by which bone senses mechanical loading is not known. Since the original (theoretical) ideas put forth by Harold Frost [3, 262, 263] regarding mechanosensation in bone, considerable effort has been spent trying to understand this process at the cellular/molecular level. Several articles and reviews exist in the literature describing what is currently known about the bone cell response to loading [264–280].

Our first encounter with the HBM kindred [211] led us to hypothesize that the mutation in this family had somehow altered the skeletal response to mechanical loading [11]. The normal shapes of the long bones, the infrequent occurrence of nerve root compression, the normal bone remodeling, and the absence of continued bone gain during adult life suggested this hypothesis to us as an explanation for how the G171V mutation in LRP5 affected bone and that the normal role of LRP5 might be an important element of the adaptive response to mechanical loading of the skeleton. The mechanosensation pathway detects and responds to changes in load environment such as those brought about by increases in bending of skeletal tissue or brought about by increases in loads. These signals result in bone-forming activity that lays down additional bone that prevents skeletal failure from the newly increased loads. Further, these signals arrange the new bone geometrically so that the adaptation to increased loads occurs with the least gain in total bone mass. Adaptation continues until the point is reached where the additional bone mass reduces the strains induced by the increased load and cancels the threat of failure from the increased load. Thus, the bone strains return to a "safe level," the adaptive signal is canceled, and further increases do not occur. A new steady state is reached. This requires that a "mechanosensor" exists in the skeleton such as described by Frost [5, 281].

Thus, the skeletal phenotypes described in the gain-of-function mutations in LRP5 are compatible with the hypothesis that the mutations cause a change (increase) in the sensitivity, or set point, of the mechanosensor, resulting in an exaggerated response of the skeleton in adapting to normal mechanical loads. The mostly normal shape of the skeleton, the rarity of neurologic abnormalities, the absence of hematologic abnormalities, and the absence of life-long continued gain in skeletal mass support this idea. The mechanisms of skeletal load adaptation remain in place in those persons with the mutations.

It must be acknowledged that a mechanosensor has never been described. Perhaps continued study of the biology of these mutations may ultimately lead to a full understanding of how the normal skeleton senses and responds to loads so accurately, faithfully, and economically. Phenotype information and *in vivo* loading data from transgenic mice containing the G171V mutation support this hypothesis [259, 282].

Work from our group using the *in vivo* tibia four-point bending model [283] demonstrated an increased sensitivity to loading in the HBMtg mice [284]. Changes in a number of Wnt/β-catenin signaling pathway target genes in MC3T3-E1 cells subjected to physical deformation have been reported, which correlate with changes observed in HBMtg mice bones and the HBM kindred bone cells [258, 259]. The fact that the HBMtg mice have an increased sensitivity to mechanical load [284] supports our original hypothesis that the LRP5G171V mutation altered the mechanosensation mechanism in our human kindred such that affected individuals build and maintain a skeleton with an inappropriate, high bone mass relative to the normal loading they encounter on their skeletons. Also in support of this hypothesis, Sawakami *et al.* [285] studied the *in vivo* bone formation response in *Lrp5* knockout mice and demonstrated that loss of both alleles of *Lrp5* resulted in an 88% reduction in the response in males and a 99% reduction in females compared to wild-type mice using the ulna loading model. Hens *et al.* [286] showed in studies using the TOPGAL mouse, which carries a reporter construct to monitor activation of the Wnt/β-catenin pathway, that in primary osteoblast cultures the pathway was activated by physical deformation as a means of mechanical loading. We have used the TOPGAL mouse and the *in vivo* ulna-loading model and demonstrated a rapid activation of the Wnt/β-catenin signaling pathway in osteocytes in response to loading (M. L. Johnson, unpublished results).

The key unanswered question at this time is how can we integrate the Wnt/β-catenin pathway along with the other responses that are known to occur in bone/bone cells as a result of changes in mechanical load? One key to answering this question is to understand the temporal order of events that occur from the perception of a load signal to the transduction of that signal to the ultimate response. Some of the earliest events that occur (within seconds to minutes) of a load being applied to bone/bone cells are the release of NO [287, 288] and PGE$_2$ [288–291], Ca^{+2} fluxes [292, 293], and the movement of ATP [294]. Li and colleagues recently proposed a model in which the release of ATP after fluid flow shear stress applied to osteoblasts binds to the P2X$_7$ receptor, resulting in prostaglandin release, and ATP binding to the P2Y receptor, resulting in intracellular calcium mobilization [295]. What is important to note is that both the intracellular prostaglandin and P2X$_7$ receptor–mediated signaling pathways involving PIP2 and IP3 have the potential to activate the Wnt signaling pathway through inhibition of GSK-3β via phosphorylation by Akt and thereby independently

of the Lrp5 coreceptor could initiate β-catenin signaling. This leads to the hypothesis that we are currently investigating that the initial rapid responses of osteocytes/osteoblasts to mechanical loading that trigger ATP and PGE$_2$ release also trigger the activation of the Wnt/β-catenin pathway (independent of Wnt) and that this activation leads to a feedback loop in which Wnt produced by this initial activation then acts in an autocrine/paracrine fashion to further amplify the mechanical load signal and results in a commitment to new bone formation.

VII. CONCLUSIONS AND FUTURE DIRECTIONS

The discovery that the Wnt/β-catenin signaling pathway is an important mediator of bone mass regulation has ushered in a new era in bone research. It is now clear that Lrp5/6 and the Wnt/β-catenin signaling pathway are major players in various aspects of skeletal development and bone mass accrual and play key roles in bone cell differentiation, proliferation, and apoptosis, and in the regulation and maintenance of adult bone mass and the ability of bone to respond to mechanical load. Where this will lead and how this will change our understanding of bone biology remain unknown. There are clearly several unanswered questions and challenges. From a translational perspective, the question is will we be able to develop agents that can modulate the Wnt/β-catenin signaling pathway selectively in bone to build bone mass and treat diseases such as osteoporosis? What contribution to variance in normal bone mass can be explained by polymorphisms in the genes encoding the different components of the Wnt pathway, and can this information enable us to predict future risk for osteoporosis and/or osteoporotic fracture? Major efforts in the pharmaceutical industry to develop anabolics based on Wnt/β-catenin pathway targets are currently under way. From a bone biology perspective, the questions are numerous. For example: What are the endogenous ligands that activate and/or modulate this pathway? Do different Wnts serve distinct or overlapping roles? Do Lrp5 and Lrp6 represent redundant receptors, or does each have unique functions in bone? How does the Wnt pathway function in the orchestration of basic bone cell activity? How does this pathway interact with other bone cell regulating pathways? Do all four Wnt signaling pathways function in bone? What is the role of Wnt/β-catenin signaling in bone mechanosensation and/or response to mechanical loading? Perhaps several of the answers to these questions and many more will be found in the fourth edition of this book.

REFERENCES

1. G. Galilei, *Dialogues Concerning Two New Sciences*. Running Press, Philadelphia (2002).
2. J. Wolff, *The Law of Bone Remodeling*. Berlin, 1892.
3. H. M. Frost, Bone "Mass" and the "mechanostat": A proposal. *Anat Record*. **219,** 1–9 (1987).
4. H. M. Frost, From Wolff's Law to the Utah Paradigm: Insights about bone physiology and its clinical applications. *Anat Record*. **262,** 398–419 (2001).
5. H. M. Frost, Bone's mechanostat: A 2003 update. *Anat Record*. **275A,** 1081–1101 (2003).
6. S. Takada, K. L. Stark, M. J. Shea, G. Vassileva, J. A. McMahon, and A. P. McMahon, Wnt-3a regulates somite and tailbud formation in the mouse embryo. *Genes Develop*. **8,** 174–189 (1994).
7. T. L. Greco, S. Takada, M. M. Newhouse, J. A. McMahon, A. P. McMahon, and S. A. Camper, Analysis of the vestigial tail mutation demonstrates that Wnt-3a gene dosage regulates mouse axial development. *Genes Develop*. **10,** 313–324 (1996).
8. Y. Yoshikawa, T. Fujimori, A. P. McMahon, and S. Takada, Evidence that absence of Wnt-3a signaling promotes neuralization instead of paraxial mesoderm development in the mouse. *Dev Biology*. **183,** 234–242 (1997).
9. Y. Gong, R. B. Slee, N. Fukai, G. Rawadi, S. Roman-Roman, A. M. Reginato, H. Wang, T. Cundy, F. H. Glorieux, D. Lev, M. Zacharin, K. Oexle, J. Marcelino, W. Suwairi, S. Heeger, G. Sabatakos, S. Apte, W. N. Adkins, J. Allgrove, M. Arsian-Kirchner, J. A. Batch, P. Beighton, G. C. M. Black, R. G. Boles, L. M. Boon, C. Borrone, H. G. Brunner, G. F. Carle, B. Dallapiccola, A. De Paepa, B. Floege, M. L. Halfide, B. Hall, R. C. Hennekam, T. Hirose, A. Jans, H. Juppner, C. A. Kim, K. Keppler-Noreuil, A. Kohlschuetter, D. LaCombe, M. Lambert, E. Lemyre, T. Letteboer, L. Peltonen, R. S. Ramesar, M. Romanengo, H. Somer, E. Steichen-Gersdorf, B. Steinmann, B. Sullivan, A. Superta-Furga, W. Swoboda, M.-J. van den Boogaard, W. Van Hul, M. Vikkula, M. Votruba, B. Zabel, T. Garcia, R. Baron, B. R. Olsen, and M. L. Warman, LDL receptor-related protein 5 (LRP5) affects bone accrual and eye development. *Cell*. **107,** 513–523 (2001).
10. R. D. Little, J. P. Carulli, R. G. Del Mastro, J. Dupuis, M. Osborne, C. Folz, S. P. Manning, P. M. Swain, S. C. Zhao, B. Eustace, M. M. Lappe, L. Spitzer, S. Zweier, K. Braunschweiger, Y. Benchekroun, X. Hu, R. Adair, L. Chee, M. G. FitzGerald, C. Tulig, A. Caruso, N. Tzellas, A. Bawa, B. Franklin, S. McGuire, X. Nogues, G. Gong, K. M. Allen, A. Anisowicz, A. J. Morales, P. T. Lomedico, S. M. Recker, P. Van Eerdewegh, R. R. Recker, and M. L. Johnson, A mutation in the LDL receptor-related protein 5 gene results in the autosomal dominant high-bone-mass trait. *Am J Hum Genet*. **70,** 11–19 (2002).
11. M. L. Johnson, J. L. Picconi, and R. R. Recker, The gene for high bone mass. *The Endocrinologist*. **12,** 445–453 (2002).
12. M. L. Johnson, K. Harnish, R. Nusse, and W. Van Hul, LRP5 and Wnt signaling: A union made for bone. *J Bone Miner Res*. **19,** 1749–1757 (2004).
13. R. Nusse and H. E. Varmus, Many tumors induced by the mouse mammary tumor virus contain a provirus integrated in the same region of the host genome. *Cell*. **31,** 99–109 (1982).
14. F. Rijsewijk, M. Schuermann, E. Wagenaar, P. Parren, D. Weigel, and R. Nusse, The *Drosphila* homology of the mouse mammary oncogen int-1 is identical to the segment polarity gene wingless. *Cell*. **50,** 649–657 (1987).
15. C. V. Cabrera, M. C. Alonso, P. Johnston, R. G. Phillips, and P. A. Lawrence, Phenocopies induced with antisense RNA identify the wingless gene. *Cell*. **50,** 659–663 (1987).
16. J. R. Miller, The Wnts. *Genome Biology*. **3,** reviews 3001.1–3001.15 (2001).
17. K. Willert, J. D. Brown, E. Danenberg, A. W. Duncan, I. L. Weissman, T. Reya, J. R. Yates, and R. Nusse, Wnt proteins are lipid-modified and can act as stem cell growth factors. *Nature*. **423,** 448–452 (2003).
18. T. Kadowaki, E. Wilder, J. Klingensmith, K. Zachary, and N. Perrimon, The segment polarity gene porcupine encodes a putative multitransmembrane protein involved in wingless processing. *Genes Develop*. **10,** 3116–3128 (1996).
19. R. Nusse, Wnts and hedgehogs: Lipid-modified proteins and similarities in signaling mechanisms at the cell surface. *Development*. **130,** 5297–5305 (2003).
20. J. T. Dunphy and M. E. Linder, Signalling functions of protein palmitoylation. *Biochim Biophys Acta*. **1436,** 245–261 (1998).
21. M. Zecca, K. Basler, and G. Struhl, Direct and long-range action of a *Wingless* morphogen gradient. *Cell*. **87,** 833–844 (1996).
22. A. Aulehla, C. Wehrle, B. Brand-Saberi, R. Kemler, A. Gossler, B. Kanzler, and B. G. Herrmann, Wnt3a plays a major role in the segmentation clock controlling somitogenesis. *Cell*. **4,** 395–406 (2003).
23. A. Aulehla and B. G. Herrmann, Segmentation in vertebrates: Clock and gradient finally joined. *Genes Develop*. **18,** 2060–2067 (2004).
24. J. O. Mason, J. Kitajewski, and H. E. Varmus, Mutational analysis of mouse Wnt-1 identifies two temperature-sensitive alleles and attributes of Wnt-1 protein essential for transformation of a mammary cell line. *Mol Biol Cell*. **3,** 521–533 (1992).
25. M. J. Seidensticker and J. Behrens, Biochemical interactions in the Wnt pathway. *Biochim Biophys Acta*. **1495,** 168–182 (2000).
26. U. Rothbacher and P. Lemaire, Creme de la Kremen of Wnt signalling inhibition. *Nature Cell Biol*. **4,** 172–173 (2002).
27. K. Willert and R. Nusse, β-catenin: A key mediator of Wnt signaling. *Development*. **8,** 95–102 (1998).
28. W. M. Jones and A. Bejsovec, Wingless signaling: An axin to grind. *Curr Biol*. **13,** 479–481 (2003).
29. H. Clevers, Wnt signaling: Ig-norrin the dogma. *Curr Biol*. **14,** 436–437 (2004).
30. M. Kuhl, L. C. Sheldahl, M. Park, J. R. Miller, and R. T. Moon, The Wnt/Ca+2 pathway: A new vertebrate Wnt signaling pathway takes shape. *Trends Genet*. **16,** 279–283 (2000).
31. C. Prunier, B. A. Hocevar, and P. H. Howe, Wnt signaling: Physiology and pathology. *Growth Factors*. **22,** 141–150 (2004).
32. A. Bejsovec, Wnt pathway activation: New relations and locations. *Cell*. **120,** 11–14 (2005).
33. M. L. Johnson and D. T. Summerfield, Parameters of LRP5 from a structural and molecular perspective. *Crit Rev Eukaryotic Gene Express*. **15,** 229–242 (2005).
34. M. Povelones and R. Nusse, Wnt signalling sees spots. *Nature Cell Biol*. **4,** E249–E250 (2002).
35. R. Nusse and H. E. Varmus, Wnt genes. *Cell*. **69,** 1073–1087 (1992).
36. H.-Y. Wang and G. C. Malbon, Wnt signaling, Ca+2, and cyclic GMP: Visualizing frizzled functions. *Science*. **300,** 1529–1530 (2003).

37. T. Quaiser, R. Anton, and M. Kuhl, Kinases and G proteins join the Wnt receptor complex. *BioEssays.* **28,** 339–343 (2006).

38. X. He, A Wnt-Wnt situation. *Develop Cell.* **4,** 791–797 (2003).

39. A. Wodarz and R. Nusse, Mechanisms of Wnt signaling in development. *Cell Dev Biol.* **14,** 59–88 (1998).

40. C. Y. Logan and R. Nusse, The Wnt signaling pathway in development and disease. *Annu Rev Cell Dev Biol.* **20,** 781–810 (2004).

41. R. Nusse, Wnt signaling in disease and development. *Cell Res.* **15,** 28–32 (2005).

42. G. Cossu and U. Borello, Wnt signaling and the activation of myogenesis in mammals. *EMBO J.* **18,** 6867–6872 (1999).

43. R. Imondi and J. B. Thomas, The ups and downs of Wnt signaling. *Science.* **302,** 1903–1904 (2003).

44. K. M. Cadigan and R. Nusse, Wnt signaling: A common theme in animal development. *Genes Develop.* **11,** 3286–3305 (1997).

45. P. Polakis, Wnt signaling and cancer. *Genes Develop.* **14,** 1837–1851 (2000).

46. M. Bienz and H. Clevers, Linking colorectal cancer to Wnt signaling. *Cell.* **103,** 311–320 (2000).

47. A. Kikuchi, Tumor formation by genetic mutations in the components of the Wnt signaling pathway. *Cancer Sci.* **94,** 225–229 (2003).

48. I. S. Nathke, The adenomatous polyposis coli protein: The Achilles heel of the gut epithelium. *Annu Rev Cell Dev Biol.* **20,** 337–366 (2004).

49. E. Sancho, E. Batlle, and H. Clevers, Signaling pathways in intestinal development and cancer. *Annu Rev Cell Dev Biol.* **20,** 695–723 (2004).

50. D. B. Chesire and W. B. Isaacs, β-catenin signaling in prostate cancer: An early perspective. *Endocrine-Related Cancer.* **10,** 537–560 (2003).

51. S. Hatsell, T. Rowlands, M. Hiremath, and P. Cowin, β-catenin and Tcfs in mammary development and cancer. *J Mammary Gland Biol Neoplasia.* **8,** 145–158 (2003).

52. N. S. Fearnhead, M. P. Britton, and W. F. Bodmer, The ABC of APC. *Hum Mole Genet.* **10,** 721–733 (2001).

53. A. Gregorieff and H. Clevers, Wnt signaling in the intestinal epithelium: From endoderm to cancer. *Genes Develop.* **19,** 877–890 (2005).

54. T. Reya and H. Clevers, Wnt signalling in stem cells and cancer. *Nature.* **434,** 843–850 (2005).

55. J. Taipale and P. A. Beachy, The hedgehog and Wnt signalling pathways in cancer. *Nature.* **411,** 349–354 (2001).

56. M. Katoh, Wnt/PCP signaling pathway and human cancer. *Oncol Rep.* **14,** 1583–1588 (2005).

57. A. M. C. Brown, Wnt signaling in breast cancer: Have we come full circle? *Breast Cancer Res.* **3,** 351–355 (2001).

58. C. L. Hall, S. Kang, O. A. MacDougald, and E. T. Keller, Role of Wnts in prostate cancer bone metastases. *J Cell Biochem.* **97,** 661–672 (2006).

59. G. V. De Ferrari and N. C. Inestrosa, Wnt signaling function in Alzheimer's disease. *Brain Res Rev.* **33,** 1–12 (2000).

60. F. J. T. Staal and H. C. Clevers, Wnt signalling and haematopoiesis: A Wnt-Wnt situation. *Nature Rev Immunol.* **5,** 21–30 (2005).

61. T. D. Gould and H. K. Manji, The Wnt signaling pathway in bipolar disorder. *Neuroscientist.* **8,** 497–511 (2002).

62. M. L. Johnson, The high bone mass family—The role of Wnt/LRP5 signaling in the regulation of bone mass. *J Mus Neuron Interact.* **4,** 135–138 (2004).

63. J. J. Westendorf, R. A. Kahler, and T. M. Schroeder, Wnt signaling in osteoblasts and bone diseases. *Gene.* **341,** 19–39 (2004).

64. S. L. Ferrari, S. Deutsch, and S. E. Antonarakis, Pathogenic mutations and polymorphisms in the lipoprotein receptor-related protein 5 reveal a new biological pathway for the control of bone mass. *Curr Opin Lipidol.* **16,** 207–213 (2005).

65. C. Hartmann, A Wnt canon orchestrating osteoblastogenesis. *Trends Cell Biol.* **16,** epub 7 February (2006).

66. R. T. Moon, B. Bowerman, M. Boutros, and N. Perrimon, The promise and perils of Wnt signaling through β-catenin. *Science.* **296,** 1644–1646 (2002).

67. N. Janssens, M. Janicot, and T. Perera, The Wnt-dependent signaling pathways as targets in oncology drug discovery. *Investigational New Drugs.* epub 28 January (2006).

68. G. Rawadi and S. Roman-Roman, Wnt signaling pathway: A new target for the treatment of osteoporosis. *Expert Opin Ther Targets.* **9,** 1063–1077 (2005).

69. P. Bhanot, M. Brink, C. Harryman Samos, J. C. Hsieh, Y. S. Wang, J. P. Macke, D. Andrew, J. Nathans, and R. Nusse, A new member of the frizzled family from *Drosophila* functions as a wingless receptor. *Nature.* **382,** 225–230 (1996).

70. H. Sawa, L. Lobel, and H. R. Horvitz, The *Caenorhabditis elegans* gene line-17, which is required for certain asymmetric cell divisions, encodes a putative seven-transmembrane protein similar to the *Drosophila* frizzled protein. *Genes Develop.* **10,** 2189–2197 (1996).

71. J. Yang-Snyder, J. R. Miller, J. D. Brown, C. J. Lai, and R. T. Moon, A frizzled homolog functions in a vertebrate Wnt signaling pathway. *Curr Biol.* **6,** 1302–1306 (1996).

72. X. He, J. P. Saint-Jeannet, Y. Wang, J. Nathans, I. Dawid, and H. Varmus, A member of the frizzled protein family mediating axis induction by Wnt-5A. *Science.* **275,** 1652–1654 (1997).

73. T. W. Sappington and A. S. Raikhel, Ligand-binding domains in vitellogenin receptors and other LDL-receptor family members share a common ancestral ordering of cysteine-rich repeats. *J Mol Evol.* **46,** 476–487 (1998).

74. D. H. Kim, Y. Inagaki, T. Suzuki, R. X. Ioka, S. Z. Yoshioka, K. Magoori, M. J. Kang, Y. Cho, A. Z. Nakano, Q. Liu, T. Fujino, H. Suzuki, H. Sasano, and T. T. Yamamoto, A new low density lipoprotein receptor related protein, LRP5, is expressed in hepatocytes and adrenal cortex, and recognizes apolipoprotein E. *Eur J Biochem.* **124,** 1072–1076 (1998).

75. Y. Dong, W. Lathrop, D. Weaver, Q. Qiu, J. Cini, D. Bertolini, and D. Chen, Molecular cloning and characterization of LR3, a novel LDL receptor family protein with mitogenic activity. *Biochem Biophys Res Commun.* **251,** 784–790 (1998).

76. S. D. Brown, R. C. Twells, P. J. Hey, R. D. Cox, E. R. Levy, A. R. Soderman, M. L. Metzker, C. T. Caskey, J. A. Todd, and J. F. Hess, Isolation and characterization of LRP6, a novel member of the low density lipoprotein receptor gene family. *Biochem Biophys Res Commun.* **248,** 879–888 (1998).

77. M. Wehrli, S. T. Dougan, K. Caldwell, L. O'Keefe, S. Schwartz, D. Vaizel-Ohayon, E. Schejter, A. Tomlinson, and S. DiNardo, Arrow encodes an LDL-receptor-related protein essential for wingless signaling. *Nature.* **407,** 527–530 (2000).

78. K. Tamai, M. Semenov, Y. Kato, R. Spokony, C. Liu, Y. Katsuyama, F. Hess, J. P. Saint-Jeannet, and X. He, LDL-receptor-related proteins in Wnt signal transduction. *Nature.* **407,** 530–535 (2000).

79. K. I. Pinson, J. Brennan, S. Monkley, B. J. Avery, and W. C. Skarnes, An LDL-receptor-related protein mediates Wnt signalling in mice. *Nature.* **407,** 535–538 (2000).

80. S. L. Holmen, S. A. Robertson, C. R. Zylstra, and B. O. Williams, Wnt-independent activation of β-catenin mediated by a Dkk-Fz5 fusion protein. *Biochem Biophys Res Commun.* **328,** 533–539 (2005).

81. G. Liu, A. Bafico, and S. A. Aaronson, The mechanism of endogenous receptor activation functionally distinguishes prototype canonical and noncanonical Wnts. *Mol Cell Biol.* **25,** 3475–3482 (2005).

82. A. Rattner, J.-C. Hsieh, P. M. Smallwood, D. J. Gilbert, N. G. Copeland, N. A. Jenkins, and J. Nathans, A family of secreted proteins contains homology to the cysteine-rich ligand-binding domain of frizzled receptors. *Proc Natl Acad Sci USA.* **94,** 2859–2863 (1997).

83. P. W. Finch, X. He, M. J. Kelley, A. Uren, R. P. Schaudies, N. C. Popescu, S. Rudikoff, S. A. Aaronson, H. E. Varmus, and J. S. Rubin, Purification and molecular cloning of a secreted, frizzled-related antagonist of Wnt action. *Proc Natl Acad Sci USA.* **94,** 6770–6775 (1997).

84. L. Leyns, T. Bouwmeester, S.-H. Kim, S. Piccolo, and E. M. De Roberts, Frzb-1 is a secreted antagonist of Wnt signaling expressed in the Spemann organizer. *Cell.* **88,** 747–756 (1997).

85. S. Wang, M. Krnks, K. Lin, F. P. Luyten, and M. Moos, Frzb, a secreted protein expressed in the Spemann organizer, binds and inhibits Wnt-8. *Cell.* **88,** 757–766 (1997).

86. M. van den Heuvel and P. W. Ingham, Smoothened encodes a receptor-like serpentine protein required for hedgehog signalling. *Nature.* **382,** 547–551 (1996).

87. J. Alcedo, M. Ayzenzon, T. Von Ohlen, M. Noll, and J. E. Hooper, The *Drosophila smoothened* gene encodes a seven-pass membrane protein, a putative receptor for the hedgehog signal. *Cell.* **86,** 221–232 (1996).

88. D. K. Strickland, S. L. Gonias, and W. S. Argraves, Diverse roles for the LDL receptor family. *Trends Endocrinol Metab.* **13,** 66–74 (2002).

89. A. Nykjaer and T. E. Willnow, The low-density lipoprotein receptor gene family: A cellular Swiss army knife? *Trends Cell Biol.* **12,** 273–280 (2002).

90. J. L. Goldstein and M. S. Brown, Binding and degradation of low density lipoproteins by cultured human fibroblasts. *J Biol Chem.* **249,** 5133–5162 (1974).

91. D. J. Figueroa, J. F. Hess, B. Ky, S. D. Brown, V. Sandig, A. Hermanowski-Vosatka, R. C. Twells, J. A. Todd, and C. P. Austin, Expression of the type I diabetes-associated gene LRP5 in macrophages, vitamin A system cells, and the Islets of Langerhans suggests multiple potential roles in diabetes. *J Histochem Cytochem.* **48,** 1357–1368 (2000).

92. L. M. Boyden, J. Mao, J. Belsky, L. Mitzner, A. Farhi, M. A. Mitnick, D. Wu, K. Insogna, and R. P. Lifton, High bone density due to a mutation in LDL-receptor-related protein 5. *N Engl J Med.* **346,** 1513–1521 (2002).

93. E. Van Wesenbeeck, E. Cleiren, J. Gram, R. Beals, O. Benichou, D. Scopelliti, L. Key, T. Renton, C. Bartles, Y. Gong, M. Warman, M. Vernejoul, J. Bollerslev, and W. Van Hul, Six novel missense mutations in the LDL receptor-related protein5 (LRP5) gene in different conditions with an increased bone density. *Am J Hum Genet.* **72,** 763–771 (2003).

94. E. A. Streeten, H. Morton, and D. J. McBride, Osteoporosis pseudoglioma syndrome: 3 siblings with a novel LRP5 mutation. *J Bone Miner Res.* **18**(Suppl 2), S35 (2003).

95. M. R. Rickels, X. Zhang, S. Mumm, and M. P. Whyte, Skeletal disease accompanying high bone mass and novel LRP5 mutation. *ASBMR Meeting on Advances in Skeletal Anabolic Agents for the Treatment of Osteoporosis.* Abstract T6 (2004).

96. M. Whyte, W. Reinus, and S. Mumm, High-bone-mass disease and LRP5. *N Engl J Med.* **350,** 2096–2098 (2004).

97. M. R. Rickels, X. Zhang, S. Mumm, and M. Whyte, Oropharyngeal skeletal disease accompanying high bone mass and novel LRP5 mutation. *J Bone Miner Res.* **20,** 878–885 (2005).

98. E. A. Streeten, E. Puffenberger, H. Morton, and D. McBride, Osteoporosis pseudoglioma syndrome: 4 siblings with a compound heterozygote LRP5 mutation. *J Bone Miner Res.* **19**(Suppl 1), S182 (2004).

99. L. Y. Jin, H. H. L. Lau, D. K. Smith, K. S. Lau, P. T. Cheung, E. Y. W. Kwan, L. Low, V. Chan, and A. W. C. Kung, A family with osteoporosis-pseudoglioma syndrome (OPG) due to compound heterozygous mutation of the LRP5 gene. *J Bone Miner Res.* **19**(Suppl 1), S129 (2004).

100. C. Toomes, H. Bottomley, R. Jackson, K. Towns, S. Scott, D. Mackey, J. E. Craig, L. Jiang, Z. Yang, R. Trembath, G. Woodruff, C. Y. Gregory-Evans, K. Gregory-Evans, M. J. Parker, G. C. M. Black, L. M. Downey, K. Zhang, and C. Inglehearn, Mutations in LRP5 or FZD4 underlie the common familial exudative vitreoretinopathy locus on chromosome 11q. *Am J Hum Genet.* **74,** 721–730 (2004).

101. X. Jiao, V. Ventruto, M. T. Trese, B. S. Shastry, and J. F. Hejtmancik, Autosomal recessive familial exudative vitreoretinopathy is associated with mutations in LRP5. *Am J Hum Genet.* **75,** 878–884 (2004).

102. U. Choudhury, M. C. Vernejoul, S. Deutsch, T. Chevalley, J. P. Bonjour, B. E. Antonarakis, R. Rizzoli, and S. L. Ferrari, Genetic variation in LDL receptor related protein 5 (LRP5) is a major risk factor for male osteoporosis: Results from a cross-sectional, longitudinal and case control study. *J Bone Miner Res.* **18,** S69(abstract F135) (2003).

103. S. Ferrari, S. Deutsch, U. Choudhury, T. Chevalley, J. Bonjour, E. Dermitzakis, R. Rizzoli, and S. Antonarakis, Polymorphisms in the low-density lipoprotein receptor-related protein 5(LRP5) gene are associated with variation in vertebral bone mass vertebral bone size, and stature in Whites. *Am J Hum Genet.* **74,** 866–875 (2004).

104. J. M. Koh, M. H. Jung, J. S. Hong, H. J. Park, J. S. Chang, H. D. Shin, S. Y. Kim, and G. S. Kim, Association between bone mineral density and LDL receptor-related protein 5 gene polymorphisms in young Korean men. *J Korean Med Sci.* **19,** 407–412 (2004).

105. T. Mizuguchi, I. Furuta, Y. Watanbe, K. Tsukamoto, H. Tomita, M. Tsujihata, T. Ohta, T. Kishino, N. Matsumoto, H. Minakami, N. Niikawa, and K. Yoshiura, LRP5, low-density-lipoprotein-receptor-related protein 5, is a determinant for bone mineral density. *J Hum Genet.* **49,** 80–86 (2004).

106. M. Okubo, A. Horinishi, D. H. Kim, T. T. Yamamoto, and T. Murase, Seven novel sequence variants in the human low density lipoprotein receptor related protein 5 (LRP5) gene. *Hum Mutat.* **19,** 186–188 (2002).

107. M. A. Koay, P. Y. Woon, Y. Zhang, L. J. Miles, E. L. Duncan, S. H. Ralston, J. E. Compston, C. Cooper, R. Keen, B. L. Langdahl, A. MacLelland, J. O'Riordan, H. A. Pols, D. M. Reid, A. G. Uitterlinden, J. A. H. Wass, and M. A. Brown, Influence of LRP5 polymorphisms on normal variation in BMD. *J Bone Miner Res.* **19,** 1619–1627 (2004).

108. H. Hartikka, O. Makitie, M. Mannikko, A. S. Doria, A. Daneman, W. G. Cole, L. Ala-Kokko, and E. B. Sochett, Heterozygous mutations in the LDL receptor-related protein 5 (LRP5) gene are associated with primary osteoporosis in children. *J Bone Miner Res.* **20,** 783–789 (2005).

109. J. Bollerslev, S. G. Wilson, I. M. Dick, F. M. Islam, T. Ueland, L. Palmer, A. Devine, and R. L. Prince, LRP5 gene polymorphisms predict bone mass and incident fractures in elderly Australian women. *Bone.* **36,** 599–606 (2005).

110. D. L. Koller, S. Ichikawa, M. L. Johnson, D. Lai, X. Xuei, H. J. Edenberg, P. M. Conneally, S. L. Hui, C. C. Johnston, M. Peacock, T. Foroud, and M. J. Econs, Contribution of the

LRP5 gene to normal variation in peak bone BMD in women. *J Bone Miner Res.* **20,** 75–80 (2005).

111. N. M. Rajamannan, M. Subramaniam, F. Caira, S. R. Stock, and T. C. Spelsberg, Atorvastatin inhibits hypercholesterolemia-induced calcification in the aortic valves via the LRP5 receptor pathway. *Circulation.* **112**(9 Suppl), I229–I234 (2005).

112. B. H. Hoang, T. Kubo, J. H. Healey, R. Sowers, B. A. Mazza, R. Yang, A. G. Huvos, P. A. Meyers, and R. Gorlick, Expression of LDL receptor-related protein 5 (LRP5) as a novel marker for disease progression in high-grade osteosarcoma. *Int J Cancer.* **109,** 106–111 (2004).

113. P. J. Hey, R. C. Twells, M. S. Phillips, N. Yusuke, S. D. Brown, Y. Kawaguchi, R. Cox, X. Guochun, V. Dugan, H. Hammond, M. L. Metzker, J. A. Todd, and J. F. Hess, Cloning of a novel member of the low-density lipoprotein receptor family. *Gene.* **216,** 103–111 (1998).

114. T. A. Springer, An extracellular β-propeller module predicted in lipoprotein and scavenger receptors, tyrosine kinases, epidermal growth factor precursor, and extracellular matrix components. *J Mol Biol.* **283,** 837–862 (1998).

115. H. Jeon, W. Meng, J. Takagi, M. J. Eck, T. A. Springer, and S. C. Blacklow, Implications for familial hypercholesterolemia from the structure of the LDL receptor YWTD-EGF domain pair. *Nat Struct Biol.* **8,** 499–504 (2001).

116. N. Itasaki, C. M. Jones, S. Mercurio, A. Rowe, P. M. Domingos, J. C. Smith, and R. Krumlauf, Wise, a context-dependent activator and inhibitor of Wnt signaling. *Development.* **130,** 4295–4305 (2003).

117. B. Mao, W. Wu, Y. Li, D. Hoppe, P. Stannek, A. Glinka, and C. Niehrs, LDL-receptor-related protein 6 is a receptor for Dickkopf proteins. *Nature.* **411,** 321–325 (2001).

118. M. V. Semenov, K. Tamai, B. K. Brott, M. Kühl, S. Sokol, and H. Xi, Head inducer Dickkopf-1 is a ligand for Wnt coreceptor LRP6. *Curr Biol.* **11,** 951–961 (2001).

119. N. S. Tolwinski, M. Wehrli, A. Rives, N. Erdeniz, S. DiNardo, and E. Wieschaus, Wg/Wnt signal can be transmitted through arrow/LRP5,6 and axin independently of Zw3/Gsk3beta activity. *Dev Cell.* **4,** 407–418 (2003).

120. A. W. Glinka, W. Wu, H. Delius, A. P. Monaghan, C. Blumenstock, and C. Niehrs, Dickkopf-1 is a member of a new family of secreted proteins and functions in head induction. *Nature.* **391,** 357–362 (1998).

121. A. P. Monaghan, P. Kioschis, W. Wu, A. Zuniga, D. Bock, A. Poustka, H. Delius, and C. Niehrs, Dickkopf genes are co-ordinately expressed in mesodermal lineages. *Mech Dev.* **87,** 45–56 (1999).

122. V. E. Krupnik, J. D. Sharp, C. Jiang, K. Robison, T. W. Chickering, L. Amaravadi, D. E. Brown, D. Guyot, G. Mays, K. Leiby, B. Chang, T. Duong, A. D. Goodearl, D. P. Gearing, S. Y. Sokol, and S. A. McCarthy, Functional and structural diversity of the human Dickkopf gene family. *Gene.* **238,** 301–313 (1999).

123. B. M. Bhat, K. M. Allen, J. Graham, W. Liu, A. Morales, A. Anisowicz, H. Lam, C. McCauley, V. Coleburn, M. Cain, W. Fortier, H. Wang, F. J. Bex, and P. Yaworsky, Functional modulation of LRP5–Wnt-Dkk1 activity by various mutations in LRP5 beta-propeller 1. *J Bone Miner Res.* **18,** S46 (2003).

124. P. Bafico, G. Liu, A. Yaniv, A. Gazit, and A. A. Aaronson, Novel mechanism of Wnt signaling inhibition mediated by Dickkopf-1 interaction with LRP6/arrow. *Nat Cell Biol.* **3,** 683–686 (2001).

125. Y. Zhang, Y. Wang, X. Li, J. Zhang, J. Mao, Z. Li, J. Zheng, L. Li, S. Harris, and D. Wu, The LRP5 high-bone-mass G171V

126. B. K. Brott and S. Y. Sokol, Regulation of Wnt/LRP signaling by distinct domains of Dickkopf proteins. *Mol Cell Biol.* **22,** 6100–6110 (2002).

127. L. Li, J. Mao, L. Sun, W. Liu, and D. Wu, Second cysteine-rich domain of Dickkopf-2 activates canonical Wnt signaling pathway via LRP-6 independently of Dishevelled. *J Biol Chem.* **277,** 5977–5981 (2002).

128. B. Mao, W. Wu, G. Davidson, J. Marhold, M. Li, B. M. Mechler, J. Delius, D. Hoppe, P. Stannek, C. Walter, A. Glinka, and C. Niehrs, Kremen proteins are Dickkopf receptors that regulate Wnt/β-catenin signalling. *Nature.* **417,** 664–667 (2002).

129. W. J. Chen, J. L. Goldstein, and M. S. Brown, NPXY, a sequence often found in cytoplasmic tails, is required for coated pit-mediated internalization of the low density lipoprotein receptor. *J Biol Chem.* **265,** 3116–3123 (1990).

130. B. Mao and C. Niehrs, Kremen2 modulates Dickkopf2 activity during Wnt/LRP6 signaling. *Gene.* **302,** 179–183 (2003).

131. J.-C. Hsieh, L. Lee, L. Zhang, S. Wefer, K. Brown, T. Rosenquist, and B. C. Holdener, Mesd encodes an LRP5/6 chaperone essential for specification of mouse embryonic polarity. *Cell.* **112,** 355–367 (2003).

132. J. Culi and R. S. Mann, Boca, an endoplasmic reticulum protein required for wingless signaling and trafficking of LDL receptor family members in *Drosophila. Cell.* **112,** 343–354 (2003).

133. M. Ai, S. Holmen, W. Van Hul, B. O. Williams, and M. L. Warman, Reduced affinity to and inhibition by Dkk1 form a common mechanism by which high bone mass-associated missense mutations in LRP5 affect canonical Wnt signaling. *Mol Cell Biol.* **25,** 4946–4955 (2005).

134. Y. Li, J. Chen, L. M. McCormick, J. Wang, and G. Bu, Mesd binds to mature LDL-receptor-related protein-6 and antagonizes ligand binding. *J Cell Science.* **118,** 5305–5314 (2005).

135. M. E. Brunkow, J. C. Gardner, J. Van-Ness, B. W. Paeper, B. R. Kovacevich, S. Proll, J. E. Skonier, L. Zhao, P. J. Sabo, Y. Fu, R. S. Alisch, L. Gillett, T. Colbert, P. Tacconi, D. Galas, H. Hamersma, P. Beighton, and J. Mulligan, Bone dysplasia sclerosteosis results from loss of the SOST gene product, a novel cystine knot-containing protein. *Am J Hum Genet.* **68,** 577–589 (2001).

136. X. Li, Y. Zhang, H. Kang, W. Liu, P. Liu, J. Zhang, S. E. Harris, and D. Wu, Sclerostin binds to LRP5/6 and antagonizes canonical Wnt signaling. *J Biol Chem.* **280,** 19883–19887 (2005).

137. M. Semenov, K. Tamai, and X. He, SOST is a ligand for LRP5/LRP6 and a Wnt signaling inhibitor. *J Biol Chem.* **280,** 26770–26775 (2005).

138. M. Ai, S. Heeger, C. F. Bartels, D. K. Schelling, and Osteoporosis-Pseudoglioma Collaborative Group, Clinical and molecular findings in osteoporosis-pseudoglioma syndrome. *Am J Hum Genet.* **77,** 741–753 (2005).

139. J. Noordermeer, J. Klingensmith, N. Perrimon, and R. Nusse, Dishevelled and armadillo act in the wingless signalling pathway in *Drosophila. Nature.* **367,** 80–83 (1994).

140. E. Siegfried, E. L. Wilder, and N. Perrimon, Component of wingless signalling in *Drosophila. Nature.* **367,** 76–80 (1994).

141. J. Klingensmith, R. Nusse, and N. Perrimon, The *Drosophila* segment polarity gene Dishevelled encodes a novel protein required for response to the wingless signal. *Genes Develop.* **8,** 118–130 (1994).

142. F. J. Diaz-Benjumea and S. M. Cohen, Wingless acts through the Shaggy/Zeste-white 3 kinase to direct dorsal-ventral axis

formation in the *Drosophila* leg. *Development.* **120,** 1661–1670 (1994).

143. X. He, M. Semenov, K. Tamai, and X. Zeng, LDL receptor-related proteins 5 and 6 in Wnt/β-catenin signaling: Arrows points the way. *Development.* **131,** 1663–1677 (2004).

144. J. M. Peters, R. M. McKay, J. P. McKay, and J. M. Graff, Casein kinase I transduces Wnt signals. *Nature.* **401,** 345–350 (1999).

145. K. Willert, M. Brink, A. Wodarz, H. Varmus, and R. Nusse, Casein kinase 2 associates with and phosphorylates Dishevelled. *EMBO J.* **16,** 3089–3096 (1997).

146. T. Q. Sun, B. Lu, J. J. Feng, C. Reinhard, Y. N. Jan, W. J. Fantl, and L. T. Williams, PAR-1 is a Dishevelled-associated kinase and a positive regulator of Wnt signaling. *Nat Cell Biol.* **3,** 628–636 (2001).

147. J. M. Gonzalez-Sancho, K. R. Brennan, L. A. Castelo-Soccio, and A. M. Brown, Wnt proteins induce Dishevelled phosphorylation via an LRP5/6 independent mechanism, irrespective of their ability to stabilize β-catenin. *Mol Cell Biol.* **24,** 4757–4768 (2004).

148. L. Zeng, F. Fagotto, T. Zhang, W. Hsu, T. J. Vasicek, W. L. Perry, B. M. Gumbiner, and F. Constantini, The mouse fused locus encodes axin, an inhibitor of the Wnt signaling pathway that regulates embryonic axis formation. *Cell.* **90,** 181–192 (1997).

149. M. Mai, C. Qian, A. Yokomizo, D. I. Smith, and W. Liu, Cloning of the human homolog of conductin (AXIN2), a gene mapping to chromosome 17q23–q24. *Genomics.* **55,** 341–344 (1999).

150. T. Nakamura, F. Hamada, T. Ihidate, K. Anai, K. Kawahara, K. Toyoshima, and T. Akiyama, Axin, an inhibitor of the Wnt signalling pathway, interacts with β-catenin, GSK-3β and APC and reduces the β-catenin level. *Genes Cells.* **3,** 395–403 (1998).

151. K. Itoh, V. E. Krupnick, and S. Y. Sokol, Axis determination in *Xenopus* involves biochemical interactions of axin, glycogen synthase kinase 3 and β-catenin. *Curr Biol.* **8,** 591–598 (1998).

152. S. Ikeda, S. Kishida, H. Yamamoto, H. Murai, S. Koyama, and A. Kikuchi, Axin, a negative regulator of the Wnt signaling pathway, forms a complex with GSK-3β and β-catenin and promotes GSK-3β-dependent phosphorylation of β-catenin. *EMBO J.* **17,** 1371–1384 (1998).

153. M. J. Hart, R. de los Santos, I. N. Albert, B. Rubinfeld, and P. Polakis, Downregulation of β-catenin by human axin and its association with the APC tumor suppressor, β-catenin. *Curr Biol.* **8,** 573–581 (1998).

154. J. Behrens, B. A. Jerchow, M. Wurtele, J. Grimm, C. Asbrand, R. Wirtz, M. Kuhl, D. Wedlich, and W. Birchmeier, Functional interaction of an axin homolog, conductin, with β-catenin, APC and GSK3β. *Science.* **280,** 596–599 (1998).

155. H. Yamamoto, S. Kishida, T. Uochi, S. Ikeda, S. Koyama, M. Asashima, and A. Kikuchi, Axil, a member of the axin family, interacts with both glycogen synthase kinase 3β and β-catenin and inhibits axis formation of *Xenopus* embryos. *Mol Cell Biol.* **18,** 2867–2875 (1998).

156. C. Sakanaka, J. B. Weiss, and L. T. Williams, Bridging of β-catenin and glycogen synthase kinase-3b by axin and inhibition of β-catenin-mediated transcription. *Proc Natl Acad Sci USA.* **95,** 3020–3030 (1998).

157. F. Fagotto, E. Jho, L. Zeng, T. Kurth, T. Joos, C. Kaufmann, and F. Constantini, Domains of axin involved in protein-protein interactions, Wnt pathway inhibition, and intracellular localization. *J Cell Biol.* **145,** 741–756 (1999).

158. G. H. Farr, D. M. Ferkey, C. Yost, S. B. Pierce, C. Weaver, and D. Kimelman, Interaction among GSK-3, GBP, axin, and APC in *Xenopus* axis specification. *J Cell Biol.* **148,** 691–701 (2000).

159. J. Mao, J. Wang, B. Liu, W. Pan, G. Farr, C. Flynn, H. Yuan, S. Takada, D. Kimelman, L. Li, and D. Wu, Low-density lipoprotein receptor-related protein-5 binds to axin and regulates the canonical Wnt signaling pathway. *Mol Cell.* **7,** 801–809 (2001).

160. G. Liu, A. Bafico, V. K. Harris, and S. A. Aaronson, A novel mechanism for Wnt activation of canonical signaling through the LRP6 receptor. *Mol Cell Biol.* **23,** 5825–5835 (2003).

161. K. Tamai, X. Zeng, C. Liu, X. Zhang, Y. Harada, Z. Chang, and X. He, A mechanism for Wnt coreceptor activation. *Mol Cell.* **13,** 149–156 (2004).

162. K. Brennan, J. M. Gonzalez-Sancho, L. A. Castelo-Soccio, L. R. Howe, and A. M. Brown, Truncated mutants of the putative Wnt receptor LRP6/arrow can stabilize β-catenin independently of frizzled proteins. *Oncogene.* **23,** 4873–4884 (2004).

163. G. Davidson, W. Wu, J. Shen, J. Bilic, U. Fenger, P. Stannek, A. Glinka, and C. Niehrs, Casein kinase 1gamma couples Wnt receptor activation to cytoplasmic signal transduction. *Nature.* **438,** 867–872 (2005).

164. X. Zeng, K. Tamai, B. Doble, S. Li, H. Huang, R. Habas, H. Okamura, J. Woodgett, and X. He, A dual-kinase mechanism for Wnt co-receptor phosphorylation and activation. *Nature.* **438,** 873–877 (2005).

165. E. Hay, C. Faucheu, I. Suc-Royer, R. Touitou, V. Stiot, B. Vayssiere, R. Baron, S. R. Roman, and G. Rawadi, Interaction between LRP5 and Frat1 mediates the activation of the Wnt canonical pathway. *J Biol Chem.* **280,** 13616–13623 (2005).

166. L. Li, H. Yuan, C. D. Weaver, J. Mao, G. H. Farr, D. J. Sussman, J. Jonkers, D. Kimelman, and D. Wu, Axin and Frat1 interact with Dvl and GSK, bridging Dvl to GSK in Wnt-mediated regulation of LEF-1. *EMBO J.* **18,** 4233–4240 (1999).

167. P. T. Rowley, Inherited susceptibility to colorectal cancer. *Annu Rev Med.* **56,** 539–554 (2005).

168. K. W. Kinzler, M. C. Nilbert, L.-K. Su, B. Vogelstein, T. M. Bryan, D. B. Levy, K. J. Smith, A. C. Preisinger, P. Hedge, D. McKechnie, R. Finniear, A. Markham, J. Groffen, M. S. Boguski, S. F. Altschul, A. Horii, H. Ando, Y. Miyoshi, Y. Miki, I. Nishisho, and Y. Nakamura, Identification of FAP locus genes from chromosome 5q21. *Science.* **253,** 661–665 (1991).

169. I. Nishisho, Y. Nakamura, Y. Miyoshi, Y. Miki, H. Ando, A. Horii, K. Koyama, J. Utsunomiya, S. Baba, P. Hedge, A. Markham, A. J. Krush, G. Petersen, S. R. Hamilton, M. C. Nilbert, D. B. Levy, T. M. Bryan, A. C. Preisinger, K. J. Smith, L.-K. Su, K. W. Kinzler, and B. Vogelstein, Mutations of chromosome 5q21 genes in FAP and colorectal cancer patients. *Science.* **253,** 665–669 (1991).

170. J. Groden, A. Thliveris, W. Samowitz, M. Carlson, L. Gelbert, H. Albertsen, G. Joslyn, J. Stevens, L. Spirio, M. Robertson, L. Sargean, K. Krapcho, E. Wolff, R. Burt, J. P. Hughes, J. Warrington, J. McPherson, J. Wasmuth, D. Le Paslier, H. Abderrahim, D. Cohen, M. Leppert, and R. White, Identification and characterization of the familial adenomatous polyposis coli gene. *Cell.* **66,** 589–600 (1991).

171. L. K. Su, B. Vogelstein, and K. W. Kinzler, Association of the APC tumor suppressor protein with catenins. *Science.* **262,** 1734–1737 (1993).

172. B. Rubinfeld, B. Souza, I. Albert, O. Muller, S. H. Chamberlain, F. R. Masiarz, S. Munemitsu, and P. Polakis, Association of the

APC gene product with β-catenin. *Science*. **262,** 1731–1734 (1993).

173. K. Vleminckx, E. Wong, K. Guger, B. Rubinfeld, P. Polakis, and B. M. Gumbiner, Adenomatous polyposis coli tumor suppressor protein has signaling activity in *Xenopus laevis* embryos resulting in the induction of an ectopic dorsoanterior axis. *J Cell Biol.* **136,** 411–420 (1997).

174. B. Rubinfeld, I. Albert, E. Porfiri, C. Fiol, S. Munemitsu, and P. Polakis, Binding of GSK3β to the APC-β-catenin complex and regulation of complex assembly. *Science.* **272,** 1023–1026 (1996).

175. J. Papkoff, B. Rubinfeld, B. Schryver, and P. Polakis, Wnt-1 regulates free pools of catenins and stabilizes APC-catenin complexes. *Mol Cell Biol.* **16,** 2128–2134 (1996).

176. S. E. Hardt and J. Sadoshima, Glycogen synthase kinase-3b: A novel regulator of cardiac hypertrophy and development. *Circulation Res.* **90,** 1055–1063 (2002).

177. C. Liu, Y. Li, M. Semenov, C. Han, G.-H. Baeg, Y. Tan, Z. Zhang, X. Lin, and X. He, Control of β-catenin phosphorylation/degradation by a dual-kinase mechanism. *Cell.* **108,** 837–847 (2002).

178. E. Jho, S. Lomvardas, and F. Costantini, A GSK3beta phosphorylation site in axin modulates interaction with beta-catenin and Tcf-mediated gene expression. *Biochem Biophys Res Commun.* **266,** 28–35 (1999).

179. V. Stambolic and J. R. Woodgett, Mitogen inactivation of glycogen synthase kinase-3β in intact cells via serine 9 phosphorylation. *Biochem J.* **303,** 701–704 (1994).

180. S. Haq, G. Choukroun, Z. B. Kang, H. Ranu, T. Matsui, A. Rosenzweig, J. D. Molkentin, A. Alessandrini, J. Woodgett, R. Hajjar, A. Michael, and T. Force, Glycogen synthase kinase-3β is a negative regulator of cardiomyocyte hypertrophy. *J Cell Biol.* **151,** 117–130 (2000).

181. M. Delcommenne, C. Tan, V. Gray, L. Rue, J. Woodgett, and S. Dedhar, Phosphoinositide-3–OH kinase-dependent regulation of glycogen synthase kinase 3 and protein kinase B/AKT by the integrin-linked kinase. *Proc Natl Acad Sci USA.* **95,** 11211–11216 (1998).

182. D. Cook, M. J. Fry, K. Hughes, R. Sumathipala, J. R. Woodgett, and T. C. Dale, Wingless inactivates glycogen synthase kinase-3 via an intracellular signalling pathway which involves a protein kinase C. *EMBO J.* **15,** 4526–4536 (1996).

183. X. Fang, S. X. Yu, Y. Lu, R. C. Bast, J. R. Woodgett, and G. B. Mills, Phosphorylation and inactivation of glycogen synthase kinase 3 by protein kinase A. *Proc Natl Acad Sci USA.* **97,** 11960–11965 (2000).

184. H. Yamamoto, T. Hinoi, T. Michiue, A. Fukui, H. Usui, V. Janssens, C. Van Hoof, J. Goris, M. Asashima, and A. Kikuchi, Inhibition of the Wnt signaling pathway by the PR61 subunit of protein phosphatase 2A. *J Biol Chem.* **276,** 26875–26882 (2001).

185. J. M. Seeling, J. R. Miller, R. Gil, R. T. Moon, R. White, and D. M. Virshup, Regulation of β-catenin signaling by the B56 subunit of protein phosphatase 2A. *Science.* **283,** 2089–2091 (1999).

186. W. Hsu, L. Zeng, and F. Costantini, Identification of a domain of axin that binds to the serine/threonine protein phosphatase 2a and a self-binding domain. *J Biol Chem.* **274,** 3439–3445 (1999).

187. Z.-H. Gao, J. M. Seeling, V. Hill, A. Yochum, and D. M. Virshup, Casein kinase I phosphorylates and destabilizes the β-catenin degradation complex. *Proc Natl Acad Sci USA.* **99,** 1182–1187 (2002).

188. P. D. McCrea, C. W. Turck, and B. Gumbiner, A homolog of the armadillo protein in *Drosophila* (plakoglobin) associated with E-cadherin. *Science.* **254,** 1359–1361 (1991).

189. C. Yost, M. Torres, J. R. Miller, E. Huang, D. Kimelman, and R. T. Moon, The axis-inducing activity, stability, and subcellular distribution of β-catenin is regulated in *Xenopus* embryos by glycogen synthase kinase 3. *Genes Develop.* **10,** 1443–1454 (1996).

190. H. Aberle, A. Bauer, J. Stappert, A. Kispert, and R. Kemler, β-catenin is a target for the ubiquitin-proteasome pathway. *EMBO J.* **16,** 3797–3804 (1997).

191. R. A. Cavallo, R. T. Cox, M. M. Moline, J. Roose, G. A. Polevoy, H. Clevers, M. Peifer, and A. Bejsovec, *Drosophila* Tcf and Groucho interact to repress wingless signalling activity. *Nature.* **395,** 604–608 (1998).

192. G. Chen, J. Fernandez, S. Mische, and A. J. Courey, A functional interaction between the histone deacetylase Rpd3 and the corepressor Groucho in *Drosophila* development. *Genes Develop.* **13,** 2218–2230 (1999).

193. A. Hecht, K. Vleminckx, M. P. Stemmler, F. van Roy, and R. Kemler, The p300/CBP acetyltransferases function as transcriptional coactivators of β-catenin in vertebrates. *EMBO J.* **19,** 1839–1850 (2000).

194. N. Barker, A. Hurlstone, H. Musisi, A. Miles, M. Bienz, and H. Clevers, The chromatin remodelling factor Brg-1 interacts with β-catenin to promote target gene activation. *EMBO J.* **20,** 4935–4943 (2001).

195. P. N. Adler, Planar signaling and morphogenesis in *Drosophila. Develop Cell.* **2,** 525–535 (2002).

196. M. Mlodzik, Planar cell polarization: Do the same mechanisms regulate *Drosophila* tissue polarity and vertebrate gastrulation? *Trends Genet.* **18,** 564–571 (2002).

197. L. C. Sheldahl, M. Park, C. C. Malbon, and R. T. Moon, Protein kinase C is differentially stimulated by Wnt and frizzled homologs in a G-protein-dependent manner. *Current Biology.* **9,** 695–698 (1999).

198. V. L. Katanaev, R. Ponzielli, M. Semeriva, and A. Tomlinson, Trimeric G protein-dependent frizzled signaling in *Drosophila. Cell.* **120,** 111–122 (2005).

199. D. G. S. Capelluto, T. G. Kutateladze, R. Habas, C. V. Finklestein, X. He, and M. Overduin, The DIX domain targets Dishevelled to actin stress fibres and vesicular membranes. *Nature.* **419,** 726–729 (2002).

200. L. C. Shedahl, D. C. Slusarski, P. Pandur, J. R. Miller, M. Kuhl, and R. T. Moon, Dishevelled activates Ca+2 flux, PKC, and CamKII in vertebrate embryos. *J Cell Biol.* **161,** 769–777 (2006).

201. M. Kuhl, The Wnt/calcium pathway: Biochemical mediators, tools and future experiments. *Front Biosci.* **9,** 967–974 (2004).

202. A. E. Chen, D. B. Ginty, and C.-M. Fan, Protein kinase A signalling via CREB controls myogenesis induced by Wnt proteins. *Nature.* **433,** 317–322 (2005).

203. O. Pourquie, A new canon. *Nature.* **433,** 208–209 (2005).

204. J. B. van Meurs, F. Rivadeneira, M. Jhamai, W. Hugens, A. Hofman, J. P. van Leeuwen, H. A. Pols, and A. G. Uitterlinden, Common genetic variation of the low-density lipoprotein receptor-related protein 5 and 6 genes determine fracture risk in elderly White men. *J Bone Miner Res.* **21,** 141–150 (2006).

205. R. K. Beals, S. W. McLoughlin, R. L. Teed, and C. McDonald, Dominant endosteal hyperostosis. Skeletal characteristics and review of the literature. *J Bone Joint Surg Am.* **83A,** 1643–1649 (2001).

206. R. K. Beals, Endosteal hyperostosis. *J Bone Joint Surg Am.* **58**, 1172–1173 (1976).

207. D. Scopelliti, R. Orsini, D. Ventucci, and D. Carratelli, Malattia di van buchem. *Minerva Stomatol.* **48**, 227–234 (1999).

208. T. Renton, E. Odell, and N. A. Drage, Differential diagnosis and treatment of autosomal dominant osteosclerosis of the mandible. *Br J Oral Max Surg.* **40**, 55–59 (2002).

209. J. Bollerslev and P. E. Anderson, Radiological, biochemical and heredity evidence of two types of autosomal dominant osteopetrosis. *Bone.* **9**, 7–13 (1988).

210. W. van Hul, J. Bollerslev, J. Gram, E. Van Hul, W. Wuyts, O. Benichou, F. Vanhoenacker, and P. J. Willems, Localization of a gene for autosomal dominant osteopetrosis (Albers-Schonberg disease) to chromosome 1p21. *Am J Hum Genet.* **61**, 363–369 (1997).

211. M. L. Johnson, G. Gong, W. J. Kimberling, S. M. Recker, D. K. Kimmel, and R. R. Recker, Linkage of a gene causing high bone mass to human chromosome 11 (11q12–13). *Am J Hum Genet.* **60**, 1326–1332 (1997).

212. M. L. Kwee, W. Balemans, E. Cleiren, J. J. P. Gille, F. van Der Blij, J. M. Sepers, and W. Van Hul, Clinical vignette: An autosomal dominant high bone mass phenotype in association with craniosynostosis in an extended family is caused by an *LRP5* missense mutation. *J Bone Miner Res.* **20**, 1254–1260 (2005).

213. P. Crabbe, W. Balemans, A. Willaert, I. Van Pottelbergh, E. Cleiren, P. J. Coucke, M. Ai, S. Goenaere, W. van Hul, A. de Paepe, and J. M. Kaufman, Missense mutations in *LRP5* are not a common cause of idiopathic osteoporosis in adult men. *J Bone Miner Res.* **20**, 1951–1959 (2005).

214. S. L. Ferrari, S. Deutsch, C. Baudoin, M. Cohen-Solal, A. Ostertag, S. E. Antonarakis, R. Rizzoli, and M. C. de Vernejoul, LRP5 gene polymorphisms and idiopathic osteoporosis in men. *Bone.* **37**, 770–775 (2006).

215. H.-W. Deng, F.-H. Xu, Q.-Y. Huang, H. Shen, H. Deng, T. Conway, Y. J. Liu, Y. Z. Liu, J. L. Li, H. T. Zhang, K. M. Davies, and R. R. Recker, A whole-genome linkage scan suggests several genomic regions potentially containing quantitative trait loci for osteoporosis. *J Clin Endocrinol Metab.* **87**, 5151–5159 (2002).

216. Y. Li, C. Fuhrmann, E. Schwinger, A. Gal, and H. Laqua, The gene for autosomal dominant exudative vitreoretinopathy (Criswick-Schepans) on the long arm of chromosome 11. *Am J Ophthalmol.* **113**, 712–713 (1992).

217. Y. Li, B. Muller, C. Fuhrmann, C. E. van Nouhuys, H. Laqua, P. Humphries, E. Schwinger, and A. Gal, The autosomal dominant familial exudative vitreoretinopathy locus maps on 11q and is closely linked to D11S533. *Am J Hum Genet.* **51**, 749–754 (1992).

218. J. Robitaille, M. L. E. MacDonald, A. Kaykas, L. C. Sheldahi, J. Zeisler, M. P. Dube, L. H. Zhang, R. R. Singaraja, D. L. Guernsey, B. Zheng, L. F. Siebert, A. Hoskin-Mott, M. T. Trese, S. N. Pimstone, B. S. Shastry, R. T. Moon, M. R. Hayden, Y. P. Goldberg, and M. E. Samuels, Mutant frizzled-4 disrupts retinal angiogenesis in familial exudative vitreo-retinopathy. *Nat Genet.* **32**, 326–330 (2002).

219. M. Kato, M. S. Patel, R. Levasseur, I. Lobov, B. H.-J. Chang, D. A. Glass, C. Hartmann, L. Li, T. H. Hwang, C. F. Brayton, R. A. Lang, G. Karsenty, and L. Chan, Cbfa 1-independent decrease in osteoblast proliferation, osteopenia, and persistent embryonic eye vascularization in mice deficient in Lrp5, a Wnt coreceptor. *J Cell Biol.* **157**, 303–314 (2002).

220. G. Karsenty, The genetic transformation of bone biology. *Genes Develop.* **13**, 3037–3051 (1999).

221. T. Gaur, C. J. Lengner, H. Hovhannisyan, R. A. Bhat, P. V. N. Bodine, B. S. Komm, A. Javed, A. J. van Wijnen, J. L. Stein, G. S. Stein, and J. B. Lian, Canonical Wnt signaling promotes osteogenesis by directly stimulating *Runx2* gene expression. *J Biol Chem.* **280**, 33132–33140 (2005).

222. T. P. Hill, D. Spater, M. M. Taketo, W. Birchmeier, and C. Hartmann, Canonical Wnt/β-catenin signaling prevents osteoblasts from differentiating into chondrocytes. *Develop Cell.* **8**, 727–738 (2005).

223. T. F. Day, X. Guo, L. Garrett-Beal, and Y. Yang, Wnt/β-catenin signaling in mesenchymal progenitors controls osteoblast and chondrocyte differentiation during vertebrate skeletogenesis. *Develop Cell.* **8**, 739–750 (2005).

224. X. Li, P. Liu, W. Liu, P. Maye, J. Zhang, Y. Zhang, M. Hurley, C. Guo, A. Boskey, L. Sun, S. E. Harris, D. W. Rowe, H. Z. Ke, and D. Wu, Dkk2 has a role in terminal osteoblast differentiation and mineralized matrix formation. *Nature Genetics.* **37**, 945–952 (2005).

225. R. L. van Bezooijen, P. ten Dijke, S. E. Papapoulos, and C. W. Lowik, SOST/sclerostin, an osteocyte-derived negative modulator of bone formation. *Cytokine Growth Factor Res.* **16**, 319–327 (2005).

226. K. E. S. Poole, R. L. van Bezooijen, N. Loveridge, H. Hamersma, S. E. Papapoulos, C. W. Lowik, and J. Reeve, Sclerostin is a delayed secreted product of osteocytes that inhibits bone formation. *FASEB J.* **19**, 1842–1844 (2005).

227. P. V. N. Bodine, J. Billiard, R. A. Moran, H. Ponce-de-Leon, S. McLarney, A. Mangine, M. J. Scrimo, R. A. Bhat, B. Stauffer, J. Green, G. S. Stein, J. B. Lian, and B. S. Komm, The Wnt antagonist secreted frizzled-related protein-1 controls osteoblast and osteocyte apoptosis. *J Cell Biochem.* **96**, 1212–1230 (2005).

228. P. V. N. Bodine, W. Zhao, Y. P. Kharode, F. J. Bex, A.-J. Lambert, M. B. Goad, T. Gaur, G. S. Stein, J. B. Lian, and B. S. Komm, The Wnt antagonist secreted frizzled-related protein-1 is a negative regulator of trabecular bone formation in adult mice. *Mole Endocrin.* **18**, 1222–1237 (2004).

229. G. Rawadi, B. Vayssiere, F. Dunn, R. Baron, and S. Roman-Roman, BMP-2 controls alkaline phosphatase expression and osteoblast mineralization by a Wnt autocrine loop. *J Bone Miner Res.* **18**, 1842–1953 (2003).

230. D. G. Winkler, M. S. K. Sutherland, E. Ojala, E. Turcott, J. C. Geoghegan, D. Shpektor, J. E. Skonier, C. Yu, and J. A. Latham, Sclerostin inhibition of Wnt-3a-induced C3H10T1/2 cell differentiation is indirect and mediated by bone morphogenetic proteins. *J Biol Chem.* **280**, 2498–2502 (2005).

231. H. Hu, M. J. Hilton, X. Tu, K. Yu, D. M. Ornitz, and F. Long, Sequential roles of hedgehog and Wnt signaling in osteoblast development. *Development.* **132**, 49–60 (2005).

232. C. N. Bennett, K. A. Longo, W. S. Wright, L. J. Suva, T. F. Lane, K. D. Hankenson, and O. A. MacDougald, Regulation of osteoblastogenesis and bone mass by Wnt10b. *Proc Natl Acad Sci USA.* **102**, 3324–3329 (2005).

233. J. A. Kennell and O. A. MacDougald, Wnt signaling inhibits adipogenesis through β-catenin-dependent and -independent mechanisms. *J Biol Chem.* **280**, 24004–24010 (2005).

234. N. H. Kulkarni, D. L. Halladay, R. R. Miles, L. M. Gilbert, C. A. Frolik, R. J. S. Galvin, T. J. Martin, M. T. Gillespie, and J. E. Onyia, Effects of parathyroid hormone on Wnt signaling pathway in bone. *J Cell Biochem.* **95**, 1178–1190 (2005).

235. P. V. N. Bodine, Y. P. Kharode, L. Seestaller-Wehr, P. Green, C. Milligan, and F. J. Bex, The bone anabolic effects of parathyroid hormone (PTH) are blunted by deletion of the Wnt

antagonist secreted frizzled-related protein (sFRP)-1. *J Bone Miner Res.* **19**(Suppl 1), S17 (abstract 1063) (2004).

236. U. T. Iwaniec, G. Liu, R. R. Arzaga, L. M. Donovan, R. Brommage, and T. J. Wronski, Lrp5 is not essential for the stimulatory effect of PTH on bone formation in mice. *J Bone Miner Res.* **19**(Suppl 1), S18 (abstract 1064) (2004).

237. Y. P. Kharode, P. V. N. Bodine, P. Green, C. Milligan, I. Li, E. Smith-Adaline, and F. J. Bex, Bone anabolic effects of PTH in LRP5 (G171V) transgenic high bone mass mice. *J Bone Miner Res.* **19**(Suppl 1), S104 (abstract F525) (2004).

238. H. Keller and M. Kneissel, SOST is a target for PTH in bone. *Bone.* **37**, 148–158 (2005).

239. T. Bellido, A. A. Ali, I. Gubrij, L. I. Plotkin, Q. Fu, C. A. O'Brien, S. C. Manolagas, and R. L. Jilka, Chronic elevation of PTH in mice reduces expression of sclerostin by osteocytes: A novel mechanism for hormonal control of osteoblastogenesis. *Endocrinology.* epub 4 August (2005).

240. M. D. Castellone, H. Teramoto, B. O. Williams, K. M. Druey, and J. S. Gutkind, Prostaglandin E$_2$ promotes colon cancer cell growth through a G$_s$-axin-β-catenin signaling axis. *Science.* **310**, 1504–1510 (2005).

241. H. S. Melkonyan, W. C. Chang, J. P. Shapiro, M. Mahadevappa, P. A. Fitzpatrick, M. C. Kiefer, L. D. Tomei, and S. R. Umansky, SARPs: A family of secreted apoptosis-related proteins. *Proc Natl Acad Sci USA.* **94**, 13636–13641 (1997).

242. Z. Zhou, J. Wang, X. Han, J. Zhou, and S. Linder, Up-regulation of human secreted frizzled homolog in apoptosis and its down-regulation in breast tumors. *Int J Cancer.* **78**, 95–99 (1998).

243. Y. Ahmed, S. Hayashi, A. Levine, and E. Wieschaus, Regulation of armadillo by a *Drosophila* APC inhibits neuronal apoptosis during retinal development. *Cell.* **93**, 1171–1182 (1998).

244. I. S. Nathke, The adenomatous polyposis coli protein. *Mol Pathol.* **52**, 169–173 (1999).

245. M. Hetman, J. E. Cavanaugh, D. Kimelman, and Z. Xia, Role of glycogen synthase kinase-3b in neuronal apoptosis induced by trophic withdrawal. *J Neurosci.* **20**, 2567–2574 (2000).

246. K. A. Longo, J. A. Kennell, M. J. Ochocinska, S. E. Ross, W. S. Wright, and O. A. MacDougald, Wnt signaling protects 3T3–L1 preadipocytes from apoptosis through induction of insulin-like growth factors. *J Biol Chem.* **277**, 38239–38244 (2002).

247. Z. You, D. Saims, S. Chen, Z. Zhang, D. C. Guttridge, K.-L. Guan, O. A. MacDougald, A. M. C. Brown, G. Evan, J. Kitajewski, and C.-Y. Wang, Wnt signaling promotes oncogenic transformation by inhibiting c-Myc-induced apoptosis. *J Cell Biol.* **157**, 429–440 (2002).

248. P. Babij, W. Zhao, C. Small, Y. Kharode, P. Yaworsky, M. Bouxsein, P. Reddy, P. Bodine, J. Robinson, B. Bhat, J. Marzolf, R. Moran, and F. Bex, High bone mass in mice expressing a mutant LRP5 gene. *J Bone Miner Res.* **18**, 960–974 (2003).

249. Y. P. Kharode, P. D. Green, J. T. Marzolf, W. Zhao, R. Askew, P. Yaworsky, and F. J. Bex, Alteration in bone density of mice due to heterozygous inactivation of Lrp6. *J Bone Miner Res.* **18**, S60 (abstract F032) (2003).

250. S. L. Holmen, T. A. Giambernardi, C. R. Zylstra, B. D. Buckner-Berghuis, J. H. Resau, J. F. Hess, V. Glatt, M. L. Bouxsein, M. Ai, M. L. Warman, and B. O. Williams, Decreased BMD and limb deformities in mice carrying mutations in both Lrp5 and Lrp6. *J Bone Miner Res.* **19**, 2033–2040 (2004).

251. H. Haegel, L. Larue, M. Ohsugi, L. Federov, K. Herrenknecht, and R. Kemler, Lack of β-catenin affects mouse development at gastrulation. *Development.* **121**, 3529–3537 (1995).

252. J. Huelsken, R. Vogel, V. Brinkmann, B. Erdmann, C. Birchmeier, and W. Birchmeier, Requirement for β-catenin in anterior-posterior axis formation in mice. *J Cell Biol.* **148**, 567–578 (2000).

253. R. Fodde, W. Edelmann, K. Yang, C. van Leeuwen, C. Carlson, B. Renault, C. Breukel, E. Alt, M. Lipkin, and P. M. Khan, A targeted chain-termination mutation in the mouse APC gene results in multiple intestinal tumors. *Proc Natl Acad Sci USA.* **91**, 8969–8973 (1994).

254. A. R. Moser, A. R. Shoemaker, C. S. Connelly, L. Clipson, K. A. Gould, C. Luongo, W. F. Dove, P. H. Siggers, and R. L. Gardner, Homozygosity for the min allele of Apc results in disruption of mouse development prior to gastrulation. *Dev Dyn.* **203**, 422–433 (1995).

255. S. L. Holmen, C. R. Zylstra, A. Mukherjee, R. E. Sigler, M.-C. Faugere, M. L. Bouxsein, L. Deng, T. L. Clemens, and B. O. Williams, Essential role of β-catenin in postnatal bone acquisition. *J Biol Chem.* **280**, 21162–21168 (2005).

256. H.-M. I. Yu, B. A. Jerchow, T.-J. Sheu, B. Liu, F. Costantini, J. E. Puzas, W. Birchmeier, and W. Hsu, The role of Axin2 in calvarial morphogenesis and craniosynostosis. *Development.* **132**, 1995–2005 (2005).

257. K. D. Hausler, N. J. Horwood, A. Uren, J. Ellis, C. Lengel, T. J. Martin, J. S. Rubin, and M. T. Gillespie, Secreted frizzled-related protein (sFRP-1) binds to RANKL to inhibit osteoclast formation. *J Bone Miner Res.* **16**, S153 (2001).

258. J. A. Robinson, M. Chatterjee-Kishore, P. Yaworsky, D. M. Cullen, W. Zhao, C. Li, Y. P. Kharode, L. Sauter, P. Babij, E. L. Brown, A. A. Hill, M. P. Akhter, M. L. Johnson, R. R. Recker, B. S. Komm, and F. J. Bex, Wnt/β-catenin signaling is a normal physiological response to mechanical loading in bone. *J Biol Chem* **281**, 31720–31728 (2006).

259. M. Chatterjee-Kishore, P. Yaworsky, W. Zhao, J. A. Robinson, C. Li, Y. P. Kharode, E. Fortier, L. Sauter, M. Cain, B. Bhat, M. Wasko, P. Babij, R. Bhat, P. V. N. Bodine, E. L. Brown, M. L. Johnson, M. Akhter, D. Cullen, R. R. Recker, and F. J. Bex, Anabolic response of bone to mechanical load involves the Wnt/β-catenin pathway. *submitted* (2006).

260. D. A. Glass, P. Bialek, J. D. Ahn, M. Starbuck, M. S. Patel, H. Clevers, M. M. Taketo, F. Long, A. P. McMahon, R. A. Lang, and G. Karsenty, Canonical Wnt signaling in differentiated osteoblasts controls osteoclast differentiation. *Develop Cell.* **8**, 751–764 (2005).

261. G. J. Spencer, J. S. Utting, S. L. Etheridge, T. R. Arnett, and P. G. Genever, Wnt signalling in osteoblasts regulates expression of the receptor activator of NFκB ligand and inhibits osteoclastogenesis in vitro. *J Cell Science.* **119**, 1283–1296 (2006).

262. H. M. Frost, *The Laws of Bone Structure*. Charles C. Thomas, Springfield (1964).

263. H. M. Frost, Perspectives: Bone's mechanical usage windows. *Bone Miner.* **19**, 257–271 (1992).

264. E. Burger and J. Veldhuijzen, Influence of mechanical factors on bone formation, resorption, and growth *in vitro*. *Bone.* **7**, 37–56 (1993).

265. C. H. Turner and F. M. Pavalko, Mechanotransduction and functional response of the skeleton to physical stress: The mechanisms and mechanics of bone adaption. *J Orthoped Sci.* **3**, 346–355 (1998).

266. H. J. Donahue, Gap junctional intercellular communication in bone: A cellular basis for the mechanostat set point. *Calcif Tissue Int.* **62**, 85–88 (1998).

267. S. Weinbaum, S. C. Cowin, and Y. Zeng, A model for the excitation of osteocytes by mechanical loading-induced bone fluid shear stresses. *J Biomech.* **27**, 339–360 (1994).

268. B. Cheng, Y. Kato, S. Zhao, J. Luo, E. Sprague, L. F. Bonewald, and J. X. Jiang, Prostaglandin E2 is essential for gap junction-mediated intercellular communication between osteocytes in response to mechanical strain. *Endocrinology.* **142,** 3464–3473 (2001).

269. J. Li, L. Duncan, R. Burr, and C. Turner, L-type calcium channels mediate mechanically induced bone formation in vivo. *J Bone Miner Res.* **17,** 1795–1800 (2002).

270. R. L. Duncan and C. H. Turner, Mechanotransduction and the functional response of bone to mechanical strain. *Calcif Tissue Int.* **57,** 344–358 (1995).

271. P. P. Cherian, X. Wang, S. Gu, L. F. Bonewald, E. Sprague, and J. X. Jiang, Mechanical strain opens connexin 43 hemichannels in osteocytes: A novel mechanism for the release of prostaglandin. *Molec Biol Cell.* **16,** 3100–3106 (2005).

272. S. C. Rawlinson, A. A. Pitsillides, and L. E. Lanyon, Involvement of different ion channels in osteoblasts' and osteocytes' early responses to mechanical strain. *Bone.* **19,** 609–614 (1996).

273. G. Zaman, A. A. Pitsillides, S. C. Rawlinson, R. F. Suswillo, J. R. Mosley, M. Z. Cheng, L. A. Platts, M. Hukkanen, J. M. Polak, and L. E. Lanyon, Mechanical strain stimulates nitric oxide production by rapid activation of endothelial nitric oxide synthase in osteocytes. *J Bone Miner Res.* **14,** 1123–1131 (1999).

274. T. M. Skerry, Identification of novel signaling pathways during functional adaptation of the skeleton to mechanical loading: The role of glutamate as a paracrine signaling agent in the skeleton. *J Bone Miner Metab.* **17,** 66–70 (1999).

275. P. P. Cherian, B. Cheng, S. Gu, E. Sprague, L. F. Bonewald, and J. X. Jiang, Effects of mechanical strain on the function of gap junctions in osteocytes are mediated through the prostaglandin EP2 receptor. *J Biol Chem.* **278,** 43146–43156 (2003).

276. E. H. Burger and J. Klein-Nulend, Mechanotransduction in bone—Role of the lacuno-canalicular network. *FASEB J.* **13,** S101–S112 (1999).

277. E. H. Burger, J. Klein-Nulend, A. Van Der Plas, and P. J. Nijweide, Function of osteocytes in bone—Their role in mechanotransduction. *J Nutr.* **125,** 2020S–2023S (1995).

278. E. J. Erlich and L. E. Lanyon, Mechanical strain and bone cell function: A review. *Osteoporos Int.* **18,** 688–700 (2002).

279. F. M. Pavalko, S. M. Norvell, D. B. Burr, C. H. Turner, R. L. Duncan, and J. P. Bidwell, A model for mechanotransduction in bone cells: The load-bearing mechanosomes. *J Cell Biochem.* **88,** 104–112 (2003).

280. A. G. Robling, A. B. Castillo, and C. H. Turner, Biomechanical and molecular regulation of bone remodeling. *Ann Rev Biomed Engineer.* **8,** 6.1–6.44 (2006).

281. H. M. Frost, From Wolff's law to the mechanostat: A new "face" of physiology. *J Orthoped Sci.* **3,** 282–286 (1998).

282. M. P. Akhter, D. J. Wells, S. J. Short, D. M. Cullen, M. L. Johnson, G. Haynatzki, P. Babij, K. M. Allen, P. J. Yaworsky, F. Bex, and R. R. Recker, Bone biomechanical properties in Lrp5 mutant mice. *Bone.* **35,** 162–169 (2004).

283. M. P. Akhter, D. M. Rabb, D. B. Kimmel, and R. R. Recker, Locating the loading region in the rat tibia bending model. *J Bone Miner Res.* **18,** S46 (1992).

284. D. M. Cullen, M. P. Akhter, D. Mace, M. L. Johnson, P. Babij, and R. R. Recker, Bone sensitivity to mechanical loads with the Lrp5 HBM mutation. *J Bone Miner Res.* **17**(Suppl 1), S332 (2002).

285. K. Sawakami, A. G. Robling, N. D. Pitner, S. J. Warden, J. Li, M. L. Warman, and C. H. Turner, Site-specific osteopenia and decreased mechanoreactivity in Lrp5–mutant mice. *J Bone Miner Res.* **19**(Suppl 1), S38 (abstract 1149) (2004).

286. J. R. Hens, K. M. Wilson, P. Dann, X. Chen, M. C. Horowitz, and J. J. Wysolmerski, TOPGAL mice show that the canonical Wnt signaling pathway is active during bone development and growth and is activated by mechanical loading in vitro. *J Bone Miner Res.* **20,** 1103–1113 (2005).

287. D. L. Johnson, T. N. McAllister, and J. A. Frangos, Fluid flow stimulates rapid and continuous release of nitric oxide in osteoblasts. *Am J Physiol.* **271,** E201–E208 (1996).

288. R. Smalt, F. T. Mitchell, R. L. Howard, and T. J. Chambers, Induction of NO and prostaglandin E2 in osteoblasts by wall stress but not mechanical strain. *Am J Physiol.* **273,** E751–E758 (1997).

289. D. W. Murray and N. Rushton, The effect of strain on bone cell prostaglandin E2 release: A new experimental method. *Calcif Tissue Int.* **47,** 35–39 (1990).

290. K. M. Reich and J. A. Frangos, Effect of flow on prostaglandin E2 and inositol trisphosphate levels in osteoblasts. *Am J Physiol.* **261,** C428–C432 (1991).

291. N. E. Ajubi, J. Klein-Nulend, M. J. Alblas, E. H. Burger, and P. J. Nijweide, Signal transduction pathways involved in fluid flow-induced PGE2 production by cultured osteocytes. *Am J Physiol.* **276,** E171–E178 (1999).

292. A. J. El-Haj, S. L. Minter, S. C. Rawlinson, R. Suswillo, and L. E. Lanyon, Cellular responses to mechanical loading in vitro. *J Bone Miner Res.* **5,** 923–932 (1990).

293. D. L. Ypey, A. F. Weidema, K. M. Hold, A. Van der Laarse, J. H. Ravesloot, A. Van Der Plas, and P. J. Nijweide, Voltage, calcium, and stretch activated ionic channels and intracellular calcium in bone cells. *J Bone Miner Res.* **7**(Suppl 2), S377 (1992).

294. K. M. Reich, C. V. Gay, and J. A. Frangos, Fluid shear stress as a mediator of osteoblast cyclic adenosine monophosphate production. *J Cell Physiol.* **143,** 100–104 (1990).

295. J. Li, D. Liu, H. Z. Ke, R. L. Duncan, and C. H. Turner, The P2X$_7$ nucleotide receptor mediates skeletal mechanotransduction. *J Biol Chem.* **280,** 42952–42959 (2005).

Cytokines and Bone Remodeling

GREGORY R. MUNDY, BABATUNDE OYAJOBI, GLORIA GUTIERREZ, JULIE STERLING, SUSAN PADALECKI, FLORENT ELEFTERIOU, AND MING ZHAO

I. INTRODUCTION

In normal individuals, bone is maintained by a process of local bone remodeling achieved via a finely regulated balance between the processes of bone formation and resorption mediated by osteoblasts and osteoclasts, respectively. Bone remodeling is regulated, in part, by local factors including cytokines generated in the bone microenvironment, and influenced by systemic hormones, including parathyroid hormone, calcitonin, sex hormones, leptin, vitamin D metabolites, and the sympathetic nervous system. The purpose of this chapter is to summarize what is currently known about the role of cytokines and their receptors in bone remodeling. In recent years, there has been an explosion of information on multiple aspects of the effects of cytokines on bone. This has become an enormous topic, and it is not possible to cover all aspects in this chapter. Rather, this current chapter focuses on selected important recent advances. In particular, this chapter covers in detail new information on the RANK-RANKL system, the BMPs, the current status of recent work on the control of bone remodeling by the sympathetic nervous system, and sclerostin. There has been important new information on the role of the Wnt signal transduction pathway on bone formation and resorption over the past 5 years, but it will not be specifically reviewed here, but rather discussed as a pathway through which some of the cytokines, and particularly the BMPs, mediate their effects. (See also Chapter 15, Johnson.) A list of the relevant local factors currently known to regulate bone remodeling is provided in Table 16-1.

Recent advances in molecular biological techniques have meant that most of the biological activities ascribed to cytokines have now been associated with specific molecules, and their receptors identified and molecularly cloned. A number of these cytokines and their cognate receptors are expressed by bone cells, marrow cells, or accessory cells in the bone microenvironment. Moreover, studies using knockout and transgenic mice have increased our understanding of the complex signal transduction mechanisms utilized by cytokines and are opening up new and exciting areas of study. Cytokines tend to be pleiotropic and multifactorial and may have overlapping and seemingly redundant biological effects. Some of this redundancy is apparent in the receptor mechanisms and signal transduction pathways used by groups of cytokines. Classic examples that illustrate this vividly are the various cytokines belonging to the interleukin (IL)-6 family, such as IL-6, leukemia inhibitory factor (LIF), oncostatin-M, and IL-11, which utilize a common signal transduction protein known as gp130. These cytokines bind to distinct membrane-associated receptors, which form hetero- or homodimers upon binding to the ligand. These dimers then complex with gp130, leading to its activation by the phosphorylation of tyrosine residues. This subsequently activates several tyrosine kinase cascades within the cells by a common tyrosine kinase, JAK2. One of these cascades involves phosphorylation of the transcription factor STAT-2. Another involves ras and MAP-2 kinase and leads to phosphorylation of the transcription factor, nuclear factor (NF-κB) [1]. The roles of these signal transduction pathways and those used by other cytokines are currently being unraveled

TABLE 16-1   Local Regulators of Bone Remodeling Reviewed in This Chapter

1. RANK ligand and its signaling receptor, RANK
2. Osteoprotegerin
3. Macrophage–colony-stimulating factor and its receptor, c-Fms
4. Vascular endothelial growth factor
5. Tumor necrosis factor
6. Interleukin-6 (IL-6)
7. Interleukin-15 (IL-15), interleukin-17 (IL-17), and interleukin-18 (IL-18)
8. Bone morphogenetic proteins
9. Hedgehog (Hh) signaling molecules
10. Sclerostin
11. Parathyroid hormone-related peptide (PTHrP)
12. Neuronal regulation of bone remodeling

in bone cells, but observations already made in other cells and tissues are holding true for bone, with just a few exceptions. The reasons that individual members of cytokine families have seemingly distinct effects on cells involved in bone remodeling remain unclear.

II. EVIDENCE FOR A ROLE OF CYTOKINES IN OSTEOCLASTIC BONE RESORPTION

A considerable amount of data has been accumulated since the mid-1970s indicating that cytokines play a role in both physiological and pathological bone remodeling. As mentioned previously, osteoclast formation and activity are regulated by factors that are generated in the bone microenvironment acting in an autocrine, paracrine, or juxtacrine fashion.

These include macrophage–colony-stimulating factor (M-CSF), also known as colony-stimulating factor-1 (CSF1), IL-6, IL-1, IL-11, tumor necrosis factor (TNF)-α, TNF-β, granulocyte macrophage–colony-stimulating factor (GM-CSF), transforming growth factor (TGF)-α, TGF-β, leukemia inhibitory factor (LIF), and bone morphogenetic proteins (BMPs). For some of these cytokines, the precise cellular source within the bone microenvironment has not been defined, although possibilities include immune cells and bone cells of either osteoblastic or osteoclastic lineages. In addition, biologically active forms of some of these cytokines may be derived from sequestered stores within the bone

matrix. However, the relative importance of cytokines in the formation of new osteoclasts and the activation of mature osteoclasts *in vivo* is still unclear. The availability of genetic mouse models has meant that the role of cytokines in bone metabolism is now increasingly being examined in an *in vivo* context. However, the functional redundancies among related cytokines or groups of cytokines that share common signaling pathways mean that the direct ablation of individual genes for these factors does not always impact bone remodeling adversely.

The most convincing evidence that cytokines are involved in control of osteoclastic bone resorption comes from genetic mouse models. The op/op mouse has a mutant form of M-CSF that causes the condition [2–7], and this is described in more detail in the section on M-CSF. The genetic mouse models that prove the role of RANKL and osteoprotegerin in control of osteoclast activity and function are also described under the relevant section.

There are other lines of evidence from studies using transgenic mice and inhibitors of cytokine activities that indicate that cytokines such as IL-1, IL-6, and TNF-α are important in disorders of bone remodeling *in vivo*. However, available data are complex to interpret, and there are often conflicting reports. IL-1 and TNF-α have consistently been shown to play a major role in the rapid bone loss associated with estrogen-depleted states such as postmenopause and after ovariectomy [8], and there is also evidence from studies with mice lacking the IL-1 type I receptor that implies that IL-1 may be an important mediator of the effects of ovariectomy on bone mass. Mice deficient in the type I IL-I receptor (IL-IR1), which is the signaling receptor for both IL-1χ and IL-1β, do not lose bone after ovariectomy [9]. The soluble p75 TNF receptor blocks the osteoclastogenic effect of TNF-α [10], and mice engineered to overexpress this soluble TNF receptor do not lose bone after ovariectomy [11]. Regarding IL-6, the increase in bone resorption observed following ovariectomy in mice can be corrected by the administration of neutralizing antibodies to IL-6, as shown by the experiments of Jilka and colleagues [12]. The same abnormality can be reversed by treatment of mice with estrogen. Also, IL-6 knockout mice are protected against bone loss induced by ovariectomy, further implicating IL-6 in bone remodeling *in vivo* [13]. Surprisingly, transgenic mice overexpressing IL-6 do not have osteopenia as would be predicted [14]. These seemingly discrepant observations are probably related to the fact that in the presence of estrogen deficiency, there is likely increased production of several cytokines by cells in the bone marrow microenvironment leading to increased osteoclastic activity. Whether production of each cytokine is restricted to specific cell types has to be determined. There is considerable evidence that IL-6 and/or TNF-α may synergize with IL-1 to enhance osteoclastic bone resorption. For example, Pacifici and colleagues suggested that the simultaneous block of IL-1 and TNF may be necessary to completely abrogate the rapid bone loss seen in the early postovariectomy period [15, 16]. However, neutralization of either of these factors may nevertheless lead to some decrease in bone resorption in certain situations.

There is also evidence that other osteotropic cytokines may be involved in other disease states. Much of this evidence comes from "gain of function" rather than "loss of function" experiments, with evidence that there is increased production of certain cytokines associated with increased bone loss. This is certainly true in myeloma, in some solid tumors, and in chronic inflammatory diseases associated with a local increase in bone loss such as rheumatoid arthritis and periodontal diseases [17]. It is also true in Paget's disease, where there is increased production of proresorptive cytokines by multinucleated osteoclasts, especially IL-6 [18–20]. Because overproduction of these cytokines in these conditions may enhance bone resorption through the stimulation of osteoclast formation and differentiation, pathologic bone lesions associated with a large increase in osteoclasts may be self-perpetuating.

III. THE OSTEOCLAST AS A CELL SOURCE OF CYTOKINES INVOLVED IN OSTEOCLASTIC RESORPTION

Abundant evidence indicates that osteoclast formation and activity are regulated by factors generated in the bone cell microenvironment. As mentioned in the preceding section, these factors may be produced by immune cells or cells in the osteoblast lineage or be derived from the bone matrix itself. However, convincing data support the notion that the osteoclast itself may also be a source of autocrine or paracrine factors, which can modulate bone remodeling. The subject of osteoclast as a secretory cell has been reviewed comprehensively elsewhere [21–24]. The osteoclast expresses IL-6 in prodigious amounts. Moreover, IL-6, at least in human systems, can stimulate the formation of cells with osteoclast characteristics [19]. Antibodies to IL-6 inhibit bone resorption by isolated human giant cells on calcified matrices, and, similarly, antisense oligonucleotides to IL-6 inhibit the capacity of human giant cells to form resorption pits on sperm whale dentine [25]. Furthermore, it appears that IL-6 may mediate

some of the effects of IL-1 and TNF on bone resorption, as an anti–IL-6 neutralizing antibody and a potent IL-6 antagonist that binds to IL-6 receptor but does not dimerize with gp130 both blocked IL-1 and TNF-induced osteoclast formation in human marrow cultures [26]. However, IL-6 is not the only cytokine that is produced by isolated osteoclasts. TGF-β, interleukin-1, annexin II (lipocortin-II), and human stem cell antigen I [24, 27, 28] have all been shown to be expressed by osteoclasts, and each of these factors may regulate osteoclasts' function. TGF-β inhibits osteoclast formation [29, 30] and is a powerful stimulator of osteoclast apoptosis [31]. These effects are probably mediated, in part, via paracrine mechanisms involving alterations in the stromal/osteoblastic cell expression of the receptor activator of NF-κB ligand (RANKL) and osteoprotegerin (OPG) [32, 33]. TGF-β may also generate prostaglandins in the microenvironment of osteoclasts [29], which can exert independent effects on osteoclast formation and activity, most probably via modulating RANKL expression [34, 35]. IL-1 and annexin-II both stimulate osteoclast formation, whereas human stem cell antigen I inhibits osteoclast formation. The relative importance of all these osteoclast products in the formation of new osteoclasts is not clear. However, one possibility is that as osteoclasts undergo apoptosis within the bone remodeling unit, at the conclusion of the remodeling sequence, some of these cytokines may be released by the dying osteoclast to produce a new generation of osteoclasts derived from their marrow precursors.

IV. THE OSTEOBLAST AS A CELL SOURCE OF CYTOKINES INVOLVED IN OSTEOCLASTIC RESORPTION

There is compelling evidence from *ex vivo* studies that the commitment of osteoclast progenitors (spleen, bone marrow, or peripheral blood derived) to differentiate to multinucleated cells with characteristics of mature osteoclasts requires direct cell–cell contact with osteoblastic or related marrow stromal cells [36]. Furthermore, it has been known for some time that almost all of the known bone-resorbing cytokines, such as IL-1, IL-6, and IL-11, as well as the systemic bone-resorbing hormones, such as parathyroid hormone (PTH), PTH-related protein (PTHrP), 1,25-dihydroxyvitamin D_3, and PGE_2 appear to exert their effect only in the presence of stromal/osteoblastic cells [36]. Because these agents activate different signal transduction pathways on osteogenic cells, it was recognized that there is a convergence in their downstream response, and the existence of a membrane-associated factor on the surface of cells of the osteoblastic lineage essential for osteoclast progenitors to proliferate and differentiate was therefore proposed. This factor, which was variously termed "osteoclast differentiation factor" (ODF) and "stromal cell-derived osteoclast formation activity" (SOFA), was postulated to be inducible by cytokines and hormones known to regulate osteoclast differentiation. Although M-CSF was known to be membrane associated and to be important for osteoclastic bone resorption, recombinant M-CSF alone could not induce osteoclast formation in the absence of stromal/osteoblastic cells. Anderson and colleagues [37] reported the molecular cloning of a novel membrane-bound member of the TNF receptor (TNFR) family from a cDNA library established from human bone marrow–derived myeloid dendritic cells. Simultaneously, they reported the cloning of the mouse orthologue of the receptor from a fetal mouse liver cDNA library. This receptor, which activated NF-κB activity, was designated receptor activator of NF-κB (RANK) and a search for its cognate ligand led to the cloning of RANK ligand (RANKL). RANKL was shown to be identical to TNF-α–related activation-induced cytokine (TRANCE), a TNF ligand family member cloned from murine thymoma EL40.5 cells and shown to activate c-jun-N-terminal kinase [38]. Subsequently, using a novel secreted TNFR homologue known as osteoprotegerin (OPG)/osteoclastogenesis-inhibitory factor (OCIF) as a probe, two groups independently reported the cloning of the same molecule, which they designated OPG ligand (OPGL) and osteoclast differentiation factor (ODF), respectively [39, 40]. As will be discussed later, RANKL/TRANCE/OPGL/ODF have now been shown to be the same molecule whose expression is obligatory for osteoclastic resorption and normal bone modeling and remodeling. As proposed by Suda and others, we refer to this cytokine hereafter as RANKL [41]. See also Chapter 5 (Blair) for detailed discussion of osteoclast biology and the RANK, RANKL, OPG system.

V. RANK LIGAND AND ITS SIGNALING RECEPTOR, RANK

RANKL (TNFSF11) is synthesized as a type II integral membrane protein with its N-terminus in the cytoplasm and a C-terminus extending extracellularly. Expression of RANKL in a human embryonic kidney fibroblast (293) cell line was reported to generate a membrane-bound as well as a secreted form of RANKL representing the extracellular C-terminal domain [42]. It has also been reported that the extracellular domain of RANKL can be cleaved by a TNF-convertase (TACE)-

like enzyme *in vitro* and that recombinant RANKL can be cleaved by purified TACE [43]. This cleaved form retains some biological activity in osteoclast assay systems [43]. Although it has been reported that T cells shed RANKL on activation *in vitro* [44], there is as yet no evidence that a soluble form of RANKL exists *in vivo* or is generated by proteolytic cleavage in the bone microenvironment. In addition, membrane-bound RANKL is more potent than an engineered soluble form [35]. Nevertheless, it is tempting to speculate that in certain pathological conditions, such as malignancy, there could be "shedding" of RANKL by tumor-associated metalloproteinases (MMPs) as has been described for other members of the TNF ligand family such as TNF and Fas. In this regard, a factor was identified from a human tumor associated with an increased number of osteoclasts in bone and hypercalcemia which appears to be a previously unidentified cytokine that stimulates osteoclast formation in the presence of M-CSF [45], and another group also identified another factor from a mouse tumor with similar biological activity [46]. More recently, MMP-7 has been shown to be capable of processing RANKL to a soluble form that promotes osteoclast activation. In addition, there is evidence of reduced RANKL shedding *in vivo* in MMP-7–deficient mice with concomitant reduction in experimental prostate cancer–induced osteolysis [47].

RANKL, in the presence of macrophage–colony-stimulating factor (M-CSF) induces osteoclast formation in all model systems presently available to study osteoclast development. For example, RANKL stimulated formation of osteoclasts from spleen-derived osteoclast progenitors in the absence of osteoblasts/stromal cells, and this was abrogated by simultaneous addition of OPG [48] or a recombinant soluble form of the extracellular domain of RANK generated on its own or as a chimeric protein fused to the Fc region of human immunoglobulin (RANK.Fc) [49]. In the presence of M-CSF, RANKL also stimulated osteoclast formation in human and murine bone marrow cultures and also in human peripheral blood monocyte cultures [50–52], and it induced formation of TRAP-positive colonies in an agar culture of bone marrow cells [42]. Treatment of stromal/osteoblastic cells of human and murine origins with known stimulators of osteoclast formation, $1\alpha,25(OH)_2D_3$, PTH, PGE2, VEGF, IL-11, IL-6, IL-1, and TNF induces or enhances RANKL messenger RNA levels [34, 35, 40, 42, 48, 53–57]. Interestingly, $1\alpha,25(OH)_2D_3$ and dexamethasone-treatment of marrow stromal ST2 cells results in an increase in PGE2 production, and it has been suggested that PGE2 initiates the cascade which leads to enhanced RANKL expression [58]. Treatment of 45Ca-labeled fetal mouse or rat long bones with a recombinant soluble form of

RANKL also stimulated the release of 45Ca from bone, which was completely inhibited by simultaneously adding OPG or RANK.Fc [48, 49]. Like OPG, polyclonal antibodies against RANKL inhibited bone resorption in organ cultures induced by soluble RANKL and by $1\alpha,25(OH)_2D_3$, PTH, PGE2, and IL-1 [40, 59]. These results clearly indicate that bone resorption induced by these osteotropic factors is mediated by RANKL. In culture systems devoid of stromal or osteoblastic cells and where there is essentially no continuing osteoclast formation such as isolated rat osteoclasts, recombinant RANKL has been shown to rapidly induce actin rings in the cells and to markedly increase their bone-resorbing activity [59, 60]. A single parenteral administration of recombinant RANKL increased blood ionized calcium within 1 hour in mice [60]. Also, systemic injection of RANKL twice daily for 3 days led to sustained hypercalcemia although the number of osteoclasts was almost identical to those of untreated mice [42]. Taken together, these data indicate that RANKL stimulates not only osteoclast differentiation but also activates mature osteoclasts, thereby directly impacting their function. Genetic studies in mice have also provided compelling evidence in support of a critical role for RANKL in bone modeling and remodeling *in vivo*. Overexpression of a soluble form of RANKL in transgenic mice recapitulates the phenotypic features of postmenopausal osteoporosis in humans with increased osteoclastic resorption, cortical porosity, skeletal fragility, and reduced biomechanical strength [61]. RANKL knockout mice have also been generated that exhibited typical osteopetrosis with total occlusion of bone marrow space within endosteal bone. The bones of these RANKL null-mutant mice lack osteoclasts although osteoclast progenitors were present that were shown to differentiate into functionally active osteoclasts when cocultured with normal osteoblasts/stromal cells from wild-type littermates [62]. In addition, RANKL (−/−) mice completely lack lymph nodes and have a defect in thymocytes and lymph node organogenesis. These results suggest that RANKL is an absolute requirement for osteoclast development, and it also plays an important role in T cell differentiation [63].

It is likely that increased expression of RANKL may play a role in pathological situations associated with bone destruction such as malignancies. Taken together, the available data strongly suggest that RANKL is not only the final common mediator of osteoclast activation in multiple myeloma and cancer-induced osteolytic bone diseases including breast cancer metastases, but also a tumor cell product and that aberrant expression of the cytokine within the tumoral marrow milieu is likely to be pivotal in initiating and maintaining the characteristic bone destruction. One of the earliest reports supporting this notion demonstrated that direct cell–cell contact between myeloma cells and marrow stroma–derived ST2 cells enhances RANKL expression in the myeloma cells and also stimulates production of a soluble factor(s) capable of enhancing RANKL expression in the marrow stromal cells [64]. It was postulated then that increased RANKL expression within the bone microenvironment cells may explain the increased osteoclastic activity and destructive bone lesions that are characteristic of multiple myeloma. This is likely to be true also for other tumors metastatic to bone such as breast cancer. In this regard, it has been reported that although RANKL expression was undetectable in either mouse breast cancer cells or bone marrow stromal cells, its expression was markedly elevated in cocultures of both cell types [65]. It has also been reported that substantial numbers of multinucleated cells with osteoclastic characteristics form in cocultures of activated human CD4+ T helper cells and adherent murine splenic osteoclast precursors in the presence of M-CSF, independent of stromal/osteoblastic cells [66]. CD4+ T cells express RANKL constitutively, and other accessory cells in the synovial pannus such as macrophages secrete M-CSF. This would explain the extensive localized bone destruction observed adjacent to chronic inflammatory tissues characterized by CD4+ T cell infiltration such as rheumatoid arthritic synovium.

As mentioned previously, RANKL was originally cloned as a ligand for the receptor, RANK. RANK (TNFRSF11A) is a type I transmembrane protein with a C-terminal cytoplasmic tail much longer than that of all known members of the TNFR superfamily. Like other members of the family, RANK has four extracellular cysteine-rich domains. However, unlike most other TNFR family members, RANK messenger RNA is ubiquitously expressed with highest levels in the skeletal muscle and thymus and in spleen- and bone-marrow–derived osteoclast precursors [37, 67]. To date, RANK has been shown to bind only to RANKL; it does not bind other members of the TNF ligand family such as lymphotoxin, TNF-α, Fas ligand, CD27 ligand, CD30 ligand, CD40 ligand, 4-1BB ligand, or TRAIL. It has also been conclusively demonstrated that the formation of mature osteoclasts from osteoclast precursors as well as activation of mature osteoclasts can only be induced via RANK signaling [59, 67, 68]. Although a number of studies have suggested that other pro-osteoclastogenic factors such as IL-1 and TNF-α and MIP-1α can directly stimulate osteoclast formation and bone resorption independent of RANK-mediated signaling, none of these factors alone or in concert stimulates osteoclast formation *in vivo* in RANK-deficient mice [64]. In the last few years, a number of signaling pathways downstream of

RANK that mediate osteoclastogenesis have been identified using a combination of genetic and biochemical studies. First, two groups independently generated mice that were lacking both the p50 and p52 subunits of NF-κB [69, 70] and reported that these double (NF-κB-1 and NF-κB-2) knockout mice developed severe osteopetrosis because of a defect in osteoclast differentiation. There was a complete absence of osteoclasts although there were osteoclast progenitors and the number of macrophages was increased. In addition to NF-κB (see above), binding of RANKL to RANK also leads to the activation of c-jun N-terminal kinase (JNK), TNF receptor associated family of adaptor molecules (TRAFs), transcription factors nuclear factor of activated T cells (NFAT) c1 and c-Fos, as well as intracellular calcium signaling pathways [71, 72]. RANK activates JNK in bone marrow– and spleen-derived hematopoietic osteoclast progenitors [43, 59, 67], and also activates JNK, but not NF-κB, in a monocyte/macrophage cell line RAW 264.7, which has been shown to differentiate into mature osteoclast-like cells when cultured in the presence of RANKL and M-CSF [67]. Consistent with this, mice genetically engineered to express a dominant-negative form of c-Jun are profoundly osteopetrotic. Interestingly, overexpression of RANK in human embryonic kidney fibroblasts (293) cells induces ligand-independent NF-κB and JNK activation [67, 73], suggesting that pathological conditions associated with RANK overexpression may result in increased osteoclast formation independent of RANKL.

As RANK has no intrinsic kinase activity, it activates NF-κB via interactions with TRAFs. Several members of the TRAF family have been implicated in regulating signals from various TNF/TNFR family members [74]. There is evidence that TRAF2, TRAF5, and TRAF6 interact with the C-terminal 85–amino acid cytoplasmic tail of RANK, and it is likely that the signals through RANK are mediated primarily through these TRAFs [67, 73, 75–78]. Of these three TRAFs, TRAF6 appears unique in several respects. First, it interacts with a novel C-terminal domain of the cytoplasmic tail of RANK distinct from the known binding motifs for TRAF1, TRAF2, TRAF3, and TRAF5 although TRAF6 also associates with a short N-terminal sequence within the cytoplasmic domain [73, 77]. Second, overexpression of an N-terminal truncated TRAF6, acting as a dominant-negative, inhibited RANKL-induced NF-κB activation in human embryonic kidney (293) cell line. Third, unlike other TRAFs, TRAF6 has also been implicated in IL-1–induced NF-κB activation [79]. Last, whereas the other TRAF null-mutant mice currently available, such as the TRAF2 and TRAF3 null-mutant mice, have a normal skeletal phenotype, TRAF6 knockout mice exhibit severe osteopetrosis with defective bone remodeling and delayed tooth eruptions [76, 80]. However, unlike in

RANKL (–/–) mice, the bones of TRAF6 (–/–) mice had a few mature osteoclasts, suggesting that there might be some redundancy in TRAF usage in osteoclast development. It seems therefore that the TRAF6 (–/–) osteoclasts are impaired either in their ability to attach to bone or are defective in some way in their ability to form ruffled borders similar to mice bearing mutations in the c-src tyrosine kinase. Although overexpression of RANK in human embryonic kidney 293 cells stimulated JNK and NF-κB, when the C-terminal cytoplasmic tail of RANK necessary for TRAF binding was deleted, the truncated RANK receptor was still capable of stimulating JNK activity, but not NF-κB. This suggests that interaction with TRAFs is critical for NF-κB activation, but not for the activation of the JNK pathway [67]. Apart from TRAFs, there is also evidence to show that signals mediated by RANK activate NF-κB in rat osteoclasts by acting through phospholipase C to release Ca^{2+} from intracellular stores [81]. This RANKL/RANK-induced increase in intracellular Ca^{2+} in turn also activates calmodulin and its downstream effectors such as the Ca^{2+}/calmodulin-dependent phosphatase calcineurin (see below). In addition to TRAF6, an AP-1 complex containing c-Fos has also been shown to play an important role in RANKL-induced osteoclast formation and activation. Mice lacking Fos (which encodes c-Fos) develop osteopetrosis due to an early differentiation block along the osteoclast lineage. RANKL induces not only transcription of Fos [82, 83], but it also induces transcription of Fosl1 (which encodes Fra-1) in a c-Fos–dependent manner, and expression of a Fosl1 transgene rescues the osteopetrosis of c-Fos null-mutant mice [83]. Finally, recent studies indicate that nuclear factor of activated T cells *NFATc1* may be the "master" transcriptional activation of osteoclastogenesis in the same way that myoD is the master transcriptional activator in muscle development since RANKL-induced TRAF6, c-fos, and calcineurin, which in turn all converge on *NFATc1*. NFATc1-null embryonic stem cells do not differentiate into osteoclasts in response to RANKL in contrast to wild-type ES cells, and ectopic expression of *NFATc1* causes RANKL-independent terminal differentiation of osteoclast precursors to mature osteoclasts [82, 84]. Moreover, RANKL induction of *NFATc1* is mediated by both TRAF6 and c-Fos, and NFATc1 rescues osteoclastogenesis in precursors lacking c-Fos [85]. RANKL-evoked calcium oscillations also lead to calcineurin-mediated activator of *NFATc1* [86]. Consistent with this, inhibition of calcineurin with either the immunosuppressant drugs cyclosporin A and FK506, or ectopic expression of a specific calcineurin inhibitory peptide, potently inhibited the RANKL-induced differentiation of RAW264.7 cells into osteoclasts, whereas ectopic expression of a constitutively active, calcineurin-independent *NFATc1* mutant was sufficient to induce these cells to differentiate into osteoclasts [84].

Taken together, all of the available data suggest that inhibitors of the biological activity of RANKL hold promise as therapeutic agents for a wide variety of conditions in which bone remodeling is dysregulated, including postmenopausal osteoporosis and cancer-induced bone diseases. In a proof-of-principle study, a genetically engineered form of RANK generated by fusing the entire extracellular domain to the Fc region of human IgG1 (RANK.Fc) was shown to efficiently block hypercalcemia induced by PTHrP-secreting human tumor xenografts in nude mice, further confirming the critical role of RANKL/RANK interaction in osteoclastic bone resorption [49]. Transgenic mice overexpressing a soluble RANK.Fc fusion protein have severe osteopetrosis because of a marked reduction in osteoclast numbers and a decrease in bone resorption indices although their teeth erupted normally [67] and systemic administration of mesenchymal stem cells retrovirally transduced with RANK.Fc prevented bone loss in ovariectomized mice [87]. RANK (–/–) mice have also been generated [88, 89], and as expected, they exhibit severe osteopetrosis due to complete absence of osteoclasts and lack of bone resorption. Although these RANK null-mutant mice formed incisors, there was failure of teeth eruption. This confirms the absolute requirement of an intact RANKL/RANK pathway for osteoclastogenesis *in vitro* and *in vivo*. The discovery that the RANKL/RANK pathway is indispensable not only for normal bone resorption but also pathological bone resorption induced by myeloma and other cancers that metastasize to the skeleton has spurred the clinical development of fully human monoclonal antibodies to human RANKL, which have the advantage over Fc fusion proteins in that they have much longer half-lives. The most advanced of this is a humanized antibody known as denosumab (AMG162). Clinical trials with this agent are ongoing in patients with osteoporosis as well as multiple myeloma and bone metastases from breast cancer, and recent data suggest that it decreases bone resorption and that, in postmenopausal women with low bone mass, it also increases bone mineral density [90–92].

VI. OSTEOPROTEGERIN

Our understanding of the biology of bone modeling and remodeling was given an impetus with the discovery of a novel secreted member of the TNFR superfamily almost a decade ago. One group isolated a heparin-binding protein from conditioned media of human fibroblast cultures that profoundly inhibited osteoclast formation. This protein was thus designated "osteoclastogenesis inhibitory factor" (OCIF) [93]. Independently, another group cloned a novel TNFR family member that constitutively lacked a transmembrane domain and was thus secreted. When expressed, the recombinant protein was shown to inhibit physiological and pathological bone resorption, and hepatic overexpression of the gene in transgenic mice resulted in severe osteopetrosis. The receptor was therefore termed osteoprotegerin (OPG) [39]. Subsequent molecular cloning of the cDNA coding for OCIF revealed its identity with OPG [40]. Other groups also independently cloned the same receptor molecule. The TNF receptor-like molecule 1 (TR1) and follicular dendritic cell-derived receptor 1 (FDCR-1) [94–96] have each been shown to have complete sequence identity to OPG/OCIF. We will hereafter refer to the protein (including OCIF, FDCR-1, and TR1) as OPG. Like other members of the TNF receptor family, OPG has four cysteine-rich domains (Dl–D4). In addition, there are two homologous death domain regions (D5 and D6) in OPG. Both D5 and D6 share structural features with other death domains previously described in other members of the TNFR family, including the TNF receptor p55, Fas, DR3, and TRAIL receptor. These death domains have been shown to mediate apoptotic signals. Although the precise role of D5 and D6 of OPG is still not known, the death domain–homologous regions are active in mediating apoptotic signals [68, 97].

OPG (TNFRSF11B) has only two known ligands, RANKL and TRAIL, both of which are type II membrane-bound TNF homologues [42, 98]. In contrast to most other TNF receptor family members, OPG is secreted and has been reported to circulate *in vivo* [99]. There is now incontrovertible evidence that it acts as a nonsignaling decoy receptor for RANKL and thereby regulates bone turnover [71]. Early studies suggested that serum concentrations of OPG increase with age in both men and women and are significantly higher in postmenopausal osteoporotic women compared to age-matched controls. It was proposed that the increased levels of OPG in the former group reflect a compensatory response to the enhanced bone resorption that occurs postmenopause rather than a cause of the osteoporosis [99]. However, the significance of changes in serum OPG levels in humans remains unclear, and there have been inconsistencies in reports in different cohorts of patients [100]. The reason for this remains unclear but may relate to differences in study design, assays, and/or methodology.

The role of OPG in normal bone remodeling has been highlighted in more detail in studies in genetic experiments with OPG-deficient mice produced by targeted disruption of the gene [101, 102]. OPG (–/–)

mice are viable and fertile, but they exhibit profound osteoporosis from birth caused by enhanced osteoclast formation and function as well as prolonged osteoclast survival, similar to transgenic mice overexpressing the soluble form of RANKL (see above). Histological analyses of skeletons of OPG-deficient mice show a destruction of growth plates and lack of trabeculae, and histomorphometrical analyses revealed an increase in bone resorption indices in long bones of adult OPG null-mutant mice. This is accompanied by a marked decrease in the biomechanical strength and mineral densities of their bones. Interestingly, osteoblast surface areas were also increased in OPG-deficient mice. Intravenous injection of recombinant OPG.Fc (OPG expressed as a fusion protein with Fc portion of IgG; AMG007) and transgenic overexpression of OPG in an OPG null background effectively rescued the severe osteoporotic phenotype caused by OPG deficiency [103]. OPG null-mutant mice also develop calcification of the aorta and renal arteries. These results indicate that OPG is a physiological humoral regulator of osteoclast-mediated bone resorption during postnatal life. They also suggest that OPG might play a role in preventing arterial calcification.

In the presence of M-CSF, RANKL induced osteoclast formation from spleen cells in the absence of osteoblasts/stromal cells, and this was abrogated by OPG. OPG also strongly inhibits osteoclast formation induced by a range of osteotropic agents including $1\alpha,25(OH)2D3$, PTH, PGE2, IL-1, and IL-11 in cocultures of osteoblasts/stromal cells and hemapoietic osteoclast progenitors. Interestingly, in contrast to their stimulatory effect on RANKL mRNA expression, M-CSF, PGE2, $1\alpha,25(OH)_2D_3$, and dexamethasone strongly inhibit OPG mRNA expression, suggesting that the regulation of OPG levels is also critical for osteoclastogenesis induced by known osteotropic factors [42, 104–107]. This has led some workers to postulate that prolonged downregulation of OPG may be one of the mechanisms involved in glucocorticoid-induced osteoporosis [108, 109]. OPG also directly inhibits the bone-resorbing activity of isolated mature osteoclasts [110]. As mentioned earlier, treatment of ^{45}Ca-radiolabeled fetal mouse long bones with a soluble form of RANKL also stimulated the release of ^{45}Ca from the bone tissues, which was completely inhibited by simultaneous addition of OPG. This effect of OPG to inhibit bone resorption is due, in part, to its ability to suppress osteoclast survival [111]. In contrast, OPG gene expression and production in marrow stromal/osteoblastic cells are markedly upregulated by TGF-β [32, 33], and this likely explains the powerful effect of TGF-β to inhibit osteoclast formation [29] and to enhance osteoclast apoptosis [31]. Furthermore, TGF-β had no

effect on OPG expression by human dental mesenchymal cells [112], raising the possibility of tissue-specific regulation of OPG expression. *In vivo*, parenteral administration of OPG results in a marked increase in bone mineral density and bone volume associated with a decrease in the number of active osteoclasts both in normal and ovariectomized rats [113]. Serum calcium concentration was also rapidly decreased by parenteral administration of OPG, independent of any changes in urinary calcium excretion, in thyroparathyroidectomized rats whose serum calcium levels were raised acutely by administration of PTH [114]. This suggests that OPG, in addition to its effect on osteoclastogenesis, also affects the function and/or survival of mature osteoclasts.

In human bone biopsies, RANKL/OPG mRNA ratios correlate positively with indices of bone remodeling, with occurrence of hip fractures in women [115], and with osteolysis due to a variety of causes [116]. A single systemic administration of recombinant OPG.Fc suppressed bone resorption in postmenopausal women for up to 6 weeks [117]. Juvenile Paget's disease, a familial disease characterized by greatly accelerated bone turnover, has been shown to be due to inactivating mutations in the gene encoding OPG, which leads to a poorly secreted protein with impaired activity [118–121]. Systemic administration of recombinant OPG.Fc reversed the effect of the disease, further confirming the critical role of OPG in bone remodeling in humans [122]. OPG.Fc also decreased osteolysis in a mouse model of myeloma bone disease [123] and decreased serum calcium levels in tumor-bearing nude mice [113, 124], suggesting that it has therapeutic potential for the treatment of cancer-induced bone disease and malignancy-induced hypercalcemia. Similarly, OPG.Fc prevented bone loss in a rat model of chronic renal insufficiency and secondary hyperparathyroidism [125].

VII.   MACROPHAGE–COLONY-STIMULATING FACTOR AND ITS RECEPTOR, C-FMS

Another cytokine that has clearly been shown to play an important role in bone resorption is monocyte-macrophage–colony-stimulating factor (M-CSF; CSF-1), which is expressed *in vivo* as three different isoforms: as a secreted glycoprotein, secreted proteoglycan, or as a biologically active cell surface expressed glycoprotein (csM-CSF). M-CSF on its own does not stimulate osteoclastic bone resorption in organ culture assays. However, it has been known for over 15 years that it is capable of stimulating the formation of cells with osteoclast characteristics in long-term human marrow cultures, as well as in murine marrow cultures [126–128]. Like IL-1, M-CSF induces fusion of preosteoclasts [59, 129] and prolongs survival of the multinucleated osteoclast-like cells [130]. However, unlike IL-1, M-CSF does not augment their pit-forming capacity when seeded on calcified matrices [59, 130 132]. M-CSF has also been implicated in the bone disease osteopetrosis [2, 3, 133].

With the exception of M-CSF as exemplified by the op/op mouse variant of osteopetrosis and the osteo-petrotic toothless (tl/tl) rat, naturally occurring models of total deficiency of cytokines involved in bone remodeling are rare. In both models of osteopetrosis, frame-shift mutations in the coding region of the csf-1 gene lead to failure of secretion of biologically active M-CSF by stromal cells, osteoblasts, or other accessory cells [134]. Consequently, mature macrophages do not survive for long, and osteoclasts fail to form during the neonatal period, resulting in inadequately remodeled bone demonstrating that M-CSF is required for normal osteoclastogenesis and bone remodeling in the mouse and rat, at least up until the late neonatal period. However, osteoclasts do form beyond the neonatal period with sufficient function to reverse the osteopetrosis by 22 weeks, indicating that M-CSF is not required for osteoclast formation beyond the first few weeks of life. These data show that, in the mouse, secretion of biologically active M-CSF is an absolute requirement for normal osteoclast formation during this period of life. The impairment in osteoclastic resorption and osteopetrosis can be rescued by exogenous administration of M-CSF during the neonatal period in both the op/op mouse and tl/tl rats [2–4]. Moreover, it has also been demonstrated that exogenous administration of vascular endothelial growth factor (VEGF) to neonatal op/op mutants also reverses the osteopetrosis, suggesting that this factor may be responsible, in part, for the spontaneous improvement observed in affected op/op mice as they mature [5]. Interestingly, GM-CSF and IL-3, which are the other major growth factors for cells of the monocyte/macrophage lineage, can also partially reverse the osteopetrosis in these mutant mice [6], implying that other factors are essential for normal bone remodeling in this form of osteopetrosis. Consistent with this hypothesis, enforced expression of bcl-2 in cells of the monocyte/macrophage lineage resulting in their prolonged survival partially rescues the osteoclast defect in csf-1 mutant op/op mice [7] More recent genetic and biochemical studies also demonstrate that treatment of op/op mice with an anti-M-CSF antibody (that neutralizes all three M-CSF isoforms) decreased osteoclast numbers and induced mild osteopetrosis [135]. However, csM-CSF alone cannot completely restore all of the *in vivo* functions of M-CSF, including its effects on bone, demonstrating that the secreted forms are nonredundant [136]. Consistent with this notion, the osteopetrotic defect in the op/op mouse was completely rescued by osteoblast-specific targeting of secreted glycoprotein form of M-CSF [137].

A number of studies have also suggested a role for M-CSF in adult bone remodeling in humans. Pacifici and colleagues provided evidence that IL-1 and TNF levels are increased *in vivo* in estrogen-deficiency states [8]. There is a substantial body of evidence to indicate that production of both the secreted and cell surface forms of M-CSF by bone marrow stromal cells (BMSC) and osteoblasts is regulated by osteotropic cytokines including IL-1 and TNF [138–140]. Furthermore, the increased ability of BMSC to support osteoclast formation in the estrogen-deficient state is via IL-1 and TNF-mediated stimulation of M-CSF production [141]. Indeed, a monoclonal antibody to the signaling receptor for M-CSF (c-Fms) ameliorated the exuberant osteoclastogenesis associated with TNF-induced inflammatory osteolysis in mice [142]. Lastly, it has recently been demonstrated that estrogen blocks M-CSF production by BMSC by directly inhibiting its gene expression [143]. It is likely that the osteoclastogenic effect of M-CSF is, in part, mediated by its ability to downregulate OPG expression [104, 144].

M-CSF mediates its direct effects on osteoclastic bone resorption through a receptor tyrosine kinase, the proto-oncogene known as c-Fms. Presumably, the presence of this receptor tyrosine kinase on osteoclast precursors is responsible for M-CSF mediating its effects on osteoclast formation. There may be a hierarchy of receptor tyrosine kinases involved in normal and pathological bone resorption. Other stimulators of bone resorption that mediate their effects on osteoclasts and receptor tyrosine kinases include epidermal growth factor (EGF), transforming growth factor-β (TGF-β), and platelet-derived growth factor (PDGF). The EGF receptor itself is a receptor tyrosine kinase. PDGF mediates its effects on osteoclast formation presumably through a receptor tyrosine kinase. Since activation of these different receptor tyrosine kinases leads to osteoclast formation and osteoclastic bone resorption, they likely play an important role in bone resorption. However, these receptor tyrosine kinases may not be the only tyrosine kinases involved in bone resorption. In this regard, there is cross-talk between c-Fms and the VEGF receptor 1 (VEGFR/Flt-1), another high-affinity tyrosine kinase, and genetic studies in which an Flt-1 signaling deficiency was introduced in the op/op background demonstrate that this interaction plays a role in osteoclastogenesis [145]. There has been an increase in our understanding of the pathway downstream of c-Fms and how that impacts osteoclast formation and bone remodeling. For example, signals from c-Fms activate Syk tyrosine kinase, which in turn activates Vav3 (a Rho family guanine nucleotide exchange factor). Consistent with that, mice null mutant for the Vav3 gene have increased bone mass and density [146].

Mice deficient in expression of the nonreceptor tyrosine kinase c-src also develop osteopetrosis with failure of osteoclastic bone resorption [147]. However, in these mice, the defect differs from that which occurs in mice with op/op osteopetrosis. In c-src–deficient osteopetrotic mice, there is a failure of ruffled border formation and polarization of the osteoclasts [148]. Nevertheless, osteoclasts form normally. It appears that this receptor tyrosine kinase may, among other things, be involved in osteoclast polarization, which is required for normal osteoclastic bone resorption. Thus, there may be a hierarchy of tyrosine kinases involved in normal osteoclastic bone resorption. Interestingly, c-src has recently been implicated in signaling by M-CSF. Treatment of normal isolated osteoclasts with M-CSF results in increased osteoclast size and cytoplasmic spreading [149–151], and this is associated with increased src kinase activity [151].

VIII. VASCULAR ENDOTHELIAL GROWTH FACTOR

VEGF and its receptors, VEGFR1 and VEGFR2, play important roles in skeletal development. They have been implicated in hypertrophic cartilage remodeling, endochondral ossification, and angiogenesis [152]. It appears that VEGF-mediated capillary invasion provides an essential signal regulating growth plate morphogenesis and triggers cartilage remodeling. Interestingly, it has been shown that, as with M-CSF, a single injection of recombinant VEGF can induce osteoclast recruitment and survival in the neonatal period in osteopetrotic (op/op) mice [5]. Also, recombinant VEGF can substitute for M-CSF in the formation of osteoclast-like cells *in vitro* in the presence of RANKL [5]. Although these cytokines are not related, these data suggest that M-CSF and VEGF have overlapping functions with regard to osteoclastic bone resorption.

Evidence also exists that VEGF may have direct effects on osteoblasts [153]. Midy and Plouet (1994) demonstrated that VEGF was able to bind to osteoblasts *in vitro* and induce both migration and alkaline phosphatase expression. More recently, it has been shown in primary calvarial cultures, that while VEGF has no effect on osteoblast proliferation, it is capable of inducing mineralization [154]. In addition, in an *in vitro* organ culture system, VEGF was able to stimulate bone growth as evidenced by calvarial thickening [154]. It has been proposed that VEGF production by endothelial cells can increase expression of BMP-2 and BMP-4, resulting in stimulation of osteoblast differentiation and bone formation [155, 156]. However, a clear role for VEGF in adult bone remodeling *in vivo* remains to be demonstrated.

IX. TUMOR NECROSIS FACTOR

TNF and TNFR-related proteins form a large family of related cytokines which share unique attributes that couple them directly to common signaling pathways, involving cell proliferation, differentiation, and survival. The wide range of biological effects associated with TNF derives from the interaction of this cytokine with its two receptors TNFR1 and TNFR2, TNFR1 being the most crucial for the majority of TNF's biological activities.

TNF is known to enhance programmed cell death (apoptosis) [157] and activation of the transcription factor nuclear factor kappa B (NF-κB). In fact, the most potent NF-κB activators are IL-1 and TNF. When NF-κB activity is decreased, there is an increase in cellular susceptibility to TNF-induced apoptosis, whereas activation of NF-κB protects against apoptosis. TNF binds to its receptor and recruits a protein called TNF-associated receptor death domain (TRADD).

TNF has been shown to stimulate osteoclastic bone resorption both *in vitro* and *in vivo* [158–160]. RANKL, the TNFR-related RANK receptor, and OPG are crucial factors in the regulation of osteoclastogenesis. Both RANKL and TNF are required for osteoclast formation, and their actions are synergistic. TNF inhibits osteoblast function by several mechanisms: decreased osteoblast differentiation [161], inhibition of the production of matrix proteins [160], and induction of osteoblast resistance to active metabolites of vitamin D [162]. TNF effects *in vivo* have been demonstrated using Chinese hamster ovarian (CHO) cells transfected with the human TNF-α gene. Nude mice bearing tumors that express TNF in large amounts develop hypercalcemia and demonstrate increased osteoclastic bone resorption [158]. TNF stimulates cells at all stages in the osteoclast lineage, in much the same way as does IL-1. TNF has been implicated in hypercalcemia in several human and animal tumors associated with the humoral hypercalcemia of malignancy [45, 163–165]. Antibodies to TNF reduce the blood-ionized calcium in these models as well as some of the other paraneoplastic syndromes associated with malignancy, including leukocytosis and cachexia. In these models, TNF is not produced by the tumor cells but rather by the host immune cells, possibly as part of the immune defense mechanism generated by the presence of the tumor [163].

It is known now that increased production of TNF in postmenopausal women [166] is a contributing factor for the bone involution observed in estrogen deficiency, as evidenced by amelioration of the bone loss caused by ovariectomy in rats and mice when TNF activity is blocked [11, 15, 16].

Lymphotoxin alpha (LTα, formerly known as tumor necrosis factor-β) is also produced in the early stages of an inflammatory reaction. The major sources of this cytokine are macrophages, monocytes, and T cells. LTα is also expressed by B lymphocytes and natural killer cells, suggesting an important role of LTα in the immune response process.

Early studies identified LTα as the osteolytic factor responsible for the osteolytic bone lesions present in myeloma. In these studies LTα was found in the supernatant fractions in cultures from myeloma cell lines and fresh myeloma bone marrow [167]. When given by injection or infusion, lymphotoxin causes hypercalcemia and increased bone resorption in rodents [167]. TNF-α and LTα are potent inducers of interleukin-6 (IL-6) production, a major growth factor for myeloma cells [168]. However, the role of LTα in myeloma bone disease has been downplayed by more recent studies failing to find significant differences in the amount of this cytokine in supernatants derived from bone marrow cultures or fresh bone marrow plasma derived from myeloma patients compared to controls. Other factors have now been identified that seem to play a bigger role in the osteolysis associated with multiple myeloma: RANKL and the chemokine macrophage inflammatory protein-1 (MIP-1) [169].

X. INTERLEUKIN-6 (IL-6)

Interleukin-6 is a multifunctional cytokine that has a number of unique effects in bone. IL-6 and its soluble receptor (sIL-6R) activate the glycoprotein IL-6(gp)-130 signaling pathway [170]. Other cytokines use the same common signal transducer, gp130 [171], and often have similar functions. Signaling by the IL-6–type cytokines involves binding to specific receptors and activation of Janus kinases and transcription factors of the STAT family, which in turn stimulate RANKL expression in stromal cells/osteoblasts [172]. Although IL-6 was originally regarded as a key pathway for the regulation of osteoclastogenesis [173], subsequent experiments in vivo with IL-6 knockout (−/−) mice failed to demonstrate an essential role of IL-6 for normal bone resorption and homeostasis [55].

IL-6 is known to exert multiple effects in the bone microenvironment and has been implicated in osteoporosis, rheumatoid arthritis, Paget's disease, and multiple myeloma. IL-6 is produced by both stromal cells and osteoblastic cells in response to systemic hormones such as PTH, PTHrP, vitamin D_3 (calcitriol), and cytokines (TGF-β, IL-1, and TNF-α) [174–176]. The stimulation of IL-6 expression in these cells by TNF occurs through a transcriptional NF-κB–dependent mechanism [177]. IL-6 stimulates osteoclast differentiation in the presence of soluble IL-6R [170], and it mediates the effects of IL-1 and TNF on osteoclast formation [26].

IL-6 can enhance the effects of other cytokines and systemic hormones on bone resorption both in vitro and in vivo. We have previously reported that IL-6 not only has synergistic effects with IL-1 and PTH in organ culture and cell culture systems for assessing bone resorption, but it also has synergistic effects on the bone resorbing capacity of PTH in vivo. This has been shown using CHO cells transfected with PTH and with CHO cells transfected with IL-6 [176]. The effects of both agents together are much greater than either agent alone.

The effects of IL-6 on bone resorption in vivo alone are modest. We have found that when IL-6 is expressed by CHO cells transfected into nude mice, there are only small effects on serum calcium, and bone resorption is not observed unless enormous amounts of circulating IL-6 are present [178]. This is in contrast to other cytokines, such as IL-1, TNF-α, and lymphotoxin.

In osteoblasts, IL-6 seems to enhance osteoblast differentiation [179]. It is possible that IL-6/sIL-6R might have an effect on osteoblast proliferation and apoptosis and that this effect is mediated through other factors. In rat calvariae, IL-6 has been shown to stimulate IGF-1 and BMP2 [80, 180].

The role of IL-6 in the pathogenesis of multiple myeloma is still not clearly understood. IL-6 levels have been associated with the presence of bone lesions and the progression of the disease by some investigators [181], while others found very little effect on tumor burden [182]. Cells of the osteoblast lineage release significant amounts of IL-6 in response to stimulation by myeloma cells [183], possibly contributing to the proliferation and survival of the tumor cells.

Cell–cell interactions between myeloma cells and marrow stromal cells result in upregulation of IL-6 production by the marrow stromal cells. In addition, IL-6 can act as an antiapoptotic factor increasing the survival of myeloma cells [184]. IL-6 has also been shown to induce expression of MIP-1α in myeloma cells and may play a role in the resistance of myeloma cells to chemotherapy [185].

IL-6 has also been implicated in the bone loss associated with postmenopausal osteoporosis. Jilka and colleagues [12] suggested that excess production of IL-6 may account in large part for the bone loss associated with ovariectomy and estrogen withdrawal. These workers have shown that neutralizing antibodies to IL-6 reduce osteoclastic resorption associated with ovariectomy in mice. They propose that IL-6 production by stromal cells and cells in the osteoblast lineage is enhanced in the presence of estrogen deficiency. There were no significant differences in osteoclast numbers between IL-6–deficient and wild-type mice. Furthermore, ovariectomy did not induce any change in osteoclast number in IL-6–deficient mice compared to wild-type mice [13]. These data suggest that IL-6 may play a more important role in osteoclast development in pathological conditions such as estrogen-depleted states rather than in physiological bone turnover. This is further emphasized by observations of morphologically normal osteoclasts in bones of gp130-deficient mice, at least in the early neonatal period [186]. IL-6 seems not to be essential for bone remodeling in physiological conditions, but plays an important role in osteoblast generation in conditions where there is high bone turnover.

Because IL-6 potentiates the bone-resorbing actions of factors such as PTH and PTHrP, its increased production may be important in some disease states where there is overproduction of these factors, which presumably lead to increased levels of IL-6 in the bone microenvironment. This may be true in some patients with severe primary hyperparathyroidism and secondary hyperparathyroidism and in some malignancies. In each of these conditions, there is excess production of peptides that induce IL-6 production in osteoblasts, and experimentally there is good evidence to believe that IL-6 production in bone may enhance the bone-resorbing effects of other factors such as PTH or PTHrP on murine osteoclasts. IL-6 knockout mice have secondary hyperparathyroidism despite reduced biochemical markers of resorption [187]. In humans, circulating levels of IL-6 and its receptor are elevated in states of PTH excess, correlate strongly with markers of bone resorption, and revert to normal following the correction of hyperparathyroidism [188].

In rodents a low-dose PTH infusion increased circulating levels of IL-6 and biochemical markers of bone resorption [189]. When a neutralizing antibody to IL-6 was used, there was a reduction in the levels of biochemical markers of bone resorption in response to PTH infusion with no change in markers of bone formation, suggesting a systemic regulation of IL-6 by PTH in mice and reinforcing the role of IL-6 as an important mediator of the bone-resorbing actions of PTH *in vivo*.

In a search for naturally occurring inhibitors of IL-6, conditioned media harvested from human and murine immune cells were examined. It was found that the monocyte-macrophage cell lines U937 and P388DI produce a biological activity that impaired the proliferative effects of IL-6 on bone [190]. These factors were purified to homogeneity, and it was found that they could be ascribed to a factor in the TGF-β superfamily, activin A. This cytokine has previously been shown to be present in considerable amounts in the bone matrix and acts as a stored, endogenous inhibitor of IL-6. More recently, other groups have shown the inhibitory effect of Activin A on IL-6 in chronic inflammatory diseases (arthritis) [191].

XI. INTERLEUKIN-15 (IL-15), INTERLEUKIN-17 (IL-17), AND INTERLEUKIN-18 (IL-18)

IL-15 is an IL-2–like cytokine produced by T cells, activated monocytes, dendritic cells, fibroblasts, and endothelial cells that binds the same receptor as IL-2. IL-15 has been reported to stimulate the formation of TRAP-positive, calcitonin receptor–positive multinucleated osteoclast-like cells in rat bone marrow cultures that resorb calcified matrices [192]. IL-15 and IL-2 also share some receptor components, but IL-2 does not stimulate the formation of osteoclast-like cells. Although IL-15 is a potent inducer of TNF, this effect to stimulate the formation of osteoclast-like cells is not blocked by a specific anti-TNF-neutralizing antibody. IL-15 levels in synovial fluids of rheumatoid arthritis patients are markedly elevated [193], raising the possibility that this cytokine may play a role in the local destruction of bone associated with chronic inflammatory disease. It has been shown that both the membrane-associated and soluble forms of IL-15 are expressed mainly in fibroblast-like synoviocytes of the lining layer. IL-1β and TNF-α induce IL-15 in fibroblast-like synoviocytes; this in turn induces the production of IL-17 in T cells, which potentiates IL-1β and TNF-α production in monocyte-macrophages, completing the proinflammatory cycle. IL-15 may be a therapeutic target in rheumatoid arthritis. A soluble fragment of the α chain of IL-15RI that antagonizes IL-15 has been reported to prevent the development of collagen-induced arthritis (CIA) in mice. In this model of rheumatoid arthritis, the IL-15RI α chain attenuated the clinical and histological abnormalities and blunted the cell-mediated and humoral responses to type II collagen [194].

IL-17 is also a product of activated T cells that induces the production of prostaglandin E2, TNF-α, IL-1β [195], IL-12, and IL-6 by bone marrow stromal cells. This cytokine has been found to act on osteoblastic cells; it stimulates cyclooxygenase-2–dependent

PGE2 synthesis, inducing RANKL gene expression in stromal/osteoblastic cells and subsequently causing differentiation of osteoclast progenitors into mature osteoclasts [196]. Furthermore, although it has no effect on either basal or IL-1–induced bone resorption in bone organ cultures, IL-17 markedly enhances TNF-α–induced osteoclastic bone resorption in fetal mouse long bones [197]. IL-17 plays a significant role in the bone destruction associated with rheumatoid arthritis. In human rheumatoid arthritis bone explants, IL-17 increased bone resorption and decreased bone formation [198]. The levels of IL-17 are also markedly elevated in rheumatoid arthritis synovial fluids compared to osteoarthritis synovial fluids from normal controls. It has also been reported that IL-17 is spontaneously produced in organ cultures of synovial tissues derived from rheumatoid arthritis patients [199].

IL-18 is a proinflammatory cytokine produced by marrow stromal/osteoblastic cells [57]. IL-18 inhibits osteoclast formation in the presence of osteoclastogenic agents including 1α,25-dihydroxyvitamin D_3, PGE2, PTH, IL-1, and IL-11. Unlike IL-15 and IL-17, IL-18 inhibits osteoclast formation in cocultures of murine spleen cells and osteoblasts, an effect likely mediated via T-cell–produced GM-CSF, as neutralizing antibodies to GM-CSF abolished osteoclast formation. IL-18, which is homologous to IL-1, binds to IL-1R–related protein 1 (IL-1RrP-1), which is in turn highly homologous to IL-1R. Both IL-1R and IL-1RrP-1 signal through IL-1R–associated kinase (IRAK) and both recruit TRAF6 and activate NF-κB [200]. Recently, it has been reported that IRAK-4 is an essential component of the IL-18 signaling cascade [201]. It was also reported that this cytokine increases the expression of OPG mRNA in stromal cells and osteoblasts [202]. IL-18 is effective in inhibiting bone destruction in murine models of breast cancer as well as lung metastasis in bone [203, 204]. These results suggest that therapeutic strategies utilizing this information may be useful for reducing pathological bone loss.

These pro- and anti-inflammatory cytokines seem to play an important part in the initiation and perpetuation of chronic inflammatory processes, which makes them potential targets for anti-inflammatory therapy in diseases like arthritis.

XII. BONE MORPHOGENETIC PROTEINS

A. Introduction

Multipotential mesenchymal cells have the capacity to undergo the commitment process to give rise to progeny with more limited or monopotential differentiation capacity, including osteoblasts and chondroblasts. Maturation of osteoblasts is important for bone formation during bone remodeling, and dysfunction of osteoblasts results in a reduction of bone formation and osteoporosis. Mechanisms responsible for commitment and specification of uncommitted mesenchymal precursor cells to the osteoblast lineage are not fully understood. Bone morphogenetic proteins (BMPs) appear to play regulatory roles in the commitment of mesenchymal precursor cells to the osteoblast lineage [205].

BMPs, structurally related to the transforming growth factor-β (TGF-β) superfamily [206, 207], were originally identified from bone matrix using an ectopic bone formation assay [208]. When BMPs are implanted subcutaneously or intramuscularly in mice or rats, they induce massive amounts of new cartilage and bone at implantation sites. BMPs have been shown to induce primary embryonic limb and mesenchymal precursor cells to differentiate into mature chondroblasts and osteoblasts [209–211]. These results indicate the regulatory roles of BMPs in the commitment of mesenchymal cells to the osteoblast and chondroblast lineages. More than 20 BMP proteins have recently been identified. These BMPs mediate their functions through type I and type II serine/threonine kinase receptors. BMPs and BMP receptors play critical roles in osteoblast differentiation and bone formation. Recently, BMP-2, one of the most important BMP family members, was recognized as an osteoporosis-associated gene [212]. In this subsection of this chapter, we review recent progress in understanding the biological functions of BMPs in bone.

B. BMP Signal Transduction

BMP signals are mediated through type I and type II BMP receptors, which are members of the TGF-β receptor superfamily. Both type I and type II BMP receptors have inducible intracellular serine/threonine kinase activity. Three type I receptors and three type II receptors have been identified. They include type IA BMP receptor (BMPR-IA), type IB BMP receptor (BMPR-IB), type I Activin receptor (ActR-I), type II BMP receptor (BMPR-II), type II Activin receptor (ActR-II), and type IIB Activin receptor (ActR-IIB) [213–218]. These receptors have different binding affinities to BMP ligands and have been found to be responsible for mediating various functions of the BMPs. Whereas BMPR-IA, BMPR-IB, and BMPR-II are specific to BMPs, ActR-IIA and ActR-IIB are also the signaling receptors for activins. ActR-IA has been shown to bind both BMP-7 and Activin when expressed in COS cells. In embryonic P19 cells and MC3T3-E1 cells, endogenous ActR-IA mediates only BMP signaling, but not Activin signaling [219]. No cross interaction between BMP ligands and TGF-β receptors has yet been demonstrated. Recently, a new coreceptor for BMPs, namely Dragon, was identified [220]. Dragon directly binds to BMP2 and BMP4, but not to BMP7 or other TGF-β ligands. Dragon also associates with type I and type II BMP receptors. The enhancing action of Dragon on BMP signaling is reduced by Noggin (a BMP antagonist discussed in Section XII.E) and dominant negative forms of Smad1, BMPR-IA, and BMPR-IB, indicating that the action of Dragon is BMP ligand and Smad dependent.

Upon binding with BMP ligands, the receptors form a heterotetrameric receptor complex composed of two pairs of type I and type II receptor complexes and transduce external BMP signal to an intracellular phosphorylation cascade. In the TGF-β receptor system, ligands bind to type II receptors in the absence of type I receptors, but type I receptors can bind ligands only in the presence of type II receptors. Similarly, in the BMP receptor complex, type II receptors are primary binding proteins for ligands, and type I receptors transduce BMP signals to Smads. However, evidence has shown that BMPs, unlike TGF-β, also bind to type I receptors in the absence of type II receptors. In the presence of type II receptors, the binding of BMPs to type I receptors is accelerated [217, 218]. There are clear domains on BMP-2, -4, and -7 ligands that bind specifically to the type I receptors [221]. After ligand binding, type II receptor kinase activity transphosphorylates the type I receptor through an SGSGS motif (GS domain) [214–218]. Phosphorylation of the GS domain is required for the activation of type I receptor serine/threonine kinase as ligand-mediated responses are impaired after mutation of serine and threonine residues in the GS domain [222, 223]. In contrast, a site mutation in the GS domain (Gln→Asp) of the type I BMP receptor results in a receptor with a constitutively activated kinase. In these mutants, signals are transduced from the type I BMP receptor in the absence of ligand and the type II BMP receptor [224].

The type I receptors act as effectors for BMP signal transduction [225]. The ligand-activated type I receptors recruit and phosphorylate BMP pathway–restricted Smads, including Smad1 [224], Smad5 [179], and Smad8 [226]. In TGF-β signaling, a TGF-β–specific Smad (Anchor for Receptor Activation [SARA]) has been identified [227]. SARA directly binds to type I receptors and Smad2/Smad3 and plays dual roles in regulating Smad phosphorylation. Recently, a BMP-specific SARA (SARAb) was reported to function in BMP signaling by regulating phosphorylation and dephosphorylation of Smads [228]. After phosphorylation, the Smads are released from the receptor complex and recruit a common mediator Smad4, forming a

nuclear translocation complex. This complex migrates into the nucleus and transactivates specific target genes. These Smad proteins share two functional regions: a conserved C-terminal domain (MH2 domain) and a conserved N-terminal domain (MH1 domain). These conserved domains are respectively responsible for the interaction with the type I receptor and the interaction with the target DNA in the transcriptional complex [229]. Smad6 and Smad7 are two inhibitory Smad proteins in the BMP signaling pathway [230]. Smad6 and Smad7 lack the C-terminal SSXS motif, and diverge from the rest of the Smads in the N-terminal region. Multiple mechanisms are involved in the inhibition of BMP signaling by Smad6. Smad6 inhibits BMPR-IB–activated phosphorylation of Smad1 [231]. Smad6 also specifically competes with Smad4 for binding to receptor-activated Smad1, yielding an inactive Smad1–Smad6 complex. Thus, Smad6 is a Smad4 decoy [232]. Recently, Smad6 was found to play a role in Smurf1-mediated Runx2 degradation [233].

In the nucleus, Smad1, Smad5, and Smad8 bind to DNA by their MH1 domain and regulate transcription through BMP-responsive elements (BRE) in the target genes. The consensus BRE, GCCGnCGC, responds only to BMP stimulation, but not to TGF-β or activin [234]. The immediate downstream target genes for BMP signaling during osteoblast and chondroblast differentiation have yet to be identified. Recent studies have suggested that several osteogenesis-related genes are BMP/Smad-responsive genes. They include Runx2, Osterix, Dlx5, ZNF450, Id1, and SOST [235–242]. To regulate gene transcription, other transcription cofactors are needed for Smad function. Numerous such DNA-binding cofactors have been found to interact with BMP-responsive Smads and coregulate gene transcription with Smads. Cofactors p300 or CBP can bind to the MH2 domain of Smad1, Smad5, or Smad4 and activate transcription through their histone acetylase activity [243, 244]. Runx2 is an osteoblast-specific transcription factor. Mice with a targeted disruption in Runx2 die after birth with complete lack of both endochondral and intramembranous ossification, caused by a maturational arrest of osteoblasts [245–247]. This demonstrates that Runx2 plays an essential role in osteogenesis. Recent evidence has indicated that the cooperation between Runx2 and BMP-activated Smads in the nucleus is required to induce expression of genes related to the osteoblast phenotype. Two mechanisms are likely responsible for the integration of Runx2 and BMP/Smad in two distinct pathways. By directly interacting with Smads [248, 249], (1) Runx2 mediates intranuclear targeting of Smad1 and Smad5 by recruiting these Smads to the nucleus [250]; (2) Smads and Runx2 form a complex with DNA through BRE and

Runx2-responsive element OSE2 on the target gene and coactivate transcription [251]. Ski and Tob are repressors of transcription [252]. Ski can bind to Smad1, Smad5, and Smad4 and represses gene transcription [253, 254]. Tob is also known to associate with these BMP-regulated Smads and negatively regulates osteoblast proliferation and differentiation by suppressing the interaction of Smad with receptor and the transcriptional activity of Smads [255, 256]. Tob deficiency also superenhances osteoblastic activity after ovariectomy [257]. Another transcription repressor that can directly bind to Smad1 is homeobox c-8 (Hoxc-8). The Smad1 interaction with Hoxc-8 dislodges Hoxc-8 inhibition on DNA, resulting in the induction of gene expression. Smad6 was shown to interact with Hoxc-8 as part of the negative feedback circuit in the BMP signaling pathway [258, 259]. Smad6 also recruits transcription corepressor C-terminal binding protein (CtBP) to repress BMP-induced transcription [260].

Studies on Smad deletion that turns off BMP signaling result in interesting findings. Smurf1 (Smad ubiquitin regulatory factor 1), a member of the Hect family of E3 ubiquitin ligases, was found to selectively interact with receptor-regulated Smad1 and Smad5 to trigger their ubiquitination and consequent proteolytic processing [261–263]. In addition, recent data have shown that Smurf1 also mediates Runx2 degradation [251]. Overexpression of a Smurf1 transgene in bone driven by Col1a1 promoter inhibits postnatal bone formation in mice [264]. Thus, Smurf1 is a specific negative regulator of BMP/Smad signaling.

C. Role of BMP Signaling in Osteogenesis and Chondrogenesis

BMPs play important roles in osteoblast and chondrocyte differentiation. BMP-2 stimulates osteoblast differentiation in primary osteoblastic cells and in cell lines derived from osteogenic tissues [265–270]. BMPs also induce nonosteogenic precursor cells to differentiate into cells with the osteoblast phenotype. For example, BMP-2 induces myoblast C2C12 cells to differentiate into osteoblasts, and this effect is mediated by Smad1 and Smad5 [271, 272]. BMPR-IB is a critical receptor component for osteoblast differentiation, while BMPR-IA is responsible for differentiation toward adipocytes. Overexpression of dominant negative BMPR-IB in osteoblast precursor 2T3 cells blocked BMP-2–induced ALP activity and mineralized matrix formation in these cells [273]. BMPs act as autocrine and paracrine factors during osteoblast differentiation [269, 270]. An accepted working model for the molecular mechanism for BMP-2–induced osteoblast differentiation that is

supported by numerous recent studies is BMP-2-Dlx5-Runx2/Osx switches. As we described in BMP signaling, Runx2 has been widely recognized as a master transcription factor that plays a pivotal role in osteoblast marker gene expression [242, 247]. Osterix (Osx) is a recently identified zinc-finger–containing transcription factor [240]. In Osx null mice, no bone formation occurs. Since Runx2 is an earlier osteogenic transcription factor and Osx mainly functions during terminal differentiation of osteoblasts, and Osx is not expressed in Runx2 null mice, Osx acts downstream of Runx2 [240]. Recent studies have found that BMPs stimulate expression of Runx2 and Osx in osteoblasts. BMP-2 and BMP-7 increase Runx2 mRNA in pluripotent mouse fibroblast C3H10T1/2 cells and in mesenchymal precursor 2T3 cells [242, 250, 273, 274]. BMP-2 increases Osx expression in primary osteoblasts that is blocked by Noggin [236, 240]. However, pretreatment with protein synthesis inhibitor cycloheximide blocks the BMP-2–induced expression of Runx2 and Osx [275, 276], suggesting that these osteogenic master genes are not the direct target of BMP signaling. Dlx5 is a homeodomain transcription factor that is co-expressed in the developing skeleton with BMP-2 and BMP-4 [277]. Forced expression of Dlx5 in cells leads to osteocalcin expression and mineralized matrix formation [278, 279]. Null mutation of Dlx5 in mice causes delayed cranial ossification and abnormal osteogenesis [280]. It is found that BMP-2 treatment, overexpression of constitutively active BMPR-IA or IB, and overexpression of Smad1 or Smad5 upregulate Dlx5 expression [277]. Thus, it is possible that BMP-2 regulates Runx2 and Osx expression through Dlx5. This model is supported by the most recent studies in which Runx2 and Osx expression are completely inhibited by Dlx5 antisense, and Runx2 expression is induced by overexpression of Dlx5 even in the absence of BMP-2 [281]. These results suggest that Dlx5, as an upstream regulator of Runx2, and Osx are indispensable mediators of BMP-2–induced osteoblast differentiation.

The skeletal system forms in large part through endochondral ossification, in which mesenchymal cells condense and differentiate into chondrocytes [282]. BMP signaling is indispensable for normal chondrogenesis [283, 284]. BMP-2, BMP-6, and BMP-7 have been shown to induce chondrogenic commitment of pluripotent mesenchymal cell lines, such as C3H10T1/2, into chondrocytes [285–288]. BMP antagonist Noggin suppresses the formation of mesenchymal condensations, whereas cartilage primordium is enlarged in Noggin knockout mice [289, 290]. BMPs are involved in the regulation of both early and terminal chondrocyte differentiation [283, 291]. BMPs promote Sox9 and type II and type X collagen expression in C3H10T1/2 and ATDC5 cells, and the implantation of BMPs and Noggin beads leads to marked reduction in these genes [292, 293]. Mice that lack GDF5, a member of the BMP family, develop brachypodism and other appendicular abnormalities that resemble the human chondrodysplasia disease Hunter-Thompson type and Grebe type, in which GDF5 null mutations were identified [294–296]. GDF5 binds to BMPR-IB with high affinity. In BMPR-IB null mice, proliferation of prechondrogenic cells and chondrocyte differentiation in the phalangeal region are markedly reduced. The appendicular skeletal defects are more severe in BMPR-IB and BMP-7 double knockout mice [297]. Recently, it was reported that BMPR-IA and BMPR-IB double deficiency in mice results in a severe generalized chondrodysplasia, suggesting that BMPR-IA and BMPR-IB have overlapping functions and are essential for chondrogenesis *in vivo* [298]. In addition to roles in early chondrogenesis, BMPs have important functions in the growth plate at later stages. Most of the BMPs (BMP-2, -3, -4, -5, and -7) are expressed in the perichondrium. BMP-7 is also expressed in proliferating chondrocytes. BMP-2 is mainly expressed in hypertrophic cartilage and overlaps with BMP-6 expression [299–301]. Moreover, BMPR-IA and BMPR-IB have a distinct and overlapping expression pattern in the growth plate. The few cartilage condensations are delayed and never form an organized growth plate in mice with double mutations in these receptors [298]. These results indicate that BMP signaling may have multiple functions in the growth plate. BMP-2 has been shown to accelerate longitudinal bone growth by stimulating growth plate chondrocyte proliferation and chondrocyte hypertrophy and cartilage matrix synthesis, which are blocked by Noggin. The stimulation of BMPs and the inhibitory effects of Noggin on type X collagen in hypertrophic differentiation at the growth plate indicate that, in addition to promoting initial hypertrophic differentiation, BMP signaling plays an essential role in regulating the most terminal stages of cell differentiation in the growth plate [301, 302]. This notion was recently confirmed in chondrocyte-targeted Smad4 null mice. The abrogation of Smad4 in chondrocytes results in dwarfism with a severe disorganized growth plate characterized by an expanded resting zone of chondrocytes, reduced chondrocyte proliferation, and accelerated hypertrophic differentiation [303]. It is well known that the Indian hedgehog (Ihh)/PTHrP pathway plays a broad role in chondrogenesis and endochondral bone formation [283, 304]. Evidence from recent studies on the cooperation between BMP and Ihh/PTHrP pathways has demonstrated that synergistic interaction of these two pathways is required for normal chondrogenesis, in which Gli transcription factors exert a key function [300, 301, 305–308].

Several pieces of evidence indicate that, in addition to skeletal development, BMP signaling is also an important mechanism for maintaining postnatal bone formation and bone remodeling. Deficiency in ActR-IIB causes multiple skeletal and tooth defects, and overexpression of dominant negative BMPR-IB in osteoblasts significantly reduces bone formation in adult mice [309, 310]. Results from a model of postnatal osteoblast-specific disruption of BMPR-IA have suggested that BMPR-IA signaling regulates postnatal osteoblast function and bone remodeling [311]. In BMP-7–deficient mice, skeletal abnormalities are identified in discrete areas: the rib cage, the skull, and the hindlimbs [312], suggesting that BMP-7 plays a role in bone development and bone formation. Preliminary data from mice in which BMP-2 and BMP-4 are conditionally deficient in osteoblasts have shown reduced bone mass in adult mice (Harris, unpublished observations). These findings demonstrate critical roles for the BMPs in both bone development and postnatal bone formation.

D.   Regulation of BMP Gene Expression

Osteoporosis results from an imbalance between bone resorption and bone formation, resulting in net bone loss. Although the mechanisms are not entirely clear, the defect in bone formation is likely related to decreased availability or effects of bone growth factors, such as BMPs. Of the more than 20 BMP members, BMP-2 has been most extensively studied. Skeletal aging studies have shown that both anabolic activity and gene expression of BMP-2 are decreased [313–317] in caged animals with osteopenia, suggesting that the decrease in BMP-2 function may be one of the molecular mechanisms responsible for osteoporosis. A recent study has found a link between osteoporosis and specific polymorphisms in the BMP-2 gene, implicating BMP-2 as an osteoporosis-associated gene [212].

BMP-2 is an autocrine and paracrine growth factor and is expressed from early stages of embryonic development through adulthood, primarily in bone-forming tissues [299]. During osteoblastic differentiation, BMP-2 mRNA is induced and maintains the sustained phenotype of mature osteoblasts [270, 318]. The mouse BMP-2 gene has been mapped to chromosome 3 and contains an 11-kb transcription unit and 3 exons [319, 320]. Our group has previously characterized the 5′ flanking region of the mouse BMP-2 gene [320, 321]. Using a mouse BMP-2 promoter–luciferase reporter gene, we have identified different groups of compounds that stimulate BMP-2 expression, including proteasome inhibitors, statins, and microtubule inhibitors. *In vitro* and *in vivo* studies have shown that proteasome inhibitors and statins not only promote osteoblast differentiation, but also increase bone formation in mice, and Noggin blocks the anabolic effects of these compounds [322, 323]. However, the precise mechanisms responsible for compound-induced BMP-2 gene transcription remain to be elucidated.

Previous studies have indicated that BMP-2 gene regulation during limb morphogenesis and osteoblast differentiation may involve multiple mechanisms and signaling pathways. Studies on mouse mesenchymal stem cells have found that estrogens activate BMP-2 transcription, which requires ERα and ERβ acting via variant estrogen-responsive element binding sites in the promoter, with ERα being the more efficacious regulator. Estrogenic compounds may enhance bone formation by increasing the transcription of the BMP-2 gene [324]. In studies on human mesenchymal stem cells, PGE2 has been shown to induce BMP-2 expression, an effect that is blocked by the PGE2 inhibitor NS-398. Stimulation of BMP-2 gene expression by COX-2–induced PGE2 is mediated via binding to the EP4 receptor [325, 326]. Analysis of homeobox a13 (HOXa13) expression reveals a pattern of localization overlapping with sites of reduced BMP-2 expression in HOXa13 mutant limbs. A novel series of BMP-2 enhancer regions has been identified to directly interact with the HOXa13 DNA–binding domain and activate gene expression [327]. Retinoic acid (RA) has been found to elevate endogenous BMP-2 transcription across species from chicken and rodents to humans. In F9 embryonal cells, retinoic acid, combined with cAMP analogs, sharply induces the BMP-2 mRNA during the differentiation. The RA-enhanced transcription is indirect, since BMP-2 promoter lacks a classical retinoic acid–responsive element. The mechanism likely involves the interaction of retinoic acid receptor and SP1 protein. An Sp1 site that is conserved between the species contributes to the retinoic acid responsiveness of the BMP-2 promoter [328–330]. In addition, interleukin-1β, -6, interferon-α, and 1,25(OH) vitamin D_3 have been reported to play a role in BMP-2 gene regulation [326, 331].

Evidence has shown that BMP-2 is an autoregulatory protein [270, 321, 332, 333]. Mouse BMP-2 gene transcription is directed by the proximal promoter element [333]. In cultures of osteoblast precursor 2T3 cells that undergo mineralized matrix formation in the presence of BMP-2, BMP-2 stimulates its own expression [334]. This autoregulation is mediated by the PI 3 kinase/Akt pathway in a BMP receptor–activated–Smad-dependent pattern [332].

Hedgehog (Hh) signaling has an essential function in osteogenesis and chondrogenesis. Gli proteins, which are zinc finger transcription factors, mediate the Hh signal to target genes. Null mutations of Gli genes cause severe skeletal abnormalities [335]. In the absence of Hh, Gli3 protein is proteolytically processed to form a C' terminal truncated Gli3 (trGli3). Gli1 and Gli3 are capable of inducing BMP-4 and BMP-7 expression [308]. Results from our recent studies reveal that truncated Gli3 formed by proteasomal processing functions as a powerful repressor of BMP-2 transcription and calvarial bone formation [322]. On the other hand, Gli2 is a potent enhancer of BMP-2 gene transcription. Overexpression of Gli2 in osteoblasts increases BMP-2 expression and osteoblast differentiation, which is attenuated by interference with Gli2 siRNA. In Gli2 deficient mice, BMP-2 expression in the growth plate is significantly reduced. Gli2 and Gli3 regulate BMP-2 transcription by interacting with the BMP-2 promoter through specific Gli responsive elements [336–338]. These findings provide further evidence for cross-talk between BMP and Hh pathways in chondrogenesis. In another recent study, we found that the transcription factor NF-κB plays an important role in BMP-2 gene regulation. Mice with double mutations in p50/p52 genes reveal decreased chondrocyte numbers in the proliferating zone of the growth plate, where BMP-2 expression is significantly reduced. The promoter assay in TMC23 chondrocytic cells demonstrated that NF-κB transactivates the BMP-2 gene through NF-κB responsive elements in the promoter [339].

E. Extracellular BMP Antagonists

Negative regulation of BMP signaling plays a critical role in governing BMP actions in skeleton. In addition to intracellular negative regulators such as Smad6, Tob, Ski, and Smurf1, as described previously, certain classes of BMP-inducible extracellular polypeptides have recently been recognized as BMP antagonists [340–343]. These secreted BMP antagonists directly bind to BMPs and prohibit BMPs from binding to their receptors. These polypeptides share a cystine-knot structure that is conserved in a superfamily including TGF-β and BMPs. Based on the size of the cystine-knot, the BMP antagonists are divided into three groups: (1) Noggin and Chordin; (2) Twisted gastrulation (Tsg); and (3) the DNA family composed of Gremlin, PRDC, Coco, Cer1, Dan, and sclerostin (SOST) [344, 345].

Noggin, a secreted homodimeric glycoprotein, was originally identified in the Spemann organizer of *Xenopus* embryos and plays a role in dorsal-ventral patterning during embryonic development [346–348]. Noggin binds to BMP-2 and BMP-4 with high affinity and to other BMP family members with varying degrees of affinity. Noggin does not bind to TGF-β [349]. By interfering with the binding between BMPs and the receptors, Noggin blocks BMP activity in undifferentiated and differentiated osteoblastic cells by inhibiting osteoblastogenesis [350–353]. Homozygous Noggin-null mice are embryonic lethals and have serious skeletal abnormalities, including excess cartilage and joint lesions. Although heterozygous null mice appear normal, heterozygous missense mutations in the human Noggin locus result in individuals with proximal symphalangism and multiple synostosis syndrome, characterized by joint fusions [290, 354, 355]. To further investigate the function of Noggin in skeletal development and postnatal bone formation, Noggin transgenic mouse models have been recently established [350, 351] using mouse or rat osteoblast-specific osteocalcin promoters. Transgenic mice overexpressing Noggin in the bone microenvironment showed dramatic decreases in bone mineral density, bone formation rate, and trabecular bone volume from ages 1 month to 8 months. These results indicate that the overproduction of Noggin inhibits osteoblast differentiation and bone formation, leading to osteopenia and fractures. Based on the crystal structure of the BMP-2–BMPR-IA ectodomain complex [356, 357], a three-dimensional crystal structure of Noggin-BMP-7 complex was recently analyzed [358, 359]. Noggin, as a cystine-knot protein, inhibits BMP signaling by blocking the molecular interfaces of the binding epitopes for both type I and type II receptors, thus preventing BMP-7 binding to the receptors. This mechanism provides an ideal molecular model to explore small molecular compounds that block interactions between Noggin and BMP ligands.

Chordin is another BMP antagonist found in the Spemann organizer [360, 361]. Chordin specifically binds to BMP-2 and BMP-4 and blocks BMP signaling [360]. It is known that Chordin plays a role in osteoblastic function and chondrocytic maturation [353, 362, 363]. Twisted gastrulation (Tsg), a secreted protein, can form a ternary complex with BMP and Chordin by directly binding. Thus, Tsg stabilizes BMP/Chordin/Tsg complex and prevents binding of BMPs to their receptors [364–366]. However, Tsg also has agonistic activity. It is known that the proteolytic activity of Tolloid metalloprotease is specific for Chordin. The presence of Tsg makes Chordin susceptible for cleavage by Tolloid, which leads to degradation of Chordin, resulting in a release of BMP from the complex and activation of BMP signaling [367–369]. Tsg overexpression inhibits BMP action in stromal and preosteoblastic cells and, accordingly, arrests their differentiation along the osteoblastic lineage [352].

Sclerostin and Gremlin are two novel members of the Dan family of BMP antagonists. Sclerosteosis is an autosomal recessive disorder caused by mutations in the SOST gene. Patients with homozygous mutations in this gene have a progressive sclerosing bone dysplasia known as sclerosteosis [370–372]. Sclerostin is abundantly expressed in long bones and cartilages. Sclerostin binds to BMP-6 and BMP-7 with high affinity and blocks the activation of BMP signaling, thus decreasing osteoblast activity and reducing the differentiation of osteoprogenitors [373–375]. Since high bone mass diseases with phenotypic similarity are caused by both gain-of-function mutations in LRP5, a Wnt receptor, and loss of SOST, it is likely that primary interactions between BMP and Wnt pathways are involved in modulating osteoblast differentiation. Results from a recent study have suggested that Wnt induces osteoblast differentiation through BMPs that are blocked by the BMP antagonist sclerostin. The expression of BMP proteins in this autocrine loop is essential for Wnt-3A–induced osteoblast differentiation [376]. Recently, sclerostin has been found to promote apoptosis of human osteoblastic cells [377], providing a novel mechanism for regulation of bone formation in which BMPs function in maintaining the survival of osteoblasts. Interestingly, Noggin and sclerostin, two BMP antagonists, have been reported to bind to each other with high affinity. The Noggin-sclerostin complex is competitive with BMP binding and attenuates the activity of each BMP antagonist [378]. The pleiotrophic nature of Noggin and sclerostin represents a novel mechanism for the fine-tuning of BMP activity in bone homeostasis. Based on its suppressive role in bone formation, sclerostin could be a therapeutic target for the treatment of osteoporosis. Gremlin, identified from a *Xenopus* ovarian library for its axial patterning activities [379], is known as Drm (down-regulated by v-mos), a homologue in rodent [380]. Gremlin knockout mice are neonatally lethal due to the absence of kidneys [381]. In transgenic mice, skeletal overexpression of gremlin impairs bone formation and causes osteopenia and spontaneous fractures [382].

F. Clinical Utilization of BMPs

BMPs were originally identified for their osteoinductive activity in ectopic locations [206, 208]. For over 40 years, the osteoinductive properties of the BMPs have been successfully used in preclinical models for accelerating fracture healing and repairing large bone defects in various mammalian animals, i.e., rat, canine, rabbit, sheep, and nonhuman primate, in which BMPs are usually applied in combination with various bioresorbable bone graft substitutes, such as collagen composites, coral, and various ceramics [383, 384]. Recombinant human BMP-2 and BMP-7 (rhBMP-2 and rhBMP-7) are the BMPs that have been most extensively used in this field. In humans, rhBMP-2 and rhBMP-7 have been used in clinical trials of orthopedic patients for spine fusion, fracture repair, and other bone defects [383, 385–393]. More than 250,000 spine fusions are performed each year in the United States. The successful fusion rate (more than 98%) using rhBMP-2/ACS (rhBMP-2 applied to absorbable collagen sponge), known as InFuse (United States) and InductOs (Europe), led to FDA approval in 2002 for its application in spine fusion in place of iliac crest bone autografts. With a fusion rate (50–75%) less than that obtained with rhBMP-2, rhBMP-7 has received a humanitarian device exemption (HDE) approval for posterolateral fusion nonunions. In a clinical trial on fracture repair, a study group released a report on the application of rhBMP-2/ACS in open tibial fractures in over 450 patients [393]. The results showed that rhBMP-2 (1.5 mg/mL) accelerates fracture and wound healing, reduces the frequency of secondary interventions and invasiveness of the procedures, resulting in a 44% reduction in the risk of failure in healing compared with the standard of care control group. Thus, rhBMP-2 has also received FDA approval for open long bone fractures.

Osteoporosis is characterized by progressive bone loss with age in both men and women, due in part to decreased bone formation. Systemic intra-peritoneal injections of rhBMP-2 reverse this phenotype in either estrogen-deficient or senile osteopenic mouse models. This change in bone mass in adults is coupled to an increase in mesenchymal stem cell numbers, osteogenic activity, and proliferation as well as a decrease in apoptosis [394]. These results and others described previously suggest that the BMP pathway could become a therapeutic target for osteoporosis, and it may be very fruitful to seek potential small molecular compounds that enhance gene expression and activity of endogenous BMPs in bones during aging. Statins, which are widely used for lowering serum cholesterol, have been found to enhance new bone formation and reduce fracture risk [323, 395]. *In vitro* studies have demonstrated that these compounds stimulate BMP-2 and BMPR-II gene expression in multiple cell lines, suggesting that statins increase bone formation by enhancing the activity of the BMP pathway.

G. Conclusion

BMPs play an important role in postnatal bone formation. Disruption of the BMP signaling pathway results in bone loss in animals. Recently, BMP-2 was recognized as an osteoporosis-associated gene. Thus, regulation of BMP signaling activity and BMP gene expression are potential therapeutic strategies for the prevention and treatment of osteoporosis.

XIII.   HEDGEHOG (HH) SIGNALING MOLECULES

Hedgehog (Hh) signaling plays a critical role during endochondral bone development. This is in part through its regulation of other genes. Especially important are the regulation of PTHrP expression and chondrocyte proliferation and differentiation by Hh signaling. However, growing evidence suggests that Hedgehog signaling plays an equally important role in the regulation of other signaling pathways important both in chondrocyte differentiation and osteoblast differentiation.

Hedgehog signaling functions as an important regulator of many developmental processes during embryogenesis. There are three different Hh proteins, Desert, Indian (Ihh), and Sonic (Shh), each of which is expressed in different tissues during embryogenesis to regulate distinct processes. Desert Hh functions mostly during spermatogenesis, but both Shh and Ihh have important roles during skeletal development. However, Ihh is the most relevant of the Hh family members in endochondral bone formation during embryonic development. Hh binds to its receptor Patched-1 (Ptc-1), leading to an activation of smoothened (Smo), which activates a signaling cascade that leads to gene activation [396]. Ihh is a master regulator of bone development, which is expressed in the developing growth plate primarily by the pre- and early hypertrophic chondrocytes, that is critical for coordinating chondrocyte proliferation and differentiation as well as osteoblast differentiation [282]. Ihh null mice die shortly after birth, displaying severe dwarfism with a reduction in chondrocyte proliferation, improper maturation of chondrocytes, and a failure of osteoblast development in endochondral bones [397].

While the Hh signaling pathway is a complex signaling network, the only known and well-described transcriptional mediator of the Hh network is the Gli family of transcription factors, Gli1, Gli2, and Gli3 [398]. Gli1 was originally isolated as a gene amplified in human glioblastoma. All three family members have since been demonstrated to be critical in many cellular processes during embryonal development [399]. While these proteins have been primarily identified in the transduction of Hh signaling, they have also been demonstrated to be important for the mediation of other signaling pathways, including the Wnt [400] and FGF pathways [401], perhaps through cross-talk with the Hh signaling pathway.

In *Drosophila*, ubiquitination plays a critical role in regulating hedgehog. In the absence of hedgehog signaling, full length (155 kDa) *cubitus interruptus* (Ci), the *Drosophila* ortholog of Gli, is processed by the proteasome to a truncated form (75 kDa), which acts as a transcriptional repressor [402]. In the presence of Hh signaling, this processing is inhibited, and the FlCi positively regulates the transcription of target genes. As with the *Drosophila* homologue Ci, processing of the Gli proteins appears to play an important role in regulating the mammalian response to hedgehog proteins [403–405]. In mammals, Gli2 and Gli3 contain both a repressor and activator region, and Gli constructs in which the activator region is deleted act as repressors of Hh signaling (Figure 3 in Sasaki *et al.*, 1997). Both Gli2 and Gli3 are processed by the proteasome upon treatment with protein kinase A, suggesting that proteasomal processing is an important step in the regulation of Gli responsive promoters. Full-length Gli3 (190 kDa) is degraded to a truncated (repressor) form (83 kDa), while Gli2 does not appear to be specifically degraded to a truncated form when transfected into tissue culture cells [403, 404].

Genetic knockout models have been generated and described for each of the three Gli family members. The phenotypes of these knockouts are distinct with little overlap, suggesting a distinct role for each of these transcriptional regulators. When mice were generated overexpressing a Gli1 construct with the zinc finger domain mutated (Gli1$^{-zfd/-zfd}$), the mice appeared normal with no obvious phenotype. On the other hand, Gli2$^{-zfd/-zfd}$ mice displayed severe defects and died shortly after birth with severe skeletal abnormalities [335]. These defects included cleft palate, absence of vertebral bodies and intervertebral discs, short limbs, and short sternum. Further examination of these mice demonstrated that these mice displayed a delay in endochondral ossification [406]. In addition there is an increase in size of the cartilaginous growth plate, a decrease in bone in the tibia and femur, and a failure to develop the primary ossification centers in the vertebral bodies. At the growth plate, these mice exhibit an increase in both the proliferating and hypertrophic chondrocytes with no change in matrix mineralization [406]. These effects are likely due to changes in expression of multiple downstream regulators. They include a reduction in expression of Ihh, Ptc, and PTHrP in the prehypertrophic chondrocytes as well as a decrease in angiogenic markers in the hypertrophic chondrocytes [406]. Additionally, a decrease in chondroclasts at the cartilage/bone interface and a reduction in osteoblasts lining the trabecular surfaces were observed [406]. These data indicate that Gli2 is an important regulator of the developing growth plate, while Gli1 alone does not seem to play a significant role. However, when the

heterozygous Gli2 mutants, which are otherwise normal, are crossed with homozygous Gli1 mutants, defects in the ventral spinal cord and lungs are observed, but when double mutants are generated, the only differences from the Gli2$^{-zfd/-zfd}$ mouse are postaxial nubbins on the limbs [407]. This suggests that Gli1 does play some role in developmental processes that Gli2 can normally compensate for the loss of Gli1. Thus, Gli1 does not seem to be a critical regulator of skeletal development.

Similar to the Gli2 null mice, the Gli3 null mice (Gli3$^{XtJ/XtJ}$) die at birth with multiple severe skeletal defects [408]; however, these are distinct from those of the Gli2 null mice, suggesting a discrete role of each of these proteins in endochondral bone formation. The craniofacial defects include enlarged maxillary region, reduced external nasal processes, failure of skull vault formation, cleft palate, and tooth defects [335, 408]. In the fore- and hindlimbs, these mice exhibit severe polydactyly and a shortening of the tibia, humerus, ulna, and radius [335]. Double homozygous mutants of Gli2 and Gli3 die before day 10.5 p.c., while heterozygous double mutants display an enhancement of many of the observed phenotypes, indicating some redundancy in function between the two transcription factors [335]. Results of these genetic models clearly suggest that Gli2 and Gli3 are major regulators of skeletal development.

Hedgehog signaling plays an important role in the regulation of several proteins that are important for normal skeletal development. Probably the most well characterized of these is the regulation of PTHrP during chondrocyte differentiation. Mice with the Ihh gene knockout die shortly after birth, do not secrete PTHrP, have an increase in hypertrophic chondrocytes, and have a decrease in proliferating chondrocytes [397]. It is well established that the effect of Ihh on delaying chondrocyte hypertrophy is mediated through an increase in PTHrP [397, 409, 410]. Since Gli2 and Gli3 expression in the growth plate overlaps with that of PTHrP [411], this suggests a direct control of PTHrP expression by the Gli family members. Furthermore, knocking out Gli3 in the Ihh −/− partially rescued the chondrocyte phenotype of Ihh −/− mice, including re-expression of PTHrP [411]. These data indicate a repressor role for Gli3 in the regulation of PTHrP. In addition, unpublished data from our laboratory have indicated that Gli2 can activate PTHrP promoter activity in the developing growth plate, suggesting a positive regulatory role for Gli2 in this regulation [412]. Finally, Gli2 null mice exhibit a decrease in PTHrP expression in the proliferating chondrocytes [406], providing further support for this notion. While both Gli2 and Gli3 appear to play important roles in skeletal development, Gli3 (in its processed truncated form) appears to primarily act as a repressor, while Gli2 acts primarily as an activator.

In addition to its role in chondrocyte differentiation, Ihh promotes osteoblast differentiation through coordination with factors such as the BMPs [307] and Wnt signaling pathways. BMP and Ihh signaling appear to act in parallel to maintain a normal chondrocyte proliferation rate [301]. BMP signaling delays the process of hypertrophic differentiation and modulates the expression of Ihh [301]. Furthermore, Ihh misexpression in a developing chick limb demonstrated an increase in the expression of BMP2 and BMP4, but not BMP5 or BMP7 [413] in the perichondrium. Other experiments suggest that Hh signaling leads to an alteration in the response of cells to BMP signaling [414]. While a complicated cross-talk appears to exist between the two pathways, it is clear that both must work together for the proper regulation of chondrocyte proliferation and differentiation.

In osteoblasts, a similar cross-talk appears to exist. Hh signaling has been clearly shown to be essential for osteoblast formation at the bone collar, as demonstrated by the lack of Runx2 expression in the perichondrium of Ihh −/− mice [397, 415]. This induction, at least in part, is in a BMP-dependent manner [416]. To further support a role for Hh signaling in osteoblast differentiation, *in vitro* data have demonstrated that Gli3 can increase BMP2 promoter activity and that TrGli3 inhibited BMP2 promoter activity in a dose-dependent manner [322]. Additionally Gli1 and Gli3 have been demonstrated to increase BMP4 and BMP7 promoter activity [308].

In addition to cross-talk between Hh and BMP signaling, it has become apparent that the Wnt and Hh signaling pathways also work in conjunction to regulate osteoblast development. It has been demonstrated that, through sequential analysis of a conditional β-catenin knockout, Wnt signaling functions downstream of Ihh signaling in osteoblast development, and the expression Wnt7b is regulated by Ihh [417]. Additionally, this same group has demonstrated that the repressor function of Gli3 is critical in the regulation of endochondral bone formation and that activators of Hh signaling (likely Gli2) can activate canonical Wnt signaling. For instance, Gli2 has been shown to regulate Wnt8 expression [400]. Since Wnt7b is regulated by Ihh signaling, and Wnt8 expression is regulated by Gli2 [400], it is likely that the increase in canonical Wnt signaling is through regulation of ligand expression.

XIV. SCLEROSTIN

Sclerostin (SOST) was discovered during studies on the disease sclerosteosis, which is a rare aggressive bone disorder associated with bony overgrowth affecting the bones of the face and with syndactyly [418]. It is very rare and has been most frequently described in Afrikaners. It is an autosomal recessive disorder associated with loss of SOST function due to variable mutations in the SOST gene [370]. So far, five distinct mutations have been described. The x-ray features are first apparent in the patients by the age of 5. Heterozygotes also show some thickening of the skull, with increases in bone mineral density and increases in bone formation rates. A related condition is Van Buchem disease, which is even rarer and less severe [419]. The mutations responsible for this condition are downstream of the SOST gene but affect secretion of the protein. It has been described mostly in people from the Netherlands.

Sclerostin is normally expressed by osteocytes, but in SOST deficiency there is no evidence for expression of this factor by osteocytes [420]. SOST has been found to inhibit bone formation *in vitro*, and in transgenic mice there is markedly decreased bone formation. SOST is therefore an osteocyte-expressed negative regulator of bone formation. The molecular mechanism of action has recently been characterized. Although it was thought initially that it worked predominantly by binding to BMPs and acting as an endogenous BMP antagonist, it now seems more likely that its mechanism of action is mediated through effects on the LRP-5 and LRP-6 receptors, and it functions by acting as a soluble antagonist in a manner similar to DKK-1 and DKK-2 [376, 421] (see Chapter 15, Johnson). *In vivo*, it is regulated by PTH, and it appears possible that this effect of PTH to inhibit SOST may be the means by which PTH exerts its anabolic effects on bone. Precisely where or how PTH acts on the SOST gene has yet to be determined.

Recently, null-mutant mice have been described that have a bone phenotype with similarities to the human disease. Monoclonal antibodies have been developed to sclerostin, and these antibodies cause a very similar effect in aged ovariectomized rats to those of PTH, with marked increases in BMD, bone formation, and osteoblast activity.

XV. PARATHYROID HORMONE-RELATED PEPTIDE (PTHRP)

PTHrP was first identified as the factor responsible in most cases for the humoral hypercalcemia of malignancy, and so thought of as a systemic circulating factor. However, it is now well known to be a local factor in physiological control of chondrocyte differentiation in the growth plate [422, 423], and in local osteolysis associated with metastatic breast cancer [424]. Thus, it can be a cytokine depending on the context in which it is produced. This may be an important concept for control of normal bone cell function and bone turnover. Horwitz and coworkers [425] showed that it is, like PTH, an anabolic factor for bone formation when it is administered as a pharmacologic agent. Miao and colleagues recently showed that PTHrP produced in the bone microenvironment may be responsible for control of osteoblast differentiation, functioning in this situation as a paracrine factor [426]. In the heterozygotes of PTHrP null-mutant mice, they found that there was a severe osteoporotic phenotype by 3 months of age characterized by decreased bone formation, as assessed histologically and by microCT. Bone formation was impaired, but in addition there was also a decrease in osteoclastic resorption surface. The bone loss was reversed by treatment with PTH, suggesting that PTHrP is a local endogenous regulator of bone turnover and bone formation.

XVI. NEURONAL REGULATION OF BONE REMODELING

Concepts on the control of bone remodeling have traditionally focused on local factors. However, these concepts have recently been modified by studies on the role of the hypothalamus in this process.

A. Regulation by Hypothalamic Neurons

1. LEPTIN

The concept of a hypothalamic regulation of bone remodeling originated from the discovery that absence of leptin or its receptor induced a high bone mass phenotype in mice in spite of hypogonadism, which suggested a major role of leptin in the regulation of bone formation [427]. Further studies using transgenic animal models characterized by high or low serum level of leptin confirmed the pronounced effect of leptin on bone formation [428, 429]. Because many physiological functions of leptin, i.e., the regulation of body weight, energy expenditure, and reproduction, are mediated by the central nervous system (CNS), these results suggested that bone remodeling was a homeostatic process regulated by the hypothalamus where the leptin receptor is highly expressed. However, other studies demonstrated expression of leptin and leptin receptor in osteoblasts or osteoblast progenitors and a positive effect of leptin on bone formation *in vitro* [430, 431], which suggested an alternative mechanism whereby

leptin controls bone formation. Administration of leptin peripherally to mice could not affect bone mass but was reported to reduce bone fragility via an unidentified mechanism [121]. Overexpression of leptin in osteoblasts *in vivo* via the 2.3-kb alpha1 collagen promoter did not affect bone mass either. Therefore, the main site of action of leptin for its regulation of bone formation appears to be central rather than peripheral, but a more subtle and direct role of leptin on osteoblast biology cannot be excluded at the present time. Furthermore, the relative contribution of these two sites of action remains unknown. Analysis of mice with selective deletion of the leptin receptor in osteoblasts versus hypothalamus will address this question.

The central nature of the regulation of bone remodeling by leptin has been further supported by studies using chemical lesioning and intracerebroventricular (icv) infusion of leptin, which defined a population of ventromedial hypothalamic (VMH) neurons as constituting a major hypothalamic center responsive to leptin for its function on bone formation [432]. Indeed, destruction of VMH neurons, which highly express the leptin receptor, recapitulated the bone phenotype of mice lacking leptin and blocked the antiosteogenic effect of leptin intracerebroventricular infusion. Therefore, leptin appears to be a central master regulator integrating the regulation of major physiological functions, including bone remodeling.

2. NPY

Neuropeptide Y (NPY) is a neuropeptide whose expression is negatively regulated by leptin and acting through multiple receptors' subtypes (Y1, Y2, Y4, and Y5). Y2 receptor is strongly expressed in neurons of the arcuate nucleus that also express the leptin receptor. Selective conditional deletion of Y2 in the hypothalamus induced a high bone mass in mice, bringing further support to the concept of a central regulation of bone remodeling [433]. Deletion of both Y2 and Y4 receptors induced an increase in bone turnover and further increase in bone mass compared to Y2 –/– mice, as well as a lean phenotype accompanied by a low level of leptin, suggesting that hypoleptinemia in these animals worsened the bone phenotype of Y2 –/– mice [434]. The downstream mediator of Y2 signaling in the brain mediating the Y2 effect on bone mass is unknown.

3. CART AND MC4R

Cocaine- and amphetamine-regulated transcript (Cart) is another neuropeptide whose expression is controlled by leptin. Cart-deficient mice display a low bone mass phenotype caused by an increase in bone resorption [435]. The mechanism of action of Cart, i.e., central or peripheral, is unknown as yet, mainly due to the fact that the receptor for Cart has not been characterized. However, the fact that hypothalamic Cart expression is increased in Melanocortin receptor 4 (Mc4r) deficient mice and that patients and mice deficient for Mc4r display a high bone mass phenotype caused by a decrease in bone resorption suggested that Mc4r and Cart function in a linear pathway. Interestingly, Cart expression was decreased in Y2Y4 −/− mice, which may possibly have triggered the increase in osteoclast surface observed in these animals [434].

4. CANNABINOIDS

The cannabinoid system, mostly known for its involvement in psychotropic, analgesic, and orectic processes, also regulates bone mass *in vivo*. The cannabinoid type 1 (CB1) receptor is expressed in the CNS and sympathetic nervous system (SNS), but also in osteoclasts, and its absence in mutant animals induced a high bone mass due to a defect that may involve osteoclast survival [436]. However, the contribution of central versus peripheral CB1 receptors in mediating this effect remains to be determined. The CB2 receptor is more specific for peripheral tissues and is notably expressed in osteoblasts and osteoclasts. Mice deficient for CB2 receptor displayed a low bone mass resulting from a high bone turnover. *In vitro* analyses demonstrated a direct effect of CB2 agonists on osteoblast proliferation and the generation of osteoclasts, suggesting a peripheral mode of action [437].

5. IL1

The proinflammatory cytokine IL1 is produced by peripheral and central tissues including bone cells, glia, and neurons, and was known for its activity to potentiate osteoclastogenesis. Surprisingly, IL1R-deficient mice displayed a low bone mass phenotype, which was recapitulated by antagonizing IL1R signaling in the CNS, thereby demonstrating a possible central origin of IL1 signaling in the regulation of bone resorption [438]. This low bone mass was accompanied by a reduced length and diameter of long bones, and was mostly caused by an increase in osteoclast numbers.

B. Regulation by the Peripheral Nervous System

1. β2 ADRENERGIC RECEPTOR (β2AR) SIGNALING IN OSTEOBLASTS REGULATES BONE FORMATION AND BONE RESORPTION

The peripheral nervous system is composed of efferent nerves, which are responsible for transmitting signals from the brain to various organs throughout the body, including bones. The autonomic part of the nervous system is further divided into the sympathetic nervous system and cholinergic nervous system, secreting the neurotransmitters norepinephrine and acetylcholine, respectively.

The initial observation of an increased bone mass and low sympathetic tone in leptin and dopamine β-hydroxylase-deficient mice (the enzyme generating norepinephrine), the presence of nerves within the bone microenvironment, and the selective detection, among all postsynaptic adrenergic receptors, of Adrβ2 in primary osteoblasts first supported the hypothesis that the sympathetic nervous system and adrenergic signaling could relay leptin signaling in the VMH to osteoblasts. Pharmacological treatment of mice *in vivo* with nonselective adrenergic agonists and antagonists, the analysis of mice deficient for Adrβ2, and bone marrow transplantation experiments between WT and Adrβ2 −/− mice all confirmed this hypothesis and demonstrated that adrenergic signaling in osteoblasts via β2adrenergic receptor (β2AR) inhibits bone formation [432, 435]. The presence of decreased bone resorption in Adrβ2-deficient mice suggested that β2AR signaling favors bone resorption. Osteoblast-osteoclast coculture assays using WT and Adrβ2 −/− cells uncovered the cellular and molecular basis of this phenotype by demonstrating that β2AR signaling in osteoblasts indirectly increases osteoclast differentiation by upregulating the expression of RANKL. Further biochemical and molecular studies demonstrated that β2AR stimulation in osteoblasts increases cAMP level and activates PKA, which eventually results in the phosphorylation of ATF4, a CREB family member previously shown to regulate osteoblast differentiation and collagen synthesis (REF). Mutation/phosphorylation analyses and promoter studies pinpointed ATF4's Serine 254 as the target of PKA activity and as a necessary phosphorylation site for ATF4 to bind to and transactivate the RANKL promoter. These studies thus characterized a new transcription factor target of β2AR signaling and identified the crucial role of β2AR signaling in the regulation of bone formation and bone resorption.

The results of these studies have generated a strong interest in groups of patients treated with β-blockers. Retrospective studies have shown either no effect [439–442] or a positive effect of β-blockers, including reduction in fracture risk and increased BMD [443, 444]. Future prospective studies will address the effect of β-blockers in homogeneous groups of patients, without concomitant treatments and at multiple skeletal sites.

2. SENSORY NEUROPEPTIDES AND BONE REMODELING

The unexpected finding that mice deficient for CT/CGRP-α displayed an increase in bone formation rather than an alteration in bone resorption provided another example of a newly identified potential neuronal system regulating bone remodeling. CGRP-α is a sensory neuropeptide generated by alternative splicing from the Calca gene. It is produced by the central and peripheral nervous systems, notably in neurons innervating bones [445–447]. CGRP-α acts directly on osteoblasts and can stimulate their proliferation and activity [448–452]. Interestingly, CGRP may also function as an autocrine factor, since it is expressed by osteoblasts [446, 453]. In agreement with this hypothesis, transgenic mice over-expressing CGRP in differentiated osteoblasts display a bone phenotype characterized by an increased bone volume caused by an increase in the rate of bone formation [454], while mice deficient for CGRP are osteopenic due to a decrease in bone formation [455]. Regardless of its origin, these results suggest that CGRP is an anabolic factor for bone acting directly on osteoblasts.

Two other sensory neuropeptides, vasoactive intestinal peptide (VIP) and substance P (SP), may have a role in bone biology and osteoclast biology more specifically. VIP belongs to a family of structurally related peptides and is a neurotrophic factor involved in neuronal growth, differentiation, survival, and transmitter synthesis. Destruction of nerves expressing VIP induced a 50% increase in osteoclast-covered surfaces in the mandible and calvariae [456]. In agreement with this result, VIP binds osteoclasts and inhibits osteoclastogenesis induced by 1,25(OH)$_2$-vitamin D$_3$ [457–459]. VIP may also control bone resorption by indirectly stimulating Pge2 expression in osteoblasts [460, 461]. SP is another neuropeptide richly expressed in small sensory neurons that innervate bones which may contribute to the maintenance of trabecular bone integrity. Capsaicin-treated rats displayed bone loss, increased bone resorption, and decreased bone formation associated with the destruction of SP and CGRP-positive unmyelinated sensory neurons [462]. These results are in agreement with the human familial dysautonomia disease characterized by the loss of unmyelinated sensory neuron, reduced bone mineral density, and frequent fractures [463, 464].

3. BIOACTIVE AMINES

The central serotoninergic system is known to modulate mood, emotion, sleep, and appetite, while the dopaminergic system is involved in processing reward information and learning. The dopamine transporter (DAT) is an important determinant of dopamine signaling activity, since it is responsible for the rapid uptake of released dopamine into presynaptic terminals, and therefore for efficient clearance of extracellular dopamine and termination of dopamine signaling. DAT-deficient mice displayed a low bone mass phenotype of

possible central origin, since DAT was not detected in bone [465].

In contrast, another member of the family of neurotransmitter transporters for bioactive amines, the serotonin transporter (5-HTT), is expressed in bones along with most of the 5-HT receptors [466, 467]. 5-HTT uptakes 5-HT from the extracellular space and therefore downregulates serotoninergic activity. *In vitro* studies demonstrated the influence of 5-HT signaling on AP1 transcription factors binding activity regulated by PTH, suggesting that 5-HT signaling is a functional component involved in osteoblast differentiation [466]. As observed in patients treated with selective serotonin-reuptake inhibitors (SSRIs) for depression [468], blocking 5-HTT activity by SSRIs in mice led to a significant decrease in bone mass due to a decrease in bone formation [469]. In agreement with these pharmacological interventions, null mutation of the gene coding for 5-HTT reduced bone formation and bone mass [469]. Among the different 5-HT receptors expressed in bones, the 5-HT receptor 2B subtype is of particular interest since it appears to be involved in mechanical sensing by osteocytes and nitric oxide release by mechanically stimulated osteoblasts [467, 470], which suggested that osteocytes are under the control of neurogenic signals for their response to mechanical stimuli. Serotoninergic signaling could also participate in the regulation osteoclastogenesis. 5-HTT and several receptor types for 5-HT (5-HT1B, 5-HT2B, and 5-HT4) are expressed by osteoclasts [471]. The 5-HTT inhibitor fluoxetine (Prozac) inhibits osteoclast differentiation, while inhibition of 5-HT intracellular transport or the addition of 5-HT stimulates osteoclast differentiation. Moreover, specific antagonists of receptor 1B and 4 inhibit the formation of differentiated osteoclasts *in vitro* [471], suggesting that elevations in cytoplasmic levels of 5-HT may be required to enhance NF-κB activation through mechanisms to be characterized. The absence of a bone resorption phenotype in mice deficient for 5-HTT, however, suggests that the net effect of the serotoninergic system on bone mass is likely to be complex *in vivo* [469].

4. GLUTAMINERGIC SIGNALING

L-glutamate is a major excitatory amino-acid neurotransmitter in the central nervous system. Bone cells, including osteoblasts, osteoclasts and osteocytes, are equipped with the molecular machinery necessary for glutamate release, extracellular recovery, and glutamate response. Osteoblasts are in close association with glutamatergic nerve endings, but also contain glutamate-filled vesicles [472], express the glutamate transporter GLAST-1 and the glutamate receptors iGluR (NMDA, AMPA, and kainite ionotropic–type glutamate receptor),

477

as well as mGluR (metabotropic-type glutamate receptor 1, 4, and 8) [473–478]. A number of signaling molecules known to associate or colocalize with iGluRs have been detected in osteoblasts as well, including Yotio, PSD95, GRIP, and SHANK [479, 480], but their role in bone biology is still speculative. Like osteoblasts, osteoclasts express functional iGluR and mGluR as well as Glu transporters with similar characteristics as neuronal cells [481–483]. The role of glutamate signaling in bone biology has been assessed mainly by *in vitro* cell-based analyses that have shown a negative effect of receptor blockade on osteoblast and osteoclast differentiation [484–488]. Conversely, activation of NMDAR in RAW264.7 cells by specific agonists induced nuclear translocation of NF-κB, a pivotal factor for osteoclast differentiation [483], which suggested that NF-κB is involved in glutamate regulation of osteoclast formation. Mice that underexpress NMDAR1 are smaller than littermates expressing normal levels of NMDAR1, which may reflect a disruption in skeletal development [489]. However, NMDA subunit NR1-deficient mice did not show any obvious bone phenotype [482], and no significant bone phenotype has been detected in GLAST-deficient mice either [490]. Interestingly, however, GLAST expression is downregulated by mechanical loading, which suggests that this glutamate transporter may be involved in coupling mechanical signals to skeletal modeling [474, 491].

Thus, accumulating evidence over the past few years has documented that both the central and peripheral nervous systems regulate bone remodeling, mostly thanks to the use of available mutant mouse models and *in vivo* studies. The intricate connections between the regulation of bone mass and body weight, energy expenditure, reproduction, and other physiological functions are still unclear. However, the characterization of the neuronal factors regulating local bone remodeling increases the list of potential therapeutic targets and may lead to new strategies for the treatment of bone diseases.

XVII. CONCLUSION

The widespread use of genetic mouse models over the past decade has increased our understanding enormously of the cytokines that influence bone remodeling and their effects on bone remodeling *in vivo*. It is likely that during the next decade, the interactions of these cytokines and the coordinated effects that they exert on their target cells will be further clarified, with the hope that this increased knowledge will improve our understanding of the mechanisms responsible for the common diseases of the bone.

REFERENCES

1. S. Akira and T. Kishimoto, NF-IL6 and NF-kappa B in cytokine gene regulation. *Adv Immunol*, **65**, 1–46 (1997).
2. R. Felix, M. G. Cecchini, and H. Fleisch, Macrophage colony stimulating factor restores in vivo bone resorption in the op/op osteopetrotic mouse. *Endocrinology*, **127**(5), 2592–2594 (1990).
3. H. Kodama, *et al.*, Congenital osteoclast deficiency in osteopetrotic (op/op) mice is cured by injections of macrophage colony-stimulating factor. *J Exp Med*, **173**(1), 269–272 (1991).
4. H. Kodama, *et al.*, Transient recruitment of osteoclasts and expression of their function in osteopetrotic (op/op) mice by a single injection of macrophage colony-stimulating factor. *J Bone Miner Res*, **8**(1), 45–50 (1993).
5. S. Niida, *et al.*, Vascular endothelial growth factor can substitute for macrophage colony-stimulating factor in the support of osteoclastic bone resorption. *J Exp Med*, **190**(2), 293–298 (1999).
6. Y. Y. Myint, *et al.*, Granulocyte/macrophage colony-stimulating factor and interleukin-3 correct osteopetrosis in mice with osteopetrosis mutation. *Am J Pathol*, **154**(2), 553–566 (1999).
7. E. Lagasse and I. L. Weissman, Enforced expression of Bcl-2 in monocytes rescues macrophages and partially reverses osteopetrosis in op/op mice. *Cell*, **89**(7), 1021–1031 (1997).
8. R. Pacifici, Cytokines, estrogen, and postmenopausal osteoporosis—The second decade. *Endocrinology*, **139**(6), 2659–2661 (1998).
9. J. A. Lorenzo, *et al.*, Mice lacking the type I interleukin-1 receptor do not lose bone mass after ovariectomy. *Endocrinology*, **139**(6), 3022–3025 (1998).
10. Y. Abu-Amer, *et al.*, Lipopolysaccharide-stimulated osteoclastogenesis is mediated by tumor necrosis factor via its P55 receptor. *J Clin Invest*, **100**(6), 1557–1565 (1997).
11. P. Ammann, *et al.*, Transgenic mice expressing soluble tumor necrosis factor-receptor are protected against bone loss caused by estrogen deficiency. *J Clin Invest*, **99**(7), 1699–1703 (1997).
12. R. L. Jilka, *et al.*, Increased osteoclast development after estrogen loss: Mediation by interleukin-6. *Science*, **257**(5066), 88–91 (1992).
13. V. Poli, *et al.*, Interleukin-6 deficient mice are protected from bone loss caused by estrogen depletion. *EMBO J*, **13**(5), 1189–1196 (1994).
14. H. Kitamura, *et al.*, Bone marrow neutrophilia and suppressed bone turnover in human interleukin-6 transgenic mice. A cellular relationship among hematopoietic cells, osteoblasts, and osteoclasts mediated by stromal cells in bone marrow. *Am J Pathol*, **147**(6), 1682–1692 (1995).
15. R. Kitazawa, *et al.*, Interleukin-1 receptor antagonist and tumor necrosis factor binding protein decrease osteoclast formation and bone resorption in ovariectomized mice. *J Clin Invest*, **94**(6), 2397–2406 (1994).
16. R. B. Kimble, *et al.*, Simultaneous block of interleukin-1 and tumor necrosis factor is required to completely prevent bone loss in the early postovariectomy period. *Endocrinology*, **136**(7), 3054–3061 (1995).
17. G. R. Mundy and T. J. Martin, Pathophysiology of skeletal complications of cancer. In *Physiology and Pharmacology of Bone: Handbook of Experimental Pharmacology*. Chap. 18, pp. 641–671. Springer, Berlin.
18. B. G. Mills and A. Frausto, Cytokines expressed in multinucleated cells: Paget's disease and giant cell tumors versus normal bone. *Calcif Tissue Int*, **61**(1), 16–21 (1997).

19. G. D. Roodman, *et al.*, Interleukin 6. A potential autocrine/paracrine factor in Paget's disease of bone. *J Clin Invest,* **89**(1), 46–52 (1992).

20. R. J. O'Keefe, *et al.*, Osteoclasts constitutively express regulators of bone resorption: An immunohistochemical and in situ hybridization study. *Lab Invest,* **76**(4), 457–465 (1997).

21. S. V. Reddy and G. D. Roodman, Control of osteoclast differentiation. *Crit Rev Eukaryot Gene Expr,* **8**(1), 1–17 (1998).

22. T. J. Martin, E. Romas, and M. T. Gillespie, Interleukins in the control of osteoclast differentiation. *Crit Rev Eukaryot Gene Expr,* **8**(2), 107–123 (1998).

23. D. Heymann, *et al.*, Cytokines, growth factors and osteoclasts. *Cytokine,* **10**(3), 155–168 (1998).

24. G. D. Roodman, Cell biology of the osteoclast. *Exp Hematol,* **27**(8), 1229–1241 (1999).

25. S. V. Reddy, *et al.*, Interleukin-6 antisense deoxyoligonucleotides inhibit bone resorption by giant cells from human giant cell tumors of bone. *J Bone Miner Res,* **9**(5), 753–757 (1994).

26. R. D. Devlin, *et al.*, IL-6 mediates the effects of IL-1 or TNF, but not PTHrP or 1,25(OH)2D3, on osteoclast-like cell formation in normal human bone marrow cultures. *J Bone Miner Res,* **13**(3), 393–399 (1998).

27. C. Menaa, *et al.*, Annexin II increases osteoclast formation by stimulating the proliferation of osteoclast precursors in human marrow cultures. *J Clin Invest,* **103**(11), 1605–1613 (1999).

28. S. J. Choi, *et al.*, Cloning and identification of human Sca as a novel inhibitor of osteoclast formation and bone resorption. *J Clin Invest,* **102**(7), 1360–1368 (1998).

29. J. Pfeilschifter, S. M. Seyedin, and G. R. Mundy, Transforming growth factor beta inhibits bone resorption in fetal rat long bone cultures. *J Clin Invest,* **82**(2), 680–685 (1988).

30. C. Chenu, *et al.*, Transforming growth factor beta inhibits formation of osteoclast-like cells in long-term human marrow cultures. *Proc Natl Acad Sci USA,* **85**(15), 5683–5687 (1988).

31. D. E. Hughes, *et al.*, Estrogen promotes apoptosis of murine osteoclasts mediated by TGF-beta. *Nat Med,* **2**(10), 1132–1136 (1996).

32. T. Murakami, *et al.*, Transforming growth factor-beta1 increases mRNA levels of osteoclastogenesis inhibitory factor in osteoblastic/stromal cells and inhibits the survival of murine osteoclast-like cells. *Biochem Biophys Res Commun,* **252**(3), 747–752 (1998).

33. H. Takai, *et al.*, Transforming growth factor-beta stimulates the production of osteoprotegerin/osteoclastogenesis inhibitory factor by bone marrow stromal cells. *J Biol Chem,* **273**(42), 27091–27096 (1998).

34. M. R. Wani, *et al.*, Prostaglandin E2 cooperates with TRANCE in osteoclast induction from hemopoietic precursors: Synergistic activation of differentiation, cell spreading, and fusion. *Endocrinology,* **140**(4), 1927–1935 (1999).

35. K. Fuller, *et al.*, TRANCE is necessary and sufficient for osteoblast-mediated activation of bone resorption in osteoclasts. *J Exp Med,* **188**(5), 997–1001 (1998).

36. T. Suda, *et al.*, Regulation of osteoclast function. *J Bone Miner Res,* **12**(6), 869–879 (1997).

37. D. M. Anderson, *et al.*, A homologue of the TNF receptor and its ligand enhance T-cell growth and dendritic-cell function. *Nature,* **390**(6656), 175–179 (1997).

38. B. R. Wong, *et al.*, TRANCE is a novel ligand of the tumor necrosis factor receptor family that activates c-Jun N-terminal kinase in T cells. *J Biol Chem,* **272**(40), 25190–25194 (1997).

39. W. S. Simonet, *et al.*, Osteoprotegerin: A novel secreted protein involved in the regulation of bone density. *Cell,* **89**(2), 309–319 (1997).

40. H. Yasuda, *et al.*, Identity of osteoclastogenesis inhibitory factor (OCIF) and osteoprotegerin (OPG): A mechanism by which OPG/OCIF inhibits osteoclastogenesis in vitro. *Endocrinology,* **139**(3), 1329–1337 (1998).

41. T. Suda, *et al.*, Modulation of osteoclast differentiation and function by the new members of the tumor necrosis factor receptor and ligand families. *Endocr Rev,* **20**(3), 345–357 (1999).

42. D. L. Lacey, *et al.*, Osteoprotegerin ligand is a cytokine that regulates osteoclast differentiation and activation. *Cell,* **93**(2), 165–176 (1998).

43. L. Lum, *et al.*, Evidence for a role of a tumor necrosis factor-alpha (TNF-alpha)-converting enzyme-like protease in shedding of TRANCE, a TNF family member involved in osteoclastogenesis and dendritic cell survival. *J Biol Chem,* **274**(19), 13613–13618 (1999).

44. B. R. Wong, R. Josien, and Y. Choi, TRANCE is a TNF family member that regulates dendritic cell and osteoclast function. *J Leukoc Biol,* **65**(6), 715–724 (1999).

45. T. Yoneda, *et al.*, A novel cytokine with osteoclastopoietic activity. *J Periodontal Res,* **28**(6 Pt 2), 521–522 (1993).

46. M. Y. Lee, D. R. Eyre, and W. R. Osborne, Isolation of a murine osteoclast colony-stimulating factor. *Proc Natl Acad Sci USA,* **88**(19), 8500–8504 (1991).

47. C. C. Lynch, *et al.*, MMP-7 promotes prostate cancer-induced osteolysis via the solubilization of RANKL. *Cancer Cell,* **7**(5), 485–496 (2005).

48. K. Tsukii, *et al.*, Osteoclast differentiation factor mediates an essential signal for bone resorption induced by 1 alpha,25-dihydroxyvitamin D3, prostaglandin E2, or parathyroid hormone in the microenvironment of bone. *Biochem Biophys Res Commun,* **246**(2), 337–341 (1998).

49. B. O. Oyajobi, P. J. Williams, K. Traianades, T. Yoneda, D. M. Anderson, and G. R. Mundy, Efficacy of a genetically-engineered soluble receptor activator of NF-κB (RANK) fusion protein on bone resorption in vitro and in vivo. *J Bone Miner Res,* **14**(Suppl 1), S163 (1999).

50. K. Matsuzaki, *et al.*, Osteoclast differentiation factor (ODF) induces osteoclast-like cell formation in human peripheral blood mononuclear cell cultures. *Biochem Biophys Res Commun,* **246**(1), 199–204 (1998).

51. J. M. Quinn, *et al.*, A combination of osteoclast differentiation factor and macrophage-colony stimulating factor is sufficient for both human and mouse osteoclast formation in vitro. *Endocrinology,* **139**(10), 4424–4427 1998).

52. V. Shalhoub, *et al.*, Osteoprotegerin and osteoprotegerin ligand effects on osteoclast formation from human peripheral blood mononuclear cell precursors. *J Cell Biochem,* **72**(2), 251–261 (1999).

53. J. K. Min, *et al.*, Vascular endothelial growth factor up-regulates expression of receptor activator of NF-kappa B (RANK) in endothelial cells. Concomitant increase of angiogenic responses to RANK ligand. *J Biol Chem,* **278**(41), 39548–39557 (2003).

54. L. C. Hofbauer, *et al.*, Interleukin-1beta and tumor necrosis factor-alpha, but not interleukin-6, stimulate osteoprotegerin ligand gene expression in human osteoblastic cells. *Bone,* **25**(3), 255–259 (1999).

55. C. A. O'Brien, *et al.*, STAT3 activation in stromal/osteoblastic cells is required for induction of the receptor activator of NF-kappaB ligand and stimulation of osteoclastogenesis by gp130-utilizing cytokines or interleukin-1 but not 1,25-dihydroxyvitamin D3 or parathyroid hormone. *J Biol Chem,* **274**(27), 19301–19308 (1999).

56. S. K. Lee and J. A. Lorenzo, Parathyroid hormone stimulates TRANCE and inhibits osteoprotegerin messenger ribonucleic acid expression in murine bone marrow cultures: Correlation with osteoclast-like cell formation. *Endocrinology,* **140**(8), 3552–3561 (1999).

57. N. J. Horwood, *et al.,* Osteotropic agents regulate the expression of osteoclast differentiation factor and osteoprotegerin in osteoblastic stromal cells. *Endocrinology,* **139**(11), 4743–4746 (1998).

58. A. E. Adams, *et al.,* 1,25 dihydroxyvitamin D3 and dexamethasone induce the cyclooxygenase 1 gene in osteoclast-supporting stromal cells. *J Cell Biochem,* **74**(4), 587–595 (1999).

59. E. Jimi, *et al.,* Osteoclast differentiation factor acts as a multifunctional regulator in murine osteoclast differentiation and function. *J Immunol,* **163**(1), 434–442 (1999).

60. T. L. Burgess, *et al.,* The ligand for osteoprotegerin (OPGL) directly activates mature osteoclasts. *J Cell Biol,* **145**(3), 527–538 (1999).

61. A. Mizuno, *et al.,* Transgenic mice overexpressing soluble osteoclast differentiation factor (sODF) exhibit severe osteoporosis. *J Bone Miner Metab,* **20**(6), 337–344 (2002).

62. Y. Y. Kong, W. J. Boyle, and J. M. Penninger, Osteoprotegerin ligand: A common link between osteoclastogenesis, lymph node formation and lymphocyte development. *Immunol Cell Biol,* **77**(2), 188–193 (1999).

63. Y. Y. Kong, *et al.,* OPGL is a key regulator of osteoclastogenesis, lymphocyte development and lymph-node organogenesis. *Nature,* **397**(6717), 315–323 (1999).

64. B. O. Oyajobi, K. Traianedes, T. Yoneda, and G. R. Mundy, Expression of RANK Ligand (RANKL) by myeloma cells requires binding to bone marrow stromal cells via a VCAM-1 interaction. *Bone,* **23**(Suppl), S180 (1998).

65. N. Chikatsu, Y. Takeuchi, T. Kikuchi, E. Ogata, and T. Fujita, Adhesion of cancer cells to bone marrow stromal cells induces ODF/OPGL expression and osteoclast-like cell formation in vitro. *2nd International Conference on Cancer-Induced Bone Disease.* Davos, Switzerland (1999).

66. N. J. Horwood, *et al.,* Activated T lymphocytes support osteoclast formation in vitro. *Biochem Biophys Res Commun,* **265**(1), 144–150 (1999).

67. H. Hsu, *et al.,* Tumor necrosis factor receptor family member RANK mediates osteoclast differentiation and activation induced by osteoprotegerin ligand. *Proc Natl Acad Sci USA,* **96**(7), 3540–3545 (1999).

68. N. Nakagawa, *et al.,* RANK is the essential signaling receptor for osteoclast differentiation factor in osteoclastogenesis. *Biochem Biophys Res Commun,* **253**(2), 395–400 (1998).

69. G. Franzoso, *et al.,* Requirement for NF-kappaB in osteoclast and B-cell development. *Genes Dev,* **11**(24), 3482–3496 (1997).

70. V. Iotsova, *et al.,* Osteopetrosis in mice lacking NF-kappaB1 and NF-kappaB2. *Nat Med,* **3**(11), 1285–1289 (1997).

71. T. Wada, *et al.,* RANKL-RANK signaling in osteoclastogenesis and bone disease. *Trends Mol Med,* **12**(1), 17–25 (2006).

72. H. Takayanagi, Mechanistic insight into osteoclast differentiation in osteoimmunology. *J Mol Med,* **83**(3), 170–179 (2005).

73. L. Galibert, *et al.,* The involvement of multiple tumor necrosis factor receptor (TNFR)-associated factors in the signaling mechanisms of receptor activator of NF kappaB, a member of the TNFR superfamily. *J Biol Chem,* **273**(51), 34120–34127 (1998).

74. R. H. Arch, R. W. Gedrich, and C. B. Thompson, Tumor necrosis factor receptor-associated factors (TRAFs)—A family of adapter proteins that regulates life and death. *Genes Dev,* **12**(18), 2821–2830 (1998).

75. H. H. Kim, *et al.,* Receptor activator of NF-kappaB recruits multiple TRAF family adaptors and activates c-Jun N-terminal kinase. *FEBS Lett,* **443**(3), 297–302 (1999).

76. M. A. Lomaga, *et al.,* TRAF6 deficiency results in osteopetrosis and defective interleukin-1, CD40, and LPS signaling. *Genes Dev,* **13**(8), 1015–1024 (1999).

77. B. R. Wong, *et al.,* The TRAF family of signal transducers mediates NF-kappaB activation by the TRANCE receptor. *J Biol Chem,* **273**(43), 28355–28359 (1998).

78. B. G. Darnay, *et al.,* Activation of NF-kappaB by RANK requires tumor necrosis factor receptor-associated factor (TRAF) 6 and NF-kappaB-inducing kinase. Identification of a novel TRAF6 interaction motif. *J Biol Chem,* **274**(12), 7724–7731 (1999).

79. Z. Cao, *et al.,* TRAF6 is a signal transducer for interleukin-1. *Nature,* **383**(6599), 443–446 (1996).

80. L. C. Yeh, M. C. Zavala, and J. C. Lee, Osteogenic protein-1 and interleukin-6 with its soluble receptor synergistically stimulate rat osteoblastic cell differentiation. *J Cell Physiol,* **190**(3), 322–331 (2002).

81. S. V. Komarova, *et al.,* RANK ligand-induced elevation of cytosolic Ca2+ accelerates nuclear translocation of nuclear factor kappa B in osteoclasts. *J Biol Chem,* **278**(10), 8286–8293 (2003).

82. H. Takayanagi, *et al.,* RANKL maintains bone homeostasis through c-Fos-dependent induction of interferon-beta. *Nature,* **416**(6882), 744–749 (2002).

83. K. Matsuo, *et al.,* Fosl1 is a transcriptional target of c-Fos during osteoclast differentiation. *Nat Genet,* **24**(2), 184–187 (2000).

84. H. Hirotani, *et al.,* The calcineurin/nuclear factor of activated T cells signaling pathway regulates osteoclastogenesis in RAW264.7 cells. *J Biol Chem,* **279**(14), 13984–13992 (2004).

85. K. Matsuo, *et al.,* Nuclear factor of activated T-cells (NFAT) rescues osteoclastogenesis in precursors lacking c-Fos. *J Biol Chem,* **279**(25), 26475–26480 (2004).

86. S. V. Komarova, *et al.,* Convergent signaling by acidosis and receptor activator of NF-kappaB ligand (RANKL) on the calcium/calcineurin/NFAT pathway in osteoclasts. *Proc Natl Acad Sci USA,* **102**(7), 2643–2648 (2005).

87. D. Kim, *et al.,* Retrovirus-mediated gene transfer of RANK-Fc prevents bone loss in ovariectomized mice. *Stem Cells,* **24**, 1798–1805 (2006).

88. W. C. Dougall, *et al.,* RANK is essential for osteoclast and lymph node development. *Genes Dev,* **13**(18), 2412–2424 (1999).

89. J. Li, I. Sarosi, X. Q. Yan, S. M. McCabe, H. L. Tan, C. Caparelli, S. Morony, R. Elliot, G. Van, and S. Kaufman, Absolute requirement for the TNFR-related protein RANK during osteoclastogenesis and in regulation of bone mass and calcium metabolism. *J Bone Miner Res,* **14**(Suppl), S149 (1999).

90. J. J. Body, *et al.,* A study of the biological receptor activator of nuclear factor-kappaB ligand inhibitor, denosumab, in patients with multiple myeloma or bone metastases from breast cancer. *Clin Cancer Res,* **12**(4), 1221–1228 (2006).

91. M. R. McClung, E. M. L., S. B. Cohen, M. A. Bolognese, G. C. Woodson, A. H. Moffett, M. Peacock, P. D. Miller, S. N. Lederman, C. H. Chesnut, D. Lain, A. J. Kivitz, D. L. Holloway, C. Zhang, M. C. Peterson, and P. J. Bekker, AMG 162 Bone Loss Study Group, Denosumab in postmenopausal women with low bone mineral density. *N Engl J Med,* **23**, 821–831 (2006).

92. P. J. Bekker, *et al.*, A single-dose placebo-controlled study of AMG 162, a fully human monoclonal antibody to RANKL, in postmenopausal women. *J Bone Miner Res,* **19**(7), 1059–1066 (2004).

93. E. Tsuda, *et al.*, Isolation of a novel cytokine from human fibroblasts that specifically inhibits osteoclastogenesis. *Biochem Biophys Res Commun,* **234**(1), 137–142 (1997).

94. T. J. Yun, *et al.*, OPG/FDCR-1, a TNF receptor family member, is expressed in lymphoid cells and is up-regulated by ligating CD40. *J Immunol,* **161**(11), 6113–6121 (1998).

95. B. S. Kwon, *et al.*, TR1, a new member of the tumor necrosis factor receptor superfamily, induces fibroblast proliferation and inhibits osteoclastogenesis and bone resorption. *FASEB J,* **12**(10), 845–854 (1998).

96. K. B. Tan, *et al.*, Characterization of a novel TNF-like ligand and recently described TNF ligand and TNF receptor superfamily genes and their constitutive and inducible expression in hematopoietic and non-hematopoietic cells. *Gene,* **204**(1–2), 35–46 (1997).

97. T. Morinaga, *et al.*, Cloning and characterization of the gene encoding human osteoprotegerin/osteoclastogenesis-inhibitory factor. *Eur J Biochem,* **254**(3), 685–691 (1998).

98. J. G. Emery, *et al.*, Osteoprotegerin is a receptor for the cytotoxic ligand TRAIL. *J Biol Chem,* **273**(23), 14363–14367 (1998).

99. K. Yano, *et al.*, Immunological characterization of circulating osteoprotegerin/osteoclastogenesis inhibitory factor: Increased serum concentrations in postmenopausal women with osteoporosis. *J Bone Miner Res,* **14**(4), 518–527 (1999).

100. A. Rogers and R. Eastell, Circulating osteoprotegerin and receptor activator for nuclear factor kappaB ligand: Clinical utility in metabolic bone disease assessment. *J Clin Endocrinol Metab,* **90**(11), 6323–6331 (2005).

101. A. Mizuno, *et al.*, Severe osteoporosis in mice lacking osteoclastogenesis inhibitory factor/osteoprotegerin. *Biochem Biophys Res Commun,* **247**(3), 610–615 (1998).

102. N. Bucay, *et al.*, Osteoprotegerin-deficient mice develop early onset osteoporosis and arterial calcification. *Genes Dev,* **12**(9), 1260–1268 (1998).

103. H. Min, *et al.*, Osteoprotegerin reverses osteoporosis by inhibiting endosteal osteoclasts and prevents vascular calcification by blocking a process resembling osteoclastogenesis. *J Exp Med,* **192**(4), 463–474 (2000).

104. G. E. Wise, *et al.*, CSF-1 regulation of osteoclastogenesis for tooth eruption. *J Dent Res,* **84**(9), 837–881 (2005).

105. O. N. Vidal, *et al.*, Osteoprotegerin mRNA is increased by interleukin-1 alpha in the human osteosarcoma cell line MG-63 and in human osteoblast-like cells. *Biochem Biophys Res Commun,* **248**(3), 696–700 (1998).

106. H. Brandstrom, *et al.*, Regulation of osteoprotegerin mRNA levels by prostaglandin E2 in human bone marrow stroma cells. *Biochem Biophys Res Commun,* **247**(2), 338–341 (1998).

107. H. Yasuda, *et al.*, Osteoclast differentiation factor is a ligand for osteoprotegerin/osteoclastogenesis-inhibitory factor and is identical to TRANCE/RANKL. *Proc Natl Acad Sci USA,* **95**(7), 3597–3602 (1998).

108. L. C. Hofbauer, B. L. Riggs, D. L. Lacey, C. R. Dunstan, T. C. Spelsberg, and S. Khosla, Glucocorticoids stimulate osteoprotegerin ligand and inhibit osteoprotegerin production in human osteoblastic lineage cells: Potential paracrine mechanisms of glucocorticoid-induced osteoporosis. *18th Endocrine Society Meeting,* San Diego, CA (1999).

109. N. O. Vidal, *et al.*, Osteoprotegerin mRNA is expressed in primary human osteoblast-like cells: Down-regulation by glucocorticoids. *J Endocrinol,* **159**(1), 191–195 (1998).

110. Y. Hadeda, *et al.*, Osteoclastogenesis inhibitory factor (OCIF) directly inhibits bone-resorbing activity of isolated mature osteoclasts. *Biochem Biophys Res Commun,* **251**(3), 796–801 (1998).

111. T. Akatsu, *et al.*, Osteoclastogenesis inhibitory factor suppresses osteoclast survival by interfering in the interaction of stromal cells with osteoclast. *Biochem Biophys Res Commun,* **250**(2), 229–234 (1998).

112. M. Sakata, *et al.*, Expression of osteoprotegerin (osteoclastogenesis inhibitory factor) in cultures of human dental mesenchymal cells and epithelial cells. *J Bone Miner Res,* **14**(9), 1486–1492 (1999).

113. T. Akatsu, *et al.*, Osteoclastogenesis inhibitory factor exhibits hypocalcemic effects in normal mice and in hypercalcemic nude mice carrying tumors associated with humoral hypercalcemia of malignancy. *Bone,* **23**(6), 495–498 (1998).

114. M. Yamamoto, *et al.*, Hypocalcemic effect of osteoclastogenesis inhibitory factor/osteoprotegerin in the thyroparathyroidectomized rat. *Endocrinology,* **139**(9), 4012–4015 (1998).

115. B. M. Abdallah, *et al.*, Increased RANKL/OPG mRNA ratio in iliac bone biopsies from women with hip fractures. *Calcif Tissue Int,* **76**(2), 90–97 (2005).

116. E. Grimaud, *et al.*, Receptor activator of nuclear factor kappaB ligand (RANKL)/osteoprotegerin (OPG) ratio is increased in severe osteolysis. *Am J Pathol,* **163**(5), 2021–2031 (2003).

117. P. J. Bekker, *et al.*, The effect of a single dose of osteoprotegerin in postmenopausal women. *J Bone Miner Res,* **16**(2), 348–360 (2001).

118. C. Middleton-Hardie, *et al.*, Deletion of aspartate 182 in OPG causes juvenile Paget's disease by impairing both protein secretion and binding to RANKL. *J Bone Miner Res,* **21**(3), 438–445 (2006).

119. B. Chong, *et al.*, Idiopathic hyperphosphatasia and TNFRSF11B mutations: Relationships between phenotype and genotype. *J Bone Miner Res,* **18**(12), 2095–2104 (2003).

120. M. P. Whyte, *et al.*, Osteoprotegerin deficiency and juvenile Paget's disease. *N Engl J Med,* **347**(3), 175–184 (2002).

121. T. Cundy, *et al.*, A mutation in the gene TNFRSF11B encoding osteoprotegerin causes an idiopathic hyperphosphatasia phenotype. *Hum Mol Genet,* **11**(18), 2119–2127 (2002).

122. T. Cundy, *et al.*, Recombinant osteoprotegerin for juvenile Paget's disease. *N Engl J Med,* **353**(9), 918–923 (2005).

123. P. I. Croucher, *et al.*, Osteoprotegerin inhibits the development of osteolytic bone disease in multiple myeloma. *Blood,* **98**(13), 3534–3540 (2001).

124. S. Morony, *et al.*, A chimeric form of osteoprotegerin inhibits hypercalcemia and bone resorption induced by IL-1beta, TNF-alpha, PTH, PTHrP, and 1,25(OH)2D3. *J Bone Miner Res,* **14**(9), 1478–1485 (1999).

125. J. Padagas, *et al.*, The receptor activator of nuclear factor-kappaB ligand inhibitor osteoprotegerin is a bone-protective agent in a rat model of chronic renal insufficiency and hyperparathyroidism. *Calcif Tissue Int,* **78**(1), 35–44 (2006).

126. C. S. Lader and A. M. Flanagan, Prostaglandin E2, interleukin 1alpha, and tumor necrosis factor-alpha increase human osteoclast formation and bone resorption in vitro. *Endocrinology,* **139**(7), 3157–3164 (1998).

127. U. Sarma and A. M. Flanagan, Macrophage colony-stimulating factor induces substantial osteoclast generation and bone resorption in human bone marrow cultures. *Blood,* **88**(7), 2531–2540 (1996).

128. S. Tanaka, *et al.*, Macrophage colony-stimulating factor is indispensable for both proliferation and differentiation of osteoclast progenitors. *J Clin Invest,* **91**(1), 257–263 (1993).

129. H. Amano, S. Yamada, and R. Felix, Colony-stimulating factor-1 stimulates the fusion process in osteoclasts. *J Bone Miner Res,* **13**(5), 846–853 (1998).

130. E. Jimi, T. Shuto, and T. Koga, Macrophage colony-stimulating factor and interleukin-1 alpha maintain the survival of osteoclast-like cells. *Endocrinology,* **136**(2), 808–811 (1995).

131. E. Jimi, *et al.*, Interleukin 1 induces multinucleation and bone-resorbing activity of osteoclasts in the absence of osteoblasts/stromal cells. *Exp Cell Res,* **247**(1), 84–93 (1999).

132. K. Fuller, *et al.*, Macrophage colony-stimulating factor stimulates survival and chemotactic behavior in isolated osteoclasts. *J Exp Med,* **178**(5), 1733–1744 (1993).

133. H. Yoshida, *et al.*, The murine mutation osteopetrosis is in the coding region of the macrophage colony stimulating factor gene. *Nature,* **345**(6274), 442–444 (1990).

134. D. E. Dobbins, *et al.*, Mutation of macrophage colony stimulating factor (Csf1) causes osteopetrosis in the tl rat. *Biochem Biophys Res Commun,* **294**(5), 1114–1120 (2002).

135. S. Wei, *et al.*, Modulation of CSF-1–regulated post-natal development with anti-CSF-1 antibody. *Immunobiology,* **210**(2–4), 109–119 (2005).

136. X. M. Dai, *et al.*, Incomplete restoration of colony-stimulating factor 1 (CSF-1) function in CSF-1–deficient Csf1op/Csf1op mice by transgenic expression of cell surface CSF-1. *Blood,* **103**(3), 1114–1123 (2004).

137. S. L. Abboud, *et al.*, Osteoblast-specific targeting of soluble colony-stimulating factor-1 increases cortical bone thickness in mice. *J Bone Miner Res,* **18**(8), 1386–1394 (2003).

138. G. Q. Yao, *et al.*, The cell-surface form of colony-stimulating factor-1 is regulated by osteotropic agents and supports formation of multinucleated osteoclast-like cells. *J Biol Chem,* **273**(7), 4119–4128 (1998).

139. J. H. Falkenburg, *et al.*, Differential transcriptional and post-transcriptional regulation of gene expression of the colony-stimulating factors by interleukin-1 and fetal bovine serum in murine fibroblasts. *Blood,* **78**(3), 658–665 (1991).

140. M. L. Sherman, *et al.*, Transcriptional and post-transcriptional regulation of c-jun expression during monocytic differentiation of human myeloid leukemic cells. *J Biol Chem,* **265**(6), 3320–3323 (1990).

141. R. B. Kimble, *et al.*, Estrogen deficiency increases the ability of stromal cells to support murine osteoclastogenesis via an interleukin-1 and tumor necrosis factor-mediated stimulation of macrophage colony-stimulating factor production. *J Biol Chem,* **271**(46), 28890–28897 (1996).

142. H. Kitaura, *et al.*, M-CSF mediates TNF-induced inflammatory osteolysis. *J Clin Invest,* **115**(12), 3418–3427 (2005).

143. S. Srivastava, *et al.*, Estrogen blocks M-CSF gene expression and osteoclast formation by regulating phosphorylation of Egr-1 and its interaction with Sp-1. *J Clin Invest,* **102**(10), 1850–1859 (1998).

144. N. Yamada, *et al.*, Down-regulation of osteoprotegerin production in bone marrow macrophages by macrophage colony-stimulating factor. *Cytokine,* **31**(4), 288–297 (2005).

145. S. Niida, *et al.*, VEGF receptor 1 signaling is essential for osteoclast development and bone marrow formation in colony-stimulating factor 1–deficient mice. *Proc Natl Acad Sci USA,* **102**(39), 14016–14021 (2005).

146. R. Faccio, *et al.*, Vav3 regulates osteoclast function and bone mass. *Nat Med,* **11**(3), 284–290 (2005).

147. P. Soriano, *et al.*, Targeted disruption of the c-src proto-oncogene leads to osteopetrosis in mice. *Cell,* **64**(4), 693–702 (1991).

148. B. F. Boyce, *et al.*, Requirement of pp60c-src expression for osteoclasts to form ruffled borders and resorb bone in mice. *J Clin Invest,* **90**(4), 1622–1627 (1992).

149. R. L. Lees and J. N. Heersche, Macrophage colony stimulating factor increases bone resorption in dispersed osteoclast cultures by increasing osteoclast size. *J Bone Miner Res,* **14**(6), 937–945 (1999).

150. A. Teti, *et al.*, Colony stimulating factor-1–induced osteoclast spreading depends on substrate and requires the vitronectin receptor and the c-src proto-oncogene. *J Bone Miner Res,* **13**(1), 50–58 (1998).

151. K. L. Insogna, *et al.*, Colony-stimulating factor-1 induces cytoskeletal reorganization and c-src–dependent tyrosine phosphorylation of selected cellular proteins in rodent osteoclasts. *J Clin Invest,* **100**(10), 2476–2485 (1997).

152. H. P. Gerber, *et al.*, VEGF couples hypertrophic cartilage remodeling, ossification and angiogenesis during endochondral bone formation. *Nat Med,* **5**(6), 623–628 (1999).

153. V. Midy and J. Plouet, Vasculotropin/vascular endothelial growth factor induces differentiation in cultured osteoblasts. *Biochem Biophys Res Commun,* **199**(1), 380–386 (1994).

154. E. Zelzer, *et al.*, Skeletal defects in VEGF(120/120) mice reveal multiple roles for VEGF in skeletogenesis. *Development,* **129**(8), 1893–1904 (2002).

155. D. Chen, M. Zhao, and G. R. Mundy, Bone morphogenetic proteins. *Growth Factors,* **22**(4), 233–241 (2004).

156. P. J. Bouletreau, *et al.*, Hypoxia and VEGF up-regulate BMP-2 mRNA and protein expression in microvascular endothelial cells: Implications for fracture healing. *Plast Reconstr Surg,* **109**(7), 2384–2397 (2002).

157. R. L. Jilka, *et al.*, Osteoblast programmed cell death (apoptosis): Modulation by growth factors and cytokines. *J Bone Miner Res,* **13**(5), 793–802 (1998).

158. R. A. Johnson, *et al.*, Tumors producing human tumor necrosis factor induced hypercalcemia and osteoclastic bone resorption in nude mice. *Endocrinology,* **124**(3), 1424–1427 (1989).

159. O. Kudo, *et al.*, Proinflammatory cytokine (TNFalpha/IL-1alpha) induction of human osteoclast formation. *J Pathol,* **198**(2), 220–227 (2002).

160. D. R. Bertolini, *et al.*, Stimulation of bone resorption and inhibition of bone formation in vitro by human tumour necrosis factors. *Nature,* **319**(6053), 516–518 (1986).

161. L. Gilbert, *et al.*, Inhibition of osteoblast differentiation by tumor necrosis factor-alpha. *Endocrinology,* **141**(11), 3956–3964 (2000).

162. N. Mayur, *et al.*, Tumor necrosis factor alpha decreases 1,25-dihydroxyvitamin D3 receptors in osteoblastic ROS 17/2.8 cells. *J Bone Miner Res,* **8**(8), 997–1003 (1993).

163. T. Yoneda, *et al.*, Evidence that tumor necrosis factor plays a pathogenetic role in the paraneoplastic syndromes of cachexia, hypercalcemia, and leukocytosis in a human tumor in nude mice. *J Clin Invest,* **87**(3), 977–985 (1991).

164. M. Sabatini, *et al.*, Increased production of tumor necrosis factor by normal immune cells in a model of the humoral hypercalcemia of malignancy. *Lab Invest,* **63**(5), 676–682 (1990).

165. M. Sabatini, *et al.*, Stimulation of tumor necrosis factor release from monocytic cells by the A375 human melanoma via granulocyte-macrophage colony-stimulating factor. *Cancer Res,* **50**(9), 2673–2678 (1990).

166. S. Srivastava, *et al.*, Estrogen decreases TNF gene expression by blocking JNK activity and the resulting production of c-Jun and JunD. *J Clin Invest,* **104**(4), 503–513 (1999).

167. I. R. Garrett, *et al.*, Production of lymphotoxin, a bone-resorbing cytokine, by cultured human myeloma cells. *N Engl J Med,* **317**(9), 526–532 (1987).

168. L. Mantovani, *et al.*, Differential regulation of interleukin-6 expression in human fibroblasts by tumor necrosis factor-alpha and lymphotoxin. *FEBS Lett,* **429**(3), 426 (1998).

169. F. P. Lai, *et al.*, Myeloma cells can directly contribute to the pool of RANKL in bone bypassing the classic stromal and osteoblast pathway of osteoclast stimulation. *Br J Haematol,* **126**(2), 192–201 (2004).

170. T. Tamura, *et al.*, Soluble interleukin-6 receptor triggers osteoclast formation by interleukin 6. *Proc Natl Acad Sci USA,* **90**(24), 11924–11928 (1993).

171. T. Kishimoto, *et al.*, Interleukin-6 family of cytokines and gp130. *Blood,* **86**(4), 1243–1254 (1995).

172. P. C. Heinrich, *et al.*, Interleukin-6–type cytokine signalling through the gp130/Jak/STAT pathway. *Biochem J,* 1998. **334**(Pt 2), 297–314 (1995).

173. G. D. Roodman, Interleukin-6: An osteotropic factor? *J Bone Miner Res,* **7**(5), 475–478 (1992).

174. J. H. Feyen, *et al.*, Interleukin-6 is produced by bone and modulated by parathyroid hormone. *J Bone Miner Res,* **4**(4), 633–638 (1989).

175. Y. Ishimi, *et al.*, IL-6 is produced by osteoblasts and induces bone resorption. *J Immunol,* **145**(10), 3297–3303 (1990).

176. J. de la Mata, *et al.*, Interleukin-6 enhances hypercalcemia and bone resorption mediated by parathyroid hormone-related protein in vivo. *J Clin Invest,* **95**(6), 2846–2852 (1995).

177. K. Kurokouchi, *et al.*, TNF-alpha increases expression of IL-6 and ICAM-1 genes through activation of NF-kappaB in osteoblast-like ROS17/2.8 cells. *J Bone Miner Res,* **13**(8), 1290–1299 (1998).

178. K. Black, I. R. Garrett, and G. R. Mundy, Chinese hamster ovarian cells transfected with the murine interleukin-6 gene cause hypercalcemia as well as cachexia, leukocytosis and thrombocytosis in tumor-bearing nude mice. *Endocrinology,* **128**(5), 2657–2659 (1991).

179. R. Nishimura, *et al.*, Smad5 and DPC4 are key molecules in mediating BMP-2–induced osteoblastic differentiation of the pluripotent mesenchymal precursor cell line C2C12. *J Biol Chem,* **273**(4), 1872–1879 (1998).

180. N. Franchimont, *et al.*, Interleukin-6 with its soluble receptor enhances the expression of insulin-like growth factor-I in osteoblasts. *Endocrinology,* **138**(12), 5248–5255 (1997).

181. T. Iwasaki, *et al.*, Clinical significance of interleukin-6 gene expression in the bone marrow of patients with multiple myeloma. *Int J Hematol,* **70**(3), 163–168 (1999).

182. O. F. Ballester, *et al.*, High levels of interleukin-6 are associated with low tumor burden and low growth fraction in multiple myeloma. *Blood,* **83**(7), 1903–1908 (1994).

183. A. Karadag, A. M. Scutt, and P. I. Croucher, Human myeloma cells promote the recruitment of osteoblast precursors: Mediation by interleukin-6 and soluble interleukin-6 receptor. *J Bone Miner Res,* **15**(10), 1935–1943 (2000).

184. G. Teoh and K. C. Anderson, Interaction of tumor and host cells with adhesion and extracellular matrix molecules in the development of multiple myeloma. *Hematol Oncol Clin North Am,* **11**(1), 27–42 (1997).

185. M. Alsina, E. G., J. Painter, and W. S. Dalton, The effects of interleukin-6 (IL-6) on myeloma cell adhesion to fibronectin (FN) are mediated by macrophage inflammatory protein-1a, (MIP-1a). *Blood,* **98**, 636a (2001).

186. K. Kawasaki, *et al.*, Osteoclasts are present in gp130–deficient mice. *Endocrinology,* **138**(11), 4959–4965 (1997).

187. B. Sun, M. Mitnick, S. Rudikoff, N. Troiana, and K. Insogna, Interleukin-6 knock-out mice have secondary hyperparathyroidism and increased bone density. *J Bone Miner Res,* **15**(Suppl 1), S215 (2002).

188. A. Grey, *et al.*, Circulating levels of interleukin-6 and tumor necrosis factor-alpha are elevated in primary hyperparathyroidism and correlate with markers of bone resorption—A clinical research center study. *J Clin Endocrinol Metab,* **81**(10), 3450–3454 (1996).

189. A. Grey, *et al.*, A role for interleukin-6 in parathyroid hormone-induced bone resorption in vivo. *Endocrinology,* **140**(10), 4683–4690 (1999).

190. G. Guiterrez, J. Rossini, J. Chavez, T. Nishihara, T. Yoneda, and G. R. Mundy, Identification, purification and expression of a naturally-occurring inhibitor of interleukin-6 (IL-6). *J Bone Miner Res,* **9**, 79 (1994).

191. E. W. Yu, *et al.*, Suppression of IL-6 biological activities by activin A and implications for inflammatory arthropathies. *Clin Exp Immunol,* **112**(1), 126–132 (1998).

192. Y. Ogata, *et al.*, A novel role of IL-15 in the development of osteoclasts: Inability to replace its activity with IL-2. *J Immunol,* **162**(5), 2754–2760 (1999).

193. I. B. McInnes, *et al.*, The role of interleukin-15 in T-cell migration and activation in rheumatoid arthritis. *Nat Med,* **2**(2), 175–182 (1996).

194. H. Ruchatz, *et al.*, Soluble IL-15 receptor alpha-chain administration prevents murine collagen-induced arthritis: A role for IL-15 in development of antigen-induced immunopathology. *J Immunol,* **160**(11), 5654–5660 (1998).

195. D. V. Jovanovic, *et al.*, IL-17 stimulates the production and expression of proinflammatory cytokines, IL-beta and TNF-alpha, by human macrophages. *J Immunol,* **160**(7), 3513–3521 (1998).

196. S. Kotake, *et al.*, IL-17 in synovial fluids from patients with rheumatoid arthritis is a potent stimulator of osteoclastogenesis. *J Clin Invest,* **103**(9), 1345–1352 (1999).

197. R. L. Van Bezooijen, *et al.*, Interleukin-17: A new bone acting cytokine in vitro. *J Bone Miner Res,* **14**(9), 1513–1521 (1999).

198. M. Chabaud, *et al.*, IL-17 derived from juxta-articular bone and synovium contributes to joint degradation in rheumatoid arthritis. *Arthritis Res,* **3**(3), 168–177 (2001).

199. M. Chabaud, *et al.*, Human interleukin-17: A T cell-derived proinflammatory cytokine produced by the rheumatoid synovium. *Arthritis Rheum,* **42**(5), 963–970 (1999).

200. E. Thomassen, *et al.*, Binding of interleukin-18 to the interleukin-1 receptor homologous receptor IL-1Rrp1 leads to activation of signaling pathways similar to those used by interleukin-1. *J Interferon Cytokine Res,* **18**(12), 1077–1088 (1998).

201. N. Suzuki, *et al.*, IL-1 receptor-associated kinase 4 is essential for IL-18–mediated NK and Th1 cell responses. *J Immunol,* **170**(8), 4031–4035 (2003).

202. C. Makiishi-Shimobayashi, *et al.*, Interleukin-18 up-regulates osteoprotegerin expression in stromal/osteoblastic cells. *Biochem Biophys Res Commun,* **281**(2), 361–366 (2001).

203. A. Nakata, *et al.*, Inhibition by interleukin 18 of osteolytic bone metastasis by human breast cancer cells. *Anticancer Res,* **19**(5B), 4131–4138 (1999).

204. T. Iwasaki, *et al.*, Interleukin-18 inhibits osteolytic bone metastasis by human lung cancer cells possibly through suppression of osteoclastic bone-resorption in nude mice. *J Immunother,* **25**(Suppl 1), S52–60 (2002).

205. J. E. Aubin, *et al.*, Osteoblast and chondroblast differentiation. *Bone,* **17**(2 Suppl), 77S–83S (1995).

206. J. M. Wozney, *et al.*, Novel regulators of bone formation: Molecular clones and activities. *Science,* **242**(4885), 1528–1534 (1988).

207. J. Massague, TGF-beta signal transduction. *Annu Rev Biochem,* **67**, 753–791 (1998).

208. M. R. Urist, Bone: Formation by autoinduction. *Science,* **150**(698), 893–899 (1965).

209. I. Asahina, T. K. Sampath, and P. V. Hauschka, Human osteogenic protein-1 induces chondroblastic, osteoblastic, and/or adipocytic differentiation of clonal murine target cells. *Exp Cell Res,* **222**(1), 38–47 (1996).

210. M. Ahrens, *et al.*, Expression of human bone morphogenetic proteins-2 or -4 in murine mesenchymal progenitor C3H10T1/2 cells induces differentiation into distinct mesenchymal cell lineages. *DNA Cell Biol,* **12**(10), 871–880 (1993).

211. V. Rosen, *et al.*, Responsiveness of clonal limb bud cell lines to bone morphogenetic protein 2 reveals a sequential relationship between cartilage and bone cell phenotypes. *J Bone Miner Res,* **9**(11), 1759–1768 (1994).

212. U. Styrkarsdottir, *et al.*, Linkage of osteoporosis to chromosome 20p12 and association to BMP2. *PLoS Biol,* **1**(3), E69 (2003).

213. H. Yamashita, *et al.*, Osteogenic protein-1 binds to activin type II receptors and induces certain activin-like effects. *J Cell Biol,* **130**(1), 217–226 (1995).

214. B. L. Rosenzweig, *et al.*, Cloning and characterization of a human type II receptor for bone morphogenetic proteins. *Proc Natl Acad Sci USA,* **92**(17), 7632–7636 (1995).

215. T. Nohno, *et al.*, Identification of a human type II receptor for bone morphogenetic protein-4 that forms differential heteromeric complexes with bone morphogenetic protein type I receptors. *J Biol Chem,* **270**(38), 22522–22526 (1995).

216. M. Kawabata, A. Chytil, and H. L. Moses, Cloning of a novel type II serine/threonine kinase receptor through interaction with the type I transforming growth factor-beta receptor. *J Biol Chem,* **270**(10), 5625–5630 (1995).

217. P. ten Dijke, *et al.*, Identification of type I receptors for osteogenic protein-1 and bone morphogenetic protein-4. *J Biol Chem,* **269**(25), 16985–16988 (1994).

218. B. B. Koenig, *et al.*, Characterization and cloning of a receptor for BMP-2 and BMP-4 from NIH 3T3 cells. *Mol Cell Biol,* **14**(9), 5961–5974 (1994).

219. M. Macias-Silva, *et al.*, Specific activation of Smad1 signaling pathways by the BMP7 type I receptor, ALK2. *J Biol Chem,* **273**(40), 25628–25636 (1998).

220. T. A. Samad, *et al.*, DRAGON, a bone morphogenetic protein co-receptor. *J Biol Chem,* **280**(14), 14122–14129 (2005).

221. J. L. Wrana, Structural biology: On the wings of inhibition. *Nature,* **420**(6916), 613–614 (2002).

222. R. Wieser, J. L. Wrana, and J. Massague, GS domain mutations that constitutively activate T beta R-I, the downstream signaling component in the TGF-beta receptor complex. *EMBO J,* **14**(10), 2199–2208 (1995).

223. J. Carcamo, A. Zentella, and J. Massague, Disruption of transforming growth factor beta signaling by a mutation that prevents transphosphorylation within the receptor complex. *Mol Cell Biol,* **15**(3), 1573–1581 (1995).

224. P. A. Hoodless, *et al.*, MADR1, a MAD-related protein that functions in BMP2 signaling pathways. *Cell,* **85**(4), 489–500 (1996).

225. S. Ebara and K. Nakayama, Mechanism for the action of bone morphogenetic proteins and regulation of their activity. *Spine,* **27**(16 Suppl 1), S10–15 (2002).

226. Y. Chen, A. Bhushan, and W. Vale, Smad8 mediates the signaling of the ALK-2 [corrected] receptor serine kinase. *Proc Natl Acad Sci USA,* **94**(24), 12938–12943 (1997).

227. F. Itoh, *et al.*, The FYVE domain in Smad anchor for receptor activation (SARA) is sufficient for localization of SARA in early endosomes and regulates TGF-beta/Smad signalling. *Genes Cells,* **7**(3), 321–331 (2002).

228. W. Shi, C. Sun, J. Wang, and X. Cao, A new Smad anchor for receptor activation (SARA) for BMP signaling. *J Bone Miner Res,* **19**(Suppl 1), S12 (2004).

229. R. S. Lo, *et al.*, The L3 loop: A structural motif determining specific interactions between SMAD proteins and TGF-beta receptors. *EMBO J,* **17**(4), 996–1005 (1998).

230. S. Bai, *et al.*, Smad6 as a transcriptional corepressor. *J Biol Chem,* **275**(12), 8267–8270 (2000).

231. T. Imamura, *et al.*, Smad6 inhibits signalling by the TGF-beta superfamily. *Nature,* **389**(6651), 622–626 (1997).

232. A. Hata, *et al.*, Smad6 inhibits BMP/Smad1 signaling by specifically competing with the Smad4 tumor suppressor. *Genes Dev,* **12**(2), 186–197 (1998).

233. R. Shen, *et al.*, Smad6 interacts with Runx2 and mediates Smad ubiquitin regulatory factor 1–induced Runx2 degradation. *J Biol Chem,* **281**(6), 3569–3576 (2006).

234. K. Kusanagi, *et al.*, Characterization of a bone morphogenetic protein-responsive Smad-binding element. *Mol Biol Cell,* **11**(2), 555–565 (2000).

235. A. J. Edgar, *et al.*, Bone morphogenetic protein-2 induces expression of murine zinc finger transcription factor ZNF450. *J Cell Biochem,* **94**(1), 202–215 (2005).

236. Y. Ohyama, *et al.*, Spatiotemporal association and bone morphogenetic protein regulation of sclerostin and osterix expression during embryonic osteogenesis. *Endocrinology,* **145**(10), 4685–4692 (2004).

237. K. Yagi, *et al.*, Bone morphogenetic protein-2 enhances osterix gene expression in chondrocytes. *J Cell Biochem,* **88**(6), 1077–1083 (2003).

238. T. Lopez-Rovira, *et al.*, Direct binding of Smad1 and Smad4 to two distinct motifs mediates bone morphogenetic protein-specific transcriptional activation of Id1 gene. *J Biol Chem,* **277**(5), 3176–3185 (2002).

239. T. Katagiri, *et al.*, Identification of a BMP-responsive element in Id1, the gene for inhibition of myogenesis. *Genes Cells,* **7**(9), 949–960 (2002).

240. K. Nakashima, *et al.*, The novel zinc finger-containing transcription factor osterix is required for osteoblast differentiation and bone formation. *Cell,* **108**(1), 17–29 (2002).

241. S. C. Xu, *et al.*, Bone morphogenetic protein-2 (BMP-2) signaling to the Col2alpha1 gene in chondroblasts requires the homeobox gene Dlx-2. *DNA Cell Biol,* **20**(6), 359–365 (2001).

242. P. Ducy, *et al.*, Osf2/Cbfa1: A transcriptional activator of osteoblast differentiation. *Cell,* **89**(5), 747–754 (1997).

243. J. N. Topper, *et al.*, CREB binding protein is a required coactivator for Smad-dependent, transforming growth factor beta transcriptional responses in endothelial cells. *Proc Natl Acad Sci USA,* **95**(16), 9506–9511 (1998).

244. C. Pouponnot, L. Jayaraman, and J. Massague, Physical and functional interaction of SMADs and p300/CBP. *J Biol Chem,* **273**(36), 22865–22868 (1998).

245. S. Mundlos, *et al.*, Mutations involving the transcription factor CBFA1 cause cleidocranial dysplasia. *Cell,* **89**(5), 773–779 (1997).

246. S. Mundlos, *et al.*, Cbfa1, a candidate gene for cleidocranial dysplasia syndrome, is essential for osteoblast differentiation and bone development. *Cell,* **89**(5), 765–771 (1997).

247. T. Komori, *et al.*, Targeted disruption of Cbfa1 results in a complete lack of bone formation owing to maturational arrest of osteoblasts. *Cell,* **89**(5), 755–764 (1997).

248. R. Nishimura, *et al.*, Core-binding factor alpha 1 (Cbfa1) induces osteoblastic differentiation of C2C12 cells without

interactions with Smad1 and Smad5. *Bone,* **31**(2), 303–312 (2002).

249. S. K. Zaidi, *et al.,* Integration of Runx and Smad regulatory signals at transcriptionally active subnuclear sites. *Proc Natl Acad Sci USA,* **99**(12), 8048–8053 (2002).

250. K. S. Lee, *et al.,* Runx2 is a common target of transforming growth factor beta1 and bone morphogenetic protein 2, and cooperation between Runx2 and Smad5 induces osteoblast-specific gene expression in the pluripotent mesenchymal precursor cell line C2C12. *Mol Cell Biol,* **20**(23), 8783–8792 (2000).

251. M. Zhao, *et al.,* E3 ubiquitin ligase Smurf1 mediates core-binding factor alpha1/Runx2 degradation and plays a specific role in osteoblast differentiation. *J Biol Chem,* **278**(30), 27939–27944 (2003).

252. A. Nohe, *et al.,* Signal transduction of bone morphogenetic protein receptors. *Cell Signal,* **16**(3), 291–299 (2004).

253. K. Luo, Negative regulation of BMP signaling by the ski oncoprotein. *J Bone Joint Surg Am,* **85**–A(Suppl 3), 39–43 (2003).

254. W. Wang, *et al.,* Ski represses bone morphogenic protein signaling in *Xenopus* and mammalian cells. *Proc Natl Acad Sci USA,* **97**(26), 14394–14399 (2000).

255. Y. Yoshida, *et al.,* Tob proteins enhance inhibitory Smad-receptor interactions to repress BMP signaling. *Mech Dev,* **120**(5), 629–637 (2003).

256. Y. Yoshida, *et al.,* Negative regulation of BMP/Smad signaling by Tob in osteoblasts. *Cell,* **103**(7), 1085–1097 (2000).

257. M. Usui, *et al.,* Tob deficiency superenhances osteoblastic activity after ovariectomy to block estrogen deficiency-induced osteoporosis. *Proc Natl Acad Sci USA,* **101**(17), 6653–6658 (2004).

258. X. Yang, *et al.,* Smad1 domains interacting with Hoxc-8 induce osteoblast differentiation. *J Biol Chem,* **275**(2), 1065–1072 (2000).

259. X. Shi, *et al.,* Smad1 interacts with homeobox DNA-binding proteins in bone morphogenetic protein signaling. *J Biol Chem,* **274**(19), 13711–13717 (1999).

260. X. Lin, *et al.,* Smad6 recruits transcription corepressor CtBP to repress bone morphogenetic protein-induced transcription. *Mol Cell Biol,* **23**(24), 9081–9093 (2003).

261. S. X. Ying, Z. J. Hussain, and Y. E. Zhang, Smurf1 facilitates myogenic differentiation and antagonizes the bone morphogenetic protein-2–induced osteoblast conversion by targeting Smad5 for degradation. *J Biol Chem,* **278**(40), 39029–39036 (2003).

262. H. Zhu, *et al.,* A SMAD ubiquitin ligase targets the BMP pathway and affects embryonic pattern formation. *Nature,* **400**(6745), 687–693 (1999).

263. C. H. Heldin and P. ten Dijke, SMAD destruction turns off signalling. *Nat Cell Biol,* **1**(8), E195–197 (1999).

264. M. Zhao, *et al.,* Smurf1 inhibits osteoblast differentiation and bone formation in vitro and in vivo. *J Biol Chem,* **279**(13), 12854–12859 (2004).

265. P. L. Kuo, *et al.,* Bone morphogenetic protein-2 and -4 (BMP-2 and -4) mediates fraxetin-induced maturation and differentiation in human osteoblast-like cell lines. *Biol Pharm Bull,* **29**(1), 119–124 (2006).

266. R. S. Thies, *et al.,* Recombinant human bone morphogenetic protein-2 induces osteoblastic differentiation in W-20-17 stromal cells. *Endocrinology,* **130**(3), 1318–1324 (1992).

267. Y. Takuwa, *et al.,* Bone morphogenetic protein-2 stimulates alkaline phosphatase activity and collagen synthesis in cultured osteoblastic cells, MC3T3–E1. *Biochem Biophys Res Commun,* **174**(1), 96–101 (1991).

268. A. Yamaguchi, *et al.,* Recombinant human bone morphogenetic protein-2 stimulates osteoblastic maturation and inhibits myogenic differentiation in vitro. *J Cell Biol,* **113**(3), 681–687 (1991).

269. D. Chen, *et al.,* Bone morphogenetic protein 2 (BMP-2) enhances BMP-3, BMP-4, and bone cell differentiation marker gene expression during the induction of mineralized bone matrix formation in cultures of fetal rat calvarial osteoblasts. *Calcif Tissue Int,* **60**(3), 283–290 (1997).

270. S. E. Harris, *et al.,* Expression of bone morphogenetic protein messenger RNA in prolonged cultures of fetal rat calvarial cells. *J Bone Miner Res,* **9**(3), 389–394 (1994).

271. A. Nakashima, T. Katagiri, and M. Tamura, Cross-talk between Wnt and bone morphogenetic protein 2 (BMP-2) signaling in differentiation pathway of C2C12 myoblasts. *J Biol Chem,* **280**(45), 37660–37668 (2005).

272. T. Katagiri, *et al.,* Bone morphogenetic protein-2 converts the differentiation pathway of C2C12 myoblasts into the osteoblast lineage. *J Cell Biol,* **127**(6 Pt 1), 1755–1766 (1994).

273. D. Chen, *et al.,* Differential roles for bone morphogenetic protein (BMP) receptor type IB and IA in differentiation and specification of mesenchymal precursor cells to osteoblast and adipocyte lineages. *J Cell Biol,* **142**(1), 295–305 (1998).

274. M. H. Lee, *et al.,* Transient upregulation of CBFA1 in response to bone morphogenetic protein-2 and transforming growth factor beta1 in C2C12 myogenic cells coincides with suppression of the myogenic phenotype but is not sufficient for osteoblast differentiation. *J Cell Biochem,* **73**(1), 114–125 (1999).

275. M. H. Lee, *et al.,* BMP-2–induced Osterix expression is mediated by Dlx5 but is independent of Runx2. *Biochem Biophys Res Commun,* **309**(3), 689–694 (2003).

276. M. H. Lee, *et al.,* BMP-2–induced Runx2 expression is mediated by Dlx5, and TGF-beta 1 opposes the BMP-2–induced osteoblast differentiation by suppression of Dlx5 expression. *J Biol Chem,* **278**(36), 34387–34394 (2003).

277. M. H. Lee, *et al.,* Dlx5 specifically regulates Runx2 type II expression by binding to homeodomain-response elements in the Runx2 distal promoter. *J Biol Chem,* **280**(42), 35579–35587 (2005).

278. T. Tadic, *et al.,* Overexpression of Dlx5 in chicken calvarial cells accelerates osteoblastic differentiation. *J Bone Miner Res,* **17**(6), 1008–1014 (2002).

279. K. Miyama, *et al.,* A BMP-inducible gene, Dlx5, regulates osteoblast differentiation and mesoderm induction. *Dev Biol,* **208**(1), 123–133 (1999).

280. D. Acampora, *et al.,* Craniofacial, vestibular and bone defects in mice lacking the Distal-less-related gene Dlx5. *Development,* **126**(17), 3795–3809 (1999).

281. H. M. Ryoo, M. H. Lee, and Y. J. Kim, Critical molecular switches involved in BMP-2–induced osteogenic differentiation of mesenchymal cells. *Gene,* **366**(1), 51–57 (2006).

282. H. M. Kronenberg, Developmental regulation of the growth plate. *Nature,* **423**(6937), 332–336 (2003).

283. B. S. Yoon and K. M. Lyons, Multiple functions of BMPs in chondrogenesis. *J Cell Biochem,* **93**(1), 93–103 (2004).

284. T. Kobayashi, *et al.,* BMP signaling stimulates cellular differentiation at multiple steps during cartilage development. *Proc Natl Acad Sci USA,* **102**(50), 18023–18027 (2005).

285. C. M. Shea, *et al.,* BMP treatment of C3H10T1/2 mesenchymal stem cells induces both chondrogenesis and osteogenesis. *J Cell Biochem,* **90**(6), 1112–1127 (2003).

286. M. K. Majumdar, E. Wang, and E. A. Morris, BMP-2 and BMP-9 promotes chondrogenic differentiation of human

multipotential mesenchymal cells and overcomes the inhibitory effect of IL-1. *J Cell Physiol*, **189**(3), 275–284 (2001).

287. J. Kramer, *et al.*, Embryonic stem cell-derived chondrogenic differentiation in vitro: Activation by BMP-2 and BMP-4. *Mech Dev*, **92**(2), 193–205 (2000).

288. W. Ju, *et al.*, The bone morphogenetic protein 2 signaling mediator Smad1 participates predominantly in osteogenic and not in chondrogenic differentiation in mesenchymal progenitors C3H10T1/2. *J Bone Miner Res*, **15**(10), 1889–1899 (2000).

289. S. Pizette and L. Niswander, BMPs are required at two steps of limb chondrogenesis: Formation of prechondrogenic condensations and their differentiation into chondrocytes. *Dev Biol*, **219**(2), 237–249 (2000).

290. L. J. Brunet, *et al.*, Noggin, cartilage morphogenesis, and joint formation in the mammalian skeleton. *Science*, **280**(5368), 1455–1457 (1998).

291. M. Wahl, *et al.*, Transcriptome analysis of early chondrogenesis in ATDC5 cells induced by bone morphogenetic protein 4. *Genomics*, **83**(1), 45–58 (2004).

292. C. Shukunami, *et al.*, Requirement of autocrine signaling by bone morphogenetic protein-4 for chondrogenic differentiation of ATDC5 cells. *FEBS Lett*, **469**(1), 83–87 (2000).

293. R. Fernandez-Lloris, *et al.*, Induction of the Sry-related factor SOX6 contributes to bone morphogenetic protein-2–induced chondroblastic differentiation of C3H10T1/2 cells. *Mol Endocrinol*, **17**(7), 1332–1343 (2003).

294. F. P. Luyten, Cartilage-derived morphogenetic protein-1. *Int J Biochem Cell Biol*, **29**(11), 1241–1244 (1997).

295. E. E. Storm, *et al.*, Limb alterations in brachypodism mice due to mutations in a new member of the TGF beta-superfamily. *Nature*, **368**(6472), 639–643 (1994).

296. J. T. Thomas, *et al.*, A human chondrodysplasia due to a mutation in a TGF-beta superfamily member. *Nat Genet*, **12**(3), 315–317 (1996).

297. S. E. Yi, *et al.*, The type I BMP receptor BMPRIB is required for chondrogenesis in the mouse limb. *Development*, **127**(3), 621–630 (2000).

298. B. S. Yoon, *et al.*, BMPr1a and BMPr1b have overlapping functions and are essential for chondrogenesis in vivo. *Proc Natl Acad Sci USA*, **102**(14), 5062–5067 (2005).

299. H. C. Anderson, *et al.*, Bone morphogenetic protein (BMP) localization in developing human and rat growth plate, metaphysis, epiphysis, and articular cartilage. *J Histochem Cytochem*, **48**(11), 1493–1502 (2000).

300. E. Minina, *et al.*, Interaction of FGF, Ihh/Pthlh, and BMP signaling integrates chondrocyte proliferation and hypertrophic differentiation. *Dev Cell*, **3**(3), 439–449 (2002).

301. E. Minina, *et al.*, BMP and Ihh/PTHrP signaling interact to coordinate chondrocyte proliferation and differentiation. *Development*, **128**(22), 4523–4534 (2001).

302. F. De Luca, *et al.*, Regulation of growth plate chondrogenesis by bone morphogenetic protein-2. *Endocrinology*, **142**(1), 430–436 (2001).

303. J. Zhang, *et al.*, Smad4 is required for the normal organization of the cartilage growth plate. *Dev Biol*, **284**(2), 311–322 (2005).

304. S. Provot and E. Schipani, Molecular mechanisms of endochondral bone development. *Biochem Biophys Res Commun*, **328**(3), 658–665 (2005).

305. M. F. Bastida, *et al.*, Levels of Gli3 repressor correlate with BMP4 expression and apoptosis during limb development. *Dev Dyn*, **231**(1), 148–160 (2004).

306. F. Liu, J. Massague, and A. Ruiz i Altaba, Carboxy-terminally truncated Gli3 proteins associate with Smads. *Nat Genet*, **20**(4), 325–326 (1998).

307. F. Long, *et al.*, Ihh signaling is directly required for the osteoblast lineage in the endochondral skeleton. *Development*, **131**(6), 1309–1318 (2004).

308. S. Kawai and T. Sugiura, Characterization of human bone morphogenetic protein (BMP)-4 and -7 gene promoters: Activation of BMP promoters by Gli, a sonic hedgehog mediator. *Bone*, **29**(1), 54–61 (2001).

309. M. Zhao, *et al.*, Bone morphogenetic protein receptor signaling is necessary for normal murine postnatal bone formation. *J Cell Biol*, **157**(6), 1049–1060 (2002).

310. S. P. Oh and E. Li, Gene-dosage-sensitive genetic interactions between inversus viscerum (iv), nodal, and activin type IIB receptor (ActRIIB) genes in asymmetrical patterning of the visceral organs along the left-right axis. *Dev Dyn*, **224**(3), 279–290 (2002).

311. Y. Mishina, *et al.*, Bone morphogenetic protein type IA receptor signaling regulates postnatal osteoblast function and bone remodeling. *J Biol Chem*, **279**(26), 27560–27566 (2004).

312. G. Luo, *et al.*, BMP-7 is an inducer of nephrogenesis, and is also required for eye development and skeletal patterning. *Genes Dev*, **9**(22), 2808–2820 (1995).

313. R. A. Meyer, Jr., *et al.*, Gene expression in older rats with delayed union of femoral fractures. *J Bone Joint Surg Am*, **85-A**(7), 1243–1254 (2003).

314. A. Matsumoto, *et al.*, Effect of aging on bone formation induced by recombinant human bone morphogenetic protein-2 combined with fibrous collagen membranes at subperiosteal sites. *J Periodontal Res*, **36**(3), 175–182 (2001).

315. R. Takae, *et al.*, Immunolocalization of bone morphogenetic protein and its receptors in degeneration of intervertebral disc. *Spine*, **24**(14), 1397–1401 (1999).

316. Y. Yazaki, *et al.*, Immunohistochemical localization of bone morphogenetic proteins and the receptors in epiphyseal growth plate. *Anticancer Res*, **18**(4A), 2339–2344 (1998).

317. J. C. Fleet, *et al.*, The effects of aging on the bone inductive activity of recombinant human bone morphogenetic protein-2. *Endocrinology*, **137**(11), 4605–4610 (1996).

318. N. Ghosh-Choudhury, *et al.*, Expression of the BMP 2 gene during bone cell differentiation. *Crit Rev Eukaryot Gene Expr*, **4**(2–3), 345–355 (1994).

319. M. Guttenbach, *et al.*, Chromosomal localization of the genes encoding ALDH, BMP-2, R-FABP, IFN-gamma, RXR-gamma, and VIM in chicken by fluorescence in situ hybridization. *Cytogenet Cell Genet*, **88**(3–4), 266–271 (2000).

320. J. Q. Feng, *et al.*, Structure and sequence of mouse bone morphogenetic protein-2 gene (BMP-2): Comparison of the structures and promoter regions of BMP-2 and BMP-4 genes. *Biochim Biophys Acta*, **1218**(2), 221–224 (1994).

321. J. Q. Feng, *et al.*, Bone morphogenetic protein 2 transcripts in rapidly developing deer antler tissue contain an extended 5 non-coding region arising from a distal promoter. *Biochim Biophys Acta*, **1350**(1), 47–52 (1997).

322. I. R. Garrett, *et al.*, Selective inhibitors of the osteoblast proteasome stimulate bone formation in vivo and in vitro. *J Clin Invest*, **111**(11), 1771–1782 (2003).

323. G. Mundy, *et al.*, Stimulation of bone formation in vitro and in rodents by statins. *Science*, **286**(5446), 1946–1949 (1999).

324. S. Zhou, *et al.*, Estrogens activate bone morphogenetic protein-2 gene transcription in mouse mesenchymal stem cells. *Mol Endocrinol*, **17**(1), 56–66 (2003).

325. T. Arikawa, K. Omura, and I. Morita, Regulation of bone morphogenetic protein-2 expression by endogenous prostaglandin E2 in human mesenchymal stem cells. *J Cell Physiol*, **200**(3), 400–406 (2004).

326. A. S. Virdi, *et al.*, Modulation of bone morphogenetic protein-2 and bone morphogenetic protein-4 gene expression in osteoblastic cell lines. *Cell Mol Biol (Noisy-le-grand)*, **44**(8), 1237–1246 (1998).

327. W. M. Knosp, *et al.*, HOXA13 regulates the expression of bone morphogenetic proteins 2 and 7 to control distal limb morphogenesis. *Development*, **131**(18), 4581–4592 (2004).

328. K. L. Abrams, *et al.*, An evolutionary and molecular analysis of BMP2 expression. *J Biol Chem*, **279**(16), 15916–15928 (2004).

329. L. M. Helvering, *et al.*, Regulation of the promoters for the human bone morphogenetic protein 2 and 4 genes. *Gene*, **256**(1–2), 123–138 (2000).

330. L. C. Heller, *et al.*, Transcriptional regulation of the BMP2 gene. Retinoic acid induction in F9 embryonal carcinoma cells and *Saccharomyces cerevisiae*. *J Biol Chem*, **274**(3), 1394–1400 (1999).

331. M. J. Fowler, Jr., *et al.*, Induction of bone morphogenetic protein-2 by interleukin-1 in human fibroblasts. *Biochem Biophys Res Commun*, **248**(3), 450–453 (1998).

332. N. Ghosh-Choudhury, *et al.*, Requirement of BMP-2–induced phosphatidylinositol 3–kinase and Akt serine/threonine kinase in osteoblast differentiation and Smad-dependent BMP-2 gene transcription. *J Biol Chem*, **277**(36), 33361–33368 (2002).

333. N. Ghosh-Choudhury, *et al.*, Autoregulation of mouse BMP-2 gene transcription is directed by the proximal promoter element. *Biochem Biophys Res Commun*, **286**(1), 101–108 (2001).

334. N. Ghosh-Choudhury, *et al.*, Immortalized murine osteoblasts derived from BMP 2–T-antigen expressing transgenic mice. *Endocrinology*, **137**(1), 331–339 (1996).

335. R. Mo, *et al.*, Specific and redundant functions of Gli2 and Gli3 zinc finger genes in skeletal patterning and development. *Development*, **124**(1), 113–123 (1997).

336. M. Zhao, J. A. Sterling, M. Qiao, B. O. Oyajobi, S. E. Harris, and G. R. Mundy, The Hedgehog signaling molecule Gli2 regulates both BMP2 and PTHrP expression in the growth plate. *J Bone Miner Res*, **20**(Suppl 1), S40 (2005).

337. M. Zhao, M. Qiao, J. A. Sterling, D. Chen, B. O. Oyajobi, I. R. Garrett, G. G. Guiterrez Rossini, and G. R. Mundy, The zinc finger transcription factor Gli2 enhances BMP-2 gene transcription, osteoblast differentiation and bone formation. *J Bone Miner Res*, **19**(Suppl 1), S46 (2004).

338. M. Zhao, G. Rossini, M. Qiao, S. E. Harris, G. R. Mundy, and D. Chen, Identification of the Gli family of zinc finger proteins as powerful regulators of BMP-2 transcription in osteoblasts. *J Bone Miner Res*, **18**(Suppl 2), S59 (2003).

339. J. Q. Feng, *et al.*, NF-kappaB specifically activates BMP-2 gene expression in growth plate chondrocytes in vivo and in a chondrocyte cell line in vitro. *J Biol Chem*, **278**(31), 29130–29135 (2003).

340. M. Yanagita, BMP antagonists: Their roles in development and involvement in pathophysiology. *Cytokine Growth Factor Rev*, **16**(3), 309–317 (2005).

341. E. Canalis, A. N. Economides, and E. Gazzerro, Bone morphogenetic proteins, their antagonists, and the skeleton. *Endocr Rev*, **24**(2), 218–235 (2003).

342. A. H. Reddi, Interplay between bone morphogenetic proteins and cognate binding proteins in bone and cartilage development: Noggin, chordin and DAN. *Arthritis Res*, **3**(1), 1–5 (2001).

343. J. Massague and Y. G. Chen, Controlling TGF-beta signaling. *Genes Dev*, **14**(6), 627–644 (2000).

344. O. Avsian-Kretchmer and A. J. Hsueh, Comparative genomic analysis of the eight-membered ring cystine knot-containing bone morphogenetic protein antagonists. *Mol Endocrinol*, **18**(1), 1–12 (2004).

345. U. A. Vitt, S. Y. Hsu, and A. J. Hsueh, Evolution and classification of cystine knot-containing hormones and related extracellular signaling molecules. *Mol Endocrinol*, **15**(5), 681–694 (2001).

346. M. Wijgerde, *et al.*, Noggin antagonism of BMP4 signaling controls development of the axial skeleton in the mouse. *Dev Biol*, **286**(1), 149–157 (2005).

347. D. M. Valenzuela, *et al.*, Identification of mammalian noggin and its expression in the adult nervous system. *J Neurosci*, **15**(9), 6077–6084 (1995).

348. W. C. Smith and R. M. Harland, Expression cloning of noggin, a new dorsalizing factor localized to the Spemann organizer in *Xenopus* embryos. *Cell*, **70**(5), 829–840 (1992).

349. L. B. Zimmerman, J. M. De Jesus-Escobar, and R. M. Harland, The Spemann organizer signal noggin binds and inactivates bone morphogenetic protein 4. *Cell*, **86**(4), 599–606 (1996).

350. X. B. Wu, *et al.*, Impaired osteoblastic differentiation, reduced bone formation, and severe osteoporosis in noggin-overexpressing mice. *J Clin Invest*, **112**(6), 924–934 (2003).

351. R. D. Devlin, *et al.*, Skeletal overexpression of noggin results in osteopenia and reduced bone formation. *Endocrinology*, **144**(5), 1972–1978 (2003).

352. E. Gazzerro, *et al.*, Overexpression of twisted gastrulation inhibits bone morphogenetic protein action and prevents osteoblast cell differentiation in vitro. *Endocrinology*, **146**(9), 3875–3882 (2005).

353. E. Gazzerro, V. Gangji, and E. Canalis, Bone morphogenetic proteins induce the expression of noggin, which limits their activity in cultured rat osteoblasts. *J Clin Invest*, **102**(12), 2106–2114 (1998).

354. J. Marcelino, *et al.*, Human disease-causing NOG missense mutations: Effects on noggin secretion, dimer formation, and bone morphogenetic protein binding. *Proc Natl Acad Sci USA*, **98**(20), 11353–11358 (2001).

355. Y. Gong, *et al.*, Heterozygous mutations in the gene encoding noggin affect human joint morphogenesis. *Nat Genet*, **21**(3), 302–304 (1999).

356. J. Nickel, *et al.*, The crystal structure of the BMP-2:BMPR-IA complex and the generation of BMP-2 antagonists. *J Bone Joint Surg Am*, **83–A**(Suppl 1, Pt 1), S7–14 (2001).

357. T. Kirsch, W. Sebald, and M. K. Dreyer, Crystal structure of the BMP-2–BRIA ectodomain complex. *Nat Struct Biol*, **7**(6), 492–496 (2000).

358. J. Groppe, *et al.*, Structural basis of BMP signaling inhibition by noggin, a novel twelve-membered cystine knot protein. *J Bone Joint Surg Am*, **85–A**(Suppl 3), 52–58 (2003).

359. J. Groppe, *et al.*, Structural basis of BMP signalling inhibition by the cystine knot protein noggin. *Nature*, **420**(6916), 636–642 (2002).

360. S. Piccolo, *et al.*, Dorsoventral patterning in *Xenopus*: Inhibition of ventral signals by direct binding of chordin to BMP-4. *Cell*, **86**(4), 589–598 (1996).

361. Y. Sasai, *et al.*, *Xenopus* chordin: A novel dorsalizing factor activated by organizer-specific homeobox genes. *Cell*, **79**(5), 779–790 (1994).

362. G. Tardif, *et al.*, Differential regulation of the bone morphogenic protein antagonist chordin in human normal and osteoarthritic chondrocytes. *Ann Rheum Dis*, **65**(2), 261–264 (2006).

363. D. Zhang, *et al.*, A role for the BMP antagonist chordin in endochondral ossification. *J Bone Miner Res*, **17**(2), 293–300 (2002).

364. I. C. Scott, *et al.*, Homologues of twisted gastrulation are extracellular cofactors in antagonism of BMP signalling. *Nature,* **410**(6827), 475–478 (2001).

365. C. Chang, *et al.*, Twisted gastrulation can function as a BMP antagonist. *Nature,* **410**(6827), 483–487 (2001).

366. J. J. Ross, *et al.*, Twisted gastrulation is a conserved extracellular BMP antagonist. *Nature,* **410**(6827), 479–483 (2001).

367. M. Oelgeschlager, *et al.*, The evolutionarily conserved BMP-binding protein twisted gastrulation promotes BMP signalling. *Nature,* **405**(6788), 757–763 (2000).

368. S. Piccolo, *et al.*, Cleavage of Chordin by Xolloid metalloprotease suggests a role for proteolytic processing in the regulation of Spemann organizer activity. *Cell,* **91**(3), 407–416 (1997).

369. P. Blader, *et al.*, Cleavage of the BMP-4 antagonist chordin by zebrafish tolloid. *Science,* **278**(5345), 1937–1940 (1997).

370. W. Balemans, *et al.*, A generalized skeletal hyperostosis in two siblings caused by a novel mutation in the SOST gene. *Bone,* **36**(6), 943–947 (2005).

371. W. Balemans, *et al.*, Increased bone density in sclerosteosis is due to the deficiency of a novel secreted protein (SOST). *Hum Mol Genet,* **10**(5), 537–543 (2001).

372. M. E. Brunkow, *et al.*, Bone dysplasia sclerosteosis results from loss of the SOST gene product, a novel cystine knot-containing protein. *Am J Hum Genet,* **68**(3), 577–589 (2001).

373. D. G. Winkler, *et al.*, Osteocyte control of bone formation via sclerostin, a novel BMP antagonist. *EMBO J,* **22**(23), 6267–6276 (2003).

374. N. Kusu, *et al.*, Sclerostin is a novel secreted osteoclast-derived bone morphogenetic protein antagonist with unique ligand specificity. *J Biol Chem,* **278**(26), 24113–24117 (2003).

375. R. L. Van Bezooijen, S. E. Papapoulos, and C. W. Lowik, Bone morphogenetic proteins and their antagonists: The sclerostin paradigm. *J Endocrinol Invest,* **28**(8 Suppl), 15–17 (2005).

376. D. G. Winkler, *et al.*, Sclerostin inhibition of Wnt-3a-induced C3H10T1/2 cell differentiation is indirect and mediated by bone morphogenetic proteins. *J Biol Chem,* **280**(4), 2498–2502 (2005).

377. M. K. Sutherland, *et al.*, Sclerostin promotes the apoptosis of human osteoblastic cells: A novel regulation of bone formation. *Bone,* **35**(4), 828–835 (2004).

378. D. G. Winkler, *et al.*, Noggin and sclerostin bone morphogenetic protein antagonists form a mutually inhibitory complex. *J Biol Chem,* **279**(35), 36293–36298 (2004).

379. D. R. Hsu, *et al.*, The *Xenopus* dorsalizing factor Gremlin identifies a novel family of secreted proteins that antagonize BMP activities. *Mol Cell,* **1**(5), 673–683 (1998).

380. L. Z. Topol, *et al.*, DRM/GREMLIN (CKTSF1B1) maps to human chromosome 15 and is highly expressed in adult and fetal brain. *Cytogenet Cell Genet,* **89**(1–2), 79–84 (2000).

381. O. Michos, *et al.*, Gremlin-mediated BMP antagonism induces the epithelial-mesenchymal feedback signaling controlling metanephric kidney and limb organogenesis. *Development,* **131**(14), 3401–3410 (2004).

382. E. Gazzerro, *et al.*, Skeletal overexpression of Gremlin impairs bone formation and causes osteopenia. *Endocrinology,* **146**(2), 655–665 (2005).

383. S. D. Boden, The ABCs of BMPs. *Orthop Nurs,* **24**(1), 49–52, quiz 53–54 (2005).

384. T. Sakou, Bone morphogenetic proteins: From basic studies to clinical approaches. *Bone,* **22**(6), 591–603 (1998).

385. J. P. Fiorellini, *et al.*, Randomized study evaluating recombinant human bone morphogenetic protein-2 for extraction socket augmentation. *J Periodontol,* **76**(4), 605–613 (2005).

386. M. Boakye, *et al.*, Anterior cervical discectomy and fusion involving a polyetheretherketone spacer and bone morphogenetic protein. *J Neurosurg Spine,* **2**(5), 521–525 (2005).

387. P. V. Mummaneni, *et al.*, Contribution of recombinant human bone morphogenetic protein-2 to the rapid creation of interbody fusion when used in transforaminal lumbar interbody fusion: A preliminary report. Invited submission from the Joint Section Meeting on Disorders of the Spine and Peripheral Nerves, March 2004. *J Neurosurg Spine,* **1**(1), 19–23 (2004).

388. T. H. Lanman and T. J. Hopkins, Lumbar interbody fusion after treatment with recombinant human bone morphogenetic protein-2 added to poly(L-lactide-co-D,L-lactide) bioresorbable implants. *Neurosurg Focus,* **16**(3), E9 (2004).

389. R. E. Jung, *et al.*, Effect of rhBMP-2 on guided bone regeneration in humans. *Clin Oral Implants Res,* **14**(5), 556–568 (2003).

390. T. H. Lanman, *et al.*, Use of rhBMP-2 in combination with structural cortical allografts: Clinical and radiographic outcomes in anterior lumbar spinal surgery. *J Bone Joint Surg Am,* **87**(6), 1205–1212 (2005).

391. T. H. Lanman, *et al.*, The effectiveness of rhBMP-2 in replacing autograft: An integrated analysis of three human spine studies. *Orthopedics,* **27**(7), 723–728 (2004).

392. T. H. Lanman, J. D. Dorchak, and D. L. Sanders, Radiographic assessment of interbody fusion using recombinant human bone morphogenetic protein type 2. *Spine,* **28**(4), 372–377 (2003).

393. S. Govender, *et al.*, Recombinant human bone morphogenetic protein-2 for treatment of open tibial fractures: A prospective, controlled, randomized study of four hundred and fifty patients. *J Bone Joint Surg Am,* **84–A**(12), 2123–2134 (2002).

394. G. Turgeman, *et al.*, Systemically administered rhBMP-2 promotes MSC activity and reverses bone and cartilage loss in osteopenic mice. *J Cell Biochem,* **86**(3), 461–474 (2002).

395. H. Oxlund and T. T. Andreassen, Simvastatin treatment partially prevents ovariectomy-induced bone loss while increasing cortical bone formation. *Bone,* **34**(4), 609–618 (2004).

396. P. W. Ingham and A. P. McMahon, Hedgehog signaling in animal development: Paradigms and principles. *Genes Dev,* **15**(23), 3059–3087 (2001).

397. B. St-Jacques, M. Hammerschmidt, and A. P. McMahon, Indian hedgehog signaling regulates proliferation and differentiation of chondrocytes and is essential for bone formation. *Genes Dev,* **13**(16), 2072–2086 (1999).

398. M. M. Cohen, Jr., The hedgehog signaling network. *Am J Med Genet A,* **123**(1), 5–28 (2003).

399. K. Koebernick and T. Pieler, Gli-type zinc finger proteins as bipotential transducers of hedgehog signaling. *Differentiation,* **70**(2–3), 69–76 (2002).

400. J. L. Mullor, *et al.*, Wnt signals are targets and mediators of Gli function. *Curr Biol,* **11**(10), 769–773 (2001).

401. R. Brewster, J. L. Mullor, and A. Ruiz i Altaba, Gli2 functions in FGF signaling during antero-posterior patterning. *Development,* **127**(20), 4395–4405 (2000).

402. P. Aza-Blanc, *et al.*, Proteolysis that is inhibited by hedgehog targets *Cubitus interruptus* protein to the nucleus and converts it to a repressor. *Cell,* **89**(7), 1043–1053 (1997).

403. A. Ruiz i Altaba, Gli proteins encode context dependent positive and negative functions: Implications for development and disease. *Development,* **126**(14), 3205–3216 (1999).

404. B. Wang, J. F. Fallon, and P. A. Beachy, Hedgehog-regulated processing of Gli3 produces an anterior/posterior repressor

gradient in the developing vertebrate limb. *Cell,* **100**(4), 423–434 (2000).

405. H. Sasaki, *et al.,* A binding site for Gli proteins is essential for HNF-3beta floor plate enhancer activity in transgenics and can respond to Shh in vitro. *Development,* **124**(7), 1313–1322 (1997).

406. D. Miao, *et al.,* Impaired endochondral bone development and osteopenia in Gli2-deficient mice. *Exp Cell Res,* **294**(1), 210–222 (2004).

407. H. L. Park, *et al.,* Mouse Gli1 mutants are viable but have defects in SHH signaling in combination with a Gli2 mutation. *Development,* **127**(8), 1593–1605 (2000).

408. C. C. Hui and A. L. Joyner, A mouse model of greig cephalopolysyndactyly syndrome: The extra-toes mutation contains an intragenic deletion of the Gli3 gene. *Nat Genet,* **3**(3), 241–246 (1993).

409. S. J. Karp, *et al.,* Indian hedgehog coordinates endochondral bone growth and morphogenesis via parathyroid hormone related-protein-dependent and -independent pathways. *Development,* **127**(3), 543–548 (2000).

410. A. Vortkamp, *et al.,* Regulation of rate of cartilage differentiation by Indian hedgehog and PTH-related protein. *Science,* **273**(5275), 613–622 (1996).

411. L. Koziel, *et al.,* Gli3 acts as a repressor downstream of Ihh in regulating two distinct steps of chondrocyte differentiation. *Development,* **132**(23), 5249–5260 (2005).

412. J. A. Sterling, B. O. Oyajobi, M. Zhao, A. Gupta, M. Banerjee, and G. R. Mundy, PTH-rP Expression in Cartilage Cells is Regulated by the Gli Family of Transcription Factors. American Society for Bone and Mineral Research 26th Annual Meeting. Seattle, Washington (2004).

413. S. Pathi, *et al.,* Interaction of Ihh and BMP/noggin signaling during cartilage differentiation. *Dev Biol,* **209**(2), 239–353 (1999).

414. L. C. Murtaugh, J. H. Chyung, and A. B. Lassar, Sonic hedgehog promotes somitic chondrogenesis by altering the cellular response to BMP signaling. *Genes Dev,* **13**(2), 225–237 (1999).

415. U. I. Chung, *et al.,* Indian hedgehog couples chondrogenesis to osteogenesis in endochondral bone development. *J Clin Invest,* **107**(3), 295–304 (2001).

416. G. van der Horst, *et al.,* Hedgehog stimulates only osteoblastic differentiation of undifferentiated KS483 cells. *Bone,* **33**(6), 899–910 (2003).

417. H. Hu, *et al.,* Sequential roles of Hedgehog and Wnt signaling in osteoblast development. *Development,* **132**(1), 49–60 (2005).

418. W. Balemans and W. Van Hul, Identification of the disease-causing gene in sclerosteosis—Discovery of a novel bone anabolic target? *J Musculoskelet Neuronal Interact,* **4**(2), 139–142 (2004).

419. G. G. Loots, *et al.,* Genomic deletion of a long-range bone enhancer misregulates sclerostin in Van Buchem disease. *Genome Res,* **15**(7), 928–935 (2005).

420. R. L. Van Bezooijen, *et al.,* Sclerostin is an osteocyte-expressed negative regulator of bone formation, but not a classical BMP antagonist. *J Exp Med,* **199**(6), 805–814 (2004).

421. X. Li, *et al.,* Sclerostin binds to LRP5/6 and antagonizes canonical Wnt signaling. *J Biol Chem,* **280**(20), 19883–19887 (2005).

422. B. Lanske, *et al.,* PTH/PTHrP receptor in early development and Indian hedgehog-regulated bone growth. *Science,* **273**(5275), 663–666 (1996).

423. A. C. Karaplis, *et al.,* Lethal skeletal dysplasia from targeted disruption of the parathyroid hormone-related peptide gene. *Genes Dev,* **8**(3), 277–289 (1994).

424. T. A. Guise, *et al.,* Evidence for a causal role of parathyroid hormone-related protein in the pathogenesis of human breast cancer-mediated osteolysis. *J Clin Invest,* **98**(7), 1544–1549 (1996).

425. M. J. Horwitz, *et al.,* Short-term, high-dose parathyroid hormone-related protein as a skeletal anabolic agent for the treatment of postmenopausal osteoporosis. *J Clin Endocrinol Metab,* **88**(2), 569–575 (2003).

426. D. Miao, *et al.,* Osteoblast-derived PTHrP is a potent endogenous bone anabolic agent that modifies the therapeutic efficacy of administered PTH 1–34. *J Clin Invest,* **115**(9), 2402–2411 (2005).

427. P. Ducy, *et al.,* Leptin inhibits bone formation through a hypothalamic relay: A central control of bone mass. *Cell,* **100**, 197–207 (2000).

428. T. A. Cock, *et al.,* Enhanced bone formation in lipodystrophic PPARgamma(hyp/hyp) mice relocates haematopoiesis to the spleen. *EMBO Rep,* **5**(10), 1007–1012 (2004).

429. F. Elefteriou, *et al.,* Serum leptin level is a regulator of bone mass. *Proc Natl Acad Sci USA,* **101**(9), 3258–3263 (2004).

430. J. E. Reseland, *et al.,* Leptin is expressed in and secreted from primary cultures of human osteoblasts and promotes bone mineralization. *J Bone Miner Res,* **16**(8), 1426–1433 (2001).

431. T. Thomas, *et al.,* Leptin acts on human marrow stromal cells to enhance differentiation to osteoblasts and to inhibit differentiation to adipocytes. *Endocrinology,* **140**(4), 1630–1638 (1999).

432. S. Takeda, *et al.,* Leptin regulates bone formation via the sympathetic nervous system. *Cell,* **111**(3), 305–317 (2002).

433. P. A. Baldock, *et al.,* Hypothalamic Y2 receptors regulate bone formation. *J Clin Invest,* **109**(7), 915–921 (2002).

434. P. A. Baldock, *et al.,* Hypothalamic control of bone formation: Distinct actions of leptin and y2 receptor pathways. *J Bone Miner Res,* **20**(10), 1851–1857 (2005).

435. F. Elefteriou, *et al.,* Leptin regulation of bone resorption by the sympathetic nervous system and CART. *Nature,* **434**, 514–520 (2005).

436. A. I. Idris., *et al.,* Regulation of bone mass, bone loss and osteoclast activity by cannabinoid receptors. *Nat Med,* **11**(7), 774–779 (2005).

437. O. Ofek, *et al.,* Peripheral cannabinoid receptor, CB2, regulates bone mass. *Proc Natl Acad Sci USA,* **103**(3), 696–701 (2006).

438. A. Bajayo, *et al.,* Central IL-1 receptor signaling regulates bone growth and mass. *Proc Natl Acad Sci USA,* **102**(36), 12956–12961 (2005).

439. L. Rejnmark, *et al.,* Fracture risk in perimenopausal women treated with beta-blockers. *Calcif Tissue Int,* **75**(5), 365–372 (2004).

440. R. Levasseur, *et al.,* Beta-blocker use, bone mineral density, and fracture risk in older women: Results from the Epidemiologie de l'Osteoporose prospective study. *J Am Geriatr Soc,* **53**(3), 550–552 (2005).

441. I. R. Reid, *et al.,* Effects of a beta-blocker on bone turnover in normal postmenopausal women: A randomized controlled trial. *J Clin Endocrinol Metab,* **90**(9), 5212–5216 (2005).

442. I. R. Reid, *et al.,* Beta-blocker use, BMD, and fractures in the study of osteoporotic fractures. *J Bone Miner Res,* **20**(4), 613–618 (2005).

443. J. Pasco, *et al.,* Beta-adrenergic blockers reduce the risk of fracture partly by increasing bone mineral density: Geelong Osteoporosis Study. *J Bone Miner Res,* **19**, 19–24 (2003).

444. R. G. Schlienger, *et al.,* Use of beta-blockers and risk of fractures. *JAMA,* **292**(11), 1326–1332 (2004).

445. S. Imai, *et al.,* Calcitonin gene-related peptide, substance P, and tyrosine hydroxylase-immunoreactive innervation of rat bone marrows: An immunohistochemical and ultrastructural

investigation on possible efferent and afferent mechanisms. *J Orthop Res,* **15**(1), 133–140 (1997).

446. S. Imai and Y. Matsusue, Neuronal regulation of bone metabolism and anabolism: Calcitonin gene-related peptide-, substance P-, and tyrosine hydroxylase-containing nerves and the bone. *Microsc Res Tech,* **58**(2), 61–69 (2002).

447. S. G. Amara, *et al.,* Alternative RNA processing in calcitonin gene expression generates mRNAs encoding different polypeptide products. *Nature,* **298**(5871), 240–244 (1982).

448. A. Vignery and T. L. McCarthy, The neuropeptide calcitonin gene-related peptide stimulates insulin-like growth factor I production by primary fetal rat osteoblasts. *Bone,* **18**(4), 331–335 (1996).

449. J. Cornish, *et al.,* Comparison of the effects of calcitonin gene-related peptide and amylin on osteoblasts. *J Bone Miner Res,* **14**(8), 1302–1309 (1999).

450. T. Kawase and D. M. Burns, Calcitonin gene-related peptide stimulates potassium efflux through adenosine triphosphate-sensitive potassium channels and produces membrane hyperpolarization in osteoblastic UMR106 cells. *Endocrinology,* **139**(8), 3492–3502 (1998).

451. A. Bjurholm, *et al.,* Neuroendocrine regulation of cyclic AMP formation in osteoblastic cell lines (UMR-106–01, ROS 17/2.8, MC3T3–E1, and Saos-2) and primary bone cells. *J Bone Miner Res,* **7**(9), 1011–1019 (1992).

452. V. P. Michelangeli, *et al.,* Effects of calcitonin gene-related peptide on cyclic AMP formation in chicken, rat, and mouse bone cells. *J Bone Miner Res,* **4**(2), 269–272 (1989).

453. H. Drissi, *et al.,* Expression of the CT/CGRP gene and its regulation by dibutyryl cyclic adenosine monophosphate in human osteoblastic cells. *J Bone Miner Res,* **12**(11), 1805–1814 (1997).

454. R. Ballica, *et al.,* Targeted expression of calcitonin gene-related peptide to osteoblasts increases bone density in mice. *J Bone Miner Res,* **14**(7), 1067–1074 (1999).

455. T. Schinke, *et al.,* Decreased bone formation and osteopenia in mice lacking alpha-calcitonin gene-related peptide. *J Bone Miner Res,* **19**(12), 2049–2056 (2004).

456. E. L. Hill and R. Elde, Distribution of CGRP-, VIP-, D beta H-, SP-, and NPY-immunoreactive nerves in the periosteum of the rat. *Cell Tissue Res,* **264**(3), 469–480 (1991).

457. H. Mukohyama, *et al.,* The inhibitory effects of vasoactive intestinal peptide and pituitary adenylate cyclase-activating polypeptide on osteoclast formation are associated with upregulation of osteoprotegerin and downregulation of RANKL and RANK. *Biochem Biophys Res Commun,* **271**(1), 158–163 (2000).

458. P. Lundberg, *et al.,* Vasoactive intestinal peptide regulates osteoclast activity via specific binding sites on both osteoclasts and osteoblasts. *Bone,* **27**(6), 803–810 (2000).

459. M. Ransjo, *et al.,* Microisolated mouse osteoclasts express VIP-1 and PACAP receptors. *Biochem Biophys Res Commun,* **274**(2), 400–404 (2000).

460. E. L. Hohmann, L. Levine, and A. H. Tashjian, Jr., Vasoactive intestinal peptide stimulates bone resorption via a cyclic adenosine 3′,5′-monophosphate-dependent mechanism. *Endocrinology,* **112**(4), 1233–1239 (1983).

461. S. Rahman, *et al.,* The regulation of connective tissue metabolism by vasoactive intestinal polypeptide. *Regul Pept,* **37**(2), 111–121 (1992).

462. S. C. Offley, *et al.,* Capsaicin-sensitive sensory neurons contribute to the maintenance of trabecular bone integrity. *J Bone Miner Res,* **20**(2), 257–267 (2005).

463. J. Pearson, *et al.,* The sural nerve in familial dysautonomia. *J Neuropathol Exp Neurol,* **34**(5), 413–424 (1975).

464. C. Maayan, *et al.,* Bone mineral density and metabolism in familial dysautonomia. *Osteoporos Int,* **13**(5), 429–433 (2002).

465. M. Bliziotes, *et al.,* Bone histomorphometric and biomechanical abnormalities in mice homozygous for deletion of the dopamine transporter gene. *Bone,* **26**(1), 15–19 (2000).

466. M. M. Bliziotes, *et al.,* Neurotransmitter action in osteoblasts: Expression of a functional system for serotonin receptor activation and reuptake. *Bone,* **29**(5), 477–486 (2001).

467. I. Westbroek, *et al.,* Expression of serotonin receptors in bone. *J Biol Chem,* **276**(31), 28961–28968 (2001).

468. N. Weintrob, *et al.,* Decreased growth during therapy with selective serotonin reuptake inhibitors. *Arch Pediatr Adolesc Med,* **156**(7), 696–701 (2002).

469. Warden, S. J., *et al.,* Inhibition of the serotonin (5-hydroxytryptamine) transporter reduces bone accrual during growth. *Endocrinology,* **146**(2), 685–693 (2005).

470. A. E. Goodship, L. E. Lanyon, and H. McFie, Functional adaptation of bone to increased stress. An experimental study. *J Bone Joint Surg Am,* **61**(4), 539–546 (1979).

471. R. Battaglino, *et al.,* Serotonin regulates osteoclast differentiation through its transporter. *J Bone Miner Res,* **19**(9), 1420–1431 (2004).

472. P. S. Bhangu, *et al.,* Evidence for targeted vesicular glutamate exocytosis in osteoblasts. *Bone,* **29**(1), 16–23 (2001).

473. Y. Gu and S. J. Publicover, Expression of functional metabotropic glutamate receptors in primary cultured rat osteoblasts. Cross-talk with N-methyl-D-aspartate receptors. *J Biol Chem,* **275**(44), 34252–34259 (2000).

474. D. J. Mason, *et al.,* Mechanically regulated expression of a neural glutamate transporter in bone: A role for excitatory amino acids as osteotropic agents? *Bone,* **20**(3), 199–205 (1997).

475. C. Chenu, *et al.,* Glutamate receptors are expressed by bone cells and are involved in bone resorption. *Bone,* **22**(4), 295–299 (1998).

476. C. Chenu, Glutamatergic regulation of bone remodeling. *J Musculoskelet Neuronal Interact,* **2**(3), 282–284 (2002).

477. C. Chenu, Glutamatergic innervation in bone. *Microsc Res Tech,* **58**(2), 70–76 (2002).

478. C. M. Serre, *et al.,* Evidence for a dense and intimate innervation of the bone tissue, including glutamate-containing fibers. *Bone,* **25**(6), 623–629 (1999).

479. T. G. Smart, Regulation of excitatory and inhibitory neurotransmitter-gated ion channels by protein phosphorylation. *Curr Opin Neurobiol,* **7**(3), 358–367 (1997).

480. A. J. Patton, *et al.,* Expression of an N-methyl-D-aspartate-type receptor by human and rat osteoblasts and osteoclasts suggests a novel glutamate signaling pathway in bone. *Bone,* **22**(6), 645–649 (1998).

481. L. Espinosa, *et al.,* Active NMDA glutamate receptors are expressed by mammalian osteoclasts. *J Physiol,* **518**(Pt 1), 47–53 (1999).

482. N. M. Peet, *et al.,* The glutamate receptor antagonist MK801 modulates bone resorption in vitro by a mechanism predominantly involving osteoclast differentiation. *FASEB J,* **13**(15), 2179–2185 (1999).

483. B. Merle, *et al.,* NMDA glutamate receptors are expressed by osteoclast precursors and involved in the regulation of osteoclastogenesis. *J Cell Biochem,* **90**(2), 424–436 (2003).

484. E. Hinoi, S. Fujimori, and Y. Yoneda, Modulation of cellular differentiation by N-methyl-D-aspartate receptors in osteoblasts. *FASEB J,* **17**(11), 1532–1534 (2003).

485. A. F. Taylor, Osteoblastic glutamate receptor function regulates bone formation and resorption. *J Musculoskelet Neuronal Interact,* **2**(3), 285–290 (2002).

486. P. G. Genever and T. M. Skerry, Regulation of spontaneous glutamate release activity in osteoblastic cells and its role in differentiation and survival: Evidence for intrinsic glutamatergic signaling in bone. *FASEB J,* **15**(9), 1586–1588 (2001).
487. T. M. Skerry and P. G. Genever, Glutamate signalling in non-neuronal tissues. *Trends Pharmacol Sci,* **22**(4), 174–181 (2001).
488. I. Laketic-Ljubojevic, *et al.,* Functional characterization of N-methyl-D-aspartic acid-gated channels in bone cells. *Bone,* **25**(6), 631–637 (1999).
489. A. R. Mohn, *et al.,* Mice with reduced NMDA receptor expression display behaviors related to schizophrenia. *Cell,* **98**(4), 427–436 (1999).
490. C. Gray, *et al.,* Glutamate does not play a major role in controlling bone growth. *J Bone Miner Res,* **16**(4), 742–749 (2001).
491. D. J. Mason, Glutamate signalling and its potential application to tissue engineering of bone. *Eur Cell Mater,* **7,** 12–25; discussion 25–26 (2004).

CHAPTER 17

Skeletal Growth Factors

ERNESTO CANALIS

I. INTRODUCTION

Bone formation and resorption are regulated by systemic and local factors acting in concert to maintain bone mass. Calciotropic and steroid hormones have been studied extensively for their effects on bone remodeling. However, there is compelling evidence to support the concept that systemic and locally produced growth factors play a central role in the regulation of bone remodeling. Growth factors regulate the replication, differentiation, and function of bone cells. This chapter will be limited to the description of factors that have a major effect on bone formation and osteoblastic function, whereas other chapters of this book will describe cytokines that regulate bone resorption. This is a somewhat arbitrary division since bone remodeling is coupled and osteoclastogenesis is highly dependent on osteoblastic signals. Furthermore, cytokines with primary effects on cells of the osteoclast lineage also play a role in the process of bone formation.

Although skeletal cells synthesize a variety of factors, some skeletal growth factors, such as platelet-derived growth factor (PDGF), vascular endothelial growth factor (VEGF), fibroblast growth factor (FGF), transforming growth factor β (TGF β), bone morphogenetic proteins (BMP), and insulin-like growth factors (IGF) have been studied in more detail. Some of these factors act as bone cell mitogens, and as such they are important in the maintenance of an adequate number of skeletal cells. They also may increase bone cell replication when additional cells are needed, such as during fracture healing and repair. A factor may play a role in the differentiation of cells and osteoblastogenesis, or may stimulate the differentiated function of mature cells. There are no growth factors specifically synthesized by skeletal cells, and those known as skeletal growth factors also are expressed in various nonskeletal tissues. However, growth factors are regulated specifically in bone at the level of synthesis or activity by agents that act primarily on this tissue. Growth factors synthesized by skeletal cells may be present in the systemic circulation, and act both as local and systemic regulators of bone remodeling. The source of circulating growth factors varies, although it is frequently the liver, the circulating platelets, or peripheral tissues. The systemic form of a factor can be regulated by agents and mechanisms different from those affecting the locally produced factor, and it is conceivable that the roles of the circulating and local form of a growth factor differ. This is not only because their synthesis is regulated by different hormones, but also because they may become available under different physiological or pathophysiological circumstances. It is tempting to believe that the local form of a growth factor plays a more immediate, and possibly important role in the control of cell function since it has a more direct access to its target cell. Locally synthesized growth factors can act either as autocrine factors and affect cells of the same class, or paracrine factors and affect different or adjacent cells.

In this chapter the function and regulation of selected growth factors will be discussed and their relevance to skeletal physiology will be considered.

II. PLATELET-DERIVED GROWTH FACTOR

PDGF was originally isolated from human platelets, and four members of the *pdgf* gene family have been identified—*pdgf a*, *pdgf b*, *pdgf c*, and *pdgf d* [1]. VEGF shares a high degree of sequence homology with PDGF and these factors are often referred to as members of the PDGF/VEGF family [2]. This family of genes encodes a highly conserved cystine knot motif. The recently discovered *pdgf c* and *pdgf d* genes have a Clr/Cls, urchin endothelial growth factor, BMP 1 (CUB) domain linked to the cystine knot core motif by a hinge domain [3–6]. Several reports have indicated that proteolytic release of the core from the CUB domain is required for the activation of PDGF C and PDGF D, although activity for the full-length peptide and independent activity for the CUB domain have been postulated [3, 5, 6]. PDGFs must form homo- or heterodimers to exhibit activity, and they can form PDGF AA, BB, AB, CC, or DD dimers [3]. The various *pdgf* genes are conspicuously expressed during development and in adult tissues. Transcripts for the *pdgf a*, *pdgf b*, and *pdgf c* genes are detected in osteoblasts, but their basal expression is relatively low [3, 7–9]. PDGF D is synthesized by myocardial and vascular cells, where it induces cell proliferation and fibrosis, but there are no reports of PDGF D synthesis by the osteoblast [10]. Since PDGF is present in circulating platelets, it can act as a local and systemic regulator of cell function [11]. The five PDGF isoforms described can interact with either one of two PDGF receptors, which have differential binding specificity for the various PDGF dimers [12–15]. PDGF receptor (PDGFR) α ligates PDGF A, B, and C chains and PDGFR β binds PDGF B and D chains [3]. The two PDGF receptors are structurally and functionally related, and PDGF binding results in receptor dimerization and the formation of PDGF αα, ββ, and αβ receptor dimers [3]. For receptor activation, PDGF AA and PDGF CC require PDGFR αα, or αβ dimers, PDGF DD requires PDGFR ββ, or αβ dimers, whereas PDGF AB and PDGF BB can activate either PDGFR αα, αβ, or ββ dimers.

PDGF AA, AB, and BB are the isoforms studied more extensively in skeletal cells, and they exert similar biological actions. However, in skeletal as well as non-skeletal cells, PDGF BB is more potent than PDGF AA, and PDGF AB has intermediate activity [16]. The primary action of PDGF in bone is the stimulation of DNA synthesis and of cell replication. Histomorphometric analysis reveals an increase in cells of the osteoblastic lineage, but the effect is not specific and PDGF causes a generalized stimulation of cell replication in bone [17]. The mitogenic effect is observed primarily in the periosteal layer, a zone rich in fibroblasts and preosteoblasts. It could be presumed that preosteoblastic cells, replicating under the influence of PDGF, eventually differentiate into mature osteoblasts. However, an inhibitory effect on the differentiation of stromal cells into cells of the osteoblastic lineage has been reported [18]. This would indicate that the cells affected could remain in a proliferative undifferentiated state. Some cells, responding to other local signals, may differentiate, and as a consequence of the increased cell number, a modest increase in collagen synthesis is observed following exposure to PDGF. It is important to note that, in accordance with the impaired cell differentiation, PDGF inhibits the expression of the mature osteoblastic

phenotype, and decreases alkaline phosphatase activity and type I collagen mRNA levels [16, 17, 19]. In calvariae, PDGF inhibits mineral apposition rates. These effects may be direct and indirect since PDGF inhibits the synthesis of IGF-I and -II in osteoblasts, and IGF-I and -II enhance osteoblastic functions [20, 21]. PDGF enhances bone resorption by increasing the number of osteoclasts, an effect that may be secondary to an increase in the expression of interleukin (IL)-6, a cytokine known to induce osteoclastogenesis [22]. In agreement with its effects on bone resorption, PDGF increases the expression of matrix metalloproteinases (MMP) by the osteoblast [23, 24]. PDGF increases the rate of transcription of the collagenase 3 (MMP-13) gene, and mRNA stability in transcriptionally arrested osteoblasts. MMP-13 is a proteinase capable of initiating the degradation of type I collagen at neutral pH, and necessary to achieve a bone resorptive response to parathyroid hormone (PTH) [25].

Cells of the osteoblastic lineage express PDGF α and β receptors. PDGF binding to its osteoblast receptor results in receptor dimerization, and activation of tyrosine kinase activity, leading to activation of protein kinase C (PKC), and intracellular calcium signaling pathways [26, 27]. In rodent, but not in human osteoblasts, IL-1 increases PDGF α receptor transcripts and the binding and mitogenic activity of PDGF AA [16, 28, 29]. TGF β decreases PDGF binding, and hormones have no effect on PDGF binding to osteoblasts [30]. Information on the activity of PDGF CC and PDGF DD in skeletal cells is limited. PDGF CC interacts with PDGF αα and αβ receptor dimers, and has potent mitogenic activity for mesenchymal cells [4]. PDGF CC induces the differentiation and regulation of endothelial cells and has potent angiogenic properties, directly or indirectly by upregulating VEGF [31]. These properties of PDGF CC could be important during the vascularization of endochondral bone formation, as it has been described for VEGF. PDGF DD interacts with PDGF β receptors, has mitogenic activity for vascular cells, and induces tissue fibrosis, but its effects on the skeleton are not known [10].

Although there is considerable knowledge about the actions of PDGF *in vitro*, information about its effects on the skeleton *in vivo* is more limited. Consistent with some of its *in vitro* effects, the systemic administration of PDGF BB to ovariectomized rats prevents bone loss, and increases the number of osteoblasts and bone formation [32]. It is likely that the mitogenic effects of PDGF on preosteoblasts result in an increased number of osteoblasts, which are capable of forming bone. PDGF does not change osteoclast number when administered systemically to ovariectomized rats, but this may be related to this particular model where the ovariectomy causes a substantial increase in bone resorption and remodeling, precluding an additional effect by PDGF. Topical application of PDGF to craniotomy defects in rodents stimulates soft tissue repair, but not osteogenesis [33]. The effects of PDGF on endothelial cell proliferation and angiogenesis are likely beneficial to the process of wound healing, and PDGF accelerates the healing response of wounds due to an increase in cellularity and in the formation of granulation tissue [34].

Genetically engineered mice with gain and loss of function mutations have provided important information on the physiological role of PDGF during development and post-natality. Null mutations of *pdgf b*, *pdgf* α and β receptors cause embryonic lethality, and *pdgf a* deletions cause prenatal and perinatal death [35]. Therefore, these models have not allowed for the study of the function of PDGF in the postnatal skeleton. *pdgf b* and *pdgf* β receptor null mice develop microvascular bleeding and absent vascular and mesangial cells [36, 37]. *pdgf a* and *pdgf* α receptor null mutants have defective alveolar formation in the lungs leading to emphysema, and have reduced intestinal villi, thin dermis, and spermatogenic arrest, but *pdgf a* null mutants do not manifest a skeletal phenotype [38, 39]. *pdgf c* null mice exhibit neonatal lethality and numerous skeletal developmental abnormalities, including cleft palate and spina bifida [40]. Similar defects are observed in *pdgf* α receptor null mice, which exhibit a phenotype characterized by embryonic lethality, cleft face, spina bifida, and vascular and skeletal defects [38, 39]. Although PDGF B can interact with the PDGF α receptor, loss of function mutations of the *pdgf b* gene do not resemble the *pdgf* α receptor null phenotype, indicating that the functions of the PDGF α and β receptors are not redundant [36, 37]. The phenotype of *pdgf d* gene deletion has not been reported, but overexpression of PDGF DD, like that of PDGF CC, results in tissue fibrosis [10, 41]. This may be secondary to the mitogenic properties of PDGF CC and PDGF DD or due to the induction of tissue inhibitors of metalloproteinases.

The major source of PDGF is the systemic circulation, and skeletal cells probably become exposed to significant concentrations of PDGF following platelet aggregation. Nevertheless, skeletal cells express the *pdgf a*, *pdgf b*, and *pdgf c* genes indicating that PDGF isoforms may act as autocrine regulators of skeletal cell function [8, 9, 42]. The expression of the *pdgf a* gene is enhanced by TGF β and by PDGF, and that of *pdgf b* enhanced by TGF β [9, 42]. Following an initial induction of PDGF A, an autoregulatory mechanism may serve to maintain local levels of the growth factor. The regulation of *pdgf* gene expression in skeletal

cells is analogous to its regulation in nonskeletal cells indicating that there are no specific transcription factors responsible for the expression of PDGF in bone. Systemic hormones are not known to regulate *pdgf* gene expression in osteoblasts. Since both TGF β and PDGF are released by platelets following platelet aggregation, the subsequent induction of PDGF by these factors in the bone microenvironment may be a mechanism to ensure adequate levels of PDGF in skeletal tissues in conditions that follow platelet aggregation, such as fracture repair. Under basal conditions there may be no need for skeletal cells to be exposed to significant concentrations of PDGF, and its levels are low [9, 42].

III. VASCULAR ENDOTHELIAL GROWTH FACTOR

VEGF shares sequence homology and angiogenic properties with PDGFs, and often they are referred to as members of the PDGF/VEGF family [3, 43]. VEGF A belongs to a gene family also composed of *placenta growth factor*, *vegf b*, *vegf c*, and *vegf d* [43]. In the mouse, differential splicing results in three VEGF A isoforms, VEGF A 120, VEGF A 164, and VEGF A 188. VEGF A is essential for angiogenesis and *vegf a* and *vegf receptor 1* and *2* genes are expressed by chondrocytes and osteoblasts [44, 45]. VEGF A is required for blood vessel formation and vessel invasion into cartilage during the process of endochondral bone formation, and for chondrocyte survival during skeletal development [46, 47]. The expression of *vegf a* by chondrocytes requires runt related transcription factor-2 (Runx-2), a transcription factor essential for normal osteogenesis [47]. VEGF A also is required for intramembranous bone formation, it enhances osteoblastic maturation *in vitro*, and VEGF receptor 1 signaling is essential for osteoclast development [47–49]. *vegf* α gene null deletions are lethal due to defective hematopoiesis and defective blood vessel formation, but mice expressing one of three VEGF A isoforms, VEGF A 120, survive and their study has documented the function of VEGF A on endochondral bone formation and osteoblastic maturation [50]. Bone histomorphometric analysis following adenoviral vector delivery of VEGF A also demonstrates a stimulatory effect of VEGF on osteoblast number, and *in vivo* VEGF A promotes angiogenesis and fracture repair [45, 51].

The expression of VEGF by osteoblasts is regulated by various growth factors. Activation of mitogen activated protein (MAP) kinases by BMP, TGF β, or FGF-2 and activation of phosphatidylinositol-3 (PI-3) kinase by IGF-I induces *vegf* expression in osteoblasts [52–55]. Following growth factor–induced synthesis, VEGF may serve to couple angiogenesis and osteoblastic differentiation and function [46,56]. The stimulatory effect of IGF-I on *vegf a* gene expression is mediated by the hypoxia inducible factor (HIF) 2 α. This may result in an increased vascular supply to the local environment, and may serve as a protective mechanism in response to changes in oxygen availability to the osteoblast [52]. It is of interest that the phenotypes of the conditional deletion of *hif-1* and of *vegf-a* in cartilage are analogous [47].

IV. FIBROBLAST GROWTH FACTOR

FGFs form a family of at least 23 structurally related polypeptides, characterized by their affinity to glycosaminoglycan heparin binding sites [57]. FGFs were initially isolated from the central nervous system and subsequently found in a variety of tissues, where they regulate cell function [58, 59]. Skeletal cells synthesize both FGF-1 and -2, the forms of FGF most extensively studied for their actions on the skeleton [60,61].

FGF has mitogenic activity in skeletal and nonskeletal cells and potent angiogenic properties [62–64]. FGF increases a population of cells of the osteoblastic lineage, which differentiate into osteoblasts [63]. However, FGF, like other potent mitogens, does not enhance the differentiated function of the osteoblast directly, and FGF-2 inhibits alkaline phosphatase activity, type I collagen, osteocalcin and osteopontin synthesis, independently of its stimulatory effects on osteoblastic cell growth [65,66]. These effects are paralleled *in vivo* and transgenic mice overexpressing FGF-2 are osteopenic, although *fgf 2* null mice exhibit impaired bone formation [67,68]. The inhibitory effect of FGF-2 on osteoblast differentiation is secondary to the induction of the transcription factor Sox 2 and the inhibition of Wnt signaling, which is essential for osteoblastogenesis [69,70]. Confirming an inhibitory effect on osteoblastic function, FGF-2 suppresses the synthesis of IGF-I, a factor that stimulates the differentiated function of the osteoblast [20]. FGF-2 increases bone resorption by favoring osteoclastogenesis, and stimulates the synthesis of collagenase 3 (MMP-13) by the osteoblast [71,72]. Marrow stromal cells from *fgf-2* null mice exhibit decreased osteoclastogenesis in response to PTH. FGF-2 induces TFG β1 transcription, and TGF β could mediate selected actions of FGF in bone [73].

The actions of FGF can be regulated by modifications in the affinity or number of FGF receptors in target cells. FGF receptors are a family of four distinct receptors, FGFR1 through 4 [74–76]. FGF receptors are transmembrane protein responsive kinases with two to three immunoglobulin-like domains. Differential splicing of the extracellular region of FGF receptors can generate receptor variants with different ligand binding specificity [74,76]. The four known receptors, and variants, bind the members of the FGF family of polypeptides with different affinity, and have different signaling and mitogenic potential. Activation of FGFR1, 2, and 3 by FGFs induces a mitogenic response, whereas activation of FGFR4 does not [75]. Accordingly, mutations of the *fgfr-1, -2,* or *-3* cause diverse skeletal syndromes including achondroplasia, a common cause of dwarfism [77]. Studies on FGF receptors in preosteoblasts and osteoblasts have been limited. Activation of the signal transducers and activators of transcription 1 regulates FGF receptor in skeletal cells, and suppresses FGFR-3 expression in osteoblasts and mediates FGF-2 actions in chondrocytes [78,79].

In vivo administration of FGF-2 promotes bone and cartilage repair and topical administration of FGF-2 accelerates fracture healing in rodents and nonhuman primates [80,81]. The mechanism is related to the mitogenic properties of FGF-2 in skeletal cells in conjunction with its angiogenic activity. Although studies in *fgf-2* null mice demonstrate that FGF-2 is required for bone formation, transgenic mice overexpressing FGF-2 in the bone environment are osteopenic [67,68]. This observation is in accordance with *in vitro* studies demonstrating an inhibitory effect of FGF-2 on Wnt signaling, IGF-I expression, and osteoblastic function [20,70]. Embryos harboring null mutations of the *fgf receptor 2* die prior to skeletogenesis, but conditional inactivation of the FGF receptor 2 results in skeletal dwarfism secondary to decreased proliferation of osteoprogenitors without affecting osteoblast differentiation [82]. These observations confirm early *in vitro* studies demonstrating a central role of FGF-2 on preosteoblastic cell proliferation, but not a stimulatory effect on osteoblastic function.

Investigations on the regulation of FGF synthesis in skeletal cells are limited to FGF-2. In fibroblasts, heat shock induces the release of FGF, suggesting that its secretion is related to cellular stress [83]. The regulation of FGF-2 expression in osteoblasts and nonskeletal fibroblasts is similar, and TGF β1 and FGF-2 increase FGF-2 synthesis in osteoblasts [61]. Sequence analysis of the FGF-2 promoter reveals AP-1 recognition sequences; therefore, signals that induce Fos and Jun, components of the AP-1 complex, have the potential to upregulate *fgf-2* gene transcription [84]. PDGF enhances the expression of FGF-2 mRNA levels in fibroblasts, and possibly in osteoblasts. Therefore, skeletal growth factors with mitogenic properties, such as FGF-2 and PDGF, are major inducers of FGF-2 synthesis. PTH causes a transient induction of FGF-2 transcripts in osteoblasts [85]. FGF is stored in the extracellular matrix, where it is complexed by extracellular matrix and cell surface–associated heparan sulfate proteoglycans, which may prolong its half-life and allow future interactions with its receptors [86]. Binding of FGF to heparan sulfate proteoglycans is a necessary step for the presentation of the factor to its receptor. The heparan sulfate proteoglycan syndecan is an integral membrane proteoglycan that binds FGF and components of the extracellular matrix, suggesting that it can regulate the effects of FGF-2 on cell growth [57].

V. TRANSFORMING GROWTH FACTOR BETA

TGF β belongs to a family of closely related polypeptides with various degrees of structural homology and important effects on cell function [87, 88]. There are five TFG β genes, and mammalian cells express TGF β1, 2, and 3. Bone matrix contains TGF β1, 2, and 3 homodimers as well as 1.2 and 2.3 heterodimers [88, 89]. TGF β1, 2, and 3 have similar effects on bone cell function although their potency differs [89]. TGF β stimulates DNA synthesis and cell replication and has a modest stimulatory effect on collagen synthesis in calvariae [90]. Bone histomorphometric analysis of intact rat calvariae exposed to TGF β demonstrates a stimulatory effect on mineral apposition rates, and *in vivo* studies have confirmed a stimulatory effect of TGF β on bone formation [91–93]. The effects of TGF β on cell replication and the differentiated function of the osteoblast are dependent on the target cell, its state of differentiation and the culture conditions used. In primary cultures of rat osteoblasts, TGF β has a biphasic stimulatory effect on DNA synthesis, whereas in rat osteosarcoma cells it inhibits cell growth [94–96]. In osteoblasts, TGF β inhibits alkaline phosphatase activity and osteocalcin synthesis, suggesting an inhibitory effect on their differentiated function. TGF β stimulates collagen synthesis by increasing *type I collagen* gene transcription [97]. In osteosarcoma cells, TGF β increases type I collagen, fibronectin, osteonectin, and osteopontin mRNA expression [98]. In stromal cell cultures, TGF β favors chondrocytic differentiation, and opposes the effect of BMP on osteoblastogenesis [99]. Consequently, whereas TGF β stimulates selected parameters of osteoblastic function, it does not direct the maturation of undifferentiated cells toward osteoblasts, and seems more relevant to chondrogenesis than to osteoblastogenesis. The actions of TGF β on bone resorption also have been a source of controversy. TGF β has a biphasic effect on osteoclastogenesis.

At low concentrations it enhances osteoclast formation, whereas at high concentrations it is inhibitory. The stimulatory effect on osteoclast formation seems to be related to the production of prostaglandins and the inhibitory effect secondary to a decrease in the differentiation of early stem cells into cells of the osteoclast lineage because of a shift toward cells of the granulocyte lineage [100, 101]. The impaired osteoclastogenesis would explain a decrease in bone resorption [102]. Osteoblasts express type I and II TGF β receptors, and PTH and glucocorticoids modify TGF β binding to its receptors on osteoblasts [103, 104]. Glucocorticoids shift the binding of TGF β from type I and II receptors to betaglycan, which is not a signal transducing molecule [105].

In accordance with its actions *in vitro*, the systemic administration of TGF β2 to experimental animals stimulates cancellous bone formation, and subperiosteal injections of TGF β1 and 2 induce osteogenesis and chondrogenesis in rat femurs [93, 106]. In contrast, transgenic mice overexpressing TGF β2, under the control of the osteoblast specific osteocalcin promoter, exhibit osteopenia [107]. Although TGF β induces chondrocyte differentiation, suppression of TGF β signaling *in vivo* also favors terminal chondrocyte differentiation, suggesting a dual role of TGF β on chondrocyte maturation [108]. Targeted gene disruption of the mouse *tgf β* gene does not result in changes in skeletal development [109]. This may be secondary to an early lethal phenotype of *tgf β* null mice, due to a severe inflammatory disease. TGF β enhances soft tissue wound healing, and like FGF-2 it may play a role in fracture repair [110]. However, its cytostatic actions and its fibrosis-inducing properties would limit its topical and systemic use in the treatment of metabolic bone disorders [111–113].

TGF β is secreted as a latent high molecular weight complex consisting of the carboxy terminal remnant of the TGF β precursor and a TGF β binding protein [114–117]. The biologically active levels of TGF β depend on changes in its synthesis and in the activation of its latent form. By inducing lysosomal proteases, bone regulatory agents increase the levels of biologically active TGF β in bone [118, 119]. FGF, TGF β itself, and estrogens increase TGF β1 synthesis in osteoblasts, and ovariectomy reduces the concentration of TGF β in rodent bone [118, 120]. The latter may suggest a role of TGF β in the abnormalities observed in the estrogen deficient state. There have been limited studies on the regulation of TGF β2 and TGF β3 synthesis in osteoblasts, but the *tgf β2* and *tgf β3* gene promoters contain cyclic AMP responsive elements, indicating a potential regulation by cyclic AMP inducers [121, 122].

VI. BONE MORPHOGENETIC PROTEIN

BMPs are members of the TGF β superfamily of polypeptides, and were originally identified because of their ability to induce endochondral bone formation [123–125]. BMPs account for most of the TGF β superfamily of peptides, and the proteins display extensive conservation among species [123]. BMP-1 is a protease, which cleaves procollagen fibrils, and is unrelated to other BMPs [126]. BMP-3, or osteogenin, is different from other BMPs since it lacks their osteogenetic properties, and it inhibits osteogenesis and opposes BMP-2 actions [127–129]. Although BMPs are synthesized by skeletal cells, their synthesis is not limited to bone, and they are expressed by a variety of extraskeletal tissues, where they play a critical role in developmental and cellular functions. BMP-1 through -6 are expressed by osteoblasts, but the degree of expression depends on the cell line examined, and its stage of differentiation [130–133]. BMP-2, -4, and -6 are the most readily detectable BMPs in osteoblast cultures and BMP-2 and -4 are 92% identical in their amino acid sequence and have virtually identical biological activities. Experiments using kinase-deficient truncated BMP receptors or using BMP antagonists have demonstrated that the locally synthesized BMPs play an autocrine role in osteoblastic differentiation and function [134].

A fundamental function of BMPs is the induction of mesenchymal cell differentiation toward cells of the osteoblastic lineage, and to the promotion of osteoblastic activity [135, 136]. As osteoblasts undergo terminal differentiation, they undergo apoptosis, an expected result of cell maturation [137–139]. Consequently, BMPs favor osteoblastic cell death. The genesis and differentiation of bone-forming osteoblasts and bone-resorbing osteoclasts are coordinated events. Receptor activator of nuclear factor-κ B-ligand (RANK-L) and colony stimulating factor 1 are osteoblast products and are major determinants of osteoclastogenesis [140]. Osteoprotegerin, a secreted receptor of the tumor necrosis factor receptor family, acts as a decoy receptor that binds RANK-L, precluding RANK-L binding to RANK and its effects on osteoclastogenesis and bone resorption. BMPs play a direct and indirect role in osteoclastogenesis. By inducing osteoblast maturation, there is increased RANK-L availability. In addition, BMPs sensitize osteoclasts to the effects of RANK-L on osteoclastogenesis and osteoclast survival [141, 142]. BMPs also induce osteoprotegerin transcription, and this may temper their effects on osteoclastogenesis [143]. BMPs inhibit collagenase 3 or MMP-13 expression in osteoblasts, a matrix metalloprotease required for normal bone resorption [144, 145].

BMPs induce endochondral ossification and chondrogenesis [146]. BMPs stimulate chondrocyte maturation and function, increasing the expression of type II and type X collagens and the incorporation of sulfate into glycosaminoglycans [147, 148]. Overexpression of BMP-2 and -4 in developing limbs results in an increase in chondrocyte number and in matrix cartilage, which may lead to joint fusions [149]. The anabolic effects of Indian and Sonic hedgehog and BMP-2 and -4 in metatarsal cultures are analogous, and BMPs mediate their actions on endochondral ossification [150]. Whereas BMPs induce osteogenesis and chondrogenesis, they prevent terminal differentiation of myogenic cells, inhibiting the transcription of the muscle-specific nuclear factors MyoD and myogenin [151, 152]. BMPs act in conjunction with other growth factors, and by inducing the differentiation of cells of the osteoblastic lineage, BMPs increase the pool of IGF-I producing and IGF-I target cells [153].

BMPs interact with type IA or activin receptor like kinase (ALK)-3 and type IB or ALK-6, and BMP type II receptors [154]. Upon ligand binding and activation of the type I receptor, dimers of the type I and type II receptor initiate a signal transduction cascade activating the signaling mothers against decapentaplegic (Smad) or the MAP kinase signaling pathways [155, 156]. Following receptor activation by BMPs, Smad 1, 5, and 8 are phosphorylated at serine residues, and translocated into the nucleus following heterodimerization with Smad 4 [157, 158]. In the nucleus, Smads can bind to DNA sequences directly, bind and cooperate with other transcription factors, or bind and displace nuclear factors from their DNA binding sites. MAP kinase signaling results in P38 MAP kinase or extracellular regulated kinase (ERK) activation by BMPs [87, 159].

The transcriptional and post-transcriptional regulation of BMP expression in osteoblasts reveals autoregulation of BMP synthesis since BMP-2 and -4 mRNA levels are BMP dependent [130, 132]. BMPs cause an early, short lived, induction of BMP-4 mRNA in osteoblasts followed by an inhibitory effect. A positive feedback loop regulating BMP-2 and -4 expression involving Runx-2 is possible since BMPs induce Runx-2 expression and the BMP-2 and -4 promoters contain Runx-2 binding sequences. BMP-6 expression in osteoblasts is steroid-dependent, and BMP-6 mRNA levels are induced by estrogens [133].

Mice deficient in BMP-2 are not viable due to placental and developmental defects, and the *bmp-4* null mutation is lethal between 6.5 and 9.5 days of gestation due to a lack of mesodermal differentiation, and patterning defects [160, 161]. Mice with disruptions of the BMP signaling *smad 5* gene also develop multiple embryonic defects, some reminiscent of those observed in *bmp-2* null mutants [162,163]. The lethality of these mutations has prevented the assessment of the impact of BMP-2 and -4 on the adult skeleton. The *bmp-6* null mutation is not lethal, and skeletogenesis is normal except for delayed ossification in the sternum [164]. *bmp* Gene inactivation results in significant developmental defects outside the skeleton, and *bmp-7* null mice display lack of eye and glomerular development, and *bmp-8* null mutations result in defective spermatogenesis [165,166].

BMP activity is regulated by a large group of secreted polypeptides that bind and limit BMP action. These extracellular BMP antagonists prevent BMP signaling by binding BMPs, and precluding their binding to cell surface receptors [167,168]. Extracellular BMP antagonists include noggin, follistatin, and follistatin related gene; ventroptin; twisted gastrulation; the chordin family, which is comprised of chordin, chordin-like, neuralin, CR rich motor neuron, BMP binding endothelial cell precursor–derived regulator, kielin, and crossveinless; and the Dan/cerberus family of genes, which is comprised of the tumor suppressor Dan, Cerberus, Cer 1, gremlin and its rat homologue drm, the protein related to Dan and Cerberus, caronte, Dante, sclerostin (the product of the *sost* gene), Wise, and Coco [167,169–171]. It is of interest that selected BMP antagonists, such as sclerostin and Coco, also block Wnt signaling [172,173]. The pattern of tissue expression of BMP antagonists is dependent on the gene studied. Sclerostin is expressed selectively in osteoblasts and osteocytes, where it is regulated by PTH, and ectodin is expressed in tooth enamel [174,175]. Overexpression of noggin and gremlin, two classic BMP antagonists, or sclerostin, in the skeletal microenvironment prevents osteoblastic differentiation *in vitro*, and *in vivo* causes osteopenia secondary to decreased bone formation [176–179]. Inactivation of the *noggin* gene causes intrauterine lethality, and articular developmental defects leading to joint fusions [180]. It is of interest that the synthesis of many BMP antagonists, such as that of noggin and gremlin, is induced by BMPs in osteoblasts, suggesting the existence of a protective mechanism to prevent skeletal cells from excessive exposure to BMPs [181,182].

VII. INSULIN-LIKE GROWTH FACTOR

IGF-I and -II have structural similarities with proinsulin and are considered essential for normal cell growth in multiple tissues including bone [183]. IGF-I and -II are present in the systemic circulation and are secreted by the osteoblast. In the circulation, IGF-I forms a large molecular weight complex with

insulin-like growth factor binding proteins (IGFBPs) and the acid labile subunit [184]. Although several IGFBPs form the complex, IGFBP-3 is the most abundant, and systemic IGF-I and IGFBP-3 levels are growth hormone (GH) dependent. IGFBPs are present in the circulation at concentrations in excess of those of IGF-I. Consequently, there is little free IGF-I in plasma. Systemic IGF-I is mostly synthesized in the liver and it is responsible for the growth-promoting effects of GH in various tissues [185–187]. It is important to note that IGF-I is synthesized by multiple peripheral tissues, where it is regulated by alternate hormones, and only to a minor extent by GH. In addition to their function as systemic regulators of growth, IGF-I and -II play an important role in the autocrine and paracrine regulation of cell metabolism in a variety of tissues, including bone [188,189]. IGF-I and -II are the most prevalent growth factors present in the skeletal tissue.

IGF-I and -II have similar effects on bone formation, although IGF-I is more potent than IGF-II [190,191]. IGFs are modest mitogens, increasing the replication of preosteoblastic cells, which presumably differentiate into mature osteoblasts. The most important function of IGFs is to enhance the differentiated function of the osteoblast. IGFs stimulate type I collagen transcription, an effect independent from their mitogenic actions, and increase mineral apposition rates [191,192]. IGF-I and -II inhibit collagenase 3 (MMP-13) synthesis by the osteoblast and as a consequence decrease bone collagen degradation [193]. IGF-I and -II are important in the maintenance of the differentiated osteoblast phenotype. IGFs not only enhance osteoblastic function, but their synthesis is differentially regulated by factors that stimulate or inhibit the differentiated expression of the osteoblastic phenotype [20]. A similar role has been suggested for IGF-II in myoblasts, where its expression follows that of genes that are determinants of myogenic differentiation [194]. It is important to note that whereas IGF-I stimulates the differentiated function of the osteoblast, it does not induce the differentiation of marrow stromal cells toward the osteoblastic pathway [195]. Indirectly, IGF-I might favor osteoblastogenesis since it stabilizes β catenin; consequently it has the potential to enhance the Wnt/β catenin signaling pathway, which is essential for osteoblastogenesis [196]. Although the primary role of IGFs is the stimulation of bone formation, IGF-I can increase the synthesis of RANK-L by osteoblasts and, in this way, enhance osteoclast recruitment, although *in vivo* experiments have demonstrated an inconsistent impact of IGF-I on osteoclastogenesis [197–199].

Skeletal cells express the IGF-I and IGF-II receptors, and the IGF-I receptor mediates their anabolic actions. This receptor is a transmembrane glycoprotein tetramer with ligand-activated tyrosine kinase activity. Insulin receptor substrate (IRS) 1 and 2 are well-characterized substrates for the IGF receptor tyrosine kinase, mediating the effects of IGF-I [200, 201]. In osteoblasts, tyrosine phosphorylation of IRS molecules by the activated receptor results in the activation of PI-3 kinase-phospho Akt signaling or in the activation of the MAP kinases, p38, Jun-N-terminal kinases, and ERK1/2 [202]. The IGF-II receptor is the same as the mannose-6-phosphate receptor, does not have a function in IGF signal transduction, and clears IGF-II, regulating its levels during fetal development [203]. IGF-I receptor number in osteoblasts can be modulated by various agents, known to regulate bone cell metabolism, including PDGF, glucocorticoids, and 1,25 dihydroxyvitamin D_3 [204–207].

IGF-I has been tested for its effects on bone metabolism *in vivo* in experimental animals and humans. IGF-I increases bone formation and prevents trabecular bone loss in experimental conditions of skeletal unloading and increases bone mineral density (BMD) in ovariectomized animals [208]. The short-term administration of IGF-I to normal humans results in an increase in serum levels of type I procollagen peptide and an increase in the excretion of collagen crosslinks, demonstrating an increase in bone turnover [209]. IGF-I increases BMD in patients with osteopenia secondary to anorexia nervosa, but the potential use of IGF-I in humans may be limited by possible side effects and the lack of skeletal specificity [210]. The content of IGF-I in human cortical bone decreases with age, a decline that parallels the one observed in serum concentrations of IGF-I with aging [211, 212]. Consequently, it is not known whether it is due to a decrease in skeletal IGF-I accumulation from the systemic circulation, or due to a decrease in the synthesis of IGF-I by the aging skeleton.

The study of genetically engineered mice has provided additional insight into the action of IGF-I *in vivo*. Transgenic mice overexpressing IGF-I under the control of the osteocalcin promoter exhibit a transient increase in trabecular bone volume secondary to an increase in osteoblast function and bone formation [197]. *igf-1* And *igf-1 receptor* gene deletions have provided valuable information on the role of IGF-I during development, and conditional gene deletions have provided information on the effects of IGF-I in the adult skeleton [186, 213]. *igf-1* Deletion causes a reduction in chondrocyte maturation, and femoral length, and osteopenia secondary to a decrease in bone formation [186, 214]. Mice carrying mutations of the GH releasing hormone receptor or of the GH receptor have absent GH secretion or action, and consequently low serum IGF-I [215, 216]. These models allow for the determination of the contribution of systemic IGF-I

to the skeleton, and the phenotype of either mutant is characterized by small growth plates, osteopenia, and reduced cortical bone, but normal trabecular bone volume. This suggests a more pronounced role of systemic IGF-I on cortical than on trabecular bone. Mice carrying a liver-specific *igf-1* deletion display a 75% reduction in total serum IGF-I levels, normal free IGF-I levels, and absent skeletal phenotype attributed to the normal serum levels of free IGF-I and the extra hepatic synthesis of IGF-I [217]. However, mice carrying dual deletions of *igf-1* and the *acid labile subunit* display marked reduction in total serum IGF-I and a significant reduction of cortical bone volume [218]. *igf-1 Receptor* null mice die after birth and demonstrate severe growth retardation, and the conditional disruption of the *igf-1 receptor* gene selectively in osteoblastic cells causes a decrease in osteoblast number, and impaired bone formation resulting in reduced trabecular bone volume [186, 198]. This observation documents the fundamental role played by IGF-I in the maintenance of bone formation and structure. Accordingly, deletion of the *irs-1* or *-2* gene causes osteopenia [219, 220]. However the phenotypes are not identical and *irs-1* null mice exhibit low bone turnover osteopenia and do not respond to PTH, whereas *igf-2* null mice exhibit increased bone resorption and respond to the anabolic effects of PTH in bone [220]. The various models described confirm the anabolic function of IGF-I on bone. Skeletal IGF-I might play a more important role in the maintenance of trabecular bone, and systemic IGF-I might be more important in the regulation of cortical bone.

The *igf-1* gene is complex, contains six exons, and has alternate promoters in exons 1 and 2 [221, 222]. The exon 1 promoter has four transcription initiation sites, and is responsible for the regulation of IGF-I expression in most extrahepatic tissues including bone [223]. The IGF-I exon 2 promoter has two transcription initiation sites and is responsible for the transcriptional regulation of IGF-I by GH in the liver [223]. IGF-I exon 2 is minimally expressed by osteoblasts, and GH is not a major inducer of IGF-I in these cells [224]. Hormones and growth factors regulate the synthesis of IGF-I in osteoblasts, and PTH, PTH related peptide, and other inducers of cyclic AMP in osteoblasts increase IGF-I synthesis [224]. IGF-I mediates selected anabolic actions of PTH in bone *in vitro* and *in vivo* [225]. The stimulatory effect of PTH on collagen synthesis *in vitro* is decreased by IGF-I neutralizing antibodies, and the stimulatory effect of PTH on bone formation *in vivo* is not observed in *igf-1* or *irs-1* null mice [220, 226]. These observations do not exclude the possibility that other factors mediate selected actions of PTH on the skeleton. For example, the effects of PTH on cell replication in the skeleton are independent of IGF-I synthesis and are dependent on Notch activation [227]. PTH also increases the levels of, and activates, skeletal TGF β, which could mediate selected effects of PTH in bone [228]. Estrogens increase and glucocorticoids inhibit IGF-I synthesis in osteoblasts [229, 230]. Selected inhibitory effects of glucocorticoids on bone metabolism can be explained by reduced IGF-I levels in the bone microenvironment. However, glucocorticoids also inhibit osteoblastic gene expression and function directly. In addition to hormones, skeletal growth factors regulate IGF-I synthesis. PDGF, FGF, and TGF β1 decrease IGF-I transcripts and polypeptide levels in osteoblasts, and this inhibition of IGF-I synthesis correlates with their inhibitory actions on osteoblastic differentiated function [20]. In contrast, BMP-2, an agent that enhances osteoblastic differentiation and function, increases IGF-I synthesis in osteoblasts [153].

The *igf-2* gene is complex and contains four promoters, and in osteoblasts, like in hepatocytes, IGF-II expression is under the control of the IGF-II P3 promoter [231, 232]. Hormones do not modify IGF-II synthesis in skeletal cells. The skeletal growth factors FGF-2, PDGF, and TGF β1 inhibit IGF-II transcription by inhibiting IGF-II P3 activity, an effect analogous to their inhibition of IGF-I expression, demonstrating a coordinated suppression of IGF-I and -II synthesis by mitogenic growth factors [231].

VIII. INSULIN-LIKE GROWTH FACTOR BINDING PROTEINS

IGFBPs are a family of related proteins known to bind specifically IGF-I and -II. There are six classic IGFBPs, termed IGFBP-1 to -6, and additional IGFBP related proteins [184]. There are significant structural similarities in IGFBPs among species, indicating a high degree of evolutionary conservation. Osteoblasts express IGFBP-1 to -6 transcripts [233]. IGFBPs regulate the bioavailability of IGFs and prevent their degradation. IGFBPs can potentiate or inhibit the effects of IGF-I and -II on cell function and vary in their affinity for IGF-I and IGF-II [183, 184]. The binding of IGFs to IGFBPs may sequester the growth factor and preclude its interactions with cell surface receptors, although IGFBPs associated with the extracellular matrix may increase the local effective concentration of the growth factor and potentiate its effects [234, 235]. In addition, IGFBPs may have IGF independent effects and regulate cellular events directly.

IGFBP-2 is important in the storage and transport of IGFs, and IGFBP-2 serum levels correlate with BMD in humans [236]. *In vitro*, IGFBP-2 prevents the effects

of IGF-I on osteoblast function, and overexpression of IGFBP-2 under the control of the cytomegalovirus (CMV) promoter leads to small mice with decreased bone mineral content and failure to respond to the anabolic effects of GH in bone [237]. IGFBP-3 is the major component of the IGF complex in serum and, like the circulating IGF-I, is GH dependent [184]. IGFBP-3 can inhibit or stimulate IGF activities, the latter by upregulating IGF-I delivery to cell surface receptors. Overexpression of IGFBP-3 under the control of the CMV or phosphoglycerate kinase promoter causes growth retardation and osteopenia [238]. IGBP-4 is an IGF-I inhibitory binding protein, but under specific conditions IGFBP-4 and IGFBP-5 were reported to stimulate bone cell function independent from their interactions with IGF-I [239, 240]. However, transgenic mice overexpressing either IGFBP-4 or IGFBP-5 under the control of the osteoblast specific osteocalcin promoter exhibit osteopenia and decreased bone formation [241, 242]. The osteopenia is probably secondary to sequestration of the IGF-I present in the bone microenvironment inhibiting its biological activity. It is possible that the different reported effects of IGFBP-4 and -5 are dependent on their interactions with extracellular matrix proteins, or on the levels of IGFBP present in a specific tissue. However, both transgenic models indicate that IGFBP-4 and IGFBP-5 are inhibitory proteins in the skeletal environment. This was documented further using retroviral vectors to overexpress IGFBP-5, in osteoblastic cells, which provoked an inhibition of osteoblastic cell function [243]. IGFBP-1 is important for glucose homeostasis, and there is limited information on the function of IGFBP-6 in skeletal tissue.

The synthesis of IGFBPs is regulated by transcriptional, post-transcriptional, and post-translational mechanisms. Although the six IGFBPs are expressed by skeletal and nonskeletal cells, their basal and regulated expression is cell specific [244, 245]. *In vitro* studies indicate that the pattern of IGFBP expression is dependent on the stages of osteoblast differentiation. IGFBP-2 and -5 expression is highest in the proliferative phase of rat osteoblastic cell cultures, and IGFBP-3, -4, and -6 expression peaks during the maturation phase [246]. The regulation of IGFBP expression during osteoblastic cell differentiation may be related to the relative levels of autocrine and paracrine factors present in the cellular environment. IGFs increase osteoblast IGFBP-5 expression, whereas growth factors with mitogenic activity inhibit IGFBP-5 and stimulate IGFBP-4 expression [247, 249].

In addition to local autocrine and paracrine factors, systemic hormones modulate IGFBP synthesis; however, the effects appear to be cell line and culture condition dependent [244]. GH increases IGFBP-3 in normal rat osteoblasts, but not in osteosarcoma cells, and cyclic AMP inducers increase the synthesis of IGFBP-3, -4, and -5 [245]. Since agents that induce cyclic AMP in osteoblasts also stimulate the synthesis of IGF-I, the induction of the binding proteins may be a mechanism to control overexposure of cells to the newly synthesized IGF-I. Conversely, glucocorticoids inhibit the synthesis of IGF-I, IGFBP-3, -4, and -5, although they increase the expression of the inhibitory IGFBP-2 in osteoblasts, leading to a marked suppression of IGF-I available to skeletal cells [229, 233]. 1,25-Dihydroxyvitamin D_3 increases osteoblast IGFBP-3 and -4 expression [250].

The abundance of IGFBPs in the extracellular space can be regulated by proteolytic degradation. IGFBP proteases have been characterized from diverse sources, including osteoblasts, which secrete MMPs and serine proteases [251]. Interestingly, the protease activity for IGFBP-4, an inhibitory IGFBP, and for IGFBP-5 is modulated by IGFs, which promote the degradation of IGFBP-4 and stabilize IGFBP-5, suggesting an alternate mechanism by which IGF activity can be modulated in bone [252]. The activity or synthesis of IGFBP proteases is modulated by agents known to regulate bone remodeling, such as IL-6 and glucocorticoids [253, 254].

IX. HEPATOCYTE GROWTH FACTOR

Hepatocyte growth factor (HGF), also known as Scatter factor, is a large molecular weight polypeptide known for its angiogenic and mitogenic properties [255]. HGF plays a role in liver and kidney repair [256, 257]. HGF signals via the product of the proto-oncogene *c-met*, a tyrosine kinase–activated receptor, and HGF and *c-met* are expressed by mesenchymal cells, osteoblasts, and osteoclasts [258, 259]. HGF is mitogenic for cells of the osteoblastic and osteoclastic lineage, and its synthesis by the osteoblast is enhanced by growth factors with a role in wound and fracture repair [260]. Therefore, HGF may play a role in bone remodeling and repair. Studies in transgenic mice overexpressing HGF under the control of the metallothionein I promoter and studies in *hgf* or *c-met* null mice have revealed that HGF is required for muscle cell migration [261, 262]. *hgf* Null mutants fail to form normal muscles, but due to embryonic lethality, the skeletal phenotype was not examined. FGF-2 and PDGF increase the synthesis of HGF by osteoblasts, whereas glucocorticoids are inhibitory [258, 262]. Since HGF plays a role in mitogenesis and tissue repair, inhibition of its synthesis by glucocorticoids may be relevant to the inhibitory effect of these steroids on wound healing

and tissue repair. Recently, HGF and c-met were found to be expressed at the site of fractured bone and HGF to induce BMP receptors [263]. Through this mechanism HGF may contribute to fracture healing and repair.

ACKNOWLEDGMENTS

The work described in this chapter was supported by grants from the National Institutes of Health AR 21707, DK 42424, and DK 45227. The author thanks Ms. Marcia Dupont and Ms. Mary Yurczak for valuable secretarial assistance.

REFERENCES

1. L. Fredriksson, H. Li, and U. Eriksson. The PDGF family: Four gene products form five dimeric isoforms. *Cytokine Growth Factor Rev,* **15**, 197–204 (2004).
2. N. Ferrara and T. Davis-Smyth. The biology of vascular endothelial growth factor. *Endocr Rev,* **18**, 4–25 (1997).
3. L. J. Reigstad, J. E. Varhaug, and J. R. Lillehaug. Structural and functional specificities of PDGF-C and PDGF-D, the novel members of the platelet-derived growth factors family. *FEBS J,* **272**, 5723–5741 (2005).
4. D. G. Gilbertson, M. E. Duff, J. W. West, J. D. Kelly, P. O. Sheppard, P. D. Hofstrand, Z. Gao, K. Shoemaker, T. R. Bukowski, M. Moore, A. L. Feldhaus, J. M. Humes, T. E. Palmer, and C. E. Hart. Platelet-derived growth factor C (PDGF-C), a novel growth factor that binds to PDGF alpha and beta receptor. *J Biol Chem,* **276**, 27406–27414 (2001).
5. E. Bergsten, M. Uutela, X. Li, K. Pietras, A. Ostman, C. H. Heldin, K. Alitalo, and U. Eriksson. PDGF-D is a specific, protease-activated ligand for the PDGF beta-receptor. *Nat Cell Biol,* **3**, 512–516 (2001).
6. W. J. LaRochelle, M. Jeffers, W. F. McDonald, R. A. Chillakuru, N. A. Giese, N. A. Lokker, C. Sullivan, F. L. Boldog, M. Yang, C. Vernet, C. E. Burgess, E. Fernandes, L. L. Deegler, B. Rittman, J. Shimkets, R. A. Shimkets, J. M. Rothberg, and H. S. Lichenstein. PDGF-D, a new protease-activated growth factor. *Nat Cell Biol,* **3**, 517–521 (2001).
7. C. Betsholtz, A. Johnsson, C. H. Heldin, B. Westermark, P. Lind, M. S. Urdea, R. Eddy, T. B. Shows, K. Philpott, and A. L. Mellor. cDNA sequence and chromosomal localization of human platelet-derived growth factor A-chain and its expression in tumour cell lines. *Nature,* **320**, 695–699 (1986).
8. H. Ding, X. Wu, I. Kim, P. P. Tam, G. Y. Koh, and A. Nagy. The mouse Pdgfc gene: Dynamic expression in embryonic tissues during organogenesis. *Mech Dev,* **96**, 209–213 (2000).
9. S. Rydziel, S. Shaikh, and E. Canalis. Platelet-derived growth factor-AA and -BB (PDGF-AA and -BB) enhance the synthesis of PDGF-AA in bone cell cultures. *Endocrinology,* **134**, 2541–2546 (1994).
10. A. Ponten, E. B. Folestad, K. Pietras, and U. Eriksson. Platelet-derived growth factor D induces cardiac fibrosis and proliferation of vascular smooth muscle cells in heart-specific transgenic mice. *Circ Res,* **97**, 1036–1045 (2005).
11. B. Westermark and C. H. Heldin. Platelet-derived growth factor. Structure, function and implications in normal and malignant cell growth. *Acta Oncol,* **32**, 101–105 (1993).

12. L. Claesson-Welsh, A. Eriksson, B. Westermark, and C. H. Heldin. cDNA cloning and expression of the human A-type platelet-derived growth factor (PDGF) receptor establishes structural similarity to the B-type PDGF receptor. *Proc Natl Acad Sci USA,* **86,** 4917–4921 (1989).

13. R. G. Gronwald, F. J. Grant, B. A. Haldeman, C. E. Hart, P. J. O'Hara, F. S. Hagen, R. Ross, D. F. Bowen-Pope, and M. J. Murray. Cloning and expression of a cDNA coding for the human platelet-derived growth factor receptor: Evidence for more than one receptor class. *Proc Natl Acad Sci USA,* **85,** 3435–3439 (1988).

14. R. A. Seifert, C. E. Hart, P. E. Phillips, J. W. Forstrom, R. Ross, M. J. Murray, and D. F. Bowen-Pope. Two different subunits associate to create isoform-specific platelet-derived growth factor receptors. *J Biol Chem,* **264,** 8771–8778 (1989).

15. Y. Yarden, J. A. Escobedo, W. J. Kuang, T. L. Yang-Feng, T. O. Daniel, P. M. Tremble, E. Y. Chen, M. E. Ando, R. N. Harkins, and U. Francke. Structure of the receptor for platelet-derived growth factor helps define a family of closely related growth factor receptors. *Nature,* **323,** 226–232 (1986).

16. M. Centrella, T. L. McCarthy, W. F. Kusmik, and E. Canalis. Isoform-specific regulation of platelet-derived growth factor activity and binding in osteoblast-enriched cultures from fetal rat bone. *J Clin Invest,* **89,** 1076–1084 (1992).

17. J. M. Hock and E. Canalis. Platelet-derived growth factor enhances bone cell replication, but not differentiated function of osteoblasts. *Endocrinology,* **134,** 1423–1428 (1994).

18. H. Tanaka and C. T. Liang. Effect of platelet-derived growth factor on DNA synthesis and gene expression in bone marrow stromal cells derived from adult and old rats. *J Cell Physiol,* **164,** 367–375 (1995).

19. E. Canalis, T. L. McCarthy, and M. Centrella. Effects of platelet-derived growth factor on bone formation in vitro. *J Cell Physiol,* **140,** 530–537 (1989).

20. E. Canalis, J. Pash, B. Gabbitas, S. Rydziel, and S. Varghese. Growth factors regulate the synthesis of insulin-like growth factor-I in bone cell cultures. *Endocrinology,* **133,** 33–38 (1993).

21. B. Gabbitas, J. Pash, and E. Canalis. Regulation of insulin-like growth factor-II synthesis in bone cell cultures by skeletal growth factors. *Endocrinology,* **135,** 284–289 (1994).

22. G. D. Roodman. Interleukin-6: An osteotropic factor? *J Bone Miner Res,* **7,** 475–478 (1992).

23. L. S. Holliday, H. G. Welgus, C. J. Fliszar, G. M. Veith, J. J. Jeffrey, and S. L. Gluck. Initiation of osteoclast bone resorption by interstitial collagenase. *J Biol Chem,* **272,** 22053–22058 (1997).

24. S. Varghese, A. M. Delany, L. Liang, B. Gabbitas, J. J. Jeffrey, and E. Canalis. Transcriptional and posttranscriptional regulation of interstitial collagenase by platelet-derived growth factor BB in bone cell cultures. *Endocrinology,* **137,** 431–437 (1996).

25. W. Zhao, M. H. Byrne, B. F. Boyce, and S. M. Krane. Bone resorption induced by parathyroid hormone is strikingly diminished in collagenase-resistant mutant mice. *J Clin Invest,* **103,** 517–524 (1999).

26. L. Claesson-Welsh. Platelet-derived growth factor receptor signals. *J Biol Chem,* **269,** 32023–32026 (1994).

27. C. H. Heldin, A. Ernlund, C. Rorsman, and L. Ronnstrand. Dimerization of B type platelet-derived growth factor receptors occurs after ligand binding and is closely associated with receptor kinase activation. *J Biol Chem,* **264,** 8905–8912 (1989).

28. R. S. Gilardetti, M. S. Chaibi, J. Stroumza, S. R. Williams, H. N. Antoniades, D. C. Carnes, and D. T. Graves. High-affinity binding of PDGF-AA and PDGF-BB to normal human osteoblastic cells and modulation by interleukin-1. *Am J Physiol,* **261,** C980–C985 (1991).

29. T. Tsukamoto, T. Matsui, H. Nakata, M. Ito, T. Natazuka, M. Fukase, and T. Fujita. Interleukin-1 enhances the response of osteoblasts to platelet-derived growth factor through the alpha receptor-specific up-regulation. *J Biol Chem,* **266,** 10143–10147 (1991).

30. K. N. Kose, J. F. Xie, D. L. Carnes, and D. T. Graves. Proinflammatory cytokines downregulate platelet derived growth factor-alpha receptor gene expression in human osteoblastic cells. *J Cell Physiol,* **166,** 188–197 (1996).

31. X. Li, M. Tjwa, L. Moons, P. Fons, A. Noel, A. Ny, J. M. Zhou, J. Lennartsson, H. Li, A. Luttun, A. Ponten, L. Devy, A. Bouche, H. Oh, A. Manderveld, S. Blacher, D. Communi, P. Savi, F. Bono, M. Dewerchin, J. M. Foidart, M. Autiero, J. M. Herbert, D. Collen, C. H. Heldin, U. Eriksson, and P. Carmeliet. Revascularization of ischemic tissues by PDGF-CC via effects on endothelial cells and their progenitors. *J Clin Invest,* **115,** 118–127 (2005).

32. B. H. Mitlak, R. D. Finkelman, E. L. Hill, J. Li, B. Martin, T. Smith, M. D'Andrea, H. N. Antoniades, and S. E. Lynch. The effect of systemically administered PDGF-BB on the rodent skeleton. *J Bone Miner Res,* **11,** 238–247 (1996).

33. L. J. Marden, R. S. Fan, G. F. Pierce, A. H. Reddi, and J. O. Hollinger. Platelet-derived growth factor inhibits bone regeneration induced by osteogenin, a bone morphogenetic protein, in rat craniotomy defects. *J Clin Invest,* **92,** 2897–2905 (1993).

34. T. F. Deuel, R. S. Kawahara, T. A. Mustoe, and A. F. Pierce. Growth factors and wound healing: Platelet-derived growth factor as a model cytokine. *Annu Rev Med,* **42,** 567–584 (1991).

35. C. Betsholtz. Insight into the physiological functions of PDGF through genetic studies in mice. *Cytokine Growth Factor Rev,* **15,** 215–228 (2004).

36. P. Leveen, M. Pekny, S. Gebre-Medhin, B. Swolin, E. Larsson, and C. Betsholtz. Mice deficient for PDGF B show renal, cardiovascular, and hematological abnormalities. *Genes Dev,* **8,** 1875–1887 (1994).

37. P. Soriano. Abnormal kidney development and hematological disorders in PDGF beta-receptor mutant mice. *Genes Dev,* **8,** 1888–1896 (1994).

38. M. Fruttiger, L. Karlsson, A. C. Hall, A. Abramsson, A. R. Calver, H. Bostrom, K. Willetts, C. H. Bertold, J. K. Heath, C. Betsholtz, and W. D. Richardson. Defective oligodendrocyte development and severe hypomyelination in PDGF-A knockout mice. *Development,* **126,** 457–467 (1999).

39. M. D. Tallquist and P. Soriano. Cell autonomous requirement for PDGFRalpha in populations of cranial and cardiac neural crest cells. *Development,* **130,** 507–518 (2003).

40. H. Ding, X. Wu, H. Bostrom, I. Kim, N. Wong, B. Tsoi, M. O'Rourke, G. Y. Koh, P. Soriano, C. Betsholtz, T. C. Hart, M. L. Marazita, L. L. Field, P. P. Tam, and A. Nagy. A specific requirement for PDGF C in palate formation and PDGFR-alpha signaling. *Nat Genet,* **36,** 1111–1116 (2004).

41. J. S. Campbell, S. D. Hughes, D. G. Gilbertson, T. E. Palmer, M. S. Holdren, A. C. Haran, M. M. Odell, R. L. Bauer, H. P. Ren, H. S. Haugen, M. M. Yeh, and N. Fausto. Platelet-derived growth factor C induces liver fibrosis, steatosis, and hepatocellular carcinoma. *Proc Natl Acad Sci USA,* **102,** 3389–3394 (2005).

42. S. Rydziel, C. Ladd, T. L. McCarthy, M. Centrella, and E. Canalis. Determination and expression of platelet-derived growth factor-AA in bone cell cultures. *Endocrinology,* **130,** 1916–1922 (1992).

43. C. J. Robinson and S. E. Stringer. The splice variants of vascular endothelial growth factor (VEGF) and their receptors. *J Cell Sci,* **114**, 853–865 (2001).

44. N. Ferrara, K. Carver-Moore, H. Chen, M. Dowd, L. Lu, K. S. O'Shea, L. Powell-Braxton, K. J. Hillan, and M. W. Moore. Heterozygous embryonic lethality induced by targeted inactivation of the VEGF gene. *Nature,* **380**, 439–442 (1996).

45. E. Zelzer and B. R. Olsen. Multiple roles of vascular endothelial growth factor (VEGF) in skeletal development, growth, and repair. *Curr Top Dev Biol,* **65**, 169–187 (2005).

46. H. P. Gerber, T. H. Vu, A. M. Ryan, J. Kowalski, Z. Werb, and N. Ferrara. VEGF couples hypertrophic cartilage remodeling, ossification and angiogenesis during endochondral bone formation. *Nat Med,* **5**, 623–628 (1999).

47. E. Zelzer, R. Mamluk, N. Ferrara, R. S. Johnson, E. Schipani, and B. R. Olsen. VEGFA is necessary for chondrocyte survival during bone development. *Development,* **131**, 2161–2171 (2004).

48. S. Niida, T. Kondo, S. Hiratsuka, S. Hayashi, N. Amizuka, T. Noda, K. Ikeda, and M. Shibuya. VEGF receptor 1 signaling is essential for osteoclast development and bone marrow formation in colony-stimulating factor 1–deficient mice. *Proc Natl Acad Sci USA,* **102**, 14016–14021 (2005).

49. J. Street, M. Bao, L. deGuzman, S. Bunting, F. V. Peale, Jr., N. Ferrara, H. Steinmetz, J. Hoeffel, J. L. Cleland, A. Daugherty, N. van Bruggen, H. P. Redmond, R. A. Carano, and E. H. Filvaroff. Vascular endothelial growth factor stimulates bone repair by promoting angiogenesis and bone turnover. *Proc Natl Acad Sci USA,* **99**, 9656–9661 (2002).

50. E. Zelzer, W. McLean, Y. S. Ng, N. Fukai, A. M. Reginato, S. Lovejoy, P. A. D'Amore, and B. R. Olsen. Skeletal defects in VEGF(120/120) mice reveal multiple roles for VEGF in skeletogenesis. *Development,* **129**, 1893–1904 (2002).

51. M. O. Hiltunen, M. Ruuskanen, J. Huuskonen, A. J. Mahonen, M. Ahonen, J. Rutanen, V. M. Kosma, A. Mahonen, H. Kroger, and S. Yla-Herttuala. Adenovirus-mediated VEGF-A gene transfer induces bone formation in vivo. *FASEB J,* **17**, 1147–1149 (2003).

52. N. Akeno, J. Robins, M. Zhang, M. F. Czyzyk-Krzeska, and T. L. Clemens. Induction of vascular endothelial growth factor by IGF-I in osteoblast-like cells is mediated by the PI3K signaling pathway through the hypoxia-inducible factor-2alpha. *Endocrinology,* **143**, 420–425 (2002).

53. Y. Kanno, A. Ishisaki, M. Yoshida, H. Tokuda, O. Numata, and O. Kozawa. SAPK/JNK plays a role in transforming growth factor-beta-induced VEGF synthesis in osteoblasts. *Horm Metab Res,* **37**, 140–145 (2005).

54. H. Tokuda, K. Hirade, X. Wang, Y. Oiso, and O. Kozawa. Involvement of SAPK/JNK in basic fibroblast growth factor-induced vascular endothelial growth factor release in osteoblasts. *J Endocrinol,* **177**, 101–107 (2003).

55. H. Tokuda, D. Hatakeyama, T. Shibata, S. Akamatsu, Y. Oiso, and O. Kozawa. p38 MAP kinase regulates BMP-4–stimulated VEGF synthesis via p70 S6 kinase in osteoblasts. *Am J Physiol Endocrinol Metab,* **284**, E1202–E1209 (2003).

56. M. M. Deckers, R. L. van Bezooijen, G. van der Horst, J. Hoogendam, C. van der Bent, S. E. Papapoulos, and C. W. Lowik. Bone morphogenetic proteins stimulate angiogenesis through osteoblast-derived vascular endothelial growth factor A. *Endocrinology,* **143**, 1545–1553 (2002).

57. N. Itoh and D. M. Ornitz. Evolution of the Fgf and Fgfr gene families. *Trends Genet,* **20**, 563–569 (2004).

58. A. Baird and D. M. Ornitz. Fibroblast growth factors and their receptors. In *Skeletal Growth Factors* (E. Canalis, ed.), pp. 167–178. Lippincott Williams & Wilkins, Philadelphia (2000).

59. F. Esch, A. Baird, N. Ling, N. Ueno, F. Hill, L. Denoroy, R. Klepper, D. Gospodarowicz, P. Bohlen, and R. Guillemin. Primary structure of bovine pituitary basic fibroblast growth factor (FGF) and comparison with the amino-terminal sequence of bovine brain acidic FGF. *Proc Natl Acad Sci USA,* **82**, 6507–6511 (1985).

60. R. K. Globus, J. Plouet, and D. Gospodarowicz. Cultured bovine bone cells synthesize basic fibroblast growth factor and store it in their extracellular matrix. *Endocrinology,* **124**, 1539–1547 (1989).

61. M. M. Hurley, C. Abreu, G. Gronowicz, H. Kawaguchi, and J. Lorenzo. Expression and regulation of basic fibroblast growth factor mRNA levels in mouse osteoblastic MC3T3–E1 cells. *J Biol Chem,* **269**, 9392–9396 (1994).

62. E. Canalis, J. Lorenzo, W. H. Burgess, and T. Maciag. Effects of endothelial cell growth factor on bone remodelling in vitro. *J Clin Invest,* **79**, 52–58 (1987).

63. E. Canalis, M. Centrella, and T. McCarthy. Effects of basic fibroblast growth factor on bone formation in vitro. *J Clin Invest,* **81**, 1572–1577 (1988).

64. R. Montesano, J. D. Vassalli, A. Baird, R. Guillemin, and L. Orci. Basic fibroblast growth factor induces angiogenesis in vitro. *Proc Natl Acad Sci USA,* **83**, 7297–7301 (1986).

65. S. B. Rodan, G. Wesolowski, K. Yoon, and G. A. Rodan. Opposing effects of fibroblast growth factor and pertussis toxin on alkaline phosphatase, osteopontin, osteocalcin, and type I collagen mRNA levels in ROS 17/2.8 cells. *J Biol Chem,* **264**, 19934–19941 (1989).

66. T. L. McCarthy, M. Centrella, and E. Canalis. Effects of fibroblast growth factors on deoxyribonucleic acid and collagen synthesis in rat parietal bone cells. *Endocrinology,* **125**, 2118–2126 (1989).

67. T. Sobue, T. Naganawa, L. Xiao, Y. Okada, Y. Tanaka, M. Ito, N. Okimoto, T. Nakamura, J. D. Coffin, and M. M. Hurley. Over-expression of fibroblast growth factor-2 causes defective bone mineralization and osteopenia in transgenic mice. *J Cell Biochem,* **95**, 83–94 (2005).

68. A. Montero, Y. Okada, M. Tomita, M. Ito, H. Tsurukami, T. Nakamura, T. Doetschman, J. D. Coffin, and M. M. Hurley. Disruption of the fibroblast growth factor-2 gene results in decreased bone mass and bone formation. *J Clin Invest,* **105**, 1085–1093 (2000).

69. M. L. Johnson, K. Harnish, R. Nusse, and W. Van Hul. LRP5 and Wnt signaling: A union made for bone. *J Bone Miner Res,* **19**, 1749–1757 (2004).

70. A. Mansukhani, D. Ambrosetti, G. Holmes, L. Cornivelli, and C. Basilico. Sox2 induction by FGF and FGFR2 activating mutations inhibits Wnt signaling and osteoblast differentiation. *J Cell Biol,* **168**, 1065–1076 (2005).

71. Y. Okada, A. Montero, X. Zhang, T. Sobue, J. Lorenzo, T. Doetschman, J. D. Coffin, and M. M. Hurley. Impaired osteoclast formation in bone marrow cultures of Fgf2 null mice in response to parathyroid hormone. *J Biol Chem,* **278**, 21258–21266 (2003).

72. S. Varghese, S. Rydziel, and E. Canalis. Basic fibroblast growth factor stimulates collagenase-3 promoter activity in osteoblasts through an activator protein-1–binding site. *Endocrinology,* **141**, 2185–2191 (2000).

73. M. Noda and R. Vogel. Fibroblast growth factor enhances type beta 1 transforming growth factor gene expression in osteoblast-like cells. *J Cell Biol,* **109**, 2529–2535 (1989).

74. D. M. Ornitz and P. Leder. Ligand specificity and heparin dependence of fibroblast growth factor receptors 1 and 3. *J Biol Chem,* **267**, 16305–16311 (1992).

75. J. K. Wang, G. Gao, and M. Goldfarb. Fibroblast growth factor receptors have different signaling and mitogenic potentials. *Mol Cell Biol,* **14**, 181–188 (1994).

76. S. Werner, D. S. Duan, C. de Vries, K. G. Peters, D. E. Johnson, and L. T. Williams. Differential splicing in the extracellular region of fibroblast growth factor receptor 1 generates receptor variants with different ligand-binding specificities. *Mol Cell Biol,* **12**, 82–88 (1992).

77. R. Shiang, L. M. Thompson, Y. Z. Zhu, D. M. Church, T. J. Fielder, M. Bocian, S. T. Winokur, and J. J. Wasmuth. Mutations in the transmembrane domain of FGFR3 cause the most common genetic form of dwarfism, achondroplasia. *Cell,* **78**, 335–342 (1994).

78. M. Sahni, D. C. Ambrosetti, A. Mansukhani, R. Gertner, D. Levy, and C. Basilico. FGF signaling inhibits chondrocyte proliferation and regulates bone development through the STAT-1 pathway. *Genes Dev,* **13**, 1361–1366 (1999).

79. L. Xiao, T. Naganawa, E. Obugunde, G. Gronowicz, D. M. Ornitz, J. D. Coffin, and M. M. Hurley. Stat1 controls postnatal bone formation by regulating fibroblast growth factor signaling in osteoblasts. *J Biol Chem,* **279**, 27743–27752 (2004).

80. H. Kawaguchi, T. Kurokawa, K. Hanada, Y. Hiyama, M. Tamura, E. Ogata, and T. Matsumoto. Stimulation of fracture repair by recombinant human basic fibroblast growth factor in normal and streptozotocin-diabetic rats. *Endocrinology,* **135**, 774–781 (1994).

81. H. Kawaguchi, K. Nakamura, Y. Tabata, Y. Ikada, I. Aoyama, J. Anzai, T. Nakamura, Y. Hiyama, and M. Tamura. Acceleration of fracture healing in nonhuman primates by fibroblast growth factor-2. *J Clin Endocrinol Metab,* **86**, 875–880 (2001).

82. K. Yu, J. Xu, Z. Liu, D. Sosic, J. Shao, E. N. Olson, D. A. Towler, and D. M. Ornitz. Conditional inactivation of FGF receptor 2 reveals an essential role for FGF signaling in the regulation of osteoblast function and bone growth. *Development,* **130**, 3063–3074 (2003).

83. A. Jackson, S. Friedman, X. Zhan, K. A. Engleka, R. Forough, and T. Maciag. Heat shock induces the release of fibroblast growth factor 1 from NIH 3T3 cells. *Proc Natl Acad Sci USA,* **89**, 10691–10695 (1992).

84. F. Shibata, A. Baird, and R. Z. Florkiewicz. Functional characterization of the human basic fibroblast growth factor gene promoter. *Growth Factors,* **4**, 277–287 (1991).

85. M. M. Hurley, S. Tetradis, Y. F. Huang, J. Hock, B. E. Kream, L. G. Raisz, and M. G. Sabbieti. Parathyroid hormone regulates the expression of fibroblast growth factor-2 mRNA and fibroblast growth factor receptor mRNA in osteoblastic cells. *J Bone Miner Res,* **14**, 776–783 (1999).

86. M. Mali, K. Elenius, H. M. Miettinen, and M. Jalkanen. Inhibition of basic fibroblast growth factor-induced growth promotion by overexpression of syndecan-1. *J Biol Chem,* **268**, 24215–24222 (1993).

87. L. Attisano and J. L. Wrana. Signal transduction by the TGF-beta superfamily. *Science,* **296**, 1646–1647 (2002).

88. J. A. Barnard, R. M. Lyons, and H. L. Moses. The cell biology of transforming growth factor beta. *Biochim Biophys Acta,* **1032**, 79–87 (1990).

89. M. Centrella, T. L. McCarthy, and E. Canalis. Transforming growth factor-beta and remodeling of bone. *J Bone Joint Surg Am,* **73**, 1418–1428 (1991).

90. M. Centrella, J. Massague, and E. Canalis. Human platelet-derived transforming growth factor-beta stimulates parameters of bone growth in fetal rat calvariae. *Endocrinology,* **119**, 2306–2312 (1986).

91. J. M. Hock, E. Canalis, and M. Centrella. Transforming growth factor-beta stimulates bone matrix apposition and bone cell replication in cultured fetal rat calvariae. *Endocrinology,* **126**, 421–426 (1990).

92. C. Marcelli, A. J. Yates, and G. R. Mundy. In vivo effects of human recombinant transforming growth factor beta on bone turnover in normal mice. *J Bone Miner Res,* **5**, 1087–1096 (1990).

93. D. Rosen, S. C. Miller, E. DeLeon, A. Y. Thompson, H. Bentz, M. Mathews, and S. Adams. Systemic administration of recombinant transforming growth factor beta 2 (rTGF-beta 2) stimulates parameters of cancellous bone formation in juvenile and adult rats. *Bone,* **15**, 355–359 (1994).

94. M. Centrella, T. L. McCarthy, and E. Canalis. Transforming growth factor beta is a bifunctional regulator of replication and collagen synthesis in osteoblast-enriched cell cultures from fetal rat bone. *J Biol Chem,* **262**, 2869–2874 (1987).

95. M. Noda and G. A. Rodan. Type beta transforming growth factor (TGF beta) regulation of alkaline phosphatase expression and other phenotype-related mRNAs in osteoblastic rat osteosarcoma cells. *J Cell Physiol,* **133**, 426–437 (1987).

96. D. M. Rosen, S. A. Stempien, A. Y. Thompson, and S. M. Seyedin. Transforming growth factor-beta modulates the expression of osteoblast and chondroblast phenotypes in vitro. *J Cell Physiol,* **134**, 337–346 (1988).

97. R. A. Ignotz and J. Massague. Transforming growth factor-beta stimulates the expression of fibronectin and collagen and their incorporation into the extracellular matrix. *J Biol Chem,* **261**, 4337–4345 (1986).

98. L. F. Bonewald, M. B. Kester, Z. Schwartz, L. D. Swain, A. Khare, T. L. Johnson, R. J. Leach, and B. D. Boyan. Effects of combining transforming growth factor beta and 1, 25–dihydroxyvitamin D3 on differentiation of a human osteosarcoma (MG-63). *J Biol Chem,* **267**, 8943–8949 (1992).

99. S. Spinella-Jaegle, S. Roman-Roman, C. Faucheu, F. W. Dunn, S. Kawai, S. Gallea, V. Stiot, A. M. Blanchet, B. Courtois, R. Baron, and G. Rawadi. Opposite effects of bone morphogenetic protein-2 and transforming growth factor-beta1 on osteoblast differentiation. *Bone,* **29**, 323–330 (2001).

100. C. Chenu, J. Pfeilschifter, G. R. Mundy, and G. D. Roodman. Transforming growth factor beta inhibits formation of osteoclast-like cells in long-term human marrow cultures. *Proc Natl Acad Sci USA,* **85**, 5683–5687 (1988).

101. D. M. Shinar and G. A. Rodan. Biphasic effects of transforming growth factor-beta on the production of osteoclast-like cells in mouse bone marrow cultures: The role of prostaglandins in the generation of these cells. *Endocrinology,* **126,** 3153–3158 (1990).

102. J. Pfeilschifter, S. M. Seyedin, and G. R. Mundy. Transforming growth factor beta inhibits bone resorption in fetal rat long bone cultures. *J Clin Invest,* **82,** 680–685 (1988).

103. M. Centrella, T. L. McCarthy, and E. Canalis. Parathyroid hormone modulates transforming growth factor beta activity and binding in osteoblast-enriched cell cultures from fetal rat parietal bone. *Proc Natl Acad Sci USA,* **85**, 5889–5893 (1988).

104. M. Centrella, T. L. McCarthy, and E. Canalis. Glucocorticoid regulation of transforming growth factor beta 1 activity and binding in osteoblast-enriched cultures from fetal rat bone. *Mol Cell Biol,* **11**, 4490–4496 (1991).

105. F. Lopez-Casillas, S. Cheifetz, J. Doody, J. L. Andres, W. S. Lane, and J. Massague. Structure and expression of the membrane proteoglycan betaglycan, a component of the TGF-beta receptor system. *Cell,* **67,** 785–795 (1991).

106. M. E. Joyce, A. B. Roberts, M. B. Sporn, and M. E. Bolander. Transforming growth factor beta and the initiation of chondrogenesis and osteogenesis in the rat femur. *J Cell Biol,* **110**, 2195–2207 (1990).

107. A. Erlebacher and R. Derynck. Increased expression of TGF-beta 2 in osteoblasts results in an osteoporosis-like phenotype. *J Cell Biol,* **132**, 195–210 (1996).

108. R. Serra, M. Johnson, E. H. Filvaroff, J. LaBorde, D. M. Sheehan, R. Derynck, and H. L. Moses. Expression of a truncated, kinase-defective TGF-beta type II receptor in mouse skeletal tissue promotes terminal chondrocyte differentiation and osteoarthritis. *J Cell Biol,* **139**, 541–552 (1997).

109. M. M. Shull, I. Ormsby, A. B. Kier, S. Pawlowski, R. J. Diebold, M. Yin, R. Allen, C. Sidman, G. Proetzel, and D. Calvin. Targeted disruption of the mouse transforming growth factor-beta 1 gene results in multifocal inflammatory disease. *Nature,* **359**, 693–699 (1992).

110. L. S. Beck, T. L. Chen, P. Mikalauski, and A. J. Ammann. Recombinant human transforming growth factor-beta 1 (rhTGF-beta 1) enhances healing and strength of granulation skin wounds. *Growth Factors,* **3**, 267–275 (1990).

111. W. A. Border and E. Ruoslahti. Transforming growth factor-beta in disease: The dark side of tissue repair. *J Clin Invest,* **90**, 1–7 (1992).

112. A. Migdalska, G. Molineux, H. Demuynck, G. S. Evans, F. Ruscetti, and T. M. Dexter. Growth inhibitory effects of transforming growth factor-beta 1 in vivo. *Growth Factors,* **4**, 239–245 (1991).

113. A. B. Roberts, M. B. Sporn, R. K. Assoian, J. M. Smith, N. S. Roche, L. M. Wakefield, U. I. Heine, L. A. Liotta, V. Falanga, and J. H. Kehrl. Transforming growth factor type beta: Rapid induction of fibrosis and angiogenesis in vivo and stimulation of collagen formation in vitro. *Proc Natl Acad Sci USA,* **83**, 4167–4171 (1986).

114. R. Derynck, J. A. Jarrett, E. Y. Chen, and D. V. Goeddel. The murine transforming growth factor-beta precursor. *J Biol Chem,* **261**, 4377–4379 (1986).

115. K. Miyazono, U. Hellman, C. Wernstedt, and C. H. Heldin. Latent high molecular weight complex of transforming growth factor beta 1. Purification from human platelets and structural characterization. *J Biol Chem,* **263**, 6407–6415 (1988).

116. K. Miyazono, A. Olofsson, P. Colosetti, and C. H. Heldin. A role of the latent TGF-beta 1–binding protein in the assembly and secretion of TGF-beta 1. *EMBO J,* **10**, 1091–1101 (1991).

117. L. M. Wakefield, D. M. Smith, K. C. Flanders, and M. B. Sporn. Latent transforming growth factor-beta from human platelets. A high molecular weight complex containing precursor sequences. *J Biol Chem,* **263**, 7646–7654 (1988).

118. M. J. Oursler, C. Cortese, P. Keeting, M. A. Anderson, S. K. Bonde, B. L. Riggs, and T. C. Spelsberg. Modulation of transforming growth factor-beta production in normal human osteoblast-like cells by 17 beta-estradiol and parathyroid hormone. *Endocrinology,* **129**, 3313–3320 (1991).

119. M. J. Oursler, B. L. Riggs, and T. C. Spelsberg. Glucocorticoid-induced activation of latent transforming growth factor-beta by normal human osteoblast-like cells. *Endocrinology,* **133**, 2187–2196 (1993).

120. R. D. Finkelman, N. H. Bell, D. D. Strong, L. M. Demers, and D. J. Baylink. Ovariectomy selectively reduces the concentration of transforming growth factor beta in rat bone: Implications for estrogen deficiency-associated bone loss. *Proc Natl Acad Sci USA,* **89**, 12190–12193 (1992).

121. R. Lafyatis, R. Lechleider, S. J. Kim, S. Jakowlew, A. B. Roberts, and M. B. Sporn. Structural and functional characterization of the transforming growth factor beta 3 promoter. A cAMP-responsive element regulates basal and induced transcription. *J Biol Chem,* **265**, 19128–19136 (1990).

122. T. Noma, A. B. Glick, A. G. Geiser, M. A. O'Reilly, J. Miller, A. B. Roberts, and M. B. Sporn. Molecular cloning and structure of the human transforming growth factor-beta 2 gene promoter. *Growth Factors,* **4**, 247–255 (1991).

123. M. Kawabata and K. Miyazono. Bone morphogenetic proteins. In *Skeletal Growth Factors* (E. Canalis, ed.), pp. 269–290. Lippincott Williams & Wilkins, Philadelphia (2000).

124. E. A. Wang, V. Rosen, J. S. D'Alessandro, M. Bauduy, P. Cordes, T. Harada, D. I. Israel, R. M. Hewick, K. M. Kerns, and P. LaPan. Recombinant human bone morphogenetic protein induces bone formation. *Proc Natl Acad Sci USA,* **87**, 2220–2224 (1990).

125. J. M. Wozney, V. Rosen, A. J. Celeste, L. M. Mitsock, M. J. Whitters, R. W. Kriz, R. M. Hewick, and E. A. Wang. Novel regulators of bone formation: Molecular clones and activities. *Science,* **242**, 1528–1534 (1988).

126. M. I. Uzel, I. C. Scott, H. Babakhanlou-Chase, A. H. Palamakumbura, W. N. Pappano, H. H. Hong, D. S. Greenspan, and P. C. Trackman. Multiple bone morphogenetic protein 1–related mammalian metalloproteinases process pro-lysyl oxidase at the correct physiological site and control lysyl oxidase activation in mouse embryo fibroblast cultures. *J Biol Chem,* **276**, 22537–22543 (2001).

127. M. E. Bahamonde and K. M. Lyons. BMP3: To be or not to be a BMP. *J Bone Joint Surg Am,* **83–A** (Suppl 1), S56–S62 (2001).

128. A. Daluiski, T. Engstrand, M. E. Bahamonde, L. W. Gamer, E. Agius, S. L. Stevenson, K. Cox, V. Rosen, and K. M. Lyons. Bone morphogenetic protein-3 is a negative regulator of bone density. *Nat Genet,* **27**, 84–88 (2001).

129. F. P. Luyten, N. S. Cunningham, S. Ma, N. Muthukumaran, R. G. Hammonds, W. B. Nevins, W. I. Woods, and A. H. Reddi. Purification and partial amino acid sequence of osteogenin, a protein initiating bone differentiation. *J Biol Chem,* **264**, 13377–13380 (1989).

130. D. Chen, M. A. Harris, G. Rossini, C. R. Dunstan, S. L. Dallas, J. Q. Feng, G. R. Mundy, and S. E. Harris. Bone morphogenetic protein 2 (BMP-2) enhances BMP-3, BMP-4, and bone cell differentiation marker gene expression during the induction of mineralized bone matrix formation in cultures of fetal rat calvarial osteoblasts. *Calcif Tissue Int,* **60**, 283–290 (1997).

131. N. Ghosh-Choudhury, M. A. Harris, J. Q. Feng, G. R. Mundy, and S. E. Harris. Expression of the BMP 2 gene during bone cell differentiation. *Crit Rev Eukaryot Gene Expr,* **4**, 345–355 (1994).

132. R. C. Pereira, S. Rydziel, and E. Canalis. Bone morphogenetic protein-4 regulates its own expression in cultured osteoblasts. *J Cell Physiol,* **182**, 239–246 (2000).

133. D. J. Rickard, L. C. Hofbauer, S. K. Bonde, F. Gori, T. C. Spelsberg, and B. L. Riggs. Bone morphogenetic protein-6 production in human osteoblastic cell lines. Selective regulation by estrogen. *J Clin Invest,* **101**, 413–422 (1998).

134. M. Suzawa, Y. Takeuchi, S. Fukumoto, S. Kato, N. Ueno, K. Miyazono, T. Matsumoto, and T. Fujita. Extracellular matrix-associated bone morphogenetic proteins are essential for differentiation of murine osteoblastic cells in vitro. *Endocrinology,* **140**, 2125–2133 (1999).

135. A. Yamaguchi, T. Ishizuya, N. Kintou, Y. Wada, T. Katagiri, J. M. Wozney, V. Rosen, and S. Yoshiki. Effects of BMP-2, BMP-4, and BMP-6 on osteoblastic differentiation of bone marrow-derived stromal cell lines, ST2 and MC3T3–G2/PA6. *Biochem Biophys Res Commun,* **220**, 366–371 (1996).

136. S. E. Gitelman, M. Kirk, J. Q. Ye, E. H. Filvaroff, A. J. Kahn, and R. Derynck. Vgr-1/BMP-6 induces osteoblastic differentiation of pluripotential mesenchymal cells. *Cell Growth Differ,* **6**, 827–836 (1995).

137. E. Hay, J. Lemonnier, O. Fromigue, and P. J. Marie. Bone morphogenetic protein-2 promotes osteoblast apoptosis

through a Smad-independent, protein kinase C-dependent signaling pathway. *J Biol Chem,* **276**, 29028–29036 (2001).

138. R. M. Pereira, A. M. Delany, and E. Canalis. Cortisol inhibits the differentiation and apoptosis of osteoblasts in culture. *Bone,* **28**, 484–490 (2001).

139. Y. Yokouchi, J. Sakiyama, T. Kameda, H. Iba, A. Suzuki, N. Ueno, and A. Kuroiwa. BMP-2/-4 mediate programmed cell death in chicken limb buds. *Development,* **122**, 3725–3734 (1996).

140. S. L. Teitelbaum. Bone resorption by osteoclasts. *Science,* **289**, 1504–1508 (2000).

141. Y. H. Gao, T. Shinki, T. Yuasa, H. Kataoka-Enomoto, T. Komori, T. Suda, and A. Yamaguchi. Potential role of cbfa1, an essential transcriptional factor for osteoblast differentiation, in osteoclastogenesis: Regulation of mRNA expression of osteoclast differentiation factor (ODF). *Biochem Biophys Res Commun,* **252**, 697–702 (1998).

142. H. Kaneko, T. Arakawa, H. Mano, T. Kaneda, A. Ogasawara, M. Nakagawa, Y. Toyama, Y. Yabe, M. Kumegawa, and Y. Hakeda. Direct stimulation of osteoclastic bone resorption by bone morphogenetic protein (BMP)-2 and expression of BMP receptors in mature osteoclasts. *Bone,* **27**, 479–486 (2000).

143. M. Wan, X. Shi, X. Feng, and X. Cao. Transcriptional mechanisms of bone morphogenetic protein-induced osteoprotegrin gene expression. *J Biol Chem,* **276**, 10119–10125 (2001).

144. S. Varghese and E. Canalis. Regulation of collagenase-3 by bone morphogenetic protein-2 in bone cell cultures. *Endocrinology,* **138**, 1035–1040 (1997).

145. S. Varghese, S. Rydziel, and E. Canalis. Bone morphogenetic protein-2 suppresses collagenase-3 promoter activity in osteoblasts through a runt domain factor 2 binding site. *J Cell Physiol,* **202**, 391–399 (2005).

146. P. Leboy, G. Grasso-Knight, M. D'Angelo, S. W. Volk, J. V. Lian, H. Drissi, G. S. Stein, and S. L. Adams. Smad-Runx interactions during chondrocyte maturation. *J Bone Joint Surg Am,* **83-A**(Suppl 1), S15–S22 (2001).

147. C. D. Grimsrud, P. R. Romano, M. D'Souza, J. E. Puzas, E. M. Schwarz, P. R. Reynolds, R. N. Roiser, and R. J. O'Keefe. BMP signaling stimulates chondrocyte maturation and the expression of Indian hedgehog. *J Orthop Res,* **19**, 18–25 (2001).

148. F. De Luca, K. M. Barnes, J. A. Uyeda, S. De Levi, V. Abad, T. Palese, V. Mericq, and J. Baron. Regulation of growth plate chondrogenesis by bone morphogenetic protein-2. *Endocrinology,* **142**, 430–436 (2001).

149. D. Duprez, E. J. Bell, M. K. Richardson, C. W. Archer, L. Wolpert, P. M. Brickell, and P. H. Francis-West. Overexpression of BMP-2 and BMP-4 alters the size and shape of developing skeletal elements in the chick limb. *Mech Dev,* **57**, 145–157 (1996).

150. V. Krishnan, Y. Ma, J. Moseley, A. Geiser, S. Friant, and C. Frolik. Bone anabolic effects of sonic/Indian hedgehog are mediated by bmp-2/4–dependent pathways in the neonatal rat metatarsal model. *Endocrinology,* **142**, 940–947 (2001).

151. T. Katagiri, A. Yamaguchi, M. Komaki, E. Abe, N. Takahashi, T. Ikeda, V. Rosen, J. M. Wozney, A. Fujisawa-Sehara, and T. Suda. Bone morphogenetic protein-2 converts the differentiation pathway of C2C12 myoblasts into the osteoblast lineage. *J Cell Biol,* **127**, 1755–1766 (1994).

152. T. Katagiri, S. Akiyama, M. Namiki, M. Komaki, A. Yamaguchi, V. Rosen, J. M. Wozney, A. Fujisawa-Sehara, and T. Suda. Bone morphogenetic protein-2 inhibits terminal differentiation of myogenic cells by suppressing the transcriptional activity of MyoD and myogenin. *Exp Cell Res,* **230**, 342–351 (1997).

153. E. Canalis and B. Gabbitas. Bone morphogenetic protein 2 increases insulin-like growth factor I and II transcripts and polypeptide levels in bone cell cultures. *J Bone Miner Res,* **9**, 1999–2005 (1994).

154. H. Yamashita, P. ten Dijke, C. H. Heldin, and K. Miyazono. Bone morphogenetic protein receptors. *Bone,* **19**, 569–574 (1996).

155. R. Derynck, Y. Zhang, and X. H. Feng. Smads: Transcriptional activators of TGF-beta responses. *Cell,* **95**, 737–740 (1998).

156. A. Nohe, S. Hassel, M. Ehrlich, F. Neubauer, W. Sebald, Y. I. Henis, and P. Knaus. The mode of bone morphogenetic protein (BMP) receptor oligomerization determines different BMP-2 signaling pathways. *J Biol Chem,* **277**, 5330–5338 (2002).

157. F. Liu, A. Hata, J. C. Baker, J. Doody, J. Carcamo, R. M. Harland, and J. Massague. A human Mad protein acting as a BMP-regulated transcriptional activator. *Nature,* **381**, 620–623 (1996).

158. K. Miyazono. Signal transduction by bone morphogenetic protein receptors: Functional roles of Smad proteins. *Bone,* **25**, 91–93 (1999).

159. C. F. Lai and S. L. Cheng. Signal transductions induced by bone morphogenetic protein-2 and transforming growth factor-beta in normal human osteoblastic cells. *J Biol Chem,* **277**, 15514–15522 (2002).

160. G. Winnier, M. Blessing, P. A. Labosky, and B. L. Hogan. Bone morphogenetic protein-4 is required for mesoderm formation and patterning in the mouse. *Genes Dev,* **9**, 2105–2116 (1995).

161. H. Zhang and A. Bradley. Mice deficient for BMP2 are nonviable and have defects in amnion/chorion and cardiac development. *Development,* **122**, 2977–2986 (1996).

162. H. Chang, D. Huylebroeck, K. Verschueren, Q. Guo, M. M. Matzuk, and A. Zwijsen. Smad5 knockout mice die at midgestation due to multiple embryonic and extraembryonic defects. *Development,* **126**, 1631–1642 (1999).

163. X. Yang, L. H. Castilla, X. Xu, C. Li, J. Gotay, M. Weinstein, P. P. Liu, and C. X. Deng. Angiogenesis defects and mesenchymal apoptosis in mice lacking SMAD5. *Development,* **126**, 1571–1580 (1999).

164. M. J. Solloway, A. T. Dudley, E. K. Bikoff, K. M. Lyons, B. L. Hogan, and E. J. Robertson. Mice lacking Bmp6 function. *Dev Genet,* **22**, 321–339 (1998).

165. G. Q. Zhao, K. Deng, P. A. Labosky, L. Liaw, and B. L. Hogan. The gene encoding bone morphogenetic protein 8B is required for the initiation and maintenance of spermatogenesis in the mouse. *Genes Dev,* **10**, 1657–1669 (1996).

166. G. Luo, C. Hofmann, A. L. Bronckers, M. Sohocki, A. Bradley, and G. Karsenty. BMP-7 is an inducer of nephrogenesis, and is also required for eye development and skeletal patterning. *Genes Dev,* **9**, 2808–2820 (1995).

167. E. Canalis, A. N. Economides, and E. Gazzerro. Bone morphogenetic proteins, their antagonists, and the skeleton. *Endocr Rev,* **24**, 218–235 (2003).

168. L. B. Zimmerman, J. M. Jesus-Escobar, and R. M. Harland. The Spemann organizer signal noggin binds and inactivates bone morphogenetic protein 4. *Cell,* **86**, 599–606 (1996).

169. D. R. Hsu, A. N. Economides, X. Wang, P. M. Eimon, and R. M. Harland. The *Xenopus* dorsalizing factor Gremlin identifies a novel family of secreted proteins that antagonize BMP activities. *Mol Cell,* **1**, 673–683 (1998).

170. J. J. Pearce, G. Penny, and J. Rossant. A mouse cerberus/Dan-related gene family. *Dev Biol,* **209**, 98–110 (1999).

171. R. P. Ray and K. A. Wharton. Twisted perspective: New insights into extracellular modulation of BMP signaling during development. *Cell,* **104**, 801–804 (2001).

172. E. Bell, I. Munoz-Sanjuan, C. R. Altmann, A. Vonica, and A. H. Brivanlou. Cell fate specification and competence by Coco, a maternal BMP, TGFbeta and Wnt inhibitor. *Development,* **130**, 1381–1389 (2003).

173. D. G. Winkler, M. S. Sutherland, E. Ojala, E. Turcott, J. C. Geoghegan, D. Shpektor, J. E. Skonier, C. Yu, and J. A. Latham. Sclerostin inhibition of Wnt-3a-induced C3H10T1/2 cell differentiation is indirect and mediated by bone morphogenetic proteins. *J Biol Chem,* **280**, 2498–2502 (2005).

174. T. Bellido, A. A. Ali, I. Gubrij, L. I. Plotkin, Q. Fu, C. A. O'Brien, S. C. Manolagas, and R. L. Jilka. Chronic elevation of parathyroid hormone in mice reduces expression of sclerostin by osteocytes: A novel mechanism for hormonal control of osteoblastogenesis. *Endocrinology,* **146**, 4577–4583 (2005).

175. J. Laurikkala, Y. Kassai, L. Pakkasjarvi, I. Thesleff, and N. Itoh. Identification of a secreted BMP antagonist, ectodin, integrating BMP, FGF, and SHH signals from the tooth enamel knot. *Dev Biol,* **264**, 91–105 (2003).

176. R. D. Devlin, Z. Du, R. C. Pereira, R. B. Kimble, A. N. Economides, V. Jorgetti, and E. Canalis. Skeletal overexpression of noggin results in osteopenia and reduced bone formation. *Endocrinology,* **144**, 1972–1978 (2003).

177. E. Gazzerro, Z. Du, R. D. Devlin, S. Rydziel, L. Priest, A. N. Economides, and E. Canalis. Noggin arrests stromal cell differentiation in vitro. *Bone,* **32**, 111–119 (2003).

178. E. Gazzerro, R. C. Pereira, V. Jorgetti, S. Olson, A. N. Economides, and E. Canalis. Skeletal overexpression of gremlin impairs bone formation and causes osteopenia. *Endocrinology,* **146**, 655–665 (2005).

179. G. G. Loots, M. Kneissel, H. Keller, M. Baptist, J. Chang, N. M. Collette, D. Ovcharenko, I. Plajzer-Frick, and E. M. Rubin. Genomic deletion of a long-range bone enhancer misregulates sclerostin in Van Buchem disease. *Genome Res,* **15**, 928–935 (2005).

180. L. J. Brunet, J. A. McMahon, A. P. McMahon, and R. M. Harland. Noggin, cartilage morphogenesis, and joint formation in the mammalian skeleton. *Science,* **280**, 1455–1457 (1998).

181. E. Gazzerro, V. Gangji, and E. Canalis. Bone morphogenetic proteins induce the expression of noggin, which limits their activity in cultured rat osteoblasts. *J Clin Invest,* **102**, 2106–2114 (1998).

182. R. C. Pereira, A. N. Economides, and E. Canalis. Bone morphogenetic proteins induce gremlin, a protein that limits their activity in osteoblasts. *Endocrinology,* **141**, 4558–4563 (2000).

183. J. I. Jones and D. R. Clemmons. Insulin-like growth factors and their binding proteins: Biological actions. *Endocr Rev,* **16**, 3–34 (1995).

184. V. Hwa, Y. Oh, and R. G. Rosenfeld. The insulin-like growth factor-binding protein (IGFBP) superfamily. *Endocr Rev,* **20**, 761–787 (1999).

185. J. Baker, J. P. Liu, E. J. Robertson, and A. Efstratiadis. Role of insulin-like growth factors in embryonic and postnatal growth. *Cell,* **75**, 73–82 (1993).

186. J. P. Liu, J. Baker, A. S. Perkins, E. J. Robertson, and A. Efstratiadis. Mice carrying null mutations of the genes encoding insulin-like growth factor I (Igf-1) and type 1 IGF receptor (Igf1r). *Cell,* **75**, 59–72 (1993).

187. L. S. Mathews, R. E. Hammer, R. L. Brinster, and R. D. Palmiter. Expression of insulin-like growth factor I in transgenic mice with elevated levels of growth hormone is correlated with growth. *Endocrinology,* **123**, 433–437 (1988).

188. E. Canalis, T. McCarthy, and M. Centrella. Isolation and characterization of insulin-like growth factor I (somatomedin-C) from cultures of fetal rat calvariae. *Endocrinology,* **122**, 22–27 (1988).

189. S. Mohan, J. C. Jennings, T. A. Linkhart, and D. J. Baylink. Primary structure of human skeletal growth factor: Homology with human insulin-like growth factor-II. *Biochim Biophys Acta,* **966**, 44–55 (1988).

190. E. Canalis. Effect of insulinlike growth factor I on DNA and protein synthesis in cultured rat calvaria. *J Clin Invest,* **66**, 709–719 (1980).

191. T. L. McCarthy, M. Centrella, and E. Canalis. Regulatory effects of insulin-like growth factors I and II on bone collagen synthesis in rat calvarial cultures. *Endocrinology,* **124**, 301–309 (1989).

192. J. M. Hock, M. Centrella, and E. Canalis. Insulin-like growth factor I has independent effects on bone matrix formation and cell replication. *Endocrinology,* **122**, 254–260 (1988).

193. E. Canalis, S. Rydziel, A. M. Delany, S. Varghese, and J. J. Jeffrey. Insulin-like growth factors inhibit interstitial collagenase synthesis in bone cell cultures. *Endocrinology,* **136**, 1348–1354 (1995).

194. K. M. Rosen, B. M. Wentworth, N. Rosenthal, and L. Villa-Komaroff. Specific, temporally regulated expression of the insulin-like growth factor II gene during muscle cell differentiation. *Endocrinology,* **133**, 474–481 (1993).

195. T. Thomas, F. Gori, T. C. Spelsberg, S. Khosla, B. L. Riggs, and C. A. Conover. Response of bipotential human marrow stromal cells to insulin-like growth factors: Effect on binding protein production, proliferation, and commitment to osteoblasts and adipocytes. *Endocrinology,* **140**, 5036–5044 (1999).

196. M. P. Playford, D. Bicknell, W. F. Bodmer, and V. M. Macaulay. Insulin-like growth factor 1 regulates the location, stability, and transcriptional activity of beta-catenin. *Proc Natl Acad Sci USA,* **97**, 12103–12108 (2000).

197. G. Zhao, M. C. Monier-Faugere, M. C. Langub, Z. Geng, T. Nakayama, J. W. Pike, S. D. Chernausek, C. J. Rosen, L. R. Donahue, H. H. Malluche, J. A. Fagin, and T. L. Clemens. Targeted overexpression of insulin-like growth factor I to osteoblasts of transgenic mice: Increased trabecular bone volume without increased osteoblast proliferation. *Endocrinology,* **141**, 2674–2682 (2000).

198. M. Zhang, S. Xuan, M. L. Bouxsein, D. von Stechow, N. Akeno, M. C. Faugere, H. Malluche, G. Zhao, C. J. Rosen, A. Efstratiadis, and T. L. Clemens. Osteoblast-specific knock-out of the insulin-like growth factor (IGF) receptor gene reveals an essential role of IGF signaling in bone matrix mineralization. *J Biol Chem,* **277**, 44005–44012 (2002).

199. H. Mochizuki, Y. Hakeda, N. Wakatsuki, N. Usui, S. Akashi, T. Sato, K. Tanaka, and M. Kumegawa. Insulin-like growth factor-I supports formation and activation of osteoclasts. *Endocrinology,* **131**, 1075–1080 (1992).

200. L. M. Chuang, M. G. Myers, Jr., G. A. Seidner, M. J. Birnbaum, M. F. White, and C. R. Kahn. Insulin receptor substrate 1 mediates insulin and insulin-like growth factor I-stimulated maturation of *Xenopus* oocytes. *Proc Natl Acad Sci USA,* **90**, 5172–5175 (1993).

201. M. G. Myers, Jr., X. J. Sun, B. Cheatham, B. R. Jachna, E. M. Glasheen, J. M. Backer, and M. F. White. IRS-1 is a common element in insulin and insulin-like growth factor-I signaling to the phosphatidylinositol 3-kinase. *Endocrinology,* **132**, 1421–1430 (1993).

202. A. Grey, Q. Chen, X. Xu, K. Callon, and J. Cornish. Parallel phosphatidylinositol-3 kinase and p42/44 mitogen-activated protein kinase signaling pathways subserve the mitogenic and antiapoptotic actions of insulin-like growth factor I in osteoblastic cells. *Endocrinology,* **144**, 4886–4893 (2003).

203. P. F. Collett-Solberg and P. Cohen. Genetics, chemistry, and function of the IGF/IGFBP system. *Endocrine,* **12**, 121–136 (2000).

204. A. Bennett, T. Chen, D. Feldman, R. L. Hintz, and R. G. Rosenfeld. Characterization of insulin-like growth factor I receptors on cultured rat bone cells: Regulation of receptor concentration by glucocorticoids. *Endocrinology,* **115**, 1577–1583 (1984).

205. Y. Hakeda, S. Harada, T. Matsumoto, K. Tezuka, K. Higashino, H. Kodama, T. Hashimoto-Goto, E. Ogata, and M. Kumegawa. Prostaglandin F2 alpha stimulates proliferation of clonal osteoblastic MC3T3–E1 cells by up-regulation of insulin-like growth factor I receptors. *J Biol Chem,* **266**, 21044–21050 (1991).

206. H. Kurose, K. Yamaoka, S. Okada, S. Nakajima, and Y. Seino. 1,25–Dihydroxyvitamin D3 [1,25–(OH)2D3] increases insulin-like growth factor I (IGF-I) receptors in clonal osteoblastic cells. Study on interaction of IGF-I and 1,25–(OH)2D3. *Endocrinology,* **126**, 2088–2094 (1990).

207. M. Rubini, H. Werner, E. Gandini, C. T. Roberts, Jr., D. LeRoith, and R. Baserga. Platelet-derived growth factor increases the activity of the promoter of the insulin-like growth factor-1 (IGF-1) receptor gene. *Exp Cell Res,* **211**, 374–379 (1994).

208. M. Machwate, E. Zerath, X. Holy, P. Pastoureau, and P. J. Marie. Insulin-like growth factor-I increases trabecular bone formation and osteoblastic cell proliferation in unloaded rats. *Endocrinology,* **134**, 1031–1038 (1994).

209. P. R. Ebeling, J. D. Jones, W. M. O'Fallon, C. H. Janes, and B. L. Riggs. Short-term effects of recombinant human insulin-like growth factor I on bone turnover in normal women. *J Clin Endocrinol Metab,* **77**, 1384–1387 (1993).

210. S. Grinspoon, L. Thomas, K. Miller, D. Herzog, and A. Klibanski. Effects of recombinant human IGF-I and oral contraceptive administration on bone density in anorexia nervosa. *J Clin Endocrinol Metab,* **87**, 2883–2891 (2002).

211. E. Canalis. Skeletal growth factors and aging. *J Clin Endocrinol Metab,* **78**, 1009–1010 (1994).

212. V. Nicolas, A. Prewett, P. Bettica, S. Mohan, R. D. Finkelman, D. J. Baylink, and J. R. Farley. Age-related decreases in insulin-like growth factor-I and transforming growth factor-beta in femoral cortical bone from both men and women: Implications for bone loss with aging. *J Clin Endocrinol Metab,* **78**, 1011–1016 (1994).

213. A. Efstratiadis. Genetics of mouse growth. *Int J Dev Biol,* **42**, 955–976 (1998).

214. D. Bikle, S. Majumdar, A. Laib, L. Powell-Braxton, C. Rosen, W. Beamer, E. Nauman, C. Leary, and B. Halloran. The skeletal structure of insulin-like growth factor I-deficient mice. *J Bone Miner Res,* **16**, 2320–2329 (2001).

215. W. H. Beamer and E. M. Eicher. Stimulation of growth in the little mouse. *J Endocrinol,* **71**, 37–45 (1976).

216. N. A. Sims, P. Clement-Lacroix, F. Da Ponte, Y. Bouali, N. Binart, R. Moriggl, V. Goffin, K. Coschigano, M. Gaillard-Kelly, J. Kopchick, R. Baron, and P. A. Kelly. Bone homeostasis in growth hormone receptor-null mice is restored by IGF-I but independent of Stat5. *J Clin Invest,* **106**, 1095–1103 (2000).

217. S. Yakar, J. J. Liu, B. Stannard, A. Butler, D. Accili, B. Sauer, and D. LeRoith. Normal growth and development in the absence of hepatic insulin-like growth factor I. *Proc Natl Acad Sci USA,* **96**, 7324–7329 (1999).

218. S. Yakar, C. J. Rosen, W. G. Beamer, C. L. Ackert-Bicknell, Y. Wu, J. L. Liu, G. T. Ooi, J. Setser, J. Frystyk, Y. R. Boisclair, and D. LeRoith. Circulating levels of IGF-1 directly regulate bone growth and density. *J Clin Invest,* **110**, 771–781 (2002).

219. N. Ogata, D. Chikazu, N. Kubota, Y. Terauchi, K. Tobe, Y. Azuma, T. Ohta, T. Kadowaki, K. Nakamura, and H. Kawaguchi. Insulin receptor substrate-1 in osteoblast is indispensable for maintaining bone turnover. *J Clin Invest,* **105**, 935–943 (2000).

220. M. Yamaguchi, N. Ogata, Y. Shinoda, T. Akune, S. Kamekura, Y. Terauchi, T. Kadowaki, K. Hoshi, U. I. Chung, K. Nakamura, and H. Kawaguchi. Insulin receptor substrate-1 is required for bone anabolic function of parathyroid hormone in mice. *Endocrinology,* **146**, 2620–2628 (2005).

221. L. J. Hall, Y. Kajimoto, D. Bichell, S. W. Kim, P. L. James, D. Counts, L. J. Nixon, G. Tobin, and P. Rotwein. Functional analysis of the rat insulin-like growth factor I gene and identification of an IGF-I gene promoter. *DNA Cell Biol,* **11**, 301–313 (1992).

222. S. W. Kim, R. Lajara, and P. Rotwein. Structure and function of a human insulin-like growth factor-I gene promoter. *Mol Endocrinol,* **5**, 1964–1972 (1991).

223. M. L. Adamo, H. Ben Hur, C. T. Roberts, Jr., and D. LeRoith. Regulation of start site usage in the leader exons of the rat insulin-like growth factor-I gene by development, fasting, and diabetes. *Mol Endocrinol,* **5**, 1677–1686 (1991).

224. T. L. McCarthy, M. Centrella, and E. Canalis. Cyclic AMP induces insulin-like growth factor I synthesis in osteoblast-enriched cultures. *J Biol Chem,* **265**, 15353–15356 (1990).

225. E. Canalis, M. Centrella, W. Burch, and T. L. McCarthy. Insulin-like growth factor I mediates selective anabolic effects of parathyroid hormone in bone cultures. *J Clin Invest,* **83**, 60–65 (1989).

226. D. D. Bikle, T. Sakata, C. Leary, H. Elalieh, D. Ginzinger, C. J. Rosen, W. Beamer, S. Majumdar, and B. P. Halloran. Insulin-like growth factor I is required for the anabolic actions of parathyroid hormone on mouse bone. *J Bone Miner Res,* **17**, 1570–1578 (2002).

227. L. M. Calvi, G. B. Adams, K. W. Weibrecht, J. M. Weber, D. P. Olson, M. C. Knight, R. P. Martin, E. Schipani, P. Divieti, F. R. Bringhurst, L. A. Milner, H. M. Kronenberg, and D. T. Scadden. Osteoblastic cells regulate the haematopoietic stem cell niche. *Nature,* **425**, 841–846 (2003).

228. J. Pfeilschifter and G. R. Mundy. Modulation of type beta transforming growth factor activity in bone cultures by osteotropic hormones. *Proc Natl Acad Sci USA,* **84**, 2024–2028 (1987).

229. A. M. Delany, D. Durant, and E. Canalis. Glucocorticoid suppression of IGF I transcription in osteoblasts. *Mol Endocrinol,* **15**, 1781–1789 (2001).

230. M. Ernst and G. A. Rodan. Estradiol regulation of insulin-like growth factor-I expression in osteoblastic cells: Evidence for transcriptional control. *Mol Endocrinol,* **5**, 1081–1089 (1991).

231. V. Gangji, S. Rydziel, B. Gabbitas, and E. Canalis. Insulin-like growth factor II promoter expression in cultured rodent osteoblasts and adult rat bone. *Endocrinology,* **139**, 2287–2292 (1998).

232. M. A. van Dijk, F. M. van Schaik, H. J. Bootsma, P. Holthuizen, and J. S. Sussenbach. Initial characterization of the four promoters of the human insulin-like growth factor II gene. *Mol Cell, Endocrinol,* **81**, 81–94 (1991).

233. R. Okazaki, B. L. Riggs, and C. A. Conover. Glucocorticoid regulation of insulin-like growth factor-binding protein expression in normal human osteoblast-like cells. *Endocrinology,* **134**, 126–132 (1994).

234. D. L. Andress and R. S. Birnbaum. Human osteoblast-derived insulin-like growth factor (IGF) binding protein-5 stimulates

osteoblast mitogenesis and potentiates IGF action. *J Biol Chem,* **267,** 22467–22472 (1992).

235. J. I. Jones, A. Gockerman, W. H. Busby, Jr., C. Camacho-Hubner, and D. R. Clemmons. Extracellular matrix contains insulin-like growth factor binding protein-5: Potentiation of the effects of IGF-I. *J Cell Biol,* **121,** 679–687 (1993).

236. S. Amin, B. L. Riggs, E. J. Atkinson, A. L. Oberg, L. J. Melton, III, and S. Khosla. A potentially deleterious role of IGFBP-2 on bone density in aging men and women. *J Bone Miner Res,* **19,** 1075–1083 (2004).

237. F. Eckstein, T. Pavicic, S. Nedbal, C. Schmidt, U. Wehr, W. Rambeck, E. Wolf, and A. Hoeflich. Insulin-like growth factor-binding protein-2 (IGFBP-2) overexpression negatively regulates bone size and mass, but not density, in the absence and presence of growth hormone/IGF-I excess in transgenic mice. *Anat Embryol (Berl),* **206,** 139–148 (2002).

238. J. V. Silha, S. Mishra, C. J. Rosen, W. G. Beamer, R. T. Turner, D. R. Powell, and L. J. Murphy. Perturbations in bone formation and resorption in insulin-like growth factor binding protein-3 transgenic mice. *J Bone Miner Res,* **18,** 1834–1841 (2003).

239. N. Miyakoshi, X. Qin, Y. Kasukawa, C. Richman, A. K. Srivastava, D. J. Baylink, and S. Mohan. Systemic administration of insulin-like growth factor (IGF)-binding protein-4 (IGFBP-4) increases bone formation parameters in mice by increasing IGF bioavailability via an IGFBP-4 protease-dependent mechanism. *Endocrinology,* **142,** 2641–2648 (2001).

240. C. Richman, D. J. Baylink, K. Lang, C. Dony, and S. Mohan. Recombinant human insulin-like growth factor-binding protein-5 stimulates bone formation parameters in vitro and in vivo. *Endocrinology,* **140,** 4699–4705 (1999).

241. R. D. Devlin, Z. Du, V. Buccilli, V. Jorgetti, and E. Canalis. Transgenic mice overexpressing insulin-like growth factor binding protein-5 display transiently decreased osteoblastic function and osteopenia. *Endocrinology,* **143,** 3955–3962 (2002).

242. M. Zhang, M. C. Faugere, H. Malluche, C. J. Rosen, S. D. Chernausek, and T. L. Clemens. Paracrine overexpression of IGFBP-4 in osteoblasts of transgenic mice decreases bone turnover and causes global growth retardation. *J Bone Miner Res,* **18,** 836–843 (2003).

243. D. Durant, R. M. Pereira, and E. Canalis. Overexpression of insulin-like growth factor binding protein-5 decreases osteoblastic function in vitro. *Bone,* **35,** 1256–1262 (2004).

244. C. Hassager, L. A. Fitzpatrick, E. M. Spencer, B. L. Riggs, and C. A. Conover. Basal and regulated secretion of insulin-like growth factor binding proteins in osteoblast-like cells is cell line specific. *J Clin Endocrinol Metab,* **75,** 228–233 (1992).

245. T. L. McCarthy, S. Casinghino, M. Centrella, and E. Canalis. Complex pattern of insulin-like growth factor binding protein expression in primary rat osteoblast enriched cultures: Regulation by prostaglandin E2, growth hormone, and the insulin-like growth factors. *J Cell Physiol,* **160,** 163–175 (1994).

246. R. S. Birnbaum and K. M. Wiren. Changes in insulin-like growth factor-binding protein expression and secretion during the proliferation, differentiation, and mineralization of primary cultures of rat osteoblasts. *Endocrinology,* **135,** 223–230 (1994).

247. E. Canalis and B. Gabbitas. Skeletal growth factors regulate the synthesis of insulin-like growth factor binding protein-5 in bone cell cultures. *J Biol Chem,* **270,** 10771–10776 (1995).

248. T. L. Chen, L. Y. Chang, D. A. DiGregorio, A. J. Perlman, and Y. F. Huang. Growth factor modulation of insulin-like growth factor-binding proteins in rat osteoblast-like cells. *Endocrinology,* **133,** 1382–1389 (1993).

249. Y. Dong and E. Canalis. Insulin-like growth factor (IGF) I and retinoic acid induce the synthesis of IGF-binding protein 5 in rat osteoblastic cells. *Endocrinology,* **136,** 2000–2006 (1995).

250. T. Moriwake, H. Tanaka, S. Kanzaki, J. Higuchi, and Y. Seino. 1,25–Dihydroxyvitamin D3 stimulates the secretion of insulin-like growth factor binding protein 3 (IGFBP-3) by cultured human osteosarcoma cells. *Endocrinology,* **130,** 1071–1073 (1992).

251. C. A. Conover. Insulin-like growth factor binding protein proteolysis in bone cell models. *Prog Growth Factor Res,* **6,** 301–309 (1995).

252. S. Kanzaki, S. Hilliker, D. J. Baylink, and S. Mohan. Evidence that human bone cells in culture produce insulin-like growth factor-binding protein-4 and -5 proteases. *Endocrinology,* **134,** 383–392 (1994).

253. N. Franchimont, D. Durant, and E. Canalis. Interleukin-6 and its soluble receptor regulate the expression of insulin-like growth factor binding protein-5 in osteoblast cultures. *Endocrinology,* **138,** 3380–3386 (1997).

254. S. Rydziel, A. M. Delany, and E. Canalis. AU-rich elements in the collagenase 3 mRNA mediate stabilization of the transcript by cortisol in osteoblasts. *J Biol Chem,* **279,** 5397–5404 (2004).

255. A. J. Strain. Hepatocyte growth factor: Another ubiquitous cytokine. *J Endocrinol,* **137,** 1–5 (1993).

256. K. Matsumoto and T. Nakamura. Emerging multipotent aspects of hepatocyte growth factor. *J Biochem (Tokyo),* **119,** 591–600 (1996).

257. J. Okano, G. Shiota, and H. Kawasaki. Protective action of hepatocyte growth factor for acute liver injury caused by D-galactosamine in transgenic mice. *Hepatology,* **26,** 1241–1249 (1997).

258. F. Blanquaert, A. M. Delany, and E. Canalis. Fibroblast growth factor-2 induces hepatocyte growth factor/scatter factor expression in osteoblasts. *Endocrinology,* **140,** 1069–1074 (1999).

259. M. Grano, F. Galimi, G. Zambonin, S. Colucci, E. Cottone, A. Z. Zallone, and P. M. Comoglio. Hepatocyte growth factor is a coupling factor for osteoclasts and osteoblasts in vitro. *Proc Natl Acad Sci USA,* **93,** 7644–7648 (1996).

260. G. Zambonin, C. Camerino, G. Greco, V. Patella, B. Moretti, and M. Grano. Hydroxyapatite coated with hepatocyte growth factor (HGF) stimulates human osteoblasts in vitro. *J Bone Joint Surg Br,* **82,** 457–460 (2000).

261. F. Bladt, D. Riethmacher, S. Isenmann, A. Aguzzi, and C. Birchmeier. Essential role for the c-met receptor in the migration of myogenic precursor cells into the limb bud. *Nature,* **376,** 768–771 (1995).

262. F. Blanquaert, R. C. Pereira, and E. Canalis. Cortisol inhibits hepatocyte growth factor/scatter factor expression and induces c-met transcripts in osteoblasts. *Am J Physiol Endocrinol Metab,* **278,** E509–E515 (2000).

263. Y. Imai, H. Terai, C. Nomura-Furuwatari, S. Mizuno, K. Matsumoto, T. Nakamura, and K. Takaoka. Hepatocyte growth factor contributes to fracture repair by upregulating the expression of BMP receptors. *J Bone Miner Res,* **20,** 1723–1730 (2005).

Intercellular Communication during Bone Remodeling

T. JOHN MARTIN AND GIDEON A. RODAN

I. INTRODUCTION

Bone remodeling refers to the renewal process whereby small pockets of old bone, disposed throughout the skeleton and separated from others geographically as well as chronologically, are replaced by new bone throughout adult life. The process is such that the entire adult human skeleton is replaced in 10 years. A major feature of bone remodeling is that it does not occur uniformly throughout the skeleton, but takes place asynchronously in focal or discrete packets known as basic multicellular units (BMUs) of bone turnover. The BMU describes a packet of bone that is being resorbed and then fully rebuilt.

In order to maintain skeletal balance, bone resorption is followed by bone formation, which restores the amount of bone removed by resorption. This tight linkage of the two processes is referred to as coupling, a concept that is supported by two main sets of observations. Kinetic studies using radiotracers of calcium or strontium to estimate the rates of bone formation and resorption in animals or humans, under physiological and pathological conditions, showed that when bone resorption increases, bone formation increases as well [1]. Hyperparathyroidism and estrogen deficiency are examples of conditions in which resorption and formation are increased. On the other hand, when bone resorption decreases, for example, during estrogen replacement therapy, bone formation does also [2]. The second type of evidence is histological. Examination of bone sections showed that osteoclastic bone resorption and osteoblastic bone formation are contiguous during bone remodeling and can be logically conceived to follow each other in the BMU [3]. The histomorpho-

metric estimation of remodeling (or turnover) rates is based on the assumption that resorption precedes formation [4, 5]. The concept that bone formation and resorption are coupled during the bone remodeling process was developed more than 30 years ago. It is based on the principle that bone resorption occurs in order to release calcium for physiological needs and to reshape the bone structure to equip it better for its mechanical function. This "coupling" has been amply confirmed and, with some exceptions discussed in the following sections, is a general characteristic of bone remodeling. Understanding the tightly controlled processes of bone resorption and formation that take place in individual BMUs throughout the skeleton requires appreciation of the pathways of control of osteoblasts and osteoclasts, how they communicate, and how they are influenced by products of cells of the immune system. This will be the subject of the discussion in this chapter, with the remodeling process as a whole considered by Parfitt (see Chapter 3).

A number of discoveries in the late 1990s revealed much of the molecular signaling processes that influence local processes in bone, including osteoblast differentiation and function and the control of osteoclast formation and activity. As yet it is unknown what controls the extent either of bone resorption or of the bone formation that replaces it, and how in particular the two are contrived to be equal. We propose to consider the processes of intercellular communication in the bone remodeling process, how cells of the osteoblast lineage influence the resorption process as well as bone formation, and consider current views of possible cellular and molecular mechanisms by which bone formation is coupled to resorption.

II. SEQUENCE OF CELLULAR EVENTS IN BONE REMODELING

Cancellous bone remodeling starts on the bone surface, the cellular sequence initiated when mechanical deformation or microcracks in old bone provoke signaling, most likely from osteocytes, that leads to osteoclast development and bone resorption. This sequence of events is initiated asynchronously throughout the skeleton, at sites that are geographically and chronologically separated from each other. Both bone resorption and bone formation occur at the same place in these BMUs, so that there is no change in the shape of the bone [4]. Until the early 1980s it was believed that bone metabolism was regulated by circulating hormones. Parathyroid hormone (PTH) and $1,25(OH)_2$ vitamin D promoted bone resorption, sex steroids had some poorly defined beneficial effects on the skeleton, and it was assumed that unknown factors promote bone formation. The discoveries of subsequent years revealed that, although circulating hormones are important, the key influences are locally generated cytokines that influence bone cell function and communication in complex ways, and often are themselves regulated in turn by the hormones (see Chapter 16, Mundy). Discoveries of the many intercellular communication and signaling pathways in bone, together with recent contributions from mouse and human genetics, have contributed greatly to the understanding of bone physiology and pathology. Indeed, many cytokines that had originally been discovered by virtue of their actions on the immune or hematopoietic systems have been revealed as fundamental local factors in the control of bone cell function. This is exemplified most directly and simply by the remarkable number of skeletal phenotypes that exist in genetically altered mice, which either under- or overexpress these cytokines or their receptors. The very nature of the remodeling process, occurring as it does in different parts of the skeleton at different times, highlights the importance of locally generated and regulated factors in the process.

III. CELL INTERACTIONS EARLY IN REMODELING

Several potential stimuli could lead to initiation of remodeling sequences. Among these are (i) pressure changes sensed by osteocytes and resulting in signals delivered to surface cells; (ii) damage in the form of microcracks in bone, leading to osteocyte stimulation or even apoptosis and the release of signals; or (iii) regulated production of local cytokines or growth factors. The first essential step in the cycle is the generation of active osteoclasts from hemopoietic precursors. These are likely derived from early and late precursors available in marrow adjacent to activation sites, or could be recruited from blood available at the bone interface through a sinus structure of bone remodeling compartments [6]. Each of these is a likely contributor, although there is no direct *in vivo* proof of either. The formation of osteoclasts in these sites needs to be regulated through direct interaction with cells of the osteoblast lineage, stimulated by cytokines and prostanoids (to be discussed in following sections). When the remodeling cycle is initiated, say by mechanical strain, which would generate cytokines or prostanoids [5], it has been proposed that the thin layer of nonmineralized matrix under these cells is initially digested by collagenase to expose the mineralized matrix that osteoclasts can resorb [7–9].

When osteoclasts start resorbing bone, the process lasts 2–4 weeks and is carried out by groups of osteoclasts to a depth of about 30 μm [10]. One of the important unanswered questions about osteoclast behavior and the control of resorption in the remodeling cycle is this: How does the osteoclast know when to stop resorbing? The process is likely to finish with osteoclast death, which has been studied *in vitro* to some extent, but its regulation *in vivo* remains obscure. Toward the end of resorption, mononuclear cells are seen at the bottom of resorption pits [11]. Macrophages had been long considered responsible for the post-resorption digestion of collagen fragments, but recent evidence has implicated bone lining cells also, with these cells identified cytologically at sites of resorption, both in calvariae and long bones, and shown engulfing the collagen fragments on the bone surface after the osteoclasts have resorbed [12]. This activity appears to be mediated by membrane matrix metalloproteinases. In the reversal phase, which is the transition from resorption to formation, the reversal plane can be identified microscopically with certain stains [11] or polarized light.

The reversal line (cement line) contains a large abundance of osteopontin [13], which is produced both by osteoclasts and osteoblasts. This is an arginine-glycine-aspartic acid (RGD), containing extracellular matrix protein that interacts with integrin receptors $\alpha_v\beta_3$ in osteoclasts and primarily $\alpha_v\beta_5$ in osteoblasts. These integrin receptors were shown not only to mediate cell attachment to the extracellular matrix, but also to act as signal transducing receptors [14]. It is not yet fully established what influence osteopontin has on osteoclast or osteoblast activity. Its presence on the reversal line raises the possibility that it may signal either cessation of osteoclast activity or initiation of osteoblastic bone formation or possibly both.

The next major stage of the remodeling process is the recruitment and differentiation of mesenchymal precursors to osteoblasts, sufficient to synthesize the amount of bone lost in the resorption process at that site. As is the case with osteoclasts, the potential sources of osteoblast precursors are several. One is that lining cells, the single layer of flattened cells that have ceased their bone-forming function, can revert to that activity. Other likely sources are adjacent marrow stromal cells and even blood-borne osteoblast precursors [15]. Once again, although clues exist in support of each of these mechanisms, direct *in vivo* evidence is lacking. During maturation osteoblasts become cuboidal, polarized cells that are rich in endoplasmic reticulum and contain a large oval nucleus. Osteoblasts are connected to each other and form a contiguous layer. They seem to cooperate in the production of the extracellular bone matrix, since the dimensions of the fibrillar organization of collagen exceed the size of single cells. Moreover, since organization of collagen is so well suited to withstand the mechanical forces exerted on bone, osteoblasts probably sense and respond to mechanical strain. Osteocytes, which are embedded in bone and connected with each other and with surface cells by canaliculi, are particularly well situated to carry out this function. The importance of osteocytes in the bone formation role, suspected for some time, has gained further support from the discovery of sclerostin, a protein product of the osteocyte that is a powerful inhibitor of bone formation [16, 17], and whose production is inhibited by PTH [18, 19].

The remodeling of cortical bone follows similar stages, triggered by cues that may start in cells lining the Haversian canals or in osteocytes [20]. Osteoclasts excavate a "cutting cone," which is refilled by osteoblast activity. Open questions in this sequence of events relate primarily to the signals that govern osteoclast and osteoblast recruitment and termination of osteoclast and osteoblast activity, the identity of cells at the reversal phase, and the precise composition of the matrix on the reversal line.

IV. INTERACTION OF OSTEOBLAST LINEAGE CELLS WITH OSTEOCLASTS

The first indications of the importance of intercellular signaling in bone came in the early 1980s, in that when osteoclasts were isolated from newborn rat or mouse bone, they required the presence of contaminating osteoblastic cells in order to be fully active and resorb bone [21]. The observations that isolated osteoblasts of various origins responded to bone-resorbing hormones and possessed receptors for these factors, in addition to the lack of evidence demonstrating receptors or direct responses to these hormones in osteoclasts, led to the concept that bone-resorbing factors must act first on osteoblasts, most likely bone lining cells. This was proposed to release factors that influence the formation and the bone-resorbing activity of osteoclasts [22]. Furthermore, since osteoclasts are derived from hemopoietic progenitors and not from a local bone cell, the case was argued that since the osteoclast derives from a "wandering" cell, it made sense to have its activity programmed by an authentic bone cell, i.e., the osteoblast [23].

In vitro studies of osteoclast formation from bone marrow cells convincingly demonstrated the requirement for osteoblasts or stromal cells [24]. This, together with the fact that actual contact between these cells and osteoclast precursors is necessary [25, 26], strongly indicated that a molecule expressed on the cell membrane of osteoblast/stromal cells is important in promoting osteoclast formation. This prediction was fulfilled with the discovery of RANK ligand (RANKL), a 316-amino-acid, type-II transmembrane protein, which is a member of the TNF ligand family [27, 28]. Produced by osteoblastic stromal cells and activated T cells, RANKL in the presence of M-CSF, but without any accompanying stromal/osteoblastic cells, promotes the formation of osteoclasts from hemopoietic cells (Figure 18-1). When RANKL (−/−) mice were generated, they were found to be osteopetrotic because of failed osteoclast formation [29]. The action of RANKL is antagonized by osteoprotegerin (OPG), a soluble member of the TNF receptor family, produced by osteoblastic stromal cells as a decoy receptor that inhibits RANKL action. Overexpression of OPG in transgenic mice results in osteopetrosis [30], and OPG (−/−) mice exhibit severe bone loss through excessive osteoclast formation and bone resorption [31]. The receptor for RANKL on hemopoietic cells is RANK (receptor activator of NF-κB), and RANK (−/−) mice are also osteopetrotic [32]. In addition to these effects on osteoclast formation, RANKL is able to activate mature osteoclasts [33] and OPG to inhibit their activity [34]. The formation of RANKL and of OPG in osteoblastic stromal cells is regulated by the hormones and cytokines that influence bone resorption [35]. The identification in the promoter region of OPG and of functional binding sites for the osteoblast master-switch transcription factor, runx2 (cbfa1) [36], provides a further molecular mechanism for the control of bone homeostasis by the osteoblast lineage. runx2, In addition to promoting and maintaining the osteoblast phenotype and thereby favoring bone formation, can drive the stromal lineage toward the capacity to inhibit bone resorption by promoting OPG formation. These mechanisms of interaction between the osteoblastic and hemopoietic lineages operate at sites of initiation

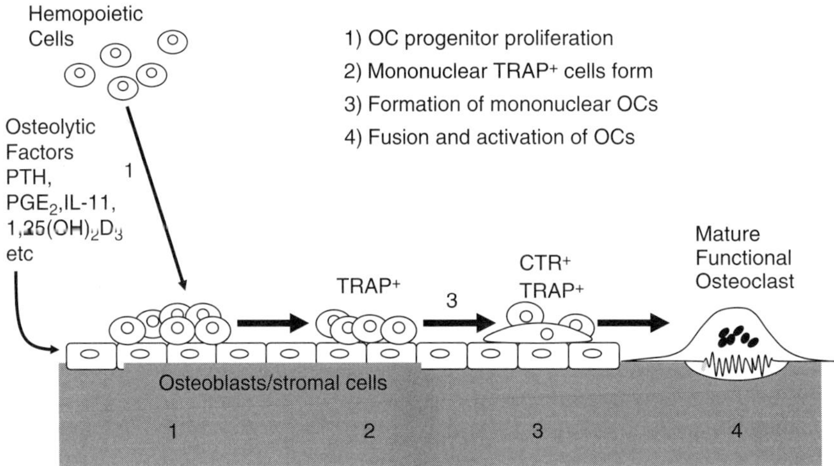

Figure 18-1   Regulation of osteoclast differentiation by the osteoblast lineage. Lining cells and osteoblast precursors produce M-CSF and respond to resorbing factors with production of RANKL, which promotes osteoclast differentiation, activity, and survival.

of remodeling to ensure the orderly progression of precursors to osteoclasts, regulate their activity, and limit their lives [37]. For further discussion of these subjects, see Chapters 4 (Lian), 5 (Blair), and 6 (Bonewald) on osteoblasts, osteoclasts, and osteocytes.

V. FACTORS PROPOSED TO MEDIATE THE COUPLING OF BONE FORMATION TO RESORPTION

An outstanding question concerning bone remodeling relates to the mechanisms by which the amount of bone formed in a BMU is linked to the amount resorbed—the coupling mechanism. Baylink and colleagues [38, 39] suggested that "coupling" is due to bone formation factors released from the bone matrix during bone resorption. Indeed, a large number of substances that are mitogenic to osteoblasts or stimulate bone formation *in vivo* could be extracted from bone matrix [40]. They include insulin-like growth factor (IGF)-I and -II [41]; acidic and basic fibroblast growth factor (FGF) [42]; transforming growth factor-β (TGF-β) 1 and 2 [43] and TGF-β heterodimers [44]; bone morphogenetic proteins (BMPs) 2, 3, 4, 6, and 7 [45–47]; platelet-derived growth factor (PDGF) [48]; and probably others (see also Chapter 17, Canalis). Several questions should be considered regarding the role of these substances in the coupling of bone formation to bone resorption: (i) Which cells produce them and under what circumstances? (ii) Do they stimulate bone formation *in vivo*? (iii) Can they be released from the matrix in active form and in controlled amounts during bone resorption? (iv) Is there evidence for an increase in the abundance of these substances at sites of bone remodeling? and (v) Are there regulated mechanisms by which they are activated?

IGF-I, IGF-II, bFGF, TGF-β, and PDGF are produced by rat osteoblastic cells. IGF-I and IGF-II production is enhanced by stimulators of bone formation, such as prostaglandin E (PGE) and PTH [49, 50]. Elevated levels of IGF-I mRNA were found in bone from estrogen-deficient rats, where bone turnover is increased [51]. During bone growth in rats, there is a close association between osteogenesis and IGF-I expression [52]. However, following marrow ablation, which causes a substantial increase in bone formation, the rise in IGF-I mRNA was seen after the histological appearance of differentiated osteoblasts, suggesting that it did not initiate bone formation in that system [53]. In human bone, the major form of IGF is IGF-II, which was also shown to be produced by human bone cells in culture [38, 54].

Bone is one of the most abundant sources of TGF-β [55]. This growth factor is produced by all osteoblastic cells examined, and its production is increased by estrogen and FGF (in osteosarcoma cells) [56, 57]. The bone morphogenetic proteins (BMPs) are members of the TGF-β superfamily. BMP-2 and BMP-4 are produced in adult bovine pre-odontoblasts [58] as well as in human fetal teeth [59]. BMP-7 (OP-1) was localized in human embryos in hypertrophied chondrocytes, osteoblasts, periosteum, as well as other tissues [65], while BMP-3 was found in human embryonic lung and kidney, in addition to perichondrium, periosteum, and osteoblasts [60]. Both bFGF [61] and PDGF [48] were shown to be produced by bone cells or bone explants in culture. These factors could thus be involved in bone remodeling, but the time and site for their synthesis and secretion *in vivo* have not yet been determined. Prostaglandin E, primarily E2, is another bone cell–produced cytokine, which *in vitro* is upregulated by mechanical strain [62] and stimulates both bone resorption and formation [63].

In remodeling, osteoblasts are recruited from a pool of committed cells and need to sustain the osteoblast phenotype. Further light on mechanisms by which new bone is formed came from the discovery of runx2 (Cbfa1), an essential transcription factor required for osteoblast differentiation [64, 65]. runx2 Not only programs the primitive mesenchymal cell to express osteoblast-specific genes, but has also been shown to be important in maintaining the osteoblast phenotype in mature bone [66]. The regulated expression of runx2 may be an important aspect of this. Study of its regulation is at an early stage, but it will be important to know whether growth factor effects proceed through the runx2 pathway in the remodeling process. Mouse genetics has uncovered a fascinating paracrine role for parathyroid hormone-related protein (PTHrP) in the communication processes essential for normal bone remodeling. Although discovered as a hormone responsible for the humoral hypercalcemia of malignancy, it appeared to have no hormonal role in the normal postnatal animal, but rather acted as a paracrine regulator in several tissues. One of these is bone, where PTHrP is produced by osteoblasts and shown in genetically manipulated mice to be a crucial local regulator of bone remodeling [67]. This new information increases understanding of the effects of the hormone, PTH, many of whose known actions, both on bone formation and resorption, are likely to be pharmacological effects, reflecting the physiological role of the local regulator, PTHrP [68] (see Chapter 10, Nissenson).

Many of the growth factors stimulate bone formation *in vivo*. IGF-I, injected into humans or rats, increases both bone resorption and bone formation [69], and reports on its effect on the bone balance are inconclusive [70]. When injected together with the IGF binding protein IGF BP-3 into rats, it was reported to increase bone volume [71]. BMPs injected into bone stimulate bone formation locally and produce a positive bone balance. TGF-β, from the same family of proteins, has a similar effect. When injected next to the periosteum or endosteum, there is a substantial augmentation in local bone formation in rats and other species [72, 73].

At the same time, following TGF-β, there is an increase in endocortical bone resorption. Thus, like IGF-I, TGF-β seems to stimulate both resorption and formation; however, the local balance is clearly positive. It was proposed that TGF-β, which is produced as an inactive precursor in bone and bone cells [74], is present in the matrix and can be activated by acidification or proteolytic cleavage, and is activated by resorbing osteoclasts [75]. It remains to be shown if the other growth factors also survive the proteolytic cleavage of the acidic hydrolases present in the resorption lacunae.

Other questions raised by this model of coupling, via growth factor release from the matrix, relate to the time course and the distance between the resorption and for-

mation processes and whether activation can be controlled with sufficient precision in this way. Osteoclastic bone resorption proceeds for about 2 weeks before formation follows and continues for 3–4 months. The osteoblast precursors, which should respond to the "coupling factors," are many microns away from where active osteoclast resorption is in progress. The osteoblastic lineage cells produce TGF-β in latent form and the IGFs as complexes bound to a family of specific, high-affinity binding proteins (IGFBPs), which regulate their bioavailability. TGF-β may be released from latent complexes at appropriate sites in bone by plasmin generated locally through the action of plasminogen activators, in a manner that is controlled temporally and spatially by hormones and cytokines [76]. A similar local control could free IGF-1 from association with its inhibitory binding protein [77]. Although there is no obvious skeletal phenotype in mice with inactivated genes for plasminogen activators, *in vivo* investigation of such possibilities would require treatment of such animals with anabolic agents such as PTH.

The preceding theory of coupling predominantly requires dissolution of growth factors from matrix and their activation by acidification and/or proteases. An extension of this, that is not necessarily exclusive, arises from the work of Boyde and colleagues [78], in which they showed *in vitro* that rat calvarial cells grown on bone slices with mechanically excavated crevices and grooves made bone in those defects, filling them exactly to a flat surface. The findings suggested that the topography of the bone can determine the timing, siting, and extent of new bone formation, and that *in vivo* this would take place in the resorbed spaces prepared by osteoclasts. Both the proposed growth factor involvement and the work of Gray *et al.* [78] imply that once the formation process is established, the participating cells themselves are able to sense spatial limits, and most likely do so by chemical communication that takes place between the developing osteoblasts. The likely mediators of these signaling processes are the same growth factors and cytokines that are proposed to be of matrix origin. In this way the bone surface left after osteoclastic resorption, the so-called reversal surface, is implicated as an initiating influence. If active growth factors are contained in this surface, they clearly could play a role either by acting on osteoblasts or intermediary cells that recruit the osteoblasts. Local matrix molecules, such as osteopontin, could also play such a role. Most of all, *in vivo* evidence is needed to show the presence by immunochemistry and the activity by bioassays, as illustrated for TGF-β *in vitro*, of specific growth factors at bone remodeling sites. The technology for such investigations may become available soon.

VI. OSTEOCLAST PRODUCTS IN THE COUPLING PROCESS

Finally, observations made in genetically manipulated mice suggest that the osteoclast itself could also be the source of an activity that contributes to the fine control of the coupling process. Generation of coupling activity was suggested by increased bone formation in OPG–/– mice [79], which are severely osteoporotic because of excessive osteoclast formation. In bone sections from mice in this high bone turnover state, sites of active bone resorption very commonly had active osteoblasts located nearby, suggesting that the coupling activity in this high turnover state was more likely derived from osteoclasts themselves.

Some indication of an osteoclast role comes also from human genetics. In individuals with the osteopetrotic syndrome, ADOII, due to inactivating mutations in the chloride-7 channel (ClC-7), bone resorption is deficient because of failure of the osteoclast acidification process. Bone formation in these patients is nevertheless normal, rather than diminished, as might be expected because of the greatly impaired resorption [80]. Furthermore, in mice deficient in either c-*src* [81], ClC-7 [82], or tyrosine phosphatase epsilon [83], bone resorption is inhibited without inhibition of formation. In these three knockout mouse lines, osteoclast resorption is greatly reduced by the mutation, although osteoclast numbers are not reduced. Indeed, osteoclast numbers are actually increased because of reduced osteoclast apoptosis. A possibility is that these osteoclasts, although unable to resorb bone, are nevertheless capable of generating a factor (or factors) contributing to bone formation. On the other hand, mice lacking c-*fos*, which are unable to generate osteoclasts, have reduced bone formation as well as resorption [84].

The cytokines that signal through gp130 play an important role in intercellular communication processes in bone, with evidence indicating that they can be involved in regulation in mice in which each of the two gp130-dependent signaling pathways was specifically attenuated. Inactivation of the SHP2/ras/MAPK signaling pathway (gp130$^{Y757F/Y757F}$ mice) yielded mice with greater osteoclast numbers and bone resorption, as well as greater bone formation than wild-type mice. This increased bone remodeling resulted in less bone because the increase in resorption was relatively greater than that in formation. In other words, the coupling process was imprecise in a way that resembles the result of estrogen withdrawal, as in ovariectomy. gp130$^{Y757F/Y757F}$ Mice crossed with IL-6 null mice had similarly high osteoclast numbers and increased bone resorption; however, these mice showed no corresponding increase in bone formation and thus had extremely low bone mass. Thus, resorption alone is insufficient to promote the coupled bone formation, but the active osteoclasts are the likely source. Furthermore, this indicated that stimulation of bone formation coupled to the high level of bone resorption in gp130$^{Y757F/Y757F}$ mice is an IL-6-dependent process, though it does not necessarily show that it is mediated by IL-6 itself [85].

The several concepts that could collectively contribute to explain the coupling mechanism are illustrated in Figure 18-2.

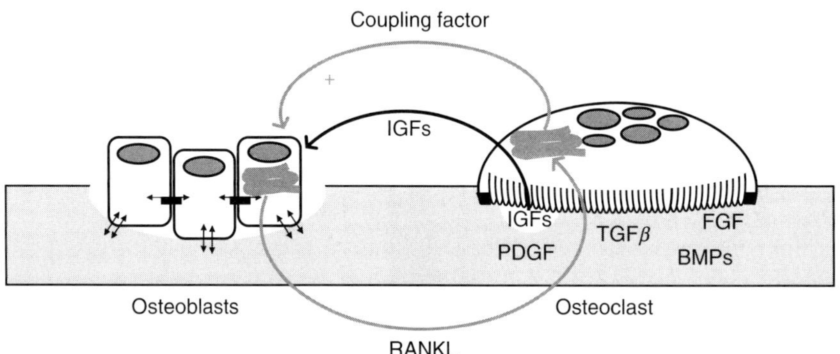

FIGURE 18-2   Three main pathways contributing to the process of coupling of bone formation to resorption. (1) Osteoclasts resorb matrix, releasing stored growth factors that are able to promote osteoblast precursor division or maturation. (2) Preosteoblasts in the resorption space divide, communicate with each other through gap junctions and paracrine signals to program differentiation, and sense spatial requirements. (3) Osteoclasts, promoted and activated by RANKL generated in osteoblasts, release coupling activity that is required for the osteoblast response.

VII. SIMILARITIES BETWEEN BONE REMODELING AND INFLAMMATION

In searching for the molecules that link bone resorption to formation, it may be useful to point to striking similarities between bone remodeling and inflammation. Inflammation starts with trauma produced by injury or by a foreign body. Bone remodeling starts with a stimulus that exposes the mineralized bone surface. In inflammation, the foreign body is recognized by white blood cells, for example, macrophages, which start secreting cytokines and growth factors. The cytokines stimulate the production and migration of other white blood cells to the site of inflammation. Bone exposed to mechanical strain, which probably initiates remodeling, attracts mononuclear cells that stain positively for nonspecific esterases. Many cytokines involved in inflammation are potent stimulators of osteoclastic bone resorption and osteoclast differentiation *in vitro* [24] (see Chapter 16, Mundy). Indeed the impact of the immune system upon bone cell function has become increasingly apparent. T cells produce many cytokines that have an impact on osteoblast or osteoclast differentiation. Although T cells represent about 2–3% of bone marrow cells, they become an abundant population in inflammatory states, e.g., periodontal disease, rheumatoid arthritis. IL-1 and TNF-α are predominantly derived from monocytes. Among the T-cell–derived cytokines, IFN-γ, GM-CSF, IL-4, and IL-13 function as negative regulators of osteoclastogenesis [86]. IL-17 is a T-cell cytokine that promotes osteoclast formation and bone resorption through a prostaglandin-dependent mechanism, similar to that with IL-1 [87]. IL-18, a stromal/osteoblastic product, inhibits osteoclast formation by acting upon T cells to promote GM-CSF production [88]. It may be that local imbalances of pro- and anti-osteoclastogenic cytokines determine whether there is a net loss of bone in inflammatory conditions affecting bone directly. These discoveries revealed much more of the cellular and molecular processes involved in the generation of resorption sites, and therefore in the bone remodeling process. There is little doubt of the importance in remodeling of the TNF and TNF receptor ligand family members. Their roles will be put into the context of hormone and drug actions upon bone, and will undoubtedly have applications for new therapeutic approaches. (See Table 18-1.)

One of the main anti-osteoporotic effects of estrogen is to inhibit proliferation and differentiation of osteoclast precursors. The precise mechanism of these effects and the cellular targets of estrogen have yet to be fully elucidated. Estrogen receptors are expressed by monocytes, osteoblasts, osteoclast precursors, as well as osteoclasts. Thus, estrogen could suppress

TABLE 18-1 Comparison of Sequence of Events and Cellular Interactions in Bone Remodeling and Inflammation

Stage	Inflammation	Bone remodeling
Injury	Tissue damage foreign body	Pressure, microfracture
Reaction	White blood cells, macrophage (local and hematogenous)	Osteoclasts
Repair	Mesenchymal cells (perivascular, fibroblasts)	Osteoblast lineage
	Fibrosis, scar formation	Bone formation

osteoclastogenesis by regulating any one or more of these cell types. Current evidence is that production of at least six factors—IL-1, TNF, IL-6 and the IL-6 receptor complex, RANKL, M-CSF, and GM-CSF—is enhanced in conditions of estrogen deficiency [86, 89]. In view of their pro-osteoclastogenic effects, all of these cytokines are considered potential mediators of the effects of estrogen on bone (see also Chapter 12, Komm).

In inflammatory states, the next phase after initiation is recruitment of fibroblasts, which produce matrix and encapsulate the foreign body. Fibroblast growth factor (FGF) and other growth factors are involved in this process [90]. The analogous phase in bone remodeling is the recruitment of osteoblasts that cover the resorption surface with mineralized matrix. PGE, IL-1, TGF-β, and FGF were all shown to stimulate bone formation *in vivo* [91–94]. An important part of inflammation is neovascularization, probably stimulated by FGF, vascular endothelial growth factor (VEGF), and other cytokines [95]. The importance of angiogenesis in osteogenesis has long been recognized, and bone-derived cells were recently shown to produce VEGF in response to PGE [96], and VEGF to promote osteoclast formation *in vitro* [97].

This analogy would suggest that, as in inflammation, T cells can substitute for osteoblastic stromal cells in promoting osteoclast formation, and evidence has been produced for that [98, 99]. It suggests further that during the resorption process or at its termination, factors released by osteoclasts or cells present on the reversal surface, for example, macrophages, attract the preosteoblasts to that surface. Interestingly, osteopontin, an extracellular molecule made by macrophages and osteoclasts and found in inflammatory and atherosclerotic lesions [100, 101], is present on the reversal surface [13], and is chemotactic. Osteopontin is also abundantly produced by macrophages found in tumors, and these are often encapsulated by fibrous tissue. Osteopontin could be one of the molecules that plays a role in the transition from resorption to formation.

VIII.   BONE MASS HOMEOSTASIS

The putative biochemical mediators of bone resorption and bone formation discussed in this chapter do not explain a major aspect of "coupling"—namely, what determines the extent of bone resorption and bone formation in each remodeling cycle. There clearly is bone mass homeostasis. All healthy individuals have a bone mineral density or bone mineral content that distributes normally around a mean with a standard deviation of about 10%. Bone mass is clearly genetically controlled [102, 103], and there is much interest in possible genes that may be involved [104]. Bone mass or bone mineral content, as measured, for example, in the lumbar spine noninvasively by DXA, is determined by the amount of both cortical and cancellous bone. The amount of cortical bone is determined by periosteal bone formation, which continues throughout life, as well as endosteal and Haversian bone remodeling. Cancellous bone volume is determined by the relative extent of bone resorption and bone formation on the cancellous bone surface. The genetic determinants of bone mass thus should control these processes. Steroid hormones and sex hormones in particular are likely to participate in the genetic determination of bone mass. Men clearly have larger and thicker bones than women. The reduction in bone mass due to estrogen or androgen deficiency is well documented. Moreover, an estrogen receptor–deficient man [105], as well as mice in which the estrogen receptor was "knocked out" [106, 107], are osteopenic. In addition, the epiphyses did not close in that ER-deficient man, suggesting an estrogen role in that function in males, as well as in females. It is not known exactly how sex steroids control bone formation or bone resorption. Receptors for sex steroids have been detected in osteoblastic cells from various species, including humans, and estrogens were shown to inhibit osteoclast activity *in vitro* [108]. The effects of estrogens and androgens on bone are discussed in detail in Chapter 12 (Komm) and Chapter 13 (Wiren). The sex steroids could have both direct and indirect effects, acting both on bone resorption and bone formation. Since these are systemic hormones and their level is most likely not determined by skeletal function, they do not generate the signals that terminate resorption or formation, but they can provide a general background for the cellular responses to such signals. Frost indeed proposed that estrogen levels determine the "set point"

for the response of the skeleton to mechanical signals [109]. The mechanical stimuli may be the most direct input in the maintenance of bone mass and thus play a central role in bone mass homeostasis and by extension in the coupling of bone resorption and bone formation.

IX.   THE ROLE OF MECHANICAL FUNCTION (STRAIN) IN THE COUPLING OF BONE RESORPTION TO BONE FORMATION

The effects of mechanical forces on bone formation and resorption, mediated by the strain in the matrix, have long been known and are very well documented. A decrease in mechanical load produced by immobilization or weightlessness causes a reduction in bone mass, which is due both to increased bone resorption, which occurs initially, and decreased bone formation, which is sustained for a longer duration [110]. Eventually, the system reaches a new steady state where the available bone mass is probably adequate for the prevailing mechanical load. The effects of weightlessness are a clear illustration of "uncoupling," bone resorption being increased and bone formation decreased, implicating mechanical load in the "coupling" phenomenon. Examination of trabeculae from human vertebrae by scanning electron microscopy provided a visual illustration of this phenomenon [111]. Trabeculae, which were not mechanically loaded since one of the extremities was loose and disconnected, showed very extensive resorption without evidence of bone formation. On the other hand, trabeculae, which were connected at both ends and thus mechanically loaded, had shallower resorption lacunae and evidence of bone formation.

The initial increase in bone turnover produced by a reduction in mechanical load, the lower bone formation rate produced by immobilization, and the stabilization of bone mass at a new steady state in effect point to mechanical strain as a factor that couples bone resorption to bone formation. In trabeculae mechanically weakened by resorption, and possibly cortical bone as well, bone formation would be stimulated until the strain is dissipated. The resulting structure would thus be ideally suited to sustain the prevailing strain. This would explain trabecular architecture, which matches the strain distribution in the bone and would explain the increase in the diameter of long bones to compensate for decreased bone mass, recently observed in mice with osteogenesis imperfecta [112]. It could explain why the gain produced by an inhibitor of bone resorption, such as estrogen or bisphosphonates, levels off after 2–3 years, possibly when the existing bone mass has maximized its resistance to the prevailing loads. Consistent with this model is the fact that in nonosteopenic minipigs [113] the potent bone resorption inhibitor alendronate did not cause any changes in bone mass. Furthermore, in osteoporotic patients treated with inhibitors of bone resorption, bone mass continues to increase for some time after the filling of the remodeling spaces, and the increment in mechanically loaded cortical bones at the hip, for example, is larger than in less loaded ones, such as the wrist.

Experimental studies suggest that relatively limited mechanical input is probably sufficient to maintain the "genetically programmed" skeletal mass [114], that short-term bone loss can be demonstrated by total rest or weightlessness (hypogravity) [115], and that very strenuous exercise, such as professional tennis playing, is necessary to produce exercise-dependent significant increases in bone mass [116]. Thus, if we accept the fact that bone mass and bone structure are controlled by mechanical strain and that bone formation is proportional to mechanical strain, we have to conclude that mechanical strain is at least one of the factors that couples bone resorption to formation. How this is brought about at the biochemical and molecular levels has not been satisfactorily elucidated and remains one of the current challenges of skeletal research.

X. INTEGRATED VIEW OF THE COUPLING OF BONE RESORPTION AND BONE FORMATION

Bone has three major functions: mechanical support, homeostasis of calcium and other ions, and housing of hemopoiesis. Bone remodeling is initiated by stimuli generated to fulfill one of these functions. Mechanical stimuli are clearly local and were recently shown to be able to initiate remodeling [117]. The change in extracellular matrix strain, perceived by lining cells or osteocytes, probably primes a specific site in bone for remodeling. Local events could include the release of arachidonic acid metabolites, probably prostaglandin E, and other cytokines plus possibly direct interaction with osteoclasts or osteoclast precursors, leading to a local round of bone remodeling. The evidence from mouse genetics ascribing a crucial role for bone-derived PTHrP in bone remodeling is an important new factor to consider [72]. It is not known what determines the extent of resorption and the depth or the size of the resorption lacunae, but bone formation is then initiated, through direct stimulation of osteoblast precursors by the initial factors, such as PGE; through

regulated release of PTHrP to promote differentiation of committed osteoblast precursors; through release of growth factors from the matrix or from other cells at the resorption site, such as vascular cells or macrophages; through interaction with matrix molecules at the resorption surfaces, such as osteopontin; or all of the above. Once in progress, bone formation probably continues as long as the bone forming cells perceive the osteogenic stimulus of mechanical strain. Likely transducers of that strain are integrins, through which cells are anchored in the matrix. Integrins were shown to act as signal transducing receptors and to affect the phosphorylation of intracellular molecules in ways similar to those produced by growth factor receptors [118], possibly leading to similar outcomes of gene expression and protein synthesis. (See Table 18-2.)

Both bone resorption and bone formation, presumably controlled by the homeostatic inputs of mechanical forces, occur in an endocrine "field." Thus, factors that suppress osteoclast activity, such as estrogens, would modulate the rate and possibly the extent of the resorptive phase, which would increase in the absence of estrogen or the presence of stimulators of osteoclast activity, such as interleukins or parathyroid hormone. The same may hold true for bone formation where factors reported to enhance osteoblast activity, such as PTHrP, IGF, androgens, TGF-β and BMPs, and others, may augment the rate and possibly the extent of the bone forming phase. If the kinetic constraints, determined primarily by the rate of bone resorption, are not rate limiting, the steady-state bone density is most likely determined by the mechanical load. On the other hand, if bone resorption proceeds at an excessive pace that becomes rate limiting, such as in estrogen deficiency, bone formation, albeit increased, will not keep pace and bone loss will occur. Once bone resorption is slowed down by estrogen or other therapy, bone mass can again reach its homeostatic level, determined by mechanical loads. Thus, the relative effects of various hormones and other factors would be to modulate the resorption or formation arm of the equation permitting or preventing the maintenance of the homeostatic bone mass and the rate at which it is reached. This is another way of expressing the "set point hypothesis" for mechanical control of bone mass [109]. For example, not all estrogen-deficient women or hyperthyroid patients with increased bone turnover lose bone to the same extent. This could also explain why exercise may be more effective in maintaining or gaining bone mass in estrogen replete postmenopausal women.

The second stimulus of bone turnover is calcium recruitment from the skeleton, initiated by PTH. Cortical bone seems to be a preferential target for PTH-stimulated bone resorption, possibly a reflection

Table 18-2   Bone Remodeling/Skeletal Homeostasis

Skeletal functions

Homeostasis of calcium and other ions

Mechanical support and levers for muscle action

Support of hemopoiesis

Participating cells

Osteoblasts (mesenchyme-derived cells)

Osteocytes (osteoblast lineage cells)

Lining cells (osteoblast lineage cells)

Marrow stromal cells (mesenchyme-derived cells)

Osteoclasts (hemopoietically derived)

B lymphocytes (hemopoietically derived)

T lymphocytes (hemopoietically derived)

Molecular mediators

Major endocrine factors

Parathyroid hormone

Sex steroids (estrogens and androgens)

Calcitonin

Glucocorticoids

Calcitriol [1,25(OH)2D]

Thyroid hormones

Paracrine/autocrine factors

Insulin-like growth factors (IGFs) and IGF-binding proteins (IGFBPs)

Transforming growth factor family, including bone morphogenetic proteins (BMPs 2, 4, 6, and others)

Fibroblast growth factor family

Prostanoids (PGE2 and others)

Interleukins (IL-1, -6, -11, -17, and others)

Colony-stimulating factors (M-CSF and GM-CSF)

Tumor necrosis factors (RANK ligand, TNF, and others)

TNF receptors (osteoprotegerin)

Parathyroid hormone-related protein

Sclerostin

Wnt signaling

Matricrine factors

Collagen (type I)

Osteopontin

Fibronectin

Vitronectin

Thrombospondin

Mechanical stimuli

---

of the distribution of PTH receptors in bone. On the other hand, elevated PTH concentrations may increase the general level of bone resorption wherever it occurs, augmenting bone loss produced by lack of mechanical function. The beneficial effects of calcium supplements and vitamin D on hip fractures are consistent with this effect, as well as the bone gain observed after parathyroidectomy in vertebral BMD, which contains a considerable amount of cancellous bone [119].

The third function of the skeleton, housing of the hemopoietic system, probably does not affect bone mass significantly under usual circumstances, but may lie at the basis of the response of the skeleton to lymphokines and other cytokines and explain the bone loss associated with inflammation in periarticular regions and the periodontium. It has been reported that increased red blood cell formation enlarges the marrow cavity [120] and malignancies of the bone marrow, such as multiple myeloma, are clearly associated with extensive bone resorption. The feedback mechanisms, which come into play for enlarging the marrow cavity when increased hemopoiesis is needed, are probably mediated by the interleukins that increase osteoclastogenesis, such as IL-1, IL-6, IL-11, and TNF-α. Production of these interleukins during inflammation or in response to local tumors, would lead to similar bone destruction. The similarity between the phases of inflammation and bone remodeling was pointed out previously, but it is not yet known if factors involved in the later steps of inflammation, probably FGF and TGF-β, which were shown to stimulate osteoblast proliferation, play a role in bone formation during normal bone remodeling.

An intriguing possibility of central control of bone remodeling and homeostasis comes from the discovery that both ob/ob mice (leptin gene mutated to inactivity) and db/db mice (leptin receptor inactive) have greatly increased bone mass despite their hypogonadism and increased circulating glucocorticoid. Strikingly, this phenotype is corrected by intracerebroventricular injection of leptin [121].

In conclusion, an integrated view of the bone remodeling process should take into account that bone mass is homeostatically controlled by mechanical function in a hormonal environment (or by hormones in a mechanical field) and that there is a close relationship between bone and hemopoiesis and a similarity between bone remodeling and the cycle of inflammation and tissue repair. Rapidly accumulating new information should test and undoubtedly modify these hypotheses.

REFERENCES

1. W. H. Harris and R. P. Heaney. Skeletal renewal and metabolic bone disease. *N Engl J Med,* **280,** 193–202 (1969).
2. K. M. Prestwood, C. C. Pilbeam, J. A. Burleson, F. N. Woodiel, P. D. Delmas, L. J. Deftos, and L. G. Raisz. The short term effects of conjugated estrogen on bone turnover in older women. *J Clin Endocrinol Metab,* **79,** 366–371 (1994).

3. H. M. Frost. Dynamics of bone remodeling. In *Bone Biodynamics*, pp. 315–333. Little, Brown, Boston (1964).

4. G. A. Rodan. Perspective: Mechanical loading, estrogen deficiency and the coupling of bone formation to bone resorption. *J Bone Miner Res*, **6**, 527–530 (1991).

5. A. M. Parfitt. Skeletal heterogeneity and the purpose of bone remodeling: Implications for the understanding of osteoporosis. In *Osteoporosis* (R. Marcus, D. Feldman, and J. Kelsey, eds.), pp. 315–319. Academic, San Diego, CA (1996).

6. M. Hauge, D. Qvesel, E. F. Eriksen, L. Mosekilde, and F. E. Melsen. Cancellous bone remodeling occurs in specialized compartments lined by cells expressing osteoblast markers. *J Bone Min Res*, **16**, 1575–1582 (2001).

7. S. Sakamoto and M. Sakamoto. Bone collagenase, osteoblasts and cell-mediated bone resorption. In *Bone and Mineral Research* (W. A. Peck, ed.), pp. 49–102. Elsevier, Amsterdam (1986).

8. T. J. Chambers, J. A. Darby, and K. Fuller. Mammalian collagenase predisposes bone surfaces to osteoclastic resorption. *Cell Tissue Res*, **241**, 671–675 (1985).

9. T. J. Chambers and K. Fuller. Bone cells predispose bone surfaces to resorption by exposure of mineral to osteoclastic contact. *J Cell Sci*, **6**, 155–165 (1985).

10. E. F. Eriksen. Normal and pathological remodeling of human trabecular bone: Three dimensional reconstruction of the remodeling sequence in normals and in metabolic disease. *Endocr Rev*, **4**, 379–408 (1986).

11. A. R. Villanueva, C. Kypikowski, and A. M. Parfitt. A new method for identification of cement lines in undecalcified, plastic embedded sections of bone. *Stain Technol*, **61**, 83–88 (1986).

12. S. Peerez-Amodio, W. Beertsen, and V. Everts. Pre-osteoclasts induce retraction of osteoblasts before their fusion to osteoclasts. *J Bone Min Res*, **19**, 1722–1728 (2004).

13. J. Chen, M. D. McKee, A. Nanci, and J. Sodek. Bone sialoprotein mRNA expression and ultrastructural localization in fetal porcine calvarial bone: Comparisons with osteopontin. *Histochem J*, **26**, 67–78 (1994).

14. R. Hynes. Integrins: Versatility, modulation, and signaling in cell adhesion. *Cell*, **69**, 11–25 (1992).

15. G. Z. Eghbali-Fatourechi, J. Lamsam, D. Fraser, D. Nagel, B. L. Riggs, and S. Khosla. Circulating osteoblast-lineage cells in humans. *N Engl J Med*, **352**, 1959–1966 (2005).

16. D. G. Winkler, M. K. Sutherland, J. C. Geoghegan, C. Yu, T. Hayes, J. E. Skonier, D. Shpektor, M. Jonas, B. R. Kovascevich, K. Staehling-Hampton, M. Appleby, M. E. Brunkow, and J. A. Latham. Osteocyte control of bone formation via sclerostin, a novel BMP antagonist. *EMBO J*, **22**, 6267–6276 (2003).

17. R. L. van Bezooijen, B. A. Roelen, A. Visser, L. van der Wee-Pals, E. de Wilt, M. Karperien, H. Hamersma, S. E. Papapoulos, P. ten Dijke, and C. W. Lowik. Sclerostin is an osteocyte-expressed negative regulator of bone formation, but not a classical BMP antagonist. *J Exp Med*, **199**, 805–814 (2004).

18. H. Keller and M. Nessel. SOST is a target gene for PTH in bone. *Bone*, **37**, 148–158 (2005).

19. T. Bellido, A. A. Ali, I. Gubrij, Q. Fu, C. A. O'Brien, S. C. Manolagas, and R. L. Jilka. Chronic elevation of parathyroid hormone in mice reduces expression of sclerostin by osteocytes: A novel mechanism for hormonal control of osteoblastogenesis. *Endocrinology*, **146**, 4577–4583 (2005).

20. T. M. Skerry, L. Bitensky, J. Chayen, and L. E. Lanyon. Early strain-related changes in enzyme activity in osteocytes following bone loading in vivo. *J Bone Miner Res*, **4**, 783–788 (1989).

21. P. M. J. McSheehy and T. J. Chambers. Osteoblastic cells mediate osteoclastic responsiveness to parathyroid hormone. *Endocrinology*, **118**, 824–828 (1986).

22. B. M. Thomson, J. Saklatvala, and T. J. Chambers. Osteoblasts mediate interleukin-1 stimulation of bone resorption by rat osteoclasts. *J Exp Med*, **164**, 104–112 (1986).

23. G. A. Rodan and T. J. Martin. Role of osteoblasts in hormonal control of bone resorption: A hypothesis. *Calcif Tissue Int*, **33**, 349–351 (1981).

24. T. J. Chambers. The cellular basis for bone resorption. *Clin Orthop Relat Res*, **151**, 283–293 (1980).

25. N. Takahashi, T. Akatsu, N. Udagawa, T. Sasaki, A. Yamaguchi, J. M. Moseley, T. J. Martin, and T. Suda. Osteoblastic cells are involved in osteoclast formation. *Endocrinology*, **123**, 2600–2602 (1988).

26. T. Suda, N. Takahashi, and T. J. Martin. Modulation of osteoclast differentiation. *Endocr Rev*, **13**, 66–80 (1992).

27. D. L. Lacey, E. Timms, H. L. Tan, M. J. Kelley, C. R. Dunstan, T. Burgess, R. Elliott, A. Colombero, G. Elliott, S. Scully, H. Hsu, J. Sullivan, N. Hawkins, E. Davy, C. Capparelli, A. Eli, Y. X, Qian, S. Kaufman, I. Sarosi, V. Shalhoub, G. Senaldi, J. Guo, J. Delaney, and W. J. Boyle. Osteoprotegerin ligand is a cytokine that regulates osteoclast differentiation and activation. *Cell*, **93**, 165–176 (1998).

28. H. Yasuda, N. Shima, N. Nakagawa, S. I. Mochizuki, K. Yano, N. Fujise, Y. Sato, M. Goto, K. Yamaguchi, M. Kuriyama, T. Kanno, A. Murakami, E. Tsuda, T. Morinaga, and K. Higashio. Identity of osteoclastogenesis inhibitory factor (OCIF) and osteoprotegerin (OPG): A mechanism by which OPG/OCIF inhibits osteoclastogenesis in vitro. *Endocrinology*, **139**, 1329–1337 (1998).

29. Y. Y. Kong, H. Hoshida, I. Sarosi, H. L. Tan, E. Timms, C. Capparelli, S. Morony, A. J. Oliveira-dos Santos, G. Van, A. Itie, W. Khoo, and A. Wakeham. OPGL is a key regulator of osteoclastogenesis, lymphocyte development and lymph-node organogenesis. *Nature*, **397**, 315–323 (1999).

30. W. S. Simonet, D. L. Lacey, C. R. Dunstan, M. Kelley, M. S. Chang, R. Luthy, H. Q. Nguyen, S. Wooden, L. Bennett, T. Boone, G. Shimamoto, M. DeRose, R. Elliott, A. Colombero, H. L. Tan, G. Trail, J. Sullivan, E. Davy, N. Bucay, L. Renshaw-Gregg, T. M. Hughes, D. Hill, W. Pattison, P. Campbell, S. Sander, G. Van, J. Tarpley, P. Derby, R. Lee, and W. J. Boyle. Osteoprotegerin: A novel secreted protein involved in the regulation of bone density. *Cell*, **89**, 309–319 (1997).

31. N. Bucay, I. Sarosi, C. R. Dunstan, S. Morony, I. Tarpley, C. Capparelli, S. Scully, H. L. Tan, W. Xu, D. L. Lacey, W. I. Boyle, and W. S. Simonet. Osteoprotegerin-deficient mice develop early onset osteoporosis and arterial calcification. *Genes Dev*, **12**, 1260–1268 (1998).

32. W. C. Dougall, M. Glaccum, K. Charrier, K. Rohirbach, K. Brasel, T. De Smedt, E. Daro, J. Smith, M. E. Tometsko, C. R. Maliszewski, A. Armstrong, V. Shen, S. Bain, D. Cosman, D. Anderson, P. J. Morrissey, J. J. Peschon, and J. Schuh. RANK is essential for osteoclast and lymph node development. *Genes Dev*, **13**, 2412–2424 (1999).

33. T. L. Burgess, Y. Qian, S. Kautman, B. D. Ring, G. Van, C. Capparelli, Mm. Kelley, H. Hsu, W. I. Boyle, C. R. Dunstan, S. Hu, and D. L. Lacey. The ligand for osteoprotegerin OPGl directly activates mature osteoclasts. *J Cell Biol*, **145**, 527–538 (1999).

34. B. J. Kwon, G. Wang, N. Udagawa, V. Haridas, Z. H. Lee, K. K. Kim, K. O. Oh, J. Greene, Y. Li, J. Su, R. Gentsz, B. B. Aggarwal, and J. Ni. TRI, a new member of the tumor necrosis factor receptor superfamily, induces fibroblast proliferation and inhibits osteoclastogenesis and bone resorption. *FASEB J*, **12**, 845–854 (1998).

35. N. J. Horwood, J. Elliott, T. J. Martin, and M. T. Gillespie. Osteotropic agents regulate the expression of osteoclast differentiation factor and osteoprotegerin in osteoblastic stromal cells. *Endocrinology,* **139,** 4743–4746 (1998).

36. K. Thirunavukkarasu, D. L. Halladay, R. R. Miles, X. Yang, R. I. S. Galvin, S. Chandrasekhar, T. J. Martin, and J. E. Onyia. The osteoblast-specific transcription factor Cbfal regulates the expression of osteoprotegrin (OPG), a potent inhibitor of osteoclast differentiation and function. *J Biol Chem,* **275,** 25163–25172 (2000).

37. T. Suda, N. Takahashi, N. Udagawa, E. Jimi, M. T. Gillespie, and T. J. Martin. Modulation of osteoclast differentiation and function by the new members of the human necrosis factor receptor and ligand families. *Endocr Rev,* **20,** 345–357 (1999).

38. S. Mohan, J. C. Jennings, T. A. Linkhart, and D. J. Baylink. Primary structure of human skeletal growth factor: Homology with human insulin-like growth factor-11. *Biochem Biophys Acta,* **966,** 44–55 (1988).

39. S. Mohan and D. J. Baylink. The role of IGF-11 in the coupling of bone formation to resorption. In *Modern Concepts of Insulin-like Growth Factors*, pp. 169–184. Elsevier, New York (1991).

40. P. V. Hauschka, A. E. Mavrakos, M. D. Iafrati, S. E. Doleman, and M. Klagsbrun. Growth factors in bone matrix: Isolation of multiple types by affinity chromatography on heparin sepharose. *J Biol Chem,* **261,** 12665–12674 (1986).

41. C. Schmid and M. Ernst. Insulin-like growth factors. In *Cytokines and Bone Metabolism*, pp. 229–265. CRC Press, Boca Raton, FL (1992).

42. S. B. Rodan and G. A. Rodan. Fibroblast growth factor and platelet-derived growth factor. In *Cytokines and Bone Metabolism*, pp. 115–145. CRC Press, Boca Raton, FL (1992).

43. M. Centrella, T. L. McCarthy, and E. Canalis. Transforming growth factor-beta and remodeling in bone. *J Bone Joint Surg,* **73–A,** 1418–1428 (1991).

44. Y. Ogawa, D. K. Schmidt, J. R. Daschy, R. J. Chang, and C. B. Glasser. Purification and characterization of transforming growth factor beta2.3 and beta1.2 heterodimers from bovine bone. *J Biol Chem,* **267,** 2325–2328 (1992).

45. S. E. Harris, M. Sabatini, M. A. Harris, J. Q. Feng, J. Wozney, and G. R. Mundy. Expression of bone morphogenetic protein messenger. RNA in prolonged cultures of fetal rat calvarial cells. *J Bone Miner Res,* **9,** 389–394 (1994).

46. S. Vukicevic, V. Latin, P. Chen, R. Batorsky, A. H. Reddi, and K. Sampath. Localization of osteogenic protein-1 (bone morphogenetic protein-7) during human embryonic development: High affinity to basement membranes. *Biochem Biophys Res Commun,* **198,** 693–700 (1994).

47. S. Vukicevic, V. M. Paralkar, N. S. Cunningham, J. S. Gutkind, and A. H. Reddi. Autoradiographic localization of osteogenin binding sites in cartilage and bone during rat embryonic development. *Dev Biol,* **140,** 209–214 (1990).

48. S. Rydziel, C. Ladd, T. L. McCarthy, M. Centrella, and E. Canalis. Determination and expression of platelet-derived growth factor-AA in bone cell cultures. *Endocrinology,* **130,** 1916–1922 (1992).

49. T. L. McCarthy, M. Centrella, and E. Canalis. Parathyroid hormone enhances the transcript and polypeptide levels of insulin-like growth factor 1 in osteoblast-enriched cultures from fetal rat bone. *Endocrinology,* **124,** 1247–1253 (1989).

50. T. L. McCarthy, M. Centrella, L. G. Raisz, and E. Canalis. Prostaglandin E2 stimulates insulin-like growth factor 1 synthesis in osteoblast-enriched cultures from fetal rat bone. *Endocrinology,* **128,** 2895–2900 (1991).

51. R. T. Turner, P. H. Backup, P. H. Sherman, E. L. Hill, G. L. Evans, and T. C. Spelsberg. Mechanism of action of estrogen on intramembranous bone formation and regulation of osteoblast differentiation and activity. *Endocrinology,* **131,** 883–889 (1992).

52. D. M. Shinar, N. Endo, D. Halperin, G. A. Rodan, and M. Weinreb. Differential expression of insulin-like growth factor-1 (IGF-1) and IGF-11 messenger ribonucleic acid in growing rat bone. *Endocrinology,* **132,** 1158–1167 (1993).

53. L. J. Suva, J. G. Seedor, N. Endo, H. A. Quartuccio, D. D. Thompson, I. Bab, and G. A. Rodan. Pattern of gene expression following rat tibial marrow ablation. *J Bone Miner Res,* **8,** 379–388 (1993).

54. S. Mohan and D. J. Baylink. Evidence that the inhibition of TE85 human bone cell proliferation by agents which stimulate camp production may in part be mediated by changes in the IGF-11 regulatory system. *Growth Regul,* **1,** 110–118 (1991).

55. M. Centrella, M. C. Horowitz, J. M. Wozney, and T. L. McCarthy. Transforming growth factor-beta gene family members and bone. *Endocr Rev,* **15,** 27–39 (1994).

56. B. S. Komm, C. M. Terpening, D. J. Benz, K. A. Graeme, A. Gallegos, M. Korc, G. L. Greene, B. W. O'Malley, and M. R. Haussler. Estrogen binding, receptor mRNA, and biologic response in osteoblast-like osteosarcoma cells. *Science,* **241,** 81–84 (1988).

57. M. Noda and R. Vogel. Fibroblast growth factor enhances type 1 transforming growth factor gene expression in osteoblast-like cells. *J Cell Biol,* **109,** 2529–2535 (1989).

58. M. Nakashima, H. Nagasawa, Y. Yamada, and A. H. Reddi. Regulatory role of transforming growth factor-beta, bone morphogenetic protein-2, and protein-4 on gene expression of extracellular matrix proteins and differentiation of dental pulp cells. *Dev Biol,* **162,** 18–28 (1994).

59. K. Heikinheimo. Stage-specific expression of decapentaplegic-Vg-related genes 2, 4 and 6 (bone morphogenetic proteins 2, 4 and 6) during human tooth morphogenesis. *J Dent Res,* **73,** 590–597 (1994).

60. S. Vukicevic, M. N. Helder, and F. P. Luyten. Developing human lung and kidney are major sites for synthesis of bone morphogenetic protein-3 (osteogenin). *J Histochem Cytochem,* **42,** 869–875 (1995).

61. R. K. Globus, J. Plouet, and D. Gospodarowicz. Cultured bovine bone cells synthesize basic fibroblast growth factor and store it in their extracellular matrix. *Endocrinology,* **124,** 1539–1547 (1989).

62. D. Somjen, I. Binderman, E. Berger, and A. Harell. Bone remodeling induced by physical stress is prostaglandin E2 mediated. *Biochim Biophys Acta,* **627,** 91–100 (1980).

63. L. G. Raisz and T. J. Martin. Prostaglandins in bone and mineral metabolism. In *Bone and Mineral Research* (W. A. Peck, ed.), pp. 286–310. Elsevier, Amsterdam (1984).

64. P. Ducy, R. Zhan, V. Geoffroy, A. L. Ridall, and G. Karsenty. Osf2/Cbfal: A transcriptional activator of osteoblast differentiation. *Cell,* **89,** 747–754 (1997).

65. T. Komori, H. Yagi, S. Nomura, A. Yamaguchi, K. Sasaki, K. Deguchi, Y. Shimizu, R. T. Bronson, Y. H. Gao, M. Inada, M. Sato, R. Okamoto, Y. Kitamura, S. Yoshiki, and T. Kishimoto. Targeted disruption of Cbfal results in a complete lack of bone formation owing to maturational arrest of osteoblasts. *Cell,* **89,** 755–764 (1997).

66. P. Ducy, M. Starbuck, M. Priemel, J. Shen, G. Pinero, V. Geoffroy, M. Amling, and G. Karsenty. A Cbfal-dependent genetic pathway controls bone formation beyond embryonic development. *Genes Dev,* **13,** 1025–1036 (1994).

67. D. Miao, B. He, Y. Jiang, T. Kobayashi, M. A. Soroceanu, J. Zhao, H. Su, X. Tong, N. Amizuka, A. Gupta, H. K. Genant, H. M. Kronenberg, D. Goltzman, and A. C. Karaplis. Osteoblast-derived PTHrP is a potent endogenous bone anabolic agent that modifies the therapeutic efficacy of administered PTH 11–34. *J Clin Invest,* **115,** 2402–2411 (2005).

68. T. J. Martin. Osteoblast-derived PTHrP is a physiological regulator of bone formation. *J Clin Invest,* **115,** 2322–2324 (2005).

69. E. M. Spencer, C. C. Liu, E. C. C. Si, and G. A. Howard. In vivo actions of insulin-like growth factor (IGF-1) on bone formation and resorption in rats. *Bone,* **12,** 21–26 (1991).

70. J. H. Tobias, J. W. M. Chow, and T. J. Chambers. Opposite effects of insulin-like growth factor-1 on the formation of trabecular and cortical bone in adult female rats. *Endocrinology,* **131,** 2387–2392 (1992).

71. C. M. Bagi, R. Brommage, L. Deleon, S. Adams, D. Rosen, and A. Sommer. Benefit of systemically administered rh1GF-1GFBP-3 on cancellous bone in ovariectomized rats. *J Bone Miner Res,* **9,** 1301–1312 (1994).

72. M. A. Critchlow, Y. S. Bland, and D. E. Ashhurst. The effects of age on the response of rabbit periosteal osteoprogenitor cells to exogeneous transforming growth factor-beta. *J Cell Sci,* **107,** 499–516 (1994).

73. M. E. Joyce, A. B. Roberts, M. B. Sportn, and M. E. Bolander. Transforming growth factor and the initiation of chondrogenesis and osteogenesis in the rat femur. *J Cell Biol,* **110,** 2195–2207 (1990).

74. L. F. Bonewald, L. Wakefield, R. O. C. Oreffo, A. Escobedo, D. R. Twardzik, and G. R. Mundy. Latent forms of transforming growth factor beta (TGFbeta) derived from bone cultures: Identification of a naturally occurring 100-kDa complex with similarity to recombinant TGFbeta. *Mol Endocrinol,* **5,** 741–745 (1991).

75. R. O. C. Oreffo, G. R. Mundy, S. M. Seyedin, and L. F. Bonewald. Activation of the bone-derived latent TGF beta complex by isolated osteoclasts. *Biochem Biophys Res Commun,* **158,** 817–823 (1989).

76. T. J. Martin, E. H. Allan, and S. Fukumoto. The plasminogen activator and inhibitor system in bone remodeling. *Growth Regul,* **3,** 209–214 (1993).

77. P. G. Campbell, J. F. Novak, T. B. Yanosick, and J. H. McMaster. Involvement of the plasmin system in dissociation of the insulin-like growth factor-binding protein complex. *Endocrinology,* **130,** 1401–1412 (1992).

78. C. Gray, S. J. Jones, and A. Boyde. Topographically induced bone formation *in vitro*: Implications for bone implants and bone grafts. *Bone,* **18,** 115–123 (1996).

79. M. Nakamura, N. Udagawa, S. Matuura, M. Mogi, *et al.* Osteoprotegerin regulates bone formation through a coupling mechanism with bone resorption. *Endocrinology,* **144,** 5441–5449 (2003).

80. E. Cleiren, O. Benichou, E. Van Hul, J. Gram, J. Bollerslev, F. R. Singer, K. Beaverson, A. Aledo, M. P. Whyte, T. Yoneyhama, M. C. de Vernejoul, and W. Van Hul. Albers-Schonberg disease (autosomal dominant osteopetrosis type II) results from mutations in the ClCN7 chloride channel gene. *Hum Mol Genet,* **10,** 2861–2867 (2001).

81. C. Lowe, T. Yoneda, B. F. Boyce, H. Chen, G. R. Mundy, and P. Soriano. Osteopetrosis in Src-deficient mice is due to an autonomous defect of osteoclasts. *Proc Nat Acad Sci USA,* **90,** 4485–4489 (1993).

82. D. Kasper, R. Planells-Cases, J. C. Fuhrmann, O. Scheel, O. Zeitz, K. Ruether, A. Schmitt, M. Poet, R. Steinfeld, M. Schweizer, U. Kornak, and T. J. Jentsch. Loss of the chloride channel ClC-7 leads to lysosomal storage disease and neurodegeneration. *EMBO J,* **24,** 1079–1091 (2005).

83. K. Aoki, E. Didomenico, N. A. Sims, K. Mukhopadhyay, L. Neff, A. Houghton, M. Amling, J. B. Levy, W. C. Horne, and R. Baron. The tyrosine phosphatase SHP-1 is a negative regulator of osteoclastogenesis and osteoclast resorbing activity: Increased resorption and osteopenia in me(v)/me(v) mutant mice. *Bone,* **25,** 261–267 (1999).

84. A. E. Grigoriadis, Z. Q. Wang, M. G. Cecchini, W. Hofstetter, R. Felix, H. A. Fleisch, and E. F. Wagner. C-Fos: A key regulator of osteoclast-macrophage lineage determination and bone remodeling. *Science,* **266,** 443–448 (1994).

85. N. A. Sims, B. F. Jenkins, J. M. Quinn, A. Nakamura, *et al.* Glycoprotein 130 regulates bone turnover and bone size by distinct downstream signaling pathways. *J Clin Invest,* **113,** 379–389 (2004).

86. T. J. Martin, E. Romas, and M. T. Gillespie. Interleukins in the control of osteoclast differentiation. *Crit Rev Euk Gene Expr,* **8,** 107–123 (1998).

87. S. Kotake, N. Udagawa, N. Takahashi, K. Matsuzaki, K. Itoh, S. Ishiyama, S. Saito, K. Inoue, N. Kamateni, M. T. Gillespie, T. J. Martin, and T. Suda. IL-17 in synovial fluids from patients with rheumatoid arthritis is a potent stimulator of osteoclastogenesis. *J Clin Invest,* **103,** 1345–1352 (1999).

88. N. Horwood, N. Udagawa, J. Elliott, D. Grail, H. Okamura, M. Kurimoto, A. R. Dunn, T. J. Martin, and M. T. Gillespie. IL-18 inhibits osteoclast formation via T-cell production of GM-CSF. *J Clin Invest,* **103,** 1345–1352 (1999).

89. E. Eghbali-Fatourechi, S. Khosla, A. Sanyal, W. J. Boyle, D. L. Lacey, and B. L. Riggs. Role of RANK ligand in mediating increased bone resorption in early post-menopausal women. *J Clin Invest,* **111,** 1221–1230 (2003).

90. D. Gospodarowicz. Fibroblast growth factor. *Clin Orthop Relat Res,* **257,** 231–248 (1990).

91. W. S. S. Jee, H. Z. Ke, and X. J. Li. Long-term anabolic effects of prostaglandin-E2 on tibial diaphyseal bone in male rats. *Bone Miner,* **15,** 33–55 (1991).

92. B. F. Boyce, T. B. Aufdemorte, I. R. Garrett, A. J. P. Yates, and G. R. Mundy. Effects of interleukin-1 on bone turnover in normal mice. *Endocrinology,* **125,** 1142–1150 (1989).

93. M. Noda and J. J. Camilliere. In vivo stimulation of bone formation by transforming growth factor. *Endocrinology,* **124,** 2991–2994 (1989).

94. T. Nakamura, K. Hanada, M. Tamura, T. Shibanushi, H. Nigi, M. Tagawa, S. Fukumoto, and T. Matsumoto. Stimulation of endosteal bone formation by systemic injections of recombinant basic fibroblast growth factor in rats. *Endocrinology,* **136,** 1276–1284 (1995).

95. J. Folkman. Angiogenesis in cancer, vascular rheumatoid and other disease. *Nature Med,* **1,** 27–31 (1995).

96. S. Harada, J. A. Nagy, K. A. Sullivan, K. A. Thomas, N. Endo, G. A. Rodan, and S. B. Rodan. Induction of vascular endothelial growth factor expression by prostaglandin E2 and E1 in osteoblasts. *J Clin Invest,* **93,** 2490–2496 (1994).

97. S. Niida, M. Kaku, H. Amano, H. Yoshida, H. Kataoka, S. Nishikawa, K. Tanne, N. Maeda, S. Nishikawa, and H. Kodama. Vascular endothelial growth factor can substitute for macrophage colony-stimulating factor in the support of osteoclastic bone resorption. *J Exp Med,* **190,** 293–298 (1999).

98. N. J. Horwood, V. Kartsogiannis, J. M. W. Quinn, E. Romas, T. J. Martin, and M. T. Gillespie. Activated T lymphocytes support osteoclast formation in vitro. *Biochem Biophys Res Commun,* **265,** 144–150 (1999).

99. Y. Y. Kong, U. Feige, U. Sarosil, B. Bolon, A. Tafuri, S. Morony, C. Capparelli, J. Li, R. Elliott, S. McCabe, T. Wong, G. Campagnuolo, E. Moran, E. R. Bogoch, G. Van, L. T. Nguyen, P. S. Ohashi, D. L. Lacey, E. Fish, W. J. Boyle, and J. M. Penninger. Activated T cells regulate bone loss and joint destruction in adjuvant arthritis through osteoprotegerin ligand. *Nature*, **402**, 304–309 (1999).

100. C. M. Giachelli, N. Bae, M. Almeida, D. T. Denhardt, C. E. Alpers, and S. M. Schwartz. Osteopontin is elevated during neointima formation in rat arteries and is a novel component of human atherosclerotic plaques. *J Clin Invest*, **92**, 1686–1696 (1993).

101. E. R. O'Brien, M. R. Garvin, D. K. Stewart, T. Hinohara, J. B. Simpson, S. M. Schwartz, and C. M. Giachelli. Osteopontin is synthesized by macrophage, smooth muscle, and endothelial cells in primary and restenotic human coronary atherosclerotic plaques. *Arterioscler Thromb*, **14**, 1648–1656 (1994).

102. J. C. Christian, C, Slemenda, and C. C. Johnston. Heritability of bone mass: A longitudinal study in aging male twins. *Am J Hum Genet*, **44**, 429–433 (1989).

103. N. A. Morrison, J. Cheng Qi, A. Tokita, P. J. H. Kelly, L. Crofts, T. V. Nguyen, P. N. Sambrook, and J. A. Eisman. Prediction of bone density from vitamin D receptor alleles. *Nature*, **367**, 284–287 (1994).

104. L. D. Spotila, C. D. Constantinou, L. Sereda, A. Ganguly, B. L. Riggs, and D. J. Prockop. Mutation in a gene for type 1 procollagen (COL1A2) in a woman with post-menopausal osteoporosis: Evidence for phenotypic and genotypic overlap with mild osteogenesis imperfecta. *Proc Natl Acad Sci USA*, **88**, 5423–5427 (1991).

105. E. P. Smith, J. Boyd, G. R. Frank, H. Takahashi, R. M. Cohen, B. Specker, T. C. Williams, D. L. B. Lubahn, and K. S. Korach. Estrogen resistance caused by mutation in the estrogen-receptor gene in a man. *N Engl J Med*, **331**, 1056–1061 (1994).

106. D. B. Lubahn, J. S. Moyer, T. S. Golding, J. F. Couse, K. S. Korach, and O. Smithies. Alteration of reproductive function but not prenatal sequal development after insertional disruption of the mouse estrogen receptor gene. *Proc Natl Acad Sci USA*, **90**, 11162–11166 (1993).

107. K. S. Korach. Insights from the study of animals lacking functional estrogen receptor. *Science*, **266**, 1524–1527 (1994).

108. M. J. Oursler, L. Pederson, L. Fitzpatrick, B. L. Riggs, and T. Spelsberg. Human giant cell tumors of the bone (osteoclastomas) are estrogen target cells. *Proc Natl Acad Sci USA*, **91**, 5227–5231 (1994).

109. H. M. Frost. The mechanosts: A proposed pathogenic mechanism of osteoporosis and the bone mass effects of mechanical and nonmechanical agents. *Bone*, **2**, 783–785 (1987).

110. D. T. Thompson and G. A. Rodan. Indomethacin inhibition of tenotomy-induced bone resorption in rats. *J Bone Miner Res*, **3,** 409–414 (1988).

111. L. Mosekilde. Consequences of the remodeling process for vertebral trabecular bone structure: A scanning electron microscopy study (uncoupling of loaded structures). *Bone Mineral*, **10**, 13–35 (1990).

112. J. Bonadio, K. J. Jepsen, M. K. Mansoura, R. Jaenisch, J. L. Huhn, and S. A. Goldstein. A murine skeletal adaptation that significantly increases cortical bone mechanical properties. *J Clin Invest*, **92**, 1697–1705 (1993).

113. M. H. Lafage, R. Balena, M. A. Battle, M. Shea, J. G. Seedor, H. Klein, W. C. Hayes, and G. A. Rodan. Comparison of alendronate and sodium fluoride effects on cancellous and cortical bone in minipigs. *J Clin Invest*, **95**, 2127–2133 (1995).

114. C. T. Rubin and L. E. Lanyon. Regulation of bone mass by mechanical strain magnitude. *Calcif Tissue Int*, **37**, 411–417 (1985).

115. V. Schneider, V. Ogancy, A. LeBlanc, A. Rakhmanov, A. Bakulin, A. Grigoriev, and L. VarRonin. Spaceflight bone loss and changes in fat and lean body mass. *J Bone Miner Res*, **7**, S122 (1992).

116. A. L. Huddleston, D. Rockwell, D. N. Kulund, and R. B. Harrison. Bone mass in lifetime tennis athletes. *JAMA*, **244**, 1107–1109 (1980).

117. S. Morii and D. B. Burr. Increased intracortical remodeling following fatigue damage. *Bone*, **14**, 103–109 (1993*)*.

118. L. J. Kornberg, S. H. Earp, C. E. Turner, C. Prockop, and R. L. Juliano. Signal transduction by integrins: Increased protein tyrosine phosphorylation caused by clustering beta-l integrins. *Proc Natl Acad Sci USA*, **88**, 8392–8396 (1991).

119. S. J. Silverberg, F. Gartenberg, D. McMahon, and J. P. Bilezikian. Parathyroidectomy improves bone density at cancellous sites in asymptomatic primary hyperparathyroidism. *J Bone Miner Res*, **8**, S169 (1993).

120. D. Shinar and G. A. Rodan. Relationships and interactions between bone and bone marrow. In *The Hematopoietic Microenvironment: The Functional and Structural Basis of Blood Cell Development* (M. W. Long and M. S. Wicha, eds.), pp. 79–109. Johns Hopkins University Press, Baltimore, MD (1993).

121. P. Ducy, M. Amling, S. Takeda, M. Priemel, A. F. Schilling, F. T. Beil, J. Shen, C. Vinson, J. M. Rueger, and G. Karsenty. Leptin inhibits bone formation through a hypothalamic relay: A central control of bone mass. *Cell*, **100**, 197–207 (2000).

Index

C

D

E

G

T

U

V